HANDBOOK of BEHAVIOR STATE CONTROL

Cellular and Molecular Mechanisms

HANDBOOK of
BEHAVIORAL STATE CONTROL
Cellular and Molecular Mechanisms

Edited by
Ralph Lydic
Helen A. Baghdoyan

CRC Press
Boca Raton   London   New York   Washington, D.C.

Library of Congress Cataloging-in-Publication Data

Handbook of behavioral state control : cellular and molecular
   mechanisms / edited by Ralph Lydic and Helen A. Baghdoyan.
      p. cm.
   Includes bibliographical references and index.
   ISBN 0-8493-3151-X (alk. paper)
      1. Arousal (Physiology).   2. Neuropsychology.   3. Molecular
neurobiology.   4. Emotions.   5. Consciousness.   I. Lydic, Ralph
II. Baghdoyan, Helen A.
QP405.H36   1998
571.7'1 — dc21   98-34858
                                                                                                      CIP

This book contains information obtained from authentic and highly regarded sources. Reprinted material is quoted with permission, and sources are indicated. A wide variety of references are listed. Reasonable efforts have been made to publish reliable data and information, but the author and the publisher cannot assume responsibility for the validity of all materials or for the consequences of their use.

Neither this book nor any part may be reproduced or transmitted in any form or by any means, electronic or mechanical, including photocopying, microfilming, and recording, or by any information storage or retrieval system, without prior permission in writing from the publisher.

All rights reserved. Authorization to photocopy items for internal or personal use, or the personal or internal use of specific clients, may be granted by CRC Press LLC, provided that $.50 per page photocopied is paid directly to Copyright Clearance Center, 222 Rosewood Drive, Danvers, MA 01923 USA. The fee code for users of the Transactional Reporting Service is ISBN 0-8493-3151-X/99/$0.00+$.50. The fee is subject to change without notice. For organizations that have been granted a photocopy license by the CCC, a separate system of payment has been arranged.

The consent of CRC Press LLC does not extend to copying for general distribution, for promotion, for creating new works, or for resale. Specific permission must be obtained in writing from CRC Press LLC for such copying.

Direct all inquiries to CRC Press LLC, 2000 Corporate Blvd., N.W., Boca Raton, Florida 33431.

Trademark Notice: Product or corporate names may be trademarks or registered trademarks and are used for identification and explanation, without intent to infringe.

© 1999 by CRC Press LLC

No claim to original U.S. Government works
International Standard Book Number 0-8493-3151-X
Library of Congress Card Number 98-34858
Printed in the United States of America 1 2 3 4 5 6 7 8 9 0
Printed on acid-free paper

Preface

All mental and behavioral states are a product of the brain. A unified view of diverse mental states such as sleep, pain, anesthesia, anxiety, and substance abuse now is emerging from the mind/body materialism of contemporary neuroscience. The rationale of this handbook is to create a unifying resource for state-of-the-art information on the cellular and molecular mechanisms generating diverse behavioral states. The need for this handbook is emphasized by the absence of any other current resource that provides a unifying synthesis for the neurobiology of behavioral states. The purpose of this handbook is to provide a working reference on: (1) the cellular and molecular mechanisms generating arousal states, (2) pharmacological and non-pharmacological methods of behavioral state control, and (3) the bidirectional interaction between arousal states and the neurobiology of pain, and between sleep and the immune system. With these purposes realized, the handbook provides a current, unified resource on the neurobiology of behavioral state control.

Mental and behavioral state disorders are common. Perturbations of mental and behavioral state control range from normal stress and sleep deprivation to debilitating neuropsychiatric disorders. A recent poll, commissioned by the Dana Alliance for Brain Initiatives, found that 86% of Americans know someone with a neurological disease, brain injury, or mental illness. Mental state disorders comprise the leading cause of all hospital admissions. Even the lives of healthy individuals are organized with a primary emphasis devoted to mental and behavioral state control. All humans, for example, daily experience the altered mental state of sleep. The single best predictor of daily performance is adequacy of the previous night's sleep. Disordered sleep powerfully disrupts the cognitive and emotional integrity of waking consciousness, and these disruptions now are recognized as a national health problem. Disorders such as anxiety are common, and 10% of the U.S. population experiences panic attacks. Psychoses and affective illnesses are serious mental state disorders, and about 6% of Americans experience clinically significant depression. Another telling statistic emphasizing the importance of behavioral states is the observation that in the U.S., in 1995, there were 31,000 suicides compared to 22,000 murders.

Drug use, abuse, and addiction make it clear that mental state control is a major social problem. Drug addiction creates unique mental state disorders, and substance abuse can be viewed as a maladaptive self-treatment of psychic pain. Drugs also provide one of the most effective and beneficial tools available for the clinical management of psychic and physical pain. In the U.S. alone, general anesthesia is produced more than 50,000 times each day. Yet, for no anesthetic agent is it presently understood how mental states are controlled and pain perception eliminated. Techniques for determining when a patient is anesthetized are imperfect, and surgery has been documented to have proceeded on paralyzed but conscious patients. The handbook describes these and other dissociated states which are of special relevance for psychiatrists, anesthesiologists, neurologists, pharmacologists, psychologists, and neuroscientists.

Two basic research paradigms guide most studies of brain and behavior. One approach studies simpler systems (fruit flies, worms, artificial membranes) in the hope of someday eventually being able to reconstruct an understanding of behavioral states from a collection of thoroughly characterized parts. A second paradigm incorporates the view that the science of behavior is part of biology and that states of consciousness, and their physiological traits, are generated as emergent processes by anatomically distributed neuronal networks. There are often tensions between these two approaches, as research programs using these two paradigms must compete for limited funds. One of the most exciting developments in contemporary neuroscience is the use of cellular and molecular techniques to elucidate complex physiological and behavioral traits comprising behavioral states. Many of the chapters comprising this handbook demonstrate the value of methodological and conceptual approaches that are successfully elucidating behavioral states and physiological traits from a cellular and molecular perspective.

Finally, the 38 chapters comprising this volume were solicited by eight Section Editors, each widely respected for their expertise and record of productivity. We thank the Section Editors and their contributing authors for their enthusiastic support of this handbook. Production of this handbook was made possible by the efforts of key individuals, in addition to the authors and section editors. Paul Petralia gave sound advice and support that were essential for initiating this project. Norina Frabotta offered editorial guidance, and Pam Myers provided expert secretarial support ensuring the timely presentation of these chapters. Our continuing efforts to elucidate the cellular and molecular mechanisms underlying the generation of behavioral states and state-dependent changes in autonomic control are supported by the Department of Anesthesia and by grants HL40881, HL57120, and MH45361.

The Editors

Ralph Lydic, Ph.D., is Director of Anesthesia Research and the Julien F. Biebuyck Professor of Anesthesia and Professor of Cellular and Molecular Physiology at The Pennsylvania State University, College of Medicine. Dr. Lydic's research career has maintained a focus on the neurobiology of sleep and breathing. In 1979, he earned his Ph.D. in Physiology from Texas Tech University, using single-cell recording techniques to test the hypothesis that the onset of rapid eye movement (REM) sleep causes diminished discharge of pontine respiratory neurons. Dr. Lydic's postdoctoral years (1979–1981) were spent in the Department of Physiology and Biophysics at Harvard Medical School. In 1981, Dr. Lydic joined the Laboratory of Neurophysiology at Harvard Medical School, where he served as Assistant Professor of Physiology. In 1986, Dr. Lydic moved his laboratory to the Pulmonary Division of The Pennsylvania State University's College of Medicine, where his research emphasized the neural control of breathing. In July 1989, Dr. Lydic moved to the Department of Anesthesia, where he was appointed Director of the Division of Anesthesia and Neuroscience Research. Since July 1991, Dr. Lydic has served as Professor in the Department of Anesthesia and in the Department of Cellular and Molecular Physiology.

Awards and honors resulting from Dr. Lydic's research include an Upjohn Pharmaceutical Scholarship (Harvard Medical School); Neurobiology Program Scholarship (Woods Hole Marine Biological Laboratory); Neurobiology Program Scholarship (Cold Spring Harbor Laboratory, NY); National Research Service Award (Harvard Medical School); William F. Milton Award (Harvard Medical School); Mentor for Scholl Fellowship, National SIDS Foundation (Pennsylvania State University); Mentor for Parker B. Francis Fellowship (Pennsylvania State University); Visiting Scientist, NASA Division of Space Life Sciences, Johnson Space Center (1994–1995); Dunaway-Burnham Visiting Scholar, Dartmouth Medical School (1995); Mentor for Proctor and Gamble Award from the American Physiological Society (1996); and Mentor for Precollege Science Education Initiative, Howard Hughes Medical Institute (1996–1998).

Dr. Lydic has a long-standing interest and commitment to the American Physiological Society (APS). He has served the APS in a variety of offices including Chairman, Central Nervous System (CNS) Section (1986–1992); Program Advisory Committee (1986–1992); CNS Section Advisory Committee (1989–1992); Long-Range Planning Committee (1989–1992); Chairman, FASEB Theme Committee: "Nervous System Function and Disorder" (1990); Nominating Committee (1990–1992); Committee on Committees (1994–1996); and Public Affairs Committee (1997–2000).

Dr. Lydic's research program ranges from the level of transmembrane cell signaling to integrative aspects of respiratory and arousal state control. Dr. Lydic's studies aim to elucidate the cellular and molecular mechanisms that cause respiratory depression during the loss of waking consciousness. These basic studies are funded by the National Heart, Lung, and Blood Institute of the National

Institutes of Health because of their potential clinical relevance for disorders such as sudden infant death syndrome, adult sleep apnea, and anesthesia-induced respiratory depression.

Helen A. Baghdoyan, Ph.D., is Professor of Anesthesia and Pharmacology at The Pennsylvania State University, College of Medicine in Hershey, PA. Dr. Baghdoyan received her Ph.D. in neuropsychopharmacology from the University of Connecticut in 1980. Her research career has maintained a focus on the neurobiology of sleep, beginning with her Ph.D. thesis which characterized the effects of systemically administered monoacylcadaverines on the sleep/wake cycle of the mouse. As a postdoctoral fellow in the Department of Psychiatry at Harvard Medical School, her research defining the cholinergic model of rapid eye movement (REM) sleep was supported by a National Research Service Award. Dr. Baghdoyan joined the faculty as Assistant Professor of Psychiatry (Neuroscience) at Harvard, and in 1987 she moved her laboratory to the Department of Anesthesia at The Pennsylvania State University, College of Medicine. Her research on brainstem cholinergic mechanisms of REM sleep generation has been supported by the National Institute of Mental Health since 1989.

Dr. Baghdoyan has served the Sleep Research Society as a member of the Executive Committee, and the American Physiological Society as a member of the Program Advisory Committee, as Councilor of the Central Nervous System (CNS) Section, and as Secretary/Treasurer of the CNS Section. Currently, Dr. Baghdoyan is serving the National Institutes of Health as a regular member of the Integrative, Functional, and Cognitive Neuroscience-3 Study Section.

The Contributors

Angel Alonso, Ph.D.
Department of Neurology and Neurosurgery
Magill University and
  Montreal Neurological Institute
Montreal, Quebec
Canada

Helen A. Baghdoyan, Ph.D.
Department of Anesthesia
Pennsylvania State College of Medicine
Hershey, Pennsylvania

Ruben Baler, Ph.D.
Section on Neuroendocrinology
Laboratory of Developmental Neurobiology
National Institute of Child Health
  and Human Development
Bethesda, Maryland

Bruno Barbagli
Department of Experimental Medicine
Laboratory of Experimental Medicine
  and Faculty of Medicine
Institut National de la Santé et de la
  Recherche Médicale
Lyon, France

Valerie Bégay, Ph.D.
Laboratory of Neurobiology and
  Cellular Neuroendocrinology
Department of Neurosciences
Paris, France

Marina Bentivoglio, M.D.
Institute of Human Anatomy
Medical Faculty
Strada le Grazie — Borgo Roma
Verona, Italy

Jean-Francois Bernard, M.D., Ph.D.
Unité de Recherches de Physiopharmacologie
  du Systéme Nerveux
Institut National de la Santé et de la
  Recherche Médicale
Paris, France

Marianne Bernard, Ph.D.
Section on Neuroendocrinology
Laboratory of Developmental Neurobiology
National Institute of Child Health
  and Human Development
Bethesda, Maryland

Jean-Marie Besson, D.Sc.
Unité de Recherches de Physiopharmacologie
  du Systéme Nerveux
Institut National de la Santé et de la
  Recherche Médicale
Paris, France

Romuald Boissard
Department of Experimental Medicine
Laboratory of Experimental Medicine
  and Faculty of Medicine
Institut National de la Santé et de la
  Recherche Médicale
Lyon, France

György Buzsáki, Ph.D.
Center for Molecular and Behavioral Neuroscience
Rutgers University
Newark, New Jersey

Greg Cahill, Ph.D.
Department of Biology
University of Houston
Houston, Texas

Brian E. Cairns, Ph.D.
Department of Pharmacology and Toxicology
Faculty of Pharmaceutical Sciences
University of British Columbia
Vancouver, British Columbia
Canada

José M. Calvo, M.D., Ph.D.
Instituto Mexicano de Psiquiatría
Mexico City, Mexico

J. Patrick Card, Ph.D.
Department of Neuroscience
University of Pittsburgh
Pittsburgh, Pennsylvania

Vincent Cassone, Ph.D.
Department of Biology
Texas A & M University
College Station, Texas

Victoria Chapman, Ph.D.
Department of Pharmacology
University College London
London, England

Boris A. Chizh, M.D., Ph.D.
Department of Physiology
School of Medical Sciences
University of Bristol
United Kingdom

James J. Chrobak, Ph.D.
Department of Psychiatry
Center for Neuroscience
University of California
Davis, California

Steven L. Coon, Ph.D.
Department of Biology
Texas A & M University
College Station, Texas

Anthony Dickenson, Ph.D.
Department of Pharmacology
University College London
London, England

Rene Drucker-Colin, M.D., Ph.D.
Department of Physiology
Faculty of Medicine
Universidad Nacional Autónoma de México
Mexico City, Mexico

Jack Falcón, Ph.D.
Laboratory of Neurobiology and
  Cellular Neuroendocrinology
Department of Neurosciences
l'UMR CNRS No. 6558, UFR Sci.
Poitiers, France

Jidong Fang, M.D., Ph.D.
Department of Veterinary and Comparative
  Anatomy, Pharmacology, and Physiology
Washington State University
Pullman, Washington

Casimir A. Fornal, Ph.D.
Department of Psychology
Program of Neuroscience
Princeton University
Princeton, New Jersey

Patrice Fort, Ph.D.
Department of Experimental Medicine
Laboratory of Experimental Medicine
  and Faculty of Medicine
Institut National de la Santé et de la
  Recherche Médicale
Lyon, France

Jonathan A. Gastel, Ph.D.
Section on Neuroendocrinology
Laboratory of Developmental Neurobiology
National Institute of Child Health
  and Human Development
Bethesda, Maryland

Damien Gervasoni
Department of Experimental Medicine
Laboratory of Experimental Medicine
  and Faculty of Medicine
Institut National de la Santé et de la
  Recherche Médicale
Lyon, France

Margarita Gómez
Department of Physiology
Faculty of Medicine
Universidad Nacional Autónoma de México
Mexico City, Mexico

Gigliola Grassi-Zucconi, Ph.D.
Department of Cell Biology
Faculty of Biological Sciences
University of Perugia
Perugia, Italy

Maria Grazia de Simoni, M.D.
Department of Neurochemistry
Instituto de Ricerche Farmacologiche
  "Mario Negri"
Milan, Italy

Robert W. Greene, M.D., Ph.D.
Neuroscience Laboratory
Harvard Medical School
Veterans Administration Medical Center
Brockton, Massachusetts

Jan Gybels, M.D., Ph.D.
Professor Emeritus of Neurosurgery
Department of Neurosciences and Psychiatry
Neurosurgery
K.U. Leuven
Leuven, Belgium

Paul E. Hardin, Ph.D.
Department of Biology
University of Houston
Houston, Texas

Nick A. Hartell, Ph.D.
Department of Physiology
School of Medical Sciences
University of Bristol
Bristol, England

P. Max Headley, Ph.D.
Department of Physiology
School of Medical Sciences
University of Bristol
Bristol, England

Mary M. Heinricher, Ph.D.
Division of Neurosurgery
Oregon Health Sciences University
Portland, Oregon

Steven J. Henriksen, Ph.D.
Department of Neuropharmacology
The Scripps Research Institute
LaJolla, California

Juan F. Herrero, M.D., Ph.D.
Department of Physiology
School of Medical Sciences
University of Bristol
Bristol, England

Luca Imeri, M.D.
Instituto di Fisiologia Umana II
Università degli Studi
Milan, Italy

P. Michael Iuvone, Ph.D.
Department of Pharmacology
Emory University School of Medicine
Atlanta, Georgia

Barry L. Jacobs, Ph.D.
Department of Psychology
Program of Neuroscience
Princeton University
Princeton, New Jersey

Anabel Jimenéz-Anguiano, Ph.D.
Department of Physiology
Faculty of Medicine
Universidad Nacional Autónoma de México
Mexico City, Mexico

Barbara E. Jones, Ph.D.
Department of Neurology and Neurosurgery
McGill University
Montreal Neurological Institute
Montreal, Quebec
Canada

David C. Klein, Ph.D.
Section on Neuroendocrinology
Laboratory of Developmental Neurobiology
National Institute of Child Health
  and Human Development
Bethesda, Maryland

George F. Koob, Ph.D.
Division of Psychopharmacology
Department of Neuropharmacology
The Scripps Research Institute
La Jolla, California

Morten P. Kristensen, Ph.D.
Department of Pharmacology and Toxicology
Faculty of Pharmaceutical Sciences
University of British Columbia
Vancouver, British Columbia
Canada

James M. Krueger, Ph.D.
Department of Veterinary and Comparative
  Anatomy, Pharmacology, and Physiology
College of Veterinary Medicine
Washington State University
Pullman, Washington

Pierre-Hervé Luppi, Ph.D.
Department of Experimental Medicine
Laboratory of Experimental Medicine
  and Faculty of Medicine
Institut National de la Santé et de la Recherche Médicale
Lyon, France

Ralph Lydic, Ph.D.
Department of Anesthesia
College of Medicine
The Pennsylvania State University
Hershey, Pennsylvania

Mark W. Mahowald, M.D.
Minnesota Regional Sleep Disorders Center
Hennepin County Medical Center
Minneapolis, Minnesota

Dolores Martínez-González, M.D.
Department of Physiology
Faculty of Medicine
Universidad Nacional Autónoma de México
Mexico City, Mexico

Steve McGaraughty, Ph.D.
Division of Neurosurgery
Oregon Health Sciences University
Portland, Oregon

Wallace B. Mendelson, M.D.
Sleep Research Laboratory
The University of Chicago
Chicago, Illinois

Emmanuel Mignot, M.D., Ph.D.
Stanford Center for Narcolepsy Research
Sleep Disorders Center
Palo Alto, California

Michel Mühlethaler, Ph.D.
Départment de Physiologie
Centre Médicale Universitaire
Université de Geneve
Geneva, Switzerland

Eric Murillo-Rodríguez
Department of Physiology
Faculty of Medicine
Universidad Nacional Autónoma de México
Mexico City, Mexico

Luz Navarro, Ph.D.
Department of Physiology
Faculty of Medicine
Universidad Nacional Autónoma de México
Mexico City, Mexico

Tore A. Nielsen, Ph.D.
Centre d'Etude du Sommeil
Hospital du Sacre-Coeur
Montreal, Quebec, Canada

Seiji Nishino, M.D., Ph.D.
Stanford Center for Narcolepsy Research
Sleep Disorders Center
Palo Alto, California

Mark R. Opp, Ph.D.
Department of Psychiatry and Behavioral Sciences
University of Texas Medical Branch
Galveston, Texas

Marcela Palomero
Department of Physiology
Faculty of Medicine
Universidad Nacional Autónoma de México
Mexico City, Mexico

Christelle Peyron, Ph.D.
Department of Experimental Medicine
Laboratory of Experimental Medicine
    and Faculty of Medicine
Institut National de la Santé et de la
    Recherche Médicale
Lyon, France

Hugh D. Piggins, Ph.D.
Anatomy and Human Biology Group
King's College
London, England

Oscar Prospéro-Garcia, M.D., Ph.D.
Department of Physiology
Faculty of Medicine
Universidad Nacional Autónoma de México
Mexico City, Mexico

Donald G. Rainnie, M.D., Ph.D.
Neuroscience Laboratory
Harvard Medical School
Veterans Administration Medical Center
Brockton, Massachusetts

Claire Rampon, Ph.D.
Department of Experimental Medicine
Laboratory of Experimental Medicine
    and Faculty of Medicine
Institut National de la Santé et de la
    Recherche Médicale
Lyon, France

Alison Reeve, Ph.D.
Department of Pharmacology
University College London
London, England

Timothy Roehrs, Ph.D.
Department of Psychiatry
Sleep Disorders and Research Center
Henry Ford Hospital
Detroit, Michigan

Patrick H. Roseboom, Ph.D.
Section on Neuroendocrinology
Laboratory of Developmental Neurobiology
National Institute of Child Health and Human Development
Bethesda, Maryland

Thomas Roth, Ph.D.
Department of Psychiatry
Sleep Disorders and Research Center
Henry Ford Hospital
Wayne State University
Detroit, Michigan

Benjamin Rusak, Ph.D.
Department of Psychology
Dalhousie University
Halifax, Nova Scotia
Canada

Manuel Sánchez, M.D.
Department of Physiology
Faculty of Medicine
Universidad Nacional Autónoma de México
Mexico City, Mexico

Clifford B. Saper, M.D., Ph.D.
Department of Neurology
Beth Israel Deaconess Medical Center
Boston, Massachusetts

Tom Scammell, M.D.
Department of Neurology
Beth Israel Deaconess Medical Center
Boston, Massachusetts

Carlos H. Schenck, M.D.
Minnesota Regional Sleep Disorders Center
Hennepin County Medical Center
Minneapolis, Minnesota

William J. Schwartz, M.D.
Department of Neurology
University of Massachusetts Medical School
Worcester, Massachusetts

Amita Sehgal, Ph.D.
Department of Neuroscience
University of Pennsylvania Medical Center
Philadelphia, Pennsylvania

Kazue Semba, Ph.D.
Department of Anatomy and Neurobiology
Dalhousie University
Halifax, Nova Scotia
Canada

Barry J. Sessle, M.D.S., Ph.D.
Faculty of Dentistry
University of Toronto
Toronto, Ontario
Canada

John E. Sherin, M.D., Ph.D.
Department of Neurology
Beth Israel Deaconess Medical Center
Boston, Massachusetts

Priyattam J. Shiromani, Ph.D.
Neuroscience Laboratory
Harvard Medical School
Veterans Administration Medical Center
Brockton, Massachusetts

Jerome M. Siegel, Ph.D.
Neurobiology Research
Department of Psychiatry and
    Brain Research Institute
University of California, Los Angeles
Sepulveda Veterans Administration Medical Center
Sepulveda, California

Karina Simón-Arceo, Ph.D.
Instituto Mexicano de Psiquiatría
Mexico City, Mexico

Peter J. Soja, Ph.D.
Faculty of Pharmaceutical Sciences
University of British Columbia
Vancouver, British Columbia
Canada

Mircea Steriade, M.D., D.Sc.
Department of Physiology
University of Laval
School of Medicine
Laval, Quebec
Canada

Joseph S. Takahashi, Ph.D.
Howard Hughes Medical Institute
Department of Neurobiology and Physiology
Northwestern University
Evanston, Illinois

Linda A. Toth, D.V.M., Ph.D.
Department of Infectious Diseases
Comparative Medicine Division
St. Jude Children's Research Hospital
Memphis, Tennessee

Luis Villanueva, D.D.S., Ph.D.
Unité de Recherches de Physiopharmacologie
  du Systéme Nerveux
Institut National de la Santé et de la
  Recherche Médicale
Paris, France

J.L. Weller
Section on Neuroendocrinology
Laboratory of Developmental Neurobiology
National Institute of Child Health
  and Human Development
Bethesda, Maryland

John T. Williams, Ph.D.
Vollum Institute of Biomedical Research
Oregon Health Science University
Portland, Oregon

William D. Willis, Jr., M.D., Ph.D.
Department of Anatomy and Neurosciences
University of Texas Medical Branch
Galveston, Texas

Lisa D. Wilsbacher
Department of Neurobiology and Physiology
Northwestern University
Evanston, Illinois

Johnathan P. Wisor, Ph.D.
Sleep Disorders and Research Center
Department of Psychiatry
Stanford University School of Medicine
Palo Alto, California

Martin Zatz, M.D., Ph.D.
Section on Biochemical Pharmacology
Laboratory of Cellular and Molecular Regulation
National Institute of Mental Health
National Institutes of Health
Bethesda, Maryland

Piotr Zlomanczuk, Ph.D.
Department of Physiology
Rydygier Medical School
Bydgoszcz, Poland

Contents

Section I. Mammalian Circadian (24-Hour) Rhythms
William J. Schwartz, Section Editor

Chapter 1. Introduction: Endogenous Pacemakers and Daily Programs 3
William J. Schwartz and Piotr Zlomanczuk

Chapter 2. Anatomy of the Mammalian Circadian Timekeeping System 13
J. Patrick Card

Chapter 3. Intercellular Interactions and the Physiology
of Circadian Rhythms in Mammals 31
Hugh D. Piggins and Benjamin Rusak

Chapter 4. The Molecular Basis of the Pineal Melatonin Rhythm:
Regulation of Serotonin *N*-Acetylation 45
*David C. Klein, Ruben Baler, Patrick H. Roseboom, J.L. Weller, Marianne Bernard,
Jonathan A. Gastel, Martin Zatz, P. Michael Iuvone, Valerie Bégay, Jack Falcón,
Greg Cahill, Vincent M. Cassone, and Steven L. Coon*

Chapter 5. Molecular Components of a Model Circadian Clock:
Lessons From *Drosophila* 61
Paul E. Hardin and Amita Sehgal

Chapter 6. Strategies for Dissecting the Molecular Mechanisms of Mammalian
Circadian Rhythmicity 75
Lisa D. Wilsbacher, Jonathan P. Wisor, and Joseph S. Takahashi

Section II. Daily Alterations In Arousal State
Mark W. Mahowald, Section Editor

Chapter 7. The Evolution of REM Sleep 87
Jerome M. Siegel

Chapter 8. Mentation During Sleep: The NREM/REM Distinction 101
Tore A. Nielsen

Chapter 9. Narcolepsy 129
Emmanuel Mignot and Seiji Nishino

Chapter 10. Dissociated States of Wakefulness and Sleep 143
Mark W. Mahowald and Carlos H. Schenck

Section III. Neuroanatomical and Neurochemical Basis of Behavioral States
Kazue Semba, Section Editor

Section Introduction 159
Kazue Semba

Chapter 11. The Mesopontine Cholinergic System: A Dual Role
in REM Sleep and Wakefulness 161
Kazue Semba

Chapter 12. An Integrative Role for Serotonin in the Central Nervous System 181
Barry L. Jacobs and Casimir A. Fornal

Chapter 13. Inhibitory Mechanisms in the Dorsal Raphe Nucleus
and Locus Coeruleus During Sleep 195
*Pierre-Hervé Luppi, Christelle Peyron, Claire Rampon, Damien Gervasoni,
Bruno Barbagli, Romuald Boissard, and Patrice Fort*

Chapter 14. Cholinergic and GABAergic Neurons of the Basal Forebrain:
Role in Cortical Activation 213
Barbara E. Jones and Michel Mühlethaler

Chapter 15. Immediate Early Gene Expression in Sleep and Wakefulness 235
Marina Bentivoglio and Gigliola Grassi-Zucconi

Section IV. Cellular and Network Mechanisms of Behavioral State Control
Robert W. Greene, Section Editor

Chapter 16. Synaptic and Intrinsic Membrane Properties Regulating
Noradrenergic and Serotonergic Neurons During Sleep/Wake Cycles 257
John T. Williams

Chapter 17. Mechanisms Affecting Neuronal Excitability in Brainstem
Cholinergic Centers and their Impact on Behavioral State 277
Robert W. Greene and Donald G. Rainnie

Chapter 18. Intrinsic Electroresponsiveness of Basal Forebrain Cholinergic
and Non-Cholinergic Neurons 297
Angel Alonso

Chapter 19. Hypothalamic Regulation of Sleep 311
Priyattam J. Shiromani, Tom Scammell, John E. Sherin, and Clifford B. Saper

Chapter 20. Cellular Substrates of Oscillations in Corticothalamic Systems During States of Vigilance 327
Mircea Steriade

Chapter 21. State-Dependent Changes in Network Activity of the Hippocampal Formation 349
James J. Chrobak and György Buzsáki

Section V. Molecules Modulating Mental States
Steven J. Henriksen, Section Editor

Chapter 22. Neuronal Mediation of Addictive Behavior 365
George F. Koob

Chapter 23. Cholinergic Enhancement of REM Sleep from Sites in the Pons and Amygdala 391
José M. Calvo and Karina Simón-Arceo

Chapter 24. State-Altering Effects of Benzodiazepines and Barbiturates 407
Wallace B. Mendelson

Chapter 25. State-Altering Actions of Ethanol, Caffeine, and Nicotine 421
Timothy Roehrs and Thomas Roth

Chapter 26. Psychomimetic Drugs, Marijuana, and 5-HT Antagonists 433
Oscar Prospéro-García, Eric Murillo Rodríguez, Anabel Jiménez-Anguiano, Luz Navarro, Manuel Sánchez, Margarita Gómez, Dolores Martínez-González, Marcela Palomero, and Rene Drucker-Colín

Section VI. State-Dependent Processing in Somatosensory Pathways
Peter J. Soja, Section Editor

Section Introduction 443
Peter J. Soja

Chapter 27. Somatosensory Transmission in the Trigeminal Brainstem Complex and its Modulation by Peripheral and Central Neural Influences 445
Barry J. Sessle

Chapter 28. Anatomy, Physiology, and Descending Control of Lumbosacral Sensory Neurons Involved in Tactile and Pain Sensations 463
William D. Willis, Jr.

Chapter 29. Pain-Modulating Neurons and Behavioral State 487
Mary M. Heinricher and Steve P. McGaraughty

Chapter 30. Electrophysiology of Spinal Sensory Processing in the Absence and Presence of Surgery and Anesthesia 505
P. Max Headley, Boris A. Chizh, Juan F. Herrero, and Nick A. Hartell

Chapter 31. Transmission Through Ascending Trigeminal and Lumbar Sensory Pathways: Dependence on Behavioral State 521
Peter J. Soja, Brian E. Cairns, and Morten P. Kristensen

Section VII. Pain and Anesthesia
Jean-Marie Besson, Section Editor

Section Introduction 545
Jean-Marie Besson

Chapter 32. Future Pharmacological Treatments of Pain 549
Anthony Henry Dickenson, Victoria Chapman, and Alison Reeve

Chapter 33. The Multiplicity of Ascending Pain Pathways 569
Luis Villanueva and Jean-François Bernard

Chapter 34. Is There Still Room in the Neurosurgical Treatment of Pain
for Making Lesions in Nociceptive Pathways? 587
Jan Gybels

Section VIII. Immunological Alterations in Arousal States
James M. Krueger, Section Editor

Chapter 35. Cytokines and Sleep Regulation 609
James M. Krueger and Jidong Fang

Chapter 36. Fever, Body Temperature, and Levels of Arousal 623
Mark R. Opp

Chapter 37. Microbial Modulation of Arousal 641
Linda A. Toth

Chapter 38. Immune Alterations in Neurotransmission 659
Luca Imeri and Maria Grazia de Simoni

Index 677

Section I

Mammalian Circadian (24-Hour) Rhythms

William J. Schwartz, Section Editor

Chapter 1

Introduction: Endogenous Pacemakers and Daily Programs

William J. Schwartz and Piotr Zlomanczuk

Contents

1.1 Some Introductory Concepts and Definitions 4
1.2 The Suprachiasmatic Nucleus as a Circadian Pacemaker 5
1.3 Analyzing Pacemaker Input and Output Pathways 7
1.4 Assembling a Multicellular Circadian Pacemaker 8
Acknowledgment 9
References 10

The Earth's daily rotation about its axis is a pervasive environmental influence on the control of behavioral state. Since the dawn of life, most organisms have had to cope with the challenge of living with alternating cycles of light and darkness. Central to this adaptation is the existence of an endogenous 24-hr clock that regulates biological processes in the temporal domain.[1,9,28,42] In mammals, circadian (in Latin, *circa* = about, *dies* = day) rhythmicity in a host of behavioral, physiological, and biochemical variables is the overt manifestation of this internal timekeeping system. But, such clocks also exist in prokaryotes, as in the autotrophic cyanobacterium *Synechococcus*, which obtains its energy both through photosynthesis and nitrogen fixation. The problem for *Synechococcus* is that nitrogenase is inactivated in the presence of oxygen; the solution is to separate nitrogenase and photosynthesis in time, so that nitrogenase activity is restricted to night (when photosynthesis and oxygen production are low).[25]

Circadian clocks provide organisms with a mechanism that can recognize local time (like a sundial) and measure its passage (like an hourglass). This allows body rhythms to be integrated for concerted action and phased to the local (geophysical) time of day, optimizing the economy of biological systems and allowing for a predictive, rather than purely reactive, homeostatic control.[27] The benefits are adaptive, allowing night-active rodents, for example, to shift their activities to the daytime, either seasonally (as in montane voles during Wyoming winters[41]) or in response to competition for a common habitat (as in golden spiny mice in the Israeli desert[13]). Circadian clocks contribute to the regulation of reproductive rhythms, seasonal behaviors, and celestial navigation and migration. The practical importance of human circadian rhythmicity, as well as its consequences for health and disease, are now being realized.

The circadian system is a unique and powerful model for investigating the cellular and molecular mechanisms that underlie behavioral state control. A natural stimulus (light) can be presented in a controlled fashion, resulting in a long-lasting neural change (in the clock's oscillation) which can be measured as an adaptive behavioral response (a permanent phase shift of overt rhythmicity). Moreover, a neural pacemaker that actually generates circadian oscillations in mammals has been localized to a discrete nucleus within the brain. As a result, there have been remarkable advances in our understanding of circadian timekeeping at multiple levels of biological organization, ranging from intracellular regulatory molecules and gene expression, to intercellular networks and multi-synaptic pathways, and finally to integrated patterns of physiology and behavior.

The chapters in this section highlight current mechanistic studies of circadian timekeeping in mammals (especially rodents) at the cellular and molecular level. These investigations owe much to insights and progress made in other organisms, including bacteria, algae, fungi, plants, mollusks, insects, frogs, and birds. While some of this work is discussed in the chapters (and especially by Hardin and Sehgal), it is mostly outside their scope, and readers are referred to excellent recent reviews.[3,5,7,14,17,23,38,40,46,57]

1.1 Some Introductory Concepts and Definitions

Ordinarily, biological activities are synchronized (entrained) to the natural day/night cycle by environmental light and darkness (Figure 1.1). However, even in aperiodic environments that lack external timing cues, some rhythms continue to oscillate ("free run") with approximately 24-hr (circadian) periods (Figure 1.2). The features of these self-sustaining rhythms have suggested the existence of an endogenous, temperature-compensated, timekeeping mechanism.[1,9,28] The fact that the clock's endogenous period is not exactly 24 hr does not mean that it is imprecise. Rather, this property allows for more stable entrainment by environmental cycles and for organisms to adapt successfully to seasonal changes in day length (photoperiod). At a minimum, the timing system consists of input (afferent) pathways for entrainment to light/dark cycles, a circadian pacemaker that generates the oscillation, and output (efferent) pathways for expression of overt, measurable rhythms. In unicellular organisms, all of these functions are performed within a single cell, but in multicellular organisms, different structures can be distinguished as performing different tasks.

The pacemaker works as a clock, because its endogenous period is accurately entrained to the external 24-hr period, primarily by light-induced phase shifts that reset the pacemaker's oscillation (Figure 1.2). Advances or delays occur because the pacemaker is differentially sensitive to light exposure at different phases of its free-running circadian cycle. This rhythm of light sensitivity can be quantified as a phase-response curve (PRC) by plotting the phase shifts that occur in a measured rhythm when light pulses are applied at different phase points across the free-running circadian cycle in constant darkness (Figure 1.3). Light presented during the early subjective night is interpreted as a late dusk and delays the succeeding rhythm, whereas light exposure during the late subjective night is interpreted as an early dawn and causes a phase advance. Light given at times other than the subjective night has little or no phase-shifting effect. Variations in photic sensitivity, in concert with changes in the pacemaker's endogenous period and the amplitude of its oscillation, can dramatically affect the temporal sequencing of various clock-controlled events.

Although the notion of a linear system with three formal elements (inputs → pacemaker → outputs) has been heuristically useful, the emerging complexity of cellular biochemistry — including networks of signaling and regulatory molecules with extensive crosstalk between metabolic cascades — obscures the definition of functional borders between elements. The existence of nested feedback loops further complicates the conceptual and experimental analysis of this problem. For

Introduction

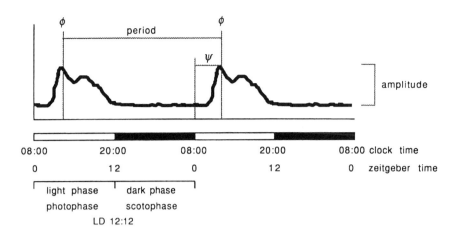

FIGURE 1.1
Description of some commonly used terms in circadian rhythm research. Amplitude: peak to trough difference in a rhythm; for ideal oscillator, mean to peak difference. Period: length of one complete cycle of a rhythm; the reciprocal of frequency. *Zeitgeber:* periodic environmental signal that entrains a rhythm; here, the light-dark (LD) cycle. *Zeitgeber* time (zt): time base under a *zeitgeber* cycle; in LD 12:12, zt 0 = lights-on at dawn, and zt 12 = lights-off at dusk. Circadian time (ct): time base under constant conditions, where 1 circadian hr = τ/24 hr; ct 12 = phase of the free-running cycle that occurs when zt 12 would have occurred on the first day following release from LD 12:12; in nocturnal rodents, it corresponds to the onset of locomotor activity. Subjective day: half cycle from ct 0 to ct 12. Subjective night: half cycle from ct 12 to ct 24 (ct 0). ϕ: phase, or reference point; the instantaneous state of the oscillation within a period. ψ: phase angle (difference) of one rhythm relative to another, especially to a *zeitgeber*. τ: "endogenous" period of a rhythm free-running in constant environmental conditions; t: period of a *zeitgeber* cycle. (Adapted from Schwartz, W.J., *Ann. Neurol.*, 41, 289, 1997.)

example, the pacemaker's photic input may be gated by a rhythm of visual sensitivity in the eyes, and feedback from some of the pacemaker's rhythmic outputs may modulate the behavior of the pacemaker itself. Especially well-studied in hamsters,[30] non-photic stimuli (e.g., benzodiazepines and other pharmacological agents, cage changing, social interactions) generate a PRC essentially opposite (180° out of phase) to the photic PRC, with phase advances during the late subjective day and small phase delays during the subjective night (Figure 1.3).

All of these points highlight some of the problems inherent in experimental investigations of clock function. Apparent arrhythmicity of a population of subjects may be attributed either to a loss of rhythmicity of each subject or to a desynchronization of rhythmicity between subjects because each of the individuals expresses a rhythm that differs in phase or period. Even when an overt rhythm is abolished, the cause might equally be inactivation of the pacemaker or merely the uncoupling of an output pathway from the still-oscillating pacemaker (loss of the clock's "hands" rather than damage to its "gears"). This difficulty emphasizes the possible confound when pacemaker activity is assessed by measurements limited to a single output. On the other hand, alterations in the free-running period of a rhythm must reflect changes in pacemaker behavior, either by a direct action on the pacemaker or by an indirect action via an input pathway.

1.2 The Suprachiasmatic Nucleus as a Circadian Pacemaker

The suprachiasmatic nucleus (SCN) in the anteroventral hypothalamus is a paired nucleus straddling the midline, bordering the third ventricle, and bounded anteroventrally by the optic chiasm. The body of evidence that identifies the SCN as a circadian pacemaker in mammals is so compelling and

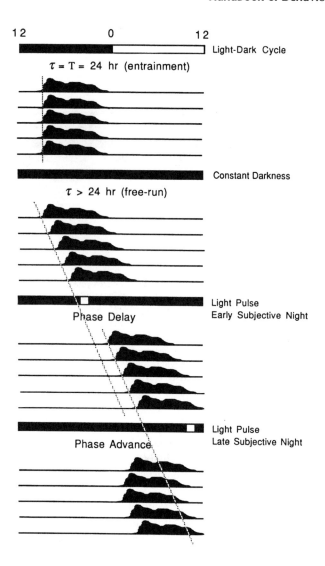

FIGURE 1.2
A nocturnal rhythm depicted as an "actogram". Activity over the course of each 24-hr interval is plotted horizontally from left to right, with succeeding days stacked vertically from top to bottom. In this case, the rhythm in a constant (time-free) environment starts later each day (i.e., the rhythm expresses a free-running circadian period [τ] of greater than 24 hr). The value of τ = the slope of a line fitted to a series of phases, e.g., onsets of activity during free-run (dotted lines). A light pulse administered during the early subjective night delays the succeeding rhythm, whereas a pulse during the late subjective night causes a phase advance (see text).

multidisciplinary in nature that the strength of this functional localization is unsurpassed by that of any other structure in the central nervous system.[18,21] The homologue of the nucleus also appears to play a crucial timekeeping role in a few species of lizards and birds that have been examined.

The functional data have been gathered mostly, but not exclusively, in rats and hamsters. Electrical or pharmacological stimulation of the nucleus causes predictable phase shifts of overt circadian rhythms, whereas destruction of the SCN results in a breakdown of the entrainment or generation of a wide array of such rhythms. More than 75% of the nucleus must be ablated in order to eliminate expressed rhythmicity. No recovery of function is found even after prolonged postoperative survival. Circadian rhythms of single or multiple unit electrical activities in the SCN have

Introduction

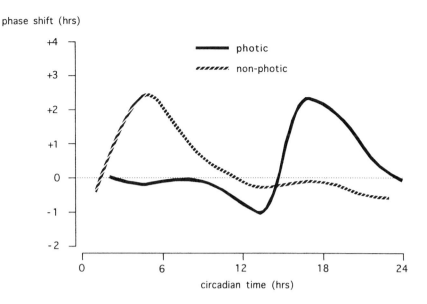

FIGURE 1.3
Phase-response curves (PRCs) to photic and non-photic stimuli in hamsters. The PRC plots the direction and amount of phase shifts against the times that pulses of the stimulus are given. When pulses span the entire free-running circadian cycle, the waveforms of the PRCs follow the lines shown. (Diagram based on Mrosovsky, N., *Biol. Rev.*, 71, 343, 1996.)

been recorded extracellularly *in vivo*[22] and *in vitro* in hypothalamic slices,[11] organotypic slice cultures,[2] and dissociated cell cultures.[56] In nocturnal rodents, the firing rate is high during the subjective light phase (subjective day) and low during the subjective dark phase (subjective night); the same is true in diurnal chipmunks. Similarly, metabolic activity (as measured by the rate of glucose utilization) is high during the subjective day and low during the subjective night in both nocturnal and diurnal animals, and this rhythm also persists *in vitro*.[33] Rhythms of neuroactive peptides synthesized in the SCN have been measured in the cerebrospinal fluid (CSF) and in tissue punches and sections at the mRNA and protein levels.[15] The most intensively studied of these molecular rhythms is a circadian rhythm of the levels of the neuropeptide arginine vasopressin.[37] CSF levels are high during the subjective day and low during the subjective night in both nocturnal and diurnal animals, and the rhythm persists *in vitro* in hypothalamic explants[8] and slices,[12] organotypic slice cultures,[47,50] and dissociated cell cultures.[31,55] Finally, neural grafts of fetal SCN tissue re-establish overt rhythmicity in arrhythmic, SCN-lesioned recipients, and the rhythms restored by the transplants display properties that are characteristic of the circadian pacemakers of the donors rather than those of the hosts.[34]

In order to develop these findings into a more complete understanding of the cellular and molecular neurobiology of the SCN, a number of laboratories have begun by analyzing the detailed anatomy of the nucleus and its connections. In Chapter 2, Card comprehensively reviews current information on the structure, neurochemistry, and connectivity of the circadian timekeeping system in mammals.

1.3 Analyzing Pacemaker Input and Output Pathways

One strategy for identifying elements of the circadian system is to trace the cascade of events that comprise the pacemaker's entrainment pathways; ultimately these pathways must converge and terminate on components of the oscillatory machinery in order to cause phase shifts of overt rhythmicity. The two families of PRCs (Figure 1.3) — the photic (light-type) and the non-photic (dark-type, for its similarity to the PRC generated by pulses of darkness administered to animals in constant light) — provide a formal framework for investigating their underlying cellular and molecular substrates.[49] In Chapter 3, Piggins and Rusak critically discuss the neurochemical interactions that may account for the shape of these PRCs and the physiology of entrainment.

The "circadian" visual system is anatomically and physiologically distinct from the visual systems responsible for reflex oculomotor function and image formation.[29] A direct retinohypothalamic tract monosynaptically innervates SCN neurons, and it appears to include a specialized photoreceptive mechanism[32] that may rely on green-sensitive cones,[16] a subset of ganglion cells,[26] and a population of SCN neurons that function as "luminance" detectors[20] (with electrophysiological responses to light that are sustained, proportional to light intensity, and elicited from large receptive fields lacking a retinotopic organization). An additional special feature in the SCN is the light-induced expression of immediate early genes that encode for transcription factors; there is now suggestive (although mostly circumstantial) evidence that such factors (especially c-Fos) are involved in the photic input pathway.[43] All of these unique properties of the circadian visual system are likely to account for the preserved circadian responses of some apparently "sightless" animals, including the blind mole rat,[53] retinally degenerate mouse mutant *rd/rd*,[10] and even some blind people who lack pupillary light reflexes and conscious light perception.[6]

Relatively less is known about the organization of SCN output pathways, which govern a program of daily rhythms exhibiting a range of waveforms and phases different from the central oscillation that drives them. The circadian pacemaker might orchestrate this via metabolic cascades, in which the formation of one control factor regulates the formation of a second, and so on. The circadian signal would undergo even further modification as the pacemaker's outputs are coupled to effector cells by multiple mechanisms (e.g., synaptic transmission, hormonal secretion, and indirectly through the rhythmic regulation of behavior). The fidelity of the circadian signal might be altered by such multistep output pathways, so overt rhythms could differ substantially from their underlying cellular and molecular oscillations.

An alternative possibility is that the multiplicity of overt rhythms reflects several circadian pacemakers, even within single cells[39] or outside the SCN,[51] each of which controls only one rhythm. In this view, the pacemaker and its "hand" would be integrally related, like the observed oscillation of a feedback circuit. In nonmammalian vertebrates, there is clear evidence for circadian oscillators in the retina[5] and pineal gland.[52] In mammals, the pineal is driven by the SCN and appears to have lost its capacity to function as an independent circadian oscillator. However, mammals do seem to have multiple oscillators, given the recent demonstration that hamster retinas *in vitro* exhibit circadian rhythms of melatonin synthesis.[51] The role of this ocular pacemaker in mammalian circadian organization is unclear (in mammals, circulating melatonin levels are derived from the pineal).

In any case, it is obvious that clock regulation of physiology and behavior must rely in some way on changes in cellular biochemistry. There are many ways by which enzymatic activities might be altered, ranging from dramatic changes in absolute levels to subtle changes in structure. Many of the control mechanisms studied thus far lie at the transcriptional level, stimulating research to identify the *cis*-acting elements and *trans*-acting factors that mediate circadian effects on the promoters of regulated genes. The circadian rhythm of the enzymatic activity of pineal *N*-acetyltransferase is probably both the most robust and, with the recent cloning of the responsible gene, the best understood biochemical rhythm in mammals. In Chapter 4, Klein and his associates clearly show why pineal melatonin has become a paradigm of clock-controlled hormonal regulation.

1.4 Assembling a Multicellular Circadian Pacemaker

The cellular and molecular basis of the actual oscillatory mechanism of the circadian pacemaker in the SCN is unknown. Whether circadian rhythmicity is a property of individual cells or instead emerges from an intercellular (network) interaction was, until recently, the focus of much debate. Accumulating evidence that SCN cells continue to oscillate after dissociation and culture[31,48,55] was elegantly verified using a system that simultaneously monitors the neuronal firing rates of multiple individual dispersed cells cultured on fixed microelectrode arrays.[56] Single cells dissociated from neonatal SCN show circadian firing rhythms with widely varying phases, in part because different cells express independent circadian periods. The genotype-specific, free-running period characteristically expressed by whole animals may represent the mean period arising from the coupling of these multiple SCN cellular oscillators.[19] These data raise two fundamental questions that remain unanswered. First, what are the mechanism(s) for intercellular communication that synchronize the disparate activities of individual cells? Second, what is the molecular nature of the intracellular circadian oscillation that actually keeps biological time?

In their chapter, Piggins and Rusak consider possible intra-SCN coupling mechanisms. Interest has focused on gamma-aminobutyric acid (GABA), particularly given the recent report[54] that bath application of GABA to rat SCN slices *in vitro* inhibited neuronal firing rate during the subjective night, while its application was excitatory during the subjective day. Both effects were mediated by a $GABA_A$ receptor; the change in sign likely followed a rhythm of the GABA equilibrium potential — positive relative to the resting membrane potential during the day, negative during the night — reflecting a circadian oscillation of intracellular chloride concentrations. Such a switching mechanism could serve a feedback function to amplify SCN activity during the day and suppress it at night. There is also evidence to suggest that synchronization of SCN cells can occur via Ca^{2+}-independent non-synaptic mechanisms.[4] Timekeeping persists when Na^+-dependent action potentials are blocked by tetrodotoxin[44,56] or the hypothermia of hibernation,[24] and in the fetal SCN before most synapses form.[36]

Current ideas about the intracellular circadian oscillatory mechanism come mainly from initial molecular genetic studies of single-gene mutants in fruit flies (*Drosophila*) and fungi (*Neurospora*). This work has led to the identification of several molecules essential to circadian clock function and to the general view that the pacemaker's core consists of autoregulatory feedback loops with oscillating levels of nuclear proteins negatively regulating the transcription of their own mRNAs. In Chapter 5, Hardin and Sehgal astutely dissect this remarkable story, primarily as developed in the *Drosophila* model.

Circadian rhythmicity in mammals is also genetically determined.[45] A spontaneous single-gene clock mutation was discovered nearly 10 years ago in the hamster (*tau*),[35] but the lack of an adequate genetic map in this species had prevented further molecular analysis. The situation is far different today, with explosive advances in the identification of putative clock genes in mammals. In the final chapter of this section, Wilsbacher, Wisor, and Takahashi provide a lucid update on the emerging genetic and molecular study of these circadian rhythm genes. Common features of these genes in *Neurospora*, *Drosophila*, and mammals hint that a structural element of circadian clocks may have been conserved for the last 900 million years. Whether this observation speaks to the evolutionary origin of all clocks or to common molecular mechanisms will be elucidated by further comparative studies. We already know that the likely molecular substrate for light's phase-shifting action is different in *Neurospora* and *Drosophila*. Future molecules undoubtedly await discovery, including those functioning to close the transcription-translation feedback loop, build the loop, or lubricate it.

All the chapters in this section amply demonstrate why the study of biological timekeeping is now at such a fertile and exciting stage, involving diverse methodologies and uniting investigators from a wide array of disciplines.

Acknowledgment

W.J.S. is supported by NINDS RO1 NS24542.

References

1. Aschoff, J., Ed., *Handbook of Behavioral Neurobiology*. Vol. 4. *Biological Rhythms,* Plenum Press, New York, 1981.
2. Belenky, M., Wagner, S., Yarom, Y., Matzner, H., Cohen, S., and Castel, M., The suprachiasmatic nucleus in stationary organotypic culture, *Neuroscience,* 70, 127, 1996.
3. Block, G.D., Khalsa, S.B.S., McMahon, D.G., Michel, S., and Guesz, M., Biological clocks in the retina: cellular mechanisms of biological timekeeping, *Int. Rev. Cytol.,* 146, 83, 1993.
4. Bouskila, Y. and Dudek, F.E., Neuronal synchronization without calcium-dependent synaptic transmission in the hypothalamus, *Proc. Natl. Acad. Sci. USA,* 90, 3207, 1993.
5. Cahill, G.M. and Besharse, J.C., Circadian rhythmicity in vertebrate retinas: regulation by a photoreceptor oscillator, *Prog. Retinal Eye Res.,* 14, 267, 1995.
6. Czeisler, C.A., Shanahan, T.L., Klerman, E.B., Martens, H., Brotman, D.J., Emens, J.S., Klein, T., and Rizzo, III, J.F., Suppression of melatonin secretion in some blind patients by exposure to bright light, *New Engl. J. Med.,* 332, 6, 1995.
7. Dunlap, J.C., Genetic and molecular analysis of circadian rhythms, *Ann. Rev. Genet.,* 30, 579, 1996.
8. Earnest, D.J. and Sladek, C.D., Circadian rhythms of vasopressin release from individual rat suprachiasmatic explants *in vitro*, *Brain Res.,* 382, 129, 1986.
9. Edmunds, Jr, L.N., *Cellular and Molecular Bases of Biological Clocks: Models and Mechanisms for Circadian Timekeeping,* Springer, New York, 1988.
10. Foster, R.G., Provencio, I., Hudson, D., Fiske, S., DeGrip, W., and Menaker, M., Circadian photoreception in the retinally degenerate mouse (*rd/rd*), *J. Comp. Physiol. A,* 169, 39, 1991.
11. Gillette, M.U., Medanic, M., McArthur, A.J., Liu, C., Ding, J.M., Faiman, L.E., Weber, E.T., Tcheng, T.K., and Gallman, E.A., Intrinsic neuronal rhythms in the suprachiasmatic nuclei and their adjustment, in *Circadian Clocks and Their Adjustment, Ciba Foundation Symposium 183,* Chadwick, D.J. and Ackrill, K., Eds., John Wiley & Sons, Chichester, 1995, 134.
12. Gillette, M.U. and Reppert, S.M., The hypothalamic suprachiasmatic nuclei: circadian patterns of vasopressin secretion and neuronal activity *in vitro*, *Brain Res. Bull.,* 19, 135, 1987.
13. Haim, A. and Rozenfeld, F.M., Temporal segregation in coexisting *Acomys* species: the role of odour, *Physiol. Behav.,* 54, 1159, 1993.
14. Hall, J.C., Tripping along the trail to the molecular mechanisms of biological clocks, *Trends Neurosci.,* 18, 230, 1995.
15. Inouye, S.I.T. and Shibata, S., Neurochemical organization of circadian rhythm in the suprachiasmatic nucleus, *Neurosci. Res.,* 20, 109, 1994.
16. Jiménez, A.J., García-Fernández, J.M., González, B., and Foster, R.G., The spatio temporal pattern of photoreceptor degeneration in the aged *rd/rd* mouse retina, *Cell Tissue Res.,* 284, 193, 1996.
17. Johnson, C.H., Golden, S.S., Ishiura, M., and Kondo, T., Circadian rhythms in prokaryotes, *Molec. Microbiol.,* 21, 5, 1996.
18. Klein, D., Moore, R.Y., and Reppert, S.M., Eds., *Suprachiasmatic Nucleus: The Mind's Clock,* Oxford University Press, New York, 1991.
19. Liu, C., Weaver, D.R., Strogatz, S.H., and Reppert, S.M., Cellular construction of a circadian clock: period determination in the suprachiasmatic nuclei, *Cell,* 91, 855, 1997.
20. Meijer, J.H., Groos, G.A., and Rusak, B., Luminance coding in a circadian pacemaker: the suprachiasmatic nucleus of the rat and the hamster, *Brain Res.,* 382, 109, 1986.
21. Meijer, J.H. and Rietveld, W.J., Neurophysiology of the suprachiasmatic circadian pacemaker in rodents, *Physiol. Rev.,* 69, 671, 1989.

Introduction

22. Meijer, J.H., Schaap, J., Watanabe, K., and Albus, H., Multiunit activity recordings in the suprachiasmatic nuclei: *in vivo* versus *in vitro* models, *Brain Res.,* 753, 322, 1997.
23. Millar, A.J. and Kay, S.A., The genetics of phototransduction and circadian rhythms in Arabidopsis, *BioEssays,* 19, 209, 1997.
24. Miller, J.D., Cao, V.H., and Heller, C., Thermal effects on neuronal activity in suprachiasmatic nuclei of hibernators and nonhibernators, *Am. J. Physiol.,* 266, R1259, 1994.
25. Mitsui, A., Kumazawa, S., Takahashi, A., Ikemoto, H., and Arai, T., Strategy by which nitrogen-fixing unicellular cyanobacteria grow photoautotrophically, *Nature,* 323, 720, 1986.
26. Moore, R.Y., Speh, J.C., and Card, J.P., The retinohypothalamic tract originates from a distinct subset of retinal ganglion cells, *J. Comp. Neurol.,* 352, 351, 1995.
27. Moore-Ede, M.C., Physiology of the circadian timing system: predictive versus reactive homeostasis, *Am. J. Physiol.,* 250, R735, 1986.
28. Moore-Ede, M.C., Sulzman, F.M., and Fuller, C.A., *The Clocks That Time Us,* Harvard University Press, Cambridge, 1982.
29. Morin, L.P., The circadian visual system, *Brain Res. Rev.,* 67, 102, 1994.
30. Mrosovsky, N., Locomotor activity and non-photic influences on circadian clocks, *Biol. Rev.,* 71, 343, 1996.
31. Murakami, N., Takamure, M., Takahashi, K., Utunomiya, K., Kuroda, H., and Etoh, T., Long-term cultured neurons from rat suprachiasmatic nucleus retain the capacity for circadian oscillation of vasopressin release, *Brain Res.,* 545, 347, 1991.
32. Nelson, D.E. and Takahashi, J.S., Sensitivity and integration in a visual pathway for circadian entrainment in the hamster (*Mesocricetus auratus*), *J. Physiol.,* 439, 115, 1991.
33. Newman, G.C., Hospod, F.E., Patlak, C.S., and Moore, R.Y., Analysis of *in vitro* glucose utilization in a circadian pacemaker model, *J. Neurosci.,* 12, 2015, 1992.
34. Ralph, M.R. and Lehman, M.N., Transplantation: a new tool in the analysis of the mammalian circadian pacemaker, *Trends Neurosci.,* 14, 362, 1991.
35. Ralph, M.R. and Menaker, M., A mutation of the circadian system in golden hamsters, *Science,* 241, 1225, 1988.
36. Reppert, S.M. and Schwartz, W.J., The suprachiasmatic nuclei of the fetal rat: characterization of a functional circadian clock using ^{14}C-labeled deoxyglucose, *J. Neurosci.,* 4, 1677, 1984.
37. Reppert, S.M., Schwartz, W.J., and Uhl, G.R., Arginine vasopressin: a novel peptide rhythm in cerebrospinal fluid, *Trends Neurosci.,* 10, 76, 1987.
38. Roenneberg, T. and Mittag, M., The circadian program of algae, *Sem. Cell Dev. Biol.,* 7, 753, 1996.
39. Roenneberg, T. and Morse, D., Two circadian oscillators in one cell, *Nature,* 362, 362, 1993.
40. Rosbash, M., Molecular control of circadian rhythms, *Curr. Opin. Genet. Dev.,* 5, 662, 1995.
41. Rowsemitt, C.N., Petterborg, L.J., Claypool, L.E., Hoppensteadt, F.C., Negus, N.C., and Berger, P.J., Photoperiodic induction of diurnal locomotor activity in *Microtus montanus,* the montane vole, *Can. J. Zool.,* 60, 2798, 1982.
42. Schwartz, W.J., Understanding circadian clocks: from c-Fos to fly balls, *Ann. Neurol.,* 41, 289, 1997.
43. Schwartz, W.J., Aronin, N., Takeuchi, J., Bennett, M.R., and Peters, R.V., Towards a molecular biology of the suprachiasmatic nucleus: photic and temporal regulation of c-*fos* gene expression, *Sem. Neurosci.,* 7, 53, 1995.
44. Schwartz, W.J., Gross, R.A., and Morton, M.T., The suprachiasmatic nuclei contain a tetrodotoxin-resistant circadian pacemaker, *Proc. Natl. Acad. Sci. USA,* 84, 1694, 1987.
45. Schwartz, W.J. and Zimmerman, P., Circadian timekeeping in BALB/c and C57BL/6 inbred mouse strains, *J. Neurosci.,* 10, 3685, 1990.

46. Sehgal, A., Ousley, A., and Hunter-Ensor, M., Control of circadian rhythms by a two-component clock, *Molec. Cell Neurosci.,* 7, 165, 1996.
47. Shinohara, K., Honma, S., Katsuno, Y., Abe, H., and Honma, K., Two distinct oscillators in the rat suprachiasmatic nucleus *in vitro, Proc. Natl. Acad. Sci. USA,* 92, 7396, 1995.
48. Silver, R., Lehman, M.N., Gibson, M., Gladstone,W.R., and Bittman, E.L., Dispersed cell suspensions of fetal SCN restore circadian rhythmicity in SCN-lesioned adult hamsters, *Brain Res.,* 525, 45, 1990.
49. Smith, R.D., Turek, F.W., and Takahashi, J.S., Two families of phase-response curves characterize the resetting of the hamster circadian clock, *Am. J. Physiol.,* 262, R1149, 1992.
50. Tominaga, K., Inouye, S.T., and Okamura, H., Organotypic slice culture of the rat suprachiasmatic nucleus: sustenance of cellular architecture and circadian rhythm, *Neuroscience,* 59, 1025, 1994.
51. Tosini, G. and Menaker, M., Circadian rhythms in cultured mammalian retina, *Science,* 272, 419, 1996.
52. Underwood, H. and Goldman, B.D., Vertebrate circadian and photoperiodic systems: role of the pineal gland and melatonin, *J. Biol. Rhythms,* 2, 279, 1987.
53. Vuillez, P., Herbin, M., Cooper, H.M., Nevo, E., and Pevet, P., Photic induction of Fos immunoreactivity in the suprachiasmatic nuclei of the blind mole rat (*Spalax ehrenbergi*), *Brain Res.,* 654, 81, 1994.
54. Wagner, S., Castel, M., Gainer, H., and Yarom, Y., GABA in the mammalian suprachiasmatic nucleus and its role in diurnal rhythmicity, *Nature,* 387, 598, 1997.
55. Watanabe, K., Koibuchi, N., Ohtake, H., and Yamaoka, S., Circadian rhythms of vasopressin release in primary cultures of rat suprachiasmatic nucleus, *Brain Res.,* 624, 115, 1993.
56. Welsh, D.K., Logothetis, D.E., Meister, M., and Reppert, S.M., Individual neurons dissociated from rat suprachiasmatic nucleus express independently phased circadian firing rhythms, *Neuron,* 14, 697, 1995.
57. Zatz, M., Melatonin rhythms: trekking toward the heart of darkness in the chick pineal, *Sem. Cell Dev. Biol.,* 7, 811, 1996.

Chapter 2

Anatomy of the Mammalian Circadian Timekeeping System

J. Patrick Card

Contents

2.1 Introduction 13
2.2 The Organization of the SCN 14
    2.2.1 Cytoarchitecture 14
    2.2.2 Neurochemical Organization 14
2.3 SCN Afferents 18
    2.3.1 The Retinohypothalamic Tract 18
    2.3.2 The Geniculohypothalamic 20
    2.3.3 Serotoninergic Afferents 20
    2.3.4 Other Afferents 20
2.4 SCN Efferents 22
    2.4.1 Preoptic Area 22
    2.4.2 The Subparaventricular Zone and the Paraventricular Nucleus 22
    2.4.3 Dorsomedial Nucleus 24
    2.4.4 Commissural Connections 24
    2.4.5 Other Efferents 24
2.5 Summary 24
References 25

2.1 Introduction

The pioneering studies that demonstrated a retinal projection to the suprachiasmatic nuclei (SCN) of the hypothalamus[42] introduced an era of anatomical and functional analysis that has firmly established these nuclei as the biological clock of the mammalian central nervous system. Creative anatomical analyses have provided considerable insight into the functional organization of this hypothalamic nucleus, as well as the means through which it imposes its temporal influences upon other systems. Indeed, the genetically determined rhythmic alterations in metabolic activity of SCN neurons have provided an important morphological index of the functional activity of SCN neurons.[56,58] Anatomical studies have also proven to be integral to the functional dissection of the SCN and have placed it within the larger context of circuitry that has come to be known as the circadian timing system (CTS). The purpose of this chapter is to provide an overview of this system.

2.2 The Organization of the SCN

The SCN have been subjected to rigorous morphological analysis in a variety of species. The most thorough characterizations have been conducted in rat and hamster, but common features have emerged from studies of a number of species, including the human. In essence, two major subdivisions of the nucleus have been defined on the basis of morphology, neurochemical phenotype, and the terminal arborization of afferents. The following sections consider that data along with recent information demonstrating functional segregation of neurons that give rise to the efferent projection of the nucleus.

2.2.1 Cytoarchitecture

At their intermediate axes, the rat SCN form distinct spherical cell groups embedded in the dorsal surface of the optic chiasm, on either side of the third ventricle (Figure 2.1). Small, densely packed bipolar neurons constitute the dorsomedial subdivision (dmSCN), while more widely dispersed multipolar neurons form the ventrolateral subfield (vmSCN). Each nucleus is approximately 1 mm in length, has a widest diameter of .5 mm, and contains approximately 10,000 neurons.

Numerous investigations employing a variety of morphological methods have provided evidence in support of the above differentiations. One of the earliest and most thorough characterizations of the intrinsic anatomy of the SCN was published by van den Pol in 1980 (see Reference 70 for a review). On the basis of Golgi impregnations and electron microscopy, he defined a number of subclasses of SCN neurons that are differentially distributed within SCN subdivisions. He also examined the synaptology of the nucleus and defined a number of mechanisms through which SCN neurons integrate information and communicate with other regions of the neuraxis. Chief among the observations derived from that study are that local synaptic communication between SCN neurons occur largely via axo-dendritic and dendro-dendritc synapses, that commissural connections exist between the two nuclei, and that subpopulations of SCN neurons can be defined by their synaptic targets. These observations provided keen insights into the cellular interactions of the SCN and also provided a strong foundation for subsequent analysis of the functional organization of the nuclei.

2.2.2 Neurochemical Organization of the SCN

Characterization of the phenotypic diversity of SCN neurons has dramatically advanced our understanding of the functional organization of these nuclei. The earliest immunohistochemical studies of peptidergic neurons in the SCN provided evidence of dmSCN and vlSCN subdivisions and also revealed neurochemical identities for neurons that have subsequently been used to probe the functional activity of the nuclei. There is now substantial evidence that peptides are differentially expressed in SCN neurons, that many of these peptidergic systems are sequestered within distinct subfields of the nucleus, and that peptide levels vary across the circadian cycle or in response to photic challenges during sensitive periods.

Localization of vasopressin (VP) and its carrier protein neurophysin provided the first demonstration that dmSCN neurons are distinguished by their peptide content (see Reference 70 for a

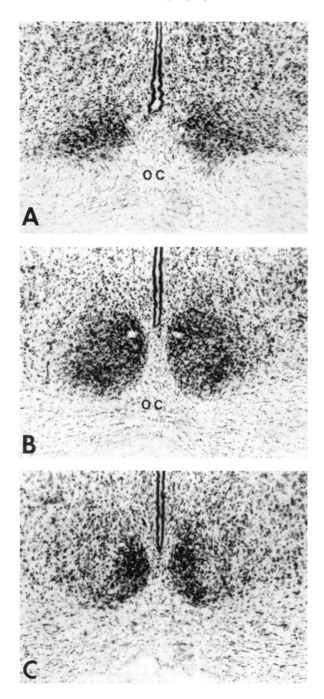

FIGURE 2.1
Cresyl violet-stained coronal sections through the rostral (**A**), intermediate (**B**), and caudal (**C**) extent of the rat SCN are illustrated. The SCN form compact cell masses that are embedded in the optic chiasm (oc) on either side of the third ventricle. At rostral levels, the nuclei present as homogeneous cell masses. At intermediate and caudal levels, differences in the cytoarchitecture of the nucleus define dorsomedial (dmSCN) and ventrolateral (vlSCN) subdivisions. The dmSCN is characterized by small, densely packed bipolar neurons, while the the vlSCN contains slightly larger cells that are more widely dispersed. The dispersed distribution of neurons in the vlSCN makes it difficult to define the lateral borders of the nuclei.

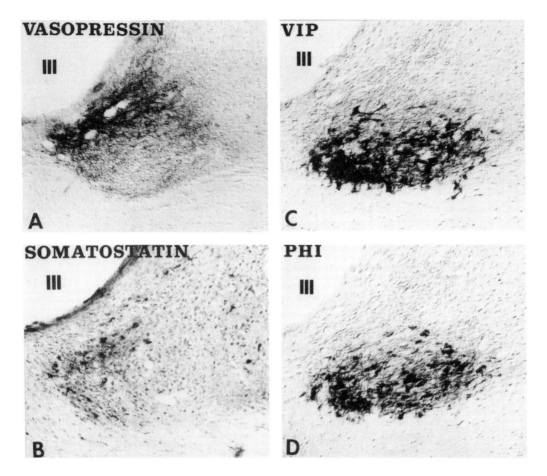

FIGURE 2.2
The differential distribution of peptidergic populations of neurons in the rat SCN are illustrated. The dmSCN contains a dense concentration of vasopressinergic neurons (**A**) and a smaller group of somatostatin-containing cells (**B**) that are most prominent along the border of the two subdivisions. The ventrolateral subdivision is dominated by neurons that contain vasoactive intestinal polypeptide (VIP; **C**) and an alternatively spliced transcript of the VIP precursor known as peptide histidine isoleucine (PHI; **D**). III = third ventricle.

review). It is now well established that VP is the principal peptide in this SCN subfield (Figure 2.2A) and that the levels of VP and its mRNA fluctuate across the circadian cycle.[20,57] The circadian fluctuation of VP is independent of photic input and exhibits highest levels during the light phase of the photoperiod. The majority of axons that arise from these cells project out of the nucleus, but synaptic contacts with dmSCN neurons have been demonstrated, and there is a small plexus of fibers in the vlSCN. A small group of somatostatin-containing neurons is also present in the dmSCN.[16] They are preferentially concentrated at the border of the dm- and vlSCN (Figure 2.2B), are distinct from other peptidergic populations in the nucleus,[66] and synapse largely within the SCN.[13] Like VP, somatostatin and its mRNA exhibit a circadian fluctuation that peaks during the subjective day.[59] Angiotensin II[74] and VGF, an NGF-inducible gene product,[70] are also prevalent in the dmSCN, but little information is available regarding the metabolism or projections of these neurons.

Vasoactive intestinal polypeptide (VIP) was the first peptide to be localized exclusively within the vlSCN.[8] These neurons fill the vlSCN, extend into the underlying optic chiasm, and co-localize an alternately spliced transcript of the VIP precursor protein known as peptide-histidine isoleucine (PHI) (Figures 2.2C,D). Several studies have shown that the levels of VIP/PHI mRNA vary across the circadian cycle, with the highest levels occurring during the dark phase of the photoperiod in the rat;[1] however, unlike VP, the fluctuation of VIP/PHI is dependent upon photic input.[65] Neurons containing corticotropin-releasing hormone (CRH) have also been reported in vlSCN, and Daikoku and coworkers identified CRH-containing cells that colocalize VIP/PHI (see Reference 74 for a review). A substantial number of vlSCN neurons also contain gastrin releasing peptide (GRP) (see Reference 70 for a review). GRP mRNA is known to colocalize with VIP/PHI, and co-injection of these peptides into the region of the SCN alters the functional activity of SCN neurons.[2] Romijn and colleagues[55] have presented evidence that these peptides are differentially regulated and suggest that there are at least three phenotypically distinct groups of cells in the vlSCN: those that contain PHI or GRP alone and those that contain VIP/PHI.

Several studies have reported species and strain differences in the phenotype of SCN neurons. For example, Wollnik and Bihler[76] reported differences in the number of VP neurons and the area of VP- and NPY-containing fibers in three strains of rat. The Brattleboro rat harbors a genetic defect that prevents expression of VP in the SCN,[48] and the mink SCN does not contain detectable VP.[29] The recent localization of calcium-binding proteins in SCN neurons has also revealed interesting differences in the phenotypic organization of the SCN. Jacobowitz and Winsky[25] reported a population of calretinin-immunoreactive neurons in rat that define the dorsolateral portion of the nucleus. Only scattered calretinin-containing neurons are found in the hamster SCN, but Silver and colleagues[60] have identified vlSCN neurons that contain calbindin immunoreactivity and exhibit light-induced Fos expression.

The human SCN also contains subdivisions comparable to those identified in lower species. In an early investigation, Stopa and colleagues[61] demonstrated segregation of VP- and VIP-containing neurons in dorsal and ventral parts of the human SCN. This observation has been confirmed and expanded in a number of subsequent studies (see References 12, 35, 41, and 63 for recent reviews). However, the organization of these cell groups differs from that of lower species, and there are additional phenotypic differences in the cells that make up the human SCN. Most notable in this respect are prominent populations of NPY- and neurotensin-containing neurons in the human SCN. A large group of NPY-immunoreactive neurons co-extensive with the VIP-containing population has been defined in the human; these cells are absent in lower species.[41] Similarly, large numbers of neurotensin-containing neurons are present in the human SCN and extend throughout both subfields rather than being confined to the vlSCN.[35]

In recent years, it has become increasingly clear that GABA is the principal small molecule neurotransmitter in the SCN. Localization of glutamic acid decarboxylase (GAD) in early studies suggested that GABA neurons are widely distributed in the SCN, and subsequent analyses have shown that GABA is present in essentially all SCN neurons (see Reference 45 for a review). Using dual labeling ultrastructural immunocytochemical localizations, Buijs and colleagues[7] demonstrated that locally arborizing axons arising from these neurons synapse upon neurons in both the ipsilateral and contralateral nucleus. However, data from that analysis also revealed that only 20 to 30% of SCN terminals contain immunohistochemically detectable levels of GABA. In discussing the potential explanation for the latter observation, Buijs and colleagues entertained the hypothesis that terminal levels of GABA fluctuate throughout the circadian cycle. This possibility is supported by the recent demonstration of differential regulation of the two isoforms of GAD mRNA in the rat dm- and vlSCN.[24] This analysis demonstrated a circadian fluctuation of GAD_{65} mRNA that peaks during the light phase of the photoperiod and is more pronounced in the dmSCN. Levels of GAD_{67} were lower and did not exhibit a statistically significant circadian variation. It remains to be determined whether this rhythm is endogenously generated or stimulated by light. However, considered with data demonstrating both GAD isoforms in human SCN,[19] these data support the conclusion that GABA is the major small molecule neurotransmitter in the mammalian SCN.

There are a number of recent observations suggesting that nitric oxide plays a role in the function of the SCN. These grew out of early observations by Amir[3] demonstrating that light-induced stimulation of heart rate in dark-adapted rats could be blocked by infusion of nitric oxide (NO) and cGMP blockers into the area of the SCN. Subsequent studies have identified subpopulations of neurons in the SCN that contain the enzymes necessary for the generation of NO. Decker and Reuss[14] identified scattered neurons in the vlSCN of hamster that colocalize NADPH-diaphorase and the neuronal isoform of nitric oxide synthase (NOS). This observation was confirmed in rat SCN, and it was further shown that NOS is found within a subset of the VIP neurons.[53] Light and electron microscopic analyses have shown that these cells elaborate a dense plexus of fibers within the vlSCN.[11,73] These and other observations support a role for nitric oxide in mediating the effects of light in the SCN.

2.3 SCN Afferents

2.3.1 The Retinohypothalamic Tract

Certainly, the best characterized projection to the SCN, both from an anatomical and functional standpoint, arises from the retina. As noted previously, demonstration of the retinohypothalamic projection in 1972 ultimately led to identification of the SCN as the circadian pacemaker, and it is now known that the projection is essential for entrainment of the circadian activity of SCN neurons.[42] Not surprisingly, considerable effort has been expended upon the characterization of this projection, and quite a bit is known about its organization, synaptology, and neurochemical phenotype in a variety of species. There is compelling evidence that the projection is glutamatergic (see Reference 17 and Chapter 3 in this volume for reviews). One of the characteristic features of the terminal arbor is that it terminates bilaterally within the vlSCN. This was apparent in the early autoradiographic studies and has been repeatedly confirmed with modern anterograde tracers. Two recent advances in tract-tracing technology have provided evidence in support of functional parcellation of the RHT. First, use of more specific anterograde tracers such as horseradish peroxidase[50] and the β-subunit of cholera toxin (CT)[26,34] has revealed that the RHT is far more extensive than previously appreciated. The increased sensitivity of CT has proven to be particularly effective in providing further insight into the extent of the terminal arbor in the SCN and has also revealed projections into other regions of hypothalamus, supporting division of the RHT into medial and lateral components. Second, the use of neurotropic alpha herpesviruses for transneuronal analysis has provided new insights into both the origin and organization of the RHT. In 1991, we identified a strain of a swine alpha herpesvirus that exhibits a preferential affinity for functionally distinct components of the visual system.[10] After intravitreal injection, this virus replicates in retinal ganglion cells (Figure 2.3A), is transported anterogradely through the centrally projecting axons of these cells, and then passes transynaptically to infect the synaptic targets of the retinal afferents (Figures 2.3B,C). Examination of retinorecipient regions of the neuraxis revealed a selective infection of the SCN and intergeniculate leaflet of thalamus along with the pretectum and a subset of the accessory optic nuclei. We subsequently defined the molecular basis for the selective tropism of this virus and used it to demonstrate that the retinal projection to the SCN arises from a subpopulation of retinal ganglion cells that correspond to type III, or W, cells.[46] Other studies have also exploited this approach to demonstrate that retinal projections to medial and lateral hypothalamus arise from different populations of retinal ganglion cells.[31,33] Collectively, these and other data support the functional division of the RHT into subdivisions that subserve both circadian and photoperiodic functions.

Guldner and colleagues[21,22] have conducted extensive analyses of the synaptology and morphological plasticity of the RHT projection to the SCN. A complete consideration of this data is beyond

Anatomy of the Mammalian Circadian Timekeeping System

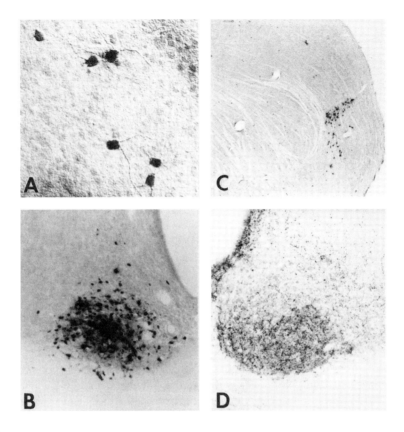

FIGURE 2.3
The origin and organization of visual inputs to the rat SCN are illustrated. Figures (A) and (B) illustrate neurons in the retina and vlSCN that are infected with a neurotropic virus that has previously been shown to have a differential affinity for the portion of the visual system involved in photic modulation of circadian function. After intravitreal infection of the virus, a distinct subset of retinal ganglion cells that conform to type III or W cells are infected by the virus (**A**). The virus replicates within these retinal ganglion cells, is transported anterogradely through their centrally projecting axons, and passes transynaptically to infect retinorecipient cells in the ventrolateral subdivision of the nucleus (**B**). The virus also infects a secondary visual projection arising from the intergeniculate leaflet of the thalamus (**C**) terminating in the region of the SCN that is coextensive with the retinorecipient neurons (**D**). This projection, known as the geniculohypothalamic tract, contains neuropeptide Y.

the scope of this short review but can be found in recent comprehensive presentations.[21,22] However, it is important to note that these studies have shown that retinal boutons exhibit a unique morphology distinguished by "lucent" mitochondria, that the boutons preferentially synapse upon the dendrites and spines of SCN neurons, and that terminals often form "complex synaptic arrangements" in which a single bouton synapses upon multiple profiles. The morphometric studies that have documented these findings have also demonstrated that optic synapses in the SCN exhibit morphological characteristics consistent with both excitatory and inhibitory synapses, with Gray type I (excitatory) synapses predominating (see Reference 22 for review). Interestingly, structural plasticity in these boutons has been demonstrated in animals exposed to different lighting regimes, leading Guldner and colleagues[22] to propose that long-term activity or disuse can alter the sign of *some* optic synapses. Electrophysiological evidence for both excitatory and inhibitory responses of SCN neurons to light or optic nerve stimulation has also been reported (see References 36 and 62 for recent reviews), but whether the inhibitory responses are the direct consequence of optic inputs or are due to local circuit GABAergic connections in the SCN remains a point of debate. Nevertheless, it is clear that there is considerable structural plasticity in SCN optic synapses, and it seems probable that these adaptive changes in morphology reflect a dynamic role for the RHT in SCN function.

2.3.2 The Geniculohypothalamic Tract

A projection from the lateral geniculate complex to the SCN was apparent in the early autoradiographic analyses of geniculate efferents and subsequently shown to arise from a distinct lamina of neurons interposed between the dorsal and ventral geniculate nuclei known as the intergeniculate leaflet (IGL; Figure 2.3C).[44] It is now well established that the IGL gives rise to a dense neuropeptide Y (NPY)-containing projection that terminates densely and bilaterally within the vlSCN (Figure 2.3D).[44] Functional analysis has implicated this projection in non-photic entrainment of the SCN (see References 40 and 44 and Chapter 3 in this volume for reviews).

Several aspects of the organization and connections of the IGL have proven to be informative regarding its function in the CTS. In particular, Pickard[49] demonstrated that at least a portion of the retinal afferents that innervate the IGL are collaterals of fibers that also innervate the SCN. These optic afferents form complex synaptic glomeruli in the IGL that are analogous to those demonstrated in the SCN[44] and are known to synapse upon the NPY neurons that give rise to the GHT.[64] Thus, photic influences of the retina act upon the SCN not only via direct projections through the RHT but also in a multi-synaptic fashion through the IGL. In this way, the IGL acts to integrate photic information with other non-photic information to exert a regulatory influence upon the SCN. It has also been shown that the IGL contains a large population of enkephalinergic neurons that are distinct from the NPY population and gives rise to a commissural projection to the contralateral IGL (see Reference 44 for a review). Although the enkephalin-containing cells do not project to the SCN in rat,[9] studies have shown that they contribute to the hamster GHT.[47] Nevertheless, a commissural connection exists between the hamster IGLs, although its peptidergic phenotype has not been established.[47] Finally, Moore and Speh[45] have shown that both NPY- and ENK-containing neurons co-localize GABA in the rat. Thus, GABA is the principal small molecule neurotransmitter in both the SCN and IGL.

2.3.3 Serotoninergic Afferents

Serotoninergic afferents innervate both the SCN and IGL; however, recent evidence indicates that different raphe nuclei give rise to each projection with the median raphe innervating the SCN and the dorsal raphe projecting to the IGL.[37] In the SCN, serotoninergic afferents are coextensive with the visual inputs and are known to synapse upon VIP neurons.[5] A large literature has documented the effects of this afferent system on circadian function. In addition to a direct effect upon SCN neurons, strong evidence also supports the conclusion that serotonin acts in the SCN to modulate the activity of retinal afferents (see Reference 52 and Chapter 3 in this volume for a comprehensive discussion). Analysis of serotonin receptor subtypes has proven to be of considerable value in defining the SCN synaptology underlying this effect. Pickard and colleagues[51] report that $5HT_{1B}$ receptor agonists block light-induced phase shifts of hamster circadian activity rhythms as well as light-induced Fos expression in all but a small population of neurons in the caudal dorsolateral portion of the nucleus. The latter observation is consistent with the functional heterogeneity in the RHT postulated by Treep and colleagues[69] and, considered with the demonstration of $5HT_{1B}$ mRNA in retinal ganglion cells, suggests that differential expression of serotonin receptor subtypes in retinal ganglion cells may underlie this functional parcellation.

2.3.4 Other Afferents

Although the aforementioned afferents constitute the best characterized projections to the SCN, a number of other regions of the neuraxis project upon this nucleus (Figure 2.4). A number of

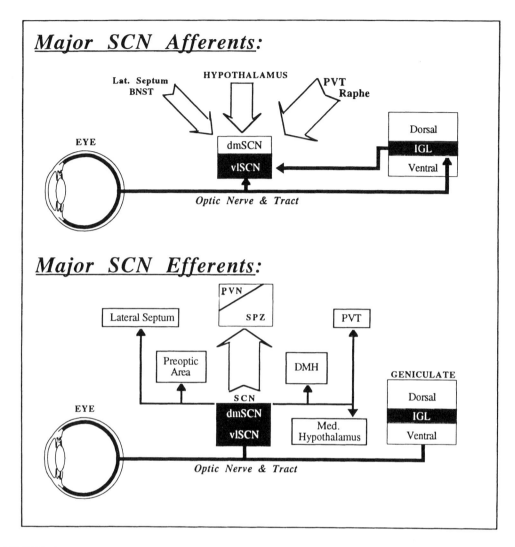

FIGURE 2.4
This schematic diagram illustrates the major cell and fiber systems that comprise the circadian timing system.

hypothalamic projections have been identified, including the preoptic, arcuate, ventromedial, and dorsomedial nuclei; the bed nucleus of the stria terminalis; lateral hypothalamic area; and caudal hypothalamus (posterior hypothalamic area and tuberomammillary nuclei). Additional extrahypothalamic projections from the zona incerta, paraventricular thalamic nuclei, the lateral septal nucleus, pretectum, ventral subiculum, and infralimbic cortex have also been demonstrated (see Reference 42 for a review). Injections of [^{3}H]D-aspartate into the SCN have provided evidence that many of these projections are excitatory.[15,38] Notable exceptions to this are GABAergic neurons of the IGL and arcuate, tuberomammillary, and pretectal nuclei.

Afferents arising in the rostral paraventricular thalamic nucleus (PVT) have been characterized by both anterograde and retrograde tracing methods.[39] Like the SCN, the rostral portion of the PVT contains a very dense concentration of melatonin binding sites, suggesting that this nucleus may be involved in entrainment of the CTS.[4,75] However, lesions of the PVT do not alter photic entrainment of activity rhythms or seasonal responses of the reproductive axis.[18] Nevertheless, it remains possible that the multiple sites of melatonin binding in the CNS create a redundancy in the system that compensates for the loss of the PVT. Additionally, it is important to note that the projection between the PVT and SCN is reciprocal, raising the possibility that the PVT is part of a group of regions responsible for distributing temporal information arising in the SCN.

The SCN also has reciprocal connections with the lateral septum (LS). Retrograde studies indicate that the afferents arise from a restricted subfield in the ventrolateral LS, which Risold and Swanson[54] have shown to be part of a topographically organized disynaptic circuit between the hypothalamus and hippocampal formation. Neurons in this area project into the periventricular hypothalamus, including the SCN, and receive projections from the ventral CA1 and subiculum. This circuitry, combined with that from the paraventricular thalamic nucleus, provides a substrate for communication of the CTS with regions of the neuraxis involved in motivated behaviors. The reciprocal nature of these circuits further suggests that the CTS not only responds to changes in behavioral state, but is also in a position to influence these changes through its efferent projections.

2.4 SCN Efferents

A number of informative studies have provided insight into the distribution of SCN efferents. Among the common principles that have emerged are that (1) the nuclei project to a restricted set of structures, primarily in the surrounding hypothalamus, (2) some areas receive a particularly dense innervation and may serve as "relays" for the distribution of temporal information arising in the SCN, and (3) with the exception of SCN commissural connections, the efferents of each nucleus are largely ipsilateral.

In reviewing work from a number of sources, Watts[74] concluded that SCN efferents can be divided into six general groups (Figure 2.4). These include a projection to the preoptic area, subparaventricular zone (SPZ), medial hypothalamus, dorsal hypothalamus and thalamus, lateral septum, and the IGL. Watts and colleagues emphasized the importance of the projection to the SPZ and developed the concept that the SPZ is an important relay that plays an integral role in the CTS. These data are considered with other literature that has clarified and expanded our understanding of the way in which the SCN functions within the CTS.

2.4.1 Preoptic Area

Watts and colleagues[74] reported a sparse SCN projection into the preoptic area that is amplified by more dense projections arising from the area surrounding the SCN and the SPZ. The importance of this projection is emphasized by the fact that this area of hypothalamus contains prominent cell groups involved in the regulation of sleep, reproduction, fluid homeostasis, and thermoregulation. The most dense input is found in the periventricular and medial preoptic areas, but scattered fibers are found throughout this region. van der Beek and colleagues[71] have demonstrated VIP afferents from the SCN synapse upon gonadotrophin-releasing hormone containing neurons. Similar monosynaptic contacts between the SCN and hypothalamic neurons projecting to the median eminence have also been reported;[23] however, the precise synaptology through which the SCN influences the other temporally dependent neuroendocrine rhythms remains to be established.

2.4.2 Subparaventricular Zone and the Paraventricular Nucleus

A very dense projection to the SPZ from both subdivisions of the SCN has been demonstrated in a number of studies. Watts and colleagues[74] estimate that this pathway constitutes three quarters of all SCN efferents. Two aspects of this projection have particularly important functional implications for the CTS. First, to what extent does the SCN innervate the paraventricular nucleus (PVN)? Well-documented influences of the SCN upon the rhythmic synthesis and release of melatonin from the pineal have been demonstrated, and the majority of evidence supports the view that this is accomplished via a multisynaptic projection involving the PVN.[43] The rhythmic release of other hormones controlled by the PVN that are dependent upon temporal cues from the SCN are also well documented. Nevertheless, the existence of an SCN projection to the PVN has been a subject of considerable debate. Second, what role does the SPZ play in the regulation of circadian function? This indistinct region had not been recognized prior to the demonstration of a dense terminal arbor of SCN efferents, yet the density of the SCN projection into this area supports the conclusion that it plays an important role in the control of circadian function.

The density of the projection of the SCN into the SPZ has naturally focused attention upon this region. However, evidence in support of a direct projection of the SCN to the PVN has been apparent from the earliest autoradiographic studies. The organization and synaptology of this input have been clarified in a number of recent investigations which have not only verified the projection but also provided evidence in support of a topographical organization in the SCN projection to the SPZ/PVN. Kalsbeek and co-workers[28] report that VP neurons of the hamster SCN project to the medial parvicellular PVN, while the anterior and dorsal parvicellular subfields receive projections from VIP neurons. Evidence in support of this view has been presented in rat by Teclemariam-Mesbah et al.,[67] who demonstrated VIP axons of SCN origin in relation to PVN neurons in both the dorsal and ventral parvicellular subfields that project to thoracic spinal cord. A slightly different segregation has been reported in rat by Vrang and colleagues.[72]. These investigators reported projections of dmSCN neurons to the medial and dorsal parvicellular PVN and vlSCN projections to the SPZ. The differences reported in these studies may be related to differences in species or injection site, but they emphasize a common theme in which the SCN provides direct input to both the SPZ and PVN.

The functional significance of a projection from the SCN to the PVN can be found in a very large literature that has examined the influence of the SCN on the rhythmic release of melatonin from the pineal gland.[43] Studies employing lesions, knife cuts, and electrical stimulation of the PVN have shown that this nucleus is necessary for the nocturnal rise in melatonin release from the pineal. The regulation of pineal activity is thought to be achieved via a multisynaptic circuit involving the SCN, PVN, preganglionic sympathetic neurons in the intermediolateral cell column (IML), and the superior cervical ganglion. Two other findings support the conclusion that the neural control of pineal activity is controlled via the aforementioned circuitry. Retrogradely labeled neurons in the dorsal, lateral, and medial parvicellular subdivisions of PVN are present after injection of tracer into the IML but are absent from the SPZ. In addition, injection of neurotropic virus into the pineal produces a retrograde transynaptic infection of the previously postulated circuit that involves the PVN but not the SPZ.[30] Temporal separation in the infection of dm- and vlSCN neurons is also consistent with a division of function in this projection, such that the circadian influences of the SCN upon melatonin secretion are mediated by the dmSCN projection to dorsal and lateral parvicellular PVN, and the inhibitory effects of light on melatonin secretion are mediated via projections of vlSCN neurons to the medial parvicellular PVN.

The function of the more prominent projection of the SCN into the SPZ is less clear. PHA-L studies indicate that the SPZ projects upon many of the same targets as the SCN and that the density of the SPZ efferents is greater than those arising from the SCN. Anterograde studies also suggest that there is a topography in the organization of SPZ efferents. On this basis, Watts and colleagues have postulated that the SPZ acts to integrate and amplify the temporal information arising from the SCN. This important observation clearly requires further attention.

2.4.3 Dorsomedial Nucleus

A projection of the SCN to the DMH is well documented.[74] The DMH gives rise to a dense and complex intrahypothalamic innervation and a smaller projection to the brainstem and telencelphalon;[68] Therefore, it is possible that the SCN influences a variety of systems through this projection. The PVN is one of the primary targets of DMH efferents, and there is reason to believe that this pathway may mediate the effects of the SCN upon the plasma corticosteriod rhythm. Although the diurnal rhythm in plasma corticosterone is dependent upon the SCN, efforts to demonstrate a direct synaptic input to CRH neurons in the PVN have been unsuccessful.[6] In contrast, Kalsbeek and co-workers[28] have demonstrated a dense plexus of VP fibers in the DMH that disappear following SCN lesions, and they have further demonstrated that microinfusion of VP into the PVN/DMV region of SCN lesioned animals suppresses plasma levels of corticosterone to basal daytime levels.[27] Thus, available evidence supports the conclusion that SCN influences upon corticosterone rhythms are mediated through the DMH.

2.4.4 Commissural Connections

There is now substantial evidence for commissural projections between the two suprachiasmatic nuclei, although the full extent and organization of these connections remain to be established. Injection of anterograde tracer into one SCN has revealed a dense projection to the contralateral dmSCN and a lesser projection to the contralateral vlSCN.[7,50] Ultrastructural analysis has shown that the fibers synapse in the contralateral nucleus and are not simply in transit to other target zones.[7] Retrograde transynaptic passage of virus into the SCN following injection of virus into the SPZ also indicates that there are strong commissural connections between dorsomedial subfields.[32] These findings provide an important substrate for the integrated activity of the SCN in the control of circadian function and also suggest that local circuit connections between nuclei, as well as subfields of the same nucleus, provide a dynamic regulatory capacity through which the SCN can modulate the activity of functionally diverse systems.

2.4.5 Other Efferents

As noted earlier, the SCN projects to a very restricted set of structures that are largely confined to the diencephalon. In addition to the major projections described above, projections have also been described to the ventromedial and arcuate nuclei, the zona incerta, and the posterior hypothalamic area. A prominent projection to the paraventricular thalamic nucleus has also been demonstrated, along with a lesser projection to the paratenial nucleus. Only a minor projection has been demonstrated to the IGL, although a much larger projection to this region is known to arise from the retrochiasmatic area.[9,44]

2.5 Summary

Taken together, studies of the structure, neurochemical phenotype, and connectivity of the SCN support the following conclusions regarding the functional organization of the circadian timing system.

1. The SCN are characterized by distinct subdivisions which differ in structure, peptidergic phenotype, and connectivity of constituent neurons.

2. Local circuit connections provide a substrate for the integrated activity of the two suprachiasmatic nuclei and their subdivisions.
3. GABA is the principal small molecule neurotransmitter in the SCN and IGL and co-localizes with essentially all neurons in these regions.
4. Visual input from the SCN and IGL terminate selectively within the vlSCN and appear to affect the activity of a subset of SCN neurons selectively.
5. The RHT and a select number of other SCN afferents contain excitatory neurotransmitters.
6. The differential distribution of serotonin receptor subtypes on SCN neurons and retinal afferents expands the way in which this SCN input can be regulated.
7. Efferent projections of the SCN terminate in a restricted set of structures that are largely confined to the diencephalon, particularly the hypothalamus.
8. A subset of SCN efferents act monosynaptically to provide temporal cues to hypothalamic neurons involved in neuroendocrine regulation.
9. SCN efferent targets such as the subparaventricular zone act as "relay centers" responsible for distribution of processed temporal information.

The above features emphasize the dynamic responsiveness and adaptability that characterize the circadian timing system. Further study of the organization of the intrinsic circuitry and efferent projections of the SCN promise to provide additional valuable insights into the way in which this biological clock functions within the CTS to impose temporal organization upon a variety of physiological and behavioral processes essential for survival of the parent organism.

References

1. Albers, H.E., Stopa, E.G., Zoeller, R.T., Kauer, J.S., King, J.C., Fink, J.S., Mobtaker, H., and Wolfe, H., Day-night variation in prepro vasoactive intestinal peptide/peptide histidine isoleucine mRNA within the rat suprachiasmatic nucleus, *Mol. Brain Res.*, 7, 85, 1990.
2. Albers, H.E., Liou, S.-Y., Stopa, E.G., and Zoeller, R.T., Interaction of colocalized neuro-peptides: functional significance in the circadian timing system, *J. Neurosci.*, 11, 846, 1991.
3. Amir, S., Blocking NMDA receptors or nitric oxide production disrupts light transmission to the suprachiasmatic nucleus, *Brain Res.*, 586, 336, 1992.
4. Bittman, E.L. and Weaver, D.R., The distribution of melatonin binding sites in neuroendocrine tissues in the ewe, *Biol. Reprod.*, 43, 986, 1990.
5. Bosler, O. and Beaudet, A., VIP neurons as prime synaptic targets for serotonin afferents in rat suprachiasmatic nucleus: a combined radioautographic and immunocytochemical study, *J. Neurocytol.*, 14, 749, 1985.
6. Buijs, R.M., Markman, M., Nunes-Cardoso, B., Hou, Y.-X., and Shinn, S., Projections of the suprachiasmatic nucleus to stress-related areas in the rat hypothalamus: a light and electron microscopic study, *J. Comp. Neurol.*, 335, 42, 1993.
7. Buijs, R.M., Hou, Y.-X., Shinn, S., and Renaud, L.P., Ultrastructural evidence for intra- and extranuclear projections of GABAergic neurons of the suprachiasmatic nucleus, *J. Comp. Neurol.*, 340, 381, 1994.
8. Card, J.P., Brecha, N., Karten, H.J., and Moore, R.Y., Immunocytochemical localization of vasoactive intestinal polypeptide-containing cells and processes in the suprachiasmatic nucleus of the rat: light and electron microscopic analysis, *J. Neurosci.*, 1, 1289, 1981.
9. Card, J.P. and Moore, R.Y., Organization of lateral geniculate-hypothalamic connections in the rat, *J. Comp. Neurol.*, 284, 135, 1989.

10. Card, J.P., Whealy, M.E., Robbins, A.K., Moore, R.Y., and Enquist, L.W., Two α-herpesvirus strains are transported differentially in the rodent visual system, *Neuron*, 6, 957, 1991.
11. Chen, D., Hurst, W.J., Ding, J.M., Faiman, L.E., Mayer, B., and Gillette, M.U., Localization and characterization of nitric oxide synthase in the rat suprachiasmatic nucleus: evidence for a nitregic plexus in the biological clock, *J. Neurochem.*, 68, 855, 1997.
12. Dai, J., Swaab, D.F., and Buijs, R., Distribution of vasopressin and vasoactive intestinal polypeptide (VIP) fibers in the human hypothalamus with special emphasis on suprachiasmatic nucleus efferent projections, *J. Comp. Neurol.*, 383, 397, 1997.
13. Daikoku, S., Hisano, S., and Kagotani, Y., Neuronal associations in the rat suprachiasmatic nucleus demonstrated by immunoelectron microscopy, *J. Comp. Neurol.*, 325, 559, 1992.
14. Decker, K., Reuss, S., Nitric oxide-synthesizing neurons in the hamster suprachiasmatic nucleus: a combined NOS- and NADPH-staining and retinohypothalamic tract staining, *Brain Res.*, 666, 284, 1994.
15. DeVries, M.J. and Lakke, J.F., Retrograde labeling of retinal ganglion cells and brain neuronal subsets by [^{3}H]D-aspartate injections in hamster hypothalamus, *Brain Res. Bull.*, 38, 349–354, 1995.
16. Dierickx, K. and Vandesande, F., Immunocytochemical localization of somatostatin-containing neurons in the rat hypothalamus, *Cell Tissue Res.*, 201, 349, 1979.
17. Ebling, F.J., The role of glutamate in the photic regulation of the suprachiasmatic nucleus, *Prog. Neurobiol.*, 50, 109, 1996.
18. Ebling, F.J., Maywood, E.S., Humby, T., and Hastings, M.H., Circadian and photoperiodic time measurement in male Syrian hamsters following lesions of the melatonin-binding sites of the paraventricular thalamus, *J. Biol. Rhythms*, 7, 241, 1992.
19. Gao, G. and Moore, R.Y., Glutamic acid decarboxylase message isoforms in human suprachiasmatic nucleus, *J. Biol. Rhythms* 11, 172, 1996.
20. Gillette, M.U. and Reppert, S.M., The hypothalamic suprachiasmatic nuclei: circadian patterns of vasopressin secretion and neuronal activity *in vitro*, *Brain Res. Bull*, 19, 135, 1987.
21. Guldner, F.H. and Wolff, J.R., Complex synaptic arrangements in the rat suprachiasmatic nucleus: a possible basis for the "Zeitgeber" and non-synaptic synchronization of neuronal activity, *Cell Tissue Res.*, 284, 203, 1996.
22. Guldner, F.H., Bahar, E., and Ingham, C.A., Structural plasticity of optic synapses in rat suprachiasmatic nucleus: adaptation to long-term influence of light and darkness, *Cell Tissue Res.*, 287, 43, 1997.
23. Horvath, T.L., Suprachiasmatic efferents avoid phenestrated capillaries but innervate neuroendocrine cells, including those producing dopamine, *Endocrinology*, 138, 1312, 1997.
24. Huhman, K.L., Hennessey, A.C., and Albers, H.E., Rhythms of glutamic acid decarboxylase mRNA in the suprachiasmatic nucleus, *J. Biol. Rhythms,* 11, 311, 1996.
25. Jacobowitz, D.M. and Winsky, L., Immunocytochemical localization of calretinin in the forebrain of the rat, *J. Comp. Neurol.*, 304, 198, 1991.
26. Johnson, R.F., Moore, R.Y., and Morin, L.P., Retinohypothalamic projections in the rat and hamster using cholera toxin, *Brain Res.*, 462, 301, 1988.
27. Kalsbeek, A., Buijs, R.M., van Heerikhuize, J.J., Arts, M., and van der Woude, T.P., Vasopressin-containing neurons of the suprachiasmatic nuclei inhibit corticosterone release, *Brain Res.*, 580, 62, 1992.
28. Kalsbeek, A., Teclemariam-Mesbah, R., and Pevet, P., Efferent projections of the suprachiasmatic nucleus in the Golden Hamster (*Mesocricetus auratus*), *J. Comp. Neurol.*, 332, 293, 1993.
29. Larsen, P.J. and Mikkelsen, J.D., The suprachiasmatic nucleus of the mink (*Mustela vision*): apparent absence of vasopressin-immunoreactive neurons, *Cell Tiss. Res.*, 273, 239, 1993.

30. Larsen, P.J., Enquist, L.W., and Card, J.P., Characterization of the multisynaptic neuronal control of the rat pineal gland using viral transneuronal tracing, *Eur. J. Neurosci.*, 10, 128, 1998.
31. Leak, R. and Moore, R.Y., Identification of retinal ganglion cells projections to the lateral hypothalamic area of the rat, *Brain Res.*, 770, 105, 1997.
32. Leak, R., Card, J.P., and Moore, R.Y., Application of the pseudorabies virus to tracing functional pathways in the brain, *Soc. Neurosci. Abstr.*, 76, 6, 1995.
33. Levine, J.D., Zhao, X.-S., and Miselis, R.R., Direct and indirect retinohypothalamic projections to the supraoptic nucleus in the female rat, *J. Comp. Neurol.*, 341, 214, 1994.
34. Levine, J.D., Weiss, M.L., Rossenwasser, A.M., and Miselis, R.R., Retinohypothalamic tract in the female albino rat: a study using horseradish peroxidase conjugated to cholera toxin, *J. Comp. Neurol.*, 306, 344, 1991.
35. Mai, J.K., Kedziora, O., Techhaus, L., and Sofroniew, M.V., Evidence for subdivisions in the human suprachiasmatic nucleus, *J. Comp. Neurol.*, 305, 508, 1991.
36. Meijer, J.H., Integration of visual information by the suprachiasmatic nucleus, in *Suprachiasmatic Nucleus: The Mind's Clock,* Klein, D.C., Moore, R.Y., and Reppert, S.M., Eds., Oxford University Press, New York, 1991, pp. 107.
37. Meyer-Bernstein, E.L. and Morin, L.P., Differential serotonergic innervation of suprachiasmatic nucleus and the intergeniculate leaflet and its role in circadian rhythm modulation, *J. Neurosci.*, 16, 2097, 1996.
38. Moga, M.M. and Moore, R.Y., Putative excitatory amino acid projections to the suprachiasmatic nucleus in the rat, *Brain Res.*, 743, 171, 1996.
39. Moga, M.M., Weis, R.P., and Moore, R.Y., Efferent projections of the paraventricular thalamic nucleus in the rat, *J. Comp. Neurol.*, 359, 221, 1995.
40. Moore, R.Y., The enigma of the geniculohypothalamic tract: why two visual entraining pathways?, *J. Interdisc. Cycle Res.*, 23, 144, 1992.
41. Moore, R.Y., The organization of the human circadian timing system, *Prog. Brain Res.*, 93, 101, 1992.
42. Moore, R.Y., Entrainment pathways and the functional organization of the circadian system, *Prog. Brain Res.*, 111, 103, 1996.
43. Moore, R.Y., Neural control of the pineal gland, *Behav. Brain Res.*, 73, 125, 1996.
44. Moore, R.Y. and Card, J.P., Intergeniculate leaflet: an anatomically and functionally distinct subdivision of the lateral geniculate complex, *J. Comp. Neurol.* 344, 403, 1994.
45. Moore, R.Y. and Speh, J.C., GABA is the principal neurotransmitter of the circadian system, *Neurosci. Lett.*, 150, 112, 1993.
46. Moore, R.Y., Speh, J.C., and Card, J.P., The retinohypothalamic tract originates from a distinct subset of retinal ganglion cells, *J. Comp. Neurol.*, 352, 351, 1995.
47. Morin, L.P. and Blanchard, J. Organization of the hamster intergeniculate leaflet: NPY and ENK projections to the suprachiasmatic nucleus, intergeniculate leaflet and posterior limitans nucleus, *Visual Neurosci.*, 12, 57, 1995.
48. Peterson, G.M., Watkins, W.B., and Moore, R.Y., The suprachiasmatic hypothalamic nuclei of the rat. VI. Vasopressin neurons and circadian rhythms, *Behav. Neural Biol.*, 29, 236, 1980.
49. Pickard, G.E., Bifurcating axons of retinal ganglion cells terminate in the hypothalamic suprachiasmatic nucleus and the intergeniculate leaflet of the thalamus, *Neurosci. Lett.*, 55, 211, 1985.
50. Pickard, G.E. and Silverman, A.-J., Direct retinal projections to the hypothalamus, piriform cortex, and accessory optic nuclei in the golden hamster as demonstrated by a sensitive anterograde horseradish peroxidase technique, *J. Comp. Neurol.*, 196, 155, 1981.

51. Pickard, G.E., Weber, E.T., Scott, P.A., Riberdy, A.F., and Rea, M.A., 5HT1B receptor agonists inhibit light-induced phase shifts of behavioral circadian rhythms and expression of the immediate-early gene c-fos in the suprachiasmatic nucleus, *J. Neurosci.* 16, 8208, 1996.

52. Rea, M.A., J.D. Glass, and Colwell, C.S., Serotonin modulates photic responses in the hamster suprachiasmatic nuclei, *J. Neurosci.*, 14, 3635, 1994.

53. Reuss, S., Decker, K., Robeler, L., Layes, E., Schollmayer, A., and Spessert, R., Nitric oxide synthase in the hypothalamic suprachiasmatic nucleus of rat: evidence from histochemistry, immunohistochemistry and Western blot; and colocalization with VIP, *Brain Res.*, 695, 257, 1995.

54. Risold, P.Y. and Swanson, L.W., Structural evidence for functional domains in the rat hippocampus. *Science*, 272, 1484, 1996.

55. Romijn, H.J., Sluiter, A.A., Pool, C.W., Wortel, J., and Buijs, R.M., Differences in colocalization between Fos and PHI, VIP and VP in neurons of the rat suprachiasmatic nucleus after a light stimulus during the phase delay versus the phase advance period of the night, *J. Comp. Neurol.*, 372, 1, 1996.

56. Schwartz, W.J., SCN metabolic activity *in vivo*, in *Suprachiasmatic Nucleus, The Mind's Clock*, Klein, D.C., Moore, R.Y., and Reppert, S.M., Eds., Oxford University Press, New York, 1991, p. 144.

57. Schwartz, W.J., Coleman, R.J., and Reppert, S.M., A daily vasopressin rhythm in rat cerebrospinal fluid, *Brain Res.*, 263, 105, 1983.

58. Schwartz, W.J., Aronin, N., Takeuchi, J., Bennett, M.R., and Peters, R.V., Towards a molecular biology of the suprachiasmatic nucleus: photic and temporal regulation of c-*fos* gene expression, *Sem. Neurosci.*, 7, 53, 1995.

59. Shinohara, K., Isobe, Y., Takeuchi, J., and Inouye, S.T., Circadian rhythms of somatostatin-immunoreactivity in the suprachiasmatic nucleus of the rat, *Neurosci. Lett.*, 129, 59, 1991.

60. Silver, R., Romero, M.-T., Besmer, H.R., Leak, R., Nunez, J.M., and LeSauter, J., Calbindin-D28K cells in the hamster SCN express light-induced Fos, *NeuroReport* 7, 1224, 1996.

61. Stopa, E.G., King, J.C., Lydic, R., and Schoene, W.C., Human brain contains vasopressin and VIP neuronal subpopulations in the suprachiasmatic region, *Brain Res.*, 297, 159, 1984.

62. Strecker, G.J., Bouskila, Y., and Dudek, F.E., Neurotransmission and electrophysiological mechanisms in the suprachiasmatic nucleus, *Sem. Neurosci.*, 7, 43, 1995.

63. Swaab, D.F., Fliers, E., and Partiman, T.S., The suprachiasmatic nucleus of the human brain in relation to sex, age and senile dementia, *Brain Res.*, 342, 37, 1985.

64. Takatsuji, K., Miguel-Hidalgo, J.-J., and Tohyama, M., Retinal fibers make synaptic contact with neuropeptide Y and enkephalin immunoreactive neurons in the intergeniculate leaflet of the rat, *Neurosci. Lett.*, 125, 73, 1991.

65. Takeuchi, J., Nagasaki, H., Shinohara, K., and Inouye, S.T., A circadian rhythm of somatostatin messenger RNA levels, but not of vasoactive intestinal polypeptide/peptide histidine isoleucine messenger RNA levels in rat suprachiasmatic nucleus, *Mol. Cell. Neurosci.*, 3, 29, 1992.

66. Tanaka, M., Okamura, H., Matsuda, T., Shigeyoshi, Y., Hisa, Y., Chihara, K., and Ibata, Y., Somatostatin neurons form a distinct peptidergic neuronal group in the rat suprachiasmatic nucleus: a double labeling *in situ* hybridization study, *Neurosci. Lett.*, 215, 119, 1996.

67. Teclemariam-Mesbah, R., Kalsbeek, A., Pevet, P., and Buijs, R.M., Direct vasoactive intestinal polypeptide-containing projection from the suprachiasmatic nucleus to spinal projecting hypothalamic paraventricular neurons, *Brain Res.*, 748, 71, 1997.

68. Thompson, R.H., Canteras, N.S., and Swanson, L.W., Organization of projections from the dorsomedial nucleus of the hypothalamus: A PHA-L study, *J. Comp. Neurol.*, 376, 143, 1996.

69. Treep, J.A., Abe, H., Rusak, B., Goguen, D.M. Two distinct retinal projections to the hamster suprachiasmatic nucleus, *J. Biol. Rhythms,* 10, 299–307, 1995.

70. van den Pol, A.N., The suprachiasmatic nucleus: morphological and cytochemical substrates for cellular interaction, in *Suprachiasmatic Nucleus, The Mind's Clock,* Klein, D.C., Moore, R.Y., and Reppert, S.M., Eds., Oxford University Press, New York, 1991, p. 17.
71. van der Beek, E.M., Wiegant, V.M., van der Donk, H.A., van den Hurk, R., and Buijs, R., Lesions of the suprachiasmatic nucleus indicate the presence of a direct vasoactive intestinal polypeptide-containing projection to gonadotrophin-releasing hormone neurons in the female rat, *J. Neuroendo.,* 5, 137, 1993.
72. Vrang, N., Larsen, P.J., Moller, M., and Mikkelsen, J.D., Topographical organization of the rat suprachiasmatic-paraventricular projection, *J. Comp. Neurol.,* 353, 585, 1995.
73. Wang, H. and Morris, J.F., Presence of neuronal nitric oxide synthase in the suprachiasmatic nuclei of mouse and rat, *Neuroscience,* 74, 1059, 1996.
74. Watts, A.G., The efferent projections of the suprachiasmatic nucleus: anatomical insights into the control of circadian rhythms, in *Suprachiasmatic Nucleus, The Mind's Clock,* Klein, D.C., Moore, R.Y., and Reppert, S.M., Eds., Oxford University Press, New York, 1991, p. 77.
75. Weaver, D.R., Rivkees, S.A., and Reppert, S.M., Localization and characterization of melatonin receptors in rodent brain by *in vitro* autoradiography, *J. Neurosci.,* 9, 2581, 1989.
76. Wollnik, F. and Bihler, S., Strain differences in the distribution of arg-vasopressin- and neuropeptide Y-immunoreactive neurons in the suprachiasmatic nucleus of rats, *Brain Res.,* 724, 191, 1996.

Chapter 3

Intercellular Interactions and the Physiology of Circadian Rhythms in Mammals

Hugh D. Piggins and Benjamin Rusak

Contents

3.1	Introduction	31
3.2	Retinal Projections to the SCN	32
3.3	Other Projections to the SCN	33
	3.3.1 The Geniculohypothalamic Tract	33
	3.3.2 The Raphe Projection	35
	3.3.3 Acetylcholine and Nerve Growth Factor	36
3.4	Intrinsic SCN Neurochemicals	36
3.5	Ionic Mechanisms of SCN Cells	37
3.6	SCN Efferents	38
3.7	Future Prospects	39
References		39

3.1 Introduction

Individual, cultured neurons of the rat suprachiasmatic nuclei (SCN) function as independent circadian oscillators,[93] and the SCN *in vivo* maintains circadian rhythmicity in the absence of sodium-dependent action potentials.[69] However, both overt expression of rhythmicity in behavior and photic regulation of the SCN depend on intercellular communication,[69] as must coordination of rhythmicity among SCN neurons. Previous chapters have outlined the basic features of circadian organization and the anatomical features of the SCN and its associated systems. The goal of this chapter is to summarize our knowledge of the functions of major neurochemical systems projecting to the SCN, and of intrinsic SCN neurochemicals in intercellular communication required for the generation of daily rhythms and their synchronization by external events.

3.2 Retinal Projections to the SCN

A monosynaptic retinal projection to the SCN (the retinohypothalamic tract, RHT) and a polysynaptic pathway via retinally innervated cells of the intergeniculate leaflet (IGL) and ventral lateral geniculate nucleus (vLGN) of the thalamus convey photic information that entrains the SCN to cycles of light and dark.[52] The RHT arises from a distinct subset of retinal ganglion cells of the gamma or Perry type III class, the axons of which innervate primarily the ventral division of the rodent SCN, as well as other hypothalamic structures. In the rat, the RHT forms mainly Gray type I synapses (presumably excitatory) with SCN cells, although an estimated 25% are Gray type II synapses.[52] At least some of the ganglion cells projecting to the SCN also project via collaterals towards the LGN and terminate in the IGL. The SCN targets of this bifurcating projection may differ physiologically from other retinorecipient SCN cells, but the functional significance of this difference remains unknown.[80]

There is considerable evidence that the RHT utilizes glutamate as its main transmitter:[16] glutamate immunoreactivity is detected in RHT terminals in the SCN; stimulating the optic nerve evokes the release of tritiated aspartate and glutamate into the SCN in a slice preparation; and *in vivo* microdialysis studies confirm that light evokes increases in extracellular glutamate levels in the SCN. The dipeptide N-acetyl-aspartyl-glutamate (NAAG) has been identified in the RHT of the rat and cat and may serve as a source of glutamate that affects SCN neurons.

Neurophysiological studies both *in vivo* and *in vitro* have shown that light, optic nerve stimulation, and glutamate application activate most SCN cells in nocturnal rodents, and these effects can be blocked by local applications of glutamate receptor antagonists.[17,32] Ionotropic and metabotropic receptors for glutamate have been identified in the SCN through detection of various receptor subunits and their mRNAs,[16,20,78] and labeled glutamatergic compounds bind with high density in the rodent SCN [42].

Antagonists which act on ionotropic glutamate receptors have been shown to reduce the phase-shifting effects of light on rodent locomotor rhythms and to attenuate light-induced immediate early gene (IEG) expression (especially c-Fos) in the SCN.[1,16] Although bolus injections of glutamate or aspartate into the SCN were not found to mimic the effects of light on circadian rhythms, microinjections of N-methyl-D-aspartatate (NMDA) into the SCN region of hamsters phase shifted activity rhythms in a manner resembling the effects of light,[49] as did electrical stimulation of the optic chiasm in rats.[43]

Despite the apparent congruence of these results implying that photic stimuli act via ionotropic glutamate receptors to induce IEG expression and phase reset the SCN, some caution is required in the interpretation of these findings. Increasing light intensity may partially overcome the blockade by ionotropic glutamate antagonists of photic phase shifts in hamsters,[16] and these antagonists do not block c-Fos expression in a distinctive region of the dorsolateral SCN.[1] Gene expression in this same region appears to be activated selectively in other experimental paradigms, implying that retinal innervation, receptor subtypes or neuronal characteristics differ between the dorsolateral and other regions of the SCN.[80] There are also regional differences in the sensitivity of SCN neurons to a metabotropic glutamate receptor agonist,[68] although the function of such receptors in this system remains unknown. A relatively small percentage of SCN neurons are suppressed by light in nocturnal rodents, and this proportion is similar to the proportion of presumably inhibitory Gray type II synapses in the the RHT innervation; however, it has not been determined whether these suppressions involve an activation of an inhibitory interneuron in the SCN. The proportion of such suppressions of SCN firing by light is apparently higher in diurnal rodents.[44] Similarly, diurnal rodents differ from nocturnal species in their phase-shift responses to light and in the regulation of IEG expression in the SCN.[2] The basis for these apparent differences between diurnal and nocturnal species remains to be established and should be of considerable interest.

One of the effects of glutamatergic activation of SCN cells is the release of the diffusible gas nitric oxide (NO).[86] Treatments which block the formation of NO in SCN slice preparations can block glutamate-induced phase shifts, while increasing NO release can cause phase shifts.[18] NO is known in other systems to diffuse rapidly and (by SCN standards) large distances from its site of release. Despite earlier uncertainties, recent studies have demonstrated widespread immunoreactivity for a specific isoform of NOS (neuronal or nNOS) in many cells of the ventral portion of the rat SCN.[19,64,89]

A main effector of NO is believed to be soluble guanylyl cyclase, which liberates cGMP. Microinjection of inhibitors of the cGMP-dependent protein kinase (PKG) can block the phase-advancing but not the phase-delaying effects of light on the pacemaker *in vivo*.[41,92] Similarly, inhibition of NOS with intracerebroventricular injections of L-NAME can selectively block phase advance shifts.[45,91] The phase-dependent effects of inhibiting NOS and PKG imply that different signal transduction pathways mediate or modulate photic effects at different phases, but it remains unclear what roles NO formation and release play with respect to the other processes that have been hypothesized to mediate photic effects on SCN cells.

One of these processes is the light-stimulated increase of the mRNAs and proteins of numerous IEGs in SCN neurons (and apparently some glia[5]) in the retinorecipient zone.[1] This activity is of interest because it is generally restricted to circadian phases at which light can shift the clock[66] and because the IEG proteins formed act as transcription factors that may play a role in triggering critical molecular events leading to shifts of circadian phase. This hypothesis is reinforced by the finding that intraventricular injection of antisense oligonucleotides to block production of the IEG proteins c-Fos and Jun-B can prevent light-induced phase shifts in rats.[94] It is clear from other studies that c-Fos production alone is not sufficient to cause rhythm phase shifts, but the production of a set of IEG proteins may be an essential step in the light-entrainment pathway.

Other neurochemicals have been suggested recently to play a role in RHT function. Immunoreactivity for the peptide Substance P (SP) has been demonstrated in the rat, human, and monkey RHT, and application of SP activates rodent SCN neurons[74] and phase-shifts the metabolic and electrical activity rhythms of rat SCN neurons with a light-type phase response curve (PRC).[74] However, microinjections of SP into the SCN region of Syrian hamsters do not evoke significant phase shifts in the locomotor rhythm at any phase tested.[61] These findings suggest that SP may function as an RHT neurotransmitter or modulator in rats but not hamsters, or that *in vivo* and slice preparations differ in responsiveness to SP.

Pituitary adenylate cyclase activating polypeptide (PACAP) has been identified in the retinal terminals innervating both the SCN and IGL[24] and possibly in cells of the ventral portion of the rat SCN.[62] PACAP application to the SCN *in vitro* resets the circadian clock with a PRC resembling that of dark pulses,[24] an action apparently mediated through adenylate cyclase and the formation of cyclic AMP. These results are consistent with earlier *in vitro* studies demonstrating that cAMP analogs phase-advance the SCN pacemaker during the projected day.[22] It remains unclear why application of a peptide found in retinal projections should have a dark-like effect on SCN phase *in vitro*.

3.3 Other Projections to the SCN

3.3.1 The Geniculohypothalamic Tract

The geniculohypothalamic tract (GHT) in several species arises from a distinctive group of neurons in the IGL and vLGN. Some of these cells are photically responsive and project monosynaptically to the SCN, overlapping the RHT innervation.[25,52] Ablation of the GHT alters entrainment, responses to constant light, and the phase-shifting effects of both light and dark pulses. The principal neurotransmitters of the GHT are thought to be neuropeptide Y (NPY) and γ-amino-butyric acid (GABA), but enkephalin is also present in some species.

Injections of NPY into the SCN *in vivo* or bath applications to a hypothalamic slice cause phase advances during the mid-subjective day, typical of a dark-type or activity-mediated PRC,[3,26] and these effects do not depend on NPY increasing activity. These findings have suggested the hypothesis that NPY release into the SCN mediates non-photic (activity- or arousal-based) phase shifts.[53] Consistent with this hypothesis, injection of NPY antisera into the SCN can block the phase-shifting effects on hamster activity rhythms of intense exercise during the middle of the subjective day.[8] This treatment can also enhance the effects of light on circadian rhythms,[6] implying that NPY and light have opposite effects on SCN pacemaker cells.[7]

Early studies on the effects of NPY on single-unit activity in the rodent SCN found diverse effects, which may be attributable to differences in the methods of drug application. Micropressure ejections of NPY directly from the recording micropipette evoked mainly activations in the spontaneous firing rates of hamster SCN cells, particularly during the subjective day;[38] however, other studies have found that NPY applied in the bath to the SCN *in vitro* evoked complex responses including primarily suppressions.[37,72] Whole-cell patch recordings and calcium imaging studies have shown that bath-applied NPY opposes the effects of excitatory inputs to the SCN,[83] suggesting that NPY released from the GHT functions presynaptically to modulate glutamate release in the SCN.

Schmahl and Böhmer[67] reported that bath applications of NPY could either suppress or increase firing of rat SCN cells *in vitro*, but that strychnine (a glycine receptor antagonist) in a small sample of cells could block the suppressive effects, leaving only activations. These complex and partially contradictory results do not present a simple picture of NPY effects at a cellular level, and it remains possible that NPY has very different effects on different SCN neurons.

Some reports suggest that NPY acts on the SCN via Y2, but not Y1 receptors,[23,28] but the nature of the action remains unclear. Tetrodotoxin (TTX) was found to block phase-shifting effects of NPY agonists on the SCN *in vivo* but not *in vitro*, implying that it acts trans-synaptically *in vivo* but directly on SCN cells in a slice preparation.[23,28] Similarly, evidence that the $GABA_A$ antagonist bicuculline blocks the phase-shifting effects of NPY on rodent behavioral rhythms[27] implies a trans-synaptic action before altering pacemaker function, which is inconsistent with evidence that NPY can phase shift *in vitro* in the presence of TTX.[23] These puzzling discrepancies require further evaluation, as they imply a radical difference in the way an important neurochemical afferent to the SCN acts in two different experimental situations. It is not clear which of these results most closely reflects the way NPY acts physiologically.

Most if not all SCN neurons and many IGL efferents to the rat SCN contain (GABA) or the GABA-synthesizing enzyme, GAD,[51] while subunits for $GABA_{A-C}$ receptors are found throughout the rat SCN.[54] These studies suggest that GABA may play an important role in circadian rhythm processes. Application of GABA to SCN cells *in vitro* and *in vivo* can inhibit spontaneous neuronal activity, and this action is blocked by the $GABA_A$ antagonist bicuculline.[36,39] Wagner et al.,[88] however, reported that GABA could either activate or suppress SCN neuronal activity, depending on the circadian phase of application. This finding was attributed to circadian changes in transmembrane Cl^- distributions, but these provocative conclusions require further assessment.

The $GABA_A$ agonist muscimol has been shown to phase advance SCN neuronal firing-rate rhythms when applied during the middle of the subjective day,[79] thus linking GABA to the phase-shifting effects of non-photic stimuli. Consistent with this interpretation, peripherally administered benzodiazepines, such as triazolam, which are believed to function by modulating the frequency of opening of the $GABA_A$ receptor-linked Cl^- channel, reset hamster behavioral rhythms with sensitivity characteristic of a dark-type PRC.[81] Benzodiazepines also suppress spontaneous activity in the SCN and potentiate the inhibitory actions of GABA on SCN neurons. These studies suggest that benzodiazepines act directly or with endogenous GABA on $GABA_A$ receptors to modulate the phase of the circadian pacemaker. Microinjections of the benzodiazepine triazolam directly into the SCN region, however, do not phase shift rodent behavioral rhythms, indicating that benzodiazepines act at a site upstream of the SCN to shift the clock. Ablation of the IGL can prevent benzodiazepine-induced phase shifts,[46] implying that these nuclei are critical for conveying the phase-shifting input to the SCN.

A likely role for GABA is the modulation of excitatory tone within the SCN. Gillespie et al.[21] have recently demonstrated that microinjections of $GABA_A$ (but not $GABA_B$) receptor antagonists into the SCN region block the effects of light on hamster wheel-running rhythms. It is not clear whether drugs acting on GABA receptors have their effects by modulating intrinsic GABAergic systems or GABAergic afferents to the SCN, such as those in the GHT. A recent SCN slice study suggested that GABA, acting via $GABA_B$ receptors, inhibited spontaneous neuronal activity and the release of glutamate from RHT terminals.[31] Results obtained with both GABA and NPY indicate that GHT afferent systems can counter the functional effects of RHT activation of SCN cells. Thus, systems mediating photic and non-photic effects (the RHT and portions of the GHT, respectively) can modulate each other's effects on the SCN pacemaker. How the photic functions of the GHT are blended into this mixture of antagonistic actions remains unclear.

3.3.2 The Raphe Projection

The SCN receive a substantial innervation from the median raphe nucleus, bringing high levels of serotonin (5-HT) primarily to the ventral and medial SCN.[47] Applications of 5-HT to SCN cells *in vitro* or *in vivo* suppress spontaneous neuronal activity and attenuate light-evoked firing and release of glutamate, acting via one or more receptors that are similar to the $5-HT_7$ receptor subtype.[70,77,95] Similarly, serotonergic agonists phase reset the electrical activity rhythm of the SCN *in vitro* with a dark-type PRC,[22] and peripheral injections of 8-OH-DPAT have similar phase-shifting effects. The shape of the PRC for serotonergic agents suggests that serotonin may play a role in conveying non-photic information to the SCN circadian clock, but some non-photic phase-shifting stimuli are still effective after destruction of the serotonergic projection to the SCN.[53]

Despite the similarity of phase shifts evoked by 5-HT agonists in an SCN slice preparation and after systemic treatments, the *in vivo* effects do not appear to be mediated by direct actions on the SCN. Microinjections of 8-OH-DPAT into the SCN region *in vivo* fail to reset hamster rhythms, whereas microinjections into the median raphe complex phase advance rhythms during the subjective day, as peripheral injections do.[50] These results suggest that in the intact rat 8-OH-DPAT acts in the raphe nuclei, possibly by activating autoreceptors that reduce 5-HT release at raphe targets. These effects may involve reduction of serotonergic activity in the SCN or, more likely, in the IGL, perhaps thereby permitting increased release of NPY in the SCN. If this is the mechanism for the effects of peripheral 8-OH-DPAT treatments, it remains unclear by what mechanism bath applications of serotonergic agonists phase shift the SCN firing-rate rhythm in a slice preparation. These observations demonstrate that effects of systemic drug treatments that are mimicked by similar treatments of the SCN *in vitro* are not necessarily achieved through the same mechanism.

Other 5-HT receptors are also involved in mediating its effects in the SCN. Activation of $5-HT_{1B}$ receptors can reduce the effects of light on the SCN, and these receptors are found presynaptically on retinal afferents to the SCN.[57] Thus, 5-HT acts through several routes to modulate photic effects on the SCN, but the functional significance of this modulatory influence remains unclear. Selective ablation of the median raphe with a specific 5-HT neurotoxin does not much alter the phase-resetting effects of light,[48] but damage to the raphe can alter the timing of activity onset and the duration of the active phase, in addition to sensitizing rats to the disruptive effects of constant light on rhythm organization.[52] Most of the available data are consistent with 5-HT acting to reduce photic effects on the circadian system via several receptor subtypes and in several sites, but it remains to be determined under what physiological conditions such actions are normally evoked.

3.3.3 Acetylcholine and Nerve Growth Factor

The SCN receive a modest innervation from neurons in the basal forebrain and brainstem which contain acetylcholine.[12] Immunocytochemical studies have demonstrated the presence of muscarinic and some types of nicotinic cholinergic receptors in and near the SCN,[84] while neurophysiological studies have found functional responses to cholinergic agonists in the SCN. The influence of cholinergic receptor activation on the phase of the SCN pacemaker varies across studies. The nonselective cholinergic agonist, carbachol, has been found to phase-shift SCN electrical activity and behavioral rhythms with a PRC that somewhat resembles that of light, but differs in some respects in most studies.[10,34,35] The phase-shifting effects of carbachol both *in vivo* and *in vitro* were blocked by muscarinic but not nicotinic antagonists.[10,34,35]

Mecamylamine, a nicotinic antagonist, blocked effects of light on the hamster SCN,[97] but, because it can also affect glutamatergic transmission in some systems, the specificity of this effect is uncertain. Cholinergic effects on circadian pineal rhythms in rats, however, were also mediated by nicotinic, but not muscarinic, receptors,[96] suggesting differences among species or among functional endpoints studied.

Tetrodotoxin did not block the phase-shifting actions of carbachol on the SCN *in vitro*, indicating a direct effect on the pacemaker in this preparation.[34] However, *in vivo* the non-competitive NMDA receptor antagonist, MK-801, attenuated the phase-advancing actions of carbachol on hamster behavioral rhythms, suggesting a presynaptic action in regulating glutamate release.[16] Interpretation is again complicated, however, because MK-801 affects calcium channels and may therefore alter activity through receptors other than the NMDA receptor. The temporal patterns and pharmacological profiles of sensitivity to carbachol differ among studies, suggesting that various neurochemical mechanisms are recruited differentially at different doses and in different species or experimental preparations, and may modify the effects of carbachol on circadian rhythms.

Immunoreactivity for the low-affinity nerve growth factor (NGF) receptor (p75-NGFR) is found on axon terminals of retinal ganglion cells and basal forebrain cholinergic neurons which project to the SCN.[11] These receptors appear functional, as NGF injections into the SCN mimic the phase-shifting effects of carbachol on hamster behavioral rhythms. The source of this NGF is unknown, but one hypothesis is that NGF released from SCN cells acts on basal forebrain and retinal afferents to modulate presynaptically the release of ACh and glutamate. Through this mechanism, NGF and/or related trophic factors (e.g., BDNF) may function as neuromodulators of the circadian pacemaker.

3.4 Intrinsic SCN Neurochemicals

The retinally innervated ventrolateral division of the SCN contains most of the vasoactive intestinal polypeptide (VIP)- and gastrin-releasing peptide (GRP)-containing neurons, and the dorsomedial division contains most of the arginine vasopressin (AVP)- and somatostatin (SS)-expressing neurons (see Chapter 2). VIP-containing neurons constitute roughly 7% of the rodent SCN neuronal population[40] and give rise to extensive intra- and inter-SCN connections, as well as extra-SCN projections.[90] Electron microscopic (EM) studies have shown VIP-ir axon terminals forming synapses on neurons in the dorsomedial SCN whose neurochemical phenotypes are unknown.[82] These VIP-innervated cell bodies are also innervated by GABA-containing axons, and GABA is also frequently found in VIP-immunoreactive neurons, implying a close functional relation (see below).

VIP-ir neurons are directly innervated by retinal ganglion cells, and some express IEGs following a light pulse delivered during the late subjective night,[65] implying some role in photic entrainment. This contention is consistent with the results of *in vivo* and *in vitro* studies demonstrating that VIP treatments mimic some of the effects of light and glutamate on the phase of the circadian pacemaker.[60,75] In constant conditions, VIP mRNA and protein do not oscillate, but levels of VIP synthesis (and presumably release) within the SCN are determined by the lighting cycle, with high levels during the dark phase and low levels during the light.[4] However, two studies have demonstrated that VIP release in the SCN can become circadian following cysteamine-induced damage to somatostatin neurons[30] or *in vitro* following chemical destruction of glial elements in the SCN.[29] The physiological implications of these findings remain unclear, but they suggest a capacity for oscillation in VIP neurons in some situations.

Neurons containing gastrin-releasing peptide (GRP), found primarily in the ventral SCN and embedded in the dorsal aspect of the optic chiasm, comprise approximately 4% of the total neuronal population of the rat SCN.[40] These neurons give rise to a variety of projections within the SCN as well as to extra-SCN sites.[90] Many GRP-containing neurons synthesize GABA and are innervated by GABA-ir axons.[82]

Neurons expressing GRP are innervated directly by the RHT, and GRP mRNA and protein levels within the SCN can be altered by retinal illumination but do not oscillate spontaneously.[4,30] Nocturnal light exposure also induces IEG expression in some GRP-containing cells.[65] GRP potently excites about half of SCN neurons tested *in vitro*[58,59] and phase shifts rodent locomotor rhythms *in vivo* and SCN electrical rhythms *in vitro* with sensitivity resembling a light-type PRC.[60,75] Whether GRP release is involved physiologically in the mediation of photic effects on rhythms remains to be assessed.

One immunocytochemical study showed an apparent overlap of VIP/PHI (peptide histidine isoleucine)-containing cell bodies with those of GRP-containing neurons within the ventral portion of the rat SCN in consecutive coronal sections.[55] This observation led to the suggestion that VIP, PHI, and GRP were co-released and that the ratios of their levels predicted effects on circadian phase.[4] More recent immunocytochemical studies have suggested that few PHI/VIP-containing cell bodies also express GRP,[56] while functional studies have demonstrated potent neurophysiological and phase-shifting effects of these peptides when they were applied alone to the SCN.[59,60] Thus, these peptides do not appear to be colocalized to any great extent, and predictions of the "ratio model" are inconsistent with numerous findings. It remains to be determined whether and how these and other peptides interact in the SCN.

AVP- and SS-containing neurons are found in the dorsal and medial regions of the SCN.[15] Both populations of neurons maintain a circadian rhythm in peptide synthesis when rodents are placed in constant conditions, and in isolated hypothalamic slice preparations.[29] AVP levels in the CSF of monkeys are also rhythmic and depend on an intact SCN. AVP activates rodent SCN neurons via an AVP1a receptor subtype; however, microinjections of AVP into the SCN region fail to shift hamster behavioral rhythms.[30] Thus, AVP release is under clear circadian control in the SCN but does not appear to play a role in mediating phase-shifting effects on the SCN. In contrast, SS mimics the phase-resetting effects of glutamate on the firing-rate rhythm of the SCN *in vitro*, although acute effects of SS on SCN neurons and behavioral effects have not been assessed.[29]

Because peptides in SCN neurons are colocalized with classical small molecule transmitters (especially GABA), it remains an important goal to elucidate how the release of these peptides is regulated physiologically and how their functions interact with the effects of classical neurotransmitters.

3.5   Ionic Mechanisms of SCN Cells

Circadian rhythms in electrical activity of the SCN have been demonstrated using both single- and multi-unit recordings from hamster and rat SCN neurons in hypothalamic slice preparations, and from the SCN of several species *in vivo*. Firing-rate rhythms typically peak during the middle of the subjective day and reach a trough during the subjective night. These rhythms persist when glutamate and GABA receptor antagonists are added to the bath, indicating that intercellular communication involving these neurotransmitters is not necessary to sustain these synchronized population rhythms. Application of TTX prevents recording of electrical activity, but when it is washed out, the rhythm returns at a phase predicted by the rhythm prior to TTX treatment. Thus, blockade of sodium-dependent action potentials neither prevents the generation nor shifts the phase of the circadian rhythm *in vitro*, a result consistent with *in vivo* studies.[73]

The intracellular mechanisms underlying the generation of the circadian rhythm in cellular firing rate are poorly understood. Circadian variations have been described in the mean input conductance and current required to hold the neurons at –60 mV, with higher mean input conductances and holding currents recorded during the middle of the subjective day than during the subjective night. These results suggest that mechanisms intrinsic to SCN neurons are responsible for generating circadian changes in the state of membrane ion channels. Jiang et al.[31] have hypothesized that these changes are probably attributable to oscillations in as yet undescribed Na^+ and K^+ channels across the circadian cycle. They concluded that the slow, inactivating, outward potassium currents ($I_{K,O}$); the calcium-activated potassium current (I_{Ca}) described previously in SCN neurons cultured from rat fetuses; and a fast-activating, outwardly rectifying K^+ current ($I_{K(FR)}$) reported by Bouskila and Dudek[13] are unlikely candidates for driving these rhythms, because they are not activated until the cell is depolarized beyond –40mV. Further research is required to establish the roles of these currents in the generation and propagation of circadian phase information within the SCN slice preparation.

Bouskila and Dudek[14] demonstrated that SCN neurons *in vitro* show a coordinated, circadian rhythm in multi-unit activity in Ca^{2+}-free solutions, indicating that Ca^{2+}-dependent synaptic transmission is not required for synchronization of the circadian activity of SCN cells. Because antagonists to GABA and glutamate neurotransmission also do not prevent such rhythmicity, non-classical neurotransmission mechanisms may be involved. Such results also suggest that circadian changes in the effects of GABA in the SCN[88] are unlikely to underlie the generation of cellular rhythmicity. Neuroanatomical evidence showing wide regions of membrane appositions among SCN cells and the close association of glial processes raises the possibility that other ionic means of intercellular communication may couple the electrical activity of SCN cells.[71,82]

3.6 SCN Efferents

Suprachiasmatic nuclei efferents have been widely mapped, but their functions are poorly understood. The use of SCN fetal transplants to restore behavioral rhythms to SCN-ablated, arrhythmic rodents has, however, provided some novel insights into the efferent regulation of rhythmicity by the SCN. Reciprocal transplant studies between different strains or species with different activity phenotypes have established that the period of the rescued circadian rhythm is that of the donor tissue and is therefore genetically determined.[63] Although early studies suggested that innervation of the host by the SCN transplant was associated with rhythm restoration, SCN transplants at sites remote from the anterior hypothalamus were also able to restore rhythmicity (with longer latency). Recent studies have shown that ventricular implants of SCN donor tissue encapsulated in a polymer coating, which prevents innervation in either direction, still restore activity rhythms to arrhythmic hamsters.[76] Apparently, a diffusible substance can convey donor-specific periodicity to an arrhythmic host, but the specific neurochemicals involved and the targets of their actions remain to be established. By contrast, endocrine rhythms, especially pineal melatonin rhythmicity, are not restored by such grafts. These results suggest that multiple efferent mechanisms may convey rhythmicity from the SCN to various target tissues in intact animals.

Suprachiasmatic nuclei transplantation studies have also addressed the role of the SCN in the deterioration of rhythmicity in aged rodents. Age-related deterioration of circadian rhythm amplitude can be reversed by implantation of fetal SCN tissue into aged, but otherwise intact, rodents.

Such transplants also restore the diminished responsiveness of the host SCN to phase-shifting stimuli such as light and triazolam.[85] Implantation of a fetal SCN of a different genotype into an animal with a damaged SCN that can still drive behavioral rhythmicity can lead to the emergence of two different rhythms expressed simultaneously. These "temporal chimeras" imply that each SCN can independently promote activity by the host.[87] SCN transplants may act through a combination of factors that directly regulate neural systems generating activation or arousal and factors that may have a trophic influence on the host's own pacemaker system.

3.7 Future Prospects

The major input pathways to the SCN have been well characterized in terms of their anatomy and some of their physiology. The specific receptor subtypes mediating their effects remains an open issue, but the available evidence indicates multiple routes by which each of these acts: retinal, raphe, and IGL projections all seem to involve effects mediated by several receptors, and even several transmitters. Blockade of activity involving IEGs, NO, or classical neurotransmitter receptors can inhibit photic phase shifts, but the precise sequence in which these mechanisms are normally activated and their functional relations remain to be analyzed.

The anatomical and neurochemical basis for intercellular communication among SCN neurons is less well understood and the neurophysiological mechanisms and neurotransmitter systems involved have hardly been explored. The evidence available, however, points to considerable complexity and the possibility of novel mechanisms of intercellular communication.

The mechanisms by which efferent projections from the SCN affect target systems also remain unclear. The evidence that a humoral factor regulates some behavioral rhythms while synaptically mediated mechanisms are required to regulate other rhythms hints at complexity in efferent mechanisms as well. Since the existence of circadian oscillators remote from the SCN is also well established, as in the case of mammalian retinal functions and in the case of a food-entrainable oscillator, future studies will also have to address the mechanisms of communication among these systems, which undoubtedly normally function in concert.

Future studies will also have to further investigate the role of glia in SCN cellular communication. Several findings point to important roles for glia, in addition to their probable functions in regulating extracellular glutamate and perhaps the levels of other transmitters in the SCN. Glia also wrap synaptic clusters in the SCN, apparently isolating these from other influences. There are reports of circadian oscillations in glial structure, of astrocytes in the SCN showing c-Fos expression in response to light, and of changes in rhythmicity after glia are destroyed by toxic agents. While no coherent picture of the role of SCN glia in circadian functions has yet emerged, future studies will undoubtedly identify new and significant functions for these components of the mammalian SCN.

References

1. Abe, H. and Rusak, B., Physiological mechanisms regulating photic induction of Fos-like protein in hamster suprachiasmatic nucleus, *Neurosci. BioBehav. Rev.*, 18, 531, 1994.
2. Abe, H., Honma, S., Shinohara, K., and Honma, K.–I., Circadian modulation in photic induction of Fos-like immunoreactivity in the suprachiasmatic nucleus of diurnal chipmunk, *Eutamias asiaticus*, *J. Comp. Physiol. A.*, 176, 159, 1995.
3. Albers, H.E. and Ferris, C.F., Neuropeptide Y: role in light-dark cycle entrainment of hamster circadian rhythms, *Neurosci. Lett.*, 50, 163, 1984.

4. Albers, H.E., Liou, S.-Y., Ferris, C.F., Stopa, E.G., and Zoeller, R.T., Neurotransmitter co-localization and circadian rhythms, *Prog. Brain Res.*, 92, 289, 1992.
5. Bennett, M.R. and Schwartz, W.J., Are glia among the cells that express immunoreactive c-Fos in the suprachiasmatic nucleus, *NeuroReport,* 5, 1737, 1994.
6. Biello, S.M., Enhanced photic phase shifting after treatment with antiserum to neuropeptide Y, *Brain Res.*, 673, 25, 1995.
7. Biello, S.M., Golombek, D.A., and Harrington, M.E., Neuropeptide Y and glutamate block each other's phase shifts in the suprachiasmatic nucleus *in vitro, Neuroscience*, 77, 1049, 1997.
8. Biello, S.M., Janik, D., and Mrsovsky, N., Neuropeptide Y and behaviorally induced phase shifts, *Neuroscience*, 62, 273, 1994.
9. Bina, K.G. and Rusak, B., Nerve growth factor phase shifts circadian activity rhythms in Syrian hamsters, *Neurosci. Lett.*, 206, 97, 1996.
10. Bina, K.G. and Rusak, B., Muscarinic receptors mediate carbachol-induced phase shifts of circadian activity rhythms in Syrian hamsters, *Brain Res.*, 743, 202, 1996.
11. Bina, K.G., Rusak, B., and Semba, K., Sources of p75-nerve growth factor receptor-like immunoreactivity in the rat suprachiasmatic nucleus, *Neuroscience*, 77, 461, 1997.
12. Bina, K.G., Rusak, B., and Semba, K., Localization of cholinergic neurons in the forebrain and brainstem that project to the suprachiasmatic nucleus of the hypothalamus of the rat, *J. Comp. Neurol.*, 335, 295, 1993.
13. Bouskila, Y. and Dudek, F.E., Neuronal synchronization without calcium-dependent synaptic transmission in the hypothalamus, *Proc. Nat. Acad. Sci. USA.*, 90, 3207, 1993.
14. Bouskila, Y. and Dudek, F.E., A rapidly activating type of outward rectifier current K^+ current and A-current in rat suprachiasmatic nucleus neurones, *J. Physiol.*, 488, 339, 1995.
15. Card, J.-P. and Moore, R.Y., The suprachiasmatic nucleus of the golden hamster: immunohistochemical analysis of cell and fiber distribution, *Neuroscience,* 13, 415, 1985.
16. Colwell, C.S. and Menaker, M., Regulation of circadian rhythms by excitatory amino acids, in *Excitating Amino Acids: Their Roles in Neuroendocrine Function,* Brann, D.W. and Mahesh, V.B., Eds., CRC Press, Boca Raton, FL, 1996.
17. Cui, L.-N. and Dyball, R.E.J., Synaptic input from the retina to the suprachiasmatic nucleus changes with the light-dark cycle in the Syrian hamster, *J. Physiol.*, 497, 483, 1996.
18. Ding, J.M., Chen, D., Weber, E.T., Faiman, L.E., Rea, M.A., and Gillette, M.U., Resetting the biological clock: mediation of nocturnal circadian shifts by glutamate and NO, *Science*, 266, 1713, 1994.
19. Ding, J.M., Faiman, L.E., Hurst, W.J., Kuriashkina, L.R., and Gillette, M.U., Resetting the biological clock: mediation of nocturnal CREB phosphorylation via light, glutamate, and nitric oxide, *J. Neurosci.*, 17, 667, 1997.
20. Ebling, F.J.P., The role of glutamate in the photic regulation of the suprachiasmatic nucleus, *Prog. Neurobiol.*, 50, 109, 1996.
21. Gillespie, C.F., Huhman, K.L., Babagbemi, T.O., and Albers, H.E., Bicuculline increases and muscimol reduces the phase-delaying effects of light and VIP/PHI/VIP in the suprachiasmatic region, *J. Biol. Rhythms*, 11, 137, 1996.
22. Gillette, M.U., Regulation of entrainment pathways by the suprachiasmatic circadian clock: sensitivities to second messengers, *Prog. Brain Res.*, 111, 121, 1996.
23. Golombek, D.A., Biello, S.M., Rendon, R.A., and Harrington, M.E., Neuropeptide Y phase shifts the circadian clock *in vitro* via a Y2 receptor, *NeuroReport*, 7, 1315, 1996.
24. Hannibal, J., Ding, J.M., Fahrenkrug, J., Larsen, P.J., Gillette, M.U., and Mikkelsen, J.D., Pituitary adenylate cyclase-activating peptide (PACAP) in the retinohypothalamic tract: a potential regulator of the biological clock, *J. Neurosci.*, 17, 2637, 1997.

25. Harrington, M.E., Nance, D.W., and Rusak, B., Double-labeling of neuropeptide Y-immunoreactive neurons which project from the geniculate to the suprachiasmatic nuclei, *Brain Res.*, 410, 275, 1987.
26. Huhman, K.L. and Albers, H.E., Neuropeptide Y microinjected into the suprachiasmatic region phase shifts circadian rhythms, *Peptides*, 15, 1475, 1994.
27. Huhman, K.L., Babagbemi, T.O., and Albers, H.E., Bicuculline blocks neuropeptide Y-induced phase advances when microinjected in the suprachiasmatic nucleus of Syrian hamsters, *Brain Res.*, 675, 333, 1995.
28. Huhman, K.L., Gillespie, C.F., Marvel, C.F., and Albers, H.E., Neuropeptide Y phase shifts circadian rhythms via Y2-like receptors, *NeuroReport*, 7, 1249, 1996.
29. Inouye, S.-I.T., Circadian rhythms of neuropeptides in the suprachiasmatic nucleus, *Prog. Brain Res.*, 111, 75, 1996.
30. Inouye, S.-I.T. and Shibata, S., Neurochemical organization of circadian rhythm in the suprachiasmatic nucleus, *Neurochem. Res.*, 20, 109, 1994.
31. Jiang, Z.-G., Allen, C.N., and and North, R.A., Presynaptic inhibition by baclofen of retinohypothalamic excitatory synaptic transmission in rat suprachiasmatic nucleus, *Neuroscience*, 64, 813, 1995.
32. Jiang, Z.-G., Yang, Y.Q., Liu, Z.-P., and Allen, C.N., Membrane properties and synaptic inputs of suprachiasmatic nucleus neurons in rat brain slices, *J. Physiol.*, 499, 141, 1997.
33. LeSauter, J., Lehman, M.N., and Silver, R., Restoration of circadian rhythmicity by transplants of SCN "micropunches", *J. Biol. Rhythms*, 11, 163, 1996.
34. Liu, C. and Gillette, M.U., Cholinergic regulation of the suprachiasmatic nucleus circadian rhythm via a muscarinic mechanism at night, *J. Neurosci.*, 16, 744, 1996.
35. Liu, C., Ding, J.M., Faiman, L.E., and Gillette, M.U., Coupling of muscarinic cholinergic receptors and cGMP in nocturnal regulation of the suprachiasmatic circadian clock, *J. Neurosci.*, 17, 659, 1997.
36. Liou, S.Y. and Albers, H.E., Single unit response of neurons within the hamster suprachiasmatic nucleus to GABA and low chloride perfusion during the day and night, *Brain Res. Bull.*, 25, 93, 1990.
37. Liou, S.Y. and Albers, H.E., Single unit response of neurons within the hamster suprachiasmatic nucleus to neuropeptide Y, *Brain Res. Bull.*, 27, 825, 1991.
38. Mason, R., Harrington, M.E., and Rusak, B., Electrophysiological responses of hamster suprachiasmatic neurons to neuropeptide Y in the hypothalamic slice preparations, *Neurosci. Lett.*, 80, 173, 1987.
39. Mason, R., Biello, S.M., and Harrington, M.E., The effects of GABA and benzodiazepines on neurones in the suprachiasmatic nucleus (SCN) of Syrian hamsters, *Brain Res.*, 552, 533, 1991.
40. Madeira, L., Andrade, J.P., Lieberman, A.R., Sousa, N., Almeida, O.F.X., and Paula-Barbosa, M.M., Chronic alcohol consumption and withdrawal do not induce cell death in the suprachiasmatic nucleus, but lead to irreversible depression of peptide immunoreactivity and mRNA levels, *J. Neurosci*, 17, 1302, 1997.
41. Mathur, A., Golombek, D.A., and Ralph, M.R., cGMP-dependent protein kinase inhibitors block light-induced phase advances of circadian rhythms *in vivo*, *Am. J. Physiol.*, 270, R1031, 1996.
42. Meeker, R.B., Greenwood, R.S., and Hayward, J.N., Glutamate receptors in the rat hypothalamus and pituitary, *Endocrinology*, 134, 621, 1994.
43. Meijer, J.H., Integration of visual information by the suprachiasmatic nucleus, in *Suprachiasmatic Nucleus: The Mind's Clock*, Klein, D.C., Moore, R.Y., and Reppert, S.M., Eds., Oxford University Press, New York, 1991, chap. 5.
44. Meijer, J.H., Watanabe, K., Dètàri, L., de Vries, M.J., Albus, H., Treep, J.A., Schaap, J., and Rietveld, W.J., Light entrainment of the mammalian biological clock, *Prog. Brain Res.*, 111, 175, 1996
45. Melo, L., Golombek, D.A., and Ralph, M.R., Regulation of circadian photic responses by nitric oxide, *J. Biol. Rhythms*, 12, 319, 1997.

46. Meyer, E.L., Harrington, M.E., and Rhamani, T., A phase-response curve to the benzodiazepine chlordiazepoxide and the effect of geniculohypothalamic tract ablation, *Physiol. Behav.*, 53, 237, 1993.
47. Meyer-Bernstein, E.L. and Morin, L.P., Differential serotonergic innervation of the suprachiasmatic nucleus and the intergeniculate leaflet and its role in circadian rhythm modulation, *J. Neurosci.*, 16, 2097, 1996.
48. Meyer-Bernstein, E.L., Blanchard, J.H., and Morin, L.P., The serotonergic projection from the median raphe nucleus modulates activity phase onset, but not other circadian rhythm parameters, *Brain Res.*, 755, 112, 1997.
49. Mintz, E.M. and Albers, H.E., Microinjection of NMDA into the SCN region mimics the phase shifting effect of light on hamsters, *Brain Res.*, 758, 245, 1997.
50. Mintz, E.M., Gillespie, C.F., Marvel, C.L., Huhman, K.L., and Albers, H.E., Serotonergic regulation of circadian rhythms, *Neuroscience,* 79, 563, 1997.
51. Moore, R.Y. and Speh, J.C., GABA is the principal neurotransmitter of the circadian system, *Neurosci. Lett.*, 150, 112, 1993.
52. Morin, L.P., The circadian visual system, *Brain Res. Rev.*, 67, 102, 1994.
53. Mrosovsky, N., Locomotor activity and non-photic influences on circadian clocks, *Biol. Rev.*, 71, 343, 1996.
54. O'Hara, B.F., Andretic, R., Heller, H.C., Carter, D.B., and Kilduff, T.S., $GABA_A$, $GABA_B$, $GABA_C$, and NMDA receptor subunit expression in the suprachiasmatic nucleus and other brain regions, *Mol. Brain Res.*, 28, 239, 1995.
55. Okamura, H., Murakami, S., Uda, K., Sugano, T., Takahashi, Y., Yanaihara, C., Yanaihara, N., and Ibata, Y., Coexistence of vasoactive intestinal polypeptide (VIP)-, peptide histidine isoleucine (PHI), and gastrin-releasing peptide (GRP)-like immunoreactivity in neurons of the rat suprachiasmatic nucleus, *Biol. Res.*, 7., 295, 1986.
56. Okamura, H. and Ibata, Y., GRP immunoreactivity shows a day-night difference in the suprachiasmatic nuclear soma and efferent fibers: comparison to VIP immunoreactivity, *Neurosci. Lett.*, 181, 165, 1994.
57. Pickard, G.E., Weber, E.T., Scott, P.A., Riberdy, A.F., and Rea, M.A., 5HT1B receptor agonist inhibit light-induced phase shifts of behavioral circadian rhythms and expression of the immediate-early gene c-*fos* in the suprachiasmatic nucleus, *J. Neurosci.*, 16, 8208, 1996.
58. Piggins, H.D. and Rusak, B., Electrophysiological effects of pressure-ejected bombesin-like peptides on hamster suprachiasmatic nucleus neurons *in vitro, J. Neuroendo.*, 5, 575, 1993.
59. Piggins, H.D., Cutler, D.J., and Rusak, B., Effects of ionophoretically applied bombesin-like peptides on hamster suprachiasmatic nucleus neurons *in vitro, Eur. J. Pharmacol.*, 271, 413, 1994.
60. Piggins, H.D., Antle, M.C., and Rusak, B., Neuropeptides phase shift the mammalian circadian pacemaker, *J. Neurosci.*, 15, 5612, 1995.
61. Piggins, H.D. and Rusak, B., Effects of microinjections of substance P into the suprachiasmatic nucleus region on hamster wheel-running rhythms, *Brain Res. Bull.*, 42, 451, 1997.
62. Piggins, H.D., Stamp, J.A., Burns, J., Rusak, B., and Semba, K., Distribution of pituitary adenylate cyclase activating polypeptide (PACAP) immunoreactivity in the hypothalamus and extended amygdala of the rat, *J. Comp. Neurol.*, 376, 278, 1996.
63. Ralph, M.R., Foster, R.G., Davis, F.C., and Menaker, M., Transplanted suprachiasmatic nucleus determines circadian pacemaker, *Science,* 247, 975, 1990.
64. Reuss, S., Components and connections of the circadian timing system in mammals, *Cell Tissue Res.*, 285, 353, 1996.
65. Romijn, H.J., Sluiter, A.A., Pool, C.W., Wortel, J., and Buijs, R.M., Difference in co-localization between Fos and PHI, GRP, VIP and VP in neurons of the rat suprachiasmatic nucleus after a light stimulus during the phase delay versus phase advance period of the night, *J. Comp. Neurol.*, 372, 1, 1996.

66. Rusak, B., Robertson, H.A., Wisden, W., and Hunt, S.P., Light pulses that shift rhythms induce gene expression in the suprachiasmatic nucleus, *Science,* 248, 1237, 1990.

67. Schmahl, C. and Böhmer, G., Effects of excitatory amino acids and neuropeptide Y on the discharge activity of suprachiasmatic neurons in rat brain slices, *Brain Res.,* 746, 151, 1997.

68. Scott, G. and Rusak, B., Activation of hamster suprachiasmatic neurones in vitro via metabotropic glutamate receptors, *Neuroscience,* 71, 533, 1996.

69. Schwartz, W.J., Further evaluation of the tetrodotoxin-resistant circadian pacemaker in the suprachiasmatic nuclei, *J. Biol. Rhythms*, 6, 149, 1991.

70. Selim, M., Glass, J.D., Hauser, U.E., and Rea, M.A., Serotonergic inhibition of light-induced fos protein expression and extracellular glutamate in the suprachiasmatic nuclei, *Brain Res.*, 621, 181, 1993.

71. Servière, J. and Lavialle, M., Astrocytes in the mammalian circadian clock: putative roles, *Prog. Brain Res.*, 111, 57, 1996.

72. Shibata, S. and Moore, R.Y., Neuropeptide Y and vasopressin effects on rat, *J. Biol. Rhythms*, 3, 265, 1988.

73. Shibata, S. and Moore, R.Y., Tetrodotoxin does not affect circadian rhythms in neuronal activity and metabolism in rodent suprachiasmatic nucleus in vitro, *Brain Res.*, 606, 259, 1993.

74. Shibata, S., Tsuneyoshi, A., Hamada, T., Tominaga, K., and Watanabe, S., Effects of substance P on circadian rhythms of firing activity and 2-deoxyglucose uptake in the rat suprachiasmatic nucleus in vitro, *Brain Res.*, 597, 257, 1992.

75. Shibata, S., Ono, M., Tominaga, K., Hamada, T., Watanabe, A., and Watanabe, S., Involvement of vasoactive intestinal polypeptide in NMDA-induced phase delay of firing activity rhythm in the suprachiasmatic nucleus in vitro, *Neurosci. Biobehav. Rev.*, 18, 591, 1994.

76. Silver, R., LeSauter, J., Tresco, P.A., and Lehman, M.N., A diffusible coupling signal from the transplanted suprachiasmatic nucleus controlling circadian locomotor rhythms, *Nature,* 382, 810, 1996.

77. Srkalovic, G., Selim, M., Rea, M.A., and Glass, J.D., Serotonergic inhibition of extracellular glutamate in the suprachiasmatic nuclear region assessed using in vivo brain microdialysis, *Brain Res.*, 656, 302, 1994.

78. Stamp, J.A., Piggins, H.D., Rusak, B., and Semba, K., Distribution of ionotropic glutamate receptor subunit immunoreactivity in the suprachiasmatic nucleus and intergeniculate leaflet of the hamster, *Brain Res.*, 756, 215, 1997.

79. Tominaga, K., Shibata, S., Hamada, T., and Watanabe, S., $GABA_A$ receptor agonist muscimol can reset the phase of neural activity rhythm in the rat suprachiasmatic nucleus in vitro, *Neurosci. Lett.*, 166, 81, 1994.

80. Treep, J.A., Abe, H., Rusak, B., and Goguen, D.M., Two distinct retinal projections to the hamster suprachiasmatic nucleus, *J. Biol. Rhythms*, 10, 299, 1995.

81. Turek, F.W. and van Reeth, O., Altering the mammalian circadian clock with the short-acting benzodiazepine, triazolam, *Trends Neurosci.*, 11, 535, 1998.

82. van den Pol, A.N., The suprachiasmatic nucleus: morphological and cytochemical substrates for cellular communication, in *Suprachiasmatic Nucleus: The Mind's Clock*, Klein, D.C., Moore, R.Y., and Reppert, S.M., Eds., Oxford University Press, New York, 1991, chap. 2.

83. van den Pol, A.N., Obrietan, K., Chen, G., and Belousov, A.B., Neuropeptide Y-mediated long-term depression of excitatory activity in suprachiasmatic nucleus neurons, *J. Neurosci.*, 16, 5883, 1996.

84. van der Zee, E.A., Streetfland, C., Strosberg, A.D., Schröder, A.D., and Luiten, P.G.M., Colocalization of muscarinic and nicotinic receptors in cholinoreceptive neurons of the suprachiasmatic region in young and aged rats, *Brain Res.*, 542, 348, 1991.

85. van Reeth, O., Zhang, Y., Zee, P.C., and Turek, F.W., Grafting fetal suprachiasmatic nuclei in the hypothalamus of old hamsters restores responsiveness of the circadian clock to a phase shifting stimulus, *Brain Res.*, 643, 338, 1994.
86. Vincent, S.R., Nitric oxide: a radical neurotransmitter in the central nervous system, *Prog. Neurobiol.*, 42, 129, 1994.
87. Vogelbaum, M.A. and Menaker, M., Temporal chimeras produced by hypothalamic transplants, *J. Neurosci.*, 12, 3619, 1992.
88. Wagner, S., Castel, M., Gainer, H., and Yarom, Y., GABA in the mammalian suprachiasmatic nucleus and its role in diurnal rhythmicity, *Nature*, 387, 598, 1997.
89. Wang, H. and Morris, J.F., Presence of neuronal nitric oxide synthase in the suprachiasmatic nuclei of mouse and rat, *Neuroscience*, 74, 1059, 1996.
90. Watts, A.G., The efferent projections of the suprachiasmatic nucleus: anatomical insights into the control circadian rhythms, in *Suprachiasmatic Nucleus:The Mind's Clock*, Klein, D.C., Moore, R.Y., and Reppert, S.M., Oxford University Press, New York, 1991, chap. 4.
91. Weber, E.T., Gannon, R.L., Michel, A.M., Gillette, M.U., and Rea, M.A., Nitric oxide synthase inhibitor blocks light-induced phase shifts of circadian activity rhythm, but not c-fos expression in the suprachiasmatic nucleus of the Syrian hamster, *Brain Res.*, 692, 137, 1995.
92. Weber, E.T., Gannon, R.L., and Rea, M.A., cGMP-dependent protein kinase inhibitor blocks light-induced phase shift advances of circadian activity *in vivo, Neurosci. Lett.*, 197, 227, 1995.
93. Welsh, D.K., Logothetis, D.E., Meister, M., and Reppert, S.M., Individual neurons dissociated from rat suprachiasmatic nucleus express independently phased circadian firing rates, *Neuron*, 14, 697, 1995.
94. Wollnik, F., Brysch, W., Uhlmann, E., Gillardon, F., Bravo, R., Zimmerman, M., Schlingensiepen, K.H., and Herdegen, T., Block of c-Fos and JunB expression by antisense oligonucleotides inhibits light-induced phase shifts of the mammalian circadian clock, *Eur. J. Neurosci.*, 7, 388, 1995.
95. Ying, S.-W. and Rusak, B., 5-HT$_7$ receptors mediate serotonergic effects on light-sensitive suprachiasmatic nucleus neurons, *Brain Res.*, 755, 246, 1997.
96. Zatz, M. and Herkenham, M.A., Injection of α-bungarotoxin near the suprachiasmatic nucleus blocks the effects of light on nocturnal pineal enzyme activity, *Brain Res.*, 213, 438, 1981.
97. Zhang, Y., Zee, P.C., Kirby, J.D., Takahashi, J.S., and Turek, F.W., A cholinergic antagonist, mecamylamine, blocks light-induced Fos immunoreactivity in specific regions of the hamster suprachiasmatic nucleus, *Brain Res.*, 615, 107, 1993.

Chapter 4

The Molecular Basis of the Pineal Melatonin Rhythm: Regulation of Serotonin N-Acetylation

David C. Klein, Ruben Baler, Patrick H. Roseboom, J.L. Weller, Marianne Bernard, Jonathan A. Gastel, Martin Zatz, P. Michael Iuvone, Valerie Bégay, Jack Falcón, Greg Cahill, Vincent M. Cassone, and Steven L. Coon

Contents

4.1	General Characteristics of Melatonin Rhythm-Generating Systems		46
	4.1.1	The Site of Melatonin Production	46
	4.1.2	Sources of ~24-Hr Signals	47
	4.1.3	Sites of Photodetection and the Effects of Light	48
4.2	AANAT — The Melatonin Rhythm-Generating Enzyme		49
	4.2.1	Tissue Distribution	49
	4.2.2	Functional Anatomy of AANAT	50
	4.2.3	Regulation of AANAT Activity by Synthesis and Proteolysis	52
4.3	Examples of Melatonin Rhythm-Generating Systems		52
	4.3.1	Regulation in Mammals	52
		4.3.1.1 Organization of the Mammalian Melatonin Rhythm-Generating System	52
		4.3.1.2 Adrenergic Signal Transduction	53
		4.3.1.3 Features of AANAT Regulation in the Rat	53
		4.3.1.4 Features of Regulation in Sheep	54
	4.3.2	Regulation in Chicken	54
		4.3.2.1 Organization of the Melatonin Rhythm-Generating System in the Chicken	54
		4.3.2.2 Cellular Regulatory Mechanisms	55
	4.3.3	Regulation in Fish	55
		4.3.3.1 Pike and Zebrafish	55
		4.3.3.2 Trout	55

| 4.4 | Final Comments | 56 |
| References | | 56 |

The day/night rhythm in circulating levels of melatonin (*N*-acetyl 5-methoxytryptamine) is a constant characteristic of vertebrate physiology (Figure 4.1);[1] circulating melatonin is always elevated about tenfold at night relative to day values. Melatonin is considered to be the hormone of the night[2] because it provides the organism with a highly reliable humoral signal that is proportional to the duration of the night phase. Changes in night length, such as those which occur on a seasonal basis, are translated into changes in the duration and/or timing of the period of elevated melatonin production. The highly reliable and accurate translation of night length into melatonin production by vertebrates reflects a set of regulatory mechanisms which limit high levels of melatonin production to the night period and minimize synthesis in the light at any time.

Seasonal lengthening and shortening of the nocturnal melatonin signal can globally change physiology in some species, altering reproduction, body weight, behavior associated with reproduction, and coat color.[1,3] In some cases, such as sheep, animals become reproductively active in response to shorter periods of melatonin production; in others, such as the Syrian hamster, this inhibits reproduction. Melatonin has a role in all vertebrates to modulate endogenous clock function; it also influences a broad range of physiological functions, including sleep.[1,3,4]

In sharp contrast to the conserved day/night pattern of melatonin production, there is remarkable species-to-species diversity in both the anatomical organization of the systems which generate this rhythm and in the molecular and cellular mechanisms which control melatonin biosynthesis. Such diversity indicates many solutions have evolved to ensure that the nocturnal increase in melatonin is a reliable indicator of the night period. This emphasizes the essential role the rhythm in melatonin plays in vertebrate physiology.

This chapter describes conserved and species-specific features of melatonin rhythm-generating systems. It is divided into three sections. The first is an overview of the basic functional components of these systems. The second is devoted to a molecule which has unique importance in vertebrate circadian biology — serotonin *N*-acetyltransferase (arylalkylamine *N*-acetyltransferase, AANAT). This enzyme is the critical interface between regulatory mechanisms and melatonin synthesis. This pivotal role has earned it the designation "the melatonin rhythm-generating enzyme". The last section emphasizes the diverse nature of melatonin rhythm-generating systems by describing examples from three vertebrate classes.

4.1 General Characteristics of Melatonin Rhythm-Generating Systems

The fundamental components of melatonin rhythm-generating systems are a site of melatonin production, a source of ~24 hour signals, and a detector through which light acts on the system (Figure 4.2).

4.1.1 The Site of Melatonin Production

The source of circulating melatonin is the pineal gland, which can be considered to be the melatonin factory. The ability of this tissue to produce high levels of melatonin reflects several biochemical features, including high levels of tryptophan hydroxylase (the first enzyme in the conversion of circulating tryptophan to serotonin), a high concentration of serotonin, and high levels of the enzymes required for the serotonin–*N*-acetylserotonin–melatonin pathway, i.e., AANAT and

The Molecular Basis of the Pineal Melatonin Rhythm

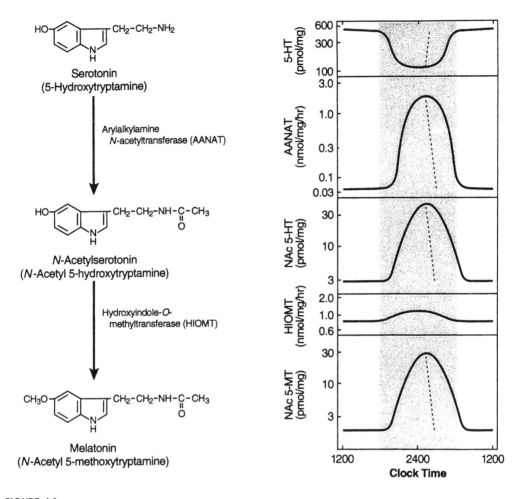

FIGURE 4.1
Rhythms in pineal indole metabolism. The dashed lines at night (shaded) represent the effects of an unexpected exposure to light at night. (From Klein, D.C. et al., *Rec. Prog. Hormone Res.*, 52, 307, 1997. With permission.)

hydroxyindole-*O*-methyltransferase (HIOMT; see Figure 4.1). Other tissues do not possess all these features, and for this reason pinealectomy causes melatonin to nearly disappear from the circulation.[1] As discussed below, the low levels of melatonin synthesized in the retina do not contribute to circulating melatonin.

As indicated above, AANAT is an interface which converts regulatory input into changes in the production of melatonin. Day/night changes in melatonin production occur in response to changes in AANAT activity, which controls *N*-acetylserotonin levels. The rate of melatonin production by HIOMT is a mass action function of *N*-acetylserotonin (Figure 4.1).

4.1.2 Sources of ~24-Hr Signals

The rhythmic ~24-hr signals which determine the pattern of melatonin production are generated by one or more clocks. In most lower vertebrates, an internal clock is typically located in the pineal gland. In mammals, the internal clock is in the suprachiasmatic nucleus (SCN) of the hypothalamus. In some vertebrates, such as the chicken, multiple internal clocks control melatonin synthesis.

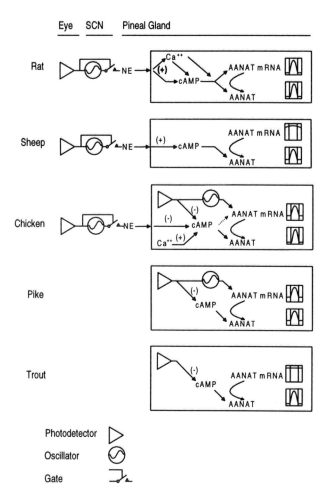

FIGURE 4.2
Melatonin rhythm-generating systems. For details see the text.

Systems that lack an internal clock have also been identified, such as the trout pineal gland. In these cases, the environmental dark/light cycle generates a diurnal rhythm in melatonin production.

One functional advantage of an endogenous clock which drives melatonin synthesis — as opposed to a simple light-off/dark-on system — is that the clock prevents high levels of melatonin synthesis from occurring during the day if animals are in darkness. Another advantage is that it allows an animal to anticipate day/night transitions. In this way, eating schedules and the sleep/wake cycle are optimally synchronized with the environmental light/dark cycle.

4.1.3 Sites of Photodetection and the Effects of Light

The third component of melatonin rhythm-generating systems is the photodetector. Light entrains and modulates the clock and also switches off melatonin production by turning off AANAT activity. The photoreceptors which act on the SCN clock are located in the lateral eyes; photoreceptors which act on pineal clocks are located in the pineal gland.

The entrainment function of light is the main mechanism which resets the clock and synchronizes it with the environmental light/dark cycle. Without this resetting influence, the internal clock free-runs, i.e., drifts out of phase with the light/dark cycle. The modulation function adjusts the period that the clock stimulates the pineal gland at night, which increases in the winter (long nights) and decreases in summer (short nights). The entrainment and modulation actions of light involve alterations in clock function. Light also acts to suppress melatonin production by interrupting clock stimulation of AANAT. This minimizes synthesis of melatonin during the day and reinforces limitations imposed by the internal clock.

4.2 AANAT — The Melatonin Rhythm-Generating Enzyme

The importance of AANAT in vertebrate circadian biology has stimulated significant interest in the regulation of enzyme activity, the links between pineal clocks and AANAT, the mechanism of enzyme action, and the basis of tissue-specific expression. Our current knowledge of this enzyme is summarized here; a more detailed description is available elsewhere.[5]

AANAT is referred to both as serotonin *N*-acetyltransferase and as arylalkylamine-*N*-acetyltransferase. The former nomenclature reflects the fact that serotonin is the best known substrate of the enzyme; the latter nomenclature recognizes the general chemical family to which serotonin belongs. A small group of other arylalkylamines are also substrates, including tryptamine, methoxytryptamine, phenylethylamine, and tyramine.[5]

Although it is clear that AANAT is the key regulator of the large day/night change in melatonin production, and that day/night changes in HIOMT protein do not appear to play a dominant role in generating this rhythm, it is important to note that melatonin production is subject to limitations imposed by the activity of tryptophan hydroxylase and the availability of serotonin and cofactors, in addition to HIOMT activity.

Two features of the pattern of the serotonin-melatonin pathway are seen in all vertebrates. One feature is the reciprocal relationship between serotonin and the *N*-acetylated derivatives, with high levels of serotonin occurring during the day and low levels at night. The second is the switch-off effect of light, which converts the nighttime pineal indole pattern to a daytime pattern (see the broken line in Figure 4.1). Although this is very rapid in the rat, the rate varies somewhat from species to species.

4.2.1 Tissue Distribution

AANAT is selectively expressed in the pineal gland and to a lesser and more variable degree in the retina.[5,8–12] Melatonin synthesis in the retina is relatively low and is thought to serve a local function.[13–16] Very low levels of AANAT expression have also been detected in brain regions, pituitary, and testes;[5] melatonin synthesis in these sites, however, has not been documented.

It is of interest to note that the retinal expression of AANAT is a reflection of a more general pineal/retinal overlap in gene expression. In some species, both tissues synthesize melatonin, detect light, and contain endogenous clocks. Pineal cells and retinal photoreceptor cells develop from adjacent areas of the roof of the diencephalon, and it is thought that both types of cells share a common ancestral photoneuroendocrine cell capable of phototransduction and circadian synthesis of melatonin.[17,18]

The pineal/retinal pattern of gene expression may be determined in part by genomic photoreceptor conserved elements (PCEs), which are short DNA sequences that identify genes for expression in these tissues.[19] Such elements occur in photoreceptor-related genes in both vertebrates and invertebrates, including opsins, and in the HIOMT and AANAT genes.[20,21] It is suspected that developmental expression of these genes is regulated by PCE binding proteins.

```
                                                       pka
                  1
Human       mstqsthpLK PeaprlppGi pespscQRRHxxLPAsEFRCL tPEDAvsaFE
Rat         ~~mlsihpLK PealhlplGt seflgcQRRHxxLPAsEFRCL tPEDAtsaFE
Sheep       mstpsvhcLK PsplhlpsGi pgspgrQRRHxxLPAnEFRCL tPEDAagvFE
Chicken     mpvlgavpfLK Ptplq...Gp rnspgrQRRHxxLPAsEFRCL sPEDAvsvFE
Consensus   --------LK P-------G- ------QRRHxxLPA-EFRCL -PEDA---FE

                  51
Human       IEREAFISVl GvCPLyLDEi rHFLTLCPEL SLGWFeEGcL VAFIIGSLWD
Rat         IEREAFISVs GtCPLhLDEi rHFLTLCPEL SLGWFeEGcL VAFIIGSLWD
Sheep       IEREAFISVs GnCPLnLDEv qHFLTLCPEL SLGWFvEGrL VAFIIGSLWD
Chicken     IEREAFISVs GdCPLhLDEi rHFLTLCPEL SLGWFeEGrL VAFIIGSLWD
Consensus   IEREAFISV- G-CPL-LDE- -HFLTLCPEL SLGWF-EG-L VAFIIGSLWD
            Region C/c-1            Region D/c-1      Region D/c-2

                  101
Human       keRLmQesLt LHrsgGhiaH lHvLAVHRaF RQQGrGpiLl WRYLhhlgsq
Rat         keRLtQesLt LHrpgGrtaH lHvLAVHRtF RQQGkGsvLl WRYLhhlgsq
Sheep       eeRLtQesLa LHrprGhsaH lHaLAVHRsF RQQGkGsvLl WRYLhhvgaq
Chicken     qdRLsQaaLt LHnprGtavH iHvLAVHRtF RQQGkGsiLm WRYLqylrcl
Consensus   --RL-Q---L- LH---G---H -H-LAVHR-F RQQG-G---L- WRYL------
                                       Motif A

                  151
Human       PavRrAaLMC EdaLVPFYer fsFhavGPCa itvGsLtFmE lhcslggHpf
Rat         PavRrAvLMC EnaLVPFYek fgFqamGPCa itmGsLtFtE lqcslrcHtf
Sheep       PavRrAvLMC EdaLVPFYqr fgFhpaGPCa ivvGsLtFtE mhcslrgHaa
Chicken     PcaRpAvLMC EdfLVPFYek cgFvavGPCq vtvGtLaFtE mqhevrgHaf
Consensus   P--R-A-LMC E--LVPFY-- --F---GPC- ---G-L-F-E -------H--
                              Motif B
                  pka
                  201
Human       lRRNSgc
Rat         lRRNSgc
Sheep       lRRNSdr
Chicken     mRRNSgc
Consensus   -RRNS--
```

FIGURE 4.3

AANAT amino acid sequences. Deduced amino acid sequences of AANAT from human (GenBank accession # U40347), rat (GenBank accession # U38306), sheep (GenBank accession # U29663), and chicken (GenBank accession # U46502). The consensus sequence identifies amino acids that are identical between all species listed. Capital letters conform to the consensus sequences. The conserved putative cyclic nucleotide-dependent protein kinase phosphorylation sites are shaded. The conserved motifs shared with other members of the A/B (GNAT)[5,8,12a] superfamily are underlined, as are the regions unique to AANATs; the nomenclature assigned to these regions reflects limited homology with C and D motifs identified within the superfamily and conserved (c) AANAT sequences.

4.2.2 Functional Anatomy of AANAT

At this writing, structure/function relationships of the AANAT molecule are under active investigation, and some of the concepts presented here should be regarded as "best guesses". AANAT is a cytosolic ~24-kDa protein (203 to 207 amino acids; see Figure 4.3).[5,8–12] The catalytic domain

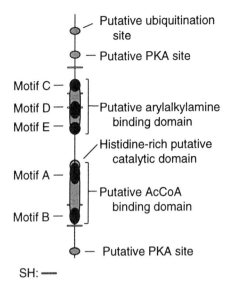

FIGURE 4.4
The functional anatomy of AANAT. The identified features are conserved in all the available deduced amino acid sequences (see Figure 4.3). PKA, cyclic AMP-dependent protein kinase; SH, cysteine. (Modified from Klein, D.C. et al., *Rec. Prog. Hormone Res.*, 52, 307, 1997. With permission.)

(~140 amino acids), in which substrate binding and acetyl transfer occur, occupies the central core of the molecule (Figure 4.4). The apparent AcCoA binding site is characterized by two motifs — motifs A and B, which occur within a 60-residue stretch (Figure 4.3). These motifs are sequences of functionally similar residues, rather than identical amino acids. Their presence in tandem is the identifying feature of the members of a superfamily of acetyltransferases which otherwise exhibit little similarity; the terms A/B and GNAT have been used to identify this superfamily.[5,8,12a] Although all members use AcCoA as an acetyl donor, each member exhibits high specificity toward a distinct narrow set of substrates, such as histones, antibiotics, diamines, or biogenic amines.

The putative arylalkylamine binding domain of AANAT (~50 amino acids) is characterized by three conserved regions (C/c-1, D/c-1, and D/c-2; Figures 4.3 and 4.4). Interestingly, another A/B superfamily arylalkylamine *N*-acetyltransferase has been identified in *Drosophila melanogaster* (DMNAT).[22] It has the very low homology to AANAT as is seen with other A/B superfamily members and does not contain these conserved regions. Accordingly, it does not belong to the AANAT gene family.

The putative catalytic domain which transfers acetyl groups from AcCoA to arylalkylamine acceptors is located between the arylalkylamine and AcCoA binding domains. Acetyl transfer is presumed to reflect the catalytic action of the imidazole moiety of histidines in this region.[5,23]

Two prominent putative regulatory features of the enzyme are the cyclic AMP-dependent protein kinase (PKA) phosphorylation sites in the C- and N-terminal regions of the protein. These sites are suspected of being important because they are conserved in all AANAT molecules and also because cyclic AMP is known to be critical for maintaining AANAT activity in all systems.

The initial 25-amino-acid portion of AANAT bounded by the N-terminal PKA phosphorylation site has a relatively high abundance of prolines and a 100% conserved lysine; otherwise, this region has relatively poor sequence conservation. The conserved lysine could play a very important regulatory role because it may be a site for ubiquitination. This is thought to target proteins for degradation by the proteasome, the macromolecular complex which contains multiple proteolytic activities.[24] The high abundance of prolines in this region may also play a role in this process. The presence of the PKA site and the lysine in this region suggests that they mediate adrenergic-cyclic AMP-regulated proteolysis, as discussed below.[24]

4.2.3 Regulation of AANAT Activity by Synthesis and Proteolysis

A very close relationship exists between AANAT protein and activity.[24] This is evident at all times and persists during the light-induced turn-off. This is of importance in understanding the regulation of AANAT because it eliminates the possibility that large changes in activity reflect posttranslational modifications that shift existing populations of AANAT molecules between active to inactive forms. The requirement for *de novo* synthesis for activity to increase makes the light-induced turn-off a one-way "off-only" switch.[24,25] This prevents spikes in melatonin synthesis at inappropriate times in response to transient, sporadic fluctuations in second messengers. Such spikes might occur if activity were only regulated by a simple posttranslational event, such as phosphorylation. This mechanism enhances the integrity of the melatonin rhythm-generating system by reducing "noise".

Cyclic AMP plays at least two roles in regulating AANAT, each of which differ in relative importance on a species-to-species basis. It appears likely that in all species cyclic AMP regulates activity by blocking AANAT proteolysis. This might function as the sole regulator of enzyme activity in species in which AANAT mRNA levels are constantly elevated. A hypothetical scenario is that AANAT protein is always made when AANAT mRNA is available and that cyclic AMP acts to inhibit proteolysis of newly synthesized molecules of AANAT. Cyclic AMP might influence degradation through a direct influence on AANAT phosphorylation or through an indirect influence via phosphorylation of a protein involved in targeting AANAT for degradation. For example, cyclic AMP might inhibit the hypothetical conjugation of AANAT to ubiquitin, or it might activate deubiquitination of ubiquitinated AANAT. In pineal glands in which AANAT mRNA is always available, cyclic AMP inhibition of proteolysis allows AANAT protein and activity to increase immediately at the start of the night — without a lag — and to be maintained at a high level throughout the night. As a result, melatonin can be produced from dusk to dawn in most natural lighting cycles.

The second mechanism through which cyclic AMP controls AANAT activity is by regulating AANAT mRNA. As described below, AANAT mRNA is nearly undetectable during the day in the rat, and cyclic AMP induces a 100- to 300-fold increase. Without this increase, AANAT activity cannot increase. The advantage of this is that it eliminates the possibility that AANAT activity and melatonin production could increase during the day. It is not unreasonable to suspect that in cases where there is a rhythm in AANAT mRNA, inappropriate production of melatonin does not favor species survival.

It should be added that in addition to cyclic AMP, other factors play a role in regulating AANAT activity. These include calcium and unidentified factors which link the pineal clock to expression of the AANAT gene.[1,2,5,10]

4.3 Examples of Melatonin Rhythm-Generating Systems

The following section covers pineal melatonin rhythm-generating systems but will not cover how retinal melatonin synthesis is controlled.[27,28] Although similarities in melatonin production exist between the pineal gland and retina, this is not always the case. For example, in the hamster, mouse, and rat, the retina differs from the pineal in that the former contains a clock, whereas the latter does not.[29] In this way, the rodent retina appears to be similar to the retinae and pineal glands of lower vertebrates.

4.3.1 Regulation in Mammals

4.3.1.1 Organization of the Mammalian Melatonin Rhythm-Generating System

The anatomical organization of the rhythm-generating system is essentially identical in all mammals. The clock that drives pineal melatonin synthesis is in the SCN,[30] which is connected to the pineal gland by a neural pathway passing through central and peripheral structures.[25] SCN cells project to cells in the paraventricular nucleus, which in turn send projections down the spinal cord to the intermediolateral cell column and synapse with preganglionic cells. These innervate superior cervical ganglia cells which send norepinephrine (NE)-containing sympathetic projections to the pineal gland. At night, stimulatory signals from the SCN cause the release of NE into the pineal extracellular space. Photic signals act via the retina and travel to the SCN via a retinal hypothalamic projection which exits the optic nerves at the optic chiasm.[25,31]

Light at night acts downstream of the SCN clock to block release of NE in the pineal gland. This effect is enhanced at the level of the pineal gland by the rapid uptake of residual extracellular NE into sympathetic nerve terminals; light is not known to act directly on the mammalian pineal gland.

4.3.1.2 Adrenergic Signal Transduction

The most important positive influence of NE is β_1-adrenergic stimulation of adenylate cyclase.[25] This is potentiated by simultaneous stimulation of α_1-adrenergic receptors which elevates intracellular Ca^{++} ($[Ca^{++}]_i$) and increases the activity of several phospholipases and activates protein kinase C.[25,32] Activation of protein kinase C is primarily responsible for sensitizing adenylate cyclase[33] to β_1-adrenergic activation. The resulting increase in cyclic AMP is essential for the increase in AANAT activity; $[Ca^{++}]_i$ also appears to act in an independent manner to enhance downstream effects of cyclic AMP.[33]

4.3.1.3 Features of AANAT Regulation in the Rat

Cyclic AMP acts to regulate AANAT protein and activity in the rat by increasing AANAT mRNA accumulation and by preventing proteolysis.[9,24,25] The time course of the nocturnal increase in melatonin production in the rat, as is true also of the hamster,[1,34] is characterized by a lag phase followed by a sigmoidal shaped response which returns to basal values prior to lights-on.

4.3.1.3.1 Transcriptional Regulation. Cyclic AMP increases AANAT mRNA 100- to 300-fold, from nearly undetectable levels through PKA phosphorylation of a cyclic AMP response element binding protein (CREB). CREB resides on the AANAT gene, bound to a cyclic AMP response element (CRE).[9,35,36] The consequent increase in AANAT mRNA is accompanied by an increase in AANAT protein and activity, provided adrenergic stimulation is maintained. In contrast to AANAT protein and activity, AANAT mRNA does not decrease rapidly when stimulation is abruptly blocked by light exposure or adrenergic blockade.

The amplitude of the increase in AANAT mRNA appears to be governed in part by the inducible cyclic AMP early repressor (ICER), a negatively acting transcription factor which competes with CREB for binding to the CRE.[37] Although the mRNA encoding this protein exhibits a dramatic rhythm in abundance similar to that of AANAT, ICER protein is relatively stable and does not undergo dramatic day/night changes. It is reasonable to suspect that ICER protein provides an integrated molecular memory of the duration of previous night periods and that this influences future patterns of response, by limiting the transcriptional response.[38]

Rat pineal AANAT mRNA gradually decreases at the end of the night. This may be due to several redundant mechanisms. First, it is generally thought that the strength of SCN stimulation starts to decrease late in the night. Second, it is likely that the cyclic AMP response of pineal cells to adrenergic stimulation gradually decreases during the course of the night, due to down-regulation and desensitization. A third possible negative influence is the induction of the early immediate gene, Fos-related antigen-2 (Fra-2). Each night Fra-2 protein increases rapidly, with dynamics similar to those of AANAT mRNA and AANAT protein.[39] Fra-2 heterodimerizes with a member of the Jun family to form a complex which is thought to bind strongly to AP-1 sites in the AANAT promoter, yet not induce transcription. As a result, it may suppress transcription.

The presence of significant levels of AANAT mRNA only at night, as is the case in the rat, essentially restricts AANAT protein synthesis to the night. The time required for AANAT mRNA to accumulate imposes a lag on the timing of the increase in AANAT protein and activity. This may be of special functional importance in fine-tuning the shape of the melatonin signal in seasonal breeders with short gestation periods, such as the hamster,[34] where very subtle changes in the duration of the night period control reproduction.[1]

4.3.1.3.2 Regulation by Inhibition of Proteolysis. The second important regulatory mechanism through which cyclic AMP acts in the rat is to prevent proteolysis of AANAT protein. As described above, when cyclic AMP is high, AANAT appears to be long-lived and not subject to significant proteolysis. However, when cyclic AMP drops, both activity and protein drop in parallel.[24] The half-life of the decrease in activity and protein in the rat is approximately 3.5 min. It is also likely that cyclic AMP-inhibition of protein degradation permits AANAT protein to accumulate when AANAT mRNA increases and also maintains elevated levels of AANAT protein at night.[24]

4.3.1.3.3 Unidentified Influences of Cyclic AMP. The current model of how cyclic AMP regulates AANAT protein and activity in the rat is based on the dynamics of mRNA and the regulation of proteolysis. It proposes that AANAT protein and activity increase when mRNA is available and proteolysis in inhibited. However, other mechanisms may play a role. For example, cyclic AMP might hypothetically enhance AANAT translation.[21]

4.3.1.4 Features of Regulation in Sheep

Sheep are representative of those species in which the melatonin rhythm has a square wave pattern, i.e., melatonin increases rapidly after lights off. Other mammals exhibiting this pattern include the human and monkey.[1,40] AANAT activity is regulated by an adrenergic-cyclic AMP mechanism in sheep[41] as it is in the rat. However, regulation in sheep differs from that in rats in that AANAT mRNA is always high — day and night (Figure 4.2). It is reasonable to speculate that the primary mechanism regulating melatonin synthesis is the cyclic AMP inhibition of AANAT proteolysis. Day levels of AANAT activity in sheep are higher than those in the rat, probably due to higher daytime values of AANAT mRNA in the daytime sheep pineal gland.

4.3.2 Regulation in Chicken

The best-studied model of regulation of pineal AANAT activity in birds is the chicken.[42] There are a number of interesting differences between regulation of melatonin production in the chicken and in mammals (Figure 4.2).

4.3.2.1 Organization of the Melatonin Rhythm-Generating System in the Chicken

Two clocks drive the melatonin rhythm in birds. One is located in the SCN and the other in the pineal gland.[43] The SCN clock appears to provide negative influences on melatonin production, mediated by the release of NE during the day. The increase in melatonin appears to reflect the combined effect of an increase in cyclic AMP and of independent influences of the pineal clock on AANAT mRNA.[10,26] Light acts on the system through two routes, the retinal-SCN system and through pineal photoreceptors.

4.3.2.2 Cellular Regulatory Mechanisms

The negative influence of the SCN clock on pineal function in the bird appears to be mediated by NE, acting through α_2-adrenergic receptors to decrease cyclic AMP levels during the day. This maintains low levels of AANAT activity and melatonin production during the day. It is interesting to note that although light suppresses AANAT activity, it does not suppress the clock-driven rhythm in AANAT mRNA.[10]

The positive influences which increase melatonin at night reflect an increase in AANAT mRNA and cyclic AMP. The increase in AANAT mRNA is driven by the pineal clock, without a strong role of cyclic AMP;[26] the link between cyclic AMP effects and clock effects on functional expression of the AANAT gene is not well understood.[2,44] However, a dark-associated increase in cyclic AMP promotes the increase in AANAT activity, and it is reasonable to suspect that this reflects an increase in AANAT protein due to cyclic AMP-dependent inhibition of AANAT proteolysis.

In addition to cyclic AMP, $[Ca^{++}]_i$ plays an important role in regulation of AANAT activity and melatonin production. $[Ca^{++}]_i$ can influence cyclic AMP production and, via this mechanism, alter AANAT activity. Furthermore, there is clear evidence that $[Ca^{++}]_i$ plays a role in phase shifting the pineal clock.[2,44] Photic regulation of chicken pinealocyte $[Ca^{++}]_i$ has not been established, although this does appear to occur in the retina.[27]

4.3.3 Regulation in Fish

As is true of the chicken, the pineal glands of most fish have an endogenous clock which drives melatonin production (Figure 4.2).[46,47] This makes the fish pineal gland an excellent, yet somewhat overlooked, model for the study of circadian mechanisms.

4.3.3.1 Pike and Zebrafish

The translation of the pineal-clock-driven rhythm in AANAT mRNA in the pike and zebrafish pineal glands[12] into changes in AANAT activity and melatonin production are controlled by light acting through cyclic AMP.[46] Although it has not yet been determined that changes in activity are due to changes in AANAT protein nor that inhibition of proteolysis is involved, these seem to be reasonable hypotheses to pursue. Cyclic AMP follows a diurnal bimodal rhythm in pike, with peaks occurring at the L/D and the D/L transitions.[47] These variations are circadian in nature, yet it is not known whether they are controlled by a clock or are part of the clock mechanism, nor is it known in which cells the increase in cyclic AMP is occurring. However, it is reasonable to propose that the increase in AANAT expression and activity is due to the increase in cyclic AMP which occurs at the L/D transition. The functional importance of the D/L associated increase in cyclic AMP is unclear.

4.3.3.2 Trout

Trout are an example of a melatonin rhythm-generating system that lacks an endogenous clock in their pineal gland.[48–51] Rather, the trout pineal gland responds to darkness and light directly without the imposition of a clock. As a result, a dark-on/light-off relationship to melatonin production can be demonstrated at all times, day and night. Other species with this type of regulation include lizards and the lamprey eel.[52,53]

Trout pineal AANAT mRNA levels are continually elevated, and it appears that AANAT activity increases in the dark when cyclic AMP is high and decreases in the light which causes a decrease in cyclic AMP. Ca^{++} also appears to play a role in the control of AANAT because $[Ca^{++}]_i$ parallels melatonin secretion. $[Ca^{++}]_i$ increases in the dark as a consequence of photoreceptor depolarization, which triggers the opening of voltage-gated Ca^{++} channels.[54] Conversely, light exposure hyperpolarizes the cells, the channels close, and $[Ca^{++}]_i$ decreases. The effects of $[Ca^{++}]_i$ are mediated by cyclic AMP-dependent and -independent mechanisms.[55]

4.4 Final Comments

This chapter has reviewed the molecular basis of melatonin rhythm-generating systems in vertebrates. The conserved and the species-specific features of these rhythms described here represent a rich body of information. The reader is encouraged to obtain a more thorough and detailed description of the unique features of these systems from the original reports and reviews cited here and to ponder the interesting question of the functional advantages of the unique features of these systems. In some cases, the species-specific differences, such as the role of transcriptional regulation, could serve primarily as a mechanism which tailors the melatonin production signal and regulates the lag period between the onset of night and the rise of serum melatonin. In other cases, the adaptive advantages are not obvious. Future research in this area should enable us to better link the diverse modes of regulation to the role of melatonin in each species; in addition, our understanding of the role of melatonin in vertebrate physiology may be improved. Finally, an understanding of the molecules and molecular mechanisms involved in the regulation of melatonin production will provide new pharmacological targets for drugs which modulate circadian rhythms.

References

1. Arendt, J., *Melatonin and the Mammalian Pineal Gland*, Chapman and Hall, London, 1995, p. 201.
2. Zatz, M., Melatonin rhythms: trekking toward the heart of darkness in the chick pineal, *Cell. Dev. Biol.*, 7, 811, 1996.
3. Karsch, F.J., Woodfill, C.J.I., Malpaux, B., Robinson, J.E., and Wayne, N.L., Melatonin and mammalian photoperiodism: synchronization of annual reproductive cycles, in *Suprachiasmatic Nucleus: The Mind's Clock*, Klein, D.C., Moore, R.Y., and Reppert, S.M., Eds., Oxford University Press, New York, 1991, p. 217.
4. Cziesler, C.A. and Turek, F.W., Eds., Melatonin, sleep, and circadian rhythms: current progress and controversies, *J. Biol. Rhythms*, 12 (special issue), 1997.
5. Klein, D.C., Coon, S.L., Roseboom, P.H., Weller, J.L., Bernard, M., Gastel, J.A., Zatz, M., Iuvone, P.M., Rodriguez, I.R., Bégay, V., Falcón, J., Cahill, G.M., Cassone, V.M., and Baler, R., The melatonin rhythm-generating enzyme: molecular regulation of serotonin N-acetyltransferase in the pineal gland, *Rec. Prog. Hormone Res.*, 52, 307, 1997.
6. Namboodiri, M.A.A., Dubbels, R., and Klein, D.C., Arylalkylamine N-acetyltransferase from mammalian pineal gland, *Methods Enzymol.*, 142, 583, 1986.
7. Sugden, D., Ceña, V., and Klein, D.C., Hydroxyindole-O-methyltransferase, *Methods Enzymol.*, 142, 590, 1986.
8. Coon, S.L., Roseboom, P.H., Baler, R., Weller, J.L., Namboodiri, M.A.A., Koonin, E.V., and Klein, D.C., Pineal serotonin N-acetyltransferase (EC 2.3.1.87): expression cloning and molecular analysis, *Science*, 270, 1681, 1995.
9. Roseboom, P.H., Coon, S.L., Baler, R., McCune, S.K., Weller, J.L., and Klein, D.C., Melatonin synthesis: analysis of the more than 150-fold nocturnal increase in serotonin N-acetyltransferase messenger ribonucleic acid in the rat pineal gland, *Endocrinology*, 137, 3033, 1996.
10. Bernard, M., Iuvone, P.M., Cassone, V.M., Roseboom, P.H., Coon, S.L., and Klein, D.C., Melatonin synthesis: photic and circadian regulation of serotonin N-acetyltransferase mRNA in the chicken pineal gland and retina, *J. Neurochem.*, 68, 213, 1997.

11. Coon, S.L., Mazuruk, K., Bernard, M., Roseboom, P.H., Klein, D.C., and Rodriguez, I.R., The human serotonin N-acetyltransferase (EC 2.3.1.87) gene (AANAT): structure, chromosomal localization, and tissue expression, *Genomics*, 34, 76, 1996.

12. Bégay, V., Falcón, J., Cahill, G., Klein, D.C., and Coon, S.L., Transcripts encoding two melatonin synthesis enzymes in the teleost pineal organ: circadian regulation in pike and zebrafish, but not in trout, *Endocrinology*, 139, 905, 1998.

12a. Neuwald, A.F. and Landsman, D., GCN5-related histone N-acetyltransferases belong to a diverse superfamily that includes the yeast SPT10 protein, *Trends Biochem. Sci.*, 22, 154, 1997.

13. Besharse, J.C., Iuvone, P.M., and Pierce, M.E., Regulation of rhythmic photoreceptor metabolism: a role for post-receptor neurons, in *Progress in Retinal Research*, Osborne, N. and Chader, G.J., Eds., Pergamon Press, Oxford, 1988, p. 21.

14. Iuvone, P.M., Circadian rhythms of melatonin biosynthesis in retinal photoreceptor cells: signal transduction, interactions with dopamine, and speculations on a role in cell survival, in *Retinal Degeneration and Regeneration*, Kato, S. Osborne, N.N., and Tamai, M., Eds., Kugler Publications, New York, 1996, p. 3.

15. Lewy, A.J., Tetsuo, M., Markey, S.P., Goodwin, F.K., and Kopin, I.J., Pinealectomy abolishes plasma melatonin in the rat, *J. Clin. Endocrinol., Metab.*, 50, 204, 1977.

16. Reppert, S.M. and Sagar, S.M., Characterization of the day-night variation of retinal melatonin content in the chick, *Invest. Ophthalmol. Vis. Sci.*, 24, 294, 1983.

17. O'Brien, P.J. and Klein, D.C., Eds., *Pineal and Retinal Relationships*, Academic Press, Orlando, FL, 1986.

18. Oksche, A., The development of the concept of photoneuroendocrine systems: historical perspective, in *Suprachiasmatic Nucleus: The Mind's Clock*, Klein, D.C., Moore, R.Y., and Reppert, S.M., Eds., Oxford University Press, New York, 1991, p. 5.

19. Kikuchi, T., Raju, K., Breitman, M.L., and Shinohara, T., The proximal promoter of the mouse arrestin gene directs expression in photoreceptor cells and contains an evolutionarily conserved retinal factor-binding site, *Mol. Cell. Biol.*, 13, 4400, 1993.

20. Rodriguez, I.R., Mazuruk, K., Schoen, T.J., and Chader, G.J., Structural analysis of the human hydroyindole-O-methyltransferase gene, *J. Biol. Chem.*, 269, 31969, 1994.

21. Baler, R. and Klein, D.C., The rat arylalkylamine N-acetyltransferase gene promoter: intronic determinants of promoter strength and tissue specificity, *J. Biol. Chem.* (submitted).

22. Hintermann, E., Grieder, N.C., Amherd, R., Brodbeck, D., and Meyer, U.A., Cloning of an arylalkylamine N-acetyltransferase (aaNAT1) from *Drosophila melanogaster* expressed in the nervous system and the gut, *Proc. Natl. Acad. Sci. USA*, 93, 12315, 1996.

23. Klein, D.C. and Kirk, K.L., 2-Fluoro-L-histidine: a histidine analog which inhibits enzyme induction, in *Symposium on Biochemistry Involving Carbon-Fluorine Bonds*, ACS Symposium Series No. 28, Washington, D.C., 1976, p. 35.

24. Gastel, J.A., Roseboom, P.H., Rinaldi, P.A., Weller, J.L., and Klein, D.C., Melatonin production: proteasomal proteolysis in serotonin N-acetyltransferase regulation, *Science*, 279, 1358, 1998.

25. Klein, D.C., Photoneural regulation of the mammalian pineal gland, in *Photoperiodism, Melatonin, and the Pineal*, Evered, D. and Clark, S., Eds., Ciba Foundation Symposium 117, Pitman Press, London, 1985, p. 38.

26. Bernard, M., Klein, D.C., and Zatz, M., Chick pineal clock regulates serotonin N-acetyltransferase mRNA rhythm in culture, *Proc. Natl. Acad. Sci. USA*, 94, 304, 1997.

27. Iuvone, P.M., Bernard, M., Alonso-Gomez, A., Greve, P., Cassone, V.M., and Klein, D.C., Cellular and molecular regulation of serotonin N-acetyltransferase activity in chicken retinal photoreceptors, *Biol. Signals*, 6, 217, 1997.

28. Cahill, G.M. and Besharse, J.C., Circadian rhythmicity in vertebrate retinas: regulation by a photoreceptor oscillator, *Prog. Retinal Eye Res.*, 14, 267, 1995.

29. Tosini, G. and Menaker, M., Circadian rhythms in cultured mammalian retina, *Science*, 272, 419, 1996.
30. Klein, D.C., Moore, R.Y., and Reppert, S.M., Eds., *Suprachiasmatic Nucleus: The Mind's Clock*, Oxford University Press, New York, 1991.
31. Illnerová, H., The suprachiasmatic nucleus and rhythmic pineal melatonin production, in *Suprachiasmatic Nucleus: The Mind's Clock*, Klein, D.C., Moore, R.Y., and Reppert, S.M., Eds., Oxford University Press, New York, 1991, p. 197.
32. Yu, L., Schaad, N., and Klein, D.C., Calcium potentiates cyclic AMP stimulation of pineal *N*-acetyltransferase (E.C. 2.3.1.87), *J. Neurochem.*, 60, 1436, 1993.
33. Sugden, D., Vanecek, J., Klein, D.C., Thomas, T.P., and Anderson, W.B., Activation of protein kinase C potentiates isoprenaline-induced cyclic AMP accumulation in rat pinealocytes, *Nature*, 314, 359, 1985.
34. Tamarkin, L., Reppert, S.M., Klein, D.C., Pratt, B., and Goldman, B.P., Studies on the daily pattern of pineal melatonin in the Syrian hamster, *Endocrinology*, 107, 1525, 1980.
35. Roseboom, P.H. and Klein, D.C., Norepinephrine stimulation of pineal cyclic AMP response element-binding protein phosphorylation: primary role of a β-adrenergic receptor/cyclic AMP mechanism, *Molec. Pharmacol.*, 47, 439, 1995.
36. Baler, R., Covington, S., and Klein, D.C., The rat arylalkylamine *N*-acetyltransferase gene promoter: cAMP activation via a cAMP-responsive element-CCAAT complex, *J. Biol. Chem.*, 272, 6979, 1997.
37. Stehle, J.H., Foulkes, N.S., Molina, C.A., Simonneaux, V., Pévet, P., and Sassone-Corsi, P., Adrenergic signals direct rhythmic expression of transcriptional repressor CREM in the pineal gland, *Nature*, 365, 314, 1993.
38. Foulkes, N.S., Borjigin, J., Snyder, S.H., and Sassone-Corsi, P., Transcriptional control of circadian hormone synthesis via the CREM feedback loop, *Proc. Natl. Acad. Sci. USA*, 93, 14140, 1996.
39. Baler, R. and Klein, D.C., Circadian expression of transcription factor Fra-2 in the rat pineal gland, *J. Biol. Chem.*, 270, 27319, 1995.
40. Reppert, S.M., Perlow, M.J., Tamarkin, L., and Klein, D.C., A diurnal melatonin rhythm in primate cerebrospinal fluid, *Endocrinology*, 104, 295, 1979.
41. Van Camp, G., Ravault, J.P., Falcón, J., Collin, J.P., and Voisin, P., Regulation of melatonin release and *N*-acetyltransferase activity in ovine pineal cells, *J. Neuroendocrinol.*, 3, 477, 1991.
42. Zatz, M. and Mullen, D.A., Two mechanisms of photoendocrine transduction in cultured chick pineal cells: pertussis toxin blocks the acute but not the phase-shifting effects of light on the melatonin rhythm, *Brain Res.*, 453, 63, 1988.
43. Cassone, V.M., Melatonin and suprachiasmatic nucleus function, in *Suprachiasmatic Nucleus: The Mind's Clock*, Klein, D.C., Moore, R.Y., and Reppert, S.M., Eds., Oxford University Press, New York, 1991, p. 309.
44. Zatz, M., Does the circadian pacemaker act through cyclic AMP to drive the melatonin rhythm in chick pineal cells?, *J. Biol. Rhythms*, 7, 301, 1992.
45. Zatz, M. and Heath, J.R., III, Calcium and photoentrainment in chick pineal cells revisited: effects of caffeine, thapsigargin, EGTA, and light on the melatonin rhythm, *J. Neurochem.*, 65, 1332, 1995.
46. Thibault, C., Collin, J.P., and Falcón, J., Intrapineal circadian oscillator(s), cyclic nucleotides and melatonin production in pike pineal photoreceptor cells, in *Melatonin and Pineal Gland: From Basic Science to Clinical Application*, Touitou, Y., Arendt, J., and Pévet, P., Eds., Elsevier, Amsterdam, 1993, p. 11.
47. Falcón, J. and Gaildrat, P., Variations in cyclic adenosine 3',5'-monophosphate and cyclic guanosine 3',5'-monophosphate content and efflux from the photosensitive pineal organ of the pike in culture, *Pflügers Arch. Eur. J. Physiol.*, 433, 336, 1997.

48. Zachmann, A., Ali, M.A., and Falcón, J., Melatonin rhythms in the pineal organ of fishes and its effects: an overview, in *Rhythms in Fishes*, Ali, M.A., Ed., NATO-ASI series A, Plenum Press, New York, 1992, p. 149.

49. Thibault, C., Falcón, J., Greenhouse, S.S., Lowery, C.A., Gern, W.A., and Collin, J.P., Regulation of melatonin production by pineal photoreceptor cells: role of cyclic nucleotides in the trout (*Oncorhynchus mykiss*), *J. Neurochem.*, 61, 332, 1993.

50. Gern, W.A. and Greenhouse, S.S., Examination of *in vitro* melatonin secretion from superfused trout (*Salmo gairdneri*) pineal organs maintained under diel illumination or continuous darkness, *Gen. Comp. Endocrinol.*, 71, 163, 1988.

51. Underwood, H., The pineal and melatonin: regulators of circadian function in lower vertebrates, *Experientia*, 45, 914, 1989.

52. Bolliet, V., Ali, M.A., Anctil, M., and Zachmann, A., Melatonin secretion *in vitro* from the pineal complex of the lamprey *Petromyzon marinus, Gen. Comp. Endocrinol.*, 89, 101, 1993.

53. Max, M. and Menaker, M., Regulation of melatonin production by light, darkness, and temperature in the trout pineal, *J. Comp. Physiol. [A]*, 170, 479, 1992.

54. Bégay, V., Bois, P., Colin, M.P., Lenfant, J., and Falcón, J., Calcium and melatonin production in dissociated trout pineal photoreceptor cells in culture, *Cell Calcium*, 16, 37, 1994.

55. Bégay, V., Collin, J.P., and Falcon, J., Calciproteins regulate cyclic AMP content and melatonin secretion in trout pineal photoreceptors, *NeuroReport*, 5, 2019, 1994.

Chapter 5

Molecular Components of a Model Circadian Clock: Lessons from *Drosophila*

Paul E. Hardin and Amita Sehgal

Contents

5.1	Clocks in *Drosophila*	61
5.2	Genetics of the *Drosophila* Clock	62
5.3	The Circadian Feedback Loop	62
5.4	Feedback Loop Synchrony and Tissue Distribution	64
5.5	Circadian mRNA Cycling	65
	5.5.1 Feedback Regulation of mRNA Cycling	65
	5.5.2 Requirements and Roles for Circadian Feedback Regulation	66
5.6	Control of Protein Expression	66
	5.6.1 Regulation of Protein Cycling	66
	5.6.2 Subcellular Localization of the PER and TIM Proteins	68
	5.6.3 Exceptions to the Rule	68
	5.6.4 Role of the PER and TIM Proteins in Entrainment to Light	69
5.7	Concluding Remarks	70
References		71

5.1 Clocks in *Drosophila*

Drosophila have been the subject of circadian clock research for several decades. Early studies by Pittendrigh and his co-workers uncovered one of the basic properties of circadian clocks — temperature compensation.[47] Though this characteristic is not of much relevance in mammals, other work in which strains of *Drosophila* were selected for early or late (emergence of adult flies from their pupal cases) indicated that the circadian oscillator had a genetic basis.[47] These early selection experiments opened the door for the now classic genetic screen of Konopka and Benzer, which resulted in the first single gene circadian clock mutant — the *period* (*per*) gene.[39] The genetic approach to identifying clock genes is now widely used, and has uncovered genes whose mutant forms alter the period of, or eliminate, circadian rhythms in cyanobacteria, *Chlamydamonas*, *Neurospora*, *Arabidopsis*, *Drosophila*, hamsters, and mice.[10,22] Among these mutants, *frequency* (*frq*) from *Neurospora* and *per* and *timeless* (*tim*) from *Drosophila* appear to encode components of circadian oscillators.[22,49,58] In this review, we will focus on recent molecular and biochemical studies of *per* and *tim*, how they have advanced our understanding of the *Drosophila* oscillator mechanism, and what they may tell us about the mammalian circadian oscillator.

5.2 Genetics of the *Drosophila* Clock

Several pieces of behavioral evidence indicate that *per* is intimately associated with the *Drosophila* circadian pacemaker. Mutations due to single amino acid changes in *per* protein (PER) can either shorten (per^S) or lengthen (per^L) the free-running (in constant darkness, or DD) circadian period, while a non-functional truncation of PER (per^{01}) abolishes free-running circadian rhythms.[3,39,78] These effects are seen in circadian locomotor activity rhythms of individual adults as well as eclosion rhythms of populations, demonstrating that *per* affects two fundamentally different types of rhythms occurring at different developmental stages.[39] PER appears to be necessary for pacemaker function, as the phase of rhythmic activity is dependent on when PER is induced (using a heat-inducible promoter) rather than the prior entrainment conditions.[16] Subsequent screens have uncovered additional rhythm mutants, of which only the *tim* locus appears to encode a second component of the circadian oscillator.[22]

Like per^{01}, the initial *tim* mutant (designated tim^{01}) is arrhythmic for both locomotor activity and eclosion rhythms.[60] The second *tim* mutant, called tim^{SL}, is an allele-specific suppressor of per^L.[52] This suppression indicates that *tim* protein (TIM) directly interacts with *per* protein (PER), a point that has been substantiated both *in vitro* (i.e., yeast two-hybrid)[18] and *in vivo*[40,81] and is discussed more fully below. Because of its interaction with *per* and its effect on both *Drosophila* rhythms, *tim* is thought to encode a second component of the *Drosophila* circadian oscillator. The cloning and sequencing of the *per* and *tim* genes revealed little in the way of potential biochemical function; however, the temporal expression, subcellular localization, and light sensitivity of the *per* and/or *tim* gene products have led to a model for how these genes contribute to circadian oscillator function.[28,62,82]

5.3 The Circadian Feedback Loop

The levels of *per* RNA and protein cycle in a circadian manner, where the *per* mRNA peak (occurring at ~ZT15) phase leads the PER peak (occurring at ~ZT21) by approximately 6 hr.[13,28,62,82] These fluctuations are seen during LD and DD conditions, indicating that these are true molecular circadian rhythms.[13,28,82] The *per* mutants influence the phase (in LD) and period (in DD) of *per* RNA and protein cycling in parallel to their effects on behavioral rhythms.[13,28,82] In addition, per^{01} mRNA, which does not cycle due to the lack of PER, can be returned to cycling by a *per* transgene that restores behavioral rhythmicity.[28] These results suggest that *per* molecular oscillations constitute a feedback loop whereby *per* mRNA is the template for PER synthesis, and PER is necessary for circadian fluctuations in *per* RNA (Figure 5.1).[28,31] This feedback loop has been the foundation for studying the *Drosophila* circadian oscillator mechanism, and, considering that a similar feedback loop in *frq* expression is present in *Neurospora*,[2] such feedback loops may be a general mechanism underlying circadian oscillator function.

Subsequent studies not only support the existence of a circadian feedback loop in *Drosophila*, but also add to its mechanistic detail in the following areas: (1) *per* RNA cycling is

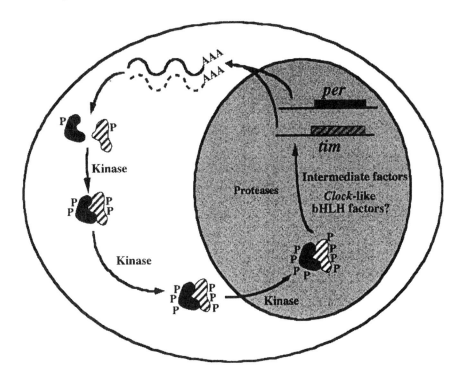

FIGURE 5.1
Circadian feedback loop. The direction of the feedback is shown by the arrows. The outer white oval represents the cytosol and the inner shaded oval represents the nucleus. The *per* (filled bar) and *tim* (hatched bar) genes give rise to cycling levels of *per* (solid curved line) and *tim* (dashed curved line) mRNA in the cytoplasm. PER (filled shape) and TIM (hatched shape) are produced in the cytoplasm and translocate into the nucleus. The proteins are progressively phosphorylated (represented by Ps) by unknown kinases in the cytoplasm and nucleus,[19] and appear to be rhythmically degraded by proteases.[65] When in the nucleus, PER and/or TIM feedback to regulate their own (and perhaps other) gene transcription through intermediate factors such as a *Clock*-like protein.

transcriptionally regulated;[31] (2) PER is nuclear in tissues relevant to behavioral rhythmicity;[42] (3) *per* RNA cycling is not found in ovaries, the only tissue in which PER is not nuclear;[27] (4) PER acts intracellularly to repress transcription of its own gene;[80] (5) PER directly binds to TIM before they are translocated to the nucleus around the middle of the night;[40,81] (6) the *tim[01]* mutant blocks nuclear localization of PER and, consequently, *per* and *tim* mRNA cycling;[60,73] (7) *tim* mRNA and protein products cycle in phase with those from the *per* gene;[36,45,61,81] and (8) light acts to reset the oscillator's phase by destabilizing TIM.[36,40,45,81] A refined view of the circadian feedback loop that takes these data into account suggests that the *per* and *tim* genes are transcribed when PER and TIM levels are low. These mRNAs then give rise to PER and TIM, which are localized to the nucleus. Nuclear PER and/or TIM then repress their own transcription, and after these nuclear proteins break down the next cycle of gene transcription begins.

Given that this feedback loop is operating, what evidence is there to indicate that it is important for circadian oscillator function? The best evidence comes from experiments using inducible *per* genes, where inappropriately timed PER production leads to a phase delay or a phase advance, depending upon when the induction occurs during the circadian cycle.[12] This experiment strongly suggests that the *per* feedback loop is a core component of the *Drosophila* circadian pacemaker. The evidence for TIM involvement in feedback loop function is equally strong, as light-dependent TIM destabilization (described more fully below) alters the phase of the oscillator, and TIM directly interacts with PER, the only other known oscillator component. Both *per* and *tim* satisfy many of the criteria that were developed, based on theoretical considerations, to define the properties of an oscillator component.[2,79] Thus, PER and TIM, along with the feedback loop they comprise, appear to be bona fide components of the circadian oscillator.

5.4 Feedback Loop Synchrony and Tissue Distribution

The *per* gene is expressed in many different neuronal and non-neuronal tissues in the head (i.e., photoreceptors, antennae, brain glia, "dorsal" and "lateral" brain neurons, proboscis) and the body (i.e., cardia, thoracic ganglion, gut, ovaries, testes, Malpighian tubules, salivary glands, fat bodies).[15,41,53,62] Because the *per* feedback loop is operating in all of these tissues except the ovary,[27] it is important to know which of these tissue(s) are necessary and/or sufficient to mediate behavioral rhythmicity. Analysis of internally marked mosaics show that *per* expression in the central brain can mediate normal circadian locomotor activity rhythms, while *per* expression in brain glia is only sufficient for weak, long-period locomotor activity rhythms.[15] Subsequent studies showed that *disconnected* (*disco*) mutants, which are arrhythmic in eclosion and locomotor activity assays, lack lateral neurons,[11,29] while transgenes that express *per* only in lateral neurons rescue moderately strong behavioral rhythms having somewhat long (~26 hr) periods.[17] Taken together, these studies make a strong case for the lateral neurons being the site of the locomotor activity clock, or the fly's version of the SCN.

The *per* feedback loop is not only operating in cells that mediate locomotor activity rhythms, but also in other head and non-ovarian body cells.[27,30] The phase of the *per* feedback loop, as measured by *per* RNA cycling, is similar in all of the head and body fractions measured. This synchrony among *per* oscillators in different body parts suggests that they are coordinated, or coupled, in some way. The mechanism underlying this coupling will be largely dependent upon the nature of these oscillators, which could range from totally autonomous pacemakers (oscillators that can be entrained by light and function independently of other oscillators) to either autonomous or non-autonomous "slave" oscillators, whose entrainment and phase are dependent upon an autonomous "master" pacemaker.[48]

There is precedent for both tissue-autonomous oscillators and master/slave oscillators. *In vitro* cultures of the rat SCN,[21] gypsy moth testes,[20] *Xenopus* eyes,[4] the chicken pineal,[66] and *Aplysia* eyes[37] show that autonomous, light-entrainable pacemakers control free-running rhythms of sperm release (moth testes), melatonin release (*Xenopus* eyes and chicken pineal), and electrical activity (*Aplysia* eyes and SCN). In *Drosophila*, the lateral brain neurons (LNs) function autonomously to control locomotor activity rhythms.[17] Explants of the prothoracic gland (which is part of the ring-gland complex) contain a light-entrainable PER-TIM circadian oscillator.[14] Likewise, PER and TIM cycle in Malpighian tubules of headless flies, indicative of an autonomous body oscillator.[19] The circadian rhythm of eclosion is presumably generated by the PER/TIM clock in the ring gland. While a circadian function has not yet been attributed to the Malpighian tubules, it is not difficult to imagine that excretion itself or some other physiological activity in this tissue is clock-controlled. These data suggest that multiple oscillators are present in the fly, a situation similar to that found in reptiles and birds where clocks have been described in the SCN, the pineal gland, and the eye.[67] Mammals, on the other hand, are thus far known to have clocks only in the SCN and the eye.[65,69]

In contrast to PER and TIM function in lateral neurons, their expression in the eye is clearly not required for free-running activity rhythms, as the eye is dispensable for this purpose.[32] As to whether or not the eye plays a role in the entrainment of the clock to light is still a debatable issue (discussed further below). Even if it were to do so, it is not clear that the processing of light cues in the eye would require PER/TIM. The prediction is that the input (or entrainment) pathway is some kind of phototransduction cascade that terminates in clock molecules, but these do not have to be in the same cells. Thus, expression of PER/TIM in the eye may reflect an eye-specific clock. Given that clocks have been described in the eyes of a variety of other organisms,[5] it is perhaps to be expected that the *Drosophila* eye will contain a clock. Studies using photoreceptor-specific expression of *per* indicate that this is indeed the case.[6]

5.5 Circadian mRNA Cycling

5.5.1 Feedback Regulation of mRNA Cycling

The mechanisms underlying circadian feedback regulation are being uncovered by identifying sequences and factors that control *per* transcription. To map circadian regulatory sequences, a series of transgenes in which *lacZ* was driven by *per* upstream fragments fused to either the P-element transposase or *hsp70* basal promoters was transformed into wild-type flies (which supply the required PER protein). All transgenic lines containing a 69-bp *per* upstream fragment from −563 to −494 (the transcription start site is +1) were capable of mediating *lacZ* mRNA cycling.[25] When *lacZ* expression was driven by this fragment alone, mRNA cycling was essentially indistinguishable from wild-type under both 12-hr light/12-hr dark and constant dark conditions. The function of this circadian regulatory fragment is also PER-dependent, as cycling is abolished in flies that lack PER function (i.e., *per^{01}* flies). In addition, this fragment drives expression throughout much of the normal *per* spatial pattern, which is important as other factors involved in circadian regulation are probably not expressed in cells lacking feedback loop function. The ability of this 69-bp fragment to activate heterologous basal promoters, operate at different distances from the transcription start site, and function in an orientation-independent manner indicates that it is a transcriptional enhancer.[25]

Because this 69-bp *per* upstream fragment can mediate high-amplitude *lacZ* mRNA cycling, its sequence was searched for common transcription factor binding sites. One of the sites identified was a consensus E-box, which is notable because it is a target for basic helix-loop-helix (bHLH) transcription factors.[44] Because a subgroup of these bHLH factors are related to PER by virtue of the PAS domain,[35] it suggested a model for PER-mediated repression: sequestration of a transcriptional activator via PAS-PAS interactions. Deletions in the E-box drive *lacZ* mRNA expression at close to the trough level of wild-type *per* mRNA, showing that this sequence is involved in transcriptional activation. However, the residual expression from both sets of mutant lines cycles with an overall low amplitude,[25] indicating that mRNA cycling is mediated by sequences outside the E-box. The sequestration of an activator by PER, however, does not solely account for *per* mRNA cycling, as the lack of functional PER or TIM leads to median levels of *per* mRNA rather than peak levels. This observation indicates that PER and/or TIM somehow regulate the activation and repression of *per* and (almost certainly) *tim* transcription.

Little is known about the factors that mediate transcriptional feedback by PER and/or TIM. It is clear that at least one, and undoubtedly several, DNA binding proteins interact with the circadian enhancer. One obvious candidate for such a factor is a bHLH-PAS transcriptional activator, because (1) disruption of the E-box target site in the *per* circadian enhancer results in a substantial drop in transcription, and (2) a bHLH-PAS protein is known to function within the clock in mice.[1,38] This mouse bHLH-PAS factor is called CLOCK, which was initially identified as a mutation that leads to period alterations as a heterozygote and arrhythmicity as a homozygote.[72] Since E-box mutants do not abolish *per* mRNA cycling, other factors apparently mediate this aspect of *per* regulation. For transcriptional feedback, a factor must also interact with PER and/or TIM. This could be the same factor which binds DNA or a separate intermediate factor. The latter possibility is supported by studies in the giant silk moth, *Antheraea pernyi*, where cycling levels of mRNA and protein are observed in brain neurons even though PER is never transported to the nucleus.[55] Searches for new clock genes in *Drosophila*, factors that interact with PER and/or TIM, and factors that bind to the *per* circadian enhancer should provide insight into the mechanisms of transcriptional feedback.

Even though the major features of mRNA cycling (i.e., amplitude, phase, and overall level) can be reproduced by the 69-bp *per* circadian enhancer, other sequences within and upstream of *per* have an effect on mRNA expression. For instance, the amplitude of mRNA cycling mediated by the circadian enhancer can be increased or decreased by its flanking sequences.[25] In addition, the first intron of *per* can drive proper spatial expression within the brain but is not capable of mediating circadian transcription.[26] When the luciferase reporter gene system is used to monitor mRNA levels, inclusion of transcribed sequences affects the phase of mRNA cycling.[63] In this case, location of the regulatory elements raises the possibility that *per* mRNA cycling may also be regulated at the post-transcriptional level.

5.5.2 Requirements and Roles for Circadian Feedback Regulation

Although *per* mRNA cycling was among the first features of the circadian feedback loop to be discovered, its requirement for oscillator function and behavioral rhythmicity has yet to be tested. However, several pieces of evidence argue that *per* mRNA cycling is not necessary for clock function. Constructs containing a 7.2-kb *per* genomic fragment, lacking the promoter, first exon, and most of the first intron, are able to rescue rhythms in per^{01} flies.[17,24] The 7.2:2 line expresses rhythms in PER abundance and nuclear localization exclusively in the LNs, but because of the low levels of *per* mRNA, it is impossible to determine whether mRNA cycles in this strain. If this same 7.2-kb fragment is driven by the heat-inducible *hsp70* promoter, it also is capable of rescuing behavioral rhythms in per^{01} flies.[16] While this promoter is thought to be constitutively active, its widespread expression pattern precludes measuring mRNA abundance in the normal *per* expressing cells. Likewise, *per* driven by the constitutively active *glass* gene promoter is capable of rescuing behavioral rhythms in per^{01} flies.[74] Although transgene mRNA cycling was not tested, these results suggest that the oscillator in these *glass-per* flies can function without mRNA cycling.

If the circadian oscillator can function without *per* mRNA cycling, then what is the role of transcriptional feedback regulation? One role would be within the oscillator itself. If *per* mRNA cycling were not required for oscillator function, another oscillator component such as *tim* could be receiving feedback information and produce high amplitude mRNA cycling. Thus, redundancy may be built in so that loss of mRNA cycling in one oscillator components does not lead to a breakdown in oscillator function. Another role for transcriptional feedback is within the entrainment mechanism. Because light acts to decrease TIM abundance, high levels of *tim* mRNA early in the dark phase are thought to replenish TIM levels — leading to a phase delay; whereas, low levels of *tim* mRNA late in the dark phase could not replenish TIM levels — leading to a phase advance.[36,45,81] A third role for circadian feedback is to control the expression of clock output genes. A group of 20 *Drosophila* rhythmically expressed genes (Dregs) has been identified,[70] the expression of which is dependent on *per* gene function,[70,71] indicating that circadian feedback loop function is important for rhythmic Dreg expression.

5.6 Control of Protein Expression

5.6.1 Regulation of Protein Cycling

On first thought it may seem absurd to devote an entire section to the regulation of protein cycling. Given that *per* and *tim* RNA cycle, it does not require a great leap of faith to believe that the proteins also cycle. In this case, however, it is by no means clear that cyclic protein expression depends upon

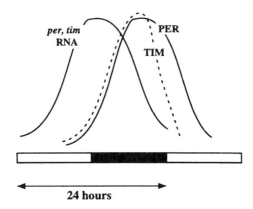

FIGURE 5.2
RNA and protein profiles of the *per* and *tim* genes. This schematic representation depicts the relative phases of the oscillations in levels of *per* and *tim* RNA and proteins. While the figure represents the profile in the presence of light/dark cycles (indicated by the bar at the bottom), a similar pattern is observed under free-running conditions.

oscillating levels of RNA. On the contrary, available data (discussed above) suggest that cyclic PER expression may be achieved even when it is encoded by a non-cycling mRNA.[6] In addition, diurnal fluctuations of TIM are observed in flies that completely lack feedback function (i.e., in per^{01} flies) due to the light responsiveness of TIM.[45,81] It is still an open question as to whether or not the two clock proteins could sustain endogenous oscillations in the absence of feedback to either gene (although this appears unlikely).

What is clear is that transcription-independent mechanisms contribute to the daily rise and fall of the PER and TIM proteins.[9] Perhaps the most striking feature of the RNA and protein profiles of these two genes is the 6-hr delay that precedes the peak of the proteins after the RNAs have peaked (Figure 5.2). This delay is critical to maintain molecular cycles, as it separates the phase of RNA synthesis from the feedback inhibitory effects of the proteins. The mechanisms that account for this delay have not been completely elucidated. In any situation where the protein profile does not parallel that of the RNA, it is usually due to translational control of the RNA or due to regulated stability of the protein. Current thinking in the field generally favors the latter possibility. Characterization of PER-beta galactosidase fusion proteins shows that a fusion protein which contains the N-terminal half of PER does not cycle, but the inclusion of additional PER sequences (including a PEST sequence) permits rhythmic expression with a phase that is indistinguishable from that of wild-type PER (i.e., the delay) is manifest.[9] Both fusion proteins produce a beta-galactosidase degradation product that does not cycle (probably due to the high stability of beta-galactosidase), and higher levels of the cleavage product are generated by the cycling protein.[9] These data are indicative of an a proteolysis-based mechanism for PER cycling. In addition, protein turnover is implicated in other aspects of clock function — PER requires TIM for stability, and entrainment of the clock to light is mediated by protein degradation (see section below on entrainment of the clock to light).

The falling phase of the protein profile must certainly depend upon regulated protein turnover. The timely disappearance of the PER and TIM proteins in a daily cycle presumably releases the restraint on transcription and allows RNA levels to accumulate once again. Consistent with this idea, an earlier decline of the PER and TIM proteins in per^s flies is followed by an earlier rise in RNA levels of the two genes.[43] The most likely explanation for the truncated phase of protein expression in these mutant flies is reduced stability of the protein(s). The per^s mutation replaces a serine residue with an asparagine and thus may eliminate a specific phosphorylation event.[3,78] Phosphorylation was proposed to target the PER protein for degradation,[13] but this does not preclude other regulatory roles for it. Both PER and TIM are heavily phosphorylated in a cyclic fashion (increased phosphorylation with the progression of the night),[13,43] and there are probably multiple phosphorylation sites that could mediate a number of different functions.

On the subject of phosphorylation, the tim^{SL} mutation, which suppresses the per^L mutation (and to a much lesser extent the per^s mutation), alters the phosphorylation of TIM.[52] Because tim^{SL} suppresses all *in vivo* phenotypes of the per^L mutation and yet does not alter the interaction between TIM and PER^L in the yeast two-hybrid system, it is thought to be a bypass suppressor. In addition to further supporting a role for phosphorylation, the tim^{SL} mutation underscores the importance of the PER/TIM complex in circadian timekeeping. Evidently these two proteins must be treated as a unit, where the ultimate output is determined by the state of each protein and by the interactions between them.

5.6.2 Subcellular Localization of the PER and TIM Proteins

It is now well established that each protein requires the other for nuclear transport;[36,45,73] however, while TIM is stably expressed in the cytoplasm in the absence of PER, PER levels are extremely low in tim^{01} (*tim* null mutant) flies.[36,45,73] Here again, protein turnover is thought to be important, the implication being that PER is unstable in the cytoplasm unless it is bound to TIM. Cell-culture studies indicate the presence of cytoplasmic localization domains in both proteins.[54] Deletion of this domain in either protein permits nuclear entry in the absence of the partner. While these studies have not yet been done in flies, it is clear that nuclear transport of these proteins is a regulated process. In the lateral neurons, nuclear entry of PER, and therefore by inference that of TIM, appears to be temporally controlled.[8] Thus, the protein is expressed in perinuclear regions at ZT17 and moves into the nucleus at ~ZT18. The per^L mutation, which produces a temperature-sensitive lengthening of circadian period, delays nuclear entry in a temperature-dependent manner.[8] This delay is caused by a reduced, also temperature-sensitive, interaction with TIM and is rescued partially by the tim^{SL} mutation (discussed above).[18,52] Thus, the timing of nuclear entry could play a critical role in determining circadian period. Since temporally controlled nuclear entry would ensure that PER and TIM are not transported to the nucleus as soon as they are synthesized, it might also contribute to the delay in feedback by the two proteins. However, temporal nuclear entry cannot be a universal mechanism, as nuclear expression of the two proteins in photoreceptor cells is not gated.[50,57] Both proteins are visible in nuclei once they attain detectable levels, and the perinuclear staining that characterizes lateral neuron expression at a specific time of the night is not observed in photoreceptor cells.[57]

5.6.3 Exceptions to the Rule

Having emphasized the general feature of PER/TIM nuclear localization, it is necessary to point out that there are exceptions to this rule. In the *Drosophila* ovary, for instance, PER expression is cytoplasmic, and, as one might expect, the RNA does not cycle.[27,42] Several neurons in the head of the beetle, *Pachmorphas exguttata*, express PER not only in the cytoplasm of the cell bodies, but also in neurites.[17] Likewise, PER and TIM expression is cytoplasmic in neurons (including axons) within the silk moth brain.[55,56] Perhaps the most surprising result, though, is that unlike the case of the beetle neurons, where expression of PER is largely noncyclic, both PER and TIM cycle in the cytoplasm of the silk moth brain. The *per* RNA also cycles, and, in addition, an antisense RNA is expressed and cycles with a reverse phase. It is speculated that the antisense RNA participates in regulating cyclic expression of the proteins. The question of how these proteins might function in the cytoplasm is still open to debate. Considering all we know about transcription-based clock mechanisms through studies in *Neurospora* and *Drosophila*, not to mention the recently isolated mammalian *Clock* gene,[38] it appears unlikely that the entire clock would be assembled in the cytoplasm. One would have to assume then that other molecules are doing the pacemaking in nuclei of these silk moth neurons, and PER and TIM function as "accessory" proteins.[23] Alternatively, but consistent with the "accessory" protein assumption, these neurons may be involved in the output pathway rather than in pacemaking itself (the only caveat to this idea being that these are the only neurons that express PER/TIM in the silk moth brain). In this context, note that PER is expressed in nuclei of photoreceptor cells and embryonic gut cells in the silk moth.[55,56] Thus, the mechanism of action in the moth cannot be totally alien to that observed in fruit flies.

5.6.4 Role of the PER and TIM Proteins in Entrainment to Light

Molecular analysis of *per* provided a basic clock mechanism (the feedback loop), but it did not elucidate the mechanisms that synchronize the clock to light. Unlike *frq*, whose transcription is activated by light,[7] neither *per* RNA nor protein displayed an acute response to light. It was not until the TIM protein was characterized that the light-responsive component of the *Drosophila* clock was identified. We now know, through work done in several labs, that light rapidly reduces levels of the TIM protein, presumably effecting its degradation.[36,40,45,81] This light-induced degradation of TIM explains how endogenous clocks which sometimes have periods very different from 24 hr (such as the *per^s* mutant with a 19-hr period) can synchronize to a 24-hr environmental cycle. Basically, because the endogenous cycle depends upon a light-sensitive protein, normal progression of the cycle is either arrested until lights-off or accelerated by lights-on. Thus, either one of those transitions (lights-off or lights-on) could conceivably lock the phase of the endogenous cycle with the phase of the environment. TIM-based models can also account for nonparametric entrainment of the clock, i.e., phase shifts in the rhythm in response to pulses of light.[59,77]

Details of the light response of TIM will not be discussed here, as they have been elaborated upon in several recent reviews.[50,59,77] Suffice to say that TIM is degraded by light in both photoreceptor cells and lateral neurons, with maximal response seen at 30 minutes and 90 minutes, respectively.[36,45] Light-induced changes in TIM are followed by corresponding changes in PER and subsequently in the two RNAs, thus shifting the phase of the entire loop. Apparently phosphorylated forms of TIM are more sensitive to light, supporting the notion that phosphorylation renders the protein more susceptible to proteolytic action.[81] In addition, phosphorylation must play a more direct role in the light response, as light also affects the phosphorylation state of the two proteins.[43] Light-induced changes in PER/TIM phosphorylation precede other events in the feedback loop, but they appear to be downstream of TIM degradation. While levels of TIM are similarly affected, at least qualitatively, by a delaying (in the early part of the night) and an advancing (in the second half of the night) light pulse, phosphorylation of PER is delayed and advanced, respectively.[40] PER itself is not required for the TIM response to light, but it is certainly critical in determining the overall effect on the clock. The *per^s* mutation, for instance, is known to augment behavioral resetting in response to light. On a molecular level, it turns out that light extended beyond ZT12 has enhanced effects on the levels and the phosphorylation of the PER^S protein compared to wild-type PER.[43] The same conditions inhibit accumulation and phosphorylation of TIM.[43]

As mentioned above, the *Neurospora* clock is reset by a light-induced increase in levels of *frq* RNA.[7] Thus, light, and presumably other zeitgebers that reset the clock, do so by changing the level of a clock component. Since *frq* cycles with a phase opposite that of *per/tim*, it follows that *frq* would be induced by light while *per/tim* are degraded. Based on available data, it may be said that the *Drosophila* clock requires two genes (*per* and *tim*) to carry out the functions that, in *Neurospora*, are performed by one gene (*frq*).

Unfortunately precious little is known about the input or entrainment pathway that transmits light to the *Drosophila* clock. The little information we have is mostly negative and indicates that external photoreceptors (eyes and ocelli) are generally dispensable for purposes of entrainment.[32]

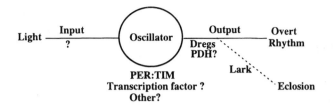

FIGURE 5.3
Molecular components of the *Drosophila* circadian system. Genes that are known to function in specific aspects of the circadian system are indicated. It is assumed that the different output pathways diverge at some point. *Lark*, for instance, is known to affect only eclosion rhythms and not locomotor activity. It is speculated that PDH is involved in output, but this has yet to be confirmed. Additional mutants that affect circadian rhythms in *Drosophila* have been isolated,[5] but because their molecular characterization is limited (or nonexistent), their role in the circadian system is not known and they are, therefore, excluded from this diagram.

Flies that lack these structures show minor deficits in their synchronization to light/dark cycles, but it is clear that they retain the ability to process light cues for circadian entrainment.[32] This naturally leads to the idea that there are extraocular photoreceptors, a candidate being the "seventh eye" described by Hofbauer and Buchner.[34] Photoreception capabilities could also exist in the lateral neurons themselves or perhaps in a subset of these. It should be noted that circadian photoreception in one cell type does not preclude its presence in another. It is likely that redundant pathways mediate entrainment of the clock to light. Finally, we know as little, or perhaps less, about the molecular components of the clock phototransduction cascade as we do about the cells involved. Again, the visual transduction cascade is not essential — visually blind *NorpA* mutants have shorter periods under free-running conditions, yet they entrain normally to light.[11,76]

5.7 Concluding Remarks

In summary, studies of *Drosophila* (as well as *Neurospora*) have provided a basic picture of how a biological clock is generated (Figure 5.3). However, the story is not over by any means. Some of the important questions to be answered include *"How is a circadian period is generated?"* What we know thus far about the regulation of *per* and *tim* does not account for the entire 24-hr period of the feedback loop. As discussed above, the lag between RNA synthesis and feedback inhibition by the proteins is a critical feature of this loop, but while there are suggestions of how this lag may be generated there is no definitive answer yet. Similarly, the length of inhibition by the proteins may be determined solely by the stability of the proteins, but to date we do not know the half-life of either protein. *"How is feedback effected by the PER and TIM proteins?"* The mechanisms that mediate transcriptional effects of these proteins have not been identified. There is a general consensus that one or more transcriptional activators must play a role, but the identity of these remains elusive, perhaps the only clue being that they will likely contain PAS domains. *"Are other clock components involved?"* The guess would be that there are other components of the *per/tim* feedback loop. In support of this, a transcriptional activator of the *frq* gene in *Neurospora*, product of the *wc2* gene, appears to be a clock component.[7]

An important issue on everybody's mind is, of course, the evolutionary conservation of the clock. Until recently it seemed that clock mechanisms would be conserved, although the molecules themselves may not be. Models for clocks usually invoke feedback loops, and oscillations in the mammalian SCN are generated intracellularly, thus generally supporting a transcription/translation feedback loop as opposed to one that would involve complex interactions between cells.[75] The recent cloning of the mouse *Clock* gene[38] and the isolation of a mammalian *per* homologue[64,68] indicate that the molecules may also be conserved.

As far as other components of the circadian system are concerned, the *Drosophila* system, sadly enough, does not contribute much. As mentioned above, there is a paucity of knowledge about the input pathway. The situation is the same for the output pathway. Although a couple of *per* downstream genes have been cloned, their function and their location within the output pathway are not known.[51,71] Pigment-dispersing hormone (PDH) is co-expressed with PER and TIM in the lateral neurons and therefore is a candidate for an output humoral factor, but its characterization in *Drosophila* is limited.[33,36] Mutations that affect only one clock output (e.g., only eclosion) are known,[46] so the output pathway must diverge at some point, but where this occurs is the question. Clearly there is enough to be done to keep *Drosophila* researchers busy for years to come.

References

1. Antoch, M.P., Song, E.-J., Chang, A.-M. et al., Functional identification of the mouse circadian *Clock* gene by transgenic BAC rescue, *Cell*, 89, 655, 1997.
2. Aronson, B.D., Johnson, K.A., Loros, J.J., and Dunlap, J.C., Negative feedback defining a circadian clock: autoregulation of the clock gene *frequency*, *Science*, 263, 1578, 1994.
3. Baylies, M.K., Bargiello, T.A., Jackson, F.R., and Young, M.W., Changes in abundance and structure of the *per* gene product can alter periodicity of the *Drosophila* clock, *Nature*, 328, 390, 1987.
4. Besharse, J.C. and Iuvone, P.M., Circadian clock in *Xenopus* eye controlling retinal serotonin *N*-acetyltranferase, *Nature*, 305, 133, 1983.
5. Block, G.D., Khalsa, S.B., McMahon, D.G., Michel, S., and Geusz, M., Biological clocks in the retina: cellular mechanisms of biological timekeeping, *Int. Rev. Cytol.*, 146, 83, 1993.
6. Cheng, Y. and Hardin, P., *Drosophila* photoreceptors contain an autonomous circadian oscillator that can function without *period* mRNA cycling, *J. Neurosci.*, 18, 741, 1998.
7. Crosthwaite, S.K., Loros, J.J., and Dunlap, J.C., Light-induced resetting of a circadian clock is mediated by a rapid increase in *frequency* transcript, *Cell*, 81, 1003, 1995.
8. Curtin, K.D., Huang, Z.J., and Rosbash, M., Temporally regulated entry of the *Drosophila period* protein contributes to the circadian clock, *Neuron*, 14, 365, 1995.
9. Dembinska, M.E., Stanewsky, R., Hall, J.C., and Rosbash, M., Circadian cycling of a *period-lacZ* fusion protein in *Drosophila*: evidence for cyclical degradation, *J. Biol. Rhythms*, 12, 157, 1997.
10. Dunlap, J.C., Genetic analysis of circadian clocks, *Ann. Rev. Physiol.*, 55, 683, 1993.
11. Dushay, M.S., Rosbash, M., and Hall, J.C., The *disconnected* visual system mutations in *Drosophila melanogaster* drastically disrupt circadian rhythms, *J. Biol. Rhythms*, 4, 1, 1989.
12. Edery, I., Rutila, J.E., and Rosbash, M., Phase shifting of the circadian clock by induction of the *Drosophila period* protein, *Science*, 263, 237, 1994.
13. Edery, I., Zwiebel, L.J., Dembinska, M.E., and Rosbash, M., Temporal phosphorylation of the *Drosophila period* protein, *Proc. Natl. Acad. Sci. USA*, 91, 2260, 1994.
14. Emery, I.F., Noveral, J.M., Jamison, C.F., and Siwicki, K.K., Rhythms of *Drosophila period* gene expression in culture, *Proc. Natl. Acad. Sci. USA*, 94, 4092, 1997.
15. Ewer, J., Frisch, B., Hamblen-Coyle, M.J., Rosbash, M., and Hall, J.C., Expression of the *period* clock gene within different cell types in the brain of *Drosophila* adults and mosaic analysis of these cells' influence on circadian behavioral rhythms, *J. Neuroscience*, 12, 3321, 1992.
16. Ewer, J., Hamblen-Coyle, M., Rosbash, M., and Hall, J.C., Requirement for *period* gene expression in the adult and not during development for locomotor activity rhythms of imaginal *Drosophila melanogaster*, *J. Neurogenetics*, 7, 31, 1990.

17. Frisch, B., Hardin, P.E., Hamblen-Coyle, M.J., Rosbash, M.R., and Hall, J.C., A promoterless *period* gene mediates behavioral rhythmicity and cyclical *per* expression in a restricted subset of the *Drosophila* nervous system, *Neuron*, 12, 555, 1994.
18. Gekakis, N., Saez, L., Delahaye-Brown, A.-M. et al., Isolation of *timeless* by PER protein interaction: defective interaction between *timeless* protein and long-period mutant PER^L, *Science*, 270, 815, 1995.
19. Giebultowicz, J.M. and Hege, D.M., Circadian clock in Malpighian tubules, *Nature*, 386, 664, 1997.
20. Giebultowicz, J.M., Riemann, J.G., Raina, A.K., and Ridgway, R.L., Circadian system controlling release of sperm in the insect testes, *Science*, 245, 1098, 1989.
21. Green, D.J. and Gillette, M.R., Circadian rhythm of firing rate recorded from single cells in the rat suprachiasmatic brain slice, *Brain Res.*, 245, 198, 1982.
22. Hall, J.C., Tripping along the trail to the molecular mechanisms of biological clocks, *Trends Neurosci.*, 18, 230, 1995.
23. Hall, J.C., Are cycling gene products as internal zeitgebers no longer the zeitgeist of chronobiology?, *Neuron*, 17, 799, 1996.
24. Hamblen, M., Zehring, W.A., Kyriacou, C.P. et al., Germ-line transformation involving DNA from the *period* locus in *Drosophila melanogaster*: overlapping genomic fragments that restore ciorcadian and ultradian rhythmicity to *per^0* and *per$^-$* mutants, *J. Neurogenetics*, 3, 249, 1986.
25. Hao, H., Allen, D.L., and Hardin, P.E., A circadian enhancer mediates PER-dependent mRNA cycling in *Drosophila*, *Mol. Cell. Biol.*, 17, 3687, 1997.
26. Hao, H., Cheng, Y., and Hardin, P. unpublished observations, 1997.
27. Hardin, P.E., Analysis of *period* mRNA cycling in *Drosophila* head and body tissues suggests that body oscillators are subservient to head oscillators, *Mol. Cell. Biol.*, 14, 7211, 1994.
28. Hardin, P.E., Hall, J.C., and Rosbash, M., Feedback of the *Drosophila period* gene product on circadian cycling of its messenger RNA levels, *Nature*, 342, 536, 1990.
29. Hardin, P.E., Hall, J.C., and Rosbash, M., Behavioral and molecular analyses suggest that circadian output is disrupted by *disconnected* mutants in *D. melanogaster*, *EMBO J.*, 11, 1, 1992.
30. Hardin, P.E., Hall, J.C., and Rosbash, M., Circadian cycling in the levels of protein and mRNA from *Drosophila melanogaste's period* gene, in *Molecular Genetics of Biological Rhythms*, Young, M.W., Ed., Marcel Dekker, New York, 1992, p. 155.
31. Hardin, P.E., Hall, J.C., and Rosbash, M., Circadian oscillations in *period* gene mRNA levels are transcriptionally regulated, *Proc. Natl. Acad. Sci. USA*, 89, 11711, 1992.
32. Helfrich, C., Role of optic lobes in the regulation of the locomotor activity rhythm of *Drosophila melanogaster*: behavioral analysis of neural mutants, *J. Neurogenet.*, 3, 321, 1986.
33. Helfrich-Forster, C., *Drosophila* rhythms: from brain to behavior, *Sem. Cell Dev. Biol.*, 7, 791, 1996.
34. Hofbauer, A. and Buchner, E., Does *Drosophila* have seven eyes?, *Naturwiss.*, 76, 335, 1989.
35. Huang, Z.J., Edery, I., and Rosbash, M., PAS is a dimerization domain common to *Drosophila* Period and several transcription factors, *Nature*, 364, 259, 1993.
36. Hunter-Ensor, M., Ousley, A., and Sehgal, A., Regulation of the *Drosophila* protein *Timeless* suggests a mechanism for resetting the circadian clock by light, *Cell*, 84, 677, 1996.
37. Jacklet, J., Circadian rhythm of optic nerve impulses recorded from the isolated eye of *Aplysia*, *Science*, 217, 562, 1969.
38. King, D.P., Zhao, Y., Sangoram, A.M. et al., Positional cloning of the mouse circadian *Clock* gene, *Cell*, 89, 641, 1997.
39. Konopka, R.J. and Benzer, S., Clock mutants of *Drosphila melanogaster*, *Proc. Natl. Acad. Sci. USA*, 68, 2112, 1971.
40. Lee, C., Parikh, V., Itsukaichi, T., Bae, K., and Edery, I., Resetting the *Drosophila* clock by photic regulation of PER and PER-TIM complex, *Science*, 271, 1740, 1996.

41. Liu, X., Lorenz, L.J., Yu, Q., Hall, J.C., and Rosbash, M., Spatial and temporal expression of the *period* gene in *Drosophila melanogaster*, *Genes Dev.*, 2, 228, 1988.
42. Liu, X., Zwiebel, L.J., Hinton, D., Benzer, S., Hall, J.C., and Rosbash, M., The *period* gene encodes a predominantly nuclear protein in adult *Drosophila*, *J. Neuroscience*, 12, 2735, 1992.
43. Marrus, S.B., Zeng, H., and Rosbash, M., Effect of constant light and circadian entrainment of *per^s* flies: evidence for light-mediated delay of the negative feedback loop in *Drosophila*, *EMBO J.*, 15, 6877, 1996.
44. Murre, C., McCaw, P.S., and Baltimore, D., A new DNA binding and dimerization motif in immunoglobulin, enhancer binding, *daughterless*, *MyoD*, and *myc* proteins, *Cell*, 56, 777, 1989.
45. Myers, M.P., Wager-Smith, K., Rothenfluh-Hilfiker, A., and Young, M.W., Light-induced degradation of TIMELESS and entrainment of the *Drosophila* circadian clock, *Science*, 271, 1736, 1996.
46. Newby, L.M. and Jackson, F.R., A new biological rhythm mutant of *Drosophila melanogaster* that identifies a gene with essential embryonic function, *Genetics*, 135, 1077, 1993.
47. Pittendrigh, C.S., On temperature independence in the clock system controlling emergence time in *Drosophila*, *Proc. Natl. Acad. Sci. USA*, 40, 1018, 1954.
48. Pittendrigh, C.S., Circadian systems: general perspective, in *Biological Rhythms*, Aschoff, J., Ed., Plenum Press, New York, 1981, p. 57.
49. Rosbash, M., Molecular control of circadian rhythms, *Curr. Opin. Genet. Dev.*, 5, 662, 1995.
50. Rosbash, M., Allada, R., Dembinska, M. et al., A *Drosophila* circadian clock, *CSHSQB*, 61, 265, 1996.
51. Rouyer, F., Rachidi, M., Pikielny, C., and Rosbash, M., A new gene encoding a putative transcription factor regulated by the *Drosophila* circadian clock, *EMBO J.*, 16, 3944, 1997.
52. Rutila, J.E., Zeng, H., Le, M., Curtin, K.D., Hall, J.C., and Rosbash, M., The *tim^{SL}* mutant of the *Drosophila* rhythm gene *timeless* manifests allele-specific interactions with *period* gene mutants, *Neuron*, 17, 921, 1996.
53. Saez, L. and Young, M.W., *In situ* localization of the *per* clock protein during development of *Drosophila melanogaster*, *Mol. Cell. Biol.*, 8, 5378, 1988.
54. Saez, L. and Young, M.W., Regulation of nuclear entry of the *Drosophila* clock proteins *period* and *timeless*, *Neuron*, 17, 911, 1996.
55. Saumann, I. and Reppert, S.M., Circadian clock neurons in the silk moth *Antheraea pernyi*: novel mechanisms of period protein regulation, *Neuron*, 17, 889, 1996.
56. Saumann, I., Tsai, T., Roca, A.L., and Reppert, S.M., Period protein is necessary for circadian control of egg hatching behavior in the silk moth *Antheraea pernyi*, *Neuron*, 17, 901, 1996.
57. Schotland, P. and Sehgal, A. unpublished observations, 1996.
58. Sehgal, A., Molecular genetic analysis of rhythms invetebrates and invertebrates, *Curr. Opin. Neurobiol.*, 5, 824, 1995.
59. Sehgal, A., Ousley, A., and Hunter-Ensor, M., Control of circadian rhythms by a two-component clock, *Mol. Cell. Neurosci.*, 7, 165, 1996.
60. Sehgal, A., Price, J.L., Man, B., and Young, M.W., Loss of circadian behavioral rhythms and *per* RNA oscillations in the *Drosophila* mutant *timeless*, *Science*, 263, 1603, 1994.
61. Sehgal, A., Rothenfluh-Hilfiker, A., Hunter-Ensor, M., Chen, Y., Myers, M.P., and Young, M.W., Rhythmic expression of *timeless*: a basis for promoting circadian cycles in *period* gene autoregulation, *Science*, 270, 808, 1995.
62. Siwicki, K.K., Eastman, C., Petersen, G., Rosbash, M., and Hall, J.C., Antibodies to the *period* gene product of *Drosophila* reveal diverse distribution and rhythmic changes in the visual system, *Neuron*, 1, 141, 1988.
63. Stanewsky, R., Jamison, C.F., Plautz, J.D., Kay, S.A., and Hall, J.C., *Period* gene expression is controlled by two circadian regulated elements, *EMBO J.*, 16, 5006, 1997.

64. Sun, Z.S., Albrecht, U., Zhuchenko, O., Bailey, J., Eichele, G., and Lee, C.C., RIGIU, a putative mammalian ortholog of the *Drosophila period* gene, *Cell*, 90, 1003, 1997.
65. Takahashi, J.S., Molecular neurobiology and genetics of circadian rhythms in mammals, *Ann. Rev. Neurosci.*, 18, 531, 1995.
66. Takahashi, J.S., Hamm, H., and Menaker, M., Circadian rhythms of melatonin release from individual superfused chicken pineal glands *in vitro*, *Proc. Natl. Acad. Sci. USA*, 77, 2319, 1980.
67. Takahashi, J.S. and Zatz, M., Regulation of circadian rhythmicity, *Science*, 217, 1104, 1982.
68. Tei, H., Okamura, H., Shigeyoshi, Y., et al., Circadian oscillation of a mammalian homologue of the *Drosophila period* gene, *Nature*, 389, 512, 1997.
69. Tosini, G. and Menaker, M., Circadian rhythms in cultured mammalian retina, *Science*, 272, 419, 1996.
70. van Gelder, R., Bae, H., Palazzolo, M., and Krasnow, M., Extent and character of circadian gene expression in *Drosophila melanogaster*: identification of twenty oscillating mRNAs in the fly head, *Curr. Biol.*, 5, 1424, 1995.
71. van Gelder, R.N. and Krasnow, M.A., A novel circadianly expressed *Drosophila melanogaster* gene dependent on the *period* gene for its rhythmic expression, *EMBO J.*, 15, 1625, 1996.
72. Vitaterna, M.H., King, D.P. and Chang, A.-M. et al., Mutagenesis and mapping of a mouse gene, *Clock*, essential for circadian behavior, *Science*, 264, 719, 1994.
73. Vosshall, L.B., Price, J.L., Sehgal, A., Saez, L., and Young, M.W., Block in nuclear localization of *period* Protein by a second clock mutation, *timeless*, *Science*, 263, 1606, 1994.
74. Vosshall, L.B. and Young, M.W., Circadian rhythms in *Drosophila* can be driven by *period* gene expression in a restricted group of central brain cells, *Neuron*, 15, 345, 1995.
75. Welsh, D.K., Logothetis, D.E., Meister, M., and Reppert, S.M., Individual neurons dissociated from rat suprachiasmatic nucleus express independently phased circadian firing patterns, *Neuron*, 14, 697, 1995.
76. Wheeler, D.A., Hamblen-Coyle, M.J., Dushay, M.S., and Hall, J.C., Behavior in light-dark cycles of *Drosophila* mutants that are arrhythmic, blind, or both, *J. Biol. Rhythms*, 8, 67, 1993.
77. Young, M.W., Wager-Smith, K., Vosshall, L., Saez, L., and Myers, M.P., Molecular anatomy of a light-sensitive circadian pacemaker in *Drosophila*, *CSHSQB*, 61, 279, 1996.
78. Yu, Q., Jacquier, A.C., Citri, Y., Hamblen, M., Hall, J.C., Rosbash, M., Molecular mapping of point mutations in the *period* gene that stop or speed up biological clocks in *Drosophila melanogaster*, *Proc. Natl. Acad. Sci. USA*, 84, 784, 1987.
79. Zatz, M., Perturbing the pacemaker of the chick pineal, *Discov. Neurosci.*, 8, 67, 1992.
80. Zeng, H., Hardin, P.E., and Rosbash, M., Constitutive overexpression of the *Drosophila period* protein inhibits *period* mRNA cycling, *EMBO J.*, 13, 3590, 1994.
81. Zeng, H., Qian, Z., Myers, M.P., and Rosbash, M., A light-entrainment mechanism for the *Drosophila* circadian clock, *Nature*, 380, 129, 1996.
82. Zerr, D.M., Hall, J.C., Rosbash, M., and Siwicki, K.K., Circadian fluctuations of *period* protein immunoreactivity in the CNS and the visual system of *Drosophila*, *J. Neurosci.*, 10, 2749, 1990.

Chapter 6

Strategies for Dissecting the Molecular Mechanisms of Mammalian Circadian Rhythmicity

Lisa D. Wilsbacher, Jonathan P. Wisor, and Joseph S. Takahashi

Contents

6.1	Introduction	75
6.2	Identification of Circadian Rhythm Genes	77
	6.2.1 Genetic Approaches	77
	6.2.2 Molecular Approaches	78
	6.2.3 Gene Expression Approaches	79
6.3	Analysis of Circadian Rhythm Genes	79
	6.3.1 Molecular Pathway Analysis	79
	6.3.2 Functional Analysis	80
6.4	Summary	81
Acknowledgments		81
References		81

6.1 Introduction

Circadian (~24-hour) rhythms regulate daily fluctuations in cellular processes, physiology, and behavior that occur in nearly all organisms, from mammals to bacteria.[50] Much has been learned about the mechanisms that underlie circadian rhythmicity through both physiological and molecular analyses.[13,50] However, until recently, the knowledge gained by these two approaches diverged in practical applicability; organisms in which circadian physiology is well understood (chick, rodents, *Aplysia*, *Bulla*) are not all genetically amenable, and orthologs of the basic circadian clock components identified through genetic analysis in *Drosophila* and *Neurospora* were only recently isolated in mammals.[13,50] The cloning of the mouse genes *Clock*, *mperiod1* (*mper1*), and *mperiod2* (*mper2*) provides examples of circadian gene discovery and identification in mammals and should allow the molecular dissection of a physiologically well-studied clock to begin in earnest.[1,2,27,44,48,52]

The transcription-translation negative feedback models of the circadian oscillator, based on the *period* (*per*) and *timeless* (*tim*) genes in *Drosophila*[21,43,59] and the *frequency* (*frq*) gene in *Neurospora*,[3] are attractive and provide a working hypothesis for the generation of circadian rhythmicity in cells. The observations in the fruit fly system which led to our current understanding of the negative feedback loop are briefly reviewed here. Konopka and Benzer[29] isolated *per* in a chemical mutagenesis screen in *Drosophila*; the three original mutations conferred short periodicity (per^S), long periodicity (per^L), or no periodicity (per^0) to eclosion and locomotor activity rhythms. The *per* gene was cloned using germline transformation to rescue the null phenotype.[6,58] Subsequently, Hardin et al.[21] demonstrated that *per* mRNA expression oscillates with a period matching that of the behavioral rhythm in both wild-type and mutant flies. Circadian rhythms of PER protein accumulation, nuclear translocation, and phosphorylation levels are also expressed in the fly head; like *per* mRNA, the *per* allele controls the period of PER protein rhythms.[12,14,30,47,61] Interestingly, these protein studies revealed ~6-hr lag in protein synthesis after mRNA synthesis.[14]

More recently, a second circadian gene, *timeless*, was isolated in *Drosophila*.[16,35] The tim^0 null mutation causes arrhythmic eclosion and locomotor behavior; in addition, all *per* mRNA and protein rhythms are abolished.[42] The expression patterns of *tim* show a striking similarity to those seen with *per*: the period and phase of *tim* and *per* mRNA expression are identical to each other in both wild-type and per^S flies, and, furthermore, no *tim* mRNA rhythm was detected in tim^0 mutant flies.[43] Like PER, TIM displays diurnal and circadian rhythms in nuclear translocation, abundance, and phosphorylation.[24,36,60] These rhythms depend on the *tim* and *per* genes, and TIM accumulation also displays a 6-hr lag behind mRNA expression.[24,36,60] The parallels in *per* and *tim* expression immediately suggested that these gene products may interact to regulate circadian rhythmicity.[43] Several important observations supported this hypothesis and provided major insights concerning the effect of light on the clock. First, PER nuclear localization is blocked in *tim* mutant flies, which suggests that TIM may stabilize PER or mediate PER nuclear entry.[41] Second, inhibition of PER abundance and phosphorylation rhythms, as seen in tim^0 mutants, can also be elicited in wild-type flies by constant light.[39] Third, PER and TIM appear to interact directly both *in vitro*[16,30,41] and *in vivo*.[60] Finally, several groups independently recognized that light (in both LD and DD) leads to the rapid disappearance of TIM, while light does not affect *tim* mRNA.[24,30,36,60]

Together these data strongly suggested that a negative regulatory loop generates circadian rhythmicity at the molecular level.[21,24,30,36,60] These results also introduced a mechanism whereby a unidirectional stimulus (light) can elicit a bidirectional response in the clock (phase delays and advances). In this model:

1. *per* and *tim* transcripts begin to increase during the subjective day and peak at about circadian time 12 (CT12, where CT0 marks the beginning of subjective day and CT12 marks the beginning of subjective night).

2. PER and TIM proteins do not accumulate in the cytoplasm until a threshold level of one or both is reached late in the day; they then interact and enter the nucleus around CT18, where PER and/or TIM act to repress *per* and *tim* transcription.

3. Maximal PER and TIM levels (CT21) strongly inhibit *per* and *tim* transcription, while protein turnover causes PER and TIM to decrease through the late subjective night and early morning.

4. The release of transcription inhibition by PER/TIM turnover allows *per* and *tim* mRNA to accumulate and begin the next cycle.

Light could exert its effect at two stages of this cycle. In the late day or early evening (stage 2), light degrades TIM and delays the entry of PER into the nucleus. During the late night or early morning (stage 3), light degrades TIM and therefore destabilizes PER to release the inhibition of transcription earlier. These two effects would cause a phase delay or a phase advance, respectively.

Implicit in this model is the activation or inhibition of other clock molecules as a direct or indirect result of the *per/tim* expression cycle. PER could mediate feedback inhibition through a conserved stretch of amino acids, the PAS domain.[23,37] This domain, named for PER, the aryl hydrocarbon receptor (AHR),[7] the aryl hydrocarbon receptor nuclear translocator (ARNT),[22] and the *Drosophila* gene *single-minded* (SIM),[37] mediates protein dimerization[23] and is proposed to increase the specificity of interaction between transcription factors containing the domain.[40] The other original members of this family are basic helix-loop-helix (bHLH) domain transcription factors; PER, however, does not contain a known DNA-binding domain and therefore is not likely to act as a transcription factor on its own.[23] Together these observations suggest that PER may act as a dominant-negative regulator of transcription by binding to other bHLH-PAS transcription factors, then preventing normal DNA binding and transcription activation.[23]

Could the model described in *Drosophila* apply to mammals as well? The identities of the cloned mammalian circadian genes suggest that the answer is yes: *Clock* is a bHLH-PAS transcription factor, and both *mper1* and *mper2* display a rhythm in mRNA expression.[1,2,27,44,48,52] Clearly, though, we are only beginning to discern the molecular mechanisms of circadian rhythmicity. To better understand the mammalian circadian system, the components of the system must be identified, molecular interactions between components must be analyzed, and the function of each component must be determined. The strategies described in this chapter illustrate the utility of genetics and molecular biology in reaching these goals.

6.2   Identification of Circadian Rhythm Genes

6.2.1   Genetic Approaches

The first step in the dissection of mammalian circadian clock mechanisms is the identification of its components. Currently, three putative mammalian circadian genes have been cloned: *Clock*, *mper1*, and *mper2*.[1,2,27,44,48,52] While the identification of these genes marks a turning point in the analysis of mammalian circadian rhythmicity, clearly more circadian genes remain unknown. Forward genetic techniques provide one of the tools to isolate these components.

The forward genetic approach (from phenotype to gene), long considered "impractical" in the mouse,[8] is in fact a feasible method of mammalian gene identification.[51,55] This method requires no prior knowledge of the mechanisms of the system being studied and is therefore appropriate for relatively unbiased identification of genes in that system. Several requirements discouraged investigators from taking this approach to study mammalian behavior in the past. First, forward genetics requires that a large number of gametes (on the order of thousands) be screened in order to scan the genome fully for mutations, and, second, a robust phenotype and suitable behavioral assay (preferably automated) must be available in order to efficiently screen the large number of animals.[51] The identification of recessive mutations requires additional breeding and testing, but recessive screens remain an important source of behavioral gene information and should not be dismissed as "impractical". Upon identification of a mutant phenotype and genetic mapping of the locus, the gene may be identified using functional rescue with transgenes, positional cloning techniques, or candidate gene analysis.[10,11,51]

The circadian gene *Clock* was the first to be cloned using forward genetics in mammals.[2,27] The *Clock* mutation was isolated using an *N*-ethyl-*N*-nitrosourea (ENU) mutagenesis screen in mice. In constant darkness, period length increases by about one hour in heterozygotes and by about 4 hr initially with eventual loss of circadian rhythmicity in homozygotes;[55] therefore, the *Clock* mutation affects both the period length and the persistence of circadian rhythmicity. The gene was localized to mouse chromosome 5 in a critical region of about 0.3 centiMorgans, but no known candidate genes had been placed in this area. A contiguous physical map of the region containing *Clock* was generated from yeast artificial chromosome (YAC) and bacterial artificial chromosome (BAC) clones,[27] and three BAC clones which spanned more than 75% of the genetic region were chosen for transgenic line generation, transcription unit identification, and genomic sequencing.

Complete functional rescue of the *Clock* mutation was achieved by one BAC clone (BAC 54) in transgenic mouse experiments: in animals that were heterozygous or homozygous for the *Clock* mutation, the 140-kilobase (kb) BAC 54 transgene restored the free-running period to values indistinguishable from wild-type.[2] Transcription unit analysis then identified a candidate expressed sequence in which mRNA levels were dramatically reduced in *Clock* homozygotes as compared to wild-type animals. The full-length cDNA of this candidate completely mapped within the BAC 54 genomic sequence; therefore, integration of these results indicated that a transcription unit which spans ~100,000 kb and encodes a bHLH-PAS domain protein is the *Clock* gene.[2,27] Finally, the original ENU-induced mutation, an A→T transversion in the third position of the 5′ splice donor site of exon 19, was discovered and shown to cause exon skipping.[27] Exon 19 encodes a portion of the putative transactivation domain of the CLOCK protein, which suggests that the mutant protein binds DNA but does not activate transcription in a normal fashion. The dominant-negative molecular nature of the *Clock* mutation is entirely consonant with genetic analysis, suggesting *Clock* is an antimorph.[26] In summary, the cloning of the *Clock* gene provides an important proof of principle that forward genetic approaches can be used to study complex behaviors such as circadian rhythmicity in mice.

6.2.2 Molecular Approaches

Obviously, not all mammalian circadian rhythm genes must be identified using forward genetics alone. Due to the interactive nature of the negative feedback loop model, molecular biology remains an important tool for the isolation of circadian genes. Methods which identify protein-protein interactions, such as the yeast two-hybrid assay,[5,32] will be extremely informative now that a molecular entry point into the circadian system is available. In addition, an appreciation of both sequence homology and functional conservation across species will accelerate the discovery of mammalian orthologs of circadian genes.

To this end, molecular techniques and sequence homology played large roles in the cloning of the putative mammalian orthologs of *Drosophila period* (*dper*) by two independent groups.[48,52] For several years, researchers were unsuccessful in cloning mammalian *per* by traditional screening methods using *dper* as probe. Tei et al.[52] finally succeeded using a technique called "intramodule scanning" (IMS) PCR, in which the human genome was scanned with *dper* PAS-specific degenerate primers. In a different approach, Sun et al.[48] identified a cDNA with sequence homology to *dper* in a search for human chromosome 17-specific transcripts. The gene these groups isolated displayed 44% overall amino acid similarity (not identity) to *dper*; importantly, *mper1** contained homology to *dper* not only in the PAS region but also in other domains. In addition, Sun et al.[48] reported the presence of a basic-helix-loop-helix domain, although only three basic residues exist in the putative basic region. The presence of a PAS domain immediately suggests that *mper1* could act as a negative regulator of *Clock*, particularly if further studies rule out a functional DNA-binding domain in *mper1*. Both groups demonstrated that *mper1* mRNA expression oscillates in a circadian manner in the SCN.[48,52] However, the nature of *mper1* expression differs from that found in *Drosophila* in

---

* The first putative human and mouse orthologs of *dper* were cloned by Sun et al.,[48] who named them *RIGUI* and *m-rigui*, respectively, and by Tei et al.,[52] who named them h*PER* and m*Per*, respectively. Subsequently, these genes were renamed h*Per1* and m*Per1* by Shearman et al.[44] and *mper1* (the mouse clone only) by Albrecht et al.[1] A second putative human ortholog was discovered in the nucleotide database (KIAA0347) and was named h*Per2*;[44] the mouse ortholog was named m*Per2*[44] or *mper2*.[1] Finally, a partial human BAC clone with homology to *dper* was found in the database and tentatively named h*Per3*.[44]

several ways. First, peak *mper1* expression occurred at CT 6 in the SCN, which supports pharmacological evidence that the mammalian oscillator is day-phased.[13,50] Second, additional mouse neural tissues exhibited diurnal (retina) or circadian (pars tuberalis, cerebellar Purkinje neurons) *mper1* mRNA oscillations which were out of phase with one another.[48] Finally, *mper1* expression responds to light: the phase of *mper1* expression can be shifted over the course of several days by a corresponding shift in the light/dark (LD) cycle,[48] and light can induce *mper1* expression during the subjective night.[1,45]

During the cloning of *mper1*, the human clones KIAA0347 and Z98884, each distinct from *hper1* yet homologous to *dper*, were detected in the database.[1,44] The full-length cDNA KIAA0347 (*hper2*) maps to human chromosome 6, while BAC genomic clone Z98884 (*hper3*) maps to human chromosome 1, which indicates that three different *per*-like loci exist in humans.[44] The *mper2* gene was isolated; like *mper1*, a circadian rhythm in *mper2* expression exists in the SCN which shifts accordingly with a shift in the LD cycle.[1,44] As of January 1998, "*mper3*" had not yet been cloned.

These examples demonstrate the use of molecular biology and homology in gene identification. As more circadian genes are isolated in *Drosophila* and other organisms, one can imagine relatively rapid identification of orthologs in mammals. Conversely, orthologs of *Clock* and other mammalian circadian genes may be found in *Drosophila* using the same approach. Importantly, this method of gene identification also lends itself well to functional studies of circadian rhythm generation, as demonstrated by the mRNA expression experiments with *mper1* and *mper2*. All told, mechanisms of circadian rhythmicity throughout evolution may involve more of the same genes than previously imagined.

6.2.3 Gene Expression Approaches

The identification of candidate mammalian circadian genes based on differential transcription, such as rhythmic expression, SCN specificity, or light responsiveness, is in theory an excellent approach. Differential display has been successfully applied towards the isolation of circadian-regulated genes in the *Xenopus* retina and in the rat retina and pineal gland.[15,17,18] Three recently developed techniques, serial analysis of gene expression (SAGE), cDNA-based amplified restriction fragment length polymorphism (AFLP) analysis, and oligonucleotide microarray hybridization, may also be useful for identification of differentially expressed transcripts.[4,31,54] These methods are more sensitive and specific than differential display; in addition, microarray analysis can be used to analyze hundreds of transcripts rapidly and simultaneously. Finally, these techniques are well suited to elucidate genes on the input pathway to, and the output pathway from, the clock, as conditions such as a phase shift-inducing light pulse may be chosen to select for these genes.

6.3 Analysis of Circadian Rhythm Genes

6.3.1 Molecular Pathway Analysis

As mentioned above, molecular biology remains an important method of circadian gene analysis. Expression studies, *in vitro* experiments and cell culture studies provide relatively rapid results in the investigation of circadian component interactions. The information gained from these experiments may then be functionally confirmed in the living, behaving animal.

The traditional molecular method of analysis has been to search for a circadian rhythm of gene expression via mRNA and protein studies. This approach, nearly dogma in the field,[19] remains a valid technique to establish the circadian control of a gene with no mutant phenotype.[1,44,48,52] However, it should be emphasized that output genes can oscillate; therefore, functional evidence such as a change in period or loss of rhythmicity is necessary in order to conclude such a gene is a clock component. Furthermore, the idea that all true circadian genes must oscillate in expression[3] may not be fully accurate; to date, there is no evidence of circadian *Clock* mRNA expression, yet this gene is clearly required for normal circadian behavior.[48,55] While the relevant oscillation "requirement" should be functional activity as opposed to expression, it is still possible that genes essential for the generation of circadian oscillations may have no intrinsic rhythms. The combination of molecular biological techniques and functional analysis through reverse genetics may soon alter the view that a gene must oscillate in expression in order to be a "true" clock component.

The identification of *Clock*, *mper1*, and *mper2* allows the isolation of additional clock genes via molecular interactions. Discovery of the CLOCK protein binding partner(s) will be an important step in the molecular dissection of the circadian clock, as *Clock* is a strong candidate for a positive element in a transcription-translation feedback loop. Dimerization is required for the activity of bHLH factors,[57] and most dimers bind the E-box (5'-CANNTG-3') promoter sequence to activate transcription.[25,49] Therefore, we predict that genes activated by CLOCK will contain an E-box within their promoters. Interestingly, the *Drosophila per* promoter contains a 69-base pair enhancer which mediates robust rhythmicity of the *per* transcript, and an E-box within this enhancer is required for high-level expression.[20] As *mper1* and *mper2* are candidate targets of CLOCK, this result is quite significant. Similarly, the mammalian ortholog of *tim* is an obvious candidate for identification and analysis. The role and mechanism of action for mammalian *tim* should prove interesting, as the role of *tim* in light-responsiveness of the night-phased *Drosophila* oscillator would not be consistent with the mammalian day-phased oscillator.

The immediate questions to address are derived from the *Drosophila* negative feedback loop model. Does mPER interact with CLOCK via the PAS domain to inhibit CLOCK activity? Does a delay in translation and nuclear localization of mPER1 or mPER2 exist? Is it required for rhythmicity, and how is it mediated? Most of these questions may be answered using protein interaction, biochemical, and cell culture techniques. Indeed, Saez and Young[41] took a cell culture approach to probe *Drosophila* PER-TIM interactions. Using *Drosophila* S2 cells, which do not contain an endogenous circadian clock, they determined that PER and TIM accumulate in the cytoplasm for 2 to 3 hr before entering the nucleus, while singly expressed proteins lacking a cytoplasmic localization domain translocated rapidly to the nucleus; these results suggested that the 6-hr transcription-translation lag *in vivo* in *Drosophila* may not be controlled by the circadian clock alone.[41] With an appropriate mammalian cell line, information from cell culture studies would be useful in evaluating the current mammalian clock genes.

6.3.2 Functional Analysis

Because circadian rhythmicity ultimately affects behavior, the conclusions drawn from *in vitro* analysis of clock components must be tested *in vivo*. Altering the activity of a putative clock component to elicit a change in the whole animal's behavior provides one of the strongest pieces of evidence available that the molecule is truly a clock component. Based on information gained from molecular analysis, rational decisions regarding reverse genetic techniques may be made to confirm the function of candidate circadian genes.

In reverse genetics (from gene to phenotype), a known locus is altered and the phenotypic effects of that alteration are measured.[51] Several elegant reverse genetic approaches have been applied to the study of mammalian behavior using transgenesis,[33,34] conventional gene knockouts,[46,51] and conditional gene knockouts.[9,46,53] Transgenic technology using dominant-negative forms of a candidate clock gene represents a straightforward method of perturbing circadian function, as this approach should prevent progression through the circadian cycle. This method is especially attractive for the study of genes such as *Clock*, where the putative basic function (i.e., transcriptional activation) is easy to target with a dominant-negative molecule. Gene knockouts are also a common targeting technique and are often used to prove the necessity of a particular locus in the normal expression of a behavior. This approach has limitations, however, in that the gene is disrupted in every cell during all stages of life; therefore, behavioral abnormalities may be secondary to developmental problems. To address this issue, researchers have recently combined transgenic and gene-targeting techniques to create conditional transgenic and knockout mice in which a gene is altered in a developmental- and tissue-specific manner; the second generation of these gene-targeting experiments has been elegantly applied in the field of learning and memory.[33,53] In addition, temporal control of exogenous gene expression would be especially desirable in circadian experiments. Several inducible systems have been created and tested in transgenic mice, including the reverse tetracycline-controlled transactivator (rtTA) system,[28] the ecdysone-inducible system,[38] and the mifepristone-inducible system.[56] The theoretical power of inducible transgenics in circadian gene analysis is unmistakable: the ability of a gene to elicit a behavioral phase-shift as a result of induced expression could demonstrate the circadian function of that gene.

These reverse genetic methods, when used on a gene-by-gene basis, will almost certainly indicate whether a candidate clock gene truly plays a role in rhythm generation or maintenance. Furthermore, breeding lines which harbor transgenic or targeted mutations together to search for genetic interactions may provide additional insights to mechanisms of circadian function. Integration of *in vivo* gene analyses with *in vitro* and cell culture information will allow us to better understand the organization of the circadian system as a whole.

6.4   Summary

The combination of genetic and molecular techniques has brought the mammalian circadian field to a period of discovery similar to that experienced recently in *Drosophila*. The first goal of mammalian clock investigation, the discovery of circadian genes, has been achieved in part and should certainly continue using both approaches. The phases of molecular and functional analyses have already begun and promise to be the most exciting, informative stages of circadian study. Clearly, much more remains to be discovered, but it appears that the overall goal will be attained: the understanding of all components and mechanisms of circadian rhythm generation.

Acknowledgments

Special thanks to R. Keith Barrett for critical review of the manuscript. Research was supported in part by the NSF Center for Biological Timing, an Unrestricted Grant in Neuroscience from Bristol-Myers Squibb, NIH grants R37 MH39593 and P01 AG11412 (J.S.T.), and NEI grant T32 EY07128 and MSTP fellowship T32 GM08152 (L.D.W.). J.S.T. is an investigator in the Howard Hughes Medical Institute.

References

1. Albrecht, U., Sun, Z.S., Eichele, G., and Lee, C.C., A differential response of two putative mammalian circadian regulators, *mper1* and *mper2*, to light, *Cell*, 91, 1055, 1997.
2. Antoch, M.P., Song, E.-J., Chang, A.-M., Vitaterna, M.H., Zhao, Y., Wilsbacher, L.D., Sangoram, A.M., King, D.P., Pinto, L.H., and Takahashi, J.S., Functional identification of the mouse circadian *Clock* gene by transgenic BAC rescue, *Cell*, 89, 655, 1997.
3. Aronson, B.D., Johnson, K.A., Loros, J.J., and Dunlap, J.C., Negative feedback defining a circadian clock: autoregulation of the clock gene *frequency*, *Science*, 263, 1578, 1994.

4. Bachem, C.W.B., van der Hoeven, R.S., de Brujin, S.M., Vreugdenhil, D., Zabeau, M., and Visser, R.G.F., Visualization of differential gene expression using a novel method of RNA fingerprinting based on AFLP: analysis of gene expression during potato tuber development, *Plant J.*, 9, 745, 1996.

5. Bai, C. and Elledge, S.J., Gene identification using the yeast two-hybrid system, *Meth. Enzymol.*, 273, 331, 1996.

6. Bargiello, T.A., Jackson, F.R., and Young, M.W., Restoration of circadian behavioural rhythms by gene transfer in *Drosophila, Nature*, 312, 752, 1984.

7. Burbach, K.M., Poland, A., and Bradfield, C.A., Cloning of the Ah-receptor cDNA reveals a distinctive ligand-activated transcription factor, *Proc. Natl. Acad. Sci. USA*, 89, 8185, 1992.

8. Capecchi, M.R., The new mouse genetics: altering the genome by gene targeting, *Trends Genet.*, 5, 70, 1989.

9. Chen, C. and Tonegawa, S., Molecular genetic analysis of synaptic plasticity, activity-dependent neural development, learning, and memory in the mammalian brain, *Ann. Rev. Neurosci.*, 20, 157, 1997.

10. Collins, F.S., Positional cloning: let's not call it reverse anymore, *Nature Genet.*, 1, 3, 1992.

11. Copeland, N.G., Jenkins, N.A., Gilbert, D.J., Eppig, J.T., Maltais, L.J., Miller, J.C., Dietrich, W.F., Weaver, A., Lincoln, S.E., Steen, R.G., Stein, L.D., Nadeau, J.H., and Lander, E.S., A genetic linkage map of the mouse: current applications and future prospects, *Science*, 262, 57, 1993.

12. Curtin, K.D., Huang, Z.J., and Rosbash, M., Temporally regulated nuclear entry of the *Drosophila period* protein contributes to the circadian clock, *Neuron*, 14, 365, 1995.

13. Dunlap, J.C., Genetic and molecular analysis of circadian rhythms, *Ann. Rev. Genet.*, 30, 579, 1996.

14. Edery, I., Zwiebel, L.J., Dembinska, M.E., and Rosbash, M., Temporal phosphorylation of the *Drosophila period* protein, *Proc. Natl. Acad. Sci. USA*, 91, 2260, 1994.

15. Gauer, F., Kedzierski, W., and Craft, C.M., Identification of circadian gene expression in the rat pineal and retina by mRNA differential display, *Neurosci. Lett.*, 187, 69, 1995.

16. Gekakis, N., Saez, L., Delahaye-Brown, A.-M., Myers, M.P., Sehgal, A., Young, M.W., and Weitz, C.J., Isolation of *timeless* by PER protein interaction: defective interaction between *timeless* protein and long-period mutant PER^L, *Science*, 270, 811, 1995.

17. Green, C.B. and Besharse, J.C., Identification of vertebrate circadian clock-regulated genes by differential display, *Meth. Mol. Biol.*, 85, 219, 1997.

18. Green, C.B. and Besharse, J.C., Use of a high stringency differential display screen for identification of retinal mRNAs that are regulated by a circadian clock, *Mol. Brain Res.*, 37, 157, 1996.

19. Hall, J.C., Are cycling gene products as internal zeitgebers no longer the zeitgeist of chronobiology?, *Neuron*, 17, 799, 1996.

20. Hao, H., Allen, D.L., and Hardin, P.E., A circadian enhancer mediates PER-dependent mRNA cycling in *Drosophila melanogaster*, *Mol. Cell. Biol.*, 17, 3687, 1997.

21. Hardin, P.E., Hall, J.C., and Rosbash, M., Feedback of the *Drosophila period* gene product on circadian cycling of its messenger RNA levels, *Nature*, 343, 536, 1990.

22. Hoffman, E.C., Reyes, H., Chu, F.-F., Sander, F., Conley, L.H., Brooks, B.A., and Hankinson, O., Cloning of a factor required for activity of the Ah (Dioxin) receptor, *Science*, 252, 954, 1991.

23. Huang, Z.J., Edery, I., and Rosbash, M., PAS is a dimerization domain common to *Drosophila period* and several transcription factors, *Nature*, 364, 259, 1993.

24. Hunter-Ensor, M., Ousley, A., and Sehgal, A., Regulation of the *Drosophila* protein Timeless suggests a mechanim for resetting the circadian clock by light, *Cell*, 84, 677, 1996.

25. Kadesch, T., Consequences of heteromeric interactions among helix-loop-helix proteins, *Cell Growth Differ.*, 4, 49, 1993.

26. King, D.P., Vitaterna, M.H., Chang, A.M., Dove, W.F., Pinto, L.H., Turek, F.W., and Takahashi, J.S., The mouse *Clock* mutation behaves as an antimorph and maps within the W^{19H} deletion, distal of *Kit*, *Genetics*, 146, 1049, 1997.
27. King, D.P., Zhao, Y., Sangoram, A.M., Wilsbacher, L.D., Tanaka, M., Antoch, M.P., Steeves, T.D.L., Vitaterna, M.H., Kornhauser, J.M., Lowrey, P.L., Turek, F.W., and Takahashi, J.S., Positional cloning of the mouse circadian *Clock* gene, *Cell*, 89, 641, 1997.
28. Kistner, A., Gossen, M., Zimmermann, F., Jerecic, J., Ullmer, C., Lubbert, H., and Bujard, H., Doxycycline-mediated quantitative and tissue-specific control of gene expression in transgenic mice, *Proc. Natl. Acad. Sci.*, 93, 10933, 1996.
29. Konopka, R.J. and Benzer, S., Clock mutants of *Drosophila melanogaster*, *Proc. Natl. Acad. Sci. USA*, 68, 2112, 1971.
30. Lee, C., Parikh, V., Itsukaichi, T., Bac, K., and Edery, I., Resetting the *Drosophila* clock by photic regulation of PER and a PER-TIM complex, *Science*, 271, 1740, 1996.
31. Lockhart, D.J., Dong, H., Byrne, M.C., Follettie, M.T., Gallo, M.V., Chee, M.S., Mittmann, M., Wang, C., Kobayashi, M., Horton, H., and Brown, E.L., Expression monitoring by hybridization to high-density oligonucleotide arrays, *Nature Biotechnol.*, 14, 1675–1680, 1996.
32. Luban, J. and Goff, S.P., The yeast two-hybrid system for studying protein-protein interactions, *Curr. Opin. Biotechnol.*, 6, 59, 1995.
33. Mayford, M., Bach, M.E., Huang, Y.Y., Wang, L., Hawkins, R.D., and Kandel, E.R., Control of memory formation through regulated expression of a CaMKII transgene, *Science*, 274, 1678, 1996.
34. Mayford, M., Wang, J., Kandel, E.R., and O'Dell, T.J., CaMKII regulates the frequency-response function of hippocampal synapses for the production of both LTD and LTP, *Cell*, 81, 891, 1995.
35. Myers, M.P., Sehgal, A., Wager-Smith, K., Wesley, C.S., and Young, M.W., Positional cloning and sequence analysis of the *Drosophila* clock gene, *timeless*, *Science*, 270, 805, 1995.
36. Myers, M.P., Wager-Smith, K., Rothenfluh-Hilfiker, A., and Young, M.W., Light-induced degradation of TIMELESS and entrainment of the *Drosophila* circadian clock, *Science*, 271, 1736, 1996.
37. Nambu, J.R., Lewis, J.O., Wharton, K.A.J., and Crews, S.T., The *Drosophila single-minded* gene encodes a helix-loop-helix protein that acts as a master regulator of CNS midline development, *Cell*, 67, 1157, 1991.
38. No, D., Tao, T.-P., and Evans, R.M., Ecdysone-inducible gene expression in mammalian cells and transgenic mice, *Proc. Natl. Acad. Sci. USA*, 93, 3346, 1996.
39. Price, J.L., Dembinska, M.E., Young, M.W., and Rosbash, M., Suppression of PERIOD protein abundance and circadian cycling by the *Drosophila* clock mutation *timeless*, *EMBO J.*, 14, 4044, 1995.
40. Reisz-Porszasz, S., Probst, M.R., Fukunaga, B.N., and Hankinson, O., Identification of functional domains of the aryl hydrocarbon receptor nuclear translocator protein (ARNT), *Mol. Cell Biol.*, 14, 6075, 1994.
41. Saez, L. and Young, M., Regulation of nuclear entry of the *Drosophila* clock proteins Period and Timeless, *Neuron*, 17, 911, 1996.
42. Sehgal, A., Price, J.L., Man, B., and Young, M.W., Loss of circadian behavioral rhythms and *per* RNA oscillations in the *Drosophila* mutant *timeless*, *Science*, 263, 1603, 1994.
43. Sehgal, A., Rothenfluh-Hilfiker, M., Hunter-Ensor, M., Chen, Y., Myers, M.P., and Young, M.W., Rhythmic expression of *timeless*: a basis for promoting circadian cycles in *period* gene autoregulation, *Science*, 270, 808, 1995.
44. Shearman, L.P., Zylka, M.J., Weaver, D.R., Kolakowski, L.F.J., and Reppert, S.M., Two *period* homologs: circadian expression and photic regulation in the suprachiasmatic nuclei, *Neuron*, 19, 1261, 1997.

45. Shigeyoshi, Y., Taguchi, K., Yamamoto, S., Takekida, S., Yan, L., Tei, H., Moriya, T., Shibata, S., Loros, J.J., Dunlap, J.C., and Okamura, H., Light-induced resetting of a mammalian circadian clock is associated with rapid induction of the *mPer1* transcript, *Cell*, 91, 1043, 1997.

46. Silva, A.J., Smith, A.M., and Giese, K.P., Gene targeting and the biology and learning and memory, *Ann. Rev. Genet.*, 31, 527, 1997.

47. Siwicki, K.K., Eastman, C., Petersen, G., Rosbash, M., and Hall, J.C., Antibodies to the period gene product of *Drosophila* reveal diverse tissue distribution and rhythmic changes in the visual system, *Neuron*, 1, 141, 1988.

48. Sun, Z.S., Albrecht, U., Zhuchenko, O., Bailey, J., Eichele, G., and Lee, C.C., *RIGUI*, a putative mammalian ortholog of the *Drospholia period* gene, *Cell*, 90, 1003, 1997.

49. Swanson, H.I., Chan, W.K., and Bradfield, C.A., DNA binding specificities and pairing rules of the Ah receptor, ARNT, and SIM proteins, *J. Biol. Chem.*, 270, 26292, 1995.

50. Takahashi, J.S., Molecular neurobiology and genetics of circadian rhythms in mammals, *Ann. Rev. Neurosci.*, 18, 531, 1995.

51. Takahashi, J.S., Pinto, L.P., and Vitaterna, M.H., Forward and reverse genetic approaches to behavior in the mouse, *Science*, 264, 1724, 1994.

52. Tei, H., Okamura, H., Shigeyoshi, Y., Fukuhara, C., Ozawa, R., Hirose, M., and Sakaki, Y., Circadian oscillation of a mammalian homologue of the *Drosophila period* gene, *Nature*, 389, 512, 1997.

53. Tsien, J.Z., Chen, D.F., Gerber, D., Tom, C., Mercer, E.H., Anderson, D.J., Mayford, M., Kandel, E.R., and Tonegawa, S., Subregion- and cell type-restricted gene knockout in mouse brain, *Cell*, 87, 1317–1326, 1996.

54. Velculescu, V.E., Zhang, L., Vogelstein, B., and Kinzler, K.W., Serial analysis of gene expression, *Science*, 270, 484, 1995.

55. Vitaterna, M.H., King, D.P., Chang, A.-M., Kornhauser, J.M., Lowrey, P.L., McDonald, J.D., Dove, W.F., Pinto, L.H., Turek, F.W., and Takahashi, J.S., Mutagenesis and mapping of a mouse gene, *Clock*, essential for circadian behavior, *Science*, 264, 719, 1994.

56. Wang, Y., DeMayo, F.J., Tsai, S.T., and O'Malley, B.W., Ligand-inducible and liver-specific target gene expression in transgenic mice, *Nature Biotechnol.*, 15, 239, 1997.

57. Weintraub, H., Davis, R., Tapscott, S., Thayer, M., Krause, M., Benezra, R., Blackwell, T.K., Turner, D., Rupp, R., and Hollenberg, S., The MyoD gene family: nodal point during specification of the muscle cell lineage, *Science*, 251, 761, 1991.

58. Zehring, W.A., Wheeler, D.A., Reddy, P., Konopka, R.J., Kyriacou, C.P., Rosbash, M., and Hall, J.C., P-element transformation with *period* locus DNA restores rhythmicity to mutant, arrhythmic *Drosophila melanogaster, Cell*, 39, 369, 1984.

59. Zeng, H., Hardin, P.E., and Rosbash, M., Constitutive overexpression of the *Drosophila period* protein inhibits *period* mRNA cycling, *EMBO J.*, 13, 3590, 1994.

60. Zeng, H., Qian, Z., Myers, M.P., and Rosbash, M., A light-entrainment mechanism for the *Drosophila* circadian clock, *Nature*, 380, 129, 1996.

61. Zerr, D.M., Hall, J.C., Rosbash, M., and Siwicki, K.K., Circadian fluctuations of *period* protein immunoreactivity in the CNS and the visual system of *Drosophila*, *J. Neurosci.*, 10, 2749, 1990.

Section II

Daily Alterations in Arousal State

Mark W. Mahowald, Section Editor

Chapter 7

The Evolution of REM Sleep

Jerome M. Siegel

Contents

7.1 Introduction 87
7.2 Definition of REM Sleep 87
7.3 REM Sleep in Mammals 90
7.4 Sleep in the Echidna 90
7.5 Sleep in the Platypus 91
7.6 How Do the Echidna and Platypus Data Fit Together? 92
7.7 Reptilian Sleep 95
7.8 Conclusions 97
Acknowledgment 97
References 97

7.1 Introduction

An understanding of the nature, amount, and distribution of rapid eye movement (REM) sleep across the animal kingdom allows one to form hypotheses about its evolutionary history and function. Attributes of sleep common to several branches of the evolutionary tree are likely to have been present in the common ancestor. Conversely, attributes present in only one branch are likely to have arisen in that branch.

Most studies of sleep have been conducted in humans, with lesser numbers in "standard" laboratory animals such as the rat, rabbit, and dog. Relatively few studies have been conducted of the more than 4000 other mammalian species. However, those studies that have been undertaken clearly show that REM sleep amounts vary enormously across the animal kingdom. REM sleep has been identified in birds, but very few avian species have been investigated.[3] The number of studies of amphibian and reptilian sleep is miniscule, with few such studies using rigorous electrophysiological and behavioral indices.

7.2 Definition of REM Sleep

The criteria for defining a state as "sleep" or as REM sleep can become a significant issue in interpreting data from animal studies. Our understanding of the nature of sleep states is largely based on studies of the cat. In this species, we know the electroencephalogram (EEG) changes correlated with sleep and their developmental history. We also know the basic parameters of the changes in neuronal activity in the cortex responsible for the EEG changes during the sleep cycle. We know much about the thalamic neurons generating cortical rhythmicity.[42] NonREM sleep promoting neurons in the basal forebrain have been identified, as have wake-inducing systems in the posterior hypothalamus and brainstem.[23,28,35] Finally, a system of neurons generating the EEG, eye movement, twitches, and underlying muscle atonia of REM sleep have been identified in the brainstem (reviewed in Reference 40). This system utilizes noradrenergic and serotonergic REM sleep-off neurons and GABAergic, cholinergic, glycinergic, and glutamatergic REM sleep-on cells, as well as other neurons. Cells that are active in both waking and REM sleep are important in generating some of the phasic motor phenomena of both of these states. The driving force responsible for the triggering of REM sleep and for controlling its duration is completely unknown.

Given what we now know about REM sleep, how can we go about defining it in newly examined animals? Ideally and ultimately, one needs to know the activity of all of the cell groups listed above to say that a state has all the characteristics of REM sleep in the cat. If a previously unexamined species were found to have two states of sleep — one with phasic motor activity and one without — one would want to know if the other aspects of REM sleep, documented in the cat, were present. Specifically, are there noradrenergic and serotonergic REM sleep-off cells and GABAergic, glutamatergic, and cholinergic REM sleep-on cells? Are there brainstem pre-motor cells firing in bursts during REM sleep? Is there active inhibition of motoneurons during this state? Are there ascending glutamatergic and cholinergic systems responsible for the low-voltage EEG activity of this state?

These relatively rigorous criteria would provide a useful description of the commonalties in the neuronal activity features that characterize REM sleep across the animal kingdom. However, we must appreciate that today these key features of REM sleep have been documented only in the adult cat. We do not know if the same pattern of REM sleep-on, REM sleep-off, and REM-waking active cells is present in humans during REM sleep. We do not know if the neurochemistry of these systems is the same in all species. The key anatomical structures responsible for REM sleep control in the cat are distributed differently in the human, rat, and other species.[4,22,26,31,39] Although these cell groups are present in other mammalian species, they are also present in modified form in fish and amphibia.[19,20,24,46,49,50] Moreover, these cell groups are present in neonatal rats and cats even though "REM sleep" characteristics are very different at these ages. Chemical microinjection studies in the rat have yet to duplicate clearly the rapid induction of REM sleep and muscle atonia reported more than 30 years ago in the cat and repeatedly confirmed by others.[14] Serotonin depletion in the cat produces insomnia, but has no such effect in the rat.[36,37]

In neonatal humans,[18,43] cats,[29] and rats,[13] the EEG voltage reduction that characterizes REM sleep in the adult human, cat, and rat does not occur. However, the phasic motor activation (twitching) that characterizes REM sleep in the adult animal is present. Because the amplitude of the twitching is more intense in neonates than in the adult, either the motor inhibition is less effective or the intensity of the phasic excitation is greater in newborns. Are these differences in motor activation and inhibition and EEG changes caused by the activity of aminergic, cholinergic, GABAergic, and glutamatergic cells? Are they a result of incomplete myelination of the axons of these systems? Are they due to immaturity of postsynaptic receptors? Are they due to neuronal differentiation and synaptogenesis? Is some combination of the above critical and, if so, what is the relative role of these changes and other changes not itemized?

I mention these points because of their relevance to the issue of state definition in non-laboratory species. We can observe a developmental continuity between "active sleep" in the neonate and REM sleep in the adult. The neonatal state of twitching gradually acquires the EEG voltage reduction and muscle atonia of REM sleep. This continuity makes it easy to accept that the neonatal "active sleep" state is closely related to the state of REM sleep (also known as paradoxical sleep) seen in the adult.

The behavioral similarity of the REM sleep state in the human and cat makes it reasonable to hypothesize that the neuronal activity changes known to underlie the state of REM sleep in the cat are occurring in the human. However, minor and major differences might be found when it becomes possible to monitor the precise behavior of chemically identified cell groups in the human. While rapid eye movements were the feature of REM sleep that gave this state its most mellifluous name, animals can have few or no eye movements in waking and REM sleep, yet still have a state of phasic motor activity (accompanied by low-voltage EEG) within sleep.[1]

Both active sleep in the neonate and REM sleep in the adult can be defined by purely behavioral criteria. We must remember that the EEG derives its value because of its correlation with behavioral measures of sleep. If animals are responsive and locomoting, we say they are awake, even if their EEG is high in voltage, a condition that can be created by certain brain lesions and by administration of the muscarinic receptor blocker atropine.[27]

There are a few ambiguous cases in which behavior alone is not sufficient to indicate waking. For example, slow swimming in circles in dolphins occurs during a state of raised arousal thresholds and unilateral EEG synchrony, indicating that it is in fact a sleep state, at least for one half of the brain.[32,33] Some birds are known to fly continuously for a period of days.[3] Must they be awake throughout these periods? We know that sleep-deprived and pathologically sleepy humans engage in automatic behaviors during which the EEG is synchronized and they are unresponsive to the environment.[16] Other "parasomnias" include the locomotion during sleep walking and the vocalizations during sleep talking. Despite their behavioral resemblance to waking, response and arousal thresholds are elevated at these times, precisely the reason these behaviors can be so dangerous. To qualify as sleep, any state must have a raised arousal threshold relative to unambiguous waking; otherwise, we are seeing a state of relaxed wakefulness. In humans, we can also add the requirement of lack of awareness of the environment during sleep; however, this is a more difficult criterion to use in animals. Even in humans, reduced awareness of certain aspects of the environment can occur while we are focused on others, without meeting any common sense definition of sleep.

What most readily distinguishes REM or paradoxical sleep from nonREM or quiet sleep is the motor activity, first observed in the extra-ocular muscles, which occurs during REM sleep. It is triggered by activation of brainstem reticulo-motor systems.[40] These systems include the paramedian pontine reticular formation, which controls eye movements, as well as the descending reticulospinal systems that cause the distal muscle twitches that characterize REM sleep. The phasic motor activation of REM sleep is usually accompanied by a simultaneous tonic inhibition of muscle tone, resulting in only occasional breakthroughs of limb movement, but frequent eye movements, irregularities in the activity of the respiratory musculature, and signs of autonomic motor activation occur.

I propose the following working definition of REM sleep: REM sleep is a sleep state in which there is repetitive phasic activation of brainstem reticulo-motor systems. The simultaneous activation of excitatory and inhibitory motor systems makes it necessary to conduct careful monitoring of motor output and perhaps the monitoring of central motor systems before concluding that no motor activation is occurring. Conversely, the presence of local motor reflexes, such as eye blinks triggered by corneal irritation, would not be sufficient for identification of REM sleep. It would be necessary to provide other evidence of brainstem motor system activation. This definition would classify "active sleep", the term some have used in newborn animals, as REM sleep. To understand fully the nature of sleep in any animal, the nature of state-specific neuronal activity must be known. To the extent that the neuronal activity patterns may vary across species, there may be corresponding variation in the functional role of sleep.

As will be described below, there are tremendous variations in the amount of sleep exhibited by different species. We must consider the possibility that equally dramatic differences in the pattern of activity of key neuronal cell groups accompany sleep and in particular REM sleep in different species. It is reasonable to hypothesize that the aminergic, GABAergic, and glutamatergic cell populations, so important in REM sleep in the cat, have the same pattern of discharge in all animals having REM sleep. This could be true even if the magnitude of activity change may differ. Similarly, one would expect that these same general patterns would hold across development and senescence. However, one can imagine that some species may have evolved qualitatively different aspects of REM sleep. It is not inconceivable that certain species may have "REM sleep" with only one of the monoaminergic cell groups turning off, say, the locus coeruleus, while the serotonergic cell group remains active. If this were the case, would it be proper to call the resulting state REM sleep or should a new name be coined for such a sleep state? This is more a semantic issue than a scientific one. Our goal should be to characterize fully the neuronal activity that underlies behavioral states and determine the similarities and differences in this activity across species.

7.3 REM Sleep in Mammals

Zepelin[51] and Zepelin and Rechtschaffen[52] compiled the work that had been done on sleep time in mammals. Some of this work was based on EEG, electromyogram (EMG), and electrooculogram (EOG) recordings with implanted electrodes in laboratory animals. However, most species were observed in zoos, using behavioral criteria to distinguish sleep from waking and REM sleep from quiet (nonREM) sleep. Zepelin's tabulation of the range of sleep times in various placental and marsupial mammal species showed that REM sleep could vary from as little as 40 min a day (e.g., in cattle) to as much as 6 hr a day in the black-footed ferret[25] and 7 hr a day in the thick-tailed oppossum.[52] Zepelin[51] sought some behavioral, ecological, or physiological correlate of this variation. One point that is obvious from the data that he compiled is that closely related animals do not have similar sleep parameters. Within the rodents, total sleep times range from 7.0 to 16.6 hr and REM sleep times from .8 to 3.4 hr. Within primates, sleep times range from 8 to 17 hr and REM sleep times from .7 to 1.9 hr. Thus, the adaptations linked to mammalian order appear to have relatively little to do with determining REM sleep time. For example, primates, with their high intelligence, manual dexterity, bipedal locomotion, long lifespan, and other features, do not as a group have higher amounts of REM sleep than rodents. Within orders, there is a tremendous variation of REM sleep time, even though the amount is relatively fixed for each species.

Zepelin[51] concluded that small animals spend more hours a day asleep. Thus, large animals such as the elephant and giraffe sleep 3.3 to 3.9 and 4.6 hr,[47] respectively, while the ground squirrel and little brown bat sleep 16.6 and 19.9 hr. Zepelin also found a small positive correlation between REM sleep and total sleep time. Animals that are small tend to have larger amounts of REM sleep. However, this relationship did not account for much of the variation in REM sleep distribution. Prior developmental studies by Jouvet-Mounier had pointed out that "altricial" animals (those born too immature to care for themselves, such as the cat, human, and rat) had much larger amounts of REM sleep at birth than "precocial" mammals (animals that are relatively independent soon after birth, such as the guinea pig and horse). REM sleep amounts decrease with age in altricial mammals and to a lesser extent in precocial mammals; however, altricial mammals continue to have much larger amounts of REM sleep than precocial mammals as adults. Zepelin showed that immaturity at birth is the single best predictor of REM sleep time throughout life. Before considering the meaning of this relation, we must examine its generality.

7.4 Sleep in the Echidna

Of the more than 4000 existing mammalian species, all but three are classified as placentals or marsupials. The marsupials are in general more altricial than placentals, and many have very large amounts of REM sleep. The third major branch of the mammalian tree is the monotremes. The monotremes are the only mammals that hatch from eggs. The three species are the short- and long-nosed echidna and the platypus. The long-nosed echidna, native to New Guinea, is considered endangered, and its sleep has not been studied. The short-nosed echidna is common in Australia. It eats ants and has a relatively long lifespan (up to 30 years).

If immaturity at birth is correlated with REM sleep time across the entire mammalian line, monotremes should have large amounts of it. After hatching, the hairless and defenseless echidna and platypus newborns climb in the mother's pouch, getting all their nutrition from their mother for a period of 4 to 6 months.

The first study of echidna sleep produced a surprising result. Allison et al.[2] concluded that the echidna had no REM sleep. They found that the echidna exhibited a high-voltage EEG during sleep. No periods of sleep with the low-voltage EEG and elevated arousal thresholds typical of REM sleep were seen. They saw no rapid eye movements or evidence of phasic motor activation during sleep. Since the echidna was the only monotreme to have had its sleep studied and the only mammal to be shown to lack REM sleep, Allison et al. hypothesized that all the monotremes lack REM sleep.

The monotremes diverged from the other mammalian lines early in the evolution of mammals. Therefore, Allison's hypothesis implies that REM sleep first evolved after the divergence of the placentals and marsupials from the monotremes. Allison concluded that the evolution of REM sleep was linked to the development of vivaparity (live birth). This hypothesis has had an enormous effect on subsequent theories of REM sleep evolution and function. It implied that quiet sleep was the original form of sleep and that REM sleep was a relatively advanced physiological trait. It was consistent with scattered but unconvincing data suggesting a role for REM sleep in learning. It also implied that the reptilian ancestors of mammals did not have REM sleep, as their earliest offspring (the monotremes) did not have this state.

We hypothesized that brainstem neuronal activity during echidna sleep might show aspects of REM sleep even though the forebrain EEG had not shown any low-voltage activity during sleep in the study of Allison et al. We implanted microwire recording electrodes in the midbrain and pons of the echidna, recording neuronal activity throughout the sleep cycle. In the cat, dog, and other laboratory mammals, unit activity in most of the units in the brainstem reticular formation is slow and regular during nonREM sleep. During REM sleep, unit activity becomes highly irregular, with periodic burst discharge. This burst discharge spreads to premotor neurons in the extraocular and spinal motor systems, producing the twitching and rapid eye movements that characterize REM sleep. When we recorded reticular neurons in the echidna during periods of high-voltage EEG activity, we did not see the nonREM sleep pattern. Figure 7.1 shows a typical example of what we saw instead. Unit discharge tended to be irregular and bursty, even though the EEG showed the pattern of nonREM sleep. Figure 7.2 presents instantaneous rate plots that allow one to compare the discharge pattern of reticular units recorded in the cat, dog, and echidna.

We quantified the amount of discharge occurring in bursts in brainstem units in the echidna and compared this data to data we collected in the same anatomical regions in the cat and dog. We found that the irregularity of discharge in the echidna was significantly greater than that seen in nonREM sleep in the cat and dog but was significantly less than that seen in REM sleep. In other words, from the standpoint of brainstem unit activity, the sleep state in the echidna appeared intermediate between REM and nonREM sleep. The echidna did not simply have a quiet sleep state as seen in the cat, dog, and rat.

7.5 Sleep in the Platypus

We next turned our attention to the platypus.[41] Because the platypus is a semi-aquatic mammal and cannot be confined without severe stress, it has been difficult to capture, maintain, and study in captivity. We dealt with these problems by utilizing an electrically shielded, artificial platypus enclosure, implanting telemetry devices to continuously monitor the platypus electroencephalogram, electrooculogram, electrocardiogram (ECG), and electromyogram while it was active and inactive, in the burrow and underwater.[41]

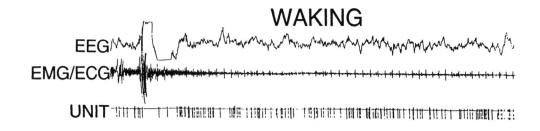

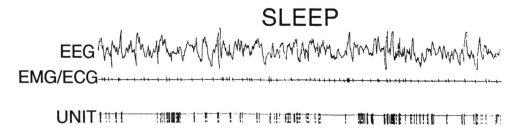

FIGURE 7.1
Unit discharge of a representative neuron recorded in nucleus reticularis pontis oralis of the echidna during waking and sleep. Note irregularity of neuronal discharge during sleep. EEG = electroencephalogram; EMG/ECG = electromyogram/electrocardiogram. Unit, pulse output of window discriminator triggered by neuron. (From Siegel, J.M. et al., *J. Neuroscience,* 15, 3500–3506, 1996. With permission.)

When the platypus was underwater and quiet, showing its typical diving response, EEG voltage was at its lowest level. A comparable low-voltage EEG was also present when the animal was awake in the burrow. At sleep onset, EEG amplitude increased, a state we termed "quiet sleep with moderate voltage EEG" (QS-M).

Phasic events began as soon as 30 to 90 sec after the onset of QS-M periods. REM sleep, as defined by muscle atonia, phasic EMG potentials, and rapid eye movements, was always accompanied by an EEG which was of moderate (REM-M) or high (REM-H) amplitude, with consistently more power in all of the frequency bands assessed than during waking states. In this respect, platypus REM sleep EEG is unlike the low-voltage REM sleep EEG seen in adult placental and marsupial mammals. We obtained confirmation of the periods of REM sleep by video recording of posture and behavior in the burrow. We found that all of the REM episodes occurred while the animal was immobile in a curled or prone sleep posture. We found that the phasic EOG and EMG potentials were correlated with rapid movements of the eyes, neck, and bill.

The platypus spends 60.1% of its sleep time (>8 hr/day) in a state with the EOG, EMG, ECG, and arousal threshold changes typical of REM sleep. This amount is greater than has been seen in any other animal.

7.6 How Do the Echidna and Platypus Data Fit Together?

There are both similarities and differences in our findings in the echidna and platypus. Both species have a relatively high-voltage EEG throughout sleep. Even during periods of phasic motor activity, the EEG of platypus did not show the low-voltage pattern seen during REM sleep in most adult mammals. However, infant placental mammals show a similar pattern of high-voltage EEG during active sleep, suggesting that in this respect ontogeny is recapitulating phylogeny.

The Evolution of REM Sleep

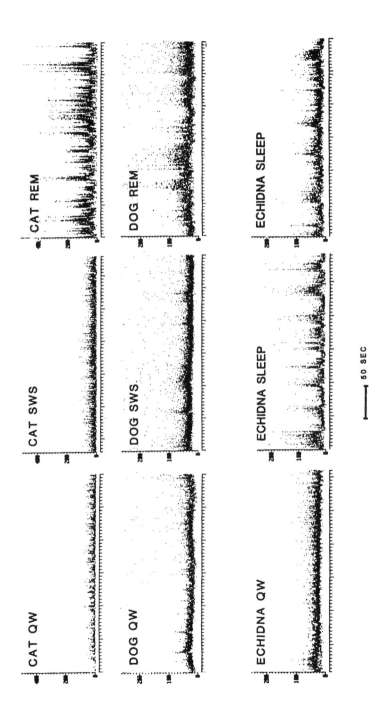

FIGURE 7.2
Instantaneous compressed rate plots of representative units recorded in nucleus reticularis pontis oralis of cat, dog, and echidna. Each point represents the discharge rate for the prior interspike interval. In cat, QW (quiet wake) and SWS (slow-wave sleep; nonREM sleep) discharge rate is low and relatively regular. The rate increases and becomes highly variable during REM sleep. A similar pattern can be seen in a unit recorded in the dog. (From Siegel, J.M. et al., *J. Neuroscience*, 15, 3500–3506, 1996. With permission.)

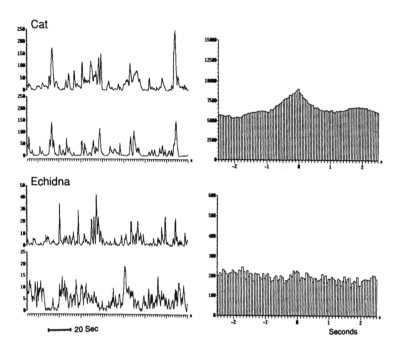

FIGURE 7.3
Rate histograms and cross-correlogram of discharge in a pair of cat reticularis pontis oralis units recorded during REM sleep (top), compared with a pair of echidna reticularis pontis oralis units recorded during sleep (bottom). Cross-correlograms of each pair computed at 50 msec binwidth are shown at right. Unit pairs in both the cat and echidna were recorded from adjacent microwires on a single bundle of 32 μ microwires. While most cat and dog units fire synchronously and are cross-correlated during REM sleep (12), none of the echidna unit pairs were cross-correlated in sleep. (From Siegel, J.M. et al., *J. Neuroscience,* 15, 3500–3506, 1996. With permission.)

The phasic motor activation seen in the extraocular, bill, and head musculature of the platypus indicates that burst discharge is occurring in its motor and premotor brainstem reticular systems. We saw that the echidna spends a large proportion of its sleep time with a burst-pause discharge pattern in brainstem reticulo-motor systems. Thus, this aspect of REM sleep neuronal activity is present to some extent in the echidna; however, we did not see twitches in neck or eye muscles in the echidna.

Another, perhaps related, difference in the sleep behavior of the echidna and platypus is in the arousal threshold. The platypus is extremely difficult to arouse from sleep and can even be picked up without awakening it. This deep sleep is consistent with its very safe sleep condition, in a burrow that few predators have been able to gain access to.[6] The echidna, in contrast has a relatively unsafe sleeping situation, often out in the open.[15] It has frequently been preyed upon by other animals. It has a relatively low arousal threshold during sleep. When disturbed, it immediately begins digging to attain a safer sleeping or hiding position. Because of its vulnerable sleeping position, any twitching of its large quills would attract attention and endanger it. Thus, the uninhibited twitching shown by the platypus would be maladaptive in the echidna. The mechanism underlying the subdued twitching in the echidna can be seen in its brainstem unit activity. Whereas we see phasic burst discharge in the echidna, the intensity of the bursts is significantly reduced relative to that in the cat and dog. We also found that the bursts during sleep were not generally cross-correlated in the reticular units of the echidna, while they were in the cat (Figure 7.3). Less synchronized burst discharge can explain the lack of twitching seen during sleep in the echidna.

The echidna is thought to have evolved from a platypus-like ancestor. But the divergence occurred over 50 million years ago, an enormous amount of time allowing for a large amount of evolution in sleep behavior, even in the relatively static monotreme line. The question is whether the phasic motor activity of the platypus or the muted brainstem "burstiness" of the echidna is the more recent development. I favor the theory that the platypus pattern represents the more primitive pattern. The behavioral aspects of REM sleep in the platypus are similar to those in placental and marsupial mammals and particularly to the vigorous phasic activity seen in neonates. The lack of EEG voltage reduction in REM sleep in the platypus is similar to the neonatal pattern in other mammals.[13,29] The large amounts of REM sleep also fit the neonatal pattern.[51] So the sleep of the platypus fits well with the general mammalian pattern, at least that of neonates. The echidna's lack of prominent motor activation during sleep is unique. The simplest conclusion is that the echidna's sleep pattern is the more recently evolved. The absence of other mammalian species without phasic motor activity during sleep suggests that the evolutionary route that allowed the echidna to reduce its brainstem activation during sleep may not have been open to other species. Once REM sleep evolved into a state that recruited neurons in the forebrain, in addition to the brainstem, a reduction in burst discharge and a reversal of forebrain changes may not have been possible.

The common element of sleep in both examined monotreme species is high-voltage EEG during phasic activation of brainstem reticulo-motor systems. This combined with the presence of a relatively high-voltage EEG in neonates suggests that the EEG voltage reduction seen in adult mammals during REM sleep is a more recently evolved feature of REM sleep. If dreaming requires such forebrain EEG "activation" one may speculate that dreaming evolved after the divergence of the monotreme line from the marsupials and placentals. Of course it has been argued that even in humans, dreaming is unrelated to EEG state and is not restricted to REM sleep.[38]

7.7 Reptilian Sleep

We find a REM sleep state in the platypus and aspects of REM sleep in the echidna. All other examined mammals have been found to have REM sleep. (Reports that the dolphin lacks REM sleep[32] are inconsistent with earlier work[8,9] and with more recent work.[34]) REM sleep is known to be present in birds. These findings suggest that reptiles, the common ancestors of birds and mammals, may have REM sleep. If REM sleep is not present in reptiles it must have evolved twice. Even if it evolved twice, one would expect some precursor state to have existed in reptiles ancestral to both mammals and birds, given the (apparently) similar form of REM sleep in birds and mammals.

The three major orders of living reptiles are (1) lizards and snakes, (2) crocodilians, and (3) chelonians (turtles and tortoises). Tauber et al.[44] reported evidence for REM sleep in the chameleon lizard. They reasoned that this animal, having very mobile active eyes in waking, would be more likely to show eye movement periods during sleep. They, like most subsequent researchers working with reptiles, found relatively little modulation of forebrain EEG across the sleep-wake cycle, compared to the dramatic modulation in mammals. Some spiking occurred in forebrain leads during sleep, but no change in EEG occurred during periods of rapid eye movement. In "REM" sleep, one eye could be open while the other remained shut. No change in muscle tone occurred during sleep. No arousal threshold testing is described in this brief report.

In a study of sleep in the lizard *Ctenosaura pectinata*, Tauber et al.[45] reported an increase in EEG voltage with arousal, accompanied by an increase in spikes recorded from the cortex. This contrasts to the blocking of spikes with arousal that they reported in the chameleon. The EEG voltage increase with arousal was seen in brainstem as well as forebrain leads. During sleep, EEG amplitude was reduced. Arousal threshold was reported to be elevated. Two to 3 hr after sleep onset, eye movements began to appear independently in the two eyes. These eye movement periods recurred at 4- to 25-min intervals throughout the sleep period. The duration of these periods is not described. Heart rate slowed during eye movement periods. Tauber et al. reported an elevation of arousal threshold throughout sleep, but did not report it separately for rapid eye movement periods or quantify arousal threshold with sleep.

Ayala-Guerrero and Huitron-Resendiz[5] conducted a similar investigation of this same species and also concluded that *Ctenosaura pectinata* had REM sleep, although the example they show looks very much like waking. However, Flanigan et al.[7] investigated this same species and also *Iguana iguana*. They saw no evidence for REM sleep in either species.

Huntley[21] reported that the desert iguana *Dipsosaurus dorsalis* has REM sleep, which was identified largely on the basis of EEG and EMG measures. Huntley found that a high-voltage EEG accompanied waking in the iguana. A low-voltage EEG pattern characterized quiescent behavioral sleep pattern. "Paradoxical sleep" periods were characterized by a return to the waking high voltage while the EMG reached minimal levels. In contrast to the high and irregular respiratory rate of waking, the paradoxical sleep state had a cessation of respiration. Heart rate was much slower than heart rate during waking and was variable as in waking. Paradoxical sleep time was greater at higher temperatures. Arousal thresholds for paradoxical and quiet sleep states were not analyzed. The percentage of times a standard shock delivered to thoracic leads aroused the animals was determined. It was found that sleep was accompanied by a reduced response percentage. Unfortunately, relatively few tests using this method were performed during "paradoxical sleep", so there is uncertainty as to whether the reduction in response frequency in this state was real. The absence of threshold tests (using ascending stimulation intensities) also makes it uncertain if the states scored as paradoxical were in fact sleep, as opposed to waking states. Huntley does not provide any evidence of phasic motor activity during this state.

A series of papers on the crocodilian *Caiman sclerops* concludes that this reptile does not have REM sleep. Flanigan et al.[12] found that caimans had small changes in the EEG across the sleep/wake continuum. Forebrain EEG spikes increased with behavioral quiescence and decreased with arousal. Sleep deprivation increased spiking during recovery sleep. These workers saw no evidence of paradoxical sleep. This same species was studies by Warner and Huggins[48] and Meglasson and Huggins,[30] who did not see the frequent spiking noted by Flanigan et al. They attribute this difference to the submersion of the nares of the animals in the Flanigan et al. study. They felt that the difference was related to cessation of respiration or to nasal stimulation. Also in contrast to the Flanigan study, they reported slow waves during quiescence in their animals. They suggest that the state with slow waves was not seen in the prior study. Flanigan et al. had reported that more than one week of adaptation was required before animals showed sleep, but Warner and Huggins saw sleep within hours of introduction into their recording situation. Warner and Huggins did not see evidence of REM sleep, but they point out that, because they did not record for 24-hr periods, they could easily have missed it. They attribute many of the differences in their findings and the findings of Flanigan et al. to the presence of other caimans in their recording situation, relaxing the recorded animal, or to the use of higher temperatures and younger animals in their study. Flanigan et al.[10] also investigated two species of chelonians (turtles and tortoises). They report no evidence for REM sleep in either species.

These data leave the question of the existence of REM sleep in reptiles unresolved. Several problems outlined above may explain "false positive" reports of REM sleep in reptiles. In particular, transient arousals might masquerade as REM sleep. Waking can only be distinguished from REM sleep through the use of arousal threshold tests.

On the other hand, it is certainly possible that "false negative" reports have overlooked periods of REM sleep in reptiles. Brief "bird-like" periods of REM sleep might easily be missed. Without behavioral observation, REM sleep periods might be mistaken for waking. It is also possible that some reptiles such as *Dipsosaurus dorsalis* have REM sleep while others such as *Caiman sclerops* do not. However, the ubiquity of REM sleep in mammals makes the search for an underlying commonality of REM and nonREM sleep states in reptiles attractive. One does not want to accept uncritically the principle that certain reptiles have REM sleep with aspects of autonomic and muscle tone control identical to mammals and that others completely lack this state, without first looking for replication of the key positive and negative findings.

A more fundamental issue is one of state definition. Since neocortex is absent in reptiles, there is no particular reason to expect a mammalian REM sleep-like voltage reduction. Indeed, the Huntley paper reports just the reverse pattern; increased EEG amplitude in REM sleep and waking. However, no other reptilian study, even those purporting to see REM sleep, report such a pattern. Cardiac variability, while correlated with REM sleep in most mammals, does not always differentiate REM sleep from waking. In the mole, heart rate is significantly less variable in REM sleep than in nonREM sleep.[1]

In keeping with the discussion at the beginning of this chapter, we propose that this uncertainty about the identification of REM sleep in reptiles can only be resolved by monitoring neuronal activity along with arousal thresholds. In particular, the activity of brainstem cholinergic, serotonergic, and noradrenergic cell groups must be sampled to better characterize the neuronal activity correlates of the observed macropotentials (EEG, ECG, EOG, EMG) characteristic of each state. Once these activity correlates are known, we will understand which if any elements of brainstem neuronal activity are correlated with the defined state. This will permit a more meaningful characterization of state. It will also allow a powerful insight into the evolution of sleep-cycle discharge patterns in aminergic and cholinergic cell groups. Because these groups are centrally involved in human psychopathology as well as state control, understanding how their state-related activity evolved could be of great clinical significance. In particular, it could lead to the identification of receptor and other rate-control mechanisms that may differ in reptiles, mammals, and birds.

7.8 Conclusions

We have found that the platypus has a sleep state with eye movements and twitching of the head and bill. The echidna has not been observed to have such phasic motor activity during sleep; however, the echidna does have a pattern of regular burst discharge in brainstem reticular cells similar to, but less intense than, that seen in REM sleep. Both monotreme species have these periods of "brainstem activation" during periods of EEG synchronization. These findings suggest that EEG voltage reduction during REM sleep is a more recently evolved feature of this state and that REM sleep in mammals originated as a brainstem state.

The presence of REM sleep in all three branches of the mammalian tree suggests that it or a very similar state was present in the earliest mammals. REM sleep is present in birds. The ubiquity of REM sleep in mammals and its presence in birds are most parsimoniously explained if one hypothesizes a single origin of this state in the common reptilian ancestors of birds and mammals. This hypothesis would predict a REM sleep state or at the least a state with many aspects of REM sleep in living reptiles. Studies at the neuronal level are necessary to test this hypothesis.

Acknowledgment

Research supported by the Medical Research Service of the Department of Veterans Affairs and USPHS grant NS32819.

References

1. Allison, T. and VanTwyver, H., Sleep in the moles, *Scalopus aquaticus* and *Condylura cristata*, *Exp. Neurol.*, 27, 564–578, 1970.

2. Allison, T., Van Twyver, H., and Goff, W.R., Electrophysiological studies of the echidna, *Tachyglossus aculeatus*. I. Waking and sleep, *Arch. Ital. Biol.*, 110, 145–184, 1972.
3. Amlaner, C.J. and Ball, N.J., Avian sleep, in *Principles and Practice of Sleep Medicine*, Kryger, M.H., Roth, T., and Dement, W.C., Eds., W.B. Saunders, Philadelphia, 1994, pp. 81–94.
4. Armstrong, D.A., Saper, C.B., Levey, A.I., Wainer, B.H., and Terry, R.D., Distribution of cholinergic neurons in the rat brain: demonstrated by the immunohistochemical localization of choline acetyltransferase, *J. Comp. Neurol.*, 216, 53–68, 1983.
5. Ayala-Guerrero, F. and Huitron-Resendiz, S., Sleep patterns in the lizard *Ctenosaura pectinata*, *Physiol. Behav.*, 49, 1305–1307, 1991.
6. Burrell, C.M.Z.S., *The Platypus*, Angus & Robinson, Sydney, 1927.
7. Flanigan, W.F., Sleep and wakefulness in iguanid lizards, *Ctenosaura pectinata* and *Iguana iguana*, *Brain Behav. Evol.*, 8, 401–436, 1973.
8. Flanigan, W.F., Nocturnal behavior of captive small cetaceans. I. The bottlenosed porpoise, *Tursiops truncatus*, *Sleep Res.*, 3, 84, 1974.
9. Flanigan, W.F., Nocturnal behavior of captive small cetaceans. II. The beluga whale, *Delphinapterus leucas*, *Sleep Res.*, 3, 85, 1974.
10. Flanigan, W.F., Sleep and wakefulness in chelonian reptiles. II. The red-footed tortoise, *Gechelone carbonaria*, *Arch. Ital. Biol.*, 112, 253–277, 1974.
11. Flanigan, W.F., Knight, C.P., Hartse, K.M., and Rechtschaffen, A., Sleep and wakefulness in chelonian reptiles. I. The box turtle, *Terrapene carolina*, *Arch. Ital. Biol.*, 112, 227–252, 1974.
12. Flanigan, W.F., Wilcox, R.H., and Rechtschaffen, A., The EEG and behavioral continuum of the crocodilian *Caiman sclerops*, *Electroencephalogr. Clin. Neurophys.*, 34, 521–538, 1973.
13. Frank, M.G. and Heller, H.C., Development of REM and slow wave sleep in the rat, *Am. J. Physiol.*, 272, R1792–R1799, 1997.
14. George, R., Haslett, W.L., and Jenden, D.J., A cholinergic mechanism in the brainstem reticular formation: induction of paradoxical sleep, *Int. J. Neuropharmacol.*, 3, 541–552, 1964.
15. Griffiths, M., *Echidnas*, Pergamon Press, New York, 1968.
16. Guilleminault, C., Billiard, M., Montplaisir, J., and Dement, W.C., Altered states of consciousness in disorders of daytime sleepiness, *J. Neurol. Sci.*, 26, 377–393, 1975.
17. Hallanger, R.M. and Wainer, B.H., Ascending projections from the pedunculopontine tegmental nucleus and the adjacent mesopontine tegmentum in the rat, *J. Comp. Neurol.*, 274, 483–15, 1988.
18. Harper, R.M., Leake, B., Miyahara, L., Hoppenbrouwers, T., Sterman, M.B., and Hodgman, J., Development of ultradian periodicity and coalescence at 1 cycle per hour in electroencephalographic activity, *Exp. Neurol.*, 73, 127–143, 1981.
19. Holmqvist, B.I., Ostholm, T., Alm, P., and Ekstrom, P., Nitric oxide synthase in the brain of a teleost, *Neurosci. Lett.*, 171, 205–208, 1994.
20. Hoogland, P.V. and Vermeulen-Vanderzee, E., Distribution of choline acetyltransferase immunoreactivity in the telencephalon of the lizard *Gekko gecko*, *Brain Behav. Evol.*, 36, 378–390, 1990.
21. Huntley, A.C., Electrophysiological and behavioral correlates of sleep in the desert iguana, *Dipsosaurus dorsalis hallowell*, *Comp. Biochem. Physiol.*, 86A, 325–330, 1987.
22. Jones, B.E. and Beaudet, A., Distribution of acetylcholine and catecholamine neurons in the cat brainstem: a choline acetyltransferase and tyrosine hydroxylase immunohistochemical study, *J. Comp. Neurol.*, 261, 15–32, 1987.
23. Lin, J.S., Sakai, K., Vanni-Mercier, G., and Jouvet, M., A critical role of the posterior hypothalamus in the mechanisms of wakefulness determined by microinjection of muscimol in freely moving cats, *Brain Res.*, 479, 225–240, 1989.

24. Luebke, J., Weider, J., McCarley, R., and Greene, R., Distribution of NADPH-daphorase positive somata in the brainstem of the monitor lizard *Varanus exanthematicus*, *Neurosci. Lett.*, 148, 129–132, 1992.
25. Marks, G.A. and Shaffery, J.P., A preliminary study of sleep in the ferret, *Mustela putorius furo*: a carnivore with an extremely high proportion of REM sleep, *Sleep*, 19, 83–93, 1996.
26. Marsel-Mesulam, M., Geula, C., Bothwell, M.A., and Hersh, L.B., Human reticular formation: cholinergic neurons of the pedunculopontine and laterodorsal tegmental nuclei and some cytochemical comparisons to forebrain cholinergic neurons, *J. Comp. Neurol.*, 281, 611–633, 1989.
27. McGinty, D.J. and Siegel, J.M., Sleep states, in *Handbook of Behavioral Neurobiology: Motivation*, Satinoff, E. and Teitelbaum, P., Eds., Plenum Press, New York, 1983, pp. 105–181.
28. McGinty, D. and Szymusiak, R., Keeping cool: a hypothesis about the mechanisms and functions of slow-wave sleep, *Trends Neurosci.*, 13, 480–487, 1990.
29. McGinty, D.J., Stevenson, M., Hoppenbrouwers, T., Harper, R.M., Sterman, M.B., and Hodgman, J., Polygraphic studies of kitten development: sleep state patterns, *Dev. Psychobiol.*, 10, 455–469, 1977.
30. Meglasson, M.D. and Huggins, S.E., Sleep in a crocodilian, *Caiman sclerops*, *Comp. Biochem. Physiol.*, 63a, 561–567, 1979.
31. Mesulam, M., Mufson, E.J., Levey, A.I., and Wainer, B.H., Atlas of cholinergic neurons in the forebrain and upper brainstem of the macaque based on monoclonal choline acetyltransferase immunohistochemistry and acetylcholinesterase histochemistry, *J. Neurosci.*, 12, 669–686, 1984.
32. Mukhametov, L.A., Supin, A.Y., and Polyakova, I.G., Interhemispheric asymmetry of the electroencephalographic sleep patterns in dolphins, *Brain Res.*, 134, 581–584, 1977.
33. Mukhametov, L.M., Unihemispheric slow-wave sleep in the amazonian dolphin, *Inia geofffrensis*, *Neurosci. Lett.*, 79, 128–132, 1987.
34. Mukhametov, L.M., Paradoxical sleep peculiarities in aquatic mammals, *Sleep Res.*, 24A, 202, 1995.
35. Nitz, D. and Siegel, J.M., GABA release in the posterior hypothalamus of the cat as a function of sleep/wake state, *Am. J. Physiol.*, 40, R1707–R1712, 1996.
36. Pujol, J.F., Buguet, A., Froment, J.L., Jones, B., and Jouvet, M., The central metabolism of serotonin in the cat during insomnia. A neurophysiological and biochemical study after administration of p-cholorphenylalanine or destruction of the raphe system, *Brain Res.*, 29, 195–212, 1971.
37. Rechtschaffen, A., Lovell, R.A., Freedman, D.X., Whitehead, W.E., and Aldrich, M., The effect of parachlorophenylalanine on sleep in the rat: some implications for the serotonin-sleep hypothesis, in *Serotonin and Behavior*, Barchas, J. and Usdin, E., Eds., 1973, pp. 1–25.
38. Rosenlicht, N., Maloney, T., and Feinberg, I., Dream report length is more dependent on arousal level than prior REM duration, *Brain Res. Bull.*, 34, 99–101, 1994.
39. Shiromani, P.J., Armstrong, D.M., Berkowitz, A., Jeste, D.V., and Gillin, J.C., Distribution of choline acetyltransferase immunoreactive somata in the feline brainstem: implications for REM sleep generation, *Sleep*, 11, 1–16, 1988.
40. Siegel, J.M., Brainstem mechanisms generating REM sleep, in *Principles and Practices of Sleep Medicine*, Kryger, M.H., Roth, T., and Dement, W.C., Eds., W.B. Saunders, Philadelphia, 1994, pp. 125–144.
41. Siegel, J.M., Manger, P.R., Nienhuis, R., Fahringer, H.M., and Pettigrew, J.D., The platypus has REM sleep, *Sleep Res.*, 26, 177, 1997.
42. Steriade, M., McCormick, D.A., and Sejnowski, T.J., Thalamocortical oscillations in the sleeping and aroused brain, *Science*, 262, 679–685, 1993.
43. Sterman, M.B., McGinty, D.J., Harper, R.M., Hoppenbrouwers, T., and Hodgman, J.E., Developmental comparison of sleep EEG power spectral patterns in infants at low and high risk for sudden death, *Electroencephalogr. Clin. Neurophysiol.*, 53, 166–181, 1982.

44. Tauber, E.S., Roffwarg, H.P., and Weitzman, E.D., Eye movements and electroencephalogram activity during sleep in diurnal lizards, *Nature,* 212, 1612–1613, 1966.
45. Tauber, E.S., Rojas-Ramirez, J., and Hernandez-Peon, R., Electrophysiological and behavioral correlates of wakefulness and sleep in the lizard, *Ctenosaura pectinata, Electroencephalogr. Clin. Neurophysiol.,* 24, 424–433, 1968.
46. Ten Donkelaar, H.J., Bangma, G.C., Barbas-Henry, H.A., de Boer-van Huizen, R., and Wolters, J.G., The brainstem in a lizard, *Varanus exanthematicus, Adv. Anat. Embryol. Cell Biol.,* 107, 1–168, 1987.
47. Tobler, I. and Schwierin, B., Behavioural sleep in the giraffe (*Giraffa camelopardalis*) in a zoological garden, *J. Sleep Res.,* 5, 21–32, 1996.
48. Warner, B.F. and Huggins, S.E., An electroencephalographic study of sleep in young caimans in a colony, *Comp. Biochem. Physiol.,* 59, 139–144, 1978.
49. Wolters, J.G., Ten Donkelaar, H.J., Steinbusch, H.W.M., and Verhofstad, A.A.J., Distribution of serotonin in the brain stem and spinal cord of the lizard *Varanus exanthematicus*: an immunohistochemical study, *J. Neurosci.,* 14, 169–193, 1985.
50. Wolters, J.G., Ten Donkelaar, H.J., and Verhofstad, A.A.J., Distribution of catecholamines in the brain stem and spinal cord of the lizard *Varanus exanthematicus*: an immunohistochemical study based on the use of antibodies to tyrosine hydroxylase, *J. Neurosci.,* 13, 469–493, 1984.
51. Zepelin, H., Mammalian sleep, in *Principles and Practice of Sleep Medicine,* Kryger, M.H., Roth, T., and Dement, W.C., Eds., W.B. Saunders, Philadelphia, 1994, pp. 69–80.
52. Zepelin, H. and Rechtschaffen, A., Mammalian sleep, longevity and energy metabolism, *Brain Behav. Evol.,* 10, 425–470, 1974.

Chapter 8

Mentation During Sleep: The NREM/REM Distinction

Tore A. Nielsen

Contents

- 8.1 Introduction 102
  - 8.1.1 The Discovery of REM and NREM Mentation 102
  - 8.1.2 Widespread Evidence for Cognitive Activity in NREM Sleep 105
    - 8.1.2.1 Discriminating "Dreaming" from "Cognitive Activity" 105
    - 8.1.2.2 Evidence for Cognitive Activity Outside of REM Sleep 105
    - 8.1.2.3 Summary 107
  - 8.1.3 The One-Generator (1-gen) Model of Sleep Mentation 108
  - 8.1.4 The Two-Generator (2-gen) Model of Sleep Mentation 109
- 8.2 Experimental Results Bearing on the Models 110
  - 8.2.1 Memory Sources 110
  - 8.2.2 Report Interrelationships 111
  - 8.2.3 Event-Related Potentials 111
  - 8.2.4 Memory Consolidation 111
  - 8.2.5 Stimulation Effects 112
  - 8.2.6 Residual Stage Differences 112
  - 8.2.7 Subject Differences 112
- 8.3 The Concept of "Activation" in Explaining Sleep Mentation Differences 113
  - 8.3.1 Are Memory Activation and Cortical Activation Isomorphic? 113
  - 8.3.2 Effects of Changes in Activation Across the Night 114
  - 8.3.3 Neurobiological Conceptions of Memory Activation 114
    - 8.3.3.1 Multiple Activation Systems 114
    - 8.3.3.2 Multiple Memory Systems 115
  - 8.3.4 Partialing Out Activation: Problems With the Use of Report Length 115
- 8.4 Summary 116
- 8.5 Toward Rapprochement: A Possible Role for Phantom REM Sleep 117
- Acknowledgments 119
- References 120

8.1 Introduction

The comparison of REM and NREM mentation reports has remained one of the principal laboratory methods for studying dreaming processes. As the influence of cognitive science on sleep research has grown, experimental methods for comparing REM and NREM mentation have also evolved. Accordingly, some researchers have employed the techniques of cognitive neuroscience and neurobiology to support neurobiological theories of dreaming,[3,73,75,77,85,107,115,172] while others, in part disillusioned with neurobiological reductionism, have applied cognitive-psychological methods to support their dream theories.[4,14,16,17,21,25,30,52,87,105,122,127,140,156,157,159,176,177] These developments have begun to reorient how dreaming is investigated, modeled, and explained. Perhaps most importantly, as the shift toward cognitive experimentation progresses, potential answers are emerging to the long-standing question of whether REM and NREM sleep mentation are qualitatively different types of phenomenon.

The present work reviews this cognitively oriented research with a specific focus on hypothesized mechanisms of REM and NREM dream generation. It is concerned primarily with the question of whether there exist quantitative and qualitative differences between the two types of mentation. To this end, two classes of dream generation model are discussed: the one generator (1-gen) model, which assumes that REM and NREM mentation arise from a common source, and the two-generator (2-gen) model, which assumes that the two types of mentation arise from different sources. The principal model within each of these classes is presented in summary form and the experimental findings pertinent to the models reviewed. Finally, the construct of *phantom REM sleep* is introduced as an example compromise position in this continuing debate.

8.1.1 The Discovery of REM and NREM Mentation

The first reports of a seemingly exclusive association between recurrent rapid eye movement (REM) episodes and vivid dreaming[10,36,37] triggered a barrage of attempts to clarify this psychophysiological relationship. A perspective quickly emerged — now referred to by many as the "REM sleep = dreaming" perspective (see References 12, 54, 103, 126, 142 for a critique and commentaries) — which considered dreaming to be an exclusive product of REM sleep. What little mentation was observed in NREM sleep was attributed to pitfalls in research design, such as the recall of mentation from previous REM episodes or the confabulation of content on awakening. Although many studies soon cast doubt on the "REM sleep = dreaming" equation,[49,64] a controversy over the *qualitative* nature of NREM and REM sleep mentation soon replaced it.

On one side of this controversy, REM and NREM sleep mentation reports were viewed as stemming from qualitatively different imagery generation systems; this was suggested by the finding that REM sleep reports were less thought-like; more elaborate; more affectively, visually, and kinesthetically involving; and more related to waking life than were NREM sleep reports.[49,64,121,145] Such differences led many to conclude that REM and NREM sleep should be contrasted with wakefulness as qualitatively distinct states of consciousness,[168] an approach I refer to as the two-generator (2-gen) approach. The best-known 2-gen model — a variant of the earlier activation-synthesis (A-S) hypothesis — has been developed principally by Hobson's group[80] and researchers inspired by them.[162] However, McCarley and colleagues[115,172] have also updated the A-S hypothesis. In addition, a psycholinguistically based 2-gen theory has been proposed.[22] Hobson's model postulates that a flow of neural information from brainstem to forebrain determines the formal properties of sleep mentation, that properties of the physiological state are isomorphic with properties of the mentation. The physiologically distinct REM and NREM stages of sleep thus necessarily produce phenomenologically distinct forms of mental experience. Dreaming is thereby explained neuro-reductionistically by reference to a sequence of neural events that is hypothesized to sustain it (see Figure 8.1).

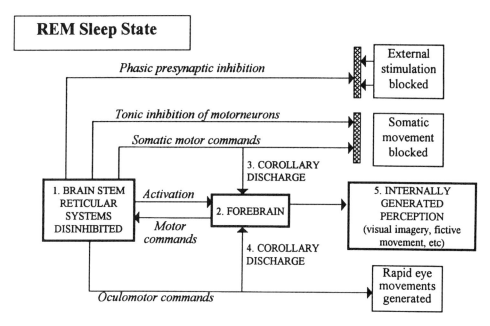

FIGURE 8.1

Two-generator model. Schematic of the activation synthesis (A-S) precursor to the AIM model of dreaming. This configuration is favored when the activation factor of AIM state space is high and the input source and modulation mode are low. A cessation of aminergic activity disinhibits the brainstem reticular formation (1). The latter activates the cortex (2) and sends it information about the somatic motor commands (3) and rapid eye movements (4) which it generates. The former results in hallucinated visual perception, the latter in fictive movements (5). External stimuli and active movements are both blocked by inhibitory processes. As these key physiological features are absent or reduced during NREM sleep, dreaming per se is not possible. Rather, NREM is characterized by nondreaming mentation. (Adapted from Hobson, J.A., *The Dreaming Brain: How the Brain Creates Both the Sense and the Nonsense of Dreams,* Basic Books, New York, 1988, pp. 208–209.)

On the other side of the controversy, some researchers have developed one-generator (1-gen) models of sleep mentation. The first step in this development was the demonstration that some form of cognitive activity can occur in all states of NREM sleep. Foulkes' implementation[49,64] of a more liberal set of criteria for defining the contents of sleep mentation reports as cognitive activity allowed him to demonstrate an incidence of cognitive activity during NREM sleep that was much higher than the near-zero levels which had been observed by some of his predecessors. Numerous investigators have replicated these findings (see later section). Although qualitative differences between REM and NREM reports were observed in many of these studies, the overall impression Foulkes derived from his work was that cognitive activity of some form continues through all stages of sleep. Further, the majority of this activity consists of dreaming.[50]

The second step toward development of 1-gen models consisted in the adoption of methods for conducting comparisons of mentation quality between reports of obviously different frequencies and lengths. Both Foulkes[60] and Antrobus[2] proposed methods for removing quantitative differences and thus permitting, presumably, fair tests of qualitative differences. Both investigators found that when length of report was controlled, qualitative differences between REM and NREM reports tended to diminish — a finding squarely supporting the notion that REM and NREM mentation derives from a common imagery source that is working at different levels of activation. Foulkes[52] subsequently elaborated a cognitive model of sleep mentation based upon the 1-gen assumption (see Figure 8.2). Other 1-gen models have been elaborated by Antrobus[2] and Feinberg and March.[46]

Both the 1-gen and 2-gen models have had a significant impact on sleep research. That Foulkes' original findings were replicated and his model tested by so many indicates that his cognitive-

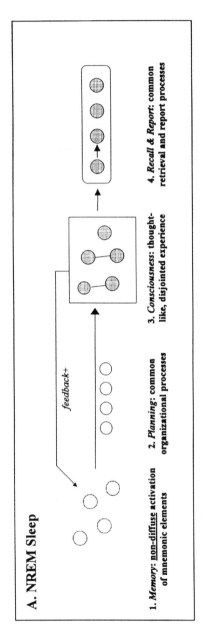
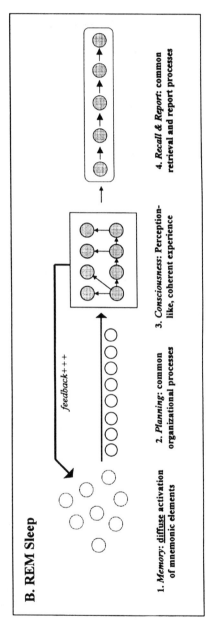

FIGURE 8.2
One-generator model. Schematic model of dream generation mechanism which is hypothesized to be common to both NREM and REM sleep. Both types of mentation are hypothesized to derive from similar memory elements (1) and to be synthesized by an identical set of planning/organizational processes (2) into similar conscious dream experiences (3). The two states differ primarily in that memory elements (1) are less diffusely activated during NREM than during REM sleep, thus being less available for further processing. As a consequence, there is both less material to recall and report (4) from a NREM sleep awakening and a weaker integration of the material recalled. Experiential feedback which may further drive mnemonic activation (1) may, too, be weaker in NREM (*feedback+*) than in REM (*feedback+++*) sleep. (Adapted from Foulkes, D., *Dreaming: A Cognitive-Psychological Analysis*, Lawrence Erlbaum Associates, Hillsdale, NJ, 1985.)

psychological experiment and his 1-gen model have had a widespread influence. Nevertheless, 2-gen thinking concerning dreaming appears to have garnered more visibility in both neuroscience and popular sectors. The merits and drawbacks of 1-gen and 2-gen models are now being scrutinized with ever more precision and vigor, even while they are only rarely compared directly one with the other (although see excellent an review in Reference 138).

8.1.2 Widespread Evidence for Cognitive Activity in NREM Sleep

8.1.2.1 Discriminating "Dreaming" from "Cognitive Activity"

Differences between how "dreaming" and "cognitive activity" are defined are critical to understanding the 1-gen and 2-gen positions. Cognitive activity — the object of study for most 1-gen theorists — is a more inclusive term than is dreaming, referring to the remembrance of any mental activity having occurred just prior to waking up. This activity may be visual imagery, thinking, reflecting, bodily feeling, or vague and fragmentary impressions. Dreaming — the object of most 2-gen theorists — is much more likely to be defined as imagery that includes a mixture of sensory hallucinations, emotions, storylikeness, dramatic progression, coherence, bizarreness, etc. and that excludes some types of cognitions such as simple thinking, reflecting, body feelings, and vague and fragmentary impressions.

Previous studies of REM and NREM mentation have employed cognitive activity and dreaming definitions inconsistently. For example, Dement and Kleitman[38] accepted only reports that were "coherent, fairly detailed description of dream content," whereas Goodenough et al.[69] permitted "a dream recalled in some detail," and Foulkes and Rechtschaffen[59] allowed "at least one item of specific content." In these three studies, the different levels of stringency varied inversely with the amount of NREM mentation attributed to the subjects, i.e., 7, 35, and 62%, respectively.

There is as yet no widely accepted standard definition of dreaming. There have been some attempts to differentiate minimal forms of dreaming from more elaborate forms and to discriminate among dreaming of different levels of complexity, such as everyday and archetypal dreaming;[20,84] mundane, transcendental, and existential dreaming;[19] lucid and nonlucid dreaming;[100] and ordinary vs. apex[74] or titanic[84] dreaming. Despite these developments, however, definitions of dreaming and its different forms still vary from study to study. Unlike the definition of cognitive activity, there is still no standard definition of dreaming which is widely implemented by the research community.

It was recently proposed[80] that six general characteristics define dreaming and differentiate it from both waking consciousness and NREM cognition, i.e., hallucinoid imagery, narrative structure, cognitive bizarreness, hyperemotionality, delusional acceptance, and deficient memory of previous mental content. A lack of all or most of these defines nondreaming mentation. However, there is as yet no demonstration of the discriminant validity of these features with samples drawn from both REM and NREM mentation.

8.1.2.2 Evidence for Cognitive Activity Outside of REM Sleep

Numerous studies have demonstrated the existence of cognitive activity during stages of sleep other than REM. How much of this activity qualifies as a form of dreaming has been less clearly shown. Some of the strongest evidence for NREM cognition is that specific NREM contents are at times closely associated with pre-awakening stimuli.[134] Such stimuli may be naturally occurring, such as sleep talking that is concordant with specific contents in the report,[7,143] or experimentally induced, such as an auditory or somatic stimulus which is incorporated into the mentation and subsequently reported.[24,50,59,102,145,146]

To illustrate such "tagging" of NREM mentation, Rechtschaffen et al.[145] cite the report of a subject who was stimulated during stage 2 sleep with a 500-Hz subwaking threshold tone (7 sec), followed by a pause (27 sec) and a second tone (7 sec), and then awakened 32 sec later:

"... a little whistling tone was going on ... and then it went off. And (the other person) said, 'Oh, you had better get things over with quickly, because you may have to wake up soon' ... I just said 'Oh!' to this, and I think I heard the whistling noise again. Then the same scene was there for some time, and I was just walking around trying to think of what was going on." (p. 412)

Similar tagging of stimuli has been demonstrated in mentation reports from all NREM sleep stages.[174] Moreover, presleep suggestions given to hypnotizable subjects produce more frequent dreaming about the suggested topics in all stages of sleep sampled (stages 2, 3, and 4 and REM) when subjects are hypnotized (44%) than when not hypnotized (25%).

Some clinical studies of NREM parasomnias also provide convincing evidence of vivid mental experiences outside of REM sleep;[48,89] sleep terrors arising from stage 3 and 4 sleep often result in reports of dramatic and frightening content. Although for some of these awakenings the content may be triggered by the arousal process itself,[18] for others there is evidence of a dramatic progression seeming to lead up to, if not to induce, the terror awakening. Fisher et al.[47,48] also found evidence of a number of stage 2 nightmares that seemed qualitatively similar to the nightmares of REM sleep.

8.1.2.2.1 Sleep Onset. Perhaps the most dream-like mentation reports have been collected from the NREM stages of sleep onset. These include the traditional Rechtschaffen and Kales[144] stages 1 and 2,[32,62,63,104,187] as well as the stages of a more detailed scoring grid.[82,124] Sleep onset is remarkable because cognitive activity reported at this time equals or surpasses in frequency (e.g., 90 to 98%) activity following awakenings from REM sleep.[63] Reports of this activity are also at times found to be as long as reports from REM sleep.[51,62,184,185] Moreover, many sleep onset reports clearly contain dreaming, defined as hallucinated dramatic episodes, not simply as isolated scenes, flashes or nonhallucinated images; from 31 to 76% of sleep onset reports are dreams in this sense, depending upon the specific EEG stage of awakening.[185]

8.1.2.2.2 NREM Sleep. Many more studies of sleep mentation have concentrated on NREM stages of sleep other than those at sleep onset. Although in many studies all NREM sleep stages (2, 3, and 4) are indiscriminately combined, stage 2 sleep is by far the most frequently examined stage. To provide an overview of the previous literature in this area, studies of REM and NREM mentation published since 1953 were consulted. Of these, 34 studies[5,10,23,35,36,38,45,58–60,64,67,69,79,86,90,92,98,108,119,120,125,131,135,139,145,155,164,167,175,189,194,195,197] were retained for the calculation of global estimates of mentation recall (see Table 8.1). Excluded were studies of patient populations whose illnesses, such as depression, anorexia, or psychosomatic problems, could have affected recall processes. To weight findings from each study equally in the global average, only one estimate of recall was included per study. When one study furnished recall values for different subgroups (e.g., young vs. old, male vs. female, high vs. low habitual recall), an average of these groups was calculated to produce a more general data point. Thus, the global estimates include results from subgroups of normal subjects who have known low rates of recall, but these contributions are offset by subjects from subgroups with high rates of recall. Estimates were also calculated separately for studies prior to and after Foulkes' work,[64] as the latter first clarified the importance of discriminating reports defined as "dreams" from those defined as "cognitive activity".

In comparing all studies (Table 8.1), the overall difference in recall rate from REM sleep (81.8 ± 8.7%) and NREM sleep (42.5 ± 21.0%) was close to 40%. However, this REM-NREM difference was much larger for the pre-1962 studies (59.1%) than it was for the post-1962 studies (33.2%). Differences in median recall for the two states parallel those for the mean — total 40%, pre-1962: 59%, post-1962: 37%.

TABLE 8.1
Rates of Recall of Cognitive Activity From
REM and NREM Sleep in 34 Studies

	No. Studies	Mean ± S.D. (%)	Median (%)	Mode (%)	Range (%)
REM (<1962)	8	75.5 ± 10.8	77	85	60–88
REM (≥1962)	21	84.1 ± 6.7	86	86	71–93
REM total	29	81.8 ± 8.7	85	86	60–93
NREM (<1962)	8	16.4 ± 12.5	18	n/a	0–35
NREM (≥1962)	25	50.9 ± 15.5	49	46	23–75
NREM total	33	42.5 ± 21.0	45	27	0–75
Stages 3 and 4	8	52.5 ± 18.6	54	n/a	19–75

The NREM estimate has not changed much in 30 years as indicated by the similarity of the present mean (42.5%) with that of 45.9 ± 15.8% calculated from the nine studies reviewed in Foulkes.[50] The present mean recall rate for REM sleep (81.8%) also compares favorably both with (1) Dement's average of 83.3% obtained by pooling 2240 REM sleep awakenings from 214 subjects (over 885 nights)[39] and (2) an average of 81.7 ± 15.0% calculated from 12 studies reviewed in Herman et al.[74]

8.1.2.2.3 Stages 3 and 4 Sleep. Some studies have found evidence of cognitive activity specifically in stages 3 and 4 sleep.[8,9,28,67,68,74,135] On average, the recall of cognition here is equal to that of stage 2 sleep. A survey of eight studies[28,45,49,108,119,132,135,155] revealed an average recall rate from stages 3 and 4 of 52.5 ± 18.6% (Table 8.1). The average stage REM recall rate estimated from the same studies was 82.2 ± 8.1% (median: 83.5%, range 71 to 91%). These values for stages 3 and 4 are consistent with the finding that stage 2 and 4 mentation differences in recall and content disappear for awakenings conducted at similar times of the night.[181] Three studies[119,132,135] found average recall rates to be higher in stage 3 (M = 56%) than in stage 4 sleep (M = 38%), a finding also true of children 9 to 11 yrs. (42 vs. 26%) and 11 to 13 yrs. (42 vs. 25%).[51] However, Pivik[132] found nearly identical levels of recall of cognitive activity in stages 3 (41 to 56%) and 4 (38 to 58%).

Some subjects appear to have little or no recall of mentation from stages 3 and 4 sleep. For example, 10 of 60 subjects (17%) in one study[28] reported *no* mentation whatsoever after several nights of 1 awakening per night from stages 3 and 4 sleep; an additional 20 subjects (33%) required from 1 to 5 additional nights before recalling some cognitive activity from these stages.

8.1.2.2.4 Wakefulness. A number of studies have examined wakefulness for evidence that hallucinatory imagery might be occurring altogether outside of the sleep state. Such imagery has been reported for quiet relaxed wakefulness.[56,60,61,147,148] Note, however, that such studies do not assess "cognitive activity" in the same general sense as do other studies of NREM sleep mentation simply because cognitive activity would be expected at virtually all times during the waking state. Rather, these studies have targeted — and found — more dream-like forms of mentation at this time.

8.1.2.3 Summary

Numerous studies have replicated the finding that mentation can occur outside of stage REM sleep as it is traditionally defined. However, in accordance with the distinction between dreaming and cognitive activity discussed earlier, the generalizations that one may draw from the literature vary depending upon which type of activity is being considered. Although most studies support the contention that cognitive activity can occur in all stages of NREM sleep, *no* research supports the claim that cognitive activity occurs continuously throughout sleep. The rate of awakenings from NREM sleep which produce no recall whatsoever is large — around 50% according to the present global estimate — and some subjects (e.g., 17%) show no NREM recall at all after repeated attempts. Thus, a representative generalization of the available findings is that about one out of two awakenings from NREM sleep produce no cognitive activity of any kind, with some subjects demonstrating little or no such activity.

Further, as dreaming is defined as a subset of cognitive activity, there is also no support for the claim that dreaming occurs continuously throughout NREM sleep. In fact, one liberal estimate is that one quarter to one half of NREM reports bearing cognitive activity fulfill a minimal definition of dreaming.[64] Thus, from 12 to 25% of NREM awakenings in susceptible subjects will produce reports of dreaming — and most of these will be rather short reports at that. As for the more elaborate forms of dreaming which involve several characters and dramatic elaborations over several scenes, their prevalence may be even lower. It has been suggested that the most vivid form of dreaming — what has been referred to as *apex* dreaming — may occupy only 7% of recalled NREM mentation.[74] Of course, sleep onset NREM reports are an unusual exception to this conclusion, a discrepancy to be discussed at greater length in a later section.

Whether and with what frequency dream-like activity occurs during relaxed wakefulness remains more problematic. Although there is evidence of non-pathological hallucinatory activity during wakefulness, it is not clear to what extent this activity differs from fantasy and other forms of respondent thought.[96] It is noteworthy that Rechtschaffen[142] has tentatively proposed that dream-like processes might be found during wakefulness when there is loss of volitional control (e.g., during the "lapses" suffered by sleep-deprived subjects). In a different vein, Mahowald[109] has proposed a continuity between the waking state hallucinations arising in diverse psychiatric conditions and vivid dreaming.

8.1.3 The One-Generator (1-gen) Model of Sleep Mentation

A single imagery generator model drawn from Foulkes[52] is expressed schematically in Figure 8.2. Both REM and NREM sleep mentation are hypothesized to derive from three component processes: (1) a memory component (or mnemonic activation), (2) a planning component (or organizational process), and (3) a conscious organization component (or conscious interpretation). This mentation is then subject to processes of recall and verbal report (4) which figure less centrally in this model than do the first three processes. Differences between dreams collected from REM and NREM sleep hinge primarily upon the first step of the sequence: different levels of memory activation. When memory activation is at its highest and most diffuse — during most REM sleep and during only some NREM sleep episodes — then planning/organizational mechanisms are more vigorously stimulated, and conscious dream experience is both more probable and more coherent. When activation is at its lowest and least diffuse — during most stage 2 NREM sleep and especially during stages 3 and 4 NREM sleep — then planning/organization is less intensely stimulated, and conscious dream experience is both less probable and less coherent. Mnemonic activation is the driving determinant of whether a dream will take place and of the specific form it will assume. And, the primary variable affected by memory activation is the diffuseness or availability of diverse memory elements.

Foulkes[54] makes a strong case that dreaming cannot be simply explained by analysis of REM sleep physiology, despite the fact that variations in brain activation associated with REM and NREM sleep seem to support this model. Rather, he argues that REM sleep is neither a necessary nor a sufficient condition for the occurrence of dreaming as suggested by two types of evidence. First, REM-NREM mentation differences are more quantitative than they are qualitative[2,60] — thus REM

TABLE 8.2
Recall of Cognitive Activity From REM and NREM Sleep in 6 Age Cohorts of Children

Age Cohort	Sleep Onset REM (%)	Other NREM (%)	NREM (%)
3–5	27	18	6
5–7	31	31	8
7–9	48	31	21
9–11	66	61	32
11–13	66	67	31
13–15	67	67	40

Source: From Foulkes, D., *Children's Dreams: Longitudinal Studies,* John Wiley & Sons, New York, 1982. With permission.

sleep is not necessary for such mentation. As will be seen below, this argument is still disputed by findings of residual qualitative differences in REM and NREM mentation reports after control for quantitative differences. Second, both REM and NREM dreaming are infrequent in children (see Table 8.2) and only emerge gradually as their (presumably) requisite waking state cognitive skills mature[51,57] — thus, REM sleep is not sufficient for sleep mentation. To illustrate the latter, Foulkes' analysis of the content of 7- to 9-year-old children's dreams reveals REM/NREM differences on some content scales, but these differences were not marked enough to suggest that the two report types were qualitatively different at this young age. Evidence for his argument is bolstered further by a replication study,[57] though an attempted replication by an independent research group has been problematic (see exchange in References 55 and 149).

8.1.4 The Two-Generator (2-gen) Model of Sleep Mentation

Hobson's[78,80,81] AIM model of sleep mentation is rooted in his and McCarley's earlier, activation-synthesis (A-S) hypothesis of dreaming (Figure 8.1).[76] The latter combines an assumption of formal mind-brain isomorphism with a detailed description of differences in REM and NREM sleep animal physiology to explain dreaming (see References 77, 88, and 116 for reviews of the physiological findings). Analogous physiological differences in human sleep determine differences in the form, but not necessarily the content, of mental experiences in REM and NREM sleep. The most global and robust level of formal isomorphism is that of the brain-mind state, with REM sleep, NREM sleep, and wakefulness comprising the three principle divisions. Each state has its particular physiology and its accompanying, isomorphic mental features. Accordingly, the authors draw a distinction between dreaming and nondreaming mentation and operationalize the former with six characteristics (listed earlier). Both of these sleep states are phenomenologically distinct from waking state consciousness.

Three global factors determine the major physiological state changes in the AIM model, thus their formal isomorphs: (1) activation level, or A, reflects midbrain neuronal firing levels and corresponds to the information processing capacity of the brain; (2), input source, or I, is the relative weighting of external vs. internal information signals; and (3) mode of information processing, or M, is the relative strength of linear or analytic processing (feature extraction) vs. parallel or synthetic processing (form recognition). Each behavioral state is determined by a different weighting of A, I, and *M*. During wakefulness, all three are high. During REM sleep, *A* is high and both *I* and *M* are low. In NREM sleep, all three are partially reduced. Other weightings of *A*, *I*, and *M* produce common and unusual behavioral states (e.g., reverie, lucid dreaming, hallucination, coma).

As spelled out by the activation-synthesis hypothesis and its more recent modifications,[113] specific mental features are assumed to be isomorphic with specific physiological processes within each state. Hobson's group has accordingly developed and validated (on non-laboratory dream reports) a variety of dream-content measures to index the mental component of these isomorphic relationships: for example, self-representation and bizarreness,[149] emotions profile,[118] visual attention continuity,[179] narrative structure,[178] thematic coherence,[173] and transformation of characters, objects, and settings.[151] These instruments add a number of novel measures of dream structure to the arsenal of existing measures;[40,192] however, their value for discriminating REM and NREM sleep mentation qualitatively, for determining their isomorphic relationships to specific physiological variables, and thus for validating the 2-gen model all remain to be demonstrated.

The A-S hypothesis precursor to the AIM model has received criticism on many fronts, primarily for its neurophysiological reductionism,[53] for its challenge to psychoanalytic assumptions,[188] for its over-reliance on mind-body isomorphism,[101] for its inconsistencies with neurological findings (e.g., patients with brainstem and forebrain lesions),[170] and for ignoring a purportedly central role for forebrain centers in the regulation of REM sleep.[3,186] More recent modifications to the model have begun to address some of these criticisms, primarily by validating measures of higher-order cognitive processes, but most criticisms have gone unanswered.

8.2 Experimental Results Bearing on the Models

A large body of research can be brought to bear on the 1-gen and 2-gen models of sleep mentation. Although a complete review of the literature is beyond the scope of this chapter, seven general classes of research were examined for evidence supporting either the 1- or the 2-gen models of mentation production. Note that some of these types of results[1-4] less directly address the question of REM/NREM content differences than do others[5-7] because they measure variables directly or indirectly correlated with sleep mentation, rather than variables indexing the content or occurrence of sleep mentation per se. In general terms, research in categories 1 and 2 support the 1-gen model of mentation to a limited extent, whereas research in categories 3 to 7 tend to support the 2-gen model:

1. Memory sources as inferred from personal associations to mentation
2. Inter-relationships between mentation contents from different reports
3. Event-related potentials (ERPs)
4. Memory consolidation studies
5. Stimulation affects on mentation
6. Residual differences in stage-related measures of mentation quality
7. Subject differences in mentation content

8.2.1 Memory Sources

The prediction that REM and NREM reports of equivalent length should derive from memory sources of equivalent character (1-gen model) is supported by some studies showing that judges categorize REM report associations to be semantic knowledge sources more often than they do

NREM report associations when length of report is not controlled, but that such differences are lost when length is controlled.[29] Other studies using similar methodologies nevertheless demonstrate differences between the sources of REM and NREM (sleep onset) reports.[27,31,32] Such differences are frequently attributed to differences in the memory processes that feed into the imagery system, rather than to dreaming processes per se (e.g., memory material that is selective during sleep onset, but undifferentiated during REM sleep):[26] "Such a system, although operating on memory sources which are sometimes different in quality and quantity, succeeds in generating products (i.e., dreams) which are very similar or, at least, not easily discriminable according to content analysis criteria." In this case, qualitative differences have been identified for a hypothesized form of nocturnal cognitive activity (memory activation), but the latter is dissociated from the dreaming mechanism by conceptualizing it as part of generalized "memory".[27] The justification for dissociating nocturnal memory activity from dreaming in this manner seems to be premised upon its identity to waking state memory, even though, as described in a later section, Foulkes[52] describes it as quite different from waking state memory.

8.2.2 Report Interrelationships

Some studies also support the 1-gen prediction that contents from different reports should be highly similar — even related — regardless of stage.[33,146] For example, one study[33] found that low-level (paradigmatic and lexical) relationships, but not high-level (syntagmatic and propositional) relationships between pairs of mentation reports were higher within the same night than they were between nights, regardless of whether the report pairs came from different stages (e.g., REM-NREM) or the same stage (e.g., REM-REM). Presumably, these similarities in content reflect the operations of a common imagery generator.

8.2.3 Event-Related Potentials

ERP studies support the 2-gen model to the extent that long-latency components, such as P300, have been shown to have higher amplitudes and/or reduced latencies in REM than in NREM sleep.[1,123,129,130,154,182] Moreover, most studies report that ERPs in REM sleep resemble those of the waking state more than they do the diminished potentials of NREM sleep. Long-latency components are widely thought to be associated with cognitive processes such as selective attention (N1 or N100), sensory mismatch (N2-P3a), orienting (N2), surprise (P3b), novelty (P3a), and semantic processing (N400),[99,158] i.e., the types of processes which might be expected to contribute to sleep mentation. Donchin's[41] proposition that P300 reflects processes of creating, maintaining, and updating an internal model of the immediate environment seems especially pertinent to the intense reality simulations of dreaming.

8.2.4 Memory Consolidation

Research on processes of memory consolidation has repeatedly found evidence of differential cognitive activity in REM and NREM sleep (see References 42, 117, and 165 for reviews). There is compelling evidence that REM sleep is more often implicated in consolidation of new skills than is NREM sleep and that REM sleep is implicated in more kinds of memory consolidation than is NREM sleep. The single memory function which has been demonstrated for NREM sleep to date (rotor pursuit) is a qualitatively different task from those that have been shown to require REM sleep (e.g., logic tasks, visual discrimination).[95,166]

8.2.5 Stimulation Effects

Experiments which have employed manipulations of stimuli either prior to or during sleep have shown that such stimuli differentially influence REM and NREM sleep mentation.[24,44,72,180] This literature does not paint an entirely clear portrait of the differences, however, with some studies suggesting superior processing of verbal materials during REM sleep[44] and others superior processing during Stage 2 sleep.[24,180]

8.2.6 Residual Stage Differences

Many studies have compared the contents of mentation reports from REM and NREM stages. And although it is true that many of the differences which were discovered early in the history of this controversy tend to diminish or disappear when report length is controlled, many qualitative differences remain after such controls are effected.[2,5,29,32,60,83,125,139] REM and NREM reports have been found to differ on self-reflectiveness,[141] bizarreness,[136] visual and verbal imagery[5] — even when report length is controlled. We demonstrated superior narrative linkage in REM vs. NREM reports.[125] Strauch and Meier[177] found fewer characters and lower levels of self-involvement in NREM than in REM dreams, again, regardless of report length. Even the rigorous length-controlled content analyses of Foulkes and Schmidt[60] revealed more per-unit self-representation in REM than in sleep onset (SO) dreams and more per-unit characterization in REM than in NREM dreams. In a recent study,[190] total word count and visual imagery count were both found to differentiate REM from stage 2 mentation reports. When total word count was controlled as a covariate, REM mentation reports were still more visually salient than NREM reports. Antrobus and colleagues[5] have recently replicated this finding, failing to replicate their own earlier study.[2] Many of the categories examined would seem to be quite fundamental to imagery production, e.g., representations of self and characters are two of the most ubiquitous constituents of dreams, occurring in over 95% of reports by some estimates.[71,169]

A study of fictive movement in dreams[139] found similar evidence of qualitative differences. Words relating to fictive movement were much more frequent for REM (68%) than for NREM reports (26%). Other findings in this study are more consistent with the 1-gen model, e.g., positive correlations between fictive imagery word count and total word count for both REM and NREM sleep reports. However, the authors suggest that the sleep of the few subjects responsible for this effect may have been atypical in some way that caused their NREM sleep to be influenced by REM sleep processes. Specifically, an increase in REM sleep "pressure" (by, for example, REM deprivation or depression) may have lengthened the REM dream reports as well as raising the odds that a NREM awakening would coincide with a pre-REM or post-REM sleep transitional window — what I refer to in a later section as phantom REM sleep.

8.2.7 Subject Differences

Some studies demonstrating subject differences in REM and NREM processes prove to be consistent with the 2-gen model. For example, Zimmerman[197] found a number of content differences between REM and NREM reports for deep sleepers but not light sleepers. We[125] found greater episodic progression in REM than in NREM reports only for high frequency recallers. Other subject variables that may be associated with stage differences include the differential association of age with late-night activation effects on REM and NREM mentation,[190] large differences in recall of REM (but not NREM) mentation between normal and insomniac patients[155] and normal and depressed patients,[150] the effects of introspective style on salience of REM and NREM content,[191] and differential correlations between laboratory measures of REM and NREM salience (recall and length) and measures of psychopathology.[59] A number of other correlates of dream recall were also recently reviewed.[161]

It should also be apparent that these subject attributes of sleep style, age, degree of insomnia and depression, habitual recall of dreaming, introspective style, and psychopathology as well as other factors will affect so-called normal populations to greater or lesser degrees and may be determinant factors in subject self-selection for participation in sleep and dream studies. The existence of so many unexplained subject-related factors could lead to the more conservative conclusion that the 1-gen and 2-gen models do not necessarily hold true as general principles of sleep mentation formation, but that they hold true only for sub-groups of subjects under specific circumstances. It is conceivable, for example, that a second imagery generator develops only in adults, in subjects so predisposed, or in certain pathological conditions. Control of subject variables would seem to be an essential step in clarifying the limits of the 1-gen and 2-gen models of sleep mentation production.

In summary, evidence from several lines of research suggest that REM and NREM cognitive processes differ in fundamental ways, even after activation and report length factors are taken into consideration. How the 1-gen model accounts for these differences remains obscure. One argument is that the observed qualitative differences are relatively minor and attributable to slight differences in the number or types of memory elements input into the planning/organizing phase of imagery generation. Although promising, this explanation identifies the source of REM and NREM report differences in memory activation processes. Not only is the distinctiveness of such processes from dreaming per se highly speculative, but the processes themselves are very difficult to access experimentally (and thus to prove or disprove). Pending further study, their existence and role in imagery production remain largely speculative.

8.3 The Concept of "Activation" in Explaining Sleep Mentation Differences

8.3.1 Are Memory Activation and Cortical Activation Isomorphic?

The cognitive-psychological 1-gen model identifies memory activation as the instigating force for dream formation (see, for example, Reference 52, p. 145). However, Foulkes[52] very carefully limits his hypothesis about activation to mental — not physiological — language, even though known relationships between cerebral activation and sleep/wake stages (e.g., lower activation in NREM, higher activation in SO and REM, highest activation in active wakefulness) might seem to be consistent with the 1-gen model. This view of Foulkes' marks a shift away from his earlier view[60] and is not shared by all 1-gen theorists.

This assertion that cortical and cognitive activation are not related would seem to fly in the face of some basic tenets of cognitive neuroscience and psychophysiology. The argument is in some respects analogous to claiming that fearful dream experiences (i.e., affective activation) are not accompanied by autonomic arousal — which is not typically the case;[47,128] however, the fact that it is *sometimes* the case[47] suggests that two such variables — one phenomenological and one physiological — may be interrelated in a manner much more complex than either strict isomorphism or strict dissociationism would suggest.

Studies bearing on the question of whether cortical activation (as measured by EEG power) is associated with cognitive activation (as measured by word count, recall, etc.) have produced mixed findings and thus do not clearly support either model (see Reference 4 for a review). With EEG slowing and an increase in voltage there is an associated decrease in mentation recall,[132,135] and there is generally more EEG slowing in NREM than in REM sleep.[43] Both delta and beta amplitude appear to be related to successful dream recall from REM sleep whether subjects are depressed or healthy.[153] If EEG-defined (delta) activation is statistically controlled, stage differences in mentation are still found.[190] On the other hand, at least one study[193] found no relationships between EEG power and word count of either REM sleep reports or waking imagery reports.

8.3.2 Effects of Changes in Activation Across the Night

Both the recall of mentation[66,69,97,133,163,183] and its salience[34,50] increase in later REM episodes, changes likely due to activation associated with circadian factors.[5] The circadian factor appears to influence REM and NREM mentation equally — a finding which tends to support the 1-gen model. On the other hand, one study examining total word count and visual imagery count[190] limits this time of night effect to younger subjects.

In perhaps the most rigorous examination of time of night activation effects,[5] both stage and diurnal activation effects were found to enhance mentation characteristics such as word count and visual clarity. However, the effect size for time of night activation (e.g., 0.23 for visual clarity) was only about 30% of the effect size for REM-NREM stage activation (e.g., 0.70). Antrobus[5] interprets the activation effect to support the early A-S hypothesis of dreaming[76] — a 2-gen model.

8.3.3 Neurobiological Conceptions of Memory Activation

If memory activation is isomorphic with brain activation in some sense, then it may be asked whether brain activation stems from one or two (or more) "activation generators" in a manner parallel to the hypothesized imagery processes in the 1-gen and 2-gen models. Two types of evidence bearing on this question tend to support the notion of 2 (or more) activational systems, although the notion of a single, unifying system cannot be excluded: (1) evidence for multiple activation systems differentially targeting brain regions across sleep states, and (2) evidence for multiple memory systems.

8.3.3.1 Multiple Activation Systems

Much evidence suggests that there exist multiple activating systems in the brain.[13,114,152] There are at least five diffuse, anatomically and neurochemically distinct ascending networks of neurons (noradrenergic, cholinergic, dopaminergic, serotonergic, histaminergic) which project widely across the cortex. All of these systems, when stimulated, lead to a desynchronization of the EEG, which justifies labeling them as "activation" or "arousal" systems. Cortical activation during REM sleep is based upon a different subset of these neural systems than it is during either NREM sleep or the waking state. Most of the multiple activation systems have been associated with distinct cognitive functions in the waking state. For example, the waking state cholinergic system is thought to enhance discrimination accuracy (by increasing the "signal" in the signal-to-noise ratio) and utilization of spatial cues.[152] In contrast, waking state noradrenergic systems appear to facilitate processing of temporal cues and reducing distraction (by reducing "noise").

Although suggestive, evidence for multiple activation systems is not undisputed proof of multiple imagery generation systems. Some interpret these arousal systems to be integrated at both early and late stages, exerting a unified influence on conscious experience.[13] For example, brainstem contributions to cortical organization are suggested by the preservation of motor coordination, directed attention, and emotional evaluation after callosotomy.[13] At later stages, the synchronous firing of cortical cells at fast frequencies (e.g., 20 to 50 Hz)[114] appears to contribute to integrated cortical activation. On the other hand, the scarce data available indicate that cortical oscillations characterize REM sleep and wakefulness, but not NREM sleep,[107] a finding tending to support the 2-gen model.

8.3.3.2 Multiple Memory Systems

The status of hypothesized memory processes during sleep remains unclear. For example, it is not apparent how the "diffuse activation of memory" posited by Foulkes[52] should be reconciled with the multiple memory systems described by cognitive scientists.[171] Foulkes[52] describes a level of cognitive activity at which elements of past memories may be dissociated and fused in novel and/or unexpected combinations.[65] However, much cognitive research suggests that memory is organized into several qualitatively distinct types, none of which corresponds readily to this diffuse level.[171] Notwithstanding these problems with the 1-gen model, proponents of the 2-gen model have also not broached the problem of reconciling memory with dream formation processes.

8.3.4 Partialing Out Activation: Problems With the Use of Report Length

The 1-gen model is supported largely by methodological and statistical procedures which are presumed to remove the effects of differential activation on mentation reports from different states and thus to level the playing field for comparing these reports on qualitative dimensions. Presumably, longer reports (reflecting greater activation) are associated with more numerous opportunities for particular elements to manifest in the mentation report.

A control for report length may be effected in several ways. Most studies estimate activation with total word count (TWC),[2] which is a tally, transformed by $\log_{10}(X+1)$ to remove positive skew, of all descriptive words in the report, excluding those pertaining to commentary, associations, hesitations, and redundancies. TWC may be partialled out of correlational relationships or its variance estimated and removed in some other way.[5,106,190,196]

It should be noted that the continuous measure of TWC is only moderately better suited to distinguish REM and NREM mentation effects than is a simple binary variant of the same variable (i.e., presence or absence of content). Although TWC clearly discriminates REM from NREM reports ($F = 33.6$, $p < .0001$), its binary variant accounts for a large proportion of the variance ($F = 18.6$, $p < .0004$).[5] The same is true of a word count of strictly visual descriptors ($F = 51.8$, $p < .0001$) and its binary variant (i.e., presence or absence of visual content; $F = 43.2$, $p < .0001$).[5]

A procedure conceptually related to TWC is to weight dependent variables using an estimate of length that is based upon the temporal progression of the experience. Foulkes and Schmidt[60] devised a method of parsing reports on the basis of events that occur contiguously, the so-called "temporal unit". In a related manner, we[125] opted to use the presence/absence of story components (character, action, setting) as a control for assessing the organization of these same components.

Hunt and colleagues[83] have challenged such methods of controlling for report length, in particular the use of TWC. They argue that variations in report length are an expected correlate of mentation that is qualitatively remarkable in some way, that "more words are necessary to describe more bizarre experiences" (p. 181). To partial out report length from a given scale may be to partial out the variable from itself (p. 181) and may "cripple our ability to study what is most distinctive about dreams by misleadingly diluting a key measure of the dreaming process" (p. 190). Worse yet, using mundane words to control for non-verbal variables such as imagery bizarreness may be an arbitrary transformation that leads to unpredictable and potentially artificial effects.[83]

Working with report length and bizarreness ratings, Hunt et al. demonstrated that (1) a bizarre pictorial stimulus does indeed require more words to describe than does a mundane stimulus; (2) a count of bizarreness words is a valid measure of bizarreness, i.e., correlates highly ($r = .92$) with standard bizarreness scales; and (3) partialing out of the residual, non-bizarreness word count does not eliminate significant correlations between bizarreness and other measures — whereas partialing out of TWC does (consistent with previous studies).[196] Thus, it is apparent that weighting dependent measures of qualitative aspects of dream mentation with a verbally based quantitative measure of activation such as TWC may introduce a significant loss of information related to the dependent variable under study. However, applying the same rules of counting words to the dependent measure and weighting it with a total count that is free of this measure will largely remove this bias.

It should also be noted that other methods of controlling for report length, such as temporal unit coding[60] or minimal representation of narrative constituents,[125] are in principle open to the same criticism dogging the TWC partialing method. That is, by weighting dependent measures relative to some other variable (e.g., element contiguity, constituent co-occurrence), one may inadvertently partial out variance which is primordial to the dependent variable itself. The use of such transformations requires careful consideration of known and potential relationships between dependent variables and the covariates used to weight them.

These considerations are particularly important for assessment of the 1-gen and 2-gen models because the former is supported largely by the finding that weighting by report length diminishes apparently qualitative differences. If this attenuation of differences is an artifact produced by inappropriate transformations of the dependent measures, then such transformations should be solidly justified, replaced by more appropriate controls or avoided altogether.

8.4 Summary

Although research can be found which supports both 1- and 2-gen models, it seems clear that much of this research tends to favor the 2-gen model: there are REM NREM differences to be found in sleep mentation and in the nocturnal physiological processes such as long-latency ERPs that are associated with complex cognitions in the waking state. The principal argument in favor of the 1-gen model, that qualitative differences are artifacts of more primary quantitative differences, has been challenged by many studies demonstrating process differences and residual differences in content qualities even after effecting controls for report length. Further, this argument has been countered on the grounds that qualitative differences appear to produce quantitative ones.[83] A second argument, that residual qualitative differences are attributable to differences in memory inputs, has merit but has not been supported by all measures attempting to quantify these inputs. There are also important theoretical questions as to whether memory indeed functions in the diffuse manner proposed by the 1-gen model and whether memory inputs should be considered to be independent of the dreaming process per se. There remains a critical argument — that REM sleep is not sufficient for dreaming — which has been largely ignored by groups other than Foulkes'; specifically, it remains to be shown whether it is cognitive competence (1-gen) or maturation of REM sleep physiological systems (2-gen) which is the predominant, necessary antecedent for dreaming.

On the other hand, the evidence taken together is not entirely consistent with the 2-gen model proposed by Hobson's group either. As Foulkes[53] has rightly indicated, evidence for neurobiological isomorphism is marginal at best, leaving most of the conclusions of this model extremely speculative. This particular 2-gen model is also quite weak in describing the nature of NREM mentation and thus has not yet provided a firm basis for predicting qualitative differences between REM and NREM cognitive activity. As a model driven by physiological analogues to cognitive activity, this 2-gen model can also be criticized on the grounds that it is too selective in the physiological processes it purports to explain. In particular, the model does not adequately account for forebrain mechanisms that would seem to be central to complex cognitive operations such as the narrative synthesis of dreaming.

8.5 Toward Rapprochement: A Possible Role for Phantom REM Sleep

The literature presents an apparent paradox. On the one hand, there is indisputable proof that cognitive activity — that at times approaches the complexity of dreaming — occurs on about 50% of awakenings from NREM sleep. On the other hand, there is much evidence that REM and NREM cognitions and their physiological correlates differ in important respects. Widespread evidence for the occurrence of dreaming in NREM sleep supports a 1-gen model, whereas numerous examples of stage-related differences support a 2-gen model. How may these paradoxical views be reconciled? One possible avenue for rapprochement is the notion of REM sleep-related cognitive processes that may under some circumstances become active in NREM sleep — covert processes I refer to as phantom REM sleep. The latter I define as an episode of NREM sleep for which some REM sleep processes are active, but for which REM sleep cannot be identified according to standard criteria. There is clinical and experimental evidence that components of sleep states may combine in atypical patterns as a consequence of illness and other unusual circumstances;[111] however, there is an increasing number of indications in the literature that dissociations and admixtures of REM and NREM sleep processes may also be found during normal sleep. Below I review some findings consistent with the possibility that REM and NREM sleep are not always clearly distinguishable phenomena, leaving open the possibility that some REM sleep processes may continue to influence NREM sleep mentation.

Phantom REM sleep may implicate the activation of subcortical REM sleep processes in the absence of complete or full EEG desynchronization, such as appears to occur just prior to and following REM sleep.[115] McCarley has shown in animals that neuronal (membrane) activity underlying REM sleep can begin well before EEG or PGO signs of REM sleep; the transition at this level is "gradual, continuous, and of long duration" (p. 375). Such activity may also continue after the offset of a REM episode. He speculates that NREM dreaming takes place during such REM-active transitions. McCarley's physiological observations are consistent with studies in humans demonstrating that heart-rate variability shifts from a predominantly parasympathetic profile in NREM sleep to a predominantly sympathetic one in REM sleep, and that this shift occurs up to 15 min prior to the EEG-defined onset of REM sleep.[160]

Phantom REM sleep may involve atypical NREM sleep episodes for which only one electrophysiological criterion earmarking it as REM sleep is lacking. REM sleep with high tonic chin muscle activity is a type of phantom REM sleep that in its most extreme form is known as REM sleep behavior disorder.[112] In non-pathological populations phantom REM sleep is well known to polysomnographers as the notorious "missing" REM sleep episode early in the night — usually the first of the night. At this time, most of the electrophysiological signs of REM sleep are present — e.g., about 90 min have elapsed since sleep onset, cessation of spindling, EEG desynchronization — but sometimes chin muscle tonus may remain high, or distinct rapid eye movements may fail to appear, or a brief waking arousal may occur. Such stages are scored as stage 1, stage 2, or, at times, indeterminate sleep, even though intuition would have it that REM sleep was present at some level. In other cases, the missing first REM period may have fewer electrophysiological signs but may nonetheless be suspected by poorly defined architectural cues and their occurrence at about 90 min after sleep onset.

In a similar vein, Karacan and colleagues[93] have reported that nocturnal penile tumescence — which is a fairly robust correlate of REM sleep[94] — will often occur at the 90-min junctures where REM sleep would be expected to occur yet where it could not be positively identified. They provide a striking hypnogram of a single subject with three consecutive nocturnal erections overlying three corresponding phantom REM episodes. Of 19 erections, 12 occurring during NREM sleep were related to this type of expected but incomplete REM sleep episode; an additional four occurred on returning to sleep immediately after an experimental REM sleep awakening.

Stickgold et al.[173] reported evidence consistent with the preceding formulation of phantom REM sleep. They exploited a two-channel (eye movement, body movement) home-monitoring device to track whether REM or NREM sleep accompanied spontaneous awakenings with mentation. Eleven subjects logged 239 reports over 110 nights, 194 of which could be classified for sleep stage. The familiar difference in recall from REM (83%) and NREM (54%) sleep awakenings was replicated, and the median word count was found to be 7 times higher for REM[148] than for NREM[21] reports. More importantly, word count revealed that NREM mentation reports were longest if they occurred within 15 min of a prior REM sleep episode, whereas REM mentation reports were longest if they occurred 30 to 45 min into a REM episode. In fact, seven of the nine longest NREM reports (word count > 100) occurred within 15 min of a REM episode.

This study replicates a similar early finding[70] that NREM reports occurring within 5 min of the last REM sleep eye movement more often produced evidence of prior cognitive activity (81.8%) than did reports occurring more than 10 min from the last eye movement (3.8%), as well as a more recent finding[6] that NREM reports occurring 5 min after the end of a REM sleep episode contain more words per report than do those occurring 15 min after the end of REM sleep. Hobson and Stickgold take this type of finding to entertain two notions: one, that "long NREM reports reflect transitional periods when some aspects of REM physiology continue to exert an influence" and, two, that "reports given early in NREM periods might reflect sleep mentation from the preceding REM period" (p. 25). The former possibility is supported by studies cited previously indicating that processes such as PGO activity and relative sympathetic tone may precede the onset of REM sleep by as much as 15 min. The latter possibility has been suggested by several authors as an explanation for dreaming during NREM sleep[91,115,139,195] and is consistent with the fact that most studies do not systematically control how close NREM awakenings are to both prior *and* subsequent REM sleep episodes. On the other hand, it should be noted that at least one study[92] has found that NREM awakenings made before any REM sleep episodes have occurred in a night (i.e., prior to the first REM sleep episode) nevertheless led to substantial recall of cognitive activity (43%), suggesting that it could not be due to REM sleep influences. A second study,[50] in which awakenings at least 30 min post-REM targeted the middle of the NREM episodes, also found a sizable recall rate of 64.6%. Both of these recall rates are within or exceed one standard deviation of the grand mean recall rate for NREM sleep presented earlier. They argue against the hypothesis that residual REM sleep processes influence NREM mentation, although they do not exclude the possibility that phantom REM sleep of some type occurred during those NREM episodes. Further, the reconsideration below of sleep onset as a possible source of phantom REM sleep also counters one of these types of evidence.

Phantom REM sleep may be manifest during some sleep onset (SO) episodes. The brief SO transitions between wakefulness and sleep demonstrate most of the electrophysiological signs of REM sleep, such as transient EMG suppressions and phasic muscle twitches and jerks, as well as extremely vivid sleep mentation. We have shown that topographic distributions of fast-frequency EEG power for SO images and REM sleep are very similar.[124] Although rapid eye movements themselves are less conspicuous at SO, they are nevertheless observed.[185] However, the slow eye movements so characteristic of SO also occur frequently in REM sleep;[137] they may constitute an unrecognized marker of REM sleep (see Reference 138 for discussion). It is thus possible that the vivid dreaming observed at SO derives from a brief, but nevertheless undetected, passage through REM sleep into descending stage 2 sleep. The sleep onset REM (SOREM) episodes observed so frequently in sleep-disordered and normal individuals[15] may simply be instances of this transition stage which happen to demonstrate all of the inclusion criteria required to score an epoch of REM sleep. Such "unmasking" of phantom REM sleep at SO is perhaps also influenced by build-up of REM pressure. For example, we found that SOREM episodes on the Multiple Sleep Latency Test (MSLT) were twice as frequent in sleepy patients (severe sleep apnea syndrome [SAS] and idiopathic hypersomnia) than they were in non-sleepy patients (mild SAS and periodic leg movements in sleep without hypersomnia; Nielsen, Montplaisir, and Gosselin, unpublished data). The fact that reports of dreaming during an MSLT nap are not good predictors of the presence of classical REM sleep[11] may also reflect the difficulty of differentiating phantom REM sleep from REM sleep as it is classically defined.

Additional evidence pointing to the possible existence of latent REM sleep at SO is the observation that a variety of sleep starts are common at sleep onset among healthy subjects. Such starts consist of abrupt motor jerks and sudden flashes of visual, auditory, and somesthetic imagery, and it has been suggested that these are intrusions of isolated REM sleep events into NREM sleep.[110]

It should be noted that polysomnographers quantifying the standard Rechtschaffen and Kales criteria have always accepted a certain degree of ambiguity in their scoring of REM sleep. This ambiguity is inherent in the notion of REM sleep "efficiency". Within the limits of a given REM sleep episode there can occur transitions into other stages — typically stage 2 or wakefulness — which reduces the efficiency of the REM episode. Within an arbitrary limit of 15 min of any "intruding" stage, the alternate activity is calculated to be a temporary aberration of an otherwise continuous REM sleep episode. Beyond the 15-min limit, alternate activity is thought to signal a transition to a new REM/NREM cycle and is no longer factored into the efficiency score of the preceding REM episode. Rather, an additional REM episode is scored, despite the fact that its periodicity falls dramatically short of the typical 90-min value. Thus, up to an arbitrary limit of 15 min, the notion of REM sleep efficiency implies that the underlying physiological state of REM sleep is not completely suspended during the intrusion of another stage. At the very least, some physiological factor determining REM sleep "propensity" continues to be active at these times. In view of the physiological research reviewed above, this 15-min limit may be an entirely appropriate window for detecting latent REM sleep processes.

Of course, the examples reviewed here do not constitute a definitive proof that REM sleep processes are latent during other stages of sleep. They might help to explain some, but certainly not all, instances of dreaming taking place during NREM sleep. The finding of cognitive activity during NREM stages 3 and 4 sleep is particularly at odds with this notion as there is less compelling evidence that phantom REM sleep might occur in these stages. Nevertheless, such events are at least feasible (e.g., stage 4 sleep terrors might involve arousal processes associated with REM sleep), and to date the findings available do converge on the possibility that REM sleep processes may account for much NREM stage 2 mentation.

As various phenomena of state overlap and intrusion among normal and sleep-disordered subjects are documented with increasing precision, new information about sleep stages and their component processes is accumulating. Future studies will undoubtedly clarify which stage-related processes respect, and which fail to respect, the scoring criteria traditionally applied in polysomnography. And although it might appear to some that a search for extra-REM physiological correlates of dreaming is an exercise in tautology,[54] mounting evidence suggests that further work in this direction may help to resolve to the current impasse over 1-gen and 2-gen models of sleep mentation. A vital part of this work may be to revisit the criteria for identifying REM and NREM sleep — much as researchers have collectively revisited the criteria for identifying dreaming and other nocturnal cognitive activities.

Acknowledgments

This work was supported by the Medical Research Council of Canada and the "Fonds de la recherche en santé du Québec". Special thanks to Dominique Petit, Ph.D., for proofreading this manuscript.

References

1. Addy, R.O., Dinner, D.S., Luders, H., Lesser, R.P., Morris, H.H., and Wyllie, E., The effects of sleep on median nerve short latency somatosensory evoked potentials, *Electroencephalogr. Clin. Neurophysiol.*, 74, 105–111, 1989.
2. Antrobus, J., REM and NREM sleep reports: comparison of word frequencies by cognitive classes, *Psychophysiology*, 20, 562–568, 1983.
3. Antrobus, J., The neurocognition of sleep mentation: rapid eye movements, visual imagery and dreaming, in *Sleep and Cognition*, Bootzin, R.R., Kihlstrom, J.F., and Schacter, D.L., Eds., American Psychological Association, Washington, D.C., 1990, pp. 3–24.
4. Antrobus, J., Dreaming: cognitive processes during cortical activation and high afferent thresholds, *Psychol. Rev.*, 98, 96–121, 1991.
5. Antrobus, J., Kondo, T., Reinsel, R., and Fein, G., Dreaming in the late morning: summation of REM and diurnal cortical activation, *Cons. Cogn.*, 4, 275–299, 1995.
6. Antrobus, J.S., Fein, G., Jordan, L., Ellman, S.J., and Arkin, A.M., Measurement and design in research on sleep reports, in *The Mind in Sleep*, 2nd ed., Ellman, S.J. and Antrobus, J., Eds., John Wiley & Sons, New York, 1991, pp. 83–122.
7. Arkin, A.M., Toth, M.F., Baker, J., and Hastey, J.M., The frequency of sleep talking in the laboratory among chronic sleep talkers and good dream recallers, *J. Nerv. Ment. Dis.*, 151, 369–374, 1970.
8. Armitage, R., Changes in Dream Content as a Function of Time of Night, Stage of Awakening and Frequency of Recall, unpublished Masters thesis, Carleton University, Ottawa, 1980.
9. Armitage, R., Hoffmann, R., and Moffitt, A., Interhemispheric EEG activity in sleep and wakefulness: individual differences in the basic rest-activity cycle (BRAC), in *The Neuropsychology of Sleep and Dreaming*, Antrobus, J.S. and Bertini, M., Eds., Lawrence Erlbaum Associates, Hillsdale, NJ, 1992, pp. 17–47.
10. Aserinsky, E. and Kleitman, N., Regularly occurring periods of eye motility, and concomitant phenomena during sleep, *Science*, 118, 273–274, 1953.
11. Benbadis, S.R., Wolgamuth, B.R., Perry, M.C., and Dinner, D.S., Dreams and rapid eye movement sleep in the multiple sleep latency test, *Sleep*, 18, 105–108, 1995.
12. Berger, M., Dreaming and REM sleep (commentary), *World Fed. Sleep Res. Soc. Newslett.*, 3, 13, 1994.
13. Berlucchi, G., One or many arousal systems? Reflections on some of Giuseppe Moruzzi's foresights and insights about the intrinsic regulation of brain activity, *Arch. Ital. Biol.*, 135, 5–14, 1997.
14. Bertini, M. and Violani, C., The postawakening testing technique in the investigation of cognitive asymmetries during sleep, in *The Neuropsychology of Sleep and Dreaming*, Antrobus, J.S. and Bertini, M., Eds., Lawrence Erlbaum Associates, Hillsdale, NJ, 1992, pp. 47–63.
15. Bishop, C., Rosenthal, L., Helmus, T., Roehrs, T., and Roth, T., The frequency of multiple sleep onset REM periods among subjects with no excessive daytime sleepiness, *Sleep*, 19, 727–730, 1996.
16. Bosinelli, M., Cavallero, C., and Cicogna, P., Self-representation in dream experiences during sleep onset and REM sleep, *Sleep*, 5, 290–299, 1982.
17. Botman, H.I. and Crovitz, H.F., Facilitating the reportage of dreams with semantic cues, *Imag. Cogn. Pers.*, 9, 115–129, 1989.
18. Broughton, R.J., Sleep disorders: disorders of arousal? Enuresis, somnambulism, and nightmares occur in confusional states of arousal, not in "dreaming sleep", *Science*, 159, 1070–1078, 1968.
19. Busink, R. and Kuiken, D., Identifying types of impactful dreams — a replication, *Dreaming*, 6, 97–119, 1996.
20. Cann, D.R. and Donderi, D., Jungian personality typology and the recall of everyday and archetypal dreams, *J. Pers. Soc. Psychol.*, 50, 1021–1030, 1986.

21. Cartwright, R., A network model of dreams, in *Sleep and Cognition,* Bootzin, R.R., Kihlstrom, J.F., and Schacter, D.L., Eds., American Psychological Association, Washington, D.C., 1990, pp. 179–189.
22. Casagrande, M., Violani, C., and Bertini, M., A psycholinguistic method for analyzing two modalities of thought in dream reports, *Dreaming,* 6, 43–55, 1996.
23. Castaldo, V. and Holzman, P., The effects of hearing one's own voice on dreaming content: a replication, *J. Nerv. Ment. Dis.,* 148, 74–82, 1969.
24. Castaldo, V. and Shevrin, H., Different effects of auditory stimulus as a function of rapid eye movement and non-rapid eye movement sleep, *J. Nerv. Ment. Dis.,* 150, 195–200, 1970.
25. Cavallero, C. and Cicogna, P., Comparing reports of the same dream: proposals for a structural analysis, *Percept. Motor Skills,* 57, 339–356, 1983.
26. Cavallero, C. and Cicogna, P., Memory and dreaming, in *Dreaming as Cognition,* Cavallero, C. and Foulkes, D., Eds., Harvester Wheatsheaf, New York, 1993, pp. 38–57.
27. Cavallero, C., Cicogna, P., and Bosinelli, M., Mnemonic activation in dream production, in *Sleep '86,* Koella, W.P., Obál, Jr., F., Schultz, H., and Visser, P., Eds., Gustav Fischer Verlag, New York, 1988, pp. 91–94.
28. Cavallero, C., Cicogna, P., Natale, V., Occhionero, M., and Zito, A., Slow wave sleep dreaming, *Sleep,* 15, 562–566, 1992.
29. Cavallero, C., Foulkes, D., Hollifield, M., and Terry, R., Memory sources of REM and NREM dreams, *Sleep,* 13, 449–455, 1990.
30. Cicogna, P., Cavallero, C., and Bosinelli, M., Analyzing modifications across dream reports, *Percept. Motor Skills,* 55, 27–44, 1982.
31. Cicogna, P., Cavallero, C., and Bosinelli, M., Differential access to memory traces in the production of mental experience, *Int. J. Psychophysiol.,* 4, 209–216, 1986.
32. Cicogna, P., Cavallero, C., and Bosinelli, M., Cognitive aspects of mental activity during sleep, *Am. J. Psychol.,* 104, 413–425, 1991.
33. Cipolli, C., Fagioli, I., Baroncini, P., Fumai, A., Marchiò, B., and Sancini, M.The thematic continuity of mental experiences in REM and NREM sleep, *Int. J. Psychophysiol.,* 6, 307–313, 1988.
34. Cohen, D.B., Changes in REM dream content during the night: implications for a hypothesis about changes in cerebral dominance across REM periods, *Percept. Motor Skills,* 44, 1267–1277, 1977.
35. Conduit, R., Bruck, D., and Coleman, G., Induction of visual imagery during NREM sleep, *Sleep,* 20, 948–956, 1997.
36. Dement, W., Dream recall and eye movements during sleep in schizophrenics and normals, *J. Nerv. Ment. Dis.,* 122, 263–269, 1955.
37. Dement, W. and Kleitman, N., Cyclic variations in EEG during sleep and their relation to eye movements, body motility, and dreaming, *Electroencephalogr. Clin. Neurophysiol.,* 9, 673–690, 1957.
38. Dement, W. and Kleitman, N., The relationship of eye movement during sleep to dream activity: an objective method for the study of dreaming, *J. Exp. Psychol.,* 53, 339–346, 1957.
39. Dement, W.C., *Some Must Watch While Some Must Sleep,* W.W. Norton, New York, 1976.
40. Domhoff, G.W., *Finding Meaning in Dreams: A Quantitative Approach,* Plenum Press, New York, 1996.
41. Donchin, E., Bashore, T.R., Coles, M.G.H., and Gratton, G., Cognitive psychophysiology and human information processing, in *Psychophysiology: Systems, Processes, and Applications,* Coles, M.G.H., Donchin, E., and Porges, S.W., Eds., Guilford Press, New York, 1986, pp. 244–267.
42. Dujardin, K., Guerrien, A., and Leconte, P., Sleep, brain activation and cognition, *Physiol. Behav.,* 47, 1271–1278, 1990.

43. Dumermuth, G., Langer, B., Lehmann, D., Meier, C.A., and Dinkelmann, R., Spectral analysis of all-night sleep EEG in healthy adults, *Eur. Neurol.,* 22, 322–339, 1983.
44. Evans, F.J., Hypnosis and sleep: techniques for exploring cognitive activity during sleep, in *Hypnosis: Research Developments and Perspectives,* Fromm, E. and Shor, R.E., Eds., Aldine/Atherton, Chicago, 1972, pp. 43–83.
45. Fein, G., Feinberg, I., Insel, T.R., Antrobus, J.S., Price, L.J., Floyd, R.C., and Nelson, M.A., Sleep mentation in the elderly, *Psychophysiology,* 22, 218–225, 1985.
46. Feinberg, I. and March, J.D., Observations on delta homeostasis, the one–stimulus model of NREM–REM alternation and the neurobiologic implications of experimental dream studies, *Behav. Brain Res.,* 69, 97–108, 1995.
47. Fisher, C., Byrne, J., Edwards, A., and Kahn, E., A psychophysiological study of nightmares, *J. Am. Psychoanalyt. Assoc.,* 18, 747–782, 1970.
48. Fisher, C., Byrne, J.V., Edwards, A., and Kahn, E., REM and NREM nightmares, *Int. Psychiatry Clin.,* 7, 183–187, 1970.
49. Foulkes, D., *The Psychology of Sleep,* Charles Scribners Sons, New York, 1966.
50. Foulkes, D., Nonrapid eye movement mentation, *Exp. Neurol.,* Suppl. 4, 28–38, 1967.
51. Foulkes, D., *Children's Dreams: Longitudinal Studies,* John Wiley & Sons, New York, 1982.
52. Foulkes, D., *Dreaming: A Cognitive-Psychological Analysis,* Lawrence Erlbaum Associates, Hillsdale, NJ, 1985.
53. Foulkes, D., Dreaming and consciousness, *Eur. J. Cognitive Psychol.,* 2, 39–55, 1990.
54. Foulkes, D., Dreaming and REM sleep, *J. Sleep Res.,* 2, 199–202, 1993; Foulkes, D., A research report — dreaming and REM sleep, *Newslett. World Fed. Sleep Res. Soc.,* 3; 11–13, 18, 28; 1994.
55. Foulkes, D., Misrepresentation of sleep-laboratory dream research with children, *Percep. Motor Skills,* 83, 205–206, 1996.
56. Foulkes, D. and Fleisher, S., Mental activity in relaxed wakefulness, *J. Abnorm. Psychol.,* 84, 66–75, 1975.
57. Foulkes, D., Hollifield, M., Sullivan, B., Bradley, L. et al., REM dreaming and cognitive skills at ages 5–8: a cross-sectional study, *Int. J. Behav. Develop.,* 13, 447–465, 1990.
58. Foulkes, D. and Pope, R., Primary visual experience and secondary cognitive elaboration in Stage R, *Percept. Motor Skills,* 37, 107–118, 1973.
59. Foulkes, D. and Rechtschaffen, A., Presleep determinants of dream content: effects of two films, *Percept. Motor Skills,* 19, 983–1005, 1964.
60. Foulkes, D. and Schmidt, M., Temporal sequence and unit composition in dream reports from different stages of sleep, *Sleep,* 6, 265–280, 1983.
61. Foulkes, D. and Scott, E., An above–zero baseline for the incidence of momentary hallucinatory mentation, *Sleep Res.,* 2, 108, 1973.
62. Foulkes, D., Spear, P.S., and Symonds, J.D., Individual differences in mental activity at sleep onset, *J. Abnorm. Psychol.,* 71, 280–286, 1966.
63. Foulkes, D. and Vogel, G., Mental activity at sleep onset, *J. Abnorm. Psychol.,* 70, 231–243, 1965.
64. Foulkes, W.D., Dream reports from different stages of sleep, *J. Abnorm. Psychol.,* 65, 14–25, 1962.
65. Freud, S., *The Interpretation of Dreams,* Basic Books, New York, 1958 (orig. 1900).
66. Goodenough, D.R., Dream recall: history and current status of the field, in *The Mind in Sleep,* Arkin, A.M., Antrobus, J.S., and Ellman, S.J., Eds., Lawrence Erlbaum Associates, Hillsdale, NJ, 1978, pp. 113–141.
67. Goodenough, D.R., Lewis, H.B., Shapiro, A., Jaret, L., and Sleser, I., Dream reporting following abrupt and gradual awakenings from different types of sleep, *J. Personality Social Psychol.,* 2, 170–179, 1965.

68. Goodenough, D.R., Lewis, H.B., Shapiro, A., and Sleser, I., Some correlates of dream reporting following laboratory awakenings, *J. Nerv. Ment. Dis.,* 140, 365–373, 1965.

69. Goodenough, D.R., Shapiro, A., Holden, M., and Steinschriber, L., A comparison of "dreamers" and "nondreamers": eye movements, electroencephalograms, and the recall of dreams, *J. Abnorm. Psychol.,* 59, 295–302, 1959.

70. Gordon, H.W., Frooman, B., and Lavie, P., Shift in cognitive asymmetries between wakings from REM and NREM sleep, *Neuropsychologia,* 20, 99–103, 1982.

71. Hall, C. and van de Castle, R.I., *The Content Analysis of Dreams,* Appleton-Century-Crofts, New York, 1966.

72. Hauri, P., Evening activity, sleep mentation, and subjective sleep quality, *J. Abnorm. Psychol.,* 76, 270–275, 1970.

73. Herman, J.H., Transmutative and reproductive properties of dreams: evidence for cortical modulation of brain–stem generators, in *The Neuropsychology of Sleep and Dreaming,* Antrobus, J.S. and Bertini, M., Eds., Lawrence Erlbaum Associates, Hillsdale, NJ, 1992, pp. 251–265.

74. Herman, J.H., Ellman, S.J., and Roffwarg, H.P., The problem of NREM recall re–examined, in *The Mind in Sleep,* Arkin, A.M., Antrobus, J.S., and Ellman, S.J., Eds., Lawrence Erlbaum Associates, Hillsdale, NJ, 1978, pp. 59–92.

75. Herman, J.H. et al., Evidence for a directional correspondence between eye movements and dream imagery in REM sleep, *Sleep,* 7, 52–63, 1984.

76. Hobson, J.A. and McCarley, R.W., The brain as a dream state generator: an activation-synthesis hypothesis of the dream process, *Am. J. Psychiatry,* 134, 1335–1348, 1977.

77. Hobson, J.A., *The Dreaming Brain: How the Brain Creates Both the Sense and the Nonsense of Dreams,* Basic Books, New York, 1988.

78. Hobson, J.A., A new model of brain-mind state: activation level, input source, and mode of processing (AIM), in *The Neuropsychology of Sleep and Dreaming,* Antrobus, J.S. and Bertini, M., Eds., Lawrence Erlbaum Associates, Hillsdale, NJ, 1992, pp. 227–247.

79. Hobson, J.A., Goldfrank, F., and Snyder, F., Respiration and mental activity in sleep, *J. Psychiatric Res.,* 3, 79–90, 1965.

80. Hobson, J.A. and Stickgold, R., Dreaming: a neurocognitive approach, *Cons. Cogn.,* 3, 1–15, 1994.

81. Hobson, J.A. and Stickgold, R., The conscious state paradigm: a neurocognitive approach to waking, sleeping, and dreaming, in *The Cognitive Neurosciences,* Gazzaniga, M.S., Ed., MIT Press, Cambridge, MA, 1995, pp. 1373–1389.

82. Hori, T., Hayashi, M., and Morikawa, T., Topography and coherence analysis of hypnagogic EEG, in *Sleep Onset: Normal and Abnormal Processes,* Ogilvie, R.D. and Harsh, J.R., Eds., American Psychological Association, Washington, D.C., 1994, pp. 237–253.

83. Hunt, H.T., Ruzycki–Hunt, K., Pariak, D., and Belicki, K., The relationship between dream bizarreness and imagination: artifact or essence?, *Dreaming,* 3, 179–199, 1993.

84. Hunt, H.T., *The Multiplicity of Dreams: Memory, Imagination, and Consciousness,* Yale University Press, New Haven, 1989.

85. Jouvet, M., *Le Sommeil et le Reve,* Editions Odile Jacob, Paris, 1992.

86. Jouvet, M., Michel, F., and Mounier, D., Analyse electroencephalographique comparee du sommeil physiologique chez le chat et chez l'homme, *Rev. Neurolog. (Paris),* 103, 189–204, 1960.

87. Kahan, T.L. and LaBerge, S., Lucid dreaming as metacognition: implications for cognitive science, *Cons. Cogn.,* 3, 246–264, 1994.

88. Kahn, A., Pace–Schott, E., Hobson, A., Kahn, D., Paceschott, E.F., and Hobson, J.A., Consciousness in waking and dreaming: the roles of neuronal oscillation and neuromodulation in determining similarities and differences, *Neuroscience,* 78, 13–38, 1997.

89. Kahn, E., Fisher, C., and Edwards, A., Night terrors and anxiety dreams, in *The Mind in Sleep*, 2nd ed., Ellman, S.J. and Antrobus, J.S., Eds., John Wiley & Sons, New-York, 1991, pp. 437–448.
90. Kales, A., Hoedemaker, F.S., Jacobson, A., Kales, J.D., Paulson, M.J., and Wilson, T.E., Mentation during sleep: REM and NREM recall reports, *Percept. Motor Skills*, 24, 555–560, 1967.
91. Kales, A., Jacobson, A., Paulson, M.J., Kales, J.D., and Walter, R.D., Somnambulism: psychophysiological correlates. I. All-night EEG studies, *Arch. Gen. Psychiatry*, 14, 586–594, 1966.
92. Kamiya, J., *Behavioral and Physiological Concomitants of Dreaming*, progress report to National Institute of Mental Health, 1962.
93. Karacan, I., Goodenough, D.R., Shapiro, A., and Starker, S., Erection cycle during sleep in relation to dream anxiety, in *The Content Analysis of Verbal Behavior*, Gottschalk, L.A., Ed., Spectrum, New York, 1979, PP. 887–893.
94. Karacan, I., Hursch, C.J., Williams, R.L., and Thornby, J.I., Some characteristics of nocturanl penile tumescence in young adults, *Arch. Gen. Psychiatry*, 26, 351–356, 1972.
95. Karni, A., Tanne, D., Rubenstein, B.S., Askenasy, J.J., and Sagi, D., Dependence on REM–sleep of overnight improvement of a perceptual skill, *Science*, 265, 679–682, 1994.
96. Klinger, E., *Meaning and Void*, University of Minnesota Press, Minneapolis, MN, 1977.
97. Kondo, T., Fein, G., and Antrobus, J.S., Late REM activation and sleep mentation, *Sleep Res.*, 18, 147, 1989.
98. Kremen, I., Dream Reports and Rapid Eye Movements, unpublished doctoral dissertation, Harvard University, Cambridge, MA, 1961.
99. Kutas, M., Event-related brain potential (ERP) studies of cognition during sleep, in *Sleep and Cognition*, Bootzin, R.R., Kihlstrom, J.F., and Schacter, D.L., Eds., American Psychological Association, Washington, D.C., 1990, pp. 43–57.
100. Laberge, S.P., Nagel, L.E., Dement, W.C., and Zarcone, V.P., Lucid dreaming verified by volitional communication during REM sleep, *Percept. Motor Skills*, 52, 727–732, 1981.
101. Labruzza, A.L., The activation-synthesis hypothesis of dreams: a theoretical note, *Am. J. Psychiatry*, 135, 1536–1538, 1978.
102. Lasaga, J.I. and Lasaga, A.M., Sleep learning and progressive blurring of perception during sleep, *Percept. Motor Skills*, 37, 51–62, 1973.
103. Lavie, P., Dreaming and REM sleep (commentary), *World Fed. Sleep Res. Soc. Newslett.*, 3, 14–15, 1994.
104. Lehmann, D., Grass, P., and Meier, B., Spontaneous conscious covert cognition states and brain electric spectral states in canonical correlations, *Int. J. Psychophysiol.*, 19, 41–52, 1995.
105. Lehmann, D. and Koukkou, M., Physiological and mental processes during sleep: a model of dreaming, in *Psychology of Dreaming*, Bosinelli, M. and Cicogna, P., Eds., Cooperativa Libraria Universitaria Editrice, Bologna, 1984, pp. 51–63.
106. Levin, R. and Livingston, G., Concordance between two measures of dream bizarreness, *Percept. Motor Skills*, 72, 837–838, 1991.
107. Llinas, R.R. and Paré, D., Of dreaming and wakefulness, *Neuroscience*, 44, 521–535, 1991.
108. Lloyd, S.R. and Cartwright, R.D., The collection of home and laboratory dreams by means of an instrumental response technique, *Dreaming*, 5, 63–73, 1995.
109. Mahowald, M.W. et al., Sleeping dreams, waking hallucinations and the central nervous system, *Dreaming*, 8, 89–102, 1998.
110. Mahowald, M.W. and Rosen, G.M., Parasomnias in children, *Pediatrician*, 17, 21–31, 1990.
111. Mahowald, M.W. and Schenck, C.H., Dissociated states of wakefulness and sleep, *Neurology*, 42(suppl. 6), 44–51, 1992.

112. Mahowald, M.W. and Schenck, C.H., REM sleep behavior disorder, in *Principles and Practice of Sleep Medicine,* 2nd ed., Kryger, M.H., Roth, T., and Dement, W.C., Eds., W.B. Saunders, Philadelphia, 1994, pp. 574–588.
113. Mamelak, A.N. and Hobson, J., Dream bizarreness as the cognitive correlate of altered neuronal behavior in REM sleep, *J. Cognitive Neurosci.,* 1, 201–222, 1989.
114. Marrocco, R.T., Witte, E.A., and Davidson, M.C., Arousal systems, *Curr. Opin. Neurobiol.,* 4, 166–170, 1994.
115. McCarley, R.W., Dreams and the biology of sleep, in *Principles and Practice of Sleep Medicine,* 2nd ed., Kryger, M.H., Roth, T., and Dement, W.C., Eds., W.B. Saunders, Philadelphia, 1994, pp. 373–383.
116. McCarley, R.W. and Hobson, J.A., The form of dreams and the biology of sleep, in *Handbook of Dreams. Research, Theory and Applications,* Wolman, B.B., Ed., Van Nostrand Reinhold, New York, 1979, pp. 76–130.
117. McGrath, M.J. and Cohen, D.B., REM sleep facilitation of adaptive waking behavior: a review of literature, *Psychol. Bull.,* 85, 24–57, 1978.
118. Merritt, J.M., Stickgold, R., Pace–Schott, E., Williams, J., and Hobson, J.A., Emotion profiles in the dreams of men and women, *Cons. Cogn.,* 3, 46–60, 1994.
119. Moffitt, A., Hoffmann, R., Wells, R., Armitage, R., Pigeau, R., and Shearer, J., Individual differences among pre– and post–awakening EEG correlates of dream reports following arousals from different stages of sleep, *Psychiatric J. Univ. Ottawa,* 7, 111–125, 1982.
120. Molinari, S. and Foulkes, D., Tonic and phasic events during sleep: psychological correlates and implications, *Percept. Motor Skills,* 29, 343–368, 1969.
121. Monroe, L.J., Rechtschaffen, A., Foulkes, D., and Jensen, J., Discriminability of REM and NREM reports, *J. Pers. Soc. Psychol.,* 2, 456–460, 1965.
122. Montangero, J., Dream production mechanisms and cognition, *New Ideas Psychol.,* 9, 353–365, 1991.
123. Nakano, S., Tsuji, S., Matsunaga, K., and Murai, Y., Effect of sleep stage on somatosensory evoked potentials by median nerve stimulation, *Electroencephalogr. Clin. Neurophysiol.,* 96, 385–389, 1995.
124. Nielsen, T.A., Germain, A., and Oullet, L., Atonia-signalled hypnagogic imagery: comparative EEG mapping of sleep onset transitions, REM sleep and wakefulness, *Sleep Res.,* 24, 133, 1995.
125. Nielsen, T.A., Kuiken, D.L., Hoffmann, R., and Moffitt, A.R., REM and NREM sleep mentation differences: a question of story structure?, *J. Sleep Res.* (submitted).
126. Nielsen, T.A. and Montplaisir, J., Dreaming and REM sleep (commentary), *World Fed. Sleep Res. Soc. Newslett.,* 3, 15–16, 1994.
127. Nielsen, T.A. and Powell, R.A., The "dream–lag" effect: a 6-day temporal delay in dream content incorporation, *Psychiatric J. Univ. Ottawa,* 14, 561–565, 1989.
128. Nielsen, T.A. and Zadra, A.L., EEG features distinguishing nightmare from non-nightmare REM episodes: elevated alpha power and right hemisphere asymmetries, *Sleep Res.,* 26, 250, 1997.
129. Noguchi, Y., Yamada, T., Yeh, M., Matsubar, M., Kokubun, Y., Kawada, J., Shiraish, G., and Kajimoto, S., Dissociated changes of frontal and parietal somatosensory-evoked potentials in sleep, *Neurology,* 45, 154–160, 1995.
130. Nordby, H., Hugdahl, K., Stickgold, R., Bronnick, K.S., and Hobson, J.A., Event-related potentials (ERPs) to deviant auditory stimuli during sleep and waking, *NeuroReport,* 7, 1082–1086, 1996.
131. Orlinsky, D., Psychodynamic and Cognitive Correlates of Dream Recall – A Study of Individual Differences, unpublished doctoral dissertation, University of Chicago, 1962.
132. Pivik, R.T., Mental Activity and Phasic Events During Sleep, unpublished Ph.D. dissertation, Stanford University, 1971.

133. Pivik, R.T., Sleep: physiology and psychophysiology, in *Psychophysiology: Systems, Processes, and Applications,* Coles, M.G.H., Donchin, E., and Porges, S.W., Eds., Elsevier, Amsterdam, 1986.
134. Pivik, R.T., Tonic states and phasic events in relation to sleep mentation, in *The Mind in Sleep. Psychology and Psychophysiology,* 2nd ed., Ellman, S. and Antrobus, J., Eds., John Wiley & Sons, New York, 1991, pp. 214–248.
135. Pivik, T. and Foulkes, D., NREM mentation: relation to personality, orientation time, and time of night, *J. Consult. Clin. Psychol.,* 32, 144–151, 1968.
136. Porte, H.S. and Hobson, J.A., Bizarreness in REM and NREM reports, *Sleep Res.,* 15, 81, 1986.
137. Porte, H.S., Slower eye movement in sleep, *Sleep Res.,* 26, 253, 1997.
138. Porte, H.S., REM dreaming versus non–REM dreaming: debate terminable or interminable?, *Dreaming* (submitted).
139. Porte, H.S. and Hobson, J.A., Physical motion in dreams — one measure of three theories, *J. Abnorm. Psychol.,* 105, 329–335, 1996.
140. Powell, R.A., Cheung, J.S., Nielsen, T.A., and Cervenka, T.M., Temporal delays in incorporation of events into dreams, *Percept. Motor Skills,* 81, 95–104, 1995.
141. Purcell, S., Mullington, J., Moffitt, A., Hoffmann, R., and Pigeau, R., Dream self-reflectiveness as a learned cognitive skill, *Sleep,* 9, 423–437, 1986.
142. Rechtschaffen, A., Dreaming and REM sleep (commentary), *World Fed. Sleep Res. Soc. Newslett.,* 3, 16–18, 1994.
143. Rechtschaffen, A., Goodenough, D., and Shapiro, A., Patterns of sleep talking, *Arch. Gen. Psychiatry,* 7, 418–426, 1962.
144. Rechtschaffen, A. and Kales, A., *A Manual of Standardized Terminology, Technique and Scoring System for Sleep Stages of Human Subjects,* HEW Neurological Information Network, Bethesda, MD, 1968.
145. Rechtschaffen, A., Verdone, P., and Wheaton, J., Reports of mental activity during sleep, *Can. Psychiatric Assoc. J.,* 8, 409–414, 1963.
146. Rechtschaffen, A., Vogel, G., and Shaikun, G., Interrelatedness of mental activity during sleep, *Arch. Gen. Psychiatry,* 9, 536–547, 1963.
147. Reinsel, R., Antrobus, J., and Wollman, M., Bizarreness in dreams and waking fantasy, in *The Neuropsychology of Sleep and Dreaming,* Antrobus, J.S. and Bertini, M., Eds., Lawrence Erlbaum Associates, Hillsdale, NJ, 1992, pp. 157–185.
148. Reinsel, R., Wollman, M., and Antrobus, J.S., Effects of environmental context and cortical activation on thought. Special issue: cognition and dream research, *J. Mind Behav.,* 7, 259–275, 1986.
149. Resnick, J., Stickgold, R., Rittenhouse, C.D., and Hobson, J.A., Self–representation and bizarreness in children's dream reports collected in the home setting, *Cons. Cogn.,* 3, 30–45, 1994.
150. Riemann, D., Löw, H., Schredl, M., Wiegand, M., Dippel, B., and Berger, M., Investigations of morning and laboratory dream recall and content in depressive patients during baseline conditions and under antidepressive treatment with trimipramine, *Psychiatric J. Univ. Ottawa,* 15, 93–99, 1990.
151. Rittenhouse, C.D., Stickgold, R., and Hobson, J.A., Constraint on the transformation of characters, objects, and settings in dream reports, *Cons. Cogn.,* 3, 100–113, 1994.
152. Robbins, T.W. and Everitt, B.J., Arousal systems and attention, in *The Cognitive Neurosciences,* Gazzaniga, M.S., Ed., MIT Press, Cambridge, 1995, pp. 703–720.
153. Rochlen, A., Hoffmann, R., and Armitage, R., EEG correlates of dream recall in depressed outpatients and healthy controls, *Dreaming,* 8, 109–123, 1998.
154. Roschke, J., Prentice-Cuntz, T., Wagner, P., Mann, K., and Frank, C., Amplitude frequency characteristics of evoked potentials during sleep: an analysis of the brain's transfer properties in depression, *Biol. Psychiatry,* 40, 736–743, 1996.

155. Rotenberg, V.S., The estimation of sleep quality in different stages and cycles of sleep, *J. Sleep Res.,* 2, 17–20, 1993.

156. Roussy, F., Camirand, C., Foulkes, D., Dekoninck, J., Loftis, M., and Kerr, N.H., Does early-night REM dream content reliably reflect presleep state of mind?, *Dreaming,* 6, 121–130, 1996.

157. Rubinstein, F., Persistent sexual symbolism: Shakespeare and Freud, *Lit. Psychol.,* 34, 1–26, 1988.

158. Salisbury, D.F., Stimulus processing awake and asleep: similarities and differences in electrical CNS responses, in *Sleep Onset: Normal and Abnormal Processes,* Ogilvie, R.D. and Harsh, J.R., Eds., American Psychological Association, Washington, D.C., 1994, pp. 289–308.

159. Salzarulo, P. and Cipolli, C., Linguistic organization and cognitive implication of REM and NREM sleep-related reports, *Percept. Motor Skills,* 49, 767–777, 1979.

160. Scholz, U.J., Bianchi, A.M., Cerutti, S., and Kubicki, S., Vegetative background of sleep — spectral analysis of the heart rate variability, *Physiol. Behav.,* 62, 1037–1043, 1997.

161. Schredl, M. and Montasser, A., Dream recall: state or trait variable? Part II. State factors, investigations and final conclusions, *Imag. Cogn. Pers.,* 16, 227–261, 1997.

162. Seligman, M.E. and Yellen, A., What is a dream?, *Behav. Res. Ther.,* 25, 1–24, 1987.

163. Shapiro, A., Goodenough, D.R., and Gryler, R.B., Dream recall as a function of method of awakening, *Psychosom. Med.,* 25, 174–180, 1963.

164. Slover, G.P.T., Morris, R.W., Stroebel, C.F., and Patel, M.K., Case study of psychophysiological diary: infradian rhythms, in *Advances in Chronobiology,* Pauley, J.E. and Scheving, L.E., Eds., Alan R. Liss, New York, 1987, pp. 439–449.

165. Smith, C., Sleep states and memory processes, *Behav. Brain Res.,* 69, 137–145, 1995.

166. Smith, C. and MacNeill, C., Impaired motor memory for a pursuit rotor task following stage 2 sleep loss in college students, *J. Sleep Res.,* 3, 206–213, 1994.

167. Snyder, F., The organismic state associated with dreaming, in *Psychoanalysis and Current Biological Thought,* Greenfield, N. and Lewis, W., Eds., University of Wisconsin Press, Madison, 1965, pp. 275–315.

168. Snyder, F., Progress in the new biology of dreaming, *Am. J. Psychiatry,* 122, 377–390, 1965.

169. Snyder, F.The phenomenology of dreaming, in *The Psychoanalytic Implications of the Psychophysiological Studies on Dreams,* Madow, L. and Snow, L.H., Eds., Charles C Thomas, Springfield, IL, 1970, pp. 124–151.

170. Solms, M., New findings on the neurological organization of dreaming: implications for psychoanalysis [review], *Psychoanalyt. Q.,* 64, 43–67, 1995.

171. Squires, N.K. and Knowlton, R., Memory systems, in *The Cognitive Neurosciences,* Gazzaniga, M.S., Ed., MIT Press, Cambridge, MA, 1995, pp. 825–837.

172. Steriade, M. and McCarley, R.W., Dreaming, in *Brainstem Control of Wakefulness and Sleep,* Steriade, M. and McCarley, R.W., Eds., Plenum Press, New York, 1990, pp. 415–433.

173. Stickgold, R., Pace–Schott, E., and Hobson, J.A., A new paradigm for dream research: mentation reports following spontaneous arousal from REM and NREM sleep recorded in a home setting, *Cons. Cogn.,* 3, 16–29, 1994.

174. Stoyva, J., The Effects of Suggested Dreams on the Length of Rapid Eye Movement Periods, unpublished doctoral dissertation, University of Chicago, 1961.

175. Stoyva, J.M., Finger electromyographic activity during sleep: its relation to dreaming in deaf and normal subjects, *J. Abnorm. Psychol.,* 70, 343–349, 1965.

176. Strauch, I., The effects of meaningful acoustic stimuli on waking mentation and dreams, in *Sleep '86,* Koella, W.P., Obál, Jr., F., Schulz, H., and Visser, P., Eds., Gustav Fischer Verlag, Stuttgart, 1988, pp. 87–90.

177. Strauch, I. and Meier, B., *In Search of Dreams: Results of Experimental Dream Research*, State University of New York Press, New York, 1996.
178. Sutton, J.P., Rittenhouse, C.D., Pace–Schott, E., Merritt, J.M., Stickgold, R., and Hobson, J.A., Emotion and visual imagery in dream reports: a narrative graphing approach, *Cons. Cogn.*, 3, 89–99, 1994.
179. Sutton, J.P., Rittenhouse, C.D., Pace–Schott, E., Stickgold, R., and Hobson, J.A., A new approach to dream bizarreness: graphing continuity and discontinuity of visual attention in narrative reports, *Cons. Cogn.*, 3, 61–88, 1994.
180. Tilley, A.J., Sleep learning during stage 2 and REM sleep, *Biol. Psychol.*, 9, 155–161, 1979.
181. Tracy, R.L. and Tracy, L.N., Reports of mental activity from sleep stages 2 and 4, *Sleep Res.*, 2, 125, 1973.
182. van Sweden, B., van Dijk, J.G., and Caekebeke, J.F.V., Auditory information procession in sleep: late cortical potentials in an oddball paradigm, *Neuropsychobiology*, 29, 152–156, 1994.
183. Verdone, P., Temporal reference of manifest dream content, *Percept. Motor Skills*, 20, 1253–1268, 1965.
184. Vogel, G., Foulkes, D., and Trosman, H., Ego functions and dreaming during sleep onset, *Arch. Gen. Psychiatry*, 14, 238–248, 1966.
185. Vogel, G.W., Sleep-onset mentation, in *The Mind in Sleep*, Arkin, A.M., Antrobus, J.S., and Ellman, S.J., Eds., Lawrence Erlbaum Associates, Hillsdale, NJ, 1978, pp. 97–108.
186. Vogel, G.W., An alternative view of the neurobiology of dreaming, *Am. J. Psychiatry*, 135, 1531–1535, 1978.
187. Vogel, G.W., Sleep-onset mentation, in *The Mind in Sleep*, 2nd ed., Ellman, S.J. and Antrobus, J.S., Eds., John Wiley & Sons, New York, 1991, pp. 125–136.
188. Wasserman, M.D., Psychoanalytic dream theory and recent neurobiological findings about REM sleep, *J. Am. Psychoanalyt. Assoc.*, 32, 831–846, 1984.
189. Waterman, D., *Rapid Eye Movement Sleep and Dreaming: Studies of Age and Activation*, University of Amsterdam, 1992.
190. Waterman, D., Elton, M., and Kenemans, J.L., Methodological issues affecting the collection of dreams, *J. Sleep Res.*, 2, 8–12, 1993.
191. Weinstein, L.N., Schwartz, D.G., and Ellman, S.J., Sleep mentation as affected by REM deprivation: A new look, in *The Mind in Sleep*, 2nd ed., Ellman, S.J. and Antrobus, J.S., Eds., John Wiley & Sons, New York, 1991, pp. 377–395.
192. Winget, C. and Kramer, M., *Dimensions of Dreams*, University Presses of Florida, Gainesville, 1979.
193. Wollman, M.C. and Antrobus, J.S., Cortical arousal and mentation in sleeping and waking subjects, *Brain Cognition*, 6, 334–346, 1987.
194. Wolpert, E.A., Studies in psychophysiology of dreams. II. An electromyographic study of dreaming, *Arch. Gen. Psychiatry*, 2, 121–131, 1960.
195. Wolpert, E.A. and Trosman, H., Studies in psychophysiology of dreams. I. Experimental evocation of sequential dream episodes, *Arch. Neurol. Psychiatry*, 79, 603–606, 1958.
196. Wood, J.M., Sebba, D., and Domino, G., Do creative people have more bizarre dreams? A reconsideration, *Imag. Cogn. Pers.*, 9, 3–16, 1989.
197. Zimmerman, W.B., Sleep mentation and auditory awakening thresholds, *Psychophysiology*, 6, 540–549, 1970.

Chapter 9

Narcolepsy

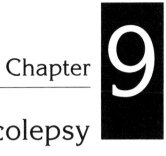

Emmanuel Mignot and Seiji Nishino

Contents

9.1	Clinical Aspects of Narcolepsy	129
	9.1.1 Prevalence of Narcolepsy and Natural History	129
	9.1.2 Excessive Daytime Sleepiness	130
	9.1.3 Cataplexy, a Pathognomonic Symptom for Narcolepsy	130
	9.1.4 Sleep Paralysis and Hypnagogic Hallucinations	130
	9.1.5 Frequently Associated Disorders	131
	9.1.6 Usual Diagnostic Procedures	131
	9.1.7 The Narcolepsy Spectrum	131
9.2	Treatment	131
9.3	Neuropharmacology and Neurochemistry of Narcolepsy	132
9.4	Genetic Aspects of Human Narcolepsy	134
9.5	HLA DQB1*0602 and DQA1*0102 Predispose to Human Narcolepsy	135
9.6	Narcolepsy and Autoimmunity	135
9.7	Non-HLA Genes and Narcolepsy	136
9.8	Conclusion	137
References		137

9.1 Clinical Aspects of Narcolepsy

Narcolepsy is a neurological disorder characterized by excessive daytime somnolence (EDS) and symptoms of abnormal REM sleep, namely cataplexy, sleep paralysis, and hypnagogic hallucinations.[1,3,16,53] Nighttime sleep is also typically disrupted in these patients, sometimes resulting in disabling insomnia. Recent well-documented clinical descriptions of the syndrome can be found in Bassetti and Aldrich.[3]

9.1.1 Prevalence of Narcolepsy and Natural History

Narcolepsy is the second most frequent cause of daytime sleepiness diagnosed in sleep disorder centers, after obstructive sleep apnea syndrome (OSAS). It is the most frequent neurological disorder primarily associated with daytime sleepiness. Narcolepsy has long been considered to be a rare condition, but population-based prevalence studies have demonstrated that narcolepsy/cataplexy affects 0.026 to 0.067% of the general population in Western Europe and the U.S.[23,42] The disorder may be less frequent in Israel (0.002%)[29] but more prevalent in Japan, where two prevalence studies have led to the conclusions that 0.16 and 0.18% of the population having narcolepsy.[19,68] A less strict definition of narcolepsy, including some cases of hypersomnia without cataplexy, would lead to a higher prevalence for all these studies. This suggests that narcolepsy is at least as prevalent as multiple sclerosis and that most patients remain undiagnosed and untreated. The symptoms usually begin to appear during adolescence but are occasionally seen in prepubertal children as young as 2 to 5 years old.[10] It is most frequently diagnosed many years after onset, i.e., during adulthood. It is a life-long condition and, because of the nature of the symptoms, has severe psychosocial consequences.[6,7]

9.1.2 Excessive Daytime Sleepiness

Sleepiness is a prerequisite for diagnosing narcolepsy in international classifications,[69] although patients occasionally experience isolated cataplexy or other abnormal symptoms of REM sleep before developing daytime sleepiness. Patients frequently nap several times per day and experience recurrent sleep attacks. Sleep attacks may occur in the middle of normal daily activities and lead to episodes of "automatic behavior" such as writing or talking in an unintelligible manner. Microsleep episodes ("blanking out") frequently occur. EDS in narcolepsy is not always distinguishable from the sleepiness caused by other sleep disorders, such as OSAS or other conditions accompanied by poor nocturnal sleep. Sleepiness in narcolepsy is relieved by short naps, but the refreshed feeling narcoleptic subjects experience after waking up typically vanishes rapidly.

9.1.3 Cataplexy, a Pathognomonic Symptom for Narcolepsy

In international classifications, cataplexy is defined as sudden episodes of bilateral loss of postural muscle tone in association with intense emotions.[69] This relatively broad definition may not be specific enough to define genuine cataplexy; in questionnaire surveys,[19,23] 6 to 30% of the general population report experiencing cataplexy-like symptoms. In cataplexy associated with narcolepsy, muscle weakness is usually brief and bilateral and occurs several times per month, unless treatment has been initiated. Knee buckling, head dropping, facial muscle jerking, or jaw sagging are the most typical cataplectic attacks. Positive emotions associated with joking, laughter, humor, and game playing are by far the most typical triggers. Anger is also often cited, but it is rarely the only trigger for cataplexy. Attacks triggered by stress or fear or those occurring in the context of athletic or sexual activities are the least specific. In severe episodes, patients will lose their muscle tone entirely, fall progressively to the ground, and remain paralyzed for a few minutes. Reminiscent of REM sleep atonia, monosynaptic reflexes are abolished during the episode.[16] Consciousness is always conserved at the onset of cataplexy, but patients may occasionally experience dream-like hallucinations during long and severe attacks. Isolated cataplexy without somnolence is rare and is most often observed in children before the development of full-blown narcolepsy. The importance of well-defined cataplexy, as described above, for the diagnosis of narcolepsy should be emphasized.

9.1.4 Sleep Paralysis and Hypnagogic Hallucinations

Sleep paralysis and hypnagogic hallucinations are frequently associated with narcolepsy, but these symptoms have poor diagnostic predictive value. During sleep paralysis, patients are unable to move for a few minutes, most often when awakening or while falling asleep. Hypnagogic hallucinations are dream-like experiences, most frequently visual or auditory, occurring at sleep onset. Both experiences occur in the normal population independently of narcolepsy. Isolated episodes of sleep paralysis are reported in 4 to 60% of the general population, these episodes being frequently associated with sleep deprivation.[12] It is thus important to distinguish muscle weakness when it occurs after emotions (cataplexy) or during sleep-stage transitions (sleep paralysis). Hypnagogic hallucinations also frequently occur in non-narcoleptic subjects. In a survey in the U.K., 37% of the general population experienced hypnagogic hallucinations at least twice a week over the past year.[51]

9.1.5 Frequently Associated Disorders

Narcolepsy often occurs in association with other sleep disorders, such as periodic leg movements,[15,72] REM behavior disorder and various other parasomnias,[32,60] and OSAS.[17] Depression, anxiety, and various other neuropsychiatric complications are also observed in many narcoleptic patients.[6,7]

9.1.6 Usual Diagnostic Procedures

The Multiple Sleep Latency Test (MSLT) is used in the U.S. or Western European countries to confirm the diagnosis. In this test, nocturnal polysomnography is first performed to exclude other possible causes of EDS and to control for nighttime sleep amount and quality. This is followed the next day by four to five naps during which sleep latency is measured and the occurrence of REM sleep is noted.[9] A mean sleep latency of 5 minutes during these naps indicates pathological sleepiness, whereas a mean sleep latency of 5 to 10 minutes is considered borderline. Narcoleptic patients also frequently exhibit REM episodes during these short naps (sleep onset REM periods, or SOREMPs), with two or more SOREMPS in four to five naps being considered diagnostic of narcolepsy.

9.1.7 The Narcolepsy Spectrum

Many patients do not experience the full-blown narcolepsy/cataplexy syndrome even though they exhibit typical MSLT results (SL ≤ 5 or 8 min and ≥2 SOREMPs), and the question of where the "narcolepsy spectrum" stops has been the object of much debate recently.[2,46] This is a very important issue, as a diagnosis of narcolepsy may lead the clinician to stop the search for other possible causes of EDS. The current consensus, based on clinical and human leukocyte antigen (HLA) genotyping data, is that narcolepsy with definite cataplexy is a homogeneous etiological entity, whereas "hypersomnias" without or with atypical/doubtful cataplexy may represent a more heterogeneous group even in the presence of abnormal MSLT results.[2,41] It is thus important to examine patients with sleepiness and no cataplexy thoroughly for any other possible cause of EDS before giving the diagnosis of narcolepsy.

9.2 Treatment

Narcolepsy is best managed using a multi-faceted, individually tailored approach. Patient education in sleep hygiene, nutrition, and scheduled napping and a good understanding of the disorder are helpful, but in most cases are not sufficient to control the problem. Pharmacological treatment is required in most cases, with large interindividual differences in treatment response and development of side effects. The pharmacological treatment of narcolepsy is beyond the scope of this chapter and has been thoroughly reviewed elsewhere.[50] Abnormal REM sleep (i.e., cataplexy) is usually treated using low doses of antidepressant compounds, most typically protriptyline, imipramine, or clomipramine, while amphetamine-like stimulants are used to control excessive daytime sleepiness.

9.3 Neuropharmacology and Neurochemistry of Narcolepsy

The mode of action of the currently available treatments for narcolepsy has been extensively studied using a canine model of narcolepsy/cataplexy.[50] All available compounds alleviate the symptoms of narcolepsy via their effects on monoaminergic transmission. Anticataplectic compounds are mostly acting on abnormal REM sleep by blocking norepinephrine uptake and thus increasing adrenergic tone in the central nervous system (CNS).[35,48,50] Stimulant compounds relieve sleepiness via their presynaptic effects on the release and reuptake of dopamine.[48,50]

Cholinergic systems are also involved in the control of abnormal REM sleep in narcolepsy. Cholinergic stimulation using either cholinesterase inhibitors or direct cholinergic agonists exacerbate canine cataplexy, while anticholinergic drugs administered at high doses have a beneficial effect in narcoleptic canines.[48,50] Anticholinergic drugs have been suggested to be helpful in some human patients, but cardiovascular side effects hamper the use of active drug dose ranges in human patients. These findings, however, have important theoretical implications, as brainstem and basal forebrain cholinergic systems are well known to be activated during natural REM sleep. It also offers the possibility that centrally active anticholinergic compounds having some therapeutic effects on narcolepsy could be discovered in the future.

In order to identify the neural structures involved in the pharmacological control of sleepiness and cataplexy, studies have been performed where small doses of pharmacological agents are injected or perfused in specific brain areas and the behavioral effects observed in narcoleptic animals *in vivo*.[48–50,54–56] Perfusion experiments using a technique called *in vivo* dialysis are also used to measure neurotransmitter levels during cataplexy or sleep or after drug administration in freely moving narcoleptic or control Dobermans.[55,56] Other experiments have used neurochemical measurements in brain regions of narcoleptic and control animals[44] or have aimed at identifying patterns of neuronal degeneration in the CNS of narcoleptic animals.[63–64] A schematic summary of the model obtained is depicted in Figure 9.1.

Results obtained to date have shown that a widespread hypersensitivity to cholinergic stimulation in the brainstem and the basal forebrain underlie abnormal REM sleep in narcolepsy.[48–50,54–55] In the pons, a region where cholinergic stimulation is known to produce REM sleep or REM sleep atonia in control animals, cholinergic agonists at doses 10 times lower than those used to induce atonia in control animals induce cataplexy in narcoleptic animals.[54] Narcoleptic animals are thus hypersensitive to cholinergic stimulation in a region often considered to be primary for the generation of REM sleep. More surprisingly, we also recently found that cataplexy can be induced by moderate levels of cholinergic stimulation in the basal forebrain (BF) area (diagonal band of Brocca);[49] this area is not known for its role in normal REM sleep control, but bilateral perfusion of very high non-physiological doses of carbachol, a cholinergic agonist, can also induce REM sleep atonia-like symptoms in control animals.[49] This suggests that cholinergic hypersensitivity is widespread in the CNS of narcoleptic animals, a result that could explain why narcoleptic subjects have abnormal REM tendencies. BF neurons are known to respond to the arousing quality of appetitive stimuli in experimental animals.[57,59] We also observed that acetylcholine release is increased in this area during food presentation in control canines.[50] Increased cholinergic release in the BF during emotions could thus more specifically trigger cataplexy in narcolepsy.

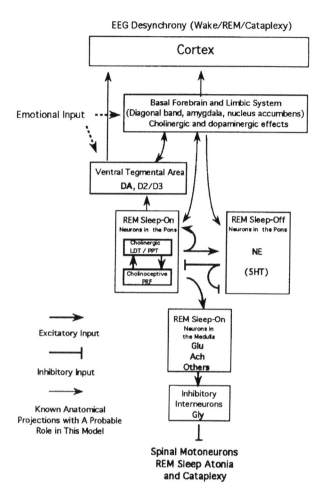

FIGURE 9.1

Neurochemical mechanisms involved in canine narcolepsy (hypothetical model). Pontine cholinergic systems are one of the last relays in the generation of cataplexy and abnormal REM sleep in narcolepsy, as they are for normal REM sleep in healthy individuals. Cataplexy or REM-like behaviors can be induced by the injection of M_2 (and M_3) agonists into the PRF in normal canines,[54] but narcoleptic animals are more sensitive to the same pharmacological manipulation. We have also found that acetylcholine release in the PRF (the cholinergic input comes from the LDT/PPT) is increased during spontaneous cataplexy in narcoleptic canines,[55] suggesting that cholinergic systems in the PRF are activated during cataplexy as they are during natural REM sleep. The model of reciprocal interaction of pontine cholinergic and aminergic systems for the regulation of REM sleep[26,33] underlies the adrenergic/cholinergic balance observed for the pharmacological control of cataplexy. It is thus believed that neuronal activity from cholinergic and cholinoceptive neurons in LDT/PPT and PRF, respectively, after synaptic relay in the medullary reticular formation, further descends through the spinal cord and inhibits the motoneurons in the spinal cord, producing muscle atonia in REM sleep or, in narcolepsy, cataplexy.[61,62] Our second major experiment finding is that carbachol injections in the BF induce cataplexy in narcoleptic dogs while inducing alertness in control dogs.[49] Carbachol injections can also induce cataplexy-like behavior in control dogs, but very high doses (about 10 times) are required to induce muscle atonia in these animals.[49] Cholinergic hypersensitivity of forebrain structures is thus also involved in narcolepsy. Because the BF is a region known to be tightly connected with the limbic system, acetylcholine release in this area could be triggered by limbic input during emotional excitement, thus resulting in cataplexy and a global REM sleep-like activation of cholinergic systems in narcoleptic canines. Midbrain dopaminergic systems are also critical to the expression of excessive daytime sleepiness and cataplexy in narcolepsy. Stimulation of dopamine release and re-uptake blockage are two key pharmacological properties mediating the EEG-activating effects of available stimulants.[50] Dopamine autoreceptor stimulation in the VTA results in increased sleepiness and cataplexy.[56] The cataplexy-inducing effect of the latter manipulation might be due to mesolimbic dopaminergic projections and their established role in mediating the expression of pleasurable stimuli.[45] Ach = acetylcholine; BF = basal forebrain; DA = dopamine; D2/D3 = dopaminergic D2/D3 receptors; NE = norepinephrine; LDT/PPT = laterodorsal tegmentum and pedunculopontine tegmentum; PRF = pontine reticular formation; VTA = ventral tegmental area.

Other experiments emphasize the importance of dopaminergic systems in the pathophysiology of narcolepsy.[50,56] As mentioned above, increased dopaminergic transmission via stimulation of release or inhibition of re-uptake likely mediates the EEG alerting effects of most currently available stimulant medications. Increased dopaminergic receptors and changes in dopamine content have been reported in various brain regions in narcoleptic canines, with special emphasis on the amygdala.[44] We also observed that dopamine autoreceptor stimulation, a manipulation that decreases dopaminergic tone in the central nervous system, increases cataplexy and sleepiness in narcoleptic animals.[56] This result was obtained after systemic administration or local perfusion of D_2/D_3 autoreceptor compounds within the ventral tegmental area (VTA), a region in which dopaminergic cell bodies are densely packed. Because midbrain dopaminergic neurons do not change their activities across the sleep/wake cycle in cats and rats,[43,66] it has long been believed that dopaminergic systems are less important than other monoaminergic systems for sleep-wake regulation. The finding that autoreceptor stimulation in the VTA produces cataplexy and sleepiness in narcoleptic canines, however, suggests that midbrain dopaminergic systems are critical for the control of EDS. Midbrain dopaminergic neurons have also recently been shown to be important for experiencing pleasurable emotions.[45] The involvement of the dopaminergic mesolimbic system in the pathophysiology of narcolepsy could thus not only explain daytime sleepiness in narcolepsy but might also explain why cataplexy is primarily associated with positive emotions.

The model is also supported by recent findings indicating neuronal degeneration in the BF area and in the amygdala,[63,64] two regions where previous neurochemical studies have found significant alterations in monoamine content in narcoleptic animals.[44]

9.4 Genetic Aspects of Human Narcolepsy

Since its description in the nineteenth century by Westphal[71] and Gelineau,[14] narcolepsy has been reported in a familial context, thus suggesting a genetic etiology for the disorder. More recent studies suggest, however, that narcolepsy is not a simple genetic disease in humans.[42] Only 8 to 10% of all narcoleptic patients can identify another member of the family with narcolepsy/cataplexy, and the familial risk of a first-degree relative of a patient with narcolepsy/cataplexy is only 1 to 2%.[42] This risk is low, and in most cases narcoleptic patients can be reassured that their relatives are unlikely to develop the illness. A risk of 1 to 2% is, however, still 10 to 40 times higher than the prevalence observed in the general population, suggesting the existence of predisposing genetic factors.

The risk for a relative to develop EDS or isolated symptoms of abnormal REM sleep is more uncertain. In a recent study, Billiard[5] found that 4.7% of first-degree relatives of patients with narcolepsy/cataplexy were excessively sleepy without any other obvious clinical reasons (e.g., OSAS, periodic leg movements). This value is remarkably similar to that reported in several other family studies for narcolepsy-like symptoms in first-degree relatives, including studies done in Japan;[20,42] however, it is difficult to conclude from these studies that genetic factors are involved in predisposing relatives of narcoleptic patients to excessive sleepiness, as the prevalence of EDS in the general population may be as high as 13%.[11,24]

Non-genetic factors are also involved in the pathophysiology of human narcolepsy. Twin studies using established registries have not been performed, with the exception of a Finnish study which identified three discordant dizygotic twins in a sample of 11,354 twin individuals.[23] Available data suggest that most but not all monozygotic twin pairs are discordant for narcolepsy. Of 16 monozygotic pairs reported in the literature, 25 to 31% are concordant for narcolepsy/cataplexy, depending on how strictly concordance is determined from available published information.[42,52] This indicates that in most human cases, narcolepsy results from the interaction of environmental factors on a specific genetic background.

9.5 HLA DQB1*0602 and DQA1*0102 Predispose to Human Narcolepsy

In 1983, narcolepsy/cataplexy was shown to be tightly associated with HLA DR2 in Japan.[21] This association was quickly confirmed in Caucasians, with 85 to 98% of all patients being DR2 positive.[4,28,47] Further studies have now established that across all ethnic groups, the association is tightest with the DQB1 allele (DQB1*0602) rather than with DR2, an allele that is also found in 12 to 40% of the control population in various ethnic groups.[22,31,36,41,58] This is especially important in the black population, for whom DR2 is a less sensitive marker for narcolepsy.[31,36,41,58]

The association is much tighter when narcolepsy is clinically defined by the presence of clear-cut cataplexy.[22,41,58] For example, in a recent, multi-center study involving the clinical trial of the compound modafinil in 509 HLA-typed narcoleptic patients, 76% vs. only 41% DQB1*0602 positivity was reported in patients with cataplexy (n = 421) vs. those without cataplexy (n = 88), respectively.[24] DQB1*0602 positivity was even higher (85 to 100%) when cataplexy was strictly defined clinically.[41] This result suggests that narcolepsy/cataplexy is a more homogeneous etiological entity than narcolepsy without cataplexy. The 41% frequency found in the narcolepsy-without-cataplexy group is, however, still higher than the corresponding control DQB1*0602 allele frequencies (approximately 24% in a matched sample). This suggests the existence of some genuine narcolepsy cases in the group without cataplexy (approximately 25% in this sample), but the prevalence of this symptomatic variant in the general population remains to be established.

Studies using additional genetic markers in the HLA DQ region have now established that HLA DQB1*0602 and another HLA gene allele located nearby, HLA DQA1*0102, are the actual narcolepsy susceptibility genes located in the HLA region.[13,40,42] Multiple CA repeat markers have been isolated and characterized in the region to define the shortest genomic segment conserved in susceptibility haplotypes across all ethnic groups.[13,40] Results demonstrate that only the areas surrounding the coding region of DQA1 and DQB1 are conserved across all susceptibility haplotypes, but polymorphisms can be observed in the area flanking DQA1*0102 and DQB1*0602 and between these two DQ genes (Figure 9.2). Furthermore, the HLA DQ region has now been entirely sequenced in control subjects and no other candidate genes have been observed.[13] The results to date thus suggest that the HLA DQA1*0102/DQB1*0602 (DQ1) heterodimer molecule is the main predisposing factor in the HLA class II region.[13,40,41]

9.6 Narcolepsy and Autoimmunity

The finding of an HLA association in narcolepsy suggests that the immune system is involved in the pathophysiology of narcolepsy, as most diseases with an HLA association are autoimmune in nature. Since 1983, however, all attempts to demonstrate that human narcolepsy is an autoimmune disease have failed.[8,38] It has not been possible to demonstrate any inflammatory process in the central nervous system of narcoleptic patients (CSF oligoclonal bands, cellular infiltration), and none of the systemic immune abnormalities usually found in autoimmune pathologies (sedimentation rate, complement and serum immunoglobulin levels, lymphocyte subsets ratios, C-reactive protein, autoantibodies) could be detected.[8,38] We also recently carried out bone marrow transplantation studies in two adult canine littermates with and without narcolepsy and could not induce disease susceptibility/resistance after adoptive transfer of the peripheral immune system, as it is usually the case in typical autoimmune diseases.[37] Finally, the brain of several human and canine patients was also systematically studied using immunocytochemical techniques to search for localized abnormal HLA expression, but this yielded no positive results.[67] Interestingly, however, HLA expression was increased diffusely in the white matter of narcoleptic canines, the increase corresponding mainly to enhanced expression by microglial cells.[67]

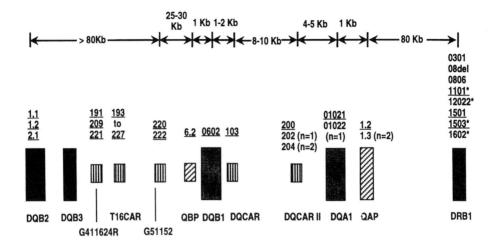

FIGURE 9.2
Schematic summary of the narcolepsy susceptibility region within the HLA region. Genes and markers are depicted by the vertical bars; alleles observed in narcoleptic patients are listed above each marker. DQB2, DQB3, DQB1, DQA1, DRB1 are HLA genes and pseudogenes. QAP and QBP are designations for the promoter region of the DQA1 and DQB1 genes, respectively. T16CAR, G411624R, G51152, DQCAR, and DQCARII are microsatellite CA repeats identified in the HLA class II region.[40] Allelic variation in the DRB1 and DQB2/DQB3 region is large, thus indicating that these genomic segments are not primarily involved in narcolepsy susceptibility. In contrast, the DQA1*0102 to DQB1*0602 DNA segment is relatively conserved across all susceptibility haplotypes. Additional diversity is also found at the level of G51152, DQCARII, and QAP, but in all cases the coding areas of DQA1*0102 and DQB1*0602 seem to be maintained. Note that both DQA1*01021 and DQA1*01022 were observed; these two HLA alleles have a silent coding difference at position 109 (ATT to ATC) and thus encode the same protein sequence. * = predominantly found in the black population. Frequently observed alleles are underlined; other variants are rare. For details, see Mignot.[40]

9.7 Non-HLA Genes and Narcolepsy

Whether or not narcolepsy is autoimmune, HLA DQA1*0102 and DQB1*0602 are neither necessary nor sufficient for the development of narcolepsy. Of the general population, 12 to 35% carry the exact same HLA alleles, but only 0.026 to 0.18% of the general population have narcolepsy. These alleles have been sequenced in narcoleptic individuals and were found to be identical to the reference sequence.[30,36,40,70] Typical cases of narcolepsy-with-cataplexy without these alleles have also been reported.[18,36,40–41,47,58,65]

Family studies in narcolepsy also underscore the role of HLA. Studies suggest that familial cases may be more frequently DQB1*0602 negative than are sporadic cases, with up to one third of narcoleptic probands of multiplex families being negative for this allele.[18,39,42,65] These non-DQB1*0602 families frequently display significant familial aggregation with no apparent linkage with HLA.[18,39,42,65] This data, together with the fact that two of the five reported concordant twin pairs are DQB1*0602 negative,[42,52] suggest increased genetic penetrance in some non-DQB1*0602 negative narcolepsy cases. Even in DQB1*0602-positive multiplex families, susceptibility does not seem to be linked to a specific HLA DQB1*0602 haplotype, further suggesting that most, if not all, DQB1*0602 haplotypes of the general population predispose equally to narcolepsy.[42] Overall, HLA contributes only minimally to the 10- to 40-fold increase in susceptibility observed in first-degree relatives (estimated lambda HLA is only 2), and other non HLA-related genetic factors are likely to be involved.

Canine narcolepsy offers a unique opportunity to identify one of these non-HLA narcolepsy genes. In this animal model, the major susceptibility gene, *canarc-1*, is unlinked to dog leucocyte antigen (DLA) polymorphisms.[34] A linkage study was therefore undertaken, and our results now indicate that *canarc-1* is tightly linked with a DNA segment with high homology to the human immunoglobulin switch gene (current LOD score 17.2 at 0% recombination).[34]

The genomic region containing the *canarc-1* marker has now been cloned and mapped in canines. Fluorescence *in situ* hybridization (FISH) indicates that *canarc-1* is located on a different canine chromosome than the canine immunoglobulin switch loci. The corresponding region syntenic in humans has been identified, and polymorphic markers are being developed in the region. This should allow us to test the involvement of the human equivalent of *canarc-1* in human narcoleptic patients (sporadic and familial cases) in the near future. A canine bacterial artificial chromosome (BAC) library has also been constructed, and the corresponding genomic segments are also being searched to identify *canarc-1* candidate genes in canines.

9.8 Conclusion

Although all attempts to date have yielded negative results, it cannot be ruled out that narcolepsy may be an autoimmune disease. It may be that the autoimmune process is difficult to detect because it involves a discrete area of the brain or only produces neuronal degeneration without any overt sign of inflammation.[63,64] Alternatively, fundamental mechanisms shared by the peripheral immune system and sleep regulatory centers may be involved. Many studies are now suggesting the existence of immune-related mechanisms specific to the brain. A key player in these mechanisms is the microglial cell population, which we have shown displays increased HLA expression in canine narcolepsy.[67] Immunoglobulin-like molecules and cytokines are also abundant in the brain, and their functions are only now starting to become understood. Cytokines also have some trophic effects in the brain and may impact neuronal differentiation at specific stages of development. These findings — together with the fact that many cytokines have sleep-inducing effects[27] and our recent finding that thalidomide, a sleep-inducing compound with immune-modulating properties dramatically aggravates cataplexy in narcoleptic animals (25) — suggest that novel immune-related CNS mechanisms might be involved in the pathophysiology of narcolepsy.

References

1. Aldrich, M.S., The neurobiology of narcolepsy/cataplexy, *Prog. Neurobiol.*, 41, 533, 1993.
2. Aldrich, M.S., The clinical spectrum of narcolepsy and idiopathic hypersomnia, *Neurology*, 46, 393, 1996.
3. Bassetti, C. and Aldrich, M.S., Narcolepsy, *Neurol. Clin.*, 14(3), 545, 1996.
4. Billiard, M. and Seignalet, J., Extraordinary association between HLA-DR2 and narcolepsy, *Lancet*, 1, 226, 1985.
5. Billard, M., Pasquié-Magnetto, V., Heckman, M., Carlander, B., Besset, A., Zachariev, Z., Eliaou, J.F., and Malafosse, A., Environmental and genetic factors family studies in narcolepsy, *Sleep*, 17, S54, 1995.
6. Broughton, R.J., Ghanem, Q., Hishikawa, Y., Surhita, Y., Nevsimalova, S., and Roth, B., Life effects of narcolepsy in 180 patients from North America, Asia and Europe compared to controls, *Le Journal Canadien des Sciences Neurologiques*, 8(4), 299, 1981.
7. Broughton, R.J., Guberman, A., and Roberts, J., Comparison of the psychosocial effect of epilepsy and narcolepsy/cataplexy: a controlled study, *Epilepsia*, 25(4), 423, 1984.
8. Carlander, B., Eliaou, J.F., and Billiard, M., Autoimmune hypothesis in narcolepsy, *Neurophysiol. Clin.*, 23, 15, 1993.

9. Carskadon, M.A., Dement, W.C., Milter, M.M., Roth, T., Westbrook, P.R., and Keenan, S., Guidelines for the multiple sleep latency test (MSLT): a standard measure of sleepiness, *Sleep*, 9(4), 519, 1986.
10. Challamel, M.J., Mazzola, M.E., Nevsimalova, S., Cannard, C., Louis, J., and Revol, M., Narcolepsy in children, *Sleep*, 17, S17, 1994.
11. D'Alessandro, R., Rinaldi, R., Cristina, E., Gamberini, G., and Lugaresi, E., Prevalence of excessive daytime sleepiness: an open epidemiological problem, *Sleep*, 18(5), 389, 1995.
12. Dahlitz, M. and Parkes, J.D., Sleep paralysis, *Lancet*, 341, 406, 1993.
13. Ellis, M., Hetisimer, A.H., Ruddy, D.A., Hansen, S.L., Kronmal, G.S., McClelland, E., Quintana, L, Drayna, D.T., Aldrich, M.S., and Mignot, E., HLA class II haplotype and sequence analysis support a role for DQ in narcolepsy, *Immunogenetics*, 46, 410, 1997.
14. Gélineau, J.B.E., De la narcolepsie, *Gazette des Hôpitaux (Paris)*, 53, 626; 54, 635; 1880.
15. Godbout, R. and Montplaisir, J., Comparisons of sleep parameters in narcoleptics with and without periodic leg movements of sleep, in *Sleep'84*, Koella, W.P., Ruther, E., and Schulz, H., Eds, Fisher Verlag, Gustav, 1985, p. 380.
16. Guilleminault, C., Dement, W.C., and Passouant, P., *Narcolepsy*, Spectrum Publications, New York, 1976, p. 689.
17. Guilleminault, C. and Dement, W.C., 235 cases of excessive daytime sleepiness: diagnosis and tentative classification, *J. Neurol. Sci.*, 31, 13, 1977.
18. Guilleminault, C., Mignot, E., and Grumet, F.C., Familial patterns in narcolepsy, *Lancet*, 2, 1376, 1989.
19. Honda, Y., Census of narcolepsy, cataplexy and sleep life among teen-agers in Fujisawa City [abstract], *Sleep Res.*, 8, 191, 1979.
20. Honda, Y., Asaka, A., Tanimura, M., and Furusho, T., A genetic study of narcolepsy and excessive daytime sleepiness in 308 families with a narcolepsy or hypersomnia proband, in *Sleep/Wake Disorders: Natural History, Epidemiology and Long Term Evolution*, Guilleminault, C. and Lugaresi, E., Eds., Raven Press, New York, 1983, p. 187.
21. Honda, Y., Asake, A., Tanaka, Y.U., and Jujui, T., Discrimination of narcolepsy by using genetic markers and HLA, *Sleep*, 12, 254, 1983.
22. Honda, Y. and Matsuki, K., Genetic aspects of narcolepsy, in *Handbook of Sleep Disorders*, Thorpy, M.J., Ed., Marcel Dekker, New York, 1990, p. 217.
23. Hublin, C., Kaprio, J., Partinen, M., Koskenvuo, M., Heikkila, K., Koskimies, S., and Guilleminault, C., The prevalence of narcolepsy: an epidemiological study of the Finnish twin cohort, *Ann. Neurol.*, 35, 709, 1994.
24. Hublin, C., Kaprio, J., Partinen, M., Heikkikla, K., and Koskenvuo, M., Daytime sleepiness in an adult, Finnish population, *J. Intern. Med.*, 239, 417, 1996.
25. Kanbayshi, T., Nishino, S., Tafti, M., Hishidawa, Y., Dement, W.C., and Mignot, E., Thalidomide, a TNF alpha inhibitor, dramatically aggravates canine cataplexy, *Neuroreport*, 7, 1881, 1996.
26. Karczmar, A.G., Longo, V.G., and De Carolis, S., A pharmacological model of paradoxical sleep: the role of cholinergic and monoaminergic systems, *Physiol. Behav.*, 5, 175, 1970.
27. Krueger, J.M. and Toth, L.A., Cytokines as regulators of sleep, *Ann. N.Y. Acad. Sci.*, 739, 299, 1994.
28. Langdon, N., Welch, K.I., Van Dam, M., Vaugham, R., and Parkes, J., Genetic markers in narcolepsy, *Lancet*, 2, 1178, 1984.
29. Lavie, P. and Peled, R., Narcolepsy is a rare disorder in Israel [letter], *Sleep*, 10(6), 608, 1987.
30. Lock, C.B., So, A.K.L., Welch, K.I., Parkes, J.D., and Trowsdale, J., MHC class II sequences of an HLA-DR2 narcoleptic, *Immunogenetics*, 27, 449, 1988.
31. Matsuki, K., Grumet, F.C., Lin, X., Guilleminault, C., and Dement, W.C., DQ rather than DR gene marks susceptibility to narcolepsy, *Lancet*, 339, 1052, 1992.

32. Mayer, G. and Meier-Ewert, K., Motor dyscontrol in sleep of narcoleptic patients (a life-long development?), *J. Sleep Res.*, 2, 143, 1993.
33. McCarley, R.W. and Hobson, J.A., Neuronal excitability modulation over the sleep cycle: a structural and mathematical model, *Science*, 189, 58, 1975.
34. Mignot, E., Wang, C., Rattazzi, C., Gaiser, C., Lovett, M., Guilleminault, C., Dement, W.C., and Grumet, F.C., Genetic linkage of autosomal recessive canine narcolepsy with an immunoglobulin µ chain switch-like segment, *Proc. Natl. Acad. Sci. USA*, 88, 3475, 1991.
35. Mignot, E., Renaud, A., Nishino, S., Arrigoni, J., Guilleminault, C., and Dement, W.C., Canine cataplexy is preferentially controlled by adrenergic mechanisms: evidence using monoamine selective uptake inhibitors and release enhancers, *Psychopharmacology*, 113, 76, 1993.
36. Mignot, E., Lin, X., Arrigoni, J., Macaubas, C., Olive, F., Hallmeyer, J., Underhill, P., Guilleminault, C., Dement, W.C., and Grumet, F.C., DQB1*0602 and DQA1*0102 (DQ1) are better markers than DR2 for narcolepsy in Caucasian and black Americans, *Sleep*, 17(8), S60, 1994.
37. Mignot, E., Lister, E., Burnett, R., Deeg, H.J., Nishino, S., Dement,W.C., and Strob, R., Bone marrow transplantation studies in canine narcolepsy [abstract], *Sleep Res.*, 24, 297, 1995.
38. Mignot, E., Tafti, M., Dement, W.C., and Grumet, F.C., Narcolepsy and immunity, *Adv. Neuroimmunol.*, 5(1), 23, 1995.
39. Mignot, E., Meehan, J., Grumet, F.C., Hallmeyer, J., Guilleminault, C., Hesla, P.E., Nevsimalova, S., Mayer, G., and Dement, W.C., HLA class II and narcolepsy in thirty-three multiplex families, [abstract], *Sleep Res.*, 25, 303, 1996.
40. Mignot, E., Kimura, A., Latterman, A., Lin, X., Yasunaga, S., Mueller-Eckardt, G., Rattazzi, C., Lin, L., Guilleminault, C., Grumet, F.C., Mayer, G., Dement, W.C., and Underhill, P., Extensive HLA class II studies in 58 non DRB1*15 (DR2) narcoleptic patients with cataplexy, *Tissue Antigens*, 49, 329, 1997.
41. Mignot, E., Hayduk, R., Black, J., Grumet, F.C., U.S. Modafinil Study Group, and Guilleminault, C., HLA DQB1*0602 is associated with cataplexy in 509 narcoleptic patients, *Sleep*, 20 (11), 1997, 1012.
42. Mignot, E., Genetic and familial aspects of narcolepsy, *Neurology*, 50(S1), 516, 1998.
43. Miller, J.D., Farber, J., Gatz, P., Roffwarg, H., and German, D.C., Activity of mesencephalic dopamine and non-dopamine neurons across stages of sleep and waking in the rat, *Brain Res.*, 273, 133, 1983.
44. Miller, J.D., Faull, K.F., Bowersox, S.S., and Dement, W.C., CNS monoamines and their metabolites in canine narcolepsy: a replication study, *Brain Res.*, 509, 169, 1990.
45. Mirenowicz, J. and Schultz, W., Preferential activation of midbrain dopamine neurons by appetitive rather than adversive stimuli, *Nature*, 379, 449, 1996.
46. Moskovitch, A., Partinen, M., and Guilleminault, C., The positive diagnosis of narcolepsy and narcolepsy's borderland, *Neurology*, 43, 55, 1993.
47. Mueller-Eckhardt, G., Meier-Ewert, K., Schendel, D., Reinecker, F., Multhoff, G., and Muller-Eckhardt, C., HLA and narcolepsy in a German population, *Tissue Antigens*, 28, 163, 1986.
48. Nishino, S., Reid, M., Dement, W.C., and Mignot, E., Neuropharmacology and neurochemistry of canine narcolepsy, *Sleep*, 17, S84, 1994.
49. Nishino, S., Tafti, M., Reid, M.S., Shelton, J., Siegel, J.M., Dement, W.C., and Mignot, E., Muscle atonia is triggered by cholinergic stimulation of the basal forebrain: implication for the pathophysiology of canine narcolepsy, *J. Neurosci.*, 15(7), 4806, 1995.
50. Nishino, S. and Mignot, E., Pharmacological aspects of human and canine narcolepsy, *Prog. Neurobiol.*, 52, 27, 1997.
51. Oyahon, M.M., Priest, R., Caulet, M., and Guilleminault, C., Hypnagogic and hypnopompic hallucinations: pathological phenomena?, *Br. J. Psychiatr.*, 169, 459, 1996.

52. Partinen, M., Hublin, C., Kaprio, J., Koskenvuo, M., and Guilleminault, C., Twin studies in narcolepsy, *Sleep,* 17, S13, 1994.
53. Rechtschaffen, A. and Dement, W.C., Studies on the relation of narcolepsy, cataplexy, and sleep with low-voltage random EEG activity, in *Sleep and Altered States of Consciounesss,* Vol. XLV, Association for Research in Nervous and Mental Disease, Eds., William & Wilkins, Baltimore, MD, 1967, p. 448.
54. Reid, M.S., Tafti, M., Geary, J., Nishino, S., Siegel, J.M., Dement, W.C., and Mignot, E., Cholinergic mechanisms in canine narcolepsy. I. Modulation of cataplexy via local drug administration into the pontine reticular formation, *Neuroscience,* 59(3), 511, 1994.
55. Reid, M.S., Tafti, M., Geary, J., Nishino, S., Siegel, J.M., Dement, W.C., and Mignot, E., Cholinergic mechanisms in canine narcolepsy. II. Acetylcholine release in the pontine reticular formation is enhanced during cataplexy, *Neuroscience,* 59(3), 523, 1994.
56. Reid, M.S., Tafti, M., Nishino, S., Sampathkumaran, R., Siegel, J., and Mignot, E., Local administration of dopaminergic drugs into the ventral tegmental area modulates cataplexy in the narcoleptic canine, *Brain Res.,* 733, 83, 1996.
57. Richardson, R.T. and Delang, M.R., Electrical studies of the function of the nucleus basils in primates, in *The Basal Forebrain: Anatomy to Function,* Napier, P.W., Kalivas, P.W., and Hanin, I., Eds., Plentum Press, New York, 1991, p. 145.
58. Rogers, A.E., Mehan, J., Guilleminault, C., Grumet, F.C., and Mignot, E., HLA DR15(DR2) and DQB1*0602 typing studies in 188 narcoleptic patients with cataplexy, *Neurology,* 48, 1550, 1997.
59. Rolls, E.T., Sanghera, M.K., and Roper-Hall, A., The latency of activation of neurons in the lateral hypothalamus and substantia innominata during feeding in the monkey, *Brain Res.,* 164, 121, 1979.
60. Schenck, C.H. and Mahowald, M.W., Motor dyscontrol in narcolepsy: rapid eye movement sleep without atonia and REM sleep behavior disorder, *Ann. Neurol.,* 32, 3, 1992.
61. Siegel, J.M., Nienhuis, R., Fahringer, H., Paul, R., Shiromani, P., Dement, W.C., Mignot, E., and Chiu, C., Neuronal activity in narcolepsy: identification of cataplexy related cells in the medial medulla, *Science,* 262, 1315, 1991.
62. Siegel, J.M., Nienhuis, R., Fahringer, H., Chiu, C., Dement, W.C., Mignot, E., and Lufkin, R., Activity of medial mesopontine units during cataplexy and sleep-waking states in the narcoleptic dog, *J. Neurosci.,* 13, 313, 1992.
63. Siegel, J.M., Fahringer, H.M., Anderson, L., Niemhuis, R., Gulyani, S., Nassiri, J., Mignot, E., and Switzer R.C., Evidence of localized neuronal degeneration in narcolepsy: studies in the narcoleptic dog, *Sleep Res.,* 24, 354, 1995.
64. Siegel, J.M., Niemhius, R., Mignot, E., Fahringer, H.M., and Jamgochian, G.M., Neuronal degeneration in the amygdala of the narcoleptic dog, *Sleep Res.,* 25, 370, 1996.
65. Singh, S.M., George, C.F.P., and Kryger, M.H., Genetic heterogeneity in narcolepsy, *Lancet,* 335, 726, 1991.
66. Steinfels, G.F., Heym, J., Strecker, R.E., and Jacobs, B.L., Response of dopaminergic neurons in cat to auditory stimuli presented across the sleep-waking cycle, *Brain Res.,* 277, 150, 1983.
67. Tafti, M., Nishino, S., Aldrich, M.S., Dement, W.C., and Mignot, E., Narcolepsy is associated with increased microglia expression of MHC class II moleucles, *J. Neurosci.,* 16(15), 4588, 1996.
68. Tashiro, T., Kanbayashi, T., and Hishikawa, Y., An epidemiological study of narcolepsy in Japanese, *Proc. 4th Int. Symp. on Narcolepsy,* Tokyo, Japan, June 16–17, 1994, p. 13.
69. Thorpy, M.J., International classification of sleep disorders, in *Diagnostic and Coding Manual,* American Sleep Disorders Association, Rochester, MN, 1990, p. 426.
70. Uryu, N., Maeda, M., Nagata, Y., Matsuki, K., Juji, T., Honda, Y., Kasai, J., Ando, A., Tsuji, K., and Inoko, H., No difference in the nucleotide sequence of the DQbeta beta1 domain between narcoleptic and healthy individulals with DR2, Dw2, *Hum. Immunol.,* 24, 175, 1989.

71. Westphal, C., Eigenthünlich mit einschlafen verbundene anfälle, *Arch. Psychiatr. Nervenkrank.*, 7, 631, 1877.
72. Wetter, T.C. and Pollmachen, T., Restless legs and periodic leg movements in sleep syndromes, *J. Neurol.*, 244(4, suppl. 1), S37, 1997.

Chapter 10

Dissociated States of Wakefulness and Sleep

Mark W. Mahowald and Carlos H. Schenck

Contents

10.1	Introduction	143
10.2	Normal State Determination	145
10.3	Evolution of Concept of State	145
10.4	Experimental State Dissociation (Animal)	146
	10.4.1 Lesion/Stimulation	146
	10.4.2 Pharmacologic	146
	10.4.3 Sleep Deprivation	146
10.5	Clinical State Dissociation-Neurologic-(Human)	147
	10.5.1 Wakefulness Variations	147
	10.5.2 REM Sleep Variations	147
	10.5.3 NREM Sleep Variations	148
	10.5.4 Miscellaneous Dissociations	149
10.6	Clinical State Dissociation-Psychogenic-(Human)	149
10.7	Clinical and Laboratory Evaluation of Neurologic and Psychogenic Dissociative Disorders	150
10.8	Comment	150
10.9	Summary	151
References		152

"The state of a system at a given instant is the set of numerical values which its variables have at that instant."[9]

10.1 Introduction

This chapter is unique among others in this handbook, as it deals with observational, experiential, and clinical phenomena in humans which for obvious reasons do not lend themselves to cellular or molecular analysis. Explanation of these phenomena is made possible by extrapolation from the basic science research data, described in other sections of this handbook, to the human clinical condition, underscoring the invaluable benefits of close collaboration between clinicians and basic science researchers.

Over the centuries, the presentations and concepts of dissociated or automatic behaviors have changed dramatically — ranging from demon possession, witchcraft, shamanism, hysteria, various psychiatric conditions, and frank malingering to the current notion of psychobiologic phenomena. The purpose of this review is to discuss such behaviors from the perspective of their state-dependent nature. To better understand these fascinating conditions, there will be a brief look at the states of human existence, with a review of the currently known spectrum of dissociated and automatic behaviors — both non-psychologic (sleep-related and neurologic) and psychologic. This will permit placing neurologic and psychiatric conditions into the more global perspective of dissociation in general.

Following is an excerpt from an interview with a 39-year-old patient with "status dissociatus" (discussed below). Prior to the development of the state dissociation concept, the only "explanation" of this phenomenon would have been (erroneously) psychiatric:

> MD: What was your sense of consciousness in the sleep lab when you appeared to be asleep and you were constantly mumbling and talking and moving around in bed?
>
> PT: I felt like I was just about awake. I felt like I was awake and talking to the person I was talking to in my dream. Say you were that person, I could talk to you and answer you back and have you answer me back with your regular tone of voice.
>
> MD: In your imagination? In your dream?
>
> PT: Yes, and I'd get mad and tell you just what I thought, word for word, just like I'm speaking right now, clear as a bell.
>
> MD: But did you have the same feeling at that time as you do right now, when you're obviously perfectly awake?
>
> PT: Yes, it's the same feeling.
>
> MD: Then how do you know you're asleep?
>
> PT: Because I wake up and I remember it.
>
> MD: But doesn't that mean that there must be some difference in how you perceive your consciousness if you can wake up from it, in contrast to right now? After all, you don't expect to wake up from this interview right here the same way you woke up from one of those episodes in the sleep lab.
>
> PT: I don't know; it's hard to say.
>
> MD: Do you feel alert in the state you're in at night in bed?
>
> PT: Yes.
>
> MD: And you hear people talking?
>
> PT: Yes.
>
> MD: But these are not people who are actually in the room, right?
>
> PT: No. They're dream characters. It's like a dream; I'm very alert in it, but it's still a dream, because I wake up *and then I know that I hadn't been awake and they're not there.* I went to bed in the sleep lab and I thought that because I dreamt and dreamt and went through a whole story, you know, I thought I was sleeping for two days. But when I woke up, only 13 minutes had gone by! It surprised the hell out of me: how could I do all that thinking in 13 minutes? *These dreams are too real for me.* When I dream of actually getting in a fight, it'll be a fight like I was fighting outside, right to the teeth.

10.2 Normal State Determination

State determination — wakefulness (W), rapid eye movement (REM) sleep, and non-rapid eye movement (NREM) sleep — may be made using various criteria: behavioral (eyes open/closed, body position, movements, reactivity to the environment), electrographic (electroencephalogram [EEG], electrooculogram [EOG], electromyogram [EMG]), or neuronal state (brain neuronal activity). The state-determining properties of each state usually cycle in a predictable and uniform manner, resulting in the behavioral appearance of a single prevailing state. However, even in normal subjects, the electrographic and neuronal activity transition among states is gradual and variable, with the simultaneous occurrence or rapid oscillation of multiple state-determining markers. Within each state, variability and fluctuation of the central nervous system (CNS) are ongoing activities.[60,142]

There is compelling evidence that there is extensive reorganization of the central nervous system activity as it moves across states of being. Factors involved in state generation are complex and include a wide variety of neurotransmitters, neuromodulators, neurohormones, and a vast array of "sleep factors" which act upon the multiple neural networks. These facts lead to the conclusion that sleep is a fundamental property of numerous neuronal groups, rather than a phenomenon that requires the whole brain.[59,77] Therefore, state determination is the result of a dynamic interaction among many variables, including circadian, neural network, neurotransmitter, and myriad sleep-promoting substances. Some identical neuronal groups are extremely active in more than one state, with differing state-dependent effects — i.e., many REM sleep phenomena are similar to the alerting response seen in wakefulness.[16,47,104] Some brainstem regions effect motor suppression in REM sleep but motor facilitation in W.[21,148]

REM is the best-studied sleep state. REM sleep is comprised of both tonic (occurring throughout the entire REM cycle) and phasic (occurring intermittently during the REM cycle) components. Each of these elements is generated, modulated, and executed by different neuronal groups located at multiple levels of the neuraxis — from cerebral cortex to spinal cord — and orchestrated by the dorsomedial pontine region.[67,133,144]

There are striking similarities neurophysiologically between wakefulness and REM sleep. The ascending brainstem reticular system potentiates thalamic and cortical responses during both states. Although the EEG is said to be "desynchronized" during wakefulness and REM sleep (in contrast to being "synchronized" during NREM sleep), there is fascinating and compelling evidence that during aroused states of the brain (wakefulness and REM sleep), activation of the mesencephalic reticular formation facilitates oscillatory activity in the gamma frequency band (>30 Hz), which enhances synchronization of cortical responses which may play a role in the processing of sensory signals. The waking and REM sleep "desynchronized" EEG pattern is more apparent than real; it is actually highly synchronized.[107,139] This synchronization may be important for information processing by facilitating the establishment of synchrony over large distances in the cortex, linking remote neurons into functional groups.[75]

10.3 Evolution of Concept of State

The clinical concept of states of being has changed dramatically over the past few decades. It was formerly thought that human existence encompassed only two states: wakefulness and sleep, with sleep being considered as simply the passive absence of wakefulness. With the discovery of REM sleep in 1953, it became apparent that sleep is not a unitary phenomenon, but rather consists of two completely different states, and each state is an active, rather than a quiescent, process.[68] Each state consists of a number of physiologic variables, which, under normal circumstances, tend to occur in concert, resulting in the appearance of one of the three conventional states of being: W, REM sleep, and NREM sleep.[60,62] Animal experiments and evaluation of humans in the sleep laboratory indicate that the "three states of being" concept must be further expanded to include the observations that the physiologic event markers of one state may intrude into other states and that the states may oscillate rapidly, resulting in the appearance of bizarre, previously difficult-to-explain, and occasionally extraordinary animal and human behaviors, which can occur in diverse naturalistic and clinical settings — with important treatment implications.[87,88]

10.4 Experimental State Dissociation (Animal)

The recurrent recruitment of state-determining parameters is amazingly consistent. However, multiple experimental examples of state component dissociation exist.[94] These fall into three categories.

10.4.1 Lesion/Stimulation

Hypothalamic, thalamic, and brainstem manipulation/stimulation induces state dissociation.[40,57,69,93,114,145,147] Recent studies in molecular biology may add a fascinating new dimension to such dissociation. For example, 6-hydroxydopamine (6-OHDA) lesions of the locus coeruleus inhibit the expected immediate early gene (c-Fos) expression (normally expected during wakefulness) in the cortex and hippocampus without changing the EEG. Therefore, following such lesions, the cortex may be at least partially functionally asleep, without a sleep EEG pattern.[23]

10.4.2 Pharmacologic

Manipulation of the cholinergic/glutamate neurotransmitter systems results in a variety of state dissociations.[11,32,33,35,53,54,66,79,98,146]

10.4.3 Sleep Deprivation

REM sleep deprivation in cats results in the appearance of ponto-geniculo-occipital (PGO) spikes during NREM sleep.[34] In addition to these experimental dissociations, there is evidence in the animal kingdom for the natural occurrence of clinically wakeful behavior during physiologic sleep. Two examples that dispel the concept of "all or none" state declaration are (1) the concurrence of swimming or flight during sleep in birds,[5] and (2) the phenomenon of unihemispheric sleep in some aquatic mammals (bottle-nosed dolphin, common porpoise, and northern fur seal) guaranteeing continued respiration while "sleeping".[149] Another naturally occurring dissociated state is seen during the arousal from torpor in hibernating ground squirrels, when there is an "uncoupling between thalamic, EMG, and cortical REM correlates."[76]

Both experimentally induced and naturally occurring state dissociations in animals serve to predict spontaneously occurring "experiments in nature" and drug-induced state dissociation in humans, which undoubtedly exist on a broad spectrum of expression. Such state dissociations are the consequence of timing or switching errors in the normal process of the dynamic reorganization of the CNS as it moves from one state (or mode) to another. Elements of one state persist, or are recruited, erroneously into another state, often with fascinating and dramatic consequences.

10.5 Clinical State Dissociation-Neurologic-(Human)

A number of well-documented state dissociations in humans occur spontaneously or as the result of neurologic dysfunction or medication administration. It may be more valid and practical to assign each as a variant of the predominant or prevailing parent state (W/NREM/REM), instead of identifying each possible dissociated state as an independent entity.

10.5.1 Wakefulness Variations

Narcolepsy is the prototypic dissociated state arising from the background of wakefulness and may best be thought of as a disorder of "state boundary control" (see Chapter 9). The symptom of cataplexy (sudden loss of muscle tone, usually in response to an emotionally laden event) is simply the isolated intrusion of REM sleep atonia into wakefulness. The element of surprise in triggering cataplexy supports the described similarity between the alerting response and REM sleep.[47] Similarly, the symptom of sleep paralysis is the persistence of REM atonia into wakefulness. The hypnagogic (occurring at sleep onset) and hypnopompic (occurring upon awakening) hallucinations are dream mentation occurring during wakefulness. These "hallucinations" (dreams) may be particularly frightening if accompanied by sleep paralysis.[143] Narcoleptic patients may experience waking dreams, particularly during drowsiness, and be misdiagnosed and even treated as having schizophrenia.[36,131] The occurrence of ambiguous or dissociated sleep is well documented in the untreated narcoleptic.[31] The induction of dissociated states in narcolepsy by tricyclic antidepressant administration indicates that genetically determined and pharmacologically potentiated state-disrupting factors may act in concert.[12,18,52] Forced awakenings in normal individuals may result in sleep paralysis.[141]

Other examples include drug-[2,43,129] and sleep deprivation-induced[8] hallucinations and a case of simultaneously occurring EEG patterns of W and NREM sleep.[108] Another likely example is "peduncular hallucinosis", which is a condition associated with deep midline intracranial lesions.[19,20,37,39,46,81,96,115,121] Interestingly, the lesions (diencephalic, hypothalamic, third ventricular region) reported to result in these hallucinations are virtually identical to those which cause "symptomatic" narcolepsy.[4,44]

10.5.2 REM Sleep Variations

The REM sleep behavior disorder (RBD) is the best-studied and perhaps most frequently documented dissociated state arising from the background of REM sleep. In retrospect, RBD was predicted in 1965 by animal experiments[69] but not formally recognized in humans until 1986.[89] During normal REM sleep, there is background atonia involving all somatic musculature (sparing the diaphragm and extraocular muscles). Although this generalized atonia may be briefly interrupted by excitatory inputs resulting in muscle jerks and twitches,[22] the prevailing atonia prevents motor activity associated with dream mentation. In RBD, motor behavior attendant with dream imagery may be vigorous, occasionally with injurious results.

There is an acute, transient form of RBD seen most frequently in the setting of drug intoxication or withdrawal states and also a chronic form of RBD, most often affecting older males. Nearly half of subjects with chronic RBD have identifiable underlying neurologic disorders.[89] The fact that over half of cases are idiopathic and tend to occur in the elderly suggests that RBD may be the reverse of sleep ontogeny (see comment section). The absence of identifiable peri-locus coeruleus lesions in the "symptomatic" subgroup is of interest and confirms animal experimental data which indicate that supra-pontine lesions may also affect REM sleep atonia.[28] In the animal model, chloramphenicol administration can reverse the peri-locus coeruleus lesion-induced REM without atonia, indicating that other structures are capable of inducing REM atonia.[1] An analogous situation during W is the fact that cortical, rubral, and pontine neurons all contribute to anterior horn cell phasic excitation, indicating that motor activity may be initiated at several levels of the CNS.[59]

Electrographic dissociation of REM sleep (absence of muscle atonia during REM sleep) and full-blown RBD may be induced in humans by the commonly prescribed tricyclic antidepressants, serotonin-specific reuptake inhibitors (SSRIs), and venlafaxine.[14,112,127,130] Spontaneously occurring RBD has been reported in dogs and cats.[56]

Another example of a mixed W/REM state is that of "lucid dreaming", during which the dreamer is aware of the fact that he/she is dreaming and has the ability to influence the course of the dream. REM sleep is the parent state during lucid dreaming, yet the subject has the facility to signal the presence of such a dream physically by means of voluntary eye and digit movements. Suppression of the H-reflex, a characteristic of REM sleep, persists during such dreaming.[17] Some "out-of-body" experiences may represent a variation on this theme.[78]

10.5.3 NREM Sleep Variations

Disorders of arousal are the most impressive and most frequent of the NREM sleep phenomena. These share common features: a positive family history, suggesting a genetic component, and they arise only from NREM sleep, particularly from slow-wave sleep (stages 3 and 4 of NREM sleep).[91] Disorders of arousal occur on a broad spectrum ranging from confusional arousals through somnambulism (sleep walking) to sleep terrors (also termed *pavor nocturnus*). Some take the form of "specialized" behaviors such as sleep-related eating and sleep-related sexual activity, without conscious awareness.[120,125]

Confusional arousals (also termed "sleep drunkenness") are often seen in children and are characterized by movements in bed, occasionally thrashing about, or inconsolable crying.[109,118] Such arousals are commonly seen in normal individuals, particularly in the setting of sleep deprivation coupled with forced awakenings and may be seen in other conditions associated with excessive daytime sleepiness, such as narcolepsy or obstructive sleep apnea. These arousals may occasionally result in injurious or violent behaviors with forensic science implications.[90]

Sleep walking is prevalent in childhood (1 to 17%), peaking at 11 to 12 years of age, and is far more common in adults (4 to 10%) than generally acknowledged.[15,48,63,74] Sleep walking may be either calm or agitated, with varying degrees of complexity and duration.

The sleep terror is the most dramatic disorder of arousal. It is frequently initiated by a loud, blood-curdling scream associated with extreme panic, followed by prominent motor activity such as hitting the wall or running around or out of the bedroom, resulting in bodily injury or property damage. A universal feature is inconsolability. Complete amnesia for the activity is typical but may be incomplete.[41,42,70] Although usually benign, these behaviors may be violent, resulting in considerable injury to the victim or others or damage to the environment, occasionally with forensic implications.[86]

The clinical features, laboratory evaluation, and treatment of disorders of arousal have been extensively reviewed elsewhere.[91] It is commonly felt that persistence of these behaviors beyond childhood or their development in adulthood is an indication of significant psychopathology.[71,135] Numerous studies have dispelled this myth, indicating that significant psychopathology in adults with disorders of arousal is usually not present.[51,84,100,124]

A recently described phenomenon, the cyclic alternating pattern (CAP), may play a role in the etiology of the disorders of arousal. The CAP is a physiological component of NREM sleep and is functionally correlated with long-lasting arousal oscillations. The CAP is a measure of NREM instability with a high level of arousal oscillation.[150] There is no difference in the macrostructural sleep parameters between patients with disorders of arousal and controls. However, patients with disorders of arousal have been found to have increases in CAP rate, in number of CAP cycles, and in arousals with EEG synchronization.

A common thread linking RBD and the disorders of arousal is the appearance of motor activity which is dissociated from waking consciousness. In RBD, the motor behavior closely correlates with dream imagery, and in disorders of arousal it often occurs in the absence of (remembered) mentation. This dissociation of behavior from consciousness may be explained by the presence of locomotor centers (LMCs), from the mesencephalon to the medulla, which are capable of generating complex behaviors without cortical input.[13,50,99,102,132] These areas project to the central pattern generator of the spinal cord, which itself is able to produce complex stepping movements in the absence of supra-spinal influence.[101] This accounts for the fact that decorticate experimental and decapitated barnyard animals are capable of performing very complex, integrated motor acts.

It is likely that during NREM sleep, the LMCs are not activated. The theory that LMCs are actively inhibited during REM sleep is supported by the observations that smaller peri-locus coeruleus lesions result in REM sleep without atonia — without any behavioral manifestations — but larger lesions are necessary to produce active motor movements. Clearly, the isolated loss of REM atonia is insufficient to explain complex REM sleep motor activity in the experimental animal,[105] suggesting a release of LMCs during the parent state, just as is seen in the loss of REM atonia. Dissociation of the LMCs from the parent state of REM or NREM sleep would explain the presence of complex motor behavior seen in RBD and disorders of arousal. It is likely that the complex motor activity associated with amnesia characteristic of alcohol-induced "black-outs"[151] and with "unconscious" behavior occurring during partial complex seizures represents dissociation between LMCs and waking consciousness and/or memory.[113] Such dissociation between behavior and consciousness may be related to inactivation of attentional or memory systems.[61]

10.5.4 Miscellaneous Dissociations

Nocturnal penile tumescence in males is one of the tonic elements of REM sleep, usually occurring with the other physiologic markers of the REM state.[73] Administration of tricyclic antidepressants[136] and mono-amine oxidase inhibitors[137,138] may selectively suppress the electrographic features of REM with persistence of at least one tonic component — penile tumescence. Similar penile tumescence/REM polygraphic dissociation has been reported in a post-traumatic pontine lesion in a man.[80]

A condition termed "status dissociatus" is the most extreme form of RBD and appears to represent a complete breakdown of state-determining boundaries. Clinically, these patients, by behavioral observation, appear to be either awake or "asleep"; however, clinically, their behavioral "sleep" is very atypical, characterized by frequent muscle twitching, vocalization, and reports of dream-like mentation upon spontaneous or forced awakening. Polygraphically, there are few, if any, features of either conventional REM or NREM sleep; rather, there is the simultaneous admixture of elements of wakefulness, REM sleep, and NREM sleep. Conditions associated with status dissociatus include protracted withdrawal from alcohol abuse, narcolepsy, olivopontocerebellar degeneration, and prior open heart surgery. Clonazepam may be effective in treating the sleep-related motor and verbal behaviors.[87,88] The clinical features of fatal familial insomnia, a prion disease closely related to Creutzfeldt-Jakob disease, are highly reminiscent of status dissociatus.[45,85]

10.6 Clinical State Dissociation-Psychogenic-(Human)

The single common feature of most automatisms or dissociations (organic or psychiatric) is lack of conscious awareness. Once one of the above-mentioned organic neurologic conditions has been ruled out, a psychiatric diagnosis may be considered in clinical dissociative disorders. A nearly ubiquitous antecedent feature of psychogenic dissociative disorder is severe psychic trauma, which is often incompletely or unremembered. There is now overwhelming neurophysiologic evidence in animal models that such physical or psychic trauma may lead to permanent alterations in functioning of the central nervous system,[87,88] predisposing to clinical dissociative disorders. This would suggest that these disorders also have a neuro(psycho)biological basis and may not be "functional" in the psychiatric sense. These conditions include psychogenic dissociative disorder and multiple personality disorder and have been reviewed elsewhere.[128]

10.7 Clinical and Laboratory Evaluation of Neurologic and Psychogenic Dissociative Disorders

The above discussion indicates that all types of automatic/dissociated behaviors may be the manifestation of either organic or psychiatric conditions. Thorough evaluation should be conducted to rule out organic conditions. Our experience with over 300 adult cases of sleep-related complex behavior has repeatedly indicated that clinical differentiation among the various dissociative disorders is often impossible.[87,88,92,119,124] Indications and techniques for formal sleep studies in these cases have been reviewed elsewhere.[91,126]

10.8 Comment

Review of the ontogeny of state appearance facilitates the analysis of observed experimental and clinical state dissociations. During embryogenesis, there are no clear-cut states but rather the simultaneous admixture of all states, which gradually coalesce to form the three recognizable states of W/REM/NREM.[24-27] This ontogeny of state development is supported by phyologenetic studies (see Chapter 7). The mechanisms of complex synchronization/recruitment of the state-specific variables are unknown. Basic science neurophysiologists have long known that state dissociation in animals occurs frequently, under many circumstances.[140] The inability of animals to report or indicate mentation and consciousness (i.e., waking hallucinations, mental imagery with disorders of arousal, dream-mentation-associated motor behavior in RBD) has been a significant limitation upon the evaluation of animal state dissociation and its application to the human clinical experience.

Many endogenous and exogenous factors can affect state cycling/synchronization. These include:[6,55,64]

1. Age
2. Sleep deprivation
3. Shift-work/rapid travel across time zones
4. Endogenous humoral factors (hormonal)
5. Drugs/medication
6. Affective disorders
7. Environmental stress

With the multiplicity of state markers and the relatively rapid normal cycling of states requiring recruitment of these numerous physiologic markers, there are innumerable theoretically possible state combinations. An overview of state determination by prevailing state is shown in Figure 10.1.

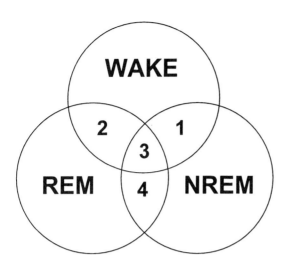

FIGURE 10.1
Overlapping states of being. (From Mahowald, M.W. and Schenck, C.H., *Neurology*, 42, 44–52, 1992. With permission.)

It is likely that major psychic or neural insults can result in an acquired functional restructuring of the CNS, which then may interfere with conventional state determination.[38] There is strong evidence that environmentally mediated events can and do affect the structure and function of the CNS[49,55,95,110] and that the CNS displays learning of new neural behaviors[65,72,116] — i.e., development of secondary epileptogenesis ("mirror foci")[103] or acquired sensory synesthesia.[29,117] Such dissociated states may play a role in the appearance of psychogenic dissociative states. Indeed, given the genetic variability of CNS development and its plasticity,[7,38,82,134] the relentless cycling, and the ever-present multiplicity of endogenous and environmental influences upon both CNS plasticity and cycling, it is actually surprising that state-component timing errors have not been identified more frequently. Truly, the drive for complete state determination must be very robust. Striking sleep abnormalities have been reported in a wide variety of degenerative[3,85,97,106,111,122,123] and acquired[10,30,58,83] neurologic conditions. This patient population should serve as a rich source of subjects at high risk for state-dissociation.

The multiple component concept of state determination must also be kept in mind when pharmacologic or lesion studies are employed to "suppress" one or another state. Such manipulation may suppress some of the commonly used markers for that state (i.e., polygraphic) without affecting other variables of that state. Recent molecular biologic studies mentioned above underscore the complexity of state determination.[23] Nielsen's concept of "phantom REM" (see Chapter 8) to explain the confusion of REM sleep vs. NREM sleep dreaming lends clinical relevance to the concept of part of one state manifesting itself during polygraphic trappings of another state.

10.9 Summary

The foregoing discussion of neurologic and psychogenic dissociative states was clearly made possible only by applying information obtained from basic science animal research studies to the human condition, without which these often dramatic and treatable conditions would have remained in the mystical or psychiatric arena, without treatment options. Continued study of state dissociation by both basic scientists and clinicians will undoubtedly identify and explain even more of these fascinating conditions, with important therapeutic implications. The reciprocal benefits of close collaboration between basic scientists and clinicians is obvious.

References

1. Aguilar-Roblero, R., Arankowsky, G., Drucker-Colin, R., Morrison, A.R., and Bayon, A., Reversal of rapid eye movement sleep without atonia by chloramphenicol, *Brain Res.*, 305, 19–26, 1984.
2. Albala A.A., Weinberg N., and Allen, S.M., Maprolitine-induced hypnopompic hallucinations, *J. Clin. Psychiatry*, 44, 149–154, 1983.
3. Aldrich, M.S., Foster, N.L., White, R.F., Bluemlein, L., and Prokopowicz, G., Sleep abnormalities in progressive supranuclear palsy, *Ann. Neurol.*, 25, 577–581, 1989.
4. Aldrich, M.S. and Naylor, M.W., Narcolepsy associated with lesions of the diencephalon, *Neurology*, 39, 1505–1508, 1989.
5. Amlander, C.J.J. and Ball, N.J., Avian sleep, in *Principles and Practice of Sleep Medicine*, Kryger, M.H., Roth, T., and Dement, W.C., Eds., W.B. Saunders, Philadelphia, 1989, pp. 30–49.
6. Anch, A.M., Browman, C.P., Mitler, M.M., and Walsh, J.K., *Sleep: A Scientific Perspective*, Prentice Hall, Englewood Cliffs, NJ, 1988.
7. Aoki, C. and Siekevitz, P., Plasticity in brain development, *Sci. Am.*, 259, 56–64, 1988.
8. Asaad, G. and Shapiro, B., Hallucinations: theoretical and clinical overview, *Am. J. Psychiatry*, 143, 1088–1097, 1986.
9. Ashby, W.R., *Design for a Brain: The Origin of Adaptive Behavior*, 2nd ed., Chapman Hall, London, 1960.
10. Autret, A., Laffont, F., de Toffol, B., and Cathala, J.-P., A syndrome of REM and non-REM sleep reduction and lateral gaze paresis after medial tegmental pontine stroke, *Arch. Neurol.*, 45, 1236–1242, 1988.
11. Baghdoyan, H.A., Rodrigo-Angulo, M.L., McCarley, R.W., and Hobson, J.A., Cholinergic induction of desynchronized sleep signs by carbachol microinjection shows intrapontine site differentiation, *Sleep Res.*, 11, 51, 1982.
12. Bental, E., Lavie, P., and Sharf, B., Severe hypermotility during sleep in treatment of cataplexy with clomipramine, *Israel J. Med. Sci.*, 15, 607–609, 1979.
13. Berntson, G.G. and Micco, D.J., Organization of brainstem behavioral systems, *Brain Res. Bull.*, 1, 471–483, 1976.
14. Besset, A., Effect of antidepressants on human sleep, *Adv. Biosci.*, 21, 141–148, 1978.
15. Bixler, E.O., Kales, A., Soldatos, C.R., and Healey, S., Prevalence of sleep disorders in the Los Angeles metropolitan area, *Am. J. Psychiatry*, 136, 1257–1262, 1979.
16. Bowker, R.M. and Morrison, A.R. The startle reflex and PGO spikes, *Brain Res.*, 102, 185–190, 1976.
17. Brylowski, A., Levitan, L., and LaBerge, S., H-reflex suppression and autonomic activation during lucid REM sleep, a case study, *Sleep*, 12, 374–378, 1989.
18. Cadilhac, J., Tricyclics and REM sleep, in *Narcolepsy: Advances in Sleep Research*, Vol. 3, Guilleminault, C., Dement, W.C., and Passouant, P., Eds., Spectrum, New York, 1976, pp. 605–623.
19. Caplan, L.R., "Top of the basilar" syndrome, *Neurology*, 30, 72–79, 1980.
20. Cascino, G.C. and Adams, R.D., Brainstem auditory hallucinations, *Neurology*, 36, 1042–1047, 1986.
21. Chase, M.H., The motor functions of the reticular formation are multifaceted and state-determined, in *The Reticular Formation Revisited*, Hobson, J.A. and Brazier, M.A.B., Eds., Raven Press, New York, 1980, pp. 449–472.
22. Chase, M.H. and Morales, F.R., Subthreshold excitatory activity and motoneurone discharge during REM periods of active sleep, *Science*, 221, 1195–1198, 1983.
23. Cirelli, C., Pompeiano, M., and Tononi, G., Neuronal gene expression in the waking state: a role for the locus coeruleus, *Science*, 274, 1211–1215, 1996.

24. Corner, M.A. Sleep and the beginnings of behavior in the animal kingdom — studies in ultradian motility cycles in early life, *Prog. Neurobiol.*, 8, 279–295, 1977.

25. Corner, M.A., Maturation of sleep mechanism in the central nervous system, *Exp. Brain Res.*, (Suppl. 8), 50–66, 1984.

26. Corner, M.A. Ontogeny of brain sleep mechanisms, in *Brain Mechanisms of Sleep*, McGinty, D.J., Drucker-Colin, R., and Morrison, A.R., and Parmeggiani, P.L., Eds., Raven Press, New York, 1985, pp. 175–197.

27. Corner, M.A. and Bour, H.L., Postnatal development of spontaneous neuronal discharges in the pontine reticular formation of free-moving rats during sleep and wakefulness, *Exp. Brain Res.*, 54, 66–72, 1984.

28. Culebras, A. and Moore, J.T., Magnetic resonace findings in REM sleep behavior disorder, *Neurology*, 39, 1519–1523, 1989.

29. Cytowic, R.E., Synesthesia an mapping of subjective sensory dimensions [editorial], *Neurology*, 39, 849–850, 1989.

30. DeBarros-Ferreira, M., Chodkiewicz, J.-P., Lairy, G.C., and Salzarulo, P., Disorganized relations of tonic and phasic events of REM sleep in a case of brain-stem tumor, *EEG Clin. Neurophysiol.*, 38, 203–207, 1975.

31. DeBarros-Ferreria, M. and Lairy, G.C., Ambiguous sleep in narcolepsy, in *Narcolepsy: Advances in Sleep Research*, Vol. 3, Guilleminault, C., Dement, W.C., and Passouant, P., Eds., Spectrum, New York, 1976, pp. 57–75.

32. Delinger, S.L., Patarca, R., and Hobson, J.A., Differential enhancement of rapid eye movement sleep signs in the cat: a comparison of microinjection of the cholinergic agonist carbachol and the beta-adrenergic antagonist propranolol on pontogeniculo-occipital wave clusters, *Brain Res.*, 473, 116–126, 1988.

33. Delorme, F., Froment, J., and Jouvet, M., Suppression du sommeil par la p-chloromethamphetamine et la — chlorophenylalanine, *C. R. Soc. Biol. (Paris)*, 160, 2347, 1966.

34. Dement, W.C., The biological role of REM sleep (circa 1968), in *Sleep Physiology and Pathology*, Kales, A., Ed., Lippincott, Philadelphia, 1969, pp. 245–265.

35. Domino, E.F., Yamamoto, K., and Dren, A.T., Role of cholinergic mechanisms in states of wakefulness and sleep, *Prog. Brain Res.*, 128, 113–133, 1968.

36. Douglass, A.B., Hays, P., Pazderka, F., and Russell, J.M., A schizophrenic variant of narcolepsy, *Sleep Res.*, 18, 173, 1989.

37. Dunn, D.W., Weisberg, L.A., and Nadell, J., Peduncular hallucinations caused by brainstem compression, *Neurology*, 33, 1360–1361, 1983.

38. Edelman, G.M., *Neural Darwinism*, Basic Books, New York, 1987.

39. Feinberg, W.M., "Peduncular hallucinosis" following paramedian thalamic infarction, *Neurology*, 39, 1535–1536, 1989.

40. Feldman, S.M. and Waller, H.J., Dissociation of electrocortical activation and behavioral arousal, *Nature*, 196, 1320–1322, 1962.

41. Fisher, C., Kahn, E., Edwards, A., and Davis, D.M., A psychophysiological study of nightmares and night terrors. I. Physiological aspects of the stage 4 night terror, *J. Nerv. Ment. Dis.*, 157, 75–98, 1973.

42. Fisher, C., Kahn, E., Edwards, A., Davis, D.M., and Fine, J., A psychophysiological study of nightmares and night terrors. III. Mental content and recall of stage 4 night terrors, *J. Nerv. Ment. Dis.*, 158, 174–188, 1974.

43. Fleminger, R., Visual hallucinations and illusions with propranolol, *Br. Med. J.*, 1, 1182, 1978.

44. Fulton, J.F. and Bailey, P., Tumors in the region of the third ventricle, their diagnosis and relation to pathological sleep, *J. Nerv. Ment. Dis.*, 69, 1–25, 1929.

45. Gambetti, P., Fatal familial insomnia and familial Creutzfeldt-Jakob disease: a tale of two diseases with the same genetic mutation, *Curr. Topics Microbiol. Immunol.*, 207, 19–25, 1996.
46. Geller, T.J. and Bellur, S.N., Peduncular hallucinosis, magnetic resonance imaging confirmation of mesencephalic infarction during life, *Ann. Neurol.*, 21, 602–604, 1987.
47. Glenn, L.L., Brainstem and spinal control of lower limb motoneurons with special reference to phasic events and startle reflexes, in *Brain Mechanisms of Sleep*, McGinty, D.J., Drucker-Colin, R., Morrison, A., and Parmeggiani, P.L., Eds., Raven Press, New York, 1985, pp. 81–95.
48. Goldin, P.R. and Rosen, R.C., Epidemiology of nine parasomnias in young adults, *Sleep Res.*, 26, 367, 1997.
49. Gorman, J.M., Liebowitz, M.R., Fyer, A.J., and Stein, J., A neuroanatomical hypothesis for panic disorder, *Am. J. Psychiatry*, 146, 148–161, 1989.
50. Grillner, S. and Dubic, R., Control of locomotion in vertebrates: spinal and supraspinal mechanisms, *Adv. Neurol.*, 47, 425–453, 1988.
51. Guilleminault, C., Moscovitch, A., and Leger, D., Forensic sleep medicine, nocturnal wandering, and violence, *Sleep*, 18, 740–748, 1995.
52. Guilleminault, C., Raynal, D., Takahashi, S., Carskadon, M., and Dement, W., Evaluation of short-term and long-term treatment of the narcolepsy syndrome with clomipramine hydrochloride, *Acta Neurol. Scand.*, 54, 71–87, 1976.
53. Harper, R.M., Relationship of neuronal activity to EEG waves during sleep and wakefulness, in *Brain Unit Activity During Behavior*, Phillips, M.I., Ed., Charles C Thomas, Springfield, IL, 1973, pp. 130–154.
54. Hazra, J., Effect of hemicholinium-3 on slow wave and paradoxical sleep of cat, *Eur. J. Pharmacol.*, 11, 395–397, 1970.
55. Hefez, A., Metz, L., and Lavie, P., Long-term effects of extreme situational stress on sleep and dreaming, *Am. J. Psychiatry*, 144, 344–347, 1987.
56. Hendricks, J.C., Lager, A., O'Brien, D., and Morrison, A.R., Movement disorders during sleep in cats and dogs, *J. Am. Vet. Med. Assoc.*, 194, 686–689, 1989.
57. Hendricks, J.C. and Mann, G.L., Different behaviors during paradoxical sleep without atonia depend upon lesion site, *Brain Res.*, 239, 81–105, 1982.
58. Hobson, J.A., Dreaming sleep attacks and desynchronized sleep enhancement, *Arch. Gen. Psychiatry*, 32, 1421–1424, 1975.
59. Hobson, J.A., Lydic, R., and Baghdoyan, H.A., Evolving concepts of sleep cycle generation: from brain centers to neuronal populations, *Behav. Brain Sci.*, 9, 371–448, 1986.
60. Hobson, J.A. and Scheibel, A.B., The brainstem core: sensorimotor integration and behavioral state control, *Neurosci. Res. Program Bull.*, 18, 1–173, 1980.
61. Hobson. J.A. and Schmajuk, N.A., Brain state and plasticity: an integration of the reciprocal interaction model of sleep cycle oscillation with attentional models of hippocampal function, *Arch. Italiennes de Biologie*, 126, 209–224, 1988.
62. Hobson, J.A. and Steriade, M., Neuronal basis of behavioral state control, in *Handbook of Physiology*, Bloom, F.E., Ed., American Physiological Society, Bethesda, MD, 1986, pp. 701–823.
63. Hublin, C., Kaprio, J., Partinen, M., Heikkila, K., and Koskenvuo, M., Prevalence and genetics of sleepwalking; a population-based twin study, *Neurology*, 48, 177–181, 1997.
64. Inoue, S., *Biology of Sleep Substances*, CRC Press, Boca Raton, FL, 1989.
65. Iriki, A., Pavlides, C., Keller, A., and Asanuma, H., Long-term potentiation in the motor cortex, *Science*, 245, 1385–1387, 1989.
66. Irmis, F., "Dissociation" between EEG and spontaneous behavior of rats after atropine, *Actvitas Nervosa Superior*, 13, 217–218, 1971.

67. Jones, B.E., Neuroanatomical and neurochemical substrates of mechanisms underlying paradoxical sleep, in *Brain Mechanisms of Sleep*, McGinty, D.J., Drucker-Colin, R., Morrison, A.R., and Parmeggiani, P.L., Eds., Raven Press, New York, 1985, pp. 139–156.
68. Jones, B.E., Basic mechanisms of sleep-wake states, in *Principles and Practice of Sleep Medicine*, 2nd ed., Kryger, M.H., Roth, T., and Dement, W.C., Eds., W.B. Saunders, Philadelphia, 1994, pp. 145–162.
69. Jouvet, M. and Delorme, F., Locus coeruleus et sommeil paradoxal, *C. R. Soc. Biol.*, 159, 895–899, 1965.
70. Kahn, E., Fisher, C., and Edwards, A., Night terrors and anxiety dreams, in *The Mind in Sleep: Psychology and Psychophysiology*, 2nd ed., Ellman, S.D. and Antrobus, J.S., Eds., John Wiley & Sons, New York, 1991, pp. 437–447.
71. Kales, J.D., Kales, A., and Soldatos, C.R., Night terrors. Clinical characteristics and personality factors, *Arch. Gen. Psychiatry*, 47, 1413–1417, 1980.
72. Kandel, E.R., Environmental determinants of brain architecture and of behavior: early experience and learning, in *Principles of Neural Science*, Kandel, E.R. and Schwartz, J.H., Eds., Elsevier/North Holland, New York, 1981, pp. 620–632.
73. Karacan, I., Salis, P.J., Thornby, J.I., and Williams, R.L., The ontogeny of nocturnal penile tumescence, *Waking Sleeping*, 1, 27–44, 1976.
74. Klackenberg, G., Somnambulism in childhood — prevalence, course and behavior correlates: a prospective longitudinal study (6–16 years), *Acta Paediatr. Scand.*, 71, 495–499, 1982.
75. Konig, P., Engel, A.K., and Singer, W., Relation between oscillatory activity and long-range synchronization in cat visual cortex, *Proc. Natl. Acad. Sci. USA*, 92, 290–294, 1995.
76. Krilowicz, B.L., Glotzbach, S.F., and Heller, H.C., Neuronal activity during sleep and completed bouts of hibernation, *Am. J. Physiol.*, 255, R1008–R1019, 1988.
77. Krueger, J.M., Toth, L.A., Floyd, R., Fang, J., Kapas, L., Bredow, S., and Obal, J.F., Sleep, microbes, and cytokines, *Neuroimmunomodulation*, 1, 100–109, 1994.
78. LaBerge, S., Levitan, L., Brylowski, A., and Dement, W.C., "Out-of body" experiences occurring in REM sleep, *Sleep Res.*, 17, 115, 1988.
79. Lai, Y.Y. and Siegel, J.M., Medullary regions mediating atonia, *J. Neurosci.*, 8, 4790–4796, 1988.
80. Lavie, P., Penile erections in a patient with nearly total absence of REM, a follow-up study, *Sleep*, 13, 276–278, 1990.
81. Lhermitte, J., Syndrome de la calotte pedonculaire, les troubles psycho-sensoriels dans les lesions du mesocephale, *Rev. Neurol.*, 38, 1359–1365, 1922.
82. Lipton, S.A., Growth factors for neuronal survival and process regeneration. Implications in the mammalian central nervous system, *Arch. Neurol.*, 46, 1241–1248, 1989.
83. Little, B.W., Brown, P.W., Rodgers-Johnson, P., Perl, D.P., and Gajdusek, D.C., Familial myoclonic dementia masquerading as Creutzfeldt-Jakob disease, *Ann. Neurol.*, 20, 231–239, 1986.
84. Llorente, M.D., Currier, M.B., Norman, S., and Mellman, T.A., Night terrors in adults: phenomenology and relationship to psychopathology, *J. Clin. Psychiatry*, 53, 392–394, 1992.
85. Lugaresi, E., Medori, R., Montagna, P., Baruzzi, A., Cortelli, P., Lugaresi, A., Tinuper, P., Zucconi, M., and Gambetti, P., Fatal familial insomnia and dysautonomia with selective degeneration of thalamic nuclei, *New Engl. J. Med.*, 315, 997–1003, 1986.
86. Mahowald, M.W., Bundlie, S.R., Hurwitz, T.D., and Schenck, C.H., Sleep violence-forensic science implications, polygraphic and video documentation, *J. Forens. Sci.*, 35, 413–432, 1990.
87. Mahowald, M.W. and Schenck, C.H., Status dissociatus – a perspective on states of being, *Sleep*, 14, 69–79, 1991.
88. Mahowald, M.W. and Schenck, C.H., Dissociated states of wakefulness and sleep, *Neurology*, 42, 44–52, 1992.

89. Mahowald, M.W. and Schenck, C.H., REM sleep behavior disorder, in *Principles and Practice of Sleep Medicine,* 2nd ed., Kryger, M.H., Dement, W., and Roth, T., Eds., W.B. Saunders, Philadelphia, 1994, pp. 574–588.
90. Mahowald, M.W. and Schenck, C.H., Complex motor behavior arising during the sleep period: forensic science implications, *Sleep,* 18, 724–727, 1995.
91. Mahowald, M.W. and Schenck, C.H., NREM parasomnias. *Neurol. Clin.,* 14, 675–696, 1996.
92. Mahowald, M.W. and Thorpy, M.J., Non-arousal parasomnias in the child, in *Principles and Practice of Sleep Medicine in the Child,* Ferber, R. and Kryger, M.H., Eds., W.B. Saunders, Philadelphia, 1995, pp. 115–123.
93. McGinty, D.J., Somnolence., recovery and hyposomnia following ventro-medial diencephalic lesions in the rat, *EEG Clin. Neurophysiol.,* 26, 70–79, 1969.
94. McGinty, D.J. and Drucker-Colin, R.R., Sleep mechanisms: biology and control of REM sleep, *Int. Rev. Neurobiol.,* 23, 391–436, 1982.
95. McKee, A.C., Kowall, N.W., and Kosik, K.S., Microtubular reorganization and dendritic growth response in Alzheimer's disease, *Ann. Neurol.,* 20, 652–659, 1989.
96. McKee, A.C., Levine, D.N., Kowall, N.W., and Richardson, E.P.J., Peduncular hallucinosis associated with isolated infarction of the substantia nigra pars reticulata, *Ann. Neurol.,* 27, 500–504, 1990.
97. Mina, M., Autret, A., Laffont, F., Beillevaire, T., and Cathala, H.P., Castaigne, P., A study on sleep in amyotrophic lateral sclerosis, *Biomedicine,* 30, 40–46, 1979.
98. Mitler, M.M. and Dement, W.C., Cataplectic-like behavior in cats after microinjection of carbachol in the pontine reticular formation, *Brain Res.,* 69, 335–343, 1974.
99. Mogenson, G.J., Limbic-motor integration. *Prog. Psychobiol. Physiol. Psychol.,* 12, 117–170, 1986.
100. Moldofsky, H., Gilbert, R., Lue, F.A., and MacLean, A.W., Sleep-related violence, *Sleep,* 18, 731–739, 1995.
101. Mori, S., Nishimura, H., and Aoki, M., Brain stem activation of the spinal stepping generator, in *The Reticular Formation Revisited,* Hobson, J.A. and Brazier, M.A.B., Eds., Raven Press, New York, 1980, pp. 241–259.
102. Mori, Sh., Integration of posture and locomotion in acute decerebrate cats and in awake, freely moving cats, *Prog. Neurobiol.,* 28, 161–195, 1987.
103. Morrell, F., Varieties of human secondary epileptogenesis, *J. Clin. Neurophysiol.,* 6, 227–275, 1989.
104. Morrison, A.R., Paradoxical sleep and alert wakefulness: variations on a theme, in *Sleep Disorders: Basic and Clinical Research,* Chase, M.H. and Weitzman, E.D., Eds., Spectrum, New York, 1983, pp. 95–122.
105. Morrison, A.R., Paradoxical sleep without atonia, *Arch Italiennes de Biologie,* 126, 275–289, 1988.
106. Mouret, J., Differences in sleep in patients with Parkinson's disease, *EEG Clin. Neurophysiol.,* 38, 653–657, 1975.
107. Munk, M.H.J., Roelfsema, P.R., Konig, P., Engel, A.K., and Singer, W., Role of reticular activation in the modulation of intracortical synchronization, *Science,* 272, 271–274, 1996.
108. Niedermeyer, E., Singer, H.S., Folstein, S.E. et al., Hypersomnia with simultaneous waking and sleep patterns in the electroencephalogram, *J. Neurol.,* 221, 1–13, 1979.
109. Nino-Murcia, G. and Dement, W.C., Psychophysiological and pharmacological aspects of somnambulism and night terrors in children, in *Psychopharmacology: The Third Generation of Progress,* Meltzer, H.Y., Ed., Raven Press, New York, 1987, pp. 873–879.
110. Oppenheim, J.S., Skerry, J.E., Tramo, M.J., and Gazzaniga, M.S., Magnetic resonance imaging morphology of the corpus collosum in monozygotic twins, *Ann. Neurol.* 26, 100–104, 1989.
111. Osorio, I. and Daroff, R.B., Absence of REM and altered NREM sleep in patients with spinocerebellar degeneration and slow saccades, *Ann. Neurol.,* 7, 277–280, 1980.

112. Passouant, P., Cadilhac, J., and Ribstein, M., Les privations de sommeil avec mouvements oculaires par les anti-depresseurs, *Rev. Neurologique*, 127, 173–192, 1972.

113. Penfield, W. and Jasper, H., *Epilepsy and the Functional Anatomy of the Human Brain*, Little, Brown and Co., Boston, 1954.

114. Pompeiano, O., *The Neurophysiological Mechanisms of the Postural and Motor Events during REM Sleep Res. Publ.*, Association for Research in Nervous and Mental Disease, 45, 351–423, 1967.

115. Reeves, A.G. and Plum, F.. Hyperphagia, rage, and dementia accompanying a ventromedial hypothalamic neoplasm, *Arch. Neurol.*, 20, 616–624, 1969.

116. Reiser, M.F., *Mind, Brain, Body — Toward a Convergence of Psychoanalysis and Neurobiology*, Basic Books, New York, 1984.

117. Rizzo, M. and Eslinger, P.J., Colored hearing synesthesia: an investigation of neural factors, *Neurology*, 39, 781–784, 1989.

118. Rosen, G.M., Ferber, R., and Mahowald, M.W., Evaluation of parasomnias in children, *Child. Adol. Psychiatric Clin. North Am.*, 5, 601–626, 1966.

119. Rosen, G.M., Mahowald, M.W., and Ferber, R., Sleepwalking: confusional arousals and sleep terrors in the child, in *Principles and Practice of Sleep Medicine in the Child*, Ferber, R. and Kryger, M., Eds., W.B. Saunders, Philadelphia, 1995, pp. 99–106.

120. Rosenfeld, D.S., Elhajjar, A.J., Sleepsex: a variant of sleepwalking, *Arch. Sexual Behav.*, 27, 269–278, 1998.

121. Rozanski, J., Peduncular hallucinosis following vertebral angiography, *Neurology*, 2, 341–349, 1952.

122. Salva, M.A.Q. and Guillemiault, C., Olivopontocerebellar degeneration, abnormal sleep, and REM sleep without atonia, *Neurology*, 36, 576–577, 1986.

123. Sasaki, H., Sudoh, K., Hamada, K., Hamada, T., and Tashiro, K., Skeletal myoclonus in olivopontocerebellar atrophy: treatment with trihexyphenidyl, *Neurology*, 37, 1258–1262, 1987.

124. Schenck, C.H., Hurwitz, T.D., Bundlie, S.R., and Mahowald, M.W., Sleep-related injury in 100 adult patients: a polysomnographic and clinical report, *Am. J. Psychiatry*, 146, 1166–1173, 1989.

125. Schenck, C.H. and Mahowald, M.W., Review of nocturnal sleep-related eating disorders, *Int. J. Eating Disorders*, 15, 343–356, 1994.

126. Schenck, C.H. and Mahowald, M.W., REM parasomnias, *Neurol. Clin.*, 14, 697–720, 1996.

127. Schenck, C.H., Mahowald, M.W., Kim, S.W., O'Connor, K.A., and Hurwitz, T.D., Prominent eye movements during NREM sleep and REM sleep behavior disorder associated with fluoxetine treatment of depression and obsessive-compulsive disorder, *Sleep*, 15, 226–235, 1992.

128. Schenck, C.H., Milner, D., Hurwitz, T.D., Bundlie, S.R., and Mahowald, M.W., Dissociative disorders presenting as somnambulism: polysomnographic, video and clinical documentation (8 cases), *Dissociation*, II(4), 194–204, 1989.

129. Schlauch, R., Hallucinations and treatment with imipramine, *Am. J. Psychiatry*, 136, 219–220, 1979.

130. Schutte, S. and Doghramji, K., REM behavior disorder seen with venlafaxine (Effexor), *Sleep Res.*, 25, 364, 1996.

131. Shapiro, B. and Spitz, H., Problems in the differential diagnosis of narcolepsy versus schizophrenia, *Am. J. Psychiatry*, 133, 1321–1323, 1976.

132. Shik, M.L. and Orlovsky, G.N., Neurophysiology of locomotor automatism, *Physiol. Rev.*, 56, 465–501, 1976.

133. Shriomani, P.J., Armstrong, D.M., Berkowitz, A., Jeste, D.V., and Gillin, J.C., Distribution of choline acetyltransferase immunoreactive somata in the feline brainstem: implication for REM sleep generation, 11, 1–16, 1988.

134. Snider, W.D. and Johnson, E.M.J., Neurotropic molecules, *Ann. Neurol.*, 26, 489–506, 1989.

135. Soldatos, C.R. and Kales, A., Sleep disorders: research in psychopathology and its practical implications, *Acta Psychiatr. Scand.*, 65, 381–387, 1982.
136. Steiger, A., Effects of clomipramine on sleep EEG and nocturnal penile tumescence: a long-term study in a healthy man, *J. Clin. Psychopharmacol.*, 8, 349–354, 1988.
137. Steiger, A., Holboer, F., and Benkert, O., Effects of brofaramine (CGP 11 305A), a short-acting, reversible, and selective inhibitor of MAO_A on sleep, nocturnal penile tumescence and nocturnal hormonal secretion in three healthy volunteers, *Psychopharmacology*, 92, 110–114, 1987.
138. Steiger, A., Holsboer, F., and Benkert, O., Dissociation of REM sleep and nocturnal penile tumescence in volunteers treated with brofaremine, *Psychiatry Res.*, 20, 177–179, 1987.
139. Steriade, M., Arousal: revisiting the reticular activating system, *Science*, 272, 225–226, 1996.
140. Steriade, M., Ropert, N., Kitsikis, A., and Oakson, G., Ascending activating neuronal core and related rostral systems, in *The Reticular Formation Revisited*, Hobson, J.A. and Brazier, M.A.B., Eds., Raven Press, New York, 1980, 125–167.
141. Takeuchi, T., Miyasita, A., Sasaki, Y., Inugami, M., and Fukuda, K., Isolated sleep paralysis elicited by sleep interruption, *Sleep*, 15, 217–225, 1992.
142. Terzano, M.G., Parrino, L., and Spaggiari, M.C., The cyclic alternating pattern sequences in the dynamic organization of sleep, *EEG Clin. Neurophysiol.*, 69, 437–447, 1988.
143. van den Hoed, J., Lucas, E.A., and Dement, W.C., Hallucinatory experiences during cataplexy in patients with narcolepsy, *Am. J. Psychiatry*, 136, 1210–1211, 1979.
144. Vertes, R.P., Brainstem control of the events of REM sleep, *Prog. Neurobiol.*, 22, 241–288, 1984.
145. Villablanca, J. and Salinas-Zeballos, M.E., Sleep-wakefulness, EEG and behavioral studies of chronic cats without the thalamus: the "athalamic" cat, *Arch Italiennes de Biologie*, 110, 383–411, 1972.
146. Vivaldi, E., McCarley, R.W., and Hobson, J.A., Evocation of desynchronized sleep signs by chemical microstimulation of the pontine brainstem, in *The Reticular Formation Revisited*, Hobson, J.A. and Brazier, M.A.B., Eds., Raven Press, New York, 1980.
147. Webster, H.H. and Jones, B.E., Neurotoxic lesions of the dorsolateral pontomesencephalic tegmentum-cholinergic cell area in the cat. II. Effects on sleep-waking states, *Brain Res.*, 458, 285–302, 1988.
148. Wills, N. and Chase, M.H., Brain stem control of masseteric reflex activity during sleep and wakefulness: mesencephalon and pons, *Exp. Neurol.*, 64, 98–117, 1979.
149. Zeplin, H., Mammalian sleep, in *Principles and Practice of Sleep Medicine,* Kryger, M.H., Roth, T., and Dement, W.C., Eds., W.B. Saunders, Philadelphia, 1989, pp. 30–49.
150. Zuccone, M., Oldani, A., Ferini-Strambi, L., and Smirne, S., Arousal fluctuations in non-rapid eye movement parasomnias: the role of cyclic alternating pattern as a measure of sleep instability, *J. Clin. Neurophysiol.*, 12, 147–154, 1995.
151. Zucker, D.K., Austin, F.M., and Branchey, L., Variables associated with alcoholic blackouts in men, *Am. J. Drug Alcohol Abuse*, 11, 295–302, 1985.

Section III

Neuroanatomical and Neurochemical Basis of Behavioral States

Kazue Semba, Section Editor

The concept of the ascending reticular activating system was originally proposed by Moruzzi and Magoun in 1949. Although its anatomical substrate remained obscure for decades, modern anatomical and physiological techniques have greatly facilitated the delineation of its components, neurochemical coding, inputs, and organization. Thus, it is now generally viewed that ascending cholinergic, monoaminergic, and glutamatergic pathways represent key components of the reticular activating system for wakefulness; that the ascending drive carried by these pathways is mostly relayed in the basal forebrain, thalamus, and hypothalamus en route to the cerebral cortex; and that the mesopontine cholinergic system has a critical role in rapid eye movement (REM) sleep. Furthermore, the mechanisms of behavioral state control via these pathways are now understood in considerable detail at synaptic levels and in cellular and molecular terms. With molecular biological techniques it is now also possible to investigate the gene expression in sleep-/wake-related neurons during different behavioral states. The chapters included in this section summarize these recent advances in the understanding of neuroanatomical and neurochemical substrates of behavioral state control.

Acetylcholine has long been thought to have a role in both REM sleep and wakefulness, and, in particular, cholinergic neurons in the mesopontine tegmentum have been implicated in both of these qualitatively different behavioral states. To obtain clues to understanding how the same population of cholinergic neurons can be involved in the control of both states, Semba discusses in her chapter evidence indicating anatomical and physiological heterogeneity of the mesopontine cholinergic neurons. It is proposed that differential projections of physiological subtypes of cholinergic neurons and differences in presynaptic mechanisms regulating acetylcholine release might explain the dual role of mesopontine cholinergic neurons in REM sleep and waking.

Serotonergic neurons in the raphe nuclei and noradrenergic neurons in the locus coeruleus have long been known for their characteristic state-dependent discharge patterns in which firing dramatically decreases from waking to slow-wave sleep to REM sleep. This has led to the widely held view that the monoaminergic neurons are involved in behavioral state control. However, reviewing recent single unit recording and *in vivo* microdialysis data from their lab for serotonergic raphe neurons, Jacobs and Fornal propose that the role of serotonergic neurons in state control might be secondary to their involvement in motor activity, particularly in tonic and repetitive activity, and its integration with sensory information processing.

Mechanisms underlying the decrease in firing rate of the monoaminergic neurons during sleep are further discussed by Luppi, Peyron, Rampon, Gervasoni, Barbagli, Boissard, and Fort in their chapter. On the basis of their pharmacoelectrophysiological data in unanesthetized animals and anatomical evidence, it is suggested that the monoaminergic neurons are under inhibitory control by GABA and glycine inputs originating from several distant sources. Of these, the glycine inputs appear to be active at similar levels across behavioral states, whereas the GABA inputs appear to vary in their magnitude and therefore might hold a key to the selective decrease in the monoaminergic neurons activity during sleep and in particular in REM sleep.

The basal forebrain is a major forebrain structure that relays ascending impulses from the reticular activating system to the cortex. The basal forebrain is known for its cholinergic neurons which supply most of acetylcholine released in the cortex. Acetylcholine is released at high levels during active states of the brain such as waking and REM sleep. However, in addition to the cholinergic neurons, a substantial number of non-cholinergic (including GABAergic) basal forebrain neurons project to the cortex. Jones and Mühlethaler review recent anatomical, electrophysiological, and pharmacological work from their labs on these two types of neurons and discuss their roles in mediating monoaminergic and glutamatergic inputs from the brainstem to the cortex to regulate the cortical EEG and behavioral arousal.

Given the time frame of behavioral states, it is likely that there are changes in gene expression across the sleep/wake cycle and that these changes are part of the regulatory mechanisms for behavioral state control. Immediate-early genes are of particular interest, because they are expressed in response to various stimuli, with their proteins acting as third messengers to mediate long-term plasticity. Bentivoglio and Grassi-Zucconi review studies using immediate-early genes as neuronal activation markers and discuss the significance and issues in the expression of these genes in the brain during normal sleep/wake states, as well as in sleep deprivation.

Chapter 11

The Mesopontine Cholinergic System: A Dual Role in REM Sleep and Wakefulness

Kazue Semba

Contents

11.1	The Role of Mesopontine Cholinergic Neurons in REM Sleep	162
11.2	The Role of Mesopontine Cholinergic Neurons in Wakefulness	163
	11.2.1 Activation of the Thalamus	163
	11.2.2 Activation of the Basal Forebrain	163
11.3	Anatomical Substrates for the Dual Role	164
	11.3.1 Efferent Projections and Axonal Collateralization	164
	11.3.2 Neurotransmitter Co-localization	165
	11.3.3 Ultrastructure	166
11.4	Physiological Substrates for the Dual Role	167
	11.4.1 *In Vivo* Studies	167
	11.4.2 *In Vitro* Studies	167
11.5	Modulatory inputs for the Dual Role	168
	11.5.1 Monoamines and Acetylcholine	168
	11.5.2 Glutamate and GABA	169
	11.5.3 Non-Classical Transmitters and Peptides	170
11.6	The Role of Non-Cholinergic Mesopontine Neurons	171
11.7	Mechanisms for the Dual Function of Mesopontine Cholinergic Neurons: Hypotheses	172
Acknowledgments		173
References		173

The mesopontine tegmentum contains cholinergic neurons that project widely to the forebrain and brainstem. These neurons are distributed in the pedunculopontine tegmental (PPT) and laterodorsal tegmental (LDT) nuclei. The mesopontine cholinergic neurons are implicated in two basic behavioral states: rapid eye movement (REM) sleep and wakefulness. These two states, in contrast to non-REM sleep, represent an active state of the brain that is characterized by the predominance of low-voltage fast activity in the cortical electroencephalogram (EEG). Except for this feature, however, the two states are rather different. REM sleep is accompanied by atonia (loss of muscle tone), respiratory depression, poikilothermia (absence of temperature regulation), and bursts of saccadic eye movements; arousal thresholds to external stimuli are raised, and most of dreaming occurs during REM sleep. Obviously, none of these phasic and tonic features is present in wakefulness. How, then, could the same population of cholinergic neurons be involved in both of these qualitatively different states and thus have a dual role in behavioral state control? In search for an answer to this question, this review will briefly summarize evidence for the involvement of mesopontine cholinergic neurons in REM sleep and wakefulness, and then discuss the anatomical and functional heterogeneity of these neurons that might provide clues to their dual role in wakefulness and REM sleep. Due to space limitations, it is not possible to cite all original papers, and the reader is referred to the cited reviews for this information.

11.1 The Role of Mesopontine Cholinergic Neurons in REM Sleep

Cholinergic neurons in the mesopontine tegmentum have long been thought to play a critical role in the generation of REM sleep (for reviews, see References 30, 78, 92, 102). Since the late 1960s, it has been known that microinjections of carbachol into the pontine reticular formation induce a REM sleep-like state. Neurons in this region start to depolarize prior to the onset of REM sleep and remain depolarized throughout a REM sleep episode. Administration of carbachol depolarizes the majority of neurons in this region through activation of non-M1, probably M2, muscarinic receptors. The cholinergic input to this region has been identified anatomically as originating in the mesopontine tegmentum.[64] More recently, chronic single-unit studies have shown that a subpopulation of neurons in the LDT and PPT become more active prior to and during REM sleep (see Section 11.4.1). Furthermore, excitotoxic lesions of the LDT and PPT, which destroyed 60% of cholinergic neurons (as well as 35% of noradrenergic neurons), virtually eliminated REM sleep without affecting wakefulness.[120] Finally, recent studies have shown that stimulation of LDT/PPT induces acetylcholine (ACh) release,[58] elicits long-latency excitatory postsynaptic potentials which are sensitive to systemic scopolamine[24] in the pontine reticular formation, and increases REM sleep time.[115]

These findings collectively suggest that REM sleep is induced by an increase in the activity of cholinergic neurons in the LDT and PPT, which in turn results in an increase in ACh release in the pontine reticular formation. ACh then depolarizes neurons in the pontine reticular formation, thus activating the efferent pathways involved in phasic (rapid eye movements, ponto-geniculo-occipital [PGO] spikes, and muscle twitches) and tonic (EEG activation, respiratory depression, and muscle atonia) events of REM sleep. Cholinergic mesopontine neurons also project to the medullary reticular formation, and this projection is thought to be involved in sensory and motor inhibition during REM sleep. The pontine reticular formation neurons might in turn provide a glutamatergic excitatory feedback to the LDT/PPT to maintain cholinergic tone and, thus, REM sleep.[36]

It should be noted that ACh is not the only input to the REM sleep induction zone of the pontine reticular formation. This region also receives monoaminergic inputs from the locus coeruleus, the A5 and A7 groups, and the raphe nuclei, as well as the serotonergic B9 group.[84] The response to the monoamine transmitters varies among reticular neurons, probably depending on the function of each neuron.[60] However, serotonin appears to have mainly inhibitory action on reticular neurons.

Consistent with this, release of serotonin in the pontine reticular formation is at the lowest levels during REM sleep, when reticular neurons fire at the highest levels.[29] In contrast, release of glutamate peaks in REM sleep, presumably contributing to the depolarization.[15] Together, these findings suggest that the activity of pontine reticular neurons are regulated through state-dependent interplay of cholinergic, serotonergic, noradrenergic, and glutamatergic inputs.

11.2 The Role of Mesopontine Cholinergic Neurons in Wakefulness

Tegmental cholinergic neurons have long been thought to have an important role in behavioral wakefulness and cortical arousal as part of the ascending reticular activating system.[91] Furthermore, chronic single-unit studies have shown that the majority of LDT and PPT neurons are active in wakefulness as well as in REM sleep, compared with slow-wave sleep (see Section 11.4.1). The activity of LDT/PPT neurons is conveyed to a number of forebrain structures including the thalamus and the basal forebrain, i.e., two structures with a critical role in cortical arousal. Both of these structures seem to be important for cortical EEG activation, because lesions of either one of the two structures reduced but did not abolish cortical EEG activation.[3,101,107]

11.2.1 Activation of the Thalamus

As a gatekeeper of the cerebral cortex, the thalamus relays sensory inputs of all modalities except olfaction to relevant regions of the cerebral cortex and, in addition, controls cortical excitability through diffuse projections to all cortical regions. Thalamocortical neurons are, in turn, constantly modulated by both sensory and non-sensory inputs of various origins, including the cholinergic tegmental input[100] (see also Chapter 20 in this book). The ascending cholinergic projections innervate the entire thalamus and represent a major brainstem input to this structure, exceeding, in terms of cell numbers, both noradrenergic and serotonergic projections from the pons.[14] Furthermore, a recent electron microscopic study indicated that cholinergic synapses are as numerous as corticothalamic synapses on lateral geniculate neurons, suggesting that the mesopontine cholinergic input plays an important role in sensory relay.[12] ACh has been shown to depolarize thalamic relay neurons, whereas it hyperpolarizes GABAergic interneurons as well as GABAergic neurons in the reticular nucleus.[2,61] Through these actions, ACh can switch the firing mode of relay neurons from the burst mode to the single spike mode and thus facilitate sensory transmission.[62,100] Stimulation of the PPT/LDT region has similar effects on thalamic neurons. These include blockade of spindles and delta oscillations, which are characteristic of slow-wave sleep,[23,103] and depolarization of relay neurons[6,22,35,44] with an increase in membrane resistance.[7] In addition, PPT stimulation also potentiates fast oscillations (20 to 40 Hz) of thalamocortical neurons, a possible correlate of higher cognitive functions.[101] Functionally, the cholinergic projections from the mesopontine tegmentum (as well as the basal forebrain) might be of particular importance during REM sleep, because in the absence of monoamine release, ACh will provide one excitatory input to thalamocortical relay neurons to keep them from hyperpolarizing into the spindle and delta oscillation modes.

11.2.2 Activation of the Basal Forebrain

There is strong evidence that basal forebrain neurons have an important role in cortical arousal (for reviews, see Reference 83 and Chapters 14 and 18 in this book). The basal forebrain contains cholinergic neurons that project widely to the cortex. There are also non-cholinergic basal forebrain neurons projecting to the cortex, some of which are GABAergic. The basal forebrain also sends projections to a number of thalamic nuclei and other subcortical structures.[1,68,85] Thus, basal forebrain neurons are positioned to influence cortical excitability both directly and indirectly through the thalamus, in each case via both cholinergic and non-cholinergic pathways. The relative importance among these basal forebrain projections in regulating cortical excitability is not clear. However, there is some evidence to suggest that the basal forebrain projection to the thalamus, rather than that to the cortex, plays a predominant role in regulating the cortical EEG.[73]

Although the basal forebrain receives a wide range of inputs from the forebrain and brainstem, the input from the PPT and LDT appears to provide one of the most powerful excitatory inputs. Many basal forebrain neurons fire in relation to the cortical EEG; the majority of them increase firing rate during EEG fast waves, while some increase activity during EEG slow waves.[8] Interestingly, train stimulation of the PPT activates fast-wave-active basal forebrain neurons, but inhibits slow-wave-active neurons,[9] consistent with the opposite roles of these neurons in cortical EEG regulation.[8] PPT stimulation also induces ACh release in the basal forebrain, and the released ACh appears to be derived from the ascending afferent terminals rather than from dendrites or local collaterals of basal forebrain neurons.[5,111] Stimulation of PPT also evokes cortical ACh release and strong EEG activation.[73] Interestingly, however, the cortical EEG activation and ACh release were blocked by infusion, in the basal forebrain, of the non-selective glutamate receptor antagonist kynurenic acid, but not by atropine.[72] These findings suggest that the effects of PPT stimulation are mediated by glutamatergic, but not cholinergic, transmission in the basal forebrain. Furthermore, α-amino-3-hydroxy-5-methylisoxazole-4-propionic acid (AMPA) and N-methyl-D-aspartate (NMDA) receptors in the basal forebrain appear to be differentially involved in cortical ACh release and EEG activation.[73] The source of glutamate in the basal forebrain is not clear, but it could be released by terminals of cholinergic neurons in the PPT because some of these neurons are known to contain glutamate.[47] Alternately, it could be released from fibers of non-cholinergic, glutamatergic PPT neurons, or glutamatergic neurons elsewhere that are activated by the latter. The functional significance of co-release of ACh and glutamate *in vivo* remains to be investigated. Released separately, glutamate would be excitatory, whereas ACh would be mostly inhibitory to cholinergic basal forebrain neurons. One consequence of co-release might be ACh-induced enhancement of bursting activity elicited by glutamate acting at NMDA receptors.[37] The transmitter content notwithstanding, mesopontine tegmental neurons appear to provide one of the most powerful excitatory drives either directly or indirectly to the thalamus, as well as to cholinergic basal forebrain neurons.

11.3 Anatomical Substrates for the Dual Role

11.3.1 Efferent Projections and Axonal Collateralization

Anatomically, cholinergic neurons in the mesopontine tegmentum are distributed in a continuum encompassing the PPT and LDT (see Reference 85 for review). The cholinergic cell population in these two nuclei are occasionally referred to as the Ch5 and Ch6 groups, respectively. These neurons give rise to both ascending and descending projections. Cholinergic projections arising from the PPT and LDT show both overlap and segregation.

The ascending projections innervate various forebrain and midbrain structures including the thalamus and basal forebrain, as discussed above, as well as the lateral preoptic area, lateral hypothalamus, substantia nigra, pretectal area, and superior colliculus.[85] These projections scarcely reach the cerebral cortex, with the exceptions of the medial prefrontal cortex in the rat[118] and the visual cortex in the cat.[17]

The thalamus is a major target of mesopontine cholinergic projections.[85] Cholinergic neurons in the PPT innervate all thalamic nuclei, whereas those in the LDT project preferentially to the

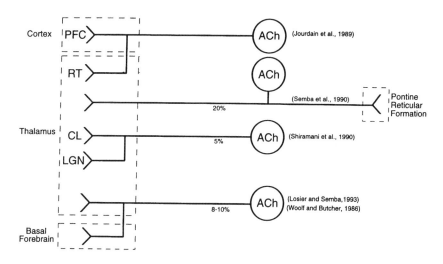

FIGURE 11.1
Examples of collateralization of cholinergic projections from the PPT and LDT.

limbic as well as certain associational and intralaminar nuclei. The density of cholinergic neuropil is not uniform; the highest densities are found in the anteroventral, reticular, lateral mediodorsal, and intralaminar nuclei, whereas all the other thalamic nuclei contain light to moderate densities.[52] Single mesopontine cholinergic neurons appear to innervate more than one thalamic nucleus in the cat,[67,105] and a subpopulation (5%) of mesopontine cholinergic neurons has been shown to innervate both the lateral geniculate nucleus and intralaminar (central-lateral) nuclei in the rat.[90] Such arrangements might concurrently facilitate sensory transmission across all modalities. In addition, about 10% of mesopontine cholinergic neurons innervating the thalamus appear to give off axon collaterals to the prefrontal cortex[31] or basal forebrain (Figure 11.1).[55,124]

The descending cholinergic projections innervate the regions of the pontine reticular formation where carbachol has been shown to induce a REM sleep-like state in animals.[64] Interestingly, the majority of cholinergic neurons projecting to the pontine reticular formation appear to innervate the thalamus, as well.[88] The significance of this collateralization is discussed in Section 11.7. The descending cholinergic projections extend more caudally, to innervate the medullary reticular formation (see Reference 85 for a review). These projections are probably involved in motor inhibition and muscle atonia during REM sleep, as well as eye movement control, acoustic startle response, and other functions.[27,39,60,74] The cholinergic component of the descending projections does not appear to reach the spinal cord, although non-cholinergic PPT neurons do innervate the spinal cord. Like the ascending projections, the descending cholinergic projections collateralize within the lower brainstem; certain cholinergic neurons project to both the vestibular nuclei and the caudal pontine reticular nucleus or to the raphe magnus and the posterior thalamus.[125]

11.3.2 Neurotransmitter Co-localization

Strong evidence indicates that cholinergic neurons contain additional neurotransmitters, raising the possibility of transmitter co-release from their terminals[85] (Table 11.1). One such transmitter is glutamate, present in a subpopulation of cholinergic neurons in the PPT and LDT. This is of particular interest because, if the detected glutamate represents a transmitter pool rather than the metabolic pool of glutamate, it could be released from cholinergic axon terminals. A number of neuropeptides are also present in mesopontine cholinergic neurons: atriopeptin in all cholinergic

TABLE 11.1
Co–localization of Neurotransmitters in Cholinergic Neurons in the PPT and LDT

Type	Transmitters	Percentage	Species	Ref.
Small–molecule transmitter	Glutamate	Subpopulation	Cat	46
		40% (PPT)	Monkey	47
Neuropeptide transmitters	Atriopeptin	100%	Rat	94
	Bombesin/GRP	30%	Rat	119
	CRF	30%	Rat	119
	Substance P	15–40%	Rat	94, 110, 118, 119
Non–classical transmitter	Nitric oxide (nitric oxide synthase)	100%	Rat	108, 117
		90–95% (PPT)	Monkey, human	76
		100%	Human	63

neurons, and bombesin, corticotropin-releasing factor (CRF), and substance P in subpopulations of cholinergic neurons. These neuropeptides might be released from cholinergic terminals in an activity-dependent manner, as demonstrated in the autonomic ganglia. These peptides could act both pre- and post-synaptically to modulate the release and the postsynaptic action of ACh.

Another substance of interest is nitric oxide synthase, which is present in virtually all mesopontine cholinergic neurons.[117] There is good evidence that nitric oxide is released by the terminals of mesopontine cholinergic neurons in the thalamus. This is based on the findings that the concentration of nitric oxide in the thalamus is correlated with the activity level of cholinergic neurons in the LDT,[123] and electrical stimulation of the LDT increases nitric oxide concentrations,[65] as well as local blood flow in the thalamus.[44] Nitric oxide has been reported to enhance ACh release in the pontine reticular formation,[51] which suggests that this might occur in other structures with cholinergic innervation. At the level of the soma, however, nitric oxide might diminish the strength of glutamatergic input, thereby reducing excitation of postsynaptic cholinergic neurons.[50]

The above findings indicate that cholinergic PPT and LDT neurons are heterogeneous with respect to co-localized neurotransmitters, in that co-localization is often seen in a subpopulation of, but not in all, cholinergic neurons. One possibility, then, is that the cholinergic projection to a given structure contains all of these neuropeptides. Alternatively, neuropeptide content might vary according to the projection, so that a given peptide might be released in one target but not in another. This possibility would have important functional implications not only for behavioral state control but also for basic understanding of transmitter systems, and remains to be investigated.

11.3.3 Ultrastructure

The ultrastructure is one anatomical feature that appears to be rather homogeneous among cholinergic neurons across the mesopontine tegmentum.[20,93,97] Cholinergic neurons are among the larger neurons in both PPT and LDT and appear to receive more synapses than the non-cholinergic neurons intermixed with them. Interestingly, about a quarter of the surface of cholinergic neuron in cross section was found to be covered with astrocytic processes, and some cholinergic neurons were directly apposed to an astrocyte. Furthermore, cholinergic dendrites as well as axons were often seen in close apposition to endothelial cells of blood capillaries or pericytes. The dendrites near capillaries might receive non-synaptic input from blood-borne substances. Most (85%) of the terminals presynaptic to cholinergic neurons contained small, round, and clear vesicles, but some contained dense-cored vesicles as well. The other (15%) presynaptic terminals contained flat vesicles.

Interestingly, in comparison with basal forebrain cholinergic neurons, tegmental cholinergic neurons appear to receive a greater number of synaptic contacts at the soma, suggesting that their activity may be under more direct synaptic control.

11.4 Physiological Substrates for the Dual Role

11.4.1 *In Vivo* Studies

Single-unit recording studies using unanesthetized animals have identified three different firing patterns among PPT/LDT neurons across the sleep/wake cycle in cats[10,66,99] and rats.[34] These studies collectively showed that more than half of the PPT/LDT neurons recorded were active during both waking and REM sleep compared with non-REM sleep (W/REM-on neurons). The majority of these neurons were fast firing (>10 spikes per sec in waking). In contrast, close to half of the recorded neurons exhibited higher firing rates in REM sleep than in non-REM sleep or wakefulness (REM-on neurons). These tended to have slower firing rates (<2 spikes per sec in waking) than W/REM-on neurons. They varied in the duration of action potentials, some with broad and others with narrow action potentials. Some of the REM-on neurons fired tonically, while others fired phasically, but all started to increase firing rate prior to the onset of REM sleep. In addition to these two main cell types, 5 to 10% of the sampled population exhibited phasic activities in relation to PGO spikes, which are prominent phasic field potentials recorded during REM sleep.[66,98] Antidromic activation from various thalamic nuclei indicated that at least some neurons in each of these three classes projected to the thalamus. Pharmacologically, a slower firing subclass of W/REM-on neurons in the PPT/LDT was consistently inhibited by carbachol, which was taken to suggest that they are probably cholinergic.[11] Responses of REM-on neurons to carbachol varied; those with broad action potentials were inhibited, whereas those with short action potential were excited.[79]

In conclusion, as a group, mesopontine tegmental neurons become active during episodes of EEG activation, and this is consistent with the notion that the ascending cholinergic pathways are a component of the reticular activating system. However, when single neurons are considered, there is significant heterogeneity in their firing patterns across the sleep/wake cycle. In particular, most tegmental neurons appear to fire either in both REM sleep and wakefulness or in REM sleep alone. Microdialysis data suggest that at least some, if not all, of these two types of neurons are cholinergic (see Section 11.7). No clear picture emerges at present with respect to the relationship between the state-dependent firing patterns, on one hand, and the shape of action potential *in vivo*, efferent projections, or pharmacological properties of these neurons, on the other.

11.4.2 *In Vitro* Studies

Electrophysiological studies using brainstem slices have identified three types of neuron in the LDT and PPT in guinea pig[49] and rat[32,33,112–114] (see also Chapter 17 in this book). Type I neurons exhibit low-threshold calcium spikes, whereas type II neurons exhibit an A current, and type III neurons display both. There has been controversy regarding the transmitter phenotype of these three cell types. Different authors identified either type II or type III neurons as cholinergic, although all agreed that type I neurons are non-cholinergic. The discrepancy might be due to differences in the region investigated (LDT vs. PPT) or the age (immature vs. adult) and/or species (rat vs. guinea pig) of the animals used. Morphologically, type II and type III neurons have larger cell bodies and more dendrites than type I neurons. Type II and type III neurons are thought to project to the thalamus on the basis of retrograde labeling[49] and appear to correspond to "broad spike" neurons described by Koyama and colleagues in their *in vivo* study.[42] "Broad spike" neurons *in vivo* have been shown to be cholinergic.[43] Type II and type III PPT neurons have been classified further by Takakusaki and colleagues[113] into short- and long-duration action potential neurons. Short-duration neurons were able to fire faster than long-duration neurons, and this and other physiological characteristics of these two types of neurons have led the authors to suggest that short-duration neurons correspond to W/REM-on neurons, whereas long-duration neurons are likely to be REM-on neurons identified *in vivo*.

In conclusion, like the *in vivo* data discussed above, the available *in vitro* data clearly show considerable heterogeneity in the membrane properties of mesopontine cholinergic neurons. It is noteworthy that this heterogeneity sharply contrasts with the homogeneity seen with monoaminergic neurons (see Chapters 12 and 16).

11.5 Modulatory Inputs for the Dual Role

The information on the modulatory inputs to cholinergic neurons is critical for understanding how the various firing patterns of cholinergic tegmental neurons are regulated in different behavioral states (for reviews, see References 50 and 60 and Chapter 17 in this book). Main afferents to both PPT and LDT arise in the brainstem reticular formation, monoaminergic nuclei, midbrain central gray, and lateral hypothalamus-zona incerta region.[86] Interestingly, the PPT and LDT are interconnected with each other, both rostrocaudally and bilaterally. In addition, the LDT and PPT receive additional, relatively selective inputs: the LDT from the medial prefrontal cortex and lateral habenula, and the PPT from the central nucleus of the amygdala, superior colliculus, and somatosensory relay nuclei in the medulla and dorsal horn. Both the LDT and PPT also receive monoaminergic inputs from the locus coeruleus and the raphe nuclei. Thus, both the PPT and LDT appear to receive limbic and autonomic, as well as sensorimotor afferents. The transmitters used by most of these inputs are yet to be identified, but many inputs probably use glutamate.

11.5.1 Monoamines and Acetylcholine

The LDT and PPT receive serotonergic raphe and noradrenergic locus coeruleus afferents.[86,95] These are of particular interest because both the monoaminergic cell populations display state-dependent firing patterns and are thought to have a permissive role in REM sleep induction by allowing cholinergic neurons to depolarize when they cease to fire.[18] Indeed, microinjections of serotonin into the LDT reduce total REM sleep time, without affecting REM sleep episode frequency, suggesting that cholinergic mechanisms have a role in the maintenance, but not the initiation, of REM sleep in unanesthetized rat.[21] Consistent with this inhibitory role, serotonin hyperpolarizes cholinergic neurons *in vitro* through activation of $5-HT_{1A}$ receptors[56] with an increase in inwardly rectifying K^+ conductances.[48] Interestingly, it was recently reported that serotonin, through $5-HT_{1A}$ receptors, inhibits REM-on neurons but not W/REM-on neurons *in vivo*.[116] Noradrenaline also has an inhibitory action on cholinergic neurons via activation of the α2 receptor and an increase in inwardly rectifying K^+ currents.[122] Electron microscopic studies have confirmed that both serotonergic[19,96] and noradrenergic axon terminals[45] synapse with cholinergic neurons. In particular, the proximal location of serotonin inputs is consistent with powerful inhibition of these neurons by serotonin. However, serotonergic synapses are not numerous, indeed representing only a small portion of synaptic inputs to cholinergic neurons. Given the powerful inhibitory effects, the paucity of serotonergic synapses might suggest a technical problem of sensitivity in detecting serotonergic terminals or an additional, paracrine action of serotonin.

Acetylcholine is likely to be a major player in the regulation of mesopontine tegmental neurons because of the dense cholinergic network interconnecting the LDT and PPT. Like serotonin and noradrenaline, ACh has an inhibitory action on mesopontine tegmental neurons through non-M1 receptors.[42,48,57] This is corroborated by the anatomical evidence that cholinergic PPT/LDT neurons express the m2 subtype of muscarinic receptors.[77] Interestingly, ACh, serotonin, and noradrenaline all appear to induce hyperpolarization by increasing the same or overlapping populations of inwardly rectifying potassium conductances. Inhibitory effects of these transmitters have also been reported *in vivo*.[42] However, it is important to note that, although all of these inputs appear to be inhibitory, functional implications are slightly different in that cholinergic inhibition is expected to be in effect in both wakefulness and REM sleep, whereas monoaminergic inhibition will be present in wakefulness only.

Histamine has been implicated in arousal, and histaminergic neurons in the tuberomammilllary nucleus show physiological properties similar to monoamines (see Chapter 19). However, unlike serotonin and noradrenaline, administration of histamine excites PPT neurons that have an A-current, a characteristic of the majority of cholinergic neurons in guinea pig brainstem slices.[38] The descending histaminergic system has been suggested to enhance wakefulness by activating ascending cholinergic pathways from the mesopontine tegmentum.[54]

11.5.2 Glutamate and GABA

Local electrical stimulation evokes both NMDA and non-NMDA receptor-mediated excitatory synaptic responses in the guinea pig LDT and PPT cholinergic neurons *in vitro*,[80,81] suggesting that these cholinergic neurons normally receive glutamatergic inputs. There is some anatomical evidence to suggest that the stoicheometry of ionotropic glutamate subunits (GluR1-7 and NMDAR1) on cholinergic neurons is non-uniform, and this might induce variability in their response to glutamate;[25] however, such physiological heterogeneity has not yet been reported. There are four possible sources of glutamatergic afferents to mesopontine cholinergic neurons. One is the medial prefrontal cortex. This is based on the report that stimulation of the medial prefrontal cortex, which is known to project to the LDT,[82] induces excitation of LDT neurons that can be blocked by non-NMDA receptor antagonists.[16] Another source of glutamatergic input is the axon terminals of cholinergic tegmental neurons which contain glutamate (see Section 11.3.2). A third source of glutamatergic input is the subthalamic nucleus, which projects to the PPT. Electrical stimulation of the subthalamic nucleus induced excitatory postsynaptic responses in "type II" (mostly cholinergic) PPT neurons.[113] Finally, the pontine and midbrain reticular formation contains numerous glutamate-immunoreactive neurons[46] and sends substantial projections to the LDT and PPT.[86,95] Functionally, the midbrain reticular formation contains many wake-active neurons,[104] whereas the pontine reticular formation contains REM-on neurons,[28] and both cell populations probably use glutamate as a transmitter.[106] Therefore, these midbrain and pontine reticular inputs may contribute significantly to the increased firing of cholinergic neurons during wakefulness and REM sleep, respectively.

The presence of GABAergic input to cholinergic mesopontine neurons is suggested by the finding that application of picrotoxin or bicuculline blocked inhibitory postsynaptic potentials (IPSPs) evoked by local electrical stimulation in cholinergic PPT/LDT neurons.[80] Through its inhibitory action at the $GABA_A$ receptor, GABA could switch the firing pattern of cholinergic neurons from the single-spike mode to the bursting mode with low-threshold calcium spikes. There are two possible sources of GABAergic input to the LDT/PPT. One is GABAergic neurons intermingled with cholinergic neurons in the same region (see Section 11.6).[30] A second source, probably selective to cholinergic neurons in the PPT, is the GABAergic neurons in the substantia nigra pars reticulata.[113] In addition to the $GABA_A$ receptor, GABA could act at the $GABA_B$ receptor to induce effects similar to those of ACh and serotonin by activating the same signaling pathways.[48]

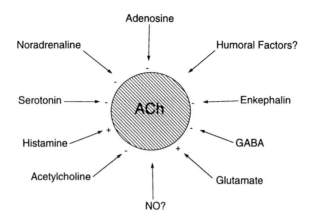

FIGURE 11.2
The modulatory inputs to mesopontine cholinergic neurons that have been identified in electrophysiological studies using brain slices *in vitro*.

11.5.3 Non-Classical Transmitters and Peptides

Adenosine is a metabolic neuromodulator suggested to promote sleep by inhibiting cholinergic neurons in the mesopontine tegmentum and basal forebrain.[71] In response to adenosine A_1 receptor antagonists, cholinergic neurons in *in vitro* slices increased firing rate, with a decrease in K^+ conductance and an increase in the hyperpolarization-activated current I_h. These findings suggest that metabolically derived extracellular adenosine tonically inhibits cholinergic neurons. This might explain an increase in the delta component in the EEG during prolonged wakefulness which is associated with an increase in extracellular adenosine levels in the basal forebrain.[70]

Recent work has investigated the role of nitric oxide synthase in mesopontine cholinergic neurons. There is evidence to suggest that activation of postsynaptic NMDA receptors at the cell body, leading to Ca^{++} entry, can produce nitric oxide, which in turn provides negative feedback onto the activated NMDA receptors in an autocrine fashion.[50]

Although the LDT/PPT region contains fiber terminals with various neuropeptides, little work has been done to investigate whether they provide synaptic inputs to cholinergic neurons. However, galanin-containing fibers, possibly originating in the hypothalamus, have been shown to form "pericellular baskets" around substance P-containing, most likely cholinergic, neurons in the human PPT.[13] Electrophysiologically, opiates inhibit possibly cholinergic PPT neurons with an A-current through activation of μ-receptors in guinea pig brainstem slices.[89]

In conclusion, tegmental cholinergic neurons appear to receive numerous synaptic and other modulatory inputs (Figure 11.2). Many of these inputs appear to be inhibitory, although a small number of excitatory inputs have also be identified, including glutamatergic and histaminergic inputs. To understand the significance of these inputs in the control of state-dependent firing patterns of mesopontine neurons, it is necessary to know the activity patterns of the input source neurons. From all we know about these inputs, it would appear that, during wakefulness, W/REM-on neurons must have powerful glutamatergic and histaminergic inputs to overcome the inhibition exerted by their monoaminergic and cholinergic inputs. The glutamate input might come from the reticular formation in the midbrain which contains many wake-active neurons. It is possible that the inhibitory inputs to W/REM-on neurons are less powerful than to REM-on neurons. For REM-on neurons, the glutamatergic input could be absent or relatively weak during wakefulness, so that they may remain hyperpolarized by monoamines and acetylcholine. During REM sleep, however, in the absence of the inhibitory monoaminergic inputs, both REM-on and W/REM-on neurons would be less hyperpolarized, and this and an additional glutamatergic input would maintain high levels of their activity. This glutamate input might arise at least in part from the pontine reticular formation as a feedback loop. Thus, it is crucial to understand the responses of REM-on and W/REM-on neurons to these different neurotransmitters originating from various sources, but very little is currently known about this key question.

11.6 The Role of Non-Cholinergic Mesopontine Neurons

Cholinergic neurons are not the only cell type found in the PPT and LDT, as non-cholinergic neurons are also present intermixed with cholinergic neurons. Because the two types of neurons probably interact with each other at local levels, it would be worth summarizing what is known about them, which is relatively little.[27,75] Non-cholinergic neurons are heterogeneous in terms of size and the number of synapses they receive. Some non-cholinergic neurons are large and receive more synapses than cholinergic neurons. However, the majority of non-cholinergic neurons appear to be smaller and receive fewer synapses than cholinergic neurons.[20,93,97] Non-cholinergic neurons are likely to contain GABA,[30] glutamate,[4,46,47] and/or neuropeptides (vasoactive intestinal peptide [VIP], CRF, neurotensin, or galanin).[109] Some non-cholinergic neurons project to the spinal cord, medullary reticular formation, and the basal ganglia, and they might represent part of the efferent systems for REM sleep phenomenology (e.g., muscle atonia, REM, PGO-spikes, EEG activation, and respiratory depression). Other non-cholinergic neurons might be interneurons and regulate the activity of cholinergic projection neurons locally. Indeed, a recent abstract reported that intracellularly labeled non-cholinergic neurons exhibited processes surrounding adjacent cholinergic neurons in the LDT.[41]

There is some evidence to suggest that non-cholinergic neurons do not necessarily receive the same synaptic inputs or show the same physiological responses as cholinergic neurons. For example, non-cholinergic, rather than cholinergic, neurons appear to be the main recipient of synapses from cholinergic terminals in the LDT and PPT. Physiologically, like cholinergic neurons, non-cholinergic neurons are probably inhibited by ACh, because 95% of cells tested in the LDT, regardless of transmitter phenotype, were inhibited.[57] Similarly, like cholinergic neurons, non-cholinergic neurons are depolarized by NMDA.[80] A small number of serotonergic synapses are made with both cholinergic and non-cholinergic neurons.[19] Serotonin hyperpolarized some non-cholinergic neurons, while eliciting no response in the others.[56] The reports on the response to noradrenaline are inconsistent; non-cholinergic neurons responded to noradrenaline with mixed responses[122] or with depolarization through activation of $\alpha 1$ adrenergic receptors.[41] Despite this inconsistency, however, the excitatory responses seen in at least some non-cholinergic neurons are in contrast to hyperpolarization consistently seen with cholinergic neurons. If some of these non-cholinergic neurons that are excited by noradrenaline turn out to be GABAergic interneurons, noradrenaline could inhibit cholinergic neurons both directly and via interneurons.

In conclusion, it is clear that non-cholinergic neurons vary in their morphology, projection, and neurotransmitter content. Non-cholinergic neurons probably respond to ACh in the same way as cholinergic neurons but differently to noradrenaline. However, most of the available data on the physiology and pharmacology of non-cholinergic neurons are based on a very small sample size, and more vigorous and detailed investigation is required. It is particularly important in future studies to investigate whether non-cholinergic neurons represent interneurons, interact with cholinergic neurons locally, and have different pharmacological responses to transmitters compared with cholinergic neurons.

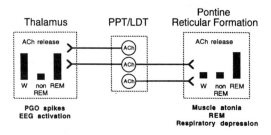

FIGURE 11.3

A schematic to summarize three sets of data with PPT/LDT neurons that are relevant to their role in behavioral state control, in particular, control of REM sleep and wakefulness: ACh release in their key target structures (the thalamus and REM sleep-inducing regions of the pontine reticular formation), efferent projections and axonal collateralization, and firing patterns across the sleep/wake cycle. Two hypotheses are proposed to explain the differential release of ACh in the two target structures. One hypothesis supposes a correlation between the firing pattern and the projections of PPT/LDT neurons. A second hypothesis proposes that presynaptic heteroreceptor activation inhibits ACh release in the reticular formation during wakefulness but not in the thalamus (see the text for detail).

11.7 Mechanisms for the Dual Function of Mesopontine Cholinergic Neurons: Hypotheses

The currently available information from the anatomical and functional data reviewed above is obviously still insufficient for presenting a clear picture about the dual role of mesopontine cholinergic neurons. However, one piece of information that brings useful insight into the dual function of mesopontine cholinergic neurons comes from recent studies using *in vivo* microdialysis to measure ACh release in unanesthetized animals. These studies have shown that ACh levels in the pontine reticular formation are elevated selectively during natural REM sleep[40] or in a carbachol-induced REM sleep-like state.[59] In contrast, in the thalamus, ACh concentrations are high not only in REM sleep but also in wakefulness, compared with slow-wave sleep.[69,121] It should be noted that in the latter study[121] microdialysis probes avoided the reticular nucleus, so that the only source of ACh measured was that released from terminals of PPT and LDT cholinergic neurons.

A paradox arises when these microdialysis data are compared with the chronic single unit data (Figure 11.3). As discussed earlier, chronic single-unit recordings from LDT and PPT neurons have shown that there are basically two types of neurons: W/REM-on and REM-on. Although the cholinergic phenotype of these neuronal types remains to be investigated, these firing patterns are, at first glance, at odds with the microdialysis data showing the different patterns of ACh release in the thalamus and pontine reticular formation. Two explanations may be offered. One is that cholinergic tegmental neurons that release ACh in the thalamus are different from those responsible for ACh release in the reticular formation. More specifically, W/REM-on cholinergic neurons and some REM-on neurons, as well, are primarily responsible for the ACh release in the thalamus (and probably in the basal forebrain), and these neurons may be responsible for cortical EEG activation during wakefulness and REM sleep. In contrast, mesopontine neurons releasing ACh in the reticular formation are primarily REM-on cholinergic neurons. Some REM-on neurons must project to both structures, because a small number of REM-on neurons were antidromically activated from the thalamus,[34,79] and more than half of cholinergic PPT/LDT neurons that innervate the carbachol-sensitive site of pontine reticular formation also project to the thalamus.[88]

A second possible explanation, not exclusive to the first, is that the differential release is due to differences in the presynaptic regulation of ACh release in the two structures, and, more specifically, to presence or absence of certain presynaptic heteroreceptors on cholinergic terminals. According to this scenario, cholinergic terminals in the thalamus would lack inhibitory adrenergic and serotonergic receptors, and, therefore, ACh release would not be affected in wakefulness when monoamines are massively released. In contrast, cholinergic terminals in the carbachol-sensitive pontine reticular formation would have presynaptic heteroreceptors, the activation of which would inhibit ACh release during wakefulness. Williams and colleagues[121] proposed that ACh release in the reticular formation is regulated presynaptically by GABA or glycine of a local origin. We have hypothesized that the heteroreceptors involved are adrenergic and serotonergic, and we tested this hypothesis with respect to noradrenaline. Preliminary data are consistent with this hypothesis.[87] The presynaptic regulation of transmitter release in behavioral state control is only beginning to be understood,[53] and it could have important functional implications.

In conclusion, the information available so far on mesopontine cholinergic neurons strongly indicates that, unlike monoaminergic neurons, mesopontine cholinergic neurons as a population have a dual role in behavioral state control, i.e., they are involved in both wakefulness and REM sleep. Furthermore, these cholinergic neurons appear to display two different firing patterns across the sleep/wake cycle: W/REM-on and REM-on. We hypothesize that these two types of neurons have different targets to innervate, such that W/REM-on neurons primarily project to the thalamus and REM-on neurons project to the reticular formation, as well as the thalamus. With their primarily ascending projections, W/REM-on neurons may be mainly responsible for cortical activation and sensory facilitation. We also hypothesize that ACh release from these two targets is regulated differently so that, in the reticular formation, it is under inhibitory control by presynaptic heteroreceptors to monoamines that are present on cholinergic terminals, but no such heteroreceptors are present on cholinergic terminals in the thalamus. In addition, it is possible that the activity of W/REM-on and REM-on neurons is regulated by different transmitters or receptors, and these neurons may also differ in their co-localized transmitters or basic membrane properties. There are many other open questions regarding the control of REM sleep, including the mechanisms of terminating a REM sleep episode, and cyclic occurrence of REM sleep. The effort to solve these questions might be aided by developing a selective neurotoxin to mesopontine cholinergic neurons, as has been the case with the study of basal forebrain cholinergic neurons.[26] Thus, combined behavioral, physiological, anatomical, and molecular approaches would be necessary to understand the biological phenomenon of sleep and wakefulness at the cellular and molecular levels.

Acknowledgments

I would like to thank Doug Rasmusson for his helpful comments on an early version of the manuscript, and Joan Burns for her technical assistance. This work was supported by the Medical Research Council (MT-14451) and the Scottish Rite Charitable Foundation of Canada.

References

1. Asanuma, C. and Porter, L.L., Light and electron microscopic evidence for a GABAergic projection from the caudal basal forebrain to the thalamic reticular nucleus in rats, *J. Comp. Neurol.*, 302, 159–172, 1990.

2. Ben-Ari, Y., Kanazawa, I., and Kelly, J.S., Exclusively inhibitory action of iontophoretic acetylcholine on single neurones of feline thalamus, *Nature*, 259, 327–329, 1976.

3. Buzsáki, G., Bickford, R.G., Ponomareff, G., Thal, L.J., Mandel, R., and Gage, F.H., Nucleus basalis and thalamic control of neocortical activity in the freely moving rat, *J. Neurosci.*, 8, 4007–4026, 1988.

4. Clements, J.R. and Grant, S., Glutamate-like immunoreactivity in neurons of the laterodorsal tegmental and pedunculopontine nuclei in the rat, *Neurosci. Lett.*, 120, 70–73, 1990.

5. Consolo, S., Bertorelli, R., Forloni, G.L., and Butcher, L.L., Cholinergic neurons of the pontomesencephalic tegmentum release acetylcholine in the basal nuclear complex of freely moving rats, *Neuroscience*, 37, 717–723, 1990.

6. Curro Dossi, R., Pare, D., and Steriade, M., Short-lasting nicotinic and long-lasting muscarinic depolarizing responses of thalamocortical neurons to stimulation of mesopontine cholinergic nuclei, *J. Neurophysiol.*, 65, 393–406, 1991.

7. Deschênes, M. and Hu, B., Membrane resistance increase induced in thalamic neurons by stimulation of brainstem cholinergic afferents, *Brain Res.*, 513, 339–342, 1990.

8. Détári, L., Rasmusson, D.D., and Semba, K., Phasic relationship between the activity of basal forebrain neurons and cortical EEG in urethane-anesthetized rat, *Brain Res.*, 759, 112–121, 1997.

9. Détári, L., Semba, K., and Rasmusson, D.D., Responses of cortical EEG-related basal forebrain neurons to brainstem and sensory stimulation in urethane-anesthetized rat, *Eur. J. Neurosci.*, 9, 1153–1161, 1997.

10. El Mansari, M., Sakai, K., and Jouvet, M., Unitary characteristics of presumptive cholinergic tegmental neurons during the sleep-waking cycle in freely moving cats, *Exp. Brain Res.*, 76, 519–529, 1989.

11. El Mansari, M., Sakai, K., and Jouvet, M., Responses of presumed cholinergic mesopontine tegmental neurons to carbachol microinjections in freely moving cats, *Exp. Brain Res.*, 83, 115–123, 1990.

12. Erisir, A., Van Horn, S.C., and Sherman, S.M., Relative numbers of cortical and brainstem inputs to the lateral geniculate nucleus, *Proc. Natl. Acad. Sci. U.S.A.*, 94, 1517–1520, 1997.

13. Gai, W.P., Blumbergs, P.C., Geffen, L.B., and Blessing, W.W., Galanin–containing fibers innervate substance P-containing neurons in the pedunculopontine tegmental nucleus in humans, *Brain Res.*, 618, 135–141, 1993.

14. Hallanger, A.E., Levey, A.I., Rye, D.B., and Wainer, B.H., The origins of cholinergic and other subcortical afferents to the thalamus in the rat, *J. Comp. Neurol.*, 262, 105–124, 1987.

15. Hasegawa, T., Azuma, S., Kimura, M., and Inoué, S., State-dependent changes of amino acids in the rat oral pontine reticular nuclei measured by microdialysis, *Soc. Neurosci. Abstr.*, 23, 2132, 1997.

16. Highfield, D. and Grant, S.J., Electrophysiological evidence for an excitatory amino acid pathway from medial prefrontal cortex to laterodorsal tegmental nucleus and rostral locus coeruleus, *Soc. Neurosci. Abstr.*, 15, 644, 1989.

17. Higo, S., Matsuyama, T., and Kawamura, S., Direct projections from the pedunculopontine and laterodorsal tegmental nuclei to area 17 of the visual cortex in the cat, *Neurosci. Res.*, 26, 109–118, 1996.

18. Hobson, J.A., McCarley, R.W., and Wyzinski, P.W., Sleep cycle oscillation: reciprocal discharge by two brainstem neuronal groups, *Science*, 189, 55–58, 1975.

19. Honda, T. and Semba, K., Serotonergic synaptic input to cholinergic neurons in the mesopontine tegmentum in rat, *Brain Res.*, 647, 299–306, 1994.

20. Honda, T. and Semba, K., An ultrastructural study of cholinergic and non-cholinergic neurons in the laterodorsal and pedunculopontine tegmental nuclei in rat, *Neuroscience*, 68, 837–853, 1995.

21. Horner, R.L., Sanford, L.D., Annis, D., Pack, A.I., and Morrison, A.R., Serotonin at the laterodorsal tegmental nucleus suppresses rapid-eye-movement sleep in freely moving rats, *J. Neurosci.*, 17, 7541–7552, 1997.
22. Hu, B., Steriade, M., and Deschênes, M., The effects of brainstem peribrachial stimulation on neurons of the lateral geniculate nucleus, *Neuroscience*, 31, 13–24, 1989.
23. Hu, B., Steriade, M., and Deschênes, M., The effects of brainstem peribrachial stimulation on perigeniculate neurons: the blockage of spindle waves, *Neuroscience*, 31, 1–12, 1989.
24. Imon, H., Ito, K., Dauphin, L., and McCarley, R.W., Electrical stimulation of the cholinergic laterodorsal tegmental nucleus elicits scopolamine-sensitive excitatory postsynaptic potentials in medial pontine reticular formation neurons, *Neuroscience*, 74, 393–401, 1996.
25. Inglis, W.L., and Semba, K., Co–localization of ionotropic glutamate receptor subunits with NADPH-diaphorase-containing neurons in the rat mesopontine tegmentum, *J. Comp. Neurol.*, 368, 17–32, 1996.
26. Inglis, W.L. and Semba, K., Discriminable excitotoxic effects of ibotenic acid, AMPA, NMDA and quinolinic acid in the rat laterodorsal tegmental nucleus, *Brain Res.*, 755, 17–27, 1997.
27. Inglis, W.L. and Winn, P., The pedunculopontine tegmental nucleus: where the striatum meets the reticular formation, *Prog. Neurobiol.*, 47, 1–29, 1995.
28. Ito, K. and McCarley, R.W., Alterations in membrane potential and excitability of cat medial pontine reticular formation neurons during changes in naturally occurring sleep-wake states, *Brain Res.*, 292, 169–175, 1984.
29. Iwakiri, H., Matsuyama, K., and Mori, S., Extracellular levels of serotonin in the medial pontine reticular formation in relation to sleep/wake cycle in cats: a microdialysis study, *Neurosci. Res.*, 18, 157–170, 1993.
30. Jones, B.E., Paradoxical sleep and its chemical/structural substrates in the brain, *Neuroscience*, 40, 637–656, 1991.
31. Jourdain, A., Semba, K., and Fibiger, H.C., Basal forebrain and mesopontine tegmental projections to the reticular thalamic nucleus: an axonal collateralization and immunohistochemical study in the rat, *Brain Res.*, 505, 55–65, 1989.
32. Kamondi, A., Williams, J.A., Hutcheon, B., and Reiner, P.B., Membrane properties of mesopontine cholinergic neurons studied with the whole-cell patch-clamp technique: implications for behavioral state control, *J. Neurophysiol.*, 68, 1359–1372, 1992.
33. Kang, Y. and Kitai, S.T., Electrophysiological properties of pedunculopontine neurons and their postsynaptic responses following stimulation of substantia nigra reticulata, *Brain Res.*, 535, 79–95, 1990.
34. Kayama, Y., Ohta, M., and Jodo, E., Firing of "possibly" cholinergic neurons in rat laterodorsal tegmental nucleus during sleep and wakefulness, *Brain Res.*, 569, 210–220, 1992.
35. Kayama, Y., Takagi, M., and Ogawa, T., Cholinergic influence of the laterodorsal tegmental nucleus on neuronal activity in the rat lateral geniculate nucleus, *J. Neurophysiol.*, 56, 1297–1309, 1986.
36. Keifer, J. C., Baghdoyan, H. A., and Lydic, R., Pontine cholinergic mechanisms modulate the cortical electroencephalographic spindles of halothane anesthesia, *Anesthesiology*, 84, 945–954, 1996.
37. Khateb, A., Fort, P., Williams, S., Serafin, M., Jones, B.E., and Mühlethaler, M., Modulation of cholinergic nucleus basalis neurons by acetylcholine and N-methyl-D-aspartate, *Neuroscience*, 81, 47–55, 1997.
38. Khateb, A., Serafin, M., and Mühlethaler, M., Histamine excites pedunculopontine neurones in guinea pig brainstem slices, *Neurosci. Lett.*, 112, 257–262, 1990.
39. Koch, M., Kungel, M., and Herbert, H., Cholinergic neurons in the pedunculopontine tegmental nucleus are involved in the mediation of prepulse inhibition of the acoustic startle response in the rat, *Exp. Brain Res.*, 97, 71–82, 1993.

40. Kodama, T., Takahashi, Y., and Honda, Y., Enhancement of acetylcholine release during paradoxical sleep in the dorsal tegmental field of the cat brain stem, *Neurosci. Lett.*, 114, 277–282, 1990.
41. Kohlmeier, K.A. and Reiner, P.B., Noradrenaline excites a subpopulation of non–cholinergic LDT neurons, *Soc. Neurosci. Abstr.*, 23, 1065, 1997.
42. Koyama, Y. and Kayama, Y., Mutual interactions among cholinergic, noradrenergic and serotonergic neurons studied by ionophoresis of these transmitters in rat brainstem nuclei, *Neuroscience*, 55, 1117–1126, 1993.
43. Koyama, Y., Honda, T., Kusakabe, M., Kayama, Y., and Sugiura, Y., *In vivo* electrophysiological distinction of histochemically identified cholinergic neurons using extracellular recording and labeling in rat laterodorsal tegmental nucleus, *Neuroscience*, 83, 1105–1112, 1998.
44. Koyama, Y., Toga, T., Kayama, Y., and Sato, A., Regulation of regional blood flow in the laterodorsal thalamus by ascending cholinergic nerve fibers from the laterodorsal tegmental nucleus., *Neurosci. Res.*, 20, 79–84, 1994.
45. Kubota, Y., Leung, E., and Vincent, S.R., Ultrastructure of cholinergic neurons in the laterodorsal tegmental nucleus of the rat: interaction with catecholamine fibers, *Brain Res. Bull.*, 29, 479–491, 1992.
46. Lai, Y.Y., Clements, J.D., and Siegel, J.M., Glutamatergic and cholinergic projections to the pontine inhibitory area identified with horseradish peroxidase retrograde transport and immunohistochemistry, *J. Comp. Neurol.*, 336, 321–330, 1993.
47. Lavoie, B. and Parent, A., Pedunculopontine nucleus in the squirrel monkey: distribution of cholinergic and monoaminergic neurons in the mesopontine tegmentum with evidence for the presence of glutamate in cholinergic neurons, *J. Comp. Neurol.*, 344, 190–209, 1994.
48. Leonard, C.L. and Llinas, R., Serotonergic and cholinergic inhibition of mesopontine cholinergic neurons controlling REM sleep: an *in vitro* electrophysiological study, *Neuroscience*, 59, 309–330, 1994.
49. Leonard, C.S. and Llinas, R., Electrophysiology of mammalian pedunculopontine and laterodorsal tegmental neurons in vitro: implications for the control of REM sleep, in *Brain Cholinergic Systems*, Steriade, M. and Biesold, D., Eds., Oxford University Press, New York, 1990, pp. 205–223.
50. Leonard, C.S., Rao, S., and Sanchez, R.M., Patterns of neuromodulation and the nitric oxide signalling pathway in mesopontine cholinergic neurons, *Semin. Neurosci.*, 7, 319–328, 1995.
51. Leonard, T.O. and Lydic, R., Pontine nitric oxide modulates acetylcholine release, rapid eye movement sleep generation, and respiratory rate, *J. Neurosci.*, 17, 774–785, 1997.
52. Levey, A.I., Hallanger, A.E., and Wainer, B.H., Choline acetyltransferase immunoreactivity in the rat thalamus, *J. Comp. Neurol.*, 257, 317–332, 1987.
53. Li, X.Y., Rainnie, D.G., McCarley, R.W., and Greene, R.W., Nicotinic pre– and post–synaptic effects in dorsal raphe nucleus, *Soc. Neurosci. Abstr.*, 22, 1740, 1996.
54. Lin, J.-S., Hou, Y., Sakai, K., and Jouvet, M., Histaminergic descending inputs to the mesopontine tegmentum and their role in the control of cortical activation and wakefulness in the cat, *J. Neurosci.*, 16, 1523–1537, 1996.
55. Losier, B. and Semba, K., Dual projections of single cholinergic and aminergic brainstem neurons to the thalamus and basal forebrain in the rat, *Brain Res.*, 604, 41–52, 1993.
56. Luebke, J., Greene, R., Semba, K., Kamondi, A., McCarley, R., and Reiner, P., Serotonin hyperpolarizes cholinergic low-threshold burst neurons in the rat laterodorsal tegmental nucleus *in vitro*, *Proc. Natl. Acad. Sci. USA*, 89, 743–747, 1992.
57. Luebke, J.I., McCarley, R.W., and Greene, R.W., Inhibitory action of muscarinic agonists on neurons in the rat laterodorsal tegmental nucleus *in vitro*, *J. Neurophysiol.*, 70, 2128–2135, 1993.
58. Lydic, R. and Baghdoyan, H.A., Pedunculopontine stimulation alters respiration and increases ACh release in the pontine reticular formation, *Am. J. Physiol.*, 264, R544–R554, 1993.

59. Lydic, R., Baghdoyan, H.A., and Lorinc, Z., Microdialysis of cat pons reveals enhanced acetylcholine release during state-dependent respiratory depression, *Am. J. Physiol.*, 261, R766–R770, 1991.
60. McCarley, R.W., Greene, R.W., Rainnie, D., and Portas, C.M., Brainstem neuromodulation and REM sleep, *Semin. Neurosci.*, 7, 341–354, 1995.
61. McCormick, D.A., Neurotransmitter actions in the thalamus and cerebral cortex and their role in neuromodulation of thalamocortical activity, *Prog. Neurobiol.*, 39, 337–388, 1992.
62. McCormick, D.A. and Prince, D.A., Actions of acetylcholine in the guinea pig and cat medial and lateral geniculate nuclei, *in vitro, J. Physiol.*, 392, 147–165, 1987.
63. Mesulam, M.-M., Geula, C., Bothwell, M.A., and Hersh, L.B., Human reticular formation: cholinergic neurons of the pedunculopontine and laterodorsal tegmental nuclei and some cytochemical comparisons to forebrain cholinergic neurons, *J. Comp. Neurol.*, 281, 611–633, 1989.
64. Mitani, A., Ito, K., Hallanger, A.H., Wainer, B.H., Kataoka, K., and McCarley, R.W., Cholinergic projections from the laterodorsal and pedunculopontine tegmental nuclei to the pontine gigantocellular tegmental field in the cat, *Brain Res.*, 451, 397–402, 1988.
65. Miyazaki, M., Kayama, Y., Kihara, T., Kawasaki, K., Yamaguchi, E., Wada, Y., and Ikeda, M., Possible release of nitric oxide from cholinergic axons in the thalamus by stimulation of the rat laterodorsal tegmental nucleus as measured with voltametry, *J. Chem. Neuroanat.*, 10, 203–207, 1996.
66. Nelson, J.P., McCarley, R.W., and Hobson, J.A., REM sleep burst neurons, PGO waves and eye movement information, *J. Neurophysiol.*, 50, 784–797, 1983.
67. Paré, D., Smith, Y., Parent, A., and Steriade, M., Projections of brainstem core cholinergic and noncholinergic neurons of cat to intralaminar and reticular thalamic nuclei, *Neuroscience*, 25, 69–86, 1988.
68. Parent, A., Pare, D., Smith, Y., and Steriade, M., Basal forebrain cholinergic and noncholinergic projections to the thalamus and brainstem in cats and monkeys, *J. Comp. Neurol.*, 277, 281–301, 1988.
69. Phillis, J.W., Tebécis, A.K., and York, D.H., Acetylcholine release from the feline thalamus, *J. Pharm. Pharmac.*, 20, 476–478, 1968.
70. Porkka-Heiskanen, T., Strecker, R.E., Thakkar, M., Bjørkum, A.A., Greene, R.W., and McCarley, R.W., Adenosine: a mediator of the sleep-inducing effects of prolonged wakefulness, *Science*, 276, 1265–1268, 1997.
71. Rainnie, D.G., Grunze, H.C.R., McCarley, R.W., and Greene, R.W., Adenosine inhibition of mesopontine cholinergic neurons: Implications for EEG arousal, *Science*, 263, 689–692, 1994.
72. Rasmusson, D.D., Clow, K., and Szerb, J.C., Modification of neocortical acetylcholine release and EEG desynchronization due to brainstem stimulation by drugs applied to the basal forebrain, *Neuroscience*, 60, 665–677, 1994.
73. Rasmusson, D.D., Szerb, J.C., and Jordan, J.L., Differential effects of α-amino-3-hydroxy-5-methyl-4-isoxazole propionic acid and N-methyl-D-aspartate receptor antagonists applied to the basal forebrain on cortical acetylcholine release and EEG desynchronization, *Neuroscience*, 72, 419–427, 1996.
74. Reese, N.B., Garcia-Rill, E., and Skinner, R.D., The pedunculopontine nucleus — auditory input, arousal and pathophysiology, *Prog. Neurobiol.*, 42, 105–133, 1995.
75. Rye, D.B., Contributions of the pedunculopontine region to normal and altered REM sleep, *Sleep*, 20, 757–788, 1997.
76. Rye, D.B., Perez, J., and Levey, A.I., Neurochemical markers of the primate pedunculopontine nucleus (PPN) and surround, *Soc. Neurosci. Abstr.*, 20, 334, 1994.
77. Rye, D.B., Thomas, J., and Levey, A.I., Distribution of molecular muscarinic (m1–m4) receptor subtypes and choline acetyltransferase in the pontine reticular formation of man and non-human primates, *Sleep Res.*, 24, 59, 1995.

78. Sakai, K., Executive mechanisms of paradoxical sleep, *Arch. Ital. Biol.*, 126, 239–257, 1988.
79. Sakai, K. and Koyama, Y., Are there cholinergic and non-cholinergic paradoxical sleep-on neurones in the pons?, *NeuroReport*, 7, 2449–2453, 1996.
80. Sanchez, R. and Leonard, C.S., NMDA receptor-mediated synaptic input to nitric oxide synthase-containing neurons of the guinea pig mesopontine tegmentum in vitro, *Neurosci. Lett.*, 179, 141–144, 1994.
81. Sanchez, R. and Leonard, C.S., NMDA-receptor-mediated synaptic currents in guinea pig laterodorsal tegmental neurons in vitro, *J. Neurophysiol.*, 76, 1101–1111, 1996.
82. Satoh, K. and Fibiger, H.C., Cholinergic neurons of the laterodorsal tegmental nucleus: efferent and afferent connections, *J. Comp. Neurol.*, 323, 387–410, 1986.
83. Semba, K., The cholinergic basal forebrain: a critical role in cortical arousal, in *The Basal Forebrain: Anatomy to Function, Adv. Exp. Med. Biol.*, Vol. 295, Napier, T.C., Kalivas, P.W., and Hanin, I., Eds., Plenum Press, New York, 1991, pp. 197–218.
84. Semba, K., Aminergic and cholinergic afferents to REM sleep induction regions of the pontine reticular formation in the rat, *J. Comp. Neurol.*, 330, 543–556, 1993.
85. Semba, K. and Fibiger, H.C., Organization of central cholinergic systems, in *Progress in Brain Research*, Vol. 79, Nordberg, A., Fuxe, K., Holmstedt, B., and Sundwall, A., Eds., Elsevier, New York, 1989, pp. 37–63.
86. Semba, K. and Fibiger, H.C., Afferent connections of the laterodorsal and the pedunculopontine tegmental nuclei in the rat: a retro- and antero-grade transport and immunohistochemical study, *J. Comp. Neurol.*, 323, 387–410, 1992.
87. Semba, K., Greene, R.W., Rasmusson, D.D., McCarley, R.W., and Weider, J., Noradrenergic presynaptic inhibition of acetylcholine release in the rat pontine reticular formation: an *in vitro* electrophysiological and *in vivo* microdialysis study, *Soc. Neurosci. Abstr.*, 23, 1065, 1997.
88. Semba, K., Reiner, P.B., and Fibiger, H.C., Single cholinergic mesopontine tegmental neurons project to both the pontine reticular formation and the thalamus in the rat, *Neuroscience*, 38, 643–654, 1990.
89. Serafin, M., Khateb, A., and Mühlethaler, M., Opiates inhibit pedunculopontine neurones in guinea pig brainstem slices, *Neurosci. Lett.*, 119, 125–128, 1990.
90. Shiromani, P., Floyd, C., and Velazquez–Moctezuma, S., Pontine cholinergic neurons simultaneously innervate two thalamic targets, *Brain Res.*, 532, 317–322, 1990.
91. Shute, C.C.D. and Lewis, P.R., The ascending cholinergic reticular system: neocortical, olfactory and subcortical projections, *Brain*, 90, 497–522, 1967.
92. Siegel, J.M., Brainstem mechanisms generating REM sleep, in *Principles and Practice of Sleep Medicine*, Kryger, M.H., Roth, T., and Dement, W.C., Eds., W.B. Saunders, Philadelphia, 1989, pp. 104–120.
93. Spann, B.M. and Grofova, I., Cholinergic and non–cholinergic neurons in the rat pedunculopontine tegmental nucleus, *Anat. Embryol.*, 186, 215–227, 1992.
94. Standaert, D.G., Saper, C.B., Rye, D.B., and Wainer, B.H., Co-localization of atriopeptin-like immunoreactivity with choline acetyltransferase- and substance P-like immunoreactivity in the pedunculopontine and laterodorsal tegmental nuclei in the rat, *Brain Res.*, 382, 163–168, 1986.
95. Steininger, T.L., Rye, D.B., and Wainer, B.H., Afferent projections to the cholinergic pedunculopontine tegmental nucleus and adjacent midbrain extrapyramidal area in the albino rat. I. Retrograde tracing studies, *J. Comp. Neurol.*, 321, 515–543, 1992.
96. Steininger, T.L., Wainer, B.H., Blakely, R.D., and Rye, D.B., Serotonergic dorsal raphe nucleus projections to the cholinergic and noncholinergic neurons of the pedunculopontine tegmental region: a light and electron microscopic anterograde tracing and immunohistochemical study, *J. Comp. Neurol.*, 382, 302–322, 1997.

97. Steininger, T.L., Wainer, B.H., and Rye, D.B., Ultrastructural study of cholinergic and noncholinergic neurons in the pars compacta of the rat pedunculopontine tegmental nucleus, *J. Comp. Neurol.*, 382, 285–301, 1997.

98. Steriade, M., Pare, D., Datta, S., Oakson, G., and Curro Dossi, R., Different cellular types in mesopontine cholinergic nuclei related to pont-geniculo-occipital waves, *J. Neurosci.*, 10, 2560–2579, 1990.

99. Steriade, M., Datta, S., Pare, D., Oakson, G., and Curro Dossi, R., Neuronal activities in brain-stem cholinergic nuclei related to tonic activation processes in thalamocortical systems, *J. Neurosci.*, 10, 2541–2559, 1990.

100. Steriade, M., Cholinergic blockade of network- and intrinsically generated slow oscillations promotes waking and REM sleep activity patterns in thalamic and cortical neurons, *Prog. Brain Res.*, 98, 345–355, 1993.

101. Steriade, M., Curro Dossi, R., Pare, D., and Oakson, G., Fast oscillations (20–40 Hz) in thalamocortical systems and their potentiation by mesopontine cholinergic nuclei in the cat, *Proc. Natl. Acad. Sci. USA*, 88, 4396–4400, 1991.

102. Steriade, M. and McCarley, R.W., *Brainstem Control of Wakefulness and Sleep*, Plenum Press, New York, 1990, 499 pp.

103. Steriade, M., McCormick, D.A., and Sejnowski, T.J., Thalamocortical oscillations in the sleeping and aroused brain, *Science*, 262, 679–685, 1993.

104. Steriade, M., Oakson, G., and Ropert, N., Firing rates and patterns of midbrain reticular neurons during steady and transitional states of the sleep-waking cycle, *Exp. Brain Res.*, 46, 37–51, 1982.

105. Steriade, M., Pare, D., Parent, A., and Smith, Y., Projections of cholinergic and non–cholinergic neurons of the brainstem core to relay and associational thalamic nuclei in the cat and macaque monkey, *Neuroscience*, 25, 47–67, 1988.

106. Stevens, D.R., McCarley, R.W., and Greene, R.W., Excitatory amino acid-mediated responses and synaptic potentials in medial pontine reticular formation neurons of the rat *in vitro*, *J. Neurosci.*, 12, 4188–4194, 1992.

107. Stewart, D.J., McFabe, D.F., and Vanderwolf, C.H., Cholinergic activation of the electrocorticogram: role of the substantia innominata and effects of atropine and quinuclidinyl benzilate, *Brain Res.*, 322, 219–232, 1984.

108. Sugaya, K. and McKinney, M., Nitric oxide synthase gene expression in cholinergic neurons in the rat brain examined by combined immunohistochemistry and *in situ* hybridization histochemistry, *Mol. Brain Res.*, 23, 111–125, 1994.

109. Sutin, E.L. and Jacobowitz, D.M., Immunocytochemical localization of peptides and other neurochemicals in the rat laterodorsal tegmental nucleus and adjacent area, *J. Comp. Neurol.*, 270, 243–270, 1988.

110. Sutin, E.L. and Jacobowitz, D.M., Localization of substance P mRNA in cholinergic cells of the rat laterodorsal tegmental nucleus: *in situ* hybridization histochemistry and immunocytochemistry, *Cell. Mol. Neurobiol.*, 10, 19–31, 1990.

111. Szerb, J.C., Clow, K., and Rasmusson, D.D., Pharmacological but not physiological modulation of cortical acetylcholine release by cholinergic mechanisms in the nucleus basalis magnocellularis, *Can. J. Physiol. Pharmacol.*, 72, 893–898, 1994.

112. Takakusaki, K. and Kitai, S.T., Ionic mechanisms involved in the spontaneous firing of tegmental pedunculopontine nucleus neurons of the rat, *Neuroscience*, 78, 771–794, 1997.

113. Takakusaki, K., Shoroyama, T., and Kitai, S.T., Two types of cholinergic neurons in the rat tegmental pedunculopontine nucleus: electrophysiological and morphological characterization, *Neuroscience*, 79, 1089–1109, 1997.

114. Takakusaki, K., Shoroyama, T., Yamamoto, T., and Kitai, S.T., Cholinergic and noncholinergic tegmental pedunculopontine projection neurons in rats revealed by intracellular labeling, *J. Comp. Neurol.*, 371, 345–361, 1996.

115. Thakkar, M., Portas, C., and McCarley, R.W., Chronic low-amplitude electrical stimulation of the laterodorsal tegmental nucleus of freely moving cats increases REM sleep, *Brain Res.*, 723, 223–227, 1996.

116. Thakkar, M.M., Strecker, R.E., and McCarley, R.W., Behavioral state control through differential serotonergic inhibition in the mesopontine cholinergic nuclei: a simultaneous unit recording and microdialysis study, *J. Neurosci.*, 18, 5490–5497, 1998.

117. Vincent, S.R., Satoh, K., Armstrong, D.M., and Fibiger, H.C., NADPH-diaphorase: a selective histochemical marker for the cholinergic neurons of the pontine reticular formation, *Neurosci. Lett.*, 43, 31–36, 1983.

118. Vincent, S.R., Satoh, K., Armstrong, D.M., and Fibiger, H.C., Substance P in the ascending cholinergic reticular system, *Nature*, 306, 688–691, 1983.

119. Vincent, S.R., Satoh, K., Armstrong, D.M., Panula, P., Vale, W., and Fibiger, H.C., Neuropeptides and NADPH–diaphorase activity in the ascending cholinergic reticular system of the rat, *Neuroscience*, 17, 167–182, 1986.

120. Webster, H.H. and Jones, B.E., Neurotoxic lesions of the dorsolateral pontomesencephalic tegmentum–cholinergic cell area in the cat. II. Effects upon sleep-waking states, *Brain Res.*, 458, 285–302, 1988.

121. Williams, J.A., Comisarow, J., Day, J., Fibiger, H.C., and Reiner, P.B., State-dependent release of acetylcholine in rat thalamus measured by *in vivo* microdialysis, *J. Neurosci.*, 14, 5236–5242, 1994.

122. Williams, J.A. and Reiner, P.B., Noradrenaline hyperpolarizes identified rat mesopontine cholinergic neurons *in vitro*, *J. Neurosci.*, 13, 3878–3883, 1993.

123. Williams, J.A., Vincent, S.R., and Reiner, P.B., Nitric oxide production in rat thalamus changes with behavioral state, local depolarization, and brainstem stimulation, *J. Neurosci.*, 17, 420–427, 1997.

124. Woolf, N.J. and Butcher, L.L., Cholinergic systems in the rat brain. III. Projections from the pontomesencephalic tegmentum to the thalamus, tectum, basal ganglia, and basal forebrain, *Brain Res. Bull.*, 16, 603–637, 1986.

125. Woolf, N.J. and Butcher, L.L., Cholinergic systems in the rat brain. IV. Descending projections of the pontomesencephalic tegmentum, *Brain Res. Bull.*, 23, 519–540, 1989.

Chapter 12

An Integrative Role for Serotonin in the Central Nervous System

Barry L. Jacobs and Casimir A. Fornal

Contents

12.1	Activity of Serotonergic Neurons in Behaving Cats	182
	12.1.1 Circadian and Sleep/Wake Cycles	182
	12.1.2 Challenges/Stressors	185
	12.1.3 Motor Activity	186
	12.1.4 PGO Waves and Hippocampal Theta	188
	12.1.5 Conclusions	189
12.2	Release of Serotonin in Forebrain Sites	189
	12.2.1 Circadian and Sleep/Wake Cycles	189
	12.2.2 Challenges/Stressors	190
	12.2.3 Conclusions	190
12.3	General Conclusions	191
References		191

Research on brain monoamine neurotransmitters began to flourish in the late 1960s due to several important advances. First, elucidation of the localization of the cell bodies, axonal pathways, and terminal arborizations of these neurons was provided by means of fluorescence histochemistry. Second, the electrophysiological identification and characterization of these neurons were accomplished through single cell recordings. Third, specific pharmacological tools for affecting these neurons and their receptors were developed.

Because of these advances, theories and hypotheses about the function of these neurons evolved and grew, becoming more specific and sophisticated. For example, we no longer believe that these neurotransmitters exert exclusively simple excitatory or inhibitory synaptic actions. This is due in large part to the discovery of the multiplicity of receptors upon which these neurotransmitters act. Similarly, we no longer conceive of individual monoamine neurotransmitters as having exclusive relationships to particular behavioral or physiological processes such as sleep, nociception, feeding, etc. And, reciprocally, we no longer believe that such behavioral and physiological processes are mediated primarily by individual neurotransmitters.

A perusal of the chapter titles in this volume exemplifies this complexity by making it clear that behavioral state is controlled by a number of different neurochemicals. This raises the obvious issue of whether the individual contributions mediating such processes can be parcelled out, or ascribed, to particular neurotransmitters. In this chapter, we attempt to do this for serotonin (5-hydroxytryptamine), within the context of a general or overarching theory of serotonergic function in the vertebrate central nervous system (CNS).

Virtually all CNS serotonin derives from a small number of neurons (tens of thousands), localized almost exclusively near the midline of the brainstem, in the raphe nuclei. Emanating from this site, serotonergic axonal projections and their terminal arborizations invade almost the entire neuraxis, from the caudal spinal cord to the frontal cortex. The fact that this is a phylogenetically old and largely conserved system, that it is among the first to develop with ontogeny, and that its cell bodies are localized in the medial aspect of the brainstem, the most primitive portion of the brain, all have important implications when considering the functional role of this system.

In the present chapter, we propose an integrative role for serotonin in the CNS. We believe that serotonin's involvement in behavioral state control is a by-product of its more fundamental role in motor control and its integration with sensory information processing and autonomic regulation. This review will focus on the two aspects of research from our laboratory that gave rise to this hypothesis: analysis of serotonergic neuronal activity in conscious animals and brain microdialysis measurements of changes in extracellular levels of serotonin in freely moving animals.

12.1 Activity of Serotonergic Neurons in Behaving Cats

For the past 15 years, we have been intensively studying the activity of brain serotonergic neurons in the domestic cat. We have carried out these experiments under a large variety of behavioral, environmental, and physiological conditions. While no series of studies such as this can be truly exhaustive, we have made a strong attempt to "cast our net widely" so as to provide a broad search for the conditions that significantly influence these cells. Our studies can be grossly divided into two groups: those of serotonergic neurons localized in the pontine/mesencephalic raphe nuclei dorsalis (DRN) and centralis superior (NCS) which primarily innervate the forebrain, and those of serotonergic neurons localized in the medullary raphe nuclei magnus (NRM), obscurus (NRO), and pallidus (NRP) which primarily innervate the brainstem and spinal cord.

12.1.1 Circadian and Sleep/Wake Cycles

One of the most fundamental variables that exerts an influence on behavioral state is the circadian cycle; therefore, as one of our first undertakings in this field, we examined whether serotonergic neurons in the DRN of the cat displayed a variation across the light/dark cycle. If behavioral state is held constant, i.e., when the activity of a given neuron is studied repeatedly during REM sleep or quiet waking at various times across a 12:12-hr light/dark cycle, the activity of these neurons was invariant.[32] Thus, to the degree that the functional activity of the serotonergic system varies across the light/dark or circadian cycle, it does so indirectly, by means of its variation across the sleep/wake cycle (see below). (It is well known that the sleep/wake cycle is strongly influenced by the light/dark and circadian cycles.) In this context, it is of some interest that the nucleus suprachiasmaticus (SCN), the site of the primary circadian clock in mammals, receives one of the most dense serotonergic inputs. When and how this input to the SCN is expressed is not known, but we speculate on this issue in a later section.

The most dramatic changes in behavioral state are seen across the sleep/wake/arousal cycle. We have examined the activity of serotonergic neurons in most of their major nuclei across this cycle. The largest cluster of brain serotonergic neurons in the mammalian brain is in the DRN. During the quiet waking state, the activity of DRN serotonergic neurons is slow and highly regular or clock-like,[31] just as it is when examined under anesthesia[1] and *in vitro*.[25] From a quiet waking rate of approximately three spikes per sec, the activity of these neurons is typically increased by approximately 10 to 30% in response to activating or arousing stimuli. Reciprocally, the activity of these neurons declines as the cat becomes drowsy and becomes even slower upon entering slow-wave sleep. Finally, this state-dependent decrease in single-unit activity culminates as the cat enters rapid-eye-movement (REM) sleep, when the activity falls virtually silent (Figure 12.1).

In general, the pattern of activity across the sleep/wake/arousal cycle in DRN serotonergic neurons is closely paralleled by serotonergic neurons in the other major groups: NCS, NRO/NRP, and NRM.[5] There are, however, some differences which may be of functional significance. For example, NRM and NRO/NRP neurons generally display a higher spontaneous firing rate than DRN or NCS neurons during comparable behavioral states. The activity of NRO/NRP neurons, as well as a subset of NCS neurons, is not as strongly related to behavioral state as DRN, NRM, and the majority of NCS cells. The former did not show as steep a decline across the sleep/wake/arousal cycle, and their activity, although significantly reduced, is not completely suppressed during REM sleep. Moreover, NRO/NRP neurons are the only ones whose activity is generally unresponsive to activating or arousing stimuli.

Thus, the activity of the brain serotonergic system as a whole appears to display a direct relationship to the level of behavioral arousal. As will be seen below, this conclusion is also supported by neurochemical evidence.

Taking these descriptive studies one step further, we investigated the neural mechanisms responsible for this decreased activity during sleep. We utilized multi-barrel microiontophoresis in head-restrained, conscious cats to address this question.[19] The iontophoretic application of the γ-aminobutyric acid (GABA) antagonist bicuculline during sleep restored the reduced activity of DRN serotonergic neurons to the waking level (the cat, of course, remained asleep). However, if the same antagonist was applied during waking, there was no change in neuronal activity. These results indicate that a GABAergic input to brain serotonergic neurons becomes activated during sleep and exerts a powerful inhibitory influence preferentially during that state. In another study in this series, we demonstrated that the excitatory response of serotonergic neurons to phasic sensory inputs (such as a click) could be blocked by iontophoretic application of the excitatory amino acid antagonist kynurenic acid. This antagonist, however, exerted no effect on the spontaneous or basal activity of these neurons. In the final study in this series, we found that the iontophoretic application of an α_1-adrenergic agonist (phenylephrine) did not alter the spontaneous waking activity of these neurons, suggesting that the noradrenergic input to these neurons already exerts its maximal effect during waking.

Our most recent experiments on this general theme explored the effects of sleep deprivation on serotonergic neurons.[13] Most, if not all, current pharmacological treatments for depression affect the serotonergic system. One intervention that at least temporarily alleviates depressive symptoms in patients is one night of total sleep deprivation. To evaluate whether there might be a serotonergic component involved in this treatment, we examined the effects of sleep deprivation on the discharge rate of serotonergic neurons in the DRN of behaving cats. Activity of single neurons during quiet waking and active waking behavior was recorded at baseline and for the duration of a 24-hr sleep deprivation period and during a 6-hr recovery sleep period. The $5-HT_{1A}$ autoreceptor agonist 8-hydroxy-2- (di-n-propylamino) tetralin (8-OH-DPAT), which reduces the firing rate of DRN serotonergic cells, was administered systemically before and after the sleep deprivation period, in order to assess whether the sensitivity of this autoreceptor changed over the course of the deprivation. Mean neuronal activity increased during the deprivation period by approximately 20% during active waking. The magnitude of this effect reached its maximum on average at the 15-hr time point. The

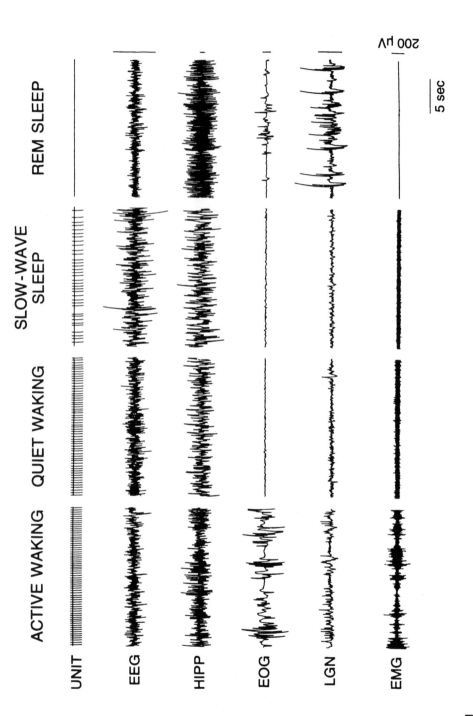

FIGURE 12.1
Polygraph records showing the activity of a typical serotonergic dorsal raphe nucleus (DRN) neuron and gross potentials across the sleep/wake/arousal cycle. Note the positive relationship between the firing rate and the level of behavioral arousal, as well as the cessation of unit activity during REM sleep, both of which are distinguishing features of DRN serotonergic neurons recorded in behaving cats. During REM sleep, ponto-geniculo-occipital (PGO) waves can be seen in the LGN trace and prominent rhythmic slow activity (theta) in the HIPP trace. EEG = cortical electroencephalogram; HIPP = hippocampal electroencephalogram; EOG = electrooculogram; LGN = lateral geniculate nucleus; EMG = nuchal electromyogram.

firing rates of DRN serotonergic neurons remained elevated above the baseline level for the entire 24-hr period in all the test conditions, but after a 6-hr recovery sleep period, neuronal activity returned to its pre-deprivation level. The ability of 8-OH-DPAT to reduce the firing rate of serotonergic cells was reduced by 14% after the 24-hr period of sleep deprivation, suggesting that the negative-feedback, somatodendritic 5-HT$_{1A}$ autoreceptor was desensitized by the sleep deprivation. Overall, these data demonstrate that sleep deprivation can increase serotonergic neurochemical system activity by causing an increase in the firing rate of individual neurons. These results, therefore, provide a plausible explanation for why, at least in part, sleep deprivation exerts an antidepressant effect.

12.1.2 Challenges/Stressors

An initial stage in a basic strategy for elucidating the behavioral and physiological roles of a brain neurochemical system might involve an examination of its response to a variety of intense conditions, especially those that are biologically and ecologically relevant. This approach should reveal at least some characteristics of the stimuli to which brain serotonergic neurons are programmed to respond. Thus, cats were exposed to the following environmental or physiological conditions while recording neuronal activity in the DRN, NCS, or NRM: a heated environment or administration of a pyrogen; drug-induced increases or decreases in arterial blood pressure; insulin-induced hypoglycemia; phasic or tonic mildly painful stimuli; loud noise; physical restraint; or a natural enemy (dog).[2,7-9,36] Despite the fact that all of these conditions evoked strong behavioral responses and/or physiological changes indicative of sympathetic activation, none of them significantly activated serotonergic neuronal activity beyond the level normally seen during an undisturbed active waking state. The following two descriptions exemplify the results from this series of experiments.

The activity of DRN serotonergic neurons was examined in response to both increased ambient temperature and pyrogen-induced fever, stimuli eliciting opposite thermoregulatory responses of heat loss and heat gain, respectively.[7] Neuronal activity remained unaffected as ambient temperature was increased from 25 to 43°C. Following prolonged heat exposure, cats displayed intense continuous panting, relaxation of posture, and a progressive rise in body/brain temperature (range 0.5 to 2.0°C), yet no change in serotonergic neuronal activity occurred. In a parallel study, a synthetic pyrogen (muramyl dipeptide) was administered, resulting in increased body/brain temperature within 30 min and lasting for approximately 6 hr. The peak elevation of body temperature was typically 1.5 to 2.5°C, yet, once again, no change in neuronal activity was observed. Consistent with these electrophysiological results, utilizing *in vivo* microdialysis, we have also found that pyrogen administration produces no change in extracellular levels of serotonin in the anterior hypothalamus, a primary thermoregulatory center of the cat.[37]

A large body of evidence implicates central serotonin in analgesia, especially those serotonergic neurons localized in the NRM and projecting to the dorsal horn of the spinal cord. Accordingly, we examined the activity of NRM serotonergic neurons in behaving cats exposed to a variety of phasic or tonic, moderately painful stimuli.[2] No change in neuronal activity was produced by these stimuli relative to the discharge rate during an undisturbed active waking baseline. There was also no change in serotonergic neuronal activity in response to the systemic administration of morphine, in a dose that produced analgesia. These results have recently been confirmed in a study reporting that identified NRM serotonergic neurons in the rat were not activated by painful stimuli eliciting the withdrawal reflex.[26]

More recently, we have begun to examine the same issue for NRO and NRP serotonergic neurons. These cells provide most of the serotonergic input to the intermediolateral column and ventral horn of the spinal cord, and also to brainstem autonomic regulatory nuclei. The activity of serotonergic NRO/NRP neurons was examined in response to cold stress, a potent activator of the sympathetic nervous system.[22] The ambient temperature in the recording chamber was lowered from 25 to 5°C and maintained at this level for 4 hr. Cats typically displayed shivering, piloerection, and a crouched or curled posture during cold exposure, but no appreciable changes in core temperature were observed. The discharge rate of approximately half of all NRO/NRP cells studied was significantly increased (~30%) during environmental cooling, whereas the remaining cells were unaffected by this manipulation. The elevation in neuronal activity seen with cold exposure was maintained throughout the entire cooling period and was unrelated to the occurrence of shivering bursts per se. Neuronal activity returned to baseline levels within 15 min after the animals were transferred back to room temperature. These results suggest that some NRO/NRP serotonergic neurons may play a role in the physiological responses underlying cold defense (e.g., increased sympathetic and/or motor output). In contrast to the increases in neuronal activity observed for many serotonergic NRO/NRP neurons, the discharge rate of serotonergic DRN neurons was not significantly altered during environmental cooling. Thus, these results do not support a specific role for DRN serotonergic neurons in thermoregulation. This is consistent with the results of our previous study, which showed that the activity of DRN neurons is unaffected by elevations in either skin temperature or body temperature.[7]

In another experiment, we examined the response of serotonergic NRO/NRP neurons to insulin administration.[18] This manipulation is of particular interest, because it is associated with increased sympathetic activation and no change, or even a decrease, in motor output. Following insulin administration, blood glucose decreased to approximately 50% of baseline levels. This was accompanied by a marked activation of the sympathoadrenal system, as indicated by increases in the plasma levels of norepinephrine and epinephrine. Surprisingly, the discharge rate of serotonergic NRO/NRP neurons was markedly reduced (~50%) after insulin. This suppression of neuronal activity was temporally correlated with the decrease in blood glucose levels. The subsequent administration of glucose, at doses sufficient to reverse hypoglycemia, promptly restored neuronal activity to baseline levels. Circulating levels of catecholamines also returned to baseline following glucose administration. The observation that serotonergic NRO/NRP neuronal activity is depressed, rather than enhanced, during the period of increased sympathetic activity following insulin administration, does not support a primary role for these neurons in sympathetic activation. Instead, the depression of serotonergic NRO/NRP neuronal activity may be directly related to the diminished muscle tone and behavioral suppression associated with insulin-induced hypoglycemia.

Overall, these results suggest that the discharge of serotonergic NRO/NRP neurons is related more closely to changes in motor output than to changes in sympathetic outflow. These neurons may still act to facilitate sympathetic activity, but only in conjunction with increased motor output, since this appears to be the primary factor associated with neuronal activation.

12.1.3 Motor Activity

Rapid-eye-movement sleep is characterized by an inhibition of the motoneurons controlling antigravity muscle tone and a resulting paralysis. Because the activity of serotonergic neurons is almost totally suppressed during REM sleep, we examined the possibility that there might be a relationship between these two phenomena. Lesions of the dorsomedial pons in cats produce a state which by all criteria appears to be REM sleep except that antigravity muscle tone is present, and the animals are thus capable of movement and even coordinated locomotion. In both waking and slow-wave sleep, the activity of DRN serotonergic neurons in these pontine-lesioned cats was similar to that of normal animals;[33] however, when these animals entered REM sleep, neuronal activity increased, instead of displaying the decrease typical of this state. Those animals displaying the greatest amount of restored muscle tone and overt behavior during REM sleep showed the highest levels of neuronal activity, with some of their serotonergic neurons discharging at a level approximating that of the waking state.

Microinjection of carbachol, a cholinomimetic agent, into this same pontine area, produces a condition somewhat reciprocal to non-atonia REM sleep. These animals are awake, as demonstrated by their ability to track visual stimuli, but are otherwise paralyzed. However, unlike the normal waking state, where serotonergic neurons are tonically active, DRN serotonergic neurons were inactive in these paralyzed animals.[30]

These data suggested that a strong positive relationship exists between tonic level of motor activity (muscle tone) and the firing rate of DRN serotonergic neurons. Recently, we have observed much more specific relationships between serotonergic neuronal activity and motor function. When cats engage in a variety of types of central pattern generator-mediated, oral-buccal activities (such as chewing/biting, licking, or grooming) approximately one fourth of DRN serotonergic neurons increase their activity by as much as two- to fivefold.[10] In contrast, the rest of the serotonergic neurons in this nucleus maintain their slow and rhythmic activity. These increases in neuronal activity invariably terminate coincident with the end of the behavioral sequence. Equally impressive is the fact that even brief (1 to 5 sec) spontaneous pauses in these behaviors are accompanied by an immediate decrease in neuronal activity to baseline levels, or below. The increased neuronal activity during these central pattern generator-mediated behaviors is typically tonic but is occasionally modulated in phase with a particular aspect of the repetitive behavior. During a variety of other non-rhythmic episodic or purposive movements, even those involving oral-buccal responses such as yawning, no increase or even a decrease in neuronal activity is seen. Reciprocally, during attentional shifts, such as those occurring during orienting movements in response to novel or imperative stimuli, the activity of DRN serotonergic neurons may fall silent for several seconds. This typically occurs in association with large eye movements, turning of the head toward the stimulus, and suppression of ongoing behaviors.

Somewhat surprisingly, most of these DRN neurons can also be activated by somatosensory stimuli applied to the head and neck region, while the same stimuli applied to the rest of the body surface are typically ineffective. The level of increased activity produced by somatosensory stimulation often approaches that seen spontaneously during oral-buccal movements (i.e., two- to fivefold).

As discussed above, serotonergic neurons in the rostral pons (DRN and NCS) provide almost the entire serotonergic innervation of the forebrain, whereas those in the caudal medulla (NRO and NRP) are the source of much of the serotonergic innervation of the spinal cord.[15] Therefore, it is interesting to compare the response properties of neurons in these two separate groups.

Contrary to pontine serotonergic neurons, where only a subgroup of neurons are activated during central pattern generator-mediated behaviors, virtually all medullary serotonergic neurons are activated under at least some of these conditions.[34] The degree of activation, however, is much less impressive, i.e., 50 to 100% above baseline vs. 100 to 400% above baseline for the DRN. In this context, it may be important to note that the basal, quiet waking discharge rate of the medullary serotonergic neurons is approximately twice that of the pontine serotonergic neurons — 5 to 6 spikes per sec vs. 2 to 3. There also appears to be at least some degree of response specificity for these neurons. Thus, virtually all medullary serotonergic neurons are activated during treadmill-induced locomotion, but only subgroups are activated during hyperpnea (induced by exposure to carbon dioxide) or during chewing/licking. Many of these individual neurons are activated in association with more than one of these motor activities. In most cases, there is a strong positive correlation between magnitude of neuronal activation and speed of locomotion and/or depth of respiration. As with pontine serotonergic neurons, the increased activity of medullary serotonergic neurons is sometimes phase-locked to the behavior (e.g., in association with the step cycle) and typically is tightly coupled to the onset and offset of the behavior. Unlike pontine serotonergic neurons, medullary serotonergic neurons are not activated by somatosensory stimuli applied to any region of the body surface. Nor is their activity significantly changed during orientation to strong or novel stimuli. Finally, when DRN neurons were examined under the identical conditions, none of them were activated during treadmill-induced locomotion, but some were activated during carbon dioxide-induced hyperpnea.[35]

12.1.4 PGO Waves and Hippocampal Theta

In addition to playing a general role in behavioral state control, the central serotonergic system has been specifically implicated in the regulation of several electrical potentials during the sleep/wake cycle, notably ponto-geniculo-occipital (PGO) waves and hippocampal rhythmic slow activity or theta. Ponto-geniculo-occipital waves are phasic, high-amplitude field potentials which can be recorded in the pons, the lateral geniculate nucleus, and the occipital cortex of many species. These waves primarily occur during the late phase of slow-wave sleep, usually as single spikes, and throughout REM sleep at a high frequency, largely in clusters that are associated with bursts of eye movements. Because PGO waves consistently precede the onset of REM sleep, the appearance of these waves during periods of slow-wave sleep usually signals the transition into REM sleep. Although the physiological significance of PGO waves is not well understood, they have been related to enhanced visual information processing and to alerting responses.[3]

Neuropharmacological studies strongly support the idea that serotonin inhibits the generation of PGO waves. Thus, systemic administration of serotonin-depleting drugs (e.g., reserpine or p-chlorophenylalanine) has been shown to release PGO waves into the waking state. The subsequent administration of the serotonin precursor 5-hydroxytryptophan restores brain serotonin levels in these animals and suppresses waking PGO wave activity.

Single-unit recordings of brain serotonergic neurons also support the hypothesis that serotonin exerts a tonic inhibitory influence on PGO wave generation. Thus, the discharge of serotonergic neurons in the DRN and NCS exhibits a strong inverse relationship to the occurrence of PGO waves.[20,23,27,32] These cells typically display a marked reduction or complete cessation of firing several seconds prior to, and during, the occurrence of PGO waves during sleep. Reciprocally, the resumption of neuronal discharge during slow-wave sleep or at the end of a REM sleep epoch is associated with a suppression of PGO wave activity. In contrast to the activity of serotonergic neurons in the midbrain, the discharge of serotonergic neurons in the medulla (i.e., NRM and NRO/NRP) was found to be unrelated to PGO waves.[6,14] This is consistent with the prominent release of PGO waves that follows destruction of the rostral but not caudal raphe nuclei.

The finding that the discharge activity of serotonergic DRN and NCS neurons is negatively correlated with the occurrence of PGO waves constitutes important physiological evidence supporting a permissive role of brain serotonin in PGO wave generation. The suppression of serotonergic neuronal activity that normally occurs during a sleep cycle may act to disinhibit cholinergic neurons in the pontine tegmentum that are believed to generate PGO waves and the REM sleep state. Finally, although PGO waves also occur during waking, usually in association with orienting or startle responses, serotonergic neurons, in general, display no strong or consistent relationship with these waking PGO waves (Fornal et al., unpublished observations).

Ascending serotonergic projections from the mesencephalic raphe nuclei have also been implicated in the control of hippocampal theta. In mammals, this rhythmic pattern of electrical activity has been related to such processes as arousal, motor behavior, orientation, attention, motivation, and learning and memory. Interestingly, the most pronounced theta activity observed under physiological conditions occurs during REM sleep, a period when serotonergic neurons are virtually silent (see Figure 12.1). Although the precise role of brain serotonin in the regulation of hippocampal theta is unclear, we have recently shown that the intravenous administration of 5-HT$_{1A}$ autoreceptor agonists (e.g., 8-OH-DPAT and ipsapirone), at doses which strongly suppress the activity of serotonergic neurons, consistently produced a rapid increase in hippocampal theta activity in awake cats.[21] These findings suggest that serotonin may tonically inhibit the appearance of hippocampal theta activity. This is consistent with the results of previous studies, which have shown that lesions of the median raphe nucleus or pharmacological depletion of brain serotonin induces continuous high-amplitude synchronized waves in the hippocampal EEG. In addition to a direct effect on the hippocampus, serotonin may influence hippocampal activity via an action on the medial septum/diagonal band, a region thought to act as a pacemaker for the hippocampal theta rhythm.

In order to examine the relationship between serotonergic neuronal activity and hippocampal theta directly, we recorded the discharge rate of serotonergic neurons in the mesencephalic raphe nuclei (i.e., DRN and NCS) together with the electrical activity in the hippocampus of freely moving cats, during a variety of waking behaviors.[11,12] A significant inverse relationship was found between the activity of a subgroup (~25%) of DRN and NCS serotonergic cells and hippocampal theta. The other serotonergic cells recorded in these nuclei displayed no obvious relationship to hippocampal theta during waking. These findings support the hypothesis that brain serotonin is involved in the modulation of hippocampal theta and that serotonin may inhibit the generation of this rhythm.

12.1.5 Conclusions

The activity of brain serotonergic neurons, regardless of the nuclei in which they are found, changes dramatically across the sleep/wake/arousal cycle. At least in part, this appears to be correlated with changes in the level of tonic motor activity that accompany these alterations in behavioral state. Somewhat surprisingly, serotonergic neurons are only minimally responsive to any of a broad range of environmental or physiological challenges/stressors. Beyond the level of neuronal activity attained during waking, many serotonergic neurons display a further, often dramatic, elevation of neuronal activity during the expression of repetitive, central pattern generator-mediated motor outputs. The particular motor pattern that is associated with increased neuronal activity varies to some extent with the serotonergic nucleus in which the neuron is localized.

We hypothesize that activation of brain serotonergic neurons facilitates motor output and, simultaneously, suppresses sensory information processing.[16,17] In addition, serotonergic neurons play an auxiliary role in coordinating appropriate autonomic and neuroendocrine outputs to the ongoing tonic or repetitive motor activity. Finally, when serotonergic neuronal activity is suppressed (for example, during orientation), motor output is disfacilitated and sensory processing is disinhibited, permitting more focused information processing to occur.

12.2 Release of Serotonin in Forebrain Sites

Traditionally, functional activity within the CNS has been measured by means of neuronal activity. As described in the previous section, this is the primary manner in which we have evaluated the brain serotonergic system. However, it is also possible that neuronal activity does not, under all conditions, directly and linearly reflect neurotransmitter release at postsynaptic target sites. Therefore, we also employ *in vivo* brain microdialysis as a method to assess neurotransmitter release more directly in behaving animals.

12.2.1 Circadian and Sleep/Wake Cycles

In the study described above, in which we examined serotonin release in the hypothalamus in response to pyrogen administration, we also assessed changes across the sleep/wake/arousal cycle.[37] Extracellular serotonin in the anterior hypothalamic/preoptic area and caudate nucleus of the freely moving cat was measured using *in vivo* microdialysis. Behavioral state was quantified based on both behavioral and polygraphic criteria. In the first phase of this study, we found that extracellular levels of serotonin in the hypothalamus and caudate nucleus were significantly increased in relation to increased levels of behavioral arousal. In a subsequent, more detailed analysis, extracellular levels of serotonin in the hypothalamus displayed a strong positive correlation with behavioral state across the entire sleep/wake/arousal continuum. Finally, consistent with electrophysiological experiments, administration of the 5-HT$_{1A}$ autoreceptor agonist 8-OH-DPAT significantly reduced extracellular levels of serotonin in both the caudate nucleus and hypothalamus. In general, these data confirm the basic validity of using single unit activity as an index of the functional activity of the brain serotonergic system.

In studies using rats we have examined changes in extracellular brain serotonin levels across the daily light/dark transition. These data support our electrophysiological data showing that behavioral state/motor activity is a more important determinant of serotonergic activity than the light/dark cycle. When we examined serotonin levels in the cerebellum across the light/dark transition we found that there was approximately a 40% increase in the dark. However, behavioral activity also increases dramatically in the dark. Thus, when we took this into account we found that there was a strong correlation between serotonin levels and behavioral activity (percent time spent in an active waking state).[24] We have also examined this issue in a number of forebrain sites in the rat.[29] Serotonin levels increased significantly during the first half hour of the dark phase in the hippocampus, corpus striatum, amygdala, and prefrontal cortex. As in the cerebellum, serotonin levels co-varied significantly with changes in time spent in active waking behaviors. When serotonin levels were compared during periods where either high levels of behavioral activity were equated in the light and dark, or where low levels of behavioral activity were equated in the light and dark, there was no significant difference. These data support our previous findings that serotonin release is importantly tied to behavioral state/behavioral activity and is only indirectly related to the light/dark cycle insofar as it underlies the circadian distribution of behavioral activity.

12.2.2 Challenges/Stressors

Employing *in vivo* microdialysis in rats, we have returned to the issue of stress and serotonergic function that we had previously examined electrophysiologically.[28] We addressed several different issues in these studies: (1) Do stressors produce a larger effect than an activating but non-stressful stimulus? (2) Do different stressors produce a different magnitude or pattern of change in levels of extracellular serotonin across various forebrain sites? (3) Is a given brain region differentially affected by different stressors? (4) Are any differences noted in the second and third issues accounted for by brainstem site of origin of the serotonergic input to these forebrain sites? To answer this, microdialysis samples were taken during the dark phase of the light/dark cycle while animals were exposed to various stressors. We found no evidence for a "stress response" in any of the forebrain sites that could not be accounted for by general activation of the animal. Thus, tail pinch, forced swimming, or exposure to a cat did not produce any larger increase (~30 to 50%) in extracellular serotonin than that seen in these same sites during a non-stressful condition: spontaneous feeding. (Increases in plasma glucocorticoid levels confirmed the "stressful" nature of the stimuli.) There was also no clearcut evidence that the different stressors, which clearly evoke different types of behavioral and physiological responses, produce a different pattern of serotonin release across the various target sites. Nor was there any compelling evidence that any one stressor had a stronger overall effect, independent of brain region, than any other one. Somewhat interesting, however, is the fact that serotonin levels in the corpus striatum were significantly less responsive overall (i.e., across all experimental manipulations), than serotonin levels in the other three sites. Thus, it is possible that serotonin which derives from the DRN exclusively (corpus striatum) is more refractory to change than serotonin which derives from the NCS alone (dorsal hippocampus) or from it and the DRN (amygdala and prefrontal cortex).

12.2.3 Conclusions

In summary, these studies employing *in vivo* microdialysis to measure changes in brain levels of serotonin in rats are consistent with our previous single unit studies in cats. The serotonergic system is not stress activatable per se. Instead, the system appears to be responsive to more general factors such as behavioral state/behavioral activity. More specifically, our single-unit studies suggest that the functional activity of the serotonin system is directly related to levels of tonic and repetitive (central pattern generator-mediated) motor activity.

12.3 General Conclusions

How do these data regarding the conditions that regulate the activity of the brain serotonergic system relate to serotonin's role in behavioral state control? We believe that the role of serotonin in state control is secondary to its involvement in motor activity and sensory information processing. The manifestations of this will be numerous and varied, depending on the particular CNS site and serotonin receptor subtype that is affected. Perhaps the clearest example of this would be the action of serotonin in the SCN (the site of the primary circadian rhythm pacemaker in mammals). It is well known that the phase of the circadian cycle can be modified by locomotion in rodents.[4] We believe that this effect is attributable to the activation of the dense serotonergic input to the SCN during running, with the magnitude and nature of the effect dependent on the duration, intensity, and timing of the locomotory episode(s). This type of action is repeated at the hundreds or thousands of serotonergic target sites within the CNS.

One of the related issues that we are currently exploring is how sustained activity (or inactivity) of the serotonergic system may alter its action, e.g., during sleep deprivation or forced locomotor activity. Preliminary evidence from our laboratory indicates that the discharge rate of medullary serotonergic neurons may decrease or "fatigue" as cats are kept running on a treadmill for extended periods of time (30 to 60 min.). This may contribute to motoric and autonomic changes typically seen in association with prolonged exercise and in this way exert a behavioral state effect.

Finally, the serotonergic system may also regulate behavioral state in a general way, due to its aforementioned widespread influence on sensory and motor systems. When neuronal activity is high, behavioral activity is facilitated and reactivity inhibited, whereas with neuronal activity low, behavioral activity is disfacilitated and reactivity disinhibited. The location of an animal's position on this sensorimotor continuum would therefore be a major contributor to the moment-to-moment determination of behavioral state.

References

1. Aghajanian, G.K., Foote, W.E., and Sheard, M.H., Lysergic acid diethylamide: sensitive neuronal units in the midbrain raphe, *Science*, 161, 706–708, 1968.
2. Auerbach, S., Fornal, C.A., and Jacobs, B.L., Response of serotonin-containing neurons in nucleus raphe magnus to morphine, noxious stimuli, and periaqueductal gray stimulation in freely moving cats, *Exp. Neurol.*, 88, 609–628, 1985.
3. Callaway, C.W., Lydic, R., Baghdoyan, H.A., and Hobson, J.A., Pontogeniculooccipital waves: spontaneous visual system activity during rapid eye movement sleep, *Cell. Mol. Neurobiol.*, 7, 105–149, 1987.
4. Edgar, D.M. and Dement, W.C., Regularly scheduled voluntary exercise synchronizes the mouse circadian clock, *Am. J. Physiol.*, 261, R928–933, 1991.
5. Fornal, C.A. and Jacobs, B.L., Physiological and behavioral correlates of serotonergic single-unit activity, in *Neuronal Serotonin*, Osborne, N.N. and Hamon, M., Eds., John Wiley & Sons, New York, 1988, pp. 305–345.

6. Fornal, C.A., Auerbach, S., and Jacobs, B.L., Activity of serotonin-containing neurons in nucleus raphe magnus in freely moving cats, *Exp. Neurol.*, 88, 590–608, 1985.
7. Fornal, C.A., Litto, W.J., Morilak, D.A., and Jacobs, B.L., Single-unit responses of serotonergic dorsal raphe nucleus neurons to environmental heating and pyrogen administration in freely moving cats, *Exp. Neurol.*, 98, 388–403, 1987.
8. Fornal, C.A., Litto, W.J., Morilak, D.A., and Jacobs, B.L., Single-unit responses of serotonergic neurons to glucose and insulin administration in behaving cats, *Am. J. Physiol.*, 257, R1345–1353, 1989.
9. Fornal, C.A., Litto, W.J., Morilak, D.A., and Jacobs, B.L., Single-unit responses of serotonergic neurons to vasoactive drug administration in behaving cats, *Am. J. Physiol.*, 259, R963–972, 1990.
10. Fornal, C.A., Metzler, C.W., Marrosu, F., Ribiero-do-Valle, L.E., and Jacobs, B.L., A subgroup of dorsal raphe serotonergic neurons in the cat is strongly activated during oral-buccal movements, *Brain Res.*, 716, 123–133, 1996.
11. Gallegos, R.A., Serotonergic Unit Activity in the Dorsal Raphe and Median Raphe Nuclei in Relation to Hippocampal Theta Activity in the Cat, Ph.D. dissertation, Princeton University, 1997.
12. Gallegos, R.A., O'Neill, P.R., and Jacobs, B.L., Relationship between serotonergic neuronal activity and hippocampal theta in the cat, *Soc. Neurosci. Abstr.*, 23, 520, 1997.
13. Gardner, J.P., Fornal, C.A., and Jacobs, B.L., Effects of sleep deprivation on serotonergic neuronal activity in the dorsal raphe nucleus of the freely moving cat, *Neuropsychopharmacology*, 17, 72–81, 1997
14. Heym, J., Steinfels, G.F., and Jacobs, B.L., Activity of serotonin-containing neurons in the nucleus raphe pallidus of freely moving cats, *Brain Res.*, 251, 259–276, 1982.
15. Jacobs, B.L. and Azmitia, E.C., Structure and function of the brain serotonergic system, *Physiol. Rev.*, 72, 165–229, 1992.
16. Jacobs, B.L. and Fornal, C.A., 5-HT and motor control: a hypothesis, *Trends Neurosci.*, 16, 346–352, 1993.
17. Jacobs, B.L. and Fornal, C.A., Activation of 5-HT neuronal activity during motor behavior, *Semin. Neurosci.*, 7, 401–408, 1995.
18. Jacobs, B.L., Martín, F.J., Fornal, C.A., and Metzler, C.W., Systemic administration of insulin decreases the activity of medullary serotonergic neurons in awake cats, *Soc. Neurosci. Abstr.*, 23, 1227, 1997.
19. Levine, E.S. and Jacobs, B.L., Neurochemical afferents controlling the activity of serotonergic neurons in the dorsal raphe nucleus: microiontophoretic studies in the awake cat, *J. Neurosci.*, 12, 4037–4044, 1992.
20. Lydic, R., McCarley, R.W., and Hobson, J.A., The time-course of dorsal raphe discharge, PGO waves, and muscle tone averaged across multiple sleep cycles, *Brain Res.*, 274, 365–370, 1983.
21. Marrosu, F., Fornal, C.A., Metzler, C.W., and Jacobs, B.L., 5-HT$_{1A}$ agonists induce hippocampal theta activity in freely moving cats: role of presynaptic 5-HT$_{1A}$ receptors, *Brain Res.*, 739, 192–200, 1996.
22. Martín, F.J., Gallegos, R.G., Fornal, C.A., Metzler, C.W., and Jacobs, B.L., Effect of environmental cooling on the activity of pontine and medullary serotonergic neurons in awake cats, *Soc. Neurosci. Abstr.*, 23, 1226, 1997.
23. McGinty, D.J. and Harper, R.M., Dorsal raphe neurons: depression of firing during sleep in cats, *Brain Res.*, 101, 569–575, 1976.
24. Mendlin, A., Martin, F.J., Rueter, L.E., and Jacobs, B.L., Neuronal release of serotonin in the cerebellum of behaving rats: an *in vivo* microdialysis study, *J. Neurochem.*, 67, 617–622, 1996.
25. Mosko, S.S. and Jacobs, B.L., Recording of dorsal raphe unit activity in vitro, *Neurosci. Lett.*, 2, 195–200, 1976.

26. Potrebic, S.B., Field, H.L., and Mason, P., Serotonin immunoreactivity is contained in one physiological cell class in the rat rostral ventromedial medulla, *J. Neurosci.*, 14, 1655–1665, 1994.
27. Rasmussen, K., Heym, J., and Jacobs, B.L., Activity of serotonin-containing neurons in nucleus centralis superior of freely moving cats, *Exp. Neurol.*, 83, 302–317, 1984.
28. Rueter, L.E. and Jacobs, B.L., A microdialysis examination of serotonin release in the rat forebrain induced by behavioral/environmental manipulations, *Brain Res.*, 739, 57–69, 1996.
29. Rueter, L.E. and Jacobs, B.L., Changes in forebrain serotonin at the light/dark transition: correlation with behavior, *NeuroReport*, 7, 1107–1111, 1996.
30. Steinfels, G.F., Heym, J., Strecker, R.E., and Jacobs, B.L., Raphe unit activity in freely moving cats is altered by manipulations of central but not peripheral motor systems, *Brain Res.*, 279, 77–84, 1983.
31. Trulson, M.E. and Jacobs, B.L., Raphe unit activity in freely moving cats: correlation with level of behavioral arousal, *Brain Res.*, 163, 135–150, 1979.
32. Trulson, M.E. and Jacobs, B.L., Raphe unit activity in freely moving cats: lack of diurnal variation, *Neurosci. Lett.*, 36, 285–290, 1983.
33. Trulson, M.E., Jacobs, B.L., and Morrison, A.R., Raphe unit activity during REM sleep in normal cats and in pontine lesioned cats displaying REM sleep without atonia, *Brain Res.*, 226, 75–91, 1981.
34. Veasey, S.C., Fornal, C.A., Metzler, C.W., and Jacobs, B.L., Response of serotonergic caudal raphe neurons in relation to specific motor activities in freely moving cats, *J. Neurosci.*, 15, 5346–5359, 1995.
35. Veasey, S.C., Fornal, C.A., Metzler, C.W., and Jacobs, B.L., Single-unit responses of serotonergic dorsal raphe neurons to specific motor challenges in freely moving cats, *Neuroscience*, 79, 161–169, 1997.
36. Wilkinson, L.O. and Jacobs, B.L., Lack of response of serotonergic neurons in the dorsal raphe nucleus of freely moving cats to stressful stimuli, *Exp. Neurol.*, 101, 445–457, 1988.
37. Wilkinson, L.O., Auerbach, S.B., and Jacobs, B.L., Extracellular serotonin levels change with behavioral state but not pyrogen-induced hyperthermia, *J. Neurosci.*, 11, 2732–2741, 1991.

Chapter 13

Inhibitory Mechanisms in the Dorsal Raphe Nucleus and Locus Coeruleus During Sleep

*Pierre-Hervé Luppi, Christelle Peyron,
Claire Rampon, Damien Gervasoni, Bruno Barbagli,
Romuald Boissard, and Patrice Fort*

Contents

13.1	Introduction		196
13.2	Effects of the Application of Glycine and GABA Antagonists on the Activity of Locus Coeruleus Neurons During Sleep		197
	13.2.1	Spontaneous Activity of Locus Coeruleus Neurons During the Sleep/Waking Cycle	197
	13.2.2	Iontophoretic Applications of Strychnine	197
	13.2.3	Iontophoretic Applications of Bicuculline	197
	13.2.4	Conclusions: Pharmacological and Physiological Significance	199
13.3	Glycinergic and GABAergic Afferent Projections to the Locus Coeruleus and Dorsal Raphe Nuclei		201
	13.3.1	Double Immunostaining Procedures	201
	13.3.2	Afferent Projections to the Locus Coeruleus	201
		13.3.2.1 Origin of the Glycinergic Input to the Locus Coeruleus	202
		13.3.2.2 Origin of the GABAergic Input to the Locus Coeruleus	203
	13.3.3	Afferent Projections to the Dorsal Raphe Nucleus	203
		13.3.3.1 Origin of the Glycinergic Input to the Dorsal Raphe Nucleus	204
		13.3.3.2 Origin of the GABAergic Input to the Dorsal Raphe Nucleus	204
	13.3.4	Physiological Role of the Glycinergic Inputs to the Dorsal Raphe Nuclei and the Locus Coeruleus	204

	13.3.5	Physiological Role of the GABAergic Inputs to the Dorsal Raphe Nuclei and the Locus Coeruleus 205
13.4		Conclusions and New Hypothesis 207
Acknowledgments 207
References 207

13.1 Introduction

In the mammalian central nervous system, main groups of noradrenergic and serotonergic neurons are found within the locus coeruleus (LC) and the dorsal raphe nucleus (DRN), respectively.[1] By means of their widespread projections throughout the entire brain, these monoaminergic neurons are thought to play crucial roles in a great variety of physiological and behavioral functions including sleep and wakefulness.[2-9] Accordingly, extracellular electrophysiological recordings in freely moving rats and cats have shown that LC noradrenergic and DRN serotonergic neurons fire tonically during wakefulness (W), decrease their activity during slow-wave sleep (SWS), and are nearly quiescent during paradoxical sleep (PS; PS-off cells).[5-9] It has been proposed that the spontaneous tonic firing of LC and DRN neurons during W, which is close to that observed in anesthetized rats,[10] is mainly due to their intrinsic pacemaker properties, as revealed by intracellular recordings from LC neurons in slices[10,11] and cultures.[12]

The mechanisms responsible for the decrease of activity of the monoaminergic neurons during SWS are not well understood. However, the classical view is that the GABAergic transmission is enhanced during SWS as supported by the strong hypnotic properties of the benzodiazepines acting on the $GABA_A$ receptors (reviewed in References 13 to 14). Moreover, Nitz and Siegel recently found an increase in the amount of GABA in the cat LC during SWS compared to W.[16] Although similar increases in GABA were not found in the DRN,[15] Levine and Jacobs showed that iontophoretic application of bicuculline, a $GABA_A$ antagonist, on DRN serotonergic neurons reversed the typical suppression of activity seen during SWS.[17] GABA-induced inhibition might therefore likely be responsible for the decrease of activity of monoaminergic neurons during SWS.

The cessation of firing of these neurons during PS, according to the classical "reciprocal interactions" models, is the result of active PS-specific inhibitory processes originating from the pontine neurons responsible for PS onset and maintenance (PS-on cells).[7,18] These neurons were first thought to be cholinergic and localized in the dorsal pons (laterodorsal tegmental, pedunculopontine, and peri-LCα nuclei). It has later been suggested that they might use GABA or glycine (GLY), rather than acetylcholine, as an inhibitory neurotransmitter.[19,20] Indeed, acetylcholine excites LC noradrenergic neurons and is only weakly inhibitory on serotonergic DRN neurons.[21,22] In contrast, in anesthetized rats, iontophoretic application of GABA or GLY strongly inhibits LC and DRN neurons, and co-iontophoresis of bicuculline or strychnine ($GABA_A$ and GLY antagonists, respectively) antagonizes these effects.[19,23-25] Furthermore, in vitro studies on slices using focal stimulation and bath-application of bicuculline and strychnine revealed GABA- and GLY-mediated inhibitory postsynaptic potentials (IPSPs) in LC neurons and GABA-mediated IPSPs in DRN cells.[26-29] In agreement with these results, GABA- and GLY-immunoreactive varicose fibers, as well as $GABA_A$ and GLY receptors, have been found in the rat LC and DRN.[19,20,30-32] Supporting the hypothesis that glycinergic neurons are responsible for the inhibition of monoaminergic neurons during PS, it has been shown that glycine was responsible for the inhibition of the somatic motoneurons during PS.[33] In contrast, supporting a role for GABA, microinjections of picrotoxin (a $GABA_A$ antagonist) in rat LC significantly reduced the duration of PS episodes.[34]

In conclusion, several lines of evidence suggested to us that GABA and/or GLY might be responsible for the inhibition of monoaminergic neurons during SWS and PS. To test this hypothesis: (1) we applied bicuculline or strychnine on LC noradrenergic cells during SWS, PS, and W using a new method which allows extracellular single-unit recordings of neurons combined with iontophoresis in the head-restrained unanesthetized rat,[35] and (2) we localized in the rat the glycinergic and GABAergic neurons potentially responsible for the inhibition of the monoaminergic neurons of the DRN and LC, by combining injections of the retrograde tracer cholera-toxin B subunit (CTb) in the DRN and LC and immunohistochemistry for GLY or glutamic acid decarboxylase (GAD; the synthetic enzyme for GABA)

13.2 Effects of the Application of Glycine and GABA Antagonists on the Activity of Locus Coeruleus Neurons During Sleep

13.2.1 Spontaneous Activity of Locus Coeruleus Neurons During the Sleep/Waking Cycle

LC neurons were recorded during at least two of the three basic vigilance stages. Noradrenergic LC neurons were identified by their typical broad action potentials (1.5 to 2 msec duration), and a phasic excitatory response to a sensory stimulus (audible click) immediately followed by an inhibition.[36–38] Their mean discharge rate was 1.39 Hz during quiet W. During SWS, LC cells showed a decrease in their firing rate (0.56 Hz), while during PS episodes they were nearly quiescent (0.01 Hz), showing only occasional single spikes.[39]

13.2.2 Iontophoretic Applications of Strychnine

Iontophoretic ejections of strychnine (50 to 250 nA, 20 to 100 sec) induced a sustained increase in the discharge rate of LC neurons regardless of the vigilance state: from 1.70 to 11.41 Hz during W, 0.39 to 9.34 Hz during SWS (Figure 13.1A), and 0.06 to 14.13 Hz during PS (Figure 13.1B).[35] Occasionally, the animal displayed successive short periods of SWS and W after strychnine administration, while the firing rate of the recorded neuron was still elevated. In these instances, we compared the neuron's increase in discharge rate during two successive SWS and W periods. We found that the increase in neuronal activity was 1.93 times as great during W as during SWS periods. This difference was statistically significant.

In all neurons tested, the strychnine effects appeared 9 to 93 sec after the onset of iontophoretic application. Recovery to baseline activity occurred several tens of seconds (range: 20 to 280 sec) after the termination of strychnine ejection. Like in anesthetized animals,[19,23] iontophoretic applications of glycine or GABA suppressed the spontaneous discharge of LC neurons during W or SWS, and glycine- but not GABA-induced inhibitions were antagonized by co-iontophoresis of strychnine.

13.2.3 Iontophoretic Applications of Bicuculline

Iontophoretic ejections of bicuculline during W, SWS, and PS (30 to 200 nA, 20 to 570 sec) induced a progressive and sustained increase in the discharge rate of LC neurons without inducing a change in vigilance state.[39] The average firing frequency increased from 1.09 to 8.72 Hz during W (Figure 13.2B) and from 0.34 to 6.91 Hz during SWS (Figure 13.2A). During PS, LC neurons were practically silent (0.01 Hz); however, following bicuculline application they showed a remarkably high mean firing rate of 6.59 Hz (Figure 13.2B). In recording some neurons, the rat displayed successive short periods of SWS and W while the effect of bicuculline on neurons was still present.

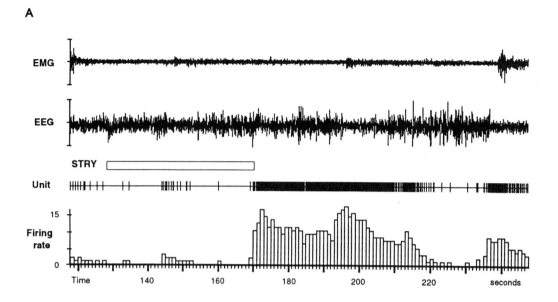

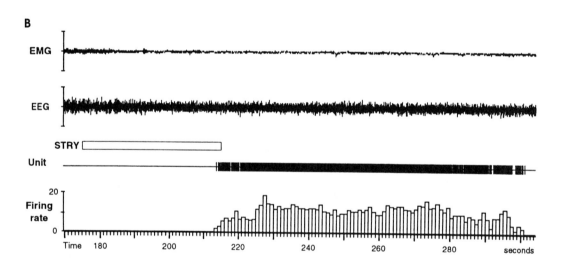

FIGURE 13.1
(A) Effects of an iontophoretic application of strychnine (90 nA, 42 sec). This application induced an activation of the neuron during SWS (the electroencephalogram, or EEG, with high voltage, slow activity, and spindles), from 0.16 to 8.9 Hz starting 42 sec after the onset of the drug application. During the following short period of W (193 to 200 sec), the discharge rate of the neuron reached 14 Hz. Then, with the return of slow waves, it strongly decreased up to the next W phase during which the discharge rate of the neuron increased again above the normal W values. (B) Effect of an iontophoretic application of strychnine during PS (80 nA, 41 sec). Note the sustained activation of the LC unit starting about 40 sec after the onset of the strychnine application.

With these neurons, we compared the increase in discharge rate during two successive SWS and W periods. We found that the increase in activity was 1.23 times as great in W as in SWS periods. In contrast with strychnine, this difference was not statistically significant. Discharge rate increases occurred 20 to 70 sec after the beginning of iontophoretic applications, while the recovery to baseline activity occurred several tens of seconds (range: 20 to 400 sec) after the end of ejection.

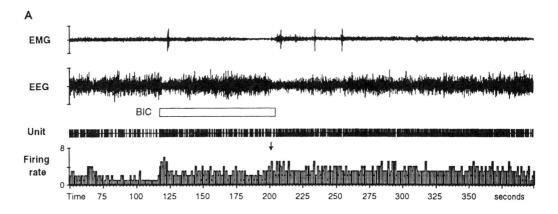

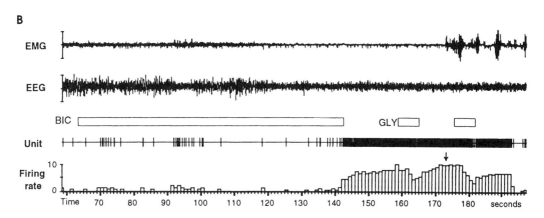

FIGURE 13.2
(A) Effects of an iontophoretic application of bicuculline (100 nA, 59 sec) during SWS. Note the strong increase in the discharge rate of the LC neuron, from 0.3 to 7.8 Hz. The effect appeared 59 sec after the onset of the ejection. Note that in contrast to strychnine, during the short period of W, the discharge rate of the neuron did not further increase (arrow). (B) Polygraphic recordings displaying the electromyogram (EMG), EEG, unit activity of a LC neuron, its firing rate, and effect of the iontophoresis of bicuculline (50 nA, 52 sec) during PS. GLY still inhibited this LC neuron during the bicuculline-induced activation.

During W or SWS, GABA-induced inhibitions were fully antagonized by bicuculline co-iontophoresis, while GLY-induced inhibitions were unaffected (Figure 13.2B).

13.2.4 Conclusions: Pharmacological and Physiological Significance

Iontophoretic applications during PS or SWS of bicuculline or strychnine induced a tonic firing in LC noradrenergic neurons. In addition, applications of bicuculline or strychnine during wakefulness induced a sustained increase in discharge rate. These results seem to indicate the existence of tonic GABA and glycinergic inputs to the LC that are active during all vigilance states. However, it is conceivable that these effects are due to non-specific excitatory effects of strychnine and bicuculline. Several arguments are against this possibility, however. With respect to bicuculline, only a slight increase in LC neuron's firing rate has been observed under anesthesia following iontophoretic or pressure applications of bicuculline.[23,40] In addition, bicuculline applications with parameters similar to those used in our studies in anesthetized rats or cats induced an increase in discharge rate of neurons from the medial hypothalamus,[41] lateral geniculate nucleus,[42] dorsal periaqueductal gray,[43] nucleus of the solitary tract,[44–46] cochlear nuclei,[47] and the spinal cord.[48,49] Finally, intracellular recordings of hippocampal or cortical neurons in slices revealed the presence of spontaneous GABA-mediated IPSPs.[50,51] Taken together, these results suggest that, like many different types of CNS neurons, the LC noradrenergic cells are tonically inhibited by GABA.

It is also unlikely that the strong and long-lasting excitations observed after strychnine applications are due to nonspecific and/or toxic effects. Indeed, if this was the case, we should have also seen a strong excitation following strychnine applications in LC neurons in anesthetized animals. In contrast, under halothane or choral hydrate anesthesia, we saw only a slight increase in the firing rate of LC neurons, even with ejection currents up to 350 nA.[19] Moreover, it has been shown that strychnine applications with similar parameters have no direct excitatory effects on other types of neurons. Thus, Bennett et al.,[44] Sun and Guyenet,[52] and Jordan et al.[46] have shown that iontophoretic applications of strychnine do not affect the spontaneous or evoked activity of cardiovascular-related neurons in the nucleus of the solitary tract or the lateral paragigantocellular nucleus. It has also been shown that applications of strychnine on inspiratory neurons from the nucleus of the solitary tract induce an activation specifically at the end of expiration.[45] Finally, following strychnine iontophoretic ejections on intracellularly recorded spinal motoneurons, no direct excitatory or toxic effects were reported.[33]

In conclusion, the effects of strychnine and bicuculline seem to indicate the existence of tonic glycinergic and GABAergic inputs to LC cells during W, SWS, and PS. Further, we found that when the same neuron was recorded during short successive periods of SWS and W while the strychnine effect on the neuron was still present, its increase in discharge rate was much greater during W than during SWS. In contrast, in the same situation, but after bicuculline administration, the increase of discharge rate of a given neuron was not statistically different between W and SWS. These results strongly suggest that release of GABA but not GLY is responsible for the inactivation of LC noradrenergic neurons during SWS. Unfortunately, due to the smaller number of LC cells recorded during PS, we were not able to make the same comparison between SWS and PS. However, with the microdialysis technique, Nitz and Siegel recently found an increase in GABA release in the cat LC during SWS as compared to waking values, a further increase during PS, and, in contrast, no detectable changes in GLY concentrations.[16] Based on these and our results, we therefore suggest that during W, the LC cells are under a tonic GABAergic inhibition which increases during SWS and even further during PS, and that the increase in GABAergic inhibition is at least partly responsible for the inactivation of these neurons during the sleep states. In contrast, the glycinergic tonic inhibition would be constant across the sleep/waking cycle and, thus, would control the general excitability of LC neurons.

It remains to be determined whether the tonic GABAergic and glycinergic inhibitions revealed in this study involve post- and/or presynaptic mechanisms. Indeed, the activation of LC cells seen after bicuculline or strychnine might be due not only to a removal of postsynaptic effects of GABA or GLY inputs but also to that of GABAergic or glycinergic presynaptic inhibition of excitatory inputs to LC cells. Such presynaptic GABAergic inhibition of excitatory terminals has been well documented for many neurons in the CNS[53] and is therefore likely to occur in LC noradrenergic neurons, as well. In contrast, glycinergic presynaptic inhibition remains to be demonstrated in CNS neurons.

Concerning the DRN serotonergic neurons, the available data are somewhat contradictory. Indeed, Nitz and Siegel[15] reported a significant increase in GABA release in the DRN during PS as compared to W and SWS, with no difference between W and SWS. In contrast, Levine and Jacobs[17] found that the iontophoretic application of bicuculline reversed the typical suppression of neuronal activity during SWS but not during PS. Additional experiments are therefore necessary to determine whether GABA or GLY plays a role in the inactivation of DRN serotonergic cells during SWS and PS.

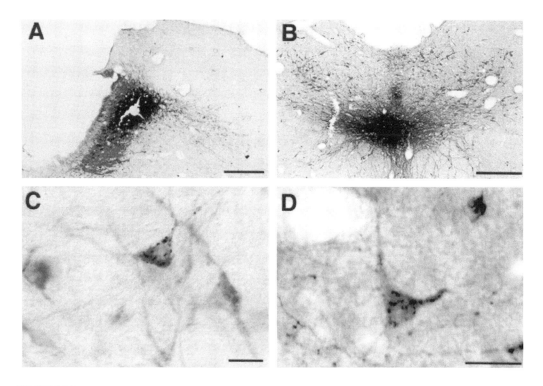

FIGURE 13.3
(**A**) and (**B**) Photomicrographs illustrating small CTb injection sites limited to the LC (A) or the ventral part of the DRN (B). (**C**) Photomicrograph showing a GAD (light color) and CTb double-labeled (black granules) neuron localized in the lateral hypothalamic area after CTb injection in the DRN. (**D**) Photomicrograph illustrating a glycine- (light color) and CTb- (black granules) labeled neuron localized in the ventrolateral part of the periaqueductal gray following CTb injection in the DRN. Bars: 250 µm (A,B) and 20 µm (C,D).

13.3 Glycinergic and GABAergic Afferent Projections to the Locus Coeruleus and Dorsal Raphe Nuclei

13.3.1 Double Immunostaining Procedures

The procedures used for retrograde tracing with cholera-toxin B subunit (CTb) and anterograde tracing with *Phaseolus vulgaris* leucoagglutinin (PHAL) have been described in detail previously.[54–56] The high sensitivity of CTb allowed us to demonstrate pathways not seen before using other tracers, including horseradish peroxidase (HRP), wheatgerm agglutinin conjugated HRP (WGA-HRP), and fluorescent tracers.

13.3.2 Afferent Projections to the Locus Coeruleus

Following iontophoretic injections of CTb into the core of the LC (Figure 13.3A), we observed a substantial number of retrogradely labeled cells in the lateral and dorsal paragigantocellular nuclei as previously described.[5] We also saw a substantial number of retrogradely labeled neurons in (1) the preoptic area dorsal to the supraoptic nucleus, (2) areas of the posterior hypothalamus, (3) the Kölliker-Fuse nucleus, and (4) the mesencephalic reticular formation. Fewer labeled cells were also observed in other regions including the hypothalamic paraventricular nucleus, the dorsal and median

A

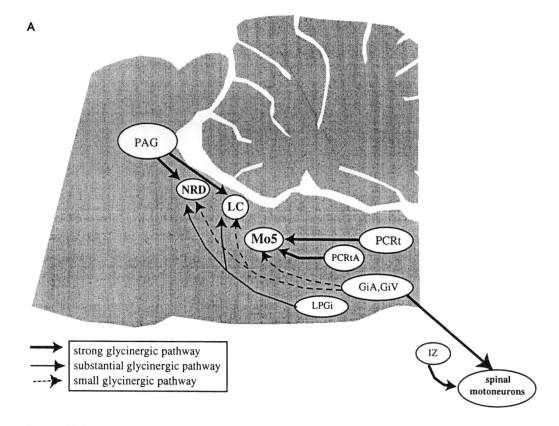

FIGURE 13.4
Schematic representation of the glycinergic (**A**) and GABAergic (**B**) afferents to the LC, dorsal raphe nucleus, motor trigeminal nucleus (Mo5), and spinal motoneurons. Abbreviations: DPGi = dorsal paragigantocellular nucleus; GiA = gigantocellular alpha nucleus; GiV = gigantocellular ventral reticular nucleus; HP = posterior hypothalamus; LC = locus coeruleus; LPGi = lateral paragigantocellular nucleus; Mo5 = trigeminal motor nucleaus; NRD = dorsal raphe nucleus; PAG = periaqueductal gray; PCRtA = nucleus parvocellular alpha; PCRt = parvocellular reticular nucleus; PoA = lateral preoptic area; SNr = substantia nigra reticulata; VTA = ventral tegmental area; IZ = intermediate zone of the spinal cord.

raphe nuclei, dorsal part of the periaqueductal gray, the area of the noradrenergic A5 group, the lateral parabrachial nucleus, and the caudoventrolateral reticular nucleus.

These results were confirmed and further extended by the anterograde transport data with CTb and PHAL. Injections of these tracers in the lateral paragigantocellular nucleus, preoptic area dorsal to the supraoptic nucleus, the posterior hypothalamic areas, the ventrolateral part of the periaqueductal gray, and the Kölliker-Fuse nucleus yielded a substantial to large number of labeled fibers in the nuclear core of the LC.

In conclusion, our results indicate that the LC receives afferents from a very large number of structures from the forebrain to the medulla. Failure to demonstrate these afferents in earlier works[5,57] was most likely due to the poor sensibility of the retrograde tracers used (HRP, WGA-HRP, and fluorogold).

13.3.2.1 Origin of the Glycinergic Input to the Locus Coeruleus

Following CTb injections centered at the LC, a large number of cells that were positive for both CTb and GLY immunoreactivity were observed in the ventrolateral and lateral parts of the periaqueductal gray, where they represented a small proportion of the CTb-positive cells (Figure 13.3D). Also at

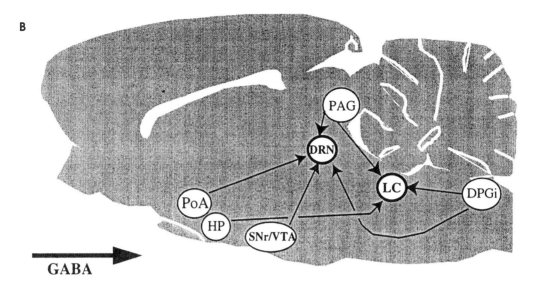

FIGURE 13.4 (continued)

the same level, a substantial number of double-labeled neurons were detected in the mesencephalic reticular formation. A moderate number of double-labeled neurons were seen in the nucleus raphe magnus, gigantocellular alpha, and lateral and dorsal paragigantocellular nuclei.[58] These results are summarized in Figure 13.4.

13.3.2.2 Origin of the GABAergic Input to the Locus Coeruleus

Following CTb injections in the LC, many brain regions contained neurons that were positive for both CTb and GAD immunoreactivity. A large number of double-labeled cells were localized in the lateral preoptic area, the lateral hypothalamic area, the lateral and ventrolateral parts of the periaqueductal gray, and the dorsal paragigantocellular nucleus. A small number of double-labeled cells were also seen in the dorsal hypothalamic area, the tuberomammillary nucleus, the mesencephalic reticular formation, the lateral parabrachial nucleus, and the nucleus raphe magnus.[59] These results are summarized in Figure 13.4.

13.3.3 Afferent Projections to the Dorsal Raphe Nucleus

We confirmed that the lateral habenula contains a large number of retrogradely labeled cells following CTb injections in the DRN (Figure 13.3B).[60] In addition, we observed a large number of retrogradely labeled cells in the orbital, cingulate, infralimbic, dorsal peduncular, and insular cortices. A substantial number of retrogradely labeled cells were also visible in the ventral pallidum, claustrum, bed nucleus of the stria terminalis, medial and lateral preoptic areas, medial preoptic nucleus, lateral, dorsal and posterior hypothalamic areas, zona incerta, subincertal nucleus, tuber cinereum, and medial tuberal nucleus.

In the brainstem, a large number of neurons were found in the ventral part of the periaqueducal gray, just above the oculomotor nuclei, and more caudally in the ventrolateral, lateral, and dorsal parts of the periaqueductal gray. At the same level, the mesencephalic and oral pontine reticular nuclei contained numerous CTb-positive cells. At pontine levels, the largest number of CTb-positive cells were seen in the lateral parabrachial nucleus. The medial parabrachial nucleus, the lateral tegmental nucleus of Castaldi, and the pontine periaqueductal gray contained a moderate number of CTb-positive cells. At medullary levels, only a small number of neurons were found in the nucleus raphe magnus, as well as parvocellular, gigantocellular alpha, lateral and dorsal paragigantocellular nuclei, and the nucleus of the solitary tract.

13.3.3.1 Origin of the Glycinergic Input to the Dorsal Raphe Nucleus

Following CTb injections in the DRN, many medium-sized cells that were positive for CTb and GLY were found bilaterally in the ventrolateral and lateral parts of the periaqueductal gray caudal to the trochlear nucleus, and in the adjacent mesencephalic reticular formation. A few small, double-labeled cells were also found in a rostral periaqueductal region dorsal to the oculomotor complex. The double-labeled cells in the periaqueductal gray represented a small proportion of CTb-labeled neurons in that region.

In the medulla, a moderate number of double-labeled cells were seen within the lateral paragigantocellular nucleus, the nucleus raphe magnus, and the gigantocellular reticular alpha nucleus.[58] These results are summarized in Figure 13.4.

13.3.3.2 Origin of the GABAergic Input to the Dorsal Raphe Nucleus

13.3.3.2.1 Neurons Immunoreactive for GAD and Serotonin in the Dorsal Raphe Nucleus. After double immunostaining of GAD (in brown) and serotonin (in black) on the same sections, GAD-immunoreactive cell bodies were observed in the lateral part of the DRN mixed with serotonergic neurons. In contrast, the medial part of the DRN contained a large number of serotonergic cells and only a few GAD-positive neurons. Double-labeled cells were not observed. These results fit perfectly with the recent findings of Stamp and Semba.[61]

13.3.3.2.2 GAD and CTb Immunoreactive Neurons. Following CTb injections in the DRN, the largest number of cells that were positive for both CTb and GAD were localized in the lateral hypothalamic area. A substantial number of double-labeled cells were also observed in the medial and lateral preoptic areas, the substantia nigra pars reticulata, the ventral tegmental area, and the ventrolateral periaqueductal gray. A moderate number of double-labeled cells were seen in the ventral pallidum, the parabrachial area, the oral pontine reticular nucleus, and the dorsal paragigantocellular nucleus. Finally, a small number of double-labeled neurons were seen in the paraventricular hypothalamic area, the lateral habenula, and the tuberomammillary, raphe magnus, and gigantocellular alpha nuclei.[59] These results are summarized in Figure 13.4.

13.3.4 Physiological Role of the Glycinergic Inputs to the Dorsal Raphe Nuclei and the Locus Coeruleus

We found a major glycinergic input to the LC and the DRN from the ventrolateral and lateral parts of the periaqueductal gray and the adjacent mesencephalic reticular formation. The lateral paragigantocellular, raphe magnus, and gigantocellular alpha nuclei provide additional small glycinergic inputs to the LC and the DRN. The LC receives a small additional projection from the dorsal paragigantocellular nucleus. These results first indicate that the glycinergic inputs to the DRN and to the LC arise from the same structures and therefore might share the same functions for monoaminergic neurons contained in these two nuclei. Of further interest regarding these results, we found a tonic glycinergic inhibition of LC neurons during the entire sleep/waking cycle. In addition, no change in glycine levels in the DRN and LC has been reported across the sleep/waking cycle by Nitz and Siegel.[15,16] It can, therefore, be proposed that glycine release onto LC and DRN cells might be constant during all vigilance states and serve to dampen their excitability generally, rather than exerting selective inhibition during sleep. This control should mainly arise from the periaqueductal gray, which is the main source of GLY to the DRN and LC. This proposal does not apparently fit with the recent results of Sastre et al.[62] in cats showing that the inactivation of the ventrolateral periaqueductal gray by muscimol (a $GABA_A$ agonist) induced a dramatic increase of PS. Such contradiction might be explained by the fact that the glycinergic neurons constitute only a minor proportion of the neurons in the ventrolateral periaqueductal gray. The effect seen by Sastre et al. could therefore be due to the inhibition of other types of neurons, in particular glutamatergic or GABAergic neurons.[63]

In addition to glycinergic neurons in the periaqueductal gray, we observed a small number of glycinergic neurons in the nucleus gigantocellular alpha and the adjacent raphe magnus projecting to the DRN and LC. Based on a number of studies in cats, we previously proposed the hypothesis that during PS, the monoaminergic neurons and the cranial and spinal somatic motoneurons might be inhibited by a single population of glycinergic neurons located in the magnocellular reticular nucleus (cat's equivalent of gigantocellular alpha and ventral nuclei). However, the present results and those we recently obtained on the glycinergic afferents to the motor trigeminal nucleus[64] indicate that at least in rats the gigantocellular alpha and ventral nuclei provide only a limited glycinergic input to the DRN, LC, and motor trigeminal nucleus and, therefore, are unlikely to contain a single population of neurons responsible for the inhibition of monoaminergic and all somatic motoneurons during PS. In fact, in rats, the only motoneurons receiving a strong glycinergic projection from the gigantocellular alpha and ventral nuclei are those of the spinal cord, as demonstrated by Holstege and Bongers.[65] In view of these results in rats, the cranial motoneurons might be inhibited during PS by neurons from the parvocellular and parvocellular alpha reticular nuclei, and glycine might not selectively be involved in the inhibition of monoaminergic neurons during sleep.[64] Additional experiments are nevertheless necessary to confirm such a conclusion, particularly in cats, in which it has been shown that the magnocellular reticular nucleus but not the parvocellular reticular nuclei contains c-Fos positive cells following long periods of PS induced by carbachol injections in the pons.[66]

13.3.5 Physiological Role of the GABAergic Inputs to the Dorsal Raphe Nuclei and the Locus Coeruleus

Our results indicate that the LC and DRN receive GABAergic inputs from neurons located in a large number of distant regions from the forebrain to the medulla (Figure 13.4). We also observed a substantial number of GAD-immunoreactive neurons in the pontine and mesencephalic periaqueductal gray that project to the LC and DRC. These results indicate that the GABA innervation of these two monoaminergic nuclei arises from multiple, distant GABAergic groups in addition to local GABAergic neurons. Such results contrast with the classical concept that GABA is mainly contained in interneurons. They suggest that the serotonergic neurons of the DRN and noradrenergic neurons of the LC could be inhibited by multiple populations of GABAergic neurons located in different structures and raises the question of the functional significance of such complexity. One possibility is that only some of these GABAergic afferents are destined to the serotonergic neurons of the DRN and the noradrenergic neurons of the LC. This seems likely for the DRN, which is a heterogeneous structure, but not for the LC, which in rats contains nearly exclusively noradrenergic cells. Another possibility is that some of these afferents are postsynaptic and the others presynaptic, but the more likely explanation is that each of these afferents is active only under specific physiological conditions.

Based on physiological and electrophysiological data (see above), we expect that one or several of these GABAergic afferents are "turned on" specifically at the onset and during SWS, while one or several others are "turned on" specifically at the onset and during PS. These afferents would thus be respectively responsible for the progressive decrease of activity of monoaminergic neurons from

W to SWS and from SWS to PS. Among the GABAergic structures revealed in our study, several are common to the DRN and the LC and are therefore good candidates for these roles.

The most likely candidate for the inhibition of the monoaminergic neurons during SWS is the lateral preoptic area. Indeed, lesion of this structure in cats and rats induced an insomnia while its stimulation induced SWS.[67-72] Neurons increasing their activity during SWS have been recorded in this area.[73-75] Moreover, c-Fos positive cells were observed in the lateral preoptic area after long periods of SWS,[76] and it has been further shown that these neurons are in part GABA- and galanin-positive and project to the tuberomammillary nucleus, which contains waking-active and presumably histaminergic neurons.[77] From these and our results, we can therefore propose that GABAergic neurons in the lateral preoptic area increase their firing just before the onset and during SWS and induce SWS via their inhibitory projections to waking-inducing structures (tuberomammillary nucleus, DRN, and LC, among others).

In addition to the lateral preoptic area, the lateral hypothalamic area, the periaqueductal gray, and the dorsal paragigantocellular nucleus provide substantial GABA inputs to the DRN and LC. The strong GABAergic projection from the lateral hypothalamic area is rather puzzling. Indeed, since the initial demonstration that the lesion of the posterior hypothalamus induces somnolence,[78] the histaminergic neurons of the posterior hypothalamus have been more specifically implicated in waking.[79] These neurons are located in the tuberomamillary nucleus, ventral and caudal to the GABAergic neurons in the lateral hypothalamic area that project to the DRN and LC. Nevertheless, muscimol injections in cats in the lateral hypothalamic area in addition to those in the tuberomammillary nucleus induced hypersomnia.[80] Additional experiments are therefore needed to determine whether the lateral hypothalamic area and its GABAergic neurons play a role in vigilance control via their projections to the DRN and LC.

The GABAergic afferents responsible for the inhibition of monoaminergic neurons during PS should be located in the brainstem. Indeed, it is well known that PS-like episodes occur in pontine or decerebrate cats.[2] Moreover, it has recently been shown that, in decerebrate animals, PS episodes induced by carbachol injections in the pons are still associated with a cessation of activity of serotonergic neurons of the raphe obscurus and pallidus nuclei.[81]

In the brainstem, we observed substantial GABAergic projections to the LC and DRN from the periaqueductal gray and the dorsal paragigantocellular nucleus. In agreement with these results, local application of bicuculline blocked the dorsal paragigantocellular-evoked inhibition of LC neurons.[23] In addition, recent findings on slices showed that focal iontophoretic application of NMDA in the ventral periaqueductal gray induced bicuculline-sensitive IPSPs in DRN serotonergic neurons.[82] The GABAergic afferents from the periaqueductal gray and the dorsal paragigantocellular nucleus could therefore be responsible for the inhibition of monoaminergic neurons during PS.

The hypothesis that this inhibition is coming from neurones located in the periaqueductal gray is further supported by two recent studies. Yamuy et al.[83] showed that after a long period of PS induced by pontine injection of carbachol, a large number of c-Fos positive cells are visible in the DRN and a region lateral to it. Moreover, Maloney and Jones[84] observed, after a PS rebound induced by deprivation, an increase in c-Fos-positive GAD immunoreactive neurons in the periaqueductal gray and the lateral tegmental nucleus.

Finally, it must be acknowledged that although a number of arguments are in favor of a role of GABA in the inhibition of monoaminergic neurons during SWS and PS, other inhibitory neuroactive substances might also participate in this inhibition. Indeed, it has for example been shown that LC and DRN cells are inhibited by local application of enkephalin.[82,84] Moreover, it has been shown that extracellular adenosine levels increase during wakefulness and decrease during SWS,[86] and that adenosine application is inhibitory on LC noradrenergic cells.[87,88]

The removal or reduction during SWS or PS of tonic excitatory inputs that may be present at high levels during W (e.g., acetylcholine or glutamate) might also contribute to the decrease of activity of LC and DRN neurons during sleep.

13.4 Conclusions and New Hypothesis

During SWS, the monoaminergic neurons would be inhibited by GABAergic neurons located in the lateral preoptic area. At the onset of and during PS, a second population of GABAergic neurons located in the periaqueductal gray or the dorsal paragigantocellular nucleus would be responsible for the complete cessation of the monoaminergic neurons during this sleep state. The cranial motoneurons would be inhibited by glycinergic neurons of the parvocellular nucleus, and those of the spinal cord by glycinergic neurons from the gigantocellular ventral nucleus and glycinergic interneurons of the intermediate zone of the spinal cord.

Acknowledgments

This work was supported by INSERM (U52), CNRS (ERS 5645), Université Claude Bernard Lyon I, and the 1996 ESRS-Synthélabo European Research Grant. The authors wish to thank C. Guillemort (GFG Co, Pierre-Bénite) for his help in designing the head-restraining system.

References

1. Dahlström, A. and Fuxe, K., Evidence for the existence of monoamine-containing neurons in the central nervous system. I. Demonstration in the cell bodies of brain stem neurons, *Acta Physiol. Scand.*, 232(Suppl.), 1, 1964.
2. Jouvet, M., The role of monoamines and acetylcholine-containing neurons in the regulation of the sleep, *Ergebn. Physiol.*, 64, 166, 1972.
3. Jacobs, B.L., Wilkinson, L.O., and Fornal, C.A., The role of brain serotonin. A neurophysiologic perspective, *Neuropsychopharmacology*, 3, 473, 1990.
4. Jacobs, B.L. and Azmitia, E.C., Structure and function of the brain serotonin system, *Physiol. Rev.*, 72, 165, 1992.
5. Aston-Jones, G., Ennis, M., Pieribone, V.A., Nickell, W.T., and Shipley, M.T., The brain nucleus locus coeruleus: restricted afferent control of a broad efferent network, *Science*, 234, 734, 1986.
6. Aston-Jones, G. and Bloom, F.E., Activity of norepinephrine-containing locus coeruleus neurons in behaving rats anticipates fluctuations in the sleep-waking cycle, *J. Neurosci.*, 1, 876, 1981.
7. Hobson, J., McCarley, R., and Wyzinski, P., Sleep cycle oscillation: reciprocal discharge by two brainstem groups, *Science*, 189, 55, 1975.
8. McGinty, D.J. and Harper, R.M., Dorsal raphe neurons: depression of firing during sleep in cats, *Brain Res.*, 101, 569, 1976.
9. Trulson, M.E. and Jacobs, B.L., Raphe unit activity in freely moving cats: correlation with level of behavioral arousal, *Brain Res.*, 163, 135, 1979.
10. Aghajanian, G.K. and VanderMaelen, C.P., Intracellular identification of central noradrenergic and serotonergic neurons by a new double labeling procedure, *J. Neurosci.*, 2, 1786, 1982.
11. Williams, J.T., North, R.A., Shefner, A., Nishi, S., and Egan, T.M., Membrane properties of rat locus coeruleus neurons, *Neuroscience*, 13, 137, 1984.
12. Masuko, S., Nakajima, Y., Nakajima, S., and Yamaguchi, K., Noradrenergic neurons from the locus ceruleus in dissociated cell culture: culture methods, morphology, and electrophysiology, *J. Neurosci.*, 6, 3229, 1986.

13. Mendelson, W.B., Neuropharmacology of sleep induction by benzodiazepines, *Crit. Rev. Neurobiol.*, 622, 1, 1992.
14. Gaillard, J.-M., *Principles and Practice of Sleep Medicine,* Kryger, M.H., Roth, T., and Dement, W.C., Eds., W.B. Saunders, Philadelphia, 1994, p. 349.
15. Nitz, D.A. and Siegel, J.M., Inhibitory amino acid neurotransmission in the dorsal raphe nucleus during sleep/wake states, *Abstr. Soc Neurosci.*, 19, 1815, 1993.
16. Nitz, D.A. and Siegel, J.M., GABA release in the locus coeruleus as a function of sleep/wake state, *Neuroscience*, 78, 795, 1997.
17. Levine, E.S. and Jacobs, B.L., Neurochemical afferents controlling the activity of serotonergic neurons in the dorsal raphe nucleus: microiontophoretic studies in the awake cat, *J. Neurosci.*, 12, 4037, 1992.
18. Sakai, K., *Sleep: Neurotransmitters and Neuromodulators,* Wauquier, A., Monti, J.M., Gaillard, J.M., and Radulovacki, M., Eds., Raven Press, New York, 1985, p. 29.
19. Luppi, P.-H, Charlety, P.J., Fort, P., Akaoka, H., Chouvet, G., and Jouvet, M., Anatomical and electrophysiological evidence for a glycinergic inhibitory innervation of the rat locus coeruleus, *Neurosci. Lett.*, 128, 33, 1991.
20. Jones, B.E., Noradrenergic locus coeruleus neurons: their distant connections and their relationship to neighboring, including cholinergic and GABAergic neurons of the central gray and reticular formation, *Prog. Brain Res.*, 88, 15, 1991.
21. Guyenet, P.G. and Aghajanian, G.K., Ach, substance P and Met-Enkephalin in the locus coeruleus: pharmacological evidence for independent sites of action, *Eur. J. Pharmacol.*, 53, 319, 1979.
22. Koyama, Y. and Kayama, Y., Mutual interactions among cholinergic, noradrenergic and serotonergic neurons studied by iontophoresis of these transmitters in rat brainstem nuclei, *Neuroscience*, 55, 117, 1993.
23. Ennis, M. and Aston-Jones, G., GABA-mediated inhibition of locus coeruleus from the dorsomedial rostral medulla, *J. Neurosci.*, 9, 2973, 1989.
24. Gallager, D.W. and Aghajanian, G.K., Effect of antipsychotic drugs on the firing of dorsal raphe cells. II. Reversal by picrotoxin, *Eur. J. Pharmacol.*, 39, 357, 1976.
25. Gallager, D.W., Benzodiazepines: potentiation of a GABA inhibitory response in the dorsal raphe nucleus, *Eur. J. Pharmacol.*, 49, 133, 1978.
26. Cherubini, E., North, R.A., and Williams, J.T., Synaptic potentials in rat locus coeruleus neurons, *J. Physiol. (Lond.)*, 406, 431, 1988.
27. Williams, J.T., Bobker, D.H., and Harris, G.C., Synaptic potentials in locus coeruleus neurons in brain slices, *Prog. Brain Res.*, 88, 167, 1991.
28. Osmanovic, S.S. and Shefner, S.A., γ-aminobutyric acid responses in rat locus coeruleus neurons in vitro: a current-clamp and voltage-clamp study, *J. Physiol. (Lond.)*, 421, 151, 1990.
29. Pan, Z.Z. and Williams, J.T., GABA- and glutamate-mediated synaptic potentials in rat dorsal raphe neurons *in vitro*, *J. Neurophysiol.*, 61, 719, 1989.
30. Wang, Q.P., Ochiai, H., and Nakai, Y., GABAergic innervation of serotonergic neurons in the dorsal raphe nucleus of the rat studied by electron microscopy double immunostaining, *Brain Res. Bull.*, 6, 943, 1992.
31. Luque, J.M., Malherbe, P., and Richards, J.G., Localization of $GABA_A$ receptor subunit mRNAs in the rat locus coeruleus, *Mol. Brain Res.*, 24, 219, 1994.
32. Zarbin, M.A., Wamsley, J.K., and Kuhar, M.J., Glycine receptors: light microscopic autoradiographic localization with [^{3}H] strychnine, *J. Neurosci.*, 5, 532, 1981.
33. Chase, M.H., Soja, P.J., and Morales, F.R., Evidence that glycine mediates the postsynaptic potentials that inhibit lumbar motoneurons during the atonia of active sleep, *J. Neurosci.*, 9, 743, 1989.

34. Kaur, S., Saxena, R.N., and Mallick, B.N., GABA in locus coeruleus regulates spontaneous rapid eye movement sleep by acting on GABA$_A$ receptors in freely moving rats, *Neurosci. Lett.*, 223, 105, 1997.
35. Darracq, L., Gervasoni, D., Soulière, F., Lin, J.S., Fort, P., Chouvet, G., and Luppi, P.-H., Effect of strychnine on rat locus coeruleus neurons during sleep and wakefulness, *NeuroReport*, 8, 351, 1996.
36. Chouvet, G., Akaoka, H., and Aston-Jones, G., Serotonin selectively decreases glutamate-induced excitation of locus coeruleus neurons, *C. R. Acad. Sci. (Paris)*, 306, 339, 1988.
37. Akaoka, H., Charlèty, P.J., Saunier, C.F., Buda, M., and Chouvet, G., Combining *in vivo* volume-controlled pressure microinjection with extracellular unit recording, *J. Neurosci. Meth.*, 42, 119, 1992.
38. Maloney, K.J., Cape, E.G., Gotman, J., and Jones, B.E., High-frequency γ-encephalogram activity in association with sleep-wake states and spontaneous behaviors in the rat, *Neuroscience*, 76, 541, 1997.
39. Gervasoni, D., Darracq, L., Fort P., Soulière F., Chouvet, G., and Luppi, P.-H., Electrophysiological evidence that noradrenergic neurones of the rat locus coeruleus are tonically inhibited by GABA during sleep, *Eur. J. Neurosci.*, 10, 964, 1998.
40. Chiang, C. and Aston-Jones, G., A 5-hydroxytryptamine2 agonist augments γ-aminobutryric acid and excitatory amino acid inputs to noradrenergic locus coeruleus neurons, *Neuroscience*, 54, 409, 1993.
41. Blume, H.W., Pittman, Q.J., and Renaud, L.P., Sensitivity of identified hypothalamic neurons to GABA, glycine and related amino acids: influence of bicuculline, picrotoxin and strychnine on synaptic inhibition, *Brain Res.*, 209, 145, 1981.
42. Murphy, P.C. and Sillito, A.M., The binocular input to cells in the feline dorsal lateral geniculate nucleus (dLGN), *J. Physiol. (Lond.)*, 415, 393, 1989.
43. Peng, Y.B., Lin, Q., and Willis, W.D., Effects of GABA and glycine receptor antagonists on the activity and PAG-induced inhibition of rat dorsal horn neurons, *Brain Res.*, 736, 189, 1996.
44. Bennett, J.A., McWilliam, P.N., and Shepheard, S.L., A gamma-aminobutyric-acid-mediated inhibition of neurons in the nucleus tractus solitarius of the cat, *J. Physiol. (Lond.)*, 392, 417, 1987.
45. Champagnat, J., Denavit-Saubiè, M., Moyanova, S., and Rondouin, G., Involvement of amino acids in periodic inhibitions of bulbar respiratory neurons, *Brain Res.*, 237, 351, 1982.
46. Jordan, D., Mifflin, S.W., and Spyer, K.M., Hypothalamic inhibition of neurons in the nucleus tractus solitarius of the cat is GABA mediated, *J. Physiol. (Lond.)*, 399, 389, 1988.
47. Palombi, P.S. and Caspary, D.M., GABA$_A$ receptor antagonist bicuculline alters response properties of posteroventral cochlear nucleus neurons, *J. Neurophysiol.*, 67, 738, 1992.
48. Curtis, D.R. and Johnston, G.A.R., Amino acid transmitters in the mammalian central nervous system, *Ergebn. Physiol.*, 69, 79, 1974.
49. Curtis, D.R., Duggan, A.W., Felix, D., and Johnston, G.A.R., Bicuculline, an antagonist of GABA and synaptic inhibition in the spinal cord of the cat, *Brain Res.*, 32, 69, 1971.
50. Otis, T.S., Staley, K.S., and Mody, I., Perpetual inhibitory activity in mammalian brain slices generated by spontaneous GABA release, *Brain Res.*, 545, 142, 1991.
51. Pitler, T.A. and Alger, B.E., Postsynaptic firing reduces synaptic GABA$_A$ responses in hippocampal pyramidal cells, *J. Neurosci.*, 12, 4122, 1992.
52. Sun, M-K. and Guyenet, P.G., GABA-mediated baroreceptor inhibition of reticulo-spinal neurons, *Am. J. Physiol.*, 249, 672, 1985.
53. McGehee, D.S. and Role, L.W., Presynaptic ionotropic receptors, *Curr. Opin. Neurobiol.*, 6, 342, 1996.
54. Luppi, P-H., Fort, P., and Jouvet, M., Iontophoretic application of unconjugated cholera toxin B subunit CTb combined with immunohistochemistry of neurochemical substances: a method for transmitter identification of retrogradely labeled neurons, *Brain Res.*, 534, 209, 1990.

55. Luppi, P.H., Aston-Jones, G., Akaoka, H., Chouvet, G., and Jouvet, M., Afferent projections to the rat locus coeruleus demonstrated by retrograde and anterograde tracing with cholera-toxin B subunit and *Phaseolus vulgaris* leucoagglutinin, *Neuroscience*, 65, 119, 1995.
56. Peyron, C., Luppi, P.-H., Fort, P., Rampon, C., and Jouvet, M., Lower brainstem catecholamine afferents to the rat dorsal raphe nucleus, *J. Comp. Neurol.*, 364, 402, 1996.
57. Cedarbaum, J.M. and Aghajanian, G.K., Afferent projections to the rat locus coeruleus as determined by a retrograde tracing technique, *J. Comp. Neurol.*, 178, 1, 1978.
58. Rampon, C., Peyron, C., Gervasoni, D., Cespuglio, R., Fort, P., and Luppi, P-H., Localization of the glycinergic neurons projecting to the rat locus coeruleus, dorsal raphe and trigeminal motor nuclei, *Soc. Neurosci. Abstr.*, 22, 1838, 1996.
59. Peyron, C., Luppi, P-H., Rampon, C., and Jouvet, M., Location of the GABAergic neurons projecting to the dorsal raphe nucleus and the locus coeruleus of the rat, *Soc. Neurosci. Abstr.*, 21, 373, 1995.
60. Aghajanian, G.K. and Wang, R.Y., Habenular and other midbrain raphe afferents demonstrated by a modified retrograde tracing technique, *Brain Res.*, 238, 463, 1977.
61. Stamp, J.A. and Semba K., Extent of colocalization of serotonin and GABA in the neurons of the rat raphe nuclei, *Brain Res.*, 677, 39, 1995.
62. Sastre, J.P., Buda, C., Kitahama, K., and Jouvet, M., Importance of the ventrolateral region of the periaqueductal gray and adjacent tegmentum in the control of paradoxical sleep as studied by muscimol microinjection in the cat, *Neuroscience*, 74, 415, 1996.
63. Beitz, A., *The Rat Nervous System*, 2nd ed., Paxinos, G., Ed., Academic Press, San Diego, 1995, p. 173.
64. Rampon, C., Peyron, C., Petit, J.M., Fort, P., Gervasoni, D., and Luppi, P.H., Origin of the glycinergic innervation of the rat trigeminal motor nucleus, *NeuroReport*, 7, 3081, 1996.
65. Holstege, J.C. and Bongers, C.M.H., A glycinergic projection from the ventromedial lower brainstem to spinal motoneurons. An ultrastructural double labeling study in rat, *Brain Res.*, 566, 308, 1991.
66. Yamuy, J., Mancillas, J.R., Morales, F.R., and Chase, M.H., C-Fos expression in the pons and medulla of the cat during carbachol-induced active sleep, *J. Neurosci.*, 13, 2703, 1993.
67. Asala, S.A., Okano, Y., Honda, K., and Inoue, S., Effects of medial preoptic area lesions on sleep and wakefulness in unrestrained rats, *Neurosci. Lett.*, 114, 300, 1990.
68. Lucas, E.A. and Sterman, M.B., Effect of a forebrain lesion on the polycyclic sleep wake patterns in the cat, *Exp. Neurol.*, 46, 368, 1975.
69. McGinty, D.J. and Stermann, M.B., Sleep suppression after basal forebrain lesions in the cat, *Science*, 160, 1253, 1968.
70. Sallanon, M., Denoyer, M., Kitahama, K., Aubert, C., Gay, N., and Jouvet, M., Long-lasting insomnia induced by preoptic neuron lesions and its transient reversal by muscimol injection into the posterior hypothalamus in the cat, *Neuroscience*, 32, 669, 1989.
71. John, J., Kumar, V.M., Gopinath, G., Ramesh, V., and Mallick, H., Changes in sleep-wakefulness after kainic acid lesion of the preoptic area in rats, *Jpn. J. Physiol.*, 44, 231, 1994.
72. Sterman, M.B. and Clemente, C.D., Forebrain inhibitory mechanisms, sleep pattern induced by basal forebrain stimulation in behaving cat, *Exp. Neurol.*, 6, 103, 1962.
73. Kaitin, K.I., Preoptic area unit activity during sleep and wakefulness in the cat, *Exp. Neurol.*, 83, 347, 1984.
74. Szymusiak, R. and McGinty, D.J., Sleep-related discharge in the basal forebrain of cats, *Brain Res.*, 370, 82, 1986.
75. Koyama, Y. and Hayaishi, O., Firing of neurons in the preoptic/anterior hypothalamic areas in rat: its possible involvement in slow wave sleep and paradoxical sleep, *Neurosci. Res.*, 19, 31, 1994.

76. Sherin, J.E., Shiromani, P.J., McCarley, R.W., and Saper, C.B., Activation of ventrolateral preoptic neurons during sleep, *Science*, 271, 216, 1996.
77. Vanni-Mercier, G., Sakai, K., Salvert, D., and Jouvet, M., Waking-state specific neurons in the caudal hypothalamus of the cat, *C. R. Acad. Sci. (Paris)*, 298, 195, 1984.
78. Von Economo, C., *Handbuch des Normalen und Patholigischen Physiologie,* Von Bethe, A., Bergman, G.V., Embden, G., and Ellinger, U.A., Eds., Springer-Verlag, Berlin, 1926, p. 291.
79. Lin, J.-S., Hou, Y., Sakai, K., and Jouvet, M., Histaminergic inputs to the mesopontine tegmentum and their role in the control of cortical activation and wakefulness in the cat, *J. Neurosci.*, 15, 1523, 1996.
80. Lin, J.S., Sakai, K., Vanni-Mercier, G., and Jouvet, M., A critical role of the posterior hypothalamus in the mechanisms of wakefulness determined by microinjections of muscimol in freely moving cats, *Brain Res.*, 429, 225, 1989.
81. Woch, G., Davies, R.O., Pack, A.I., and Kubin, L., Behavior of raphe cells projecting to the dorsomedial medulla during carbachol-induced atonia in the cat, *J. Physiol. (Lond.)*, 490, 745, 1996.
82. Jolas, T. and Aghajanian, G.K., Opioids suppress spontaneous and NMDA-induced inhibitory postsynaptic currents in the dorsal raphe nucleus of the rat in vitro, *Brain Res.*, 755, 229, 1997.
83. Yamuy, J., Sampagna, S., Lopez-Rodriguez, F., Luppi, P.-H., Morales, F.R., and Chase, M.H., Fos and serotonin immunoreactivity in the raphe nuclei of the cat during carbachol-induced active sleep: a double-labeling study, *Neuroscience*, 67, 211, 1995.
84. Maloney K.J. and Jones B.E., c-Fos expression in cholinergic, GABAergic and monoaminergic cell groups during paradoxical sleep deprivation and recovery, *Abstr. Soc. Neurosci.*, 23, 2131, 1997.
85. Bird, S.J. and Kuhar, M.J., Iontophoretic applications of opiates to the locus coeruleus, *Brain Res.*, 122, 523, 1977.
86. Porkka-Heiskanen, T., Strecker, R.E., Thakkar, M., Bjorkum, A.A., Greene, R.W., and McCarley, R.W., Adenosine: a mediator of the sleep-inducing effects of prolonged wakefulness, *Science*, 276, 1265, 1997.
87. Pan, W.J., Osmanovic, S.S., and Shefner, S.A., Characterization of the adenosine A1 receptor-activated potassium current in rat locus ceruleus neurons, *J. Pharmacol. Exp. Ther.*, 273, 537, 1995.
88. Shefner, S.A. and Chiu, T.H., Adenosine inhibits locus coeruleus neurons: an intracellular study in a rat brain slice preparation, *Brain Res.*, 366, 364, 1986.

Chapter 14

Cholinergic and GABAergic Neurons of the Basal Forebrain: Role in Cortical Activation

Barbara E. Jones and Michel Mühlethaler

Contents

14.1	Introduction	214
14.2	Neuroanatomical and Immunohistochemical Study of Basal Forebrain Neurons	214
	14.2.1 Distribution of Cholinergic and GABAergic Neurons	215
	14.2.2 Projections of Cholinergic and GABAergic Neurons	215
	14.2.3 Brainstem Afferents to Basal Forebrain Neurons	217
14.3	*In Vitro* Electrophysiological and Pharmacological Study of Basal Forebrain Neurons	218
	14.3.1 Intrinsic Properties of Identified Cholinergic and Non-Cholinergic Neurons	218
	14.3.2 Pharmacological Modulation of Cholinergic and Non-Cholinergic Neurons	220
14.4	*In Vivo* Electrophysiological and Pharmacological Study of the Role of Basal Forebrain Neurons in Cortical Activation	222
	14.4.1 Discharge Properties of Possible Cholinergic and Non-Cholinergic Neurons In Relation to Cortical Activity	223
	14.4.2 Modulation of Cortical Activity by Stimulation or Inactivation of the Basal Forebrain	223
	14.4.2.1 Electrical Stimulation or Lesion	224
	14.4.2.2 Chemical Stimulation or Inactivation	224
14.5	Summary and Conclusions	228
References		230

14.1 Introduction

The basal forebrain has long been known to represent an important relay of the brainstem reticular activating system to the cerebral cortex. Thus, from the early studies of Dempsey and Morison[15,52] and subsequently Moruzzi et al.,[53,67] it was known that ascending impulses from the reticular formation relayed both dorsally through the thalamus, via the nonspecific thalamo-cortical projection system, and ventrally through the hypothalamus and basal forebrain to reach widespread areas of the cerebral cortex and therein evoke widespread activation. Such activation accompanies, yet outlasts, the regionally limited and specific cortical response to sensory stimulation in the lightly anesthetized or sleeping animal. Tonic cortical activation accompanies and underlies the state of wakefulness, when animals and humans are conscious. It also accompanies the state of rapid eye movement (REM) or paradoxical sleep (PS), when animals and humans dream.[13,14,35] Following ablation of the thalamus, cortical activation can still be evoked from the reticular formation[15,67] and persists along with the states of waking and REM sleep,[34,74,75] indicating that the ventral extra-thalamic relay through the basal forebrain is sufficient for at least the rudimentary state modulation of cortical activity.

From early histochemical studies employing staining for acetylcholinesterase (AChE), the catabolic enzyme for acetylcholine (ACh), Shute and Lewis[65] realized that the projection from the basal forebrain to the cerebral cortex may be cholinergic. In addition, they visualized cholinesterase-stained fibers ascending from the brainstem into the basal forebrain and accordingly delineated an ascending cholinergic reticular system, which they postulated carries the activating influence from the brainstem to the cerebral cortex through a cholino-cholinergic relay.[66] Although this schema has proven to be oversimplified, it nonetheless appropriately represented the potency and importance of cholinergic neurons of the basal forebrain and brainstem in mechanisms of cortical activation.

Physiological studies by Krnjevic[42] showed that ACh had prominent excitatory effects on cortical neurons. Early pharmacological studies also indicated that muscarinic and nicotinic blockers could virtually eliminate cortical activation of both waking and REM sleep states.[18,45] Early biochemical studies by Jasper and colleagues[12,29] showed that ACh was released in the cortex maximally in association with cortical activation and the natural states of waking and REM sleep. Recent physiological, pharmacological, and biochemical studies have confirmed these early observations and conclusions.

In subsequent studies aimed at better understanding the way in which the basal forebrain neurons influence the cerebral cortex in a state-dependent manner and the way in which they in turn are influenced by afferents from the brainstem reticular activating system, the full complexity of the constituent elements of this system has come to light. Thus, as will be the topic of this chapter, it has become apparent that basalo-cortical projection neurons are comprised by cholinergic and other non-cholinergic neurons, particularly GABAergic neurons, and that these neurons are collectively but differentially influenced by multiple afferents from the brainstem.

14.2 Neuroanatomical and Immunohistochemical Study of Basal Forebrain Neurons

The basal forebrain refers generally to the subcortical gray matter at the base of the telencephalon but here, used in the more limited sense now often employed, it refers to the region of the nucleus basalis of Meynert or the magnocellular basal nucleus. The latter collection of cells corresponds in fact to those neurons which form the relay between the ascending reticular activating system and the cerebral cortex. Originally revealed by electrophysiological means in the 1940s and 50s (see above), the cortically projecting neurons of that relay were not identified by neuroanatomical studies until 1975 by Kievet and Kuypers[40] using newly available retrograde transport techniques.

It was subsequently established that the basalo-cortical projection neurons corresponded in large part to the basalo-cortical cholinesterase-stained fiber system originally visualized by Shute and Lewis[65] and also Krnjevic and Silver.[8,43,50] Since these early studies, multiple studies have confirmed with immunohistochemistry for choline acetyltransferase (ChAT), the synthetic enzyme for ACh, that a major proportion, though not all, of the cortically projecting neurons of the magnocellular basal complex are cholinergic.

14.2.1 Distribution of Cholinergic and GABAergic Neurons

The cholinergic neurons of the basal forebrain are distributed in a continuum extending from the septum, diagonal band of Broca, magnocellular preoptic nucleus, and substantia innominata into the globus pallidus.[25,49,63] The cells along this continuum have been shown to project in a topographically organized manner to the hippocampus and neocortex.[46,61,62] Thus, the most medial and rostral neurons within the septum and diagonal band were shown to project to the hippocampus, whereas the more lateral and caudal cells located through the magnocellular preoptic nucleus, substantia innominata, and globus pallidus were shown to project to the neocortex. Cells in this latter collection are often referred to as the nucleus basalis.

Through this full cell continuum, the cholinergic neurons are co-distributed with other neurons of different sizes and containing different neurotransmitters. In the septum and diagonal band, they were found to be co-distributed with a large number of GABAergic neurons, which project to the hippocampus and olfactory bulb.[7,41,76] These findings evoked the possibility that GABAergic neurons might also comprise a contingent of the nucleus basalis and relay to the neocortex.

Gritti et al.[25] examined the distribution, number, and size of GABAergic neurons in relation to the cholinergic cells in the basal forebrain of the rat, using single and dual immunostaining for glutamic acid decarboxylase (GAD) and ChAT. Through the entire basal cell complex extending from the septum, diagonal band nuclei, magnocellular preoptic nucleus, substantia innominata, and globus pallidus, GABAergic cells are co-distributed with cholinergic cells. Although intermingled with each other, the GABAergic and cholinergic cells nonetheless often form segregated clusters of cells through the basal forebrain. In all these subregions, GABAergic neurons outnumber cholinergic neurons by at least 2:1. In the nucleus basalis region (Figure 14.1A), the total estimated number of GABAergic neurons is 22,600 and of cholinergic, 11,800. In cell size, the GABAergic neurons are on average significantly smaller than the cholinergic; however, among this large population of GABAergic cells, there are neurons which are as large as the cholinergic cells and would thus form part of the magnocellular basal cell complex.

14.2.2 Projections of Cholinergic and GABAergic Neurons

Whereas it had been established that GABAergic neurons within the septum and diagonal band project in similar numbers as cholinergic neurons to the hippocampus and also olfactory bulb (see above), it was not believed that neurons other than the cholinergic neurons projected to the neocortex from the nucleus basalis;[61] yet, in the cat, evidence was presented that such projections exist from GABAergic neurons of the basal forebrain.[19]

Gritti et al.[27] investigated neocortical projections from GABAergic and cholinergic basal forebrain neurons in the rat using retrograde transport of the sensitive tracer, cholera toxin, together with immunohistochemistry for GAD and ChAT. Because we thought that perhaps GABAergic neurons contributed particularly to projections to limbic neocortex or meso-cortex but little, if at all,

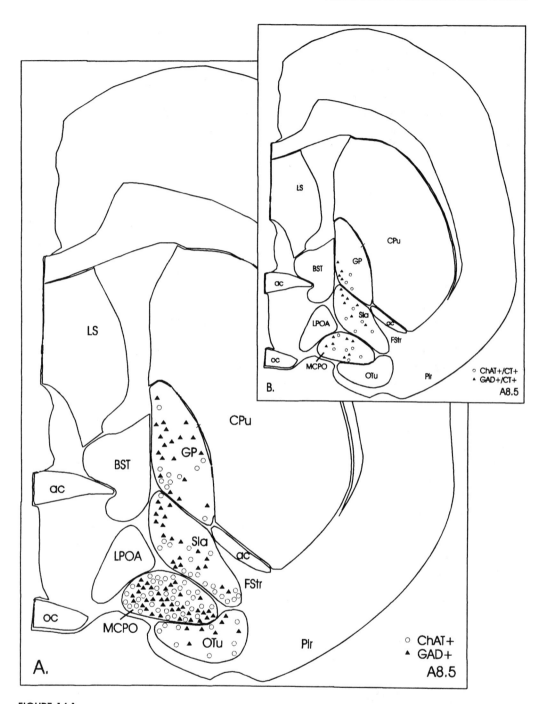

FIGURE 14.1
(A) Distribution of ChAT-positive (ChAT+) and GAD-positive (GAD+) neurons in the basal forebrain of the rat. (B) Plotted on the same figure is the distribution of cortically projecting ChAT+ and GAD+ neurons, which were retrogradely labeled with cholera toxin (CT+) injected into the orbitofrontal cortex. Cells are plotted on a computerized atlas template,[25] which corresponds approximately to level A8.5 in stereotaxic coordinates.[56] (Results and figures adapted from previously published work by Gritti et al.[25,27])

to projections to non-limbic or iso-cortex, we investigated retrograde transport from orbito-frontal cortex as meso-cortex and parietal cortex as iso-cortex. From the orbito-frontal cortex, we found that retrogradely labeled GABAergic neurons were distributed alongside retrogradely labeled cholinergic neurons through the basal forebrain, including the nucleus basalis (Figure 14.1B). An equivalent number and proportion of GABAergic to cholinergic neurons were retrogradely labeled, indicating that, as is the case for the projection to the hippocampus (archi-cortex), GABAergic neurons project in parallel with cholinergic neurons to the limbic (meso-) cortex. To our great surprise, we found that the same was true for the projection to the iso-cortex. Indeed, equivalent numbers of GABAergic and cholinergic neurons were retrogradely labeled from the parietal cortex. That GABAergic basal forebrain neurons contribute to projections to both hippocampus and neocortex has also been revealed by anterograde transport techniques by Freund and colleagues.[23,24]

One other surprising finding emerged from the studies involving retrograde transport with cholera toxin from the neocortex,[27] and that was a wider distribution of both cholinergic and GABAergic retrogradely labeled cells than that previously published for cholinergic neurons using less-sensitive retrograde tracers. Indeed, although darkly retrogradely labeled cells were concentrated within subregions where cells had previously been located in restricted distributions according to a topographical distribution,[61] lightly retrogradely labeled cells were distributed in a much wider manner without restriction to subregion. These results indicate that, as had also been found to be the case in the cat,[1,6] both GABAergic and cholinergic, cortically projecting basal forebrain neurons provide a very widespread innervation to the cortex. Thus, despite what may be a focused, high-density projection to one cortical area, the cholinergic and GABAergic basalis neurons would also provide dispersed collaterals to other cortical regions and thus have the capacity to influence the cortex in a widespread manner.

Like the cholinergic cells, the GABAergic cells that are retrogradely labeled from the cerebral cortex are relatively large-sized neurons and significantly larger than the GABAergic cells in the same region that give rise to descending projections to the posterior hypothalamus.[25-27] Thus, the GABAergic cell population is heterogeneous and probably comprised of small local projection neurons; small, caudally projecting neurons; and larger, cortically projecting cells that are unequivocally part of the magnocellular basal cell complex, both according to their size and their projections to the neocortex.

In addition to the cholinergic and GABAergic neurons in the basal cell complex, there are many other neurons which give rise to descending projections on the one hand and also ascending projections to the cerebral cortex, on the other, which cannot be accounted for as either cholinergic or GABAergic by the quantitative studies performed.[26,27] Since many cells are darkly stained for glutaminase, it is possible that glutamatergic neurons represent a significant proportion of these as yet unidentified cells (Jones, unpublished observations).

14.2.3 Brainstem Afferents to Basal Forebrain Neurons

The basal cell complex lies within the ascending, as well as descending, fibers of the medial forebrain bundle. Accordingly, the large cholinergic and GABAergic neurons with long dendrites sit in a position to receive inputs from the passing fibers originating in the brainstem. From anatomical studies employing anterograde transport of radiolabeled proteins, one of us,[33] using degeneration techniques, had confirmed and extended the earlier demonstrations by Nauta and Kuypers[54] that a major extra-thalamic projection from the brainstem reticular formation passes ventrally through the hypothalamus coursing in the medial forebrain bundle to reach the basal forebrain. Few fibers from the reticular formation reach beyond the basal forebrain to go directly to the cerebral cortex. Accordingly, through this ventral extra-thalamic route, the reticular formation influences the cerebral cortex via the cortically projecting neurons of the basal forebrain (see Reference 31 for review). The neurons of the reticular formation which project rostrally are medium-to large-sized cells which contain high concentrations of glutamate[20,31] and glutaminase (Jones, unpublished observations).

In addition to the large population of reticular neurons which project to the basal forebrain, monoaminergic neurons also project into the region of the basal cell complex, as we and others have shown.[32,64,77] Thus, relatively dense catecholaminergic fibers arising from both the noradrenergic locus coeruleus cells and the dopaminergic ventral tegmental cells collateralize within the area of the nucleus basalis. Serotonergic neurons of the dorsal and midbrain raphe also provide a dense innervation to the nucleus basalis region. These monoaminergic fibers could be collaterals of fibers continuing directly on to the cerebral cortex. Finally, there are also a small number of cholinergic neurons in the pontomesencephalic tegmentum (laterodorsal and pedunculopontine tegmental nuclei) which project to the region of the basal forebrain cholinergic cells, as Shute and Lewis had originally proposed.[32] However, the number of cholinergic cells that project up to the basal forebrain was found to be very small, revealing that the prominent fascicles of cholinesterase-stained fibers visualized by Shute and Lewis were actually comprised predominantly of monoaminergic fibers which also stain intensely for this enzyme. Thus, what appears as a relatively dense cholinergic fiber plexus, evident in ChAT-immunostained material around the basalis nucleus, may originate predominantly from collaterals of the local cholinergic basalis cells.

14.3 *In Vitro* Electrophysiological and Pharmacological Study of Basal Forebrain Neurons

Several investigations of the particular activity of individual basal forebrain neurons have been performed *in vivo*, in both anesthetized or unanesthetized animals, as will be reviewed below. However, from such studies it has not been possible to infer the particular activity of cholinergic vs. non-cholinergic and potentially GABAergic neurons. Such characterization depends upon labeling of the recorded neuron and subsequent dual staining of the labeled cell for ChAT, in the case of the cholinergic cells.

14.3.1 Intrinsic Properties of Identified Cholinergic and Non-Cholinergic Neurons

We undertook with Khateb et al.[2,39] (see also Chapter 18, this volume) the identification of cholinergic neurons *in vitro* in guinea pig basal forebrain slices. ChAT-immunoreactive neurons are distributed through the basal forebrain of the guinea pig (Figure 14.2A) in a manner similar to that in the rat (Figure 14.1A). GAD-immunoreactive neurons are also visible in the same region in the guinea pig brain (Jones, unpublished observations). In the absence of colchicine pretreatment, often necessary to augment levels of GAD in nerve cell bodies, large, prominent GAD-immunoreactive cells are evident and co-distributed with the cholinergic cells in the region, yet clustered more ventrally (Figure 14.2A).

With intracellular recordings combined with biocytin-filling of the neurons in the slice, we established that the ChAT-positive (ChAT+) neurons are distinctive in their properties within the basal forebrain.[39] From a hyperpolarized membrane potential, cholinergic neurons discharge in low-threshold bursts (Figure 14.3A1), which are subtended by low-threshold calcium spikes. The frequency of spikes within each burst is ~150 to 200 Hz, and the frequency of the bursts is ~1.5 to 5.0 Hz in the slice (at 32°C). When more depolarized, the cholinergic cells fire in a repetitive single spike mode (Figures 14.3A2,A3), the frequency of which depends upon the degree of depolarization, although it almost never exceeds 15 Hz. Thus, the cholinergic neurons display two modes of firing: one phasic in a burst mode and one tonic in a single spike mode, both slow.

Cholinergic and GABAergic Neurons of the Basal Forebrain: Role in Cortical Activation 219

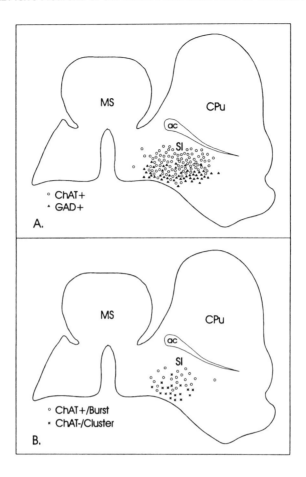

FIGURE 14.2
(A) Distribution of ChAT-positive (ChAT+) and GAD-positive (GAD+) neurons in the basal forebrain of the guinea pig, as revealed in a brain slice used for *in vitro* electrophysiological studies. (B) Plotted on the same figure is the approximate distribution of neurons which were recorded and filled with biocytin by intracellular technique and subsequently found to be ChAT+ or ChAT-negative (ChAT−) with dual staining procedures. ChAT+ cells displayed low-threshold bursts, whereas ChAT− cells displayed spike clusters (see Figure 14.3). Note that the distributions of ChAT+/burst cells and ChAT−/cluster cells are similar to the respective distributions of the ChAT+ and GAD+ neurons in (A), evoking the possibility that the non-cholinergic cells could in part be GABAergic. (Based upon work published by Khateb et al.[2,39])

In the same preparation, non-cholinergic, or ChAT-negative (ChAT−) cells were found to have distinctly different intrinsic properties from ChAT+ cells.[2] Non-cholinergic cells did not display high frequency bursts, but instead clusters of spikes (Figure 14.3B2). The spikes and clusters of spikes appeared to be subtended by subthreshold membrane potential oscillations (Figures 14.3B1,B2), which occurred at approximately the same frequency as the spikes, ~20 to 70Hz and often around 40 Hz. The spike clusters occurred in a rhythmic manner at a frequency of 4 to 10 Hz. Across different membrane potentials, these non-cholinergic cells could discharge in single spikes interposed by subthreshold membrane potential oscillations (Figures 14.3B1), in clusters of spikes also interposed by subthreshold membrane potential oscillations (Figure 14.3B2), or in continuous repetitive spikes (Figure 14.3B3). Thus, like the cholinergic cells, they displayed both phasic and tonic modes of firing, which depended, however, upon an entirely different mechanism. The phasic and rhythmic, clustered discharge occurred spontaneously in many of these neurons in the slice. Given the interspike interval and intercluster interval of their clustered discharge, these neurons could modulate both high and low frequency rhythmic activity.

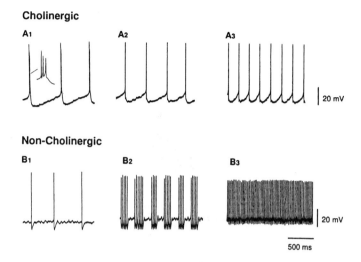

FIGURE 14.3
Intracellular recordings of cholinergic (**A**) and non-cholinergic cells (**B**). Cholinergic cells display low-threshold bursts (enlarged in inset) at a relatively hyperpolarized level of the membrane (A1) and single spikes at increasing frequencies at more depolarized levels (A2 and A3, with DC current injection). Non-cholinergic cells display single spikes (B1) or clusters of spikes (B2) interposed by subthreshold membrane potential oscillations at a frequency similar to the clustered spikes. With increasing depolarization of the membrane, the non-cholinergic cells can also fire repetitively (B3). In contrast to the cholinergic (A3), the non-cholinergic cells discharge at a relatively high frequency when strongly depolarized (B3). Both cells shown were spontaneously active. (Based upon work published by Khateb et al.[2,39])

Both ChAT+ and ChAT– neurons recorded and identified in the slice were distributed through the substantia innominata and region delineated as the magnocellular preoptic nucleus in the rat (Figure 14.2B). Both cell types were relatively large in size and could be cortically projecting cells. According to their distribution and size, the non-cholinergic neurons could also, in part at least, be GABAergic neurons. We are currently seeking unequivocal identification of GAD-immunoreactive cells.

Given the properties of the cholinergic and non-cholinergic, possibly cortically projecting, cells, the basalis neurons could influence activity of the cerebral cortex in a rhythmic manner at both low and high frequencies or, as defined in EEG activity, delta-theta and beta-gamma frequencies.

14.3.2 Pharmacological Modulation of Cholinergic and Non-Cholinergic Neurons

In order to determine the modulation of the cholinergic and non-cholinergic neurons by brainstem afferents and accordingly assess the way in which these neurons may in turn modulate cortical activity in a state-dependent manner, we examined with Khateb et al.[21,22,36–38] the pharmacological effects of neurotransmitters upon these cells in the guinea pig slice. Cholinergic cells are depolarized by glutamate and all the excitatory amino acid receptor agonists.[37] Thus t-ACPD ([±]-1-aminocyclopentane-trans-1,3-dicarboxylic acid), AMPA ([RS]-a-amino-3-hydroxy-5-methyl-4-isoxazolepropionic acid), kainate, and NMDA (*N*-methyl-D-aspartate) all depolarize the cholinergic neurons and drive them into a tonic mode of firing up to their maximal tonic frequency of 15 Hz. Tonically firing, they could stimulate high-frequency discharge in cortical neurons and thus high-frequency EEG activity typical of cortical activation. In addition, NMDA can stimulate rhythmic, low-threshold bursting (up to a frequency of 6 Hz) in the cholinergic cells, when they are held at

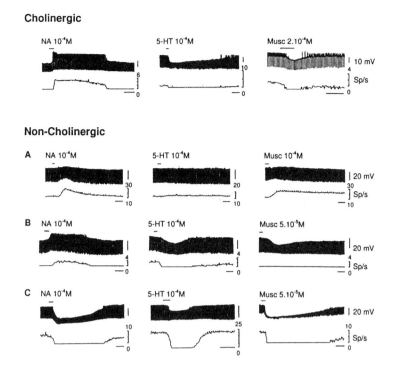

FIGURE 14.4
Pharmacological profiles of cholinergic cells (top panel) and three types (A, B, and C) of non-cholinergic cells (bottom panels). Whereas cholinergic cells respond uniformly to the major neurotransmitters, non-cholinergic cells respond differentially, thus distinguishing three different types of cells (type A, ~44% of non-cholinergic cells tested; type B, ~23%; and type C, ~15%). The upper trace in each panel shows the effect of the neurotransmitter or agonist (applied during the time indicated by horizontal bar) on the membrane potential (mV, vertical scale bars at right) and resistance of the cell (evident by changes in voltage response to short, hyperpolarizing current pulses applied during recordings, as seen in upper right trace at slow speed). The lower trace in each panel shows the effect on the rate of spontaneous firing (spikes per second, or Sp/s). (Based upon work published by Fort et al.[22])

a hyperpolarized level. These results would suggest that input from the brainstem reticular formation, which is presumed to be glutamatergic (above), could stimulate either tonic or rhythmic bursting discharge in the cholinergic basalis cells depending upon the type of receptor activated and the influence of other contingent inputs.

Cholinergic neurons are modulated in different ways by noradrenaline and serotonin, contained in brainstem afferents from the locus coeruleus and raphe, respectively. Noradrenaline (NA) depolarizes and excites the cholinergic neurons, driving them into a tonic mode of firing (Figure 14.4). These results suggest that noradrenergic locus coeruleus neurons that fire during the waking state may recruit the cholinergic neurons in tandem to stimulate cortical activation. In contrast, serotonin (5-hydroxytryptamine, or 5-HT) hyperpolarizes and inhibits the cholinergic neurons, decreasing their tonic discharge (Figure 14.4). The serotonergic raphe neurons could thus act in a different manner than the noradrenergic neurons in modulating the activity of the cholinergic basalis cells and that of cortical neurons in turn. Indeed, presumed noradrenergic locus coeruleus neurons have been found to behave differently from presumed serotonergic raphe neurons during waking behaviors, during which noradrenergic cells turn on during orientation,[4,57] whereas serotonergic cells turn off during orientation[28] to sensory stimulation. These reciprocal changes in excitatory, noradrenergic, and inhibitory serotonergic inputs would perhaps allow maximal excitation of the cholinergic basalis cells and thus maximal cholinergic excitatory input to the cerebral cortex during such behaviors. ACh, which is probably contained predominantly in local collaterals from the basalis cholinergic neurons (above), has both excitatory and inhibitory effects on the cholinergic neurons through nicotinic and muscarinic receptors, respectively.[38] Muscarine hyperpolarizes the membrane and inhibits spontaneous discharge (Figure 14.4), through what may thus be a muscarinic autoreceptor on the cholinergic basalis neurons. In the presence of muscarine or carbachol, NMDA evokes a rhythmic discharge (as above) comprised of prolonged, low-threshold bursts.[38] The significance of such rhythmic bursting *in vivo* could be to modulate rhythmic slow activity, which could potentially be within a theta range.

Like cholinergic cells, the non-cholinergic neurons were depolarized by glutamate and excitatory amino acid agonists (Khateb, Fort, Jones, and Mühlethaler, unpublished observations). It has not yet been determined whether any of these agonists, such as NMDA, might stimulate or favor rhythmic cluster discharge in these neurons. Depolarization of the non-cholinergic cells is often associated with cluster discharge (as in Figure 14.3B2), although with very strong depolarization the cells can also be made to fire tonically (Figure 14.3B3). Thus input from the reticular formation, which is presumed to be predominantly glutamatergic (above), could stimulate rhythmic cluster discharge or tonic discharge *in vivo*, depending upon the membrane potential of the cells.

The non-cholinergic neurons were differentiated into three major types according to their responses to the monoamines and to muscarine.[22] Type A cells (~44% of the non-cholinergic cells) were depolarized by noradrenaline (NA), unaffected or slightly hyperpolarized by serotonin (5-HT), and depolarized by muscarine. Type B cells (~23%) were depolarized by NA, hyperpolarized by serotonin, and hyperpolarized by muscarine. Finally, type C cells (~15%) were hyperpolarized by NA, serotonin, and muscarine. Accordingly, the non-cholinergic neurons, which share certain discharge properties, appear nonetheless to be comprised of distinct types which respond differentially to the major modulatory neurotransmitters and thus potentially play different roles in cortical modulation across the sleep/waking cycle. As the locus coeruleus is active during waking,[3] it is presumed that those cells, types A and B, which are depolarized by NA are likely to be active during the waking state, whereas those which are hyperpolarized by NA, type C, could be active during sleep.

According to the *in vitro* pharmacology, it would thus appear that during the waking state, cholinergic and the majority of non-cholinergic (including potentially GABAergic) cells may be active in parallel under the influence of ascending brainstem glutamatergic and noradrenergic inputs, which are maximally active during waking.[3,68] Among the non-cholinergic cells excited by noradrenergic input, some would be excited by cholinergic input from neighboring neurons, whereas others would be inhibited by this input, thus potentially eliciting alternate discharge patterns from the different cell types. Given the tendency for the non-cholinergic cells to fire in rhythmic clusters of spikes and the cholinergic in rhythmic bursts of spikes, these two cell populations could fire in concert in a manner promoting rhythmic activity in the cerebral cortex. Such rhythmic activity would tend to be within a theta frequency and modulated by a higher frequency, beta-gamma rhythm carried by the non-cholinergic neurons during the cluster discharge.

14.4 *In Vivo* Electrophysiological and Pharmacological Study of the Role of Basal Forebrain Neurons in Cortical Activation

The hypotheses formulated on the basis of *in vitro* physiology and pharmacology concerning the role of the basal forebrain neurons in cortical activation await testing and confirmation by *in vivo* experimentation.

14.4.1 Discharge Properties of Possible Cholinergic and Non-Cholinergic Neurons in Relation to Cortical Activity

Multiple recording studies have been performed *in vivo* to examine the discharge properties of neurons in the basal forebrain in relation to EEG activity in anesthetized animals or EEG activity and sleep/wake state in unanesthetized animals. In none of these studies to date, however, has recording allowed labeling of the recorded cells and thus identification of recorded cells as cholinergic or non-cholinergic. Such efforts are currently ongoing in anesthetized animals by Manns and colleagues (Manns, Alonso, and Jones, unpublished observations).[47a]

In anesthetized animals, stimulation of the reticular formation produces an increase in discharge in the majority of basalo-cortical projection neurons in parallel with evoked low-voltage fast activity in the EEG.[17] In this and other *in vivo* recording studies,[5,60] the basalo-cortical relay neurons were found to be heterogeneous in their discharge properties and thus thought to represent more than one cell type. More recently, Nunez[55] identified basalo-cortical neurons with different properties that he compared to those characterized by us *in vitro*. One cell type, which increased discharge rate with stimulation of the reticular formation, resembles the cholinergic cells in that it displays high-frequency bursts. Such cells show both a phasic and tonic discharge mode (Manns et al., unpublished observations).[47a] Other cells that also increase their discharge with stimulation of the reticular formation show a tendency to discharge in clusters of spikes, in addition to a tonic mode (Manns et al., unpublished observations).[47a] The identity of these different cell types awaits immunostaining for ChAT and GAD, and the relationship of their discharge pattern to EEG activity necessitates further quantitative analysis.

Also in anesthetized animals, another cell type ceases discharge upon stimulation of the reticular formation and increases its discharge rate in association with high-voltage, irregular, slow activity in the EEG[16,17,55] (also, Manns et al., unpublished observations).[47a] These neurons, which represent a minority in the basalis region in the rat, could not be antidromically activated from the cerebral cortex and thus may not be cortically projecting neurons. They may, in part at least, represent local projection neurons which could be responsible for inhibiting the cortically projecting cells.

In unanesthetized, freely moving, and naturally waking/sleeping cats, Szymusiak and McGinty[72] found cells in the basal forebrain that discharge at their highest rate in association with cortical activation and the waking state. They also identified cells that increase their rate of discharge in association with high-voltage, irregular, slow-wave activity and slow-wave sleep. Such "sleep-active" cells are less numerous than the "wake-active" cells in the basal forebrain. A certain number of "sleep-active" cells, like the large numbers of the "wake-active" cells, could be antidromically activated from the cerebral cortex and could thus also be basalo-cortical projection neurons.[73]

The *in vivo* unit recording data would thus suggest that as the neuroanatomical and *in vitro* electrophysiological results indicate, basalo-cortical projection neurons are heterogeneous and may be comprised of multiple cell types, including the cholinergic that are "wake-active" and other cells that are "sleep-active". The basal forebrain could accordingly influence the cerebral cortex in different ways across sleep/wake states.

14.4.2 Modulation of Cortical Activity by Stimulation or Inactivation of the Basal Forebrain

Determination of the role of the basal forebrain and its constituent neurons in cortical activity and sleep/wake state has also been approached by examining the effects of modulating the basal forebrain: first by electrical stimulation or lesions, which would affect all cell types, and, second, by chemical stimulation which would differentially modulate the different cell types according to their different receptors for the major neurotransmitters contained in brainstem afferents.

14.4.2.1 Electrical Stimulation or Lesion

Electrical stimulation of the nucleus basalis region increases ACh release in the neocortex[11,58] and elicits cortical activation, which is associated with a muscarinic-induced depolarization of cortical neurons and shift from slow (1 to 5 Hz) to fast (20 to 40 Hz) activity in these cells.[51] Stimulation has, however, also been associated in some cases with inhibition of cortical neuronal activity and in such cases with a lack of enhanced ACh release.[30] The latter results suggest that the basal forebrain may, via different cell types which may be differentially concentrated in different regions, either excite or inhibit cortical neuronal activity. Moreover, it was also shown many years ago in cats[69] that electrical stimulation in the basal forebrain could elicit cortical slow waves and slow-wave sleep.

Lesions of the basal forebrain are associated with decreases in ACh release from the cortex and parallel decreases in low-voltage fast cortical activity.[44,70] The loss of fast cortical activity and replacement by irregular slow activity in the cortex after basalis lesions resembles similar EEG changes that occur after administration of atropine, a muscarinic receptor antagonist.[70] On the other hand, large electrolytic and neurotoxic lesions of the basal forebrain have also been reported to be associated with a loss of slow wave sleep.[48,71] The latter lesions were very large, and included the lateral preoptic region, ventral substantia innominata, and region of the magnocellular preoptic nucleus.

The results from electrical stimulation and lesions of the basal forebrain thus indicate that the primary influence of the basal forebrain and particularly the cholinergic basalo-cortical neurons is to facilitate cortical fast activity and thus cortical activation. In addition, however, results also indicate that via different, presumably non-cholinergic, cells a secondary influence of the basal forebrain would be inhibitory to cortical neuronal activity and cortical activation and thus potentially associated with facilitation of cortical irregular slow-wave activity and slow-wave sleep.

14.4.2.2 Chemical Stimulation or Inactivation

As studied in anesthetized animals, chemical stimulation of the basalis region with glutamate can evoke cortical activation in a manner similar to that of electrical stimulation.[51] It has also been established in anesthetized animals that chemical inactivation of the basalis region decreases evoked cortical release of ACh and low-voltage fast cortical activity.[11,59] Interestingly, in the latter studies, it was found that the increase in ACh release and fast activity in the cortex, which was evoked by brainstem stimulation, was blocked by glutamate antagonists but not by cholinergic antagonists, indicating, as the neuroanatomical studies had suggested (above), that the major excitatory input from the brainstem reticular formation to the basal forebrain is via glutamatergic neurons.[31]

Cape and Jones[9] have examined the effects of chemical stimulation of the basalis region in freely moving, naturally waking/sleeping rats using a remote microinjection procedure (Figure 14.5). In these studies, EEG activity was measured by spectral analysis, and high-frequency gamma EEG activity (30 to 60 Hz) was quantified, as it was shown by Maloney et al.[47] to reflect cortical and behavioral arousal in the rat. In these experiments, stimulation with excitatory amino acid agonists (AMPA or NMDA) leads to cortical activation marked by high-frequency gamma activity and behavioral waking during the day when the animal is normally asleep the majority of the time. Conversely, chemical inactivation with procaine microinjections leads to a decrease in gamma activity and presence of delta activity, associated with a state of slow-wave sleep (Cape and Jones, in preparation).

Cape and Jones[10] have also studied the effect of monomines microinjected into the region of the cholinergic basalis neurons (Figure 14.5). Because noradrenaline and serotonin act selectively and differentially on different cholinergic and non-cholinergic cell types, it is probable that they would produce selective effects upon EEG and sleep/wake state. NA, which depolarizes the cholinergic

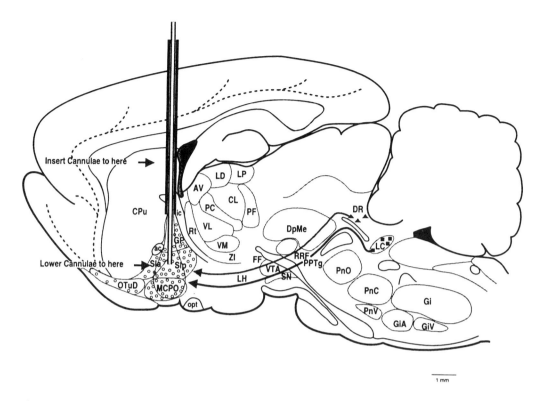

FIGURE 14.5
Schematic diagram of sagittal view of rat brain depicting microinjections of neurotransmitters into the basal forebrain. Shown are the nuclei within the brainstem reticular formation and monoaminergic and cholinergic cell groups which comprise the brainstem activating system and give rise to ascending projections up to the basal forebrain, as depicted here for fibers ascending from the noradrenergic locus coeruleus (LC) and serotonergic dorsal raphe (DR) neurons. Fiber systems terminate or leave collaterals in the basal forebrain in the region of the cholinergic basalis neurons (open circles), which in turn give rise to widespread projections to the cerebral cortex (dashed line). Bilateral microinjections of chemicals were performed in freely moving, naturally waking/sleeping rats, as here represented for noradrenaline and serotonin. Microinjections were performed by insertion of the chemical-loaded inner injection cannulae into chronically indwelling guide cannulae and subsequent lowering of the cannulae by remote control in the naturally sleeping animal. Abbreviations: ac, anterior commissure; AV, anteroventral thalamic nucleus; CL, centrolateral thalamic nucleus; CPu, caudate putamen; DpMe, deep mesencephalic reticular field; FF, fields of Forel; Gi, gigantocellular reticular field; GiA, gigantocellular reticular field, alpha part; GiV, gigantocellular reticular field, ventral part; GP, globus pallidus; ic, internal capsule; LC, locus coeruleus; LD, laterodorsal thalamic nucleus; LH, lateral hypothalamic area; LP, lateral posterior thalamic nucleus; MCPO, magnocellular preoptic nucleus; opt, optic tract; OTuD, olfactory tubercle, deep layer; PC, paracentral thalamic nucleus; PF, parafascicular thalamic nucleus; PnC, pontine reticular field, caudal part; PnO, pontine reticular field, oral part; PnV, pontine reticular field, ventral part; PPTg, pedunculopontine tegmental nucleus; R, red nucleus; RRF, retrorubral field; Rt, reticular thalamic nucleus; SIa, substantia innominata, anterior part; SIp, substantia innominata, posterior part; SN, substantia nigra; VL, ventrolateral thalamic nucleus; VM, ventromedial thalamic nucleus; VTA, ventral tegmental area; ZI, zona incerta. (From Cape, E.G. and Jones, B.E., *J. Neurosci.*, 18, 2653–2666, 1998. With permission.)

neurons and the majority of the non-cholinergic cells, produced an increase in cortical gamma activity and a decrease in delta activity (Figures 14.6 and 14.7). These EEG changes were associated with a behaviorally awake and often attentive state. In contrast, serotonin, which hyperpolarizes the cholinergic neurons and has no or a hyperpolarizing effect on non-cholinergic cells, was associated with a decrease in gamma activity and with the presence of delta activity (Figures 14.6 and 14.8), during the day when the animals are normally asleep the majority of the time. These EEG characteristics were associated with behaviorally quiet sleep or quiet waking. The results indicate that selective modulation of cholinergic and non-cholinergic neurons in the basal forebrain by

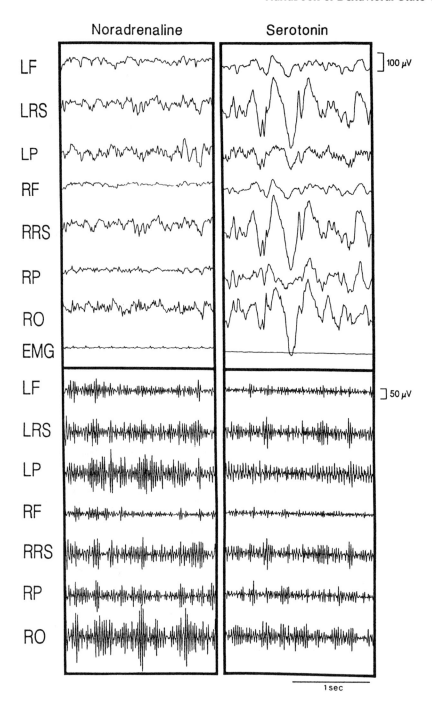

FIGURE 14.6

EEG samples following noradrenaline and serotonin microinjections. Unfiltered (above) and high-frequency, gamma (30.5 to 58.0 Hz), filtered (below) EEG samples (2 sec each) illustrating the consistent EEG patterns that occurred during the respective post-injection recording periods. Noradrenaline produced a low-voltage, fast EEG pattern (above), in association with relatively high gamma activity (below), similar to normal waking, whereas serotonin produced a high-voltage, irregular, slow EEG pattern (above), in association with relatively low gamma activity (below), similar to normal slow-wave sleep. The samples were taken ~2 to 3 min after the injection was stopped. The EEG was recorded by reference to an electrode in the rostral frontal bone from the left and right frontal (LF and RF), retrosplenial (LRS and RRS), parietal (LP and RP), and occipital (LO and RO) cortical regions. Voltage scales are the same for all cortical leads. (From Cape, E.G. and Jones, B.E., *J. Neurosci.*, 18, 2653–2666, 1998. With permission.)

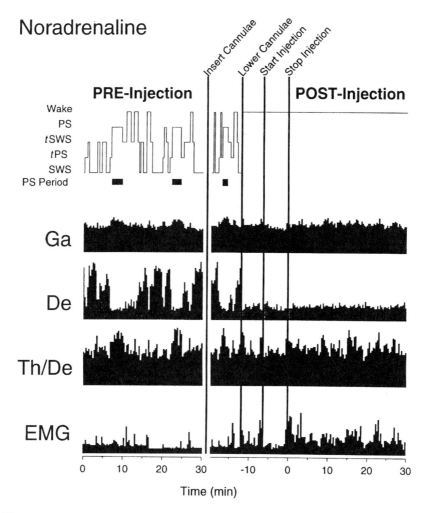

FIGURE 14.7
Hypnogram and EEG and EMG activities (per 20-sec epoch) during noradrenaline pre- and post-injection recording periods. Note the immediate occurrence of wake once the filled cannulae are inserted and the maintenance of a wake state in association with moderately high gamma EEG activity and low delta EEG activity during the entire post-injection period. Both Th/De and EMG remain relatively high. EEG frequency band activity is from the right retrosplenial lead and, together with EMG, is displayed as amplitude units or ratio scaled to maximum activity. Abbreviations: (above) PS, paradoxical sleep; SWS, slow-wave sleep; tPS, transition into paradoxical sleep; tSWS, transition into slow-wave sleep. (below) Ga, gamma; De, delta; Th, theta; EMG, electromyogram. (From Cape, E.G. and Jones, B.E., *J. Neurosci.*, 18, 2653–2666, 1998. With permission.)

natural neurotransmitters and thus the brainstem afferents from which they are normally released can be associated with selective changes in cortical activity and associated sleep/wake states. They also show that the basal forebrain cholinergic, together with the non-cholinergic, neurons have the capacity to stimulate high-frequency gamma EEG activity which reflects cortical and behavioral arousal in the rat.

In order to determine the potential influence of rhythmical burst discharge by the cholinergic neurons, Cape and Jones (manuscript in preparation) have also examined the effect of NMDA alone and in the presence of carbachol, drugs which in the slice were shown to produce rhythmic, low-threshold bursting in the cholinergic neurons. Microinjections of NMDA alone or together with carbachol significantly increased rhythmic theta activity in the limbic neocortex in association with an aroused wake state.

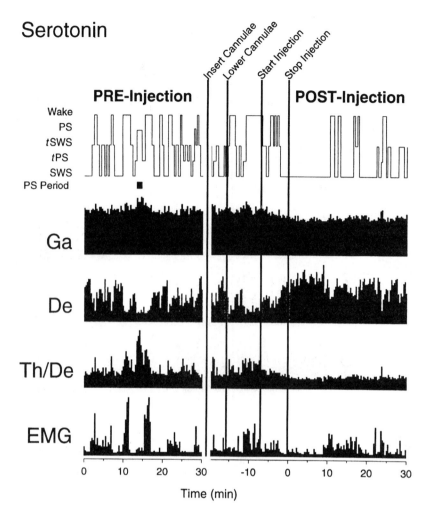

FIGURE 14.8

Hypnogram and EEG and EMG activities (per 20-sec epoch) during serotonin pre- and post-injection recording periods. Note the continuity of SWS during and following the injection in association with a decrease in gamma activity and the persistence of delta activity. No PS occurs in the post-injection period, and Th/De ratio remains low. Moderate EMG activity is present. EEG frequency band activity is from the right retrosplenial lead and, together with EMG, is displayed as amplitude units or ratio scaled to maximum activity. Abbreviations: (above) PS, paradoxical sleep; SWS, slow-wave sleep; tPS, transition into paradoxical sleep; tSWS, transition into slow-wave sleep. (below) Ga, gamma; De, delta; Th, theta; EMG, electromyogram. (From Cape, E.G. and Jones, B.E., *J. Neurosci.*, 18, 2653–2666, 1998. With permission.)

14.5 Summary and Conclusions

Recent anatomical, physiological, and pharmacological results have substantiated an important role of cholinergic nucleus basalis neurons in eliciting and maintaining cortical activation. The activation is mediated by widespread projections to the neocortex which relay ascending input from the brainstem activating system. Cholinergic neurons are excited by glutamate, which is presumably released from terminals of the reticular neurons and stimulated to discharge tonically at rates up to 15 Hz. Resulting increased release of ACh in the cortex would produce an increase in cortical neuronal activity and a shift from irregular slow-wave activity to fast activity in the EEG and particularly gamma activity (30 to 60 Hz) which reflects cortical activation and behavioral arousal. Release of noradrenaline by locus coeruleus afferents during the waking state and during particular waking behaviors would also depolarize and excite cholinergic neurons and thereby stimulate, via the basalis relay, high-frequency gamma activity in the cerebral cortex. Release of serotonin by raphe afferents would, in contrast, hyperpolarize and inhibit cholinergic neurons and thereby attenuate gamma activity in the cerebral cortex. Given their intrinsic properties, the cholinergic cells also have the capacity to discharge rhythmically in low-threshold bursts and thus to modulate cortical neurons in a slow, rhythmic manner. Rhythmic bursting in the cholinergic neurons — which can be driven by glutamate released from reticular neurons acting upon NMDA receptors in combination with cholinergic modulation from basalis collaterals acting upon muscarinic autoreceptors — could stimulate rhythmic slow activity in a theta range (4 to 8 Hz) in limbic neocortex. Such cortical activation modulated at a theta frequency from the basalo-cortical cholinergic neurons would occur during aroused waking and REM sleep.

GABAergic neurons also comprise a portion of the nucleus basalis and project in parallel with and in equal proportion to the cholinergic cells to widespread areas of the neocortex. Non-cholinergic basalis neurons recorded in the slice, which may in part correspond to GABAergic neurons and to basalo-cortical projection neurons, often discharge rhythmically in spike clusters. The spike frequency is fast and could pace cortical activity in a gamma range, and the cluster frequency is slow and could pace cortical activity in a theta range. Submitted to excitatory influences from the brainstem afferents, including glutamatergic reticular inputs, non-cholinergic, possibly GABAergic, basalis neurons could in turn drive gamma and theta activity in neocortex. Of the non-cholinergic cells, the majority are depolarized by noradrenaline and thus also stimulated to discharge presumably during waking when noradrenergic neurons are active. Many of the non-cholinergic cells are hyperpolarized by serotonin released by raphe inputs which would thus tend to attenuate this non-cholinergic input and thereby decrease both gamma and theta cortical activity from this source. Of the two non-cholinergic cell types depolarized by noradrenaline, one type is excited by muscarine, whereas another is inhibited by muscarine, suggesting perhaps different patterns of alternate discharge with the cholinergic basalis neurons of these two types. Whether one of these two is GABAergic and the other glutamatergic is a possibility that remains to be determined. These potentially wake-active neurons could in any case both be active in concert with the cholinergic cells in a manner that could facilitate both high-frequency gamma and theta activity.

Another minor type of non-cholinergic cell in the basal forebrain, which could also potentially be GABAergic, comprises those neurons which are hyperpolarized and inhibited by noradrenaline as well as serotonin and muscarine. Such cells could be sleep-active neurons that by projections to the cerebral cortex, locally to other basal forebrain neurons, and/or caudally to posterior hypothalamus and brainstem, depending in each case upon the target cells, could inhibit discharge in other activating neurons. Released from the depolarizing and driving force of cholinergic and other non-cholinergic neurons of the activating system at different levels, cortical neurons would be disfacilitated and assume the irregular discharge pattern that underlies irregular slow wave activity of slow wave sleep.

By a heterogeneous population of cells containing different neurotransmitters, invested with different receptors and endowed with different intrinsic membrane properties, the basal forebrain may thus modulate cortical activity across the sleep/waking cycle and across a spectrum of frequencies. Its primary influence on cortical activity is nonetheless to provide an activating influence through cholinergic and possibly GABAergic or other unidentified basalo-cortical projection neurons as a relay from the brainstem-activating system and potentially as a pacemaker, modulating rhythmic cortical activity in a gamma and theta range during waking and REM sleep and thus providing a substrate for widely distributed coherent cortical activity during these states when waking conscious processes or dreams occur.

References

1. Adams, C.E. et al., Basal forebrain neurons have axon collaterals that project to widely divergent cortical areas in the cat, *Brain Res.*, 397, 365–371, 1986.
2. Alonso, A., Khateb, A., Fort, P., Jones, B.E., and Mühlethaler, M., Differential oscillatory properties of cholinergic and non-cholinergic nucleus basalis neurons in guinea pig brain slice, *Eur. J. Neurosci.*, 8, 169–182, 1996.
3. Aston-Jones, G. and Bloom, F.E., Activity of norepinephrine-containing locus coeruleus neurons in behaving rats anticipates fluctuations in the sleep-waking cycle, *J. Neurosci.*, 1, 876–886, 1981.
4. Aston-Jones, G. and Bloom, F.E., Norepinephrine-containing locus coeruleus neurons in behaving rats exhibit pronounced responses to non-noxious environmental stimuli, *J. Neurosci.*, 1, 887–900, 1981.
5. Aston-Jones, G., Shaver, R., and Dinan, T., Cortically projecting nucleus basalis neurons in rat are physiologically heterogeneous, *Neurosci. Lett.*, 46, 19–24, 1984.
6. Boylan, M.K., Fisher, R.S., Hull, C.D., Buchwald, N.A., and Levine, M.S., Axonal branching of basal forebrain projections to the neocortex: a double–labeling study in the cat, *Brain Res.*, 375, 176–181, 1986.
7. Brashear, H.R., Zaborszky, L., and Heimer, L., Distribution of GABAergic and cholinergic neurons in the rat diagonal band, *Neuroscience*, 17, 439–451, 1986.
8. Butcher, L.L. and Woolf, N.J., Histochemical distribution of acetylcholinesterase in the central nervous system: clues to the localization of cholinergic neurons, in *Handbook of Chemical Neuroanatomy. Vol. 3. Classical Transmitters and Transmitter Receptors in the CNS, Part II*, Bjorklund, A., Hokfelt, T., and Kuhar, M.J., Eds., Elsevier, Amsterdam, 1984, pp. 1–50.
9. Cape, E.G. and Jones, B.E., Modulation of sleep-wake state and cortical activity following injection of agonists into the region of cholinergic basal forebrain neurons, *Soc. Neurosci. Abst.*, 20, 156, 1994.
10. Cape, E.G. and Jones, B.E., Differential modulation of high frequency gamma EEG activity and sleep-wake state by noradrenaline and serotonin microinjections into the region of cholinergic basalis neurons, *J. Neurosci.*, 18, 2653–2666, 1998.
11. Casamenti, F., Deffenu, G., Abbamondi, A.L., and Pepeu, G., Changes in cortical acetylcholine output induced by modulation of the nucleus basalis, *Brain Res. Bull.*, 16, 689–695, 1986.
12. Celesia, G.G. and Jasper, H.H., Acetylcholine released from cerebral cortex in relation to state of activation, *Neurology*, 16, 1053–1064, 1966.
13. Dement, W. and Kleitman, N., Cyclic variations in EEG during sleep and their relation to eye movements, body motility and dreaming, *Electroenceph. Clin. Neurophysiol.*, 9, 673–690, 1957.
14. Dement, W. and Kleitman, N., The relation of eye movements during sleep to dream activity: an objective method for the study of dreaming, *J. Exp. Psychol.*, 53, 339–346, 1957.
15. Dempsey, E.W., Morison, R.S., and Morison, B.R., Some afferent diencephalic pathways related to cortical potentials in the cat, *Am. J. Physiol.*, 131, 718–731, 1941.
16. Detari, L., Semba, K., and Rasmusson, D.D., Responses of cortical EEG-related basal forebrain neurons to brainstem and sensory stimulation in urethane-anaesthetized rats, *Eur. J. Neurosci.*, 9, 1153–1161, 1997.
17. Detari, L. and Vanderwolf, C.H., Activity of identified cortically projecting and other basal forebrain neurones during large slow waves and cortical activation, *Brain Res.*, 437, 1–8, 1987.
18. Domino, E.F., Yamamoto, K., and Dren, A.T., Role of cholinergic mechanisms in states of wakefulness and sleep, *Prog. Brain Res.*, 28, 113–133, 1968.
19. Fisher, R.S., Buchwald, N.A., Hull, C.D., and Levine, M.S., GABAergic basal forebrain neurons project to the neocortex: the localization of glutamic acid decarboxylase and choline acetyltransferase in feline corticopetal neurons, *J. Comp. Neurol.*, 272, 489–502, 1988.

20. Ford, B., Holmes, C., Mainville, L., and Jones, B.E., GABAergic neurons in the rat pontomesencephalic tegmentum. Codistribution with cholinergic and other tegmental neurons projecting to the posterior lateral hypothalamus, *J. Comp. Neurol.*, 363, 177–196, 1995.

21. Fort, P., Khateb, A., Pegna, A., Mühlethaler, M., and Jones, B.E., Noradrenergic modulation of cholinergic nucleus basalis neurons demonstrated by *in vitro* pharmacological and immunohistochemical evidence in the guinea pig brain, *Eur. J. Neurosci.*, 7, 1502–1511, 1995.

22. Fort, P., Khateb, A., Serafin, M., Mühlethaler, M., and Jones, B.E., Pharmacological characterization and differentiation of non-cholinergic nucleus basalis neurons *in vitro*, *NeuroReport*, 9, 1–5, 1998.

23. Freund, T.F. and Antal, M., GABA-containing neurons in the septum control inhibitory interneurons in the hippocampus, *Nature*, 336, 170–173, 1988.

24. Freund, T.F. and Meskenaite, V., Gamma-aminobutyric acid-containing basal forebrain neurons innervate inhibitory interneurons in the neocortex, *Proc. Natl. Acad. Sci. USA*, 89, 738–742, 1992.

25. Gritti, I., Mainville, L., and Jones, B.E., Codistribution of GABA- with acetylcholine-synthesizing neurons in the basal forebrain of the rat, *J. Comp. Neurol.*, 329, 438–457, 1993.

26. Gritti, I., Mainville, L., and Jones, B.E., Projections of GABAergic and cholinergic basal forebrain and GABAergic preoptic-anterior hypothalamic neurons to the posterior lateral hypothalamus of the rat, *J. Comp. Neurol.*, 339, 251–268, 1994.

27. Gritti, I., Mainville, L., Mancia, M., and Jones, B.E., GABAergic and other non-cholinergic basal forebrain neurons project together with cholinergic neurons to meso- and iso-cortex in the rat, *J. Comp. Neurol.*, 383, 163–177, 1997.

28. Jacobs, B.L. and Fornal, C.A., Activity of brain serotonergic neurons in the behaving animal, *Pharmacol. Rev.*, 43, 563–578, 1991.

29. Jasper, H.H. and Tessier, J., Acetylcholine liberation from cerebral cortex during paradoxical (REM) sleep, *Science*, 172, 601–602, 1971.

30. Jimenez-Capdeville, M.E., Dykes, R.W., and Myasnikov, A.A., Differential control of cortical activity by the basal forebrain in rats: a role for both cholinergic and inhibitory influences, *J. Comp. Neurol.*, 381, 53–67, 1997.

31. Jones, B.E., Reticular formation. Cytoarchitecture, transmitters and projections, in *The Rat Nervous System*, Paxinos, G., Ed., Academic Press Australia, New South Wales, 1995, pp. 155–171.

32. Jones, B.E. and Cuello, A.C., Afferents to the basal forebrain cholinergic cell area from pontomesencephalic — catecholamine, serotonin, and acetylcholine — neurons, *Neuroscience*, 31, 37–61, 1989.

33. Jones, B.E. and Yang, T.-Z., The efferent projections from the reticular formation and the locus coeruleus studied by anterograde and retrograde axonal transport in the rat, *J. Comp. Neurol.*, 242, 56–92, 1985.

34. Jouvet, M., Recherches sur les structures nerveuses et les mecanismes responsables des differentes phases du sommeil physiologique, *Arch. Ital. Biol.*, 100, 125–206, 1962.

35. Jouvet, M., Michel, F., and Courjon, J., Sur un stade d'activite electrique cerebrale rapide au cours du sommeil physiologique, *C. R. Soc. Biol.*, 153, 1024–1028, 1959.

36. Khateb, A., Fort, P., Alonso, A., Jones, B.E., and Mühlethaler, M., Pharmacological and immunohistochemical evidence for a serotonergic input to cholinergic nucleus basalis neurons, *Eur. J. Neurosci.*, 5, 541–547, 1993.

37. Khateb, A., Fort, P., Serafin, M., Jones, B.E., and Mühlethaler, M., Rhythmical bursts induced by NMDA in cholinergic nucleus basalis neurones *in vitro*, *J. Physiol. (Lond.)*, 487.3, 623–638, 1995.

38. Khateb, A., Fort, P., Williams, S., Serafin, M., Jones, B.E., and Mühlethaler, M., Modulation of cholinergic nucleus basalis neurons by acetylcholine and N–methyl–D–aspartate, *Neuroscience*, 81, 47–55, 1997.

39. Khateb, A., Mühlethaler, M., Alonso, A., Serafin, M., Mainville, L., and Jones, B.E., Cholinergic nucleus basalis neurons display the capacity for rhythmic bursting activity mediated by low threshold calcium spikes, *Neuroscience*, 51, 489–494, 1992.

40. Kievit, J. and Kuypers, H.G.J.M., Basal forebrain and hypothalamic connections to frontal and parietal cortex in the rhesus monkey, *Science*, 187, 660–662, 1975.

41. Kohler, C., Chan-Palay, V., and Wu, J.-Y., Septal neurons containing glutamic acid decarboxylase immunoreactivity project to the hippocampal region in the rat brain, *Anat. Embryol.*, 169, 41–44, 1984.

42. Krnjevic, K., Chemical transmission and cortical arousal, *Anesthesiology*, 28, 100–104, 1967.

43. Krnjevic, K. and Silver, A., A histochemical study of cholinergic fibres in the cerebral cortex, *J. Anat.*, 99, 711–759, 1965.

44. Lo Conte, G., Casamenti, F., Bigi, V., Milaneschi, E., and Pepeu, G., Effect of magnocellular forebrain nuclei lesions on acetylcholine output from the cerebral cortex, electrocorticogram and behaviour, *Arch. Ital. Biol.*, 120, 176–188, 1982.

45. Longo, V.G., Behavioral and electroencephalographic effects of atropine and related compounds, *Pharamacol. Rev.*, 18, 965–996, 1966.

46. Luiten, P.G.M., Gaykema, R.P.A., Traber, J., and Spencer, D.G., Cortical projection patterns of magnocellular basal nucleus subdivisions as revealed by anterogradely transported *Phaseolus vulgaris* leucoagglutinin, *Brain Res.*, 413, 229–250, 1987.

47. Maloney, K.J., Cape, E.G., Gotman, J., and Jones, B.E., High frequency gamma electroencephalogram activity in association with sleep-wake states and spontaneous behaviors in the rat, *Neuroscience*, 76, 541–555, 1997.

47a. Manns, I.D., Alonso, A., and Jones, B.E., Characterization of juxtacellularly recorded and labelled basal forebrain units in relation to cortical EEG activation, *Soc. Neurosci. Abstr.*, in press.

48. McGinty, D.J. and Sterman, M.B., Sleep suppression after basal forebrain lesions in the cat, *Science*, 160, 1253–1255, 1968.

49. Mesulam, M.-M., Mufson, E.J., Wainer, B.H., and Levey, A.I., Central cholinergic pathways in the rat: an overview based on an alternative nomenclature (Ch1–Ch6), *Neuroscience*, 10, 1185–1201, 1983.

50. Mesulam, M.-M. and Van Hoesen, G.W., Acetylcholinesterase-rich projections from the basal forebrain of the rhesus monkey to neocortex, *Brain Res.*, 109, 152–157, 1976.

51. Metherate, R., Cox, C.L., and Ashe, J.H., Cellular bases of neocortical activation: modulation of neural oscillations by the nucleus basalis and endogenous acetylcholine, *J. Neurosci.*, 12, 4701–4711, 1992.

52. Morison, R.S., Dempsey, E.W., and Morison, B.R., Cortical responses from electrical stimulation of the brain stem, *Am. J. Physiol.*, 131, 732–743, 1941.

53. Moruzzi, G. and Magoun, H.W., Brain stem reticular formation and activation of the EEG, *Electroenceph. Clin. Neurophysiol.*, 1, 455–473, 1949.

54. Nauta, W.J.H. and Kuypers, H.G.J.M., Some ascending pathways in the brain stem reticular formation, in *Reticular Formation of the Brain*, Jasper, H.H. et al., Eds., Little, Brown & Co., Boston, 1958, pp. 3–30.

55. Nunez, A., Unit activity of rat basal forebrain neurons: relationship to cortical activity, *Neuroscience*, 72, 757–766, 1996.

56. Paxinos, G. and Watson, C., *The Rat Brain in Stereotaxic Coordinates*, Academic Press, Sydney, 1986.

57. Rasmussen, K., Morilak, D.A., and Jacobs, B.L., Single unit activity of locus coeruleus neurons in the freely moving cat. I. During naturalistic behaviors and in response to simple and complex stimuli, *Brain Res.*, 371, 324–334, 1986.

58. Rasmusson, D.D., Clow, K., and Szerb, J.C., Frequency-dependent increase in cortical acetylcholine release evoked by stimulation of the nucleus basalis magnocellularis in the rat, *Brain Res.*, 594, 150–154, 1992.

59. Rasmusson, D.D., Clow, K., and Szerb, J.C., Modification of neocortical acetylcholine release and electroencephalogram desynchronization due to brainstem stimulation by drugs applied to the basal forebrain, *Neuroscience*, 60, 665–677, 1994.
60. Reiner, P.B. and McGeer, E.G., Electrophysiological properties of cortically projecting histamine neurons of the rat hypothalamus, *Neurosci. Lett.*, 73, 43–47, 1987.
61. Rye, D.B., Wainer, B.H., Mesulam, M.-M., Mufson, E.J., and Saper, C.B., Cortical projections arising from the basal forebrain: a study of cholinergic and noncholinergic components employing combined retrograde tracing and immunohistochemical localization of choline acetyltransferase, *Neuroscience*, 13, 627–643, 1984.
62. Saper, C.B., Organization of cerebral cortical afferent systems in the rat. I. Magnocellular basal nucleus, *J. Comp. Neurol.*, 222, 313–342, 1984.
63. Schwaber, J.S., Rogers, W.T., Satoh, K., and Fibiger, H.C., Distribution and organization of cholinergic neurons in the rat forebrain demonstrated by computer-aided data acquisition and three-dimensional reconstruction, *J. Comp. Neurol.*, 325, 309–325, 1987.
64. Semba, K., Reiner, P.B., McGeer, E.G., and Fibiger, H.C., Brainstem afferents to the magnocellular basal forebrain studied by axonal transport, immunohistochemistry and electrophysiology in the rat, *J. Comp. Neurol.*, 267, 433–453, 1988.
65. Shute, C.C.D. and Lewis, P.R., Cholinesterase-containing systems of the brain of the rat, *Nature*, 199, 1160–1164, 1963.
66. Shute, C.C.D. and Lewis, P.R., The ascending cholinergic reticular system: neocortical, olfactory and subcortical projections, *Brain*, 90, 497–520, 1967.
67. Starzl, T.E., Taylor, C.W., and Magoun, H.W., Ascending conduction in reticular activating system, with special reference to the diencephalon, *J. Neurophysiol.*, 14, 461–477, 1951.
68. Steriade, M., Oakson, G., and Ropert, N., Firing rates and patterns of midbrain reticular neurons during steady and transitional states of the sleep-waking cycle, *Exp. Brain Res.*, 46, 37–51, 1982.
69. Sterman, M.B. and Clemente, C.D., Forebrain inhibitory mechanisms: sleep patterns induced by basal forebrain stimulation in the behaving cat, *Exp. Neurol.*, 6, 103–117, 1962.
70. Stewart, D.J., Macfabe, D.F., and Vanderwolf, C.H., Cholinergic activation of the electrocorticogram: role of the substantia innominata and effects of atropine and quinuclidinyl benzilate, *Brain Res.*, 322, 219–232, 1984.
71. Szymusiak, R. and McGinty, D., Sleep suppression following kainic acid-induced lesions of the basal forebrain, *Exp. Neurol.*, 94, 598–614, 1986.
72. Szymusiak, R. and McGinty, D., Sleep-related neuronal discharge in the basal forebrain of cats, *Brain Res.*, 370, 82–92, 1986.
73. Szymusiak, R. and McGinty, D., Sleep-waking discharge of basal forebrain projection neurons in cats, *Brain Res. Bull.*, 22, 423–430, 1989.
74. Vanderwolf, C.H. and Stewart, D.J., Thalamic control of neocortical activation: a critical re-evaluation, *Brain Res. Bull.*, 20, 529–538, 1988.
75. Villablanca, J. and Salinas-Zeballos, M.E., Sleep-wakefulness, EEG and behavioral studies of chronic cats without the thalamus: the "athalamic" cat, *Arch. Ital. Biol.*, 110, 383–411, 1972.
76. Zaborszky, L., Carlsen, J., Brashear, H.R., and Heimer, L., Cholinergic and GABAergic afferents to the olfactory bulb in the rat with special emphasis on the projection neurons in the nucleus of the horizontal limb of the diagonal band, *J. Comp. Neurol.*, 243, 488–509, 1986.
77. Zaborszky, L., Cullinan, W.E., and Braun, A., Afferents to basal forebrain cholinergic projection neurons: an update, in *The Basal Forebrain*, Napier, T.C., Kalivas, P.W., and Hanin, I., Eds., Plenum Press, New York, 1991, pp. 43–100.

Chapter 15

Immediate Early Gene Expression in Sleep and Wakefulness

Marina Bentivoglio and Gigliola Grassi-Zucconi

Contents

15.1	Immediate Early Genes as Markers of Neuronal Activation 236		
15.2	Basal Expression of Immediate Early Genes 237		
	15.2.1	What Do We Mean by Immediate Early Gene "Basal Expression"? 237	
	15.2.2	Immediate Early Genes Are Expressed in Brain Cell Subsets in Basal Conditions 238	
		15.2.2.1	Immediate Early Gene Neuronal Basal Expression in Different Species 239
	15.2.3	Immediate Early Gene Expression in the Brain Fluctuates During 24 Hours 239	
		15.2.3.1	Immediate Early Gene Expression in the Biological Clock 241
15.3	Expression of Immediate Early Genes Oscillates in the Brain During the Sleep/Wake Cycle 241		
	15.3.1	Immediate Early Gene Expression During Spontaneous Wakefulness 241	
		15.3.1.1	Fos Expression in the Cerebral Cortex 243
		15.3.1.2	Fos Expression in the Diencephalon 245
		15.3.1.3	Fos Expression in the Brain Stem and the Role of Noradrenergic Innervation in Immediate Early Gene Expression 246
		15.3.1.4	*NGFI*-A Expression in Spontaneous Wakefulness 246
	15.3.2	Immediate Early Genes and Sleep Deprivation 247	
	15.3.3	c-*fos* Expression During Sleep 247	
15.4	Open Questions 249		
Note	 249		
References	 250		

15.1 Immediate Early Genes as Markers of Neuronal Activation

Cellular immediate early genes (IEGs) are characterized by their rapid and transient inducibility by stimuli acting on the cell surface, and they are transcribed even in the absence of protein synthesis. Some of their protein products can bind to specific DNA sequences, thus regulating the transcription of downstream genes. These proteins act as third messengers, i.e., as transducers of the nuclear signal, as within a few minutes they can transform the primary short-term signals into long-term cell responses lasting hours or days.[42] The present overview will deal with the *fos*, *jun*, and *NGFI* families of IEGs, whose expression during sleep and wake has been investigated in experimental studies.

A wealth of data has pointed out a pivotal role of the *fos* and *jun* gene families in the coupling of neuronal activity to the structural and functional changes involved in neural plasticity. c-*fos* is the IEG prototype in the nervous system. The role of *fos* as master switch — the sensor that detects incoming signals at the cell membrane to convert them into gene transcription and, therefore, in long-lasting responses through the regulation of downstream genes — has been pointed out since the mid-1980s.[35] The Fos (c-Fos, Fos B, and Fos-related antigens, or Fras) and Jun (c-Jun, Jun B, and Jun D) proteins encoded by these IEGs are transcription factors sharing the common features of a leucine-zipper region and an adjacent basic-rich domain involved in DNA binding.[54] The leucine-zipper region represents the part of the molecule that allows the dimerization between the Jun family of proteins (both homo- and heterodimers), as well as between proteins of the Jun and Fos families (Jun-Fos heterodimers). Transcription factors belonging to the Jun and Fos families can associate in dimers to constitute the transcription factor activator protein-1 (AP-1) DNA-binding complex. The AP-1 recognition sequence is found in a variety of promoter regions of target genes and modulates their transcription, thus mediating neuronal responses to many different extracellular signals. All the dimers forming the Fos and Jun proteins contribute to the AP-1 complex activity, and the resulting combinations have different regulatory effects on transcription.[31]

The IEG nerve growth factor-induced A (*NGFI*-A) — deriving its name from the initial finding that it is induced by nerve growth factor in PC12 cells and also known as *zif*/268, *egr*-1, *krox* 24, or *tis*-8 — interacts with DNA using the zinc finger binding domain; also, the protein product of this gene acts as a transcription factor regulating the expression of target genes.[39] *NGFI*-B codes for a protein that shows homology to the steroid receptor family of transcription factors.[24,40] Experimental data on the different inducibility and persistence of IEGs indicated that transcription factors may subserve different responses in neurons, but very few potential target genes are known at present.[43]

IEGs, and in particular c-*fos*, have been introduced in experimental studies as functional markers of neuronal activity (see Reference 27 for a recent review). *In situ* hybridization of c-*fos* mRNA and/or the immunohistochemical revelation of its protein product Fos provide tools to map functional neuronal activation at a single-cell resolution. In addition, Fos revelation is compatible with the chemical characterization of neurons (e.g., by means of double immunohistochemistry) and allows the study of neuronal circuits in which such cells are inserted through the combination with tract-tracing techniques. After 10 years of use, it is now clear that, as for every experimental approach, functional mapping with c-*fos* has pitfalls and limitations. For example, the gene is not a universal marker of neuronal activity, and the threshold for its induction may vary in different neuronal cell types[16] (see also Sections 15.3.1.2 and 15.3.3). However, c-*fos* has offered to neuroanatomists a sensitive and effective tool for dynamic studies and has provided intriguing data on gene expression in brain neurons during sleep and wake.

15.2 The Basal Expression of Immediate Early Genes

A number of studies have unequivocally indicated that the expression of IEGs is not only induced by several different experimental paradigms, but is also constitutive and/or elicited by physiological processes which spontaneously occur during the normal animal's life, commonly grouped under the definition of "basal conditions".

15.2.1 What Do We Mean by Immediate Early Gene "Basal Expression"?

This question may not be relevant to experiments controlling the gene induction consequent to intentionally applied stimuli, but it is certainly relevant to studies focused only on the IEG "basal expression", i.e., without a comparison with stimulated cases. "Basal conditions" are generally considered those in which no exogenous stimulation is intentionally applied and would thus represent the physiological state of the animal. While we should obviously keep this concept as such, we should also be aware that this reflects a laboratory jargon to define that in "basal conditions" the experimenters "do nothing" to the animals but (1) the animals actually *do* do something, and (2) conditions related to the normal animal's life may vary according to the animal's age, gender, hormonal influence, social interactions, and environment.

The first of the above assumptions brings about the obvious consequence that there is an inter-individual variability in the study of the basal expression of genes which may also be transiently induced in behaviors that represent the animal's normal activity. Sensory and social stimuli that may contribute to the IEG basal expression should reflect "the physiological context of daily experience";[25] however, these parameters may be relatively difficult to control and standardize, and the animal's behavior should be carefully observed and monitored throughout the period considered for the experiment. In addition, common qualitative findings (such as brain structures in which IEG expression is consistently induced in all animals in basal conditions) and their quantitative variability should be carefully evaluated.

The IEG expression can vary considerably during a lifetime, so the animal's age can represent a crucial parameter in the study of basal gene expression. For example, both the *fos* and *jun* gene families have been found to vary during the postnatal development of the rat visual cortex (see Reference 29 for review). The Fos and EGR-1 proteins are much more expressed in the visual cortex of kittens than of adult cats,[30] suggesting that these proteins could be related to plasticity during cortical neuronal maturation. Another example is provided by the variation of Fos and Fras expression during the postnatal development of the rat cerebral cortex, striatum, hippocampus, and cerebellum.[1] As for gender, the pattern of basal expression of IEGs in the brain of female rats was found to be congruent with that observed in males.[25,26] However, physiological conditions such as pregnancy, parturition, maternal experience, and maternal behavior in general have been reported to influence considerably the expression of IEGs in the brain and, in particular, in the hypothalamus (see, for example, References 10 and 34).

For the sake of simplicity and to find a common language, we will here refer to "basal conditions" as those in which young adult males of the same species and strain are used, the experiment is performed in their home cages and in an environment and with experimenters to whom the animals have been familiarized for at least one week, and housing conditions have been kept constant under a 12-hr/12-hr light/dark cycle unless specified otherwise. These protocols may be routinely applied in laboratories in which the animal's behavior or sleep is studied, but may be new, for example, to neuroanatomists, who are not trained to pay attention to the fact that one animal might have just arrived in the animal facility and another may have already been there for some time, and that the animals may or may not be familiarized with, for example, the odors of the environment (including the perfume or shaving creme of the experimenter).

Of course, not all these details are in general specified in the studies dealing with the basal expression of IEGs, and we should therefore be aware that experimental parameters may complicate the comparison among different studies, which is anyhow often hampered by the use of different probes and/or antibodies. It should also be considered that experimental studies are mostly performed in nocturnal rodents, whose period of rest corresponds to the lights-on time, when the environmental light may affect IEG expression in visual-related brain structures.

Basal conditions obviously include also sleep and wake, and it can be assumed that the animals are sleeping during the period of rest and they are awake during the period of activity. The definition of sleep and wake should, however, be based upon electrographic correlates, which also allow determination of the relative proportion of sleep and wake in the hours preceding the animal's sacrifice, thus providing precious parameters for the interpretation of data and the evaluation of inter-individual variability. We will therefore refer to sleep and wake only when the electroencephalogram (EEG) has been recorded during the experimental session. This implies that the animal should have been habituated also to the EEG recording equipment and procedure prior to the study of IEG expression during sleep and wake. The brain structures which express IEGs in these two physiological states are the focus of the present overview, and the data which presented the evidence of basal IEG expression will be first briefly dealt with.

15.2.2 Immediate Early Genes Are Expressed in Brain Cell Subsets in Basal Conditions

Since the detection of c-*fos* expression in the brain,[17,44] several findings have pointed out a constitutive expression of IEGs and the evidence of IEG expression in basal conditions. Most of these studies have only taken into account unstimulated conditions as a control of paradigms in which stimuli are applied, and a complete review of all these "control" data is beyond the scope of the present chapter. When the time of the animal's sacrifice is not specified in these reports, in all likelihood animals were killed during daytime. Altogether, these studies have pointed out that there are considerable variations in the basal expression of the different IEGs.

The protein products of the *fos, jun,* and *krox* gene families were found to display different patterns of basal expression in the rat hippocampus:[28] Jun was found to be high in granule cells of the dentate gyrus, whereas Fos-related antigens, but not Fos, were found to be expressed in some granule cells of the dentate gyrus, and Krox 24 was especially high in CA1 and in the subiculum. A low basal level of Fos in the hippocampus is also supported by the findings obtained with a *fos-lacZ* gene construct in mice,[58] which allowed researchers to distinguish Fos expression from that of Fras, unlike many antibodies which recognize both Fos and Fos-related antigens.

In the rat cortex, sampled 2 hr after the lights-on time, Fos and Jun B were found to be expressed by a restricted number of neurons with a prevalence in layer VI, whereas the positivity to the c-Jun protein was widely distributed in the neocortex.[26] In the same study,[26] the transcription factor cyclic AMP response element-binding protein (CREB), which can act on the promoters for Fos and NGFI-A, was highly expressed in neurons of all cortical areas and layers. The basal expression of proteins encoded by genes of the *fos* and *jun* families and by *krox*-24 was also examined in detail in the brain of rats perfused 2 hr before the lights-on time.[25] In this study, Jun D immunoreactivity was detected in almost all neurons in the brain, including areas which contained c-Fos-, Jun B-, and c-Jun-immunostained cells: telencephalic structures (the dorsal striatum and lateral septum), thalamic nuclei (midline nuclei and the intergeniculate leaflet, which is a component of the lateral geniculate nucleus), hypothalamic structures (the dorsomedial, paraventricular, and supraoptic nuclei), and brainstem regions (superior and inferior colliculi, central gray, locus coeruleus); c-Jun was also found to be expressed in medullary motoneurons, whereas positivity to Fos B was restricted to a low number of neurons in the brain. On the other hand, Krox-24, although not ubiquitous, was found to be intensely expressed by many neuronal populations throughout the brain.[25] These latter findings support a high constitutive expression of NGFI-A in the rat brain.[53]

In the rat periaqueductal gray, Fos positivity was detected in basal conditions in clusters of neurons grouped in distinct dorsolateral, lateral, and ventrolateral regions,[63] thus supporting a modular functional organization of this structure.[3]

As for the *fos* basal expression in the brainstem, it is noteworthy that in a study on transgenic mice carrying the *fos-lacZ* fusion gene, in which beta-galactosidase activity allowed unambiguous determination of *fos* expression,[58] the Fos-lacZ expression in the brain of unstimulated animals was highest in the raphe nuclei.

In relation to the basal expression of IEGs, it is also interesting to recall that in cortical cultured mature neurons, in which IEGs are induced by spontaneous synaptic activity, the basal expression of c-*fos*, *jun* B, *zif* 268, and *fos* B is rapidly suppressed by antagonists to L-type, voltage-sensitive calcium channels, indicating that these channels mediate the coupling of synaptic excitation to the activation of transcriptional events.[45]

15.2.2.1 Immediate Early Gene Neuronal Basal Expression in Different Species

The vast majority of the experiments exploiting IEG induction and describing IEG basal expression have been performed in rodents, and the expression of IEGs during sleep and wake has been studied mostly in rats, while a few such studies have been pursued in cats. However, it is important to note that reports on the IEG basal expression range from observations in invertebrates to humans. For example, basal Fos protein expression, also described in the rat olfactory bulb[52] (see also Section 15.3.1), was reported to occur in the olfactory system of unstimulated bees, in which Fos immunoreactivity displayed an age-dependent variation.[19] Studies on the basal expression of genes of the *fos* and *jun* families and of NGFI-A have also been performed in the cat and in the monkey (see Reference 29 for review). Fos and Jun immunoreactivities have been observed in surgical specimens resected from patients for intractable epileptic seizures,[18] and in postmortem human brain specimens which served as control for pathological changes observed in Alzheimer's disease.[2,65] In these studies, and especially in those performed in lower species such as invertebrates, or in the human brain, the technical problems represented by the specificity and sensitivity of the different commercially available antibodies are even more delicate and critical for the interpretation of data than in experimental studies in rodents. However, altogether these reports indicate that the basal expression and the induction of IEGs are highly conserved features in phylogenesis.

15.2.3 Immediate Early Gene Expression in the Brain Fluctuates During 24 Hours

Initial observations on a circadian oscillation of Fos in the rat brain pointed out a marked increase of Fos immunoreactivity after the onset of darkness in the hippocampus and striatum.[33] The occurrence of a circadian fluctuation of IEGs in the brain was extensively investigated in our laboratories.[21,22,36] We aimed in these studies at verifying the hypothesis that if the induction of IEGs, and in particular of c-*fos*, reflects neuronal activation, their expression during the activity period should differ from that occurring during the period of rest. The results of northern blot hybridization from different regions of the rat brain, sampled every 4 hr during 24 hr, revealed that the levels of c-*fos* mRNA were just detectable during the rest hours, but increased remarkably during the hours

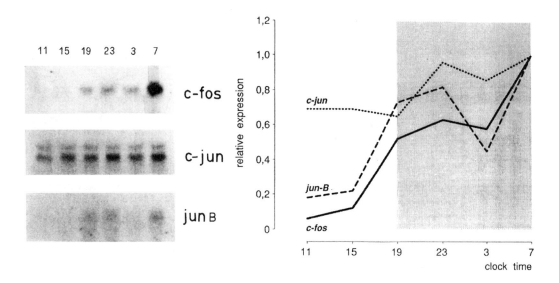

FIGURE 15.1
Spontaneous variation of the expression of three immediate early genes in the rat brain during a 24-hr cycle in basal conditions. Northern blot analysis of the mRNAs in the cerebral cortex at six time points (three rats were pooled per time point) in standard lighting conditions (12 hr light/12 hr dark) is shown on the left, and the level of their basal expression (evaluated as optical density of northern blots) during the 24-hr cycle is shown on the right. The findings show that c-*fos* expression is induced at the lights-off time (19 hr), remains high throughout nighttime, and peaks at the end of the dark period, whereas c-*jun* expression is consistently high, and the pattern of *jun* B transcript parallels that of c-*fos*.

of activity.[21,22] When such analysis was extended to the *jun* family of genes, the levels of *jun* B mRNA expression were found to be low across time points in the hippocampus, diencephalon, brainstem, and cerebellum, but displayed in the cerebral cortex and striatum an oscillation that paralleled that of c-*fos*, with the highest level of expression during the hours of darkness.[36] At variance with the fluctuation of c-*fos* and *jun* B mRNAs, the levels of c-*jun* transcript were found to be consistently high in different brain regions during 24 hr (Figure 15.1).[36]

In order to verify whether the oscillation of c-*fos* and *jun* B was endogenously determined, the study of the basal expression of these IEGs was repeated by maintaining the animals in constant light or darkness. The abolition of the light/dark cycle confirmed that the expression of c-*fos*,[21] as well as that of *jun* B in the cortex and striatum,[36] was high during the animal's subjective night (night and day are defined as subjective when not entrained to a light/dark cycle). In order to ascertain a relationship between c-*fos* expression and the state of the animal's vigilance, the experiments were also performed on animals kept awake with gentle handling during the day and allowed to sleep during the night.[21] Such manipulation of the rest/activity cycle fully supported a variation of gene expression related to the animal's physiological state and, in particular, an increase of c-*fos* expression that paralleled the animal's activity. These data are stimulating, as they showed that, at variance with c-*jun*, the expression of c-*fos* oscillates concomitantly with that of *jun* B in the cortex and striatum. Therefore, in the Fos-Jun cross-talk mechanisms at the nuclear level, c-*fos* and *jun* B may be critical for a circadian variation of genomic expression and for the level of transcription factors associated with different states of alertness. Recent data based on the study of IEG expression in nocturnal vs. diurnally active species, different seasonal and hibernation phases in squirrels, and during postnatal development in the rat strongly support a link between the expression of IEGs, and in particular of c-*fos*, with neuronal activity as well as locomotor activity.[46a]

15.2.3.1 Immediate Early Gene Expression in the Biological Clock

The role of the suprachiasmatic nucleus of the hypothalamus (SCN) as circadian pacemaker of endogenous biological rhythms is extensively dealt with in other chapters of this handbook (Chapter 1–3, 6), in which the involvement of IEGs in the molecular mechanisms underlying the photic entrainment of the SCN is also discussed. As reviewed in these chapters, light regulates the expression of many IEGs in the SCN, particularly in the retinorecipient portion of the nucleus. It will only be briefly mentioned here that a basal IEG expression, i.e., in the absence of a photic stimulation, has also been reported in the SCN. Fos expression in basal conditions was reported to be low in the rat SCN during nighttime and highest in the morning.[33] Although the expression of this gene may be driven by the light/dark cycle, c-*fos* mRNA was reported to increase in the late subjective night in the retinorecipient portion of the rat SCN,[61] and Fos was found to be expressed in the rostral portion of the SCN in the hamster during the mid-subjective day.[15] In the same species, *jun* B mRNA was detected in the late subjective night in the SCN, where it persisted until the early subjective day.[23]

The SCN is involved in the regulation of the sleep/wake cycle,[32] and changes in gene expression in the SCN may reveal the neural correlates of sleep changes. In the rat model of a sleep disorder caused by systemic infection with the parasite *Trypanosoma brucei*, which is the causative agent of African trypanosomiasis or sleeping sickness in humans, an alteration of Fos expression during the light period has provided us a clue for detecting a neural dysfunction in the otherwise normal brain.[6]

15.3 The Expression of Immediate Early Genes Oscillates in the Brain During the Sleep/Wake Cycle

Having ascertained an oscillation of IEG expression in the brain during 24 hr, with an increase during the activity period,[21] we verified with EEG monitoring whether such variation could reflect physiological states corresponding to sleep and wakefulness. To this purpose, EEGs were recorded in rats for several hours during daytime and nighttime. Northern blot analysis confirmed that c-*fos* mRNA level was highest during wake and lowest during sleep (Figure 15.2).[21] Immunohistochemistry was then used to investigate the neural substrate of Fos protein expression during sleep and wake. These experiments confirmed a considerable quantitative increase of Fos-positivity during wakefulness in respect to sleep and revealed the induction of Fos in neuronal subsets.

15.3.1 Immediate Early Gene Expression During Spontaneous Wakefulness

In the experiments performed in our laboratories,[8,20,22] rats were killed during spontaneous wake in the period of darkness of the light/dark cycle. We observed that Fos immunoreactivity increased in the brain in parallel with the increase in the proportion of wake in the 1.5 hr preceding sacrifice. In other studies,[11,50] rats that had been awake in the dark were compared to animals that had been spontaneously awake during the light period for about 30 min before sacrifice, and c-*fos* expression was found to be higher in the former group than in the latter. These studies[11,50] confirmed an increase of Fos expression after a sustained (1.5-hr) period of spontaneous wakefulness in respect to a short (0.5-hr) period, which, however, induced c-*fos* mRNA expression but was not sufficient for adequate protein synthesis.

In the cases we examined after a period of spontaneous wake, we consistently detected Fos-immunostained neurons in the cerebral cortex, septum and diagonal band region, diencephalon, superior colliculi, and periaqueductal gray, whereas labeling in the hippocampus was absent in most

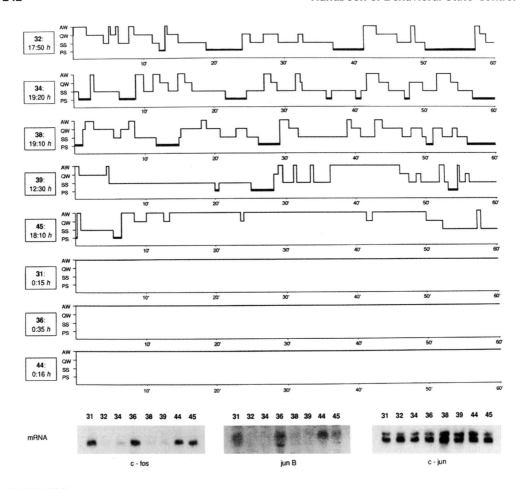

FIGURE 15.2
Spontaneous variation of immediate early gene expression in the rat brain during sleep and wake. The hypnogram derived from electroencephalographic monitoring in the hour preceding sacrifice of eight rats (the case code and time of sacrifice of each animal are indicated on the left) entrained to a standard 12-hr/12-hr light/dark cycle (lights-on at 19 hr) is shown on top. Abbreviations: AW, active wake; QW, quiet wake; SS, synchronized sleep; PS, paradoxical sleep. Northern blot analysis of the mRNAs of three immediate early genes in the cerebral cortex of each animal is shown at the bottom. Note that c-*fos* and *jun* B expression is low in rats 32, 34, 38, and 39 which had been sleeping and is high in rats 45, 31, 36, and 44 which had been awake, whereas c-*jun* expression is high in all cases.

cases and restricted to very few positive cells in some cases (Figure 15.3). These latter findings are at variance with the report of Fos expression in many regions of the hippocampal formation during spontaneous wake, in which marked labeling in the dorsomedial striatum as well as in the nucleus accumbens was also described.[50] In our experimental series, Fos-immunoreactive neurons were seen in the dorsal part of the striatum only after a sustained period of wake, when the animals' behavior had been characterized by a pronounced motor activity. High Fos expression during spontaneous wake was reported to occur also in the olfactory bulb and anterior olfactory nucleus.[49,50]

As for the Fos expression in the basal forebrain, the phenotype (see Chapter 14) of the neurons in which Fos is spontaneously induced remains to be identified. It is interesting, however, to recall that basal forebrain neurons exert a powerful effect on cortical activation both directly and through the suppression of thalamic synchronizing mechanisms,[9] and Fos induction in these cells during wake could be related to these mechanisms.

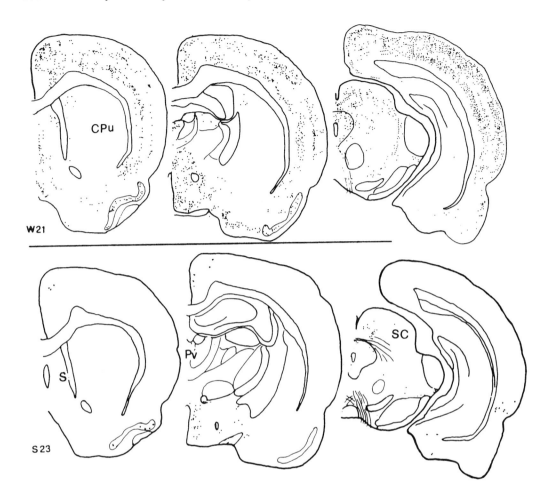

FIGURE 15.3
Distribution of Fos-immunoreactive cells in three representative sections through the brain, arranged in anteroposterior order, of rats in which the electroencephalogram had been recorded for several hours before perfusion: the animal W21 (top) had been uninterruptedly awake for 2 hr prior perfusion during the night, whereas case S23 had been perfused during daytime after a period of prevalent sleep (with a proportion of about 70% of synchronized sleep in the 2 hr preceding sacrifice). Note the remarkable increase in the number of Fos-immunostained cells in the brain after wake in respect to the rat that had been sleeping. Abbreviations: CPu, caudate-putamen; Pv, thalamic paraventricular nucleus; S, septum; SC, superior colliculus.

A comparison between c-*fos* mRNA and Fos protein expression goes beyond the scope of the present chapter, but it is important to note that dissociation between a high mRNA level and an absent or low level of Fos protein, even after a period sufficient for adequate protein synthesis, may occur during spontaneous wake, e.g., in the cerebellar cortex.[50] This mismatch could be ascribed to lack of mRNA translation or to a low level or post-translational changes of Fos protein.[16,50]

15.3.1.1 Fos Expression in the Cerebral Cortex

During wake, Fos-immunoreactive neurons have been consistently detected in many areas of the cerebral cortex,[20,50] and their number displayed an inter-individual variability. Fos-immunostained cells were mostly detected in the cingulate cortex, frontoparietal (sensorimotor), occipital, piriform, and rhinal cortices (Figures 15.3 and 15.4C,D). A marked increase of Fos immunoreactivity was

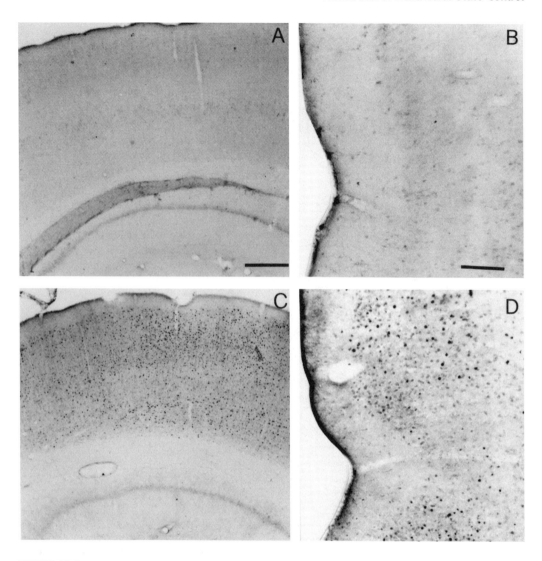

FIGURE 15.4
Microphotographs of Fos-immunoreactive cells, in which labeling of the nucleus is evident, in the parietal cortex (C) and perirhinal cortex (D) of rats that had been awake before sacrifice, compared to the absence of immunoreactivity in the same areas of rats that had been sleeping (A, B). Scale bars are equivalent to 600 µm in A (C is at the same magnification as A) and 100 µm in B (D is at the same magnification as B).

detected in the perirhinal and piriform cortices following spontaneous wake in the dark in respect to that observed after spontaneous wake during the light period.[50] We observed that neurons which displayed Fos positivity in the neocortex during wake were concentrated in two bands, corresponding to the middle and deep layers (Figure 15.3).[20] It remains to be ascertained whether Fos is expressed during spontaneous wake by either or both of the two categories of neurons of the cerebral cortex, i.e., projection and local circuit neurons. Preliminary data based on double immunocytochemical staining[47] have indicated that during spontaneous wake Fos is rarely expressed by interneurons of the cerebral cortex (characterized by their positivity to the calcium-binding proteins calbindin D28k and parvalbumin), thus suggesting that Fos is induced during wakefulness in pyramidal projection neurons.

The laminar distribution of cortical cells that express Fos during spontaneous wake indicates that both corticocortical projection neurons and neurons of the deep layers that innervate subcortical targets could be "activated" when the animals are active. On the other hand, a considerable Fos induction was observed in neurons of layer IV and of the supragranular layers of the primary visual and somatosensory cortices of rats exploring a novel environment, indicating that the activation of the granular and supragranular layers may be specifically related to the attention paid to environmental stimuli.[41] In our experience, after a sustained wake in basal conditions, Fos is markedly induced in the supragranular and deep layers but not in layer IV. IEG induction in layer IV of sensory cortices could, therefore, be selectively related to recognition mechanisms.

15.3.1.2 Fos Expression in the Diencephalon

Fos immunoreactivity has been found to be very pronounced, though selective, in the diencephalon during wake. In agreement with other reports,[11,50] we have detected diencephalic Fos-immunoreactive cell populations in several hypothalamic regions, as well as in the thalamus (Figure 15.3), where they were predominantly located in the midline nuclei and concentrated in the thalamic paraventricular nucleus.[48] Fos was found to be increased also in the intralaminar nuclei during spontaneous wake.[8,11,50] However, in our experimental series of animals perfused during nighttime after a period of spontaneous wakefulness, labeling in the midline nuclei was found to be a consistent feature,[48] whereas the anterior intralaminar domains, such as the central lateral and paracentral nuclei, contained Fos-immunostained neurons only after a sustained wake, when the labeling in the brain was overall very conspicuous.

The midline nuclei are composed of cell populations projecting to limbic and limbic-related targets, such as the prefrontal and rhinal cortices and the nucleus accumbens. The ventral midline component, the nucleus reuniens, is connected with the hippocampus, and the thalamic paraventricular nucleus, which is the most dorsal component of the midline group, is a main source of innervation of the amygdala. As for the other domains of the rat thalamus, the midline nuclei do not contain interneurons (see Reference 5 for review). We have determined that during spontaneous wake Fos-immunoreactive neurons include cell populations projecting to the amygdala, which prevailed in the anterior part of the thalamic paraventricular nucleus, and neurons projecting to the nucleus accumbens, which prevailed in the posterior part of this nucleus.[48] The latter data indicate not only that Fos is increased during wake in selected populations of thalamic projection neurons, but also whatever information is transduced through these changes in genomic expression is conveyed to the limbic system after thalamic processing.

Two other peculiarities of the thalamic midline are noteworthy in this context. First, the thalamic paraventricular nucleus is a main target of innervation of the SCN[32] and could thus represent a component of the circadian timing system. Second, neurotrophins may modulate behavioral states, and, although in general thalamus and neurotrophins have not been related, the midline is a selective site of expression of neurotrophin high-affinity receptors in the thalamus.[5] Whether the selective behavioral state-associated Fos induction in the thalamic midline could be related to the above features remains to be verified.

Both the intralaminar and midline nuclei, and especially the former, are main components of the so-called "nonspecific" thalamus, traditionally considered as the rostral pole of the brainstem reticular activating systems (see References 4 and 60 for reviews). The role of the ascending brainstem pathways and of the medial thalamus in the regulation of cortical activity and arousal is reviewed in Chapters 11 and 20. In relation to behavioral state-related gene expression, it is interesting to note that Fos is induced during wake in thalamic "nonspecific" structures, in which it increases after a sustained wake, thus paralleling changes in the physiological activity of these neurons.

Although c-*fos* mRNA was detected in the thalamic ventral complex and medial geniculate nucleus during spontaneous wake,[48] it is interesting to note that in our experience Fos protein expression does not occur in the awake animals in the principal (traditionally defined as "specific") thalamic nuclei and in the thalamic reticular nucleus. As is well known, the former nuclei are the relays which transfer information to their respective cortical target, whereas the reticular nucleus exerts a fundamental intrathalamic role as a pacemaker of thalamic activity (reviewed in Chapter 20). The lack of state-dependent Fos expression in these structures could be due to a high threshold for gene induction in their neurons and/or it may reflect their "habituation" to normally ongoing stimuli. Such selectivity of thalamic labeling indicates, however, that spontaneous Fos expression cannot be considered a universal marker for behavioral state-related neurons.

As for the hypothalamus, special emphasis was devoted to the occurrence of Fos immunoreactivity in the medial preoptic area, posterior hypothalamic area, and supramammillary nuclei after a period of wake in rats.[11,50] The hypothalamic structures involved in behavioral state control are discussed in Chapter 19 and will not be dealt with further here. It should, however, be mentioned that the medial preoptic area is considered to play a critical role in the homeostatic processes regulating sleep and wake, and the block of wake-induced Fos protein expression in this area through local bilateral injections of c-*fos* antisense oligonucleotides was found to increase the relative proportion of wakefulness the following day.[12] This elegant paradigm provided additional support to the specificity of a behavioral state-related role of c-*fos* expression in hypothalamic structures.

15.3.1.3 Fos Expression in the Brainstem and the Role of Noradrenergic Innervation in Immediate Early Gene Expression

In the brainstem, Fos-immunostained neurons were found after a period of wake in the dorsal raphe, locus coeruleus and mesopontine nuclei, central gray, and superior and inferior colliculi.[50,62] Such activation of tectal neurons could be related to their role in sensory processing and in the integration of multi-sensory and sensorimotor information.[59] On the other hand, the central gray is involved in emotional behavior-related sensory processing.[3] The role in state-dependent behavior of the monoaminergic brainstem structures, which contain neurons in which Fos is induced during wake, is extensively discussed in this handbook (see Chapters 12 and 16). The mesopontine cell groups, namely the pedunculopontine and laterodorsal tegmental nuclei, are the main source of cholinergic innervation of the diencephalon, and their role as substrate of EEG desynchronized states is discussed by Semba in Chapter 11 (see also Chapter 17).

The role of the locus coeruleus and of the noradrenergic innervation in the increase of IEG expression during wake has been investigated with elegant paradigms in the rat.[13] Based on the data indicating that the firing of neurons of the locus coeruleus increases during wake and that these cells may exert a wide influence on brain neurons through their diffuse innervation, the locus coeruleus has been destroyed unilaterally with the local administration of neurotoxin, or bilaterally through systemic neurotoxin injection, thus causing a degeneration of noradrenergic fibers. Fos expression was found to be abolished in the cerebral cortex and hippocampus on the side(s) of the lesion, as compared to the unlesioned side or intact animals, after both sleep deprivation in the light and spontaneous wake in the dark. In addition, NGFI-A expression decreased concomitantly with the degeneration of noradrenergic fibers. CREB expression was also considerably reduced in the cortical areas depleted of noradrenergic innervation. Interestingly, a dissociation between EEG activity and Fos expression was also found in these experiments, as the low-voltage, fast-activity pattern that characterizes wake was maintained after lesion of the locus coeruleus and noradrenergic fiber destruction, despite the decrease of IEG expression in these conditions.

15.3.1.4 *NGFI*-A Expression in Spontaneous Wakefulness

In situ hybridization of *NGFI*-A mRNA and immunoreactivity to its protein product indicated that the expression of this gene is increased in rats spontaneously awake both in the dark and in the light in respect to those that had been sleeping during the hours of light.[50,51] *NGFI*-A expression was augmented during wake in many neocortical and allocortical areas — in the olfactory bulb, striatum, claustrum, amygdala, hippocampal formation, anterior thalamic nuclei, hypothalamus (medial preoptic area, supraoptic, ventromedial and dorsomedial nuclei, and throughout the SCN), superior and inferior colliculi, central gray, parabrachial nuclei, and locus coeruleus.[50,62] Both NGFI-A mRNA and protein increased in some brain regions very rapidly, after only 30 min of spontaneous wake, but such variation displayed a temporal regulation and topographical selectivity: for example, after 6 hr of spontaneous wake, *NGFI*-A increased in the cortex, striatum, and claustrum, but decreased in the anterior thalamic nuclei.[50,51]

15.3.2 Immediate Early Genes and Sleep Deprivation

Sleep deprivation paradigms may imply a generalized stress response and should be evaluated with caution. For the study of IEG expression, sleep deprivation has mostly been performed with gentle handling, a procedure in which stress is minimized. The investigations based on forced wakefulness strongly supported a relationship between IEG induction and wakefulness. Northern blot analysis of IEG mRNAs displayed different changes in several brain regions of sleep-deprived rats:[46] c-*fos* underwent the greatest increase after sleep deprivation; in contrast, c-*jun* mRNA levels remained largely invariant across time points and sampled brain regions; *jun* B showed changes in the cortex comparable to those of c-*fos*, and *NGFI*-A and *NGFI*-B mRNAs also displayed changes consistent with those of c-*fos* mRNA. These data, therefore, supported our findings[21,22,36] indicating a parallel increase of c-*fos* and *jun* B expression during wake (Figure 15.1).

After sleep deprivation of up to 24 hr, many regions of the rat brain — in particular, the cerebral cortex, medial preoptic area, tectum and central gray — showed an increase of c-*fos* mRNA and Fos immunoreactivity, whereas Fos labeling was marked also in other regions, such as the striatum, amygdala, septal region, dorsal raphe, and mesopontine nuclei, in which it was scarce or absent in control animals.[12,49] The magnitude of the expression of c-*fos* mRNA and Fos protein displayed a complex temporal regulation, indicating that the induction of this gene is not strictly proportional to the amount of prior wakefulness,[12] at least after sleep deprivation. In agreement with the data observed after a period of spontaneous wakefulness, Fos and NGFI-A were also found to be induced in noradrenergic cells of the locus coeruleus of sleep-deprived rats.[62]

The induction of *NGFI*-A mRNA has been investigated in detail by means of *in situ* hybridization after a sleep deprivation up to 24 hr in the rat.[51] *NGFI*-A mRNA levels were in general much less affected by sleep deprivation performed during the hours of darkness (when the basal *NGFI*-A expression was already high) than during the light period. This study provided the interesting finding that the *NGFI*-A mRNA level can increase or decrease, depending upon the area, during forced wakefulness. For example, in the cerebral cortex and striatum the increase was highest after 6 hr of sleep deprivation, but such induction decreased after 12 hr of forced wake; in the anterior thalamic nuclei the increase was highest after 3 hr and decreased after 6 hr of sleep deprivation. Forced wakefulness also resulted in a decreased *NGFI*-A expression in the hippocampal formation but did not modify the *NGFI*-A expression in the medial preoptic area. In general, the different topographical and temporal features of c-*fos* and *NGFI*-A expression during forced wakefulness suggest that these two IEGs may be sensitive to different stimuli.[51]

15.3.3 c-*fos* Expression During Sleep

In agreement with the data provided by northern blot hybridization,[21] the cerebral cortex was found to be largely devoid of Fos-immunoreactive neurons after a period of sleep (Figures 15.3 and 15.4A,B).[20,50] When present after a period of sleep, Fos-immunostained neurons were mostly concentrated in the cingulate and rhinal cortices (Figure 15.3).[20] Such selectivity of distribution is intriguing in view of the role that neuronal activity in cell subsets of the limbic cortex could play during sleep. In addition, we have observed that the paucity of Fos-immunostained neurons during sleep paralleled the proportion of slow-wave sleep in the 2 hr preceding sacrifice, which increased as immunostained cell density decreased.[20]

The number of diencephalic Fos-immunopositive cells after a period of sleep was also found to be remarkably lower than that observed after spontaneous or forced wakefulness.[8,11,50] Accordingly, very few, if any, Fos-immunoreactive neurons were detected in the thalamic paraventricular nucleus during sleep.[48] Thus, sleep, and in particular slow-wave sleep, seems to reverse, at least in the cerebral cortex and in the thalamic midline, the c-*fos*-mediated up-regulation of gene expression that occurs during wake.

In addition, the pattern of Fos expression in the diencephalon during sleep was found to differ from that observed after a period of wakefulness: Fos-immunoreactive neurons were found in the thalamus in the intergeniculate leaflet, which is a component of the circadian timing system connected with the SCN (see Reference 32 for review). At variance with the pattern observed during wake, Fos was detected in cell subsets of the SCN.[8,50] Labeling in the intergeniculate leaflet and in the SCN could, however, follow the light/dark cycle rather than the sleep/wake cycle (see Section 15.2.3.1 and Reference 57).

A selective Fos induction was also detected during spontaneous sleep in the ventrolateral preoptic area, and sleep deprivation experiments confirmed that Fos induction in this hypothalamic cell subset was related to a recent sleep experience.[57] About half of the neurons of the ventrolateral preoptic area that displayed Fos during sleep were found to innervate the tuberomammillary nucleus of the posterior hypothalamus.[57] This latter structure is composed by histaminergic neurons that are tonically active during wake and provoke arousal (see Reference 60 for review). On this basis, the hypothesis was put forth that tuberomammillary histamine neurons may be inhibited by GABAergic cells of the ventrolateral preoptic area "activated" during sleep[57] (see Chapter 19 in this handbook for an overview on the role of hypothalamic centers during sleep).

Fos expression was also used to identify neuronal populations that could be implicated in REM sleep. This is a difficult task, as spontaneous desynchronized sleep occurs in brief bouts lasting only a few minutes within epochs of synchronized sleep, and such short REM sleep episodes may not enable a correlation with Fos induction. Therefore, REM sleep has been experimentally manipulated in order to increase its duration. After REM periods extended by means of auditory stimulation or as a recovery from sleep deprivation in the rat, Fos-immunoreactive neurons were found in the lateral hypothalamic area, basolateral amygdala, and brainstem regions which are part of the neuronal network involved in REM sleep mechanisms.[37,38] In particular, an increase of Fos immunoreactivity after a relative increase of REM sleep was detected in brainstem structures that contain REM-on cells, such as the mesopontine, parabrachial, and subcoeruleus nuclei, but not in the locus coeruleus and medial and dorsal raphe which contain REM-off neurons.

On the other hand, after a sustained REM sleep induced in the cat by the injection of the cholinergic agonist carbachol in the medial pontine reticular formation, an increase of Fos-immunoreactive neurons in respect to vehicle-treated controls was identified in the locus coeruleus, dorsal raphe, medial pontine reticular formation, and mesopontine nuclei,[55,56,64] i.e., in cell groups containing REM-on and REM-off neurons, whose Fos induction was ascribed to changes of their firing during REM sleep.[56] In addition, Fos expression during cholinergically induced REM sleep was detected in the medullary raphe nuclei and in cell groups of the dorsolateral pontine reticular formation and ventral region of the medial medullary reticular formation, involved in the generation of atonia during REM sleep.[64] Fos labeling was also detected in the lateral and medial vestibular nuclei and in neurons of the nuclei prepositus hypoglossi and intercalatus, as well as in motoneurons of the nucleus abducens, whose activation could be related to rapid eye movements and the generation of ponto-geniculo-occipital (PGO) waves during paradoxical sleep.[64] Although the complex pattern of Fos expression in these studies may reflect the neural networks involved in the generation and maintenance of desynchronized sleep, during carbachol-induced REM sleep Fos expression was not detected in the thalamic primary visual relay, the dorsal lateral geniculate nucleus, in which PGO waves are elicited in relation to REM sleep.[56]

15.4 Open Questions

The study of IEGs indicates that sleep and wake, as well as synchronized and desynchronized sleep, are characterized by different genomic expressions, the level of IEGs being high during wake and low during sleep. Such fluctuation of gene expression is not ubiquitous but occurs in certain cell populations in the brain. Thus, even taking into account the limitations discussed above, IEG induction may reveal the activation of neural networks in different behavioral states. Although stimulating, these findings leave unanswered a number of questions: Do the areas in which IEGs oscillate during sleep and wake subserve specific roles in the regulation of these physiological states and in a general "resetting" of behavioral state? Is gene induction a clue to the understanding of the alternation of sleep and wake, and of REM and non-REM sleep?

The inducibility of transcription factors indicates that external cues can modulate cell function through the regulation of gene expression. The variation of IEG expression during sleep and wake seems to indicate that this could also be true for internal cues. The high expression of IEGs during wake could be related to the animal's activity, to "a momentary excitation of single neurons in the course of transferring physiological 'everyday' information".[25] This assumption, however, does not explain why the phenomenon of "spontaneous" IEG induction is not ubiquitous, being instead restricted to cell populations in the brain: Would the neurons in which c-*fos* is not induced during wake represent those "habituated" to the ongoing stimuli? Would they represent neurons with a higher threshold for IEG induction? Which information would be processed in brain neurons through Fos expression during sleep? In view of the role played by transcription factors in neural plasticity, could behavioral state-related IEG induction underlie, at least in part, learning mechanisms?

The oscillation of IEGs affects the expression of target genes, and this brings about other questions: May the transcriptional cascade explain the biological need and the significance of sleep? Does this explain the molecular and cellular correlates of arousal, alertness, and, more in general, of consciousness?

We do not have final answers to these questions. However, the data seem to indicate that synchronized sleep "wipes out" in many brain structures the expression of some IEGs and "turns on" c-*fos* expression in other neurons: Would sleep represent a "refractory period" after IEG induction during wake? Could the alternation of IEG induction and down-regulation in selected brain subsets even represent the trigger for sleep and wake? These questions are, of course, provocative and extreme, but they highlight the complexity of an integrated signal transduction network which underlies the anatomical and functional heterogeneity of neuronal networks.

Note

The following recent article would also be of interest to readers: Basheer, R., Sherin, J.E., Saper, C.B., Morgan, J.I., McCarley, R.W., and Shiromani, P.J., Effects of sleep on wake-induced c-*fos* expression, *J. Neurosci.*, 17, 9746–9750, 1997. This study confirmed that c-*fos* induction in the cerebral cortex declines after increasing amounts of sleep.

References

1. Alcantara, A.A. and Greenough W.T., Developmental regulation of Fos and Fos-related antigens in cerebral cortex, striatum, hippocampus, and cerebellum of the rat, *J. Comp. Neurol.*, 334, 75–85, 1993.
2. Anderson, A.J., Cummings, B.J., and Cotman, C.W., Increased immunoreactivity for Jun- and Fos-related proteins in Alzheimer's disease: association with pathology, *Exp. Neurol.*, 125, 286–295, 1994.
3. Bandler, R. and Shipley, M.T., Columnar organization in the midbrain periaqueductal gray: modules for emotional expression?, *Trends Neurosci.*, 17, 379–389, 1994.
4. Bentivoglio, M. and Steriade, M., Brainstem-diencephalic circuits as a structural substrate of the ascending reticular activation concept, in *The Diencephalon and Sleep*, Mancia, M. and Marini, G., Eds., Raven Press, New York, 1990, pp. 7–29.
5. Bentivoglio, M., Chen, S., Peng, Z.-C., Bertini, G., Ringstedt, T., and Persson H., Thalamus, neurotrophins and their receptors, in *Thalamic Networks for Relay and Modulation*, Minciacchi, D., Molinari, M., Macchi, G. and Jones, E.G., Eds., Pergamon Press, New York, 1993, pp. 309–320.
6. Bentivoglio, M., Grassi-Zucconi, G., Peng, Z.-C., Bassetti, A., Edlund, C., and Kristensson, K., Trypanosomes cause dysregulation of c-*fos* expression in the rat suprachiasmatic nucleus, *NeuroReport*, 5, 712–714, 1994.
7. Bentivoglio, M., Kultas-Ilinsky, K., and Ilinsky I.A., The limbic thalamus: structure, intrinsic organization, and connections, in *Neurobiology of the Cingulate Cortex and Limbic Thalamus*, Vogt, B.A. and Gabriel, M., Eds., Birkhäuser, Boston, 1993, pp. 71–122.
8. Bentivoglio, M., Peng, Z.-C., Chen, S., Montagnese, P., Mandile, P., Vescia, S., and Grassi-Zucconi, G., Induction of c-Fos during spontaneous wakefulness in the forebrain and diencephalon: an immunocytochemical and *in situ* hybridization study in the rat, *Soc. Neurosci. Abstr.*, 18, 2, 1992.
9. Buzsaki, G. and Gage, F.H., The cholinergic nucleus basalis: a key structure in neocortical arousal, *EXS*, 57, 159–171, 1989.
10. Calamandrei, G. and Keverne, E.B., Differential expression of Fos protein in the brain of female mice dependent on pup sensory cues and maternal experience, *Behav. Neurosci.*, 108, 113–120, 1994.
11. Cirelli, C., Pompeiano, M, and Tononi, G., Fos-like immunoreactivity in the rat brain in spontaneous wakefulness and sleep, *Arch. Ital. Biol*, 131, 327–330, 1993.
12. Cirelli, C., Pompeiano, M, and Tononi G., Sleep deprivation and c-*fos* expression in the rat brain, *J. Sleep Res.*, 4, 92–106, 1995.
13. Cirelli, C., Pompeiano, M., and Tononi G., Neuronal gene expression in the waking state: a role for the locus coeruleus, *Science*, 274, 1211–1215, 1996.
14. Cirelli, C., Pompeiano, M., Arrighi, P., and Tononi, G., Sleep-waking changes after c-*fos* antisense injections in the medial preoptic area, *NeuroReport*, 6, 801–805, 1995.
15. Chambille, I., Doyle, S., and Servière, J., Photic induction and circadian expression of Fos-like protein. Immunohistochemical study in the retina and suprachiasmatic nuclei of hamster, *Brain Res.*, 612, 138–150, 1993.
16. Dragunow, M., Faull and R.L.M., The use of c-*fos* as a metabolic marker in neuronal pathway tracing, *J. Neurosci. Meth.*, 29, 261–265, 1989.
17. Dragunow, M., Peterson, M.R., and Robertson, H.A., Presence of c-*fos*-like immunoreactivity in the adult rat brain, *Eur. J. Pharmacol.*, 135, 113–114, 1987.
18. Dragunow, M., Goulding, M., Faull, R.L.M., Ralph, R., Mee, E., and Frith, R., Induction of c-*fos* mRNA and protein in neurons and glia after traumatic brain injury: pharmacological characterization. *Exp. Neurol.*, 107, 236–248, 1990.
19. Fonta, C., Gascuel, J., and Masson, C., Brain Fos-like expression in developing and adult honeybees, *NeuroReport* 6, 745–749, 1995.

20. Grassi-Zucconi, G., Giuditta, A., Mandile, P., Chen, S., Vescia, S., and Bentivoglio, M., c-*fos* spontaneous expression during wakefulness is reversed during sleep in neuronal subsets of the rat cortex, *J. Physiol. (Paris)*, 88, 91–93, 1994.

21. Grassi-Zucconi, G., Menegazzi, M., Carcereri, A., Bassetti, A., Montagnese, P., Cosi, C., and Bentivoglio, M., c-*fos* mRNA is spontaneously induced in the rat brain during the activity period of the circadian cycle, *Eur. J. Neurosci.*, 5, 1071–1078, 1993.

22. Grassi-Zucconi, G., Menegazzi, M., Carcereri, A., Vescia, S., and Bentivoglio, M., Different programs of gene expression are associated with different phases of the 24 hours and sleep-wake cycles, *Chronobiologia*, 21, 93–97, 1994.

23. Guido, M.E., Rusak, B., and Robertson, H.A., Spontaneous circadian and light-induced expression of *jun B* mRNA in the hamster suprachiasmatic nucleus, *Brain Res.*, 732, 215–222, 1996.

24. Hazel, T.G., Nathans D., and Lau, L.F., A gene inducible by serum growth factors encodes a member of the steroid and thyroid hormone receptor superfamily, *Proc. Natl. Acad. Sci. USA*, 85, 8444–8448, 1988.

25. Herdegen, T., Kovary, K., Buhl, A., Bravo, R., Zimmermann M., and Gass, P., Basal expression of the inducible transcription factors c-Jun, Jun B, Jun D, c-Fos, Fos B, and Krox-24 in the adult rat brain, *J. Comp. Neurol.*, 354, 39–56, 1995.

26. Herdegen, T., Sandkuhler, J., Gass, P., Kiessling, M., Bravo, R., and Zimmermann, M., Jun, Fos, Krox, and CREB transcription factor proteins in the rat cortex: basal expression and induction by spreading depression and epileptic seizures, *J. Comp. Neurol.*, 333, 271–288, 1993.

27. Herrera, D.G. and Robertson, H.A., Activation of c-*fos* in the brain, *Progr. Neurobiol.*, 50, 83–107, 1996.

28. Hughes, P., Lawlor, P., and Dragunow, M., Basal expression of Fos, Fos-related, Jun, and Krox 24 proteins in rat hippocampus, *Brain Res.*, 13, 355–357, 1992.

29. Kaczmarek, L. and Chaudhuri, A., Sensory regulation of immediate-early gene expression in mammalian visual cortex: implications for functional mapping and neural plasticity, *Brain Res. Rev.*, 23, 237–256, 1997.

30. Kaplan, I.V., Guo, Y., and Mower, G.D., Immediate early gene expression in cat visual cortex during and after the critical period: differences between EGR-1 and Fos proteins, *Mol. Brain Res.*, 36, 12–22, 1996.

31. Karin, M. and Smeal, T., Control of transcription factors by signal transduction pathways: the beginning of the end, *Trends Biochem.* 17, 418–422, 1992.

32. Klein, D.C., Moore R.Y., and Reppert, S.M., Eds., *Suprachiasmatic Nucleus: The Mind's Clock*, Oxford University Press, New York, 1991.

33. Kononen, J., Koistinaho J., and Alho, H., Circadian rhythm in c-*fos*-like immunoreactivity in rat brain, *Neurosci. Lett.*, 120, 105–108, 1990.

34. Luckman, S.M., Fos expression within regions of the preoptic area, hypothalamus and brainstem during pregnancy and parturition, *Brain Res.*, 669, 115–124, 1995.

35. Marx, J.L., The *fos* gene as "master switch", *Science*, 237, 854–856, 1987.

36. Menegazzi, A., Carcereri de Prati, A., and Grassi-Zucconi, G., Differential expression pattern of *jun*B and c-*jun* in the rat brain during the 24-h cycle, *Neurosci. Lett.*, 182, 295–298, 1994

37. Merchant-Nancy, H., Vazquez, J., Aguilar-Roblero, R., and Drucker-Colin, R., c-*fos* proto-oncogene changes in relation to REM sleep duration, *Brain Res.*, 579, 342–346, 1992.

38. Merchant-Nancy, H., Vazquez, J., Garcia, F., and Drucker-Colin, R., Brain distribution of c-*fos* expression as a result of prolonged rapid eye movement (REM) sleep period duration, *Brain Res.*, 681, 15–22, 1995.

39. Milbrandt, J., A nerve growth factor-induced gene encodes a possible transcriptional regulatory factor, *Science*, 238, 797–799, 1987.

40. Milbrandt, J., Nerve growth factor induces a gene homologous to the glucocorticoid receptor gene, *Neuron*, 1, 183–188, 1988.
41. Montero, V. M., c-*fos* induction in sensory pathways of rats exploring a novel complex environment: shifts of active thalamic reticular sectors by predominant sensory cues, *Neuroscience*, 76, 1069–1081, 1997.
42. Morgan, J.I. and Curran, T., Stimulus-transcription coupling in the nervous system: involvement of the inducible proto-oncogenes *fos* and *jun*, *Ann. Rev. Neurosci.*, 14, 421–451, 1991.
43. Morgan, J.I. and Curran, T., Immediate-early genes: ten years on, *Trends. Neurosci.* 18, 66–67, 1995.
44. Morgan, J.I., Cohen, D.R., Hempstead, J.L., and Curran, T., Mapping patterns of c-*fos* expression in the central nervous system after seizure, *Science*, 237, 192–197, 1987.
45. Murphy, T.H., Worley, P.F., and Baraban, J.M., L-type voltage-sensitive calcium channels mediate synaptic activation of immediate early genes, *Neuron*, 7, 625–635, 1991.
46. O'Hara, B.F., Young, K.A., Watson, F.L., and Heller, H.C., Kilduff, T.S, Immediate early gene expression in brain during sleep deprivation: preliminary observations, *Sleep*, 16, 1–17, 1993.
46a. O'Hara, B.F., Watson, F.L., Andretic, R., Wiler, S.W., Young, K.A., Bitting, L., Heller, H.C., and Kilduff, T.S., Daily variation of CNS gene expression in nocturnal versus diurnal rodents and in the developing rat brain, *Mol. Brain Res.*, 48, 73–86, 1997.
47. Peng, Z.-C., Chen S., and Bentivoglio M., A sensitive double immunostaining protocol for Fos-immunoreactive neurons, *Brain Res. Bull.*, 36, 101–105, 1994.
48. Peng, Z.-C., Grassi-Zucconi, G., and Bentivoglio, M., Fos-related protein expression in the midline paraventricular nucleus of the rat thalamus: basal oscillation and relationship with limbic efferents, *Exp. Brain Res.*, 104, 21–29, 1995.
49. Pompeiano, M., Cirelli, C., and Tononi, G., Effects of sleep deprivation on Fos-like immunoreactivity in the rat brain, *Arch. Ital. Biol.*, 130, 325–335, 1992.
50. Pompeiano, M., Cirelli, C., and Tononi, G., Immediate-early genes in spontaneous wakefulness and sleep: expression of c-*fos* and NGFI-A mRNA and protein, *J. Sleep Res.* 3, 80–96, 1994.
51. Pompeiano, M., Cirelli, C., Ronca-Testoni, S., and Tononi, G., NGFI-A expression in the rat brain after sleep deprivation, *Mol. Brain Res.*, 46, 143–153, 1997.
52. Sallaz, M. and Jourdan, F., c-*fos* expression and 2-deoxyglucose uptake in the olfactory bulb of odour-stimulated awake rats, *NeuroReport*, 4, 55–58, 1993.
53. Schlingenspien, K.-H., Luno, K., and Brysch, W., High basal expression of the zif/268 immediate early gene in cortical layers IV and VI, in CA1 and in the corpus striatum — an *in situ* hybridization study, *Neurosci. Lett.* 122, 67–70, 1991.
54. Schutte, J., Viallet, J., Nau, M., Segal, S., Fedorko, J., and Minna, J., *jun*B inhibits and c-*fos* stimulates the transforming and trans-activating activities of c-*jun*, *Cell*, 59, 987–997, 1989.
55. Shiromani, P.J., Kilduff, T.S., Bloom, F.E., and McCarley, R.W., Cholinergically induced REM sleep triggers Fos-like immunoreactivity in dorsolateral pontine regions associated with REM sleep, *Brain Res.*, 580, 351–357, 1992.
56. Shiromani, P.J., Malik, M., Winston, S., and McCarley, R.W., Time course of Fos-like immunoreactivity associated with cholinergically induced REM sleep, *J. Neurosci.*, 15, 3500–3508, 1995.
57. Sherin, J.E., Shiromani, P.J., McCarley, R.W., and Saper, C.B., Activation of ventrolateral preoptic neurons during sleep, *Science*, 271, 216–219, 1996.
58. Smeyne, R.J., Schilling, K., Robertson, L., Luk, D., Oberdick, J., Curran, T., and Morgan, J.L., Fos-lacZ transgenic mice: mapping sites of gene induction in the central nervous system, *Neuron*, 8, 13–23, 1992.
59. Stein, B.E. and Meredith, M.A., *The Merging of the Senses,* MIT Press, Cambridge, MA, 1993.

60. Steriade, M. and McCarley, R., *Brainstem Control of Wakefulness and Sleep*, Plenum Press, New York, 1990.
61. Sutin, E.L. and Kilduff, T.S., Circadian and light-induced expression of immediate-early gene mRNAs in the rat suprachiasmatic nucleus, *Mol. Brain Res.*, 15, 281–290, 1992.
62. Tononi, G., Pompeiano, M., and Cirelli, C., The locus coeruleus and immediate-early genes in spontaneous and forced wakefulness, *Brain Res. Bull.*, 15, 589–596, 1994.
63. Valverde-Navarro, A.A., Olucha F.E., Garcia-Verdugo, J.M., Hernandez-Gil, T., Ruiz-Torner, A., and Martinez-Soriano, F., Distribution of basal-expressed c-*fos*-like immunoreactive cells of the periaqueductal grey matter of the rat, *NeuroReport*, 7, 2749–2752, 1996.
64. Yamuy, J., Mancillas J.R., Morales F.R., and Chase, M.H., c-*fos* expression in the pons and medulla of the cat during carbachol-induced active sleep, *J. Neurosci.*, 13, 2703–2718, 1993.
65. Zhang, P., Hirsch, E.C., Damier, P., and Duyckaerts, C., and Javoy-Agid, F., c-*fos* protein-like immunoreactivity: distribution in the human brain and overexpression in the hippocampus of patients with Alzheimer's disease, *Neuroscience*, 46, 9–21, 1992.

Section IV

Cellular and Network Mechanisms of Behavioral State Control

Robert W. Greene, Section Editor

Chapter 16

Synaptic and Intrinsic Membrane Properties Regulating Noradrenergic and Serotonergic Neurons During Sleep/Wake Cycles

John T. Williams

Contents

16.1	Introduction		258
16.2	Locus Coeruleus		258
	16.2.1	Structure	258
	16.2.2	Intrinsic Membrane Properties	259
		16.2.2.1 Action Potentials	259
		16.2.2.2 Afterhyperpolarizations	259
		16.2.2.3 Voltage Clamp Experiments	260
	16.2.3	Synaptic Potentials	262
	16.2.4	Electrotonic Coupling	263
	16.2.5	Activity of LC Cells and Sleep	264
16.3	Raphe Nuclei		265
	16.3.1	Structure	265
	16.3.2	Intrinisic Membrane Properties	265
		16.3.2.1 Action Potentials — Afterhyperpolarizations	265
		16.3.2.2 Voltage Clamp Experiments	266
	16.2.3	Synaptic Potentials	267
		16.2.3.1 Glutamate	267
		16.2.3.2 GABA	268
		16.2.3.3 5-HT	268
		16.2.3.4 Noradrenaline	269
	16.2.4	Activity of 5-HT Cells and Sleep	270
16.4	Summary and Conclusions		271
References			272

16.1 Introduction

From studies using extracellular recording in behaving animals, it has been known for some time that the activity of monoamine cells of the locus coeruleus (LC) and dorsal raphe (DR) vary with the waking state of the animal.[1-3] During the sleep/wake cycle, the activity patterns of neurons in the locus coeruleus and dorsal raphe seem to be similarly regulated. In each case, cells fire at a slow and regular rate during slow-wave sleep or quiet resting, are silent during REM sleep, and are most active and erratic during active waking, particularly with the presentation of novel stimuli. The similarities between the firing of neurons in the locus coeruleus and dorsal raphe are also remarkable in that the overall firing rate is very low relative to many areas of the central nervous system (CNS). Since the neurons in both these nuclei release transmitter, noradrenaline, and 5-HT, over wide areas the the central nervous system, their activity is thought to have a global influence on CNS function. The intrinsic membrane properties of, and afferent synaptic input onto neurons in the locus coeruleus and dorsal raphe that shape the firing patterns of these neurons will be described based on work in isolated cells and brain slice preparations. From these fundamental properties the regulation of neuronal firing in the LC and DR during the sleep/wake cycle will be discussed.

16.2 Locus Coeruleus

16.2.1 Structure

The locus coeruleus in rat is made up of approximately 1400 neurons on each side located just lateral to the edge of the fourth ventricle.[4-6] In the rat, as in primates, this nucleus is a densely packed area containing a homogeneous population of noradrenergic neurons. This group of cells is responsible for about 50% of the noradrenaline content in the CNS. One of the distinctive aspects of this small nucleus is the enormous area of the brain over which it sends projections, both ascending and descending. As a result, the functional consequences of the activity of cells in the locus coeruleus have been linked with generalized behaviors such as the sleep/waking cycle, vigilance, and attentiveness to novel stimuli.[4-6]

Although the projections of the locus coeruleus have been known since the advent of formaldehyde-induced fluorescence for catecholamines in the 1970s, knowledge of afferent input to the LC continues to evolve. Early tract tracing studies suggested that neurons of the LC received afferent input from widespread central areas.[7] With the development of tracers where diffusion from the site of injection was more limited, afferents to the cell body region were found to be restricted to two major inputs, the paragigantocellularis and prepositus hypoglossi, and a smattering of minor inputs.[8,9] It has recently been realized that the dendritic arbor of LC neurons extends well beyond the densely packed cell body region of the nucleus in both the caudal and rostro-medial directions.[10,11] Afferents to these pericoeruleur regions have to be examined, and the number of projections onto LC cells has to be reassessed.

In spite of relatively few projections into the cell body region of the LC, numerous neurotransmitters, modulators, and peptides have been found in the cell body regions using immunohistochemical methods (see Reference 1 for review). Exogenous application of these agents has been used to determine potential physiological actions using both *in vivo* and *in vitro* preparations (Table 16.1). Many of these agents have powerful effects on the firing rate of LC cells. Unfortunately, however, almost nothing is known about the endogenous transmitters or the circumstances under which these compounds can be released synaptically.

TABLE 16.1
Transmitter Candidates in Locus Coeruleus Active When Applied Exogenously

Compound	Effect on Excitability	Preparation	Ref.
ACh nicotinic	Increase	Slice	12
ACh muscarinic	Increase	Slice	13
Adenosine	Decrease	Slice	14–17
Angiotensin	Decrease	Slice	18
ATP	Increase	Slice	19
CRF	Increase	In vivo	20
$GABA_A$	Decrease	Slice	21
$GABA_B$	Decrease	Slice	22
Galanin	Decrease	Slice	23
Glycine	Decrease	Slice	24
Glutamate	Increase	Slice	25
Enkephalin	Decrease	In vivo/slice	26–29
NPY	Decrease	Slice	30
Noradrenaline	Decrease	Slice	31
OFQ	Decrease	Slice	32
5-HT	Decrease	In vivo/slice	33, 34
Somatostatin	Decrease	Slice/culture	35, 36
Substance P	Increase	In vivo/slice/culture	37–39
Vasopressin	Increase	In vivo	40, reviewed in 41
VIP	Increase	Slice	42

16.2.2 Intrinsic Membrane Properties

16.2.2.1 Action Potentials

The properties of LC cells has been most extensively examined in brain slice preparations where they invariably exhibit spontaneous activity that ranges from 0.2 to 3 Hz.[43] The threshold for action potential generation is –55 mV, the peak is approximately +30 mV, and the afterhyperpolarization following a single action potential reaches about –75 mV. Both sodium and calcium currents are responsible for the depolarizing phase of the action potential, and spontaneous activity continues in the form of calcium spikes after blocking sodium currents with tetrodotoxin. Following a single action potential is a rapid afterhyperpolarization that decays and brings the membrane potential back towards the threshold for the next action potential (Figure 16.1).

16.2.2.2 Afterhyperpolarizations

The repolarization of the action potential that continues into the afterhyperpolarization is made up of several potassium conductances as determined from the use of potassium channel blockers. For example 4-aminopyridine (500 μM) caused a small increase in the amplitude and doubled the duration of the action potential, suggesting that IK-A plays a role in the repolarization.[44] In addition, the rate of depolarization leading to an action potential is highly dependent on the potential from

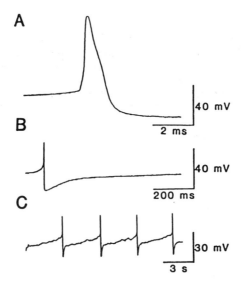

FIGURE 16.1
Waveform of the action potential, afterhyperpolarization, and slow depolarization between spontaneous action potentials in a typical locus coeruleus neuron — an intracellular recording from a locus coeruleus neuron in a coronal rat brain slice. Shown are the different components of activity: **(A)** Action potential with two components of the repolarization. **(B)** Afterhyperpolarization following a single action potential. **(C)** Slow depolarization leading to the generation of an action potential. (From Williams, J.T. et al., *Neuroscience*, 13, 137, 1984. With permission.)

which a depolarizing step in made.[43] Finally voltage clamp experiments from neurons in both slices and culture indicate the presence of a large transient potassium conductance that is activated upon depolarization from negative potentials.[43,45] It therefore appears that an IK-A-like current can limit action potential frequency and duration.

An early component of the afterhyperpolarization following a single action potential was selectively blocked by apamin, a toxin that blocks one of the small conductance calcium-activated potassium channels.[46] Following a burst of action potentials, two hyperpolarizing responses were identified, both of which required calcium entry.[46,47] The early component was apamin sensitive, and the late one was blocked by dantrolene and ryanodine. Thus, calcium entry during the action potential is an important factor in determining the rate of depolarization leading to the next action potential.

16.2.2.3 Voltage Clamp Experiments

The most dramatic component of the current/voltage plot obtained with voltage clamp experiments was the inward rectification at negative potentials (Figure 16.2).[29,43,48] Under normal recording conditions (2.5 m*M* K) the conductance of LC cells measured between –60 and –90 mV was roughly 5 nS (200 MΩ), whereas at potentials more negative than the potassium equilibrium potential (–110 mV) the conductance increased to about 15 nS. As expected, increasing the extracellular potassium increased the conductance throughout the voltage range tested and shifted the potassium equilibrium potential to less negative values.[29,48] Thus, the resting conductance was highly dependent on the extracellular potassium content. This inwardly rectifying conductance appeared to dominate the conductance of LC cells, and the reduced conductance at less negative potential can account for the high input resistance observed in these cells in spite of their very large size (25 to 50 μm diameter). The high input resistance of these cells facilitates changes in membrane potential caused by relatively small currents.

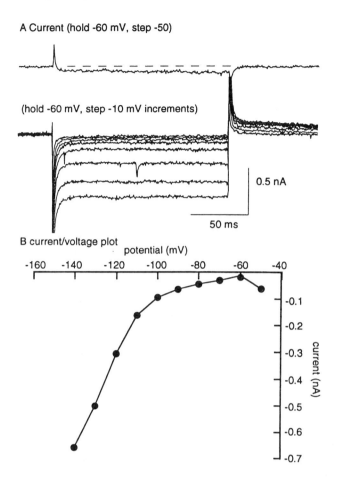

FIGURE 16.2
Inward rectification at negative potentials and a persistant inward current at −50 mV are the two components of the steady-state current voltage plot. (**A**) Top: Current trace shows the persistant inward current observed when the potential was stepped to −50 mV. Below: Superimposed current traces illustrating the currents evoked when the membrane potential was stepped from −60 mV to −140 mV in −10-mV increments. (**B**) Current/voltage plot resulting from the traces shown in (A). This is a whole cell recording from a LC neuron in a coronally sectioned slice.

The next obvious component of the current/voltage plot is the persistent inward current that was observed between −60 and −50 mV (Figure 16.2).[29,43] This inward current is presumably critically important for the regulation of spontaneous activity. This suggestion is supported by the voltage range over which it activates relative to the threshold for action potential generation. This persistent current was completely blocked by tetrodotoxin, indicating that voltage-dependent sodium channels were responsible.

A pacemaker current regulated by a cAMP-dependent mechanism has been suggested to play a role in the firing rate of LC cells.[49] Whole cell electrodes containing cAMP or superfusion of cAMP analogs were found to increase firing, whereas depletion of cAMP from the cell with the use of whole cell electrodes decreased the firing rate of LC cells in slices. The significance of this current has yet to be determined, primarily because difficulties in its isolation have limited its full characterization. Forskolin and cAMP analogs have been shown to increase the frequency of spontaneous excitatory postsynaptic potentials which would indirectly increase excitability.[50] In addition, cAMP and cAMP analogs that do not activate protein kinase A decreased the amplitude and duration of the afterhyperpolarization of LC cells, an action that would also indirectly increase excitability.[51]

Finally, forskolin in concentrations above 10 μM caused an inhibition rather than excitation, suggesting that the role of the cAMP cascade on spontaneous firing in the LC is complex.[50]

Calcium currents in LC neurons were indicated by the presence of slowly rising action potentials after treatment with tetrodotoxin.[43] These action potentials were blocked by removing the calcium from the extracellular solution or by the addition of calcium channel blockers to the extracellular solution. The amplitude and duration of the calcium-dependent action potentials were augmented with the addition of agents that block potassium conductance, such as TEA, 4-AP, and barium. With the use of acutely isolated cells, calcium currents have been measured directly.[52] All of the current found in LC cells was of the high-threshold subtype, that is there was no evidence for a low-threshold (T) type of current. The threshold for the calcium current was about –40 mV, and the peak current was observed at about 0 mV. In addition, based on pharmacological studies, there are at least L and N subtypes of high-threshold current.[53]

The membrane properties of LC cells can be summarized by considering the balance of inward and outward currents. A persistent, tetrodotoxin-sensitive, inward sodium current appears to drive action potentials that are made up inward sodium and calcium currents. The duration of the action potential is determined by a voltage-dependent potassium currents, some of which are sensitive to 4-AP and others that are blocked by TEA. Calcium entry during the action potential activates at least two potassium conductances that mediate a large (20 mV) and long-lasting (10 sec) afterhyperpolarization that limits subsequent action potential generation. Thus, there are several conductances present in LC cells that limit the frequency of action potentials, and it comes as no surprise that the frequency of LC firing measured *in vivo* is considerably lower than that observed in many other central neurons.

16.2.3 Synaptic Potentials

Electrical stimulation in the area of the locus coeruleus in brain slice preparations evokes a series of synaptic potentials. There are three short-latency synaptic responses mediated by glutamate, GABA, and glycine, followed by a slow inhibitory postsynaptic potential mediated by noradrenaline.[24,25]

Fast transmission mediated by glutamate and GABA is ubiquitous in the CNS to regulate moment-to-moment activity of most neurons. In LC cells, as elsewhere, glutamate-mediated excitatory synaptic potentials are made up of both AMPA ([RS]-a-amino-3-hydroxy-5-methyl-4-isoxazolepropionic acid) and NMDA (*N*-methyl-D-aspartate) components.[24,25] The paragigantocellularis (PGi) is at least one site that sends glutamate a projection to the locus coeruleus. Electrical stimulation in the PGi or the sciatic nerve or activation of a variety of sensory stimuli results in a short-latency excitation that is blocked by local injection of glutamate receptor antagonists into the LC.[54,55] There are two important aspects of these excitations. First is that the excitation is short lived, and any single cell fires no more than 2 to 4 action potentials following a single stimulus. Second, it appears that many if not all LC cells respond to a given stimulus.

There are at least two mechanisms that limit the firing rate of LC cells, including the afterhyperpolarization following a burst of action potentials and noradrenergic synaptic inhibition arising from projections from both the PGi and dendritic release from LC cells themselves.[31,56-59] The large afterhyperpolarization following a single action potential not only brings the membrane potential to roughly 20 mV hyperpolarized from threshold, but it also removes the inactivation of the transient potassium current (IK-A). A rapid depolarization during the afterhyperpolarization would activate an opposing outward potassium current to slow the depolarization toward threshold. Thus, this would be an effective mechanism to limit the maximum rate of firing and/or mediate an early inhibition following a strong excitatory stimulus. In fact, LC cells do not show burst activity as cells in many areas of the central nervous system do. A more prolonged inhibition is thought to be mediated by the synaptic activation of α_2-adrenoceptors.[31,56-59] In slice recordings, electrical

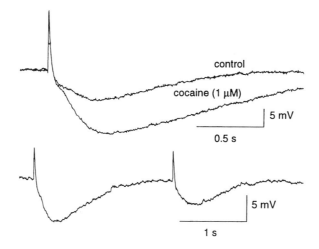

FIGURE 16.3
Postsynaptic potentials in an LC neuron. (**A**) Two superimposed potential traces. Electrical stimulation in the area of the LC was used to evoke transmitter release. A short-latency, rapid depolarization followed the stimulus. This potential was made up of glutamate (excitatory) and $GABA_A$ (inhibitiory) postsynaptic potentials. A slow hyperpolarizing synaptic potential followed the short-latency synaptic potentials. Cocaine selectively augmented and prolonged the slow hyperpolarizing synaptic potential. (**B**) Two stimuli applied at an interval of 2 sec evoked inhibitory postsynaptic potentials that showed marked paired pulse depression. This depression is thought to result from presynaptic inhibition mediated by α_2-adrenoceptors.

stimulation within the cell body region of the LC evoked an inhibitory postsynaptic potential (IPSP) that resulted from an increase in potassium conductance (Figure 16.3).[24,31,58] There was a latency of about 50 msec between the stimulus and the onset of the IPSP, the peak occurred after about 400 msec, and the total duration was about 2 sec. This IPSP was blocked by α_2-adrenoceptor antagonists. The inhibition caused by activation of α_2-adrenoceptor antagonists was sensitive to pertussis toxin, indicating a g-protein-linked activation of the underlying inwardly rectifying potassium conductance.[29,60] An inhibition with a similar time course and sensitivity to α_2-antagonists was also observed *in vivo* following electrical stimulation in the PGi.[59] Inhibition mediated by α_2-receptors is therefore slow relative to the excitatory synaptic input as well as the inhibition caused by the afterhyperpolarization following action potentials and can be mediated by afferent input as well as local transmission.

In summary, it appears that (1) excitatory afferent input is highly convergent before projecting to the LC, (2) the small response observed in individual cells is amplified in the sense that each stimulus affects a large population of cells, and (3) following a strong excitation there is an inhibitory period that results from both intrinsic membrane properties (afterhyperpolarization) and slow synaptic inhibition (dendritic release of noradrenaline). The concept that LC cells respond to both excitatory and inhibitory inputs as a group may be central to the understanding of the function of this nucleus.

16.2.4 Electrotonic Coupling

Slow synchronous oscillations in membrane potential have been observed in neurons throughout the locus coeruleus slice preparations from neonatal and often adult animals.[11,38,42,61–65] This synchronous activity was not dependent on synaptic transmission but was regulated by the excitability of locus coeruleus neurons as a group.[11,62] Superfusion with agents that caused hyperpolarization of LC neurons, such as opioids, slowed and then completely blocked synchronous activity.

The synchronous activity was thought to arise from electrotonic coupling between spontaneously active cells. The evidence for electrotonic coupling in neonates is strong in that both dye and electrical coupling between cells has been observed.[62,63] In addition, the glycyrrhetinic acid derivative, carbenoxolone, disrupted both the electrical coupling and the synchronous activity.[11,65] In slices from adult animals, synchronous activity has been reported occasionally but was reliably observed under conditions where the action potential was prolonged and the input resistance was increased, such as with the combination of TEA and/or Ba.[11,43,65] As was found in neonatal slices, carbenoxolone blocked this activity. One other potentially important observation was that the synchronous activity was not present after sectioning the brain slice rostral and caudal to the cell LC cell body region.[11] The disruption of synchrony by this procedure was taken to indicate that the coupling between LC cells was between dendrites in the pericoeruleur region outside the cell body region of the nucleus. This observation further supported the previous suggestion that the extended dendritic arbor was important for the inhibition caused by opioids, the excitation by glutamate, and the interaction between opioid and glutamate agonists.[66-68]

The significance, site, and role of electrotonic coupling in regulating central noradrenergic tone remain to be determined. What is clear, however, is that the neurons over large regions of the nucleus receive excitatory synaptic input and are thus brought to threshold at roughly at the same time. Electrotonic coupling between neurons may help regulate activity of cells in the nucleus to facilitate this synchronous activity and thus release noradrenaline in widespread areas of the CNS. The activity of a single neuron does not strongly influence other cells in the nucleus, so that in the resting state output from the nucleus would be dependent on spontaneous activity of individual cells. With the activation of a population of neurons, however, electrotonic coupling would serve to recruit others and enhance the overall output from the nucleus. The synchronous firing of LC neurons may be an efficient mechanism to augment noradrenergic tone. Such an increase is thought to be an important component of the alerting response that follows a novel or particularly salient stimulus to the animal.[69,70]

16.2.5 Activity of LC Cells and Sleep

Although the LC has been proposed to play a principal role in the various stages of the sleep/waking cycle, it now appears that control of LC cell firing is determined by intrinsic properties during only certain parts of the cycle. That is, the firing rate and pattern during slow-wave sleep and certain waking behaviors such as grooming or feeding are most like those observed in a brain slice preparation. This activity is primarily driven by the membrane properties as determined from single cell recordings in the brain slice. Presumably the firing of individual cells under these conditions is randomly distributed throughout the nucleus. During REM sleep, LC cells were generally silent, indicating an active inhibition during this period. The source and mechanism of this inhibition have been suggested to be GABA-mediated, as determined from dialysis experiments done in freely behaving cats.[71] Neurons of the prepositus hypoglossi (GABA- and enkephalin-containing) and PGi (adrenaline-containing) are all potential candidates for inhibition during this period. The firing of LC cells is the highest during waking periods. A broad range of sensory stimuli, including auditory, visual, somatosensory, and olfactory, evoke short-latency transient bursts of activity that are followed by a silent period.[1] Interestingly, the burst of activity is largest when the stimulus is applied in the transition between slow-wave sleep and waking, or under circumstances where the stimulus is particularly meaningful and results in an orienting response of the animals toward the stimulus.[69,70] The increased activity found in the LC also correlates with the onset of cortical desynchronization, further suggesting a role of the LC in cortical activation.[72] Thus, the activity in the LC has been hypothesized to play a role early in the alerting response to particularly meaningful stimuli. Based on this hypothesis, one would predict that the central noradrenaline tone would be low during REM and slow-wave sleep. During the transition between sleep and waking, sensory stimuli increase noradrenergic tone to aid in cortical desynchronization to foster the transition into waking. It therefore appears that the LC may be a component of the sleep/wake cycle.

16.3 Raphe Nuclei

16.3.1 Structure

The 5-HT-containing cells in the brain are found in a series of midline nuclei that have been numbered starting in the caudal ventral medulla (B1–B3) and then moving rostrally and dorsally into the midbrain (B6–9).[73] These subnuclei contain a heterogeneous population of neurons, and the anatomical definition of the borders of these nuclei is not distinct. The projections from the raphe nuclei are roughly arranged by the location in the brainstem; that is, neurons in the caudal nuclei project caudally and rostral projections arise from the more rostral subgroups. Each group also receives collateral innervation by other 5-HT-containing nuclei. Within each group, there are at least two neurochemically defined cell types. The primary projection cell contains 5-HT, and a second group thought to be interneurons contains GABA.

The dorsal raphe (B6/7) is the largest of the raphe nuclei, accounts for about 50% of the 5-HT found in the forebrain, and is the best characterized at a cellular and synaptic level. The lateral habenula, hypothalamus, pontine reticular formation, and other raphe nuclei are among the afferent projection areas to the dorsal raphe.[73] A series of synaptic potentials has been observed on 5-HT-containing cells within the dorsal raphe, and they are mediated by glutamate, GABA, 5-HT, and noradrenaline.[74-79] The afferent input, particularly the noradrenergic, appear to be required for slow regular firing of 5-HT neurons *in vivo*.[80] In addition, with the use of glutamate and GABA antagonists, the regulation of state-dependent activity of neurons in the dorsal raphe has been defined in behaving animals during the sleep/wake cycle.[81]

16.3.2 Intrinisic Membrane Properties

There are at least two electrophysiologically distinct groups of neurons in the raphe: primary cells, which have been identified as 5-HT containing, and interneurons, which have been presumed to be GABA containing.[82,83] Most studies have focused on the properties and synaptic regulation of primary cells. Secondary cells have been distinguished based on the action potential waveform, rate of spontaneous activity, and pharmacological experiments and will not be considered further here.

16.3.2.1 Action Potentials — Afterhyperpolarizations

The waveform of the action potential has been observed *in vivo* in brain slices and acutely dissociated neurons.[84,85] In anesthetized animals, the firing pattern of dorsal raphe was slow and regular, whereas cells were generally not spontaneously active in brain slice experiments. Based on extracellular recordings, the duration was estimated to be about 2 msec, which was very close to the duration observed in brain slice experiments using intracellular recordings. The action potential was between 70 and 80 mV in amplitude, and there were two components of the repolarizing phase similar to the the action potential observed in the locus coeruleus (Figure 16.4). When the membrane potential was depolarized from a potential more negative than –70 mV, repetitive, low-threshold, calcium-dependent depolarizations (low-threshold spikes) were often observed. The amplitude of low-threshold spikes declined over several seconds until a steady membrane potential was reached.

Using acutely dissociated neurons, Pennington et al.[85] demonstrated that the action potential consisted of both sodium and calcium components. Following the action potential, there was a

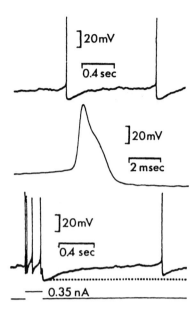

FIGURE 16.4
Waveform of the afterhyperpolarization, action potentials, and slow afterhyperpolarization following a burst of action potentials — an intracellular recording from a 5-HT-containing neuron in the dorsal raphe in a rat brain slice. Top trace shows the afterhyperpolarization following a single action potential. Middle trace shows an action potential with two components of the repolarization. Bottom trace shows the slow afterhyperpolarization resulting from a burst of action potentials that leads to the next action potential. (From Vandermaelen, C.P. and Aghajanian, G.K., *Brain Res.*, 289, 109, 1983. With permission.)

prominent afterhyperpolarization that was often separated into early and late components.[77,86] The early component was about 10 mV hyperpolarized from threshold, peaked within 10 to 30 msec after action potential, and decayed within 100 msec (Figure 16.5A). The second component was more variable in amplitude, peaked between 50 and 100 msec, and decayed over a period of 0.5 to 1 sec. The late afterhyperpolarization was blocked completely by apamin, indicating that a calcium-dependent potassium conductance was the primary current involved.[77,86] The modulation of this component of the afterhyperpolarization seems to be an important factor in regulating the firing rate and pattern of these neurons (see below).

16.3.2.2 Voltage Clamp Experiments

Voltage clamp experiments indicate that primary cells have several conductances that will play important roles in regulating activity.[87,88] Two conductances are found at potentials hyperpolarized from rest (−65 mV), Ih and IK-ir.[87] The amplitude of each of these conductances was determined by the sequential block of IK-ir with Ba and Ih with Cs (Figure 16.6). Although these conductances varied considerably from cell to cell, the overall conductance (4.5 ± 0.3 nS, n = 38) of these large cells was lower than many cells of comparable size, including LC neurons (8.3 ± 0.5 nS, n = 38). The steady-state current/voltage did not show any persistent inward current in the potential range near threshold (Figure 16.6). When voltage steps were applied from a negative holding potential, two transient currents were observed: one outward (IK-A) and a second inward (ICa-T). A complex current waveform was often observed when the membrane potential was stepped into a range where these currents activated, primarily because these two currents have striking similarities in voltage dependence and time-course.[89] In some cells, one current seemed to be more effective in the regulation of firing. For example, in cells where the calcium current dominated, low-threshold calcium spikes associated with a burst of sodium-dependent action potentials were evoked upon

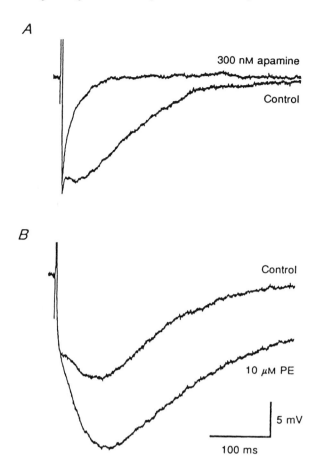

FIGURE 16.5
The slow afterhyperpolarization following a single action potential was mediated by apamine-sensitive potassium conductance and was augmented in amplitude and duration by activation of α_1-adrenoceptors. (**A**) Two superimposed voltage traces illustrating the afterhyperpolarization (control) and the selective inhibition of the late component of the afterhyperpolarization by apamine (300 n*M*). The full action potential (not shown) was evoked with a 2-msec depolarizing pulse applied through the recording electrode. (**B**) Superimposed voltage traces from another cell showing the increase in the afterhyperpolarization caused by phenylephrine (PE, 10 μ*M*). Both recordings were made with intracellular electrodes. (From Pan, Z.Z. et al., *J. Physiol.*, 478, 431, 1994.

depolarization under current clamp, whereas the excitability of cells was reduced in cells that had large transient potassium currents. In most cells, however, both currents were present such that transmitter-mediated control of these two currents would have an obvious role in determining the activity of these cells (see below).

16.2.3 Synaptic Potentials

16.2.3.1 Glutamate

Electrical stimulation in the area of the dorsal raphe in brain slice preparations evokes a series of synaptic potentials. There are two short-latency synaptic responses mediated by glutamate and GABA, followed by a slow inhibitory postsynaptic potential mediated by 5-HT that preceded an even slower excitatory postsynaptic potential caused by noradrenaline. Bursts of one to four action

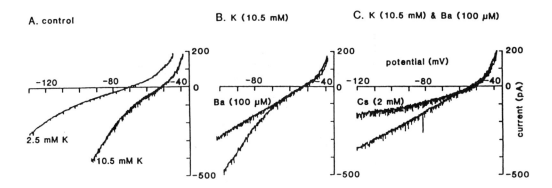

FIGURE 16.6
Currents that make up the steady-state properties of dorsal raphe neurons — a voltage clamp experiment using the switch-clamp method with an intracelluIar electrode. Slow voltate ramps were applied to observe the steady-state current voltage characteristics. (**A**) Two superimposed traces showing the increase in conductance that occurred when the extracellular solution was increased from control (2.5 mM) to 10.5 mM. The zero-current level shifted from about –70 mV to about –50 mV, and more rectification was observed at negative potentials. (**B**) Addition of BaCl$_2$ (100 μM), in the continued presence of 10.5 mM K, decreased the conductance at potentials more negative than –70 mV. This decrease in conductance resulted from the selective block of IK-ir. (**C**) Addition of CsCl (2 mM) to the solution containing both high potassium and BaCl$_2$ caused a further reduction in conductance at potentials more negative than –65 mV. This effect resulted from the blockade of Ih. (From Williams, J.T. et al., *J. Neurosci.*, 8, 3499, 1988. With permission.)

potentials that disrupt the regular activity occur upon sensory stimulation and are probably mediated by the release of glutamate.[81] In brain slice experiments, electrical stimulation evoked both NMDA- and non-NMDA-mediated excitatory postsynaptic potentials.[74] The glutamate-releasing terminals are thought to arise from neurons in areas including the habenula, hypothalamus, and pontine reticular formation.[73] In microcultures, glutamate-mediated excitatory postsynaptic potentials were evoked by stimulation of identified 5-HT cells.[90–91] Such excitatory potentials were observed even in cultures where there was only a single 5-HT-containing cell that synapsed on itself. This result suggests that under some circumstances glutamate as well as 5-HT can be released from a single cell.

16.2.32 GABA

In brain slices, GABA$_A$-mediated inhibitory postsynaptic potentials were found to have the same latency and time course as the glutamate-mediated excitatory postsynaptic potentials.[74] Inhibitory potentials mediated by GABA$_A$ receptors are thought to be mediated by both long-distance afferents as well as local interneurons. The role of the GABA-mediated inhibition during the wake/sleep cycle is discussed below.

16.2.3.3 5-HT

A slow, inhibitory, postsynaptic potential mediated by 5-HT has been described in several of the raphe nuclei as well as microcultures of 5-HT neurons (Figure 16.7).[75,76,78] This synaptic potential can be as large as 20 mV, lasts for about 1 to 2 sec, and is mediated by an increase in potassium conductance that is pertussis toxin sensitive and rectifies inwardly.[87] Antagonists at the 5-HT-1A receptor (spiperone, NAN-190, pindobind 5-HT-1A) block the inhibitory postsynaptic potential, and agents that inhibit the reuptake of 5-HT, such as cocaine and fluoxetine, prolong its duration.[87,88] Whole cell and single channel recordings from acutely isolated dorsal raphe neurons have further characterized the potassium conductance induced by 5-HT.[92,93]

Activation of the 5-HT-1A receptor has also been shown to decrease the high threshold calcium current in acutely isolated dorsal raphe neurons.[94] When the action potential waveform was used as

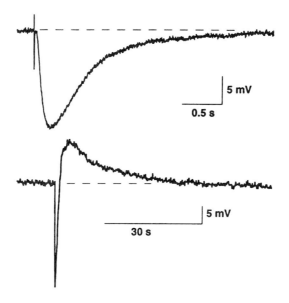

FIGURE 16.7
Slow synaptic potentials in dorsal raphe. (A) A single electrical stimulus evoked an inhibitory postsynaptic potential that was about 15 mV in amplitude and lasted about 1.5 sec. (B) Another cell, in which five stimuli were applied at 10 Hz to evoke a large inhibitory postsynaptic potential that was followed by a slow excitatory postsynaptic potential. The inhibitory postsynaptic potential was mediated by 5-HT, and the slow excitatory postsynaptic potential was caused by noradrenaline.

the voltage protocol to measure calcium currents, the inhibition of calcium entry by 5-HT was much larger (>80%) than when the more usual voltage step protocol was used.[85] This observation indicates that the inhibition of calcium entry under physiological conditions can be far greater than once thought. The combination of the increase in potassium conductance and inhibition of calcium entry by 5-HT can, therefore, result in a profound inhibition.

The release of 5-HT in the raphe nuclei as well as projection areas is also potently regulated by the 5-HT-1B/D autoreceptor. Low concentrations of the relatively selective 5-HT-1B agonist, TFMPP, caused a potent inhibition of the 5-HT-mediated IPSP in both the dorsal raphe and the nucleus prepositus hypoglossi.[95,96] In addition, there was a long-lasting (10 to 20 sec) synaptic inhibition following a single inhibitory postsynaptic potential in both nuclei. This inhibition was thought to result from the feedback activation of 5-HT-1B/1D receptors on the terminals releasing 5-HT. This feedback inhibition is also thought to play an important role in the decrease in 5-HT release found with higher concentrations of cocaine and fluoxetine.[75,76] The source of 5-HT in the dorsal raphe is from both recurrent collaterals from cells within the nucleus as well as innervation from other raphe nuclei.[97,98] Thus, it appears that there are synaptic mechanisms in place to coordinate the activity with the various raphe nuclei.

16.2.3.4 Noradrenaline

The noradrenergic input is thought to be critically involved in the "clocklike" firing of 5-HT neurons *in vivo*. Depletion of noradrenaline and α_1-antagonists decreased or abolished spontaneous firing recorded in anesthetized animals.[80] In addition, electrical stimulation in brain slice preparations evoked a slow depolarizing synaptic potential that was abolished by prazosin (Figure 16.7).[77,79] Superfusion with noradrenaline or phenylephrine caused a depolarization and evoked repetitive activity. In fact, experiments measuring the extracellular activity of 5-HT neurons in brain slices almost always included phenylephrine in the superfusion solution to cause "spontaneous" activity.[86]

Activation of α_1-adrenoceptors had two effects that were seemingly functionally antagonistic. The most obvious effect was a sustained depolarization leading to repetitive activity. Voltage clamp experiments showed that activation of α_1-adrenoceptors caused an inward current over a wide range of potentials.[77] The voltage dependence of the current was dependent on the conditions of the experiment. With whole cell recordings with patch pipettes, the inward current reversed polarity at the potassium equilibrium potential, indicating closure of a standing potassium current. With intracellular recordings, however, the current never reversed polarity, even at very negative membrane potentials. The recordings with intracellular electrodes were not technically flawed, because the potassium conductance increase caused by 5-HT did reverse polarity at the potassium equilibrium potential. Taken together these results suggest that α_1-adrenoceptor activation decreased a potassium conductance and, in addition, had another action that was disturbed by recording with whole cell electrodes.[77]

The second effect of α_1-adrenoceptor activation was an augmentation in amplitude and increase in the duration of the late component of the afterhyperpolarization following the action potential (Figure 16.5B).[77] This augmentation was observed following a single action potential,[77] as well as a burst of activity.[86] Apamin blocked the late component of the afterhyperpolarization as well as the augmentation by phenylephrine, indicating a role of intracellular calcium for this effect. Thapsigargin, an agent that depletes calcium stores, decreased the afterhyperpolarization as well as the augmentation by phenylephrine. This result suggested that release of calcium from internal stores was augmented by α_1-adrenoceptor stimulation. Caffeine increased the amplitude of the afterhyperpolarization without prolonging the duration, and this effect was blocked by ryanodine. The effect of phenylephrine on the afterhyperpolarization was not changed by either ryanodine or caffeine. The proposed mechanism for this α_1-adrenoceptor-mediated action was calcium release from ryanodine-insensitive internal stores by a rise in IP3. Further support for this mechanism came from the observation that an inhibitor of phospholipase C, manoalide, blocked the augmentation of the afterhyperpolarization caused by phenylephrine. Manoalide did not, however, affect the depolarization following the activation of phospholipase C, suggesting that the two actions of noradrenaline were mediated by separate mechanisms.

The effect of the other metabolite resulting from the breakdown of phosphoinositide by phospholipase C — DAG — was tested using phorbol esters on both the the afterhyperpolarization and the α_1-mediated receptor response.[77,86] The effect of PDBU was qualitatively different then that of phenylephrine in that it increased the amplitude of the afterhyperpolarization but did not affect the duration. In addition phorbolesters, in low concentrations, blocked both the depolarization and the increase in the afterhyperpolarization caused by activation of α_1-adrenoceptors. It is known that phosphorylation of the α_1-adrenoceptor by protein kinase C at other sites results in an inactivation of the receptor, and this mechanism has been proposed as a regulatory mechanism to control receptor activation. Application of high concentrations of phenylephrine for a period of 30 min did not result in any apparent decline of the depolarizing response, suggesting that the α_1-adrenoceptor activation of protein kinase C was not efficiently coupled to a desensensitization mechanism in dorsal raphe neurons.

Although the two effects of α_1-adrenoceptor activation on the 5-HT neurons in the dorsal raphe are very different, together these two effects can account for the very regular pattern of activity observed under resting conditions *in vivo*. The inward current drives the cell toward threshold for action potential generation and the prolonged afterhyperpolarization acts as a break to prevent bursting activity such that a regular rhythmic pattern of firing is maintained.

16.2.4 Activity of 5-HT Cells and Sleep

What intrinsic and synaptic properties regulate the activity of dorsal raphe neurons during the sleep/wake cycle *in vivo*? This question has been most directly addressed by Levine and Jacobs[81] using extracellular recordings and iontophoretic application of transmitter antagonists in behaving cats. In that study, application of bicuculline, the $GABA_A$ antagonist, reversed the suppression of activity observed during slow-wave sleep and had no effect on the activity during wakefulness, nor did it reverse the inhibition of activity found during REM sleep. The excitatory amino acid antagonist, kynurenate, reduced the sensory evoked increase in activity but had no effect on the repetitive spontaneous activity.

Although α_1-adrenoceptor antagonists were not iontophoresed in this study, local application of phenylephrine did not cause an increase in firing in waking animals, where the firing rate was already about 3 spikes per sec.[81] The interpretation of this experiment was that endogenous noradrenaline had already increased the firing to a maximal rate such that exogenous application of phenylephrine had no further action. Local applications of α_1-adrenoceptor antagonist in anesthetized rats did inhibit firing.[80] Thus, it appears that during quiet waking and under anesthesia, tonic release of noradrenaline is responsible for the repetitive activity of 5-HT neurons. It is tempting to speculate that the inhibition of activity during REM sleep results from a decrease in noradrenergic tone. The inhibition of activity in the LC during REM sleep is consistent with this hypothesis.

The role that the 5-HT innervation plays on firing of 5-HT neurons has also been examined in behaving cat with the use of systemic injections of both agonists and antagonists at the 5-HT-1A receptor.[99] In animals that were awake and alert, blockade of 5-HT-1A receptors caused an increase in firing rate, suggesting the presence of 5-HT tone that dampened the activity in the awake state. Interestingly, the firing rate was not affected in animals which were drowsy or asleep. This result suggested that 5-HT tone was elevated during waking when the firing rate was higher and low during sleep when the firing rate was lower. Thus, the 5-HT tone appeared as a negative feedback system within the raphe system.

16.4 Summary and Conclusions

The activity patterns of amine-containing neurons in the locus coeruleus and dorsal raphe during the sleep/wake cycle are similar in several ways. Both are silent during REM sleep; fire in a slow, steady pattern during slow-wave sleep or quite resting; and are most active during waking. In the waking state, cells are activated by multi-modal stimuli, indicating a strong convergence of sensory afferents in the pathways leading to each nucleus. Probably the most striking similarity between these two cell types is the expression of multiple mechanisms to limit the rate of firing. Both cell types have intrinsic membrane properties that result in relatively long duration action potentials (1.5 to 2 msec) with substantial calcium entry that results in calcium-dependent afterhyperpolarizations that are 15 to 20 mV in amplitude. Part of the reason for the large amplitude of the afterhyperpolarization is the high input resistance of these cells. Other voltage-dependent currents such as the transient outward current found in each cell type also limit excitability because this current will be de-inactivated at membrane potentials reached during the afterhyperpolarization. In addition, recurrent synaptic inhibition mediated by the monoamine transmitter produced by cells within each nucleus limit activity following a strong excitatory stimulus.

In spite of many similarities, the mechanisms that control activity of cells in each nucleus are distinctly different. Locus coeruleus neurons are spontaneously active in the absence of synaptic input such that the silence during REM sleep must be mediated by active inhibition. The regular activity of 5-HT cells in the dorsal raphe, however, was dependent on noradrenergic tone; therefore, the silence during REM sleep could result from a decline in noradrenaline afferent input. The correlation of activity of cells in both the locus coeruleus and dorsal raphe with the sleep/wake cycle suggests that each may play a role in the transitions between waking states. It is clear, however, that the regulation of activity in each is dependent on both intrinsic membrane properties and, more importantly, afferent input involving many different nuclei.

References

1. Aston-Jones, G. and Bloom, F.E., Activity of norepinephrine-containing locus coeruleus neurons in behaving rats anticipates fluctuations in the sleep-waking cycle, *J. Neurosci.*, 1, 876, 1981.
2. McGinty, D. and Harper, R.M., Dorsal raphe neurons: depression of firing during sleep in cats, *Brain Res.*, 101, 569, 1976.
3. Trulson, M.E. and Jacobs, B.L., Raphe unit activity in freely moving cats: correlation with level of behavioural arousal, *Brain Res.*, 163, 135, 1979.
4. Aston-Jones, G., Shipley, M.T., and Grzanna, R., The locus coeruleus, A5 and A7 noradrenergic cell groups, in *The Rat Nervous System*, Paxinos, G., Ed., Academic Press, San Diego, 1995, pp. 183–213.
5. Aston-Jones, G., Shipley, M.T., Chouvey, G., Ennis, M., VanBockstaele, E., Pieribone, V., Shiekhattar, R., Charlety, P., Astier, B., Valentino, R.J., and Williams, J.T., Afferent regulation of the nucleus locus coeruleus: anatomy, physiology and pharmacology, in *Progress in Brain Research Neurobiology of Locus Coeruleus*, Barns, C.D. and Pompeiano, O., Eds., Elsevier, Amsterdam, 1991, pp. 47–75.
6. Holets, V.R., The anatomy and function of noradrenaline in the mammalian brain, in *The Pharmacology of Noradrenaline in the Central Nervous System*, Heal, D.J. and Marsden, C.A., Eds., Oxford University Press, New York, 1990, pp. 1–40.
7. Cedarbaum, J.M. and Aghajanian, G.K., Afferent projections to the rat locus coeruleus as determined by a retrograde tracing technique, *J. Comp. Neurol.*, 178, 1, 1978.
8. Aston-Jones, G., Ennis, M., Pieribone, V., Nickell, T.W., and Shipley, M.T., The brain nucleus locus coeruleus: restricted afferent control of a broad efferent network, *Science* 234, 734, 1986.
9. Aston-Jones, G., Shipley, M.T., Ennis, M., Williams, J.T., and Pieribone, V.A., Restricted afferent control of locus coeruleus neurons revealed by anatomic, physiologic and pharmacologic studies, in *The Pharmacology of Noradrenaline in the Central Nervous System*, Marsden, C.A. and Heal, D.J., Eds., Oxford University Press, London, 1990, pp. 187–247.
10. Shipley, M.T., Fu, L., Ennis, M., Liu, W.L., and Aston-Jones, G., Dendrites of locus coeruleus neurons extend preferentially into two pericoerulear zones, *J. Comp. Neurol.*, 365, 56, 1996.
11. Ishimatsu, M. and Williams, J.T., Synchronous activity in locus coeruleus results from dendritic interactions in pericoerulear regions, *J. Neurosci.*, 16, 5196, 1996.
12. Egan, T.M. and North, R.A., Actions of acetylcholine and nicotine on rat locus coeruleus neurons in vitro, *Neuroscience*, 19, 565, 1986.
13. Egan, T.M. and North, R.A., Acetylcholine acts on m2-muscarinic receptors to excite rat locus coeruleus neurones, *Br. J. Phamracol.*, 85, 733, 1985.
14. Shefner, S.A. and Chiu, T.H., Adenosine inhibits locus coeruleus neurons: an intracellular study in a rat brain slice preparation, *Brain Res.*, 366, 364, 1986.
15. Regenold, J.T. and Illes, P., Inhibitory adenosine A1-receptors on rat locus coeruleus neurones: an intracellular electrophysiological study, *Nauyn-Schmiedeberg's Arch. Pharmacol.*, 341, 225, 1990.
16. Pan, W.J., Osmanovic S.S., and Shefner, S.A., Characterization of the adenosine A1 receptor-activated potassium current in rat locus coeruleus neurons, *J. Pharmacol. Exp. Ther.*, 273, 537, 1995.
17. Pan, W.J., Osmanovic S.S., and Shefner, S.A., Adenosine decreases action potential duration by modulation of A-current in rat locus coeruleus neurons, *J. Neurosci.*, 14, 1114, 1994.
18. Xiong H. and Marshall, K.C., Angiotensin II depresses glutamate depolarizations and excitatory postdynaptic potentials in locus coeruleus through angiotensin II subtype 2 receptors, *Neuroscience*, 62, 163, 1994.
19. Shen, K.Z. and North, R.A., Excitation of rat locus coeruleus neurons by adenosine 5′-triphosphate: ionic mechanism and receptor characterization, *J. Neurosci.*, 13, 894, 1993.

20. Curtis, A.L., Lechner, S.A., Pavovich, L.A., and Valentino, R.J., Activation of the locus coeruleus noradrenergic system by intracoerulear microinfusion of corticotropin-releasing factor: effects on discharge rate, cortical norepinephrine levels and cortical electroencephalographic activity, *J. Pharmacol. Exp. Ther.*, 281, 163, 1997.

21. Osmanovic, S.S. and Shefner, S.A., γ-aminobutyric acid responses in rat locus coeruleus neurons *in vitro*: a current-clamp and voltage-clamp study, *J. Physiol.*, 421, 151, 1990.

22. Osmanovic, S.S. and Shefner, S.A., Baclofen increases the potassium conductance of rat locus coeruleus neurons recorded in brain slices, *Brain Res.*, 438, 124, 1988.

23. Pierebone, V.A., Xu, Z.Q., Zhang, X., Grillner, S., Barfai, T., and Hokfelt, T., Galanin induces a hyperpolarization of norepinephrine-containing locus coeruleus neurons in the brainstem slice, *Neuroscience*, 64, 861, 1995.

24. Williams, J.T., Bobker, D.H., and Harris, G.C., Synaptic potentials in locus coeruleus neurons in brain slices, in *Progress in Brain Research, Neurobiology of Locus Coeruleus*, Barns, C.D. and Pompeiano, O., Eds., Elsevier, Amsterdam, 1991, pp. 167–172.

25. Cherubini, E., North, R.A., and Williams, J.T., Synaptic potentials in locus coeruleus neurones, *J. Physiol.*, 406, 431, 1988.

26. Bird S.J. and Kuhar, M.J., Iontophoertic application of opiates to the locus coeruleus, *Brain Res.*, 122, 523, 1977.

27. Pepper, C.M. and Henderson, G., Opiates and opioid peptides hyperpolarize locus coeruleus neurons *in vitro*, *Science*, 209, 394, 1980.

28. Williams, J.T., Egan, T.M., and North, R.A., Enkephalin opens potassium channels in mammalian central neurones, *Nature* 299, 74, 1982.

29. Williams, J.T., North, R.A., and Tokimasa, T., Inward rectification of resting and opioid-activated potassium currents in rat locus coeruleus neurons, *J. Neurosci.*, 8, 4299, 1988.

30. Illes, P. and Regenold, J.T., Interaction between neuropeptide Y and noradrenaline in central catecholamine neurons, *Nature*, 433, 62, 1990.

31. Egan, T.M., Henderson, G., North, R.A., and Williams, J.T., Noradrenaline mediated synaptic inhibition in locus coeruleus neurones, *J. Physiol.* 345, 477, 1983.

32. Connor, M., Vaughan, C.W., Chieng, B., and Christie, M.J., Nociceptin receptor coupling to a potassium conductacne in rat locus coeruleus neurones *in vitro*, *Br. J. Pharmacol.*, 119, 1614, 1996.

33. Shiekhattar, R. and Aston-Jones, G., Sensory reponsiveness of brain noradrenergic neurons is modulated by endogenous bran serotonin, *Brain Res.*, 623, 72, 1993.

34. Bobker, D.H. and Williams, J.T., Serotonin agonists inhibit synaptic potentials in the rat locus coeruleus *in vitro* via 5-HT1A and 5-HT1B receptors, *J. Pharmacol. Exp. Ther.*, 250, 37, 1989.

35. Inoue, M., Nakajima, S., and Nakajima, Y., Somatostatin induces an inward rectification in rat locus coeruleus neurons through a pertussis toxin-sensitive mechanism, *J. Physiol.*, 407, 117, 1988.

36. Chieng, B., Conner, M., and Christie M.J., The mu-opioid receptor antagonist D-Phe-Cys-Tyr-D-Trp-Orn-Thr-Pen-Thr-NH2 (CTOP) [but not D-Phe-Cys-Tyr-D-Trp-Arg-Thr-Pen-Thr-NH2 (CTAP)] produced a nonopioid receptor-mediated increase in K conductance of rat locus coeruleus neurons, *Mol. Pharmacol.*, 50, 650, 1996.

37. Guyenet, P.G. and Aghajanian, G.K., Excitation of neurons in the nucleus locus coeruleus by substance P and related peptides, *Brain Res.*, 136, 178, 1977.

38. Shen, K.Z. and North, R.A., Substance P opens cation channels and closes potassium channels in rat locus coeruleus neurons, *Neuroscience*, 50, 345, 1992.

39. Velimirovic, B.M., Koyano K., Nakajima, S., and Nakajima, Y., Opposing mechanisms of regulation of a G-protein-coupled inward rectifier K channel in rat brain neurons, *Proc. Natl. Adac. Sci. USA*, 92, 1590, 1995.

40. Olpe, H.R. and Baltzer, V., Vasopressin activated noradrenergic neurons in the rat locus coeruleus: a microiontophoretic investigation, *Eur. J. Pharmacol.,* 73, 377, 1981.
41. Olpe, H.R. and Steinmann M., Responses of locus coeruleus neurons to neuropeptides, in *Progress in Brain Research, Neurobiology of Locus Coeruleus,* Barns, C.D. and Pompeiano, O., Eds., Elsevier, Amsterdam, 1991, pp. 241–248.
42. Wang, Y.Y. and Aghajanian, G.K., Excitation of locus coeruleus neurons by vasoactive intestinal peptied: role of cAMP and protein kinase A, *J. Neurosci.,* 10, 3335, 1990.
43. Williams, J.T., North, R.A., Shefner, S.A., Nishi, S., and Egan, T.M., Membrane properties of rat locus coeruleus neurones, *Neuroscience,* 13, 137, 1984.
44. Osmanovic, S.S. and Shefner, S.A., Calcium-activated hyperpolarization in rat locus coeruleus neurons *in vitro, J. Physiol.,* 469, 89, 1993.
45. Forsythe, I.D., Linsdell, P., and Stanfield, P.R., Unitary A-currents of rat locus coeruleus neurones grown in cell culture: rectification caused by internal Mg and Na, *J. Physiol.,* 451, 553, 1992.
46. Osmanovic, S.S. and Shefner, S.A., Functional significance of the apamin-sensitive conductance in rat locus coeruleus, *Brain Res.,* 530, 283, 1990.
47. Aghajanian, G.K., Vandermaelen, C.P., and Andrade, R., Intracellular studies on the role of calcium in regulating the activity and reactivity of locuc coeruleus neurons *in vivo, Brain Res.,* 273, 237, 1983.
48. Osmanovic, S.S. and Shefner, S.A., Anomalous rectiification in rat locus coeruleus neurons, *Brain Res.,* 417, 161, 1987.
49. Alreja, M. and Aghajanian, G.K., Pacemaker activity of locus coeruleus neurons: whole-cell recordings in brain slices dependence on cAMP and protein kinase A, *Brain Res.,* 556, 339, 1991.
50. Osborne, P.B. and Williams, J.T., Forskolin enhancement of opioid currents in rat locus coeruleus neurons, *J. Neurophysiol.,* 76, 1559, 1996.
51. Shiekhattar, R. and Aston-Jones, G., Activation of adenylate cyclase attenutes the hyperpolarization following single action potentials in brain noradrenergic neurons independently of protein kinase A, *Neuroscience,* 62, 523, 1994.
52. Ingram, S., Wilding, T.J., McCleskey, E.W., and Williams, J.T., Efficacy and kinetics of opioid action on acutely dissociated neurons, *Mol. Pharm.,* 52, 136, 1997.
53. Illes, P. and Regenold, J.T., ω-Conotoxin GVIA and nifedipine inhibit Ca action potentials in rat locus coeruleus neurons, *Acta Physiol. Scand.,* 137, 459, 1989.
54. Ennis, M. and Aston-Jones, G., Activation of locus coeruleus from nucleus paragigantocellularis: a new excitatory amine acid pathway in brain, *J. Neurosci.,* 8, 3644, 1988.
55. Chiang, C. and Aston-Jones, G., Response of locus coeruleus neurons to footshock stimulation is mediated by neurons in the rostral ventral medulla, *Neuroscience,* 53, 705, 1993.
56. Ennis, M. and Aston-Jones, G., Evidence for self- and neighbor-mediated postactivation inhibition of locus coeruleus neurons, *Brain Res.,* 374, 299, 1986.
57. Harris, G.C., Hausken, Z., and Williams, J.T., Cocaine induced synchronous oscillations in central noradrenergic neurons *in vitro, Neuroscience,* 50, 253, 1992.
58. Surprenant, A.M. and Williams, J.T., Inhibitory synaptic potentials recorded from mammalian neurons prolonged by blockade of noradrenaline uptake, *J. Physiol.,* 382, 87, 1987.
59. Aston-Jones, G., Astier, B., and Ennis, M., Inhibition of noradrenergic locus coeruleus neurons by C1 adrenergic cells in the rostral ventral medualla, *Neuroscience,* 48, 371, 1992.
60. Aghajanian, G.K. and Wang, Y.Y., Pertussis toxin blocks the outward currents evoked by opiate and α_2-agonists in locus coeruleus neurons, *Brain Res.,* 371, 390, 1986.
61. Williams, J.T. and Marshall, K.C., Membrane properties and adrenergic responses in locus coeruleus of young rats, *J. Neurosci.,* 7, 3687, 1987.

62. Christie, M.J., Williams, J.T., and North, R.A., Electrical coupling synchronizes subthreshold activity in locus coeruleus neurons *in vitro* from neonatal rats, *J. Neurosci.*, 9, 3584, 1989.
63. Christie, M.J. and Jelinek, H.F., Dye-coupling among neurons of the rat locus coeruleus during postnatal development, *Neuroscience*, 56, 129, 1993.
64. Marshall, K.C., Christie, M.J., Finlayson, P.G., and Williams, J.T., Developmental aspects of the locus coeruleus-noradrenaline system, in *Progress in Brain Research, Neurobiology of Locus Coeruleus*, Barns, C.D. and Pompeiano, O., Eds., Elsevier, Amsterdam, 1991, pp.173–185.
65. Travagli, R.A., Dunwiddie, T.V., and Williams, J.T., Opioid inhibition in locus coeruleus, *J. Neurophysiol.*, 74, 519, 1995.
66. Oleskevich, S., Clements, J.D., and Williams, J.T., Opioid-glutamate interactions in rat locus coeruleus neurons, *J. Neurophysiol.*, 70, 931, 1993.
67. Travagli, R.A., Wessendorf, M., and Williams, J.T., The dendritic arbor of locus coeruleus neurons contributes to opioid inhibition, *J. Neurophysiol.*, 75, 2029, 1996.
68. Ivanov, A. and Aston-Jones, G., Extranuclear dendrites of locus coeruleus neurons: activation by glutamate and modulation of activity by alpha adrenoceptors, *J. Neurophysiol.*, 74, 2427, 1995.
69. Aston-Jones, G., Rajkowski, J., Kubiak, P., and Alexinsky, T., Locus coeruleus neuons in monkey are selectively activated by attended cues in a vigilance task, *J. Neurosci.*, 14, 4467, 1994.
70. Rajkowski, J., Kubiak, P., and Aston-Jones, G., Locus coeruleus activity in monkey: phasic and tonic changes are associated with altered vigilance, *Brain Res. Bull.*, 35, 607, 1994.
71. Nitz D. and Siegel J.M., GABA release in the locus coeruleus as a function of the sleep/wake state, *Neuroscience*, 78, 795, 1997.
72. Berridge, C.W. and Foote, S.L., Effects of locus coeruleus activation on electroencophalographic activity in neocortex and hippocampus, *J. Neurosci.*, 11, 3135, 1991.
73. Halliday, G., Harding, A., and Paxinos, G., Serotonin and tachykinin systems, in *The Rat Nervous System*, Paxinos, G., Ed., Academic Press, San Diego, 1995, pp. 929–974.
74. Pan, Z.Z. and Williams, J.T., GABA and glutamate mediated synaptic potentials in the rat dorsal raphe neurones *in vitro*, *J. Neurophysiol.*, 61, 719, 1989.
75. Pan, Z.Z., Colmers, W.F., and Williams, J.T., 5-HT-mediated synaptic potentials in the dorsal raphe nucleus: interactions with excitatory amino acid and GABA neurotransmission, *J. Neurophysiol.*, 62, 481, 1989.
76. Pan, Z.Z. and Williams, J.T., Differential actions of cocaine and amphetamine on dorsal raphe neurons *in vitro*, *J. Pharmacol. Exp. Ther.*, 251, 56, 1989.
77. Pan, Z.Z., Grudt, T.J. and Williams, J.T., α_1-adrenoceptors in rat dorsal raphe neurons: depolarization and regulation of activity, *J. Physiol.*, 478, 431, 1994.
78. Yoshimura, M. and Higashi, H., 5-Hydroxytryptamine mediates inhibitory postsynaptic potentials in rat dorsal raphe neurons, *Neurosci. Lett.*, 53, 69, 1985.
79. Yoshimura, M., Higashi, H., and Nishi, S., Noradrenaline mediates slow excitatory synaptic potentials in rat dorsal raphe neurons *in vitro*, *Neurosci. Lett.*, 61, 305, 1985.
80. Baraban, J.M. and Aghajanian, G.K., Suppression of firing activity of 5-HT neurons in the dorsal raphe by α-adrenoceptor agonists, *Neuropharmacology*, 19, 355, 1980.
81. Levine, E.S. and Jacobs, B.L., Neurochemical afferents controlling the activity of serotonergic neurons in the dorsal raphe nucleus: microiontophoretic studies in the awake cat, *J. Neurosci.*, 12, 4037, 1992.
82. Pan, Z.Z., Wessendorf, M.W., and Williams, J.T., Modulation by serotonin of the neurons in rat nucleus raphe magnus *in vitro*, *Neuroscience*, 54, 421, 1993.
83. Pan, Z.Z., Williams, J.T., and Osborne, P., Opioid actions on single nucleus raphe magnus neurons from rat and guinea pig *in vitro*, *J. Physiol.*, 427, 519, 1990.

84. Vandermaelen, C.P. and Aghajanian, G.K., Electrophysiological and pharmacolgical characterization of serotonergic dorsal raphe neurons recorded extracellularly and intracellularly in rat brain slices, *Brain Res.*, 289, 109, 1983.
85. Penington, N.J., Kelly, J.S., and Fox, A.P., Action potential waveforms reveal simultaneous changes in ICa and IK produced by 5-HT in rat dorsal raphe neurons, *Proc. Roy. Soc. Lond. B.*, 248, 171, 1992.
86. Freedman, J.E. and Aghajanian, G.K., Role of phosphoinositide metabolites in the prolongation of afterhyperpolarizations by α_1-adrenoceptors in rat dorsal raphe neurons, *J. Neurosci.*, 7, 3897, 1987.
87. Williams, J.T., Colmers, W.F., and Pan, Z.Z., Voltage and ligand activated inwardly rectifying currents in rat dorsal raphe neurons *in vitro, J. Neurosci.*, 8, 3499, 1988.
88. Bayliss, D.A., Li, Y.W., and Talley, E.M., Effects of serotonin on caudal raphe neurons: activation of an inwardly rectifying potassium conductance, *J. Neurophysiol.*, 77, 1349, 1997.
89. Aghajanian, G.K., Modulation of a transient outward current in serotonergic neurones by α_1-adrenoceptors, *Nature*, 315, 501, 1985.
90. Johnson, M.D., Synaptic glutamate release by postnatal rat serotonergic neurons in microculture, *Neuron*, 12, 433, 1994.
91. Johnson, M.D., Electrophysiological and histochemical properties of postnatal rat serotonergic neurons in dissociated cell culture, *Neuroscience*, 63, 775, 1994.
92. Penington, N.J., Kelly, J.S., and Fox, A.P., Whole-cell recordings of inwardly rectifying K currents activated by 5-HT-1A receptors on dorsal raphe neurones of the adult rat, *J. Physiol.*, 469, 387, 1993.
93. Penington, N.J., Kelly, J.S., and Fox, A.P., Unitary properties of potassium channels activated by 5-HT in acutely isolated rat dorsal raphe neurones, *J. Physiol.*, 469, 407, 1993.
94. Penington, N.J. and Kelly, J.S., Serotonin receptor activation reduces calcium current in an acutely dissociated adult central neuron, *Neuron*, 4, 751, 1990.
95. Bobker, D.H. and Williams, J.T., 5-HT mediated inhibitory postsynaptic potentials in guinea-pig prepositus hypoglossi and feedback inhibition by 5-HT, *J. Physiol.*, 422, 447, 1990.
96. Bobker, D.H. and Williams, J.T., Cocaine and amphetamine interact at 5-hydroxytryptamine synapses through distinct mechanism in guinea-pig prepositus hypoglossi, *J. Neurosci.*, 11, 2151, 1991.
97. Fornal, C.A., Litto, W.J., Metzler, C.W., Marrosu, K.T., and Jacobs, B., Single-unit reponses of serotonergic dorsal raphe neurons to 5-HT1A agonists and antagonist drug administration in behaving cats, *J. Pharmacol. Exp. Ther.*, 270, 1345, 1994.
98. Wang, R.Y. and Aghajanian, G.K., Antodromically identified serotonergic neurons in the rat midbrain raphe: evidence for collateral inhibition, *Brain Res.*, 132, 186, 1977.
99. Portas, C.M. and McCarley, R.W., Behavioral state-related changes of extracellular serotonin concentration in the dorsal raphe nucleus: a microdialysis study in freely moving cat, *Brain Res.*, 648, 306, 1994.

Chapter 17

Mechanisms Affecting Neuronal Excitability in Brainstem Cholinergic Centers and their Impact on Behavioral State

Robert W. Greene and Donald G. Rainnie

Contents

17.1	Introduction: Cholinergic Influences on Behavioral State	278
17.2	Cellular Factors Affecting Neuronal Excitability in the LDT/PPT	279
	17.2.1 Voltage-Gated Currents	279
	17.2.1.1 Outward Currents	279
	17.2.1.2 Inward Currents	280
	17.2.2 Ligand-Gated Currents	282
	17.2.2.1 Ionotropic Effectors	282
	17.2.2.2 Metabotropic Effectors	282
	17.2.3 Functional Consequences of Activating the Ligand-Gated, Inwardly Rectifying Potassium Conductance in LDT/PPT Neurons	285
17.3	Intrinsic and Extrinsic Properties of LDT/PPT Neurons and their Impact on Behavioral State	286
	17.3.1 Low-Threshold Bursts	286
	17.3.2 State-Related Monoaminergic Input	289
	17.3.2.1 Waking	289
	17.3.2.2 REM/SWS	290
	17.3.3 The Transition from Waking to Sleep	290
References		292

17.1 Introduction: Cholinergic Influences on Behavioral State

There are at least three ascending brainstem neuromodulatory systems capable of generating a level of thalamocortical activation necessary for arousal. These brainstem systems include the serotonergic neurons of the raphe nuclei, the noradrenergic neurons of the locus coeruleus, and the pontine cholinergic neurons of the laterodorsal and pedunculopontine tegmental nuclei (LDT/PPT). Activation of their target postsynaptic receptors on thalamocortical neurons, by any one of these three systems, in most cases results in a membrane potential depolarization. This depolarization moves the membrane potential out of the range which would favor delta wave oscillations ("burst mode", see below) and into a potential range of greater responsiveness to either descending cortical input or ascending input ("tonic mode").[37,65]

Unlike the commonality of effect that these neuromodulatory systems have on thalamocortical neurons, the effects of cholinergic inputs onto inhibitory thalamic interneurons are clearly discernible from the effects of the other neuromodulatory systems. Hence, a muscarinic receptor-mediated hyperpolarization which is associated with an increased conductance, most probably an inwardly rectifying potassium current,[37] is observed in thalamic interneurons. This inhibition of inhibitory thalamic interneurons would result in a disinhibition of thalamocortical relay cells. In contrast, the monoaminergic input onto inhibitory thalamic interneurons results in a membrane depolarization, which is mediated by a reduction of a potassium conductance[40] and results in increased inhibition of thalamocortical relay cells. The inhibitory control is likely to reduce thalamocortical excitability in a functionally specific manner. Thus, its reduction by cholinergic input may result in a nonspecific increase in thalamocortical excitability. Consequently, the cholinergic disinhibition is more consistent with a generalized behavioral arousal, in the sense that inhibitory control of thalamocortical neurons is reduced.

Having established a mechanism by which the pontine cholinergic nuclei may specifically influence thalamocortical activation, it becomes necessary to demonstrate state-related changes in neuronal activity. Hence, a clear correlation has been demonstrated between the firing rates of the majority of cholinergic LDT/PPT neurons and the level of thalamocortical activation, as measured by cortical EEG, which is normally associated with either waking or rapid-eye-movement (REM) sleep states.[67] In addition, during slow-wave sleep, in which the cortical EEG normally displays synchronized slow-wave activity, stimulation of the LDT/PPT can induce a level thalamocortical activation that is associated with a depolarization of thalamic relay neurons towards a "tonic" mode and is normally associated with arousal.[65] Moreover, the EEG activation that is associated with the REM state provides a clear case of generalized thalamocortical activation in the complete absence of monoaminergic influence (see below). Thus, under physiological conditions it appears that an ascending cholinergic influence is sufficient for thalamocortical activation, but not arousal.

The integral role of pontine cholinergic nuclei in waking and thalamocortical activation is further emphasized by recent studies examining the behavioral state-related consequences of their inhibition, excitation or disinhibition. Hence, when pontine cholinergic neurons are directly inhibited, waking is significantly reduced.[48,50] In contrast, electrical stimulation of the LDT significantly increases REM sleep expression.[70] Furthermore, dorsal raphe projections to the LDT/PPT can cause an inhibition of cholinergic neurons via activation of postsynaptic 5-HT receptors.[19,28,31] When serotonergic neuronal activity in the dorsal raphe nucleus is inhibited, pontine cholinergic neurons become disinhibited, and a significant increase in the time spent in the REM state is observed.[49] Recently, microdialysis of 5-HT into the LDT was shown to inhibit REM sleep behavior,[20] a finding consistent with an inhibition of LDT neurons that results in the inhibition of REM sleep behavior. These studies serve to indicate the potential importance of the control of LDT/PPT neuronal excitability in behavioral state and also in the transition between states.

At the cellular level, control of neuronal excitability is most directly mediated by two distinct families of membrane-spanning ion channels that can be distinguished according to their mechanism of gating: namely, those ion channels gated by voltage and/or calcium and those ion channels that are ligand-gated. It is within the limits of neuronal excitability set by the activity of these channels, and their intrinsic conductances, that the brainstem cholinergic arousal centers can ultimately influence behavioral state.

Despite the fact that there have been some differences in both the age and species used (juvenile vs. adult; rat vs. guinea pig) and the nuclei studied (LDT vs. PPT), there is a general agreement from electrophysiological studies that the brainstem cholinergic nuclei contain three neuronal populations.[25,26,29,31,73] These populations have been designated as type I, type II, and type III. The basis of this differentiation has been the relative expression of several time- and voltage-dependent conductances. Of particular relevance to this discussion are type III neurons, which are believed to represent the largest population of cholinergic neurons in the LDT/PPT.

17.2 Cellular Factors Affecting Neuronal Excitability in the LDT/PPT

17.2.1 Voltage-Gated Currents

Because of the importance of three specific voltage-gated conductances — namely, I_T, I_A, and I_H — on the patterning of discharge responses in type III LDT/PPT neurons, it is on these three conductances that we will focus our attention.

17.2.1.1 Outward Currents

17.2.1.1.1 I_A. The most extensively studied outward current in LDT/PPT neurons is a transient, voltage-gated, outward potassium current, usually called I_A.[25,29] This current is responsible for a relative refractory period following a membrane hyperpolarization beyond –65 mV for more than 15 msec (Figure 17.1A,B). The processes responsible for this behavior are voltage-dependent inactivation and activation of the channel conductance. In the inactivated state, the channel will not open irrespective of changes in the membrane potential. Activation is the process whereby changes in membrane potential cause the channel to open. At the resting membrane potential of LDT/PPT neurons (~–63 mV), the I_A channel is in the inactivated state. In order for activation to be expressed, the inactivation process must be reversed or de-inactivated by membrane hyperpolarization. The de-inactivation is a voltage-dependent process whereby the more hyperpolarized the membrane potential becomes, the greater the number of I_A channels de-inactivated, with a $V_{1/2}$ of about –70 mV. The de-inactivation process is also time dependent and may be described by a single exponential time course with a time constant of about 17 msec at –80 mV.[25] Consequently, neurons have to be hyperpolarized beyond –70 mV for more than 17 msec in order for more than 60% of the I_A channels to become de-inactivated (Figure 17.1C). For the same reason that steady-state de-inactivation is voltage sensitive, the kinetics of de-inactivation are also voltage dependent, being quicker at more hyperpolarized potentials. Once inactivation is removed, the I_A channel may be activated by membrane depolarization with a $V_{1/2}$ of about –35 mV. Most type III cholinergic neurons of rat LDT have a pronounced I_A current together with I_T, described below. It is the relative expression of these two conductances that determines the burst-firing characteristics of these neurons. Increased expression of I_A, by increasing periods of membrane hyperpolarization, would tend to inhibit burst firing, whereas increased expression of I_T would tend to increase burst firing following periods of membrane hyperpolarization. Moreover, Kamondi et al.[25] note that the afterhyperpolarization (AHP)

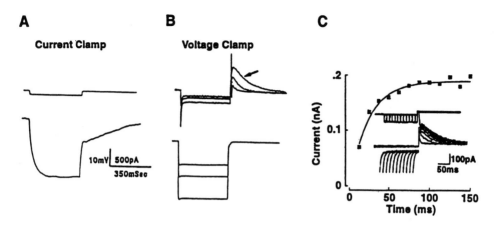

FIGURE 17.1
Transient outward current, I_A, causes a relative refractory period following a hyperpolarization. **(A)** A current clamp recording of an LDT neuron, showing a hyperpolarization (bottom trace) in response to hyperpolarizing current injection (top trace) with a delayed return to resting membrane potential. **(B)** A voltage clamp recording from the same neuron is shown to illustrate the current response (top trace) to a hyperpolarizing voltage step (bottom trace) with a transient outward tail current (arrow) that is responsible for the delay in the return to resting membrane potential seen in (A). **(C)** A graph showing the time course for the removal of inactivation by a curve of the maximal outward current vs. the duration of the preceding hyperpolarization (the overlapping voltage and current traces are inset in the graph). (Parts A and B adapted from Luebke, J.I. et al., *J. Neurophysiol.*, 70(5), 2128, 1993; Part C adapted from Kamondi, A. et al., *J. Neurophysiol.*, 68(4), 1359, 1992. With permission.)

following an action potential normally pushes the membrane potential to about –70 mV for sufficient time to remove some inactivation of the I_A channel and thus contribute to a delay in the firing of the next action potential.

17.2.1.1.2 Delayed Outward Currents. There are other outward currents in LDT/PPT neurons that contribute to neuronal excitability and which mediate responses such as action potential repolarization and the post-spike afterhyperpolarization. Typically, these delayed, outwardly rectifying currents result from a combination of voltage-gated I_K-type currents and calcium-dependent, I_C-like currents. However, the relative contributions and specific characteristics of these currents in LDT/PPT neurons have not yet been examined. With regard to I_M and I_{AHP}, they probably contribute less to tonic LDT/PPT neuronal firing behavior than that which has been observed with pyramidal neurons, because little or no spike adaptation is observed in LDT/PPT neurons. The role of I_{AHP} in burst firing of these neurons remains to be determined.

17.2.1.2 Inward Currents

Current clamp records from type III LTD/PPT neurons demonstrate the presence of all-or-nothing action potentials, rebound bursts of action potentials following transient hyperpolarizing voltage excursions, and a pronounced, slowly developing, depolarizing "sag" in the voltage response to hyperpolarizing current input. All three responses are mediated by the activation of voltage-gated inward currents. As with all central nervous system (CNS) neurons, the action potentials are dependent on a rapid, TTX-sensitive, inward sodium current. The presence of a steady-state or "persistent" inward sodium current in the potential range slightly depolarized to the resting membrane potential has yet to be established firmly. The rebound burst potential following periods of membrane hyperpolarization is mediated by an inward calcium current, I_T, and the depolarizing "sag" by an anomalous rectifying non-specific inward cation current, I_H.

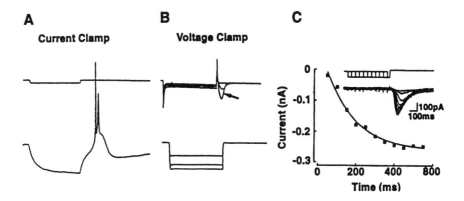

FIGURE 17.2
A voltage-sensitive, low-threshold calcium current is responsible for the low-threshold burst. (A) A current clamp recording of an LDT neuron, showing a low-threshold burst response following a hyperpolarization (top trace, current; bottom trace, voltage). (B) A voltage clamp recording from the same neuron showing the low-threshold calcium current (top traces, arrow) generated in response to hyperpolarizing voltage steps (bottom traces) that remove inactivation. The removal of inactivation is time dependent, as illustrated by the graph in (C). The peak amplitude of inward current that generates the low-threshold burst is plotted relative to the duration of the preceding hyperpolarization that is needed to remove inactivation (see inset of overlapping voltage and current traces). (Parts A and B adapted from Luebke, J.I. et al., *J. Neurophysiol.*, 70(5), 2128, 1993; part C adapted from Kamondi, A. et al., *J. Neurophysiol.*, 68(4), 1359, 1992. With permission.)

17.2.1.2.1 I_T. Perhaps the most dramatic of the firing pattern characteristics of LDT/PPT neurons is the rebound burst of action potentials resulting from activation of a low-threshold calcium current, I_T (Figure 17.2A,B).[31] Like the outward I_A current, expression of the inward I_T current is voltage dependent and is determined by the ratio of inactive and active I_T channels. Moreover, the voltage sensitivity of this current is almost identical to that of I_A with respect to steady-state activation and inactivation;[25] however, the kinetics of de-inactivation are much slower than those of I_A (Figure 17.2C). Thus, a hyperpolarization lasting several hundred milliseconds to –80 mV is required to remove sufficient inactivation of I_T such that subsequent depolarization beyond –65 mV will activate this regenerative current and evoke a low threshold burst potential.

The amplitude and duration of the rebound burst potential and the number of action potentials evoked by each potential vary from cell to cell. Hence, one may observe large bursts of five or more action potentials (AP), smaller bursts evoking only two or three APs, or a single action potential that, in the absence of I_T, would not have been evoked. This variation depends on several factors, including (1) the total number of I_T channels expressed in a given cell, (2) the degree of the removal of inactivation, and (3) the presence of other currents that might shunt I_T and thus prevent burst expression. In type III neurons, the current that is most likely to shunt I_T is I_A. As mentioned above, I_A has a voltage-sensitivity in its gating similar to I_T, so changes in membrane potential will have parallel effects on these two currents. Because one current will shunt the other, a small change in the expression of either could have a profound effect on the firing pattern of the neuron. Hence, any neurotransmitter that modulates the expression of either one of these currents could potentially change the response of a neuron from one of initial refractoriness to burst generation and vice versa.

17.2.1.2.2 I_H. Most LDT/PPT neurons display, to a varying degree, a time-dependent, depolarizing "sag" in the voltage response to a transient hyperpolarizing current step of more than 200 msec. In voltage clamp, this response is seen to be mediated by the activation of a time-dependent inward current. The voltage dependency and ionic sensitivity of the current responsible for the rectification are characteristic of the hyperpolarization-activated, non-specific cation current, I_H.[53] Because the reversal potential for this current is more depolarized than the resting membrane potential of these neurons, and its activation range (−65 to −90 mV) is more hyperpolarized than the action potential threshold, I_H is always inward. Moreover, the slow kinetics of I_H activation (tens of milliseconds) indicate that this conductance is more likely to impact the de-inactivation of I_T channels than that of I_A channels.

I_H, through an interaction with I_T, can increase the tendency of thalamic neurons to oscillate at a slow frequency (<4 Hz).[39] Hence, following a hyperpolarization of sufficient amplitude and duration to remove I_T inactivation, activation of I_H would depolarize the membrane potential close to that of threshold for the I_T-dependent, low-threshold burst. With the expression of the burst, I_T inactivates, and voltage- and calcium-sensitive outward rectifiers are activated, causing the neuron to hyperpolarize and thereby completing the cycle. Thus, the I_T/I_H interaction may induce membrane potential oscillations in thalamic neurons. As pontine cholinergic neurons also express I_H and I_T, similar oscillations may also occur in the LDT/PPT. However, increasing I_H can also prevent oscillations by preventing the hyperpolarization that is necessary to de-inactivate I_T. Thus, I_H can either facilitate or prevent slow oscillations, depending on the amplitude of the current activated. This current is itself affected by the total number of I_H channels expressed in a particular neuron and also on the voltage sensitivity ($V_{1/2}$) of the I_H which, in LDT/PPT neurons, may be altered by neuromodulators such as adenosine.[53]

17.2.2 Ligand-Gated Currents

Ligand-gated ion channels can be further subdivided into two distinct categories — namely, those channels in which the neurotransmitter recognition site (receptor) and the ion channel are an integral unit (ionotropic effectors) and those in which receptor activation induces a second messenger cascade, via activation of a GTP-dependent protein, which results in the subsequent activation of an independent ion channel (metabotropic effectors).

17.2.2.1 Ionotropic Effectors

At least three postsynaptic responses mediated by activation of ionotropic receptors have been identified pharmacologically in the LDT/PPT. These include two excitatory glutamate responses mediated by activation of α-amino-3-hydroxy-5-methylisoxazole-4-propionic acid (AMPA) receptors and N-methyl-D-aspartate (NMDA) receptors, and an inhibitory γ-amino-butyric acid (GABA) response mediated by activation of $GABA_A$ receptors.[59] Local stimulation of afferent inputs to the LDT/PPT can elicit postsynaptic potentials mediated by these three receptor subtypes in identified cholinergic and non-cholinergic neurons of albino guinea pigs. Postsynaptic responses were typical of fast AMPA-, NMDA-, and $GABA_A$-mediated potentials observed throughout the CNS. Hence, local electrical stimulation evokes an excitatory postsynaptic potential (EPSP)-inhibitory postsynaptic potential (IPSP) sequence of less than about 75 msec, with the EPSP mediated primarily by AMPA receptor activation and to a lesser extent by NMDA receptor activation. The IPSP is mediated by activation of $GABA_A$ receptors as suggested by its sensitivity to picrotoxin (10 μM).[59]

17.2.2.2 Metabotropic Effectors

The actions of several neurotransmitters appear to converge at the level of the LDT/PPT by activating a common metabotropic effector system.

17.2.2.2.1 Serotonin. In 60% of LDT/PPT neurons, 5-HT induces a TTX-resistant hyperpolarization that is associated with a decreased membrane input resistance, *in vitro*[28,31] (Figure 17.3A). About 66% of all identified cholinergic neurons responded in this manner to 5-HT, whereas only 25% of the non-cholinergic neurons responded in a similar fashion. Further, about 73% of type III cholinergic neurons were hyperpolarized by 5-HT.

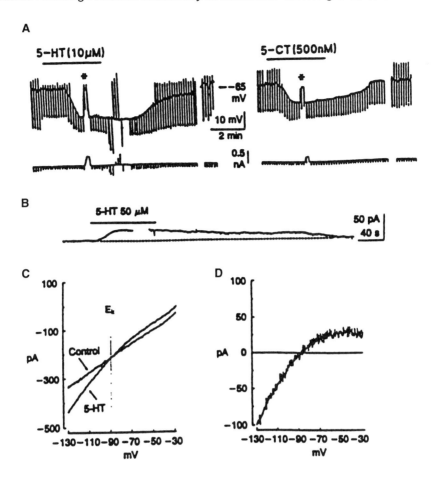

FIGURE 17.3
Serotonin induces a hyperpolarization in LDT/PPT neurons by increasing an inwardly rectifying potassium conductance. (**A**) A voltage record under current clamp shows a hyperpolarization induced by 5-HT or the 5-HT-1a agonist 5-carboxytryptamine (8-OH-DPAT induces a similar response, not shown). Downward deflections are the voltage response to injected hyperpolarizing current, used to test the input resistance that decreased during the 5-HT hyperpolarization. (**B**) In voltage clamp, 5-HT induced an outward current. (**C**) A voltage ramp was used to generate the current response I/V curves, before and during 5-HT application. (**D**) Subtraction of the I/V curve during control from the I/V curve obtained during 5-HT gave the I/V curve of the 5-HT-generated current. The greater slope conductance at more hyperpolarized potentials is characteristic of the inward rectifier potassium current. (Adapted from Luebke, J.I. et al., *Proc. Natl. Acad. Sci. USA*, 89, 743–747, 1992.)

The current mediating the 5-HT-induced hyperpolarization has a reversal potential close to the potassium equilibrium potential and increases in amplitude with membrane hyperpolarization. These properties are characteristic of an inwardly rectifying potassium current (I_{IR}; Figure 17.3B,C,D). The 5-HT response was mimicked by 5-carboxamidotryptamine (5-CT; Figure 17.3A) and 8-hydroxy-dipropylaminotetralin (8-OH-DPAT) and blocked by spiperone, and is, therefore, thought to be mediated by activation of $5HT1_A$ receptors.[28,31] In other regions of the CNS, the $5-HT1_A$ receptor-mediated increase in an inwardly rectifying potassium conductance is associated with activation of a G-protein cascade.[2] Although this has not been tested in the LTD/PPT, it is likely to be the case here, as well.

Immunohistochemical studies have demonstrated that cholinergic neurons of the rat LDT/PPT are also closely associated with $5-HT_2$ receptor immunoreactivity.[45] To date, no postsynaptic electrophysiological correlates of $5-HT_2$ receptor activation have been observed;[31] however, 5-HT-

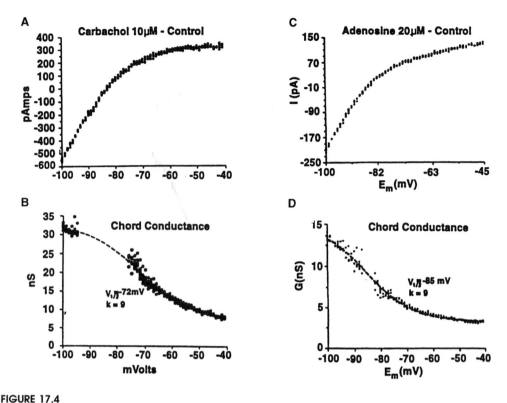

FIGURE 17.4
Metabotrobic transmitters that elicit inhibition in LTD/PPT neurons converge on a common effector mechanism, namely, an increase in inwardly rectifying potassium conductance. (A) An I/V curve of the muscarinic carbachol response shows the characteristic inward rectification. (B) A curve of the chord conductance (calculated from I/[$V_m - V_{rev}$]) shows a voltage sensitivity characteristic of the inward rectifier conductance. (C) and (D) The curves illustrated in these two graphs are similar to those in (A) and (B), except that they were elicited by exposure to adenosine acting at the A1 receptor. (Parts A and B adapted from Luebke, J.I. et al., *J. Neurophysiol.*, 70(5), 2128, 1993; parts C and D adapted from Rainnie, D.G. et al., *Science*, 263, 689, 1994. With permission.)

dependent effects that are mediated by presynaptic receptor activation have not been examined. Alternatively, it is also possible that the postsynaptic 5-HT_2 receptor is not coupled to an electrophysiologically responsive receptor/effector system.

17.2.2.2.2 Norepinephrine. The effects of norepinephrine (NE) on cholinergic neurons of the LDT/PPT appear to be even more homogenous than those of 5-HT. Williams and Reiner[74] report that over 90% of identified cholinergic neurons in these nuclei respond to NE with a TTX-resistant hyperpolarization that is also associated with a decreased membrane input resistance. As with the 5-HT response, the current mediating this response results from an increase of an inwardly rectifying potassium conductance. A similar NE conductance in other areas of the CNS is associated with activation of a G-protein-dependent, inwardly rectifying potassium channel.[46] The NE response is most probably mediated by activation of postsynaptic α_2 receptors, as it is mimicked by the specific α_2 agonist, UK-14304, and blocked by the α_2 antagonist, idazoxan.[74] Additional NE responses were observed in non-cholinergic neurons, and it is possible that these are mediated by receptors other than α_2 receptors but this remains to be examined.

17.2.2.2.3 Acetylcholine. About 90% of all rat LDT/PPT neurons recorded *in vitro* responded to the cholinergic agonist, carbachol, with a membrane hyperpolarization that was associated with a decreased membrane input resistance.[28,32] However, electrophysiological and histochemical identification of cell types revealed that the carbachol-induced hyperpolarization was not restricted to any one particular cell type. As with 5-HT and NE, the current responsible for this hyperpolarization resulted from an increase of an inwardly rectifying potassium conductance (Figure 17.4A,B). Moreover, the receptors mediating this response are probably muscarinic, as the response is mimicked by the muscarinic agonist, methacholine, and blocked by atropine as well as high concentrations of the muscarinic antagonist, pirenzepine.[28,32] The relative lack of sensitivity to prienzepine blockade (IC_{50} = 580 nM), together with an inwardly rectifying potassium conductance as an effector mechanism, indicates that a muscarinic M_2 receptor subtype probably subserves this response.

17.2.2.2.4 Adenosine. When the network excitability of the rat LDT/PPT was measured *in vitro*, using extracellular single unit recording, the firing rates of all neurons tested were reduced when exposed to bath application of adenosine (AD).[53] The reduction of network excitability resulted from both pre- and postsynaptic effects of AD. Hence, whole-cell patch clamp recording demonstrated that more than 70% of identified cholinergic neurons responded to AD with a TTX-insensitive hyperpolarization associated with a decrease in membrane input resistance. The current mediating this hyperpolarization results, in part, from an increase in an inwardly rectifying potassium conductance as observed with the other metabotropic effector mechanisms (Figure 17.4C,D).

In addition, AD acts at postsynaptic receptors to reduce the expression of the voltage-gated, non-specific cation conductance, I_H.[53] A similar effect of AD on I_H has been demonstrated in thalamic neurons.[47] Because activation of I_H channels results in a depolarizing inward current (see above), the AD-mediated reduction of I_H would appear as an outward current associated with a decreased membrane conductance. Therefore, any IPSC that would normally hyperpolarize an LDT/PPT neuron into the range of activation of I_H would now have a greater effect following the reduction of I_H by adenosine. A predictable consequence of this action would be that de-inactivation of I_T and I_A is more likely to occur, thus promoting the expression of these two currents.

The postsynaptic AD effects are probably mediated by activation of A_1 receptors, as the response to AD was mimicked by the A_1-specific agonist, N^6-cyclohexyladenosine, but not by the A_{2a} agonist, CGS21480, and was blocked by the A_1 receptor antagonist, cyclopentyltheophylline.[53] Activation of AD receptors on presynaptic terminals in the LDT/PPT results in a reduction of synaptic transmission within the nucleus;[54] however, the AD receptor subtype subserving this response has yet to be identified.

17.2.2.2.5 Histamine. Unlike most neurotransmitters studied in the LDT/PPT *in vitro*, histamine (HA) has been shown to be one of the few neurotransmitters that excite neurons of the guinea pig PPT.[27] Although morphological analysis was not attempted in this study, over 90% of the neurons responded to HA with a depolarization. It is likely, therefore, that cholinergic and non-cholinergic neurons are affected in the same manner by HA. The effect was not blocked by TTX, indicating a postsynaptic site of action, but was blocked by the H_1 antagonist, mepyramine but not by the H_2 antagonist, cimetidine, indicating activation of H_1 receptors.

17.2.2.3 Functional Consequences of Activating the Ligand-Gated, Inwardly Rectifying Potassium Conductance in LDT/PPT Neurons

Activation of serotonin $5-HT1_A$ receptors, norepinephrine α_2 receptors, acetylcholine M_2 receptors, or adenosine A_1 receptors results in the activation of an inwardly rectifying potassium conductance in LDT/PPT neurons. This remarkable convergence of different modulatory receptors onto a single effector mechanism raises the question of the functional significance of the activation of this inwardly rectifying potassium conductance on the firing pattern of LDT/PPT neurons.

The inward rectification of the ligand-gated potassium conductance is due to a hyperpolarization-activated voltage sensitivity that can be well described by a sigmoid curve with a $V_{1/2}$ of about –73 mV.[32] Because the resting membrane potential of LDT/PPT neurons is about –60 mV and the reversal potential of this current is about –90 mV, as the membrane potential hyperpolarizes towards the reversal potential, the conductance of the channel will increase. Consequently, any inhibitory input, such as $GABA_A$ IPSPs, which occur coincident with a ligand-gated increase in the inwardly rectifying potassium conductance, will be potentiated. Conversely, once the cell is hyperpolarized by this current, any depolarizing current will decrease the conductance of the inward rectifier as the cell is depolarized. Hence, activation of the inwardly rectifying potassium conductance will act to potentiate inhibitory synaptic activity but not at the expense of an excitatory drive, if it is of sufficient magnitude to overcome the inhibition. In other words, the firing of the neuron takes on a more "all-or-nothing" character. This has a permissive effect on low-threshold burst generation, such that for a given degree of hyperpolarization induced by the ligand-gated inward rectifier current, an EPSC of a given amplitude may elicit a low-threshold burst that would not have occurred if the potassium conductance were voltage insensitive.

Figure 17.5 illustrates a computer model of a typical type III cholinergic neuron from the LDT/PPT that displays a low-threshold burst. This computer model is modified from a single compartment model of a thalamic relay neuron using Hodgkin-Huxley style of mathematical equations to describe the voltage-dependence and kinetics of the various currents.[22,38] Under control conditions, the neuron responds to a transient hyperpolarizing current step with a rebound burst of two action potentials (Figure 17.5A). However, in this instance, the hyperpolarization was induced in the absence of any change in the neuron's conductance, such as would occur had the hyperpolarization been induced by an inhibitory synaptic input. In Figure 17.5B, the neuron has been hyperpolarized to a new resting membrane potential of –78 mV by an increase in a voltage-insensitive potassium conductance. Despite the intrinsic ability of this neuron to generate a low-threshold burst, a depolarizing current step has little effect on this cell because the increased, voltage-insensitive, potassium conductance shunts any depolarizing currents. However, when the same neuron is hyperpolarized to a similar resting membrane potential of –78 mV by an inwardly rectifying potassium conductance, an identical depolarizing current step will now generate a low-threshold burst response (Figure 17.5C). This permissive effect of the inward rectifier results from its voltage sensitivity.

17.3 Intrinsic and Extrinsic Properties of LDT/PPT Neurons and their Impact on Behavioral State

17.3.1 Low-Threshold Bursts

Ponto-geniculo-occipital (PGO) waves are field potentials that reflect synchronous activity of specific populations of central neurons. The appearance of PGO waves heralds the onset of REM sleep and is believed to reflect cellular mechanisms that contribute to the initiation of this behavioral state. During REM sleep, PGO waves originate in the pontine tegmentum and subsequently spread to the lateral geniculate bodies of the thalamus, and to a lesser extent into other thalamic nuclei, and then into the occipital cortex. There is considerable evidence consistent with a role of burst-firing LDT/PPT cholinergic neurons in the generation of the PGO waves. Anatomical studies have demonstrated that these neurons send significant projections to the thalamus, including the lateral geniculate nuclei.[17,60] Extracellular single-unit recordings have revealed a population of LDT/PPT neurons that fire bursts of two to six action potentials that are tightly coupled to the occurrence of PGO waves (Figure 17.6A,B).[36,58,66,68] Electrical stimulation in the region of the LDT/PPT can evoke PGO waves in the lateral geniculate nucleus,[57] and excitotoxic lesions of the LDT/PPT can reduce the occurrence of PGO waves during REM sleep by up to 75%.[72] Finally, unilateral application of nicotinic antagonists to the lateral geniculate nucleus abolished PGO wave generation on the side of application only.[21] Thus, these studies suggest that PGO waves may originate from synchronous activity of cholinergic neurons of the brainstem pontine tegmentum.

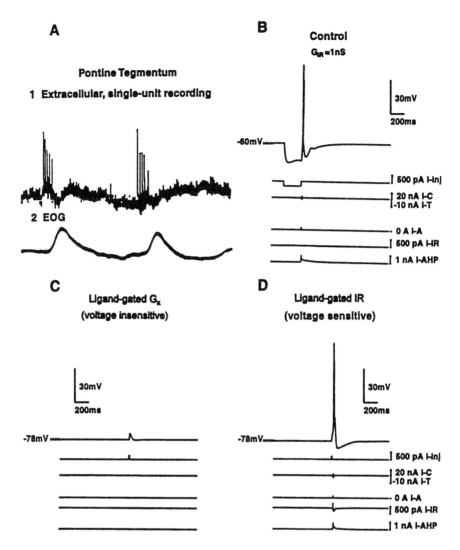

FIGURE 17.5
(A) A low-threshold burst can generate a "burst" of action potentials similar to those observed in extracellular single unit recordings. (B), (C), (D) A computer-generated model of LDT neuronal membrane potential shows a comparison between the effect of a non-voltage-sensitive potassium conductance and the same amplitude of an inwardly rectifying potassium conductance on low-threshold burst firing. In (B), the membrane is hyperpolarized by current injection in the absence of a direct effect on the conductance of the the neuron. A rebound, low-threshold burst is evoked by this current. In (C), the membrane of an identical neuron is hyperpolarized to –78 mV by a voltage-insensitive potassium conductance, and then a short-duration, depolarizing current is injected to mimic an EPSC. In (D), the neuron is identical and the membrane potential is hyperpolarized similarly to –78 mV by an increase in potassium conductance; however, the hyperpolarizing conductance is an inwardly rectifying potassium conductance. Now, when an identical, short-duration, EPSC-like current is applied, a low-threshold response is elicited. Note that the trace labeled "I-IR" shows a fast increase in current due to the action potential which provides a transient increase in potassium ion driving force. More slowly, the I_{IR} (inwardly rectifying potassium current) deactivates and becomes smaller due to the depolarization, thus allowing expression of the low-threshold burst. (Part A from McCarley, R.W. and Massaquoi, S.G., *Am. J. Physiol.*, 251, R1011, 1986. With permission.)

The cellular mechanisms that trigger the origination of PGO waves are likely to be dependent, to a large degree, upon those mechanisms that contribute to the burst-generating properties of the pontine cholinergic neurons. The presence of an intrinsic, low-threshold calcium current (I_T) that can induce burst generation in identified cholinergic LDT/PPT neurons provides such a mechanism, as

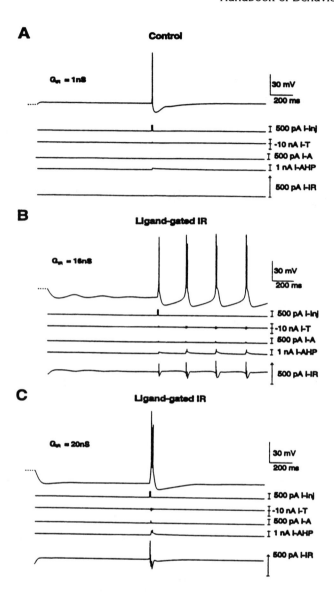

FIGURE 17.6
A computer model of an LTD neuronal membrane potential showing bursting and oscillating properties that result from an interaction of I-IR (the inwardly rectifying current) and I-T (the low-threshold calcium current). (**A**) The response of a model LTD neuron to an EPSC-like depolarizing current in control conditions with only 1 nS of IR conductance. (**B**) An increase in G_{IR}, similar to that evoked by the metabotrobic inhibitory neurotransmitters, can elicit oscillations of bursting activity. Note that a small amount of I_H is present in the model, as is observed in our recordings, but the oscillations can still occur in the complete absence of I_H. The oscillations result only from the interaction of I_{IR}, I_T, and a resting membrane potential of −58 mV. (**C**) At higher amplitudes of I_{IR}, oscillations are inhibited because the membrane potential remains hyperpolarized relative to the threshold for low-threshold burst generation.

the bursts recorded intracellularly *in vitro* would produce the same extracellular properties as those observed *in vivo* in conjunction with PGO waves (Figure 17.6). In order for the low-threshold burst to be expressed, the cholinergic neuron must receive an inhibitory input of sufficient amplitude to remove inactivation of I_T, which must then be followed by sufficient excitatory input to depolarize the membrane potential into the range of I_T activation.

Ponto-geniculo-occipital waves often appear as oscillations of bursts of activity. A similar pattern of activity can be modeled for LDT/PPT neurons based primarily on (1) a resting membrane potential of –60 mV, (2) the presence of I_T, and (3) the presence of a calcium-dependent potassium current or some other slowly activating, voltage-sensitive, outwardly rectifying potassium current (Figure 17.7B). During a burst induced by I_T, calcium fluxes into the cytosol through voltage-gated ion channels and activates calcium-dependent potassium currents. In association with inactivation of I_T, the calcium-dependent potassium conductance acts to terminate the burst. On termination of the burst, the long time constant for the decay of the calcium-dependent potassium conductance ensures that the cell hyperpolarizes towards potassium equilibrium potential, which removes inactivation of I_T. The calcium-dependent potassium current then subsides, and the cell depolarizes towards resting membrane potential and into the range of I_T activation, starting the cycle anew.

As mentioned above, within conductance limits the ligand-gated inward rectifier current has a positive modulatory effect on the expression of I_T, and, consequently it also has a positive modulatory effect on oscillations; however, with large increases of current, the tendency toward oscillation is decreased. The positive effects on oscillations result from the characteristics of the voltage sensitivity of the inward rectifier, which induces a negative slope region in the neuron's current-voltage relationship that would contribute to the "all-or-nothing" character of burst expression. Thus, with the addition of a limited amount of I_{IR}, the afterhyperpolarization induced by the delayed rectifier potassium current(s) will be enhanced, further removing inactivation from I_T. As long as the resting membrane potential in the presence of the additional I_{IR} is within the range of I_T activation, then as the hyperpolarization decays, an enhanced low-threshold burst may be generated as a result of the greater removal of inactivation. Additionally, as the neuron depolarizes, the I_{IR} decreases, thus avoiding a shunt of the I_T.

The effects of increasing the expression of an inward rectifier conductance (G_{IR}) on a model LDT/PPT neuron is illustrated in Figure 17.2. At rest, under control conditions, an EPSC-like current injection will elicit only a single action potential. A burst response is not forthcoming due to inactivation of I_T at the resting membrane potential of –58 mV. If the G_{IR} is increased a moderate amount, and then a similar EPSC-like current injection is added, the neuron responds with a burst-hyperpolarization oscillation (Figure 17.2B). A further increase in G_{IR} would still allow strong bursting behavior, but there is insufficient tendency to depolarize into the range of I_T activation resulting from the stronger hyperpolarizing influence of I_{IR} (Figure 17.2C).

17.3.2 State-Related Monoaminergic Input

17.3.2.1 Waking

All of the monoaminergic inputs to the LDT/PPT converge at the level of activation of I_{IR}; consequently, coincident activity in the separate monoaminergic centers will dictate the magnitude of I_{IR}. Having established the postsynaptic effects of monoaminergic input onto LDT/PPT neurons, as outlined above, predictions can be made as to the functional interactions of the neuromodulatory centers across behavioral states based on their known anatomical relationships and their state-related firing patterns. The major caveat for these predictions is that the presynaptic actions of the neuromodulatory inputs are not known, and even if they were, one would need to know the state-related activity of these presynaptic terminals to fully understand their effects on firing patterns of the target LDT/PPT neurons. Nevertheless, postsynaptic modulation may serve as a starting point. The ascending cholinergic activating system is most active during two distinct behavioral states: waking and REM sleep. In contrast, the monoaminergic systems, including NE, 5-HT, and HA, are most active during waking. Their firing rates then reduce during slow-wave sleep (SWS), and they become silent during REM sleep.[3,41,55,64,71] During waking, when pontine cholinergic neurons are active, brainstem noradrenergic and serotonergic systems are also active. Their inputs to the LDT/

PPT are inhibitory and appear to target primarily, but not exclusively, cholinergic neurons. The activity of the monoaminergic systems during waking would, therefore, cause an increased activation of I_{IR} which would dampen the output of the pontine cholinergic system but still allow burst responses in the face of strong excitatory synaptic input. However, the inhibition does not appear to be reciprocal — that is, cholinergic neurons do not send inhibitory efferents to monoaminergic neurons (see Chapter 16). Accordingly, the inhibition must be more complex in function than simply to prevent co-activation by these ascending activating systems. One of the possibilities includes a bias of the output of cholinergic neurons towards co-release of co-localized peptides by the monoaminergic input that would be absent during REM sleep. The bias would derive from an increased tendency to burst in the presence of increased I_{IR}.

17.3.2.2 REM/SWS

During REM sleep, the activity of the monoaminergic systems is negligible. Thus, the activation of the REM sleep state may be primarily, if not entirely, dependent on LDT/PPT cholinergic discharge. Furthermore, microinjection of cholinergic agonists into LDT/PPT target zones of the medial pontine reticular formation induces a behavioral state indistinguishable in most respects from natural REM.[1,4,5,11,44] This pharmacological model of REM provides further evidence consistent with a central role of pontine cholinergic neurons in REM state generation.

Do brainstem monoaminergic inputs to the LDT/PPT contribute to REM state generation? It is probable that the decrease in brainstem monoaminergic activity prior to REM, in and of itself, contributes to REM state generation. During the transition from SWS to REM, the firing rates of the monoaminergic neurons decrease dramatically. The postsynaptic effect of this decreased activity on LDT/PPT neurons would be a marked disinhibition due to a reduction of I_{IR}. These effects were predicted by the REM-SWS oscillation model proposed by Hobson et al.[18] and McCarley and Hobson[34] and further developed by McCarley and Massaquoi.[35] Hence, a reciprocal interaction between brainstem monoaminergic and cholinergic systems was proposed which involved an inhibition of the cholinergic neurons by the monoaminergic neurons. As demonstrated above, the predictions of monoaminergic inhibitory input have been substantiated at the cellular level by observations *in vitro*. Thus, as the activity of the monoaminergic neurons subsides during the transition to REM, the cholinergic neurons of the LDT/PPT would become disinhibited and begin to fire at the high rates which are characteristic of REM.[67,68]

There remains the question of the sufficiency of monoaminergic disinhibition to account for the increased cholinergic firing rates. In the *in vitro* slice preparation, in which cholinergic LDT/PPT neurons are practically deafferented, spontaneous firing rates are low or absent. This would suggest that an excitatory input, in addition to disinhibition, is necessary to increase LDT/PPT firing activity. Moreover, in the Hobson and McCarley model,[34] a recurrent excitatory cholinergic input was proposed. However, the presence of powerful, inhibitory, M_2-type autoreceptors on the cholinergic neurons[32] indicates that, if present, any excitatory recurrent cholinergic input to the LDT/PPT would have to be indirect. One neurotransmitter candidate that may mediate the increased excitatory drive onto cholinergic neurons is glutamate.[59] Indeed, the LDT/PPT receives a prominent input from the medial prefrontal cortex,[60,62] stimulation of which evokes a glutamate-induced excitation of LDT neurons.[13] In addition, increased excitatory drive from histaminergic neurons of the tuberomamiliary nucleus may also be a candidate. However, this circuit is thought to be more involved in cortical activation during waking.[30]

17.3.3 The Transition from Waking to Sleep

Adenosine has been suggested as an endogenous sleep factor based on its somnogenic properties when administered systemically or intraventricularly.[51,52] Caffeine induces EEG arousal,[76] in addition to its well-known stimulant properties, at doses most consistent with its action as an adenosine receptor antagonist, as both the propensity to sleep and the intensity of EEG delta wave activity are proportional to the duration of prior wakefulness.[8,9] Pape's suggestion that his observation of an AD-mediated decrease of I_H in the thalamus would enhance delta wave oscillations is of particular interest.[47] Furthermore, central administration of AD and the specific A_1 receptor agonist, N^6-cyclopentyladenosine (CPA), can selectively increase delta power.[6,61] However, these effects are due to administration of exogenous AD and hence may be categorized as only pharmacological in nature.

It is difficult to reconcile the localized control of the electrophysiological actions of AD with its putative actions as a sleep factor. Although AD is ubiquitous in cells, and A1 receptors that mediate AD-dependent inhibition are found on cortical, thalamic, and brainstem neuronal elements, the control of AD release is dependent on local factors. Thus, a global alteration of neuronal activity consistent with the onset of sleep would require a global organization of the control of AD release throughout the cortex and thalamus. This does not appear consistent with the control of AD metabolism that occurs at the level of the micro-environment. A potential solution to the apparent conflict between local control and global response was provided by Rainnie et al.,[53] who showed that endogenous AD exerts a powerful, tonic, inhibitory action on cholinergic neurons in both the basal forebrain and in the LDT/PPT. It was proposed that localized increases in extracellular AD in these two particular regions could induce sleep through a global reduction in cholinergic tone. The latter would result from an AD-induced reduction of cholinergic neuronal activity that itself would reduce the global cholinergic tone due to the widespread projections of these neurons.

In support of the AD/cholinergic hypothesis, recent *in vivo* studies have demonstrated that local infusion of AD promotes sleep,[49] and sleep deprivation is accompanied by an increase in extracellular AD in these two regions.[48] Hence, local changes in AD could inhibit the cholinergic neurons in these regions and thus reduce cholinergic tone at their target sites. To the extent that cholinergic activity is responsible for arousal, a local increase in AD in these cholinergic nuclei might be expected to decrease arousal and promote sleep.

Why does the extracellular AD concentration change across behavioral states and can this provide a clue for the function of SWS? There is a large body of evidence that AD release into the CNS extracellular space is related to the metabolic state of neuronal tissue.[7,42,43] Hence, AD might provide a negative feedback link between electrophysiological and metabolic state.[14,15,42] This same relationship for AD could be extended to global arousal and metabolic state through the inhibitory actions of AD on cholinergic neurons[53] and possibly other sleep or arousal centers. However, the link between extracellular AD levels and physiological metabolic states has not been clearly established. To date, only two non-pathological AD release mechanisms have been described: (1) by repetitive activation of excitatory inputs[33] and (2) through stimulation of intracellular cAMP.[12,56] In contrast, several pathological states that all involve increased metabolic demand relative to metabolite availability can cause an increase in electrophysiologically active AD release;[15] however, these mechanisms are not consistent with the action of AD as a physiological sleep factor that relates CNS metabolism to behavioral state. In the case of the stimulus-evoked AD release, the time course of the effects is too rapid. In the case of the slower, metabolically challenged release, the release is not related to metabolism under physiological conditions. Nevertheless, endogenous AD does exert a tonic inhibitory tone on cholinergic neurons of the basal forebrain and LDT/PPT.[53] This action is similar to an AD-mediated tonic inhibition in the hippocampus that is Ca^{2+}-independent and controlled, at least in part, by a bidirectional facilitated transport system.[10,75] The flux of AD through this transport system is primarily inward, as blockade of the transport with nitrobenzylthioionosine (NBTI) results in an increase of AD-mediated inhibition. Recently we have observed a similar effect of NBTI in the LTD/PPT *in vitro*.[16] Moreover, when NBTI was locally perfused into the basal forebrain, the extracellular AD concentration was increased in conjunction with an increase in SWS.[48] More importantly, NBTI microperfusion into the thalamus mimicked the effect on AD concentration but not the effect on behavioral state, a fact that serves to emphasize the importance of the regional localization of the AD increase in the regulation of behavioral state.

Behavioral state control evolves from the interactions of several parallel systems including, but not exclusively, three ascending activating systems of the brainstem. These are the noradrenergic, the serotonergic, and the cholinergic systems. Thus, the absence of an essential role for any one of these systems in the generation of thalamocortical activation is not unexpected. Additionally, an experimentally induced, prolonged reduction in the activity of one system may be confounded by the activation of compensatory mechanisms in a parallel system. Lesion experiments targeted at any one particular system largely confirm these expectations.[23,24,63,69] However, under physiological conditions, the excitability of a particular activating system may have a substantial effect on behavioral state, as has been demonstrated for the state of arousal and the cholinergic system.[48,50]

Here we have demonstrated that the excitability of LDT/PPT neurons of the brainstem cholinergic system results from an interplay of the activity of their voltage and ligand-gated channels. The activity of these channels is, in turn, largely controlled by the release of monoaminergic and fast amino-acid neurotransmitters from other systems. In addition, excitability is under the local control of adenosine, a neuromodulator with many characteristics indicative of its role in the LDT/PPT and basal forebrain as a sleep factor. For LDT/PPT cholinergic neurons, the convergence of monoaminergic, cholinergic, and adenosinergic receptor activation onto an inhibitory inward rectifier conductance and its interaction with those channels mediating the low-threshold burst conductance provides a context for an understanding of how their excitability may be controlled and related to other systems.

References

1. Amatruda, T., Black, D., McKenna, T., McCarley, R. W., and Hobson, J.A., Sleep cycle control and cholinergic mechanisms: differential effects of carbachol injections at pontine brain stem sites, *Brain Res.*, 98, 501, 1975
2. Andrade, R., Malenka, R.C., and Nicoll, R.A., A G-protein couples serotonin and GABA$_B$ receptors to the same channels in hippocampus, *Science*, 234, 1261, 1986.
3. Aston-Jones, G. and Bloom, F.E., Activity of norepinephrine-containing locus coeruleus neurons in behaving rats anticipates fluctuations in the sleep-waking cycle, *J. Neurosci.*, 1, 876, 1981.
4. Baghdoyan, H.A., Monaco, A.P., Rodrigo-Angulo, M.L., Assens, F., McCarley, R.W., and Hobson, J.A., Microinjection of neostigmine into the pontine reticular formation of cats enhances desynchronized sleep signs, *J. Pharmacol. Exp. Ther.*, 231, 173, 1984.
5. Baxter, B.L., Induction of both emotional behavior and a novel form of REM sleep by chemical stimulation applied to cat mesencephalon, *Exp. Neurol.*, 23, 220, 1969.
6. Benington, J.H. and Heller, H.C., Restoration of brain energy metabolism as the function of sleep, *Prog. Neurobiol.*, 45, 347, 1995.
7. Berne, M., Winn, H.R., and Rubio, R., The effect of inadequate oxygen supply to the brain on cerebral adenosine levels, in *Cerebral Hypoxia in the Pathogenesis of Migraine*, Rose, F.C. and Amery, W.K., Eds., Pitman, London, 1982, p. 89.
8. Borbély, A.A., Baumann, F., Brandeis, D., Strauch, I., and Lehman, D., Sleep deprivation: effect on sleep and EEG power density in man, *Electroencephalogr. Clin. Neurophysiol.*, 51, 483, 1981.
9. Borbély, A.A., Tobler, I., and Hanagasioglu, M., Effect of sleep deprivation on sleep and EEG power spectra in the rat, *Behav. Brain Res.*, 14, 171, 1984.
10. Geiger, J.D. and Nagy, J.I., Adenosine deaminase and [^{3}H]nitrobenzylthioinosine as markers of adenosine metabolism and transport in central purinergic systems, in *Adenosine and Adenosine Receptors*, Williams, M., Ed., Humana Press, Clifton, NJ, 1990, p. 225.
11. George R., Haslett W.L., and Jenden D.J., A cholinergic mechanism in the brainstem reticular formation: Induction of paradoxical sleep, *Int. J. Neuropharmacol.*, 3, 541, 1964.

12. Gereau, IV, R.W. and Conn, P.J., A cyclic AMP-dependent form of associative synaptic plasticity induced by coactivation of beta-adrenergic receptors and metabotropic glutamate receptors in rat hippocampus, *J. Neurosci.*, 14, 3310, 1994.
13. Grant, S.J. and Highfield, D.A., Extracellular characteristics of putative cholinergic neurons in the rat laterodorsal tegmental nucleus, *Brain Res.*, 559, 64, 1991.
14. Greene, R.W. and Haas, H.L., Endogenous adenosine inhibits hippocampal CA1 neurones: further evidence from extra- and intracellular recording, *Naunyn Schmiedebergs Arch. Pharmacol.*, 337, 561, 1988.
15. Greene, R.W. and Haas, H.L., The electrophysiology of adenosine in the mammalian central nervous system, *Prog. Neurobiol.*, 36, 329, 1991.
16. Grunze, H.C. and Greene, R.W., unpublished observations, 1997.
17. Hallanger, A.E., Levey, A.I., Lee, H.J., Rye, D.B., and Wainer, B.H., The origins of cholinergic and other subcortical afferents to the thalamus in the rat, *J. Comp. Neurol.*, 262, 105, 1987.
18. Hobson, J.A., McCarley, R.W., and Wyzinski, P.W., Sleep cycle oscillation: reciprocal discharge by two brainstem neuronal groups, *Science*, 189, 55, 1975.
19. Honda, T. and Semba, K., Serotonergic synaptic input to cholinergic neurons in the rat mesopontine tegmentum, *Brain Res.*, 647, 299, 1994.
20. Horner, R.L., Sanford, L.D., Annis, D., Pack, A.I., and Morrison, A.R., Serotonin at the laterodorsal tegmental nucleus suppresses rapid-eye-movement sleep in freely behaving rats, *J. Neurosci.*, 17, 7541, 1997.
21. Hu, B., Steriade, M., and Deschênes, M., The effects of brainstem peribrachial stimulation on neurons of the lateral geniculate nucleus, *Neuroscience*, 31, 13, 1989.
22. Huguenard, J.R. and McCormick, D.A., Simulation of the currents involved in rhythmic oscillations in thalamic relay neurons, *J. Neurophysiol.*, 68, 1373, 1992.
23. Jones, B.E., The role of noradrenergic locus coeruleus neurons and neighboring cholinergic neurons of the pontomesencephalic tegmentum in sleep-wake states, *Prog. Brain Res.*, 88, 533, 1991.
24. Jones, B.E., The organization of central cholinergic systems and their functional importance in sleep-waking states, *Prog. Brain Res.*, 98, 61, 1993.
25. Kamondi, A., Williams, J.A., Hutcheon, B., and Reiner, P.B., Membrane properties of mesopontine cholinergic neurons studied with the whole-cell patch-clamp technique: implications for behavioral state control, *J. Neurophysiol.*, 68, 1359, 1992.
26. Kang, Y. and Kitai, S., Electrophysiological properties of pedunculopontine neurons and their postsynaptic responses following stimulation of substantia nigra reticulata, *Brain Res.*, 535, 79, 1990.
27. Khateb, A., Serafin, M., and Mühlethaler, M., Histamine excites pedunculopontine neurones in guinea pig brainstem slices, *Neurosci. Lett.*, 112, 257, 1990.
28. Leonard, C.S. and Llinas, R., Serotonergic and cholinergic inhibition of mesopontine cholinergic neurons controlling REM sleep: an *in vitro* electrophysiological study, *Neuroscience*, 59, 309, 1994.
29. Leonard, C.S. and Llinas, R., Electrophysiology of mammalian pedunculopontine and laterodorsal tegmental neurons *in vitro*: implications for the control of REM sleep, in *Brain Cholinergic Systems*, Steriade M. and Biesold D., Eds., Oxford University Press, New York, 1990, p. 205.
30. Lin, J.S., Hou, Y., Sakai, K., and Jouvet, M., Histaminergic descending inputs to the mesopontine tegmentum and their role in the control of cortical activation and wakefulness in the cat, *J. Neurosci.*, 16, 1523, 1996.
31. Luebke, J.I., Greene, R.W., Semba, K., Kamondi, A., McCarley, R.W., and Reiner, P.B., Serotonin hyperpolarizes cholinergic low-threshold burst neurons in the rat laterodorsal tegmental nucleus *in vitro*, *Proc. Natl. Acad. Sci. USA*, 89, 743, 1992.

32. Luebke, J.I., McCarley, R.W., and Greene, R.W., Inhibitory action of muscarinic agonists on neurons in the rat laterodorsal tegmental nucleus *in vitro, J. Neurophysiol.*, 70, 2128, 1993.
33. Manzoni, O.J., Manabe, T., and Nicoll, R.A., Release of adenosine by activation of NMDA receptors in the hippocampus, *Science*, 265, 2098, 1994.
34. McCarley, R.W. and Hobson, J.A., Neuronal excitability modulation over the sleep cycle: a structural and mathematical model, *Sci. Washington, D.C.*, 189, 58, 1975.
35. McCarley, R.W. and Massaquoi, S.G., A limit cycle mathematical model of the REM sleep oscillator system, *Am. J. Physiol.*, 251, R1011, 1986.
36. McCarley, R.W., Nelson, J.P., and Hobson, J.A., PGO burst neurons: correlative evidence for neuronal generators of PGO waves, *Science*, 201, 209, 1978.
37. McCormick, D.A., Neurotransmitter actions in the thalamus and cerebral cortex and their role in neuromodulation of thalamocortical activity, *Prog. Neurobiol.*, 39, 337, 1992.
38. McCormick, D.A. and Huguenard, J.R., A model of the electrophysiological properties of thalamocortical relay neurons, *J. Neurophysiol.*, 68, 1384, 1992.
39. McCormick, D.A. and Pape, H.C., Properties of a hyperpolarization-activated cation current and its role in rhythmic oscillation in thalamic relay neurons, *J. Physiol. (Lond.)*, 431, 291, 1990.
40. McCormick, D.A. and Wang, Z., Serotonin and noradrenaline excite GABAergic neurones of the guinea-pig and cat nucleus reticularis thalami, *J. Physiol.*, 442, 235, 1991.
41. McGinty, D.J. and Harper, R.M., Dorsal raphe neurons: depression of firing during sleep in cats, *Brain Res.*, 101, 569, 1976.
42. McIlwain, H. and Poll, J.D., Adenosine in cerebral homeostatic role: appraisal through actions of homocysteine, colchicine, and dipyridamole, *J. Neurobiol.*, 17, 39, 1986.
43. McIlwain, H. and Pull, I., Release of adenine derivatives on electrical stimulation of superfused tissues from the brain, *J. Physiol. (Lond.)*, 221, 9P, 1972.
44. Mitler, M.M. and Dement, W.C., Cataplectic-like behavior in cats after microinjections of carbachol in pontine reticular formation, *Brain Res.*, 68, 335, 1974.
45. Morilak, D.A. and Ciaranello, R.D., 5-HT_2 receptor immunoreactivity on cholinergic neurons of the pontomesencephalic tegmentum shown by double immunofluorescence, *Brain Res.*, 627, 49, 1993.
46. North, R.A., Williams, J.T., Surprenant, A., and Christie, M.J., μ and δ receptors belong to a family of receptors that are coupled to potassium channels, *Proc. Natl. Acad. Sci. USA*, 84, 5487, 1987.
47. Pape, H.C., Adenosine promotes burst activity in guinea-pig geniculocortical neurones through two different ionic mechanisms, *J. Physiol.*, 447, 729, 1992.
48. Porkka-Heiskanen, T., Strecker, R.E., Thakkar, M., Bjørkum, A.A., Greene, R.W., and McCarley, R.W., Adenosine: a mediator of the sleep-inducing effects of prolonged wakefulness, *Science*, 276, 1265, 1997.
49. Portas, C.M., Thakkar, M., Rainnie, D.G., and McCarley, R.W., Microdialysis perfusion of 8-OH-DPAT in the dorsal raphe nucleus decreases serotonin release and increases REM sleep in freely moving cat, *J. Neurosci.*, 16, 2820, 1996.
50. Portas, C.M., Thakkar, M., Rainnie, D.G., Greene, R.W., and McCarley, R.W., Role of adenosine in behavioral state modulation: a microdialysis study in the freely moving cat, *Neuroscience*, 79, 225, 1997.
51. Radulovacki, M., Virus, R.M., Djuricic-Nedelson, M., and Green, R.D., Adenosine analogs and sleep in rats, *J. Pharmacol. Exp. Ther.*, 228, 268, 1984.
52. Radulovacki, M., Role of adenosine in sleep in rats, *Rev. Clin. Bas. Pharmac.*, 5, 327, 1985.
53. Rainnie, D.G., Grunze, H.C., McCarley, R.W., and Greene, R.W., Adenosine inhibition of mesopontine cholinergic neurons: implications for EEG arousal, *Science*, 263, 689, 1994.

54. Rainnie, D.G., McCarley, R.W., and Greene, R.W., Adenosine modulation of glutamatergic transmission in the laterodorsal tegmentum via an action at presynaptic receptors (abstract), *Sleep Res.*, 26, 39, 1997.

55. Reiner, P.B., Clonidine inhibits central noradrenergic neurons in unanesthetized cats, *Eur. J. Pharmacol.*, 115, 249, 1985.

56. Rosenberg, P.A., Knowles, R., Knowles, K.P., and Li, Y., β-adrenergic receptor-mediated regulation of extracellular adenosine in cerebral cortex in culture, *J. Neurosci.*, 14, 2953, 1994.

57. Sakai, K., Petitjean, F., and Jouvet, M., Effects of pontomesencephalic lesions and electrical stimulation upon PGO waves and EMPs in unanesthetized cats, *Electroencephalogr. Clin. Neurophysiol.*, 41, 49, 1976.

58. Sakai, K. and Jouvet, M., Brainstem PGO-on cells projecting directly to the cat lateral geniculate nucleus, *Brain Res.*, 194, 500, 1980.

59. Sanchez, R. and Leonard, C.S., NMDA receptor- mediated synaptic input to nitric oxide synthase-containing neurons of the guinea pig mesopontine tegmentum *in vitro, Neurosci. Lett.*, 179, 141, 1994.

60. Satoh, K. and Fibriger, H.C., Cholinergic neurones of the laterodorsal tegmental nucleus: efferent and afferent connections, *J. Comp. Neurol.*, 253, 277, 1986.

61. Schwierin, B., Borbély, A.A., and Tobler, I., Effects of N^6-cyclopentyladenosine and caffeine on sleep regulation in the rat, *Eur. J. Pharmacol.*, 300, 163, 1996.

62. Semba, K., Reiner, P.B., McGeer, E.G., and Fibiger, H.C., Brainstem projecting neurons in the rat basal forebrain: neurochemical, topographical, and physiological distinctions from cortically projecting cholinergic neurons, *Brain Res. Bull.*, 22, 501, 1989.

63. Siegel, J.M. Brainstem mechanisms generating REM sleep, in *Principles and Practice of Sleep Medicine,* Kryger M.H., Roth T., and Dement W.C., Eds., W.B. Saunders, Philadelphia, 1989, p. 104.

64. Steineger, T.L., Alam, N., Szymusiak, R., and McGinty, D., State-dependent discharge of neurons in the rat posterior hypothalamus, in *Annual Meeting of the Society for Neuroscience*, Washington, D.C., 1996.

65. Steriade, M., McCormick, D.A., and Sejnowski, T.J., Thalamocortical oscillations in the sleeping and aroused brain, *Science*, 262, 679, 1993.

66. Steriade, M., Paré, D., Bouhassira, D., Deschênes, M., and Oakson, G., Phasic activation of lateral geniculate and perigeniculate thalamic neurons during sleep with ponto-geniculo-occipital waves, *J. Neurosci.*, 9, 2215, 1989.

67. Steriade, M., Paré, D., Datta, S., Oakson, G., and Curró Dossi, R., Neuronal activities in brain-stem cholinergic nuclei related to tonic activation processes in thalamocortical systems, *J. Neurosci.*, 10, 2541, 1990.

68. Steriade, M., Paré, D., Datta, S., Oakson, G., and Curró Dossi, R., Different cellular types in mesopontine cholinergic nuclei related to ponto-geniculo-occipital waves, *J. Neurosci.*, 10, 2560, 1990.

69. Szymusiak, R., Magnocellular nuclei of the basal forebrain: substrates of sleep and arousal regulation, *Sleep*, 18, 478, 1995.

70. Thakkar, M., Portas, C., and McCarley, R.W., Chronic low-amplitude electrical stimulation of the laterodorsal tegmental nucleus of freely moving cats increases REM sleep, *Brain Res.*, 723, 223, 1996.

71. Trulson, M.E. and Jacobs, B.L., Raphe unit activity in freely moving cats: correlation with level of behavioral arousal, *Brain Res.*, 163, 135, 1979.

72. Webster, H.H. and Jones, B.E., Neurotoxic lesions of the dorsolateral pontomesencephalic tegmentum-cholinergic cell area in the cat. II. Effects upon sleep-waking states, *Brain Res.*, 458, 285, 1988.

73. Wilcox, K.S., Grant, S.J., Burkhart, B.A., and Christoph, G.R., *In vitro* electrophysiology of neurons in the lateral dorsal tegmental nucleus, *Brain Res. Bull.*, 22, 557, 1989.
74. Williams, J.A. and Reiner, P.B., Noradrenaline hyperpolarizes identified rat mesopontine cholinergic neurons *in vitro, J. Neurosci.*, 13, 3878, 1993.
75. Wu, P.H. and Phillis, J.W., Uptake by central nervous tissues as a mechanism for the regulation of extracellular adenosine concentrations, *Neurochem. Int.*, 6, 613, 1984.
76. Yanik, G., Glaum, S., and Radulovacki, M., The dose-response effects of caffeine on sleep in rats, *Brain Res.*, 403, 177, 1987.

Chapter 18

Intrinsic Electroresponsiviness of Basal Forebrain Cholinergic and Non-Cholinergic Neurons

Angel Alonso

Contents

18.1 Introduction 297
18.2 Electrophysiological properties 299
    18.2.1 Cholinergic Neurons 299
    18.2.2 Non Cholinergic Neurons 302
18.3 Discussion 302
Acknowledgments 305
References 305

18.1 Introduction

The basal forebrain (BF) constitutes a contiguous set of nuclei that envelops the rostral and ventral aspects of the basal ganglia. These include the medial septal nucleus, the vertical and horizontal limbs of the diagonal band of Broca, and the nucleus basalis magnocellularis.[1,2] It has been known for quite a long time that populations of cholinergic neurons within the BF give rise to a rather topographically organized projection which innervates the entire cortical mantle;[3,4] however, the majority of BF neurons are non-cholinergic, and most of these non-cholinergic cells are GABAergic.[2,5] In addition, recent studies have demonstrated that these BF GABAergic neurons also give rise, similarly to the cholinergic cells, to significant projections to most, if not all, cortical areas.[6-11]

An important role of the BF appears to relate to the functional organization of population dynamics in its target neuronal networks across the sleep-waking cycle. In this respect, it is known that the BF constitutes a final relay of the ventral extra-thalamic pathway of the reticular activating system (for review, see References 12 and 13). Indeed, numerous pharmacological, behavioral, and lesion studies indicate that the BF plays a critical role in cortical activation which characterizes the waking state as well as the paradoxical phase of sleep (REM sleep; for review, see Reference 14). In the neocortex, cortical arousal is primarily reflected by low-voltage fast electrographic activity, once considered an index of cortical "desynchronization", though now known to be caused by fast neuronal rhythms in the "gamma" range.[15-19] On the other hand, in limbic cortices, particularly in the entorhinal cortex and hippocampus, cortical arousal is primarily manifested by very robust rhythmic slow activity in the theta range (4to 12 Hz; "theta rhythm").[20-22] Limbic theta is always nevertheless accompanied by gamma activity.[23-26]

The topographical organization of the BF cortical projections appears to be such that neurons within the medial septal nucleus as well as within the horizontal and vertical limb of the diagonal band of Broca (BF subnuclei; hereafter commonly refer to as the medial septum, or MS) project primarily to the hippocampus and associated limbic structures, while neurons in the nucleus basalis (NB) project primarily, though not exclusively,[6] to the neocortex. In this respect, lesion experiments have demonstrated that the development of theta rhythmicity in the limbic network depends on the integrity of the MS.[20,27] MS cholinergic and GABAergic afferents probably play distinct roles: cholinergic inputs tonically modulating the excitability of cortical neurons and fast GABAergic inputs providing rhythmic pacemaking. This scenario is supported, for example, by the observation that after selective damage of BF cholinergic neurons the frequency of theta remains unchanged, but the power is drastically reduced.[28] Unitary recordings from MS neurons *in vivo* have demonstrated that the vast majority of MS neurons discharge rhythmically phase-locked to the theta rhythm,[27,29-31] although different subtypes of rhythmic cells have been described according to their discharge pattern and theta phase relations.[30] In addition, medial septal neurons have been shown to continue rhythmical discharge *in vivo* after anatomical disconnection of the MS.[32] Furthermore, recent *in vitro* intracellular studies have revealed that BF non-cholinergic (putative GABAergic) neurons in the BF display robust intrinsic autorhythmicity at theta, and also gamma, frequencies.[33,34]

With respect to the NB, lesions of this BF structure have been shown to cause the development of slow electrographic waves (in the delta range) in the neocortex (as opposed to LVFA or cortical activation). While some NB neurons increase their firing rate during waking,[12,35-37] still others increase their firing rates and/or switch into a bursting firing pattern during slow waves.[35,38,39] *In vitro* intracellular studies have also revealed that, similarly to putative GABAergic MS neurons, putative GABAergic NB neurons also display intrinsic autorhythmicity at theta and gamma frequencies.[33]

In order to understand how one particular brain structure (in this case the BF) influences the population behavior of complex neuronal networks, one needs to consider the intrinsic biophysical/molecular properties of the neurons involved, as well as synaptic properties of the related neuronal circuits and the actions of neuromodulatory influences. As demonstrated in the thalamo-cortical system, the above three levels of organization interact dynamically to generate population behaviors such as spindle oscillations or delta rhythm.[40-43] As compared to the wealth of information with respect to the thalamo-cortical network, much less is known in the case of the BF, though considerable information has been accumulating over the last few years. In this regard, this chapter reviews recent findings with respect to the electroresponsive properties of BF neurons. A picture is emerging in which the chemomorphological heterogeneity of BF neurons is also reflected in a diversity of electrophysiological properties. Clear-cut differences have been shown to exist between the intrinsic electroresponsiveness of cholinergic and GABAergic BF neurons, and also between cholinergic neurons in different BF subregions. Interestingly, though not surprisingly, the biophysical properties of BF non-cholinergic neurons endow these cells with intrinsic oscillatory dynamics which, in some cases, clearly relate to the genesis of some forebrain rhythms, such as the "theta" and, possibly, "gamma" rhythms. In addition, the electrophysiological and pharmacological complexity of NB cholinergic neurons suggests that these cells not only participate in the "tuning" of cortical activation but may also play an important role in the control of forebrain activities across the sleep/waking cycle.

18.2 Electrophysiological properties

18.2.1 Cholinergic Neurons

The electrophysiological properties of cholinergic neurons (see Chapter 14) in the region of the basal forebrain comprising the substantia-innominata-magnocellular preoptic area (for simplicity hereafter referred to as nucleus basalis, NB) has been described in detail in References 33, 44, and 45. The distinctive electrophysiological feature of these cells is reminiscent of thalamic neurons[46,47] in their capacity to discharge a low-threshold burst (LTB) of spikes when depolarized from their typically negative (~–70mV) resting membrane potential (Figure 18.1). In addition, LTBs evoked from the resting level (Figure 18.1A4) always emerge with a pronounced delay, characteristic of transient outward rectification. Consistent with this interpretation, when the cells are depolarized from more positive potentials (e.g., ~–65mV), LTBs develop without marked delay (Figure 18.1A3). Further depolarization decreases the robustness of the bursts (which became spike doublets; Figure 18.1A2), and finally at potentials positive to ~–55mV the bursting mechanism becomes inactivated and the cells discharge tonically with single spikes (Figure 18.1A). LTBs of BF-ACh neurons are always followed by large-amplitude (10 to 20 mV) and long-lasting afterhyperpolarizations (sAHPs) (Figures 18.1A3 and 18.2). The mechanism undelying the LTBs of NB-ACh neurons has been shown to be a very robust, low-voltage-activated (T-type) Ca^{++} current (Figures 18.1B1 through Figure 18.3),[44,48–50] which in combination with the strong, high-voltage-activated Ca^{++} currents that the cells also possess (Figure 18.1B1), trigger the sAHP.[50]

Interestingly, a conspicuous feature of BF-ACh cells is their ability to discharge with a rhythmic series of LTBs at a slow frequency (0.5 to 3Hz), both in response to a square, long-lasting current pulse and, most distinctively, when slowly depolarized from their resting membrane potential (Figure 18.2). This rhythmic bursting ability in response to membrane depolarization clearly separates the intrinsic electroresponsiveness of the NB-ACh cells from that of the thalamocortical neurons, although it bears some resemblance to the typical bursting cells of the nucleus reticularis. Indeed, the T-type, Ca^{++} current of NB neurons displays a relatively high threshold for activation (~–55 mV) and slow inactivation kinetics (Figure 18.1B2) (t_d of ~–80 msec at –50 mV),[48,51] characteristics that are similar to the I_t present in nucleus reticularis neurons.[52]

Cholinergic cells along the BF are electrophysiologically heterogeneous. While the above description refers to the cholinergic neurons in the NB, cholinergic cells in the MS do not display the intrinsic ability to burst. As described by Griffith[53] and others,[45,54] MS-ACh neurons respond to membrane depolarization with broad single spikes, never a burst. Other aspects of their electrophysiology are, however, similar to those of NB-ACh neurons. That is, MS-ACh neurons also display robust transient outward rectification, and single spikes are followed by a pronounced sAHP.

That cholinergic neurons of the NB switch their firing pattern from a bursting mode to a slow tonic mode as a function of membrane depolarization is particularly interesting, given the potential role of these neurons in the control of cortical operations during the sleep/waking cycle. It is thus of particular interest to study the action of neuromodulatory influences upon these neurons. A potentially important one is that of neurotensin (NT), as the receptors to this neuropeptide have been shown by anatomical techniques to be selectively associated to cholinergic neurons[55,56] within the BF. In addition, NT has been shown to produce awakening effects and associated actions on the EEG when injected intraventricularly. For these reasons, and in collaboration with Beaudet, we recently investigated in BF slices the neuromodulatory actions of NT on cholinergic neurons.[45] We provided unequivocal demonstration of BF cholinergic neurons as direct NT targets by confocal laser scanning microscopic demonstration of internalization of fluo-NT within biocytin-filled, electrophysiologically identified cholinergic cells. To our surprise, NT produced extremely robust neuromodulatory actions on the intrinsic electroresponsiveness of NB cholinergic neurons. As illustrated in Figure 18.3, the most significant effect of NT was to engender an intrinsically

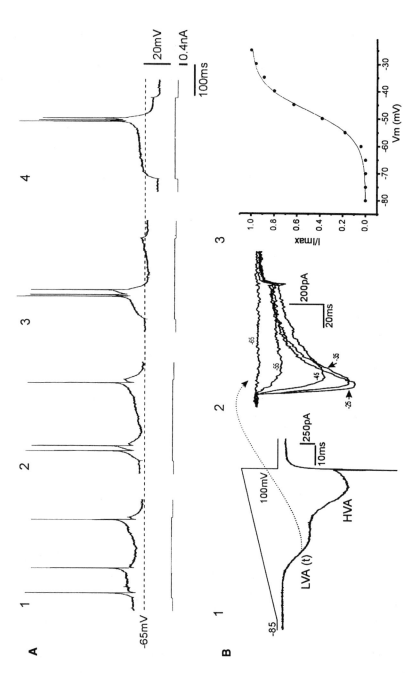

FIGURE 18.1

Low-threshold bursting discharge of cholinergic neurons. **(A)** Discharge patterns of a cholinergic neuron in response to constant current-pulse injection at the resting level (3), at a hyperpolarized level (4), and at depolarized levels (1 and 2). **(B)** Whole-cell calcium current in an acutely dissociated cholinergic neuron in 3-mM extracellular calcium. Note in panel 1 that a 100-mV ramp depolarization (1 mV/1 msec) from a holding potential of −85 mV evokes a robust current with two clear peaks at low and high voltages, respectively, indicating the presence of robust low-voltage activated (LVA; t) and high-voltage activated (HVA) calcium currents. In panel 2, the LVA current is selectively activated by step depolarizations to the indicated potentials, and panel 3 shows the activation curve constructed from this data, the continuous line corresponding to a Boltzmann fit. (Part A from Alonso, A. et al. *Eur. J. Neurosci.*, 8, 169, 1996; with permission. Part B corresponds to preliminary observations reported in Klink, R. et al., *Soc. Neurosci. Abstr.*, 19, 1993.)

Intrinsic Electroresponsiveness of Basal Forebrain Neurons

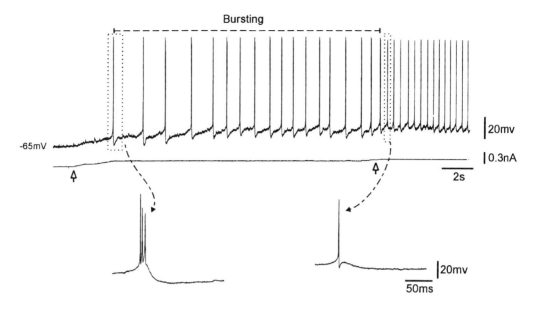

FIGURE 18.2
Firing profile of cholinergic neurons in response to D.C. depolarization. A slow ramp depolarization (3 mV/sec) triggers a rhythmic bursting discharge (dashed line) comprising LTBs at −55 mV (enlargement at bottom left). A further minor D.C. depolarization above −55 mV subsequently triggers a sustained single spike discharge (enlargement at bottom right). (From Alonso, A. et al., *Eur. J. Neurosci.*, 8, 169, 1996. With permission.)

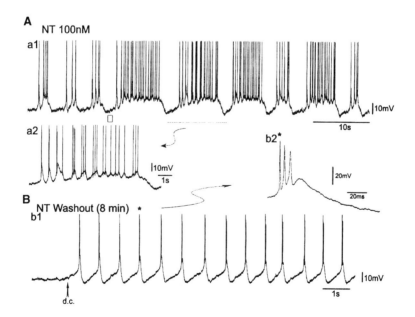

FIGURE 18.3
Neurotensin induces complex, spindle-like bursting sequences in cholinergic neurons. (**A**) Complex bursting behavior immediately following 4-min superfusion 100 n*M* neurotensin. The burst sequence signaled by a dotted line in (a1) is illustrated at an expanded time base in (a2). (**B**) Rhythmic bursting pattern triggered by D.C. depolarization from the resting level (−67 mV) after 8 min of NT washout. The burst event signaled by an asterisk in (b1) is illustrated at an expanded time base in (b2). (From Alonso, A. et al., *J. Neurosci.*, 14, 5778, 1994. With permission.)

generated, prominent, slow, rhythmic bursting pattern that could be shaped into complex spindle-like sequences. Clearly, the rich electroresponsiveness demonstrated by NB cholinergic neurons in normal conditions can be molded into complex firing behaviors under neuromodulatory influences, a statement well in line with the idea that NB cholinergic neurons play a major role in the control of cortical dynamics. NT was also found to depolarize MS cholinergic neurons strongly but never to induce a bursting pattern.[45]

18.2.2 Non-Cholinergic Neurons

The intrinsic electroresponsiveness of immunohistochemically characterized ChAT-negative neurons of the NB was recently described in detail by Alonso et al.[33] These neurons were clearly differentiated from the cholinergic cells by (1) the absence of LTB discharge, (2) a much faster spike followed by a fast afterhyperpolarization (fAHP), and (3) the ability to discharge at high firing rates. These properties had been described previously as being characteristic of MS non-cholinergic neurons.[54,57] However, we found that the most salient electrophysiological property of most non-cholinergic NB neurons was their ability to display an intrinsically generated rhythmic bursting discharge at 2 to 10 Hz (Figure 18.4A,B) composed of non-adapting spike clusters (Figure 18.4B,C,D) interspersed by rhythmic subthreshold oscillations at ~20 to 70Hz (Figure 18.4E,F). Importantly, the frequency of the subthreshold oscillations was found to be approximately the same as that of the intra-cluster firing frequency, and because spikes arise at the peak of the oscillations, spike timing and frequency might be also largely determined by the mechanism underlying the oscillations. Importantly, NB non-cholinergic cells also displayed robust transient outward rectification (note in Figures 18.4A and E that each spike cluster is terminated by a sudden break of the fAHP repolarizing phase followed by a slowly rising membrane depolarization and oscillations). Following the above observations, Serafin et al.[34] re-examined the firing pattern of MS non-cholinergic (potentially GABAergic) neurons to find that these also display a rhythmic clustered discharge equivalent to that of the non-cholinergic cells in the NB. Thus, different from cholinergic cells, most non-cholinergic cells across the BF display a very similar rhythmic firing pattern.

With regard to the mechanism underlying the rhythmic subthreshold oscillations and rhythmic spike-cluster discharge of BF non-cholinergic cells, preliminary experimental evidence[34] suggests that the subthreshold oscillations are dependent on the activation of a persistent Na current.[58] In the basalis cells, the progressive decline of the peak voltages during the spike cluster (Figure 18.4C) and the progressive increase of the membrane potential during the inter-cluster interval (Figure 18.4A,C) suggest that the transient outward rectification displayed by the cells may be an important factor in determining the spike clustering, particularly for the duration of the spike cluster and inter-cluster interval. Indeed, modeling studies carried out by Wang[59] on the issue of 40-Hz neuronal oscillations have demonstrated that the interplay of a persistent Na current and a transient outward rectifier is not only able to produce the intrinsic subthreshold oscillations but also a spike-cluster discharge.

18.3 Discussion

I have summarized above a series of recent studies on the intrinsic electroresponsiveness of BF neurons that have brought about several important new findings. On the one hand, non-cholinergic (presumably GABAergic) neurons throughout the BF show a similar electrophysiological identity, displaying robust intrinsic pacemaker mechanisms which allow them to discharge spontaneously in rhythmic trains (spike clusters) of action potentials in the theta range of frequencies. Moreover, the spike clusters of these GABAergic cells are interspersed by rhythmic subthreshold oscillations, and the frequency of these subthreshold oscillations overlaps with the intra-cluster firing frequency and

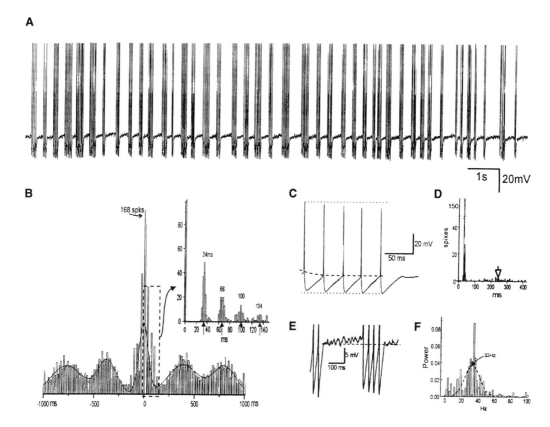

FIGURE 18.4
Basal forebrain non-cholinergic neurons fire in rhythmic clusters of spikes. (**A**) Spontaneous activity of a typical cluster cell. (**B**) Autocorrelation histogram of the discharge in (A). The inset to the right demonstrates the rhythmic character of the intra-cluster firing at an frequency of 30 Hz. (**C**) An individual spike-cluster from (A) at expanded time and voltage scales reveals that as the cluster proceeds the peak amplitude of the action potential decreases, while the peak of the AHP reaches more negative levels. (**D**) First-order inter-spike interval from (A) demonstrating (as inset in B) the regularity of the intra-cluster intervals. (**E**) A short trace from (A) at expanded time and voltage scales reveals clear subthreshold oscillations during the inter-cluster interval. Note that the oscillations evolve during a ramp depolarization leading to a new spike cluster. (**F**) Power spectrum of the membrane potential oscillations. A Gaussian fit between 10 and 60 Hz indicates a frequency peak at 33 Hz. (Part F adapted from Alonso, A. et al., *Eur. J. Neurosci.*, 8, 169, 1996.)

falls within the electrographic gamma range (40 Hz). On the other hand, the intrinsic electroresponsiveness of BF cholinergic neurons is clearly distinct from that of the GABAergic cells, although there are some heterogeneities within the population of cholinergic cells themselves (i.e., between the NB and MS). Mainly, NB cholinergic cells display a voltage-dependent switch in firing pattern that allows them to discharge rhythmically (in the delta range of frequencies: 1 to 4 Hz) with a LTB of action potentials when relatively hyperpolarized (–65 to –70 MV) and tonically with single spikes at low frequencies at more depolarized levels. In contrast, MS cholinergic neurons always discharge with single, broad-action potentials at a slow frequency.[45,53,54]

That presumptive GABAergic cells within the MS display intrinsic autorhythmicity at theta frequencies is not surprising given the important role of MS GABAergic cells in the genesis of the theta rhythm.[27–31] Indeed, the currently most supported hypothesis is that theta emerges in the hippocampus largely as a result of rhythmic pacemaker input of MS GABAergic cells onto hippocampal GABAergic interneurons. Upon disinhibition, these interneurons would then discharge to rhythmically pace the pyramidal cells with IPSPs. Numerous lines of evidence support this hypothesis. First, MS GABAergic cells project upon hippocampal GABAergic interneurons.[7,60] Second, hippocampal interneurons discharge rhythmically at theta frequencies.[61–63] Third, during theta rhythm, the membrane potential of hippocampal pyramidal cells displays rhythmic oscillations primarily driven by IPSPs.[63] Finally, selective lesion of MS cholinergic cells decreased, but failed to abolish, hippocampal theta rhythm.[28] A recent study by Toth et al.[64] has tested this hypothesis in a septo-hippocampal slice preparation, and they did, indeed, observe rhythmic firing of hippocampal interneurons and correlated rhythmic inhibition of hippocampal pyramidal cells following rhythmic stimulation of the medial septum. The fact that the oscillations of the GABAergic cells and the intracluster firing frequency are in the gamma range suggest that these neurons may also participate in the emergence of gamma oscillations which are known to occur simultaneously with the theta rhythm.[15,23–26]

With regard to the emergence of hippocampal theta, several important issues remain. For example, what is the role of the MS cholinergic cells? Cholinergic actions in the hippocampus are known to affect very substantially the excitability of the pyramidal cells[65,66] and the local network population dynamics.[67–70] These actions most likely facilitate the GABAergic pacemaker drive to enable the full emergence of theta. It is also unclear whether rhythmic activity in an isolated septal preparation can engage into synchronized population activity. Local MS connectivity appears to exist between GABAergic cells as well as between GABAergic and cholinergic neurons.[71] It might be that local MS connectivity is enough for the emergence of synchronized population activity or, perhaps, that a hippocampal feedback is necessary.[72,73]

A parallel issue to the above is the emergence of theta rhythm in other limbic cortices such as the entorhinal cortex (EC). The EC provides the major cortical afferent input to the hippocampus, and an extremely robust theta rhythm is generated by neurons in the superficial layers of the EC and, in particular, neurons in layer II.[22,74,75] EC layer II cells give rise to the major component of the perforant path.[76,77] Interestingly, these neurons also display a prominent intrinsic autorhythmicity consisting of very rhythmic subthreshold oscillations at theta frequency.[58,78] This autorhythmicity bears some similarities to that present in BF GABAergic cells and is not observed with such robustness in other limbic neurons. MS GABAergic cells also innervate the EC. This suggests that mechanisms similar to those that operate in the hippocampus for the emergence of theta may be also operative in the EC. However, great differences exist between the electrophysiology of EC and hippocampal neurons, as well as in the actions of acetylcholine upon these two neuronal populations.[79,80] While theta may have a global functional role, in different neuronal networks theta oscillations and their underlying mechanisms may affect the processing capabilities of these networks differently.

Somewhat surprising is the fact that non-cholinergic (presumptive GABAergic) cells in the NB have equivalent autorhythmic properties to those in the MS. It is not clear whether MS and NB GABAergic cells are interconnected and may form a functional system. However, recent studies[8–11] have shown that NB GABAergic cells also project quite heavily upon the neocortex and, similarly to the hippocampus, contact neocortical GABAergic interneurons.[10] It has been proposed that resonant cortico-thalamic activity may be set in the gamma range, in part due to the presence of gamma autorhythmicity in cortical interneurons[81] and other cell types.[82,83] Activation of the NB can stimulate gamma activity in the neocortex.[84] It thus seems likely that NB GABAergic neurons form part of a more extensive forebrain network that participates in the control/modulation of neocortical gamma rhythms. Equally, they may also contribute to the presence of a background, slower oscillation in the theta range to coordinate large populations of neurons, including limbic-neocortical interactions.[85]

The presence of an intrinsic rhythmic bursting mechanism in NB cholinergic neurons and not in MS cholinergic cells also came as somewhat of a surprise. On the one hand, this finding points out the heterogeneity of the BF cholinergic system and suggests some different functional role of NB vs. MS cholinergic neurons. In this respect, there is substantial evidence that NB cholinergic neurons promote cortical arousal.[38] It is to be expected, therefore, that these cells would increase their firing rates during cortical activation. Indeed, several studies have identified a group of BF projection neurons that fire at low rates during slow-wave sleep and increase their firing rates during waking and paradoxical sleep.[12,35-37] Interestingly, many of these cells switched to a bursting mode during slow-wave sleep.[35] This observation may help to interpret other somewhat contrasting studies implicating the BF in the control of sleep.[38,86-88] It seems thus reasonable to hypothesize that NB neurons may play a major role in the control of cortical dynamics across the sleep/waking cycle, manifesting a tonic discharge during the waking state and switching under neuromodulatory influences into a slow bursting pattern during slow-wave sleep. At this point, discussion about what the actions of these two distinct firing modes would be upon the target neuronal elements would be highly speculative, but it is certainly an extremely interesting subject for future research.

Acknowledgments

The author wishes to thank Dr. C. Dickson for his criticisms and suggestions on the manuscript. The work was supported by grants from the Medical Research Council of Canada. Dr. A. Alonso is an MNI Killam Scholar.

References

1. Mesulam, M.-M., Mufson, E.J., Levey, A.I., and Wainer, B.H., Cholinergic innervation of cortex by the basal forebrain: cytochemistry and cortical connections of the septal area, diagonal band nuclei, nucleus basalis (substantia innominata) and hypothalamus in the rhesus monkey, *J. Comp. Neurol.*, 214, 170, 1983.
2. Gritti, I., Mainville, L., and Jones, B.E., Codistribution of GABA- with acetylcholine-synthesizing neurons in the basal forebrain of the rat, *J. Comp. Neurol.*, 329, 438, 1993.
3. Rye, D.B., Wainer, B.H., Mesulam, M.-M., Mufson, E.J., and Saper, C.B., Cortical projections arising from the basal forebrain: a study of cholinergic and noncholinergic components employing combined retrograde tracing and immunohistochemical localization of choline acetyltransferase, *Neuroscience*, 13, 627, 1984.
4. Saper, C.B., Organization of cerebral cortical afferent systems in the rat. I. Magnocellular basal nucleus., *J. Comp. Neurol.*, 222, 313, 1984.
5. Onteniente, B., Tago, H., Kimura, H., and Maeda, T., Distribution of gamma-aminobutyric acid-immunoreactive neurons in the septal region of the rat brain., *J. Comp. Neurol.*, 248, 422, 1986.
6. Alonso, A. and Köhler, C., A study of the reciprocal connections between the septum and the entorhinal area using anterograde and retrograde axonal transport methods in the rat brain, *J. Comp. Neurol.*, 225, 327, 1984.
7. Freund, T.F. and Antal, M., GABA-containing neurons in the septum control inhibitory interneurons in the hippocampus, *Nature*, 336, 170, 1988.
8. Zaborszky, L., Carslen, J., Brashear, H.R., and Heimer, L., Cholinergic and GABAergic afferents to the olfactory bulb in the rat with special enphasis on the projection neurons in the nucleus of the horizontal limb of the diagonal band, *J. Comp. Neurol.*, 243, 488, 1986.
9. Fisher, R.S., Buchwald, N.A., Hull, C.D., and Levine, M.S., GABAergic basal forebrain neurons project to the neocortex: the localization of glutamic acid decarboxylase and choline acetyltransferase in feline corticopetal neurons, *J. Comp. Neurol.*, 272, 489, 1988.
10. Freund, T.F. and Meskenaite, V., γ-aminobutyric acid-containing basal forebrain neurons innervate inhibitory interneurons in the neocortex, *Proc. Natl. Acad. Sci. USA*, 89, 738, 1992.

11. Jones, B.E., Mainville, L., Mancia, M., and Critti, I., GABAergic and other non-cholinergic basal forebrain neurons project to both meso-iso-cortical regions in the rat brain, 21, 1617, 1995.
12. Buzsáki, G., Bickford, R.G., Ponomareff, G., Thal, L.J., Mandel, R., and Gage, F.H., Nucleus basalis and thalamic control of neocortical activity in the freely moving rat, *J. Neuroscience*, 8, 4007, 1988.
13. Jones, B.E., The organization of central cholinergic systems and their functional importance in sleep/waking states, in *Progress in Brain Res.*, Cuello, A.C., Ed., Elsevier, Amsterdam, 1993, p. 61.
14. Vanderwolf, C.H., Cerebral activity and behavior: control by central cholinergic and serotoninergic systems, *Int. Rev. Neurobiol.*, 30, 225, 1988.
15. Maloney, K.J., Cape, E.G., Gotman, J., and Jones, B.E., High-frequency electroencephalogram activity in association with sleep/wake states and spontaneous behaviors in the rat, *Neuroscience*, 76, 541, 1997.
16. Bressler, S.L., The gamma wave: a cortical information carrier, *Trends Neurosci.*, 13, 161, 1990.
17. Singer, W., Synchronization of cortical activity and its putative role in information processing and learning, *Ann. Rev. Physiol..*, 55, 349, 1993.
18. Steriade, M., Amzica, F., and Contreras, D., Synchronization of fast (30–40Hz) spontaneous cortical rhythms during brain activation, *J. Neurosci.*, 16, 392, 1996.
19. Llinas, R. and Ribary, U., Coherent 40-Hz oscillation characterizes dream state in humans, *Proc. Natl. Acad. Sci. USA*, 90, 2078, 1993.
20. Green, J.D. and Arduini, A., Hippocampal electrical activity in arousal, *J. Neurophysiol.*, 17, 533, 1954.
21. Bland, B.H., The physiology and pharmachology of hippocampal formation theta rhythms, *Prog. Neurobiol.*, 26, 1, 1986.
22. Alonso, A. and García-Austt, E., Neuronal sources of theta rhythm in the entorhinal cortex. I. Laminar distribution of theta field potentials, *Exp. Brain Res.*, 67, 493, 1987.
23. Stumpf, C., The fast component in the electrical activity of rabbit's hippocampus, *Electroencephalogr. Clin. Neurophysiol.*, 18, 477, 1965.
24. Chrobak, J.J. and Buzsaki, G., Gamma oscillations in the input network of the entorhinal-hippocampal axis of the freely-behaving rat, *J. Neurosci.*, 1997.
25. Bragin, A., Jando, G., Nadasdy, Z., Hetke, J., Wise, K., and Buzsáki, G., Gamma (40–100 Hz) oscillation in the hippocampus of the behaving rat, *J. Neurosci.*, 15, 47, 1995.
26. Soltesz, I. and Deschênes, M., Low- and high-frequency membrane potential oscillations during theta activity in CA1 and CA3 pyramidal neurons of the rat hippocampus under ketamine-xylazine anesthesia, *J. Neurophysiol.*, 70, 97, 1993.
27. Stewart, M. and Fox, S.E., Do septal neurons pace the hippocampal theta rhythm?, *Trends Neurosci.*, 13, 163, 1990.
28. Lee, M.G., Chrobak, J.J., Sik, A., Wiley, R.G., and Buzsaki, G., Hippocampal theta activity following selective lesion of the septal cholinergic system, *Neuroscience*, 62, 1033, 1994.
29. Alonso, A., Gaztelu, J.M., Buño, J.W., and García-Austt, E., Cross-correlation analysis of septo-hippocampal neurons during theta rhythm, *Brain Res.*, 413, 135, 1987.
30. Gaztelu, J.M. and Buño, W., Septo-hippocampal relationships during EEG theta rhythm, *Electroencephalogr. Clin. Neurophysiol.*, 54, 375, 1982.
31. Petsche, H., Stumpf, C., and Gogolak, G., The significance of the rabbit's septum as a rely station between the midbrain and the hippocampus. I. The control of hippocampus arousal activity by the septum cells, *Electroencephalogr. Clin. Neurophysiol.*, 19, 25, 1962.
32. Vinogradova, O.S., Brazhnik, E.S., Karanov, A.M., and Zhadina, S.D., Neuronal activity of the septum following various types of deafferentation, *Brain Res.*, 187, 353, 1980.

33. Alonso, A., Khateb, A., Fort, P., Jones, B.E., and Muhlethaler, M., Differential oscillatory properties of cholinergic and non-cholinergic neurons in the nucleus basalis, *Eur. J. Neurosci.*, 8, 169, 1996.
34. Serafin, M., Williams, S., Khateb, A., Fort, P., and Mühlethaler, M., Rhythmic firing of medial septum non-cholinergic neurons, *Neuroscience*, 75, 671, 1996.
35. Détári, L. and Vanderwolf, C.H., Activity of identified cortically projecting and other basal forebrain neurones during large slow waves and cortical activation in anaesthetized rats, *Brain Res.*, 437, 1, 1987.
36. Détári, L., Juhász, G., and Kukorelli, T., Firing properties of cat basal forebrain neurones during sleep/wakefulness cycle, *Electroencphalogr. Clin. Neurophysiol.*, 58, 362, 1984.
37. Detari, L., Juhasz, G., and Kukorelli, T., Neuronal firing in the pallidal region: firing patterns during sleep/wakefulness cycle in cats, *Electroencephalogr. Clin. Neurophysiol.*, 67, 159, 1987.
38. Szymusiak, R. and McGinty, D., Sleep-waking discharge of basal forebrain projection neurons in cats, *Brain Res. Bull.*, 22, 423, 1989.
39. Nuñez, A., Unit activity of rat basal forebrain neurons: relationship to cortical activity, *Neuroscience*, 72, 757, 1996.
40. Llinas, R.R., The intrinsic electrophysiological properties of mammalian neurons: insights into central nervous system function, *Science*, 242, 1654, 1988.
41. Steriade, M., McCormicK, D.A., and Sejnowski, T.J., Thalamocortical oscillations in the sleeping and aroused brain, *Science*, 262, 679, 1993.
42. von Krosigk, M., Bal, T., and McCormick, D.A., Cellular mechanisms of a synchronized oscillation in the thalamus, *Science*, 261, 361, 1993.
43. McCormick, D.A., Neurotransmitter actions in the thalamus and cerebral cortex and their role in neuromodulation of thalamocortical activity, *Prog. Neurobiol.*, 39, 337, 1992.
44. Khateb, A., Muhlethaler, M., Alonso, A., Serafin, M., Mainville, L., and Jones, B.E., Cholinergic nucleus basalis neurons display the capacity for rhythmic bursting activity mediated by low threshold calcium spikes, *Neuroscience*, 51, 489, 1992.
45. Alonso, A., Faure, M.-P., and Beaudet, A., Neurotensin promotes oscillatory bursting behavior and is internalized in basal forebrain cholinergic neurons, *J. Neurosci.*, 14, 5778, 1994.
46. Llinas, R. and Jahnsen, H., Electrophysiology of mammalian thalamic neurones *in vitro*, *Nature*, 297, 406, 1982.
47. Steriade, M. and Llinás, R.R., The functional states of the thalamus and the associated neuronal interplay, *Physiol. Rev.*, 68, 649, 1988.
48. Allen, T.G.J., Sim, J.A., and Brown, D.A., The whole-cell calcium current in acutely dissociated magnocellular cholinergic basal forebrain neurones of the rat, *J. Physiol.*, 460, 91, 1993.
49. Griffith, W.H., Taylor, L., and Davis, M.J., Whole-cell and single-channel calcium currents in guinea pig basal forebrain neurons, *J. Neurophysiol.*, 71, 2359, 1994.
50. Williams, S., Serafin, M., Muhlethaler, M., and Bernhein, L., Distinct contributions of high- and low-voltage-activated calcium currents to afterhyperpolarizations in cholinergic nucleus basalis neurons of the guinea pig, *J. Neurosci.*, 17, 7307, 1997.
51. Klink, R., Faure, M.P., Kay, A.R., and Alonso, A., Diversity of Na and Ca currents in cholinergic nucleus basalis neurons, *Soc. Neurosci. Abstr.*, 19, 1993.
52. Huguenard, J.R. and Prince, D.A., A novel T-type current underlies prolonged Ca^{2+}-dependent burst firing in GABAergic neurons of rat thalamic reticular nucleus, *J. Neurosci.*, 12, 3804, 1992.
53. Griffith, W.H., Membrane properties of cell types within guinea pig basal forebrain nuclei *in vitro*, *J. Neurophysiol.*, 59, 1590, 1988.
54. Markram, H. and Segal, M., Electrophysiological characteristics of cholinergic and non-cholinergic neurons in the rat medial septum-diagonal band complex., *Brain Res.*, 513, 171, 1990.

55. Szigethy, E., Wenk, G.L., and Beaudet, A., Anatomical substrate for neurotensin-acetylcholine interactions in the rat basal forebrain, *Peptides*, 9, 1227, 1989.
56. Szigethy, E., Leonard, K., and Beaudet, A., Ultrastructural localization of ^{125}I neurotensin binding sites to cholinergic neurons of the rat nucleus basalis magnocellularis, *Neuroscience*, 36, 377, 1990.
57. Griffith, W.H. and Mathews, R.T., Electrophysiology of AChE-positive neurons in basal forebrain slices, *Neurosci. Lett.*, 71, 169, 1986.
58. Alonso, A. and Llinas, R.R., Subthreshold Na$^+$-dependent theta-like rhythmicity in stellate cells of entorhinal cortex layer II, *Nature*, 342, 175, 1989.
59. Wang, X.-J., Ionic basis for intrinsic 40 Hz neuronal oscillations, *NeuroReport*, 5, 221, 1993.
60. Freund, T. and Buzsaki, G., Interneurons of the hippocampus, *Hippocampus*, 6, 347, 1996.
61. Buzsáki, G., Leung, L.-W., and Vanderwolf, C.H., Cellular bases of hippocampal EEG in the behaving rat, *Brain Res. Rev.*, 6, 139, 1983.
62. Buzsáki, G. and Chrobak, J.J., Temporal structure in spatially organized neuronal ensembles: a role for interneuronal networks, *Curr. Opin. Neurobiol.*, 5, 504, 1995.
63. Ylinen, A., Soltesz, I., Bragin, A., Penttonen, M., Sik, A., and Buzsáki, G., Intracellular correlates of hippocampal theta rhythm in identified pyramidal cells, granule cells, and basket cells, *Hippocampus*, 5, 78, 1995.
64. Tóth, K., Freund, T., and Miles, R., Disinhibition of rat hippocampal pyramidal cells by GABAergic afferents from the septum, *J. Physiol.*, 500, 463, 1997.
65. Madison, D.V., Lancaster, B., and Nicoll, R.A., Voltage clamp analysis of cholinergic action in the hippocampus, *J. Neurosci.*, 7, 733, 1987.
66. Benardo, L.S. and Prince, D.A., Cholinergic excitation of mammalian hippocampal pyramidal cells, *Brain Res.*, 249, 315, 1982.
67. Konopacki, J., Mac Iver, M.B., Roth, S.H., and Bland, B.H., Carbachol-induced EEG "theta" activity in hippocampal brain slices, *Brain Res.*, 405, 196, 1987.
68. Colom, L.V., Nassif-Caudarella, S., Dickson, C.T., Smythe, J.W., and Bland, B.H., In vivo intrahippocampal microinfusion of cabachol and bicuculline induces theta-like oscillations in the septally deafferented hippocampus, *Hippocampus*, 1, 381, 1991.
69. Krnjevic, K. and Ropert, N., Electrophysiological and pharmacological characteristics of facilitation of hippocampal population spikes by stimulation of the medial septum, *Neuroscience*, 7, 2165, 1982.
70. Krnjevic, K., Central cholinergic mechanisms and function, 1993, 285, 1993.
71. Leranth, C. and Frotscher, M., Organization of the septal region in the rat brain: cholinergic-GABAergic interconnections and the termination of hippocampo-septal fibers, *J. Comp. Neurol.*, 289, 304, 1989.
72. Alonso, A. and Köhler, C., Evidence for separate projections of hippocampal pyramidal and non-pyramidal neurons to different parts of the septum in the rat brain, *Neurosci. Lett.*, 31, 209, 1982.
73. Tóth, K., Borthegyi, Z., and Freund, T.F., Postsynaptic targets of GABAergic hippocampal neurons in the medial septum-diagonal band of Broca complex, *J. Neurosci.*, 13, 3712, 1993.
74. Alonso, A. and García-Austt, E., Neuronal sources of theta rhythm in the entorhinal cortex of the rat. II. Phase relations between unit discharges and theta field potentials, *Exp. Brain Res.*, 67, 502, 1987.
75. Dickson, C.T., Kirk, I.J., Oddie, S.D., and Bland, B.H., Classification of theta-related cells in the entorhinal cortex: cell discharges are controlled by the ascending brainstem synchronizing pathway in parallel with hippocampal theta-related cells, *Hippocampus*, 5, 306, 1995.
76. Steward, O. and Scoville, S.A., The cells of origin of entorhinal afferents to the hippocampus and fascia dentata of the rat., *J. Comp. Neurol.*, 169, 347, 1976.
77. Schwartz, S.P. and Coleman, P.D., Neurons of origin of the perforant path, *Exp. Neurol.*, 74, 305, 1981.

78. Alonso, A. and Klink, R., Differential electroresponsiveness of stellate and pyramidal-like cells of medial entorhinal cortex layer II, *J. Neurophysiol.*, 70, 128, 1993.
79. Klink, R. and Alonso, A., Muscarinic modulation of the oscillatory and repetitive firing properties of entorhinal cortex layer II neurons, *J. Neurophysiol.*, 77, 1813, 1997.
80. Klink, R. and Alonso, A., Ionic mechanisms of muscarinic depolarization in entorhinal cortex layer II neurons, *J. Neurophysiol.*, 77, 1829, 1997.
81. Llinas, R.R., Grace, A.A., and Yarom, Y., *In vitro* neurons in mammalian cortical layer 4 exhibit intrinsic oscillatory activity in the 10- to 50-Hz frequency range, *Proc. Natl. Acad. Sci. USA*, 88, 897, 1991.
82. Gutfreund, Y., Yarom, Y., and Segev, I., Subthreshold oscillations and resonant frequency in guinea-pig cortical neurons: physiology and modelling, *J. Physiol. (Lond.)*, 483, 621, 1995.
83. Gray, C.M. and McCormick, D.A., Chattering cells: superficial pyramidal neurons contributing to the generation of synchronous oscillations in the visual cortex, *Science*, 274, 109, 1996.
84. Cape, E. and Jones, B., Modulation of sleep/wake state and cortical activity following injections of agonists into the region of cholinergic basal forebrain neurons, *J. Neurosci.*, 18, 2653–2666, 1998.
85. Buzsáki, G., The hippocampo-neocortical dialogue, *Cerebral Cortex*, 6, 81, 1996.
86. Sterman, M.B. and Clemente, C.D., Forebrain inhibitory mechanisms: sleep patterns induced by basal forebrain stimulation in the behaving cat, *Exp. Neurol.*, 6, 103, 1962.
87. Szymusiak, R. and McGinty, D., Sleep-related neuronal discharge in the basal forebrain of cats, *Brain Res.*, 370, 82, 1986.
88. Szymusiak, R. and McGinty, D., Sleep suppression following kainic acid-induced lesions of the basalforebrain, *Exp. Neurol.*, 94, 598, 1986.

Chapter 19

Hypothalamic Regulation of Sleep

*Priyattam J. Shiromani, Tom Scammell,
John E. Sherin, and Clifford B. Saper*

Contents

19.1	Introduction	311
19.2	Brain Regions Responsible for Waking	312
19.3	Brain Regions Responsible for Non-REM Sleep	313
19.4	Hypothalamic Mechanisms of Sleep/Wake Control	314
19.5	Evidence that AD Is an Endogenous Somnogen	316
Acknowledgments		318
References		318

"No small art is it to sleep;
It is necessary for that purpose to keep awake all day."
Zarathustra

19.1 Introduction

Seminal studies by von Economo and Nauta indicated that the hypothalamus plays an important role in sleep; however, since those pioneering studies it has been difficult to identify discrete sleep-producing neurons. One factor hindering progress is that the hypothalamus contains neurons regulating a variety of functions such as temperature regulation, appetite control, circadian rhythms, sex drive, and energy balance. Thus, how to separate neurons governing sleep from the other neurons? The discovery of the immediate-early gene c-*fos* has been very helpful in this regard. Combined evidence from lesion, electrophysiological, pharmacological, and now c-Fos studies show that sleep-active neurons interact with wake-active cells in the hypothalamus to govern sleep/wakefulness. While this network may produce sleep/wakefulness, other factors are also likely to be involved. Adenosine is one such agent that is likely to accumulate with wakefulness and influence sleepiness. The induction of c-*fos* in sleep-active neurons also indicates that molecular events are involved. We hypothesize that a coordinated interaction between somnogens, such as adenosine, and intracellular cascades involving transcription factors governs the waxing and waning of the sleep/wake process.

In this chapter, we summarize progress in the study of hypothalamic circuitry underlying sleep, especially recent work in which the immediate-early gene, c-*fos*, has been used to define neuronal populations related to behavioral state. Combining Fos immunohistochemistry with classical tract tracing techniques has proved to be a powerful approach that permits localization of state-dependent neurons as well as identification of their neurotransmitters and projections.

19.2 Brain Regions Responsible for Waking

Brainstem and forebrain mechanisms are implicated in waking. In the brainstem, Moruzzi and Magoun[70] showed that electrical stimulation of the brainstem reticular core produces arousal in an otherwise sedated animal. Steriade[108] has identified this arousing region as the laterodorsal tegmental (LDT) and pedunculo-pontine tegmental (PPT) nuclei. These neurons are active when the EEG is desynchronized (during waking and REM sleep). They innervate the thalamus and are responsible for switching thalamic neurons from an inherent oscillatory pattern (seen during sleep) to a fast pattern (seen during waking and REM sleep). The pontine LDT/PPT cholinergic neurons, however, represent only one portion of the overall influence of the cholinergic system on waking; there is also a forebrain component.

Cholinergic neurons in the basal forebrain also are implicated in waking and EEG desynchronization.[113] The basal forebrain (BF) includes the vertical and horizontal limbs of the diagonal band of Broca, the substantia inominata, and the magnocellular preoptic area where sleep-active cells have been identified.[2,113,116,117] Putative cholinergic basal forebrain neurons have increased activity during waking and REM sleep and project to the thalamus and to cortex. These wake- and REM-active basal forebrain cells decrease their firing in response to local warming which is somnogenic.[2] Selective lesions of the BF cholinergic neurons using the toxin 192 IgG-saporin do not change total amounts of wakefulness,[9] indicating this basal forebrain cholinergic system is not in itself critical to influence wakefulness.

In the posterior hypothalamus, histaminergic neurons of the tuberomammillary nucleus (TMN) also promote wakefulness. Lesions of the posterior hypothalamus including the TMN produce hypersomnolence.[66,105] TMN neurons have their highest discharge rate during waking and are virtually silent during NREM and REM sleep.[89,115,124] These wake-active TMN neurons are hypothesized to inhibit reciprocally basal forebrain-pre-optic (BF-POA) cells implicated in sleep. For instance, the wake-active TMN neurons are inhibited by warming of the BF-POA.[53] Conversely, histamine microinjections into BF-POA produce a dose-dependent increase in wake.[59] Blockade of histamine synthesis in the POA increases sleep and decreases wake.[59] Histamine H1 and H2 receptors are postulated to mediate the arousal.[59]

Norepinephrine[82] and serotonin[42] have also been implicated in arousal. Dorsal raphe and LC neurons have their highest discharge rates during waking, then decrease their firing during non-REM (NREM) sleep, and become virtually silent during REM sleep (summarized in Reference 109). Dorsal raphe[47] and LC[46] lesions produce transient hypersomnia. Applications of beta-receptor agonists to LC targets such as the medial septal nucleus elicit arousal.[12] The role of serotonin in arousal is less clear because of the somnogenic influence of the serotonin precursors, tryptophan and 5-HTP,[47] and the ability of 5-HTP to reverse parachlorophenylalanine (PCPA)-induced insomnia.[23] The differential action of serotonin in modulating behavioral state might be related to its action on specific serotonin receptors.[13]

The current consensus is that forebrain and brainstem regions are involved in waking. The neurotransmitters involved include acetylcholine (basal forebrain and pons), histamine (TMN), norepinephrine (locus coeruleus), and serotonin (dorsal raphe).

Using c-Fos, we and others have found that following wakefulness (either natural or forced), numerous Fos-ir neurons are found in many brain regions.[8,18,35,57,74,76–78,120] Wake-active Fos-ir neurons are prominent in cholinergic areas, such as the basal forebrain and the pons (LDT/PPT). With sleep, the number of c-Fos-ir cells decreases dramatically.[17,34,98] We have concentrated on identifying c-Fos cells in the basal forebrain, where wake-active cells are hypothesized to be cholinergic, as cortical release of acetylcholine is increased during waking and REM sleep.[44,112] Electrical stimulation of the region containing the LDT/PPT also drives cortical acetylcholine release and induces cortical desynchronization.[84,85] Lesions of the basal forebrain produce slow waves in the EEG[15,86] and also decrease cortical acetylcholine levels.[22] We have found c-Fos neurons following natural waking and following PPT electrical stimulation in three basal forebrain regions: the nucleus basalis of Meynert (NBM), the substantia innominata (SI), and the magnocellular preoptic area (MCPO). The distribution of c-Fos in the NBM and SI is similar to that of choline acetyltransferase, and we predict that during wakefulness the BF cholinergic cells will be c-Fos positive. It will be important to determine the distribution Fos-ir cells in the various BF cell populations.

19.3 Brain Regions Responsible for Non-REM Sleep

Converging evidence from lesion,[4,5,45,61,90,104,116,117] stimulation (electrical and pharmacological),[23,63,67,110,111,123] and electrophysiological (detailed below) studies indicates that neurons in the BF-POA play an important role in triggering NREM sleep. For the purposes of brevity, we are referring to the BF-POA area as collectively representing the subpopulation of non-cholinergic sleep-promoting and sleep-active cells. The sleep-active cells are presumed to be GABAergic and are nested among wake-active, presumably cholinergic, cells (see Reference 113 for review).

The BF-POA and medulla[26] are the only two areas in the brain where neuronal activity increases during sleep. In virtually every brain region that has been examined, sleep is associated with a decline in neuronal discharge and metabolic activity. Sleep-active neurons are found in the POA and adjacent basal forebrain in rats, cats, and rabbits.[31,48,52,116,117] These neurons begin to fire during drowsiness, and peak activity is seen during NREM sleep. The sleep-active cells comprise about 25% of the recorded cells in the BF-POA and are mixed with the more common wake-active cells.[52,113] Whole-body warming or local warming of the POA hypothalamus induces sleep (summarized in References 64, 65, and 114). This warming increases the activity of many sleep-active neurons and suppresses activity of wake-active neurons in the POA and adjacent basal forebrain.[3,64] McGinty and Szymusiak[64] postulate that the POA monitors temperature, and sleep (primarily NREM sleep) might be induced in response to a rise in temperature.

The transmitter identity of the sleep-active neurons is unknown, but there is strong evidence that these could contain GABA. For instance, GABAergic neurons are prevalent in the BF-POA and have projections to the wake-active TMN.[37] Consistent with the descending GABAergic innervation model, Nitz and Siegel[73] recently reported increased GABA release in PH during natural sleep. Microinjection of muscimol (a GABA agonist) into the posterior hypothalamus induces sleep.[60] The muscimol induction of sleep is seen in PCPA-treated insomniac cats[60] and in POA-lesioned insomniac cats.[90] BF-POA warming (which is somnogenic) decreases firing of wake-active PH[53] and basal forebrain neurons.[2]

Because, NREM-active neurons are mixed with wake- and REM-active cells within the BF/POA, we used c-Fos to identify the sleep-active population. During sleep, a distinct cluster of Fos-ir neurons is seen in the ventral lateral preoptic area (VLPO).[95] Rats were examined following natural periods of sleep. Rats that mostly slept (>80%) had many Fos-ir cells in the VLPO, whereas rats that slept very little (<30 %) had very few Fos-ir cells in the VLPO. To rule out circadian influences on Fos-ir, rats were sleep deprived for 9 to 12 hr during the day. Some rats were killed immediately at the end of the deprivation, while others were allowed varying amounts of recovery sleep. The rats were killed at 4:00 p.m. (9-hr deprivation), 7:00 p.m. (12-hr deprivation), or 10:00

p.m. (12-hr deprivation followed by 3 hr of recovery sleep). We hypothesized that rats killed at 4:00 p.m. without any sleep should have no Fos-ir cells in the VLPO. We further hypothesized that the VLPO should demonstrate Fos-ir cells no matter when sleep occurred, including the night cycle. Our hypotheses were confirmed in that Fos-ir neurons were seen in the VLPO in association with sleep but not circadian time. Rats killed without sleep at all of the time periods showed little or no Fos-ir in the VLPO. In contrast, rats allowed recovery sleep had many Fos-ir neurons in the VLPO.

Are VLPO cells electrophysiologically active during sleep? Recently, McGinty's group[1] has found a higher percentage of sleep-active cells in the VLPO area (44%) compared to the adjacent preoptic area. This reinforces our observation of c-Fos cells during natural sleep in the VLPO.

An important feature of VLPO neurons is that they are active only in association with sleep, rather than sleep need. For instance, with sleep deprivation there is increased sleepiness. However, in our studies Fos-ir was not seen in animals that were sleepy and killed without an opportunity to sleep. Fos-ir was seen only after the animals had slept. This finding has been confirmed by Szymusiak using single-cell recording of VLPO neurons (paper presented at the APSS meeting, June 1997). He found that VLPO neurons were electrophysiologically active only during sleep (SWS and REM sleep). In sleep-deprived rats, VLPO neurons did not fire unless the animal slept, further indicating that these cells are not responding to sleep pressure. If the VLPO cells were responding to sleep need or pressure, then they should have begun to discharge during wakefulness of sleep deprivation.

These findings indicate that the VLPO may be an output element in sleep generation, influencing the production of sleep. If VLPO activation is associated with sleep, then neuro-anatomical evidence should document contact with an arousal population. The VLPO Fos-ir cells project to the histaminergic neurons of the TMN. As noted previously, TMN neurons discharge most rapidly during waking and are virtually silent during sleep.[89,115,124] VLPO neurons are GABAergic, and increased GABA has been noted during sleep in the posterior hypothalamus.[73] Sleep-active VLPO neurons usually contain galanin. In fact, galanin immunoreactivity identifies VLPO neurons better than GABA, which is ubiquitous in the POA. From a functional perspective, the fact that VLPO cells are galanin positive is extremely exciting. Galanin stimulates the release of GHRH from the hypothalamus[51] and growth hormone (GH) from the pituitary.[75] The significance of this is that sleep is marked by a pulsatile increase in GH, and this is independent of the circadian cycle (see Reference 55 for review). The GH secretion is correlated with sleep and delta sleep in humans, rats, and other species. GH promotes sleep, but galanin has not been shown to induce sleep.[121] The relationship between sleep and GH secretion is generally regarded as the best-documented example of the association between sleep and hormone secretion. The finding that VLPO Fos-ir cells which are sleep-active and also galanin positive may link the hormonal changes seen in sleep to activity in specific cell groups.

The neurochemical phenotype of VLPO coupled with its connectivity and firing profile suggests that VLPO neurons provide an important inhibitory influence on wake-active TMN cells. Neurons in the posterior hypothalamus and the tuberomammillary region, like the LC and dorsal raphe neurons, decrease firing during sleep.[115,124] Our overall hypothesis is that the VLPO cells are part of a group of sleep-active cells in the BF-POA with inhibitory projections to wake-active populations in the histaminergic tuberomammillary nucleus (TM) and the pons (norepinephrine, serotonin, and acetylcholine).

19.4 Hypothalamic Mechanisms of Sleep/Wake Control

Based on the converging lines of evidence presented above, it is possible to derive a model of hypothalamic regions involved in sleep/wake control (Figure 19.1). The major components of this model involve interactions among GABAergic, NREM-active neurons in the BF-POA, and wake-

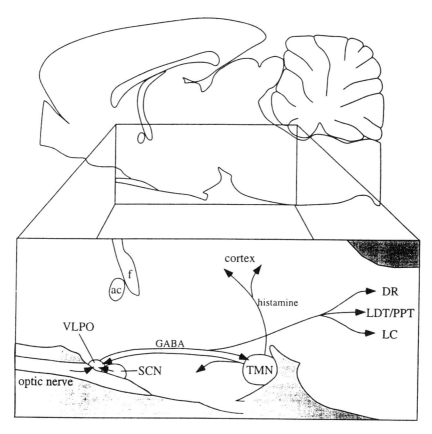

FIGURE 19.1
Summary diagram of hypothalamic mechanisms of sleep. Wake-active neurons of the tuberomammillary nucleus (TMN) send diffuse histaminergic projections to the cortex, hypothalamus, and many other regions. Sleep-active neurons of the ventrolateral pre-optic area (VLPO) receive afferents from the retina and suprachiasmatic nucleus (SCN), allowing modulation of VLPO activity by light and circadian influences. GABAergic VLPO neurons project heavily to the TMN and to a lesser degree to the dorsal raphe (DR), laterodorsal tegmental and pedunculopontine tegmental (LDT/PPT) nuclei, and locus coeruleus (LC). Through these inhibitory projections, the VLPO may promote sleep by coordinating the inhibition of wake-active neurons.

active cholinergic and monoaminergic neurons. VLPO neurons constitute a portion of the sleep-active neurons of the BF-POA that inhibit the major monoaminergic and cholinergic arousal systems in the caudal hypothalamus and pons. The wake-on cells, in turn, may inhibit the sleep-active GABA cells. The inhibitory projections from the sleep-active VLPO neurons to the TMN may promote NREM sleep and thereby set the stage for REM sleep. Once NREM sleep is established, the pontine REM sleep generator would trigger REM sleep, as outlined in the reciprocal interaction model.[109]

This network undoubtedly is influenced by the SCN. The SCN is the master clock in the brain, and one of the cycles it regulates is the circadian expression of sleep/wake rhythms. In the absence of the SCN, sleep/wake episodes continue to occur (the total amount of sleep during a 24-hr period is unchanged), but there is no day/night variation in sleep.[39] In monkeys, however, the SCN also may promote wakefulness.[25]

While such a network can generate sleep and wakefulness, the mechanisms regulating each state are unknown. It is known that the alternation between states is controlled by both circadian and homeostatic factors.[14,24] Circadian drive from the SCN regulates sleep timing, but there is also a homeostatic component: longer periods of wakefulness result in an increased propensity to sleep.

This sleep drive dissipates only with cumulative sleep[58,107] and appears specific for individual sleep stages because selective deprivation of REM sleep results in subsequent rebound of REM sleep.

At the electrophysiological level, EEG slow-wave activity (delta activity; 0.1 to 4 Hz) during sleep is proportionate to the preceding length of wakefulness. Delta activity during the initial recovery sleep is maximal after about 12 hr of continuous waking in rats.[119] As recovery sleep progresses, delta activity exponentially declines. Firing rates of most VLPO neurons are low during wakefulness and twice as high during NREM sleep.[1] During prolonged wakefulness, the firing rates of these neurons are unchanged but increase fourfold in the subsequent recovery NREM sleep (Szymusiak, R., personal communication). This increase in firing rate is tightly correlated with delta, suggesting that sleep-promoting VLPO activity may influence NREM sleep intensity. Circadian factors also modulate the expression of delta in sleep deprived rats[122] and in humans.[24]

One mechanism that could account for the phenomena of increased sleepiness and delta activity with wakefulness would be an endogenous sleep factor that accumulates during wakefulness and dissipates during sleep. At the biochemical level, a number of endogenous substances increase with sleep and may serve as sleep factors. These putative somnogens include cytokines,[54] prostaglandin D2,[63] growth hormone, and adenosine.[118] Other compounds such as delta sleep-inducing peptide, muramyl peptides, and sleep lipid[21] have been added to this list, but the site of action and receptor mechanisms of all these somnogens remain to be defined. Although systemic and ICV injections of tumor necrosis factor and interleukin 1 increase sleep,[54] their site of action is unknown. Prostaglandin D2 may most effectively promote sleep when dialyzed into the subarachniod space just anterior to the VLPO, but its effect may not occur in all species. Because sleep is ubiquitous, it is reasonable to assume that somnogens and their neural effects are similar across species.

19.5 Evidence that AD Is an Endogenous Somnogen

Compared to other potential sleep factors, the site of action and potential receptor mechanisms for adenosine (AD) are better understood. The ubiquitous stimulant caffeine is an AD receptor antagonist, suggesting that endogenous AD may promote sleep. The extracellular concentration of AD gradually rises during wakefulness and falls during sleep.[79] This AD may influence neuronal activity through membrane-bound receptors on target cells.

Several AD receptors have been pharmacologically characterized and sequenced. All of these are G protein-coupled receptors for which selective agonists and antagonists have been developed. The A1 receptor couples to an inhibitory G protein (Gi) which inhibits adenylyl cyclase and can activate voltage-sensitive potassium and calcium channels.[87] In neurons, A1 receptor activation can produce hyperpolarization, reducing the excitability of neurons.[36] In contrast, the A2 receptors have a lower affinity for AD, couple to Gs, activate adenylyl cyclase, and increase intracellular cAMP. Two functionally distinct subtypes have been sequenced: the A2a receptor is expressed in neurons,[127] and the A2b receptor is produced by glia and fibroblasts.[106] The A2b receptor also has a low affinity for AD and is unlikely to be occupied under physiologic conditions. The A3 receptor is not blocked by xanthine derivatives, is coupled to Gi, may produce hyperpolarization, and is expressed at very low levels in brain. Thus, caffeine most likely inhibits sleep via the A1 or A2a receptors, as the A2b has little affinity for AD at concentrations typically found in brain, and the A3 receptor has little affinity for caffeine or other methylxanthines.

The AD receptors are widely distributed in brain. Radiolabeled A1 agonists and *in situ* hybridization with an A1 probe labels many regions of the rat brain including potentially sleep-regulating sites such as the cortex, diagonal band nuclei, magnocellular preoptic nucleus, anterior hypothalamic area, lateral hypothalamus, mamillary and premamillary nuclei, midbrain tegmentum, pontine nuclei, reticular formation, and raphe nuclei.[28,29,33] In contrast, the A2a receptor appears restricted to striatum and olfactory tubercle in rat brain.[32,43] The A2b and A3 receptors may be expressed in brain at low levels, but their contribution to brain adenosine responses appears small,[94] and their distributions are not yet described. All of these AD receptors appear rare in the preoptic area, indicating that if AD influences VLPO activity it must do so through neurons outside of the preoptic area.

Several lines of evidence suggest that AD may be a natural somnogen, regulating everyday sleep. In rats, extracellular AD levels in brain rise to their highest at the end of the animal's nocturnal activity period (deSanchez, 1993) and fall during the light period, when most sleep occurs, in parallel with exponentially declining EEG delta power.[14,30] In cats, AD levels are higher during wakefulness than during NREM sleep, and double during 6 hr of forced wakefulness.[79] Intraperitoneal administration of A1 agonists to rats increases NREM sleep EEG delta power[10,11] and the time spent in deep NREM sleep.[81] Time spent in NREM sleep increases at the expense of waking and REM sleep for up to 6 hr following these injections. These effects are probably centrally mediated as intracerebroventricular injection of AD also increases EEG delta power.[10,11] Most importantly, caffeine and theophylline, the most widely used stimulants, competitively inhibit both the A1 and A2 receptors, suggesting that endogenous AD may be associated with sleep drive. Caffeine administered to humans prolongs sleep latency and decreases NREM sleep EEG delta power.[56] Benington and Heller[10] thoughtfully argue that the effects of AD are similar to those seen with total sleep deprivation, a manipulation that also increases sleep drive, EEG delta power, and NREM sleep.

During wakefulness, extracellular AD concentrations may rise throughout the brain, but to coordinate the production of sleep AD must act in specific sleep- or wake-regulatory regions. Ticho and Radulovacki[118] found that BF-POA injection of an A1 agonist increased the time spent in deep NREM sleep. Rainnie and colleagues[83] showed that, in slice preparations, extracellularly recorded diagonal band of Broca neurons roughly doubled their firing rates when a selective A1 antagonist was added to the medium, and firing nearly ceased when AD was added. Using *in vivo* microdialysis in the diagonal band region, Greene and others[79,80] recently demonstrated that both exogenous AD and elevations in endogenous AD (produced by blocking AD reuptake) increase NREM and REM sleep. Considered together, these results suggest that physiologic concentrations of AD may act via A1 receptors to inhibit wake-active basal forebrain neurons.

In contrast to many studies demonstrating that A1 agonists induce sleep, Satoh and colleagues[91] have just shown that a highly selective A2a agonist microdialyzed into the POA or the subarachnoid space just below the POA nearly doubled NREM and REM sleep in a dose-dependent fashion. However, treatment with an A2a antagonist did not decrease sleep, and, as yet, A2a receptors have not been identified in any brain regions traditionally implicated in sleep.

Two models have been proposed to account for the somnogenic effects of AD. Benington and Heller[10] suggest that AD may induce sleep by globally inhibiting neurons throughout the brain; however, if this were the case, focal brain lesions such as those in the BF-POA would not prevent sleep. Also, AD receptors are heterogeneously distributed in brain, and some regions such as the VLPO are more active during sleep. Rainnie, Porkka-Heiskanen, and others[79,80,83] argue that AD induces sleep by inhibiting cholinergic neurons of the basal forebrain, but lesions of basal forebrain cholinergic neurons with 192-IgG-saporin do not alter the amount of sleep,[49] and these cholinergic neurons are not well positioned to coordinate the inhibition of the histaminergic, serotonergic, and noradrenergic systems that occurs during sleep. In addition, AD promotes both NREM and REM sleep, yet cholinergic cells are relatively inactive during NREM sleep but active during REM sleep. Both the global and cholinergic models may account for some aspects of sleep, but the REM-promoting effects of AD and the coordinated inhibition of arousal systems cannot be fully accounted for by either model.

We hypothesize that the VLPO plays an essential role in the generation of sleep. High physiologic concentrations of AD are sufficient to induce sleep,[79] yet animals with lesions of the VLPO and adjacent preoptic area are profoundly insomniac.[71,90,116,117] Assuming that these animals also experience an increase in AD during wakefulness, we conclude that the VLPO and adjacent preoptic regions are required for natural, AD-induced sleep.

We propose a new model in which AD induces sleep by disinhibiting the VLPO. After prolonged wakefulness, high AD levels may inhibit neurons in many brain regions via A1 receptors. AD would also inhibit specific GABAergic neurons of the basal forebrain and anterior hypothalamic area which may inhibit the VLPO during wakefulness. Thus, AD would disinhibit the VLPO, contributing to the generation of NREM and REM sleep. In support of this model, we have found that microinjection of the nonspecific AD agonist 5AD-N-ethylcarboxamidoadenosine (NECA) into the subarachnoid space just anterior to the VLPO produced a marked increase in NREM sleep for over 3 hr.[92] This increase in NREM sleep was accompanied by intense Fos expression in the VLPO. Vehicle microinjection did not alter sleep/wake behavior, and the Fos expression in the VLPO was comparable to controls.

We have recently completed a series of experiments to define carefully the neurotransmitters and targets of these VLPO neurons. By combining immunohistochemistry for glutamic acid decarboxylase (GAD-67) and galanin with injections of a retrograde tracer, we demonstrated a selective GABAergic and galanin-ergic projection from the VLPO to the core of the tuberomammillary nucleus, the source of the ascending histaminergic arousal system. Occasional neurons in regions adjacent to the VLPO, such as the ventromedial and dorsolateral preoptic areas, also contained GABA and galanin and projected to the TMN. Histaminergic neurons of the TMN are inhibited by GABA during sleep, and these observations demonstrate that the VLPO may inhibit histaminergic activity during sleep. Using anterograde tracing, we have found that the VLPO also projects to other major arousal systems, including the dorsal and median raphe nuclei, the LDT/PPT, and the locus coeruleus. We suspect that these projections are also inhibitory, and we are further investigating the anatomic details of this system. Afferents to the VLPO are largely unknown, but Meibach and Siegel[68] have described projections from the diagonal band to the lateral POA. Preliminary work in our lab has also revealed definite projections to the VLPO from the suprachiasmatic nucleus and retina, pathways through which the VLPO may interact with the circadian and arousal systems, respectively. Thus, the VLPO is well positioned to influence and be influenced by the major arousal systems. These observations, coupled with the activation of VLPO neurons during sleep, argue strongly that the VLPO may coordinate the inhibition of multiple arousal systems during sleep.

c-fos has proved to be very helpful in identifying neurons underlying sleep or wakefulness. However, its activation must serve an important function. Therefore, we examined sleep in mice lacking the *c-fos* gene.[102] Null *c-fos* mice had about 30% less non-REM sleep; REM sleep was not changed. The null *c-fos* mice took longer to fall asleep even when they were deprived of sleep for 6 hr. Cirelli et al.[18] have found that injections of *c-fos* antisense oligonucleotides (which block c-Fos protein synthesis) also decrease sleep. These studies suggest that transcription factors, such as *c-fos*, are involved in sleep, and loss of the gene decreases sleep.

Acknowledgments

This work was supported by the DVA Medical Research Service, NIH NS30140.

References

1. Alam, M.N., Szymusiak, R., Steininger, T.L., and McGinty, D., Sleep-wake activity and thermosensitivity of neurons in the ventral lateral preoptic area of rats (abstract), *Soc. Neurosci.*, 22(1), 26, 1996.
2. Alam, M.N., McGinty, D., and Szymusiak, R., Neuronal discharge of preoptic/anterior hypothalamic thermosensitive neurons: relation to NREM sleep, *Am. J. Physiol.*, 269, R1240–R1249, 1995.

3. Alam, M.N., Szymusiak, R., and McGinty, D., Local preoptic/anterior hypothalamic warming alters spontaneous and evoked neuronal activity in the magno-cellular basal forebrain, *Brain Res.*, 696, 221–230, 1995.

4. Alam, M.N. and Mallick, B.N., Differential acute influence of medial and lateral preoptic areas on sleep/wakefulness in freely moving rats, *Brain Res.*, 525, 242–248, 1990.

5. Asala, S.A., Okano, Y., Honda, K., and Inoue, S., Effects of medial preoptic area lesions on sleep and wakefulness in unrestrained rats, *Neurosci. Lett.*, 114, 300–304, 1990.

6. Barraco, R.A., Campbell, W.R., Schoener, E.P., Shehin, S.E., and Parizon, M., Cardiovascular effects of microinjections of adenosine analogs into the fourth ventricle of rats, *Brain Res.*, 424(1), 17–25, 1987.

7. Barraco, R.A., Aggarwal, A.K., Phillis, J.W., Moron, M.A., and Wu, P.H., Dissociation of the locomotor and hypotensive effects of adenosine analogues in the rat, *Neurosci. Lett.*, 48(2), 139-44, 1984.

8. Basheer, R., Sherin, J., Morgan, J., Saper, C., McCarley, R.W., and Shiromani, P.J., Effects of sleep on wake-induced c-Fos, *J. Neurosci.*, 17, 9746–9750, 1997

9. Bassant, M.H., Apartis, E., Jazat-Poindessous, F.R., Wiley, R.G., and Lamour, Y.A., Selective immunolesion of the basal forebrain cholinergic neurons: effects on hippocampal activity during sleep and wakefulness in the rat, *Neurodegeneration*, 4, 61–70, 1995.

10. Benington, J.H. and Heller, H.C., Restoration of brain energy metabolism as the function of sleep, *Prog. Neurobiol.*, 45(4), 347–360, 1995.

11. Benington, J.H., Kodali, S.K., and Heller, H.C., Stimulation of A1 adenosine receptors mimics the electroencephalographic effects of sleep deprivation, *Brain Res.*, 6921–6922, 79–85, 1995.

12. Berridge, C.W. and Foote, S.L., Enhancement of behavioral and electroencephalographic indices of waking following stimulation of noradrenergic beta-receptors within the medial septal region of the basal forebrain, *J. Neurosci.*, 16, 6999–7009, 1996.

13. Bjorvatn, B., Bjorkum, A.A., Neckelmann, D., and Ursin, R., Sleep/waking and EEG power spectrum effects of a nonselective serotonin (5-HT) antagonist and a selective 5-HT reuptake inhibitor given alone and in combination, *Sleep*, 18, 451–462, 1995.

14. Borbely, A.A., A two-process model of sleep regulation, *Hum. Neurobiol.*, 1, 195–204, 1982.

15. Buzsaki, G., Bickford, R., Ponomareff, G., Thal, L.J., Mandel, R., and Gage, F.H., Nucleus basalis and thalamic control of neocortical activity in the freely moving rat, *J. Neurosci.*, 8, 4007–4026, 1988.

16. Chagoya de Sanchez, V., Hernandez, M.R., Suarez, J., Vidrio, S., Yanez, L., and Diaz, M.M., Day-night variations of adenosine and its metabolizing enzymes in the brain cortex of the rat-possible physiological significance for the energetic homeostasis and the sleep/wake cycle, *Brain Res.*, 6121–6122, 115–121, 1993.

17. Cirelli, C., Pompeiano, M., and Tononi, G., Fos-like immunoreactivity in the rat brain in spontaneous wakefulness and sleep, *Arch. Ital. Biol.*, 131, 327–330, 1993.

18. Cirelli, C., Pompeiano, M., and Tononi, G., Sleep deprivation and c-*fos* expression in the rat brain, *J. Sleep Res.*, 4, 92–106, 1995.

19. Cirelli, C., Pompeiano, M., Arrighi, P., and Tononi, G., Sleep-waking changes after c-*fos* antisense injections in the medial preoptic area, *NeuroReport*, 6, 801–805, 1995.

20. Cornfield, L.J., Hu, S., Hurt, S.D., and Sills, M.A., [^{3}H]2-phenylaminoadenosine ([^{3}H]CV 1808) labels a novel adenosine receptor in rat brain, *J. Pharmacol. Exp. Ther.*, 263(2) 552–561, 1992.

21. Cravatt, B.F., Prospero-Garcia, O., Siuzdak, G., Gilula, N.B., Henriksen, S.J., Boger, D.L., and Lerner, R.A., Chemical characterization of a family of brain lipids that induce sleep, *Science*, 268, 1506–1509, 1995.

22. Dekker, A.J. and Thal, L.J., Independent effects of cholinergic and serotonergic lesions on acetylcholine and serotonin release in the neocortex of the rat, *Neurochem. Res.,* 18, 277–283, 1993.
23. Denoyer, M., Sallanon, M., Kitahama, K., Aubert, C., and Jouvet, M., Reversibility of parachlorophenylalanine-induced insomnia by intrahypothalamic microinjection of L-5-hydroxytryptophan, *Neuroscience,* 28, 83–94, 1989.
24. Dijk, D.J. and Czeisler, C.A., Contribution of the circadian pacemaker and the sleep homeostat to sleep propensity, sleep structure, electroencephalographic slow waves, and sleep spindle activity in humans, *J. Neurosci.,* 15, 3526–3538, 1995.
25. Edgar, D.M., Dement, W.C., and Fuller, C.A., Effect of SCN lesions on sleep in squirrel monkeys: evidence for opponent processes in sleep/wake regulation, *J. Neurosci.,* 13, 1065–1079, 1993.
26. Eguchi, K. and Satoh, T., Characterization of the neurons in the region of solitary tract nucleus during sleep, *Physiol. Behav.,* 24, 99–102, 1980.
27. El Mansari, M., Sakai, K., and Jouvet, M., Unitary characteristics of presumptive cholinergic tegmental neurons during the sleep-waking cycle in freely moving cats, *Exp. Brain Res.,* 76, 519–529, 1989.
28. Fastbom, J., Pazos, A., and Palacios, J.M., The distribution of adenosine A1 receptors and 5'-nucleotidase in the brain of some commonly used experimental animals, *Neuroscience,.* 22(3), 813–826, 1987.
29. Fastbom, J., Pazos, A., Probst, A., and Palacios, J.M., Adenosine A1 receptors in the human brain: a quantitative autoradiographic study, *Neuroscience,.* 22(3), 827–839, 1987.
30. Feinberg, I., March, J.D, Floyd, T.C., Jimison, R., Bossom, D.L., and Katz, P.H., Homeostatic changes during post-nap sleep maintain baseline levels of delta EEG, *Electroencephalogr. Clin. Neurophysiol.,* 61(2), 134–137, 1985.
31. Findlay, A.L.R. and Hayward, J.N., Spontaneous activity of single neurons in the hypothalamus of rabbits during sleep and waking, *J. Physiol.,* 201, 237–258, 1969.
32. Fink, J.S., Weaver, D.R., Rivkees, S.A., Peterfreund, R.A., Pollack, A.E., Adler, E.M., and Reppert, S.M., Molecular cloning of the rat A2 adenosine receptor, selective co-expression with D2 dopamine receptors in rat striatum, *Brain Res. Mol. Brain Res.,* 14(3), 186–195, 1992.
33. Goodman, R.R. and Synder, S.H., Autoradiographic localization of adenosine receptors in rat brain using [^{3}H]cyclohexyladenosine, *J. Neurosci.,* 2(9), 1230–1241, 1982.
34. Grassi-Zucconi, G., Giuditta, A., Mandile, P., Chen, S., Vescia, S., and Bentivoglia, M., c-*fos* spontaneous expression during wakefulness is reversed during sleep in neuronal subsets of the rat cortex, *J. Physiol.,* 88, 91–93, 1994.
35. Grassi-Zucconi, G., Menegazzi, M., De Prati, A.C., Bassetti, A., Montagnese, P., Mandile, P., Cosi, C., and Bentivoglio, M., c-*fos* mRNA is spontaneously induced in the rat brain during the activity period of the circadian cycle, *Eur. J. Neurosci.,* 5, 1071–1078, 1993.
36. Greene, R.W. and Haas, H.L., The electrophysiology of adenosine in the mammalian central nervous system, *Prog. Neurobiol.,* 36(4), 329–341, 1991.
37. Gritti, I., Mainville, L., and Jones, B.E., Projections of GABAergic and cholinergic basal forebrain and GABAergic preoptic-anterior hypothalamic neurons to the posterior lateral hypothalamus of the rat, *J. Comp. Neurol.,* 339, 251–268, 1994.
38. Hayaishi, O. and Matsumura, H., Prostaglandins and sleep, *Adv. Neuroimmunol.,* 5(2), 211–216, 1995.
39. Ibuka, N. and Kawamura, H., Loss of circadian rhythm in sleep/wakefulness cycle in the rat by suprachiasmatic nucleus lesions, *Brain Res.,* 96, 76–81, 1975.
40. Imeri, L., Bianchi, S., Angeli, P., and Mancia, M., Differential effects of M2 and M3 muscarinic antagonists on the sleep/wake cycle, *NeuroReport,* 2, 383–385, 1991.

41. Imeri, L., Bianchi, S., Angeli, P., and Mancia, M., M1 and M3 muscarinic receptors: specific roles in sleep regulation, *NeuroReport,* 3, 276–278, 1992.
42. Jacobs, B.L. and Azmitia, E.C., Structure and function of the brain serotonin system, *Physiol. Rev.,* 72, 165–229, 1992.
43. Jarvis, M.F., Schulz, R., Hutchison, A.J., Do, U.H., Sills, M.A., and Williams, M., [^{3}H]CGS 21680, a selective A2 adenosine receptor agonist directly labels A2 receptors in rat brain, *J. Pharmacol. Exp. Ther.,* 251(3), 888–893, 1989.
44. Jasper, H.H. and Tessier, J., Acetylcholine liberation from cerbral cortex during paradoxical (REM) sleep, *Science,* 172, 601–602, 1971.
45. John, J., Kumar, V.M., Gopinath, G., Ramesh, V., and Mallick, H., Changes in sleep/wakefulness after kainic acid lesion of the preoptic area in rats, *Jpn. J. Physiol.,* 44, 231–242, 1994.
46. Jones, B.E., Bobillier, P., Pin, C., and Jouvet, M., The effects of lesions of catecholamine-containing neurons upon monoamine content of the brain and EEG and behavioral waking in the cat, *Brain Res.,* 58, 157–168, 1973.
47. Jouvet, M., The role of monoamines and acetylcholine-containing neurons in the regulation of the sleep-waking cycle, *Ergebnisse Physiol.,* 64, 166–308, 1972.
48. Kaitin, K., Preoptic area unit activity during sleep and wakefulness in the cat, *Exp. Neurol.,* 83, 347–351, 1984.
49. Kapas, L., Obal, Jr., F., Book, A.A., Schweitzer, J.B., Wiley, R.G., and Krueger, J.M., The effects of immunolesions of nerve growth factor-receptive neurons by 192 IgG-saporin on sleep, *Brain Res.,* 712(1), 53–59, 1996.
50. Kayama, Y., Ohta, M., and Jodo, E., Firing of "possibly" cholinergic neurons in the rat laterodorsal tegmental nucleus during sleep and wakefulness, *Brain Res.,* 569, 210–220, 1992.
51. Kitajima, N., Chihara, K., Abe, H., Okimura, Y., and Shakutsui, S., Galanin stimulates immunoreactive growth hormone releasing-factor secretion from rat hypothalamic slice perifused *in vitro, Life Sci.,* 47, 2371–2376, 1990.
52. Koyama, Y. and Hayaishi, O., Modulation by prostaglandins of activity of sleep-related neurons in the preoptic/anterior hypothalamic areas in rats, *Brain Res. Bull.,* 33, 367–372, 1994.
53. Krilowicz, B.L., Szymusiak, R., and McGinty, D., Regulation of posterior lateral hypothalamic arousal related neuronal discharge by preoptic anterior hypothalamic warming, *Brain Res.,* 668, 30–38, 1994.
54. Krueger, J.M. and Majde, J.A., Microbial products and cytokines in sleep and fever regulation [review], *Crit. Rev. Immunol.,* 14, 355–379, 1994.
55. Krueger, J.M. and Obal, Jr., F., Growth hormone-releasing hormone and interleukin-1 in sleep regulation [review], *FASEB J.,* 7, 645–652, 1993.
56. Landolt, H.P., Dijk, D.J., Gaus, S.E., and Borbely, A.A., Caffeine reduces low-frequency delta activity in the human sleep EEG, *Neuropsychopharmacology,* 12(3), 229–238, 1995.
57. Landis, C., Collins, B., Cribbs, L., Smalheiser, N., Sukhatne, V., Bergmann, B., and Rechtschaffen, A., Expression of EGR-1-like immunoreactivity (EGR-1LI) is altered in specific areas in brain and spinal cord of sleep-deprived rats [abstract], *Sleep Res.,* 21, 321, 1992.
58. Levine, B., Roehrs, T., Steanski, E., Zorick, F., and Roth, T., Fragmenting sleep diminishes its recuperative value, *Sleep,* 10, 590–599, 1987.
59. Lin, J.S., Sakai, K., and Jouvet, M., Hypothalamo-preoptic histaminergic projections in sleep/wake control in the cat, *Eur. J. Neurosci.,* 6, 618–625, 1994.
60. Lin, J.S., Sakai, K., Vanni-Mercier, G., and Jouvet, M., A critical role of the posterior hypothalamus in the mechanisms of wakefulness determined by micro-injections of muscimol in freely moving cats, *Brain Res.,* 479, 225–240, 1989.

61. Lucas, E.A. and Sterman, M.B., Effect of a forebrain lesion on the polycyclic sleep/wake cycle and sleep/wake patterns in the cat, *Exp. Neurol.,* 46, 368–388, 1975.
62. Martinez-Mir, M.I., Probst, A., and Palacios, J.M., Adenosine A2 receptors: selective localization in the human basal ganglia and alterations with disease, *Neuroscience,* 42(3), 697–706, 1991.
63. Matsumura, H., Nakajima, T., Osaka, T., Satoh, S., Kawase, K., Kubo, E., Kantha, S.S., Kasahara, K., and Hayaishi, O., Prostaglandin D2-sensitive, sleep-promoting zone defined in the ventral surface of the rostral basal forebrain, *Proc. Natl. Acad. Sci. USA,* 91, 11998–12002, 1994.
64. McGinty, D. and Szymusiak, R., Keeping cool: a hypothesis about the mechanisms and functions of slow-wave sleep [review], *Trends Neurosci.,* 13, 480–487, 1990.
65. McGinty, D., Szymusiak, R., and Thompson, D., Preoptic/anterior hypothalamic warming increases EEG delta frequency activity within non-rapid eye movement sleep, *Brain Res.* 667, 273–277, 1994.
66. McGinty, D.J., Somnolence, recovery and hyposomnia following ventromedial diencephalic lesions in the rat, *Electroencephalogr. Clin. Neurophysiol.,* 26, 70–79, 1969.
67. Mendelson, W.B., Martin, J.V., Perlis, M., and Wagner, R., Enhancement of sleep by microinjection of triazolam into the medial preoptic area, *Neuropsychopharmacology,* 2, 61–66, 1989.
68. Meibach, R.C. and Siegel, A., Efferent connections of the septal area in the rat: an analysis utilizing retrograde and anterograde transport methods, *Brain Res.,* 119(1), 1–20, 1977.
69. Morairty, S., Thomson, D., Szymusiak, R., Hays, T., and McGinty, D., The somnogenic effects of prostaglandin D2 infusion in rats with preoptic/anterior hypothalamic lesions, in *Proc. Annual Meeting of the Association of Professional Sleep Societies,* 1996.
70. Moruzzi, G. and Magoun, H.W., Brainstem reticular formation and activation of the EEG, *Electroencephalogr. Clin. Neurophysiol.,* 1, 455–473, 1949.
71. Nauta, W.J.H., Hypothalamic regulation of sleep: an experimental study, *J. Neurophysiol.,* 9, 285–316, 1946.
72. Nicoll, R.A., Malenka, R.C., and Kauer, J.A., Functional comparison of neurotransmitter receptor subtypes in mammalian central nervous system, *Physiol. Rev.,* 70(2), 513–565, 1990.
73. Nitz, D.N. and Siegel, J.M., GABA, glutamate, and glycine release in the posterior hypothalamus across the sleep/wake cycle, *Sleep Res.,* 24, 12, 1995.
74. O'Hara, B.F., Young, K., Watson, F., Heller, H.C., and Kilduff, T.S., Immediate early gene expression in brain during sleep deprivation: preliminary observations, *Sleep,* 16, 1–7, 1993.
75. Ottlecz, A., Snyder, G.D., and McCann, S.M., Regulatory role of galanin in control of hypothalamic-anterior pituitary function, *Proc. Natl. Acad. Sci. USA,* 85, 9861–9865, 1988.
76. Peng, Z.C., Grassi-Zucconi, G., and Bentivoglio, M., Fos-related protein expression in the midline paraventricular nucleus of the rat thalamus, basal oscillation and relationship with limbic efferents, *Exp. Brain Res.,* 104, 21–29, 1995.
77. Pompeiano, M., Cirelli, C., and Tonoli, G., Effects of sleep deprivation on Fos-like immunoreactivity in the rat brain, *Arch. Ital. Biol.,* 130, 325–335, 1992.
78. Pompeiano, M., Cirelli, C., and Tonoli, G., Immediate-early genes in spontaneous wakefulness and sleep: expression of c-*fos* and NGFI-A mRNA and protein, *J. Sleep Res.,* 3, 80–96, 1994.
79. Porkka-Heiskanen, T., Strecker, R.E., Thakkar, M., Bjorkum, A.A., Greene, R.W., and McCarley, R.W., Adenosine: a mediator of the sleep-inducing effects of prolonged wakefulness, *Science,* 276, 1265–1266, 1997.
80. Portas, C.M., Thakkar, M., Rainnie, D.J., Greene, R., and McCarley, R., Role of adenosine in behavioral state regulation: a microdialysis study in the freely moving cat, *Neuroscience,* 79, 225–235, 1997.
81. Radulovacki, M., Virus, R.M., Djuricic, N.M., and Green, R.D., Adenosine analogs and sleep in rats, *J. Pharmacol. Exp. Ther.,* 228(2), 268–274, 1984.

82. Rajkowski, J., Kubiak, P., and Aston-Jones, G., Locus coeruleus activity in monkey: phasic and tonic changes are associated with altered vigilance, *Brain Res. Bull.*, 35, 607–616, 1994.

83. Rainnie, D.G., Grunze, H.C., McCarley, R.W., and Greene, R.W., Adenosine inhibition of mesopontine cholinergic neurons, implications for EEG arousal, *Science*, 263(5147), 689–692, 1994 (published *erratum* appears in *Science*, 265(5168), 16, 1994).

84. Rasmusson, D.D., Clow, K., and Szerb, J.C., Frequency-dependent increase in cortical acetylcholine release evoked by stimulation of the nucleus basalis magnocellularis in the rat, *Brain Res.*, 594, 150–154, 1992.

85. Rasmusson, D.D., Clow, K., and Szerb, J.C., Modification of neocortical acetylcholine release and electroencephalogram desynchronization due to brainstem stimulation by drugs applied to the basal forebrain, *Neuroscience*, 60, 665–677, 1994.

86. Ray, P. and Jackson, W., Lesions of nucleus basalis alter ChAT activity and EEG in rat frontal neocortex, *Electroencephalogr. Clin. Neurophysiol.*, 79, 62–68, 1991.

87. Reppert, S.M., Weaver, D.R., Stehle, J.H., and Rivkees, S.A., Molecular cloning and characterization of a rat A1-adenosine receptor that is widely expressed in brain and spinal cord, *Mol. Endocrinol.*, 5(8), 1037–1048, 1991.

88. Risold, P.Y., Canteras, N.S., and Swanson, L.W., Organization of projections from the anterior hypothalamic nucleus: a *Phaseolus vulgaris*-leucoagglutinin study in the rat, *J. Comp. Neurol.*, 348(1), 1–40, 1994.

89. Sakai, K., El Mansari, M., Lin, J.S., Zhang, G., and Vanni-Mercier, G., The posterior hypothalamus in the regulation of wakefulness and paradoxical sleep, in *The Diencephalon and Sleep*, Mancia, M. and Marini, G., Eds., Raven Press, New York, 1990, pp. 171–198.

90. Sallanon, M., Denoyer, M., Kitahama, K., Aubert, C., Gay, N., and Jouvet, M., Long-lasting insomnia induced by preoptic neuron lesions and its transient reversal by muscimol injection into the posterior hypothalamus in the cat, *Neuroscience*, 32, 669–683, 1989.

91. Satoh, S., Matsumura, H., Suzuki, F., and Hayaishi, O., Promotion of sleep mediated by the A2a-adenosine receptor and possible involvement of this receptor in the sleep induced by prostaglandin D2 in rats, *Proc. Natl. Acad. Sci. USA*, 931(2), 5980–5984, 1996.

92. Scammel, T., Chou, T.C., Shiromani, P.J., and Saper, C.B., Adenosine in the hypothalamus and basal forebrain regulates sleep [abstract], *Soc. Neurosci.*, 23(1), 1067, 1997.

93. Schiffmann, S.N., Libert, F., Vassart, G., and Vanderhaeghen, J.J., Distribution of adenosine A2 receptor mRNA in the human brain, *Neurosci. Lett.*, 130(2), 177–181, 1991.

94. Shearman, L.P. and D.R. Weaver, [^{125}I]4-aminobenzyl-5'-N-methylcarboxamidoadenosine (^{125}I.AB-MECA) labels multiple adenosine receptor subtypes in rat brain, *Brain Res.*, 745(1-2), 10–20, 1997.

95. Sherin, J.E., Shiromani, P.J., McCarley, R.W., and Saper, C.B., Activation of ventrolateral preoptic neurons during sleep, *Science*, 271, 216–219, 1996.

96. Shiromani, P., Armstrong, D.M., Bruce, G., Hersh, L.B., Groves, P.M., and Gillin, J.C., Relation of choline acetyltransferase immunoreactive neurons with cells which increase discharge during REM sleep, *Brain Res. Bull.*, 18, 447–455, 1987.

97. Shiromani, P., Gillin, J.C., and Henriksen, S.J., Acetylcholine and the regulation of REM sleep: basic mechanisms and clinical implication for affective illness and narcolepsy, *Ann. Rev. Pharmacol. Toxicol.*, 27, 137–156, 1987.

98. Shiromani, P., Winston, S., Mallik, M., and McCarley, R.W., Rapid decline in Fos-LI in association with recovery sleep that follows total sleep deprivation, *Soc. Neurosci. Abstr.*, 19(1), 572, 1993.

99. Shiromani, P.J., Armstrong, D.M., Berkowitz, A., Jeste, D.V., and Gillin, J.C., Distribution of choline acetyltransferase immunoreactive somata in the feline brainstem: implications for REM sleep generation, *Sleep*, 11, 1–16, 1988.

100. Shiromani, P.J., Kilduff, T.S., Bloom, F.E., and McCarley, R.W., Cholinergically induced REM sleep triggers Fos-like immunoreactivity in dorsolateral pontine regions associated with REM sleep, *Brain Res.*, 580, 351–357, 1992.

101. Shiromani, P.J., Mallik, M., Winston, S., and McCarley, R.W., Time course of Fos-LI associated with carbachol induced REM sleep, *J. Neurosci.*, 15, 3500–3508, 1995.

102. Shiromani, P.J., Thakkar, J., Thakkar, M., Greco, M., Basheer, R., and McCarley, R., Sleep in c-*fos* gene knockout mice, *Soc. Neurosci. Abstr.*, 23(2), 1846, 1997.

103. Shiromani, P.J., Winston, S., and McCarley, R.W., Pontine cholinergic neurons show Fos-like immunoreactivity associated with cholinergically-induced REM sleep, *Mol. Brain Res.*, 38, 77–84, 1996.

104. Shoham, S., Blatteis, C.M., and Krueger, J.M., Effects of preoptic area lesions on muramyl dipeptide-induced sleep, *Brain Res.*, 476, 396–399, 1989.

105. Shoham, S. and Teitelbaum, P., Subcortical waking and sleep during lateral hypothalamic "somnolence" in rats, *Physiol. Behav.*, 28, 323–333, 1982.

106. Stehle, J.H., Rivkees, S.A., Lee, J.J., Weaver, D.R., Deeds, J.D., and Reppert, S.M., Molecular cloning and expression of the cDNA for a novel A2-adenosine receptor subtype, *Mol. Endocrinol.*, 6(3), 384–393, 1992.

107. Stepanski, E., Lamphere, J., Roehrs, T., Zorick, F., and Roth, T., Experimental sleep fragmentation in normal subjects, *Int. J. Neurosci.*, 33, 207–214, 1987.

108. Steriade, M., Datta, S., Pare, D., Oakson, G., and Curr-Dossi, R., Neuronal activities in brain-stem cholinergic nuclei related to tonic activation processes in thalamocortical systems, *J. Neurosci.*, 10, 2541–2559, 1990.

109. Steriade, M. and McCarley, R.W., *Brainstem Control of Wakefulness and Sleep*, Plenum Press, New York, 1990.

110. Sterman, M.B. and Clemente, C., Forebrain inhibitory mechanisms: sleep patterns induced by basal forebrain stimulation in the behaving cat, *Exp. Neurol.*, 6, 103–117, 1962.

111. Sterman, M.B. and Clemente, C., Forebrain inhibitory mechanisms: cortical synchronization induced by basal forebrain stimulation, *Exp. Neurol.*, 6, 9–102, 1962.

112. Szerb, J.C., Cortical acetylcholine release and electroencephalographic arousal, *J. Physiol. (Lond.)*, 192, 329–345, 1967.

113. Szymusiak, R., Magnocellular nuclei of the basal forebrain: substrates of sleep and arousal regulation, *Sleep*, 18, 478–500, 1995.

114. Szymusiak, R., Danowski, J., and McGinty, D., Exposure to heat restores sleep in cats with preoptic/anterior hypothalamic cell loss, *Brain Res.* 541, 134–138, 1991.

115. Szymusiak, R., Iriye, T., and McGinty, D., Sleep-waking discharge of neurons in the posterior lateral hypothalamic area of cats, *Brain Res. Bull*, 23, 111–120, 1989.

116. Szymusiak, R. and McGinty, D., Sleep-related neuronal discharge in the basal forebrain of cats, *Brain Res.*, 370, 82–92, 1986.

117. Szymusiak, R. and McGinty, D., Sleep suppression following kainic acid-induced lesions of the basal forebrain, *Exp. Neurol.*, 94, 598–614, 1986.

118. Ticho, S.R. and Radulovacki, M., Role of adenosine in sleep and temperature regulation in the preoptic area of rats, *Pharmacol. Biochem. Behav.*, 40, 33-40, 1991.

119. Tobler, I. and Borbely, A., Sleep EEG in the rat as a function of prior waking, *Electroencephalogr. Clin. Neurophysiol.*, 64, 74–76, 1986.

120. Tononi, G., Pompeiano, M., and Cirelli, C., The locus coeruleus and immediate-early genes in spontaneous and forced wakefulness, *Brain Res. Bull.*, 35, 589–596, 1994.

121. Toppila, J., Stenberg, D., Alanko, L., Asikainen, M., Urban, J.H., Turek, F.W., and Porkka-Heiskanen, T., REM sleep deprivation induces galanin gene expression in the rat brain, *Neurosci. Lett.*, 183, 171–174, 1995.
122. Trachsel, L., Tobler, I., and Borbely, A., Sleep regulation in rats: effects of sleep deprivation, light, and circadian phase, *Am. J. Physiol.*, 251, R1037–R1044, 1986.
123. Ueno, R., Ishikawa, Y., Nakayama, T., and Hayaishi, O., Prostaglandin D_2 induces sleep when microinjected into the preoptic area of the conscious rat, *Biochem. Biophys. Res. Comm.*, 109, 576–582, 1982.
124. Vanni-Mercier, G., Sakai, K., and Jouvet, M., Neurones specifiques de l'eveil dans l'hypothalamus posterieur, *C. R. Acad. Sci.*, 298, 195–200, 1984.
125. Velazquez-Moctezuma, J., Gillin, J.C., and Shiromani, P.J., Effect of specific M1, M2 muscarinic receptor agonists on REM sleep generation, *Brain Res.*, 503, 128–131, 1989.
126. Velazquez-Moctezuma, J., Shalauta, M., Gillin, J.C., and Shiromani, P.J., Cholinergic antagonists and REM sleep generation, *Brain Res.*, 543, 175–179, 1991.
127. Weaver, D.R., A2a adenosine receptor gene expression in developing rat brain, *Brain Res. Mol Brain Res.*, 20(4), 313–327, 1993.
128. Yamuy, J., Mancillas, J.R., Morales, F.R., and Chase, M.H., C-fos expression in the pons and medulla of the cat during carbachol-induced active sleep, *J. Neurosci.*, 13, 2703–2718, 1993.
129. Yergey, J.A. and Heyes, M.P., Brain eicosanoid formation following acute penetration injury as studied by *in vivo* microdialysis, *J. Cerebral Blood Flow Metab.*, 101, 143–146, 1990.
130. Zhou, Q.Y., Li, C., Olah, M.E., Johnson, R.A., Stiles, G.L., and Civelli, O., Molecular cloning and characterization of an adenosine receptor: the A3 adenosine receptor, *Proc. Natl. Acad. Sci. USA*, 89(16), 7432–7436, 1992.

Chapter 20

Cellular Substrates of Oscillations in Corticothalamic Systems During States of Vigilance

Mircea Steriade

Contents

20.1 The How and Why of Sleep and Some Methodological Issues 327
20.2 The Cortically Generated Slow Oscillation Groups Other
 Sleep Rhythms into Complex Wave Sequences 329
20.3 Fast Oscillations During Brain-Activated State 340
20.4 Concluding Remarks 342
References 344

20.1 The How and Why of Sleep and Some Methodological Issues

This chapter is a brief account about the cellular substrates of brain oscillations that characterize wake and sleep states. As such, it may answer questions concerning basic mechanisms underlying the states of vigilance but will provide only very few insights on the *why* of sleep because, among the many hypotheses that have been proposed, none of them was seriously tested.

The fact is that there are two categories of sleep investigators: some look at this global behavior from outside the brain, while others enter the brain. The former approach is used by sleep psychologists and physiologists who are interested in the states of mind and the functions of different vital systems during the wake/sleep cycle, but do not manipulate the necessary tools that would allow them to reveal the neuronal processes which control, or are associated with, the states of vigilance. Instead, they evaluate sleep functions in ontogeny and phylogeny by recording a series of macrophysiological variables (body temperature, energy expenditure, weight) and by comparing normal with sleep-deprived subjects. By contrast, neurophysiologists look at the brain from inside and sometimes from within the neurons, and it is hoped that they know what type of neuron from which they are recording and the input-output organization of various cellular classes, but often they only use sleep as a reproducible behavioral condition to reveal mechanisms underlying changes in excitability of brainstem, thalamic, and cortical networks. In other words, electrophysiologists are so overwhelmed by technical difficulties and by putting together cellular data to make sense of operations in complex networks that they do not succeed in testing hypotheses about the function(s) of sleep; at best, they offer a few embellishing ideas on the behavioral significance of their data at the end of articles or chapters (as is the case here). Neither psychologists nor neuroscientists have provided a clear picture as to the *why* of resting (quiescent) sleep or rapid-eye-movement (REM) sleep. Even the conventional wisdom that the resting slow-wave sleep (SWS) has a restorative function (in view of the fact that SWS is enhanced by extending the prior waking period) should be regarded with caution because SWS is exclusively defined on the basis of the electroencephalogram (EEG), but there is yet no sound evidence to correlate sleep quality with physiological processes (e.g., growth hormone secretion and body temperature) other than the EEG.

Nonetheless, despite the fact that sleep functions are not yet elucidated, the study of spontaneous brain oscillations and changes in neuronal excitability during sleep states is fundamental because it reveals the reorganization of thalamic and cortical networks during fluctuations in global behavioral states and raises intriguing questions that may eventually lead to the disclosure of sleep functions. As I will show in this chapter, the idea that, during quiet sleep, the brain lies in total darkness and is associated with annihilation of consciousness became obsolete in view of data showing that neurons in upper nervous instances display unexpectedly high levels of spontaneous activity, and, although the thalamic gates are closed for signals from the outside world, the intracortical dialogue is maintained. These experimental results suggest that SWS, which is commonly regarded as a state of complete rest and unconsciousness, may subserve important cerebral functions.

During the past decade, the apparent chaos of EEG waves was reduced to a few basic cellular operations in the thalamus and cortex. We are witnessing a true renaissance in this field which creates bridges between biophysicists, who recently became interested in events related to behavioral states of vigilance, and clinical researchers, who have begun to use the language of neurophysiology and to test hypotheses advanced by basic scientists. The most interesting aspects of brain oscillations — namely, their expression in large populations of neurons and their changes in relation with shifts in the state of vigilance — depend on local or global synchronizing devices and on modulatory systems with generalized actions. Because brain oscillations are expressed at the macroscopic EEG level, they should arise from the coherent activity of neurons in the cerebral cortex and thalamus. The state-dependent changes in oscillatory patterns are attributable to modulatory influences from brainstem, hypothalamus, and basal forebrain. These aspects justify my choice to study the synchronized brain oscillations *in vivo* by means of multi-site, extra- and intracellular recordings in preparations with preserved thalamocortical connectivity and intact modulatory systems. As shown below, cortically and thalamically generated rhythms interact to produce complex patterns of activity. No simple circuit can account for the full expression of different oscillations during natural states of vigilance in living animals. The complexity of brain rhythms requires not only a permanent interaction between the cortex and thalamus, but also the presence of brainstem and other modulatory systems. Extra- and intracellular recordings, combined with dual impalements of cortical and thalamic neurons *in vivo*, are performed under different anesthetics because of stability problems, but the results obtained in these acute conditions should be validated by multi-site recordings of field potentials and extracellular discharges in chronically implanted, naturally sleeping and awakening animals.

First, I will present the neuronal substrates of low-frequency oscillations (less than 15 Hz) in thalamocortical systems which characterize SWS and distinguish this deafferented state from both brain-activated states of waking and REM sleep. The conversion of slow sleep oscillations into paroxysmal discharges will be analyzed in relation to the preferential occurrence of some seizures during resting sleep. Finally, I will expose the cellular processes that account for the shift from low-frequency to fast (20 to 50 Hz) rhythms that appear during activated states.

20.2 The Cortically Generated Slow Oscillation Groups Other Sleep Rhythms into Complex Wave Sequences

Three types of oscillations characterize the state of resting sleep (Figure 20.1): spindles (7 to 14 Hz), delta (1 to 4 Hz), and slow oscillation (less than 1 Hz).[1] Each of these rhythms originate through either network or intrinsic properties of different thalamic and neocortical neurons, even after their disconnection from related structures.

The GABAergic thalamic reticular (RE) neurons can generate spindle rhythmicity after deafferentation from dorsal thalamic nuclei and cerebral cortex (Figure 20.2A)[2] due to axo-dendritic and dendrodendritic connections among RE neurons. The pacemaking role of the RE nucleus in the generation of sleep spindles was also demonstrated by loss of spindling in target thalamocortical systems after disconnection from the RE nucleus.[3] Spindles result from repetitive spike-bursts in RE cells that produce rhythmic inhibitory postsynaptic potentials (IPSPs) in thalamocortical (TC) neurons, leading to postinhibitory rebound spike-bursts that are transferred to cortex and produce excitatory postsynaptic potentials (EPSPs) in cortical cells, occasionally leading to action potentials (Figure 20.2B,C).[4]

Spindles appear during early stages of sleep, sometimes preceding overt behavioral manifestations of sleep (Figure 20.3A). As TC neurons spend much of their sleep time during spindle-related IPSPs, there is a powerful inhibition of incoming messages in their route to the cerebral cortex. Recording field potentials evoked by stimulation of prethalamic axons (a method that permits the monitoring of the presynaptic deflection reflecting the magnitude of the afferent volley, together with the synaptically relayed, thalamically generated waves) revealed that the thalamus is the first station where afferent signals are completely blocked from the very onset of sleep (Figure 20.3B).[5,6] This obliteration of synaptic transmission in the thalamus leads to the deafferentation of the cerebral cortex, a prerequisite for the process of falling asleep. More recently, intracellular recordings from thalamic and cortical neurons have shown that, because of their hyperpolarization during sleep, TC cells do not transfer to cortex signals from prethalamic relay station, whereas the internal (corticocortical and corticothalamic) dialogue of the brain may be maintained during sleep.[7]

The delta oscillation has two components. The cortical one survives complete thalamectomy,[8,9] and the spectral content in the delta band (1 to 4 Hz) results, at least partially, from the shape and duration of the K-complex,[10] a major electrographic sign of sleep (see below). The other component of delta oscillation is stereotyped, clock-like (see bottom right trace in Figure 20.1 and Figure 20.4A), and generated in the thalamus through the hyperpolarization-activated interplay between two intrinsic currents of TC neurons. Although this rhythm is seen in thalamic slices after blockage of synaptic transmission,[11,12] its appearance at the level of local field potentials and cortical EEG (Figure 20.4B) is possible because corticothalamic volleys synchronize TC cells through the action of inhibitory RE neurons,[13] which fulfill two requirements: they set the membrane potential of TC cells at the required level of hyperpolarization for the generation of delta oscillation, and they have distributed projections to the dorsal thalamus, thus synchronizing various TC cells.

As delta oscillation appears in TC neurons at a more hyperpolarized level of membrane potential than spindles,[13] the two sleep oscillations are incompatible in single TC cells.[14] These intracellular data from anesthetized preparations (Figure 20.5)[15] are supported by results obtained in naturally sleeping animals, showing that thalamic spindles are maximal at sleep onset and decrease thereafter, whereas thalamic delta waves increase gradually during resting sleep.[16] Thus,

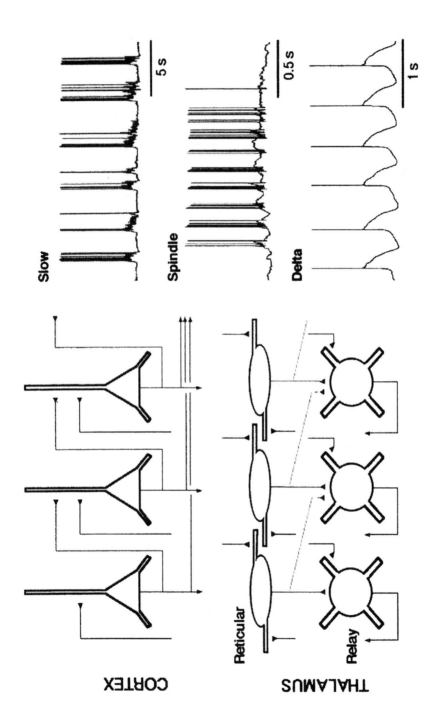

FIGURE 20.1
Building blocks of thalamic and neocortical networks, with different types of sleep oscillations generated by excitatory glutamatergic cortical neurons (top), inhibitory GABAergic thalamic reticular neurons (middle), and excitatory glutamatergic thalamocortical cells (bottom). The direction of axons is indicated by arrows. Short- and long-scale intracortical pathways are illustrated. Divergent thalamic reticular axons are shown as broken lines. Note different time calibrations in intracellular traces showing the slow cortical oscillation (less than 1 Hz), the spindles (7 to 14 Hz) in thalamic reticular neurons, and the intrinsic (clock-like) delta oscillation (1 to 4 Hz) in thalamocortical cells. These oscillations can be generated at each of these levels, even after disconnection from afferent inputs; however, in the intact brain, these structures are interacting and their rhythms are combined within complex wave sequences. (Adapted from Steriade, M. et al., *Trends Neurosci.,* 17, 199–208, 1994.)

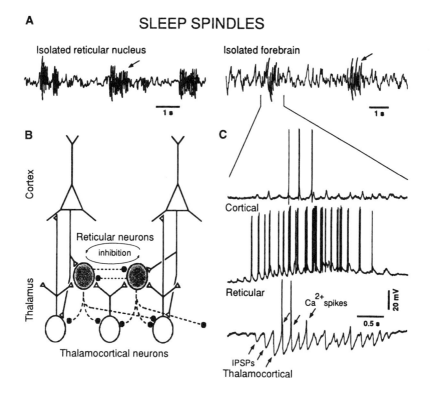

FIGURE 20.2
Sleep spindle oscillations as synaptically generated in the thalamus. (**A, left**) Field potentials recorded *in vivo* through a microelectrode inserted in the deafferented reticular nucleus (rostral pole) of a cat. Arrow indicates one spindle sequence. (**A, right**) Spindle oscillations recorded *in vivo* through a microelectrode inserted in the intralaminar centrolateral nucleus of a cat with an upper brainstem transection, creating an isolated forebrain preparation. Note the two spindle sequences (the second marked by an arrow) and, between them, lower frequency waves. (**B**) Schematic diagram of neuronal connections involved in the generation of spindle oscillation. (**C**) Intracellular recordings of one spindle sequence (see A, right) in three neuronal types (cortical, thalamic reticular, and thalamocortical) of cats *in vivo*. (Adapted from References 2 and 4.)

with increasing hyperpolarization of TC cells during resting sleep, due to the progressive reduction in firing rates of cholinergic and other types of brainstem-thalamic activating neurons,[17] the incidence of spindles is diminished while the incidence of delta waves is largely increased during deep sleep stages. On the other hand, the reappearance of spindles toward the very end of resting sleep[18] is attributable to a relative depolarization of TC cells, due to the increased firing rates of brainstem-thalamic reticular neurons that display precursor increased rates of spontaneous firing, 30 to 60 sec before the onset of REM sleep.[19]

The slow oscillation has a frequency peak at 0.7 to 0.8 Hz in cats under ketamine-xylazine anesthesia[18,20] and in the natural sleep of cats[21,22] and humans.[10,23] The cortical nature of the slow oscillation was demonstrated by its survival in the cerebral cortex after thalamectomy,[9] its absence in the thalamus of decorticated animals,[24] and the disruption of its long-range synchronization after disconnection of intracortical synaptic linkages (Figure 20.6).[25]

Intracellular analyses of the slow oscillation showed that cortical neurons throughout layers II to VI (many of them with physiologically identified callosal or thalamic projections) displayed a spontaneous oscillation consisting of prolonged depolarizing and hyperpolarizing components (Figure 20.7).[26] All major cellular classes in the cerebral cortex, as identified by electrophysiological characteristics and intracellular staining, display the slow oscillation: regular-spiking and intrinsically bursting cells, as well as local-circuit inhibitory basket cells.[20] Both long-axoned and short-

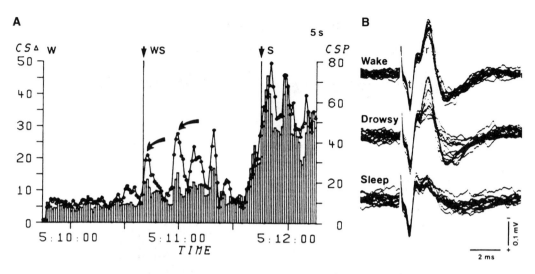

FIGURE 20.3
Spindle oscillations appear during the early stage of sleep and are associated with decreased responsiveness in the thalamus; chronically implanted behaving cats. (A) Normalized amplitudes of cortical spindle waves (CSP, filtered to 7–14 Hz; line-circle trace) and slow as well as delta waves (CSδ, filtered to 0.5–4 Hz; bar graph) during waking (W), transition from W to resting sleep (WS), and slow-wave sleep (S). Abscissa indicates real time. Note spindle sequences during WS (arrows). (B) Blockade of synaptic transmission in the thalamus at sleep onset. Field potentials evoked in the VL nucleus by stimulation of cerebello-thalamic axons. Evoked responses consisted of a presynaptic (tract, *t*) component and a postsynaptic (relayed, *r*) component. Note progressively diminished amplitude of *r* component during drowsiness, up to its disappearance during sleep, despite lack of changes in the afferent volley monitored by the *t* component. (Adapted from Steriade, M., in *Cerebral Cortex*. Vol. 9. *Normal and Altered States of Function*, Peters, A. and Jones, E.G., Eds., Plenum Press, New York, 1991, pp. 279–357.)

axoned cells exhibit similar relations with the EEG components of the slow oscillation: during the depth-positive EEG wave, cortical neurons are hyperpolarized, whereas during the sharp, depth-negative EEG deflection cortical neurons are depolarized. The long-lasting depolarization of the slow oscillation consists of EPSPs, fast prepotentials (FPPs), and fast IPSPs, reflecting the action of synaptically coupled GABAergic local-circuit cortical cells.[26] Data also indicate that the depolarizing component is made up of both *N*-methyl-D-aspartatate (NMDA)-mediated synaptic excitatory events and a voltage-dependent, persistent sodium current $g_{Na(p)}$.[26] The long-lasting hyperpolarization interrupting the depolarizing envelopes is a combination of a $g_{K(Ca)}$ and disfacilitation in the corticothalamic network. The disfacilitation mechanism is supported by measuring the membrane input resistance (R_{in}), showing that R_{in} is highest during the long-lasting hyperpolarizing component of the slow oscillation.[27]

The spectacular similarity between all types of cortical neurons and EEG wave-forms raised the question of mechanisms underlying these synchronization processes. Dual intracellular recordings *in vivo* revealed that the overt synchronization of EEG patterns is associated with the simultaneous hyperpolarizations in cortical neurons.[15,20,25]

The neuronal synchronization also implicates thalamic neurons. GABAergic RE thalamic cells are identified by their peculiar bursting pattern and cortically elicited, spindle-like, depolarizing oscillations.[20,28] Remarkably, although the intrinsic properties of RE neurons are quite different from those of neocortical cells, RE neurons exhibit patterns of the slow sleep oscillation, with prolonged depolarizations interrupted by prolonged hyperpolarizations which are very similar to those of cortical neurons (see intracellular recordings of RE cell and the second trace of cortical cells in Figure 20.7).[29] The depolarizing component of the cortically generated slow oscillation is transmitted to RE thalamic neurons, at which level it triggers rhythmic spike bursts (Figure 20.8).[24]

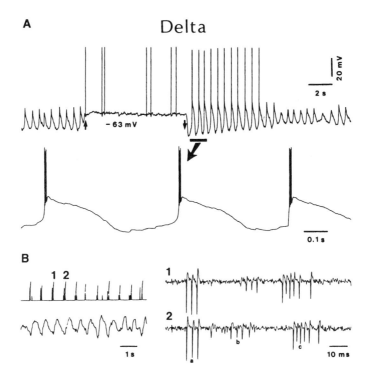

FIGURE 20.4
Intrinsic (clock-like) delta oscillation in thalamocortical neurons *in vivo*. (**A**) Delta oscillation in intracellularly recorded lateroposterior (LP) neuron after decortication. The cell oscillated spontaneously at 1.7 Hz. A 0.5-nA depolarizing current (between arrows) prevented the oscillation, and its removal set the cell back in the oscillatory mode. Three cycles are expanded below (arrow) to show high-frequency spike bursts crowning the low-threshold Ca^{++} spike. (**B**) Extracellular recording of neuronal discharges and field potentials in LP nucleus. Delta oscillation in field potentials (lower trace) is associated with spike bursts in three simultaneously recorded neurons (upper trace; 1 and 2 are expanded at right; note regular sequences in neurons a to c). (Adapted from Steriade, M. et al., *J. Neurosci.*, 11, 3200–3217, 1991.)

These spike bursts of thalamic GABAergic RE neurons impose onto TC cells rhythmic IPSPs leading to rebound spike bursts.[20,24,29,30] This is the mechanism underlying the brief sequence of EEG spindles that follows every cycle of the slow oscillation (see top trace in Figure 20.8). Distinct from the prolonged, waxing-and-waning pattern of spindle oscillations in decorticated animals or those under barbiturate anesthesia when cortical neurons display reduced spontaneous activities, the spindles triggered by the synchronous corticothalamic volleys of the slow oscillation are much shorter (Figure 20.8). This is due to the fact that the excitation of large numbers of corticothalamic neurons during the slow oscillation entrain, right from the start, a great population of thalamic neurons implicated in the genesis of spindles, thus explaining the absence of an initial waxing process[30].

Thus, although the systematization of sleep rhythms into three categories (spindles, delta, slow) may be useful for didactic purposes, in brain-intact animals and humans, the sleep oscillations are not seen in isolation but are grouped by the cortically generated slow oscillation. The coalescence of the slow and spindle oscillations is especially visible during light sleep; however, spindling is not the only sleep rhythm that is modulated and grouped by the cortical slow oscillation. The intrinsically generated delta rhythm of TC cells is influenced by the slow oscillation because the rhythmic depolarizing corticothalamic volleys increase the membrane conductance of TC cells and prevent the interplay between I_h and I_t, thus periodically dampening the clock-like delta oscillation (see bottom right trace in Figure 20.7).[29] As to the other component of delta waves, which is generated

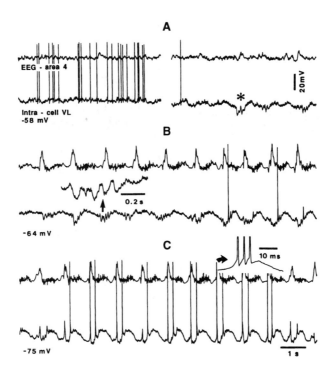

FIGURE 20.5
The dependency of transition from spindle to delta oscillation on the membrane potential of thalamocortical neurons *in vivo*. Intracellular recording of thalamic VL neuron, simultaneously with surface-EEG from area 4. (**A, left**) Tonic firing during activated EEG pattern. (**A, right**) Appearance of membrane potential oscillations in association with slight changes in EEG activity (asterisk marks the first hyperpolarizing sequence in association with the initial signs of EEG synchronization). (**B**) Hyperpolarizing spindle oscillations (8 to 10 Hz) are grouped within sequences recurring with a slow rhythm (~0.8 Hz) in close time relation with the cortical EEG rhythm. One spindle sequence is expanded above. (**C**) Hyperpolarizing current step (–0.2 nA) transformed spindles into delta potentials (3 to 4 Hz), also grouped by the same (~0.8 Hz) cortical rhythm. One spike burst crowning the low-threshold Ca^{2+} is expanded above. (Adapted from Steriade, M. et al., *Trends Neurosci.*, 17, 199–208, 1994.)

intracortically after thalamectomy, the frequency band of 1 to 4 Hz in the power spectrum during late stages of resting sleep may result from the duration (0.3 to 0.4 sec) of the depth-negative (depolarizing) component of the slow oscillation, which represents the K-complex.[10] Typical delta waves, at a frequency of 2 to 4 Hz, generated by both regular-spiking and intrinsically bursting cortical neurons, are grouped within sequences recurring with the slow rhythm.[9] And, in human sleep EEG, sequential mean amplitudes of delta waves show their periodic recurrence with the rhythm of slow oscillation.[26] That delta and slow oscillation represent two distinct phenomena in human sleep EEG was recently demonstrated by showing differences in the dynamics between the slow and the delta oscillations, as the latter declines in activity from the first to the second non-REM sleep episode, whereas the former does not.[23]

As the slow oscillation was first described intracellularly under different anesthetics, the similarity between these cellular patterns and those observed during natural sleep was validated in extracellular recordings from chronically implanted, unanesthetized animals. Under these conditions, the depth-positive waves are accompanied by silenced firing (due to cells' hyperpolarizations), while depth-negative sharp deflections are associated with brisk firing (Figure 20.9).[21]

Recent, independent experiments from two laboratories[10,23] have demonstrated the presence of the slow oscillation in human sleep. In our study,[10] during stage 2, scalp recordings showed a prevalent peak (0.8 Hz) within the frequency range of the slow oscillation as well as a minor mode around 15 Hz, reflecting spindle waves. The power spectrum revealed a major peak around 1 Hz,

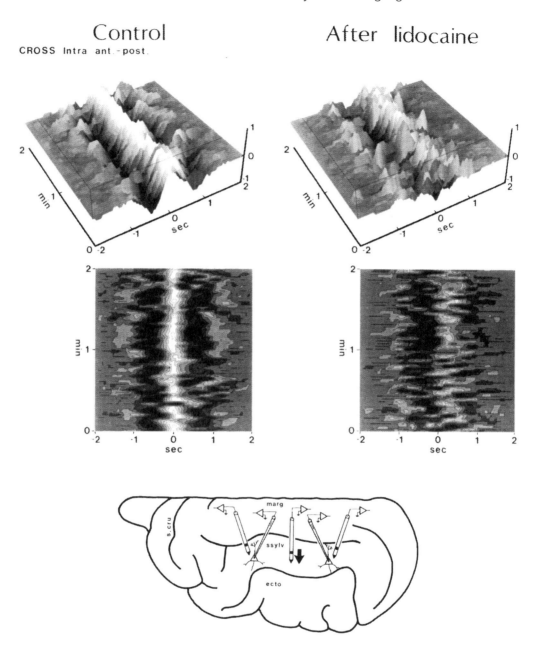

FIGURE 20.6

Disruption of synchronization of slow oscillation by intracortical disconnection of synaptic linkages. Dual intracellular recordings from the anterior and posterior parts of the suprasylvian gyrus in cat; lidocaine injection (40 µl, 20%) between the two micropipettes (see brain figure). The synchrony and its disruption after lidocaine injections are represented by sequential field analyses. The control synchrony was characterized by well-aligned, high, central peaks. After lidocaine injection, the previous pattern was replaced by a blurred sequence of lower peaks and lower valleys deviating from the central plane. (Adapted from Amzica, F. and Steriade, M., *J. Neurosci.*, 15, 4658–4677, 1995.)

which became evident from stage 2 and continued throughout resting sleep. These data invite human sleep researchers to consider the two types of oscillatory activities below 4 Hz (delta, 1 to 4 Hz; slow, <1 Hz) and, accordingly, to analyze their results by taking into account the distinctness of these two oscillations.[23]

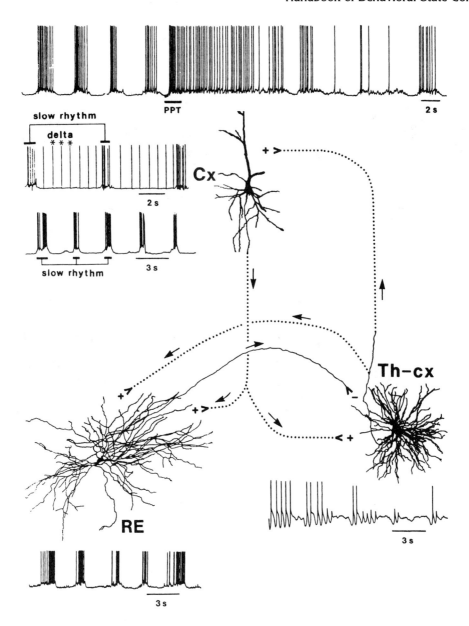

FIGURE 20.7
Low-frequency sleep oscillations and their modulation by brainstem core cholinergic system; intracellular recordings *in vivo*. Top trace shows the slow oscillation (<1 Hz) in a cortical cell recorded from cat association area 5. A brief pulse-train to the cholinergic pedunculopontine tegmental (PPT) nucleus (horizontal bar) transiently suppressed the slow cortical oscillation and replaced it by tonic firing. This effect was associated with an EEG-activated response that had a similar time course (and was blocked by a muscarinic antagonist, scopolamine; not shown). Below, summary diagram depicting several aspects of sleep oscillations in interconnected cortical and thalamic networks. The cortical neuron (Cx) was stained intracellularly with Lucifer yellow, and the thalamic reticular (RE) and thalamocortical (Th-cx) neurons were intracellularly stained with horseradish peroxidase. The direction of axons is indicated by arrows, and excitatory or inhibitory signs are indicated by + or –. The intracellular recordings of cortical neurons (from association area 5) show a slow rhythm (0.17 in upper trace, 0.3 Hz in bottom trace). In the first trace, note the appearance, between the depolarizing envelopes of the slow oscillation, of clock-like action potentials (asterisks) recurring at the delta frequency (1.5 Hz) and arising in thalamocortical cells. The thalamic reticular neuron displays the slow oscillation (0.3 Hz). The clock-like delta oscillation in the thalamocortical neuron (2.5 Hz) tends to dampen periodically within the frequency range of the slow rhythm (0.2 to 0.3 Hz), due to an increase in membrane conductance resulting from converging excitatory inputs arising in cortical neurons and inhibitory inputs arising in GABAergic thalamic reticular neurons. (Adapted from References 9, 29, 36, and 38.)

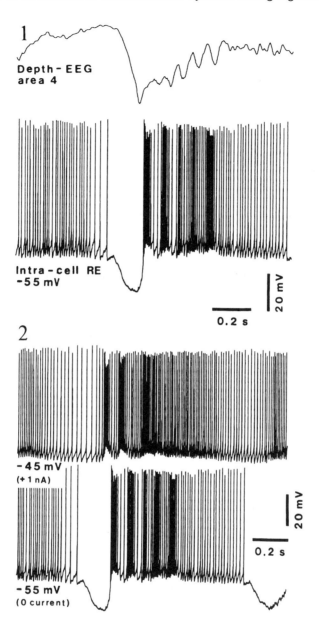

FIGURE 20.8
Effect of sharp corticothalamic volleys during the depolarizing phases of the slow sleep oscillation upon thalamic reticular cells; depth-EEG from area 4 and intracellular recording from thalamic reticular (RE) neuron. Shown in 1 is the resting membrane potential (–55 mV). Note spindle-related spike bursts in association with EEG spindles that follow the hyperpolarizing phase of the slow oscillation. Shown in 2 are traces at the resting level and under depolarizing current. (Adapted from Timofeev, I. and Steriade, M., *J. Neurophysiol.*, 76, 4152–4168, 1996.)

Naturally synchronized sleep oscillations develop, under special circumstances, into paroxysmal oscillations resembling spike-wave (SW) seizures or more complex seizure patterns. The incidence of SW seizures in humans increases during the spindle stage of EEG-synchronized sleep and is attenuated or blocked upon arousal.[31] Self-sustained SW cortical complexes at 2 to 3 Hz, lasting for 10 to 15 seconds, may occur spontaneously or follow prolonged thalamic stimulation

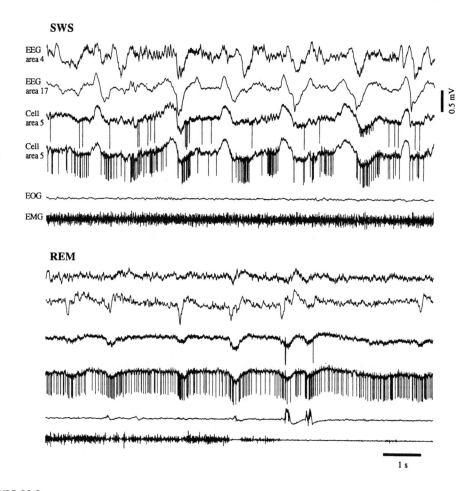

FIGURE 20.9
Patterns of slow oscillation during natural sleep of cat are similar to those recorded intracellularly under ketamine-xylazine anesthesia; chronically implanted, naturally sleeping cat. (SWS = slow-wave sleep; REM = rapid-eye-movement sleep.) SWS is separated from REM by a non-depicted period of 34 sec. Six traces represent: depth-EEG from motor (precruciate) area 4; depth-EEG from visual area 17; unit discharges and slow focal potentials from association area 5 in the anterior suprasylvian gyrus; similar recording from an adjacent focus (2 mm apart) in area 5; electrooculogram (EOG); and electromyogram (EMG). Compare the patterns of the slow oscillation during natural SWS to those in the preceding figure under ketamine-xylazine anesthesia. (Adapted from Steriade, M. et al., *J. Neurosci.*, 16, 392–417, 1996.)

during behavioral state of drowsiness or light sleep in chronically implanted monkeys.[32] Recent data strengthen the idea that the neocortex is preferentially implicated in the generation of seizures with SW complexes at 2 to 4 Hz or with poly-SW patterns interspersed with runs of fast EEG spikes at 10 to 20 Hz,[33,34] as in the Lennox-Gastaut syndrome. The results supporting this hypothesis basically consist of the fact that, with multi-site cortical and thalamic recordings, cortical processes are recruited well before any sign of thalamic entrainment. Dual simultaneous intracellular recordings from cortical and TC neurons demonstrate that, surprisingly, an important proportion (60%) of TC cells remained silent during cortically generated SW seizures, displaying a tonic hyperpolarization with phasic IPSPs in close time-relation with paroxysmal depolarizing shifts in neocortical neurons.[34] This result was explained by the fact that GABAergic RE neurons faithfully follow the spike bursts of cortical neurons and impose IPSPs onto TC cells, thus reinforcing through their exceedingly long spike bursts the hyperpolarization of TC cells. The RE neurons also contribute to the coalescence of IPSPs in TC cells, thus preventing the latter to fire rebound spike bursts.

Cellular Substrates of Oscillations in Corticothalamic Systems During Vigilance

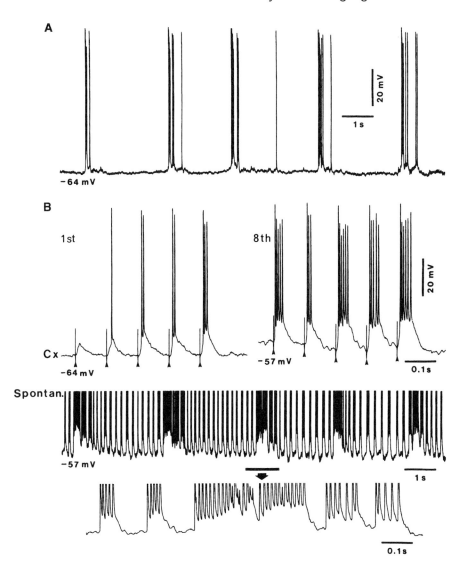

FIGURE 20.10
Self-sustained oscillation after repetitive cortical stimulation in an athalamic animal. Intracellular recording of an intrinsically bursting neuron, recorded at 1.5 mm in area 7. (**A**) Slow rhythm (0.4 Hz). Bursts marked by one and two asterisks are expanded below (spikes truncated). (**B**) Responses of the same cell to repetitive stimulation (five-shock trains at 10 Hz, repeated every 3 Hz) of the contralateral area 7. The augmenting responses to the 1st and 8th trains are illustrated. Self-sustained (Spontan.) seizure activity is illustrated after 15 such cortical pulse-trains. (Adapted from Steriade, M. et al., *J. Neurosci.*, 13, 3266–3283, 1993.)

These data have been reproduced in computer network models of SW seizures.[35] The role of neocortical neurons in generating complex seizures is also demonstrated by self-sustained discharges, after repetitive stimulation of callosal pathways, in intrinsically bursting cortical cells recorded from athalamic animals (Figure 20.10).[9] Such paroxysmal events occur on the background of a progressive depolarization of cortical neurons induced by rhythmic stimuli applied to the homotopic point in the contralateral cortex (Figure 20.10B), eventually leading to self-sustained seizures whose components partially reproduce the sleep slow oscillation from which the seizures emerged (Figure 20.10).[9]

20.3 Fast Oscillations During Brain-Activated States

Low-frequency oscillations of electrical activity, which characterize the behavioral state of resting sleep, are blocked upon brain activation, during both waking and REM sleep. Experimentally, this effect is produced by stimulation of the upper brainstem reticular core where neurons with projections to the thalamus and nucleus basalis are present.[17] The abolition of the slow sleep oscillation and its replacement by tonic firing, as a consequence of the suppression of prolonged hyperpolarizing components of slow oscillation,[36] is illustrated in the top trace of Figure 20.7. This aspect reproduces the patterns of brain activation during transition from resting sleep to either waking and REM sleep.

The blockage of the slow oscillation and other sleep rhythms is associated, upon arousal, by the appearance of fast oscillations (20 to 60 Hz, mainly 30 to 40 Hz). The recent interest in fast brain oscillations was aroused by hypotheses on their possible significance in focused attention and in phenomena of binding different features of an object into a global percept.[37] Spatially localized, coherent fast oscillations are present on the background electrical activity, without necessarily requiring optimal sensory stimuli, during all states of vigilance, including resting sleep.[21,22] The fast oscillations depend upon the depolarization of thalamic and cortical neurons[38] which occurs transiently (but predictably) during given phases of slow sleep oscillation and continuously during behavioral states with tonically increased brain excitability, waking and REM sleep. Far from being exclusively generated in the cerebral cortex, the fast oscillations also occur in thalamic neurons and are synchronized within corticothalamic networks.

Fast rhythms (25 to 45 Hz) of EEG activity appear in the occipital cortex when a dog pays intense attention to a visual stimulus;[39] when a cat is watching, during behavioral immobility, a visible but unseizable mouse;[40] and when monkeys retrieve raisins from unseen sites.[41] These oscillations are not simply related to active movements, but they are better understood within the concept of "set" cells, with discharge patterns modulated by movement preparation during conditions of increased alertness and interest for a given target. The 35- to 45-Hz oscillations also appear in humans during selective somatosensory attention, phase-locked between contralateral prefrontal and parietal cortical areas.[42]

In addition to states of attentive behavior[39-42] and sensory-elicited responses,[37] fast oscillations also appear spontaneously. The fast oscillations are produced by intrinsic properties of cortical and thalamic neurons, in conjunction with synaptic activities in reciprocal corticothalamic loops. These two, non-exclusive factors are discussed below.

Fast oscillations of cortical neurons, elicited by depolarizing current pulses, have been described both *in vitro*[43-46] and *in vivo*.[21,47-50] It was shown that these fast oscillations in cortical neurons are generated by a voltage-dependent, persistent Na^+ current, with the involvement of a delayed rectifier.[43] The impact exerted by fast-oscillating neurons on target brain structures is enhanced when, instead of single action potentials, neurons fire spike bursts at high frequencies. Fast-rhythmic-bursting cells have been described in the rostral part of thalamic intralaminar nuclei projecting to association cortex,[51] in superficial layers of visual cortex,[48] in motor cortical areas 4 and 6, and in association with cortical areas 5 and 7 (Figure 20.11).[21,49,50] The connectivity of these bursting neurons is important because it may reveal the role of cortex and thalamus in the generation of fast oscillations. Whereas early studies have emphasized that fast oscillations are exclusively generated intracortically, the inclusion of the thalamus in the synchronization process provides the advantage of linking morphologically distant and functionally different fields, because cortical areas project to association and intralaminar thalamic nuclei, which, in turn, feed back to cortical areas that may be different from the input sources.[52]

The intrinsic discharge pattern of fast-rhythmic-bursting corticothalamic neurons, as identified by antidromic invasion from appropriate thalamic nuclei, does not constitute an invariant stigma of these neurons. Indeed, after a passive response or single spikes in response to a subthreshold or

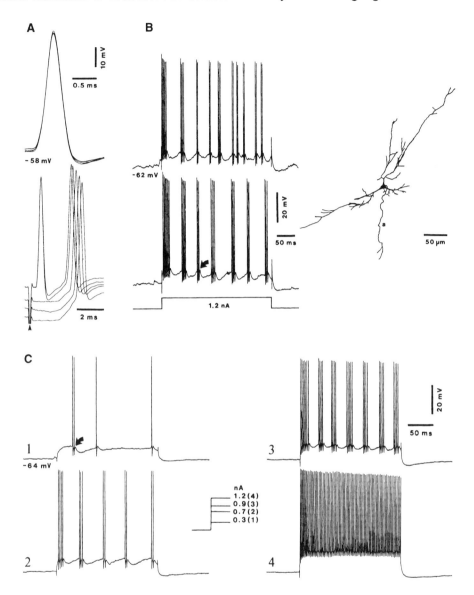

FIGURE 20.11

Corticothalamic cells display a continuum of discharge patterns, from single spikes to rhythmic (20 to 40 Hz) high-frequency (300 to 500 Hz) spike bursts and eventually to fast tonic firing without adaptation (~400 Hz), elicited by depolarizing current pulses with different intensities. (**A**) Intracellular recordings (cats under ketamine-xylazine anesthesia) of corticothalamic neuron located in layer VI of cat suprasylvian area 5. Spontaneous action potentials of a corticothalamic cell had duration of 0.35 msec at half amplitude. Below, electrophysiological identification of thalamic projection and input of the same neuron. Stimulus to thalamic lateral posterior (LP) nucleus (arrowhead) elicited antidromic activation (0.5-msec latency) and orthodromic spikes (2-msec latency). Five superimposed traces; the two top traces (at a V_m of –58 mV) depict both antidromic and synaptic responses, whereas antidromic spikes failed at more hyperpolarized levels. (**B**) Another corticothalamic neuron, antidromically invaded from central lateral (CL) rostral intralaminar nucleus. Intracellular staining (at right) showed its location in lower layer V (a, axon). Depolarizing current pulses elicited spike bursts recurring rhythmically at ~25 Hz. (**C**) Another corticothalamic cell, antidromically activated from the LP nucleus. Depolarizing current pulses with different intensitites (0.3, 0.7, 0.9, and 1.2 nA in 1 to 4, respectively) elicited changing discharge patterns, from single spikes to spike bursts (~35 Hz) and, eventually, fast-firing (~400 Hz) without frequency adaptation. (Adapted from References 49 and 50.)

a slightly suprathreshold depolarizing current pulse, corticothalamic fast-rhythmic-bursting neurons discharge rhythmic spike bursts when the stimulus intensity is raised; eventually, with a further increase in stimulus intensity, these neurons reach the pattern of tonic firing (300 to 600 Hz), without frequency adaptation (Figure 20.11C). The latter pattern is also observed in neurons which are conventionally thought to be local GABAergic neurons. Similar changes in discharge patterns occur as an effect of synaptic activity in thalamocortical networks.[50] These data indicate that the distinctions between the intrinsic electrophysiological properties of neocortical cell-classes are much more labile than conventionally thought. The hypothesis was then advanced[50] that the same network, submitted to repetitive inputs leading to cells' depolarization, would favor the transformation of regular-spiking into fast-rhythmic-bursting cortical cells. We also suggested that corticothalamic neurons are among the best candidates in the synchronizing process in view of their propensity to develop fast rhythmic bursts of action potentials. The role of reciprocal thalamocortical loops in the generation of fast oscillations supports data from magnetoencephalographic recordings in humans.[53] Although the above results on fast-rhythmic-bursting corticothalamic neurons[49,50] were obtained by applying depolarizing current pulses, the same changes in membrane potential of cortical neurons may occur during natural shifts in vigilance, with transition from functionally deafferented to brain-active states. Indeed, stimulation of upper brainstem glutamatergic and cholinergic nuclei leads to depolarization of target thalamocortical neurons, which, in turn, depolarize cortical neurons.[54]

Fast oscillations are synchronized within intracortical and corticothalamic systems. The synchronization is robust within a cortical column and among closely located cortical foci (1 to 5 mm) but becomes weaker or negligible between distant areas.[21] Interestingly, the fast waves (30 to 40 Hz) are in-phase, displaying the same polarity from the surface to deepest cortical layers (Figure 20.12).[21,55] Several findings indicate that this is not due to volume conduction from sources extrinsic to the explored area: the negative components of fast field potentials are crowned by action potentials at all cortical depths, cross-correlations between surface- and depth-recorded demonstrate time lags of 2 to 3 msec, and the fast waves undergo a reduction in amplitude, up to 70%, when the micro-electrode passes from deep cortical layers to the underlying white matter. The in-phase fast oscillation across the cortex is probably due to myriads of cortical and thalamic inputs exciting the neurons throughout the cortex and resulting in transmembrane current flow, with negligible internal longitudinal current components.[55] The spatially restricted intracortical synchronization of the fast oscillation stands in contrast with the long-range synchronization of slow rhythms.[25] The participation of thalamic nuclei in the synchronization of spontaneous fast oscillations becomes most evident when care is taken to record from directly related cortical areas and dorsal thalamic nuclei, as identified by standard physiological procedures (antidromic and orthodromic responses). In such reciprocally connected systems, the fast oscillations systematically occur with short time lags between the cortex and thalamus.[21,22] Thus, the intrathalamic and corticothalamic synchronization primarily depends on topographically organized projections, though the involvement of more circuitous or diffuse pathways involving rostral intralaminar and other dorsal thalamic nuclei should also be envisaged.

20.4 Concluding Remarks

What could be the role of spontaneous slow oscillations in resting sleep and fast oscillation during brain-active states? During resting sleep, the hyperpolarization of thalamic and neocortical neurons is effective in preventing the transfer of incoming signals from the outside world and to ensure a safe sleep. Despite the diversity of sleep oscillations, their functional outcome may be similar; however, the deafferentation process that occurs from the very onset of sleep may be just the tip of the iceberg. The rhythmic spike bursts, resulting from the de-inactivation of a transient low-threshold Ca^{++}

Cellular Substrates of Oscillations in Corticothalamic Systems During Vigilance

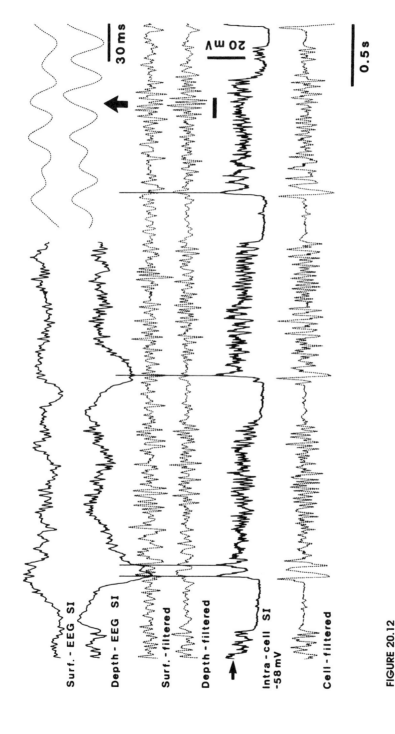

FIGURE 20.12
In-phase, surface and depth, fast oscillations (30 to 40 Hz) in cerebral cortex during the depolarizing phase of the slow oscillation. Intracellular recording of cortical neuron and EEG from surface and depth of somatosensory cortex. Dotted traces are filtered activities (10 to 100 Hz). Note fast oscillations during the depolarizing phase of the slow oscillation (0.7 Hz), their suppression during the hyperpolarizing phase, and in-phase fast oscillations (~35 Hz) in surface and depth EEG (see expanded traces and arrow at right). (Adapted from Steriade, M. et al., *J. Neurosci.*, 16, 2788–2808, 1996.)

conductance during the hyperpolarization of thalamic and neocortical neurons, may prevent the metabolic inertia that would result from complete absence of discharges and would thus favor a quick passage from quiet sleep to a brain-active state. The possibility is open that sleep rhythms re-orchestrate the intracellular processes of neurons to perform tasks best done during quiet sleep. At variance with previous assumptions, many cells recorded from neocortical areas were found to be firing as actively when the animal is asleep as when the animal is alert, although their firing patterns are changed from one state to the other.[1,21]

Why are thalamic and cortical cells so busy when the brain is supposed to rest? One of the sleep mysteries lies in this question. Moruzzi[56] suggested that sleep function may concern those neurons that during wakefulness are related to conscious behavior. During the state of resting sleep, the rhythmic spike bursts may reorganize and/or specify the circuitry and also may consolidate memory traces acquired during wakefulness.[6,29] The study of intracortical and corticothalamic synchrony during periods of sleep following conditioning sessions has revealed that changes in the coherence of rhythmic activities during the behavioral task are also expressed during sleep.[57] The persistence of plasticity during a behavioral state that is fundamentally different from that in which conditioning took place suggests, as in the hippocampus,[58,59] that the re-expression, during sleep, of information acquired in wakefulness may be related to memory consolidation.

As to the spontaneously occurring fast rhythms, they mainly characterize waking and dreaming sleep[1,21,22,53] but also appear in resting sleep or deep anesthesia when they are periodically obliterated during the hyperpolarizing episodes of the slow oscillation.[21,22] The fact that fast rhythms also appear during states of unconsciousness does not exclude the possibility that such oscillations may underlie binding operations, significant for cognitive experiences during adaptive states, open to the external world. If excitatory inputs are coincident with the depolarizing phase of the subthreshold fast oscillations, they are likely to cause the neurons to fire.[60] The impact on target neurons would increase if fast-oscillating neurons were capable of discharging high-frequency spike bursts (see Figure 20.11).[48-50] Data show that brainstem-thalamic volleys, which induce responses resembling the ponto-geniculo-occipital potentials of dreaming sleep, are followed by a dramatic synchronization of spontaneously occurring fast cortical oscillations.[21,55] Similar results may be elicited by using relevant signals during the state of wakefulness.

References

1. Steriade, M., Cellular substrates of brain rhythms, in *Electroencephalography: Basic Principles, Clinical Applications and Related Fields*, Niedermeyer, E. and Lopes da Silva, F., Eds., Williams & Wilkins, Baltimore, MD, 1993, pp. 27–62.
2. Steriade, M., Domich, L., Oakson, G., and Deschênes, M., The deafferented reticularis thalami nucleus generates spindle rhythmicity, *J. Neurophysiol.*, 57, 260–273, 1987.
3. Steriade, M., Deschênes, M., Domich, L., and Mulle, C., Abolition of spindle oscillations in thalamic neurons disconnected from nucleus reticularis thalami, *J. Neurophysiol.*, 54, 1472–1497, 1985.
4. Steriade, M., McCormick, D.A., and Sejnowski, T.J., Thalamocortical oscillations in the sleeping and aroused brain, *Science*, 262, 679–685, 1993.
5. Steriade, M., Iosif, G., and Apostol, V. Responsiveness of thalamic and cortical motor relays during arousal and various stages of sleep, *J. Neurophysiol.*, 32, 251–265, 1969.
6. Steriade, M., Alertness, quiet sleep, dreaming, in *Cerebral Cortex. Vol. 9. Normal and Altered States of Function*, Peters, A. and Jones, E.G., Eds., Plenum Press, New York, 1991, pp. 279–357.
7. Timofeev, I., Contreras, D., and Steriade, M., Synaptic responsiveness of cortical and thalamic neurones during various phases of slow oscillation in cat, *J. Physiol. (Lond.)*, 494, 265–278, 1996.

8. Villablanca, J., Role of the thalamus in sleep control: sleep/wakefulness studies in chronic diencephalic and athalamic cats, in *Basic Sleep Mechanisms*, Petre-Quadens, O. and Schlag, J., Eds., Academic Press, New York, 1974, pp. 51–81.
9. Steriade, M., Nuñez, A., and Amzica, F., Intracellular analysis of relations between the slow (<1 Hz) neocortical oscillation and other sleep rhythms of the electroencephalogram, *J. Neurosci.*, 13, 3266–3283, 1993.
10. Amzica, F. and Steriade, M. The K-complex: its slow (<1 Hz) rhythmicity and relation to delta waves, *Neurology*, 49, 952–959, 1997.
11. McCormick, D.A. and Pape, H.-C., Properties of a hyperpolarization-activated cation current and its role in rhythmic oscillation in thalamic relay neurones, *J. Physiol. (Lond.)*, 431, 319–342, 1990.
12. Leresche, N., Lightowler, S., Solesz, I., Jassik-Gerschenfeld, D., and Crunelli, V., Low-frequency oscillatory activities intrinsic to rat and cat thalamocortical cells, *J. Physiol. (Lond.)*, 441, 155–174, 1991.
13. Steriade, M., Curró Dossi, R., and Nuñez, A., Network modulation of a slow intrinsic oscillation of cat thalamocortical neurons implicated in sleep delta waves: cortical potentiation and brainstem cholinergic suppression, *J. Neurosci.*, 11, 3200–3217, 1991.
14. Nuñez, A., Curró Dossi, R., Contreras, D., and Steriade, M., Intracellular evidence for incompatibility between spindle and delta oscillation in thalamocortical neurons of cat, *Neuroscience*, 48, 75–85, 1992.
15. Steriade, M., Contreras, D., and Amzica, F., Synchronized sleep oscillations and their paroxysmal developments, *Trends Neurosci.*, 17, 199–208, 1994.
16. Lancel, M., van Riezen, H., and Glatt, A., The time course of sigma activity and slow wave activity during NREMs in cortical and thalamic EEG of the cat during baseline and after 12 hours of wakefulness, *Brain Res.*, 596, 286–295, 1992.
17. Steriade, M. and McCarley, R.W., *Brainstem Control of Wakefulness and Sleep*, Plenum Press, New York, 1990.
18. Steriade, M. and Amzica, F., Coalescence of sleep rhythms and their chronology in corticothalamic networks, *Sleep Res. Online*, 1, 1–10, 1998.
19. Steriade, M., Datta, S., Paré, D., Oakson, G., and Curró Dossi, R., Neuronal activity in brainstem cholinergic nuclei related to tonic activation processes in thalamocortical systems, *J. Neurosci.*, 10, 2541–2559, 1990.
20. Contreras, D. and Steriade, M., Cellular basis of sleep rhythms: a study of dynamic corticothalamic relationships, *J. Neurosci.*, 15, 604–622, 1995.
21. Steriade, M., Amzica, F., and Contreras, D., Synchronization of fast (30 to 40 Hz) spontaneous cortical rhythms during brain activation, *J. Neurosci.*, 16, 392–417, 1996.
22. Steriade, M., Contreras, D., Amzica, F., and Timofeev, I., Synchronization of fast (30 to 40 Hz) spontaneous oscillations in intrathalamic and thalamocortical networks, *J. Neurosci.*, 16, 2788–2808, 1996.
23. Achermann, P. and Borbély, A., Low-frequency (<1 Hz) oscillation in the human sleep electroencephalogram, *Neuroscience*, 81, 213–222, 1997.
24. Timofeev, I. and Steriade, M., Low-frequency rhythms in the thalamus of intact-cortex and decorticated cats, *J. Neurophysiol.*, 76, 4152–4168, 1996.
25. Amzica, F. and Steriade, M., Disconnection of intracortical synaptic linkages disrupts synchronization of a slow oscillation, *J. Neurosci.*, 15, 4658–4677, 1995.
26. Steriade, M., Nuñez, A., and Amzica, F., A novel slow (<1 Hz) oscillation of neocortical neurons *in vivo*: depolarizing and hyperpolarizing components, *J. Neurosci.*, 13, 3252–3265, 1993.

27. Contreras, D., Timofeeev, I., and Steriade, M., Mechanisms of long-lasting hyperpolarizations underlying slow sleep oscillations in cat corticothalamic networks, *J. Physiol. (Lond.)*, 494, 251–264, 1996.
28. Steriade, M., Domich, L., and Oakson, G., Reticularis thalami neurons revisited: activity changes during shifts in states of vigilance, *J. Neurosci.*, 6, 68–81, 1986.
29. Steriade, M., Contreras, D., Curró Dossi, R., and Nuñez, A., The slow (<1 Hz) oscillation in reticular thalamic and thalamocortical neurons: scenario of sleep rhythm generation in interacting thalamic and neocortical networks, *J. Neurosci.*, 13, 3284–3299, 1993.
30. Contreras, D. and Steriade, M., Spindle oscillations in cats: the role of corticothalamic feedback in a thalamically generated rhythm, *J. Physiol. (Lond.)*, 490, 159–180, 1996.
31. Kellaway, P., Sleep and epilepsy, *Epilepsia*, 26(suppl. 1), 15–30, 1985.
32. Steriade, M., Interneuronal epileptic discharges related to spike-and-wave cortical seizures in behaving monkeys, *Electroencephalogr. Clin. Neurophysiol.*, 37, 247–263, 1974.
33. Steriade, M. and Amzica, F., Dynamic coupling among cortical neurons during evoked and spontaneous spike-wave seizure activity, *J. Neurophysiol.*, 72, 2051–2069, 1994.
34. Steriade, M. and Contreras, D., Relations between cortical and thalamic cellular events during transition from sleep patterns to paroxysmal activity, *J. Neurosci.*, 15, 623–642, 1995.
35. Lytton, W.W., Contreras, D., Destexhe, A., and Steriade, M., Dynamic interactions determine partial thalamic quiescence in a computer network model of spike-wave seizures, *J. Neurophysiol.*, 77, 1679–1696, 1997.
36. Steriade, M., Amzica, F., and Nuñez, A., Cholinergic and noradrenergic modulation of the slow (~0.3 Hz) oscillation in neocortical cells, *J. Neurophysiol.*, 70, 1385–1400, 1993.
37. Singer, W. and Gray, C.M., Visual feature integration and the temporal correlation hypothesis, *Ann. Rev. Neurosci.*, 18, 555–586, 1995.
38. Steriade, M., Central core modulation of spontaneous oscillations and sensory transmission in thalamocortical systems, *Curr. Opin. Neurobiol.*, 3, 619–625, 1993.
39. Lopes da Silva, F.H., van Rotterdam, A., Storm ven Leeuwen, W., and Tielen, A.M., Dynamic characteristics of visual evoked potentials in the dog. II. Beta frequency selectivity in evoked potentials and background activity, *Electroencephalogr. Clin. Neurophysiol.*, 29, 260–268, 1970.
40. Bouyer, J.J., Monraron, M.F., and Rougeul, A., Fast frontoparietal rhythms during combined focused attentive behaviour and immobility in cat: cortical and thalamic localization, *Electroencephalogr. Clin. Neurophysiol.*, 51, 244–252, 1992.
41. Murthy, V.N. and Fetz, E.E. Oscillatory activity in sensorimotor cortex of awake monkeys: synchronization of local field potentials and relation to behavior, *J. Neurophysiol.*, 76, 3949–3967, 1996.
42. Desmedt, J.E. and Tomberg, C., Transient phase-locking of 40-Hz electrical oscillations in prefrontal and parietal human cortex reflects the processs of conscious somatic perception, *Neurosci. Lett.*, 168, 126–129, 1994.
43. Llinás, R., Grace, A.A., and Yarom, Y., *In vitro* neurons in mammalian cortical layer 4 exhibit intrinsic oscillatory activity in the 10- to 50-Hz frequency range, *Proc. Natl. Acad. Sci. USA*, 88, 897–901, 1991.
44. Amitai, Y., Membrane potential oscillations underlying firing patterns in neocortical neurons, *Neuroscience*, 63, 151–161, 1994.
45. Gutfreund, Y., Yarom, Y., and Segev, I., Subthreshold oscillations and resonant frequency in guinea-pig cortical neurons: physiology and modelling, *J. Physiol. (Lond.)*, 483, 621–640, 1995.
46. Plenz, D. and Kitai, S.T., Generation of high-frequency oscillations in local circuits of rat somatosensory cortex cultures, *J. Neurophysiol.*, 76, 4180–4184, 1996.

47. Nuñez, A., Amzica, F., and Steriade, M., Voltage-dependent fast (20–40 Hz) oscillations in long-axoned neocortical neurons, *Neuroscience,* 51, 7–10, 1992.
48. Gray, C.M. and McCormick, D.A., Chattering cells: superficial pyramidal neurons contributing to the generation of synchronous oscillations in the visual cortex, *Science,* 274, 109–113, 1996.
49. Steriade, M., Synchronized activities of coupled oscillators in the cerebral cortex and thalamus at different levels of vigilance, *Cerebral Cortex,* 7, 583–604, 1997.
50. Steriade, M., Timofeev, I., Dürmüller, N., and Grenier, F., Dynamic properties of corticothalamic neurons and local interneurons generating fast rhythmic (30-40 Hz) spike bursts, *J. Neurophysiol.,* 79, 483–490, 1998.
51. Steriade, M., Curró Dossi, R., and Contreras, D., Electrophysiological properties of intralaminar thalamocortical cells discharging rhythmic (~40 Hz) spike bursts at ~1000 Hz during waking and rapid-eye-movement sleep, *Neuroscience,* 56, 1–9, 1993.
52. Kato, N., Cortico-thalamo-cortical projection between visual cortices, *Brain Res.,* 509, 150–152, 1990.
53. Llinás, R. and Ribary, U., Coherent 40-Hz oscillation characterizes dream state in humans, *Proc. Natl. Acad. Sci. USA,* 90, 2078–2081, 1993.
54. Curró Dossi, R., Paré, D., and Steriade, M., Short-lasting nicotinic and long-lasting muscarinic depolarizing responses of thalamocortical neurons to stimulation of mesopontine cholinergic nuclei, *J. Neurophysiol.,* 65, 393–406, 1991.
55. Steriade, M. and Amzica, F., Intracortical and corticothalamic coherency of fast spontaneous oscillations, *Proc. Natl. Acad. Sci. USA,* 93, 2533–2538, 1996.
56. Moruzzi, G., The functional significance of sleep with particular regard to the brain mechanisms underlying consciousness, in *Brain and Conscious Experience*, Eccles, J.C., Ed., Springer, New York, 1066, pp. 345–379.
57. Amzica, F., Neckelmann, D., and Steriade, M., Instrumental conditioning of fast oscillations (20–50 Hz) in corticothalamic networks, *Proc. Natl. Acad. Sci. USA,* 94, 1985–1989, 1997.
58. Buzsáki, G., Two-stage model of memory trace formation: a role of "noisy" brain states, *Neuroscience,* 31, 551–570, 1989.
59. Wilson, M.A. and McNaughton, B.L., Reactivation of hippocampal ensemble memories during sleep, *Science,* 265, 676–679, 1994.
60. Lampl, I. and Yarom, Y., Subthreshold oscillations of the membrane potential: a functional synchronizing and timing device, *J. Neurophysiol.,* 70, 2181–2186, 1993.

Chapter 21

State-Dependent Changes in Network Activity of the Hippocampal Formation

James J. Chrobak and György Buzsáki

Contents

21.1	Theta/Gamma Dynamics Is One of the Temporally Defined Repertoires of Entorhinal-Hippocampal Circuits	350
21.2	How Is This Dynamic Property of Entorhinal-Hippocampal Circuits Modulated by the Behavioral State of the Animal?	352
21.3	The Alternative Dynamic: In the Absence of Theta/Gamma, Punctate Bursts Are Evident in the Hippocampal Formation	353
21.3.1	Hippocampal Sharp Waves	353
21.3.2	Entorhinal Sharp Waves	353
21.3.3	Dentate Spikes	355
21.4	Network Rules: State-Dependent Population Dynamics Constrain Cellular Neurophysiology	355
21.5	Local Variation: State-Dependent Population Dynamics and Local Phase Shifts in the Interneuron(s)-Innervated Subnet	357
21.6	State-Dependent Population Dynamics: A Two-State Model of Memory Formation Within the Hippocampal Formation	358
21.6	Conclusion	358
Acknowledgments		358
References		359

Distributed populations of hippocampal formation (hippocampus, subiculum, and entorhinal cortex) neurons support memory formation and consolidation in the mammalian brain. The superficial layers of the entorhinal cortex are the major entry point for input from polymodal associative cortices into the hippocampal formation.[1,2] Entorhinal circuits process and relay this input to the dentate gyrus, CA1, and the subiculum. The latter two structures project back to the deep layers of the entorhinal cortex, and from this lamina the product of hippocampal processing is conveyed to neocortical circuits. The biophysical means by which distributed neurons within this circuitry effectively interact and accomplish memory formation is not well understood. It is widely assumed that alterations in the synaptic connectivity of hippocampal networks support memory processes. Interactions among neurons in distributed networks are bound by anatomical constraints on synaptic connectivity and temporal constraints that govern their discharge charcteristics.[3] The discharge of subcortical modulatory input (i.e., cholinergic, monoaminergic, serotonergic, histaminergic) varies as a function of state and alters biochemical events and the spiking patterns of individual cortical target neurons (see previous chapters this book). This input, in turn, also alters the dynamics of larger circuits. These state-dependent dynamics alter the way large populations of hippocampal formation neurons interact and thus have profound implications for memory processes as well as the neurochemical events that underlie synaptic plasticity.[4-9]

Entorhinal-hippocampal circuits are organized into two distinct macrostates defined by the slow micro-EEG events prominent in extracellular field recordings. These macrostates are theta and sharp wave.[5,10] Theta is a rhythmic field oscillation between 4 and 12 Hz which occurs during exploratory behavior and rapid-eye-movement (REM) sleep.[11-14] Sharp waves refer to an aperiodic, large amplitude field event evident in stratum radiatum of CA1 that occurs during consummatory behavior and slow-wave sleep.[4,15]

What is important about these macrostates of the network? The discharge characteristic of whole populations of neurons varies systematically in relationship to these states. During theta and sharp waves, distinct populations of neurons are entrained into two prominent, fast-frequency rhythms: gamma (40 to 100 Hz), which occurs in relation to theta, and ripples (~200 Hz), which occur in relation to sharp waves. The fast-frequency dynamics that develop in association with theta and sharp wave allow large numbers of slow-firing neurons to be players in distributed, fast-frequency, population volleys. What is the utility of distributed, fast-frequency, population volleys? The entrainment of large populations of principle (pyramidal) neurons into temporally organized population ensembles is an intrinsic property of hippocampal-entorhinal networks and appears to be a necessary condition for communication between populations of neurons. Temporal entrainment in anatomically interconnected groups of neurons would allow for coherent ensembles throughout the brain to participate in distributed network events. In the current chapter, we describe the state-dependent micro-EEG events observed (theta and hippocampal sharp waves, as well as entorhinal sharp waves and dentate spikes) and their associated fast-frequency oscillations. We suggest that understanding the biophysical means by which distributed neurons in the hippocampal formation accomplish memory formation necessitates understanding state-dependent, neuronal dynamics.

21.1 Theta/Gamma Dynamics Is One of the Temporally Defined Repertoires of Entorhinal-Hippocampal Circuits

Whenever the rat moves, or attends to sensory stimuli, or is in REM sleep, theta waves dominate the hippocampal micro-EEG. The precise laminar organization of rodent hippocampal neurons and their dendrites allow for the observation of rhythmic field potentials at theta frequency 4 to 12 Hz. These potentials reflect extracellular currents generated by synchronized synaptic (and intrinsic membrane) potentials in pyramidal and granule cells.[16-18] The rhythmic activation of synaptic and intrinsic membrane currents, which are reflected in the extracellular current flow, determines the ensemble output of a very large number of hippocampal neurons.[19]

Theta rhythmicity can be seen in the phase-related discharge of neurons in the entorhinal cortex,[20-23] pontine nuclei,[24,25] hypothalamic nuclei,[26] septum and other areas of the basal forebrain,[27-31] and the amygdala.[32] Neurons in some of these structures exhibit intrinsic oscillatory behavior in the 4- to 12-Hz frequency range, while others exhibit resonant behavior in response to

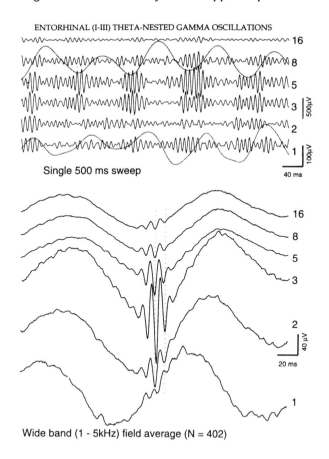

FIGURE 21.1
Simultaneous recording of theta and gamma field activity from multi-site probe (16 recording sites: 100-μm spacing) along the laminar axis of the entorhinal cortex. For clarity, only select traces are displayed. Top traces illustrate a single sweep showing the relationship between theta and gamma field potentials. Trace 1 was located superficially in layer I, while trace 16 was located near the layer III–V border. Six black traces are gamma filtered (50 to 150 Hz), two gray traces show concurrent theta waves (1 to 20 Hz). Theta records are from positions 1 and 8 as in black (gamma) traces. Note amplitude variation of gamma oscillation at different recording sites and the prominent relation between the phase of theta and amplitude of gamma waves. Bottom traces illustrate averaged (n = 402) extracellular field potentials (wide band) as triggered from negative peak of local gamma oscillation (trace 3). Note sudden phase reversal of gamma between sites 1 and 2, in contrast to gradual shift of theta waves.

synaptic input at this frequency. The rhythmic output of a vast neuronal orchestra at theta frequency reflects a consortium of mechanisms — at the cellular, local circuit, and neural systems level — for bringing neurons throughout the limbic forebrain together in time within a window of 80 to 200 msec. The ensemble output of a large population of neurons within this time window is the ostensible feature of theta. The elegance of this event is cut short only by recent evidence demonstrating that theta is coupled to a faster frequency dynamic, the gamma rhythm (see Figure 21.1). These coupled oscillatory rhythms result in the precision timing of vast ensembles of neurons during exploratory activity and REM sleep.

In association with theta, there is increased gamma frequency activity predominantly in the dentate/hilar and CA1 regions of the hippocampus,[17,33] and the superficial layers of the entorhinal cortex.[34–36] Gamma oscillations reflect synchronous membrane oscillations in pyramidal, granule, and stellate neurons caused by rhythmic fast-frequency inhibitory postsynaptic potentials (IPSPs).[32,37,38]

Within the hippocampus, select GABAergic interneurons discharge relatively synchronous trains of action potentials at the gamma and theta frequency. This coherent network output imposes a hyperpolarizing oscillation on the membrane potential of principal neurons.[17,33,36,39] Such oscillations, occurring in the context of a depolarizing input, impose a periodic fluctuation in the pyramidal cell membrane, close to, but below, discharge threshold. This network output constrains the time window during which principle neurons (pyramidal/granule) can initiate somatic spikes. A number of sparsely firing, principal cells (including layer II–III entorhinal cortical neurons, as well as granule and pyramidal cells in the hippocampus) discharge during theta, always in phase relation to locally developing gamma oscillations.[36,38] The net yield of a large number of slowly firing principle cells discharging on the same phase of the coupled theta/gamma population rhythm provides a fast-frequency network output at theta/gamma frequency to their efferent targets.

A distributed population output, entrained into local gamma rhythms, is thus fed into the circuits of the hippocampus. The entorhinal output is entrained into the theta/gamma dynamic and impinges upon granule and pyramidal cell targets that concurrently develop gamma-frequency dynamics.[33] In this manner, entorhinal afferents and their hippocampal targets develop a transient interaction at gamma frequency on each theta wave. Transiently synchronized neuronal discharges emerge within varying groups of entorhinal and hippocampal neurons, with varying subpopulations of interconnected neurons becoming entrained into a rapidly evolving/devolving range of frequencies.[38,40]

21.2 How Is This Dynamic Property of Entorhinal-Hippocampal Circuits Modulated by the Behavioral State of the Animal?

In simplest form, one predominant influence of subcortical modulatory inputs (i.e., acetylcholine, norepinephrine) is to depolarize the membrane potential of pyramidal and interneuronal populations into a voltage range that enhances the membrane currents driving theta/gamma dynamics. For example, muscarinic receptor activation promotes oscillatory activity in entorhinal stellate cells by depolarizing the membrane to the voltage range at which Na^+-dependent subthreshold membrane potential oscillations dominate the discharge character of these neurons (see Chapter 18).[31,41] Membrane depolarization alone is typically sufficient to promote this behavior. However, the particular dynamic output of a neuron varies depending upon the non-linear summation of time- and voltage-dependent currents engaged or disengaged by, for instance, muscarinic receptor activation, and, most importantly, synchronous synaptic input from a pacemaker input or an orchestrated network.

The intrinsic membrane currents of some neurons allow for the generation of theta frequency oscillations or gamma-frequency oscillations, while a few neurons will discharge gamma trains (spike clusters) at theta frequency.[31,39,42,43] Most neurons will not, however, exhibit spike output at these frequencies; rather, a membrane oscillation emerges that allows a neuron to yield spikes entrained to the frequency of synaptic input.[44–46] In CA1 neurons, for example, theta frequency membrane oscillations are voltage independent, demonstrating that the principle source of rhythmicity in these neurons is synaptic input from the network.[47] Rhythmicity results from both intrinsic membrane dynamics and rhythmic synaptic input via the network. The influence of cholinergic and noradrenergic input is to depolarize the membrane potentials of many neurons in the septum, hippocampus, and entorhinal cortex into a voltage range where the non-linear summation of intrinisic and synaptic currents promote theta/gamma dynamics. Concurrently, this dynamic acts to inhibit the membrane currents that promote dendritic Ca^{2+} bursts, particularly in CA3 neurons. The inhibition of these currents in CA3 neurons prevents the occurrence of spontaneous bursting behavior in the auto-associative network of CA3.[48] What is important here is that the effect of subcortical modulatory input is to push a heterogenous group of neurons into exhibiting specific dynamical behaviors. The product of these behaviors, as articulated by the predominant actions of the network collective, is either theta/gamma or punctate bursts. These punctate bursts are evident in the CA1–CA3 region as hippocampal sharp waves; other circuit bursts include entorhinal sharp waves and dentate spikes.

21.3 The Alternative Dynamic: In the Absence of Theta/Gamma, Punctate Bursts Are Evident in the Hippocampal Formation

21.3.1 Hippocampal Sharp Waves

Within a few seconds of sitting still, eating, or drinking, punctate sharp waves are evident in the micro-EEG of the hippocampus. The sharp wave refers to slow (50 to 150 msec), large-amplitude field potentials observed in the stratum radiatum of CA1. The specific laminar profile of this event reflects a massive depolarization (synchronous field excitatory postsynaptic potentials, or EPSPs) in the apical and basal dendritic field of CA1 pyramidal neurons.[4,15] Sharp waves originate in the population discharge of CA3 neurons, whose Schaffer collaterals impinge upon the dendrites of CA1. When free from subcortical inputs that inhibit the bursting currents in these neurons[48] and free from the dominant influence of the hilar inhibitory projections, these neurons get together and exhibit large population bursts.[49,50] Both of these inhibitory influences predominate during naturally occurring theta/gamma in the awake rat or during REM sleep; however, during consummatory behaviors and slow-wave sleep, sharp waves dominate the physiological activity of the hippocampus.

During sharp waves, the large depolarization of the CA1 neurons, by the CA3 network, sets into motion a short-lived interaction between CA1 pyramidal cells and interneurons that produces an oscillatory population discharge of the pyramidal network at a frequency near 200 Hz.[51] The specific synaptic currents mediating the high-frequency oscillation are largely mediated by rhythmic, synchronized IPSPs near the soma of CA1 neurons.[49] A high-frequency discharge of CA1 basket cells and other interneurons phase-related to the extracellular 200-Hz ripple indicates that these neurons are principally responsible for the synchronized IPSPs in pyramidal cells. During the 200-Hz oscillation associated with sharp waves, these interneurons may discharge on each wave of the ripple, while a given CA1 pyramidal cell, constrained by the rhythmic barrage of the interneurons, typically fires once or not at all. The massive depolarization of interneurons and pyramidal cells brought about by the CA3 Schaffer collaterals sets into motion an orchestrated discharge of the CA1 pyramidal cell network.

21.3.2 Entorhinal Sharp Waves

CA1 neurons innervate the subiculum and deep layer (V–VI) neurons within the entorhinal cortex.[1,2] A concurrent discharge of subicular and deep-layer entorhinal neurons occurs in conjunction with hippocampal sharp waves.[52] The output of these neurons is also entrained into local fast-frequency ripples.[53] We have also observed a slow (50 to 150 msec), large-amplitude field potential in the dendritic field of deep-layer (V–VI) entorhinal neurons during immobility and slow-wave sleep.[53,54] This entorhinal sharp wave (see Figure 21.2) likely reflects the sequential activation of the deep layer that originates in the discharge of CA1 and subicular neurons during hippocampal sharp waves. Paré and colleagues have indicated that neurons within the basolateral nucleus of the cat amygdala

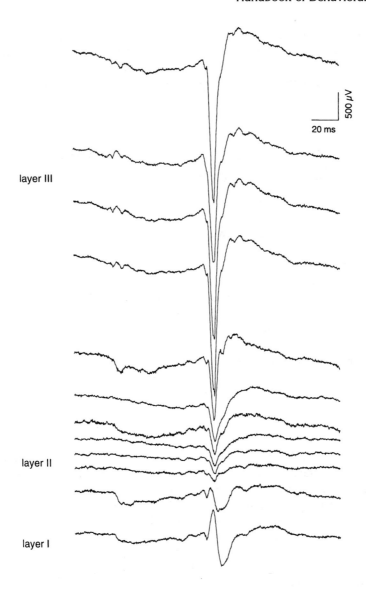

FIGURE 21.2
Simultaneous recording of entorhinal sharp wave from multi-site probe (16 recording sites; 100-μm spacing) along the laminar axis of the entorhinal cortex. For clarity, only select traces are displayed. Top trace is near layer II–V border, while bottom trace is located in layer I. These events tend to occur within tens of milliseconds of hippocampal SPWs and occur within the axonal domain of hippocampal and subicular inputs to the entorhinal cortex. Deep layer (V–VI) neurons discharge a 200-Hz population volley in relation to these events, while the output of superficial layer (II–III) neurons is extremely limited (see Reference 53 for details).

also discharge as early as 40 msec prior to entorhinal sharp potentials and that amygdala lesions abolish these events.[55] Thus, a confluence of afferents that innervate the dendritic fields of both deep and superficial layer entorhinal neurons may contribute to the genesis and amplitude of entorhinal sharp waves.

Similar to the observations in the cat entorhinal cortex,[55] we have observed that the field potential reverses in polarity near layer II of the entorhinal cortex. It should be noted that the stratification of dendritic fields in the entorhinal cortex is much less laminarly organized then that found in the hippocampus. The dendritic field of many layer V–VI neurons ramify throughout layers

I–VI, and the dendritic fields of layer II–III also ramify throughout layer I–IV.[56] While layer V–VI neurons discharge regularly with entorhinal sharp waves and exhibit ripple frequency synchronization, only a few layer II–III neurons discharge in relation to entorhinal sharp waves.[54,55] Superficial layer neurons are strongly inhibited by a dense network of basket cells that limit the output of these neurons.[52,57–60] We have observed fast-frequency oscillations within the superficial lamina of the entorhinal cortex in association with entorhinal sharp waves, but have found a limited relationship between the discharge of layer II–III neurons and this event.[54] The limited output of these neurons in association with entorhinal sharp waves may have a relation to another punctate burst evident in the dentate/hilar region during the sharp wave state, referred to as dentate spikes.

21.3.3 Dentate Spikes

Bragin and colleagues[50] describe two population events in the dentate/hilar region that emerge in conjuction with a synchronized discharge of hilar inhibitory interneurons. Dentate spikes are large-amplitude (2- to 4-mV), short-duration (<50 msec) field potentials that occur during immobility and slow-wave sleep. Based on differences in current source density analysis, two dentate spikes can be distinquished. Dentate spike one (DS1) is defined by a large current sink in the outer molecular layer (the target of lateral entorhinal cortical inputs), while DS2 is defined by a large current sink in the inner molecular layer (the target of medial entorhinal cortical inputs). The CSD analysis and the fact that dentate spikes are abolished by entorhinal lesions indicate that entorhinal inputs support the genesis of dentate spikes. Both types of dentate spikes are related to a synchronous discharge of hilar interneurons and the suppression of CA3 pyramidal neurons. It appears that the dentate spikes act to suppress the discharge of the CA3–CA1 network and thus inhibit the genesis of sharp waves. In contrast, a small percentage (<20%) of DS1 were preceded by sharp wave bursts (see Figure 21.3).

Clearly, the anatomical paths exist for the sequential interaction of hippocampal and entorhinal sharp waves and dentate spikes. In the absensce of certain subcortical modulatory inputs that are engaged during exploratory activity and REM sleep in the rat, hippocampal and entorhinal circuits engage in aperiodic population bursts: hippocampal sharp waves, entorhinal sharp waves, and dentate spikes. Although lesions of the amygdala have been shown to abolish entorhinal sharp waves[55] and lesions of the entorhinal cortex abolish one type of dentate spike,[33] a clearer relation between these three events is wanting.[50,55] Further, experiments assessing the relation of these three events, particularly in light of their potential relation to temporal lobe epilepsy,[61] is warranted.

21.4 Network Rules: State-Dependent Population Dynamics Constrain Cellular Neurophysiology

What is the functional significance of fast-frequency population dynamics? Fast-frequency population dynamics appears to be a necessary means for effective neuronal interaction among the associational pathways that connect cortical neurons across/within structures. At least within entorhinal and hippocampal circuits there are strong constraints, imposed by the network, on the time window within which temporal summation of EPSPs can occur. Rhythmic hyperpolarization of the somatic compartment of pyramidal cells, as well as rhythmic shunting of dendritic excitation, are critical structural elements in defining that time window.[62] The rhythmic modulation of principle cell membranes is an inescapable product, *in vivo*, of activating the network in which it is embedded.[6,9,63–65] For granule and CA1 neurons discharging during theta, gamma frequency (10 to 25 msec) defines that time window. For neurons in the hippocampal-entorhinal output pathway, ripple frequency (~5 msec) defines the time window. The impact of an input on the postsynaptic target is limited by the rhythmic cycle of inhibition imposed by interneuronal nets. Neurons discharging such

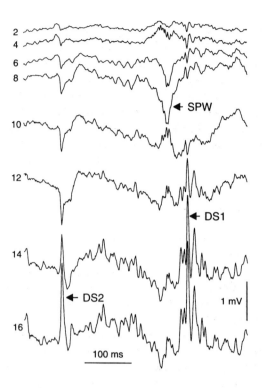

FIGURE 21.3
Simultaneous recording of field activity from multi-site probe (16 recording sites; 100-μm spacing) along the CA1-dentate axis of the dorsal hippocampus during awake immobility. For clarity, only every second trace is displayed. Sharp wave (SPW) in the CA1 region and type 1 and type 2 dentate spikes (DS1 and DS2, respectively) are indicated by arrows. Double arrow along trace 4 indicates SPW-associated, high-frequency (200-Hz) field oscillation (ripple) in the CA1 pyramidal layer. Trace 14 is in the granule cell layer. Temporal proximity of all three transients (SPW, DS1, and DS2) as shown here is very rare. SPWs were never observed after either dentate spikes (<200 msec), nor were they followed by DS2. In contrast, 5 to 15% of DS1 were preceded by SPW events. (See Reference 50 for details.)

that their EPSPs tag dendritic targets at the wrong phase of the rhythmic cycle may have no impact on the discharge of the postsynaptic neuron (see References 6, 9, 62 to 65 for further discussion).

Fortunately, the dynamics of intrinsic membrane currents and network interactions appears to be built so that principle cells resonate to hyperpolarizing inputs with depolarizing currents,[59,63,66] and excitatory volleys from associational inputs arrive in highly orchestrated, high-frequency barrages.[6,53] The dynamics built into entorhinal-hippocampal network neurons allow ensembles to act as an orchestra performing transient temporal symphonies. Importantly, the specific dynamics that are allowed are promoted/modulated by behavioral state, specifically by the activity of several subcortical modulatory inputs. The subtleties of subcortical modulation of these dynamics warrants serious investigation.[9]

We have suggested some means by which behavioral states acting via the differential engagement of subcortical modulatory inputs can direct the dynamic behavior of network neurons. Subcortical modulatory inputs, by acting on specific membrane bound receptors, are also engaging a vast repetoire of biochemical events (see other chapters this book) in hippocampal neurons. The biochemical cascades and, particularly, the calcium dynamics that evolve from the interaction of modulatory receptor activation and synaptic input are powerfully constrained by membrane dynamics that evolve as a function of the network and, more precisely, the state-dependent activity of the network. The modulation of calcium dynamics are critical to the biochemical cascades that produce long-term potentiation (LTP) and LTD.[67,68] We have suggested that by regulating the excitability of distinct dendritic domains and the spike output of neurons the rhythm imposed on a neuron by the network determines calcium flux in specific neuronal compartments.[6,69] This has direct implications for synaptic plasticity.

21.5 Local Variation: State-Dependent Population Dynamics and Local Phase Shifts in the Interneuron(s)-Innervated Subnet

Bringing anatomically distributed populations of neurons together in the time domain is a fundamental function subserved by entrainment in the nervous system. As discussed above, such "groupings" may increase the impact of the neurons on their postsynaptic targets, leading to more efficient transmission and perhaps plastic changes. Within the hippocampal formation, principle neurons discharge in phase with local fast-frequency field oscillations. It must be emphasized, though, that most judgments on correlations are based on statistical averages. The typical case of phase relationship between a particular pyramidal cell and the negative peak of the ensemble oscillatory rhythm reflects the mean behavior over a relatively long time period. However, one might take the view that deviations from the average behavior is the relevant neural information.[65,70]

O'Keefe and Recce[71] have discovered that the action potentials of "space-coding" pyramidal cells undergo a systematic phase precession while the rat crosses the field of the recorded unit. Importantly, the phase of theta at which the cell fired was a better predictor of the animal's position than the firing rate of the neuron. Several computational models attempt to explain this important physiological phenomenon.[71-73] What would be the impact of spikes occurring slightly out of phase in pyramidal cells on the neuronal ensemble? It may be argued their effect may not be efficiently "heard" by their postsynaptic principal cell targets because the depolarizing effects of the converging presynaptic action potentials are counteracted by rhythmic hyperpolarization. However, if the phase shift occurs coherently in a sufficiently large group of neurons and results in an increased cycle-by-cycle discharge of these cells, then the cooperative action of such a cell assembly may be particularly efficient to convey specific information.[7,72,74]

How do members of these cell assemblies strengthen their relationship so that they represent the same information when reactivated by the same input vectors on future occurrences? The problem one has to address is that if pyramidal cells discharge on virtually every phase of the theta cycle when the rat passes through the cell's place field then synaptic connections of the neuron may get both potentiated and depotentiated. Stimulation delivered at the time of maximum population discharge (i.e., on the negative phase of cell body layer theta or of "theta-like" oscillation in the slice) induced better long-term potentiation than stimulation presented on the opposite phase.[75,76] In fact, when the stimulus trains were delivered during the phase with the least probability of population firing (i.e., at the time when place cells discharge with their highest rate), it depotentiated the previously enhanced synaptic weights. Thus, these general rules might suggest that spike phase precession is not a particularly good condition for strengthening the connections among similarly behaving neurons.

A different scenario can be envisioned if the strongly activated pyramidal neurons can entrain their feedback interneurons. A concerted discharge of concurrently active pyramidal neurons may activate their basket cells at a short latency; therefore, basket cell discharge would show a similar phase precession as their presynaptic principal neurons. The hypothesized result is a local phase shift in the interneuron(s)-innervated subnet. Within these subnets the strongly activated neurons discharge coherently. Because their firing activity shifts in phase relative to their numerous slow discharging presynaptic partners (i.e., the average principal cell population), the probability of synaptic strengthening will be considerably higher within the subnet than the connectivity between the subnet and the "average" population.

21.6 State-Dependent Population Dynamics: A Two-State Model of Memory Formation Within the Hippocampal Formation

We have suggested that the orchestration of distributed neurons into fast-frequency ensembles is a cardinal feature of how hippocampal and entorhinal neurons interact. The requirements of principle cells (pyramidal and granule) discharging during these events necessitate synchronized excitatory synaptic volleys arriving within the dendrites on the order of 5 to 20 msec (the frequency of the rhythmic inhibition). Gamma emerges in relation to the theta rhythm during exploratory activity and REM sleep, while ripples emerge in relation to sharp waves during consummatory behavior and slow-wave sleep. Thus, the fast-frequency patterns emerge in relation to the behavioral state of the animal. Do these distinct population dynamics have something to do with the cognitive/computation operations of the hippocampal formation? What do synchronized population ensembles in the hippocampus do? We have already suggested that entorhinal and hippocampal networks use these population dynamics to interact and modify their synaptic connectivity. We presume that alterations in the synaptic connectivity of hippocampal networks support memory formation. Based on the patterns of amnestic deficts observed in humans and other mammals, it is quite clear that hippocampal circuits do not serve as the ultimate site for memory storage. Rather, hippocampal circuits participate in a process by which cortical representations become independent of hippocampal input. We suggested that the theta/gamma dynamic that develops during exploratory activity modifies the synaptic connectivity of the circuits conveying afferent input to the hippocampus, while modifications in the synaptic connectivity of the output network are modified during sharp wave-related ripples.

Whenever rats explore their environment, hippocampal circuits receive rhythmic input from neurons with the superficial layers of the entorhinal cortex.[17,20,23] During sharp waves, the output neurons of the hippocampus and entorhinal cortex participate in organized population bursts.[5,47,52,53] These patterns seem to serve companion processes. Theta synchronizes the input pathway into the hippocampus, whereas sharp waves synchronize the output pathway from the hippocampus back to neocortical structures.[52,53]

21.6 Conclusion

The non-linear interplay between intrinsic membrane currents, network connectivity, associational synaptic input, and the actions of subcortical modulatory inputs allows a higher order dimension to the behavioral state-dependent symphonies played by hippocampal circuits. The biophysical means by which distributed neurons within these circuits interact and accomplish memory formation is embedded within these state-dependent symphonies. We believe that alterations in the synaptic connectivity of hippocampal circuits support memory and that understanding how this complex adaptive system supports memory requires linking population dynamics to the biophysical cascades underlying synaptic plasticity.

Acknowledgments

This work was supported by the National Institutes of Health (NS34994), the National Science Foundation, the Human Frontiers Science Program, and Whitehall Foundation (GB), as well as the Alzheimer's Association (JJC).

References

1. Amaral, D.G. and Witter, M.P., The three-dimensional organization of the hippocampal formation: a review of anatomical data, *Neuroscience*, 31, 571, 1989.
2. Amaral, D.G. and Witter, M.P., The hippocampal formation, in *The Rat Nervous System*, Second ed., Paxinos, G., Ed., Academic Press, New York, 1995, p. 443.
3. Llinás, R.R., The intrinsic electrophysiological properties of mammalian neurons: insight into the central nervous system, *Science*, 242, 1654, 1989.
4. Buzsáki, G., Hippocampal sharp waves: their origin and significance, *Brain Res.*, 398, 242, 1986.
5. Buzsáki, G., A two-stage model of memory trace formation: a role for "noisy" brain states, *Neuroscience*, 31, 551, 1989.
6. Chrobak, J. and Buzsáki, G., Operational dynamics in the hippocampal-entorhinal axis, *Neurosci. Biobehav. Rev.*, 22, 303, 1988.
7. Lisman, J.E. and Idiart, M.A.P., A mechanism for storing 7±2 short-term memories in oscillatory subcycles, *Science*, 267, 1512, 1995.
8. Jensen, O., Idiart, M.A.P., and Lisman, J.E., Physiologically realistic formation of autoassociative memory in networks with theta/gamma oscillations: role of fast NMDA channels, *Learning Memory*, 3, 243, 1996.
9. Freund, T. and Buzsáki, G., Interneurons of the hippocampus, *Hippocampus*, 6, 345, 1996.
10. Buzáki, G., The hippocampal-neocortical dialogue, *Cerebral Cortex*, 6, 81, 1996.
11. Green, J.D. and Arduini, A.A., Hippocampal electrical activity in arousal, *J. Neurophysiol.*, 17, 533, 1954.
12. Grastyan, E., Lissak, K., Madarasz, I., and Donhoffer, H., The hippocampal electrical activity during the development of conditioned reflexes, *Electroencephalogr. Clin. Neurophysiol.*, 11, 409, 1959.
13. Vanderwolf, C.H., Hippocampal electrical activity and voluntary movement in the rat, *Electroencephalogr. Clin. Neurophysiol.*, 26, 407, 1969.
14. Arnolds, E.D.A.T., Lopes da Silva, F.H., Aitink, J.W., Kamp, A., and Boeijinga, P., The spectral properties of hippocampal EEG related to behavior in man, in *Developments in Neuroscience*, Vol. 10, Pfurtscheller, G., Buser, P., Lopes da Silva, F.H., and Petsche, H., Eds., Elsevier/North-Holland Biomedical Press, Amersterdam, 1979, p. 91.
15. Suzuki, S.S. and Smith, G.K., Spontaneous EEG spikes in the normal hippocampus. I. Behavioral correlates, laminar profiles and bilateral synchrony, *Electroencephalogr. Clin. Neurophysiol.*, 67, 438, 1987.
16. Brankack, J., Stewart, M., and Fox, S.E., Current source density analysis of the hippocampal theta rhythm: associated sustained potentials and candidate synaptic generators, *Brain Res.*, 615, 310, 1993.
17. Buzsáki, G., Leung, L., and Vanderwolf, C.H., Cellular bases of hippocampal EEG in the behaving rat, *Brain Res. Rev.*, 6, 139, 1983.
18. Lee, M.G., Chrobak, J.J., Sik, A., Wiley, R.G., and Buzsáki, G., Hippocampal theta activity following selective lesion of the septal cholinergic system, *Neuroscience*, 62, 1033, 1994..
19. Bland, B.H., Physiology and pharmacology of hippocampal formation theta rhythms, *Prog. Neurobiol.*, 26, 1, 1990.
20. Mitchell, S.J. and Ranck, Jr, J.B., Generation of theta rhythm in medial entorhinal cortex of freely moving rats, *Brain Res.*, 189, 49, 1980.
21. Alonso, A. and Garcia-Austt, E., Neuronal sources of theta rhythm in the entorhinal cortex of the rat. I. Laminar distribution of theta field potentials, *Exp. Brain Res.*, 67, 493, 1987.
22. Alonso, A. and Garcia-Austt, E., Neuronal sources of theta rhythm in the entorhinal cortex of the rat. II. Phase relations between unit discharges and theta field potentials, *Exp. Brain Res.*, 67, 502, 1987.

23. Stewart, M., Quirk, G.J., Barry, M., and Fox, S.E., Firing relation of medial entorhinal neurons to the hippocampal theta rhythm in urethane anesthetized and walking rats, *Exp. Brain Res.*, 90, 21, 1992..

24. Vertes, R.P., Brainstem modulation of the hippocampus: anatomy, physiology and significance, *The Hippocampus*, Vol. 4, Isaacson, R.L. and Pribram, K., Eds., Plenum Press, New York, 1986, p. 41.

25. Kocsis, B. and Vertes, R.P., Dorsal raphe neurons: synchronous discharge with the theta rhythm of the hippocampus in the freely behaving rat, *J. Neurophysiol.*, 68, 1463, 1992.

26. Kirk, I.J. and McNaughton, N., Supramammillary cell firing and hippocampal rhythmical slow activity, *NeuroReport*, 2, 723–235, 1991

27. Petsche, H., Stumpf, C., and Gogolak, G., The significance of the rabbit's septum as a relay station between midbrain and the hippocampus. I. The control of hippocampus arousal activity by the septum cells, *Electoencephalogr. Clin. Neurophysiol.*, 14, 202–211, 1962.

28. Stewart, M., and Fox, S.E., Two populations of rhythmically bursting neurons in rat medial septum are revealed by atropine, *J. Neurophysiol.*, 61, 982–992, 1989.

29. Stewart, M. and Fox, S.E., Do septal neurons pace the hippocampal theta rhythm?, *Trends Neurosci.*, 13, 163–168, 1990.

30. Barrenechea, C., Pedemonte, M., Nunez, A., and Garcia-Austt, E., *In vivo* intracellular recordings of medial septal and diagonal band of Broca neurons: relations with theta rhythm, *Exp. Brain Res.*, 103, 31–40, 1995.

31. Alonso, A., Khateb, A., Fort, P., Jones, B.E., and Muhlethaler, M., Differential oscillatory properties of cholinergic and non-cholinergic nucleus basalis neurons in guinea pig brain slices, *Eur. J. Neurosci.*, 8, 169, 1996.

32. Paré, D., Ong, J., and Gaudreau, H., Projection cells and interneurons of the lateral and basolateral amygdala: distinct firing patterns and differential relation to theta and delta rhythms in conscious cats, *J. Neurosci.*, 16, 3334, 1996.

33. Bragin, A., Jando, G., Nadasdy, Z., Hetke, J., Wise, K., and Buzsáki, G., Gamma (40–100 Hz) oscillation in the hippocampus of the behaving rat, *J. Neurosci.*, 15, 47, 1995.

34. Charpak, S., Pare, D., and Llinás, R.R., The entorhinal cortex entrains fast CA1 hippocampal oscillations in the anesthetized guinea pig: role of the monosynaptic component of the perforant path, *Eur. J. Neurosci.*, 7, 1548, 1995.

35. Chrobak, J.J. and Buzsáki, G., Distinct, behaviorally regulated, high-frequency oscillation synchronous input and output neurons of the hippocampal-entorhinal axis, *Soc. Neurosci. Abstr.*, 20, 357, 1994.

36. Chrobak, J.J. and Buzsáki, G., Entorhinal-hippocampal network dynamics constrain synaptic potentiation and memory formation, in *Long-Term Potentiation: Current Issues,* Baudry, M. and Davies J.L., Eds., MIT Press, Cambridge, 1997, p. 213.

37. Whittington, M.A., Traub, R.D., and Jeffreys, J.G.R., Metabotropic receptor activation drive synchronized 40 Hz oscillations in networks of inhibitory neurons, *Nature*, 373, 612, 1995.

38. Traub, R.D., Whittington, M.A., Colling, S.B., Buzsáki, G., and Jefferys, J.G.R., Analysis of gamma rhythms in the rat hippocampus *in vitro* and *in vivo*, *J. Physiol. (Lond.)*, 493, 471, 1996.

39. Soltész, I. and Deschénes, M., Low- and high-frequency membrane potential oscillations during theta activity in CA1 and CA3 pyramidal neurons of the rat hippocampus under ketamine-xylazine anesthesia, *J. Neurophysiol.*, 70, 97, 1993.

40. Wang, X-J. and Buzsáki, G., Gamma oscillation by synaptic inhibition in a hippocampal interneuronal network model, *J. Neurosci.*, 16, 6402–6413, 1996.

41. Klink, R. and Alonso, A., Muscarinic modulation of the oscillatory and repetitive firing properties of entorhinal cortex layer II neurons, *J. Neurophysiol.*, 77, 1813, 1997.

42. Alonso A. and Llinás, R.R., Subthreshold Na⁺-dependent theta-like rhythmicity in stellate cells of entorhinal cortex layer II, *Nature*, 342, 175, 1989.
43. Llinas, R., Grace, A.A., and Yarom, Y., *In vitro* neurons in mammalian cortical layer 4 exhibit intrinsic oscillatory activity in the 10- to 50-Hz frequency range, *Proc. Natl. Acad. Sci. USA*, 88, 897, 1991.
44. Wang, X.J., Multiple dynamical modes of thalamic relay neurons: rhythmic bursting and intermittent phase-locking, *Neuroscience*, 39, 21, 1994.
45. Read, H.L. and Siegel, R.M., The origins of aperiodicities in sensory neuron entrainment, *Neuroscience*, 75, 301, 1996.
46. Lampl, I. and Yarom, Y., Subthreshold oscillations and resonant behavior: two manifestations of the same mechanism, *Neuroscience*, 78, 325, 1997.
47. Ylinen, A., Soltesz, I., Bragin, A., Penttonen, M., Sik, A., and Buzsáki, G., Intracellular correlates of hippocampal theta rhythm in identified pyramidal cells, granule cells, and basket cells, *Hippocampus*, 5, 78, 1995.
48. Traub, R.D. and Miles, R., *Neuronal Networks of the Hippocampus*, Cambridge University Press, New York, 1991.
49. Ylinen, A., Sik, A., Bragin, A., Nadásdy, Z., Jandó, G., Szabó, I., and Buzsáki, G., Sharp wave-associated high-frequency oscillation (200 Hz) in the intact hippocampus: network and intracellular mechanisms, *J. Neurosci.*, 15, 30, 1995.
50. Bragin, A., Jando, G., Nadasdy, Z., van Landeghem, M., and Buzsáki, G., Dentate EEG spikes and associated interneuronal population bursts in the hippocampal hilar region of the rat, *J. Neurophysiol.*, 73, 1691, 1995.
51. Buzsáki, G., Horvath, Z., Urioste, R., Hetke, J., and Wise, K., High-frequency network oscillation in the hippocampus, *Science*, 256, 1025, 1992.
52. Chrobak, J.J. and Buzsáki, G., Selective activation of deep layer retrohippocampal neurons during hippocampal sharp waves, *J. Neurosci.*, 14, 6160, 1994.
53. Chrobak, J.J. and Buzsáki, G., High-frequency oscillations in the output networks of the hippocampal-entorhinal axis of the freely behaving rat, *J. Neuroscience*, 16, 3056, 1996.
54. Chrobak. J.J. and Buzsáki, G., Temporal structure in the discharge of superficial layer (II–III) entorhinal cortical neurons, *Soc. Neurosci. Abstr.*, 21, 1206, 1995.
55. Paré, D., Dong, J., and Gaudreau, H., Amygdalo-entorhinal relations and their reflection in the hippocampal formation: generation of sharp sleep potentials, *J. Neurosci.*, 15, 2482, 1995.
56. Llingehohl, K. and Finch, D.M., Morphological characterization of rat entorhinal neurons *in vivo*: soma-dendritic structure and anxonal domains, *Exp. Brain Res.*, 84, 57, 1991.
57. Finch, D.M., Wong, E.E., Derian, E.L., and Babb, T.L., Neurophysiology of limbic system pathways in the rat: projections from the subicular complex and hippocampus to the entorhinal cortex, *Brain Res.*, 397, 205, 1986
58. Wouterlood, F.G., Hartig, W., Bruckner, G., and Witter, M.P., Paravalbumin-immunoreactive neurons in the entorhinal cortex of the rat: localization, morphology, connectivity and ultrastructure, *J. Neurocytol.*, 24, 135, 1995.
59. Jones, R.S.G., Synaptic and intrinsic properties of neurons of origin of the perforant path in layer II of the rat entorhinal cortex *in vitro*, *Hippocampus*, 4, 335, 1994..
60. Jones, R.S.G., Entorhinal-hippocampal connections: a speculative view of their function, *Trends Neurosci.*, 16, 58, 1993.
61. Bragin, A., Csicsvari, J., Penttonen, M., and Buzsáki, G., Epileptic afterdischarge in the hippocampo-entorhinal system: current source density and unit studies, *Neuroscience*, 76, 1187, 1987.
62. Miles, R., Toth, K., Gulyas, A.E., Hajos, N., and Freund, T.F., Differences between somatic and dendritic inhibition in the hippocampus, *Neuron*, 16, 815, 1996.

63. Cobb, S.R., Buhl, E.H., Halasy, K., Paulsen, O., and Somogyi, P., Synchronization of neuronal actiivty in hippocampus by individual GABAergic interneurons, *Nature*, 378, 75, 1995.
64. Cobb, S.R., Halasy, K., Vida, I., Nyiri, G., Tamás, G., Buhl, E.H., and Somogyi, P., Synaptic effects of identified interneurons innervating both interneurons and pyramidal cells in the rat hippocampus, *Neuroscience*, 79, 629, 1997.
65. Buzsáki, G. and Chrobak, J.J., Temporal structure in spatially organized neuronal ensembles: a role for interneuronal networks, *Curr. Opin. Neurobiol.*, 5, 504, 1995.
66. Kandel, E.R., Specer, W.A., and Brinley, F.J., Electrophysiology of hippocampal neurons. 1. Sequential invasion and synaptic organization, *J. Neurophysiol.*, 24, 225, 1961.
67. Lisman, J., A mechanism for the Hebb and the anti™-Hebb processes underlying learning and memory, *Proc. Natl. Acad. Sci. USA*, 86, 9574, 1989.
68. Artola, A. and Singer, W., Long-term depression of excitatory synaptic transmission and its relationship to long-term potentiation, *Trends Neurosci.*, 16, 480, 1993.
69. Buzsáki, G., Penttonen, M., Nadasdy, Z., and Bragin, A., Pattern and inhibition-dependent invasion of pyramidal cell dendrites by fast spikes in the hippocampus *in vivo*, *Proc. Natl. Acad. Sci. USA*, 93, 9921, 1996.
70. Hopfield, J.J., Pattern recognition computation using action potential timing for stimulus representation, *Nature*, 376, 33, 1995.
71. O'Keefe, J. and Recce, M.L., Phase relationship between hippocampal place units and the EEG theta rhythm, *Hippocampus*, 3, 317, 1993.
72. Jensen, O., Idiart, M.A.P., and Lisman, J.E., Physiologically realistic formation of autoassociative memory in networks with theta/gamma oscillations: role of fast NMDA channels, *Learning Memory*, 3, 243, 1996.
73. Tsodyks, M.V., Skaggs, W.E., Sejnowski, T.J., and McNaughton, B.L., Population dynamics and theta rhythm phase precession of hippocampal place cell firing: a apiking neuron model, *Hippocampus*, 6, 271, 1996.
74. Jensen, O. and Lisman, J.E., Hippocampal CA3 region predicts memory sequences: accounting for the phase precession of place cells, *Learning Memory*, 3, 279, 1996.
75. Heurta, P.T. and Lisman, J.E., Bidirectional synaptic plasticity induced by a single burst during cholinergic theta oscillation in CA1 *in vitro*, *Neuron*, 15, 1053, 1996.

Section V

Molecules Modulating Mental States

Steven J. Henriksen, Section Editor

Chapter 22

Neuronal Mediation of Addictive Behavior

George F. Koob

Contents

22.1	Drug Addiction and Animal Models	365
22.2	Motivational View of Drug Dependence	366
22.3	Neuroadaptation, Sensitization, and Counteradaptation	367
22.4	Neuronal Mechanisms for the Positive Reinforcing Effects of Drugs	367
	22.4.1 Psychomotor Stimulants	369
	22.4.2 Opiates	371
	22.4.3 Sedative-Hypnotics	372
	22.4.4 Nicotine and THC	373
22.5	Neuronal Substrates for Negative Reinforcement Associated With Drug Addiction	374
	22.5.1 Neuronal Elements of Reward	374
	22.5.2 Drug Dependence: Neurochemical Substrates	374
	22.5.3 Molecular and Cellular Adaptations	376
	22.5.4 Neural Substrates for Drug Tolerance	377
	22.5.5 Neuronal Substrates for Drug Sensitization	377
22.6	Neuronal Substrates for Secondary Sources of Reinforcement in Drug Addiction	378
	22.6.1 Protracted Abstinence	378
	22.6.2 Neural Substrates of Relapse	378
22.7	Extended Amygdala: A Common Substrate for Drug Reinforcement	378
Acknowledgments		379
References		380

22.1 Drug Addiction and Animal Models

Drug addiction is a chronically relapsing disorder defined as a compulsion to take a drug with loss of control over drug intake. The term "substance dependence" is used to describe a syndrome basically equivalent to addiction, and the diagnostic criteria used describe symptoms that lead to loss of control in drug intake.[1] However, substance use, substance abuse, and substance dependence are separate, definable entities in most formulations,[1] and an important challenge for neurobiological research is to understand how the transition occurs between controlled substance or drug use and the

TABLE 22.1
Animal Models of Drug Addiction

Animal models for the positive reinforcing properties of drugs
    Operant drug self-administration
    Place preference
    Drug discrimination

Animal models of the negative reinforcing properties of drug withdrawal
    Operant drug self-administration in drug-dependent animals
    Operant schedules for nondrug reinforcers in dependent animals
    Place aversion
    Brain stimulation reward
    Drug discrimination

Animal models of the conditioned reinforcing properties of drugs
    Extinction with and without cues associated with drug self-administration
    Positive reinforcing properties of cues associated with drug self-administration
    Conditioned negative reinforcing effects of withdrawal — cues conditioned to the motivational effects of drug abstinence

loss of control that defines addiction or substance dependence, and what molecular, cellular, and system processes contribute to the development of substance dependence.

Progress in the neurobiology of drug dependence has depended not only on the development of molecular, neurobiological, and neuropharmacological tools for understanding the neuropharmacological mechanisms of action of drugs of abuse, but also the development of animal models of drug dependence that allows interpretation of neuropharmacological advances in the context of the disorder under study. An important issue in such a multi-disciplinary pursuit is that there are no complete animal models of addiction. Animal models exist for many elements of the syndrome of drug addiction, and these elements can be constructed according to conceptual frameworks such as different sources of reinforcement or by symptoms or diagnostic criteria for addiction.[1,2] In addition, while face validity may be useful as a starting point for the development of animal models, it is predictive validity that conveys the power to link the animal work to the human condition (Table 22.1).

22.2 Motivational View of Drug Dependence

The motivating factors for the development, maintenance, and persistence of drug addiction can be broken down into four major sources of reinforcement in drug dependence: positive reinforcement, negative reinforcement, conditioned positive reinforcement, and conditioned negative reinforcement (Table 22.2).[3] Clearly, positive reinforcing effects are critical for establishing self-administration behavior, and some have argued the hypothesis that positive reinforcement is the key to drug dependence.[4] However, while alleviation of withdrawal symptoms (negative reinforcement) may not be a major motivating factor in the *initiation* of compulsive drug use, a compelling case can be made for compromises in hedonic processing associated with drug abstinence as a driving force of addiction.[5,6] Clearly, the construct of negative reinforcement plays an important role in the *maintenance* of drug use after the development of dependence. Thus, while initial drug use may be motivated by the positive affective state produced by the drug, continued use leads to neuroadaptation

TABLE 22.2
Relationship of Addiction Components and Behavioral Constructs of Reinforcement

Addiction Component	Behavioral Construct
Pleasure	Positive reinforcement
Self-medication	Negative reinforcement
Habit	Conditioned positive reinforcement
Habit	Conditioned negative reinforcement

to the presence of drug and to another source of reinforcement, the negative reinforcement associated with relieving negative affective consequences of drug termination. Indeed, the defining feature of drug dependence has been argued to be the establishment of a negative affective state.[7] However, an even more compelling motivational force is the hypothesis that negative affective states, even during protracted abstinence, can contribute to the reinforcement associated with drug-taking by changing the "set point" for hedonic processing.[6] Much progress has been made in identifying the neuronal substrates for the acute positive reinforcing effects of drugs of abuse. A more recent focus has been on the neuronal substrates for negative reinforcement and the conditioned reinforcing effects that contribute to relapse.

22.3 Neuroadaptation, Sensitization, and Counteradaptation

Historically, neuronal systems have been hypothesized to respond to chronic drug insult by counteradaptive responses.[8] Here, the body's attempt to counter the acute effects of the drug included the development of tolerance and the manifestation of a withdrawal syndrome in the absence of drug. Modern formulations have focused on changes in hedonic processing including the manifestation of hedonic withdrawal with a minimal contribution of physical signs of withdrawal.[6,9] Another adaptive process, the phenomenon of sensitization, also has been conceptualized as a critical neuroadaptive process.[10] Such adaptations have been explored at all levels of drug dependence research from the behavioral to the molecular.[9]

If a compromised reward state (sensitized or deficit or both) defines addiction, it could be produced by several mechanisms. First, genetic or environmental factors could produce an increased sensitivity to the reinforcing effects of drugs or some neurobiological deficit that requires reversal, perhaps by drug initiation. Alternatively, chronic drug-taking itself could produce an increased sensitivity to the reinforcing effects of drugs or a hedonic deficit state that requires self-medication to reverse. These changes have been hypothesized to exist at the molecular, cellular, and system levels and combined together form a powerful motivation for drug-seeking behavior (Figure 22.1).[6]

22.4 Neuronal Mechanisms for the Positive Reinforcing Effects of Drugs

The neuronal circuits forming the medial forebrain bundle have long been hypothesized to be the neural substrates of reward. The medial forebrain bundle contains both ascending and descending

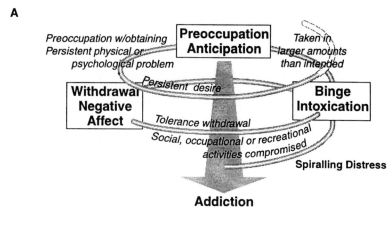

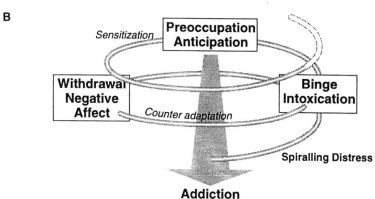

FIGURE 22.1
Diagram describing the spiralling distress/addiction cycle from two conceptual perspectives: psychiatric and dysadaptational. (A) shows the three major components of the addiction cycle with the different criteria for substance dependence for DSM-IV incorporated. (B) incorporates the places of emphasis for the theoretical constructs of sensitization and counteradaptation. (From Koob, G.F. and Le Moal, M., *Science*, 278, 52, 1997. With permission.)

pathways that include most of the brain's monoamine systems,[11–13] and the structures involved include those that support intracranial self-stimulation: the ventral tegmental area, the basal forebrain, and the medial forebrain bundle which connects these two areas.[11,12,14,15] Significant insights into the neurochemical and neuroanatomical components of the medial forebrain bundle have provided the key not only to drug reward but also to natural rewards.

The principle focus of research on the neurobiology of drug addiction has been the origins and terminal areas of the mesocorticolimbic dopamine system, and there is now compelling evidence for the importance of this system in drug reward. The major components of this drug reward circuit are the ventral tegmental area (the site of dopaminergic cell bodies), the basal forebrain (the nucleus accumbens, olfactory tubercle, frontal cortex, and amygdala), and the dopaminergic connection between the ventral tegmental area and the basal forebrain. Other components are the opioid peptide, GABA, glutamate, and serotonin systems, and presumably many other neural inputs that interact with the ventral tegmental area and the basal forebrain (see Figure 22.2).[16] The neuronal components of this circuitry for different types of drug reward will be discussed in the following sections, and a construct called the extended amygdala will be introduced which provides important insights into the relationship of drug reward to natural reward systems.

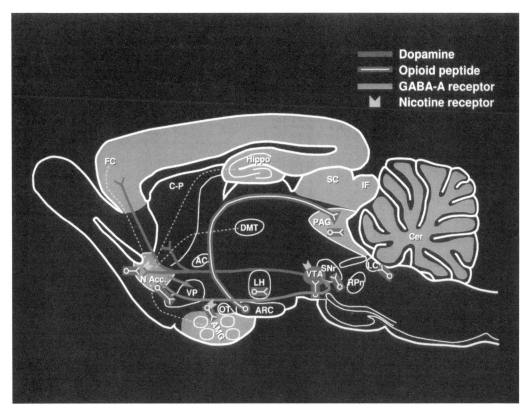

FIGURE 22.2
Sagittal rat brain section illustrating a drug (cocaine, amphetamine, opiate, nicotine, and alcohol) neural reward circuit that includes a limbic-extrapyramidal motor interface. Dotted lines indicate limbic afferents to the nucleus accumbens and lines with arrows from the nucleus accumbens to the ventral pallidum and ventral tegmental area represent efferents thought to be involved in psychomotor stimulant reward. The mesocorticolimbic dopamine system, thought to be a critical substrate for psychomotor stimulant reward, originates in the A10 cell group of the ventral tegmental area and projects to the nucleus accumbens, olfactory tubercle, ventral striatal domains of the caudate-putamen, and amygdala. Segments with the symbol ( ) indicate opioid peptide-containing neurons, systems that may be involved in opiate, ethanol, and possibly nicotine reward. These opioid peptide systems include the local enkephalin circuits (short segments) and the hypothalamic midbrain beta-endorphin circuit (long segment). Shaded regions of the frontal cortex, hippocampus, superior colliculus, inferior colliculus, periaqueductal gray, cerebellum, nucleus accumbens, and amygdala indicate the approximate distribution of $GABA_A$ receptor complexes, some of which may mediate sedative/hypnotic (ethanol) reward, determined by both tritiated flumazenil binding and expression of the alpha, beta and gamma subunits of the $GABA_A$ receptor. Notched symbols indicate nicotinic acetylcholine receptors hypothesized to be located on dopaminergic and opioid peptidergic systems. NAcc, nucleus accumbens; VP, ventral pallidum; VTA, ventral tegmental area; LH, lateral hypothalamus; SNr, substantia nigra pars reticulata; DMT, dorsomedial thalamus; PAG, periaqueductal gray; OT, olfactory tract; AC, anterior commissure; LC, locus coeruleus; AMG, amygdala; Hippo, hippocampus; Cer, cerebellum; FC, frontal cortex; SC, superior colliculus; IF, inferior colliculus. (Adapted from Koob, G.F. and Le Moal, M., *Science*, 278, 52, 1997. With permission.)

22.4.1 Psychomotor Stimulants

At the neuropharmacological level of analysis, psychomotor stimulants of high abuse potential interact initially with monoamine transporter proteins, which have been cloned and characterized,[17-19] are located on monoaminergic nerve terminals and terminate a monoamine signal by transporting the monoamine from the synaptic cleft back into the terminals. Cocaine inhibits all three monoamine transporters — dopamine, serotonin, and norepinephrine — thereby potentiating monoaminergic transmission. Amphetamine and its derivatives also potentiate monoaminergic transmission, but by

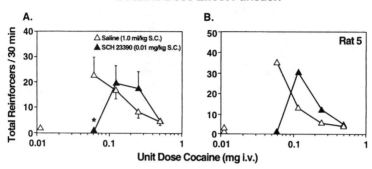

FIGURE 22.3
Shift of the cocaine dose-effect function to the right following pretreatment with the dopamine D1 antagonist, SCH-23390. (A) Effects of pretreatment with the D1 dopamine receptor antagonist, SCH-23390 (0.01 mg/kg s.c.), on the cocaine (0.06 to 0.5 mg) self-administered dose-effect function measured using the within-session dose-effect paradigm (n = 4). (B) Same as (A) but for an individual rat. (From Caine, S.B. and Koob, G.F., *Behav. Pharmacol.*, 6, 333, 1995. With permission.)

increasing monoamine release. Amphetamine itself is transported into monoaminergic nerve terminals by all three transporters, where it disrupts the storage of the monoamine transmitters. This leads to an increase in extravascular levels of the monoamines and to the reverse transport of the monoamine into the synaptic cleft via the monoamine transporters.[20] Amphetamine and cocaine are psychomotor stimulants and as such have behavioral effects such as psychomotor activation, suppression of hunger and fatigue,[21] and induction of euphoria in humans.[22] In animals, these drugs increase motor activity,[23] have psychomotor stimulant actions on operant behavior,[24] decrease food intake,[23] enhance conditioned responding,[24] decrease thresholds for reinforcing brain stimulation,[25,26] produce preferences for environments where they have been previously experienced (place preferences),[27,28] and readily act as reinforcers for drug self-administration.[29]

The mesocorticolimbic dopamine system appears to be the critical substrate for both the psychomotor stimulant effects of amphetamine and cocaine[30–32] and their reinforcing actions.[33,34] The most direct evidence implicating dopamine in the reinforcing actions of cocaine comes from studies of intravenous self-administration and studies of conditioned place preference. Dopamine receptor antagonists, when injected systemically, reliably decrease the reinforcing effects of cocaine and amphetamine self-administration in rats and block conditioned place preferences for these drugs.[33–38]

Dopamine antagonists actually shift the cocaine dose-effect function for cocaine self-administration to the right (see Figure 22.3).[39,40] Neurotoxin-induced lesions of the mesocorticolimbic dopamine system with 6-hydroxydopamine (6-OHDA) in the nucleus accumbens or ventral tegmental area produce extinction-like responding and significant, long-lasting decreases in self-administration of cocaine and amphetamine over days.[41,42] These decreases in cocaine self-administration following dopamine-selective lesions of the nucleus accumbens have now been observed in a variety of different tests and conditions, including situations where animals show a decrease in the amount of work they would perform for cocaine[29] and situations where other reinforcers such as food were unaffected but cocaine self-administration was abolished.[43]

All three dopamine receptor subtypes have been implicated in the reinforcing actions of cocaine as measured by intravenous self-administration, including the D1,[44] D2,[40,45] and D3 receptors (Table 22.3).[46] Dopamine D1 and D2 antagonists also block the place conditioning produced by amphetamine.[47,48]

Electrophysiological recordings in animals during intravenous cocaine self-administration have identified several types of neurons that respond in the nucleus accumbens in a manner time-

TABLE 21.3
Functional Classification of Dopamine Receptor Subtypes

Adenylate Cyclase Linkage	D1 Stimulatory	D2 Unlinked or Inhibitory	D3 Unlinked
Brain localization	Nucleus accumbens	Nucleus accumbens	Shell of nucleus accumbens
	Corpus striatum	Corpus striatum	
Cocaine self-administration	Agonists ↑	Agonists ↑	Agonists ↑
	Antagonists ↓	Antagonists ↓	Antagonists ?

locked to drug infusion and reinforcement, suggesting that the integration of various components of cocaine reinforcement are indeed occurring at the level of the nucleus accumbens.[49–53] One group of neurons may be involved in anticipatory neuronal responses and may be an initiation mechanism. A second group of neurons fires postcocaine and may represent a direct reinforcement effect.[50] A third group of neurons, termed progressive reversal neurons, appears to have its firing positively correlated with the inter-infusion interval linking it to the initiation of the next response.[52] The neurons that fire both before and after the cocaine-reinforced response (cocaine delivery after a response) do not fire to cocaine-only probes (noncontingent delivery), but do fire to stimuli paired with cocaine delivery (tone-houselight probes).[54] These results suggest that a subset of nucleus accumbens neurons may be involved in mediating conditioned responses associated with cocaine reinforcement.

22.4.2 Opiates

Animals will also readily self-administer opiate drugs such as heroin,[55] and, if provided in limited access situations, rats and primates[56] will maintain stable levels of opiate intake on a daily basis without obvious signs of physical dependence.[57] Systemic and central administration of competitive opiate antagonists decreases this acute opiate reinforcement as measured by an increase in the number of infusions for opiate drugs (decrease in the inter-injection interval) in intravenous self-administration studies.[34,57–60] This appears to result from a competitive interaction between the antagonist and agonist at opioid receptors.

The mu opioid receptor subtype appears to be particularly important for the reinforcing actions of heroin and morphine. Mu opioid receptor agonists produce dose-dependent decreases in heroin self-administration (increases in the inter-injection interval), and irreversible mu-selective antagonists dose-dependently increase heroin self-administration (decrease the inter-injection interval).[61] These changes reflect an increase and decrease, respectively, in opiate reward by virtue of the shift of the dose-effect function to the right and left, similar to that described for dopamine antagonists and cocaine (see Figure 22.3). Also, knockout mice without the mu receptor fail to show morphine-induced analgesia or morphine-induced conditioned place preferences.[62]

The sites of action for opioid antagonists to block the reinforcing effects of opiates appear to be associated with the same neural circuitry associated with psychomotor stimulant reward (see Figure 22.2). Intracerebral injection of quaternary derivatives of opiate antagonists — charged, hydrophilic compounds that do not readily spread from the sites in the brain at which they are injected — dose-dependently block heroin self-administration in non-dependent rats.[60,63–65] This antagonism was observed when the antagonists were injected into the ventral tegmental area[63] or the region of the nucleus accumbens.[64] The dopamine-dependent action for opiates in the ventral tegmental area is supported by neurochemical and neuropharmacological studies. Opiates can produce an increase in dopamine release in the nucleus accumbens similar to that of cocaine and ethanol[66] (also see Hemby et al.[67]). Opioid peptides are self-administered into the ventral tegmental area,[68] and microinjections of opioids into the ventral tegmental area will lower brain-stimulation reward thresholds and produce robust place preferences.[69] However, rats also will self-administer opioid peptides directly in the region of the nucleus accumbens,[70] and heroin self-administration is not blocked by cocaine-blocking doses of dopamine antagonists[71] nor by dopamine-selective lesions of the mesocorticolimbic dopamine system.[72] Nor does chronic dopamine receptor blockade alter heroin self-administration, which further demonstrates dopamine-independent mechanisms.[73] Altogether, these results suggest that neural elements in the region of the ventral tegmental area *and* the nucleus accumbens are responsible for the reinforcing properties of opiates and implicates both dopamine-dependent and dopamine-independent mechanisms of opiate action.[72,74–76]

22.4.3 Sedative-Hypnotics

Sedative-hypnotics, such as barbiturates, benzodiazepines, and ethanol, all produce a characteristic euphoria, disinhibition, anxiety reduction, sedation, and hypnosis. These drugs exert anti-anxiety effects that are reflected in a reduction of behavior suppressed by punishment in conflict situations in laboratory animals. This anti-conflict effect correlates well with their ability to act as anxiolytics in the clinic[77] and may be a major component of the reinforcing actions of these drugs.

The sedative and anti-punishment (anxiolytic) effects of sedative-hypnotics are associated with facilitation of the $GABA_A$ receptor,[78] but the actions of sedative-hypnotics on this receptor are complex. Sedative-hypnotics drugs do not bind the GABA receptor directly at the GABA-binding site. Instead, they appear to bind to other sites on the receptor complex and thereby facilitate, via allosteric effects, activation of the receptor by GABA. The net result is potentiation of GABA-induced Cl^- flux through the receptor ionophore.[79] The $GABA_A$ receptor is a heteromeric complex, and the ability of sedative-hypnotics to facilitate receptor function depends on the actual subunit composition of the receptor, which differs markedly throughout the brain. For example, recent work using chimeras and site-directed mutagenesis has revealed that ethanol requires, to produce its neuropharmacological effects, a specific site on the $GABA_A$ receptor and glycine receptor associated with amino acids serine and alanine.[80] These two amino acids have been hypothesized to be part of a binding pocket formed between two transmembrane domains. Even though benzodiazepines, barbiturates, and ethanol interact with the $GABA_A$ receptor at distinct sites, the convergence of their action on the functioning of the same protein complex may explain the long-appreciated cross-tolerance and cross-dependence exhibited by these drugs.

Studies of the neuropharmacological basis for the anxiolytic properties of sedative-hypnotics provided some of the first clues to their reinforcing properties and abuse potential.[81] GABAergic antagonists reverse many of the behavioral effects of ethanol, which led to the hypothesis that GABA has a role in the intoxicating effects of ethanol.[82,83] The partial inverse benzodiazepine agonist, R015-4513, which has been shown to reverse some of the behavioral effects of ethanol,[79] produces a dose-dependent reduction of oral ethanol (10%) self-administration in rats.[84–86] More recent studies have shown similar effects with potent GABA antagonists microinjected into the brain, with the most effective site to date being the central nucleus of the amygdala.[87]

Ethanol antagonism of the *N*-methyl-D-aspartatate (NMDA) receptor also appears to contribute to the intoxicating effects of ethanol,[88,89] and perhaps to the dissociative effects seen in people with high ethanol blood levels.[90] Ethanol inhibits the functioning of the receptor — again, not by blocking the glutamate binding site, but via a more complex allosteric effect on the receptor complex, which results in decreased glutamate-induced Na^+ and Ca^{++} flux through the receptor ionophore.[91] Whether ethanol antagonism of the NMDA receptor also contributes to its reinforcing effects remains to be established. At still higher doses, ethanol can exert more general inhibitory effects on voltage-gated ion channels, particularly Na^+ and Ca^{++} channels.[91] These actions occur only with extreme concentrations seen clinically and would, therefore, not appear to be involved in the reinforcing actions of ethanol, although they may contribute to the severe nervous system depression, even coma, seen at these blood levels.

Possibly via its initial effects on the $GABA_A$ and NMDA glutamate receptors, ethanol influences several additional neurotransmitter systems in the brain which are believed to be involved in its reinforcing properties. Again, there is much evidence implicating dopamine in the reinforcing actions of low, non-dependence-inducing doses of ethanol. Dopamine receptor antagonists have been shown to reduce lever-pressing for ethanol in non-deprived rats,[92] and extracellular dopamine levels also have been shown to increase in non-dependent rats orally self-administering low doses of ethanol.[93] However, virtually complete 6-hydroxydopamine denervation of the nucleus accumbens failed to alter voluntary responding for ethanol.[94] Thus, as with opiates, these results suggest that activation of mesocorticolimbic dopamine transmission may be associated with important aspects of ethanol reinforcement, but this activation may not be critical. In fact, evidence suggests that other sedative hypnotics such as benzodiazepines decrease mesocorticolimbic activity.[95]

Thus, other dopamine-independent, neurochemical systems likely contribute to the mediation of the reinforcing actions of ethanol with a view emerging that multiple neurotransmitters combine to "orchestrate" the reward profile of ethanol.[96] These include actions on serotonergic and opioid peptide systems.

Modulation of various aspects of serotonergic transmission, including increases in the synaptic availability of serotonin with precursor loading, blockade of serotonin reuptake, or blockade of certain serotonin receptor subtypes, can decrease ethanol intake.[97] Serotonin-3 antagonists decrease self-administration of ethanol.[98,99] More recently, serotonin-2 receptor antagonists, including some with both serotonin-2 antagonist action and serotonin-1a agonist activity, can selectively decrease acute ethanol reinforcement.[100] Several double-blind, placebo-controlled clinical studies have reported that serotonin reuptake inhibitors produced mild decreases in ethanol consumption in humans.[101] Fluoxetine has been shown to reduce depressive symptoms and ethanol consumption in depressed alcoholics,[102] but may be of limited use in nondepressed alcoholics in preventing relapse.[103,104]

The opioid receptor antagonists, naloxone and naltrexone, reduce ethanol self-administration in several animal models, implicating opioid peptide systems in acute ethanol reinforcement.[105] Opioid antagonists dose-dependently decrease consumption of sweet solutions of water as well as ethanol in operant, free-choice tests;[105] however, recent results suggest that antagonism of specific opioid receptor subtypes in specific brain regions might reveal more selective effects.[106] Two double-blind, placebo-controlled clinical trials demonstrated that naltrexone significantly reduces ethanol consumption, frequency of relapse, and "craving" for ethanol in humans.[107,108] Thus, ethanol interactions with opioid neurotransmission may contribute to certain aspects of ethanol reinforcement that may be of particular importance to the motivation associated with relapse.

22.4.4 Nicotine and THC

Nicotine has an initial molecular site of action as a direct agonist action at nicotinic acetylcholine receptors. Although brain nicotinic acetylcholine receptors are widely distributed throughout the brain, to date it is the receptors specifically in the brain mesolimbic dopamine system that have been implicated in its reinforcing actions. Nicotine self-administration in animals with limited access is blocked by nicotinic acetylcholine antagonists and dopamine antagonists.[109,110] Interestingly, nicotinic acetylcholine receptor antagonists and opioid peptide antagonists can precipitate a physical withdrawal syndrome in rats.[109,110] Thus, nicotine may alter function in both the mesocorticolimbic dopamine system and opioid peptide systems in the same neural circuitry associated with other drugs of abuse (see Figure 22.2).[111]

Tetrahydrocannabinol (THC) is a drug of abuse and dependence.[112] THC binds to the cannabinoid-1 receptor in the brain, which is widely distributed throughout the brain but particularly in the extrapyramidal motor system of the rat.[113] THC has been shown also to decrease reward thresholds in rats upon acute administration[114,115] and to produce a place preference.[116] In addition, a recent study in mice has shown intravenous self-administration of a synthetic THC analog.[117] THC activates the mesocorticolimbic dopamine system,[118] and recent data suggest that THC can selectively increase the release of dopamine in the shell of the nucleus accumbens as do other drugs of abuse.[119]

22.5 Neuronal Substrates for Negative Reinforcement Associated With Drug Addiction

Drugs of abuse have several common actions associated with acute withdrawal that appear independent of physical signs of withdrawal. Acute withdrawal is associated with a negative affective state including various negative emotions such as dysphoria, depression, irritability, and anxiety. For example, cocaine withdrawal in humans in the outpatient setting is characterized by severe depressive symptoms combined with irritability, anxiety, and anhedonia lasting several hours to several days (i.e., the "crash") and may be one of the motivating factors in the maintenance of the cocaine-dependence cycle.[120] Inpatient studies have shown similar changes in mood and anxiety states, but they generally are much less severe.[121] Opiate withdrawal is characterized by severe dysphoria, and ethanol withdrawal produces dysphoria and anxiety.

22.5.1 Neuronal Elements of Reward

Recent neurobiological studies have focused on the neural substrates and neuropharmacological mechanisms of the motivational effects of drug withdrawal, effects that may contribute to the negative reinforcement associated with drug dependence. The same neural systems implicated in the positive reinforcing effects of drugs of abuse have begun to be shown to be involved in these aversive motivational effects of drug withdrawal. Using the technique of intracranial self-stimulation to measure reward thresholds throughout the course of drug dependence, recent studies have shown that reward thresholds are increased (reflecting a decrease in reward) following chronic administration of all major drugs of abuse, including opiates, psychostimulants, alcohol, and nicotine. These effects reflect changes in the activity of the same mesocorticolimbic system (midbrain-forebrain system) implicated in the positive reinforcing effects of drugs and can last up to 72 hours depending on the drug and dose administered (see Figure 22.4).[122–127]

22.5.2 Drug Dependence: Neurochemical Substrates

Two neuroadaptive models have been conceptualized to be associated with the change in reward function accompanying the development of drug dependence: sensitization and a homeostatic adaptive mechanism.[128] With repeated drug administration, sensitization is more likely to occur with intermittent exposure to a drug, in contrast to tolerance which is more likely to occur with continuous exposure. In a recent formulation of the role of sensitization in drug dependence, a shift in an incentive-salience state described as "wanting" was hypothesized to be progressively increased by repeated exposure to drugs of abuse,[10] and the transition to pathologically strong wanting or craving was defined as compulsive use. In contrast, in a homeostatic adaptive formulation, the initial acute effect of the drug is opposed or counteracted by homeostatic changes in systems that mediate the primary drug effects.[129–131] In opponent process theory, both tolerance and dependence were inextricably linked.[129] Affective states, pleasant or aversive, were hypothesized to be opposed by centrally mediated mechanisms that reduce the intensity of these affective states.

Neuronal Mediation of Addictive Behavior

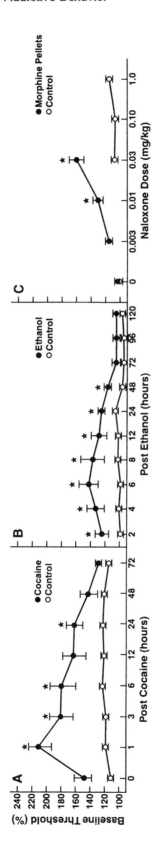

FIGURE 22.4

Changes in reward threshold associated with chronic administration of three major drugs of abuse. Reward thresholds were determined using a rate-independent discrete-trials threshold procedure for intracranial self-stimulation (ICSS) of the medial forebrain bundle. (A) Rats equipped with intravenous catheters were allowed to self-administer cocaine for 12 straight hours prior to withdrawal and reward threshold determinations. Elevations in threshold were dose-dependent with longer bouts of cocaine self-administration, yielding larger and longer-lasting elevations in reward thresholds. (Redrawn from Markou, A. and Koob, G.F., *Neuropsychopharmacology*, 4, 17, 1991. With permission.) (B) Elevations in reward thresholds with the same ICSS technique following chronic exposure to ethanol of approximately 200 mg% in ethanol vapor chambers. (Redrawn from Schulteis, G. et al., *Proc. Natl. Acad. Sci. USA*, 92, 5880, 1995. With permission.) (C) Elevations in reward thresholds using the same ICSS technique following administration of very low doses of the opiate antagonist, naloxone, to animals made dependent on morphine using two 75-mg morphine (base) pellets implanted subcutaneously. (Redrawn from Schulteis, G. et al., *J. Pharmacol. Exp. Ther.*, 271, 1391, 1994. With permission.) Asterisks refer to significant differences between treatment and control values. Values are mean ± S.E.M.

TABLE 22.4
Neurotransmitters Implicated
in the Motivational Effects of
Withdrawal From Drugs of Abuse

Neurotransmitter	Effect(s)
↓ Dopamine	"Dysphoria"
↓ Opioid peptides	Pain, "dysphoria"
↓ Serotonin	Pain, "dysphoria", depression
↓ GABA	Anxiety, panic attacks
↑ Corticotropin-releasing factor	Stress

The neuronal basis for neuroadaptive mechanisms described by both theoretical positions can be envisioned at the molecular, cellular, and system levels. Within-system adaptations have been hypothesized wherein neurochemical changes associated with the same neurotransmitters implicated in the acute reinforcing effects of drugs are altered during the development of dependence.[9] Examples of such homeostatic, within-system adaptive neurochemical events include decreases in dopaminergic and serotonergic transmission in the nucleus accumbens during drug withdrawal as measured by *in vivo* microdialysis,[125,132] increased sensitivity of opioid receptor transduction mechanisms in the nucleus accumbens during opiate withdrawal,[133] decreased GABAergic and increased NMDA glutamatergic transmission during ethanol withdrawal,[91,134,135] and differential regional changes in nicotine receptor function (Table 22.4).[136,137]

Recruitment of other neurotransmitter systems in the adaptive responses to drugs of abuse neurotransmitter systems not linked to the acute reinforcing effects of the drug has been termed a "between-system" adaptation.[9] A common between-system adaptation to repeated administration of drugs of abuse may be activation of brain and pituitary stress systems (Table 22.4). Pituitary adrenal function is activated during drug dependence, and acute withdrawal from drugs of abuse in humans and dysregulation can persist even past acute withdrawal.[138,139] Both stress and repeated administration of glucocorticoids can increase behavioral effects of psychostimulants, and it has been hypothesized that circulating glucocorticoids can function to maintain the sensitized state,[140,141] and as such, may represent a within-system adaptation.[141] Corticotropin-releasing factor (CRF) function, outside of the pituitary adrenal axis, also appears to be activated during acute withdrawal from cocaine, alcohol, opiates, and THC and thus may mediate behavioral aspects of stress associated with abstinence.[100,142-144] Rats treated repeatedly with cocaine, nicotine, and ethanol show significant anxiogenic-like responses following cessation of chronic drug administration which are reversed with intracerebroventricular administration of a CRF antagonist.[145,146] Microinjections into the central nucleus of the amygdala of lower doses of the CRF antagonist also reversed the anxiogenic-like effects of ethanol withdrawal,[145] and similar doses of the CRF antagonist injected into the amygdala were active in reversing the aversive effects of opiate withdrawal.[143]

22.5.3 Molecular and Cellular Adaptations

Insights into the molecular and cellular mechanisms of drug dependence have begun to be focused on mechanisms of motivational aspects dependence. The cAMP second messenger pathway in the nucleus accumbens is up-regulated after chronic administration of several drugs of abuse.[147-149] This up-regulation could mediate some of the documented electrophysiological changes in the nucleus accumbens associated with chronic drug exposure, such as enhanced responsiveness of D1 dopamine receptors after chronic cocaine treatment.[150] Studies involving direct administration of activators or inhibitors of the cAMP pathway into the nucleus accumbens are consistent with the interpretation that up-regulation of the cAMP pathway in this brain region may contribute to an aversive state during drug withdrawal.[148] A major task now will be to explore these mechanisms directly in animal models of dependence that reflect both the positive and negative reinforcing effects of dependence.

22.5.4 Neural Substrates for Drug Tolerance

Tolerance to the reinforcing actions of drugs of abuse may also be an important mechanism for drug addiction; however, the measurement of such phenomena in animal models such as intravenous self-administration has been limited.[151] One would hypothesize that the neural substrates for drug tolerance would overlap significantly with those associated with acute withdrawal because tolerance and withdrawal appear to be components of the same neuroadaptive process. However, tolerance also depends on learning processes, and this has been most explored in the context of opiate drugs and sedative-hypnotics such as ethanol.[152] Mechanisms for these associative processes may involve several neurotransmitters independent of their role in acute withdrawal. Norepinephrine and serotonin have long been shown to be involved in the development of tolerance to ethanol and barbiturates.[153] More recently, co-administration of glutamate receptor antagonists and opiates has been shown to block the development of tolerance to the opiate alone but not to the combination of the drugs.[154] This is again consistent with an associative component of tolerance.

Mechanisms at the molecular level for tolerance also probably overlap with those of dependence.[147] For example, up-regulation of the cAMP pathway could be a mechanism of tolerance — the changes would be expected to oppose the acute actions of opiates of inhibiting adenylyl cyclase. In addition, tolerance appears to involve the functional uncoupling of opioid receptors from their G proteins. The mechanisms underlying this uncoupling remain unknown but could involve drug-induced changes in the phosphorylation state of the receptors or G proteins that reduce their affinity for each other. Such phosphorylation of the receptor could occur via cAMP- or Ca^{++}-dependent protein kinases known to be regulated by opiate exposure (see above) or by other types of protein kinases (termed G protein receptor kinases, or GRKs) that phosphorylate and desensitize receptors only when they are in their ligand-bound conformation (see Nestler[155]). Another possible mechanism of tolerance is drug-induced changes in the ion channels that mediate the acute effects of the drugs. For example, alterations in the phosphorylation state, amount or even type of channel could conceivably contribute to drug tolerance.[156]

22.5.5 Neuronal Substrates for Drug Sensitization

The repeated administration of psychomotor stimulants can result in an enhancement of their behavioral effects, particularly if the treatment regimen involves intermittent, noncontinuous administration.[157] Usually sensitization has been observed with psychomotor stimulants and measures of motor behavior but also has been observed with opiates and other nonpsychomotor stimulant drugs.[158,159] Sensitization to the reinforcing effects of cocaine as measured by self-administration[160] and place preference[161] also has been observed.

The neurochemical substrates for sensitization have largely focused on increased activity in the mesocorticolimbic dopamine system.[162] There appears to be a time-dependent chain of neurobiological changes within the mesocorticolimbic dopamine system that lead to sensitization.[150,163,164] Repeated administration of stimulants and opiates results in increasingly larger activating effects. Systemic administration of opiates or injections of opiates directly into the ventral tegmental area produce sensitization. Also, repeated microinjections of amphetamine into the somatodendritic region of the ventral tegmental area dopamine cells, at doses that do not cause behavioral activation, are sufficient to produce sensitization to later systemic injections of amphetamine or morphine, suggesting changes at the level of the ventral tegmental area associated with sensitization.

Consistent with this hypothesis, repeated administration of cocaine produces a decrease in the sensitivity of impulse-regulating somatodendritic dopamine D2 autoreceptors which could translate into enhanced dopaminergic function with subsequent injections.[164] However, the time course of dopaminergic subsensitivity is only 4 to 8 days, and behavioral sensitization can persist for weeks. Increased release of dopamine in the nucleus accumbens accompanies the increased behavioral responsivity to psychomotor stimulants.[163]

Stressors can also cause sensitization to stimulant drugs, and there appears to be an important role for both the hypothalamic pituitary adrenal stress axis and extrahypothalamic CRF system in stress-induced sensitization.[165] In addition, a role for brain glutamate systems in sensitization has been hypothesized as a result of studies showing that administration of N-methyl-D-aspartatate glutamate receptor antagonists block the development of sensitization to psychomotor stimulants.[4,166]

22.6 Neuronal Substrates for Secondary Sources of Reinforcement in Drug Addiction

22.6.1 Protracted Abstinence

The persistence of changes in drug reinforcement mechanisms that characterize drug addiction suggests that the underlying molecular mechanisms are long-lasting, and indeed considerable attention is being directed at drug regulation of gene expression. Two types of transcription factors, CREB and novel Fos-like proteins (termed chronic FRAs or Fos-related antigens), have been hypothesized to be possible mediators of chronic drug action.[167–169] The challenge for the future will be to relate regulation of a specific transcription factor to specific features of drug reinforcement associated with specific histories of drug administration (sensitization of acute challenges vs. changes in set point associated with protracted abstinence).

22.6.2 Neural Substrates of Relapse

Animal models for the study of relapse have been limited and are currently under development.[170] Neuropharmacological probes that activate the mesocorticolimbic dopamine system have been shown to reinstate self-administration rapidly in animals trained and then extinguished on intravenous drug self-administration.[171,172] There are a limited number of observations using other models. Acamprosate, a drug being marketed in Europe to prevent relapse in alcoholics and which has potential glutamate modulatory action,[173] blocks the increase in drinking observed in rodents after a forced abstinence, again in non-dependent rats.[174] Similarly, opioid antagonists were shown to prevent the increase in drinking of ethanol in animals post-stress,[175] and, subsequently, naltrexone was shown to have efficacy in preventing relapse in detoxified human alcoholics as mentioned above.[107,108] Also, a recent study reports that agonists selective for D1 dopamine receptors, but not for D2-like receptors, can block reinstatement of lever-pressing inferred to represent cocaine-seeking behavior.[176]

22.7 Extended Amygdala: A Common Substrate for Drug Reinforcement

Recent neuroanatomical data and new functional observations have provided support for the hypothesis that the neuroanatomical substrates for the reinforcing actions of drugs may involve a common neural circuitry that forms a separate entity within the basal forebrain, termed the "extended amygdala".[177] Originally described by Johnston,[178] the term "extended amygdala" represents a macrostructure that is composed of several basal forebrain structures: the bed nucleus of the stria terminalis, the central medial amygdala, the medial part of the nucleus accumbens (e.g., shell),[179] and the area termed the sublenticular substantia innominata. There are similarities in morphology, immunohistochemistry, and connectivity in these structures.[177] They receive afferent connections from limbic cortices, hippocampus, basolateral amygdala, midbrain, and lateral hypothalamus. The efferent connections from this complex include the posterior medial (sublenticular) ventral pallidum, medial ventral tegmental area, various brainstem projections, and perhaps most intriguing from a functional point of view, a considerable projection to the lateral hypothalamus.[180]

Recent studies have demonstrated selective neurochemical and neuropharmacological actions in specific components of the extended amygdala both in the acute reinforcing effects of drugs of abuse and in the negative reinforcement associated with drug dependence. D1 dopamine antagonists are effective in blocking cocaine self-administration when the antagonist is administered directly into the shell of the nucleus accumbens, the central nucleus of the amygdala,[181] and the bed nucleus of the stria terminalis.[182] Moreover, selective activation of dopaminergic transmission occurs in the shell of the nucleus accumbens in response to acute administration of virtually all major drugs of abuse.[119,183,184] In addition, the central nucleus of the amygdala has been implicated in the GABAergic and opioidergic influences on the acute reinforcing effects of ethanol.[185,186]

Evidence for parts of the extended amygdala being involved in the aversive stimulus effects of drug withdrawal includes activation of CRF systems in the central nucleus of the amygdala.[187] Also, there are changes in sensitivity to opiate antagonists in opiate-dependent rats in the nucleus accumbens and central nucleus of the amygdala.[133,188] This concept of the extended amygdala may ultimately link the recent developments in the neurobiology of drug reward with existing knowledge of the substrates for emotional behavior,[189] essentially bridging what have been largely independent research pursuits. Perhaps more importantly, this neuronal circuit is well situated to form a heuristic model for exploring the mechanisms associated with vulnerability to relapse and concepts such as craving, both of which may involve secondary conditioned reinforcement constructs.

Drug dependence not only involves acquisition of drug taking and maintenance of drug taking, but also functions as a chronic relapsing disorder with reinstatement of drug taking after detoxification and abstinence. Both the positive and negative affective states can become associated with stimuli in the drug-taking environment or even internal cues through classical conditioning processes.[3] Re-exposure to these conditioned stimuli can provide the motivation for continued drug use and relapse after abstention. There is evidence in humans that the positive reinforcing effects of drugs such as heroin and cocaine, as measured by subjective reports of euphoria or "high", can become conditioned to previously neutral stimuli. Patients being treated for heroin addiction and allowed to self-administer either saline or heroin reported that both saline and heroin injections were pleasurable, particularly in the patient's usual injection environment.[190] Alternatively, patients, even detoxified subjects, can report negative affective symptoms like those associated with drug abstinence when returning to environments similar to those associated with drug dependence.[191] The neurobiological bases for the syndrome of protracted abstinence may involve subtle molecular and cellular changes in the circuitry associated with the extended amygdala. Elucidation of these changes will be the challenge of future research on the neurobiology of addiction.

Acknowledgments

This is publication number 11293-NP from The Scripps Research Institute. This work was supported by National Institutes of Health grants AA06420 and AA08459 from the National Institute on Alcohol Abuse and Alcoholism and DA04043, DA04398, and DA08467 from the National Institute on Drug Abuse. The author would like to thank Mike Arends for assistance with manuscript preparation.

References

1. American Psychiatric Association, *Diagnostic and Statistical Manual of Mental Disorders*, 4th ed., American Psychiatric Press, Washington, D.C., 1994.
2. World Health Organization, *International Statistical Classification of Diseases and Related Health Problems*, 10th rev. ed., World Health Organization, Geneva, 1992.
3. Wikler, A., Dynamics of drug dependence: implications of a conditioning theory for research and treatment, *Arch. Gen. Psychiatry*, 28, 611, 1973.
4. Wise, R.A., The neurobiology of craving: implications for the understanding and treatment of addiction, *J. Abnorm. Psychol.*, 97, 118, 1988.
5. Solomon, R.L., The opponent-process theory of acquired motivation: the affective dynamics of addiction, in *Psychopathology: Experimental Models*, Maser, J.D. and Seligman, M.E.P., Eds., W.H. Freeman, San Francisco, 1977, p. 124.
6. Koob, G.F. and Le Moal, M., Drug abuse: hedonic homeostatic dysregulation, *Science*, 278, 52, 1997.
7. Russell, M.A.H., What is dependence?, in *Drugs and Drug Dependence*, Edwards, G., Russell, M.A.H., Hawks, D., and MacCafferty, M., Eds., Lexington Books, Lexington, MA, 1976, p. 182.
8. Himmelsbach, C.K., Clinical studies of drug addiction: physical dependence, withdrawal and recovery, *Arch. Intern. Med.*, 69, 766, 1942.
9. Koob, G.F. and Bloom, F.E., Cellular and molecular mechanisms of drug dependence, *Science*, 242, 715, 1988.
10. Robinson, T.E. and Berridge, K.C., The neural basis of drug craving: an incentive-sensitization theory of addiction, *Brain Res. Rev.*, 18, 247, 1993.
11. Olds, J. and Milner, P., Positive reinforcement produced by electrical stimulation of septal area and other regions of rat brain, *J. Comp. Physiol. Psychol.*, 47, 419, 1954.
12. Stein, L., Chemistry of reward and punishment, in *Psychopharmacology, A Review of Progress (1957–1967)*, Efron, D.H., Ed., U.S. Government Printing Office, Washington, D.C., 1968, p. 105.
13. Nauta, J.H. and Haymaker, W., Hypothalamic nuclei and fiber connections, in *The Hypothalamus*, Haymaker, W., Anderson, E., and Nauta, W.J.H., Eds., Charles C Thomas, Springfield, IL, 1969, p. 136.
14. Liebman, J.L. and Cooper, S.J., *The Neuropharmacological Basis of Reward*, Clarendon Press, Oxford, 1989.
15. Valenstein, E.S. and Campbell, J.F., Medial forebrain bundle-lateral hypothlamic area and reinforcing brain stimulation, *Am. J. Physiol.*, 210, 270, 1966.
16. Koob, G.F., Drugs of abuse: anatomy, pharmacology, and function of reward pathways, *Trends Pharmacol. Sci.*, 13, 177, 1992.
17. Kilty, J.E., Lorang, D., and Amara, S.G., Cloning and expression of a cocaine-sensitive rat dopamine transporter, *Science*, 254, 578, 1991.
18. Blakely, R.D., Berson, H.E., Fremeau, R.T., Jr., Caron, M.G., Peek, M.M., Prince, H.K., and Bradley, C.C., Cloning and expression of a functional serotonin transporter from rat brain, *Nature*, 354, 66, 1991.
19. Giros, B., El Mestikawy, S., Bertrand, L., and Caron, M.G., Cloning and functional characterization of a cocaine-sensitive dopamine transporter, *FEBS Lett.*, 295, 149, 1991.
20. Rudnick, G. and Clark, J., From synapse to vesicle: the reuptake and storage of biogenic amine neurotransmitters, *Biochim. Biophys. Acta*, 1144, 249, 1993.
21. Angrist, B. and Sudilovsky, A., Central nervous system stimulants: historical aspects and clinical effects, in *Handbook of Psychopharmacology*, Vol. 11, Iversen, L.L., Iversen, S.D., and Snyder, S.H., Eds., Plenum Press, New York, 1976, p. 99.

22. Fischman, M.W., Schuster, C.R., and Hatano, Y., A comparison of the subjective and cardiovascular effects of cocaine and lidocaine in humans, *Pharmacol. Biochem. Behav.*, 18, 123, 1983.
23. Groppetti, A., Zambotti, F., Biazzi, A., and Mantegazza, P., Amphetamine and cocaine on amine turnover, in *Frontiers in Catecholamine Research*, Usdin, E. and Snyder, S.H., Eds., Pergamon Press, New York, 1973, p. 917.
24. Spealman, R.D., Goldberg, S.R., Kelleher, R.T., Goldberg, D.M., and Charlton, J.P., Some effects of cocaine and two cocaine analogs on schedule-controlled behavior of squirrel monkeys, *J. Pharmacol. Exp. Ther.*, 202, 500, 1977.
25. Kornetsky, C. and Esposito, R.U., Reward and detection thresholds for brain stimulation: dissociative effects of cocaine, *Brain Res.*, 209, 496, 1981.
26. Kornetsky, C. and Esposito, R.U., Euphorigenic drugs: effects on the reward pathways of the brain, *Fed. Proc.*, 38, 2473, 1979.
27. Mucha, R.F., van der Kooy, D., O'Shaughnessy, M., and Bucenieks, P., Drug reinforcement studied by the use of place conditioning in rat, *Brain Res.*, 243, 91, 1982.
28. Carr, G.D., Fibiger, H.C., and Phillips, A.G., Conditioned place preference as a measure of drug reward, in *The Neuropharmacological Basis of Reward*, Liebman, J.M. and Cooper, S.J., Eds., Oxford University Press, New York, 1989, p. 264.
29. Koob, G.F., Vaccarino, F.J., Amalric, M., and Bloom, F.E., Positive reinforcement properties of drugs: search for neural substrates, in *Brain Reward Systems and Abuse*, Engel, J. and Oreland, L., Eds., Raven Press, New York, 1987, p. 35.
30. Kelly, P.H., Seviour, P.W., and Iversen, S.D., Amphetamine and apomorphine responses in the rat following 6-OHDA lesions of the nucleus accumbens septi and corpus striatum, *Brain Res.*, 94, 507, 1975.
31. Kelly, P.H. and Iversen, S.D., Selective 6OHDA-induced destruction of mesolimbic dopamine neurons: abolition of psychostimulant-induced locomotor activity in rats, *Eur. J. Pharmacol.*, 40, 45, 1976.
32. Pijnenburg, A.J.J., Honig, W.M.M., and Van Rossum, J.M., Inhibition of d-amphetamine-induced locomotor activity by injection of haloperidol into the nucleus accumbens of the rat, *Psychopharmacologia*, 41, 87, 1975.
33. Yokel, R.A. and Wise, R.A., Increased lever pressing for amphetamine after pimozide in rats: Implications for a dopamine theory of reward, *Science*, 187, 547, 1975.
34. Ettenberg, A., Pettit, H.O., Bloom, F.E., and Koob, G.F., Heroin and cocaine intravenous self-administration in rats: mediation by separate neural systems, *Psychopharmacology*, 78, 204, 1982.
35. Phillips, A.G. and Fibiger, H.C., Anatomical and neurochemical substrates of drug reward determined by the conditioned place preference technique, in *Methods of Assessing the Reinforcing Properties of Abused Drugs*, Bozarth, M.A., Ed., Springer-Verlag, New York, 1987, p. 275.
36. Beninger, R.J. and Hahn, B.L., Pimozide blocks establishment but not expression of amphetamine-produced environment-specific conditioning, *Science*, 220, 1304, 1983.
37. Beninger, R.J. and Herz, R.S., Pimozide blocks establishment but not expression of cocaine-produced environment-specific conditioning, *Life Sci.*, 38, 1425, 1986.
38. Morency, M.A. and Beninger, R.J., Dopaminergic substrates of cocaine-induced placed conditioning, *Brain Res.*, 399, 33, 1986.
39. Caine, S.B. and Koob, G.F., Pretreatment with the dopamine agonist 7-OH-DPAT shifts the cocaine self-administration dose-effect function to the left under different schedules in the rat, *Behav. Pharmacol.*, 6, 333, 1995.
40. Bergman, J., Kamien, J.B., and Spealman, R.D., Antagonism of cocaine self-administration by selective dopamine D1 and D2 antagonists, *Behav. Pharmacol.*, 1, 355, 1990.

41. Roberts, D.C.S., Koob, G.F., Klonoff, P., and Fibiger, H.C., Extinction and recovery of cocaine self-administration following 6-hydroxydopamine lesions of the nucleus accumbens, *Pharmacol. Biochem. Behav.*, 12, 781, 1980.

42. Lyness, W.H., Friedle, N.M., and Moore, K.E., Destruction of dopaminergic nerve terminals in nucleus accumbens: effect of d-amphetamine self-administration, *Pharmacol. Biochem. Behav.*, 11, 553, 1979.

43. Caine, S.B. and Koob, G.F., Effects of mesolimbic dopamine depletion on responding maintained by cocaine and food, *J. Exp. Anal. Behav.*, 61, 213, 1994.

44. Koob, G.F., Le, H.T., and Creese, I., The D-1 dopamine receptor antagonist SCH 23390 increases cocaine self-administration in the rat, *Neurosci. Lett.*, 79, 315, 1987.

45. Woolverton, W.L. and Virus, R.M., The effects of a D1 and a D2 dopamine antagonist on behavior maintained by cocaine or food, *Pharmacol. Biochem. Behav.*, 32, 691, 1989.

46. Caine, S.B. and Koob, G.F., Modulation of cocaine self-administration in the rat through D-3 dopamine receptors, *Science*, 260, 1814, 1993.

47. Leone, P. and Di Chiara, G., Blockade of D_1 receptors by SCH 23390 antagonizes morphine- and amphetamine-induced place preference conditioning, *Eur. J. Pharmacol.*, 135, 251, 1987.

48. Beninger, R.J., Hoffman, D.C., and Mazurski, E.J., Receptor subtype-specific dopaminergic agents and conditioned behavior, *Neurosci. Biobehav. Rev.*, 13, 113, 1989.

49. Carelli, R.M., King, V.C., Hampson, R.E., and Deadwyler, S.A., Firing patterns of nucleus accumbens neurons during cocaine self-administration in rats, *Brain Res.*, 626, 14, 1993.

50. Carelli, R.M. and Deadwyler, S.A., A comparison of nucleus accumbens neuronal firing patterns during cocaine self-administration and water reinforcement in rats, *J. Neurosci.*, 14, 7735, 1994.

51. Chang, J.Y., Sawyer, S.F., Lee, R.-S., and Woodward, D.J., Electrophysiological and pharmacological evidence for the role of the nucleus accumbens in cocaine self-administration in freely moving rats, *J. Neurosci.*, 14, 1224, 1994.

52. Peoples, L.L. and West, M.O., Phasic firing of single neurons in the rat nucleus accumbens correlated with the timing of intravenous cocaine self-administration, *J. Neurosci.*, 16, 3459, 1996.

53. Peoples, L.L., Uzwiak, A.J., Gee, F., and West, M.O., Operant behavior during sessions of intravenous cocaine infusion is necessary and sufficient for phasic firing of single nucleus accumbens neurons, *Brain Res.*, 757, 280, 1997.

54. Carelli, R.M. and Deadwyler, S.A., Dual factors controlling activity of nucleus accumbens cell-firing during cocaine self-administration, *Synapse*, 24, 308, 1996.

55. Schuster, C.R. and Thompson, T., Self administration of and behavioral dependence on drugs, *Ann. Rev. Pharmacol. Toxicol.*, 9, 483, 1969.

56. Deneau, G., Yanagita, T., and Seevers, M.H., Self-administration of psychoactive substances by the monkey: a measure of psychological dependence, *Psychopharmacologia*, 16, 30, 1969.

57. Koob, G.F., Pettit, H.O., Ettenberg, A., and Bloom, F.E., Effects of opiate antagonists and their quaternary derivatives on heroin self-administration in the rat, *J. Pharmacol. Exp. Ther.*, 229, 481, 1984.

58. Goldberg, S.R., Woods, J.H., and Schuster, C.R., Nalorphine-induced changes in morphine self-administration in rhesus monkeys, *J. Pharmacol. Exp. Ther.*, 176, 464, 1971.

59. Weeks, J.R. and Collins, R.J., Changes in morphine self-administration in rats induced by prostaglandin E1 and naloxone, *Prostaglandins*, 12, 11, 1976.

60. Vaccarino, F.J., Pettit, H.O., Bloom, F.E., and Koob, G.F., Effects of intracerebroventricular administration of methyl naloxonium chloride on heroin self-administration in the rat, *Pharmacol. Biochem. Behav.*, 23, 495, 1985.

61. Negus, S.S., Weinger, N.B., Henriksen, S.J., and Koob, G.F., Effect of mu, delta, and kappa opioid agonists on heroin-maintained responding in the rat, *NIDA Res. Monogr.*, 119, 410, 1992.

62. Matthes, H.W.D., Maldonado, R., Simonin, F., Valverde, O., Slowe, S., Kitchen, I., Befort, K., Dierich, A., Le Meur, M., Dolle, P., Tzavara, E., Hanoune, J., Roques, B.P., and Kieffer, B.L., Loss of morphine-induced analgesia, reward effect and withdrawal symptoms in mice lacking the mu-opioid-receptor gene, *Nature*, 383, 819, 1996.
63. Britt, M.D. and Wise, R.A., Ventral tegmental site of opiate reward: antagonism by a hydrophilic opiate receptor blocker, *Brain Res.*, 258, 105, 1983.
64. Vaccarino, F.J., Bloom, F.E., and Koob, G.F., Blockade of nucleus accumbens opiate receptors attenuates intravenous heroin reward in the rat, *Psychopharmacology*, 86, 37, 1985.
65. Schroeder, R.L., Weinger, M.B., Vakassian, L., and Koob, G.F., Methylnaloxonium diffuses out of the rat brain more slowly than naloxone after direct intracerebral injection, *Neurosci. Lett.*, 121, 173, 1991.
66. Di Chiara, G. and Imperato, A., Opposite effects of mu and kappa opiate agonists on dopamine release in the nucleus accumbens and in the dorsal caudate of freely moving rats, *J. Pharmacol. Exp. Ther.*, 244, 1067, 1988.
67. Hemby, S.E., Martin, T.J., Co, C., Dworkin, S.I., and Smith, J.E., The effects of intravenous heroin administration on extracellular nucleus accumbens dopamine concentrations as determined by *in vivo* microdialysis, *J. Pharmacol. Exp. Ther.*, 273, 591, 1995.
68. Bozarth, M.A. and Wise, R.A., Intracranial self-administration of morphine into the ventral tegmental area in rats, *Life Sci.*, 28, 551, 1981.
69. Di Chiara, G. and North, R.A., Neurobiology of opiate abuse, *Trends Pharmacol. Sci.*, 13, 185, 1992.
70. Goeders, N.E., Lane, J.D., and Smith, J.E., Self-administration of methionine enkephalin into the nucleus accumbens, *Pharmacol. Biochem. Behav.*, 20, 451, 1984.
71. Ettenberg, A., Pettit, H.O., Bloom, F.E., and Koob, G.F., Heroin and cocaine intravenous self-administration in rats: mediation by separate neural systems, *Psychopharmacology*, 78, 204, 1982.
72. Pettit, H.O., Ettenberg, A., Bloom, F.E., and Koob, G.F., Destruction of dopamine in the nucleus accumbens selectively attenuates cocaine but not heroin self-administration in rats, *Psychopharmacology*, 84, 167, 1984.
73. Stinus, L., Cador, M., and Le Moal, M., Interaction between endogenous opioids and dopamine within the nucleus accumbens, *Ann. N.Y. Acad. Sci.*, 654, 254, 1992.
74. Stinus, L., Nadaud, D., Deminiere, J.M., Jauregui, J., Hand, T.T., and Le Moal, M., Chronic flupentixol treatment potentiates the reinforcing properties of systemic heroin administration, *Biol. Psychiatry*, 26, 363, 1989.
75. Spyraki, C., Fibiger, H.C., and Phillips, A.G., Attenuation of heroin reward in rats by disruption of the mesolimbic dopamine system, *Psychopharmacology*, 79, 278, 1983.
76. Shippenberg, T.S., Herz, A., Spanagel, R., Bals-Kubik, R., and Stein, C., Conditioning of opioid reinforcement: neuroanatomical and neurochemical substrates, *Ann. N.Y. Acad. Sci.*, 654, 347, 1992.
77. Sepinwall, J. and Cook, L., Behavioral pharmacology of anti-anxiety drugs, in *Handbook of Psychopharmacology*, Vol. 13, Iversen, L.L., Iversen, S.D., and Snyder, S.H., Eds., Plenum Press, New York, 1978, p. 345.
78. Richards, G., Schoch, P., and Haefely, W., Benzodiazepine receptors: new vistas, *Semin. Neurosci.*, 3, 191, 1991.
79. Suzdak, P.D., Glowa, J.R., Crawley, J.N., Schwartz, R.D., Skolnick, P., and Paul, S.M., A selective imidazobenzodiazepine antagonist of ethanol in the rat, *Science*, 234, 1243, 1986.
80. Mihic, S.J., Ye, Q., Wick, M.J., Koltchine, V.V., Krasowski, M.D., Finn, S.E., Mascia, M.P., Valenzuela, C.F., Hanson, K.K., Greenblatt, E.P., Harris, R.A., and Harrison, N.L., Sites of alcohol and volatile anaesthetic action on $GABA_A$ and glycine receptors, *Nature*, 389, 385, 1997.
81. Koob, G.F. and Britton, K.T., Neurobiological substrates for the anti-anxiety effects of ethanol, in *The Pharmacology of Alcohol and Alcohol Dependence*, Begleiter, H. and Kissin, B., Eds., Oxford University Press, New York, 1996, p. 477.

82. Frye, G.D. and Breese, G.R., GABAergic modulation of ethanol-induced motor impairment, *J. Pharmacol. Exp. Ther.*, 223, 750, 1982.
83. Liljequist, S. and Engel, J., Effects of GABAergic agonists and antagonists on various ethanol-induced behavioral changes, *Psychopharmacology*, 78, 71, 1982.
84. Samson, H.H., Tolliver, G.A., Pfeffer, A.O., Sadeghi, K.G., and Mills, F.G., Oral ethanol reinforcement in the rat: effect of the partial inverse benzodiazepine agonist R015-4513, *Pharmacol. Biochem. Behav.*, 27, 517, 1987.
85. Rassnick, S., D'Amico, E., Riley, E., Pulvirenti, L., Zieglgansberger, W., and Koob, G.F., GABA and nucleus accumbens glutamate neurotransmission modulate ethanol self-administration in rats, *Ann. N.Y. Acad. Sci.*, 654, 502, 1992.
86. June, H.L., Colker, R.E., Domangue, K.R., Perry, L.E., Hicks, L.H., June, P.L., and Lewis, M.J., Ethanol self-administration in deprived rats: effects of R015-4513 alone, and in combination with flumazenil (R015-1788), *Alcohol. Clin. Exp. Res.*, 16, 11, 1992.
87. Institute of Medicine, *Pathways of Addiction*, National Academy Press, Washington, D.C., 1996, p. 1.
88. Hoffman, P.L., Rabe, C., Moses, F., and Tabakoff, B., *N*-methyl-D-aspartate receptors and ethanol: inhibition of calcium flux and cyclic GMP production, *J. Neurochem.*, 52, 1937, 1989.
89. Lovinger, D.M., White, G., and Weight, F.F., Ethanol inhibits NMDA-activated ion current in hippocampal neurons, *Science*, 243, 1721, 1989.
90. Tsai, G., Gastfriend, D.R., and Coyle, J.T., The glutametergic basis of human alcoholism, *Am. J. Psychiatry*, 152, 332, 1995.
91. Fitzgerald, L.W. and Nestler, E.J., Molecular and cellular adaptations in signal transduction pathways following ethanol exposure, *Clin. Neurosci.*, 3, 165, 1995.
92. Pfeffer, A.O. and Samson, H.H., Haloperidol and apomorphine effects on ethanol reinforcement in free-feeding rats, *Pharmacol. Biochem. Behav.*, 29, 343, 1988.
93. Weiss, F., Hurd, Y.L., Ungerstedt, U., Markou, A., Plotsky, P.M., and Koob, G.F., Neurochemical correlates of cocaine and ethanol self-administration, *Ann. N.Y. Acad. Sci.*, 654, 220, 1992.
94. Rassnick, S., Stinus, L., and Koob, G.F., The effects of 6-hydroxydopamine lesions of the nucleus accumbens and the mesolimbic dopamine system on oral self-administration of ethanol in the rat, *Brain Res.*, 623, 16, 1993.
95. Murai, T., Koshikawa, N., Kanayama, T., Takada, K., Tomiyama, K., and Kobayashi, M., Opposite effects of midazolam and beta-carboline-3-carboxylate ethyl ester on the release of dopamine from rat nucleus accumbens measured by *in vivo* microdialysis, *Eur. J. Pharmacol.*, 261, 65, 1994.
96. Engel, J.A., Enerback, C., Fahlke, C., Hulthe, P., Hard, E., Johannessen, K., Svensson, L., and Soderpalm, B., Serotonergic and dopaminergic involvement in ethanol intake, in *Novel Pharmacological Interventions for Alcoholism*, Naranjo, C.A. and Sellers, E.M., Eds., Springer, New York, 1992, p. 68.
97. Sellers, E.M., Higgins, G.A., and Sobell, M.B., 5-HT and alcohol abuse, *Trends Pharmacol. Sci.*, 13, 69, 1992.
98. Fadda, F., Garau, B., Marchei, F., Colombo, G., and Gessa, G.L., MDL 72222, a selective 5-HT3 receptor antagonist, suppresses voluntary ethanol consumption in alcohol-preferring rats, *Alcohol Alcohol.*, 26, 107, 1991.
99. Hodge, C.W., Samson, H.H., Lewis, R.S., and Erickson, H.L., Specific decreases in ethanol- but not water-reinforced responding produced by the 5-HT3 antagonist ICS 205-930, *Alcohol*, 10, 191, 1993.
100. Richter, R.M. and Weiss, F., *In vivo* CRF release in rat amygdala is increased during withdrawal after cocaine self-administration with unlimited access, 1997 (unpublished).
101. Naranjo, C., Kadlec, K.E., Sanhueza, P., Woodley-Remus, D., and Sellers, E.M., Fluoxetine differentially alters alcohol intake and other consummatory behaviors in problem drinkers, *Clin. Pharmacol. Ther.*, 47, 490, 1990.

102. Cornelius, J.R., Salloum, I.M., Ehler, J.G., Jarrett, P.J., Cornelius, M.D., Perel, J.M., Thase, M.E., and Black, A., Fluoxetine in depressed alcoholics: a double-blind, placebo-controlled trial, *Arch. Gen. Psychiatry*, 54, 700, 1997.

103. Kranzler, H.R., Burleson, J.A., Korner, P., Del Boca, F.K., Bohn, M.J., Brown, J., and Liebowitz, N., Placebo-controlled trial of fluoxetine as an adjunct to relapse prevention in alcoholics, *Am. J. Psychiatry*, 152, 391, 1995.

104. Janiri, L., Gobbi, G., Mannelli, P., Pozzi, G., Serretti, A., and Tempesta, E., Effects of fluoxetine at antidepressant doses on short-term outcome of detoxified alcoholics, *Int. Clin. Psychopharmacol.*, 11, 109, 1996.

105. Hubbell, C.L., Marglin, S.H., Spitalnic, S.J., Abelson, M.L., Wild, K.D., and Reid, L.D., Opioidergic, serotonergic, and dopaminergic manipulations and rats' intake of a sweetened alcoholic beverage, *Alcohol*, 8, 355, 1991.

106. Hyytia, P., Involvement of μ-opioid receptors in alcohol drinking by alcohol-preferring AA rats, *Pharmacol. Biochem. Behav.*, 45, 697, 1993.

107. O'Malley, S.S., Jaffe, A.J., Chang, G., Schottenfeld, R.S., Meyer, R.E., and Rounsaville, B., Naltrexone and coping skills therapy for alcohol dependence: a controlled study, *Arch. Gen. Psychiatry*, 49, 881, 1992.

108. Volpicelli, J.R., Alterman, A.I., Hayashida, M., and O'Brien, C.P., Naltrexone in the treatment of alcohol dependence, *Arch. Gen. Psychiatry*, 49, 876, 1992.

109. Malin, D.H., Lake, J.R., Carter, V.A., Cunningham, J.S., and Wilson, O.B., Naloxone precipitates nicotine abstinence syndrome in the rat, *Psychopharmacology*, 112, 339, 1993.

110. Malin, D.H., Lake, J.R., Carter, V.A., Cunningham, J.S., Hebert, K.M., Conrad, D.L., and Wilson, O.B., The nicotinic antagonist mecamylamine precipitates nicotine abstinence syndrome in the rat, *Psychopharmacology*, 115, 180, 1994.

111. Corrigall, W.A., Franklin, K.B.J., Coen, K.M., and Clarke, P.B.S., The mesolimbic dopaminergic system is implicated in the reinforcing effects of nicotine, *Psychopharmacology*, 107, 285, 1992.

112. Anthony, J.C., Warner, L.A., and Kessler, R.C., Comparative epidemiology of dependence on tobacco, alcohol, controlled substances, and inhalants: basic findings from the National Comorbidity Survey, *Exp. Clin. Psychopharmacol.*, 2, 244, 1994.

113. Herkenham, M., Lynn, A.B., Little, M.D., Johnson, M.R., Melvin, L.S., de Costa, B.R., and Rice, K.C., Cannabinoid receptor localization in brain, *Proc. Natl. Acad. Sci. USA*, 87, 1932, 1990.

114. Lepore, M., Liu, X., Savage, V., Matalon, D., and Gardner, E.L., Genetic differences in delta 9-tetrahydrocannabinol-induced facilitation of brain stimulation reward as measured by a rate-frequency curve-shift electrical brain stimulation paradigm in three different rat strains, *Life Sci.*, 58, PL365, 1996.

115. Gardner, E.L., Paredes, W., Smith, D., Donner, A., Milling, C., Cohen, D., and Morrison, D., Facilitation of brain stimulation reward by delta-9-tetrahydrocannabinol, *Psychopharmacology*, 96, 142, 1988.

116. Lepore, M., Vorel, S.R., Lowinson, J., and Gardner, E.L., Conditioned place preference induced by delta 9-tetrahydrocannabinol: comparison with cocaine, morphine, and food reward, *Life Sci.*, 56, 2073, 1995.

117. Martellotta, M.C., Cossu, G., Fattore, L., Gessa, G.L., and Fratta, W., Self-administration of the cannbinoid receptor agonist WIN 55,212-2 in drug-naive mice, *Neuroscience*, 85, 327, 1998.

118. Chen, J.P., Paredes, W., Lowinson, J.H., and Gardner, E.L., Strain-specific facilitation of dopamine efflux by delta 9-tetrahydrocannabinol in the nucleus accumbens of rat: an *in vivo* microdialysis study, *Neurosci. Lett.*, 129, 136, 1991.

119. Tanda, G., Pontieri, F.E., and Di Chiara, G., Cannabinoid and heroin activation of mesolimbic dopamine transmission by a common mu1 opioid receptor mechanism, *Science*, 276, 2048, 1997.

120. Gawin, F.H. and Kleber, H.D., Abstinence symptomatology and psychiatric diagnosis in cocaine abusers: clinical observations, *Arch. Gen. Psychiatry*, 43, 107, 1986.
121. Weddington, Jr., W.W., Brown, B.S., Haertzen, C.A., Hess, J.M., Mahaffey, J.R., Kolar, A.F., and Jaffe, J.H., Comparison of amantadine and desipramine combined with psychotherapy for treatment of cocaine dependence, *Am. J. Drug Alcohol Abuse*, 17, 137, 1991.
122. Markou, A. and Koob, G.F., Postcocaine anhedonia: an animal model of cocaine withdrawal, *Neuropsychopharmacology*, 4, 17, 1991.
123. Schulteis, G., Markou, A., Gold, L.H., Stinus, L., and Koob, G.F., Relative sensitivity to naloxone of multiple indices of opiate withdrawal: a quantitative dose-response analysis, *J. Pharmacol. Exp. Ther.*, 271, 1391, 1994.
124. Leith, N.J. and Barrett, R.J., Amphetamine and the reward system: evidence for tolerance and post-drug depression, *Psychopharmacologia*, 46, 19, 1976.
125. Parsons, L.H., Koob, G.F., and Weiss, F., Serotonin dysfunction in the nucleus accumbens of rats during withdrawal after unlimited access to intravenous cocaine, *J. Pharmacol. Exp. Ther.*, 274, 1182, 1995.
126. Markou, A. and Koob, G.F., Construct validity of a self-stimulation threshold paradigm: effects of reward and performance manipulations, *Physiol. Behav.*, 51, 111, 1992.
127. Legault, M. and Wise, R.A., Effects of withdrawal from nicotine on intracranial self-stimulation, *Neurosci. Abstr.*, 20, 1032, 1994.
128. Koob, G.F., Drug addiction: the yin and yang of hedonic homeostasis, *Neuron*, 16, 893, 1996.
129. Solomon, R.L. and Corbit, J.D., An opponent-process theory of motivation. 1. Temporal dynamics of affect, *Psychol. Rev.*, 81, 119, 1974.
130. Siegel, S., Evidence from rats that morphine tolerance is a learned response, *J. Comp. Physiol. Psychol.*, 89, 498, 1975.
131. Poulos, C.X. and Cappell, H., Homeostatic theory of drug tolerance: a general model of physiological adaptation, *Psychol. Rev.*, 98, 390, 1991.
132. Weiss, F., Markou, A., Lorang, M.T., and Koob, G.F., Basal extracellular dopamine levels in the nucleus accumbens are decreased during cocaine withdrawal after unlimited-access self-administration, *Brain Res.*, 593, 314, 1992.
133. Stinus, L., Le Moal, M., and Koob, G.F., Nucleus accumbens and amygdala are possible substrates for the aversive stimulus effects of opiate withdrawal, *Neuroscience*, 37, 767, 1990.
134. Roberts, A.J., Cole, M., and Koob, G.F., Intra-amygdala muscimol decreases operant ethanol self-administration in dependent rats, *Alcohol. Clin. Exp. Res.*, 20, 1289, 1996.
135. Weiss, F., Parsons, L.H., Schulteis, G., Hyytia, P., Lorang, M.T., Bloom, F.E., and Koob, G.F., Ethanol self-administration restores withdrawal-associated deficiencies in accumbal dopamine and 5-hydroxytryptamine release in dependent rats, *J. Neurosci.*, 16, 3474, 1996.
136. Collins, A.C., Bhat, R.V., Pauly, J.R., and Marks, M.J., Modulation of nicotine receptors by chronic exposure to nicotinic agonists and antagonists, in *The Biology of Nicotine Dependence*, Bock, G. and Marsh, J., Eds., John Wiley & Sons, New York, 1990, p. 87.
137. Dani, J.A. and Heinemann, S., Molecular and cellular aspects of nicotine abuse, *Neuron*, 16, 905, 1996.
138. Kreek, M.J., Multiple drug abuse patterns and medical consequences, in *Psychopharmacology: The Third Generation of Progress*, Meltzer, H.Y., Ed., Raven Press, New York, 1987, p. 1597.
139. Kreek, M.J., Ragunath, J., Plevy, S., Hamer, D., Schneider, B., and Hartman, N., ACTH, cortisol and beta-endorphin response to metyrapone testing during chronic methadone maintenance treatment in humans, *Neuropeptides*, 5, 277, 1984.

140. Piazza, P.V. and Le Moal, M., Glucocorticoids as a biological substrate of reward: physiological and pathophysiological implications, *Brain Res. Rev.*, 25, 359, 1997.
141. Piazza, P.V. and Le Moal, M.L., Pathophysiological basis of vulnerability to drug abuse: role of an interaction between stress, glucocorticoids, and dopaminergic neurons, *Ann. Rev. Pharmacol. Toxicol.*, 36, 359, 1996.
142. Koob, G.F., Heinrichs, S.C., Menzaghi, F., Merlo-Pich, E., and Britton, K.T., Corticotropin releasing factor, stress and behavior, *Semin. Neurosci.*, 6, 221, 1994.
143. Heinrichs, S.C., Menzaghi, F., Schulteis, G., Koob, G.F., and Stinus, L., Suppression of corticotropin-releasing factor in the amygdala attenuates aversive consequences of morphine withdrawal, *Behav. Pharmacol.*, 6, 74, 1995.
144. Rodriguez de Fonseca, F., Carrera, M.R.A., Navarro, M., Koob, G.F., and Weiss, F., Activation of corticotropin-releasing factor in the limbic system during cannabinoid withdrawal, *Science*, 276, 2050, 1997.
145. Rassnick, S., Heinrichs, S.C., Britton, K.T., and Koob, G.F., Microinjection of a corticotropin-releasing factor antagonist into the central nucleus of the amygdala reverses anxiogenic-like effects of ethanol withdrawal, *Brain Res.*, 605, 25, 1993.
146. Sarnyai, Z., Biro, E., Gardi, J., Vecsernyes, M., Julesz, J., and Telegdy, G., Brain corticotropin-releasing factor mediates "anxiety-like" behavior induced by cocaine withdrawal in rats, *Brain Res.*, 675, 89, 1995.
147. Nestler, E.J., Hope, B.T., and Widnell, K.L., Drug addiction: a model for the molecular basis of neural plasticity, *Neuron*, 11, 995, 1993.
148. Self, D.W. and Nestler, E.J., Molecular mechanisms of drug reinforcement and addiction, *Ann. Rev. Neurosci.*, 18, 463, 1995.
149. Nestler, E.J., Molecular neurobiology of drug addiction, *Neuropsychopharmacology*, 11, 77, 1994.
150. Henry, D.J. and White, F.J., Repeated cocaine administration causes persistent enhancement of D1 dopamine receptor sensitivity within the rat nucleus accumbens, *J. Pharmacol. Exp. Ther.*, 258, 882, 1991.
151. Li, D.H., Depoortere, R.Y., and Emmett-Oglesby, M.W., Tolerance to the reinforcing effects of cocaine in a progressive ratio paradigm, *Psychopharmacology*, 116, 326, 1994.
152. Young, A.M. and Goudie, A.J., Adaptive processes regulating tolerance to the behavioral effects of drugs, in *Psychopharmacology: The Fourth Generation of Progress*, Bloom, F.E. and Kupfer, D.J., Eds., Raven Press, New York, 1995, p. 733.
153. Tabakoff, B. and Hoffman, P.L., Alcohol: neurobiology, in *Substance Abuse. A Comprehensive Textbook*, Lowinson, J.H., Ruiz, P. and Millman, R.B., Eds., Williams & Wilkins, Baltimore, MD, 1992, p. 152.
154. Trujillo, K.A. and Akil, H., Inhibition of morphine tolerance and dependence by the NMDA receptor antagonist MK-801, *Science*, 251, 85, 1991.
155. Nestler, E.J., Under siege: the brain on opiates, *Neuron*, 16, 897, 1996.
156. Nestler, E.J., Molecular mechanisms of drug addiction, *J. Neurosci.*, 12, 2439, 1992.
157. Stewart, J. and Badiani, A., Tolerance and sensitization to the behavioral effects of drugs, *Behav. Pharmacol.*, 4, 289, 1993.
158. Bartoletti, M., Gaiardi, M., Gubellini, C., Bacchi, A., and Babbini, M., Previous treatment with morphine and sensitization to the excitatory actions of opiates: dose-effect relationship, *Neuropharmacology*, 26, 115, 1987.
159. Phillips, T.J., Huson, M., Gwiazdon, C., Burkhart-Kasch, S., and Shen, E.H., Effects of acute and repeated ethanol exposures on the locomotor activity of BXD recombinant inbred mice, *Alcohol. Clin. Exp. Res.*, 19, 269, 1995.

160. Horger, B.A., Giles, M.K., and Schenk, S., Pre-exposure to amphetamine and nicotine predisposes rats to self-administer a low-dose of cocaine, *Psychopharmacology*, 107, 271, 1992.

161. Shippenberg, T.S. and Heidbreder, C., Sensitization to the conditioned rewarding effects of cocaine: pharmacological and temporal characteristics, *J. Pharmacol. Exp. Ther.*, 273, 808, 1995.

162. Wise, R.A. and Leeb, K., Psychomotor-stimulant sensitization: a unitary phenomenon?, *Behav. Pharmacol.*, 4, 339, 1993.

163. Kalivas, P.W. and Stewart, J., Dopamine transmission in the initiation and expression of drug- and stress-induced sensitization of motor activity, *Brain Res. Rev.*, 16, 223, 1991.

164. White, F.J. and Wolf, M.E., Psychomotor stimulants, in *The Biological Bases of Drug Tolerance and Dependence*, Pratt, J.A., Ed., Academic Press, London, 1991, p. 153.

165. Koob, G.F. and Cador, M., Psychomotor stimulant sensitization: the corticotropin-releasing factor-steroid connection, *Behav. Pharmacol.*, 4, 351, 1993.

166. Karler, R., Calder, L.D., Chaudhry, I.A., and Turkanis, S.A., Blockade of "reverse tolerance" to cocaine and amphetamine by MK-801, *Life Sci.*, 45, 599, 1989.

167. Hope, B.T., Nye, H.E., Kelz, M.B., Self, D.W., Iadarola, M.J., Nakabeppu, Y., Duman, R.S., and Nestler, E.J., Induction of a long-lasting AP-1 complex composed of altered Fos-like proteins in brain by chronic cocaine and other chronic treatments, *Neuron*, 13, 1235, 1994.

168. Hyman, S.E., Addiction to cocaine and amphetamine, *Neuron*, 16, 901, 1996.

169. Widnell, K., Self, D.W., Lane, S.B., Russell, D.S., Vaidya, V., Miserendino, M.J.D., Rubin, C.S., Duman, R.S., and Nestler, E.J., Regulation of CREB expression: *in vivo* evidence for a functional role in morphine action in the nucleus accumbens, *J. Pharmacol. Exp. Ther.*, 276, 306, 1996.

170. Koob, G.F., Animal models of drug addiction, in *Psychopharmacology: The Fourth Generation of Progress*, Bloom, F.E. and Kupfer, D.J., Eds., Raven Press, New York, 1995, p. 759.

171. deWit, H. and Stewart, J., Reinstatement of cocaine-reinforced responding in the rat, *Psychopharmacology*, 75, 134, 1981.

172. Stewart, J. and deWit, H., Reinstatement of drug-taking behavior as a method of assessing incentive motivational properties of drugs, in *Methods of Assessing the Reinforcing Properties of Abused Drugs*, Bozarth, M.A., Ed., Springer-Verlag, New York, 1987, p. 211.

173. O'Brien, C.P., Eckardt, M.J., and Linnoila, V.M.I., Pharmacotherapy of alcoholism, in *Psychopharmacology: The Fourth Generation of Progress*, Bloom, F.E. and Kupfer, D.J., Eds., Raven Press, New York, 1995, p. 1745.

174. Heyser, C.J., Schulteis, G., Durbin, P., and Koob, G.F., Chronic acamprosate eliminates the alcohol deprivation effect while having limited effects on baseline responding for ethanol in rats, *Neuropsychopharmacology*, 18, 125, 1998.

175. Volpicelli, J.R., Davis, M.A., and Olgin, J.E., Naltrexone blocks the post-shock increase of ethanol consumption, *Life Sci.*, 38, 841, 1986.

176. Self, D.W., Barnhart, W.J., Lehman, D.A., and Nestler, E.J., Opposite modulation of cocaine-seeking behavior by D1- and D2-like dopamine receptor agonists, *Science*, 271, 1586, 1996.

177. Alheid, G.F. and Heimer, L., New perspectives in basal forebrain organization of special relevance for neuropsychiatric disorders: the striatopallidal, amygdaloid, and corticopetal components of substantia innominata, *Neuroscience*, 27, 1, 1988.

178. Johnston, J.B., Further contributions to the study of the evolution of the forebrain, *J. Comp. Neurol.*, 35, 337, 1923.

179. Heimer, L. and Alheid, G., Piecing together the puzzle of basal forebrain anatomy, in *The Basal Forebrain: Anatomy to Function*, Napier, T.C., Kalivas, P.W. and Hanin, I., Eds., Plenum Press, New York, 1991, p. 1.

180. Heimer, L., Zahm, D.S., Churchill, L., Kalivas, P.W., and Wohltmann, C., Specificity in the projection patterns of accumbal core and shell in the rat, *Neuroscience*, 41, 89, 1991.

181. Caine, S.B., Heinrichs, S.C., Coffin, V.L., and Koob, G.F., Effects of the dopamine D-1 antagonist SCH 23390 microinjected into the accumbens, amygdala or striatum on cocaine self-administration in the rat, *Brain Res.*, 692, 47, 1995.

182. Epping-Jordan, M.P., Markou, A., and Koob, G.F., The dopamine D-1 receptor antagonist SCH 23390 injected into the dorsolateral bed nucleus of the stria terminalis decreased cocaine reinforcement in the rat, *Brain Res.*, 784, 105, 1998.

183. Pontieri, F.E., Tanda, G., and Di Chiara, G., Intravenous cocaine, morphine, and amphetamine preferentially increase extracellular dopamine in the "shell" as compared with the "core" of the rat nucleus accumbens, *Proc. Natl. Acad. Sci. USA*, 92, 12304, 1995.

184. Pontieri, F.E., Tanda, G., Orzi, F., and Di Chiara, G., Effects of nicotine on the nucleus accumbens and similarity to those of addictive drugs, *Nature*, 382, 255, 1996.

185. Hyytia, P. and Koob, G.F., $GABA_A$ receptor antagonism in the extended amygdala decreases ethanol self-administration in rats, *Eur. J. Pharmacol.*, 283, 151, 1995.

186. Heyser, C.J., Roberts, A.J., Schulteis, G., Hyytia, P., and Koob, G.F., Central administration of an opiate antagonist decreases oral ethanol self-administration in rats, *Neurosci. Abstr.*, 21, 1698, 1995.

187. Merlo-Pich, E., Lorang, M., Yeganeh, M., Rodriguez de Fonseca, F., Raber, J., Koob, G.F., and Weiss, F., Increase of extracellular corticotropin-releasing factor-like immunoreactivity levels in the amygdala of awake rats during restraint stress and ethanol withdrawal as measured by microdialysis, *J. Neurosci.*, 15, 5439, 1995.

188. Koob, G.F., Wall, T.L., and Bloom, F.E., Nucleus accumbens as a substrate for the aversive stimulus effects of opiate withdrawal, *Psychopharmacology*, 98, 530, 1989.

189. Davis, M., Neurobiology of fear responses: the role of the amygdala, *J. Neuropsychiatry Clin. Neurosci.*, 9, 382, 1997.

190. Levin, E.G. and Santell, L., Association of a plasminogen activator inhibitor (PAI-1) with the growth substratum and membrane of human endothelial cells, *J. Cell. Biol.*, 105, 2543, 1987.

191. O'Brien, C.P., Experimental analysis of conditioning factors in human narcotic addiction, *Pharmacol. Rev.*, 27, 533, 1975.

192. Schulteis, G., Markou, A., Cole, M., and Koob, G.F., Decreased brain reward produced by ethanol withdrawal, *Proc. Natl. Acad. Sci. USA*, 92, 5880, 1995.

Chapter 23

Cholinergic Enhancement of REM Sleep from Sites in the Pons and Amygdala

José M. Calvo and Karina Simón-Arceo

Contents

23.1 Cholinergic Regulation of REM Sleep 392
23.2 Cholinergic Evocation of REM Sleep from the Lateral Pons 392
23.3 Anatomical and Physiological Relationships Between the Pons and Amygdala 393
23.4 Cholinergic Modulation of REM Sleep from the Amygdala 396
23.5 Possible Mechanisms Underlying Long-Term REM Sleep Enhancement 398
23.6 Converging Evidence: Cholinergic Modulation of REM Sleep from Lateral Pons and Amygdala 400
References 401

Rapid-eye-movement (REM) sleep is a distinctive behavioral state that alternates with episodes of slow-wave sleep (SWS) and always is preceded and accompanied by the occurrence of ponto-geniculo-occipital (PGO) waves. A widely distributed pontine neuronal generator network has been postulated to generate REM sleep. The paramedian zone of the anterodorsal pontine tegmentum and the pontine tegmental nuclei which play a major role in PGO wave generation have been considered as major components of the neuronal network for REM sleep induction. Because PGO wave activity always precedes REM onset and PGO wave frequency increases during REM sleep, several investigators have proposed that PGO wave mechanisms constitute a key component for the triggering and maintenance of REM sleep (see Reference 1 for a review). PGO waves propagate from the pons to the visual, oculomotor, and auditory systems, and PGO wave-related potentials and neuronal firing changes can be recorded in many forebrain structures (see Reference 1 for a review). There also is a correlation between PGO waves and various peripheral phasic events such as the rapid eye movements (REMs)[2,3] and transient modifications of heart rate and respiration[4,5] (see Reference 6 for a review). Evidence also exists suggesting the participation of PGO waves in the hyperpolarization of motoneurones that modulate the muscle atonia of REM sleep.[7] Besides the data obtained from animal studies, the existence of PGO-like waves in man has also been documented.[8] Thus, these facts are consistent with the view that PGO waves are one of the most important phasic phenomenon of the REM sleep.

23.1 Cholinergic Regulation of REM Sleep

Pharmacological and physiological studies strongly suggest that serotonin (5-HT), norepinephrine (NE), acetylcholine, vasoactive intestinal polypeptide (VIP), and the corticotropin-like intermediate lobe peptide (CLIP) play a major role in REM sleep onset and maintenance. While 5-HT and NE exert an inhibitory influence over REM sleep and PGO waves,[9–11] VIP and CLIP promote the onset of REM sleep,[12,13] and acetylcholine also induces the REM sleep onset.[14,15]

The hypothesis that acetylcholine plays a major role in REM sleep induction was first proposed by Hernández-Peón[16] and Jouvet.[17] More recently, the search of specific brain sites for cholinergically eliciting REM sleep has been the aim of several studies, which have strongly supported the role of acetylcholine in REM sleep induction. Cholinergic microstimulation has been used to activate specific brainstem neuronal populations. This approach permits the induction of long-lasting REM sleep episodes at short latency, by microinjecting cholinergic drugs into a restricted paramedian zone of the anterodorsal pontine tegmentum.[18–21] The cholinergic drugs carbachol and neostigmine enhance REM sleep, and muscarinic receptor blockers[22–24] antagonize cholinergic REM sleep enhancement. Additionally, the fact that the acetylcholinesterase inhibitor neostigmine is as potent as carbachol for enhancing REM sleep[15,18] further shows that endogenous acetylcholine release is crucial in physiological REM sleep induction. This idea is supported by the increased acetylcholine release in the pontine tegmentum detected during naturally occurring REM sleep.[25,26]

Because the paramedian reticular site for the cholinergic induction of REM sleep is lacking cholinergic neurons,[27] it is likely that the sources of cholinergic input to this site during the physiological onset of REM sleep are the dorsolateral tegmentum (LDT) or the pedunculopontine tegmentum (PPT) cholinergic nuclei. Many neurons in these nuclei show specific increases in firing during REM sleep;[28-31] in addition, LDT/PPT neurons project to the most effective pontine reticular sites upon REM sleep induction by carbachol.[32] Electrical stimulation of LDT and PPT nuclei induces the acetylcholine release in the medial pontine reticular formation.[33] Moreover, the LDT and PPT nuclei, have been identified as part of a neuronal network where PGO waves are generated.[31,34–37] The fact that the lesion of the caudo-lateral pontine region suppresses PGO wave occurrence in the ipsilateral lateral geniculate body (LGB) also supports the role of lateral pontine regions in PGO wave generation.[37]

23.2 Cholinergic Evocation of REM Sleep from the Lateral Pons

The lateral pontine region is of particular interest not only because of its role in PGO wave generation, but also because of its cholinergic neurons that project to the cholinoceptive reticular formation zone,[38] where REM sleep can be cholinergically induced. Thus, the lateral pontine region mediates two major phenomena for the onset of REM sleep, namely, the PGO waves that always precede and accompany this sleep stage (see Reference 1 for a review) and the release of acetylcholine into the cholinoceptive reticular zone for REM sleep induction (see Reference 24 for review). Recent studies have shown that blocking muscarinic autoreceptors, presumed to reside on cholinergic LDT or PPT terminals in the medial reticular formation, produce an increased release of acetylcholine in this region.[39]

Physiological and pharmacological studies have shown that the lateral pontine region ($P = 4.0$, $L = 4.5$, $H = -2.5$) participates in the long-term regulation of PGO waves and of REM sleep. We found that a single unilateral carbachol microinjection ($4\ \mu g/0.25\ \mu l$ saline) in close proximity to the cholinergic neurons of the lateral pons induces the prompt occurrence of state-independent PGO waves recorded in the LGB. These cholinergically induced PGO waves appear as clusters in the ipsilateral LGB and persist for 3 days. On subsequent days, they are confined to the PGO-related states, namely, the episodes of SWS with PGO waves (*sommeil phasique a ondes lentes*, SPHOL[40]) and REM sleep. Quantitatively, the total PGO wave amount shows a significant and long-term enhancement compared with control. Carbachol injection results in a significant increase of PGO waves, reaching a mean number of PGO waves tenfold higher than control. PGO counts progressively decline compared to the day of carbachol administration, but even 6 days after there are still twice as many waves.[41,42]

While PGO waves occur continuously on the first day of carbachol administration, no changes are observed in the amount of REM sleep. Surprisingly, 24 hr after carbachol microinjection, REM% rises to more than three times that of baseline levels. Although REM% declined progressively on the following 5 days, it remained higher than control even on day 6 after a single, unilateral microinjection. It is worth noting that this prolonged enhancement of REM% is due to a significant increase in the number of episodes without any significant change in the mean duration of REM sleep episodes.[43] This fact strongly suggests that lateral pontine regions participate primarily in the onset mechanisms rather than the maintenance mechanisms of REM sleep.

The percentage of time spent in SPHOL episodes (which are transitional from SWS to REM sleep) is also significantly increased by a single carbachol microinjection. SPHOL% also shows a prolonged increase peaking at over sixfold on day 4 after carbachol administration. A decreasing trend is observed afterwards, but the SPHOL% is still over four times that of baseline control 6 days after carbachol. In contrast with the REM sleep increase, the enhancement of SPHOL episodes is due to a significant increase in both mean duration and number of episodes.[43] The latter indicates that cholinergic activation of lateral pontine regions in cat has a robust effect on the transitional stages to REM sleep and on the possibility that lateral pontine regions contribute to the onset of REM sleep.

As it will be considered later in detail, the prolonged enhancement of PGO waves and REM sleep cannot be ascribed to a permanency of carbachol within nervous tissue. It is more likely that carbachol activation of M2 muscarinic receptors metabolically stimulates cholinergic neurons, via second messengers, resulting in an enhanced acetylcholine synthesis and release. The possibility that the prolonged enhancement of PGO waves and REM sleep are mediated by the activation of M2 muscarinic receptors is supported by studies in which the microinjection of carbachol in the lateral pontine region is preceded by a microinjection of an M2 receptor blocker (methoctramine), and no changes in PGO waves or in REM sleep were detected.[44] Taken together, these findings strengthen the participation of lateral pontine regions in PGO wave generation and the onset of REM sleep and has prompted us to explore further the participation in REM sleep induction of forebrain structures anatomically and physiologically related to the lateral pontine region.

23.3 Anatomical and Physiological Relationships Between the Pons and Amygdala

Several anatomical studies have demonstrated reciprocal projections between the central amygdaloid nucleus (CN) and pontine nuclei involved in PGO wave and REM sleep generation, such as the lateral pontine region, LDT, PPT, the brachium conjuntivum (BC) area (referred to as "X" area[45]), the locus coeruleus, and the gigantocellular tegmental field[46–49] (see Reference 50 for a review). Furthermore, the existence of cholinergic neurons[51] and cholinergic projections from the LDT nucleus to the CN have been well documented.[52,53] It is of interest that neurons containing neuropeptides related with sleep and REM sleep induction have also been found in the CN. Cells in the CN contain cholecystokinin, somatostatin, and VIP,[54,55] and some of these neurons project to the lateral parabrachial region.[51,56–58]

In humans, visual and auditory hallucinations comprised of reminiscent phenomena and emotional and autonomic changes take place concomitantly with REM sleep phasic events.[59-63] In cat, elaborate behavior (orientation, predatory and aggressive attack, rage, and flight) during REM sleep without atonia has also been demonstrated.[64] Although an accurate description of this REM sleep phenomenology is now available, its brain neurobiological integration has been little explored. There is strong evidence that functions related to emotional and autonomic changes are integrated in the limbic system. The electrical stimulation in these structures, including the temporal lobe amygdala (AMG), elicits emotional and autonomic responses in cats[65-67] and monkeys.[68] In man, elaborate hallucinatory and reminiscent phenomena and the sensation of dreaming are also produced.[69-73] Thus, the limbic structures may constitute an important component of the forebrain neuronal network of REM sleep.

The AMG is a key component of the limbic system. It has extensive anatomical connections with the neocortex and with the visceral brainstem that may provide a link between them and participate in emotional integration. Accordingly, dramatic effects on emotional behavior have been observed following AMG lesions or stimulation in both humans and experimental animals.[74-76] The AMG also plays a major role in memory processing and consolidation.[77-80] On the other hand, it is well known that affective and emotional disorders are associated with sleep disruptions, particularly with REM sleep temporal distribution.[81-84] However, little is known about the neurobiological interactions between the limbic system in which emotions are integrated and brainstem structures in which REM sleep is generated. Furthermore, the precise nature of the neurochemical events regulating them is still to be elucidated.

In early studies it was shown that cells of the lateral amygdaloid nucleus exhibit bursts of increased discharge frequency during REM sleep.[85] It was also shown that, in man, the hippocampus and AMG unit activity shows increased discharge frequency during REM sleep, compared with that observed during SWS or wakefulness.[86]

To analyze the limbic changes related to the phasic phenomena of REM sleep, we have averaged the EEG activity of limbic regions such as the hippocampus, cingulate gyrus, and the AMG during the REMs or PGO waves. We found that PGO-propagated field potentials can also be recorded in those limbic structures (Figure 23.1A).[87] These findings suggest that the limbic system, including the AMG, is phasically influenced by PGO waves during REM sleep. Thus, this activation of limbic structures may account for the emotional, autonomic, and hallucinatory phenomena of REM sleep.

The rate and pattern of occurrence of PGO waves can be modulated by different midbrain and forebrain structures (see Reference 1 for review). To explore the role of the AMG in PGO wave modulation, we analyzed the effect of amygdaloid stimulation or lesion on PGO wave occurrence. Electrical stimulation of the AMG during REM sleep enhances the PGO wave density (Figure 23.1B),[88] and electrolytic lesions of the CN decreases PGO wave density.[89] The amygdaloid PGO-propagated potentials and the influence of the AMG upon PGO wave activity strongly suggest a facilitatory interaction between the AMG and pontine structures responsible for PGO waves generation.

The analysis of the neuronal unit activity of the CN and basal amygdaloid nucleus across the sleep/waking cycle reveal discharge patterns consistent with regulating PGO waves. It was found that the 65% of the neurons recorded in the CN began to increase their discharge frequency during SPHOL episodes, showing still higher values during REM sleep, and reaching their highest frequency during periods of REM sleep with PGO waves (Figure 23.2).[90] The neurons recorded in the basal amygdaloid nucleus behaved similarly to those recorded in the CN, but the percentage is lower (35%).

These findings support the existence of a physiological interaction between the AMG and pontine structures responsible for PGO wave generation. As mentioned above, the lateral pontine region is connected with the amygdaloid CN and plays a major role in PGO wave generation and consequently in the cholinergic drive for REM sleep induction. Hence, we predicted that the AMG

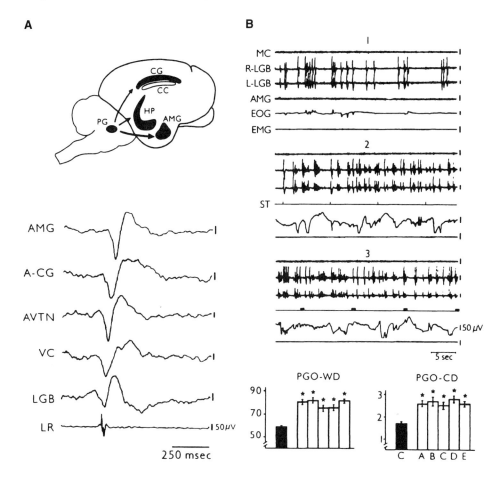

FIGURE 23.1

(A) Schematic representation (upper panel) of PGO waves propagation from the pontine generator (PG) network to the cyngulate gyrus (CG), hippocampus (HP), and the temporal lobe amygdala (AMG). CC = corpus callosum. PGO propagation (lower panel) to different structures of the limbic system and thalamus. Traces represent the average of 256 PGO waves and related thalamic and limbic potentials during REM sleep. A-CG = anterior cyngulate gyrus; AVTN = anterior ventral thalamic nucleus; VC = visual cortex; LGB = lateral geniculate body; LR = lateral rectus muscle of the eyeball. Note that A-CG and AMG display the longest latency. (B) Polygraphic recordings (upper panels) of REM sleep during control (1), during single pulses (2), and trains of pulses (3) electrical stimulation in the AMG. Note the increases of PGO waves and the number and amplitude of REMs during amygdaloid stimulation. MC = motor cortex; R-LGB and L-LGB = right and left lateral geniculate, respectively; EOG = electrooculogram; EMG = electromyogram; ST = stimulus artifact. The graphics (lower panels) show the effect of different modalities of amygdaline electrical stimulation (A, B, C, D, and E) during REM sleep on PGO wave density (PGO-WD) and on PGO wave cluster density (PGO-CD). The bars represent mean values ± SEM; $*p < 0.001$, as compared with baseline control (C, black bars).

may exert through its CN an excitatory control over the lateral pontine region, facilitating the occurrence of PGO waves and REM sleep. Because both CN and lateral pontine nuclei contain cholinergic cells, the participation of a cholinergic mechanism in this hypothetical phenomenon was explored by delivering single carbachol microinjections into the CN and different amygdaloid nuclei. Carbachol administration was preceded by three 8-hr periods of control sleep recording on consecutive days, and recordings of the same duration were carried out the day of carbachol microinjection and on the following 4 days. Different carbachol doses (4 μg/0.25 μl, 8 μg/0.5 μl, and 16 μg/1.0 μl saline) were applied into each amygdaloid nucleus.

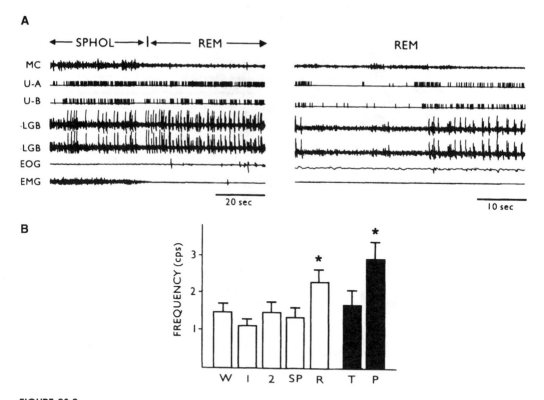

FIGURE 23.2
(A) Polygraphic recordings showing a transition from an SPHOL episode to REM sleep (left) and a REM sleep episode where the presence and absence of PGO waves can be observed (right). Both recordings show the digitalized pulses obtained from two neurons (U-A and U-B) simultaneously recorded in the amygdaloid CN. Note the relationship between increases in the unit discharge frequency and the occurrence of PGO waves. Other abbreviations are the same as in Figure 23.1. (B) Frequency mean values (± SEM) of neurons recorded in the CN during waking (W), SWS-1 (1), SWS-2 (2), SPHOL episodes (SP), REM sleep (R), tonic periods of REM sleep without PGO waves (T), and phasic periods of REM sleep with PGO waves (P). *$p < 0.01$ as compared with W, 1, 2, and SP.

23.4 Cholinergic Modulation of REM Sleep from the Amygdala

Carbachol administration into the lateral (LN), basal (BN), or basolateral (BLN) amygdaloid nuclei did not produce any change in the REM sleep amount, nor in any other sleep or waking stages. Conversely, carbachol application into the CN induced a significant and progressive enhancement of the SPHOL%, peaking on the third day after carbachol administration. A decreasing trend in SPHOL% is observed on the last 2 days, but it remains significantly higher than baseline control (Figure 23.3). REM% is also significantly and progressively increased, beginning on the first day and peaking on the fourth day after carbachol microinjection. On the last recording day, REM% declines, but it remains significantly higher than baseline control (Figure 22.3).[91] We have recently followed the long-term effect of carbachol and found that SPHOL% and REM% return to baseline control values 7 days after a single carbachol microinjection. Concomitantly with the SPHOL and REM sleep augmentation, the SWS-2 stage percentage is mildly increased, and the SWS-1 and waking stages percentage are significantly reduced.

It is worth noting the similarity between the SPHOL and REM sleep enhancement produced by carbachol microinjection into the CN and that produced by carbachol administered into the lateral pontine region. In both cases, the enhancement in REM% is due to an increase of the number of

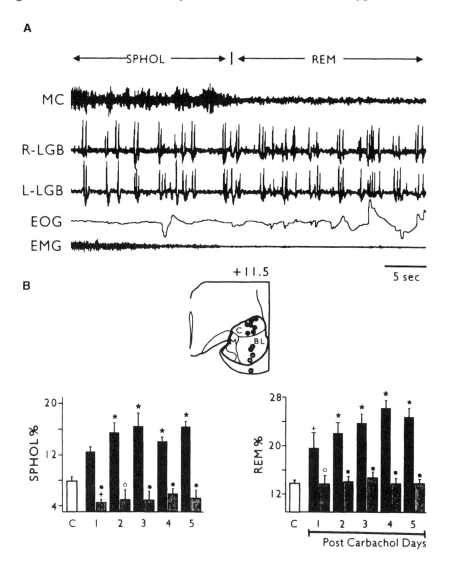

FIGURE 23.3
(A) Polygraphic recording showing the transition from an SPHOL episode to REM sleep. Abbreviations same as in Figure 23.1. (B) Schematic representation of the effective (black dots) and ineffective (gray dots) injection sites into different amygdaloid nuclei. C = central nucleus; BL = basolateral nucleus; M = medial nucleus. The graphics show the effect on SPHOL episodes and REM sleep percentage when carbachol (8 µg/0.50 µl) was applied in the effective (black bars) and ineffective (gray bars) sites. Note that effective microinjection sites are localized into the CN. C = saline control; 1 to 5 days after a single carbachol microinjection. Data are expressed as mean ± SEM; $+p < 0.001$; $*p < 0.0001$ as compared with C; $•p < 0.0001$ as compared with effective sites of injection.

episodes, whereas the enhancement in SPHOL% is due to increases in both mean duration and number of episodes.[43,91] So, the cholinergic activation of the CN also has a substantial effect on the transitional stages to REM sleep and supports the hypothesis that the CN also participates in the onset mechanisms of REM sleep.

Another observation supporting the view that the CN plays a role in the onset of REM sleep is the cholinergic activation of PGO waves from CN. Recently, we found that carbachol administration into the CN produces a significant and progressive long-term enhancement of PGO wave density (PGO waves per minute) during SPHOL episodes and REM sleep.[92] From the first day of carbachol application, PGO waves density shows a significant increase that progressively rises on the following 4 days. Additionally, the cholinergic activation of the CN provokes significant changes on the pattern of PGO wave occurrence. In control conditions, PGO waves during SPHOL episodes appear as single or double waves. After carbachol, PGO waves appear in three wave clusters, the density of which significantly increases on the days following a single carbachol microinjection. The PGO waves also show significant changes in its pattern of occurrence during REM sleep. The occurrence of single PGO waves significantly decreases, whereas that of PGO wave clusters show a significant progressive and long-term increase, reaching mean values threefold higher than control (Figure 23.4).

Taken together, these findings indicate that the AMG participates in REM sleep induction through the CN and that a cholinergic mechanism is involved in the amygdalar control of the REM sleep. In summary, cholinergic activation of the CN increases the number and mean duration of the transitional SPHOL episodes from SWS to REM sleep; it also increases the PGO waves occurring during these transitional stages. These changes may explain the increase in the number of REM sleep episodes.[91] Thus, the CN can be considered as a limbic-forebrain component of the neuronal network controlling the occurrence of REM sleep. Carbachol application into the lateral pontine region and into the CN causes similar effects on PGO wave density, SPHOL episodes, and REM sleep. Therefore, a possible mechanism for explaining the effect of carbachol applied to CN is the activation of the lateral pontine region through the efferent fibers of the CN, resulting in a long-term facilitation of PGO waves, SPHOL episodes, and REM sleep.

23.5 Possible Mechanisms Underlying Long-Term REM Sleep Enhancement

A notable feature of carbachol microinjections into the lateral pontine regions or into the CN is the long-term duration of its effect. From a pharmacological point of view the explanation of these lasting effects remains to be determined. Immediate effects of carbachol microinjected into lateral pontine regions, such as the prompt (7 to 9 min) onset of PGO waves, may be explained in terms of direct muscarinic receptor activation by the drug. In fact, it has been shown that the immediate triggering of PGO waves induced by carbachol can be blocked with an M-2 receptor antagonist (methoctramine).[44]

The long-term effects of carbachol, however, are not likely to be due only to this muscarinic activation. These prolonged effects suggest that carbachol may also induce a change in the synthesis of proteins and enzymes that respond, via second messengers, to transduction signals that follow the muscarinic receptors activation. The interaction of carbachol with muscarinic receptors has been well documented. Binding of carbachol to these receptors activates phospholipase C (PLC) via G protein.[93] The activation of PLC generates second messengers as diacylglycerol (DAG) and inositol tri-phosphate. Signaling pathways that increase the intracellular levels of Ca^{++} and DAG have been shown to modulate diverse cellular functions through the activation of protein kinase C (PKC).[94,95] In the nervous system, activation of PKC has been related to enhancement of neurotransmitter release,[96,97] modification of neuronal plasticity,[98,99] and gene expression,[100] which may take several hours or days to develop. Therefore, carbachol in lateral pontine sites and in CN may produce a long-lasting change in neuronal activity that may somehow account for prolonged REM sleep.

Recently, a modulatory role of G protein, adenylate cyclase, adenosine 3',5'-cyclic monophosphate, and protein kinase A in the cholinergic REM sleep induction from the medial pontine reticular formation has been shown.[101] These findings support the participation of intracellular signal transduction mechanisms in regulating the cholinergic induction of REM sleep and strengthen the hypothesis that such mechanisms contribute to the long-term effect of CN and lateral pontine carbachol on REM sleep.

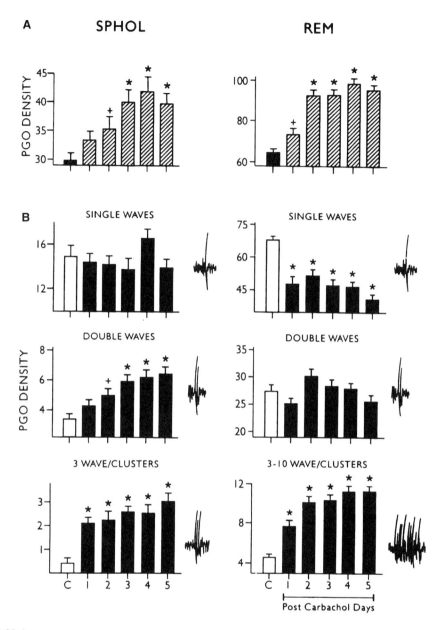

FIGURE 23.4
(A) Effect of a single carbachol (4 μg/0.25 μl) microinjection into the CN on the PGO waves density (PGO waves per min) during SPHOL episodes and REM sleep. Data are expressed as mean ± SEM; $+p < 0.05$; $*p < 0.0001$ as compared with saline control (C). (B) Effect of CN carbachol (4 μg/0.25 μl) on the density of single, double, and clusters of PGO waves occurring during SPHOL episodes (left) and REM sleep (right). Abbreviations same as in Figure 23.3. $+p < 0.002$, $*p < 0.0001$ as compared with C.

The hypothesis that cholinergic neurons play a major role in the onset of REM sleep is supported by the enhancement of REM sleep after carbachol in the cholinergic cells of the lateral pons.[43] This hypothesis is also strengthened by the lasting enhancement of REM sleep after carbachol in the CN, which also contains cholinergic neurons[50] and receives cholinergic projections from the LDT.[52,53] REM sleep episodes can also be increased by administering carbachol into the anterodorsal pontine tegmentum,[21] which lacks cholinergic neurons,[27] but this increase lasts 20 hr after carbachol administration.[43] In this case, the increase of REM% is due to a lengthening of the REM episodes mean duration. These facts suggest that cholinergic neurons are involved in the long-term regulation of REM sleep, whereas cholinoceptive neurons participate in the onset and maintenance of this sleep stage.

The long-term enhancement of REM sleep has been presumed to be the consequence of activating the cholinergic cells of the lateral pons or the CN. However, both structures, particularly the CN, circumscribe a variety of cells and fibers which contain one or more neuropeptides, such as somatostatin and VIP,[54,55] that are related to REM sleep induction. It is worth mentioning that the CN does contain cholinergic neurons and that abundant CN efferent projections to the lateral pons have been demonstrated. Nevertheless, the existence of efferent fibers from the cholinergic cells in the CN to the lateral pontine region has not yet been determined. Conversely, the peptidergic efferent fibers from the CN to the lateral pons have been well delineated.[51,56-58] Thus, it could be proposed that the CN neurons containing somatostatin and VIP may also participate in the long-term enhancement of REM sleep. Consistent with this idea is the fact that increased REM sleep can also be induced for several days by microinjecting VIP into the oral pontine tegmentum in rats.[102] Therefore, the interaction between cholinergic cells and neurons containing VIP should be defined to understand better the neurochemical mechanisms of REM sleep.

23.6 Converging Evidence: Cholinergic Modulation of REM Sleep from Lateral Pons and Amygdala

All the neurophysiological and pharmacological data obtained in the studies cited above strongly support the AMG as critically involved in the neurobiological features of REM sleep. With respect to emotional and autonomic considerations, the AMG may also represent an important brain structure for integrating the hallucinatory and reminiscent phenomena, and the emotional and autonomic changes that occur concomitantly with REM sleep phasic events. The AMG is critically involved in emotional and memory processing (see Reference 76 for a review). Recently, it has been shown in a positron emission tomography study that regional cerebral blood flow is positively correlated with REM sleep in pontine tegmentum and both amygdaloid complexes, in healthy subjects who recalled dreams upon awakening.[103] Because several limbic structures, including the AMG, are located in the temporal lobe, we have analyzed the REMs density and the emotional content of dreams reported by temporal lobe epileptic patients. Because of their focal and secondarily propagated epileptic seizures during sleep, it can be assumed that their limbic structures are under an increased state of excitability. Our results show that these patients have a higher density of REMs and report more unpleasant emotions in their dreams than healthy control subjects. Moreover, all epileptic patients report fewer different emotions per dream, but with a higher intensity than healthy subjects.[104]

Finally, it is of interest that some regions involved in PGO wave generation, such as the lateral pons, the BC area and the locus coeruleus area are also involved in the control of autonomic functions. The parabrachial nuclei comprise an important component of the respiratory control circuit.[105,106] The area surrounding the BC plays a role in a variety of visceral functions such as those related to taste[107] and central cardiovascular control,[108] and the locus coeruleus region participates in parasympathetic visceromotor functions.[109] It has been well documented that the AMG plays an important role in a variety of autonomic and somatomotor responses associated with emotion such as flight, defense, fear, and attack[65,110-115] by acting directly on the brainstem through the ventral amygdalofugal pathway.[116,117] During REM sleep, transient modifications of heart rate and respiration have been correlated with PGO waves.[4,6] Moreover, several autonomic changes are concomitant to the REMs of REM sleep.[5,63] The fact that amygdaloid electrical stimulation and carbachol application into the amygdaloid CN increases PGO activity leads to the possibility that the AMG is also involved in the autonomic changes associated with emotional phenomena of REM sleep episodes.

References

1. Callaway, C.W., Lydic, R., Baghdoyan, H.A., and Hobson, J.A., Pontogeniculooccipital waves: spontaneous visual system activity during rapid eye movement sleep, *Cell. Mol. Neurobiol.*, 7, 105, 1987.
2. Brooks, D.C., Waves associated with eye movements in the awake and sleeping cat, *Electroencephalogr. Clin. Neurophysiol.*, 24, 532, 1968.
3. Cespuglio, R., Laurent, J.P., and Jouvet, M., Etude des relations entre l'activite ponto-geniculo-occipitale (PGO) et la motricité oculaire chez le chat sous réserpine, *Brain Res.*, 83, 319, 1975.
4. Baust, W., Holzbach, E., and Zechlin, O., Phasic changes in heart rate and respiration correlated with PGO-spikes activity during REM sleep, *Pflügers Arch.*, 331, 113, 1972.
5. Orem, J. and Barnes, C.D., *Physiology in Sleep*, Academic Press, New York, 1980, p. 1.
6. Orem, J., Control of the upper airways during the sleep and the hypersomnia — sleep apnea syndrome, in *Physiology in Sleep*, Orem, J. and Barnes, C.D. Eds., Academic Press, New York, 1980, p. 273.
7. López-Rodríguez, F., Chase, M.H., and Morales, F.R., PGO-Related potentials in lumbar motoneurons during active sleep, *J. Neurophysiol.*, 68, 109, 1992.
8. Miyauchi, S., Takino, R., and Azakamki, M., Evoked potentials during REM sleep reflect dreaming, *Electroencephalogr. Clin. Neurophysiol.*, 76, 19, 1990.
9. Simon, R.P., Michel, D.G., and Brooks, D.C., The role of the raphe nuclei in the regulation of pontogeniculo-occipital waves activity, *Brain Res.*, 58, 313, 1973.
10. Cespuglio, R., Gomez, M.E., Walker, E., and Jouvet, M., Effets du refroidissement et de la stimulation des noyaux du systeme du raphé sur les etats de vigilance chez le chat, *Electroencephalogr. Clin. Neurophysiol.*, 47, 289, 1979.
11. Aston-Jones, G. and Bloom, F.E., Activity of norepinephrine-containing locus coeruleus neurons in behaving rats anticipates fluctuations in the sleep-waking cycle, *J. Neurosci.*, 876, 1981.
12. Riou, F., Cespuglio, R., and Jouvet, M., Endogenous peptides and sleep in the rat. III. The hypnogenic properties of vasoactive intestinal polypeptide, *Neuropeptides*, 2, 265, 1982.
13. El Kafi, B., Cespuglio, R., Léger, L., Marinesco, S., and Jouvet, M., Is the nucleus raphe dorsalis a target for the peptides possessing hypnogenic properties?, *Brain Res.*, 637, 211, 1994.
14. Vivaldi, E., McCarley, R.W., and Hobson, J.A., Evocation of desynchronized sleep signs by chemical microstimulation of the pontine brainstem, in *The Reticular Formation Revisited*, Hobson, J.A. and Brazier, M.A.B., Eds., Raven Press, New York, 1980, p. 513.
15. Baghdoyan, H.A., Rodrigo-Angulo, M.L., McCarley, R.W., and Hobson, J.A., Site-specific enhancement and suppression of desynchronized sleep signs following cholinergic stimulation of three brainstem regions, *Brain Res.*, 306, 39, 1984.
16. Hernández-Peón, R., Chavez-Ibarra, G., Morgane, P.J., and Timo-Iaria, C., Cholinergic pathways for sleep, alertness and rage in the limbic midbrain circuit, *Acta Neurol. Latinoamer.*, 8, 93, 1962.
17. Jouvet, M., Recherches sur les structures nerveuses et les mécanismes responsables des différentes phases du sommeil physiologique, *Arch. Ital. Biol.*, 100, 125, 1962.
18. Baghdoyan, H.A., Rodrigo-Angulo, M.L., McCarley, R.W., and Hobson, J.A., A neurochemical gradient in the pontine tegmentum for the cholinoceptive induction of desynchronized sleep signs, *Brain Res.*, 414, 245, 1987.
19. Baghdoyan, H.A., Lydic, R., Callaway, C.W., and Hobson, J.A., The carbachol-induced enhancement of desynchronized sleep signs is dose dependent and antagonized by centrally administered atropine, *NPF*, 2, 67, 1989.

20. Vanni-Mercier, G., Sakai, K., Lin, J.S., and Jouvet, M., Mapping of cholinoceptive brainstem structures responsible for the generation of paradoxical sleep in the cat, *Arch. Ital. Biol.,* 127, 133, 1989.
21. Yamamoto, K., Mamelak, A.N., Quattrochi, J., and Hobson, J.A., A cholinoceptive desynchronized sleep induction zone in the anterodorsal pontine tegmentum: locus of sensitive region, *Neuroscience,* 39, 279, 1990.
22. Velázquez-Moctezuma, J., Guillin, J. C., and Shiromani, P.J., Effect of specific M1, M2 muscarinic receptor agonists on REM sleep generation, *Brain Res.,* 128, 1989.
23. Velázquez-Moctezuma, J., Shalauta, M., Guillin, J.C., and Shiromani, P.J., Cholinergic antagonists and REM sleep generation, *Brain Res.,* 543, 175, 1991.
24. Lydic, R. and Baghdoyan, H.A., The neurobiology of rapid eye movement sleep, in *Sleep and Breathing,* Saunders, N.A. and Sullivan, C.E., Eds., Marcel Dekker, New York, 1994, p. 47.
25. Kodama, T., Takahashi, Y., and Honda, Y., Enhancement of acetylcholine release during paradoxical sleep in the dorsal tegmental field of the cat brain stem, *Neurosci. Lett.,* 144, 277, 1990.
26. Lydic, R., Baghdoyan, H.A., and Lorinc, Z., Microdialysis of cat pons reveals enhanced acetylcholine release during state-dependent respiratory depression, *Am. J. Physiol.,* 261, 766, 1991.
27. Shiromani, P.J., Armstrong, D.M., Berkowitz, A., Jeste, D.V., and Gillin, J.C., Distribution of choline acetyltransferase immunoreactive somata in the feline brainstem: implications for REM sleep generation, *Sleep,* 11, 1, 1988.
28. Datta, S., Paré, D., Oakson, G., and Steriade, M., Thalamic-projecting neurons in brainstem cholinergic nuclei increases their firing rates one minute in advance of EEG desynchronization associated with REM sleep [abstract], *Soc. Neurosci. Abstr.,* 15, 452, 1989.
29. Steriade, M., Datta, S., Paré, D., Oakson, G., and Curró-Dossi, R., Neuronal activities in brain stem cholinergic nuclei related to tonic activation processes in thalamocortical systems, *J. Neurosci.,* 10, 2541, 1990.
30. Steriade, M., Paré, D., Datta, S., Oakson, G., and Curró-Dossi, R., Different cellular types in mesopontine cholinergic nuclei related to ponto-geniculo-occipital waves, *J. Neurosci.,* 10, 2560, 1990.
31. Sakai, K., El Mansari, M., and Jouvet, M., Inhibition by carbachol microinjections of presumptive cholinergic PGO-on neurons in freely moving cats, *Brain Res.,* 213, 1990.
32. Quattrochi, J., Mamelak, A.N., Macklis, J.D., Madison, R., and Hobson, J.A., Mapping neuronal inputs to REM sleep induction sites with carbachol fluorescent microspheres, *Science.,* 245, 984, 1989.
33. Lydic, R., Keifer, J.C., Baghdoyan, H.A., and Becker, L., Microdialysis of the pontine reticular formation reveals inhibition of acetylcholine release by morphine, *Anesthesiology,* 79, 1003, 1993.
34. Saito, H., Sakai, K., and Jouvet, M., Discharge pattern of the nucleus parabrachialis lateralis nuerons of the cat during sleep and waking, *Brain Res.,* 134, 59, 1977.
35. Sakai, K. and Jouvet, M., Brain stem PGO-ON cells projecting directly to the cat dorsal lateral geniculate nucleus, *Brain Res.,* 194, 1980, 1980.
36. Datta, S. and Hobson, J.A., Neuronal activity in the caudolateral peribrachial pons-relationship to PGO waves and rapid eye movements, *J. Neurophysiol.,* 71, 95, 1994.
37. Datta, S. and Hobson, J.A., Suppression of ponto-geniculo-occipital waves by neurotoxic lesions of pontine caudo-lateral peribrachial cells, *Neuroscience,* 67, 703, 1995.
38. Jones, B.E., Paradoxical sleep and its chemical/structural substrates in the brain, *Neuroscience,* 40, 637, 1991.
39. Roth, M.T., Fleegal, M.A., Lydic, R., and Baghdoyan, H.A., Pontine acetylcholine release is regulated by muscarinic autoreceptors, *NeuroReport,* 7, 3069, 1996.
40. Thomas, J. and Benoit, O., Individualisation d'un sommeil a ondes lentes et activité phasique, *Brain Res.,* 5, 221, 1967.

41. Datta, S., Calvo, J.M., Quattrochi, J., and Hobson, J.A., Long-term enhancement of REM sleep following cholinergic stimulation, *NeuroReport,* 2, 619, 1991.
42. Datta, S., Calvo, J.M., Quattrochi, J., and Hobson, J.A., Cholinergic microstimulation of the peribrachial nucleus in the cat. 1. Immediate and prolonged increases in ponto-geniculo-occipital waves, *Arch. Ital. Biol.,* 130, 263, 1992.
43. Calvo, J.M., Datta, S., Quattrochi, J., and Hobson, J.A., Cholinergic microstimulation of the peribrachial nucleus in the cat. 2. Delayed and prolonged increases in REM sleep, *Arch. Ital. Biol.,* 130, 285, 1992.
44. Datta, S., Quattrochi, J., and Hobson, J.A., Effect of specific muscarinic M2-receptor antagonist on carbachol induced long-term REM sleep, *Sleep,* 16, 8, 1993.
45. Sakai, K., Some anatomical and physiological properties of pontomesencephalic tegmental neurons with special reference to the PGO waves and postural atonia during paradoxical sleep in the cat, in *The Reticular Formation Revisited,* Hobson, J.A. and Brazier, M.A.B., Eds., Raven Press, New York, 1980, p. 427.
46. Takeuchi, Y., McLean, J.H., and Hopkins, D.A., Reciprocal connections between the amygdala and parabrachial nuclei: ultrastructural demonstration by degeneration and axonal transport of horseradish peroxidase in the cat, *Brain Res.,* 239, 583, 1982.
47. De Olmos, J., Alheid, G.F., and Beltramino, C.A., Amygdala, in *The Rat Nervous System,* Paxinos, G., Ed., Academic Press, Australia, 1985, p. 223.
48. Semba, K. and Fibiger, H.C., Afferent connections of the laterodorsal and the pedunculopontine tegmental nuclei in the rat — a retrograde and anterograde transport and immunohistochemical study, *J. Comp. Neurol.,* 323, 387, 1992.
49. Bernard, J.F., Bester, H., and Besson, J.M., Involvement of the spino-parabrachio-amygdaloid and hypothalamic pathways in the autonomic and affective emotional aspects of pain, *Prog. Brain Res.,* 107, 243, 1996.
50. Amaral, D.G., Price, J.L., Pitkanen, A., and Carmichael, S.T., Anatomical organization of the primate amygdaloid complex, in *The Amygdala: Neurobiological Aspects of Emotion, Memory, and Mental Dysfunction,* Aggleton, J.P., Ed., Wiley-Liss, New York, 1992, p. 1.
51. Price, J.L., Russchen, F.T., and Amaral, D.G., The limbic region. II. The amygdaloid complex, in *Handbook of Chemical Neuroanatomy. Vol. 5. Integrated Systems of the CNS. Part I. Hypothalamus, Hippocampus, Amygdala, Retina,* Bjorklund, A., Hokfelt, T., and Swanson, L.W. Eds., Elsevier, Amsterdam, 1987, pp. 3, 279.
52. Petrov, T., Krukoff, T.L., and Jhamandas, J.H., Chemically defined collateral projections from the pons to the central nucleus of the amygdala and hypothalamic paraventricular nucleus in the rat, *Cell Tissue Res.,* 277, 289, 1994.
53. Kaada, B.R., Cingulate, posterior orbital, anterior insular and temporal pole cortex, *Handbook of Physiology,* Field, J., Magoun, H.W., and Hall, V.E., Eds., American Physiological Society, Washington, D.C., 1960, p. 1346.
54. Steiger, A., Guldner, J., Hemmeter, U., Rothe, B., Wiedemann, K., and Holsboer, F., Effects of growth hormone-releasing hormone and somatostatin on sleep EEG and nocturnal hormone secretion in male controls, *Neuroendocrinology,* 56, 566, 1992.
55. Honkaniemi, J., Colocalization of peptide-like and tyrosine hydroxylase-like immunoreactivities with fos-immunoreactive neurons in rat central amygdaloid nucleus after immobilization stress, *Brain Res.,* 598, 107, 1992.
56. Kawai, Y., Inagaki, S., Shiosaka, S., Senba, E., Hara, Y., Sakanaka, M., Takatsuki, K., and Tohyama, M., Long descending projections from amygdaloid somatostatin-containing cells to the lower brain stem, *Brain Res.,* 239, 603, 1982.

57. Veening, J.G., Swanson, L.W., and Sawchenko, P.E., The organization of projections from the central nucleus of the amygdala to brainstem sites involved in autonomic regulation: a combined retrograde transport-immunohistochemical study, *Brain Res.*, 303, 337, 1984.
58. Moga, M.M. and Gray, T.S., Evidence of corticotropin-releasing factor, neurotensine, and somatostatin in the neural pathway from the central nucleus of the amygdala to the parabrachial nucleus, *J. Comp. Neurol.*, 241, 275, 1985.
59. Ramsey, G., Studies of dreaming, *Psychol. Bull.*, 50, 432, 1953.
60. Shapiro, A., Goodenough, D., Biederman, I., and Sleser, I., Dream recall and the physiology of sleep, *J. Appl. Physiol.*, 19, 778, 1964.
61. Fisher, C., Gross, J., and Zuch, J., Cycle of penile erection synchronous with dreaming (REM) sleep, *Arch. Gen. Psychiatry*, 12, 29, 1965.
62. Snyder, F., The orgasmic state associated with dreaming, in *Psychoanalysis and Current Biological Thought*, Greenfield, N. and Lewis, W. Eds., University of Wisconsin Press, Madison, 1965, p. 275.
63. Taylor, W.B., Moldofsky, H., and Furedy, J.J., Heart rate deceleration in REM sleep: an orientating reaction interpretation, *Psychophysiology*, 22, 110, 1985.
64. Sastre, J.P. and Jouvet, M., Le comportement onirique du chat, *Physiol. Behav.*, 22, 279, 1979.
65. Kaada, B.R., Stimulation and regional ablation of the amygdaloid complex with reference to functional representation, in *The Neurobiology of the Amygdala*, Eleftheriou, B.E., Ed., Plenum Press, New York, 1972, p. 205.
66. McLean, P.D. and Delgado, J.M.R., Electrical and chemical stimulation of fronto-temporal portion of limbic system in the waking animal, *Electroencephalogr. Clin. Neurophysiol.*, 5, 91, 1953.
67. Ursin, H. and Kaada, B.R., Functional localization within the amygdaloid complex in the cat, *Electroencephalogr. Clin. Neurophysiol.*, 12, 1, 1960.
68. Reis, D.J. and Oliphant, M.C., Bradycardia and tachycardia following electrical stimulation of the amygdaloid region in monkey, *J. Neurophysiol.*, 27, 893, 1964.
69. Penfield, W.P. and Perot, P., The brain's record of auditory and visual experience. A final summary and discussion, *Brain*, 86, 595, 1963.
70. Heat, R.G., Pleasure response of human subjects to direct stimulation of the brain: physiologic and psychodynamic considerations, in *The Role of Pleasure in Behavior*, Heat, R.G., Ed., Harper and Row, New York, 1964, p. 219.
71. Brazier, M.A.B., Stimulation of the hippocampus in man using implanted electrodes, in *RNA and Brain Function, Memory and Learning*, Brazier, M.A.B., Ed., University of California Press, Berkley, 1966, p. 299.
72. Fernández-Guardiola, A., Reminiscences elicited by electrical stimulation of the temporal lobe in humans, in *Neurobiology of Sleep and Memory*, Drucker-Colin, R. and McGaugh, J.L. Eds., Academic Press, New York, 1977, p. 273.
73. Halgren, E., Walter, R.D., Cherlow, D.G., and Crandall, P.H., Mental phenomena evoked by electrical stimulation of the human hippocampal formation and amygdala, *Brain*, 101, 83, 1978.
74. Halgren, E., Emotional neurophysiology of the amygdala within the context of human cognition, in *The Amygdala: Neurobiological Aspects of Emotion, Memory, and Mental Dysfunction*, Aggleton, J.P., Ed., Wiley-Liss, New York, 1992, pp. 7, 191.
75. LeDoux, J.E., Emotion and the amygdala, in *The Amygdala: Neurobiological Aspects of Emotion, Memory, and Mental Dysfunction*, Aggleton, J.P., Ed., Wiley-Liss, New York, 1992, pp. 12, 339.
76. Aggleton, J.P., Ed., *The Amygdala: Neurobiological Aspects of Emotion, Memory, and Mental Disfunction*, Wiley-Liss, New York, 1992, p. 1.

77. Kesner, R.P., Learning and memory in rats with an emphasis on the role of the amygdala, in *The Amygdala: Neurobiological Aspects of Emotion, Memory, and Mental Disfunction*, Aggleton, J.P., Ed., Wiley-Liss, New York, 1992, p. 379.
78. Smith, C. and Lapp, L., Increases in number of REMs and REM density in humans following an intensive learning period, *Sleep,* 14, 325, 1991.
79. Challamel, M.J., Functional role of REM sleep during ontogenesis, *Neurophysiol. Clin.,* 22, 117, 1992.
80. Hassard, A., Reverse learning and the physiological basis of eye movement desensitization, *Med. Hypotheses,* 47, 277, 1996.
81. Taylor, S.F., Goldman, R.S., Tandon, R., and Shipley, J.E., Neuropsychological function and REM sleep in schizophrenic patients, *Biol. Psychiatry,* 32, 529, 1992.
82. Benca, R.M., Obermeyer, W.H., Thisted, R.A., and Gillin, J.C., Sleep and psychiatric disorders. A meta-analysis, *Arch. Gen. Psychiatry,* 49, 651, 1992.
83. Benson, K.L. and Zarcone, V.P., Rapid eye movement sleep eye movements in schizophrenia and depression, *Arch. Gen. Psychiatry,* 50, 474, 1993.
84. Armitage, R., Effects of antidepressant treatment on sleep EEG in depression, *J. Psychopharmacol.,* 10, 22, 1996.
85. White, T.J. and Jacobs, B.L., Single unit activity in the lateral amygdala of the cat during sleep and waking, *Electroencephalogr. Clin. Neurophysiol.,* 38, 331, 1975.
86. Ravagnati, L., Halgren, E., Babb T.L., and Crandall, P.H., Activity of the human hippocampal formation and amygdala neurons during sleep, *Sleep,* 2, 161, 1979.
87. Calvo, J.M. and Fernández-Guardiola, A., Phasic activity of basolateral amygdala, cingulate gyrus, and hippocampus during REM sleep in the cat, *Sleep,* 7, 202, 1984.
88. Calvo, J.M., Badillo, S., Morales-Ramírez, M., and Palacios-Salas, P., The role of temporal lobe amygdala in ponto-geniculo-occipital activity and sleep organization in cats, *Brain Res.,* 403, 22, 1987.
89. Calvo, J.M., Badillo, S., and Palacios-Salas, P., Participación del sistema límbico en la regulación de los fenómenos fásicos del sueño paradójico, *Anales del Instituto Mexicano de Psiquiatría,* 1, 101, 1988.
90. Calvo, J.M., El sistema límbico y el sueño, *Anales del Instituto Mexicano de Psiquiatría,* 4, 47, 1993.
91. Calvo, J.M., Simón-Arceo, K., and Fernández-Mas, R., Prolonged enhancement of REM sleep produced by carbachol microinjection into the amygdala, *NeuroReport,* 7, 577, 1996.
92. Simón-Arceo, K. and Calvo, J.M., Long term increase of REM sleep and ponto-geniculo-occipital potentials (PGO) provoked by cholinergic activation of the temporal lobe amygdala in the cat, *Salud. Ment.,* 20, 12, 1997.
93. Nathanson, N., Molecular properties of the muscarinic acetylcholine receptor, *Ann. Rev. Neurosci.,* 10, 195, 1987.
94. Berridge, M.J., Inositol triphosphate and diacylglycerol: two interacting second messengers, *Ann. Rev. Biochem.,* 56, 159, 1987.
95. Nishizuka, Y., The molecular heterogenicity of protein kinase C and its implications for cellular regulation, *Nature,* 334, 661, 1988.
96. Malenka, R.C., Ayoub, G.S., and Nicoll, R.A., Phorbol esters enhance transmitter release in rat hippocampal slices, *Brain Res.,* 403, 198, 1987.
97. Malenka, R.C., Madison, D.V., and Nicoll, R.A., Potentiation of synaptic transmission in the hippocampus by phorbol esters, *Nature,* 321, 175, 1986.
98. Huang, K.P., The mechanisms of protein kinase C activation, *Trends Neurosci.,* 12, 425, 1989.
99. Routtenberg, A., Protein kinase C activation leading to protein F1 phosphorylation may regulate synaptic plasticity by presynaptic terminal growth, *Behav. Neuronal. Biol.,* 44, 186, 1985.

100. Lui, J.P., Protein kinase C and its substrates, *Mol. Cell. Endocrinol.,* 116, 1, 1996.
101. Capece, M.L. and Lydic, R., cAMP and protein kinase A modulate cholinergic rapid eye movement sleep generation, *Am. J. Physiol.,* 273, R1430, 1997.
102. Bourgin, P., Lebrand, C., Escourrou, P., Gaultier, C., Franc, B., Hamon, M., and Adrien, J., Vasoactive intestinal polypeptide microinjections into the oral pontine tegmentum enhance rapid eye movement sleep in the rat, *Neuroscience,* 77, 351, 1997.
103. Maquet, P., Peters, J.M., Aerts, J., Delfiore, G., Degueldre, C., Luxen, A., and Franck, G., Functional neuroanatomy of human rapid-eye-movement sleep and dreaming, *Nature,* 383, 163, 1996.
104. Gruen, I., Martinez, A., Cruz-Ulloa, C., Aranday, F., and Calvo, J.M., Characteristics of the emotional phenomena in the dreams of patients with temporal lobe epilepsy, *Salud Ment.,* 20, 8, 1997.
105. Bertrand, F. and Hugelin, A., Respiratory synchronizing function of nucleus parabrachialis medialis: pneumotaxic mechanisms, *J. Neurophysiol.,* 34, 189, 1971.
106. Von Euler, C., Marttila, I., Remmers, J.E., and Trippenbach, J., Effects of lesions in the parabrachial nucleus on the mechanisms for central and reflex termination of inspiration in the cat, *Acta Physiol. Scand.,* 96, 324, 1976.
107. Norgren, R., Taste pathways to hypothalamus and amygdala, *J. Comp. Neurol.,* 166, 17, 1976.
108. Coote, J.H., Hilton, S.M., and Zbrozyna, A.W., The ponto-medullary area integrating the defense reaction in the cat and its influence on muscle blood flow, *J. Physiol. (Lond).,* 229, 257, 1973.
109. Westlund, K.D. and Coulter, J.D., Descending projection of the locus coeruleus and subcoeruleus, medial parabrachial nuclei in the monkey: axonal transport studies and dopamine-B-hydroxylase immunocytochemistry, *Brain Res. Rev.,* 2, 235, 1980.
110. Applegate, C.D., Kapp, B.S., Underwood, M.D., and McNall, C.L., Autonomic and somatomotor effects of amygdala central nucleus stimulation in awake rabbits, *Physiol. Behav.,* 31, 353, 1983.
111. Gloor, P., Amygdala, in *Handbook of Physiology,* Field, J., Magoun, H.W., and Hall, V.E., Eds., American Physiological Society, Washington, D.C., 1960, p. 1395.
112. Pascoe, J.P. and Kapp, B.S., Electrophysiological characteristics of amygdala central nucleus neurons in the awake rabbit, *Brain Res.,* 14, 331, 1985.
113. Roldán, E.R., Alvarez-Peláez, R., and Fernández de Molina, E., Electrographic study of the amygdaloid defense response, *Physiol. Behav.,* 13, 779, 1974.
114. Seggie, J., Corticomedial amygdala lesions, behavior, corticosterone, and prolactin: unexpected separation of effects, *Psychiatry Res.,* 10, 139, 1983.
115. Stock, G., Schlor, Y., Heidt, H., and Buss, J., Psychomotor behavior and cardiovascular patterns during stimulation of the amygdala, *Pflügers Arch.,* 376, 177, 1978.
116. Hilton, S.M. and Zbrozyna, A.W., Amygdaloid region for defense reactions and its efferent pathway to the brain stem, *J. Physiol.,* 165, 160, 1963.
117. Hopkins, D.A., Amygdalotegmental projections in the rat, cat and rhesus monkey, *Neurosci. Lett.,* 1, 263, 1975.

Chapter 24

State-Altering Effects of Benzodiazepines and Barbiturates

Wallace B. Mendelson

Contents

- 24.1 Abstract 408
- 24.2 Introduction 408
- 24.3 The Molecular Level 408
- 24.4 The Neuroanatomic Level — Where Do Benzodiazepines Act? 409
  - 24.4.1 Sites at which Triazolam Microinjections Alter Sleep 409
    - 24.4.1.1 Medial Preoptic Area 410
    - 24.4.1.2 Benzodiazepines 410
    - 24.4.1.3 Barbiturates 410
    - 24.4.1.4 Dorsal Raphe Nuclei 411
  - 24.4.2 Anatomic Specificity 411
    - 24.4.2.1 Locus Coeruleus 411
    - 24.4.2.2 Basomedial Nucleus of the Amygdala 411
    - 24.4.2.3 Gigantocellular Tegmental (FTG) Field 411
    - 24.4.2.4 Lateral Preoptic Area 412
    - 24.4.2.5 Horizontal Limb of the Diagonal Band of Broca 412
  - 24.4.3 Summary: Hypnotics, the MPA, and Brainstem Structures 412
- 24.5 Is an Intact Preoptic Area Necessary for Pharmacologically Induced Sleep? 413
  - 24.5.1 Recovery Sleep Following Ibotenic Acid Lesions 413
  - 24.5.2 Benzodiazepines 414
  - 24.5.3 Barbiturates 414
- 24.6 Conclusions 414
- References 416

24.1 Abstract

One of the classic puzzles in psychopharmacology is how diverse substances from such a wide range of pharmacologic classes can have the relatively similar effect of inducing sleep. In this chapter, I will suggest that a common element for a variety of such compounds is that they interact with various moieties of the $GABA_A$-benzodiazepine receptor complex. In terms of neuroanatomic site(s) of action, microinjection studies indicate that the preoptic area of the anterior hypothalamus as well as brainstem structures including the dorsal raphe nuclei are sensitive to hypnotics, and it is hypothesized that such compounds act by altering function of these reciprocally innervated areas. I will propose that knowledge of the structure of the $GABA_A$-benzodiazepine receptor complex can be used to develop other classes of agents, including analeptics and treatments for sleep-disordered breathing.

24.2 Introduction

The modern history of sleep-inducing agents began with the introduction of the intermediate-duration barbiturates in the 1920s and 1930s. These compounds, such as pentobarbital, amobarbital, and secobarbital, were the most widely used hypnotics until the introduction of the benzodiazepines (BZs), first as anxiolytics in the 1960s and then for sleep in the early 1970s. Only recently has the non-benzodiazepine zolpidem supplanted individual benzodiazepines as the most widely used hypnotic in the U.S. All of these compounds, representing diverse pharmacological classes, have one feature in common: they bind to various moieties of the $GABA_A$-benzodiazepine receptor complex. In this review, we will summarize what is known about this interaction, then move to a neuroanatomic level to examine possible site(s) at which this may take place, and close by putting these observations in the context of studies of sleep regulation.

24.3 The Molecular Level

A major step in understanding the mechanism of action of benzodiazepines (BZs) came with the discovery that labeled diazepam binds to high-affinity (K_d ~10^{-9} M) saturable, stereo-specific sites in the central nervous system.[1,2] These central receptors are found in greatest density in the synaptosomal fraction of neurons, perhaps implying a role in neurotransmission. They are thought to be macromolecular complexes functionally comprised of three distinct but interacting entities: a BZ recognition site, a $GABA_A$ recognition site, and a chloride ionophore.[3] They are primarily postsynaptic structures located on dendrites, the somatic membrane, and the initial segments of axons[4] and include binding sites for not only benzodiazepines but also barbiturates and some neurosteroids. Structurally they are comprised of at least five types of subunits, most of which are present in a variety of isoforms. The alpha and beta subunits each have four membrane-spanning sequences that contribute to the chloride ion channel and provide extracellular sites at which benzodiazepines and barbiturates may act. Alpha and gamma subunits are necessary for modulation by benzodiazepines, while the presence of alpha, gamma, and beta subunits are needed for a fully functioning receptor complex.[5] Of particular interest is the benzodiazepine type I receptor (sometimes referred to as omega 1), which is comprised of alpha-1, beta-2, and gamma-2 subunits; the type II receptor is derived from alpha-3, beta-2, and gamma-2 subunits. The former are particularly rich in the cortex and cerebellum, while the latter are more abundant in the hippocampus and spinal cord.

The traditional understanding of the function of this receptor complex is that binding by BZ receptor agonists leads to postsynaptic inhibition. This results from enhanced chloride ion flux, stabilizing the postsynaptic membrane at a level below that necessary for spike generation. Benzodiazepines and barbiturates do this through slightly different mechanisms, the former by increasing the frequency of channel opening[6] and the latter by prolonging the duration of opening.[7] The interaction of barbiturates with the receptor complex is described in more detail in a recent review.[8] An alternative approach which may explain some aspects of benzodiazepine effects is that there is also a presynaptic action on potential-dependent calcium channel function.[9-11]

The original description of the central BZ receptor complex indicated a close correlation between the affinities of various BZs for the central recognition site and their potencies as anxiolytics, anticonvulsants, and muscle relaxants.[1] (Two other types of receptors, a "peripheral receptor"[12] and a diazepam-insensitive receptor [alpha 6, beta 2, gamma 2] have not been associated with such properties and will not be considered here.) The possible role of the receptor complex in sleep induction was less clear. Subsequently, several studies in rats indicated that indeed the central BZ recognition site does mediate sleep-inducing effects of flurazepam, one of the first clinically used benzodiazepine hypnotics. The possible role of the receptor complex in pharmacologic alterations in sleep was further clarified by studies of inverse agonists, compounds that bind at the BZ recognition site but which have opposite pharmacologic effects as clinically used benzodiazepine agonists. The inverse agonist 3-hydroxymethyl-β-carboline (3-HMC), for instance, induced a powerful awakening effect, which was prevented by the BZ receptor blocker CGS 8216.[13] Similarly, a dose of 3-HMC which was so low it had minimal effects on sleep by itself was capable of blocking sleep induction by systemically administered flurazepam.[13] In another series of studies, we showed that the effects of BZs on sleep are stereospecific, in so far as the B_{10} benzodiazepine enantiomers have opposite effects on sleep. The (+) compound induced sleep while the (−) had awakening properties, which could be prevented by the benzodiazepine receptor antagonist CGS 8216.[9,14]

In summary, earlier work has indicated that benzodiazepines bind to the $GABA_A$-benzodiazepine receptor complex in a high-affinity, saturable, stereospecific manner.[3] Subsequent studies indicated that inverse agonists for the BZ recognition site increase wakefulness, low doses of inverse agonists block sleep induction by benzodiazepines, and some (+) and (−) benzodiazepine enantiomers induce sleep and wakefulness, respectively.[9] This body of data made it clear that interaction of BZs with the $GABA_A$-benzodiazepine receptor complex mediates their effects on sleep. It has not been certain, however, where that interaction might take place neuroanatomically. Let us then move from the molecular to the anatomical level and examine possible site(s) of action.

24.4 The Neuroanatomic Level — Where Do Benzodiazepines Act?

24.4.1 Sites at which Triazolam Microinjections Alter Sleep

The approach we have used to determine sites of action of hypnotics has been to microinject them into nuclei that have been associated with sleep regulation in lesion or stimulation studies. As a starting point, we chose sites from the classical studies of Hernandez-Peon (e.g., see Reference 15). Much of this work involved injections of triazolam into the medial preoptic area (MPA) of the anterior hypothalamus and into the dorsal raphe (DR) nuclei of the brainstem. We found that both pentobarbital and triazolam enhanced sleep when injected into the MPA,[10,16] and triazolam decreased sleep when administered into the DR.[9] Let us begin by briefly summarizing the role of these areas in sleep regulation, and then examine the results of microinjections of hypnotics into them.

24.4.1.1 Medial Preoptic Area

Since the studies of von Economo[17] in the wake of the encephalitis lethargica epidemic in the 1920s, there has been growing recognition that sites in the hypothalamus or nearby areas are involved in the regulation of sleep and waking. This view was strengthened by the lesion studies of Hess[18] and Nauta,[19] which indicated that the anterior hypothalamus is involved in sleep maintenance, while a more caudal area appears to have a major role in wakefulness. Later studies reported that in a basal forebrain area including the MPA, stimulation enhanced[20] and lesions decreased[21] sleep in cats. Medial preoptic area lesions in rats acutely decreased sleep, although in a manner highly dependent on ambient temperature.[22] The presence of significant concentrations of some forms of the $GABA_A$-benzodiazepine receptor complex in the MPA[23] suggests that benzodiazepine hypnotic compounds might alter sleep and waking by acting there. The projections of the MPA, which travel widely throughout the forebrain and brainstem,[24] are consistent with this view. Particularly relevant output pathways include those which go ventrally to the median and dorsal raphe nuclei and possibly to the locus coeruleus.[25] Stimulation of the MPA has long been thought to suppress neural activity in the midbrain reticular formation[26] or possibly to induce initial excitation followed by post-excitatory discharge suppression.[27] As the midbrain reticular formation and specific aminergic structures such as the dorsal raphe nuclei have long been associated with the regulation of sleep and waking, these observations are consistent with a role of the preoptic area in the process.

24.4.1.2 Benzodiazepines

Before beginning the studies of microinjection of triazolam into sites such as the MPA, it was necessary to find a dose that was unlikely to induce sleep if it should diffuse into the ventricular system. In direct intraventricular injection studies, we found that doses as high as 1.0 μg of triazolam did not alter sleep in 10 rats, and we chose to study 0.25 and 0.5 μg in the studies we will describe here. Because the lesion-induced alterations in sleep reported by Szymusiak and Satinoff[22] were dependent on ambient temperature, we maintained temperature in the chambers at approximately 26°C, well inside the range characterized in that study.

A total of 15 rats were injected with vehicle, 0.25 μg triazolam, or 0.5 μg triazolam into the MPA, in random sequence, in studies separated by one week each. The EEG and EMG were recorded for 2 hours following the 10:00 a.m. injections (lights were on from 8:00 a.m. until 8:00 p.m.), and the state of consciousness (waking, NREM, and REM sleep) was determined for each 30-sec epoch.[28] Both doses of triazolam significantly decreased sleep latency and increased total sleep time, primarily by increasing NREM sleep. REM sleep and intermittent waking time were not altered.* Thus, the MPA appears to be a possible site which might mediate the hypnotic properties of triazolam.

24.4.1.3 Barbiturates

In an analogous study, pentobarbital (1 and 100 μg) and vehicle were microinjected into the MPA in 16 rats in random sequence, and sleep studies were performed.[29] As in the case of triazolam, there was a potent hypnotic effect. Both doses significantly reduced sleep latency and increased total sleep time, primarily by increasing NREM sleep. REM sleep time, REM latency, and intermittent waking time were not affected. This suggested, then, that the MPA may mediate the hypnotic properties of barbiturates as well as benzodiazepines.

---

* In this study, as in a report on microinjecting triazolam into the dorsal raphe nuclei,[11] co-microinjection of the dihydropyridine calcium channel blocker nifedipine blocked the effects of triazolam on sleep latency. These and other studies seem to suggest that one aspect of benzodiazepine effects on sleep may be mediated by altering potential-dependent calcium channel function, a hypothesis described more fully in Reference 9.

24.4.1.4 Dorsal Raphe Nuclei

The dorsal raphe nuclei of the brainstem seemed an appropriate candidate site, as lesions produce a large, albeit transient, reduction in sleep.[30-32] The neuronal firing rate in this serotonergic center drops as an animal moves from waking to NREM to REM sleep.[33] Ascending fibers run via the midbrain tegmentum and medial forebrain bundle through the lateral hypothalamus, preoptic area, basal forebrain, and other areas to the cerebral cortex, where they interact with most cortical neurons.[34,35] When 0.5 µg of triazolam was administered into the dorsal raphe of 10 rats, sleep, surprisingly, was reduced.[36] There was a significant increase in sleep latency and also a decrease in total sleep, achieved primarily by reducing NREM sleep. Triazolam did not significantly alter intermittent waking time or amounts of REM sleep. As another kind of control, we examined sleep in "near miss" animals, in which our injection had been into surrounding tissue; there was no significant effect on sleep in this group. The awakening effect of triazolam, then, was very specific for injection into the dorsal raphe.

Our interpretation of this observation is that it is consistent with earlier studies indicating that systemically or iontophoretically applied benzodiazepines potentiate inhibition of neuronal firing in the dorsal raphe by GABA.[37] As mentioned earlier, both electrolytic lesions[30] and pharmacologically induced serotonin depletion greatly increase wakefulness, at least transiently, so it seems likely to us that triazolam produced a similar effect by inhibiting dorsal raphe activity.

24.4.2 Anatomic Specificity

It should be noted that the effects of triazolam microinjections are very specific for certain locations. We have also injected into a variety of areas traditionally thought to be involved in sleep regulation on the basis of lesion or stimulation studies, with no effect on sleep (see following).

24.4.2.1 Locus Coeruleus

This noradrenergic brainstem center, has been of interest in terms of sleep regulation since Jouvet and Delorme[38] demonstrated that lesions in cats acutely suppressed REM sleep and depleted norepinephrine in the rostral part of the brain. As with the dorsal raphe nuclei, cellular firing rates are highest in waking, declining progressively in NREM and REM sleep in restrained animals.[39] The locus coeruleus receives input from the dorsal raphe nuclei[40] and sends projections widely throughout the nervous system in a pattern that overlaps, but is not identical to, that of the dorsal raphe. We performed daytime microinjections of vehicle and 0.25 and 0.5 µg triazolam into the locus coeruleus of 13 rats. Sleep latencies were 18.3 ± 2.5 min, 18.6 ± 3.5 min, and 16.7 ± 5.3 min, respectively (df = 2, 24; F = 0.074; $p < 0.9$). Other measures including total sleep, NREM, and REM sleep were also not significantly affected. We also compared the sleep of these 13 rats to that of 10 "near miss" animals (analogously to the procedure in the dorsal raphe study) and found no differences in sleep between the two groups.

24.4.2.2 Basomedial Nucleus of the Amygdala

This area was chosen because Hernandez-Peon proposed it as a sleep regulatory site.[15] In this case, sleep latencies in 7 rats after vehicle and 0.25 and 0.5 mg triazolam were 30.5 ± 3.9, 22.0 ± 1.8, and 33.0 ± 4.9 min, respectively. Similarly, total sleep and NREM and REM sleep were not significantly affected.

24.4.2.3 Gigantocellular Tegmental (FTG) Field

This area was featured prominently in the Hobson-McCarley reciprocal interaction model of sleep regulation,[41] which suggests that, at least in restrained cats, the firing rate of its cholinoceptive cells is lowest in waking and highest in REM sleep. Sleep latencies in 11 rats given vehicle and 0.25 and 0.5 μg microinjection were 27.4 ± 5.0, 36.9 ± 8.4, and 32.6 ± 5.0 min, respectively (NS). There were no changes in total sleep, NREM, or REM sleep.

24.4.2.4 Lateral Preoptic Area

This area* was selected because of reports that electrical stimulation there in cats enhances[15] and lesions acutely decrease[42] sleep. Injections of vehicle and 0.25 and 0.5 μg triazolam into the lateral preoptic area (LPA) in 10 rats did not alter sleep latencies, which were 26.2 ± 3.2 min., 26.8 ± 3.7 min., and 29.4 ± 4.4 min, respectively (df = 2, 18; F = 0.34407; $p < 0.7$). There was also no significant effect on total sleep or NREM or REM sleep. (Ambient temperatures were held constant, at 26.8 ± 0.1°C.) Due to the recent interest in the ventrolateral preoptic area and sleep,[43] it seems likely that it would be worthwhile to explore microinjections into this slightly more ventral structure in future studies.

24.4.2.5 Horizontal Limb of the Diagonal Band of Broca

The diagonal band of Broca is in the area described by McGinty and Sterman,[21] in which lesions in cats greatly decreased sleep. Moreover, its horizontal limb is one area in which cell firing increases in NREM sleep compared to waking in cats. We administered vehicle and 0.25 and 0.5 μg triazolam into the horizontal limb of the diagonal band of Broca in 12 rats, resulting in sleep latencies of 21.3 ± 2.6 min., 19.9 ± 3.5 min., and 22.4 ± 4.4 min, respectively (df = 2, 22; F = 0.307; $p < 0.7$). Once again, there were no significant drug effects on any other sleep measure. The enhancement of sleep after injection into the MPA, then, was not seen after injection into nearby structures (the lateral preoptic area and the diagonal band of Broca). This suggests that the drug effect is very specific to the MPA.

In summary, we have given 0.5 μg triazolam intraventricularly as well as into five structures thought to be involved in sleep regulation (lateral preoptic area, horizontal limb of diagonal band of Broca, locus coeruleus, basomedial nucleus of the amygdala, FTG field), without effect on sleep. These negative studies lead us to believe that the alterations in sleep we observed after microinjection into the MPA and DR may lead to insights into the mechanism of action of benzodiazepine hypnotics. Let us then review the functions of these latter structures and examine how they might be involved in pharmacologically induced perturbations of sleep.

24.4.3 Summary: Hypnotics, the MPA, and Brainstem Structures

The MPA is a complex, sexually dimorphic structure. It contains cell bodies and fibers that cross-react in immunohistochemical studies with a wide variety of neurotransmitters, including substance P and neuropeptide Y.[44] Push-pull cannula studies indicate that catecholamines, GABA, and glutamate are released.[45] GABA is fairly uniformly distributed throughout the hypothalamus; its synthetic enzyme GAD is found in particularly high concentrations in the preoptic area,[46] as are benzodiazepine receptors.[47] Thus, it seems an appropriate area for benzodiazepine ligands to exert pharmacologic actions.

The MPA receives input from a number of forebrain and brainstem areas, including other parts of the hypothalamus as well as serotonergic fibers from the dorsal raphe nucleus and noradrenergic projections.[35] One major output descends to the median and dorsal raphe nuclei, and possibly to the

---

* The anatomic relationship of the lateral preoptic area and the horizontal limb of the diagonal band of Broca may be seen in Figure 24.1, which is derived from another study described later in the text but which also serves to illustrate the proximity of these structures to the MPA.

locus coeruleus.[25] The preoptic area contains cells that are sensitive to temperature, osmolarity, glucose, and steroids[48] and receives projections from visual, auditory, and tactile sensory systems.[35] The basal forebrain contains cells which display increased firing during NREM sleep compared to waking ("sleep-active neurons"), although there is greater density in the LPA and diagonal band of Broca than the MPA.[49] The MPA also contains neurons which become more active in the course of NREM and REM sleep.[50] Thus, the preoptic area, in general, and the MPA, in particular, appear to be involved in coordinating a variety of systems involved in reproductive and homeostatic activity.[35] Given the bidirectional interactions of sleep with cardiovascular, thermoregulatory, endocrine, and sensory systems,[51] it seems likely that the preoptic area might be an important site for the coordination of sleep with other processes. As it receives ascending brainstem input and has descending pathways to brainstem structures including the dorsal raphe, it seems possible that one mechanism by which compounds such as benzodiazepines and barbiturates act is by altering the interaction of these brainstem and anterior hypothalamic structures.

24.5 Is an Intact Preoptic Area Necessary for Pharmacologically Induced Sleep?

In view of the finding that microinjections of triazolam and pentobarbital into the MPA enhance sleep, a central question that arises is whether an intact preoptic area is necessary for the hypnotic action when these agents are given systemically. We have examined this by making ibotenic acid lesions of the preoptic area acid (0.4 μl of a solution of 2.5 μg/μl in phosphate-buffered saline, pH 7.4), allowing sleep to return to approximately baseline levels over the course of one week, and then recording sleep following intraperitoneal (IP) injections of triazolam and pentobarbital. Let us first examine the consequences of ibotenic acid lesions of the MPA and then review the effects on sleep of systemic drug injections in lesioned animals.

24.5.1 Recovery Sleep Following Ibotenic Acid Lesions

Prior to lesioning, and then on days 1, 3, 5, and 8 post-lesion, we took 2-hr samples of sleep at 10:00 a.m. (lights on from 8:00 a.m. to 8:00 p.m.) and followed the course of recovery sleep. Total sleep time at baseline had been 55.7 ± 2.4 min. It dropped to a low of 46.6 ± 3.0 min on post-lesion day 1 and returned to baseline levels on days 5 and 8 (51.8 ± 2.9 and 55.4 ± 2.5 min, respectively). Total sleep time values for sham-lesioned animals were unchanged. An analysis of variance revealed significant effects for lesioning, the day sleep was measured, and an interaction between the two. An analogous set of findings occurred for other measures such as sleep latency. In the pre-lesioned rats, baseline values were 22.5 ± 2.3 min, rising to 43.2 ± 4.4 minutes 1 day post-lesion and returning to baseline values on days 5 and 8 (26.5 ± 1.1 and 18.1 ± 1.3 min, respectively). Thus, with this particular lesion size and ambient temperature (~26°C), recovery to baseline sleep was rapid, and IP triazolam injections performed on days 7 and 9 were done in the context of relatively normal sleep.

It should be noted that, in keeping with previous studies from this laboratory, even the acute disturbances in sleep were modest,[52] lending support to the view that larger decrements occur only when lesions are large, and the ambient temperature is low (e.g., see Reference 22). For example, on post-lesion day 1, we found total sleep time to be reduced to 83.7% of typical sleep, compared to a reduction to 15.2% in animals with large lesions kept at a cooler ambient temperature (20°C) in the Szymusiak and Satinoff study.[22] Perhaps most importantly, sleep had largely returned to baseline values by post-lesion day 5, so that subsequent studies of triazolam were done in the context of relatively normal sleep.

24.5.2 Benzodiazepines

On post-lesion days 7 and 9, we administered triazolam 0.8 mg/kg or vehicle IP at 10:00 a.m. in random sequence to 10 rats, and performed 2-hr sleep studies. Lighting conditions were similar to those of the studies presented earlier. Following the study, the accurate placement of the lesions was confirmed histologically. It was found that triazolam retained a potent hypnotic effect. Sleep latency dropped from 29.3 ± 3.7 min on vehicle to 6.0 ± 1.2 min on triazolam ($p < .001$). Similarly, total sleep time increased from 44.1 ± 4.7 min on vehicle to 70.6 ± 6.6 min on triazolam ($p < .001$). Total amounts of REM sleep and intermittent waking time were unchanged. Thus, at least at the post-lesion time point at which sleep was allowed to return to normal values, an intact preoptic area did not appear to be necessary for systemically administered triazolam to produce a hypnotic effect.

24.5.3 Barbiturates

In an analogous study, pentobarbital (50 mg/kg IP) was given to 10 rats one week after histologically confirmed ibotenic acid lesions of the preoptic area. Lesion sites are depicted schematically in Figure 24.1. Pentobarbital displayed a potent hypnotic effect, as seen in Table 24.1. In summary, parenterally administered triazolam and pentobarbital were found to have hypnotic effects in rats that were 1-wk post-preoptic area lesion. One interpretation of these results would be that an intact preoptic area is not necessary for the hypnotic effects of these compounds. An alternative view would be that this interpretation is dependent on the timing (relative to making the lesion) of drug administration. Because sleep values had returned to normal levels at the point at which the study was done, it is possible that redundant sleep-regulating systems were active at that point. To put this notion in context, the history of exploring the anatomy of sleep regulation is replete with examples of the redundancy of the mechanism, which emphasizes the importance of sleep to the organism. One of the best examples has been the study of lesions of the dorsal raphe nuclei (see Reference 53 for a review); although anatomic or pharmacologic lesions of the dorsal raphe acutely induce large reductions in sleep, within a short period of time the total amount of sleep returns to normal, as other systems assume control. It is possible that we are seeing a similar process in our preoptic-lesioned rats.

24.6 Conclusions

In summary, a substantial body of evidence has accumulated indicating that both benzodiazepines and barbiturates induce sleep by acting at various moieties of the $GABA_A$-benzodiazepine receptor complex. This also appears to be true for the non-benzodiazepine hypnotic zolpidem, which binds primarily to the Type I receptors.[54] Although it is not as clearly established that ethanol's hypnotic actions are mediated by the receptor complex, it is known that ethanol alters function of the chloride ionophore (e.g., see Reference 55), and at least some of its pharmacologic properties may be affected by compounds which bind to the ionophore.[56,57*] Similarly, the anesthetics etomidate and propofol interact with the receptor complex (Ruigt, G.S.F., personal communication). As we continue to learn more about the workings of the receptor complex, it seems likely that it will be found to be the common site of action of many agents. If so, we will have begun to find the answer to an old dilemma: how to explain the observation that compounds from a variety of pharmacological classes may have relatively similar effects in inducing sleep.

---

* In the case of ethanol and pentobarbital, and possibly other agents, the specificity of the hypnotic effect may be dose dependent. One might speculate, for instance, that at low doses sleep is induced by acting at the GABAa-benzodiazepine receptor complex, while at higher doses anesthesia results from nonspecific depression of a wide range of CNS sites.

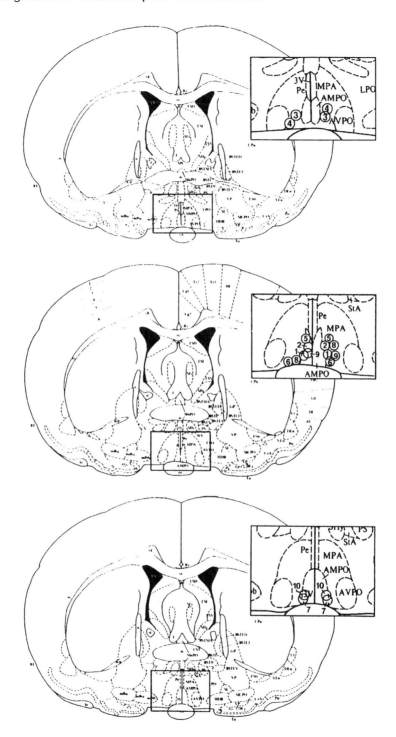

FIGURE 24.1
Ibotenic acid lesion sites in study of parenterally administered pentobarbital. The three sections (from top to bottom) are at –.26, –.30, and –.4 mm posterior to bregma. Circled numbers refer to histologically confirmed lesion site for the rat identified by that number. Abbreviations: MPA = medial preoptic area; AMPO = antero-medial preoptic area; AVPO = antero-ventral preoptic area, OX = optic chiasm, LPO = lateral preoptic area; AC = anterior commissure; BSTMA = bed nucleus strias terminalis, med. div., anterior; CC = corpus callosum; CPU = caudate putamen (striatum), CTX = cerebral cortex; F = fornix; HDB = horizontal limb of the diagonal band of Broca; MnPO = median preoptic nucleus, VP = ventral pallidum.

TABLE 24.1
Effects of Pentobarbital on Sleep in Preoptic Area-Lesioned Rats

	Vehicle	Pentobarbital	p <
Total sleep time	53.5 ± 1.8	111.6 ± 3.0	.0001
Sleep latency	19.7 ± 1.3	2.9 ± 1.0	.0001
NREM sleep	49.6 ± 1.5	111.6 ± 3.0	.0001
REM sleep	3.8 ± 1.2	0.0 ± 0.0	.01
Intermittent waking	48.6 ± 3.2	5.4 ± 3.0	.0001
REM latency	70.9 ± 9.7	—	—

Note: All values represent mean ± SEM minutes.

It should be mentioned in passing that knowledge of ligand interaction with this receptor complex may potentially be used to develop a wide range of psychoactive compounds, including analeptics. We have shown, for instance, that the benzodiazepine $B_{10}(-)$ enantiomer increases wakefulness in rats in a dose-dependent manner. It might be possible to use analogous compounds to enhance wakefulness with an agent that has the relatively benign side effect profile of benzodiazepines, as compared to that of amphetamines. Similarly, although some beta-carbolines have unacceptable side effects, others do not and indeed might potentially be useful as analeptics. The compound B-CCT, for instance, does not have proconvulsant or proconflict ("anxiogenic") properties, yet it potently increases wakefulness in rats.[58] Because benzodiazepines have mild respiratory depressant qualities, it is also possible that receptor inverse agonists might be used to treat sleep-disturbed respiration.[59]

At least one site of the hypnotic action of benzodiazepines and barbiturates is the MPA, although we have demonstrated that there are likely to be redundant systems involved. Interestingly, both ethanol[60] and adenosine[61] also enhance sleep when microinjected into the MPA. Thus, it appears to be a common neuroanatomic site for a variety of agents. Taking a broader view of sleep regulation, one can summarize the history of the field by tracing an original interest in the basal forebrain and hypothalamus in the classical studies of von Economo and Hess in the 1920s, then a subsequent ventral progression to the reticular activating system of the brainstem in the 1940s,[62] along with studies of specific brainstem nuclei such as the dorsal raphe in the 1960s,[38] and finally more recent interest once again in the basal forebrain and hypothalamus.[21,22,43] Given the bidirectional flow of outputs between brainstem structures such as the dorsal raphe nuclei and the anterior hypothalamus, it seems likely to me that we will continue to find that an important aspect of sleep regulation stems from their interaction. The studies we have presented here suggest that the effects of benzodiazepines and barbiturates on sleep may involve drug-induced perturbations of this system.

References

1. Squires, R.F. and Braestrup, C., Benzodiazepine receptors in rat brain, *Nature,* 266, 732–734, 1977.
2. Mohler, H. and Okada, T., Benzodiazepine receptor: demonstration in the central nervous system, *Science,* 198, 849–851, 1977.
3. Skolnick, P., Mendelson, W.B., and Paul, S.B., Benzodiazepine receptors in the central nervous system, in *Psychopharmacology of Sleep,* Wheatley, D., Ed., Raven Press, New York, 1981, pp. 117–134.

4. Schofield, P.R., Darlison, M.G., Fujita, N., Burt. D.R., Stephenson, F.A., Rodriguez, H., Rhee, L.M., Ramachandran, J., Reale, V., Glencorse, T.A., Seeburg, P.H., and Barnard, E.A., Sequence and functional expression of the $GABA_A$ receptor shows a ligand-gated superfamily, *Nature,* 328, 221–227, 1987.

5. Hadingham, K.L., Wingrove, P., le-Bourdelles, B.B., Plamer, K.J., Gagan, C.L., and Whiting, P.J., Cloning of cDNA sequences encoding human alpha 2 and alpha 3 gamma-aminobutyric acid-A receptor subunits and characterization of the benzodiazepine pharmacology of alpha 5-containing human gamma-aminobutyric acid-A receptors, *Mol. Pharmacol.,* 43, 970–975, 1993.

6. Twyman, R.E., Rogers, C.J., and Macdonald, R.L., Differential regulation of gamma-aminobutyric acid receptor channels by diazepam and pentobarbital, *Ann. Neurol.,* 25, 312–220, 1989.

7. Macdonald, R.L., Rogers, C.J., and Twyman, R.E., Barbiturate regulation of kinetic properties of the $GABA_A$ receptor channel of mouse spinal neurones in culture, *J. Physiol.,* 417, 483–500, 1989.

8. Harrison, N., Mendelson, W., and de Wit, H., Barbiturates, in *Psychopharmacology, Fourth Generation of Progress,* American College of Neuropsychopharmacology, CD-ROM, in press.

9. Mendelson, W.B., Neuropharmacology of sleep induction by benzodiazepines, *Crit. Rev. Neurobiol.,* 16, 221–232, 1992.

10. Mendelson, W.B. and Monti, D., Effects on sleep of triazolam and nifedipine injections into the medial preoptic area, *Neuropsychopharmacology,* 8, 227–232, 1993.

11. Mendelson, W.B., Nifedipine blocks effects of triazolam injected into the dorsal raphe nucleus on sleep, *Neuropsychopharmacology,* 10, 151–155, 1994.

12. Benavides, T., Quarteronet, D., Imbault, F., Malgouris, C., Uzan, A., Renault, C., and Dubroeucg, M.C., Labeling of "peripheral-type" benzodiazepine bindings sites in the rat brain by using [^{3}H]PK 11195; an isoquinoline carboxamide derivative: kinetic studies and autoradiographic localization, *J. Neurochem.,* 41, 1744–1750, 1983.

13. Mendelson, W.B., Cain, M., Cook, J.M., Paul, S.M., and Skolnick, P., A benzodiazepine receptor antagonist decreases sleep and reverses the hypnotic actions of flurazepam, *Science,* 219, 414–416, 1983.

14. Mendelson, W.B., Paul, S.M., and Skolnick, P., Does the benzodiazepine receptor play a role in sleep? Studies of stereospecificity, *Sleep Res.,* 11, 65, 1982.

15. Hernandez Peon, R. and Chavez Ibarra, G., Sleep induced by electrical or chemical stimulation of the forebrain, *Electroencephalogr. Clin. Neurophysiol.,* 24, 188–198, 1963.

16. Mendelson, W.B., Sleep induction by microinjection of pentobarbital into the medial preoptic area in rats, *Life Sci.,* 59, 1821–1828, 1996.

17. von Economo, C., *Schlaftheorie, Ergeb. Physiol.,* 28, 312–339, 1929.

18. Hess, W.R., Hirnreizversuche uber den mechanismus des Schlafes, *Arch. Psychiatr. Nervenkr.,* 86, 287–292, 1929.

19. Nauta, W.J.H., Hypothalamic regulation of sleep in rats: an experimental study, *J. Neurophysiol.,* 9, 285–316, 1946.

20. Sterman, M.B. and Clemente, C.D., Forebrain inhibitory mechanisms: sleep patterns induced by basal forebrain stimulation in the behaving cat, *Exp. Neurol.,* 6, 103–117, 1962.

21. McGinty, D.J. and Sterman, M.B., Sleep suppression after basal forebrain lesions in the cat, *Science,* 160, 1253–1255, 1968.

22. Syzmusiak, R. and Satinoff, E., Ambient temperature-dependence of sleep disturbances produced by basal forebrain damage in rats, *Brain Res. Bull.,* 12, 295–305, 1984.

23. Fritschy J.-M. and Mohler, H.. GABA-a receptor heterogeneity in the adult rat brain: differential regional and cellular distribution of seven major subunits, *J. Comp. Neurol.,* 359, 154–194, 1995.

24. Simerly, R.B. and Swanson, L.W., Projections of the medial preoptic nucleus: a *Phaseolus vulgaris* leucagglutinin anterograde tract-tracing study in the rat, *J. Comp. Neurol.*, 270, 209–242, 1988.
25. Swanson, L.W., in *Handbook of Chemical Neuroanatomy,* Vol. 5, Bjorklund, S., Hokfelt, T., and Swanson, L.W., Eds., Elsevier, New York, 1987, pp. 1–124.
26. Bremer, F., Preoptic hypnogenic area and the reticular activating system, *Arch. Ital. Biol.*, 111, 85–111, 1973.
27. Szymusiak, R. and McGinty, D., Responses evoked in mid brain reticular formation neurons by stimulation of the medial and lateral basal forebrain, *Sleep Res.*, 17, 18, 1988.
28. Mendelson, W.B., Majchrowicz, E., Mirmirani, N., Dawson, S., Gillin, J.C., and Wyatt, R.J., Sleep during chronic ethanol administration and withdrawal in rats, *J. Stud. Alcohol,* 39, 1213–1223, 1978.
29. Mendelson, W.B., Sleep induction by microinjection of pentobarbital into the medial preoptic area in rats, *Life Sci.*, 59, 1821–1828, 1996.
30. Jouvet, M. and Renault, J., Insomnie persistente apres lesions des noyaux raphe chez le chat, *Compt. Rend. Soc. Biol. (Paris),* 160, 1461–1465, 1966.
31. Pujol, J.F., Buguet, A., Froment, J.L., Jones, B., and Jouvet, M., The central metabolism of serotonin in the cat during insomnia: a neurophysiological and biochemical study after pchlorophenylalanine or destruction of the raphe system, *Brain Res.*, 29, 195–212, 1971.
32. Morgane, P.J. and Stern, W.C., Relationship of sleep to neuroanatomical circuits, biochemistry, and behavior, *Ann. N.Y. Acad. Sci. USA,* 193, 95–111, 1972.
33. Jacobs, B.L., Overview of the activity of brain monoaminergic neurons across the sleep/wake cycle, in *Sleep: Neurotransmitters and Neuromodulators,* Wauquier, A., Gaillard, J.M., Monti, J.M., and Radulovacki, M., Eds., Raven Press, New York, 1985, pp. 114.
34. Molliver, M.E., Serotonergic neuronal systems. What their anatomic organization tells us about function, *J. Clin. Psychopharmacol.*, 7, 3S–23S, 1987.
35. Simerly, R.B. and Swanson, L.W., The organization of neural inputs to the medial preoptic nucleus of the rat, *J. Comp. Neurol.*, 246, 312–342, 1985.
36. Mendelson, W.B., Martin, J.V., Perlis, M., and Wagner, R., Arousal induced by injection of triazolam into the dorsal raphe nucleus of rats, *Neuropsychopharmacology,* 1, 85–88, 1987.
37. Gallager, D.W., Benzodiazepines: potentiation of a GABA inhibitory response in the dorsal raphe nucleus, *Eur. J. Pharmacol.*, 49, 133–143, 1978.
38. Jouvet, M. and Delorme, F., Locus ceruleus et sommeil paradoxal, *Compt. Rend. Soc. Biol. (Paris),* 159, 895–899, 1965.
39. Jacobs, B.L., Overview of the activity of brain monoaminergic neurons across the sleep-wake cycle, in *Sleep: Neurotransmitters and Neuromodulators,* Wauquier, A., Gaillard, J.M., Monti, J.M., and Radulovacki, M., Eds., Raven Press, New York, 1985, p. 1.
40. McRae-DeGueurce, A. and Milon, H., Serotonin and dopamine afferents to the rat locus coeruleus: a biochemical study after lesioning of the ventral mesencephalic tegmental A10 region and the raphe dorsalis, *Brain Res.*, 263, 344–347, 1983.
41. Hobson, J.A., Lydic, R., and Baghdoyan, N., Evolving concepts of sleep cycle generation: from brain centers to neuronal populations, *Behav. Brain Sci.*, 9, 371–448, 1986.
42. Lucas, E.A. and Sterman, M.B., Effect of a forebrain lesion on the polycyclic sleep/wake cycle and sleep/wake patterns in the cat, *Exp. Neurol.*, 46, 368–388, 1975.
43. Sherin, J.E., Shiromani, P.J., McCarley, R.W., and Saper, C.B., Activation of ventrolateral preoptic neurons during sleep, *Science,* 271, 216–219, 1996.
44. Simerly, R.B., Gorski, R.A., and Swanson, L.W., Neurotransmitter specificity of cell and fibers in the medial preoptic nucleus: an immunohistochemical study in the rat, *J. Comp. Neurol.*, 246, 343–363, 1986.

45. Demling, J., Fuchs, E., Baumert, M., and Wuttke, W., Preoptic catecholamine, GABA, and glutamate release in ovariectomized estrogen-primed rats utilizing a push-pull cannula technique, *Neuroendocrinology*, 41, 212–218, 1985.

46. Tappaz, M.L., Brownstein, M.J., and Kopin, I.J., Glutamate decarboxylase (GAD) and gamma-aminobutyric acid (GABA) in discrete nuclei of hypothalamus and substantia nigra, *Brain Res.*, 125, 109–121, 1977.

47. Unnerstall, J.R., Niehoff, D.L., Kuhar, M.J., and Palacios, J.M., Quantitative receptor autoradiography using [^{3}H]ultrofilm: application to multiple benzodiazepine receptors, *J. Neurosci. Meth.*, 6, 59–73, 1982.

48. Boulant, J.A. and Silva, N.L., Neuronal sensitivities in preoptic tissue slices: interactions among homeostasis systeme, *Brain Res. Bull.*, 20, 871–878, 1988.

49. Szymusiak, R. and McGinty, D.J. Sleep-related neuronal discharge in the basal forebrain of cats, *Brain Res. Bull.*, 370, 82–92, 1986.

50. Koyama, Y. and Hayaishi, O., Firing of neurons in the preoptic/anterior hypothalamic areas in rat: its possible involvement in slow wave sleep and paradoxical sleep, *Neurosci. Res.*, 19, 31–38, 1994.

51. McGinty, D.J., Physiological equilibrium and the control of sleep states, in *Brain Mechanisms of Sleep*, McGinty, D.J., Morrison, A., Drucker-Colin, R., and Parmeggiani, J., Eds., Raven Press, New York, 1985, pp. 361–384.

52. Feng, P.-F., Bergmann, B.M., and Rechtachaffen, A., Sleep deprivation in rats with preoptic/anterior hyothalamic lesions, *Brain Res.*, 703, 93–99, 1995.

53. Mendelson, W.B., *Human Sleep: Research and Clinical Care*, Plenum Press, New York, 1987, pp. 14–25.

54. Depoortere, H., Zivkovic, B., Lloyd, K.G., Sanger, D.J., Perrrault, G., Langer, S.Z., and Bartholini, G., Zolpidem, a novel nonbenzodiazepine hypnotic. I. Neuropharmacological and behavioral effects, *J. Pharmacol. Exp. Therap.*, 237, 649–658, 1986.

55. Suzdak, P.D., Schwartz, R.D., Skolnick, P., and Paul, S.M., Ethanol stimulates gamma-aminobutyric acid receptor-mediated chloride transport in rat brain synaptoneurosomes, *Proc. Natl. Acad. Sci.*, 83, 4071–4075, 1986.

56. Mendelson, W.B., Martin, J.V., Wagner, R., Roseberry, C., Skolnick, P., Weissman, B.A., and Squires, R., Are the toxicities of pentobarbital and ethanol mediated by the GABA-benzodiazepine receptor-chloride ionophore complex?, *Eur. J. Pharmacol.*, 108, 63–68, 1985.

57. Koob, G.F., Mendelson, W.B., Schafer, J., Wall, T.L., Britton, K.T., and Bloom, F.E., Picrotoxinin receptor ligand blocks anti-punishment effects of ethanol, *Alcohol*, 5, 437–443, 1989.

58. Martin, J.V., Skolnick, P., Cook, J.M., and Mendelson, W.B., Inhibition of sleep and benzodiazepine receptor binding by a B-carboline derivative, *Pharmacol. Biochem. Behav.*, 34, 37–42, 1989.

59. Mendelson, W.B., Drugs which alter sleep and sleep-related respiration, in *Sleep and Respiration in Aging Adults*, Kuna, S.T. et al., Eds., Elsevier, New York, 1991, pp. 49–54.

60. Ticho, S.R., Stojanovic, M., Lekovic, G., and Radulovacki, M., Effects of ethanol injection to the preoptic area on sleep and temperature in rats, *Alcohol*, 9, 275–278, 1992.

61. Ticho, S.R. and Radulovacki, M., Role of adenosine in sleep and temperature regulation in the preoptic area of rats, *Pharmacol. Biochem. Behav.*, 40, 33–40, 1991.

62. Moruzzi, G. and Magoun, H.W., Brain stem reticular formation and activation of the EEG, *Electroencephalogr. Clin. Neurophysiol.*, 1, 455–473, 1949.

Chapter 25

State-Altering Actions of Ethanol, Caffeine, and Nicotine

Timothy Roehrs and Thomas Roth

Contents

- 25.1 Ethanol ... 422
  - 25.1.1 Effects on Sleep ... 422
  - 25.1.2 Effects on Alertness ... 423
  - 25.1.3 Interactions With State ... 423
- 25.2 Caffeine ... 424
  - 25.2.1 Effects on Sleep ... 424
  - 25.2.2 Effects on Alertness ... 424
  - 25.2.3 Interactions With State ... 425
- 25.3 Nicotine ... 425
  - 25.3.1 Effects on Sleep ... 426
  - 25.3.2 Effects on Alertness ... 426
  - 25.3.3 Interactions With State ... 426
- 25.4 State-Altering and Reinforcing Effects ... 427
  - 25.4.1 Behavioral Analysis ... 427
  - 25.4.2 Neurobiological Analysis ... 428
- 25.5 Conclusions ... 429
- References ... 429

Ethanol, caffeine, and nicotine share the distinction of being commonly used legal drugs that have profound state-altering characteristics. They are present in a variety of beverages, foods, and tobacco products. They are drugs that are recognized for their addictive liability, although caffeine is less so recognized than ethanol and nicotine. Intriguing questions for the neuroscientist include what are the mechanisms for their state-altering characteristics and how do these contribute to their wide use and/

or abuse. This chapter will examine the state-altering characteristics of each drug and attempt to relate their patterns of use to their state-altering properties and mechanisms of action.

In attempting to understand the state-altering actions of these three drugs, it should be noted that the basal state of the individual is an important consideration. In other words, the action of the drug, or its discontinuation effects, may differ in the wake vs. sleep state, or may differ depending on the level of sleepiness or alertness during the wake state. Some of the contradictory data regarding the state-altering actions of these drugs can be interpreted when the individual's basal state is considered. Further, the patterns of use/abuse may depend on the basal state of the individual.

25.1 Ethanol

Ethanol is a highly water- and lipid-soluble molecule that follows a single compartment pharmacokinetic model. Peak plasma concentrations are achieved within 30 to 60 min after consumption.[1] Thereafter, a constant 10 to 20 mg/100 ml plasma concentration are metabolized per hour. These pharmacokinetic characteristics are important in understanding the effects of ethanol on sleep, as discussed below. The primary mechanisms of action for the state-altering effects of ethanol are considered to be facilitation of the inhibitory effects of the neurotransmitter gamma-amino-butyric acid (GABA) and inhibition of excitatory effects at N-methyl-D-aspartate receptors.[2] The sedative effect of other GABA facilitating drugs, such as benzodiazepines, is well documented. Yet, there also is an extensive literature supporting the involvement of other neurotransmitter systems, such as serotonin, that also are known to have a role in modulation of sleep and wakefulness.

25.1.1 Effects on Sleep

Studies of the effects of ethanol on the sleep of healthy normals have used doses from 0.16 to 1.0 g/kg, yielding breath ethanol concentrations (BrEC) as high as .105%.[3–12] A few studies have found reduced sleep latency,[6,7,10,11] and a single study found increased sleep time at the low dose, 0.16 g/kg, but no effect at 0.32 and 0.64 g/kg.[8] When analyzing the sleep period by halves, some studies reported increased wake or light stage 1 sleep in the second half of the sleep period.[11,12] This result is interpreted to be a "rebound" effect following the completed metabolism of ethanol. Given ethanol metabolism of 10 to 20 mg/100 ml plasma concentration per hour (i.e., about .01 to .02% BrEC) and the typical peak BrEC measured before sleep of .06 to .08%, ethanol metabolism would be completed within 4 to 5 hr.

In contrast to these variable effects on sleep induction and sleep maintenance, consistent effects of ethanol on sleep staging are found. Most studies find suppression of REM sleep at least in the first half of the sleep period[3,4,9,11,12] and increased stage 3/4 sleep in the first half of the sleep period.[7,10,11,12] Several studies have assessed ethanol effects over repeated nights of administration, and clear tolerance to sedative and sleep stage effects develops.[6,9] This occurs rapidly within a week of nightly administration.

The effects of ethanol on the sleep of alcoholics have been assessed in a number of studies. The sleep of sober alcoholics is extremely disturbed; sleep latency is prolonged, sleep is fragmented, light, and shortened overall.[7,13,14] The administration of ethanol improves the sleep of the alcoholic, but only initially.[15,16] Thereafter, ethanol in the alcoholic is associated with fragmented and disturbed sleep. Thus, some have suggested that alcoholics, in part, drink to improve their sleep, and some data support this; the likelihood of relapse among alcoholic patients was associated with the severity of the individuals' sleep disturbance after detoxification.[14]

25.1.2 Effects on Alertness

The primary effect of ethanol on alertness and waking function is one of sedation and disturbed performance. The sedative effect of ethanol has been assessed directly using the Multiple Sleep Latency Test (MSLT). On the MSLT, individuals are given repeated opportunities throughout the day to fall asleep.[17] The sleep latency testing of a MSLT is conducted in a sleep-conducive environment with the person lying on a bed. Standard sleep recordings are used to document latency to sleep onset, defined as the time to the first 30-sec epoch scored as sleep. Typically four to five tests at 2-hour intervals are conducted, and the latency is averaged over the tests.[18] When sleep is totally deprived, reduced relative to one's biological need, or is disturbed by a sleep disorder or the environment, one falls asleep more rapidly on the MSLT (i.e., one becomes sleepier). And, when sleep time is increased or a patient's sleep disorder is successfully treated, sleep latency is increased (i.e., one becomes alert) on the MSLT.[17] The sedative effects of ethanol as measured by the MSLT are dose related over the dose range of 0.4 to 0.8 g/kg. The data show reduced average sleep latencies with increasing doses.[19] Importantly, as discussed later, the sedative effects continue for at least 2 hr beyond the point that breath ethanol concentration has reached zero.[20]

Laboratory performance assessment has traditionally been used to evaluate the functional-behavioral effects of drugs. State-altering effects are inferred from alterations of laboratory performance. However, one has to be cautious in that a drug can alter one specific type of performance, alter different aspects of performance in different directions (i.e., improve one type, but impair others), or even not impair performance, but rather the motivation to perform. The performance-disruptive effects of alcohol, coincident with its sedating effects as measured by the MSLT, have been studied extensively.[21] At this laboratory, we have used a divided attention task, in which subjects must follow a moving target on a video screen with a cursor while also responding to stimuli appearing in the center or the periphery of the screen, and an auditory vigilance task, in which subjects must detect long tones against the background of shorter tones. The effect of ethanol on divided attention and vigilance performance is characterized by a slowing of responses and an increase of missed stimuli.[21] Others have examined simulated automobile driving and cognitive performance, similarly finding disturbed functioning.[22-24]

25.1.3 Interactions With State

The effects following complete metabolism of ethanol vary as a function of the wake vs. sleep state. As noted above, the daytime studies in waking individuals have reported a continued sedative effect after BrEC has reached zero. On the other hand, the nighttime studies in sleeping individuals report a "rebound" wakefulness and disruption of sleep during the second half of the night, presumably after ethanol has been metabolized.

Ethanol effects also vary as a function of the waking state of the individual — that is, the level of sleepiness or alertness. The tactic used in such studies is to directly manipulate the subjects' level of sleepiness or alertness prior to, or immediately after, administering ethanol and then assessing its sedating and performance disruptive effects. Reducing the time spent in bed at night (TIB) increased the level of sleepiness the following day, and that increased sleepiness was associated with an increased sedative and performance-disruptive effect of ethanol, despite no change in the BrECs.[19,24] In contrast, increasing TIB reduces level of sleepiness the following day — in other words, increases alertness — and this increased alertness was associated with an attenuation of ethanol effects.[21] Again, importantly, the BrECs achieved were not diminished. The sedative and performance disruptive effects of ethanol can be attenuated by alerting operations after having consumed the ethanol. Daytime naps, even in people who have no complaints of sleepiness, increase basal levels of alertness. A nap completely reversed the sedative effect and attenuated the performance-disruptive effects of ethanol.[25]

Finally, there is some suggestion that the reinforcing effects of ethanol in social drinkers might vary as a function of basal waking state. Ethanol's reinforcing effects have been studied by giving social drinkers an opportunity to choose between a previously experienced color-coded ethanol or placebo beverage.[26–28] When given the opportunity to self administer multiple doses of ethanol, consistent individual differences are found. Those differences relate to the individuals' drinking histories, the subjective basal state of the individual, and the "mood-altering" effect of ethanol. Those with higher intakes were basally more sleepy and experienced ethanol as increasing vigor, elation, and positive mood, while those with lower intakes were not sleepy and experienced ethanol as increasing fatigue and reducing vigor.

25.2 Caffeine

Caffeine, a methylxanthine present in a variety of foods and beverages, is a central nervous system stimulant. It reaches peak plasma concentration within 30 to 60 min, and its duration of activity is typically 5 to 6 hr (its elimination half-life is 3 to 10 hr). The mechanism by which caffeine is thought to produce its stimulatory effect is through blockade of adenosine receptors. It has been hypothesized that adenosine has a role in the promotion of sleep.[29,30] Adenosine has an inhibitory effect on the electrophysiological activity of neurons in a number of central nervous system regions, some known to be involved in sleep, and the intravenous administration of adenosine in animals induces sleep.[30]

25.2.1 Effects on Sleep

As might be expected of a stimulant, caffeine in doses of 150 to 400 mg administered before or near bedtime prolongs the onset of sleep and reduces total sleep time in healthy normals.[31–35] To appreciate the magnitude of caffeine's disruptive effect on sleep, in one study 300 mg caffeine reduced sleep efficiency from 89 to 74% compared to the effect of 20 mg of methylphenidate and 40 mg of pemoline, which reduced sleep efficiency to 61 and 80%, respectively.[33] As to sleep-stage effects, some studies report a reduction in stage 3/4 sleep,[32,33] and a recent study which used spectral power analyses of the EEG during sleep found a reduction of slow-wave EEG activity.[36] Tolerance to the sleep-disruptive effects has not been reported, although the longest period of nightly administration has been 7 nights.[37] In that study, upon discontinuation after seven nights of caffeine (that is, a placebo on night eight), most measures returned to the baseline level with no evidence of "rebound" effects. The predicted "rebound" effects following chronic caffeine use would be increased sleep efficiency and enhanced slow-wave EEG activity. However, "rebound" effects, if present, may occur during wakefulness on the first day following the discontinuation (given its half-life and duration of activity) or not until the second discontinuation night. As discussed below, there is some evidence to suggest "rebound" in reports of waking fatigue and sleepiness during the discontinuation of chronic caffeine use.

25.2.2 Effects on Alertness

The effects of caffeine on direct measures of sleepiness or alertness (i.e., MSLT) have not been extensively studied. In one study using the MSLT, normal subjects given 250 mg caffeine (bid) at 9:00 a.m. and at 1:00 p.m. had an increased average daily sleep latency compared to placebo (i.e., decaffeinated coffee).[37] The authors also reported in that study a developing tolerance to the alerting effects of caffeine over four administrations of 250 mg. As to subjective alerting effects, a classical review of caffeine's (100 to 500 mg) effects on mood cites studies that showed increased vigor and reduced fatigue.[38] In many of those studies, subjects were undergoing sleep deprivation and/or extended military operations, and it is argued that caffeine's alerting effects are only evident in sleep-deprived and fatigued individuals. However, several recent studies have reported increased alerting effects under normal circumstances and even at low caffeine doses (< 100 mg).[39]

The performance and mood-enhancing effects of caffeine are discussed in the review cited above and in other more recent reviews.[38,40] Early studies assessing a wide range of functional capacities showed performance enhancement over the dose range of 100 to 500 mg.[38] Recent studies have demonstrated the performance-enhancing effects of caffeine at the range of low doses (32 to 64 mg) that is found in foods and over-the-counter drugs.[40] Systematic studies of the subjective effects of caffeine clearly indicate that caffeine at low doses (20 to 200 mg) increases rating of alertness, energy/activity, and concentration, while decreasing ratings of sleepiness and muddled thinking.[41]

Discontinuing the chronic use of caffeine is associated with mood and performance disturbance.[42] Studies have documented the presence of a withdrawal syndrome after a double-blind, placebo-controlled cessation of chronic, but moderate (235 mg on average), daily caffeine consumption.[42,43] On the second day of caffeine cessation (20 hr after the last caffeine), in addition to the ubiquitous headache, the symptoms recorded included reduced vigor and increased sleepiness, fatigue, and drowsiness. Further, performance was disrupted in the chronic caffeine drinker. The daily morning cup of coffee immediately after arising, which typically may be after an 8- to 12-hr abstinence beginning the previous evening, has been interpreted as restoration of plasma caffeine levels to avoid caffeine withdrawal. The studies cited above would support such an interpretation.

25.2.3 Interactions With State

No data suggest that the effects of caffeine differ in the wake vs. sleep state, although no study has directly compared effects; however, a study of another methylxanthine, theophylline, directly compared the dose-dependent sleep vs. waking effects of this drug.[44] The results showed similar but subtle differences. During the day, sleep onset was delayed (the wake state was maintained), while during the night sleep was lightened, but wake was not produced. As to interactions as a function of level of sleepiness in the wake state, a MSLT study compared the alerting effects of caffeine after 8 vs. 4 hr of bedtime the previous night.[45] Caffeine, 75 and 150 mg, produced the same 2- and 4-min increase, respectively, in average daily sleep latency, regardless of the amount of sleep the previous night. But, differential effects may occur at the other extreme of the sleepiness/alertness spectrum. In some studies, disruption of performance and anxiogenic-like mood effects are reported at higher doses.[41] One might speculate that after sleep-satiation and with full daytime alertness the effects of caffeine may be more consistently disruptive and may even be disruptive at lower doses. This would support the notion that the performance-enhancing effects of caffeine can be attributed to the chronic, mild sleep debt that most people have.

25.3 Nicotine

Nicotine is a tertiary amine with central nervous system stimulatory effects. It is present in various tobacco products. When tobacco is smoked, nicotine is readily absorbed through the lungs into arterial circulation, taking 10 to 20 sec to reach the brain.[46] Its elimination half-life is reported to be 2 to 3 hr. In the regular smoker, having a cigarette every 2 to 3 hr or more frequently, the plasma concentration accumulates over 6 to 8 hr and then declines during the nighttime hours when smoking has ceased. This pharmacokinetic information is important in appreciating its potential effects on sleep. As to the pharmacodynamics of nicotine, it binds to the nicotinic acetylcholine receptors, facilitating the release of acetylcholine, norepinephrine, dopamine, and other neurotransmitters.[46] These transmitters have all been implicated in the control of sleep and wake.

25.3.1 Effects on Sleep

The study of the effects of nicotine on the sleep of healthy non-smokers was facilitated with the development of nicotine delivery systems (i.e., nicotine gum and patches) for use in clinical smoking cessation programs. Transdermal nicotine (7 and 14 mg) in normals produced a dose-dependent increase in wakefulness and a reduction in REM percent relative to a placebo patch.[47] A study in normals and depressed patients of transdermal nicotine (17.5 mg) showed an increase in wake time and a decrease in REM percent in the normals, but increased REM percent in the patients and no alteration of their already abnormally increased wakefulness.[48] Finally, in obese non-smoking normals 15 mg transdermal nicotine reduced total sleep time and REM percent; it had no beneficial effect on sleep-disordered breathing and snoring of these obese individuals.[49]

The relation of nicotine dependence and sleep disturbance can be inferred from results of questionnaire studies of current cigarette smokers and polysomnographic studies of smokers discontinuing. A comprehensive sleep and health questionnaire found that smokers were more likely than non-smokers to report problems falling and staying asleep and remaining alert during the day.[50] Several polysomnographic studies of smokers during cessation have suggested the presence of disturbed sleep during the cessation from chronic smoking. Sixteen smokers were studied during the week before and following smoking cessation.[51] The number of arousals, awakenings, and sleep-stage changes was increased during the cessation week. In a double-blind, placebo-controlled study of heavy smokers (30 cigarettes per day on average), arousal frequency increased relative to baseline in the placebo patch group, while decreasing in the nicotine (22 mg) patch group.[52] The nicotine patch group also had an increase in percent stage 3/4 sleep relative to baseline.

25.3.2 Effects on Alertness

The effect of nicotine on direct measures of sleepiness or alertness (i.e., MSLT) has not been studied. Other indices of physiological arousal, including heart rate, blood pressure, and skin temperature, in normal non-smokers indicate that nicotine gum has dose-related (0 to 4 mg) arousal effects.[53] Cortical arousal is suggested by studies in which subcutaneous injections yielding 8.5 ng/ml peak nicotine concentrations in non-smokers increased waking EEG alpha frequency, and cigarettes with 1.2 mg nicotine increased alpha and beta EEG frequencies.[54,55] As to whether nicotine enhances the performance of non-smokers is less clear. Some studies indicate that motor responding is enhanced by 1.5 to 4 mg nicotine, while attention and cognitive processes are not affected.[53,56-58] On the other hand, in smokers the data strongly indicate that nicotine cigarettes (0.8 mg) or nicotine gum (0 to 4 mg) improve most types of performance.[59-62] In most of these studies, the smokers had abstained for 4 to 12 hr prior to assessment, suggesting a possible performance disruption associated with the cigarette abstinence. In those few studies that included a non-smoker control group, performance of the abstinent smokers at baseline was disturbed compared to that of the controls.[58]

25.3.3 Interactions With State

No data suggest that the effects of nicotine or its discontinuation differ in the wake vs. sleep state, but this question has not been directly assessed, nor have the interactive effects of the basal level of sleepiness or alertness with nicotine's effects been studied directly. As noted above, though, the data do suggest that nicotine in the doses studied does not have a performance-enhancing effect unless performance is suboptimal, as in the case of abstinent smokers. In support of this interpretation are data on the combined effects of smoking and ethanol. The sedating and performance-disruptive effects of ethanol were discussed above. A study of the combined effects of cigarette smoking and ethanol consumption found that smoking reversed the performance decline associated with ethanol.[63]

25.4 State-Altering and Reinforcing Effects

The profound state-altering actions of each of these commonly used drugs has been extensively reviewed. Each drug is also well known for it chronic use/abuse liability. How the state-altering actions of each of these drugs contributes to its reinforcing effects is of interest. The state-altering actions of the drug may contribute to either or both the initiation and the maintenance of the drug use/abuse. Discussion of the relation of state-altering and reinforcing effects can proceed at two levels of analysis, the behavioral and the neurobiological.

25.4.1 Behavioral Analysis

Use and abuse of these three drugs can be analyzed as operant behaviors. As such, the drug is viewed as a reinforcer that increases drug-seeking and self-administration behavior based on its consequences. Among important consequences are the various pharmacological effects of the drug. Parenthetically, drug reinforcement can also be associated with a variety of other non-pharmacological consequences; however, pharmacological consequences are the focus of this discussion. An important pharmacological property of both caffeine and nicotine, as reviewed above, is their alerting effect. For "sleepy" people experiencing fatigue and possibly even difficulty functioning, the alert state produced by the drug can be reinforcing.

Two models of drug reinforcement have been hypothesized. The first views the drug as a positive reinforcer, meaning that the drug produces what is inferred to be a "mood-elevating" or a "euphorogenic" effect. Subjective assessments in humans using validated, self-rated mood and drug effect questionnaires support these inferences. The second model views the drug as a negative reinforcer in that it reverses an inferred "aversive" state. Whether producing alertness in a "sleepy" individual is positive or negative reinforcement or the combination of both is difficult to determine. The answer may depend on the etiology, the level of sleepiness, and the other accompanying symptoms.

The sleepy state of the individual can result from a number of factors. Two very likely factors in caffeine and nicotine use are chronic insufficient sleep and abstinence after chronic use of caffeine or nicotine. Sleepiness due to insufficient sleep can develop in healthy people in two ways. One way in which it can develop is when individuals spend an insufficient time in bed to meet their biological sleep need, which is a hypothetical construct referring to the daily amount of sleep each individual needs to remain fully alert throughout the day. Epidemiological and laboratory data now indicate that a proportion of the general population is chronically sleepy.[64,65] Typically, the insufficient sleep is due to insufficient time in bed, rather than disturbed sleep. These people sleep well, and when their bedtime is increased they have increased sleep time and become alert during the day.[66] Sleepiness due to restricted bedtime may enhance the reinforcing efficacy of stimulants. Data from this laboratory indicate that restriction of time in bed from 8 to 4 hours in healthy normals without previous stimulant use shifts their preference for methylphenidate (10 mg) vs. placebo from about 20 to 80%.[67]

Sleepiness can also develop when time in bed is sufficient, but the continuity of sleep is disturbed. The common example of this problem in the general population would be that of rotating and shift workers. As indicated above, disturbed sleep also is clearly associated with acute caffeine use, and it also appears to be associated with nicotine use. But, whether one's sleep remains disturbed with chronic stimulant use (i.e., whether tolerance develops) and whether that disturbed sleep in turn produces subsequent sleepiness are not clear. It is known that sleep disrupted by acute ethanol consumed at bedtime leads to increased sleepiness and disturbed performance the next day.[68] Of interest, rotating and shift workers report the disproportionate use of sedating drugs, including alcohol, to improve sleep and alerting drugs to improve alertness.

Finally, abstinence from the chronic use of either of these stimulants is probably associated with increased sleepiness. Increased sleepiness and fatigue subsequent to caffeine discontinuation are clearly supported by the data, as noted above. Sleepiness has not been systematically assessed in nicotine abstinence; however, the apparently disturbed performance during nicotine abstinence, relative to non-smokers, suggests that there may be an increase of sleepiness and fatigue during nicotine abstinence. Resumption of nicotine or caffeine use then reverses the performance impairment in the chronic user and probably also sleepiness.

As to the reinforcing effect of ethanol relative to its sedating effects, it was noted above that its reinforcing effect in social drinkers varies as a function of subjective state of sleepiness or alertness of the individual. That is, "sleepy" individuals experience ethanol as increasing vigor and positive mood, while alert persons do not; they find it sedating. In the abstinent alcoholic, disturbed sleep was noted, and it has been hypothesized that resumption of ethanol use by the alcoholic may be, in part, due to the fact that ethanol improves their disturbed sleep, at least initially. The sedating effects of ethanol may be reinforcing to non-alcoholic persons who have insomnia. A recent study from this laboratory found that persons with insomnia were more likely to self administer ethanol in average doses of 0.45 g/kg before sleep than were non-insomniacs who reported similar levels of social drinking (i.e., non-hypnotic ethanol use) as the insomniacs.[69]

25.4.2 Neurobiological Analysis

The generally accepted positive-reinforcement neurobiology of drug self administration involves mesocorticolimbic projections originating in the ventral tegmental area of the rostral reticular activating system and terminating in the nucleus accumbens. It is primarily a dopamine system.[70] Citation of the large amount of evidence supporting this hypothesis is beyond the scope of this chapter. But, most drugs which are abused in society, including those discussed in this chapter, interact in some way with this system. This reinforcement system is one part of a broader dopaminergic system that projects into several other forebrain regions and as a whole is considered to have executive and integrative functions.[70] The dopaminergic neurons of the ventral tegmental area are modulated by a number of other neurotransmitter systems. It is through this modulation that the sleep/wake state and levels of sleepiness during wake could have an impact on drug reinforcement. However, the specifics of this possible modulation have not been elucidated.

The difficulty in relating reinforcing and sleep/wake mechanisms is that the neural mechanisms generating sleep and wake are complex and interrelated, involving systems that extend from the medulla through the brainstem and hypothalamus to the basal forebrain.[71] They certainly overlap at points with the classical reinforcement system described above. The neurochemicals involved are multiple and include the catecholamines, acetylcholine, histamine, and glutamate during wakefulness and serotonin, GABA, and adenosine in sleep. Modulation of the reinforcing effects of the drugs discussed in this chapter could occur through any of these systems or combinations thereof.

In the chronic use of drugs with abuse liability it is hypothesized by Koob that the neurobiological systems involved in acute positive reinforcement adapt by establishing "opponent" processes.[72] These hypothesized neurobiological opponent processes "neutralize" the acute reinforcing effects of the drug and during the abstinence are unopposed. Consequently, they produce the abstinence syndrome, an inverse state to the drug state, which becomes the basis of negative reinforcement (i.e., reversal of abstinence symptoms). It was noted in the earlier review that the drugs discussed in this chapter are associated with profound sleep/wake alterations during drug abstinence. The opponent sleep/wake adaptations could occur through modulatory mechanisms similar to those of the positive reinforcement system. However, Koob does not hypothesize that same system opponent processes are necessary.[71] The opponent processes may develop through other neurobiological systems, including the possibility of sleep/wake systems, the point being that sleep/wake state alterations are one of a number of symptoms expressed during abstinence from these drugs.

25.5 Conclusions

The state-altering actions of ethanol, caffeine, and nicotine have been well documented and are consistent with the known pharmacokinetics of the drugs. Ethanol intially enhances sleep and later in the night disrupts sleep, while caffeine and nicotine consistently disrupt sleep. During wake, ethanol is sedating and caffeine and nicotine are alerting. Interactions of basal state (i.e., sleep vs. wake and levels of sleepiness during wake) with drug effects have been documented for ethanol, but in the few studies done with caffeine interactive effects have not been observed. No studies of possible nicotine basal state interactions have been done. The chronic use of these drugs typically produces tolerance to their state-altering effects, and during abstinence a withdrawal syndrome is found that includes disturbance of sleep and of alertness during wake, which is generally the inverse of their acute effect, and is often hypothesized to be critical to and predictive of their continued use.

Behavioral and neurobiological analyses of the chronic use or abuse of these drugs can be applied to hypothesize that their state-altering action is an important factor in their chronic use and abuse. The state-altering actions of ethanol, caffeine, and nicotine may play a role in both the initiation and maintenance of chronic use or abuse. Much data indirectly support this hypothesis. A few studies have directly tested this hypothesis and have shown that basal state is related to the reinforcing effects of these drugs and that manipulations of basal state can alter their reinforcing effects. Further study of the relation of the state-alerting and reinforcing effects of these drugs will be important in both prevention and treatment of their chronic use and abuse.

References

1. Jones, A., Physiological aspects of breath-alcohol measurement, *Alcohol Drugs Driving*, 6, 1, 1991.
2. Hoffman, P.L. and Tabakoff, B., Ethanol's action on brain biochemistry, in *Alcohol and the Brain*, Tarter, R.E. and van Thiel, D.H., Eds., Plenum Press, New York, 1988, p. 19.
3. Gresham, S.C., Webb, W.B., and Williams, R.L., Alcohol and caffeine: effect on inferred visual dreaming, *Science*, 140, 1226, 1963.
4. Yules, R.B., Freedman, D.X., and Chandler, K.A., The effects of ethyl alcohol on man's electroencephalographic sleep cycle, *Electroencephalogr. Clin. Neurophysiol.*, 20, 109, 1966.
5. Yules, R.B., Lippman, M.E., and Freedman, D.X., Alcohol administration prior to sleep, *Arch. Gen. Psychiatry*, 16, 94, 1967.
6. Rundell, J.B., Lester, B.K., Griffiths, W.J., and Williams, H.L., Alcohol and sleep in young adults, *Psychopharmacology*, 26, 201, 1972.
7. Williams, H. and Salamy, A., Alcohol and sleep, in *The Biology of Alcoholism*, Kissin, A. and Begleiter, H., Eds., Plenum Press, New York, 1972, p. 475.

8. Stone, B.M., Sleep and low doses of alcohol, *Electroencephalogr. Clin. Neurophysiol.*, 48, 706, 1980.
9. Prinz, P., Roehrs, S.T., Vitaliano, P., Linnoila, M., and Weitzman, E., Effect of alcohol on sleep and nighttime plasma growth hormone and cortisol concentrations, *J. Clin. Endocrin. Metab.*, 51, 759, 1980.
10. MacLean, A. and Cairns, J., Dose-response effects of ethanol on the sleep of young men, *J. Stud. Alcohol*, 43, 434, 1982.
11. Williams, D., MacLean, A., and Cairns, J., Dose-response effects of ethanol on the sleep of young women, *J. Stud. Alcohol*, 44, 515, 1983.
12. Roehrs, S.T., Yoon, J., and Roth, T., Nocturnal and next-day effects of ethanol and basal level of sleepiness, *Hum. Psychopharm. Clin. Exp.*, 6, 307, 1991.
13. Zarcone, V.P., Sleep and alcoholism, in *Sleep Disorders Basic and Clinical Research*, Chase, M. and Weitzman, E.D., Eds., SP Medical, New York, 1983, p. 319.
14. Gillin, J.C., Smith, T.L., Irwin, M., Kripke, D.F., and Schuckit, M., EEG sleep studies in "pure" alcoholism during subacute withdrawal: relationships to normal controls, age, and other clinical variables, *Biol. Psychiatry*, 27, 477, 1990.
15. Allen, R.P., Wagman, A.M., Funderburk, F.R., and Wells, D.T., Slow wave sleep: a predictor of individual differences in response to drinking?, *Biol. Psychiatry*, 15, 345, 1980.
16. Wagman, A.M. and Allen, R.P., Effects of alcohol ingestion and abstinence on slow wave sleep of alcoholics, *Adv. Exp. Med. Biol.*, 59, 453, 1975.
17. Roth, T., Roehrs, T.A., Carskadon, M.A., and Dement, W.C., Daytime sleepiness and alertness, in *Principles and Practice of Sleep Medicine*, 2nd ed., Kryger, M.H., Roth, T., and Dement, W.C., Eds., W.B. Sauders, Philadelphia, 1994, p. 40.
18. Carskadon, M.A., Dement, W.C., Mitler, M.M., Roth, T., Westbrook, P.R., and Keenan, S., Guidelines for Multiple Sleep Latency Test (MSLT): a standard measure of sleepiness, *Sleep*, 9, 519, 1986.
19. Zwyghuizen-Doorenbos, A., Roehrs, T., Lamphere, J., Zorick, F., and Roth, T. Increased daytime sleepiness enhances ethanol's sedative effects, *Neuropsychopharmacology*, 1, 279, 1988.
20. Roth, T., Roehrs, T., and Merlotti, L., Ethanol and daytime sleepiness, *Alcohol, Drugs Driving*, 5, 357, 1990.
21. Roehrs, T., Zwyghuizen-Doorenbos, A., Timms, V., Zorick, F., and Roth, T., Sleep extension, enhanced alertness and the sedating effects of ethanol, *Pharmacol. Biol. Behav.*, 34, 321, 1989.
22. Mungas, D., Ehlers, C.L., and Wall, T.L., Effects of acute alcohol administration on verbal and spatial learning, *Alcohol Alcohol.*, 29, 163, 1994.
23. Rabbitt, P.M.A. and Maylor, E.A., Investigating models of human performance, *Br. J. Psychol.*, 82, 259, 1991.
24. Roehrs, T., Beare, D., Zorick, F., and Roth, T., Sleepiness and ethanol effects on simulated driving, *Alcohol Clin. Exp. Res.*, 18, 154, 1994.
25. Roehrs, T., Zwyghuizen-Doorenbos, A., Zwyghuizen, H., and Roth, T., Sedating effects of ethanol after a nap, *Alcohol Drugs Driving*, 5, 351, 1990.
26. de Wit, H. and Johanson, C.E., Choice procedures in human drug self-administration, in *Methods of Assessing the Reinforcing Properties of Abused Drugs*, Bozarth, M.A., Ed., Springer-Verlag, Berlin, 1987, p. 559.
27. de Wit, H., Uhlenhuth, E.H., Pierri, J., and Johanson, C.E., Individual differences in behavioral and subjective responses to alcohol, *Alcohol Clin. Exp. Res.*, 11, 52, 1987.
28. de Wit, H., Pierri, J., and Johanson, C.E., Assessing individual differences in ETOH preference using a cumulative dosing procedure, *Psychopharmacology*, 98, 113, 1989.
29. Radulovacki, M., Adenosine and sleep, in *Adenosine and the Adenine Nucleotides as Regulators of Cellular Function*, Phillips, W.C., Ed., CRC Press, Boca Raton, FL, 1991, p. 381.

30. Benington, J.H. and Heller, H.C., Restoration of brain energy metabolism as the function of sleep, *Prog. Neurobiol.*, 45, 347, 1995.
31. Brezinova, V., Effect of caffeine on sleep: EEG study in late middle age people, *Br. J. Clin. Pharmacol.*, 1, 203, 1974.
32. Karacan, I., Thornby, J.I., Anch, M., Booth, G.H, Williams, R.L, and Salis, P.J., Dose-related sleep disturbances induced by coffee and caffeine, *Clin. Pharmacol. Ther.*, 20, 682, 1977.
33. Nicholson, A.N. and Stone, B.M., Heterocyclic amphetamine derivatives and caffeine on sleep, *Br. J. Clin. Pharmacol.*, 9, 195, 1980.
34. Okuma, T., Matsuoka, H., Matuse, Y., and Toyomura, K., Model insomnia by methylphenidate and caffeine and use in the evaluation of temazepam, *Psychopharmacology*, 76, 201, 1982.
35. Bonnet, M.H. and Arand, D.L., Caffeine use as a model of acute and chronic insomnia, *Sleep*, 15, 526, 1992.
36. Landolt, H.P., Dijk, D.J., Gaus, S.E., and Borbely, A.A., Caffeine reduces low-frequency delta activity in the human sleep EEG, *Neuropsychopharmacology*, 12, 229, 1995.
37. Zwyghuizen-Dorrenbos, A., Roehrs, T.A., Lipschutz, L., Timms, V., and Roth, T., Effects of caffeine on alertness, *Psychopharmacology*, 100, 39, 1990.
38. Weiss, B. and Laties, V.G., Enhancement of human performance by caffeine and the amphetamines, *Pharmacol. Rev.*, 14, 1, 1962.
39. Sawyer, D.A., Julia, H.L., and Turin A.C., Caffeine and human behavior: arousal, anxiety, and performance effects, *J. Behav. Med.*, 5, 415, 1982.
40. Lieberman, H.R., Wurtman R.J., Emde, G.G., and Coviella, I.L.G., The effects of caffeine and aspirin on mood and performance, *J. Clin. Psychopharmacol.*, 7, 315, 1987.
41. Griffiths, R.R. and Mumford, G.K., Caffeine: a drug of abuse, in *Psychopharmacology: The Fourth Generation of Progress*, Bloom, F.E. and Kupfer, D.J., Eds., Raven Press, New York, 1995, p. 1699.
42. Griffiths, R.R. and Woodson, P.P., Caffeine physical dependence: a review of human and laboratory animal studies, *Psychopharmacology*, 94, 437, 1988.
43. Silverman, K., Evans, S.M., Strain, E.C., and Griffiths, R.R., Withdrawal syndrome after the double-blind cessation of caffeine consumption, *New Engl. J. Med.*, 327, 1109, 1992.
44. Roehrs, T., Merlotti, L., Halpin, D., Rosenthal, L., and Roth, T., Effects of theophylline on nocturnal sleep and daytime sleepiness/alertness, *Chest*, 108, 382, 1995.
45. Rosenthal, L., Roehrs, T.A., Zwyghuizen-Doorenbos, A., Plath, D., and Roth, T., Alerting effects of caffeine anfter normal and restricted sleep, *Neuropsychopharmacology*, 4, 103, 1991.
46. Benowitz, N.L., Pharmacology of nicotine: addiction and therapeutics, *Ann. Rev. Pharmacol. Toxicol.*, 36, 597, 1996.
47. Gillin, L.C., Lardon, M., Ruiz, C., Golshan, S., and Salin-Pascual, R.J., Dose-dependent effects of transdermal nicotine on early morning awakening and rapid eye movement sleep time in non-smoking normal volunteers, *J. Clin. Psychopharmacol.*, 14, 264, 1994.
48. Salin-Pascual, R.J., de la Fuente, J.R., Galicia-Polo, L., and Drucker-Colin, R., Effects of transdermal nicotine on mood and sleep in nonsmoking major depressed patients, *Psychopharmacology*, 121, 476, 1995.
49. Davila, D.G., Hurt, R.D., Offord, K.P., Harris, C.D., and Shepard, J.W., Acute effects of transdermal nicotine on sleep architecture, snoring, and sleep-disordered breathing in nonsmokers, *Am. J. Respir. Crit. Care Med.*, 150, 469, 1994.
50. Phillips, B.A. and Danner, F.J., Cigarette smoking and sleep disturbance, *Arch. Intern. Med.*, 155, 734, 1995.
51. Prosise, G.L., Bonnet, M.H., Berry, R.B., and Dickel, M.L., Effects of abstinence from smoking on sleep and daytime sleepiness, *Chest*, 105, 1136, 1994.

52. Wetter, D.W., Fiore, M.C., Baker, T.B., and Young, T.B., Tobacco withdrawal and nicotine replacement influence objective measures of sleep, *J. Cont. Clin. Psychol.*, 63, 658, 1995.
53. Heishman, S.J., Snyder, F.R., and Henningfield, J.E., Performance, subjective, and physiological effects of nicotine in non-smokers, *Drug Alcohol Depend.*, 34, 11, 1993.
54. Pritchard, W.S., Robinson, J.H., deBethizy, J.D., Davis, R.A., and Stiles, M.F., Caffeine and smoking: subjective, performance, and psychophysiological effects, *Psychophysiology*, 32, 19, 1995.
55. Foulds, J., McSorely, K., Sneddon, J., Feyerabend, C., Jarvis, M.J., and Russell, M.A., Effect of subcutaneous nicotine injections on EEG alpha frequency in non-smokers: a placebo-controlled pilot study, *Psychopharmacology*, 115, 163, 1994.
56. Wesnes, K. and Revell, A., The separate and combined effects of scopolamine and nicotine on human information processing, *Psychopharmocology*, 89, 55, 1983.
57. West, R.J. and Jarvis, M.J., Effects of nicotine on finger tapping rate in non-smokers, *Pharmacol. Biochem. Behav.*, 25, 727, 1986.
58. Foulds, J., Stapleton, J., Swettenham, J., Bell, K., McSorley, K., and Russell, M.H., Cognitive performance effects of subcutaneous nicotine in smokers and never-smokers, *Psychopharmocology*, 127, 31, 1996.
59. Houlihan, M.E., Pritchard, W.S., Krieble, K.K., Robinson, J.H., and Duke, D.W., Effects of cigarette smoking on EEG spectral-band power, dimensional complexity, and nonlinearity during reaction-time task performance, *Psychophysiology*, 33, 740, 1996.
60. Strough, C., Mangan, G., Bates, T., Frank, N., Kerkin, B., and Pellett, O., Effects of nicotine on perceptual speed, *Psychopharmocology*, 119, 305, 1995.
61. Baldinger, B., Hasenfratz, M., and Battig, K., Comparison of the effects of nicotine on a fixed rate and subject-paced version of the rapid information processing task, *Psychopharmocology*, 121, 396, 1995.
62. Parrott, A.C. and Craig, D., Cigarette smoking and nicotine gum (0, 2, and 4 mg): effects upon four visual attention tasks, *Neuropsychobiology*, 25, 34, 1992.
63. Michel, Ch. and Battig, K., Separate and combined psychophysiological effects of cigarette smoking and alcohol consumption, *Psychopharmocology*, 97, 65, 1989.
64. Breslau, N., Roth, T., Rosenthal, L., Andreski, P., Daytime sleepiness: an epidemiological study of young adults, *Am. J. Pub. Health*, 87, 1649–1653, 1997.
65. Levine, B., Roehrs, T., Zorick, F., and Roth, T., Daytime sleepiness in young adults, *Sleep*, 11, 39, 1988.
66. Roehrs, T., Shore, E., Papineau, K., Rosenthal, L., and Roth, T., A two-week sleep extension in sleepy normals, *Sleep*, 19, 576, 1996.
67. Papineau, K., Roehrs, T.A., Rosenthal, L., and Roth, T., The self-administration of methylphenidate as a function of time in bed, *Sleep Res.*, 25, 66, 1996.
68. Roehrs, T., Yoon, J., and Roth, T., Nocturnal and next-day effects of ethanol and basal level of sleepiness, *Hum. Psychopharmacol. Clin. Exp.*, 6, 307, 1991.
69. Roehrs, T.A., Papineau, K., Fortier, J., Rosenthal, L., and Roth, T., The reinforcing and subjective effects of ethanol as a hypnotic, *Sleep Res.*, 25, 70, 1996.
70. Le Moal, M., Mesocorticolimbic dopaminergic neurons: functional and regulatory roles, in *Psychopharmacology: The Fourth Generation of Progress,* Bloom, F.E. and Kupfer, D.J., Eds., Raven Press, New York, 1995, p. 283.
71. Jones, B.E., Basic mechanisms of sleep-wake states, in *Principles and Practice of Sleep Medicine,* 2nd ed., Kryger, M.H., Roth, T., and Dement, W.C., Eds., W.B. Saunders, Philadelphia, 1994, p. 145.
72. Koob, G.F., Stinus, L., LeMoal, M., and Bloom, F.E., Opponent process theory of motivation: neurobiological evidence from studies of opiate dependence, *Neurosci. Biobehav. Rev.*, 13, 135, 1989.

Chapter

Psychomimetic Drugs, Marijuana, and 5-HT Antagonists

Oscar Prospéro-García, Eric Murillo-Rodríguez, Anabel Jiménez-Anguiano, Luz Navarro, Manuel Sánchez, Margarita Gómez, Dolores Martínez-González, Marcela Palomero, and Rene Drucker-Colín

Contents

26.1	Introduction	434
26.2	Serotonin	434
26.3	Marijuana	435
	26.3.1 Effects in Humans	435
	26.3.2 Effects in Animals	435
	26.3.3 Marijuana Receptors	436
	26.3.3.1 Cannabinoids	436
	26.3.3.2 Arachidonylethanolamide (Anandamide)	436
	26.3.3.3 Cis-9-10-Octadecenoamide (Oleamide)	437
26.4	Other Psychedelic Drugs	437
	26.4.1 d-Lysergic Acid Diethylamide (LSD)	438
	26.4.2 Effects in Humans	438
	26.4.3 Effects in Animals	438
26.5	Discussion	438
Acknowledgment		439
References		439

26.1 Introduction

Many drugs have the capacity to modify brain physiology. In doing so, these psychoactive agents alter consciousness, alertness, perception, and behavioral performance. Repeated self-administration of some of these psychoactive drugs can lead to addiction to these pharmacological effects. In this chapter, we describe the action of a selected example of these drugs, including marijuana; mushroom and cactus extracts with hallucinogenic properties, i.e., psilocybin, psilocin, and mescaline; and the most astonishing drug of this group, LSD. All have an impact on conscious behavior and sleep. It is noteworthy to mention that in the 1960s the widespread use of psychogenic drugs led to what has been termed the psychedelic era. Yet, the essence of this sociological movement is not new. For centuries these drugs have been used for religious or mystical purposes, placing them in a class of drugs that might be best described as "drugs of the spirit".[54] For example, the earliest reference to marijuana comes from the Chinese emperor Sheng Nung in the year 2737 B.C., in which marijuana was termed "the liberator of sin".[52] Possibly influenced by this association, a very recently discovered endogenous molecule that binds to cannabinoid receptors in the brain of mammals has been called anandamide,[12] a name coined from the Sanskrit word *ananda* which means bliss. In part, because most of the psychoactive effects caused by these drugs result from a probable interaction with serotonin (5-HT) processes in the brain, we present a brief description of the role of this neurotransmitter in brain function.

26.2 Serotonin

Serotonin is one of the major monoaminergic neurotransmitters in the brain. Although its action in the CNS is not completely understood, it has been implicated in the regulation of both sensory alertness and sleep. In addition, serotonin's role in the modulation of mnemonic processes and in the pathophysiology of several forms of mental disease recently has been supported.[37] In addition, despite the fact that the monoaminergic theory of sleep is no longer accepted as originally proposed by Jouvet in 1969,[34] there is still experimental evidence suggesting a role for serotonin in slow-wave sleep (SWS) processes. Some evidence indicates that serotonin is involved in the maintenance of sensory consciousness.[33] The precise way in which serotonin affects consciousness is, of course, unclear; however, recent research has indicated that a major role could be in regulating the perception of environmental stimuli.[33] Studies in which unit activity of serotonin-containing cells of the dorsal raphe nucleus were monitored in freely moving cats revealed a close association between motor and sensory processes believed to be under the control of serotonin, i.e., motor activity is associated with a suppression of sensory transmission.[9] Somatosensory-evoked potential amplitude was reduced when experimental subjects executed a voluntary movement.[9] Reciprocally, sensory transmission is facilitated when serotonin-containing cells are inhibited during sensory orientation.[33] It has been shown that when a cat is in quiet, non-motoric waking, neurons containing serotonin exhibit a typical slow and regular activity. As soon as the animal begins a grooming behavior, activation of these neurons reaches 2 to 5 times the activity of quiet wakefulness.[43] In contrast, when subjects are exposed to strong stimuli causing an orienting response, activity of these neurons falls silent. This observation suggests that serotonin plays a pivotal role in the regulation of perception of environmental stimuli and motor expression.

It is very well accepted that sleep-related, phasic ponto-geniculate-occipital (PGO) cortex waves are examples of endogenously driven stimuli impinging on the cortex as well as other brain structures during rapid-eye-movement (REM) sleep. Indeed, serotonin-containing cells are exclusively silent during this sleep stage and particularly during these phasic events.[43] Also, the inhibition of serotonin synthesis produced by systemic administration of parachlorophenylalanine (PCPA) "releases" PGO waves during slow-wave sleep and wakefulness, while transiently suppressing the appearance of both sleep states. Cats treated with PCPA in a chronic manner become hyperreactive to external stimuli. For example, auditory stimulation evokes an enhanced startle reflex and simultaneously triggers bursts of PGO waves. Although the acoustic stimulation is repeated frequently, animals do not habituate to these stimuli (non-published observations from our laboratory).

These facts suggest that serotonin can regulate neural responses to stimuli, either exogenous or endogenous in origin. They also imply that impairment of serotoninergic neurotransmission can alter the coding and potential response to specific stimuli. It is interesting to note that most psychedelic drugs have the capacity of suppressing the action of endogenously released serotonin. It is now clear that psychedelic drugs can powerfully block serotoninergic receptors, and that this can result in perceptual hallucinations. Glennon[22,23] has suggested that the potency to induce hallucinations is directly proportional to the affinity for the serotonin receptor. In the next section, we will discuss the action of drugs that produce such hallucinations, as well as other effects.

26.3 Marijuana

Marijuana, grass, weed, pot, reefer, hashish, charas, bhang, ganja, dagga — all are names referring to the derivatives of the plant *Cannabis sativa*. This is a hemp plant consisting of two varieties — *indica* and *americana*. The entire plant harbors cannabinoids, the active pharmacological agent; however, the highest concentrations are found in the flowering tops. Although the hemp plant synthesizes at least 400 chemicals, only 60 are cannabinoid-like in structure. The representative active molecule is Δ^9-tetrahydrocannabinol (Δ^9-THC).

26.3.1 Effects in Humans

Marijuana is by far the illicit drug most commonly used worldwide. In the U.S., about 55% of young adults admit some experience with this drug.[52] Δ^9-THC induces several psychoactive effects after an oral dose of 20 mg/kg, or after smoking a cigarette containing 2% of Δ^9-THC.[13] Among other effects observed are impairment of short-term memory, motor coordination, cognitive ability, attention, and general and self perception, along with an increase in hunger and vividness of visual imagery. Additional effects of Δ^9-THC include decreased skin and core temperature and increased subjective sleepiness, as well as polygraphically defined alpha waves and an increase in SWS.[18,50] At higher doses, Δ^9-THC can elicit hallucinations associated with delusional and paranoid feelings. Thinking can be confused and disorganized with a profound loss of the sense of time.[19,45] Although the acute administration of Δ^9-THC can enhance sexual drive, the levels of testosterone are lowered and spermatogenesis is inhibited. Actually, Δ^9-THC may induce impotence in man.[28,35] Hormonal changes also have been described in women following both acute and chronic use of Δ^9-THC, such as the suppression of luteinizing hormone during the luteal phase of the menstrual cycle. Δ^9-THC may also have fetal teratogenic actions.[17]

Despite all these adverse effects, marijuana possesses potential therapeutic effects. The drug has been used in China as a pain reliever and to control gout, malaria, and other diseases.[52] Several effects are of contemporary interest. Beneficial effects of Δ^9-THC include attenuation of the nausea and vomiting associated with cancer chemotherapy, decreased bronchial constriction with asthma, decreased intraocular pressure from glaucoma, antipyretic actions, treatment of convulsant disorders, appetite stimulation, and decreased intestinal motility in patients suffering from diarrhea.[28]

26.3.2 Effects in Animals

Δ^9-THC produces several effects in rats, including short-term and spatial memory impairment and aversion in both place and taste preference tasks.[37,48,49] These effects can be elicited by either systemic or intrahippocampal administration of this drug.[24,39,44] Other effects include hypermotility and hyperreactivity at low doses, while hypomotility, hypothermia, rigid immobility, and antinociception are observed at higher doses.

26.3.3 Marijuana Receptors

Interestingly enough, most of all the symptoms caused by marijuana are mediated by at least two subtypes of cannabinoid receptors, CB1 and CB2. Receptor CB1 has a preferential distribution in the brain, with the highest concentration in the hippocampus,[26,27,29,41] although its mRNA has also been found in testis.[21] In contrast, CB2 has a more peripheral distribution with no apparent expression of either protein or mRNA in the brain.[46] The action of cannabinoid molecules in the brain via the CB1 receptor is apparently mediated by G-proteins to inhibit adenylate cyclase and Ca^{++} channels.[10,16,40] It is not known which transduction mechanisms mediate the activation of CB2 receptors. These receptors are regulated by at least three classes of ligands: cannabinoids, eicosanoid, and aminoalkylindol derivatives.[29] We will discuss the activity of cannabinoids and eicosanoids in the next sections.

26.3.3.1 Cannabinoids

As mentioned before, the hemp plant synthesizes about 60 cannabinoids. Most of these have no psychomimetic actions; however, those that are active mimic to a degree the action of Δ^9-THC.[47] The great extent of cannabinoid pharmacology is mediated by CBI and CB2 receptors;[13] however, some of these effects may be non-receptor specific and may be the result of cannabinoid binding in a non-saturable fashion to the membrane. Also, some of the cannabinoid effects may be mediated by interactions with other neurotransmitters and their receptors. For example, the recent description of a putative endogenous cannabinoid, anandamide,[12] suggests that some arachidonic acid derivatives may be the endogenous ligands of the cannabinoid receptor. One group of these derivatives is the prostaglandin family (PGE, PGE_2, PGE_4). Several *in vitro* studies have suggested that Δ^9-THC hampers the synthesis of PGs in the rodent brain, e.g., PGE and PGF_4. In this context, there is an extensive literature supporting the fact that PGE_2 modulates wakefulness, while PGD_2 facilitates SWS.[25] Therefore, cannabinoid sedation may be a result of a blockade of PGE_2 synthesis. It is also possible that cannabinoids bind to PGE_2 receptors, due to the great homology between the two molecules. Because PGE_2 levels increase after the performance of a learning task, the deleterious effect on memory caused by cannabinoids may be partially explained by a blockade of PGE_2 receptors.[53]

Cannabinoids also interact directly with monoamine systems. For example, Δ^9-THC binds to serotonin (5-HT) receptors,[15] resulting in inhibition of serotonin activity.[1,2] Other additional effects are an enhancement of norepinephrine synthesis[40] and the inhibition of DNA, RNA, and protein synthesis.[40]

26.3.3.2 Arachidonylethanolamide (Anandamide)

This lipid is the biosynthetic product of arachidonic acid and ethanolamine and chemically is termed an eicosanoid.[12] The original studies indicated that anandamide exists naturally in the porcine brain and binds to cannabinoid receptors;[12,20,31] however, more recent work has shown that anandamide can also be isolated from human and rat brains as well as peripheral tissues.[41] It is noteworthy to mention that anandamide binds to cannabinoid receptors at all stages of development. In addition, binding sites for anandamide have also been found in invertebrate immunocytes and microglia.[57] The presence of cannabinoid receptors among invertebrates suggests that this kind of signaling system has been conserved for about 50 million years.[57] Degradation of anandamide depends on the activity of an enzyme originally called anandamide-amidase.[11,42,58] Recent studies have characterized this enzyme, and it is now referred to as fatty acid amide hydrolase (FAAH).[42,58] This enzyme also degrades another fatty-amide with potential cannabinoid activity — oleamide.[8] Interestingly, the highest enzymatic activity of FAAH is observed in the cerebral cortex and hippocampus.[42] This activity correlates with the CB1 receptor distribution. Anandamide cellular effects are similar to those induced by Δ^9-THC, such as inhibition of adenylate cyclase activity and inhibition of N-type Ca^{++} channels.[10] In addition, anandamide activates the MAP kinase signal transduction pathway and increases arachidonic acid release and PGE$_2$ synthesis in fetal lung fibroblasts.[6,59]

Anandamide effects on behavior are similar to those induced by Δ^9-THC, although with shorter duration.[1] For example, anandamide produces analgesia, hypothermia, hypomotility, and catalepsy.[56,60] Moreover, anandamide may substitute for Δ^9-THC and CP 55,940 (a cannabimimetic drug) in a discriminative stimulus task.[49,61] Due to all these effects, and the capacity to bind the cannabinoid receptor, anandamide is definitively considered the first endogenous marijuana derivative described.

26.3.3.3 Cis-9-10-Octadecenoamide (Oleamide)

This lipid was detected in cerebrospinal fluid obtained from partially sleep-deprived cats.[7,38] The original report indicated that this fatty-amide was able to induce sleep in rats when administered systemically.[51] As yet, neither its synthetic pathway nor the regional and cellular distribution in the brain has been described. However, the enzyme that degrades anandamide — the fatty acid amide hydrolase — also degrades oleamide.[8,42] This observation suggests that oleamide may be part of an extended cannabinoid family. In fact, oleamide exhibits some effects similar to those caused by cannabinoids and anandamide. For example, in our laboratory, we have observed that oleamide produces somnolence, memory impairment, hypomotility, analgesia, and a decrease in body temperature. We have also observed that oleamide decreases c-Fos-positive neuron expression in the hippocampus.

Other groups have documented an immunosuppression effect, indicating that oleamide, like cannabinoids, possesses the capacity to affect the immune system.[36] Despite all these effects, oleamide does not bind to the cannabinoid receptors.[5] In contrast, oleamide binds to 5-HT receptors,[5] but unlike cannabinoids and anandamide, oleamide enhances serotonin activity.[30] Oleamide has also been tested to determine its interaction with glutamate receptors, which was not found.[30] Although we do not postulate oleamide as another endogenous marijuana, we believe it shares several cannabinoid properties.

In summary, cannabinoids and their receptors were only recently described, but they are well characterized in the brain of several mammals, including man. Logically, the existence of a receptor for an exogenous substance implies the existence of an endogenous ligand, which is the case for cannabinoid receptors as well. Moreover, cannabinoids not only activate unique receptors to elicit pharmacological effects, but they also interact with several other neurotransmitter systems. One of the most important interactions appears to involve serotonin receptors and their extended circuits. This observation may be crucial in explaining the hallucinogenic effects of these drug. Serotoninergic systems seem to be pivotal for the drugs with hallucinogenic properties.[22,23]

26.4 Other Psychedelic Drugs

Psychedelic mushrooms are, in the Mexican pre-Columbian culture, described as deities or gods.[3,4] These mushrooms are still in use in religious and ceremonial rituals of Mexican-Indian communities.[52,54] Teonanacatl, ololiuqui, and peyote have existed for centuries and are still employed as drugs used by Mexican-Indians to commune with their gods. It is very interesting that these drugs continue to impact modern culture, where they have been searched for recreational uses instead of for ritual or ceremonial purposes. The best example is the use of LSD, a synthetic drug, that shares its pharmacological mechanism of action with the natural product of botanical mushrooms and peyote.[52]

26.4.1 d-Lysergic Acid Diethylamide (LSD)

When Dr. Albert Hoffman synthesized LSD in the Sandoz Laboratories in Basel, Switzerland, in 1938, he never thought he was creating a new era in psychopharmacology.[52] Although it is a synthetic drug, LSD became the main drug employed for its psychedelic properties.[52]

26.4.2 Effects in Humans

LSD, psylocibin, psilocin, and mescaline cause similar psychogenic effects, with LSD being the most potent. In general, these drugs induce hallucinations, alterations of mood and thinking which are similar to those observed in spontaneously occurring psychotic states.[32] Actually, due to the observation of these effects, in the 1950s it was proposed that the pathogenesis of schizophrenia could be based on altered metabolism of serotonin.[32,52,54] Serotonin seems to be the main neurotransmitter affected by these psychogenic drugs. On the other hand, one of the most important effects caused by psychedelic drugs is an alteration of sensory perception. This effect consists of a heightened awareness of sensory stimuli often accompanied by an enhanced sense of clarity, but with poor control over what is experienced.[32] Commonly, there is a feeling of being a passive observer, similar to a dream. Frequently the attention of the user is turned inward, thereby fascinated by the perceived quality of their own thinking processes.[54] In addition, most users agree that visual and auditory perception is distorted. For example, they can see music and hear colors.[32] In this state, there is scarce capacity for differentiating the boundaries of the self from the environment, creating the sense of union with "mankind" and "cosmos".[19] This becomes the spiritual experience that gives the mystical sense to these drugs. Other physiological effects caused by these psychomimetic drugs are an increase in body temperature and blood pressure, tachycardia, hyperreflexia, tremor, muscular weakness, nausea, piloerection, and pupillary dilatation.[21]

26.4.3 Effects in Animals

Numerous studies in animals have shown that psychedelic drugs produce hyperthermia and complex behavioral responses. Although these are very predictable effects in animals, there is no universally accepted animal model to test psychedelic drugs.

26.5 Discussion

The continuous use of psychedelic drugs since ancient times suggests that there is a human desire to self-impose drug-induced altered states of existence. This desire prompts some individuals to seek methods to achieve such a state. Yet, it is clear that these altered conscious states depend on specific brain molecules, receptors, and interconnected brain circuits. REM sleep may be the closest physiological state mimicking the experiences reported by users of these drugs.[19,32,52] Indeed, by understanding the potential common circuits employed in the generation of REM sleep and those neuronal mechanisms activated by specific classes of psychedelic drugs, a better understanding of the brain mechanisms underlying these two "states" may be achieved.

Acknowledgment

This work was supported by a FUNSALUD Grant given to OPG.

References

1. Adams, I.B., Ryan, W., Singer, M., Thomas, B.F., Compton, D.R., Razdan, R.K., and Martin, B.R., Evaluation of cannabinoid receptor binding and *in vivo* activities for anandamide analogs, *J. Pharmacol. Exp. Ther.*, 273, 1172–1181, 1995.
2. Aghajanian, G.K. and Haigler, H.J., Direct and indirect actions of LSD serotonin and related compounds on serotonin-containing neurons, in *Serotonin and Behavior*, Barchas, J. and Usdin, E., Eds., Academic Press, New York, 1973, pp. 263–266.
3. Benítez, F., *Los Indios de México: Los Hongos Alucinantes*, Serie Popular, Ediciones Era, Mexico, 1972.
4. Benítez, F., *Los Indios de México: En la Tierra Mágica del Peyote*, Serie Popular, Ediciones Era, Mexico, 1972.
5. Boring, D.L., Berglund, B.A., and Howett, A.C., Cerebrodiene, arachidonyl-ethanolamide, and hybrid structures: potential for interaction with brain cannabinoid receptors, *Prostaglandins Leukot. Essent. Fatty Acids, 55,* 207–210, 1996.
6. Burnstein, S. and Hunter, S.A., Prostaglandins and cannabis. VI. Release of arachidonic acid from HeLa cells by 1 tetrahydrocannabinol and other cannabinoids, *Biochem. Pharmacol.*, 27, 1275–1280, 1978.
7. Cravatt, B.F., Próspero-García, O., Siuzdak, G., Gilula, N.B., Henriksen, S.J., Boger, D.L., and Lerner, R.A., Chemical characterization of a family of brain lipids that induce sleep, *Science,* 268, 1506–1509, 1995.
8. Cravatt, B.F., Giang, D.K., Mayfield, S.P., Boger, D.L., Lerner, R.A., and Gilula, N.B., Molecular characterization of an enzyme that degrades neuromodulatory fatty-acid amides, *Nature,* 384, 83–87, 1996.
9. Chapman, C.E., Jiang, W., and Lamarre, Y., Modulation of lemniscal input during conditioned arm movements in the monkey, *Exp. Brain Res.*, 72, 316–334, 1989.
10. Childers, S.R. and Deadwyler, S.A., Role of cyclic AMP in the actions of cannabinoid receptors, *Biochem. Pharmacol.*, 52, 819–827, 1996.
11. Desrnaud, F., Cadas, H., and Piomelli, D., Anandamide amidohydrolase activity in rat brain microsomes, *J. Biol. Chem.*, 270, 6030–6035, 1995.
12. Devane, W.A., Hanus, L., Breuer, A., Pertwee, R.G., Stevenson, L.A., Griffin, G., Gibson, D., Mandelbaum, A., Etinger, A., and Mechoulam, R., Isolation and structure of a brain constituent that binds to the cannabinoid receptor, *Science,* 258, 1945–1949, 1992.
13. Dewey, W.L., Cannabinoid pharmacology, *Pharmacol. Rev.*, 38, 151–1781 1986.
14. Díaz, J.L., *El Ábaco, la Lira y la Rosa, las Regiones del Conocimiento*, Fondo de Cultura Económica, Mexico, D.F., 1997.
15. Fan, P., Cannabinoid agonists inhibit the activation of 5-HT3 receptors in rat nodose ganglion neurons, *J. Neurophysiol.*, 73, 907–910, 1995.
16. Felder, C.C., Joyce, K.E., Briley, E.M., Mansouri, J., Mackie, K., Blond, O., Lai, Y., Ma, A.L., and Mitchell, R.L., Comparison of the pharmacology and signal transduction of the human cannabinoid CBI and CB2 receptor, *Mol. Pharmacol.*, 48, 443–450, 1995.
17. Fleishman, R.W., Hayden, D.W., Rosenkrantz, H., and Braude, M.C., Teratologic evaluation of delta-9-tetrahydrocannabinol in mice, including a review of the literature, *Teratology,* 12, 47–50, 1975.

18. Freemon, F.R., The effect of delta-9-tetrahydrocannabinol on sleep, *Psychopharmacology,* 35, 39–44, 1974.
19. Freedman, D.X., The use and abuse of LSD, *Arch. Gen. Psychiatry,* 18, 300–347, 1968.
20. Fride, E. and Mechulam, R., Pharmacological activity of the cannabinoid receptor agonist, anandamide, a brain constituent, *Eur. J. Pharmacol.,* 231, 313–314, 1993.
21. Gerard, C.M., Mollereau, C., Vassart, G., and Parmentier, M., Molecular cloning of a human cannabinoid receptor which is also expressed in testis, *Biochem. J.,* 279, 129–134, 1991.
22. Glennon, R.A., Young, R., Rosecrans, J.A., Kallman, M.J.I., Hallucinogenc agents as discriminative stimuli: a correlation with serotonin receptor affinities, *Psychopharmacology,* 68, 155–158 1980.
23. Glennon, R.A., Young, R., Bennington., F., and Morin, R.D., Behavioral and serotonin receptor properties of 4-substituted derivatives of the hallucinogenic I-(2,5-dimethoxyphenyl)-2 aminopropane, *J. Med. Chem.,* 25, 1163–1168, 1982.
24. Hampson, R.E., Foster, T.C., and Deadwyler, S.A., Effects of delta-9-tetrahydrocannabinol on sensory evoked hippocampal activity in the rat: principal components analysis and sequential dependency, *J. Pharmacol. Exp. Ther.,* 251, 870–877, 1989.
25. Hayaishi, O., Molecular mechanisms of sleep-wake regulation: roles of prostaglandins D2 and E2, *FASEB J.,* 5, 2575–2581, 1991.
26. Herkenham, M., Lynn, A.B., Johnson, M.R., Melvin, L.S., de Costa, B.R. et al., Characterization and localization of cannabinoid receptors in rat brain: a quantitative *in vitro* autoradiographic study, *J. Neurosci.,* 11, 563–583, 1991.
27. Herkerham, M., Lynn, A.B., Little, M.D., Johnson, M.R., Melvin, L.S. et al., Cannabinoid receptor localization in brain, *Proc. Natl. Acad. Sci. USA,* 87, 1932–1936, 1990.
28. Hollister, L.E., Health aspects of cannabis, *Pharmacol. Rev.,* 38, 1–20, 1986.
29. Howlett, A.C., Pharmacology of cannabinoid receptors, *Ann. Rev. Pharmacol, Toxicol.,* 35, 607-634,1995.
30. Huidobro-Toro, J.P. and Harris, R.A., Brain lipids that induce sleep are novel modulators of 5-hydroxytryptamine receptors, *Proc. Natl. Acad. Sci. USA,* 93, 8078–8082, 1996.
31. Iversen, L., Endogenous cannabinoids, *Nature,* 372, 619, 1994.
32. Jacobs, B.L., *Hallucinogens: Neurochemical, Behavioral and Clinical Perspectivas,* Raven Press, New York, 1984.
33. Jacobs, B.L. and Fornal, C.A., 5-HT and motor control: a hypothesis, *Trends Neurosci.,* 16, 346–352, 1993,
34. Jouvet, M., Biogenic amines and the states of sleep, *Science,* 163, 32–41, 1969.
35. Kolodny, R.C., Masters, W.H., Kolodny, R.M., and Toro, G., Depression of plasma testosterone levels after chronic intensive marihuana use, *N. Engl. J. Med.,* 290, 872–874, 1974.
36. Langstein, J., Hofstadter, F., and Schwarz, H., Cis-9, 10-octadecenoamide, an endogenous sleep-inducing CNS compound, inhibits lymphocyte proliferation, *Res. Immunol.,* 147, 389–396, 1996.
37. Lepore, M., Vorel, S.R., Lowinson, J., and Gardner, E.L., Conditioned place preference induced by delta-9-tetrahydrocannabinol: comparison with cocaine, morphine and food reward, *Life Sci.,* 56, 2073–2080, 1995.
38. Lerner, R.A., Siuzdak, G., Prospéro-García, O., Henriksen, S. J., Boger, D.L., and Cravatt, B.F., Cerebrodiene: a brain lipid isolated from sleep-deprived cats, *Proc. Natl. Acad. Sci. USA,* 91, 9505–9508, 1994.
39. Lichtman, A.H., Dimen, K.R., and Martin, B.K., Systematic or intrahippocampal cannabinoid administration impairs spatial memory in rats, *Psychopharmacol.,* 119, 289–290,1995.
40. Martin, B.R., Cellular effects of cannabinoids, *Pharmacol. Rev.,* 38, 45–74, 1986.

41. Matsuda, L.A., Lolait, S.J., Brownstein, M.J., Young, A.C., and Bonner, T.I., Structure of a cannabinoid receptor and functional expression of the cloned CDNA, *Nature,* 346, 561–564, 1990.
42. Maureli, S., Bisogno, T., De Petrocellis, L., Di Luccia, A., Marino, G., and Di Marzo, V., Two novel classes of neuroactive fatty acid amides are substrates for mouse neuroblastoma "anandamide amidohydrolase", *FEBS Lett.,* 377, 82–86, 1995.
43. McGinty, D. and Szymusiak, R., Neuronal unit activity patterns in behaving animals: brainstem and limbic system, *Ann. Rev. Psychol.,* 39, 135–168, 1988,
44. Molina-Holgado, V., González, M.I., and Leret, M.C., Effect of delta-9-tetrahydrocannabinol on short-term memory in the rat, *Physiol. Behav.,* 57, 172–179, 1995.
45. Moreau, J.J., *Hashish and Mental Disease,* Raven Press, New York, 1973.
46. Munro, S., Thomas, K.L., and Abu-Shaar, M., Molecular characterization of a peripheral receptor for cannabinoids, *Nature,* 365, 61–65, 1993,
47. Musty, R.E., Reggio, P., and Consroe, P., A review of recent advances in cannabinoid research and the 1994 International Symposium on Cannabis and the Cannabinoids, *Life Sci.,* 56, 1933–1944, 1995.
48. Parker, L.A. and Gillies, T., THC-induced place and taste aversions in Lewis and Sprangue-Dawley rats, *Behav. Neurosci.,* 109, 71–78, 1995.
49. Pertwee, R.G., Stevenson, L.A., and Griffin, G., Cross-tolerance between delta-9-tetrahydrocannabinol and the cannabimimetic agents, CP 55,940, WIN 55,212-2 and anandamide, *Br. J. Pharmacol.,* 110, 1483–1490, 1993.
50. Pivik, R.T., Zarcone, V., Dement, W.C., and Hollister, L.E., Delta-9-tetrahydrocannabinol and synhexyl: effects on human sleep patterns. *Clin. Pharmacol. Ther.,* 13, 426–435, 1972.
51. Prospéro-García, O, Cravatt, B.F, Siuzdak, G., Boger, D.L., Lerner, R.A., and Henriksen, S.J., cis-9, 10 octadecenoamide: a novel natural lipid isolated from cat CSF with potential sleep-modulating properties, *Sleep Res.,* 24, 50, 1995.
52. Ray, O. and Ksir,C., *Drugs, Society, and Human Behavior,* 7th ed., Mosby, Boston, 1996, chaps. 16 and 17.
53. Reichman, M., Nen, W., and Hokin, L.E., Effects of Δ^9-tetrahydrocannabinol in prostaglandin formation in the brain, *Mol. Pharmacol.,* 32, 686–690, 1987.
54. Schultes, R.E. and Hofmann, A., *Plants of the Gods,* McGraw-Hill, New York, 1979.
55. Skaper, S.D., Buriani, A., Dal Toso, R., Petrelli, L., Romanello, S., Facci, L., and Leon, A., The alimide palmitoylethanolamide and cannabinoids, but not anandamide, are protective in a delayed postglutamate paradigm of excitotoxic death in cerebellar granule neurons, *Proc. Natl. Acad. Sci. USA,* 93, 3984–3989, 1996.
56. Smith, P.B., Compton, D.R., Welch, S.P., Razdan, R.K., Mechoulam, R., and Martin, B.R., The pharmacological activity of anandamide, a putative endogenous cannabinoid, in mice, *J. Pharmacol. Ther.,* 270, 219–227, 1994.
57. Stefan, G.B., Liu, P., and Goligorsky, M.S., Cannabinoid receptors are coupled to nitric oxide in invertebrates immunocytes, microglia and human monocytes, *J. Biol. Chem.,* 271, 19238–19242, 1996.
58. Ueda, N., Kurahashi, Y., Yamamoto, S., and Tokunaga, T., Partial purification and characterization of the porcine brain enzyme hydrolyzing and synthesizing anandamide, *J. Biol. Chem.,* 270, 23823–23827, 1995.
59. Wartman, M., Campbell, D., Subramanian, A., Burstein, S.H., and Davis, R.J., The MAP-kinase signal transduction pathway is activated by endogenous cannabinoid anandamide, *FEBS Lett.,* 359C, 133–136, 1995.

60. Welch, S.P., Dunlow, L.D., Patrick, G.S., and Razdan, R.K., Characterization of anandamide- and fluoroanandamide-induced antinociception and cross-tolerance to delta 9-THC after intrathecal administration to mice: blockade of delta-9-THC-induced antinociception, *J. Pharmacol. Exp. Ther.*, 273, 1235–1244, 1995.
61. Wiley, J., Balster, R., and Martin, B., Discriminative stimulus of anandamide in rats, *Eur. J. Phamacol.*, 276, 49–54, 1995.

Section VI

State-Dependent Processing in Somatosensory Pathways

Peter J. Soja, Section Editor

Introduction and Overview

An organism's behavioral state is characterized by a multi-faceted display of physiological phenomena and is controlled by a host of complex molecules, neuronal ensembles, and highly regulated pathways. One of the defining features of sleep is the reduced responsiveness to external stimuli, which suggests that modulation occurs along sensory pathways to the cerebral cortex. Sensory information is conveyed to the cortex by a vast array of peripheral receptors, second-order tract neurons, and (pre)thalamic relay nuclei. Important challenges intersecting the fields of sensory physiology and basic sleep research are to understand how, to what extent, and where ascending sensory information is regulated during sleep.

Synaptic transmission of sensory information through thalamocortical pathways is well known to be impeded during sleep when compared to wakefulness (see Chapter 20). Thus, reduced responsiveness to external stimuli during sleep can partly be accounted for by a reduction in thalamic sensory throughput. On the other hand, suppression of prethalamic sensory transmission may also occur during sleep, but this possibility has not been rigorously investigated. Indeed, prethalamic regulation of ascending sensory transmission during wakefulness is not well understood. This lack of knowledge regarding state dependency of sensory processing exists even though a wealth of knowledge has been obtained over the past 35 years about the anatomical, electrophysiological, and pharmacological bases underlying the transmission of tactile and nociceptive information to the brain.

This paradox is primarily due to the favored use of acute animal preparations. The vast majority of scientific literature that describes the physiological organization and the afferent or descending regulation of individual trigeminal and lumbar sensory neurons derive from experiments performed under artificial conditions, i.e., anesthesia, invasive preparative surgery, artificial respiration, etc. These experimental conditions are very useful in that they permit investigators to exert substantial experimental control and optimize data yields, which have led to firm understanding of organizational and functional principles of ascending sensory systems. A major limitation of experiments conducted under artificial conditions is that they preclude analyses of the activity of sensory neurons and their regulation during naturally occurring behavioral states such as wakefulness and sleep.

Earlier investigators recognized this limitation of acute experiments and made several attempts at developing approaches for recording sensory neuron activity in conscious unanesthetized, intact animal preparations. However, compared to the multitude of studies performed in acute preparations, only a few studies have documented sensory neuron activity in the conscious animal. This is not surprising because, as anyone who has tried their hand at chronic unit recording will certainly attest, such efforts are technically demanding and often require long time periods for accumulating, analyzing, and interpreting adequate amounts of data. Nevertheless, a solid foundation has been laid for further investigations of how activity in sensory systems is affected by changes in behavioral state, e.g. arousal, attention, drowsiness, sleep, general anesthesia, etc.

Progress has recently been made been made toward elucidating state-specific activity changes of individual units in sensory relay nuclei as well as in primary and second-order neurons comprising identified ascending sensory tracts. The sensory systems of the trigeminal brainstem and lumbar spinal cord have been foci for such investigations. This section will review our current understanding of the characteristics and state-specific activity patterns of neurons constituting these regions and their ascending tracts.

This section consists of five chapters which aim to combine studies performed on sensory neurons in acute anesthetized preparations with those performed in chronic animal preparations. In Chapter 27, B.J. Sessle reviews the basic characteristics of the trigeminal sensory system, including anatomical organization, physiological mechanisms, sensory modalities relayed and regulatory influences from ascending and descending sources. Chapter 28, by W.D. Willis, surveys the corresponding characteristics of a number of ascending sensory tracts of the lumbar spinal cord. The role of the rostral ventromedial medulla in regulating ascending sensory transmission is then considered in Chapter 29 by M.M. Heinricher and S.P. McGaraughty, with particular emphasis on pain sensation and the possible involvement of the neurotransmitter serotonin. Chapter 30, by P.M. Headley, B.A. Chihz, J.F. Herrero, and N.A. Hartell, describes the effects of anesthetic agents and surgical procedures on the functioning of neurons in the spinal cord dorsal horn. Finally, in Chapter 31, P.J. Soja, B.E. Cairns, and M.P. Kristensen summarize studies of the dynamic modulation of sensory neuron functioning across the sleep/wakefulness cycle with a focus on changes in the activity of ascending tract cells and other sensory neurons in the trigeminal brainstem and lumbar spinal cord during wakefulness, quiet sleep, and active sleep.

Important technical considerations for future studies of the state dependency of ascending sensory transmission follow from the overviews. First of all, sensory cell activity is powerfully influenced by anesthesia and recent surgery. These factors render data obtained under such conditions of limited use as a baseline for comparison with sleep or other states of consciousness. Moreover, acute studies implicitly do not permit normal states of sleep and wakefulness to progress. Furthermore, the cellular composition of a given sensory relay region is highly diverse, often encompassing both local interneurons and pre-motor neurons in addition to ascending tract cells. For future studies of state-dependent modulation of sensory transmission, it is therefore essential to identify sensory cells with respect to their efferent projection to ensure that observed functional changes apply to ascending tract cells. It is hoped that research efforts along these lines will lead to a clearer understanding of how ascending sensory information is regulated before entering the thalamus and higher brain centers as a function of behavioral state.

Chapter

Somatosensory Transmission in the Trigeminal Brainstem Complex and its Modulation by Peripheral and Central Neural Influences

Barry J. Sessle

Contents

27.1	Introduction	446
27.2	Primary Afferent Inputs and Organization of the Trigeminal Brainstem Sensory Nuclear Complex	446
27.3	Trigeminal Brainstem Processing of Non-Nociceptive Afferent Inputs	449
	27.3.1 Spatiotemporal Properties of Low-Threshold Mechanoreceptive Neurons	449
	27.3.2 Neuroplastic Changes in Low-Threshold Mechanoreceptive Neurons	449
	27.3.3 Processing of Other Somatosensory Inputs	450
27.4	Trigeminal Brainstem Processing of Nociceptive Afferent Inputs	451
	27.4.1 Properties of Neurons in Subnucleus Caudalis	451
	27.4.2 Properties of Other Trigeminal Brainstem Neurons	452
	27.4.3 Convergence	452
	27.4.4. Neuroplastic Changes in Nociceptive Neurons	453
27.5	Modulation of Somatosensory Transmission	454
	27.5.1 Descending Modulation	454
	27.5.2 Afferent-Induced Modulation	456
27.6	Concluding Remarks	456
Acknowledgments		457
References		457

27.1 Introduction

This chapter will outline trigeminal (V) brainstem mechanisms related to orofacial somatosensory transmission and its modulation by segmental influences and by descending influences from higher brainstem regions. Because many of these mechanisms contribute to the processes underlying changes in behavioral state and some indeed are themselves altered by the state of the organism, an additional aim of this chapter is to provide a basis for subsequent chapters that deal specifically with state-dependent features of these processes. The chapter itself builds upon some of the material on spinal nociceptive mechanisms and their modulation presented elsewhere in this section (see Chapters 28 and 29). Given the focus of this chapter on brainstem processes, little attention is given to peripheral V somatosensory mechanisms; the interested reader can access this information from recent reviews.[13,29,30,52]

27.2 Primary Afferent Inputs and Organization of the Trigeminal Brainstem Sensory Nuclear Complex

In the spinal somatosensory system, all spinal afferents supplying deep as well as superficial tissues have their primary afferent cell bodies in the dorsal root ganglia. In the V system, in contrast, jaw muscle spindle afferents and some mechanosensitive afferents supplying periodontal tissues have their primary afferent cell bodies within the central nervous system (CNS), in the V mesencephalic nucleus. From here they project to the V motor nucleus or to adjacent nuclei (e.g., supratrigeminal nucleus, V subnucleus oralis) where they excite interneurons involved in craniofacial reflex function.[8,19,78] Nearly all the remaining primary afferents supplying craniofacial tissues have peripheral cell bodies principally in the gasserian (semilunar) ganglion. The central projections of these primary afferent cell bodies enter the brainstem and may ascend or descend in the V spinal tract from which they give off collaterals that terminate in one or more subdivisions of the V brainstem sensory nuclear complex.

The V primary afferent inputs activate second-order neurons within the V brainstem sensory nuclear complex, which can be subdivided into the principal or main sensory nucleus and the spinal tract nucleus (Figure 27.1). The latter comprises three subnuclei (oralis, interpolaris, caudalis), of which the most caudal component, subnucleus caudalis, resembles the spinal dorsal horn in that it is a laminated structure; it extends into the cervical spinal cord where it merges with the spinal dorsal horn. The subdivisions and detailed anatomical organization of each component have been extensively described.[16,27,33,41,71] As far as their axonal projections, many neurons in the four components of the V brainstem complex contribute to ascending nociceptive or non-nociceptive pathways involved in somatosensory function or modulation by virtue of their projection to areas such as the ventrobasal thalamus, pontine parabrachial area, periaqueductal gray, or brainstem reticular formation.[23,28,29,59] Some of the connections to the reticular formation and other parts of the brainstem are utilized in autonomic reflex responses to orofacial stimuli. Some neurons in the V complex and adjacent to it (e.g., supratrigeminal nucleus) also serve as interneurons in craniofacial and cervical muscle reflex pathways, and there are also intrinsic connections between neurons in different components of the complex (e.g., caudalis axonal projection to subnucleus oralis and vice versa) that underlie the modulatory influences between rostral and caudal V neurons that are described below. It is to be noted that there is a differential contribution from each nucleus/subnucleus to each of these projections — e.g., the main sensory nucleus is the principal direct brainstem relay to ventrobasal thalamus of mechanosensitive afferent input from vibrissae and most other parts of the orofacial region, whereas many subnucleus oralis neurons directly project to other brainstem structures.

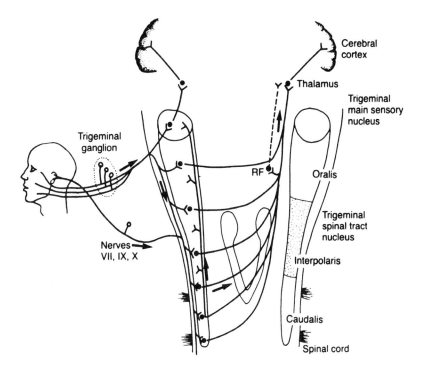

FIGURE 27.1
Major pathway transmitting sensory information from the face and mouth. Trigeminal primary afferents project via the trigeminal ganglion to second-order neurons in the trigeminal brainstem sensory complex which may project to neurons in higher levels of the brain (for example, in thalamus) or in brainstem regions such as the cranial nerve motor nuclei or reticular formation (RF). Some afferents from cranial nerves VII, IX, and X may also synapse in the trigeminal complex; not shown are the projections of some cervical nerve and XII afferents to the trigeminal complex and the projection of many VII, IX, and X afferents to the solitary tract nucleus, the horseshoe-shaped structure in the middle of the diagram. (From Sessle, B.J., in *Oral Biology*, Roth, G.I. and Calmes, R., Eds., Mosby, St. Louis, 1981. With permission.)

The low-threshold mechanosensitive primary afferents traveling in the V nerve primarily terminate in the more rostral components of the V brainstem complex and in laminae III–VI of subnucleus caudalis; some V nociceptive cutaneous afferents and dental pulp afferents may also terminate in some of these rostral components, but most of the small-diameter V nociceptive afferents terminate in subnucleus caudalis, in its laminae I, II, V, and VI.[39,50,51,70,72] In addition to its predominant input from V nerve afferents, the V brainstem complex (especially its subnucleus caudalis) may also receive afferent inputs from other cranial nerves such as VII, IX, X, and XII, as well as from upper cervical nerves.[3,41,42,54]

The V brainstem complex is somatotopically arranged in such a way that those neurons receiving mandibular afferent input are located in the dorsal part of each nucleus or subnucleus of the V brainstem complex, and those with ophthalmic afferent input in the ventral part. The maxillary region is represented in between, and oral and perioral structures are represented medially. In subnucleus caudalis, this inverted, medially facing somatotopic or topographic pattern of the face and mouth may shift, with perioral regions represented in the rostral part of the subnucleus and more lateral regions of the face more caudally; this somatotopic pattern in subnucleus caudalis has sometimes been referred to as an "onion-skin" arrangement. A particularly interesting and extensively studied somatotopic pattern in the V brainstem complex is that of the vibrissae or facial whiskers.[33,76] The brainstem projections of single vibrissal afferents are organized into columns or cylinders that run rostrocaudally and are somatotopically parcellated in the transverse plane, in the main sensory nucleus, and in the subnuclei interpolaris and caudalis; subnucleus oralis does not reveal such clear-cut spatial patterning. This patterning of the vibrissal afferent endings and the second-order neurons upon which they terminate is especially evident in rodents and appears as an array of cell patches or aggregates termed "barrelettes", reflecting the pattern of vibrissae on the ipsilateral face. The vibrissal afferents and most neurons associated with such barrelettes are optimally responsive to tactile stimulation of a single vibrissa, and each neighboring barrelette similarly "represents" the neighboring whisker on the face.

In the spinal cord and dorsal column nuclei, primary afferent structure-function relationships have been well documented, but in the V brainstem complex, structure-function correlations are usually less clear cut and are often variable between types of afferents in the four different components of the V brainstem complex. For example, there is evidence from intra-axonal labeling of functionally identified afferent fibers that some different types of afferents in subnucleus caudalis may be morphologically distinguished from each other (e.g., low-threshold vibrissal vs. low-threshold mucosal vs. nociceptive afferents), whereas in all four components of the V brainstem complex, the different functional classes of low-threshold vibrissal afferents are characterized by their morphological homogeneity. While there is this variability and in many cases an absence of structure-function correlations of V primary afferent projections to the V brainstem complex, it is nonetheless clear that the spatial distribution of the collateral endings of an afferent is influenced by the location in the orofacial region of the mechanoreceptive field of the afferent, thus accounting for the somatotopic pattern mentioned earlier.[33,39]

There are several morphologically distinct neuronal types in the V brainstem complex but, as for the V primary afferent endings, there appear to be fundamental differences between V brainstem and spinal somatosensory systems in the extent of structure-function correlations.[27,33,40] For example, some morphological differences occur between different classes of neurons in subnucleus caudalis, but structure-function relationships are less prominent than in the subnucleus interpolaris, where neurons projecting to certain brain regions differ from interpolaris "local circuit" interneurons in some receptive field and morphological features, and within each of these two neuron groups there are morphological distinguishing features correlating with projection site, receptive field size, or receptor type (e.g., mechanoreceptor vs. nociceptor). Structure-function relationships have also been extensively investigated in vibrissa-sensitive neurons, and correlations have been shown to be a strong feature in some parts of the V brainstem complex but not in other parts (e.g., main sensory vs. oralis). Henderson and Jacquin[33] have recently outlined these characteristics and the factors that may account for the structure-function differences within and between V nuclei/subnuclei in the properties of vibrissa-sensitive neurons.

The neurochemical features of some of the different neuronal types, especially in caudalis, have also been detailed.[26,27,45,50,51] For example, many of the neurons projecting out of the V brainstem complex that are found mainly in caudalis laminae I and III to VI express glutamate receptor subtypes. Some of the neurons in these laminae, and in more rostral components of the V brainstem complex, may contain other neurochemicals that contribute to the modulatory mechanisms in the V brainstem complex described below. The lamina II of caudalis, the so-called substantia gelatinosa (SG), is a case in point, as it represents an important interneuronal system contributing to the powerful segmental and descending modulation of somatosensory transmission that occurs in subnucleus caudalis and more rostral components of the V brainstem complex (see below). The several morphologically distinct cell types in the SG contain neuromodulatory chemicals such as GABA or endogenous opioids and receive a mix of low- and high-threshold peripheral afferent inputs, as well as inputs from higher brain centers. Through their arborizations within the V brainstem complex, the SG neurons relay modulatory influences from these inputs onto V brainstem neurons.[18,25–27,45,50]

27.3 Trigeminal Brainstem Processing of Non-Nociceptive Afferent Inputs

27.3.1 Spatiotemporal Properties of Low-Threshold Mechanoreceptive Neurons

The rostral components of the V brainstem complex have long been considered to be the essential brainstem regions relaying orofacial mechanosensory information related to so-called fine touch and thereby analogous to the dorsal column nuclei of the spinal somatosensory system. This view has been reinforced by evidence that the mechanoreceptive field properties of some of the neurons that relay such mechanosensory information — low-threshold mechanoreceptive (LTM) neurons — may allow for a more precise coding of orofacial mechanical stimuli, e.g., the single-whisker receptive field of most main sensory neurons vs. the multi-whisker receptive field of most more caudally located V brainstem neurons in the rat[33,47] and the finding that many caudalis LTM neurons may have less capacity than more rostral V brainstem neurons for the faithful transmission of detailed information about the spatiotemporal qualities of low-intensity mechanical stimuli[17] (although this latter conclusion has been disputed by others; see Reference 43). Nonetheless, there is now abundant evidence that at least some mechanosensitive neurons at all levels of the V brainstem complex are capable of receiving and faithfully transmitting detailed somatosensory information about light tactile stimuli delivered to localized regions of the face and mouth,[23,29] in keeping with the evidence mentioned above that V low-threshold mechanosensitive afferents project to all levels of the V brainstem complex.

The LTM neurons are abundant in all four nuclei/subnuclei of the V brainstem complex, including subnucleus caudalis where they predominate in laminae III and IV. In accordance with analogous neurons in the dorsal column nuclei and spinal dorsal horn[74] (see Chapter 28), they are activated by light tactile stimuli applied to a localized mechanoreceptive field, show graded responsiveness as stimulus intensity or area of receptive field stimulation is gradually increased, and receive low-threshold A-fiber afferent inputs that appear to release glutamate as the principal excitatory neurotransmitter. In general, as noted above, they have functional properties ensuring that a secure transmission of tactile information from orofacial mechanoreceptors through the V brainstem complex can take place. The LTM neurons thereby provide the higher levels of the brain with detailed information of the modality and spatiotemporal features (e.g., location, intensity) of an orofacial tactile stimulus.

27.3.2 Neuroplastic Changes in Low-Threshold Mechanoreceptive Neurons

The spatiotemporal coding characteristics of the LTM neurons can, however, be modified by anesthesia or behavioral state (see Chapter 31) and by other modulatory processes. Segmental and descending modulatory influences are described below, but another form of modulation is reflected in the neuroplasticity of the V low-threshold somatosensory system. The extensive investigations of the organization and neuronal properties of the whisker barrelettes in the V brainstem complex (and analogous structures in ventrobasal thalamus and cerebral cortex) that were mentioned above have also revealed the plasticity and susceptibility of the LTM neuronal properties to peripheral manipulations such as deafferentation in neonatal animals and even in adults.[33,76] Deafferentation of other orofacial tissues such as the tooth pulp in adult animals[36,46] can also lead to profound changes in the mechanoreceptive field and response properties of LTM neurons at the different levels of the V

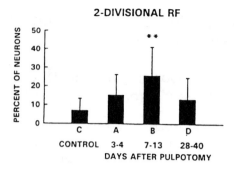

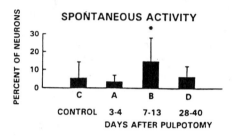

FIGURE 27.2
Time course of effects of mandibular molar pulp deafferentation on the mechanoreceptive field loci (top) and spontaneous activity (bottom) of subnucleus oralis LTM neurons in the rat. Note the significantly higher proportion of neurons in group B with spontaneous activity and a mechanoreceptive field involving both V2 and V3 or both V1 and V2 divisions. $*p < 0.05$; $**p < 0.01$; RF = mechanoreceptive field. (From Kwan, C.L. et al., *Somatosensory Motor Res.*, 10, 115, 1993. With permission.)

brainstem complex (Figure 27.2). The modulatory mechanisms underlying these expressions of neuroplasticity are still unclear but do not appear to be accounted for by morphological changes (e.g., collateral sprouting) in the uninjured low-threshold mechanosensitive afferent endings in the V brainstem complex or by alterations in certain central inhibitory circuits or presynaptic regulatory processes.[46,59,66] Interestingly, studies of the effects of the neonatal depletion of C-fibers reveal that C-fiber afferents appear to play an important role in shaping the adult mechanoreceptive field properties of the LTM neurons receiving large-diameter mechanosensitive afferent inputs.[47]

27.3.3 Processing of Other Somatosensory Inputs

A small number of low-threshold neurons in the V brainstem complex can be excited by afferent inputs other than those derived from orofacial mechanoreceptors in skin, mucosa, or periodontal tissues.[8,23] In the rostral components, in particular, some neurons can be activated by jaw movements, especially jaw opening, presumably through inputs from low-threshold mechanoreceptors in the temporomandibular joint (TMJ) or stretch-sensitive receptors in masticatory muscles. It is thought that these neurons contribute to ascending sensory pathways (e.g., to thalamus and cortex) or to brainstem circuits underlying the reflex regulation of masticatory muscle activity.

Some LTM neurons respond also to cooling of their orofacial mechanoreceptive field, apparently through inputs from some slowly adapting mechanoreceptive afferents that are sensitive to innocuous cooling. Other neurons, especially in V subnucleus caudalis, respond exclusively and with much greater sensitivity to innocuous cooling or warming.[23,29] These are specific thermosensitive neurons which appear to receive their afferent input from specific thermoreceptors and which are important for the relay to the higher brain centers of detailed information (e.g., intensity, duration, location) about orofacial thermal stimuli.

27.4 Trigeminal Brainstem Processing of Nociceptive Afferent Inputs

27.4.1 Properties of Neurons in Subnucleus Caudalis

It has been noted above that most small-diameter afferents terminate in V subnucleus caudalis, and that intra-axonal labeling studies have shown that many of these afferents are nociceptive in character. The central endings within subnucleus caudalis of small-diameter axons in the V spinal tract, presumed to represent nociceptive primary afferents, also contain several neuropeptides and excitatory amino acids.[22,34,51,56,58,69] As in the spinal dorsal horn (see Chapter 28), one of these neuropeptides is the neurokinin substance P, which is found in small-diameter somatosensory afferents and in their ganglion cell bodies and which has also been implicated in peripheral injury and inflammation. Substance P is also concentrated in the central endings of these afferents in the superficial and deep laminae of V subnucleus caudalis where the nociceptive neurons predominate (see below). Noxious stimulation of peripheral orofacial tissues results in the release from these central endings of substance P which produces a long-latency, sustained excitation of these neurons.

Several other endogenous neuropeptides, such as calcitonin gene-related peptide (CGRP) and somatostatin, have also been implicated in the excitatory processes underlying nociceptive transmission. A role for excitatory amino acids such as glutamate in V as well as spinal nociceptive mechanisms is also indicated[22] (see Chapter 28) — for example, high concentrations of glutamate receptors and localization of glutamate in neurons and afferent endings in the V brainstem complex, including subnucleus caudalis;[12,45] release of glutamate from subnucleus caudalis by noxious stimulation of craniofacial tissues;[4] and increased activity in identified V brainstem nociceptive neurons by local application of glutamate and decreased activity by glutamate antagonists.[34,56] Peripheral glutamate receptors may also be involved, as there is recent evidence of glutamate localization in tooth pulp afferents[2] and of jaw muscle activity that can be reflexively activated by glutamate injection into TMJ tissues and blocked by glutamate receptor antagonists.[9] The N-methyl-D-aspartate (NMDA) receptor, an ionotropic receptor subtype of the excitatory amino acid receptor family, as well as non-NMDA receptor mechanisms, have been specifically implicated in these various central and peripheral effects.[4,22,37]

The recent documentation of neuropeptide and excitatory amino acid involvement in nociceptive processing in V subnucleus caudalis lends support to clinical, anatomical, and electrophysiological findings indicating that subnucleus caudalis serves as the principal brainstem relay site of V nociceptive information. Clinical and related experimental observations of the effects of V tractotomy near the obex have revealed that this neurosurgical procedure (used to relieve V neuralgia in humans) produces a profound orofacial analgesia (and thermanesthesia), with only a very limited loss of tactile sensibility. Anatomically, there are parallels in structure, afferent inputs, cell types, and projection sites between subnucleus caudalis and the spinal cord dorsal horn[22,27,74] (see Chapter 28), and, as in the spinal dorsal horn, noxious stimulation induces increased *c-fos* expression in caudalis neurons.[22,24,31,67,73] There are also electrophysiological parallels, as neurons responding to cutaneous or deep noxious stimuli have been documented in subnucleus caudalis of anesthetized, decerebrate, or unanesthetized experimental animals.[21,22,29,59,77] Moreover, as in the spinal dorsal horn, these caudalis nociceptive neurons can be classified into two main groups on the basis of their cutaneous (or mucosal) receptive field properties: nociceptive-specific (NS) neurons, which receive small-diameter afferent inputs from A-delta and/or C fibers and which respond only to noxious stimuli (e.g., pinch, heat) applied to a localized craniofacial receptive field, and wide dynamic range (WDR) or convergent neurons, which may receive large-diameter and small-diameter A-fiber inputs as well as C-fiber inputs and are excited by non-noxious (e.g., tactile) stimuli as well as by noxious stimuli (see Figure 27.4).

The spatiotemporal properties of the NS and WDR neurons in subnucleus caudalis have been extensively studied, and both types of neurons have been shown in anesthetized preparations to possess a graded response as the intensity of noxious stimulation is progressively increased or as more of the receptive field is stimulated. Studies in unanesthetized or awake, behaving animals confirm their graded responsiveness and, in addressing other spatial and temporal coding features, indicate that both types of nociceptive neurons are critical elements in our ability to localize and discriminate superficial pain in the craniofacial region (see Chapter 31).[21] The NS and WDR neurons are concentrated in the superficial (I and II) and deep (V and VI) laminae of caudalis, in a somatotopic arrangement consistent with that noted above for LTM caudalis neurons. It is also a feature of the caudalis NS and WDR neurons that they project to areas (e.g., ventrobasal thalamus, pontine parabrachial area, reticular formation) implicated in nociceptive transmission or its modulation. Because comparable neurons occur in the spinal dorsal horn,[22] the close functional as well as morphological similarities between V subnucleus caudalis and the spinal dorsal horn have led to the former's designation as the medullary dorsal horn. Furthermore, it should be noted that many of the properties of these nociceptive neurons are retained in different behavioral states, although state of consciousness can influence the excitability of the neurons.[21]

27.4.2 Properties of Other Trigeminal Brainstem Neurons

Although it is clear that subnucleus caudalis is very important in V nociceptive processing, electrophysiological and behavioral studies have also implicated more rostral components of the V brainstem complex in orofacial nociceptive mechanisms. Caudalis lesions may not completely eliminate all reflex or behavioral responses to noxious orofacial stimuli, whereas rostral lesions may interfere with pain behavior evoked by noxious stimuli applied to intraoral or perioral tissues.[5,14,79] The rostral components (subnuclei interpolaris and oralis) of the V spinal tract nucleus also project to some of the same regions that are the projection sites of caudalis nociceptive neurons and that are implicated in nociceptive transmission or its modulation, and stimulation of these modulatory sites can suppress rostral V neuron activity.[61] Furthermore, consistent with the documentation of opioid-containing terminals and opiate receptors in these rostral components, injection of analgesic chemicals such as morphine into the rostral components can suppress orofacial nociceptive behavior.[48] In addition, muscle, TMJ, tooth pulp, and cranial vessel afferents, as well as cutaneous nociceptive afferents, may terminate in the rostral components;[39,51,70,72] indeed, nociceptive neurons activated by noxious stimulation of intraoral or perioral regions have been documented in subnuclei interpolaris and oralis (Figure 27.3).[15,32,55] While the spatiotemporal coding features of these rostral V nociceptive neurons have not yet been described in detail, these findings do suggest that the rostral components of the V spinal tract nucleus may represent important elements in craniofacial pain processes, especially those related to perioral and intraoral nociceptive mechanisms.

27.4.3 Convergence

Some of the NS and WDR neurons in subnucleus caudalis of cats and rats receive exclusively craniofacial cutaneous (or mucosal) afferent inputs.[1,6,35,63,64] However, the majority of the NS and WDR neurons can also be excited by electrical or natural stimulation of afferents supplying the dental pulp, jaw and tongue muscles, cranial vessels, or the TMJ, as well as by cutaneous noxious stimulation; in contrast, LTM neurons typically respond only to cutaneous stimuli.[1,6,20,35,44,60,63,68,82]

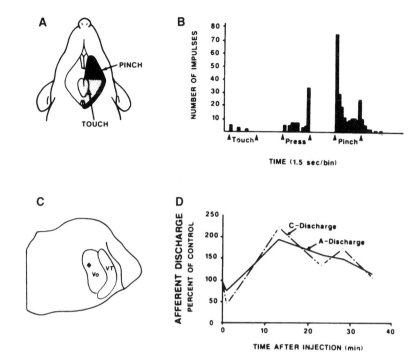

FIGURE 27.3
Response properties of a wide dynamic range (WDR) neuron and effects of 5% mustard oil injected into the deep masseter muscle. (**A**) The neuronal mechanoreceptive field is shown in the face. (**B**) Histogram shows its response to touch, pressure, and pinch stimuli applied to its mechanoreceptive field. (**C**) Its histologically confirmed recording locus within subnucleus oralis (Vo). VT = trigeminal spinal tract. (**D**) The effect of mustard oil application on the A- and C-fiber discharges evoked from the cutaneous mechanoreceptive field of the neuron. Note the time course of the facilitatory effect which affected both the A- and C-fiber cutaneous afferent inputs. (From Hu, J.W. et al, *Pain*, 48, 53, 1992. With permission.)

Very few caudalis neurons respond only to stimulation of TMJ, muscle, pulp, or cranial vessel afferents. Thus, the extensive convergent inputs from these afferents to cutaneous nociceptive neurons in subnucleus caudalis can explain the poor localization, spread, and referral of pain that are typical of pain conditions involving deep tissues such as the TMJ and associated musculature, or cranial vessel or pulpal tissues.[20,60,65,68] There is also extensive convergence in subnucleus caudalis from non-V afferents, e.g., 50% of the cutaneous nociceptive neurons can be excited by other cranial nerve afferents and by upper cervical nerve afferents.[63,64] This feature thus may provide a physiological explanation for the referral of pain between sites innervated by the V nerve or these non-V afferents.

27.4.4 Neuroplastic Changes in Nociceptive Neurons

These convergent afferent inputs, some of which can be documented only by electrical stimulation of peripheral sites, may also be involved in V brainstem neuroplastic changes that can be induced by injury, inflammation, or deafferentation. Neuroplasticity in central somatosensory pathways may be induced either by an enhanced nociceptive afferent input, as in direct stimulation of peripheral nerves by an injury and in inflammation, or by a decreased afferent input, as may occur through nerve damage resulting in deafferentation. The prolonged receptive field expansions and heightened excitability of central nociceptive neurons may be accompanied by pain behavior and have been viewed as reflections of a centrally based "functional plasticity" or "central sensitization" resulting from an "unmasking" or "strengthening" centrally of the convergent afferent inputs to the nociceptive neurons.[13,22]

Specifically in the V system, application of algesic chemicals and inflammatory irritants such as formalin or mustard oil into craniofacial tissues can markedly increase neuronal responses of NS and WDR neurons in subnuclei caudalis or oralis (see Figure 27.3).[37,38,55,63] The deep receptive field properties of nociceptive neurons are particularly expressive of this functional plasticity, and deep nociceptive inputs may be more effective than cutaneous inputs in inducing these effects. The mustard oil-induced nociceptive neuronal changes in the V brainstem complex may be accompanied by reflexively induced EMG activity in jaw-opening and jaw-closing muscles that is dependent upon a relay in subnucleus caudalis for its elicitation.[37,81] Several neurochemical processes (NMDA, neurokinins, opioids) appear involved in modulating these expressions of neuroplasticity in V brainstem nociceptive neurons.[37] As noted earlier, V brainstem neuroplasticity can also be induced by orofacial deafferentation, but these changes appear to take several days to develop and may be particularly directed at the LTM neurons.

27.5 Modulation of Somatosensory Transmission

The V brainstem neuroplasticity noted above is, in a sense, a form of modulation of somatosensory transmission. More traditional concepts of somatosensory modulation center around afferent-induced or descending modulation. While such modification of somatosensory transmission can occur at thalamic and cortical neuronal levels, the modification of ascending somatosensory information may largely occur earlier in the somatosensory pathways and involve presynaptic and postsynaptic regulatory processes, as pointed out in Chapter 28. It should be noted that these modulatory mechanisms may also be operational in varying degrees during different behavioral states and indeed contribute to specific features of these states, e.g., the atonia of active sleep (see Chapter 31).

In the V system, the intricate organization of each subdivision of the V brainstem complex and the variety of inputs and interconnections provide the basis for considerable interaction between the various inputs derived from peripheral tissues (e.g., so-called segmental or afferent inhibition) or from intrinsic brain regions (e.g., descending inhibition). Examples include the interneuronal system within the SG of subnucleus caudalis, the ascending modulatory influence of subnucleus caudalis on more rostral V brainstem neurons, and the descending inputs to rostral and caudal components of the V brainstem complex from the periaqueductal gray (PAG), nucleus raphe magnus (NRM), cerebral cortex, and several other brain centers. These modulatory processes involve a variety of endogenous neurochemical substances and receptor mechanisms,[22,25,59,61,62] some of which underlie the facilitatory influences on nociceptive transmission (e.g., substance P, NMDA), whereas others primarily exert inhibitory influences (e.g., enkephalin, GABA, 5-HT) that may involve presynaptic or postsynaptic regulatory mechanisms.

27.5.1 Descending Modulation

Neurons in the spinal somatosensory system are subject to powerful descending modulatory influences (see Chapters 28 and 29). In the V brainstem complex, the responses of V nociceptive neurons to afferent inputs can also be modulated by a number of central neural structures implicated in endogenous analgesic mechanisms, such as the PAG, NRM, anterior pretectal nucleus, and parabrachial area of the pons, just to name a few. Electrical or chemical stimulation of these structures can inhibit V brainstem neuronal and related reflex and behavioral responses to noxious stimulation of cutaneous, tooth pulp, or deep tissues.[11,23,25,29,61] Figure 27.4 shows examples of such

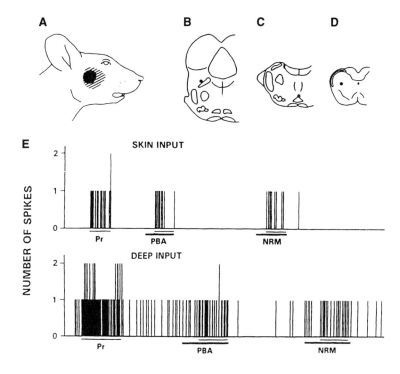

FIGURE 27.4
Effects of electrical stimulation of parabrachial area (PBA) or nucleus raphe magnus (NRM) on nociceptive responses of a caudalis nociceptive neuron. This neuron had no spontaneous activity and was activated by application of a force (200 g) applied either to its cutaneous receptive field (striped area) or its deep receptive field (dark area), as depicted in (**A**). (**B**) and (**C**) show the stimulation sites in PBA and NRM, respectively, and (**D**) the caudalis recording site. Note that PBA stimulation appeared to produce inhibition of nociceptive responses as potent as NRM stimulation (**E**). Thin and thick bars represent test (Pr = pressure) and conditioning stimulations, respectively. Binwidth: 20 msec. (From Chiang, C.Y. et al, *J. Neurophysiol.*, 71, 2430, 1994. With permission.)

descending modulation. Although these particular descending influences appear to be preferentially directed at nociceptive neurons, non-nociceptive transmission may also be affected. The LTM neurons instead are especially susceptible to modulation by corticofugal influences which may be somewhat less effective in suppressing V nociceptive transmission.

In the V brainstem complex, these descending pathways exert their modulatory effects by the release of certain of the neurochemicals mentioned above. For example, enkephalin or 5-HT may be released from the terminals of descending pathways from the PAG or NRM, respectively. Some neurochemicals also (e.g., enkephalin) or instead (e.g., GABA) appear to be released from the endings of neurons intrinsic to the V brainstem complex (e.g., the SG of subnucleus caudalis). Descending influences on V brainstem nociceptive transmission have been implicated as intrinsic mechanisms contributing to the analgesic effects of such therapeutic approaches as deep brain stimulation, acupuncture, and opiate-related drugs.[25,59,75,80] The modulatory influences on LTM neurons in the V brainstem complex are thought to contribute to the spatial coding properties of these tactile-transmission neurons,[10,16] and those from some specific brain regions also appear to play a role in the gating of V somatosensory transmission by behavioral state. For example, just as pain may be modified by behavioral factors, the receptive field and response properties of V (and spinal) nociceptive neurons are subject to behavioral variables that operate, at least in part, by utilizing some of the pathways and neurochemical mechanisms outlined above.[21] Furthermore, alterations in the efficacy of these various modulatory influences must also be considered as possible factors contributing to the neuroplastic changes of V brainstem neurons as a result of damage or inflammation of peripheral tissues (see above).

27.5.2 Afferent-Induced Modulation

The features and implications for orofacial somatosensation of so-called afferent or segmental inhibition have been well documented in the V literature, and indeed some of the first studies showing segmental inhibition of tactile-transmission neurons were demonstrated in LTM neurons at various levels of the V brainstem complex.[16,23] As in the spinal somatosensory system, these modulatory effects are thought to represent mechanisms for "sharpening" the coding properties of LTM neurons and so contribute to, for example, spatial acuity in orofacial touch sensation. In the case of nociceptive transmission in the V brainstem complex, the efficacy of noxious or non-noxious stimuli or therapeutic procedures such as acupuncture and transcutaneous electrical nerve stimulation (TENS) in suppressing craniofacial reflexes or neuronal responses has also been documented. The responses of V nociceptive neurons to small-diameter afferent inputs evoked by noxious orofacial stimuli can be suppressed by non-noxious somatosensory stimuli (e.g., vibratory or tactile) that excite large afferent nerve fibers and by stimulation of non-somatosensory afferents such as vagal nerve fibers.[49,59] Furthermore, diffuse noxious inhibitory controls (DNIC), whereby small-fiber afferent inputs from various spatially dispersed regions of the body may inhibit V and spinal neuronal responses to other small-fiber afferent inputs, are thought to contribute to the therapeutic effectiveness of procedures such as acupuncture, TENS, and counter-irritation.[35,55,75]

These afferent-induced effects can be explained by the involvement of segmental mechanisms (e.g., GABA-containing interneurons in the SG of subnucleus caudalis or more rostral components of the V brainstem complex; see above) but could also involve the recruitment of descending influences from higher brain regions. Neurochemicals implicated in these afferent-induced effects in the V brainstem complex include, in addition to GABA, the endogenous opioids, 5-HT, noradrenaline, adenosine and ATP.[18,25,57,59,75]

27.6 Concluding Remarks

It is clear from the foregoing that remarkable advances in knowledge of V brainstem somatosensory processing and its modulation have been gained over the last 30 years. This information has been important not only for our improved understanding of how touch and pain are transmitted, or modified in their transmission, through the V brainstem complex, but also for providing insights into the mechanisms and pathways contributing to the state-dependent processing of orofacial somatosensory information. There are nonetheless many issues that remain unresolved and that require much more research.[7,53] Research priorities include: (1) the processes by which the brainstem mechanisms change in, and contribute to, different behavioral states; (2) V brainstem processing of thermosensitive information and its modulation, as the focus in recent years has been on tactile and especially nociceptive brainstem mechanisms; (3) the possible processing of deep nociceptive inputs in the rostral components of the V brainstem complex, and indeed, the relative importance of the rostral and caudal components in different types of craniofacial pain; (4) the relative contribution to craniofacial pain of WDR compared to NS neurons, and the contribution of the different ascending projection pathways; (5) the mechanisms and roles of the various components of the V brainstem complex and adjacent structures in ascending somatosensory pathways and in muscle and autonomic reflex circuits; (6) the neurochemical processes underlying both nociceptive and non-nociceptive transmission in the different components of the V brainstem complex, including definition of these processes in different behavioral states and the sites within or outside the V brainstem complex where these influences originate; and (7) the development of appropriate chronic pain models and behavioral models analogous to those established in the spinal somatosensory system, in order to clarify V neuroplasticity and the underlying mechanisms of longer term alterations in afferent inputs (e.g., inflammation, nerve injury, deafferentation) so as to provide important insights into the pathogenesis of chronic craniofacial pain conditions and improved therapeutic approaches.

Acknowledgments

The studies of the author cited herein were supported by grant MT-4918 from the Medical Research Council of Canada and grant DE04786 from the U.S. National Institutes of Health. The secretarial assistance of Ms. Fong Yuen is gratefully acknowledged.

References

1. Amano, N., Hu, J.W., and Sessle, B.J., Responses of neurons in feline trigeminal subnucleus caudalis (medullary dorsal horn) to cutaneous, intraoral and muscle afferent stimuli, *J. Neurophysiol.*, 55, 227, 1986.
2. Azerad, J., Boucher, Y., and Pollin, B., Occurrence of glutamate in primary sensory trigeminal neurons innervating the rat dental pulp, *Comptes Rendus de l'Academie des Sciences Paris, Serie III*, 314, 469, 1992.
3. Beckstead, R.M. and Norgren, R., An autoradiographic examination of the central distribution of the trigeminal, facial, glossopharyngeal, and vagal nerves in the monkey, *J. Comp. Neurol.*, 184, 455, 1979.
4. Bereiter, D.A. and Benetti, A.P., Excitatory amino release within spinal trigeminal nucleus after mustard oil injection into the temporomandibular joint region of the rat, *Pain*, 67, 451, 1996.
5. Broton, J.G. and Rosenfeld, P., Effects of trigeminal tractotomy on facial thermal nociception in the rat, *Brain Res.*, 333, 63, 1985.
6. Broton, J.G, Hu, J.W., and Sessle, B.J., Effects of temporomandibular joint stimulation on nociceptive and nonnociceptive neurons of the cat's trigeminal subnucleus caudalis (medullary dorsal horn), *J. Neurophysiol.*, 59, 1575, 1988.
7. Bryant, P.S. and Sessle, B.J., Workshop recommendations on research needs and directions, in *Temporomandibular Disorders and Related Pain Conditions, Progress in Pain Research and Management*, Vol. 4, Sessle, B.J., Bryant, P.S., and Dionne, R.A., Eds., IASP Press, Seattle, 1995, p. 467.
8. Capra, N.F. and Dessem, D., Central connections of trigeminal primary afferent neurons: topographical and functional considerations, *Crit. Rev. Oral Biol. Med.*, 4, 1, 1992.
9. Cairns, B.E., Sessle, B.J., and Hu, J.W., Local application of glutamate to the rat temporomandibular joint induces increases in jaw muscle activity through activation of peripheral NMDA and non-NMDA receptors, *Neurosci. Abstr.*, 23, 1256, 1997.
10. Chiang, C.Y., Hu, J.W., Dostrovsky, J.O., and Sessle, B.J., Changes in mechanoreceptive field properties of trigeminal somatosensory brainstem neurons induced by stimulation of nucleus raphe magnus in cats, *Brain Res.*, 485, 371, 1989.
11. Chiang, C.Y, Hu, J.W, and Sessle, B.J., Parabrachial area and nucleus raphe magnus-induced modulation of nociceptive and nonnociceptive trigeminal subnucleus caudalis neurons activated by cutaneous or deep inputs, *J. Neurophysiol.*, 71, 2430, 1994.

12. Clements, J.R., Magusson, K.R., Hautman, J., and Beitz, A.J., Rat tooth pulp projections to spinal trigeminal subnucleus caudalis are glutamate-like immunoreactive, *J. Comp. Neurol.,* 309, 281, 1991.
13. Cooper, B.Y. and Sessle, B.J., Anatomy, physiology, and pathophysiology of trigeminal system paresthesias and dysesthesias, in *Oral and Maxillofacial Surgery Clinics of North America — Trigeminal Nerve Injury: Diagnosis and Management,* Vol. 4/No. 2, LaBanc, J.P. and Gregg, J.M., Eds., W.B. Saunders, Philadelphia, 1992, p. 297.
14. Dallel, R., Clavelou, P., and Woda, A., Effects of tractotomy on nociceptive reactions induced by tooth pulp stimulation in the rat, *Exp. Neurol.,* 106, 78, 1989.
15. Dallel, R., Raboisson, P., Woda, A., and Sessle, B.J., Properties of nociceptive and nonnociceptive brainstem neurons in trigeminal subnucleus oralis of the rat, *Brain Res.,* 521, 95, 1990.
16. Darian-Smith, I., Neural mechanisms of facial sensation, *Int. Rev. Neurobiol.,* 9, 301, 1966.
17. Darian-Smith, I., Rowe, M.J., and Sessle, B.J., "Tactile" stimulus intensity: information transmission by relay neurons in different trigeminal nuclei, *Science,* 160, 791, 1968.
18. Dickenson, A.H., Where and how do opioids act?, in *Proc. 7th World Congress on Pain, Progress in Pain Research and Management,* Vol. 2, Gebhart, G.F., Hammond, D.L., and Jensen, T.S., Eds., IASP Press, Seattle, 1994, p. 525.
19. Donga, R. and Lund, J.P., Discharge patterns of trigeminal commissural last-order interneurons during fictive mastication in the rabbit, *J. Neurophysiol.,* 66, 1564, 1991.
20. Dostrovsky, J.O., Davis, K.D., and Kawakita, K., Central mechanisms of vascular headaches, *Can. J. Physiol. Pharmacol.,* 69, 652, 1991.
21. Dubner, R., Recent advances in our understanding of pain, in *Oro-Facial Pain and Neuromuscular Dysfunction: Mechanisms and Clinical Correlates,* Klineberg, I. and Sessle, B., Eds, Pergamon Press, Oxford, 1985, p. 3.
22. Dubner, R. and Basbaum, A.I., Spinal dorsal horn plasticity following tissue or nerve injury, in *Textbook of Pain,* 3rd ed., Wall, P.D. and Melzack, R., Eds, Churchill Livingstone, London, 1994, p. 225.
23. Dubner, R., Sessle, B.J., and Storey, A.T., *The Neural Basis of Oral and Facial Function,* Plenum Press, New York, 1978, p. 483.
24. Eberberger, A., Anton, F., Tölle, T.R., and Zieglgänsberger, W., Morphine, 5-HT2 and 5-HT3 receptor antagonists reduce *c-fos* expression in the trigeminal nuclear complex following noxious chemical stimulation of the rat nasal mucosa, *Brain Res.,* 676, 336, 1995.
25. Fields, H.L. and Basbaum, A.I., Central nervous system mechanisms of pain modulation, in *Textbook of Pain,* 3rd ed., Wall, P.D. and Melzack, R., Eds, Churchill Livingstone, London, 1994, p. 243.
26. Gobel, S., Bennett, G.J., Allen, B., Humphrey, E., Seltzer, Z., Abdel-Moumene, M., Hayashi, H., and Hoffert, M.J., Synaptic connectivity of substantia gelatinosa neurons with reference to potential termination sites of descending axons, in *Brain Stem Control of Spinal Mechanisms,* Sjolund, B. and Bjorklund, A., Eds., Elsevier, Amsterdam, 1982, p. 135.
27. Gobel, S., Hockfield, S., and Ruda, M.A., Anatomical similarities between medullary and spinal dorsal horns, in *Oral-Facial Sensory and Motor Functions,* Kawamura, Y. and Dubner, R., Eds., Quintessence, Tokyo, 1981, p. 211.
28. Guilbaud, G., Bernard, J.F., and Besson, J.M., Brain areas involved in nociception and pain, in *Textbook of Pain,* 3rd ed., Wall, P.D. and Melzack, R., Eds., Churchill Livingstone, London, 1994, p. 113.
29. Hannam, A.G. and Sessle, B.J., Temporomandibular neurosensory and neuromuscular physiology, in *Temporomandibular Joint and Masticatory Muscle Disorders,* Zarb, G., Carlsson, G., Sessle, B., and Mohl, N., Eds., Munksgaard, Copenhagen, 1994, p. 67.

30. Hargreaves, K.M., Roszkowski, M.T., Jackson, D.L., Bowles, W., Richardson, J.D., and Swift, J.Q., Neuroendocrine and immune response to injury, degeneration, and repair, in *Temporomandibular Disorders and Related Pain Conditions, Progress in Pain Research and Management*, Vol. 4, Sessle, B.J., Bryant, P.S., and Dionne, R.A., Eds., IASP Press, Seattle, 1995, p. 273.

31. Hathaway, C.B., Hu, J.W., and Bereiter, D.A., Distribution of fos-like immunoreactivity in the caudal brainstem of the rat following noxious chemical stimulation of the temporomandibular joint, *J. Comp. Neurol.*, 356, 444, 1995.

32. Hayashi, H., Sumino, R., and Sessle, B.J., Functional organization of trigeminal subnucleus interpolaris: nociceptive and innocuous afferent inputs, projections to thalamus, cerebellum and spinal cord, and descending modulation from the periaqueductal gray, *J. Neurophysiol.*, 51, 890, 1984.

33. Henderson, T.A. and Jacquin, M.F., What makes subcortical barrels? Requisite trigeminal circuitry and developmental mechanisms, in *Cerebral Cortex*, Jones, E.G. and Diamond, I.T., Eds., Plenum Press, New York, 1995, p. 123.

34. Henry, J.L., Sessle, B.J., Lucier, G.E., and Hu, J.W., Effects of substance P on nociceptive and non-nociceptive trigeminal brain stem neurons, *Pain*, 8, 33, 1980.

35. Hu, J.W., Response properties of nociceptive and non-nociceptive neurons in the rat's trigeminal subnucleus caudalis (medullary dorsal horn) related to cutaneous and deep craniofacial afferent stimulation and modulation by diffuse noxious inhibitory controls, *Pain*, 41, 331, 1990.

36. Hu, J.W., Dostrovsky, J., Lenz, Y., Ball, G., and Sessle, B.J., Tooth pulp deafferentation is associated with functional alterations in the properties of neurons in the trigeminal spinal tract nucleus, *J. Neurophysiol.*, 56, 1650, 1986.

37. Hu, J.W., Tsai, C.-M., Bakke, M., Seo, K., Tambeli, C.H., Vernon, H., Bereiter, D.A., and Sessle, B.J., Deep craniofacial pain: involvement of trigeminal subnucleus caudalis and its modulation, in *Proc. 8th World Congress on Pain, Progress in Pain Research and Management*, Vol. 8, Jensen, T.S., Turner, J.A., and Wiesenfeld-Hallin, Z., Eds., IASP Press, Seattle, 1997, p. 497.

38. Hu, J.W., Sessle, B.J., Raboisson, P., Dallel, R., and Woda, A., Stimulation of craniofacial muscle afferents induces prolonged facilitatory effects in trigeminal nociceptive brainstem neurones, *Pain*, 48, 53, 1992.

39. Jacquin, M.F., Woerner, D., Szczepanik, A.M., Riecker, V., Mooney, R.D., and Rhoades, R.W., Structure-function relationships in rat brainstem subnucleus interpolaris. I. Vibrissa primary afferents, *J. Comp. Neurol.*, 243, 266, 1986.

40. Jacquin, M.F., Golden, J., and Rhoades, R., Structure-function relationships in rat brainstem subnucleus interpolaris. III. Local circuit neurons, *J. Comp. Neurol.*, 282, 24, 1989.

41. Johnson, L.R., Westrum, L.E., and Henry, M.A., Anatomic organization of the trigeminal system and the effects of deafferentation, in *Trigeminal Neuralgia. Current Concepts Regarding Pathogenesis and Treatment*, Fromm, G.H. and Sessle, B.J., Eds., Butterworth-Heinemann, Stoneham, 1991, p. 27.

42. Kerr, F.W.L., Craniofacial neuralgias, in *Advances in Pain Research and Therapy*, Vol. 3, Bonica, J.J., Liebeskind, J.C., and Albe-Fessard, D.G., Eds., Raven Press, New York, 1979, p. 283.

43. Kirkpatrick, D.B. and Kruger, L., Physiological properties of neurons in the principal sensory trigeminal nucleus of the cat, *Exp. Neurol.*, 48, 664, 1975.

44. Kojima, Y., Convergence patterns of afferent information from the temporomandibular joint and masseter muscle in the trigeminal subnucleus caudalis, *Brain Res. Bull.*, 24, 609, 1990.

45. Kondo, E., Kiyama, H., Yamano, M., Shida, T., Ueda, Y., and Tohyama, M., Expression of glutamate (AMPA) and gamma-aminobutyric acid (GABA)$_A$ receptors in the rat caudal trigeminal spinal nucleus, *Neurosci. Lett.*, 186, 169, 1995.

46. Kwan, C.L., Hu, J.W., and Sessle, B.J., Effects of tooth pulp deafferentation on brainstem neurons of the rat trigeminal subnucleus oralis, *Somatosensory Motor Res.*, 10, 115, 1993.

47. Kwan, C.L., Hu, J.W., and Sessle, B.J., Neuroplastic effects of neonatal capsaicin treatment on neurons in adult rat trigeminal nuclei principalis and subnucleus oralis, *J. Neurophysiol.,* 75, 298, 1996.
48. Luccarini, P., Cadet, R., Saade, M., and Woda, A., Antinociceptive effect of morphine microinjections into the spinal trigeminal subnucleus oralis, *NeuroReport,* 6, 365, 1995.
49. Maixner, W., Sigurdsson, A., Fillingim, R.B., Lundeen, T., and Booker, D.K., Regulation of acute and chronic orofacial pain, in *Orofacial Pain and Temporomandibular Disorders, Advances in Pain Research and Therapy,* Vol. 21, Fricton, J.R. and Dubner, R., Eds., Raven Press, New York, 1995, p. 85.
50. Matthews, M.A., Hernandez, T.V., and Liles, S.L., Immunocytochemistry of enkephalin and serotonin distribution in restricted zones of the rostral trigeminal spinal subnuclei: comparisons with subnucleus caudalis, *Synapse,* 1, 512, 1987.
51. Matthews, M.A., Hoffman, K.D., and Hernandez, T.V., Ulex europaeus agglutinin-I binding to dental primary afferent projections in the spinal trigeminal complex combined with double immunolabeling of substance P and GABA elements using peroxidase and colloidal gold, *Somatosensory Motor Res.,* 6, 513, 1989.
52. Narhi, M., Yamamoto, H., Ngassapa, D., and Hirvonen, T., The neurophysiological basis and the role of inflammatory reactions in dentine hypersensitivity, *Archs. Oral Biol.,* 39(Suppl.), 23S, 1994.
53. National Institute of Dental Research, *Long-Range Research Plan for the Nineties,* National Institutes of Health, Bethesda, MD, 1990.
54. Pfaller, K. and Arvidsson, J., Central distribution of trigeminal and upper cervical primary afferents in the rat studied by anterograde transport of horseradish peroxidase conjugated to wheat germ agglutinin, *J. Comp. Neurol.,* 268, 91, 1988.
55. Raboisson, P., Dallel, R., Clavelou, P., Sessle, B.J., and Woda, A., Effects of subcutaneous formalin on the activity of trigeminal brain stem nociceptive neurons in the rat, *J. Neurophysiol.,* 73, 496, 1995.
56. Salt, T.E. and Hill, R.G., Neurotransmitter candidates of somatosensory primary afferent fibres, *Neuroscience,* 10, 1083, 1983.
57. Salter, M.W., De Koninck, Y., and Henry, J.L., Physiological roles for adenosine and ATP in synaptic transmission in the spinal dorsal horn, *Prog. Neurobiol.,* 41, 125, 1993.
58. Schaible, H.-G., Ebersberger, A., Peppel, P., Beck, U., and Meßlinger, K., Release of immunoreactive substance P in the trigeminal brain stem nuclear complex evoked by chemical stimulation of the nasal mucosa and dura mater encephali — a study with antibody microprobes, *Neuroscience,* 76, 273, 1997.
59. Sessle, B.J., The neurobiology of orofacial and dental pain, *J. Dental Res.,* 66, 962, 1987.
60. Sessle, B.J., Masticatory muscle disorders: basic science perspectives, in *Temporomandibular Disorders and Related Pain Conditions, Progress in Pain Research and Management,* Vol. 4, Sessle, B.J., Bryant, P.S., and Dionne, R.A., Eds., IASP Press, Seattle, 1995, p. 47.
61. Sessle, B.J., Chiang, C.Y., and Dostrovsky, J.O., Interrelationships between sensorimotor cortex, anterior pretectal nucleus and periaqueductal gray in modulation of trigeminal sensorimotor function in the rat, in *Processing and Inhibition of Nociceptive Information,* Inoki, R., Shigenaga, Y., and Tohyama, M., Eds., Elsevier, Amsterdam, 1992, p. 77.
62. Sessle, B.J. and Hu, J.W., Raphe-induced suppression of the jaw-opening reflex and single neurons in trigeminal subnucleus oralis, and influence of naloxone and subnucleus caudalis, *Pain,* 10, 19, 1981.
63. Sessle, B.J. and Hu, J.W., Mechanisms of pain arising from articular tissues, *Can. J. Physiol. Pharmacol.,* 69, 617, 1991.
64. Sessle, B.J., Hu, J.W., Amano, N., and Zhong, G., Convergence of cutaneous, tooth pulp, visceral, neck and muscle afferents onto nociceptive and nonnociceptive neurones in trigeminal subnucleus caudalis (medullary dorsal horn) and its implications for referred pain, *Pain,* 27, 219, 1986.

65. Sessle, B.J., Hu, J.W., and Yu, X.-M., Brainstem mechanisms of referred pain and hyperalgesia in the orofacial and temporomandibular region, in *New Trends in Referred Pain and Hyperalgesia, Pain Research and Clinical Management,* Vol. 7, Vecchiet, L., Albe-Fessard, D., and Lindblom U., Eds., Elsevier, Amsterdam, 1993, p. 59.
66. Shortland, P.J., Jacquin, M.F., DeMaro, J.A., Kwan L., Hu, J.W., and Sessle, B.J., Central projections of identified trigeminal primary afferents after molar pulp deafferentation in adult rats, *Somatosensory Motor Res.,* 12, 277, 1995.
67. Strassman, A.M., Mineta, Y., and Vos, B.P., Distribution of fos-like immunoreactivity in the medullary and upper cervical dorsal horn produced by stimulation of dural blood vessels in the rat, *J. Neurosci.,* 14, 3725, 1994.
68. Strassman, A.M., Potrebic, S., and Maciewicz, R.J., Anatomical properties of brainstem trigeminal neurons that respond to electrical stimulation of dural blood vessels, *J. Comp. Neurol.,* 346, 349, 1994.
69. Suarez-Roca, H. and Maixner, W., Activation of kappa opioid receptors by U50488H and morphine enhances the release of substance P from rat trigeminal nucleus slices, *J. Pharmacol. Exp. Ther.,* 264, 648, 1993.
70. Takemura, M., Nagase, Y., Yoshida, A., Yasuda, K., Kitamura, S., Shigenaga, Y., and Matano, S., The central projections of the monkey tooth pulp afferent neurons, *Somatosensory Motor Res.,* 10, 217, 1993.
71. Torvik, A., Afferent connections to the sensory trigeminal nuclei, the nucleus of the solitary tract and adjacent structures, *J. Comp. Neurol.,* 106, 51, 1956.
72. Tsuru, K., Otani, K., Kajiyama, K., Suemune, S., and Shigenaga, Y., Central terminations of periodontal mechanoreceptive and tooth pulp afferents in the trigeminal principal and oral nuclei of the cat, *Brain Res.,* 485, 29, 1989.
73. Wakisaki, S., Sasaki, Y., Ichikawa, H., and Matsuo, S., Increase in *c-fos*-like immunoreactivity in the trigeminal nucleus complex after dental treatment, *Proc. Finn. Dent. Soc.,* 88 (suppl. 1), 551, 1992.
74. Willis, W.D. and Coggeshall, R.E., *Sensory Mechanisms of the Spinal Cord,* Raven Press, New York, 1991.
75. Woolf, C.J. and Thompson, J.W., Stimulation fibre-induced analgesia: transcutaneous electrical nerve stimulation (TENS) and vibration, in *Textbook of Pain,* 3rd ed., Wall, P.D. and Melzack, R., Eds., Churchill Livingstone, London, 1994, p. 1191.
76. Woolsey, T.A., Peripheral alteration and somatosensory development, in *Development of Sensory Systems in Mammals,* Coleman, J.R., Ed., Wiley, New York, 1990, p. 461.
77. Yokota, T., Neural mechanisms of trigeminal pain, in *Advances in Pain Research and Therapy,* Vol. 9, Fields, H.L., Dubner, R., and Cervero, F., Eds., Raven Press, New York, 1985, p. 221.
78. Yoshida, A., Yasuda, K., Dostrovsky, J.O., Bae, Y.C., Takemura, M., Shigenaga, Y., and Sessle, B.J., Two major types of premotoneurons in the feline trigeminal nucleus oralis as demonstrated by staining with HRP, *J. Comp. Neurol.,* 347, 495, 1994.
79. Young, R.F. and Perryman, K.M., Pathways for orofacial pain sensation in the trigeminal brain-stem nuclear complex of the Macaque monkey, *J. Neurosurg.,* 61, 563, 1984.
80. Young, R.F. and Rinaldi, P.C., Brain stimulation for relief of chronic pain, in *Textbook of Pain,* 3rd ed., Wall, P.D. and Melzack, R., Eds., Churchill Livingstone, London, 1994, p. 1225.
81. Yu, X.-M., Sessle, B.J., Haas, D.A., Izzo, A., Vernon, H., and Hu, J.W., Involvement of NMDA receptor mechanisms in jaw electromyographic activity and plasma extravasation induced by inflammatory irritant application to temporomandibular joint region of rats, *Pain,* 68, 169, 1996.
82. Zagami, A.S. and Lambert, G.A., Central projections from cranial blood vessels: emphasis on spinal cord and thalamus, in *Pathophysiological Mechanisms of Migraine,* Olesen J. and Schmidt, R.F., Eds., VCH, Weinheim, 1993, p. 221.

Chapter 28

Anatomy, Physiology, and Descending Control of Lumbosacral Sensory Neurons Involved in Tactile and Pain Sensations

William D. Willis, Jr.

Contents

28.1	Introduction		464
28.2	Postsynaptic Dorsal Column Pathway		464
	28.2.1	Anatomical Organization	464
	28.2.2	Response Properties	465
	28.2.3	Sensory Functions	466
	28.2.4	Descending Control	467
	28.2.5	Primary Afferent Control	467
28.3	Spinocervical Tract		468
	28.3.1	Anatomical Organization	468
	28.3.2	Response Properties	469
	28.3.3	Sensory Functions	469
	28.3.4	Descending Control	469
	28.3.5	Primary Afferent Control	470
28.4	Spinothalamic Tract		470
	28.4.1	Anatomical Organization	470
	28.4.2	Response Properties: Lamina I STT Cells	471
	28.4.3	Response Properties: STT Cells in Laminae IV–VI	472
	28.4.4	Responses of STT Cells Projecting to Intralaminar Nuclei	474
	28.4.5	Sensory Functions	474
	28.4.6	Descending Control	475
	28.4.7	Primary Afferent Control	476

28.5	Spinoreticular and Spinomesencephalic Tracts		476
	28.5.1	Anatomical Organization	476
	28.5.2	Response Properties	477
	28.5.3	Sensory Functions	477
	28.5.4	Descending Controls	477
	28.5.5	Primary Afferent Control	477
28.6	Summary		477
Acknowledgments			478
References			478

28.1 Introduction

This chapter will be concerned with neurons whose cell bodies are located in the lumbosacral spinal cord and which contribute to the processing of tactile and pain sensations. These include neurons whose axons project in the postsynaptic dorsal column (PSDC) pathway and the spinocervical (SCT), spinothalamic (STT), spinoreticular (SRT), and spinomesencephalic (SMT) tracts.[9,115,116] A number of other tracts, including those belonging to the trigeminal system, will be considered elsewhere in this section (see Chapter 27). In addition, numerous interneurons, especially in the dorsal horn and intermediate region of the spinal cord gray matter, are involved in sensory (as well as in reflex) processing. It should be noted that the most important contribution to discriminative tactile sensation is made by axons of dorsal root ganglia that enter the spinal cord through dorsal roots, ascend in the dorsal column, and synapse in the dorsal column nuclei.[115] However, consideration here will be limited to the PSDC pathway, SCT, STT, SRT, and SMT. These ascending sensory pathways are all under the control of pathways that descend from the brain.

Most of our knowledge concerning these sensory pathways has been obtained in anatomical experiments or in neurophysiological and neuropharmacological investigations on anesthetized animals. However, for some topics, psychophysical studies have been done on human subjects that allow correlations between the activity of neurons in reduced preparations and human sensation. These studies are important for interpretations of changes in somatosensory processing in altered states (see Chapter 31).

28.2 Postsynaptic Dorsal Column Pathway

28.2.1 Anatomical Organization

Some ascending axons in the dorsal columns arise from neurons of the spinal cord gray matter, in addition to those originating from dorsal root ganglion cells. These axons form the PSDC pathway. Cell bodies of neurons that give rise to the PSDC pathway in rats, cats, and monkeys are concentrated in lamina IV of the dorsal horn, although some are also in other laminae, including III, V–VII, and X.[8,115] These neurons project to the dorsal column nuclei in a somatotopically organized fashion, mostly by way of the dorsal funiculus, although a small portion of the projection is actually in the dorsal lateral funiculus.

It has been estimated that in rats there are about 750 to 1000 PSDC neurons in the cervical enlargement and 500 to 700 in the lumbar enlargement.[51] The number of these neurons is about one third of the total number of neurons, including dorsal root ganglion cells, which project to the dorsal column nuclei from the spinal cord. Bennett et al.[8] estimate that in cats and monkeys there are about

1000 PSDC neurons in the lumbar enlargement. Enevoldson and Gordon[41] provide estimates of 1700 to 2000 PSDC cells in the cat cervical enlargement and 800 to 1200 in the lumbosacral enlargement. Thus, the PSDC pathway is a substantial projection.

Postsynaptic dorsal column cells in laminae III and IV of the dorsal horn can have dorsally directed dendrites that penetrate into lamina II and that may even reach lamina I.[14] Such neurons could receive direct synaptic connections from nociceptive primary afferent fibers that terminate in laminae I and II, in addition to synapses from tactile afferents that end in laminae III and IV.[110,115] It is noteworthy that neurokinin receptors in the substantia gelatinosa are located on dendrites that ascend from neurons in deeper laminae, rather than on the intrinsic interneurons of lamina II;[10,17,79,81] however, it appears that the dendrites of PSDC cells in laminae III and IV do not contain neurokinin receptors,[83] although it is known that similarly disposed dendrites of STT cells do.[83,88] Synapses have been described on PSDC cells that contain immunoreactivity for glutamate, glycine, and GABA.[85,86] The glutamate-containing synapses accounted for 53% of the synaptic profiles contacting the cell bodies of the PSDC cells examined and are presumably excitatory. Forty percent of the synaptic profiles contained GABA immunoreactivity; however, the same terminals usually also showed co-localization of glycine. The glycine- and GABA-containing boutons are undoubtedly inhibitory synapses. Some glutamate-containing boutons were found to be postsynaptic to GABA-containing boutons, suggesting that PSDC neurons are controlled by presynaptic inhibition.[85] Only about 20% of the terminals of PSDC neurons in the dorsal column nuclei seem to be glutamatergic; the transmitter utilized by most of the PSDC neurons is unknown.[32]

28.2.2 Response Properties

Postsynaptic dorsal column neurons in the dorsal horn respond to stimuli applied to the skin. Many of the neurons appear to have a tactile function, as they can be activated by sensitive mechanoreceptors in the skin, such as hair follicle afferents, Pacinian corpuscles, or Merkel cell endings; however, more than half are excited by both innocuous and noxious cutaneous stimuli (see Reference 115 for a review). Very few are selectively excited by nociceptors, and none so far tested has responded to stimulation of visceral afferents. The receptive fields of dorsal horn PSDC cells are fairly small on the distal part of the limb, but they are large on the proximal limb; the excitatory receptive fields can also be on the hairy or glabrous skin, or both, and there are often inhibitory receptive fields, as well. PSDC cells in rats may respond just to weak or to weak and strong mechanical stimuli, but they generally do not respond to noxious heat stimuli, even when these are repeated and presumably sensitize the skin.[48] However, PSDC neurons in cats do respond to noxious heat stimuli, especially when the stimuli are repeated.[59,92]

Recently, a concentration of PSDC neurons has been described in the gray matter around the central canal (laminae X and VII) of the sacral spinal cord.[56] These neurons can be activated both by cutaneous and visceral stimuli. For example, in addition to discharging when the skin in the peroneal region is stimulated, many of these neurons show a graded excitation following graded noxious colorectal distention (CRD; Figure 28.1E), and they also show enhanced responses following acute inflammation of the colon.[1] The responses of these neurons as well as of neurons in the nucleus gracilis to noxious CRD are blocked when morphine or the non-NMDA glutamate receptor antagonist, CNQX, is infused into the gray matter of the sacral spinal cord by microdialysis (Figure 1D,E),[1] indicating that the visceral responses of gracile neurons depend on a synaptic relay in the sacral cord rather than on directly projecting axons of dorsal root ganglion cells. It is likely that interruption of the visceroreceptive component of the PSDC pathway is responsible for the reduction in pelvic cancer pain that has been reported following limited midline myelotomy at T10 in human patients (Figure 28.2).[56,89] This would imply that the PSDC pathway forms a major route by which visceral nociceptive information reaches the thalamus.

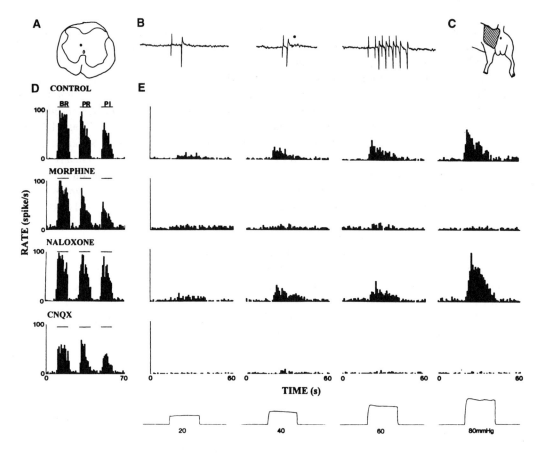

FIGURE 28.1
Responses of a postsynaptic dorsal column (PSDC) neuron in the sacral cord to mechanical stimulation of the skin and noxious colorectal distention. The recording site is shown in (**A**). The neuron was identified by antidromic activation from the dorsal column near the nucleus gracilis; the antidromic action potential, the collision test, and high frequency following are shown in (**B**). The receptive field is indicated in (**C**). The histograms in (**D**) show the responses to brush (BR), pressure (PR), and pinch (PI) stimuli applied to the skin in the receptive field. The responses in the top histogram served as a control for the effects of morphine, naloxone, and CNQX, which are shown in the lower three histograms. In (**E**) are the responses of the PSDC neuron to graded colorectal distention; the pressures attained are shown in the monitor records at the bottom to be 20, 40, 60, and 80 mmHg. The top row of histograms are the control responses, the second row the responses after morphine was administered in the sacral cord by microdialysis, the third row the reversal of the morphine action by systemic naloxone, and the fourth row the effects of microdialysis administration of CNQX. (From Al-Chaer, E.D. et al., *J. Neurophysiol.*, 76, 2675, 1996. With permission.)

28.2.3 Sensory Functions

Evidently, PSDC neurons in the dorsal horn of cats can contribute to the processing of both tactile and nociceptive information from the skin, although a nociceptive function is controversial for rats. Presumably, some of the neurons in the dorsal column nuclei that have been described respond to noxious stimulation of the skin and derive their nociceptive input from the PSDC pathway.[24,44,115] However, unmyelinated peptidergic afferent fibers that arise from dorsal root ganglion cells and that project directly to the dorsal column nuclei have also been described.[99,100] These could be the axons of primary afferent nociceptors and could explain at least some of the nociceptive responses of neurons of the dorsal column nuclei. The PSDC pathway has now been shown to include a major visceral nociceptive component.

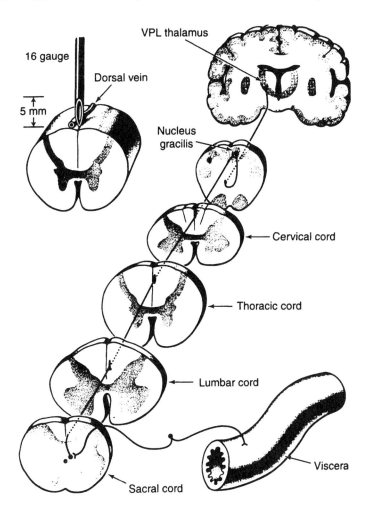

FIGURE 28.2
Drawing showing the course of the visceral pain pathway from the colon to the ventral posterior lateral (VPL) nucleus of the thalamus. A visceral primary afferent fiber is shown to innevate the colon. The afferent projects to the sacral spinal cord, where it synapses on a postsynaptic dorsal column neuron, which in turn projects to the nucleus gracilis. A neuron in the nucleus gracilis is then shown to project to the contralateral VPL nucleus. The inset at the upper left shows the neurosurgical procedure of limited midline myelotomy at T10 which was used to interrupt this pathway to relieve pelvic visceral pain. (From Nauta, H.J.W. et al., *J. Neurosurg.*, 86, 538, 1997. With permission.)

28.2.4 Descending Control

The PSDC pathway has been shown to be controlled by pathways that descend from the brain in experiments in which conduction in these pathways was blocked by cold.[93] PSDC neurons receive serotonergic synapses, presumably originating from neurons of the brainstem raphe nuclei;[90] however, the descending control systems that regulate the activity of PSDC neurons have not been examined in detail.

28.2.5 Primary Afferent Control

In addition to inhibitory controls from pathways that descend from the brain, PSDC neurons have been shown to be inhibited following volleys in primary afferent fibers; some of these inhibitory

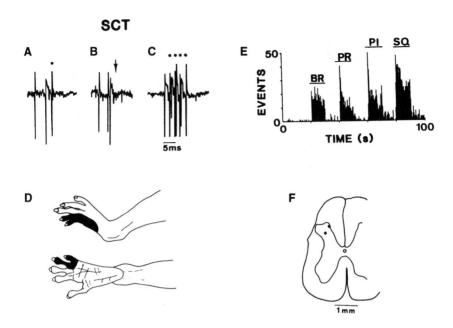

FIGURE 28.3
Responses of a neuron belonging to the primate spinocervical tract (SCT). Antidromic activation of the SCT cell from the dorsolateral funiculus at C3 just caudal to the lateral cervical nucleus is shown in (**A**). Collision is illustrated in (**B**) and high frequency following in (**C**). The receptive field is indicated in (**D**). The histogram in (**E**) shows the responses of the cell to brush (BR), pressure (PR), pinch (PI), and squeeze (SQ) stimuli applied to the receptive field. The locations of two SCT cells are shown in (**F**).

actions are due to inhibitory postsynaptic potentials,[58a] but there is also a long-lasting inhibition with the time course expected for presynaptic inhibition.[14a]

28.3 Spinocervical Tract

28.3.1 Anatomical Organization

The spinocervical tract is another pathway that originates in the spinal cord gray matter and provides somatosensory information to the thalamus after a synaptic relay. The name is derived from the fact that this pathway relays in the lateral cervical nucleus, which is a collection of neurons in the lateral funiculus at the C1–C2 level just ventrolateral to the dorsal horn (see Reference 115 for a review). Neurons giving rise to the spinocervical tract (SCT) are concentrated in laminae III and IV of the dorsal horn, although a few SCT neurons may be found in other laminae; however, many of the latter may actually be propriospinal neurons that project to the spinal gray matter at an upper cervical level. SCT neurons project to the lateral cervical nucleus by way of the dorsal part of the lateral funiculus. A lateral cervical nucleus has been identified in rats, cats, dogs, raccoons, and monkeys.[115] It has also been observed in at least some human spinal cords, although it is unclear if the SCT is an important entity in humans. The lateral spinal nucleus of the rodent spinal cord is a different structure from the lateral cervical nucleus;[4] the lateral spinal nucleus extends the length of the spinal cord, whereas the lateral cervical nucleus is a separate nucleus that is found only in the upper cervical spinal cord.[52]

The number of SCT neurons in rats and cats has been estimated. There is a total of only a few hundred SCT cells in the cervical and lumbar enlargements of rats.[4] Many more are found in cats:

550 to 800 in the lumbosacral enlargement, a total of about 2200 on each side.[15] In another study, 1100 were found in the cat cervical enlargement, 450 in the thoracic cord, and 700 in the lumbosacral enlargement.[42]

Neurons of the SCT are distributed in the dorsal horn in a somatotopically organized manner.[115] At least in cats, the dendrites of SCT cells tend to extend dorsally to the border between laminae II and III and then to turn to pass longitudinally in a rostrocaudal direction. The dendrites form longitudinal columns that are aligned with the terminations of hair follicle afferents in the dorsal horn. The neurotransmitter that is used by SCT cells is likely to be glutamate.[60]

28.3.2 Response Properties

Neurons of the SCT typically respond vigorously to weak mechanical stimuli, such as hair movement, although many also have a convergent nociceptive input (Figure 28.3)[39,115] Some SCT cells are activated by fine muscle afferents.[58] Although early investigations in cats did not reveal receptive fields for SCT neurons on the glabrous skin, such fields were identified for at least a few cat SCT neurons in later work.[62] Glabrous skin receptive fields have also been described for monkey SCT neurons (Figure 28.3).[39] Although the receptive fields of SCT neurons can be quite circumscribed (Figure 28.3), those of neurons in the lateral cervical nucleus are often relatively large (see Reference 115 for a review).

28.3.3 Sensory Functions

Behavioral studies indicate that tactile deficits in animals are more profound following lesions of the dorsal quadrant of the spinal cord than after a lesion that is limited to either the dorsal column or the dorsal lateral funiculus (see Reference 115 for a review). This supports a role for the SCT in tactile discrimination, although it is also possible that interruption of PSDC axons in the dorsal quadrant helps explain the behavioral changes. It is unclear to what extent the SCT contributes to nociceptive processing. Ventral quadrant lesions in cats are insufficient to produce analgesia,[22] and dorsal quadrant lesions are sometimes required to eliminate the nociceptive responses of neurons in the monkey ventral posterior lateral thalamic nucleus,[23] implying a possible role for the SCT in nociception. However, again axons of the PSDC in the dorsal quadrant could also be involved. As many as 20% of the SCT neurons on one side of the spinal cord in cats can be excited by stimulation of the toes,[16] suggesting that the SCT may form an early warning system that alerts the animal to tactile or other stimuli applied to certain key regions of the hairy skin.

28.3.4 Descending Control

The SCT is under the control of descending projection systems, as can be demonstrated by increases in responses of these neurons after cold block of the cord. Stimulation of axons in different parts of the spinal cord white matter (dorsal lateral funiculus, ventral funiculus, and dorsal funiculus) can result in the inhibition of SCT cells. Some of these effects are consistent with the observation that stimulation of the periaqueductal gray, nucleus cuneiformis, nucleus raphe magnus, or the medullary reticular formation results in the inhibition of SCT cells. The activity of SCT cells can also be modulated by stimulation of the somatosensory cerebral cortex. The lateral cervical nucleus is also under descending control, but much of this may be secondary to the descending modulation of SCT neurons. However, nerve fibers containing serotonin- or tyrosine hydroxylase- immunoreactivity are found within the lateral cervical nucleus, suggesting that this nucleus is directly affected by several of the descending control systems.[13,40]

28.3.5 Primary Afferent Control

Spinocervical tract cells can be inhibited by volleys in primary afferent fibers, including both cutaneous and muscle afferents.[13a,15a,55a,103a] At least part of the inhibition can be attributed to inhibitory postsynaptic potentials, as shown by intracellular recordings.[13a,55a] Inhibition from stimulation within the excitatory receptive field could last as long as 1 sec.[103a]

28.4 Spinothalamic Tract

28.4.1 Anatomical Organization

The spinothalamic tract (STT) originates in the spinal cord and terminates largely in the contralateral thalamus, although there are some ipsilateral projections as well. The presence of an STT has been confirmed in a variety of mammals, including marsupial phalangers, opossums, hedgehogs, rats, rabbits, pigs, sheep, cats, dogs, raccoons, and a number of primates, including humans (see Reference 115 for a review). STT neurons are concentrated in laminae I and IV–VI, and there are also STT cells in the ventral horn and around the central canal in rats and monkeys. This is also true for the cervical spinal cord in cats, although there are few STT cells in laminae IV–VI of the cat lumbar enlargement.

A case can be made for subdividing the STT according to the thalamic targets of these neurons. A major component of the STT projects to the ventral posterior lateral (VPL) nucleus in rats, monkeys, and humans.[53,115] However, in cats, there are few STT terminations in the VPL nucleus. Instead, much of the projection of the STT to the lateral thalamus is to the boundary region between the VPL and ventrolateral (VL) nuclei. This latter observation may be related to the presence of only a few STT cells in laminae IV–VI of the cat lumbar spinal cord. The projection in cats to the VPL/VL border may correspond to the STT projection to the VPLo nucleus in monkeys. Other important targets of the STT in the lateral thalamus include the ventral posterior inferior nucleus and the posterior complex.[109,115]

The STT also projects to several nuclei in the medial thalamus, including the intralaminar complex and parts of the medial dorsal nucleus, the nucleus submedius,[115] and the posterior part of the ventral medial nucleus (VMpo).[29] STT cells that project to the intralaminar nuclei are concentrated in the intermediate region and ventral horn, although there is also evidence for a projection from laminae I and V. The latter STT cells are likely to give collateral projections to both the lateral and medial thalamus.[28,50,108] In cats, the main projection to the nucleus submedius is by cells of lamina I;[26,108] however, this is not the case in rats, where the projection to the nucleus submedius is from deeper layers.[30] Recently, axons from lamina I in monkeys have been traced to the posterior part of the ventral medial nucleus.[29]

The number of STT cells has been estimated in rats, cats, and monkeys. In rats, 9500 cells in the entire spinal cord were labeled from one side of the thalamus.[18] In cats, Craig et al.[28] estimated that 5000 STT cells could be labeled from one side of the thalamus. For monkeys, the comparable figure is over 18,000.[3]

The dendrites of many of the cells in lamina I are oriented longitudinally, although some have a transverse disposition; based on dendritic structure, lamina I neurons could be classified as fusiform, multipolar, flattened, or pyramidal.[70] In rats, the lamina I neurons that project to the thalamus have either pyramidal or fusiform shapes, corresponding to two of the morphological classes of cells in lamina I.[71] However, in cats and monkeys, lamina I STT cells include pyramidal, fusiform, and multipolar cells.[124,125] At least some of these neurons receive primary afferent synapses containing the neuropeptides, calcitonin gene-related peptide,[20] and substance P,[88] and substance P receptors have been reported to occur on lamina I STT cells.[68] Of the synapses that contact rat lamina

I STT cell bodies and dendrites, 37% are immunoreactive for glutamate and 20% for GABA; this is true for both pyramidal and flattened cell types.[67]

Dendrites of many STT neurons in laminae IV and V of monkeys can project as far dorsally as lamina I.[115] Many of these dendrites contain NK1 receptors and so should respond to substance P.[35,83,88] It is noteworthy that few NK1 receptors occur on the intrinsic interneurons of laminae II,[10,17,79,81] so STT neurons and perhaps other types of ascending tract cells provide the dendritic substrate for much of the peptidergic synaptic transmission in the superficial dorsal horn and presumably form a major output for the transmission of nociceptive information that relays in the superficial dorsal horn.[33,80]

Dendrites of STT cells in the deep layers of the monkey dorsal horn may extend into the lateral funiculus or ventromedially into laminae VII and X (see Reference 115 for a review). Dendrites of STT cells in the ventral horn of cats penetrate the lateral white matter or extend medially into lamina X. Some of these neurons receive synaptic inputs from muscle and joint receptors.

About half of the synaptic profiles that contact the cell bodies and proximal dendrites of STT cells in laminae IV–VI contain glutamate,[112] and about one fifth contain gamma-aminobutyric acid.[21] Glutamate receptors of the NMDA and AMPA types have been demonstrated on retrogradely labeled STT cells.[119] Glutamate has been proposed as a transmitter in the synaptic terminals of STT cells in the thalamus.[11] There is also evidence that STT terminals in the thalamus contain substance P.[5,6,91]

28.4.2 Response Properties: Lamina I STT Cells

A number of observations suggest that there are distinct functional populations of STT cells. For this discussion, the STT population will be subdivided into those STT cells found in lamina I, those in laminae IV–VI, and those in deeper laminae. However, the basis for a functional distinction may relate to organizational features other than laminar distribution, so this grouping is tentative.

Recordings from STT cells in lamina I in cats have shown that these neurons are often selectively responsive to noxious cutaneous mechanical stimuli or to innocuous cold stimuli, although some STT cells were responsive to both or to stimulation of deep receptors.[27] Based on measurements of conduction velocities, some of these STT cells in cats have unmyelinated axons, although many are myelinated. On the other hand, 60% of the STT cells recorded in the superficial dorsal horn of rats were found to be of the wide dynamic range (WDR) type, responding to innocuous mechanical stimuli but more vigorously to noxious stimuli; 42% of the lamina I STT neurons were of the high-threshold (HT) type.[31]

Lamina I STT cells in monkeys appear to respond more like those in rats than in cats. Most are of the WDR type; however, others are of the HT type.[43] The proportions are 69% WDR and 29% HT.[96] Most lamina I STT cells also respond to noxious heat stimuli, and some to innocuous and noxious cold, as well.[43] The receptive fields of primate lamina I STT cells are relatively small and may be on either the glabrous or hairy skin.[43,113] There is a somatotopic relationship between the location of the receptive field on the limb and the position of the STT cell in the dorsal horn.[115]

Recently, the response properties of lamina I STT cells that project to the VMpo nucleus of the monkey thalamus have been described.[34] Many of these neurons respond to stimulation of cold thermoreceptors; others are nociceptive specific or respond to noxious heat, pinch, and cold stimuli.

High-threshold lamina I STT cells are generally activated by intradermal injection of capsaicin into the receptive field;[104] however, their discharge does not last more than a few minutes, in contrast to the duration of pain observed in human subjects, which lasts about 10 to 30 minutes. The responses of these cells to punctate mechanical stimuli increase, suggesting that they may play a role in the secondary mechanical hyperalgesia that is produced by intradermal injection of capsaicin. However, they generally do not show changes in their heat thresholds and so may not contribute to primary heat hyperalgesia. Experimental activation of the protein kinase G signal transduction pathway results in a reduction in the responsiveness of lamina I STT cells (and also of high-threshold STT cells in deeper laminae), whereas the responses of STT cells in laminae IV–VI become sensitized (see below).[78] Thus, it appears that high-threshold lamina I STT cells may be affected differently than wide dynamic-range STT cells by manipulations that result in the plastic changes in the response properties of these cells that are described as central sensitization and that are thought to underlie allodynia and secondary hyperalgesia.[118]

28.4.3 Response Properties: STT Cells in Laminae IV–VI

Some STT cells in laminae IV–VI in monkeys are of the high-threshold type, responding selectively to noxious mechanical stimulation of the skin; others are of the low-threshold (LT) type, preferring innocuous mechanical stimuli; however, most are of the wide dynamic range type; the proportions have been estimated as 20% LT, 59% WDR, and 22% HT.[96] STT cells in laminae IV–VI generally also respond to noxious thermal stimulation of the skin, especially to noxious heat.[115]

Many STT cells in the deep dorsal horn are excited by stimulation of receptors in skeletal muscle (see Reference 115 for a review). The same STT cells also respond to cutaneous stimuli, so it was suggested that these cells may contribute to the referral of muscle pain. A special group of STT cells located in Stilling's nucleus of the sacral spinal cord appear to be selectively activated by afferents from muscle spindles of the tail muscles.

Many STT cells, especially those in the sacral and the thoracic spinal cord, respond to visceral stimuli, as well as to cutaneous stimuli, and may contribute to the referral of visceral pain (see References 45 and 115 for a review). Interestingly, STT cells at a particular level of the spinal cord that are activated by visceral stimuli are often inhibited by visceral stimuli applied to an organ whose input reaches the spinal cord at a different level. For example, STT cells in the upper thoracic spinal cord that are excited by stimulation of cardiopulmonary afferents can often be inhibited by bladder distention.[46] Conversely, STT cells in the sacral spinal cord that are activated by bladder distention are often inhibited by stimulation of cardiopulmonary afferents.[126] The inhibition depends on a relay in the upper cervical segments.[57]

Repeated noxious heat stimulation of the skin results in an enhancement of later responses to noxious heat pulses applied to the same site.[115] This change in responsiveness of STT cells may contribute to the primary thermal hyperalgesia that results from noxious heat stimuli.[63,64,87] Presumably the main event responsible for this is the sensitization of primary afferent nociceptors by the exposure to noxious heat stimuli.

After repeated noxious heat stimuli, STT cells in laminae IV–VI also show increased responses to innocuous mechanical stimuli applied to the adjacent undamaged skin (see Reference 115 for a review). Because it is unlikely that the previous heat stimuli had affected sensitive mechanoreceptors, this change in the responses may help explain mechanical allodynia.

Similar conclusions have been reached in experiments in which intradermal injections of capsaicin were used to produce central sensitization of STT cells.[36,104] Following a capsaicin injection, the background discharge rate of STT cells in laminae IV–VI increases dramatically (Figure 28.4). Then, the discharge rate gradually diminishes over a period of minutes, approaching the original level after about 15 min. This time course parallels that of the pain that is experienced by human subjects given a comparable injection.[65,66] After this time, the responses of the neuron to graded noxious heat stimuli applied near the injection site are enhanced, and the threshold for activation of the cell is lowered from 48°C to 32°C. This change resembles that expected for the primary thermal hyperalgesia that occurs in human subjects after an intradermal injection of capsaicin.[65] There can also be increased responses to mechanical stimuli applied near the injection site. These help explain primary mechanical hyperalgesia, although reduced responses are found when the mechanical stimuli are applied directly on the injection site, presumably due to the desensitization of nociceptor terminals by the capsaicin. The responses to innocuous mechanical

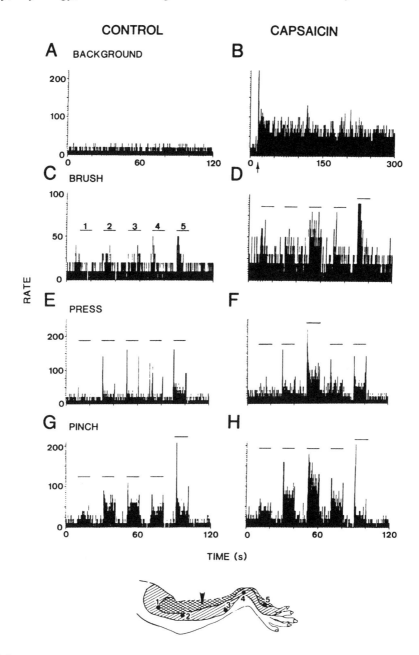

FIGURE 28.4
Changes in the responses of a wide dynamic range spinothalamic tract (STT) neuron following an intradermal injection of capsaicin. The histograms in (**A**), (**C**), (**E**), and (**G**) show the background activity of the neuron and its responses to brush, pressure (press), and pinch stimuli applied at five different locations in the receptive field (see numbers in the diagram at the bottom). Capsaicin was injected at the site indicated by the arrowhead, and the elevation in discharge rate is shown in (**B**). The histograms in (**D**), (**F**), and (**H**) show the enhancement in the responses to mechanical stimuli, as tested beginning 15 min after the injection. The size of the receptive field was also increased (compare single- and double-hatched areas in the drawing of the hindlimb). (From Dougherty, P.M. and Willis, W.D., *J. Neurosci.*, 12, 883, 1992. With permission.)

stimuli applied within several centimeters around the injection site become progressively enhanced over about 15 min and then diminish over a period of about one and a half hours. Presumably this change reflects the time course and spatial distribution of the mechanical allodynia that is seen in human subjects.[65,66] Responses to weak noxious stimuli also become enhanced, reflective of the development of secondary hyperalgesia (Figure 28.4).[36,104] Interestingly, the responses to strong noxious mechanical stimuli are not significantly changed,[36] and the responses to noxious heat are reduced.[105] The absence of secondary heat hyperalgesia is also found in human subjects given an intradermal capsaicin injection.[2]

The central sensitization that follows intradermal injection of capsaicin appears to depend on the release of excitatory amino acids and peptides, including substance P, in the dorsal horn and by the discharges of nociceptive afferent fibers activated by the capsaicin, as the sensitization can be blocked by administration of antagonists of N-methyl-D-asparate receptors or of neurokinin 1 receptors directly into the dorsal horn.[37,38] However, the enhanced responses of the STT cells to weak mechanical stimuli and to iontophoretic application of excitatory amino acids[36] long outlast the expected direct effects of excitatory amino acids and neurokinins on the STT cells. Presumably, some mechanism other than the immediate interaction of neurotransmitters and their receptors is responsible for changes that last on the order of hours. Recent experiments suggest that central sensitization depends on the activation of several signal transduction cascades, including those involving protein kinase C, protein kinase G, and protein kinase A.[25,76,82,105] In addition to enhancing excitatory responses, central sensitization is also accompanied by a reduction in the ability of inhibitory amino acids to reduce the discharges of STT cells in laminae IV–VI.[77] Apparently, during central sensitization, the responses of STT cells to peripheral input are increased because of both an enhanced excitation and a reduced inhibition.

Another circumstance in which the responses of STT cells of laminae IV–VI to weak mechanical and thermal stimuli become enhanced is in neuropathic pain. Experiments in which neuropathic pain was produced in rats using the chronic constriction injury model of Bennett and Xie[7] revealed that the discharges of STT cells in response to mechanical and thermal stimuli were often prolonged beyond the duration of the stimuli.[98] In monkeys with neuropathic pain due to tight ligation of the L7 or S1 spinal nerve (the neuropathic pain model of Kim and Chung[61]), STT cells at the segmental level just above the ligated level showed greatly enhanced responses to innocuous mechanical and thermal stimuli.[97] As allodynia and hyperalgesia are features of neuropathic pain, it was suggested that these exaggerated responses of STT cells would have contributed to this abnormal pain state.

28.4.4 Responses of STT Cells Projecting to Intralaminar Nuclei

There is a group of STT cells concentrated in the deep dorsal horn and in the ventral horn which project to the intralaminar complex of the thalamus but not to the lateral thalamus (see Reference 115 for a review). These STT cells often respond to noxious stimuli applied anywhere on the body and even face. At least some have a convergent input from skin and viscera. The responses of these neurons to cutaneous stimuli do not encode stimulus intensity well, and they clearly do not encode stimulus location. Furthermore, they can be excited for long period of time following stimulation in the brainstem reticular formation.

28.4.5 Sensory Functions

It is known from clinical observations that axons in the anterolateral quadrant of the human spinal cord can mediate touch, although not as effectively as can the pathways in the dorsal part of the spinal cord.[94] Therefore, it is possible that the small proportion of STT cells in laminae IV–VI that have low-threshold response properties are involved in the processing of touch. Although WDR STT cells respond to innocuous mechanical stimuli, it is unclear if these cells can contribute to tactile sensation. More likely, cells of this type signal pain. Normally, their responses to innocuous mechanical stimuli are small, but when the responses to innocuous mechanical stimulation are enhanced, as during central sensitization, such cells are responsible for allodynia.[114]

Most STT cells both in lamina I and in laminae IV–VI respond to noxious stimuli and presumably contribute to pain sensation. The lamina I STT cells seem best suited for providing information about the location of noxious stimuli, and the high-threshold cells in this lamina would give an unambiguous signal that a noxious stimulus had occurred. However, STT cells of the WDR type are also likely to contribute to pain sensation, although in general they would be less effective in providing information about stimulus location because their receptive fields tend to be larger than those of lamina I STT cells. If these neurons were relatively unresponsive to tactile stimuli under normal conditions, they could provide an unambiguous nociceptive signal. When their responses become sensitized following injury, the enhanced responses to innocuous mechanical stimuli might now produce allodynia.[114]

Some of the STT cells in lamina I appear to be responsible for signaling cold and warm. A special group of STT cells located in Stilling's nucleus appear to be proprioceptive, signaling tail position. Thus, the STT has a number of subpopulations of neurons that appear to play a role in tactile sensation, pain (as well as allodynia and hyperalgesia), cold and warm sensations, and perhaps proprioception.

28.4.6 Descending Control

Cells of the STT are subjected to powerful descending controls. These include both inhibitory and excitatory modulation. Most of the emphasis in the investigations of the descending control of STT cells has been on inhibitory actions, as such effects can be interpreted in terms of the descending analgesia systems. However, excitatory modulation could form a basis for central pain states such as those seen after damage to certain regions of the brain or of the spinal cord.

The first description of a brain structure that participates in the descending analgesia systems was that of Reynolds,[103] who found that abdominal surgery could be done in rats during electrical stimulation of the periaqueductal gray (PAG) without chemical anesthesia. Since then, many studies have confirmed that electrical or chemical stimulation of the PAG causes analgesia, as well as other effects (see References 9 and 115 for a review). PAG stimulation results in the inhibition of STT cells, an action that is consistent with a role of the STT in pain transmission and of the PAG in analgesia. The inhibitory action of the PAG on STT cells is mediated in part by serotonin release, as it can be partially blocked by administration of antagonists on serotonin receptors. It is also mediated in part by the inhibitory amino acids, GABA and glycine, as PAG inhibition of STT cells can be partially blocked by antagonists of $GABA_A$ (but not $GABA_B$) and glycine receptors administered directly into the spinal cord dorsal horn by microdialysis.[74,75] It is likely that norepinephrine is also involved, as the PAG inhibition of nociceptive dorsal horn neurons in rats is partially blocked by local administration of antagonists of (α-2 adrenoreceptors).[101] The effects of stimulation of the PAG are apparently transmitted by way of several descending pathways originating in the serotonergic brainstem raphe nuclei and the noradrenergic cell groups in the locus coeruleus, subcoeruleus parabrachial complex.[12,111] In many STT cells, PAG inhibition is associated with an inhibitory postsynaptic potential;[123] however, another mechanism for PAG inhibition of some STT cells is disfacilitation; that is, a tonic excitation of STT cells is removed in response to PAG stimulation, presumably because of the inhibition of excitatory interneurons.[123]

STT neurons are also inhibited following electrical stimulation in the nucleus raphe magnus (see Reference 115 for a review). Studies by several groups have demonstrated that stimulation of the nucleus raphe magnus in cats and rats produces analgesia. This inhibition is presumably mediated in part by release of serotonin in the spinal cord (see Reference 9 for a review), although it is difficult to block the inhibition with serotonin receptor antagonists. Stimulation in the nucleus raphe magnus does cause the release of serotonin in the spinal cord, but it also results in the release of amino acids, as well,[107] so the inhibition may be due to the action of more than one neurotransmitter. The cellular mechanisms of the inhibition produced in STT cells by stimulation in the nucleus raphe magnus are also complex. These include the generation of inhibitory postsynaptic potentials,[49] primary afferent depolarization, and presumably presynaptic inhibition,[84] and perhaps disfacilitation, although the latter possibility has not been investigated.[123]

Stimulation in the medullary reticular formation can also result in the inhibition of STT cells, although certain sites elicit an excitation instead.[55] The inhibition may be due in part to primary afferent depolarization.[84]

Other sources of descending modulation of STT cells include the anterior pretectal nucleus, the ventrobasal thalamus, and the cerebral cortex. Preliminary experiments indicate that stimulation in the anterior pretectal nucleus in monkeys results in the inhibition of STT cells (Rees et al., unpublished observations). This observation is of interest, as it has been shown that stimulation in the anterior pretectal nucleus in rats produces analgesia without other side effects (see Reference 102 for a review). Stimulation in the ventrobasal complex in monkeys produces inhibition of STT cells.[47] This inhibition is associated with the excitation of raphe-spinal neurons[117] and with the release of serotonin in the spinal cord.[106] One option for neurosurgical treatment of pain is to implant a stimulating electrode in the ventrobasal complex.[54] Stimulation of the somatosensory cortex in monkeys can also produce an inhibition of STT cells.[122,123] This can be attributed in part to the generation of inhibitory postsynaptic potentials and in part to disfacilitation.[123] Stimulation of the primary motor cortex, on the other hand, can result in the excitation of STT cells.[122] It is unclear as to what extent the activity of STT cells is regulated by cortical activity, but behavioral studies of medullary dorsal horn neurons in monkeys suggest that cortical activity helps determine the responsiveness of somatosensory neurons that are involved in nociceptive transmission.[19] Removal of cortical inhibition or accentuation of cortical excitation of STT cells could play a role in central pain states.

28.4.7 Primary Afferent Control

Like PSDC and SCT neurons, STT cells can be inhibited by volleys in primary afferent fibers. One form of inhibition results from innocuous stimulation of the skin;[116a] however, a more powerful inhibition is produced by noxious mechanical or thermal stimulation of the skin.[46b] This type of inhibition resembles the diffuse noxious inhibitory controls described by Le Bars and his colleagues[66a,b] and may at least in part involve supraspinal connections.[46a] STT cells can also be inhibited by stimulation of the dorsal columns[46a] or peripheral nerves, especially of Aδ and C-fibers.[22a] These effects may relate to the clinical effectiveness of dorsal column and peripheral nerve stimulation in relieving pain. Inhibition of STT cells can also be observed following stimulation of visceral afferent fibers.[12a,87a]

28.5 Spinoreticular and Spinomesencephalic Tracts

28.5.1 Anatomical Organization

The cells of the SRT in rats, cats, and monkeys are concentrated in laminae V, VII, and VIII (see Reference 115 for a review), although there is a component of the SRT that projects to the dorsal reticular nucleus of the medulla that originates largely from lamina I and IV–VI.[69,73] Most of the lamina I cells that project to the dorsal reticular nucleus in rats are of the multipolar type, although some are pyramidal or flattened.[73] On the other hand, the SMT originates chiefly from cells in laminae I and V, although there are SMT cells in other laminae, including lamina X.[115,121] The SMT neurons in lamina I in rats are of the fusiform and pyramidal types.[72]

28.5.2 Response Properties

In general, SRT and SMT cells often have convergent inputs from a number of types of sensory receptors, and their receptive fields tend to be complex; however, some SMT cells, particularly those projecting to the parabrachial region, have restricted receptive fields (see Reference 115 for a review).

28.5.3 Sensory Functions

The properties of neurons belonging to the SRT and SMT do not seem to be suited to a role in the discriminative aspects of somatic sensation, including pain sensation, although many of these neurons do respond to intense stimuli. It seems more likely that the SRT and SMT are involved in more general functions, including arousal, attentional mechanisms, the motivational-affective dimension of pain, somatic and autonomic reflex responses, endocrine responses, and feedback control of nociceptive input through engagement of the analgesia systems.

28.5.4 Descending Controls

Neurons of the SRT are influenced by both inhibitory and excitatory descending controls. Inhibition or excitation can be produced by stimulation in the medullary reticular formation.[115] Inhibition of SRT neurons results from electrical or chemical stimulation of the PAG.[115] SRT and SMT neurons in rats are inhibited following stimulation in the nucleus raphe magnus. Yezierski[120] described complex excitatory and inhibitory actions on SMT neurons following stimulation in the nucleus raphe magnus and medullary reticular formation.

28.5.5 Primary Afferent Control

Spinoreticular tract and SMT neurons can also be inhibited by primary afferent volleys.[44a,121a]

28.6 Summary

In addition to the dorsal column pathway formed by collaterals of primary afferent neurons that project directly to the dorsal column nuclei, several pathways originating from neurons of the spinal cord dorsal horn appear to contribute to the processing of tactile information. These include the postsynaptic dorsal column pathway, the spinocervical tract, and the spinothalamic tract. These same pathways are also involved in the processing of nociceptive information and so are all probably involved in pain mechanisms. However, the postsynaptic dorsal column pathway may have an especially important role in visceral pain. The spinothalamic tract transmits not only nociceptive but also thermal information to the thalamus. Part of the spinothalamic tract but also the spinoreticular and spinomesencephalic tracts are likely to have a diversity of functional roles that include arousal, attention, motivational-affective responses, somatic and autonomic reflex, endocrine responses, and engagement of the endogenous analgesia system.

Acknowledgments

The authors thank Griselda Gonzales for help with the artwork. The experiments done in the author's laboratory were supported by NIH grants NS 09743 and NS 11255.

References

1. Al-Chaer, E.D., Lawand, N.B., Westlund, K.N., and Willis, W.D., Pelvic visceral input into the nucleus gracilis is largely mediated by the postsynaptic dorsal column pathway, *J. Neurophysiol.*, 76, 2675, 1996.
2. Ali, Z., Meyer, R.A., and Campbell, J.N., Secondary hyperalgesia to mechanical but not heat stimuli following a capsaicin injection in hairy skin, *Pain*, 68, 401, 1996.
3. Apkarian, A.V. and Hodge, C.J., The primate spinothalamic pathways. I. A quantitative study of the cells of origin of the spinothalamic pathway, *J. Comp. Neurol.*, 288, 447, 1989.
4. Baker, M.. and Giesler, G.J., Anatomical studies of the spinocervical tract of the rat, *Somatosensory Res.*, 2, 1, 1984.
5. Battaglia, G. and Rustioni, A., Substance P innervation of the rat and cat thalamus. II. Cells of origin in the spinal cord, *J. Comp. Neurol.*, 315, 473, 1992.
6. Battaglia, G., Spreafico, R., and Rustioni, A., Substance P innervation of the rat and cat thalamus. I. Distribution and relation to ascending spinal pathways, *J. Comp. Neurol.*, 315, 457, 1992.
7. Bennett, G.J. and Xie, Y.K., A peripheral mononeuropathy in rat that produces disorders of pain sensation like those seen in man, *Pain*, 33, 87, 1988.
8. Bennett, G.J., Seltzer, Z., Lu, G.W., Nishikawa, N., and Dubner, R., The cells of origin of the dorsal column postsynaptic projection in the lumbosacral enlargements of cats and monkeys, *Somatosensory Res.*, 1, 131, 1983.
9. Besson, J.M. and Chaouch, A., Peripheral and spinal mechanisms of nociception, *Physiol. Rev.*, 67, 67, 1987.
10. Bleazard, L., Hill, R.G., and Morris, R., The correlation between the distribution of the NK1 receptor and the actions of tachykinin agonists in the dorsal horn of the rat indicates that substance P does not have a functional role on substantia gelatinosa (lamina II) neurons, *J. Neurosci.*, 14, 7655, 1994.
11. Blomqvist, A., Ericson, A.C., Craig, A.D., and Broman, J., Evidence for glutamate as a neurotransmitter in spinothalamic tract terminals in the posterior region of owl monkeys, *Exp. Brain Res.*, 108, 33, 1996.
12. Bowker, R.M., Westlund, K.N., and Coulter, J.D., Origins of serotonergic projections to the spinal cord in rat: an immunocytochemical-retrograde transport study, *Brain Res.*, 226, 187, 1981.
12a. Brennan, T.J., Oh, U.T., Hobbs, S.F., Garrison, D.W., and Foreman, R.D., Urinary bladder and hindlimb afferent input inhibits activity of primate T2-T5 spinothalamic tract neurons, *J. Neurophysiol.*, 61, 573, 1989.
13. Broman, J. and Pubols, B.H., Substance P-like and serotonin-like immunoreactivity in the lateral cervical nucleus of the raccoon, *J. Comp. Neurol.*, 329, 354, 1993.
13a. Brown, A.G. and Franz, D.N. Responses of spinocervical tract neurones to natural stimulation of identified cutaneous receptors, *Exp. Brain Res.*, 7, 231, 1969.
14. Brown, A.G. and Fyffe, R.E.W., Form and function of dorsal horn neurones with axons ascending the dorsal columns in cat, *J. Physiol.*, 321, 31, 1981.
14a. Brown, A.G., Brown, P.B., Fyffe, R.E.W., and Pubols, L.M., Receptive field organization and response properties of spinal neurones with axons ascending the dorsal columns in the cat, *J. Physiol.*, 337, 575, 1983.

15. Brown, A.G., Fyffe, R.E.W., Noble, R., Rose, P.K., and Snow, P.J., The density, distribution and topographical organization of spinocervical tract neurones in the cat, *J. Physiol.*, 300, 409, 1980.

15a. Brown, A.G., Koerber, H.R., and Noble, R. An intracellular study of spinocervical tract cell responses to natural stimuli and single hair afferent fibres in cats, *J. Physiol.*, 382, 331, 1987.

16. Brown, A.G., Rose, P.K., and Snow, P.J., Dendritic trees and cutaneous receptive fields of adjacent spinocervical tract neurones in the cat, *J. Physiol.*, 300, 429, 1980.

17. Brown, J.L., Liu, H., Maggio, J.E., Vigna, S.R., Mantyh, P.W., and Basbaum, A.I., Morphological characterization of substance P receptor-immunoreactive neurons in rat spinal cord and trigeminal nucleus caudalis, *J. Comp. Neurol.*, 356, 327, 1995.

18. Burstein, R., Dado, R.J., and Giesler, G.J., The cells of origin of the spinothalamic tract of the rat: a quantitative reexamination, *Brain Res.*, 511, 329, 1990.

19. Bushnell, M.C., Duncan, G.H., Dubner, R., and He, L.F., Activity of trigeminothalamic neurons in medullary dorsal horn of awake monkeys trained in a thermal discrimination task, *J. Neurophysiol.*, 52, 170, 1984.

20. Carlton, S.M., Westlund, K.N., Zhang, D., Sorkin, L.S., and Willis, W.D., Calcitonin gene-related peptide containing primary afferent fibers synapse on primate spinothalamic tract cells, *Neurosci. Lett.*, 109, 76, 1990.

21. Carlton, S.M., Westlund, K.N., Zhang, D., and Willis, W.D., GABA-immunoreactive terminals synapse on primate spinothalamic tract cells, *J. Comp. Neurol.*, 322, 528, 1992.

22. Casey, K.L. and Morrow, T.J., Supraspinal nocifensive responses of cats: spinal cord pathways, monoamines, and modulation, *J. Comp. Neurol.*, 270, 591, 1988.

22a. Chung, J.M., Fang, Z.R., Hori, Y., Lee, K.H., and Willis, W.D., Prolonged inhibition of primate spinothalamic tract cells by peripheral nerve stimulation. *Pain*, 19, 259, 1984.

23. Chung, J.M., Lee, K.H., Surmeier, D.J., Sorkin, L.S., Kim, J., and Willis, W.D., Response characteristics of neurons in the ventral posterior lateral nucleus of the monkey thalamus, *J. Neurophysiol.*, 56, 370, 1986.

24. Cliffer, K.D., Hasegawa, T., and Willis, W.D., Responses of neurons in the gracile nucleus of cats to innocuous and noxious stimuli: basic characterization and antidromic activation from the thalamus, *J. Neurophysiol.*, 68, 818, 1992.

25. Coderre, T.J., Contribution of protein kinase C to central sensitization and persistent pain following tissue injury, *Neurosci. Lett.*, 140, 181, 1992.

26. Craig, A.D. and Burton, H., The distribution and topographical organization in the thalamus of anterogradely-transported horseradish peroxidase after spinal injections in cat and raccoon, *Exp. Brain Res.*, 58, 227, 1985.

27. Craig, A.D. and Kniffki, K.D., Spinothalamic lumbosacral lamina I cells responsive to skin and muscle stimulation in the cat, *J. Physiol.*, 365, 197, 1985.

28. Craig, A.D., Linington, A.J., and Kniffki, K.D., Cells of origin of spinothalamic tract projections to the medial and lateral thalamus in the cat, *J. Comp. Neurol.*, 289, 568, 1989.

29. Craig, A.D., Bushnell, M.C., Zhang, E.T. and Blomqvist, A., A thalamic nucleus specific for pain and temperature sensation, *Nature*, 373, 19, 1994.

30. Dado, R.J. and Giesler, G.J., Afferent input to nucleus submedius in rats: retrograde labeling of neurons in the spinal cord and caudal medulla, *J. Neurosci.*, 10, 2672, 1990.

31. Dado, R.J., Katter, J.T., and Giesler, G.J., Spinothalamic and spinohypothalamic tract neurons in the cervical enlargement of rats. II. Responses to innocuous and noxious mechanical and thermal stimuli, *J. Neurophysiol.*, 71, 981, 1994.

32. De Biasi, S., Vitellaro-Zuccarello, L., Bernardi, P., Valtschanoff, J.G., and Weinberg, R.J., Ultrastructural and immunocytochemical characterization of terminals of postsynaptic ascending dorsal column fibers in the rat cuneate nucleus, *J. Comp. Neurol.*, 353, 109, 1995.

33. De Koninck, Y., Ribeiro-da-Silva, A., Henry, J.L., and Cuello, A.C., Spinal neurons exhibiting a specific nociceptive response receive abundant substance P-containing synaptic contacts, *Proc. Natl. Acad. Sci. USA,* 89, 5073, 1992.
34. Dostrovsky, J.O. and Craig, A.D., Cooling-specific spinothalamic neurons in the monkey, *J. Neurophysiol.,* 76, 3656, 1996.
35. Dougherty, P.M. and Willis, W.D., Enhancement of spinothalamic neuron responses to chemical and mechanical stimuli following combined micro-iontophoretic application of N-methyl-D-aspartic acid and substance P, *Pain,* 47, 85, 1991.
36. Dougherty, P.M. and Willis, W.D., Enhanced responses of spinothalamic tract neurons to excitatory amino acids accompany capsaicin-induced sensitization in the monkey, *J. Neurosci.,* 12, 883, 1992.
37. Dougherty, P.M., Palecek, J., Paleckova, V., Sorkin, L.S., and Willis, W.D., The role of NMDA and non-NMDA excitatory amino acid receptors in the excitation of primate spinothalamic tract neurons by mechanical, chemical, thermal, and electrical stimuli, *J. Neurosci.,* 12, 3025, 1992.
38. Dougherty, P.M., Palecek, J., Paleckova, V., and Willis, W.D., Neurokinin 1 and 2 antagonists attenuate the responses and NK1 antagonists prevent the sensitization of primate spinothalamic tract neurons after intradermal capsaicin, *J. Neurophysiol.,* 72, 1464, 1994.
39. Downie, J.W., Ferrington, D.G., Sorkin, L.S., and Willis, W.D., The primate spinocervicothalamic pathway: responses of cells of the lateral cervical nucleus and spinocervical tract to innocuous and noxious stimuli, *J. Neurophysiol.,* 59, 861, 1988.
40. Doyle, C.A. and Maxwell, D.J., Catecholaminergic innervation of the lateral cervical nucleus: a correlated light and electron microscopic analysis of tyrosine hydroxylase-immunoreactive axons in the cat, *Neuroscience,* 61, 381, 1994.
41. Enevoldson, T.P. and Gordon, G., Postsynaptic dorsal column neurons in the cat: a study with retrograde transport of horseradish peroxidase, *Exp. Brain Res.,* 75, 611, 1989.
42. Enevoldson, T.P. and Gordon, G., Spinocervical neurons and dorsal horn neurons projecting to the dorsal column nuclei through the dorsolateral fascicle: a retrograde HRP study in the cat, *Exp. Brain Res.,* 75, 621, 1989.
43. Ferrington, D.G., Sorkin, L.S., and Willis, W.D., Responses of spinothalamic tract cells in the superficial dorsal horn of the primate lumbar spinal cord, *J. Physiol.,* 388, 681, 1987.
44. Ferrington, D.G., Downie, J.W., and Willis, W.D., Primate nucleus gracilis neurons: responses to innocuous and noxious stimuli, *J. Neurophysiol.,* 59, 886, 1988.
44a. Fields, H.L., Clanton, C.H., and Anderson, S.D., Somatosensory properties of spinoreticular neurons in the cat, *Brain Res.,* 120, 49, 1977.
45. Foreman, R.D., Organization of the spinothalamic tract as a relay for cardiopulmonary sympathetic afferent fiber activity, in *Progress in Sensory Physiology,* Vol. 9, Ottoson, D., Ed., Springer-Verlag, New York, 1989, p. 1.
46. Foreman, R.D., Intraspinal modulation of visceral transmission, in *Visceral Pain, Progress in Pain Research and Management,* Vol. 5, Gebhart, G.F., Ed., IASP Press, Seattle, 1995, p. 291.
46a. Foreman, R.D., Beall, J.E., Applebaum, A.E., Coulter, J.D., and Willis, W.D., Effects of dorsal column stimulation on primate spinothalamic tract neurons, *J. Neurophysiol.,* 39, 534, 1976.
46b. Gerhart, K.D., Yezierski, R.P., Giesler, G.J., and Willis, W.D., Inhibitory receptive fields of primate spinothalamic tract cells, *J. Neurophysiol.,* 46, 1309, 1981.
47. Gerhart, K.D., Yezierski, R.P., Fang, Z.R., and Willis, W.D., Inhibition of primate spinothalamic tract neurons by stimulation in ventral posterior lateral (VPLc) thalamic nucleus: possible mechanisms, *J. Neurophysiol.,* 49, 406, 1983.
48. Giesler, G.J. and Cliffer, K.D., Postsynaptic dorsal column pathway of the rat. II. Evidence against an important role in nociception, *Brain Res.,* 326, 347, 1985.

49. Giesler, G.J., Gerhart, K.D., Yezierski, R.P., Wilcox, T.K., and Willis, W.D., Postsynaptic inhibition of primate spinothalamic neurons by stimulation in nucleus raphe magnus, *Brain Res.,* 204, 184, 1981.
50. Giesler, G.J., Yezierski, R.P., Gerhart, K.D., and Willis, W.D., Spinothalamic tract neurons that project to medial and/or lateral thalamic nuclei: evidence for a physiologically novel population of spinal cord neurons, *J. Neurophysiol.,* 46, 1285, 1981.
51. Giesler, G.J., Nahin, R.L., and Madsen, A.M., Postsynaptic dorsal column pathway of the rat. I. Anatomical studies, *J. Neurophysiol.,* 51, 260, 1984.
52. Giesler, G.J., Miller, L.R., Madsen, A.M., and Katter, J.T., Evidence for the existence of a lateral cervical nucleus in mice, guinea pigs, and rabbits, *J. Comp. Neurol.,* 263, 106, 1987.
53. Gingold, S.I., Greenspan, J.D., and Apkarian, A.V., Anatomic evidence of nociceptive inputs to primary somatosensory cortex: relationship between spinothalamic terminals and thalamocortical cells in squirrel monkeys, *J. Comp. Neurol.,* 308, 467, 1991.
54. Gybels, J.M. and Sweet, W.H., *Neurosurgical Treatment of Persistent Pain,* Karger, Basel, 1989.
55. Haber, L.H., Martin, R.F., Chung, J.M., and Willis, W.D., Inhibition and excitation of primate spinothalamic tract neurons by stimulation in region of nucleus reticularis gigantocellularis, *J. Neurophysiol.,* 43, 1578, 1980.
55a. Harrison, P.J. and Jankowska, E., An intracellular study of descending and non-cutaneous afferent input to spinocervical tract neurones in the cat, *J. Physiol.,* 356, 245, 1984.
56. Hirshberg, R.M., Al-Chaer, E.D., Lawand, N.B., Westlund, K.N., and Willis, W.D., Is there a pathway in the posterior funiculus that signals visceral pain?, *Pain,* 67, 291, 1996.
57. Hobbs, S.F., Oh, U.T., Chandler, M.J., Fu, Q.G., Bolser, D.C., and Foreman, R.D., Evidence that C1 and C2 propriospinal neurons mediate the inhibitory effects of viscerosomatic spinal afferent input on primate spinothalamic tract neurons, *J. Neurophysiol.,* 67, 852, 1992.
58. Hong, S.K., Kniffki, K.D., Mense, S., Schmidt, R.F., and Wendisch, M., Descending influences on the responses of spinocervical tract neurones to chemical stimulation of fine muscle afferents, *J. Physiol.,* 290, 129, 1979.
58a. Jankowska, E., Rastad, J., and Zarzecki, P., Segmental and supraspinal input to cells of origin of nonprimary fibres in the feline dorsal columns, *J. Physiol.,* 290, 185, 1979.
59. Kamogawa, H. and Bennett, G.J., Dorsal column postsynaptic neurons in the cat are excited by myelinated nociceptors, *Brain Res.,* 364, 386, 1986.
60. Kechagias, S. and Broman, J., Compartmentation of glutamate and glutamine in the lateral cervical nucleus: further evidence for glutamate as a spinocervical tract neurotransmitter, *J. Comp. Neurol.,* 340, 531, 1994.
61. Kim, S.H. and Chung, J.M., An experimental model for peripheral neuropathy produced by segmental spinal nerve ligation in the rat, *Pain,* 50, 355, 1992.
62. Kunze, W.A.A., Wilson, P., and Snow, P.J., Response of lumbar spinocervical tract cells to natural and electrical stimulation of the hindlimb footpads in cats, *Neurosci. Lett.,* 75, 253, 1987.
63. LaMotte, R.H., Thalhammer, J.G., Torebjörk, H.E., and Robinson, C.J., Peripheral neural mechanisms of cutaneous hyperalgesia following mild injury by heat, *J. Neurosci.,* 2, 765, 1982.
64. LaMotte, R.H., Thalhammer, J.G., and Robinson, C.J., Peripheral neural correlates of magnitude of cutaneous pain and hyperalgesia: a comparison of neural events in monkey with sensory judgments in human, *J. Neurophysiol.,* 50, 1, 1983.
65. LaMotte, R.H., Shain, C.N., Simone, D.A., and Tsai, E.F.P., Neurogenic hyperalgesia: psychophysical studies of underlying mechanisms, *J. Neurophysiol.,* 66, 190, 1991.
66. LaMotte, R.H., Lundberg, L.E.R., and Torebjörk, H.E., Pain, hyperalgesia and activity in nociceptive C units in humans after intradermal injection of capsaicin, *J. Physiol.,* 448, 749, 1992.

66a. Le Bars, D., Dickenson, A.H., and Besson, J.M., Diffuse noxious inhibitory controls (DNIC). I. Effects on dorsal horn convergent neurons in the rat, *Pain,* 6, 283, 1979.

66b. Le Bars, D., Dickenson, A.H., and Bedsson, J.M., Diffuse noxious inhibitory controls (DNIC). II. Lack of effect on non-convergent neurones, supraspinal involvement and theoretical implications, *Pain,* 6, 305, 1979.

67. Lekan, H.A. and Carlton, S.M., Glutamatergic and GABAergic input to rat spinothalamic tract cells in the superficial dorsal horn, *J. Comp. Neurol.,* 361, 417, 1995.

68. Li, J.L., Ding, Y.Q., Shigemoto, R., and Mizuno, N., Distribution of trigeminothalamic and spinothalamic-tract neurons showing substance P receptor-like immunoreactivity in the rat, *Brain Res.,* 719, 207, 1996.

69. Lima, D., A spinomedullary projection terminating in the dorsal reticular nucleus of the rat, *Neuroscience,* 34, 577, 1990.

70. Lima, D. and Coimbra, A., A Golgi study of the neuronal population of the marginal zone (lamina I) of the rat spinal cord, *J. Comp. Neurol.,* 244, 53, 1986.

71. Lima, D. and Coimbra, A., The spinothalamic system of the rat: structural types of retrogradely labeled neurons in the marginal zone (lamina I), *Neuroscience,* 27, 215, 1988.

72. Lima, D. and Coimbra, A., Morphological types of spinomesencephalic neurons in the marginal zone (lamina I) of the rat spinal cord, as shown after retrograde labeling with cholera toxin subunit B, *J. Comp. Neurol.,* 279, 327, 1989.

73. Lima, D. and Coimbra, A., Structural types of marginal (lamina I) neurons projecting to the dorsal reticular nucleus of the medulla oblongata, *Neuroscience,* 34, 591, 1990.

74. Lin, Q., Peng, Y.B., and Willis, W.D., Glycine and $GABA_A$ antagonists reduce the inhibition of primate spinothalamic tract neurons produced by stimulation in periaqueductal gray, *Brain Res.,* 654, 286, 1994.

75. Lin, Q., Peng, Y.B., and Willis, W.D., Role of GABA receptor subtypes in inhibition of primate spinothalamic tract neurons: difference between spinal and periaqueductal gray inhibition, *J. Neurophysiol.,* 75, 109, 1996.

76. Lin, Q., Peng, Y.B., and Willis, W.D., Possible role of protein kinase C in the sensitization of primate spinothalamic tract neurons, *J. Neurosci.,* 16, 3026, 1996.

77. Lin, Q., Peng, Y.B., and Willis, W.D., Inhibition of primate spinothalamic tract neurons by spinal glycine and GABA is reduced during central sensitization, *J. Neurophysiol.,* 76, 1005, 1996.

78. Lin, Q., Peng, Y.B., Wu, J., and Willis, W.D., Involvement of cGMP in nociceptive processing by and sensitization of spinothalamic neurons in primates, *J. Neurosci.,* 17, 3293, 1997.

79. Liu, H., Brown, J.L., Maggio, J.E., Vigna, S.R., Mantyh, P.W., and Basbaum, A.I., Synaptic relationship between substance P and the substance P receptor: light and electron microscopic characterization of the mismatch between neuropeptides and their receptors, *Proc. Natl. Acad. Sci. USA,* 91, 1009, 1994.

80. Ma, W., Ribeiro-da-Silva, A., De Koninck, Y., Radhakrishnan, V., Henry, J.L., and Cuello, A.C., Quantitative analysis of substance P-immunoreactive boutons on physiologically characterized dorsal horn neurons in the cat lumbar spinal cord, *J. Comp. Neurol.,* 376, 45, 1980.

81. Mantyh, P.W., DeMaster, E., Malhotra, A., Ghilardi, J.R., Rogers, S.D., Mantyh, C.R., Liu, H., Basbaum, A.I., Vigna, S.R., Maggio, J.E., and Simone, D.A., Receptor endocytosis and dendrite reshaping in spinal neurons after somatosensory stimulation, *Science,* 268, 1629, 1995.

82. Mao, J., Price, D.D., Mayer, D.J., and Hayes, R.L., Pain-related increases in spinal cord membrane-bound protein kinase C following peripheral nerve injury, *Brain Res.,* 588, 144, 1992.

83. Marshall, G.E., Shebab, S.A.S., Spike, R.C., and Todd, A.J., Neurokinin-1 receptors on lumbar spinothalamic neurons in the rat, *Neuroscience,* 72, 255, 1996.

84. Martin, R.F., Haber, L.H., and Willis, W.D., Primary afferent depolarization of identified cutaneous fibers following stimulation in medial brain stem, *J. Neurophysiol.,* 42, 779, 1979.

85. Maxwell, D.J., Ottersen, O.P., and Storm-Mathisen, J., Synaptic organization of excitatory and inhibitory boutons associated with spinal neurons which project through the dorsal columns of the cat, *Brain Res.,* 676, 103, 1995.

86. Maxwell, D.J., Todd, A.J., and Kerr, R., Colocalization of glycine and GABA in synapses on spinomedullary neurons, *Brain Res.,* 690, 127, 1995.

87. Meyer, R.A. and Campbell, J.N., Myelinated nociceptive afferents account for the hyperalgesia that follows a burn to the hand, *Science,* 213, 1527, 1981.

87a. Milne, R.J., Foreman, R.D., Giesler, G.J., and Willis, W.D., Convergence of cutaneous and pelvic visceral nociceptive inputs onto primate spinothalamic neurons, *Pain,* 11, 163, 1981.

88. Naim, M., Spike, R.C., Watt, C., Shehab, S.A.S., and Todd, A.J., Cells in laminae III and IV of the rat spinal cord that possess the neurokinin-1 receptor and have dorsally directed dendrites receive a major synaptic input from tachykinin-containing primary afferents, *J. Neurosci.,* 17, 5536, 1997.

89. Nauta, H.J.W., Hewitt, E., Westlund, K.N., and Willis, W.D., Surgical interruption of a midline dorsal column pain pathway, *J. Neurosurg.,* 86, 538, 1997.

90. Nishikawa, N., Bennett, G.J., Ruda, M.A., Lu, G.W., and Dubner, R., Immunocytochemical evidence for a serotoninergic innervation of dorsal column postsynaptic neurons in cat and monkey: light- and electron-microscopic observations, *Neuroscience,* 10, 1333, 1983.

91. Nishiyama, K., Kwak, S., Murayama, S., and Kanazawa, I., Substance P is a possible neurotransmitter in the rat spinothalamic tract, *Neurosci. Res.,* 21, 261, 1995.

92. Noble, C.R. and Riddell, J.S., Cutaneous excitatory and inhibitory input to neurones of the postsynaptic dorsal column system in the cat, *J. Physiol.,* 396, 497, 1988.

93. Noble, R. and Riddell, J.S., Descending influences on the cutaneous receptive fields of postsynaptic dorsal column neurones in the cat, *J. Physiol.,* 408, 167, 1989.

94. Noordenbos, W. and Wall, P.D., Diverse sensory functions with an almost totally divided spinal cord. A case of spinal cord transection with preservation of part of one anterolateral quadrant, *Pain,* 2, 185, 1976.

95. Oliveras, J.L., Redjemi, F., Guilbaud, G., and Besson, J.M., Analgesia induced by electrical stimulation of the inferior centralis nucleus of the raphe in the cat, *Pain,* 1, 139, 1975.

96. Owens, C.M., Zhang, D., and Willis, W.D., Changes in the response states of primate spinothalamic tract cells caused by mechanical damage of the skin or activation of descending controls, *J. Neurophysiol.,* 67, 1509, 1992.

97. Palecek, J., Dougherty, P.M., Kim, S.H., Paleckova, V., Lekan, H., Chung, J.M., Carlton, S.M., and Willis, W.D., Responses of spinothalamic tract neurons to mechanical and thermal stimuli in an experiment model of peripheral neuropathy in primates, *J. Neurophysiol.,* 68, 1951, 1992.

98. Palecek, J., Paleckova, V., Dougherty, P., Carlton, S.M., and Willis, W.D., Responses of spinothalamic tract cells to mechanical and thermal stimulation of skin in rats with experimental peripheral neuropathy, *J. Neurophysiol.,* 67, 1562, 1992.

99. Patterson, J.T., Head, P.A., McNeill, D.L., Chung, K., and Coggeshall, R.E., Ascending unmyelinated primary afferent fibers in the dorsal funiculus, *J. Comp. Neurol.,* 290, 384, 1989.

100. Patterson, J.T., Coggeshall, R.E., Lee, W.T., and Chung, K., Long ascending unmyelinated primary afferent axons in the rat dorsal column: immunohistochemical localizations, *Neurosci. Lett.,* 108, 6, 1990.

101. Peng, Y., Lin, Q., and Willis, W.D., Involvement of alpha-2 adrenoreceptors in the periaqueductal gray-induced inhibition of dorsal horn cell activity in rats, *J. Pharm. Exp. Ther.,* 278, 125, 1996.

102. Rees, H. and Roberts, M.H.T., The anterior pretectal nucleus: a proposed role in sensory processing, *Pain,* 53, 121, 1993.

103. Reynolds, D.V., Surgery in the rat during electrical analgesia induced by focal brain stimulation, *Science,* 164, 444, 1969.

103a. Short, A.D., Brown, A.G., and Maxwell, D.J., Afferent inhibition and facilitation of transmission through the spinocervical tract in the anaesthetized cat, *J. Physiol.,* 429, 511, 1990.

104. Simone, D.A., Sorkin, L.S., Oh, U., Chung, J.M., Owens, C., LaMotte, R.H. and Willis, W.D., Neurogenic hyperalgesia: central neural correlates in responses of spinothalamic tract neurons, *J. Neurophysiol.,* 66, 228, 1991.

105. Sluka, K.A., Rees, H., Chen, P.S., Tsuruoka, M., and Willis, W.D., Inhibitors of G-proteins and protein kinases reduce the sensitization to mechanical stimulation and the desensitization to heat of spinothalamic tract neurons induced by intradermal injection of capsaicin in the primate, *Exp. Brain Res.,* 115, 15, 1997.

106. Sorkin, L.S., McAdoo, D.J., and Willis, W.D., Stimulation in the ventral posterior lateral nucleus of the primate thalamus leads to release of serotonin in the lumbar spinal cord, *Brain Res.,* 581, 307, 1992.

107. Sorkin, L.S., McAdoo, D.J., and Willis, W.D., Raphe magnus stimulation-induced antinociception in the cat is associated with release of amino acids as well as serotonin in the lumbar dorsal horn, *Brain Res.,* 618, 95, 1993.

108. Stevens, R.T., Hodge, C.J., and Apkarian, A.V., Medial, intralaminar and lateral terminations of lumbar spinothalamic tract neurons: a fluorescent double-label study, *Somatosensory Motor Res.,* 6, 285, 1989.

109. Stevens, R.T., London, S.M., and Apkarian, A.V., Spinothalamocortical projections to the secondary somatosensory cortex (SII) in squirrel monkey, *Brain Res.,* 631, 241, 1993.

110. Todd, A.J., Cells in laminae III and IV of rat spinal dorsal horn receive monosynaptic primary afferent input in lamina II, *J. Comp. Neurol.,* 289, 676, 1989.

111. Westlund, K.N. and Coulter, J.D., Descending projections of the locus coeruleus and subcoeruleus/medial parabrachial nuclei in monkey: axonal transport studies and dopamine-(-hydroxylase immunocytochemistry, *Brain Res. Rev.,* 2, 235, 1980.

112. Westlund, K.N., Carlton, S.M., Zhang, D., and Willis, W.D., Glutamate-immunoreactive terminals synapse on primate spinothalamic tract cells, *J. Comp. Neurol.,* 322, 519, 1992.

113. Willis, W.D., Neural mechanisms of pain discrimination, in *Sensory Processing in the Mammalian Brain,* Lund, J.S., Ed., Oxford University Press, New York, 1989, p. 130.

114. Willis, W.D., Central plastic responses to pain, in *Progress in Pain Research and Management,* Vol. 2, Gebhart, G.F., Hammond, D.L., and Jensen, T.S., Eds., IASP Press, Seattle, 1994, p. 301.

115. Willis, W.D. and Coggeshall, R.E., *Sensory Mechanisms of the Spinal Cord,* Plenum Press, New York, 1991.

116. Willis, W.D. and Westlund, K.N., Neuroanatomy of the pain system and of the pathways that modulate pain, *J. Clin. Neurophysiol.,* 14, 2, 1997.

116a. Willis, W.D., Trevino, D.L., Coulter, J.D., and Maunz, R.A., Responses of primate spinothalamic tract neurons to natural stimulation of hindlimb, *J. Neurophysiol.,* 37, 358, 1974.

117. Willis, W.D., Gerhart, K.D., Willcockson, W.S., Yezierski, R.P., Wilcox, T.K., and Cargill, C.L., Primate raphe- and reticulospinal neurons: effects of stimulation in periaqueductal gray or VPLc thalamic nucleus, *J. Neurophysiol.,* 51, 467, 1984.

118. Willis, W.D., Sluka, K.A., Rees, H., and Westlund, K.N., Cooperative mechanisms of neurotransmitter action in central nervous sensitization, in *Toward the Neurobiology of Chronic Pain, Progress in Brain Research,* Vol. 110, Carli, G. and Zimmermann, M., Eds., Elsevier Science, Amsterdam, 1996, p. 151.

119. Ye, Z. and Westlund, K.N., Ultrastructural localization of glutamate receptor subunits (NMDAR1, AMPA GluR1 and GluR2/3) and spinothalamic tract cells, *NeuroReport*, 7, 2581, 1996.
120. Yezierski, R.P., The effects of midbrain and medullary stimulation on spinomesecephalic tract cells in the cat, *J. Neurophysiol.*, 63, 240, 1990.
121. Yezierski, R.P. and Mendez, C.M., Spinal distribution and collateral projections of rat spinomesenvcephalic tract cells, *Neuroscience*, 44, 113, 1991.
121a. Yezierski, R.P. and Schwartz, R.H. Response and receptive-field properties of spinomesencephalic tract cells in the cat. *J. Neurophysiol.*, 55, 76, 1986.
122. Yezierski, R.P., Gerhart, K.D., Schrock, B.J., and Willis, W.D., A further examination of effects of cortical stimulation on primate spinothalamic tract cells, *J. Neurophysiol.*, 49, 424, 1983.
123. Zhang, D., Owens, C.M., and Willis, W.D., Two forms of inhibition of spinothalamic tract neurons produced by stimulation of the periaqueductal gray and the cerebral cortex, *J. Neurophysiol.*, 65, 1567, 1991.
124. Zhang, E.T. and Craig, A.D., Morphology and distribution of spinothalamic lamina I neurons in the monkey, *J. Neurosci.*, 17, 3274, 1997.
125. Zhang, E.T., Han, Z.S., and Craig, A.D., Morphological classes of spinothalamic lamina I neurons in the cat, *J. Comp. Neurol.*, 367, 537, 1996.
126. Zhang, J., Chandler, M.J., and Foreman, R.D., Thoracic visceral inputs use upper cervical segments to inhibit lumbar spinal neurons in rats, *Brain Res.*, 709, 337, 1996.

Chapter 29

Pain-Modulating Neurons and Behavioral State

Mary M. Heinricher and Steve McGaraughty

Contents

29.1 Descending Nociceptive Modulatory Systems: The Rostral
   Ventromedial Medulla 487
29.2 "Building Blocks" of Descending Modulation in the
   Rostral Ventromedial Medulla 489
29.3 Functional Relationships Between On- and Off-Cells in the
   Rostral Ventromedial Medulla 492
29.4 Analysis of the Role of Serotonin in Descending Nociceptive Modulation 493
29.5 Recruitment of the Rostral Ventromedial Medulla Modulatory
   Neurons in Different Behavioral States 495
29.6 Integration of Somatomotor and Autonomic Control
   With Nociceptive Modulation 496
29.7 Conclusions 497
Acknowledgments 497
References 497

29.1 Descending Nociceptive Modulatory Systems: The Rostral Ventromedial Medulla

The sensation of pain is highly dynamic. The magnitude of the response to a given damaging or potentially damaging (i.e., noxious) stimulus can vary widely with a host of behavioral state variables. Arousal, attention, learning, fear, and stress have all been shown to influence pain sensation in humans and nociceptive responding in animals. Human subjects, for example, report increases in pain when their attention is directed towards a noxious stimulus and a decrease in both the ability to discriminate pain intensity and perceived unpleasantness of the stimulus when attention is focused in another direction.[14,21,52,59,97] Similarly, changes in nociceptive responses are seen in monkeys performing vigilance tasks[21] or in hungry cats given access to food.[23,24] For rodents, introduction of a predator, a biologically relevant fear stimulus, results in a substantial reduction in

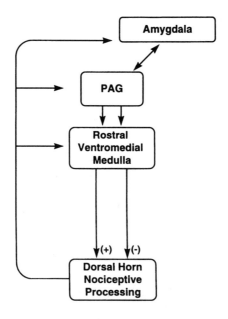

FIGURE 29.1

Brainstem nociceptive modulatory network with links in the midbrain periaqueductal gray (PAG) and rostral ventromedial medulla (RVM). The RVM receives a large input from the PAG and projects to the spinal and medullary dorsal horns to modulate processing of nociceptive information. Ascending information can influence activity in this system both directly and indirectly. The PAG-RVM axis receives inputs from spinoreticular and spinomesencephalic systems, and both feedforward and feedback processes can be activated by afferent input. Moreover, processes organized in limbic forebrain structures can gain access to nociceptive modulating circuits via reciprocal connections with the PAG. (Adapted from Heinricher, M.M., in *Messenger Molecules in Headache Pathogenesis: Monoamines, Purines, Neuropeptides and Nitric Oxide,* Olesen, J. and Edvinsson, L., Eds., Raven Press, New York, 1997. With permission.)

nociceptive responsiveness. More generally, elements of the environment that signal a potential threat can profoundly alter nociceptive thresholds.[31,93] In contrast, illness, a state that requires profound adjustments in both physiology and behavior, is associated with a state of increased nociception.[95]

Our understanding of the numerous neural systems likely to contribute to behavioral modulation of the link between noxious stimulation and behaviorally measured pain responses is still primitive. However, behavioral state variables are known to influence nociceptive processing at the first central synapse in the dorsal horn.[20,42] This indicates that pathways descending from the brain to the dorsal horn contribute to the influence of behavioral state on nociceptive processing. The best studied and probably functionally most significant of the descending pathways that modulate dorsal horn nociception originates in the rostral ventromedial medulla (RVM), a region which includes the nucleus raphe magnus and adjacent reticular formation (Figure 29.1). The RVM sends a large, diffuse projection through the dorsolateral funiculus to the dorsal horn at all levels, terminating most densely in laminae implicated in nociception, laminae I and IIo, and lamina V.[10,55] Electrical stimulation in the RVM produces a behaviorally measurable antinociception, and can inhibit dorsal horn nociceptive neurons (see Willis[98] for review). Microinjection studies demonstrate that this region is also an important substrate for opioid analgesia.[102] Indeed, the RVM is a critical link in an opioid-sensitive pain modulating network and receives a large input from the midbrain periaqueductal gray (PAG), a region also known to be involved in pain modulation and opioid analgesia (see Basbaum and Fields,[11] Mayer and Price,[68] and Yaksh et al.[102] for reviews). The PAG itself does not have a significant projection to the dorsal horn, and the RVM is a critical relay for the descending influences of the PAG on nociception.[85] The PAG-RVM system receives inputs from ascending spinal afferent systems (including spinoreticular and spinomesencephalic tracts, as well as collaterals from spinothalamic axons), and these converge with afferents from a wide range of supraspinal sites, including forebrain structures implicated in behavioral state. This system is thus ideally situated to mediate a complex integration of sensory traffic with higher order, state-related, influences.

This chapter will review our current knowledge of the "building blocks" of the descending modulatory system within the RVM, focusing on information gained from recent work using pharmacological tools and single unit recording in this region. The ultimate goal is to place this system into a behavioral context, with an understanding of how the different elements of the RVM might be recruited to regulate nociception in different behavioral states.

29.2 "Building Blocks" of Descending Modulation in Rostral Ventromedial Medulla

The recognition that electrical or glutamate stimulation or opioid microinjection into the RVM produces a potent analgesia led to the view that this region constituted an "analgesia center", distinct from the less differentiated reticular core.[9] This perspective prompted a number of straightforward predictions concerning the physiology and pharmacology of RVM neurons. As diagrammed in Figure 29.1, RVM neurons should be activated by stimulation in the PAG and project to the spinal cord. These neurons should also be activated by opioid analgesics. Indeed, many early studies confirmed these predictions, at least to a first approximation. Many RVM neurons were excited by PAG stimulation,[13,61,78,90,99] and a substantial number were found to project to the spinal cord, as shown using both anatomical and antidromic activation techniques.[11,35] However, only a minority were activated when morphine was given systemically. Many were inhibited or unresponsive (see Barbaro et al.[5] for review), demonstrating that the RVM was not, at least in terms of pharmacology, homogenous. Moreover, the responses of RVM neurons to cutaneous stimuli, noxious and non-noxious, were also mixed (see Fields and Heinricher[33] for review).

Thus, although it was possible to focus on RVM cell properties that were consistent with an analgesia center view, ignoring the full range of responses within the region was ultimately unsatisfying and, moreover, inconsistent with existing information demonstrating that the region was not homogeneous in terms of either morphology and neurochemistry (see Fields et al.[34] for review). A step toward resolving this dilemma was provided by the introduction of a lightly anesthetized rat preparation, which provided a more powerful conceptual framework for understanding the circuitry of the RVM. Spinal nocifensive reflexes are preserved in this preparation, in which rats are maintained in a lightly anesthetized state by a continuous infusion of either a short-acting barbiturate or, in some cases, a low level of a volatile anesthetic such as halothane (1% or less). Such reflexes are an indication of nociceptive responsiveness and are widely used in awake behaving animals.[22] For example, the tail flick reflex, which is a withdrawal response generally elicited by application of noxious heat to the tail surface, is evoked in these lightly anesthetized animals, as in awake animals, at a tail surface temperature of approximately 42 to 43°C. It is thus possible to monitor nociceptive responsiveness during single-cell recording experiments using standard electrophysiological techniques and thus place neurophysiological findings into a behavioral context.

With this approach, it became clear that RVM neurons fall into three classes based on: (1) how cell firing changes in association with nocifensive reflexes (the most commonly used being the tail flick, Figure 29.2),[32] and (2) how cell firing changes following administration of morphine.[5,36] Cells of one class — "off-cells" — show an abrupt cessation of activity beginning just prior to the occurrence of the tail flick and similar reflexes. This pause in firing is evident only if a reflex is evoked at a time when the cell is spontaneously active and can last anywhere from less than 200 msec to tens of minutes. These neurons become continuously active following systemically administered morphine.[36] Cells of a second class — "on-cells" — are defined by a burst of activity that

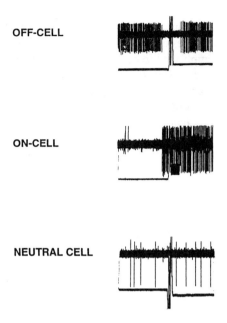

FIGURE 29.2
Characteristic tail flick-related changes in activity of on- and off-cells, and lack of response by neutral cell. Single 10-sec oscilloscope sweeps show cell activity (upper trace) and tail flick (deflection in lower trace is tail flick). Heat onset is at beginning of trace. Note abrupt pause (off-cell) and sudden burst (on-cell) associated with the tail flick.

begins just before the tail flick and similar reflexes. As is the case with the off-cell pause, the reflex-related burst of activity that characterizes on-cells varies widely in duration and magnitude, from just a few spikes closely linked to the reflex to a burst of activity lasting tens of minutes. On-cells are uniformly inhibited by opioids.[5,48,49] Both on- and off-cells exhibit similar changes in firing associated with other nocifensive reflexes such as limb withdrawal or cough. Neutral cells exhibit a wide range of spontaneous firing patterns but have in common a lack of response during the tail flick or other nocifensive reflexes as well as following opioid administration.[5,32] A substantial proportion of all three cell classes project to the spinal cord,[91] and identified on- and off-cells have recently been shown to project into the dorsal horn.[35]

In themselves, the properties described above suggest distinct roles for both the off- and on-cell classes in nociceptive modulation.[33] (A possible role for neutral cells in nociceptive modulation will be discussed below.) First, since nocifensive reflexes are suppressed when RVM neurons are activated non-selectively using direct electrical stimulation or glutamate microinjection, activation of some class of RVM neuron is sufficient to produce antinociception. The fact that off-cell activity ceases abruptly just prior to the execution of nocifensive reflexes suggests that these neurons function to inhibit dorsal horn nociceptive processing, and that the pause in firing permits responses to occur. Indeed, this pause is blocked by opioid analgesics given systemically or by microinjection into the RVM.[36,49]

Because on-cells are highly active just prior to and during execution of nocifensive reflexes, it seems unlikely that these neurons exert a significant net inhibitory effect on nociception. Indeed, the fact that the reflex occurs during a burst of on-cell firing would be more consistent with a permissive or even pro-nociceptive role for these neurons, as would the demonstration that on-cell firing is inhibited, directly, by opioids. On-cell firing is suppressed when opioids are applied to individual neurons using iontophoresis[48] or are microinjected into the RVM.[49]

The existence of pro-nociceptive neurons within the RVM may, on the surface, seem inconsistent with the fact that electrical stimulation within the RVM generally inhibits nociception.

However, electrical stimulation does not mimic the natural inputs to the region and may co-activate on- and off-cell populations artificially. If the influence of the off-cell population were more potent than that of on-cells, the net effect observed would be antinociception. Indeed, in a recent series of experiments, Zhuo and Gebhart[105,107] showed that stimulation within the RVM can produce facilitating as well as inhibiting effects on spinal nociception. The RVM has also been shown to play a critical role in the hyperalgesia associated with administration of illness-inducing agents[94] and that following injection of formalin under the skin of the paw.[96] A further demonstration of a facilitating outflow from the RVM was provided in an experiment that took advantage of the hyperalgesia induced as part of an acute opioid abstinence syndrome. When morphine-induced antinociception is reversed by naloxone, there follows a period of hyperalgesia, in which there is a significant decrease in nociceptive threshold relative to pre-morphine baseline levels. Inactivation of the RVM using a local anesthetic microinjection attenuates this hyperalgesia, demonstrating that some outflow from this region is able to exert a net pro-nociceptive effect.[56] The observation that on-cells are activated but off-cells inactive under these condition points to activation of on-cells as the underlying mechanism.[12]

The majority of work focused on defining the roles of on- and off-cells in nociceptive modulation has been performed in barbiturate-anesthetized rats maintained in a state in which it is possible to elicit a tail flick reflex at 42 to 43°C. Under these conditions, the great majority of on-cells are excited by noxious pinch as well as by non-noxious cutaneous stimulation such as firm stroking of the hairy skin of the back. A substantial number of off-cells also respond to both noxious and innocuous stimulation. In the absence of imposed stimulation, the ongoing activity of both on- and off-cells spontaneously alternates between periods of relatively high firing rate and periods or silence (or more rarely, very low activity, less than 1 Hz).[6] Simultaneous recordings from pairs of RVM neurons under these conditions indicate that all cells of the same class are active during the same periods (which can range from several seconds to tens of minutes, but are on the order of 80 sec, on average); neurons of the other class are active during alternate periods. These cycles of silence and activity may be related to EEG state.[39] Periods in which off-cells are silent and on-cells are active are associated with increased nociceptive responsiveness (i.e., a decrease in tail flick latency); periods in which on-cells are silent and off-cells are active are associated with decreased nociceptive responsiveness.[43] Nociceptive responsiveness is also increased if on-cells are induced to burst and off-cells to pause by delivery of a noxious conditioning stimulus.[82]

Anesthetic agents are known to alter the responsiveness of bulbospinal neurons in the RVM.[15] Nevertheless, cells having the key properties of on- and off-cells have also been identified in both halothane- and isoflurane-anesthetized rats.[58,73] There have been only a small number of studies in awake, freely moving animals. These have generally taken a "sensory" approach, identifying neuronal responses to stimuli delivered by the experimenter, so it can be difficult to compare the results of these experiments with on- and off-cell classes, which are defined by relationships to nociceptive responses rather than to noxious stimuli. Indeed, it is not *a priori* obvious how a neuron whose role is to modulate pain should respond to a peripheral stimulus unless that stimulus could be shown to induce an increase or decrease in nociceptive responsiveness.

The above caveat notwithstanding, there are important parallels between neurophysiological findings in awake animals and those obtained in anesthetized animals. In a series of studies in awake cats, Jacobs and colleagues[2] investigated the activity of putative serotonergic neurons and found that activity was related to arousal but not specifically with noxious stimulation; morphine administration had no effect on these neurons (see Chapter 12). Although cross-species comparisons may not be valid, the lack of response to morphine, in particular, suggests that these neurons, which displayed a slow, regular firing pattern considered indicative of serotonergic neurons, would be equivalent to neutral cells recorded in barbiturate-anesthetized rats. If so, one might expect that in awake animals, the activity of at least some neutral cells would be related to the presence of arousing stimuli, which would presumably include noxious stimuli. Oliveras et al.[76,77] observed some RVM neurons that were unaffected by noxious stimulation but many more that were excited. The excitatory response to noxious stimulation was attenuated by systemically administered morphine.[64] These neurons thus likely correspond to neutral and/or on-cell classes identified in barbiturate-anesthetized rats. Cells inhibited by noxious stimulation were not observed by these authors. Using a pharmacological strategy, McGaraughty et al.[71] observed two groups of RVM neurons, one activated and the other suppressed by systemically administered morphine, and inferred that these populations corresponded to the off- and on-cell populations identified in barbiturate-anesthetized rats. These authors suggested that the lack of off-cells in the studies by Oliveras and colleagues was related to the repeated use of noxious stimulation, which should suppress off-cell firing, as a search stimulus. Nonetheless, finding "off-like" responses to morphine in awake animals does not firmly consolidate a role for an off-cell class of neurons under physiological conditions. There is still a need to show that there exists a cell class that is inhibited in association with nociceptive responses. To this end, Clarke et al.[26] have reported typical off-cell responses in unanesthetized-decerebrated rats. Although this finding is not in intact animals, it does indicate that off-cells are not an artifact of anesthesia.

Thus, although studies carried out in anesthetized animals face significant difficulties in explaining behavior in awake animals, and anesthetics may alter the inputs to the RVM as well as the excitability or responsiveness of the RVM cells themselves, the use of the lightly anesthetized rat has provided a simplified system in which it has been possible to delineate within the RVM two cell classes that constitute functionally distinct "building blocks" for modulation of nociception. The defining characteristics of the two classes are the nocifensive reflex-related changes in activity and their opposite correlations with the analgesic actions of opioid drugs. Although each class to date has shown uniform pharmacology,[44–48,50] future work may indicate the existence of subclasses on the basis of inputs, targets, or neurochemistry. Nevertheless, recruitment of one class, the off-cells, should give rise to a net decrease in nociceptive responses. Activation of the other class, the on-cells, should produce a net increase in nociceptive responses.

29.3 Functional Relationships Between On- and Off-Cells in the Rostral Ventromedial Medulla

The function of a modulatory network depends not only on the intrinsic properties of the constituent neurons, but also on the anatomical and functional connections among these neurons and how these combine to give rise to measurable effects on behavior. Interactions among the cell classes within the RVM have so far been analyzed primarily in terms of how they relate to opioid analgesia. In as much as opioid microinjection within the RVM produces an antinociception,[102] at least a subset of nociceptive modulatory neurons within the region must be responsive to opioids. Off-cells are activated when opioids are microinjected into the RVM, but do not respond when iontophoretic methods are used to apply the drug directly to the off-cell under study.[48,49] The activation following microinjection must therefore be indirect, via disinhibition. In contrast, on-cell firing is suppressed, directly, by opioids.[5,48,49] However, non-selective inactivation of all neurons in the RVM does not yield an antinociception and has even been reported to facilitate behavioral responses to noxious stimulation in some experiments.[24,80,81] This leads to the conclusion that, without activation of off-cells, suppression of on-cell firing is not by itself sufficient to account for the antinociceptive actions of opioids.

Although a significant proportion of on-cells project to the spinal cord,[91] the line of reasoning outlined above suggested the hypothesis that some on-cells might serve as inhibitory interneurons within the RVM, linking an inhibitory action of opioids with activation of off-cells.[34] However, on-cells do not generally show extensive axonal arborizations within the RVM,[66] and blocking the on-cell burst does not interfere with the off-cell pause.[47] This demonstrates, first, that the on-cell is not likely to serve as an inhibitory interneuron within the RVM, and more generally, that on- and off-cells constitute parallel, albeit related, opioid-sensitive outflows from the RVM. The mechanism

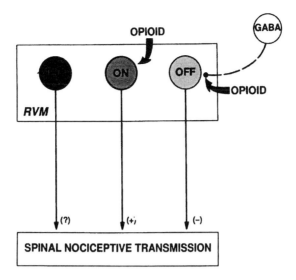

FIGURE 29.3
Three classes of neurons identified physiologically in RVM have distinct roles in nociceptive modulation. All three project to the dorsal horn. On-cells are likely to exert a net facilitating effect on nociception and are the only RVM neurons directly sensitive to opioids. Off-cells, likely to exert a net inhibitory effects on nociception, are activated by opioids. This activation is indirect, and most likely due to presynaptic suppression of GABA-mediated inhibition. Neutral cells are not responsive to opioids, and their role in nociceptive modulation and relationship to other RVM cell classes is unknown. However, a subset of neutral cells contain serotonin, and are thus likely to have some role in nociceptive modulation. (Adapted from Heinricher, M.M., in *Messenger Molecules in Headache Pathogenesis: Monoamines, Purines, Neuropeptides and Nitric Oxide*, Olesen, J. and Edvinsson, L., Eds., Raven Press, New York, 1997. With permission.)

through which opioid infusion leads to activation of off-cells thus remains an open question but presumably involves a presynaptic action on an inhibitory terminal entering the RVM from some other region (Figure 29.3).

29.4 Analysis of the Role of Serotonin in Descending Nociceptive Modulation

Since the earliest demonstrations that manipulations of the RVM could produce a behaviorally measurable analgesia, investigators have been interested in the identity of the neurotransmitter(s) mediating descending antinociceptive effects from this region. A number of neurotransmitters have been identified within RVM neurons, among them serotonin, substance P, enkephalin, thyrotropin releasing hormone, somatostatin, cholecystokinin, excitatory amino acids, and GABA, with most of these found in neurons projecting to the spinal cord.[1,17,86] Despite this neurochemical diversity, serotonin has received the largest share of attention.

The focus on serotonin has been justified by several lines of evidence. Foremost is the existence of serotonergic neurons within the RVM. In this context, it is worth noting that less than 25% of the neurons within the nucleus raphe magnus contain serotonin,[72] that the RVM extends beyond the boundaries of the nucleus raphe magnus to areas in which serotonergic neurons are not concentrated, and that there has been considerable controversy concerning the size of the descending serotonergic outflow from the RVM. Nevertheless, functional studies clearly indicate that this neurotransmitter is in some way involved in descending control. Electrical stimulation in the RVM can cause release of serotonin in the spinal cord,[41,84,87] and exogenous serotonin applied directly to the spinal cord can have a hypoalgesic effect. The hypoalgesia is dose related and was shown in some experiments to be blocked by serotonin antagonists (see Bardin et al.[7] for references). Finally, behavioral antinociception produced by activation of PAG or RVM using electrical stimulation or morphine or neuroexcitant microinjection is often attenuated by intrathecal administration of serotonergic antagonists (see Chapter 28).[16,30,40,54,60,100,101] Taken together, these observations point to an important role for serotonin in pain suppression. Indeed, it is frequently assumed that the role of 5-HT in descending control is well understood, with activation of serotonergic neurons by direct electrical stimulation or by opioids given as a textbook explanation for the analgesic actions of descending systems.

Unfortunately, this seemingly straightforward link between increased discharge of serotonergic neurons and antinociception does not withstand closer scrutiny. As reviewed in Chapter 28, the spinal actions of serotonin are not uniformly antinociceptive. Moreover, spinal administration of serotonergic antagonists attenuates, but does not block, the antinociceptive effects of manipulating the PAG or RVM. As an additional complicating factor, serotonergic antagonists were more effective in blocking analgesia as measured by the tail flick reflex than hot-plate or thermal or mechanical paw-withdrawal tests.[4,30,53] Although this could be due to the fact that the hot plate test is organized supraspinally whereas the tail flick is a spinal reflex, this is unlikely to be a complete explanation, as descending inhibition of both spinally and supraspinally organized responses evoked by tail stimulation was attenuated by spinal administration of serotonin antagonists.[16] As an alternative, it has recently been proposed that the neurochemistry of descending inhibition of responses evoked by delivering a noxious stimulus to the paws is distinct from that exerted on responses evoked by stimulating the tail.[30] This latter possibility is supported by other recent work demonstrating that the nocifensor reflexes evoked by paw and tail stimulation are under distinct controls,[74] but in any case, one would question the generality of serotonin in mediating descending antinociception. A further issue for consideration is the fact that spinal serotonin transmission is also necessary for some of the descending pro-nociceptive actions of the RVM,[105,107] which indicates that the role of serotonin in descending nociceptive modulation is not exclusively antinociceptive.

Investigations of serotonin release and metabolism also provide little support for the idea that the antinociceptive effects of RVM are necessarily associated with an increase in serotonergic cell activity above ongoing levels. Although electrical stimulation in the RVM does evoke spinal release of serotonin,[18,41,84] inhibition of nociceptive dorsal horn neurons can occur with stimulation parameters that do not cause serotonin release.[87] Moreover although early studies suggested increased release or turnover of spinal serotonin following administration of morphine either by microinjection in PAG or RVM or systemically,[92,103] a more recent study using microdialysis methods saw no increase in dorsal horn serotonin release following systemic morphine administration.[67] This microdialysis study is also consistent with findings from single unit studies of identified and presumed serotonergic neurons. In both cat and rat, neurons considered to be serotonergic (based on firing pattern) were found to be unresponsive to opioids.[2,25] In two recent single unit studies using intracellular labeling in rat, serotonin has been found only in neutral cells,[65,79] which are not activated by opioids.[5,48] Electrophysiological and release studies taken together therefore strongly argue against the idea that opioids cause an increase in activity of serotonergic neurons in the RVM.

In sum, although there is a large body of evidence indicating that serotonin can inhibit spinal nociceptive processing, there is little support for the view that increased discharge of serotonergic neurons is responsible for the hypoalgesic effects mediated through the RVM. Although pharmacological studies clearly demonstrate that serotonergic transmission is necessary for the full expression of descending modulation, what remains uncertain is the precise role of this neurotransmitter and how it interacts with relevant circuits in the dorsal horn. One speculative possibility is that serotonin has a more general role in supporting other descending controls, both pro- and antinociceptive, possibly regulating the potency of other modulatory inputs in controlling dorsal horn nociceptive circuits. This would be consistent with the fact that although intrathecal administration of serotonergic antagonists interferes with descending modulatory influences, these compounds by themselves generally have no effect on nociceptive responsiveness.[40]

The emphasis on serotonin has, to some extent, directed attention away from the numerous other neurotransmitters identified in RVM neurons. Whether pro- and antinociceptive outflows from the RVM are associated with distinct sets of neurotransmitters is obviously an intriguing question, with significant clinical relevance.

29.5 Recruitment of the Rostral Ventromedial Medulla Modulatory Neurons in Different Behavioral States

As reviewed above, investigations of the roles of the different neuronal classes within the RVM in descending control of nociception are relatively well advanced. In contrast, questions relating to how or even whether the different RVM cell classes are brought into play in different behavioral states are far from being answered. Ongoing activity in on- and off-cell populations is correlated with small fluctuations in nociceptive responsiveness,[43] but the inputs controlling "spontaneous" discharges of these neurons are not known. For example, fluctuations in on- and off-cell firing may be related to EEG state[39] and could thus in principle mediate shifts in nociceptive responsiveness associated with changes in attention and arousal, but the mechanism linking EEG state to RVM activity remains a complete mystery.

Does the RVM play an important role in the analgesic component of general anesthesia? The activity of both on- and off-cells is suppressed when anesthesia is increased to surgical levels,[6,58] so the RVM would not be expected to exert a significant descending control in that state (see Chapter 30). Thus, other mechanisms, including direct spinal actions and other descending systems, must be invoked to explain changes in somatomotor processing in anesthesia.[27]

Somatosensory processing has long been known to vary with the sleep/waking cycle. Early work pointed to an almost uniform suppression of somatosensory and motor systems during active sleep, but it has since become clear that the situation is more complex. A number of groups have recently looked at changes in responsiveness of spinal and trigeminal neurons and report a broad spectrum of effects associated with active sleep (see Chapter 31). Presumably these disparities are due to a number of factors, including cell populations recorded from, sampling issues, and the use of different species and stimuli. If so, it would imply that the role of the different neurons studied, and how each is involved in the circuits mediating both motor responses and sensation, must be known if we are to determine the significance of state-related changes in activity in the dorsal horn.

Nevertheless, the question of whether the RVM contributes to changes in somatosensory processing with the sleep/wake cycle (whatever those might be), remains of interest. Some RVM cell population displays an increase in *c-fos* expression during an active sleep-like state induced by carbachol;[104] unfortunately, identified RVM neurons have not been recorded through the sleep/wake cycle, so the identity of these neurons is unknown. Although there is no direct evidence, serotonin could provide a bridge between the sleep/waking cycle and nociceptive modulation. Activity of serotonergic neurons in the RVM varies across the sleep/wake-arousal continuum.[38] If serotonin, which is found in at least some neutral cells, supports the more specific regulatory processes represented by off-cell (and possibly on-cell) activity, one could speculate that descending modulatory influences (both inhibitory and facilitatory) from this region would be relatively ineffective during active sleep, when serotonergic neurons are inactive, and most potent during the active waking state.

One state in which the antinociceptive outflow from the RVM has been specifically implicated is as part of defense responses, i.e., in a state of stress or fear. Brainstem systems that suppress nociceptive transmission are known to be activated by a variety of biologically relevant threat stimuli, such as an attacking conspecific or potential predator. Antinociception can also be elicited as a conditioned response to previously neutral cues that have been paired in a classical conditioning paradigm with noxious or aversive events.[31,93] In such situations, hypoalgesia is recruited in concert with other behaviors and autonomic adjustments as part of a defensive reaction.[29] Fear-related responses to threat are organized in the amygdala,[28] and apparently produce hypoalgesia via connections with the PAG-RVM axis (Figure 29.1). Indeed, infusions of morphine into the amygdala produce an antinociception and, at the same time, activate off-cells and suppress on-cell firing.[51,70] Thus, one way in which the modulatory circuitry of the RVM is known to be engaged physiologically in behaving animals is via activation of defense-related processes organized in the amygdala.

29.6 Integration of Somatomotor and Autonomic Control With Nociceptive Modulation

The RVM is embedded in the reticular formation, once viewed as a rather mysterious, undifferentiated brainstem core, presenting a daunting lack of specificity in function and connectivity. This concept has since undergone considerable revision, in part due to the advent of modern anatomical techniques, which demonstrated that connections of different elements of the reticular formation are in fact highly organized.[19]

Despite this new knowledge of reticular organization, caution is required if functional inferences are to be derived from physiological analyses of neurons in this region. Barman and Gebber[8] and Koepchen et al.[57] have reviewed the conceptual issues confronting such investigations. Because the functions subserved by the reticular core are often closely intertwined (e.g., respiratory and cardiovascular variables must be co-regulated; defense responses include coordinated somatomotor, cardiovascular, and antinociceptive components, etc.), two kinds of information are required: (1) whether manipulations of the region in question affects the relevant functional variables (e.g., nociceptive responsiveness, blood pressure, motoneuron excitability), and (2) whether the region has appropriate connectivity. It is only in this context that the physiology of constituent neurons can be interpreted.

Our focus on the role of on- and off-cells and neutral cells in pain modulation was grounded in such knowledge. Indeed, the term "rostral ventromedial medulla" does not respect cytoarchitecturally defined nuclear boundaries and was instead coined to correspond to a region where: (1) bulbospinal neurons project densely to dorsal horn laminae implicated in nociception, and (2) electrical and chemical stimulation alter nociceptive responses, both spinally and supraspinally organized. In our early studies, recordings were performed only at sites at which low current (< 10 µA) stimulation produced inhibition of the tail flick reflex.[5,32] In adjacent regions (nucleus reticular gigantocellularis and on the midline caudal to the facial nucleus in raphe pallidus and raphe obscurus), stimulation did not produce antinociception, and neurons with reflex-related changes in activity were not identified.

An important question which arises then is whether the role of on- and off-cells is limited to pain modulation. The RVM does not project to the ventral horn motoneuron pool, and stimulation in the RVM does not alter monosynaptic reflexes.[37] Thus, although immediately adjacent areas, including nucleus reticularis gigantocellularis and caudal raphe nuclei, raphe pallidus, and raphe obscurus, are strongly implicated in motor control, it seems unlikely that a specific role for these neurons in motor control will be identified.

There are, however, a number of recent findings which point to interactions between cardiovascular regulation and pain modulating systems in the RVM.[63] A full discussion of this literature is beyond the scope of this review, but a series of recent studies has identified a subset of neurons identified as on- and off-cells which also exhibit some measure of correlation with blood pressure.[89] Might this indicate a role for these neurons in cardiovascular control? The argument for such a conclusion is not compelling. The projection from the RVM to the spinal intermediolateral zone is not dense, and, unlike that to the dorsal horn, has not been shown to include identified on- and off-cell projections.[35,88] Although stimulation studies have found pressor and depressor sites in the

RVM, it is not clear that this stimulation would have altered nociception.[69] The most useful studies in this regard are probably those of Zhuo and Gebhart,[105,106] who assessed effects of RVM stimulation on tail flick latency and blood pressure in the same experiment. Most animals showed no change in blood pressure or heart rate following stimuli that reliably altered the tail flick. In lesion studies, infusions of lidocaine that blocked the pain-modulating functions of the RVM did not alter blood pressure.[83] Still, the idea that a subset of RVM on- and off-cells provides an interface between pain modulation and cardiovascular control remains an intriguing possibility.

In contrast to what is seen in reduced experimental preparations using unnatural forms of stimulation, adaptive behavior in intact organisms normally requires co-regulation of motor and autonomic systems with sensory processing, including descending modulatory systems. For example, in awake, freely moving animals, RVM-stimulation-produced antinociception is accompanied by an immobility or hyporeactivity, which in rats is thought to be a component of defense responses.[75] The question of how these components are integrated is an exciting area of ongoing and future research.[3,62]

29.7 Conclusions

Both functional and anatomical studies support an important role for the RVM in descending nociceptive modulation. By adopting an approach in which the physiology and pharmacology of RVM neurons can be investigated in a behavioral context, we have made substantial advances in our understanding of the properties of its constituent neurons, the relationships between different classes of neurons, and how these properties and connections combine to give rise to variations in nociceptive processing. This information provides identified targets for molecular and cellular analyses and gives insights useful in interpreting results of behavioral studies.

Important challenges remain. As discussed at length above, serotonin may provide a bridge between behavioral state and nociceptive modulation, and a more penetrating analysis of the role of serotonin in descending control is likely to provide an important key to the understanding of the system as a whole. Questions related to whether and how this system is brought into play in different behavioral states are critical, but efforts to delineate neural mechanisms through which these components are recruited in different states are just beginning. Continuing investigations of how somatosensory processing in the spinal and trigeminal dorsal horns is altered with behavioral state should complement information gained from analyses at supraspinal levels (see Chapters 30 and 31). Finally, it must be emphasized that the RVM is just one of many descending systems controlling sensorimotor processing. The interactions between the RVM other descending systems, those involved in motor and cardiovascular control, as well as other descending nociceptive modulatory systems (such as the pontine noradrenergic cell groups), are likely to be a key factor in understanding how all of these systems relate to behavioral state.

Acknowledgment

This work was supported by a grant from NIDA (DA05608).

References

1. Antal, M., Petko, M., Polgar, E., Heizmann, C.W., and Storm-Mathisen, J., Direct evidence of an extensive GABAergic innervation of the spinal dorsal horn by fibres descending from the rostral ventromedial medulla, *Neuroscience,* 73, 509, 1996.

2. Auerbach, S., Fornal, C., and Jacobs, B.L., Response of serotonin-containing neurons in nucleus raphe magnus to morphine, noxious stimuli, and periaqueductal gray stimulation in freely moving cats, *Exp. Neurol.,* 88, 609, 1985.

3. Bandler, R. and Shipley, M.T., Columnar organization in the midbrain periaqueductal gray: modules for emotional expression?, *Trends Neurosci.,* 17, 379, 1994.

4. Barbaro, N.M., Hammond, D.L., and Fields, H.L., Effects of intrathecally administered methysergide and yohimbine on microstimulation-produced antinociception in the rat, *Brain Res.,* 343, 223, 1985.

5. Barbaro, N.M., Heinricher, M.M., and Fields, H.L., Putative pain modulating neurons in the rostral ventral medulla: reflex-related activity predicts effects of morphine, *Brain Res.,* 366, 203, 1986.

6. Barbaro, N.M., Heinricher, M.M., and Fields, H.L., Putative nociceptive modulatory neurons in the rostral ventromedial medulla of the rat display highly correlated firing patterns, *Somatosensory Motor Res.,* 6, 413, 1989.

7. Bardin, L., Bardin, M., Lavarenne, J., and Eschalier, A., Effect of intrathecal serotonin on nociception in rats: influence of the pain test used, *Exp. Brain Res.,* 113, 81, 1997.

8. Barman, S.M. and Gebber, G.L., Problems associated with the identification of brain stem neurons responsible for sympathetic nerve discharge, *J. Auton. Nerv. Syst.,* 3, 369, 1981.

9. Basbaum, A.I., in *The Reticular Formation Revisited,* Hobson, J.A. and Brazier, M.A.B., Eds., Raven Press, New York, 1980.

10. Basbaum, A.I., Clanton, C.H., and Fields, H.L., Three bulbospinal pathways from the rostral medulla of the cat: an autoradiographic study of pain modulating systems, *J. Comp. Neurol.,* 178, 209, 1978.

11. Basbaum, A.I. and Fields, H.L., Endogenous pain control systems: brainstem spinal pathways and endorphin circuitry, *Ann. Rev. Neurosci.,* 7, 309, 1984.

12. Bederson, J.B., Fields, H.L., and Barbaro, N.M., Hyperalgesia during naloxone-precipitated withdrawal from morphine is associated with increased on-cell activity in the rostral ventromedial medulla, *Somatosensory Motor Res.,* 7, 185, 1990.

13. Behbehani, M.M. and Fields, H.L., Evidence that an excitatory connection between the periaqueductal gray and nucleus raphe magnus mediates stimulation produced analgesia, *Brain Res.,* 170, 85, 1979.

14. Beydoun, A., Morrow, T.J., Shen, J.F., and Casey, K.L., Variability of laser-evoked potentials: attention, arousal and lateralized differences, *Electroencephalographr. Clin. Neurophysiol.,* 88, 173, 1993.

15. Blair, R.W. and Evans, A.R., Effects of anesthetic agents on somatosensory responses of raphespinal neurons in the cat, *Neurosci. Lett.,* 162, 133, 1993.

16. Borszcz, G.S., Johnson, C.P., and Thorp, M.V., The differential contribution of spinopetal projections to increases in vocalization and motor reflex thresholds generated by the microinjection of morphine into the periaqueductal gray, *Behav. Neurosci.,* 110, 368, 1996.

17. Bowker, R.M., Abbott, L.C., and Dilts, R.P., Peptidergic neurons in the nucleus raphe magnus and the nucleus gigantocellularis: their distributions, interrelationships, and projections to the spinal cord, *Prog. Brain Res.,* 77, 95, 1988.

18. Bowker, R.M. and Abhold, R.H., Evoked changes in 5-hydroxytryptamine and norepinephrine release: *in vivo* dialysis of the rat dorsal horn, *Eur. J. Pharmacol.,* 175, 101, 1990.

19. Brazier, M.A.B. and Hobson, J.A., Eds., *The Reticular Formation Revisited,* Raven Press, New York, 1980.

20. Bushnell, M.C., Duncan, G.H., Dubner, R., and He, L.F., Activity of trigeminothalamic neurons in medullary dorsal horn of awake monkeys trained in a thermal discrimination task, *J. Neurophysiol.,* 52, 170, 1984.

21. Bushnell, M.C., Duncan, G.H., Dubner, R., Jones, R.L., and Maixner, W., Attentional influences on noxious and innocuous cutaneous heat detection in humans and monkeys, *J. Neurosci.,* 5, 1103, 1985.

22. Carstens, E., No withdrawal from reflexes to assess pain, *Pain Forum,* 6, 119, 1997.
23. Casey, K.L. and Morrow, T.J., Supraspinal nocifensive responses of cats: spinal cord pathways, monoamines, and modulation, *J. Comp. Neurol.,* 270, 591, 1988.
24. Casey, K.L. and Morrow, T.J., Effect of medial bulboreticular and raphe nuclear lesions on the excitation and modulation of supraspinal nocifensive behaviors in the cat, *Brain Res.,* 501, 150, 1989.
25. Chiang, C.Y. and Pan, Z.Z., Differential responses of serotonergic and non-serotonergic neurons in nucleus raphe magnus to systemic morphine in rats, *Brain Res.,* 337, 146, 1985.
26. Clarke, R.W., Morgan, M.M., and Heinricher, M.M., Identification of nocifensor reflex-related neurons in the rostroventromedial medulla of decerebrated rats, *Brain Res.,* 636, 169, 1994.
27. Collins, J.G., Kendig, J.J., and Mason, P., Anesthetic actions within the spinal cord: contributions to the state of general anesthesia, *Trends Neurosci.,* 18, 549, 1995.
28. Davis, M., The role of the amygdala in fear and anxiety, *Ann. Rev. Neurosci.,* 15, 353, 1992.
29. Depaulis, A. and Bandler, R., Eds., *The Midbrain Periaqueductal Gray Matter: Functional, Anatomical, and Neurochemical Organization,* Plenum Press, New York, 1991.
30. Fang, F. and Proudfit, H.K., Spinal cholinergic and monoamine receptors mediate the antinociceptive effect of morphine microinjected in the periaqueductal gray on the rat tail, but not the feet, *Brain Res.,* 722, 95, 1996.
31. Fanselow, M.S., Conditioned fear-induced opiate analgesia: a competing motivational state theory of stress analgesia, *Ann. N.Y. Acad. Sci.,* 467, 40, 1986.
32. Fields, H.L., Bry, J., Hentall, I., and Zorman, G., The activity of neurons in the rostral medulla of the rat during withdrawal from noxious heat, *J. Neurosci.,* 3, 2545, 1983.
33. Fields, H.L. and Heinricher, M.M., Anatomy and physiology of a nociceptive modulatory system, *Phil. Trans. Roy. Soc. Lond. B,* 308, 361, 1985.
34. Fields, H.L., Heinricher, M.M., and Mason, P., Neurotransmitters in nociceptive modulatory circuits, *Ann. Rev. Neurosci.,* 14, 219, 1991.
35. Fields, H.L., Malick, A., and Burstein, R., Dorsal horn projection targets of ON and OFF cells in the rostral ventromedial medulla, *J. Neurophysiol.,* 74, 1742, 1995.
36. Fields, H.L., Vanegas, H., Hentall, I.D., and Zorman, G., Evidence that disinhibition of brain stem neurones contributes to morphine analgesia, *Nature,* 306, 684, 1983.
37. Floeter, M.K. and Fields, H.L., Evidence that inhibition of a nociceptive flexion reflex by stimulation in the rostroventromedial medulla in rats occurs at a premotoneuronal level, *Brain Res.,* 538, 340, 1991.
38. Fornal, C., Auerbach, S., and Jacobs, B.L., Activity of serotonin-containing neurons in nucleus raphe magnus in freely moving cats, *Exp. Neurol.,* 88, 590, 1985.
39. Grahn, D.A. and Heller, H.C., Activity of most rostral ventromedial medulla neurons reflect EEG/EMG pattern changes, *Am. J. Physiol.,* 257, R1496, 1989.
40. Hammond, D.L., Control systems for nociceptive afferent processing: the descending inhibitory pathways, in *Spinal Afferent Processing,* Yaksh, T.L., Ed., Plenum Press, New York, 1986, p. 363.
41. Hammond, D.L., Tyce, G.M., and Yaksh, T.L., Efflux of 5-hydroxytryptamine and noradrenaline into spinal cord superfusates during stimulation of the rat medulla, *J. Physiol.,* 359, 151, 1985.
42. Hayes, R.L., Dubner, R., and Hoffman, D.S., Neuronal activity in medullary dorsal horn of awake monkeys trained in a thermal discrimination task. II. Behavioral modulation of responses to thermal and mechanical stimuli, *J. Neurophysiol.,* 46, 428, 1981.
43. Heinricher, M.M., Barbaro, N.M., and Fields, H.L., Putative nociceptive modulating neurons in the rostral ventromedial medulla of the rat: firing of on- and off-cells is related to nociceptive responsiveness, *Somatosensory Motor Res.,* 6, 427, 1989.

44. Heinricher, M.M., Haws, C.M., and Fields, H.L., Opposing actions of norepinephrine and clonidine on single pain-modulating neurons in rostral ventromedial medulla, in *Pain Research and Clinical Management*, Dubner, R., Gebhart, G.F., and Bond, M.R., Eds., Elsevier, Amsterdam, 1988, p. 590.
45. Heinricher, M.M., Haws, C.M., and Fields, H.L., Evidence for GABA-mediated control of putative nociceptive modulating neurons in the rostral ventromedial medulla: iontophoresis of bicuculline eliminates the off-cell pause, *Somatosensory Motor Res.*, 8, 215, 1991.
46. Heinricher, M.M. and McGaraughty, S., CCK Modulates the Antinociceptive Actions of Opioids by an Action Within the Rostral Ventromedial Medulla: A Combined Electrophysiological and Behavioral Study, presented at the International Association for the Study of Pain, 8th World Congress, Vancouver, B.C., Canada, 1996.
47. Heinricher, M.M. and McGaraughty, S., Analysis of excitatory amino acid transmission within the rostral ventromedial medulla: implications for circuitry, *Pain*, 75, 247–255, 1988.
48. Heinricher, M.M., Morgan, M.M., and Fields, H.L., Direct and indirect actions of morphine on medullary neurons that modulate nociception, *Neuroscience*, 48, 533, 1992.
49. Heinricher, M.M., Morgan, M.M., Tortorici, V., and Fields, H.L., Disinhibition of off-cells and antinociception produced by an opioid action within the rostral ventromedial medulla, *Neuroscience*, 63, 279, 1994.
50. Heinricher, M.M. and Roychowdhury, S., Reflex-related activation of putative pain facilitating neurons in rostral ventromedial medulla (RVM) depends upon excitatory amino acid transmission, *Neuroscience*, 78, 1159, 1997.
51. Helmstetter, F.J., Bellgowan, P.S., and Tershner, S.A., Inhibition of the tail flick reflex following microinjection of morphine into the amygdala, *NeuroReport*, 4, 471, 1993.
52. Janssen, S.A. and Arntz, A., Anxiety and pain: attentional and endorphinergic influences, *Pain*, 66, 145, 1996.
53. Jensen, T.S. and Yaksh, T.L., Spinal monoamine and opiate systems partly mediate the antinociceptive effects produced by glutamate at brainstem sites, *Brain Res.*, 321, 287, 1984.
54. Jensen, T.S. and Yaksh, T.L., Examination of spinal monoamine receptors through which brainstem opiate-sensitive systems act in the rat, *Brain Res.*, 363, 114, 1986.
55. Jones, S.L. and Light, A.R., Termination patterns of serotoninergic medullary raphespinal fibers in the rat lumbar spinal cord: an anterograde immunohistochemical study, *J. Comp. Neurol.*, 297, 267, 1990.
56. Kaplan, H. and Fields, H.L., Hyperalgesia during acute opioid abstinence: evidence for a nociceptive facilitating function of the rostral ventromedial medulla, *J. Neurosci.*, 11, 1433, 1991.
57. Koepchen, H.P., Langhorst, P., and Seller, H., The problem of identification of autonomic neurons in the lower brain stem, *Brain Res.*, 87, 375, 1975.
58. Leung, C.G. and Mason, P., Effects of isoflurane concentration on the activity of pontomedullary raphe and medial reticular neurons in the rat, *Brain Res.*, 699, 71, 1995.
59. Leventhal, H., Brown, D., Shacham, S., and Engquist, G., Effects of preparatory information about sensations, threat of pain, and attention on cold pressor distress, *J. Pers. Soc. Psychol.*, 37, 688, 1979.
60. Lin, Q., Peng, Y.B., and Willis, W.D., Antinociception and inhibition from the periaqueductal gray are mediated in part by spinal 5-hydroxytryptamine(1A) receptors, *J. Pharmacol. Exp. Ther.*, 276, 958, 1996.
61. Lovick, T.A., Midbrain influences on ventrolateral medullo-spinal neurones in the rat, *Exp. Brain Res.*, 90, 147, 1992.
62. Lovick, T.A., The medullary raphe nuclei: a system for integration and gain control in autonomic and somatomotor responsiveness?, *Exp. Physiol.*, 82, 31, 1997.
63. Maixner, W., Interactions between cardiovascular and pain modulatory systems: physiological and pathophysiological implications, *J. Cardiovasc. Electrophysiol.*, 2, S3, 1991.

64. Martin, G., Montagne-Clavel, J., and Oliveras, J.L., Involvement of ventromedial medulla "multimodal, multireceptive" neurons in opiate spinal descending control system: a single-unit study of the effect of morphine in the awake, freely moving rat, *J. Neurosci.*, 12, 1511, 1992.
65. Mason, P., Physiological identification of pontomedullary serotonergic neurons in the rat, *J. Neurophysiol.*, 77, 1087, 1997.
66. Mason, P. and Fields, H.L., Axonal trajectories and terminations of on- and off-cells in the cat lower brainstem, *J. Comp. Neurol.*, 288, 185, 1989.
67. Matos, F.F., Rollema, H., Brown, J.L., and Basbaum, A.I., Do opioids evoke the release of serotonin in the spinal cord? An *in vivo* microdialysis study of the regulation of extracellular serotonin in the rat, *Pain*, 48, 439, 1992.
68. Mayer, D.J. and Price, D.D., Central nervous system mechanisms of analgesia, *Pain*, 2, 379, 1976.
69. McCall, R.B., Evidence for a serotonergically mediated sympathoexcitatory response to stimulation of medullary raphe nuclei, *Brain Res.*, 311, 131, 1984.
70. McGaraughty, S. and Heinricher, M.M., The effects of morphine in the basolateral nucleus of the amygdala on the ON, OFF, and neutral cells in the rostral ventromedial medulla of anesthetized rats, *Soc. Neurosci. Abstr.*, 22, 114, 1996.
71. McGaraughty, S., Reinis, S., and Tsoukatos, J., Two distinct unit activity responses to morphine in the rostral ventromedial medulla of awake rats, *Brain Res.*, 604, 331, 1993.
72. Moore, R.Y., The anatomy of central serotonin neuron systems in the rat brain, in *Serotonin Neurotransmission and Behavior*, Jacobs, B.L. and Gelperin, A., Eds., MIT Press, Cambridge, MA, 1981, p. 35.
73. Morgan, M.M. and Heinricher, M.M., Activity of neurons in the rostral medulla of the halothane-anesthetized rat during withdrawal from noxious heat, *Brain Res.*, 582, 154, 1992.
74. Morgan, M.M., Heinricher, M.M., and Fields, H.L., Inhibition and facilitation of different nocifensor reflexes by spatially remote noxious stimuli, *J. Neurophysiol.*, 72, 1152, 1994.
75. Morgan, M.M., Whitney, P.K., Boyer, J.S., and Strack, A.M., Chemical stimulation of the rostral ventromedial medulla: analysis of conditioned place avoidance and stress hormone release, *Soc. Neurosci. Abstr.*, 22, 113, 1996.
76. Oliveras, J.L., Martin, G., Montagne, J., and Vos, B., Single unit activity at ventromedial medulla level in the awake, freely moving rat: effects of noxious heat and light tactile stimuli onto convergent neurons, *Brain Res.*, 506, 19, 1990.
77. Oliveras, J.L., Vos, B., Martin, G., and Montagne, J., Electrophysiological properties of ventromedial medulla neurons in response to noxious and non-noxious stimuli in the awake, freely moving rat: a single-unit study, *Brain Res.*, 486, 1, 1989.
78. Pomeroy, S.L. and Behbehani, M.M., Physiologic evidence for a projection from periaqueductal gray to nucleus raphe magnus in the rat, *Brain Res.*, 176, 143, 1979.
79. Potrebic, S.B., Fields, H.L., and Mason, P., Serotonin immunoreactivity is contained in one physiological cell class in the rat rostral ventromedial medulla, *J. Neurosci.*, 14, 1655, 1994.
80. Proudfit, H.K., Effects of raphe magnus and raphe pallidus lesions on morphine-induced analgesia and spinal cord monoamines, *Pharmacol. Biochem. Behav.*, 13, 705, 1980.
81. Proudfit, H.K., Reversible inactivation of raphe magnus neurons: effects on nociceptive threshold and morphine-induced analgesia, *Brain Res.*, 201, 459, 1980.
82. Ramirez, F. and Vanegas, H., Tooth pulp stimulation advances both medullary off-cell pause and tail flick, *Neurosci. Lett.*, 100, 153, 1989.
83. Randich, A., Aimone, L.D., and Gebhart, G.F., Medullary substrates of descending spinal inhibition activated by intravenous administration of [D-Ala^2]methionine enkephalinamide in the rat, *Brain Res.*, 411, 236, 1987.

84. Rivot, J.P., Chiang, C.Y., and Besson, J.M., Increase of serotonin metabolism within the dorsal horn of the spinal cord during nucleus raphe magnus stimulation, as revealed by *in vivo* electrochemical detection, *Brain Res.*, 238, 117, 1982.
85. Sandkühler, J. and Gebhart, G.F., Relative contributions of the nucleus raphe magnus and adjacent medullary reticular formation to the inhibition by stimulation in the periaqueductal gray of a spinal nociceptive reflex in the pentobarbital-anesthetized rat, *Brain Res.*, 305, 77, 1984.
86. Skagerberg, G. and Björklund, A., Topographic principles in the spinal projections of serotonergic and non-serotonergic brainstem neurons in the rat, *Neuroscience*, 15, 445, 1985.
87. Sorkin, L.S., McAdoo, D.J., and Willis, W.D., Raphe magnus stimulation-induced antinociception in the cat is associated with release of amino acids as well as serotonin in the lumbar dorsal horn, *Brain Res.*, 618, 95, 1993.
88. Strack, A.M., Sawyer, W.B., Platt, K.B., and Loewy, A.D., CNS cell groups regulating the sympathetic outflow to adrenal gland as revealed by transneuronal cell body labeling with pseudorabies virus, *Brain Res.*, 491, 274, 1989.
89. Thurston, C.L. and Randich, A., Responses of on and off cells in the rostral ventral medulla to stimulation of vagal afferents and changes in mean arterial blood pressure in intact and cardiopulmonary deafferented rats, *Pain*, 62, 19, 1995.
90. Vanegas, H., Barbaro, N.M., and Fields, H.L., Midbrain stimulation inhibits tail-flick only at currents sufficient to excite rostral medullary neurons, *Brain Res.*, 321, 127, 1984.
91. Vanegas, H., Barbaro, N.M., and Fields, H. L., Tail-flick related activity in medullospinal neurons, *Brain Res.*, 321, 135, 1984.
92. Vasko, M.R., Pang, I.H., and Vogt, M., Involvement of 5-hydroxytryptamine-containing neurons in antinociception produced by injection of morphine into nucleus raphe magnus or onto spinal cord, *Brain Res.*, 306, 341, 1984.
93. Watkins, L.R., Cobelli, D.A., and Mayer, D.J., Classical conditioning of front paw and hind paw footshock induced analgesia (FSIA): naloxone reversibility and descending pathways, *Brain Res.*, 243, 119, 1982.
94. Watkins, L.R., Wiertelak, E.P., Goehler, L.E., Mooney-Heiberger, K., Martinez, J., Furness, L., Smith, K.P., and Maier, S.F., Neurocircuitry of illness-induced hyperalgesia, *Brain Res.*, 639, 283, 1994.
95. Watkins, L.R., Wiertelak, E.P., Goehler, L.E., Smith, K.P., Martin, D., and Maier, S.F., Characterization of cytokine-induced hyperalgesia, *Brain Res.*, 654, 15, 1994.
96. Wiertelak, E.P., Roemer, B., Maier, S.F., and Watkins, L.R., Comparison of the effects of nucleus tractus solitarius and ventral medial medulla lesions on illness-induced and subcutaneous formalin-induced hyperalgesias, *Brain Res.*, 748, 143, 1997.
97. Willer, J.C., Boureau, F., and Albe-Fessard, D., Supraspinal influences on nociceptive flexion reflex and pain sensation in man, *Brain Res.*, 179, 61, 1979.
98. Willis, Jr., W.D., Anatomy and physiology of descending control of nociceptive responses of dorsal horn neurons: comprehensive review, *Prog. Brain Res.*, 77, 1, 1988.
99. Willis, W.D., Gerhart, K.D., Willcockson, W.S., Yezierski, R.P., Wilcox, T.K., and Cargill, C.L., Primate raphe- and reticulospinal neurons: effects of stimulation in periaqueductal gray or VPLc thalamic nucleus, *J. Neurophysiol.*, 51, 467, 1984.
100. Xu, W., Cui, X., and Han, J.S., Spinal serotonin IA and IC/2 receptors mediate supraspinal mu opioid-induced analgesia, *NeuroReport*, 5, 2665, 1994.
101. Yaksh, T.L., Direct evidence that spinal serotonin and noradrenaline terminals mediate the spinal antinociceptive effects of morphine in the periaqueductal gray, *Brain Res.*, 160, 180, 1979.
102. Yaksh, T.L., al-Rodhan, N.R., and Jensen, T.S., Sites of action of opiates in production of analgesia, *Prog. Brain Res.*, 77, 371, 1988.

103. Yaksh, T.L. and Tyce, G.M., Microinjection of morphine into the periaqueductal gray evokes the release of serotonin from spinal cord, *Brain Res.,* 171, 176, 1979.
104. Yamuy, J., Sampogna, S., Lopez-Rodriguez, F., Luppi, P.H., Morales, F.R., and Chase, M.H., Fos and serotonin immunoreactivity in the raphe nuclei of the cat during carbachol-induced active sleep: a double-labeling study, *Neuroscience,* 67, 211, 1995.
105. Zhuo, M. and Gebhart, G.F., Characterization of descending inhibition and facilitation from the nuclei reticularis gigantocellularis and gigantocellularis pars alpha in the rat, *Pain,* 42, 337, 1990.
106. Zhuo, M. and Gebhart, G.F., Spinal serotonin receptors mediate descending facilitation of a nociceptive reflex from the nuclei reticularis gigantocellularis and gigantocellularis pars alpha in the rat, *Brain Res.,* 550, 35, 1991.
107. Zhuo, M. and Gebhart, G.F., Characterization of descending facilitation and inhibition of spinal nociceptive transmission from the nuclei reticularis gigantocellularis and gigantocellularis pars alpha in the rat, *J. Neurophysiol.,* 67, 1599, 1992.

Chapter 30

Electrophysiology of Spinal Sensory Processing in the Absence and Presence of Surgery and Anesthesia

P. Max Headley, Boris A. Chizh, Juan F. Herrero, and Nick A. Hartell

Contents

30.1	Introduction	506
30.2	Methods	507
	30.2.1 Single Motor Unit Recording in Rats Subjected to Minimum Surgery	507
	30.2.2 Recording of Dorsal Horn Activity in Anesthetized Rats	508
	30.2.3 Single Spinal Neuron Recording in Awake or Anesthetized Sheep	508
	30.2.4 *In Situ, In Vitro* Recording of Spinal Function in Mouse	509
30.3	Results and Discussion	509
	30.3.1 Effects of Surgery on Spinal Neuronal Responsivity	509
	30.3.1.1 Single Motor Unit Recordings	509
	30.3.1.2 Dorsal Horn Neurons in Surgically Prepared, Anesthetized Rats	510
	30.3.1.3 Single Spinal Neuron Responses in Awake Sheep	511
	30.3.2 Effects of Anesthetics on Spinal Neuronal Responsivity	511
	30.3.2.1 Single Motor Unit Recordings	511
	30.3.2.2 Single Neuron Recording in Conscious vs. Halothane-Anesthetized Sheep	512
	30.3.3 Sympathetic Responsiveness and Motor Activity	514
30.4	Conclusions	516
References		517

30.1 Introduction

One of the earlier suggestions that the state of spinal sensory processing could be altered by prior sensory inputs was by Livingston,[44] who postulated that "reverberating circuits" could be set up by prior nociceptive afferent inputs and that these might explain the persistence of pain states beyond the time course of the originating stimulus. Subsequent behavioral and electrophysiological investigations have confirmed that spinal processing can be markedly altered by the prior sensory experience of the spinal cord, as reviewed by Woolf and Chong.[67] These considerations have progressively raised the issue as to whether the preparative surgery required in the great majority of cases to permit access to CNS tissue for electrophysiological recording does not itself bias the recordings obtained. Findings that central processing can be altered within minutes of a peripheral noxious stimulus[66] confirmed the likelihood that sensory processing will be different after preparative surgery as compared with the normal unoperated state. Although there is long-standing evidence that this occurs (e.g., see References 2 and 63), there have been rather few specific investigations of the effects of preparative surgery. One such, by Clarke and Matthews,[9,10] provided clear evidence that certain stages of surgical preparation and mounting for electrophysiological recording were responsible for the previously noted differences in the threshold of jaw opening reflexes in behavioral studies as compared with electrophysiological studies. The brainstem site of recording did, however, preclude investigation of those effects occurring within the trigeminal system as opposed to those imposed by descending controls from "higher" centers.

One means for avoiding the complexities arising from surgical preparation is to study chronically prepared animals in which sufficient time has elapsed since surgery for normal processing to become re-established. There have been numerous attempts at developing chronic implant systems for recording from spinal neurons, but, because of formidable technical difficulties, few groups have maintained the initial impetus. One highly productive group is that of Collins,[11] who has developed a preparation permitting recordings of spinal dorsal horn neurons in conscious cats and has thereby made a substantial contribution to the understanding of neuronal responsiveness under conscious conditions.[12–18,55] There is, however, one potential limitation of this type of preparation, namely that the animals require considerable training to withstand the investigations of receptive field properties of the neuron under investigation, without making responsive movements against the fixation of the spine. This training may itself be a problem. With training, by definition, some aspect of the sensorimotor processing cascade is altered. There is no definitive information as to where this happens, but a likely consequence is altered descending control of spinal processes. It follows that highly trained animals may not reflect accurately the normal degree of spinal neuronal responsiveness. There are therefore potential (though to date unproven) advantages in recording from less trained and/or restrained animals, though inevitably some degree of habituation to handling is required for stable recordings to be feasible.

The neurotransmitter mechanisms mediating the sensitization process underlying hyperalgesia and allodynia have come under close scrutiny over the last decade. Whilst these processes are clearly multi-factorial, one transmitter receptor has received particular attention — the receptor that recognizes the glutamate analogue N-methyl-D-aspartate (NMDA).[23,31,46,67] The process of wind-up,[47] which has been taken to reflect the changes underlying hyperalgesia,[23] has been shown to be mediated by NMDA receptors.[21,24] These receptors are particularly suited to triggering longer term changes in response to barrages of sensory information. First, the voltage dependence of the Mg^{++} block of the glutamate-gated channel results in a positive feedback of inward current generated by activation of glutamate receptors; as the membrane depolarizes, Mg^{++} leaves the channel, and glutamate generates progressively more inward current. Second, the channel is permeable to Ca^{++}, which can in turn result in longer term changes of intracellular biochemistry. Consequently, strong noxious stimuli that trigger barrages of sensory traffic will elicit sufficient release of L-glutamate in the cord to overcome the Mg^{++} block and will result in increased intracellular Ca^{++} in those cells receiving nociceptive synaptic inputs. This cascade provides a neat explanation for the onset of hyperalgesic states. However, not all electrophysiological or indeed clinical data are consistent with such a simple story.

Since the turn of the century, it has been clear that exogenously administered compounds can affect spinal transmission. Since then evidence has mushroomed to indicate that a wide variety of chemicals can influence spinal sensory (and motor) processing, whether reaching the spinal cord following direct application or by vascular routes. One such group of agents are the anesthetics. Two aspects of the anesthetic triad of loss of consciousness, analgesia, and motor relaxation could have spinal components: clearly loss of consciousness cannot be mediated at this level, but both sensory inputs and motor outputs need to be mediated via the spinal cord (or brainstem equivalents) and so have the potential to be modulated by anesthetics at this level. More emphasis has been placed on the effects on sensory transmission.

Kitahata's group in Seattle in the 1960s (reviewed in Reference 41) was instrumental in showing that various anesthetics could affect different classes of spinal dorsal horn neuron at concentrations relevant to a surgical plane of anesthesia. Subsequent work has elaborated these actions and has confirmed that anesthetics do alter sensory processing at the spinal cord level.[15,20,33,41] Just as with the considerations above of preparative surgery, this finding causes difficulties for the electrophysiologist working with *in vivo* preparations; the surgery requires, on ethical grounds, that animals be rendered non-sentient during recording sessions, and the most common means for achieving this is with anesthesia. Decerebration is an alternative, but destruction of major areas of forebrain does not leave a normal CNS for investigation; in particular, activity in descending pathways is known to change with time following decerebration. The presence of anesthetics at surgically relevant concentrations does not, of course, dampen all spinal activity; nociceptive reflexes can still be elicited. Moreover, some processes appear to be resistant to the presence of anesthetics, such as the induction of proto-oncogenes, including *c-fos*.[65]

The uncertainties generated by such considerations have triggered us to develop various preparations that have allowed us to start untangling some of the components involved in the actions of surgery and anesthesia. Such findings have relevance not only to the interpretation of electrophysiological data, but also to those clinical situations in which anesthesia and surgery go hand in hand.

To investigate the effects of preparative surgery one requires a starting point involving less or no surgery. In our first such investigation, we recorded single motoneuron activity in the form of single motor unit spikes recorded via percutaneous needle electrodes; this permitted us to avoid the normal levels of surgery to the spinal column, but still required the animals to be immobilized. The next stage was to develop a preparation that permitted recordings from animals that were conscious. To avoid the requirements applying to the use of cats in such experiments (namely, that lumbar vertebrae must be fused to reduce mobility and that animals require extensive training; see above), we used sheep. These preparations did not, however, permit either stable intracellular recording or control of the extracellular medium, aspects that are necessary to investigate the molecular events underlying sensory processing. We have therefore recently developed a preparation of trunk and hindquarters of adult mouse that can be perfused with all the advantages of an *in vitro* preparation but that retains the integrated and functional afferent and efferent connections of an *in situ* preparation.

30.2 Methods

30.2.1 Single Motor Unit Recording in Rats Subjected to Minimum Surgery

Under halothane anesthesia, male Wistar rats were implanted with arterial, venous, and tracheal cannulae. The spinal cord and column were left intact, or the cord was exposed at low thoracic level via a small laminectomy performed under local anesthetic infiltration (sham spinalization), or the animals were spinalized (again with local anesthetic infiltration). Under subsequent immobilization either with an anesthetic or following midcollicular decerebration, single motor unit (SMU) spikes were recorded using bipolar tungsten wire electrodes housed in 27-gauge hypodermic needles inserted percutaneously into hindlimb flexor muscles. Different groups of animals were investigated with either different anesthetics or different degrees of surgery to the spinal column and cord. Nociceptive reflex responses were elicited with computer-controlled devices that permitted either noxious pinch or thermal stimuli, or trains of electrical stimuli, to be applied in regular cycles of constant and defined intensity and duration. Counts of actions potentials elicited during phases of each stimulus were logged onto computer and were analyzed on line in terms of percentage changes from the mean of the last three pre-test control cycles. Full details are given by Hartell and Headley.[27,29]

30.2.2 Recording of Dorsal Horn Activity in Anesthetized Rats

As a baseline comparison for data in the other preparations, studies have been performed in more traditional preparations of α-chloralose-anesthetized male Wistar rats. After insertion of tracheal, jugular vein, and carotid artery cannulae, an extensive laminectomy is performed between thoracic segment 9 and lumbar segment 4. The spinal cord is either left intact or spinalized at low thoracic level. Seven-barrel glass micropipettes are used for extracellular recording and iontophoretic administration of neurotransmitter analogues and antagonists into the region of the cell being recorded. Responses can be elicited in regular cycles by admixtures of iontophoretic administration of excitatory agents and by activation of peripheral receptive fields with noxious thermal and mechanical or non-noxious tap or vibration stimuli. Evoked responses are quantified and analyzed as above. Full details are given in References 5 and 32.

30.2.3 Single Spinal Neuron Recording in Awake or Anesthetized Sheep

Mature sheep were implanted in an initial surgical operation under halothane anesthesia. A titanium ring baseplate was fixed to one or two lumbar vertebrae, using dental acrylic fixed with bone screws, and with no immobilization of other intervertebral joints. A 20-mm diameter laminectomy was performed through the acrylic. Onto this baseplate was screwed a tube that permanently penetrated the skin. The laminectomy could be sealed within the tube, and an outer cap closed the tube. This permitted a tight double seal between recording session but also ready access to the intact dura for recording sessions. Following surgery, animals were treated with analgesic (buprenorphine and flunixin) for as long as was deemed appropriate by standard veterinary criteria (usually less than a week). Full details are given by Herrero et al.[34]

No overt training was required; however, animals were habituated over a few days to standing in a crate and to being stroked. For recording, they were restrained in a sling in the crate, with their feet just touching or just off the floor; this was done so as to prevent them from lying down during recording sessions. As with all animals, individual temperament varied, but most animals habituated to this procedure within a very few sessions. Food and water were available *ad libitum* at all times. Animals frequently ruminated during recording sessions, an important indicator of lack of stress. The cord was exposed, and a manipulator assembly inserted into the tube and fixed to the baseplate. A stepper microdrive was used to drive a glass-coated tungsten microelectrode through the dura mater and into the cord. The manipulator could be angled and rotated so that access was available over the whole exposed 20-mm length of cord. Penetrations were performed under direct visualization using an arthroscope probe inserted through a guide hole in the microelectrode drive. The electrodes were necessarily rather blunt to be strong enough to penetrate the dura. While this will have biased the units recorded, single-unit activity could readily be picked up. Recording stability depended on the temperament of the animal and on the vigor and nature of search stimuli applied in investigating receptive fields. Unit responsiveness was determined using a range both of calibrated von Frey filaments and of forceps that applied calibrated intensities of non-noxious to noxious squeeze. Spike firing rate and receptive field size were mapped. The same animals were used for conscious recording sessions and for sessions in which the animal was anesthetized with halothane in N_2O/O_2. Penetration of the dura remained possible for up to 4 months after the initial implantation.

30.2.4 *In Situ, In Vitro* Recording of Spinal Function in Adult Mouse

We have adapted a preparation of working heart and brainstem recently developed by Paton[52,53] to provide an isolated trunk-hindquarters preparation. Adult mice were deeply anesthetized with urethane i.p., were eviscerated, and were then bisected through the thorax. Upon immediate transfer of the hind part of the animal to the recording chamber, the descending aorta was cannulated and perfusion established with standard ACSF with added oncotic agent (dextran). Recordings have to date been made from afferent and efferent somatic and sympathetic nerves using suction electrodes, as well as from single neurons within the cord using sharp electrodes for extracellular or intracellular recording.[7]

30.3 Results and Discussion

30.3.1 Effects of Surgery on Spinal Neuronal Responsivity

Comparisons of results from three of the preparations described above permits us to deduce some aspects that are, and others that are not, appreciably affected by the preparative surgery.

30.3.1.1 Single Motor Unit Recordings

Comparison of results from animals prepared with the minimum surgery consistent with a low thoracic spinalization (performed under local anesthetic infiltration) with animals on which an extensive laminectomy had been performed permits certain conclusions. At relatively constant levels of intensity of pinch stimuli (applied with a pneumatic device to one toe) the greater level of surgery resulted in small (but nonetheless significant) increases of the evoked motoneuron discharge rate.[29] Despite this rather small change, there was an appreciable increase in the sensitivity of the nociceptive reflexes to pharmacological agents of various classes. These included the injectable anesthetics methohexitone, α-chloralose, and the steroid mixture of alphaxalone/alphadolone (Saffan; Mallinckrodt Veterinary),[29] the dissociative anesthetic ketamine (which is also an uncompetitive NMDA antagonist[1]),[29,30] and the μ-opioid analgesics fentanyl and morphine.[26,35] These results clearly showed that an increase in preparative surgery above a baseline level resulted in changes in the sensitivity of spinal reflex processes to a variety of agents that are specific (ketamine, fentanyl) or less specific (barbiturates, α-chloralose) in affecting neurotransmitter systems. The simplest (though incomplete) conceptual explanation for this increased sensitivity to agents of diverse pharmacology is that there is an increase in the activity level of spinal interneurons bearing the appropriate receptors; if the activity is increased, so is the potential for suppression.

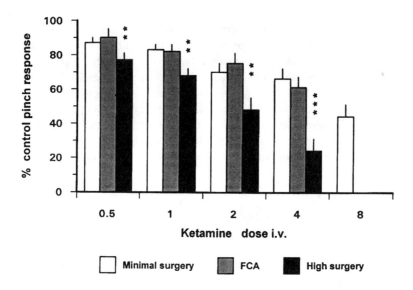

FIGURE 30.1
The contribution of NMDA receptors to spinal nociceptive reflexes varies with the degree of preceding surgery. Single motor unit activity, recorded in α-chloralose anesthetized rats, was evoked by controlled noxious pinch stimuli applied to one paw at 3-min intervals. Tests were performed in three groups of rats: those prepared with a minimal degree of surgery (n = 19, open bars), those which in addition had an inflamed paw caused by Freund's Complete Adjuvant (FCA) injection 4 days earlier (n = 5, gray bars), and those with normal paws but an extensive laminectomy (n = 13, black bars). The NMDA antagonist ketamine was administered intravenously in cumulative doses. Ketamine was significantly more effective in those animals with a greater degree of preparative surgery, whereas inflammation of one paw did not have this effect. Asterisks indicate significant difference between the two surgical groups: **$p < 0.01$, ***$p < 0.001$. (See references 28 and 30.)

An example of the changes seen is shown in Figure 30.1, which shows pooled data taken from three groups of rats: one prepared with minimal surgery to the spine, one with a extensive lumbar laminectomy as used for acute dorsal horn recordings, and one with inflammation of one hind paw induced by Freund's Complete Adjuvant injected locally 4 days previously. Cumulative doses of ketamine were given i.v., and its effects on responses were monitored as a percentage of the control response, which in this case was evoked by a pneumatically controlled pincher device. It is apparent that in animals with the greater surgery there is a marked increase (about fourfold) in the efficacy of the NMDA antagonist. Perhaps surprisingly, however, 4 days of inflammation of the paw on which the receptive field was sited did not have the same effect.[28]

30.3.1.2 Dorsal Horn Neurons in Surgically Prepared, Anesthetized Rats

The greater effectiveness of ketamine in the above experiments mirrors current hypotheses, mentioned above, that NMDA receptors become progressively more effective following barrages in nociceptive afferent nerves. As intimated above, these considerations imply that responses elicited by higher levels of afferent input would display a greater NMDA receptor-mediated component than responses to lower intensity stimuli. We have recently examined this in standard spinal cord preparations (i.e., high surgery preparation and anesthesia, in this case with α-chloralose) by testing the sensitivity to NMDA antagonists of responses elicited by alternating high and low intensities of noxious pinch or thermal stimuli. The firing rates evoked by the higher intensity stimulus were roughly double those elicited by the lower intensity. The results of this study[5] were surprising. Systemically administered NMDA antagonists reduced the low-frequency background discharge to a greater degree than the high firing rate discharges evoked by noxious stimuli applied to the receptive field, and this held whether the antagonist was uncompetitive (ketamine) or competitive

(CPP, 3-((R))-2-carboxypiperazine-4-yl-propyl-1-phosphonic acid). Moreover, when the background activity level was maintained constant by iontophoretic ejection of an excitatory amino acid from one barrel of the multi-barrel pipette, the NMDA antagonists did not alter evoked responses at all. Wind-up[47] elicited by electrical stimulation of the receptive field was, however, still reduced by NMDA antagonists. These results indicate that, under these conditions of surgery and anesthesia, NMDA receptors mediate the low level background discharge of neurons but not the much higher frequency discharges elicited by phasic noxious stimuli. Moreover, wind-up and strong nociceptive responses, despite driving neurons to similar firing rates, function differently with respect to NMDA receptors. A definitive explanation for these findings is so far elusive, and a plausible explanation may involve more than one mechanism. One possibility is that in different pathways there is differential NMDA receptor subunit composition with resulting differences in Mg^{++}-sensitivity and hence voltage dependence (see References 48 and 60). Another possibility is that prolonged nociceptive inputs (here, from the areas of surgery) cause the release of neuropeptides and that the latter subsequently upregulate the properties of NMDA receptors/channels[6,19,54,61,62] that are involved in the synaptic pathways from the areas of damage. Other pathways would not be affected, including those activated in our experiments by brief noxious stimuli (here, the test stimuli) applied to receptive fields well away from the area of surgery. These reflex and dorsal horn results therefore indicate that sensitization processes can be rather specific to particular pathways and may depend on the pre-existing state of the relevant area of CNS. Consequently, changes in NDMA receptor function are more complex than is often assumed.

30.3.1.3 Single Spinal Neuron Responses in Awake Sheep

Following the single motor unit rat study, there remained uncertainties concerning the spinal effects of anesthesia and surgery. First, these recordings were of motor outputs, not of sensory pathways, and, second, there was still a basal level of surgical intervention and in most cases presence of basal anesthetic. The results could therefore not predict the behavior of spinal neurons under normal, intact conditions. The results obtained in conscious sheep addressed some of these aspects, as recordings were obtained at least one week, and up to 4 months, after the initial implant operation.

We were somewhat surprised how similar the properties of spinal neurons appeared to be in these preparations of conscious sheep as compared with standard, high-surgery, anesthetized rat preparations.[36,37] The types of cells recorded (balance of WDR and low-threshold mechanoreceptive neurons) and the nature of their receptive fields reflected rather closely what is seen in standard acute preparations. This reflects the similar neuronal responsiveness seen in our previous reflex recordings independent of the degree of preparative surgery, but contrasts with the differences we had seen when testing anesthetic actions in acute experiments with different degrees of surgery (see also below).

30.3.2 Effects of Anesthetics on Spinal Neuronal Responsivity

Comparison of data gathered in these various preparations also permits conclusions on the manner in which anesthetics, at clinically relevant doses, may alter spinal function.

30.3.2.1 Single Motor Unit Recordings

We initially made a comparison of the potency of four injectable anesthetics (methohexitone, α-chloralose, alphaxalone/alphadolone, and ketamine) in terms of behavioral effects in intact rats with their effects on spinal nociceptive reflexes in spinalized rats.[27] It was clear that all agents tested had clear effects on nociceptive reflexes in spinalized animals at doses at or close to those causing sedation/anesthesia in behavioral experiments. This is not surprising, as many general anesthetics are known to interact with $GABA_A$ mechanisms,[40,43,45,56,57] and GABA is an important inhibitory mediator in the spinal cord; alternatively, dissociative anesthetics such as ketamine are NMDA antagonists,[1] and NMDA is known to affect various sensory and motor processes in the spinal cord.

The question arises as to whether any one anesthetic is preferable to any other as a baseline anesthetic for spinal neuronal recording. Some conclusions can be drawn. Ketamine is not a suitable primary anesthetic. In lower vertebrates, the anesthetic dose is very much higher than the NMDA blocking dose,[27] so NMDA function will be totally suppressed for hours following induction with high doses of this agent; its half-life is between about 10 min and an hour in different laboratory species (see Headley et al.[32] for references). Barbiturate and steroid anesthetics are effective enhancers of $GABA_A$ mediated inhibitions,[45,57] and this will distort the balance of spinal processing. We have used α-chloralose. While this agent, like others, is capable of enhancing GABA actions[49] and causes marked suppression in decerebrate animals,[27] it had minimal effects at the doses used as supplements in animals themselves anesthetized with α-chloralose; this, in our experience, is unlike barbiturates. It therefore results in a stable level of suppression over a range of doses, which provides some safeguard against spurious observations arising from variations in anesthetic depth. The influence of surgical trauma on the degree of suppression of spinal reflexes caused by several anesthetics has already been addressed above.

These results clearly showed that anesthetics affect spinal reflex processing at behaviorally relevant doses and that these effects will distort observations based on electrophysiological studies designed to assess spinal physiological function. Similar data have been obtained with dorsal horn recordings in acute experiments on α-chloralose-anesthetized rats. Comparison between different cell types does, however, show that anesthetic actions can be rather specific; for instance, non-nociceptive responses of dorsal horn convergent neurons are more reduced than are similar responses of neurons receiving only low threshold inputs (Dong, X.-W. and Headley, P.M.[25] and unpublished data).

30.3.2.2 Single Neuron Recording in Conscious vs. Halothane-Anesthetized Sheep

To assess more directly the effects of anesthesia on spinal function, we used our sheep preparation to compare the responsiveness of spinal neurons in the fully awake state with neurons recorded in the same animals under conditions of stable halothane anesthesia. The ideal protocol would be to follow the same neuron between conscious and anesthetized states. This is technically extremely difficult with inhalational anesthetics; even with injectable anesthetics, it has been performed rarely,[51] and we have succeeded on rather few occasions. Our report[36] was therefore a population comparison between the two states. Two interesting observations resulted.

First, the balance between cell types (and the apparent basic responsiveness of cell types) was very similar in awake and halothane-anesthetized sheep. Of particular note was that wide dynamic range (WDR, or convergent) neurons were as prevalent under awake conditions as under halothane anesthesia. This finding is at variance with the chronic cat study of Collins and Ren[16] in which WDR cell types were rather seldom encountered in the absence of anesthesia. It seems likely that this difference between the two chronic preparations results from different levels of tonic activity in descending pathways, which could in turn be accounted for by several mechanisms. One possibility is that the difference reflects the much greater degree of training required for the cat preparations; another that the cats tend towards sleep during recordings, a state that Soja[58] has shown can markedly affect activity in ascending tracts (see Chapter 31). In our study we did, however, observe some differences between awake and halothane anesthetized animals. Halothane anesthesia resulted in a small reduction of ongoing activity and an increased threshold of WDR units. Perhaps unexpectedly, the receptive field (RF) size was larger under anesthesia, an observation opposite to that reported for the effects of halothane in acute experiments.[22] It is likely that

Electrophysiology of Spinal Sensory Processing

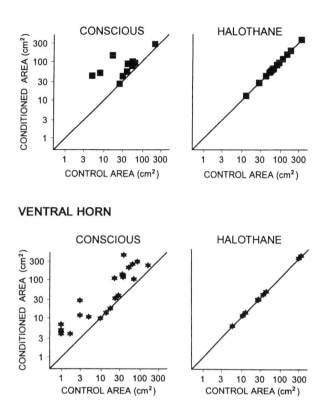

FIGURE 30.2
Dynamic changes of receptive field (RF) size can be triggered within seconds by non-noxious stimuli in conscious animals, but this effect is suppressed under conditions of anesthesia. Recordings were made from unidentified dorsal and ventral horn neurons in the lumbar spinal cord of chronically prepared sheep. The receptive field on the hindquarters was mapped for each neuron with von Frey filaments. Following measurement of the RF area under control conditions, conditioning was applied in the form of 30 to 60 sec of gentle manual rubbing or scratching of the RF. In awake animals (lefthand panels) this non-noxious conditioning elicited appreciable increases in RF area with most neurons. The charts plot the control RF area (x-axis) against the RF area following conditioning (y-axis) for dorsal horn (upper) and ventral horn neurons (lower panels). Note the logarithmic scale; in percentage terms, RF area was increased in awake animals to up to 1000% control. When neurons were recorded in the same animals under temporary halothane anesthesia (righthand panels), no such sensitization occurred (all points fall close to the line of unity slope). In both states, similar effects were seen on dorsal and ventral horn neurons. (See also Reference 38.)

in our spinally intact animals the increased RF size resulted from a reduction of descending inhibitory controls as compared to the awake state.

Second, a potentially very important observation was that under conscious conditions RF sizes were readily modified by low-threshold afferent inputs.[38] It is well established from recordings in acute preparations that noxious stimulation of peripheral tissues results in sensitization and enlargement of receptive field size of spinal neurons.[67] In previous studies, non-noxious stimuli failed to have any such effect. In conscious sheep, however, a few seconds of stroking or mild scratching of the skin caused a substantial enlargement of RF size. The magnitude of this effect is illustrated in Figure 30.2, which compares the pooled data from neurons tested with this conditioning stimulation in conscious vs. halothane anesthetized sheep. The magnitude of the effect in conscious animals and the total lack of this effect under anesthesia are impressive. Presumably this phenomenon has not been seen before simply because it is totally suppressed by anesthesia. The fact that it can occur in awake animals indicates that RF size and therefore neuronal sensitivity can be modulated dynamically depending on the level and nature of sensory input arriving at the spinal cord. This input does not need to be nociceptive. It may therefore be that altered RF size is more an indication of attentional state than of pathological change. Not all sensitization, however, was blocked by halothane; sensitization could still be elicited by noxious levels of stimulation, as would be expected. For ethical reasons, however, such conditioning stimuli could not be tested under conscious conditions, so the effects of the anesthetic on the magnitude of this effect could not be tested.

30.3.3 Sympathetic Responsiveness and Motor Activity

Undoubtedly sensory processing in the spinal cord can be better understood if there is related information on patterns of activity in the three different outflows — somatic motor, sympathetic, and ascending. While segmental motor reflexes and ascending activity have been studied extensively, the spinal sympathetic outflow in response to nociceptive inputs has received rather less attention. It seems likely, although there is little direct information, that preparative surgery would affect sympathetic responses to noxious stimuli, just as we have seen with somatic motor outputs. There is, however, a belief that much of the sympathetic response to noxious stimulation is mediated supraspinally, as suggested by reports that spinalization dramatically reduces sympathetic activity.[39] Other studies indicate that the spinal cord is itself capable of generating sympathetic activity,[50,64] although this activity can be severely affected by spinal transection. The depression of sympathetic function seen in spinalized animals has been attributed to "spinal shock",[39] during which the fall in blood pressure and secondary changes in the hemodynamics of spinal cord perfusion create a positive feedback loop. Furthermore, anesthetics are known to depress sympathetic activity.[39]

We have recently devised an arterially perfused preparation of the trunk and hindquarters of adult mouse.[7] This preparation has several advantages over *in vivo* preparations that make it particularly suitable for the study of neural transmission in the spinal cord. First, it provides mechanical stability for intracellular recording; second, it permits control over the extracellular medium, thus allowing additions of test agents that would be toxic *in vivo*; third, as a result of constant perfusion pressure, it provides stability of circulation in the spinal cord; fourth, it avoids the complexities of supraspinal loops, descending modulatory influences, and depressant effects of anesthesia.[59,64] The considerable advantages over other *in vitro* preparations are, first, that it is of adult tissue and, second, that it still permits as much control of sensory stimulation and monitoring of segmental outputs as an *in vivo* preparation. Altogether, the preparation shows a remarkable degree of responsiveness to various peripheral cutaneous and visceral stimuli. Mechanical, thermal, or electrical stimulation of the tail or hindlimb evokes brisk motor reflexes which with more intense stimulation show integrated escape responses. Under conditions of muscle paralysis, such activity can be recorded in neurograms (Figure 30.3A–C).

We have now used this preparation to make direct comparisons of activity patterns in somatic motor and sympathetic outflows and have found that there is coupling between the two segmental outflows.[8] Parallel recording of somatic motor and sympathetic nerves indicated that noxious somatic stimuli, as well as visceral stimuli such as bladder distention, elicit coupled responses in somatic motor and in sympathetic outputs (Figure 30.3B,C). The quantification of this coupling required analysis over many seconds. While such prolonged activity could be achieved by repeated and intense sensory stimulation, such inputs would feed similarly into the different outflows, obscuring any other mechanisms that might underlie coupling. In view of the fictive locomotion that can be generated by NMDA in neonatal preparations in the absence of sensory inputs,[4] we tested NMDA by perfusion. In our preparation, it evoked stepping movements of the hindlimbs or, after

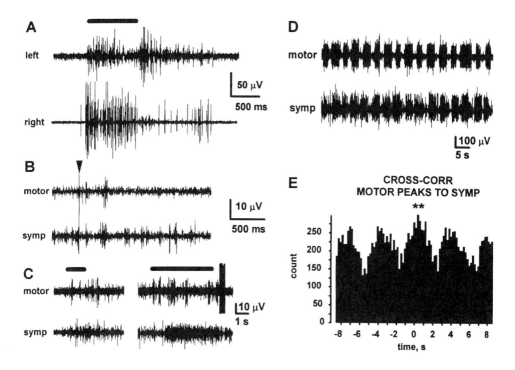

FIGURE 30.3
Brisk somatic and sympathetic responses can be evoked by peripheral stimulation *in vitro* in a new arterially perfused trunk-hindquarters preparation of adult mouse. Under some conditions the somatic motor and sympathetic outflows become coupled. (**A**) Mechanical stimulation of the tail (thick horizontal bar above traces) resulted in bilateral reflexes in flexor muscles, recorded as mulit-unit EMG. (**B**) Electrical stimulation of the tail (arrowhead) elicited responses in somatic motor (upper trace) and also in concurrently recorded sympathetic nerves (lower trace; multi-unit recordings from femoral nerve and abdominal sympathetic chain respectively). (**C**) Parallel recording of motor (femoral, upper trace) and sympathetic nerves (abdominal sympathetic trunk, lower trace) shows similar responses to noxious pinch of the tail (left) and bladder distension stimuli (20 mmHg above resting pressure, right). The timing of the stimuli is shown by the thick horizontal lines. (B) and (C) are from the same preparation. (**D**) Arterial perfusion with ACSF containing NMDA 40 (µM) resulted in regular stepping movements; following muscle paralysis, rhythmic multi-unit activity can be recorded in motor nerves (upper trace), and this activity is intermittently closely reflected in sympathetic nerves (lower trace). (**E**) Cross-correlation analysis (using the peaks of somatic motor nerve bursts as the trigger point for a histogram of sympathetic nerve activity) shows temporal coupling of the rhythm in the two outflows shown in (D). The asterisks indicate that the peak at time 0 reached 2 S.D.s above the mean level of sympathetic activity during this recording segment. (Traces in A through C reprinted from Chizh, B.A. et al., *J. Neurosci. Meth.*, 76, 177, 1998. With permission.)

paralysis, concentration-dependent rhythmic discharges in somatic motor nerves (Figure 30.3D). It is most likely that this activity is generated in motor areas of the cord, as has been established for fictive locomotion in neonatal preparations.[3,42] Parallel recordings of the two outlows indicated close similarlities in discharge patterns (Figure 30.3D); cross-correlation analysis confirmed that sympathetic activity was coupled to somatic motor activity (Figure 30.3E). This cross-talk is unlikely to be due to proprioceptive feedback because for these parallel recordings a paralyzing agent was used to prevent gross movements. Interestingly, in the absence of strong motor rhythm, the cross-correlation between the sympathetic and somatic motor activity was principally unidirectional, with a higher proportion of motor bursts having correlated sympathetic bursts than vice versa. These results have therefore uncovered a modulation of some proportion (but not all) of sympathetic outflow by motor activity. Moreover, the coupling is evidently under dynamic control, as periods of strong coupling (such as that shown in Figure 30.3D,E) alternate with periods of rather weaker coupling. Stronger coupling could be re-established by a burst of afferent sensory input.

These data indicate that the spinal cord is capaple of coordinating and modulating activity in sympathetic and somatic motor outflows independently of supraspinal mechanisms. The substantial degree of coupling between somatic motor and sympathetic outflows is under dynamic control and can be influenced by segmental inputs. This coupling of the two outputs may provide a spinally located mechanism for state-dependent modulation of sympathetic function. This would be relevant, for example, in exercise, where altered cardiovascular control is required in parallel with altered motor activity. It has long been known that some aspects of the cardiovascular response to exercise start within seconds of the onset of exercise; it may be that spinal mechanisms contribute to this. It is also notable that motor and sympathetic abnormalities are associated in some forms of "sympathetically maintained pain" or "reflex sympathetic dystrophy". It will be fascinating if the sort of coupling unearthed here could contribute to such pathology.

30.4 Conclusions

The studies summarized above address some of the issues surrounding the contribution of surgery and anesthesia to the processing of spinal sensory inputs and of motor and sympathetic outputs. There is a plentiful literature, much deriving from the 1960s, that anesthetics can influence spinal neurons; however, little of this older data was quantified, and it was certainly not compared to the situation in chronically prepared awake animals. Some of the conclusions remain the same but some are altered by the more recent results.

There is now a very large literature on the sensitization processes that can be triggered by peripheral noxious stimuli; it is therefore not surpising that surgery affects spinal function to a substantial degree. Equally there is a large literature on the role of NMDA receptors in this sensitization process, and increasingly there are clinical reports on the effects of NMDA antagonists in the prevention or treatment of hyperalgesic states. Just how and where the NMDA receptors are involved in the sensitization process, is, however, less clear. Our data are in some respects counterintuitive; one would have predicted that NMDA receptors would contribute increasingly as response amplitude increases and neuronal depolarization is greater, simply due to relief of the voltage-dependent block by Mg^{++} of NMDA channels. That NMDA receptors mediate low-level background activity but not the much more vigorous discharges elicited by intense stimuli appears inconsistent with known mechanisms. While there are various possible explanations in terms of NMDA subunit composition, differences in membrane voltage in different cellular compartments, etc., the most plausible explanation seems to us to involve the co-release of neuropeptides under conditions of continued, but not with phasic, noxious stimuli. Such peptides, the precise identity of which remains undefined, could modulate NMDA receptor/channel function such as to increase the NMDA contribution in selected pathways.

Anesthetics of several classes clearly have direct actions at the level of the spinal cord; this is a logical finding considering their known interactions with neurotransmitter systems that are known to operate in the spinal cord. Those anesthetics that enhance $GABA_A$ receptor-mediated mechanisms (barbiturates, steroids) have very clear depressant effects on spinal neuronal responsiveness. The dissociative anesthetics that are NMDA antagonists cause a variety of effects at NMDA blocking doses, but these are well below the anesthetic dose in lower mammals such as rodents. Gaseous anesthetics, the neurotransmitter interactions of which are rather less certain, appear to have relatively little action on the basal properties of spinal neurons (no change in the balance of cell types and only small changes in sensitivity to sensory stimulation). However, more subtle functions can be much more seriously affected. One example that we studied is the dynamic change in receptive field size following different intensities of sensory stimulation. Halothane effectively abolished the dynamic control of RF size that was triggered by non-noxious conditioning stimulation in awake animals; this dynamic control over RF size seems likely to reflect altered attentional state following unusual (but not necessarily unpleasant) stimulation. Whether such attention is at conscious or subconscious levels remains to be determined, although anecdotal observation indicates that sheep did not alter their feeding or cudding behavior when this effect was seen.

The effectiveness and significance of spinal neuronal processing can only really be interpreted by observation of outputs, whether in ascending tracts, as behavioral responsiveness, or as motor or sympathetic reactions. The study of sympathetic responsiveness to sensory inputs has been hampered by difficulties in distinguishing spinal and supraspinal components of such responses, following from the substantial loss of sympathetic function following spinal section in the species most used in sympathetic studies, the cat. Rodents do not display such profound "spinal shock" and therefore may be better suited to such studies. The isolated yet *in situ* trunk-hindquarters preparation of mouse that we have recently developed has several extra benefits of permitting stable intracellular recording and in controlling the extracellular medium. Most important so far, however, has been the ability to generate somatic motor activity independent of sensory inputs, and thereby to investigate the possible links between somatic and sympathetic outflows. The coupling of activity seen from somatic to sympathetic under conditions of enhanced somatic activity ("fictive" locomotion) suggests that sympathetic activity may be influenced during the state of exercise by activity in spinal motor pools, as well as by sensory and descending influences. This teleologically attractive observation opens up new concepts of the states that can control sympathetic outflow. Given the known links between sympathetic function and some chronic pain states, the role of motor activity in affecting sympathetic function may have important sensory as well as autonomic implications.

These studies serve to emphasize that if the fluctuations of spinal neuronal activity that must underlie various functional, attentional, and clinical states are to be understood, a variety of approaches is required, and the results from such different techniques must be integrated. The mechanisms underlying these fluctuations are likely to be multiple, varying from rapid modifications by physiological control mechanisms, through less rapid changes triggered by complex neuromediator release, to overt pathological changes under conditions of tissue damage; moreover, these processes are modulated by the direct spinal actions of a wide variety of pharmaceutical agents. There also appears to be dynamic control of the balance of activity in the three classes of spinal output — ascending, motor, and sympathetic — and this is likely to have important functional consequences under different behavioral states. These observations pose as many questions as they answer. To understand these relationships more fully we shall need to define not only the anatomical and synaptic mechanisms linking spinal neurons, and the inherent cell properties that influence the ability to integrate synaptic inputs onto neurons of different functional groups, but also the network properties that influence the distribution of spinally generated, afferent and descending activity. This will be a formidable challenge.

References

1. Anis, N.A., Berry, S.C., Burton, N.R., and Lodge, D., The dissociative anaesthetics, ketamine and phencyclidine, selectively reduce excitation of central mammalian neurones by *N*-methyl-D-aspartate, *Br. J. Pharmacol.*, 79, 565, 1983.
2. Brown, A.G., Effects of descending impulses on transmission through the spinocervical tract, *J. Physiol.*, 219, 103, 1971.
3. Cazalets, J-R., Borde, M., and Clarac, F., The synaptic drive from the spinal locomotor network to motoneurons in the newborn rat, *J. Neurosci.*, 16, 298, 1996.
4. Cazalets, J.R., Sqalli-Houssaini, Y., and Clarac, F., Activation of the central pattern generators for locomotion by serotonin and excitatory amino acids in neonatal rat, *J. Physiol.*, 455, 187, 1992.

5. Chizh, B.A., Cumberbatch, M J., Herrero, J.F., Stirk, G.C., and Headley, P.M., Stimulus intensity, cell excitation and the N-methyl-D-aspartate receptor component of sensory responses in the rat spinal cord in vivo, Neuroscience, 80, 251, 1997.

6. Chizh, B.A. and Headley, P.M., Thyrotropin-releasing hormone facilitates spinal nociceptive responses by potentiating NMDA receptor mediated transmission, Eur. J. Pharmacol., 300, 183, 1996.

7. Chizh, B.A., Headley, P.M., and Paton, J.F.R., An arterially perfused trunk-hindquarters preparation of adult mouse in vivo, J. Neurosci. Meth., 76, 177, 1997.

8. Chizh, B.A., Headley, P.M., and Paton, J.F.R., Coupling of sympathetic and somatic motor outflows from the spinal cord in a perfused preparation of adult mouse in vitro, J. Physiol., 508, 907, 1998.

9. Clarke, R.W. and Matthews, B., The effects of anaesthetics and remote noxious stimuli on the jaw-opening reflex evoked by tooth-pulp stimulation in the cat, Brain Res., 327, 105, 1985.

10. Clarke, R.W. and Matthews, B., The thresholds of the jaw opening reflex and trigeminal brainstem neurons to tooth pulp stimulation in acutely and chronically prepared cats, Neuroscience, 36, 105, 1990.

11. Collins, J.G., Neuronal activity recorded from the spinal dorsal horn of physiologically intact, awake, drug-free, restrained cats: a preliminary report, Brain Res., 322, 301, 1984.

12. Collins, J.G., Effects of ketamine on low intensity tactile sensory input are not dependent upon a spinal site of action, Anesth. Analg., 65, 1123, 1986.

13. Collins, J.G., Inhibition of spontaneous activity of spinal dorsal horn neurons in the intact cat is naloxone-insensitive, Brain Res., 401, 95, 1987.

14. Collins, J. G., A descriptive study of spinal dorsal horn neurons in the physiologically intact, awake, drug-free cat, Brain Res. 416, 34, 1987.

15. Collins, J.G., Kendig, J.J., and Mason, P., Anesthetic actions within the spinal cord: contributions to the state of general anesthesia, Trends Neurosci., 18, 549, 1995.

16. Collins, J.G. and Ren, K., WDR response profiles of spinal dorsal horn neurones may be unmasked by barbiturate anaesthesia, Pain, 28, 369, 1987.

17. Collins, J.G., Ren, K., Saito, Y., Iwasaki, H., and Tang, J., Plasticity of some spinal dorsal horn neurons as revealed by pentobarbital-induced disinhibition, Brain Res., 525, 189, 1990.

18. Collins, J.G., Ren, K., and Tang, J., Lack of spontaneous activity of cutaneous spinal dorsal horn neurones in awake, drug free, spinally transected cats, Exp. Neurol., 96, 299, 1987.

19. Cumberbatch, M.J., Chizh, B.A., and Headley, P.M., Modulation of excitatory amino acid responses by tachykinins and selective tachykinin agonists in the rat spinal cord, Br. J. Pharmacol., 115, 1005, 1995.

20. Davidoff, R.A. and Hackman, J.C., Drugs, chemicals, and toxins: their effects on the spinal cord, in Handbook of the Spinal Cord. Vol. 1. Pharmacology, Davidoff, R.A., Ed., Marcel Dekker, Basel, 1983, p. 409.

21. Davies, S.N. and Lodge, D., Evidence for involvement of N-methylaspartate receptors in "wind-up" of class 2 neurones in the dorsal horn of the rat, Brain Res., 424, 402, 1987.

22. de Jong, R.H., Robles, R., and Morikawa, K.-I., Actions of halothane and nitrous oxide on dorsal horn neurons ("the spinal gate"), Anesthesiology, 31, 205, 1969.

23. Dickenson, A.H., A cure for wind-up: NMDA receptor antagonists as potential analgesics, Trends Pharmacol. Sci., 11, 307, 1990.

24. Dickenson, A.H. and Sullivan, A.F., Evidence for the role of the NMDA receptor in the frequency dependent potentiation of deep rat dorsal horn nociceptive neurones following C fibre stimulation, Neuropharmacology, 26, 1235, 1987.

25. Dong, X.-W. and Headley, P.M., How selective are anaesthetics between sensory responses of dorsal horn neurones of the convergent and low threshold types?, Neurosci. Lett. (Suppl.), 38, S28, 1990.

26. Hartell, N.A. and Headley, P.M., Effects of surgical trauma on the direct spinal action of anaesthetic and analgesic agents, *Pain,* 5(Suppl.), S203, 1990.
27. Hartell, N.A. and Headley, P.M., Spinal effects of four injectable anaesthetics on nociceptive reflexes in rats: a comparison of electrophysiological and behavioural measurements, *Br. J. Pharmacol.,* 101, 563, 1990.
28. Hartell, N.A. and Headley, P.M., NMDA receptor involvement in mechanically and electrically evoked spinal reflexes in normal and mono-arthritic rats, *Abstr. Soc. Neurosci.,* 21, 214.11, 1991.
29. Hartell, N.A. and Headley, P.M., Preparative surgery enhances the direct spinal actions of three injectable anaesthetics in the anaesthetized rat, *Pain,* 46, 75, 1991.
30. Hartell, N.A. and Headley, P.M., NMDA receptor contribution to spinal nociceptive reflexes: influence of stimulus parameters and of preparative surgery, *Neuropharmacology,* 35, 1567, 1996.
31. Headley, P.M. and Grillner, S., Excitatory amino acids and synaptic transmission: the evidence for a physiological function, *Trends Pharmacol. Sci.,* 11, 205, 1990.
32. Headley, P.M., Parsons, C.G., and West, D.C., The role of *N*-methylaspartate receptors in mediating responses of rat and cat spinal neurones to defined sensory stimuli, *J. Physiol.,* 385, 169, 1987.
33. Heavner, J.E., Jamming spinal sensory input: effects of anesthetic and analgesic drugs in the spinal cord dorsal horn, *Pain,* 1, 239, 1975.
34. Herrero, J.F., Coates, T.W., Higgins, M., Livingston, A., Waterman, A.E., and Headley, P.M., A technique for recording from spinal neurones in awake sheep, *J. Neurosci. Meth.,* 46, 225, 1993.
35. Herrero, J.F. and Headley, P.M., The effects of sham and full spinalization on the systemic potency of μ- and κ-opioids on spinal nociceptive reflexes in rats, *Br. J. Pharmacol.,* 104, 166–170, 1991.
36. Herrero, J.F. and Headley, P.M., Cutaneous responsiveness of lumbar spinal neurons in awake and halothane-anesthetized sheep, *J. Neurophysiol.,* 74, 1549, 1995.
37. Herrero, J.F. and Headley, P.M., The dominant class of somatosensory neurone recorded in the spinal dorsal horn of awake sheep has wide dynamic range properties, *Pain,* 61, 133, 1995.
38. Herrero, J.F. and Headley, P.M., Sensitization of spinal neurons by non-noxious stimuli in the awake but not anesthetized state, *Anesthesiology,* 82, 267, 1995.
39. Jänig, W., Organization of the lumbar sympathetic outflow to skeletal muscle and skin of the cat hindlimb and tail, *Rev. Physiol. Biochem. Pharmacol.,* 102, 121, 1985.
40. Kirkness, E.F., Steroid modulation reveals further complexity of $GABA_A$ receptors, *Trends Pharmacol. Sci.,* 10, 6, 1989.
41. Kitahata, L.M., Modes and sites of "analgesic" action of anesthetics on the spinal cord, *Int. Anesthesiol. Clin.,* 13, 149, 1975.
42. Kjaerulff, O., Barajon, I., and Kiehn, O., Sulphorhodamine-labelled cells in the neonatal rat spinal cord following chemically induced locomotor activity *in vitro, J. Physiol.,* 478, 265, 1994.
43. Lambert, J.J., Peters, J.A., and Cottrell, G.A., Actions of synthetic and endogenous steroids on the $GABA_A$ receptor, *Trends Pharmacol. Sci.,* 8, 224, 1987.
44. Livingston, W.K., *Pain Mechanisms,* Macmillan, New York, 1943.
45. Macdonald, R.L. and Olsen, R.W., $GABA_A$ receptor channels, *Ann. Rev. Neurosci.,* 17, 569, 1994.
46. Meller, S.T. and Gebhart, G.F., Nitric oxide (NO) and nociceptive processing in the spinal cord, *Pain,* 52, 127, 1993.
47. Mendell, L.M., Physiological properties of unmyelinated fibre projection to the spinal cord, *Exp. Neurol.,* 16, 316, 1966.
48. Momiyama, A., Feldmeyer, D., and Cull-Candy, S.G., Identification of a native low-conductance NMDA channel with reduced sensitivity to Mg^{2+} in rat central neurones, *J. Physiol.,* 494, 479, 1996.
49. Nicoll, R.A. and Wojtowicz, J.M., The effects of pentobarbital and related compounds on frog motoneurons, *Brain Res.,* 191, 225, 1980.

50. Osborn, J.W., Livingstone, R.H., and Schramm, L.P., Elevated renal nerve activity after spinal transection: effects on renal function, *Am. J. Physiol.*, 253, R619, 1987.
51. Ota, K., Yanagidani, T., Hinds, A., Cannan, S., Kishikawa, K., and Collins, J.G., Halothane reduces spinal dorsal horn neuronal response to low-threshold receptive field stimulation, *Abstr. Soc. Neurosci.*, 19, 1197, 1993.
52. Paton, J.F.R., A working heart-brainstem preparation of the mouse, *J. Neurosci. Meth.*, 65, 63, 1996.
53. Paton, J.F.R., The ventral medullary respiratory network of the mature mouse studied in a working heart-brainstem preparation, *J. Physiol.*, 493, 819, 1996.
54. Randic, M., Kolaj, M., Kojic, Lj., Cerne, R., Cheng, G., and Wang, R.A., Interaction of neuropeptides and excitatory amino acids in the rat superficial dorsal horn, *Prog. Brain Res.*, 104, 225, 1995.
55. Saito, Y., Collins, J.G., and Iwasaki, H., Tonic 5-HT modulation of spinal dorsal horn neuron activity evoked by both noxious and non-noxious stimuli: a source of neuronal plasticity, *Pain*, 40, 205, 1990.
56. Sieghart, W., $GABA_A$ receptors: ligand-gated Cl-ion channels modulated by multiple drug-binding sites, *Trends Pharmacol. Sci.*, 13, 446, 1992.
57. Sieghart, W., Structure and pharmacology of gamma-aminobutyric acid A receptor subtypes, *Pharmacol. Rev.*, 47, 181, 1995.
58. Soja, P.J., Oka, J.-I., and Fragoso, M., Synaptic transmission through cat lumbar ascending sensory pathways is suppressed during active sleep, *J. Neurophysiol.*, 70, 1708, 1993.
59. Taylor, R.F. and Schramm, L.P., Differential effects of spinal transection on sympathetic nerve activities in rats, *Am. J. Physiol.*, 253, R611, 1987.
60. Tölle, T.R., Berthele, A., Zieglgänsberger, W., Seeburg, P.H., and Wisden, W., The differential expression of 16 NMDA and non-NMDA receptor subunits in the rat spinal cord and in periaqueductal gray, *J. Neurosci.*, 13, 5009, 1993.
61. Urban, L., Thompson, S.W.N., and Dray, A., Modulation of spinal excitability: co-operation between neurokinin and excitatory amino acid neurotransmitters, *Trends Pharmacol. Sci.*, 17, 432, 1994.
62. Urban, L., Thompson, S.W.N., Fox, A.J., Jeftinija, S., and Dray, A., Peptidergic afferents: physiological aspects, *Prog. Brain Res.*, 104, 255, 1995.
63. Wall, P.D. and Werman, R., The physiology and anatomy of long-ranging afferent fibres within the spinal cord, *J. Physiol.*, 255, 321, 1976.
64. Weaver, L.C. and Stein, R.D., Effects of spinal cord transection on sympathetic discharge in decerebrate-unanesthetized cats, *Am. J. Physiol.*, 257, R1506, 1989.
65. Williams, S., Evan, G.I., and Hunt, S.P., Changing patterns of c-*fos* induction in spinal neurons following thermal cutaneous stimulation in the rat, *Neuroscience*, 36, 73, 1990.
66. Woolf, C.J., Evidence for a central component of post-injury pain hypersensitivity, *Nature*, 306, 686, 1983.
67. Woolf, C.J. and Chong, M.-S., Pre-emptive analgesia — treating postoperative pain by preventing the establishment of central sensitization, *Anesth. Analg.*, 77, 362, 1993.

Chapter 31

Transmission Through Ascending Trigeminal and Lumbar Sensory Pathways: Dependence on Behavioral State

Peter J. Soja, Brian E. Cairns, and Morten P. Kristensen

Contents

31.1	Introduction		522
31.2	Studies of Neurons in the Trigeminal Sensory Nuclear Complex		522
	31.2.1	Sensory Neurons of the Trigeminal Sensory Nuclear Complex in Acute Animal Preparations	522
	31.2.2	Chronically Instrumented Animal Preparations: Methodological Approaches to Brainstem Unit Recording	523
	31.2.3	Activity of Trigeminal Sensory Nuclear Complex Neurons in Awake Animals	524
	31.2.4	Afferent and Descending Modulation of Trigeminal Sensory Neurons in Chronically Instrumented Animal Preparations	525
	31.2.5	Modulation of Trigeminal Neurons Across the Sleep/Wakefulness Cycle	525
	31.2.6	Mechanisms Modulating Trigeminal Sensory Nuclear Complex Neuron Function During Active Sleep	528
		31.2.6.1 Presynaptic Inhibition of Primary Afferent Terminals	528
		31.2.6.2 Postsynaptic Inhibition	529
		31.2.6.3 Presynaptic Control of Trigemino-Thalamic Tract Afferent Terminals	529
31.3	Studies of Spinal Cord Sensory Neurons		530
	31.3.1	Characteristics of Dorsal Spinocerebellar Tract Neurons	530
	31.3.2	Chronically Instrumented Animal Preparations: Methodological Approaches to Spinal Cord Unit Recording	531
	31.3.3	Characteristics of Lumbar Sensory Neurons in Awake Intact Preparations	532

	31.3.4	Modulation of Lumbar Sensory Neurons in Chronically Instrumented Animal Preparations 533
	31.3.5	Activity of Lumbar Sensory Neurons Across the Sleep/Wakefulness Cycle 534
	31.3.6	Mechanisms Modulating Lumbar Sensory Neuron Function During Active Sleep 534
31.4	Conclusions	 536
Acknowledgments		 538
References		 538

31.1 Introduction

Detailed overviews of the basic anatomical and physiological characteristics of the major ascending sensory pathways in the trigeminal sensory nuclei and spinal cord and their respective forms of descending control have been presented in Chapters 27 and 28. The majority of the surveyed studies of sensory transmission have been carried out in anesthetized preparations that preclude analysis of modulation under natural physiological and behavioral conditions. Despite a wealth of knowledge in the fields of sensory physiology and pain research, an important and challenging task is to understand how sensory neurons, including those located within the trigeminal sensory nuclear complex (TSNC) and lumbar spinal cord, are modulated under specific behavioral circumstances.

Activity of CNS neurons during sleep states has been intensely investigated using extra- and intracellular recording techniques (see Hobson and Steriade[56] and Steriade and McCarley[114]). For this purpose, chronically instrumented, unanesthetized animal preparations have been used to assess the activity of a variety of cell types, such as monoaminergic,[61,87] thalamic,[114] and cortical[30,114] neurons, as well as somatic motoneurons.[21] Compared to these cellular systems, lumbar and trigeminal sensory neurons, per se, have been the subject of very few electrophysiological investigations during wakefulness and sleep.

This chapter will describe what is currently known regarding state-specific modulation of ascending sensory transmission by outlining electrophysiological recording studies in the TSNC (Section 31.2.) and spinal cord (Section 31.3.) of chronically instrumented, unanesthetized animal preparations. Sensory tracts that arise from these regions receive a multitude of ascending and descending input and convey an array of sensory modalities. Thus, comparison of mechanisms that modulate activity of trigeminal and lumbar cells will afford an initial evaluation of the existence of general principles for regulating ascending sensory transmission. Analysis of the neural circuits activated during specific behavioral elements and the role of these pathways in modulating the organism's own sensory inputs are only beginning to be unraveled (see Chapter 29). Progress in this research area will require the continued use of chronic, undrugged intact animal preparations to obviate the need for anesthetic agents which inherently distort multiple aspects of sensory neuron processing (see Chapter 30).

31.2 Studies of Neurons in the Trigeminal Sensory Nuclear Complex

31.2.1 Sensory Neurons of the Trigeminal Sensory Nuclear Complex in Acute Animal Preparations

Early studies performed in acutely prepared, anesthetized animal preparations indicate that the subnuclei of the rostral TSNC, i.e., main sensory nucleus, nucleus oralis, and anterior portion of nucleus interpolaris, are principally involved in transmission of light mechanical and tactile stimuli from facial receptive fields.[34,38,42,46,70,74] In addition, there are several lines of evidence that support a role for the rostral TSNC in transmission of nociceptive information, particularly from the oral cavity.[124,126] Thus, neurons in the rostral TSNC are capable of responding not only to mechanoreceptor input, e.g., hair movement, but also to afferent input from the tooth pulp and facial cutaneous receptive fields at stimulus intensities considered to be nociceptive (see Chapter 27).

The teeth are a common source of pain in the oral cavity.[100] Anatomical studies as well as experiments utilizing antidromic stimulation in the brainstem indicate that small-diameter afferent fibers from the tooth pulp terminate at all levels of the ipsilateral TSNC in the cat; however, rostral areas receive the densest projections.[36,78] Furthermore, a significant proportion of the neurons in nucleus oralis and the main sensory nucleus, which respond to electrical stimulation of tooth pulp afferents, project to the contralateral ventrobasal thalamus.[3,103,104,115] This projection pattern suggests that the rostral TSNC serves as an important relay of ascending somatosensory information emanating from the teeth.

It is important to emphasize that much of what is known about the activity of TSNC neurons has been garnered from studies performed in anesthetized animals. However, studies by Matthews and colleagues[8,23] indicate that anesthetics and/or trauma induced by surgical procedures which are necessary to perform such experiments may distort the responses of most rostral trigeminal neurons to perioral and facial inputs (see also Chapter 30). Additionally, it is clearly not possible to investigate behavior-related modulation of trigeminal pathways in acute, anesthetized animals. These limitations provide a rationale for performing single-unit recording studies in unanesthetized, intact animal preparations.

31.2.2 Chronically Instrumented Animal Preparations: Methodological Approaches to Brainstem Unit Recording

In contrast to investigations of individual TSNC neurons in acute, anesthetized animal preparations, corresponding studies performed in chronic, intact animal preparations are relatively few in number. Nevertheless, studies have been performed in both the awake cat[7,8,99] and monkey,[11,39,40,50,57,82] and different types of stimuli, including electrical, thermal, or mechanical stimulation of peripheral receptive fields, have been used to evoke responses in TSNC cells. Evoked responses have been tested during various sub-states of wakefulness as a baseline and compared with responses obtained during specific behavioral states or elements, e.g., feeding, attention, sleep, or anesthesia. Studies performed in cats have focused principally on rostral trigeminal sensory neurons, whereas those in monkeys have been directed at caudal trigeminal sensory neurons.

Several techniques have been used to record activity of TSNC neurons in unanesthetized preparations. Banks et al.[5] as well as Boissonade and Matthews[7] reported on their procedures for recording single-unit activity of trigeminal sensory neurons in the chronic, freely moving animal. Cats were implanted with a chamber on the skull overlying the cerebellar surface. Electrode leads were also implanted into the canine tooth pulps to permit orthodromic stimulation of the cells under investigation during recording sessions.[6] In our laboratory, we have utilized cats that additionally were implanted with head-restraining devices and electrodes for recording behavioral indices of wakefulness and sleep.[14,108] The animal's head is painlessly restrained in the original stereotaxic position assumed during the surgical implant procedure, and the cats readily sleep in the experimental setup after a short adaptation period. Guided placement of a recording electrode into the TSNC in the chronically instrumented cat preparation is achieved by recording characteristic, peripherally evoked field potentials.[13,14,16]

Dubner and colleagues[39,40,50,57] utilized rhesus monkeys (*Macaca mulatta*) that were seated in chairs and had been trained to perform tasks involving detection of thermal and visual stimuli. Once training had been accomplished, a series of surgical implant procedures were performed, including fusing the first four cervical vertebrae, implanting stimulation electrodes in the ventro-posterio-medial thalamus (VPM), and attaching a recording chamber to the occipital cranium to provide head restraint and to allow lowering of recording electrodes to the medullary dorsal horn. Following recovery from these procedures, the animals were placed in the restraint apparatus, and neurons located in the medullary trigeminal nucleus caudalis were recorded before and during thermal and visual detection tasks. As the experimental approaches summarized above indicate, there are presently several techniques and species available for investigating TSNC neuron activity in awake animals.

31.2.3 Activity of Trigeminal Sensory Nuclear Complex Neurons in Awake Animals

Characterization of neurons located in the rostral TSNC of unanesthetized cats has so far been limited principally to those neurons that are activated by pulpal afferents.[8,13,14] In each of these studies, orthodromic input was provided by low-intensity electrical stimuli ($\leq$100 µA; duration $\leq$ 0.2 ms) applied to pulpal afferents. Such stimuli often evoke one or more short-latency ($\geq$3 ms) action potentials and occasionally also trigger a second, long-latency (25 ms) spike discharge in these cells. In the absence of pulpal stimuli, the majority of these neurons are silent.[8,14,99]

As has previously been described for the anesthetized cat,[3] rostral TSNC neurons activated by tooth pulp afferents in the chronically instrumented cat preparation fall into two general categories: stimulus intensity-dependent and -independent cells.[14] Stimulus intensity-dependent neurons respond with a graded spike discharge to increasing stimulus intensities. Conversely, stimulus intensity-independent neurons do not demonstrate graded responses to supra-threshold stimuli. Thus, many of the properties of these rostral TSNC neurons are similar to those found for neurons recorded in the acute cat preparation.[11,23,33,46,60,104,121] It is not clear, however, whether the stimulus-dependent neurons found in studies analyzing electrically evoked responses are equivalent to the wide-dynamic-range cells observed in recordings of responses to natural stimulus modalities.[11,49]

Neurons comprising ascending sensory pathways such as the trigemino-thalamic tract (TGT) and reflex pathways to trigeminal motoneurons are located within the TSNC. TGT neurons have been antidromically identified by applying low-intensity stimuli to the contralateral VPM of both awake cats[16] and monkeys.[11] The estimated mean axonal conduction velocity of these cells is relatively slow (cat, 12.9 m/s;[16] monkey, 14.7 m/s[11]) and of similar magnitude to conduction velocities measured in anesthetized animals.[10,41]

The majority of TGT neurons, in contrast to most other units recorded in the rostral TSNC, display ongoing spontaneous activity (mean rate: 12 ± 1 spikes per s) in the awake cat in the absence of any experimentally applied stimulus.[16] Nevertheless, TGT neurons that lack ongoing spike discharge and respond to stimulation of pulpal afferents are also found in the rostral TSNC.[16] About one third of TGT neurons in the rostral TSNC respond vigorously to hair-mechanoreceptor input elicited by air puffs directed toward orofacial receptive fields.[16] The latter finding supports the suggestion that the rostral TSNC plays an important role in the transmission of low-threshold mechanical input from facial receptive fields.

In the awake monkey, TGT neurons within subnucleus caudalis respond to low-threshold mechanical stimuli applied to facial receptive fields.[11] Additionally, TGT neurons which show either graded responses to increasingly intense mechanical stimuli or respond only to high-intensity mechanical stimuli applied to the face have also been found.[11] A portion of the latter type of TGT neurons also respond to noxious thermal stimuli (>45°C).[11]

These investigations provide quintessential confirmation of findings in acute studies of TSNC modality specificity, as the response profiles of caudal TSNC neurons in monkeys and rostral TSNC neurons recorded in the cat do not differ qualitatively when results obtained from awake and anesthetized preparations are compared. Moreover, the results emphasize the diversity of TSNC neurons, including within the TGT cell sub-population, and of their responses to peripheral stimuli. This inhomogeneity of TSNC cells must be taken into account when conducting experiments and interpreting data from studies of state-specific modulation of this region and its efferent pathways.

31.2.4 Afferent and Descending Modulation of Trigeminal Sensory Neurons in Chronically Instrumented Animal Preparations

A considerable body of literature describes afferent modulation of trigeminal sensory neurons in anesthetized animals, as established in Chapter 27, but very little parallel work has been undertaken in awake animals. With respect to afferent modulation, it is well documented in anesthetized cats that TSNC neuronal activity evoked by electrical stimulation of the pulp of one tooth can be suppressed or occasionally enhanced when a conditioning stimulus is applied to the pulp of another ipsi- or contralateral tooth, referred to here as homosynaptic conditioning (HC).[35,69,90,106] In chronically instrumented, awake cats, HC stimuli exert similar effects on TSNC neuron activity (Figure 31.1A).

The time course of suppression and facilitation of tooth pulp-evoked TSNC neuronal activity in chronic, awake cats as a result of afferent conditioning is illustrated in Figure 31.1C. Increases in tooth pulp-evoked activity over baseline appear for conditioning test intervals ranging from 10 to 80 ms with the peak effect of conditioning occurring at 30 ms. Decreases below baseline in the tooth pulp-evoked activity appear in other cells for conditioning test intervals ranging from 10 to 50 ms, again with the peak effect of conditioning occurring at 30 ms. A significant inverse correlation exists between latency-to-onset and HC-induced modulation of tooth pulp-evoked spike activity (Figure 31.1B; see also Reference 15). This result indicates that during HC-induced suppression and enhancement there is a lengthening and shortening, respectively, of the latency-to-onset of tooth pulp-driven TSNC neuron spike activity. It has been shown that HC stimuli depolarize tooth pulp afferent terminals in the anesthetized cat,[37,77] as well as in the awake cat,[13] with maximum effects appearing around 30 ms conditioning test intervals. This similarity of time courses is consistent with presynaptic inhibitory processes mediating (portions of) the HC-induced suppression of TSNC neurons; however, the pharmacological basis for this form of afferent modulation is not known.

Many acute studies have demonstrated that trigeminal sensory neuron activity can be potently suppressed by electrical stimulation of various brainstem sites (see Chapter 27). In contrast, phasic modulation of trigeminal sensory neurons induced by brainstem stimulation has not been reported for unanesthetized animals. It is important to investigate whether these types of regulatory influences impinge on TGT or TSNC neurons in a state-specific manner, as such influences can contribute to modulation of ascending sensory transmission during sleep.

31.2.5 Modulation of Trigeminal Neurons Across the Sleep/Wakefulness Cycle

A simple method for assessing aggregate TSNC excitability changes across behavioral state is to measure alterations of electrical field potentials within the region. Field potentials evoked by electrical stimulation of nerve trunks or receptive fields represent the summed activity of a population of trigeminal sensory neurons in the vicinity of the electrode tip. Decreases in the peak amplitude of the field potential (e.g., as a result of conditioning stimuli applied to peripheral or central sites or caused by alterations in behavioral state) may reflect a decreased excitability of individual neurons. Hence, monitoring field potentials can provide useful information on overall, state-dependent activity changes of trigeminal sensory neurons. Inferior alveolar nerve- or tooth

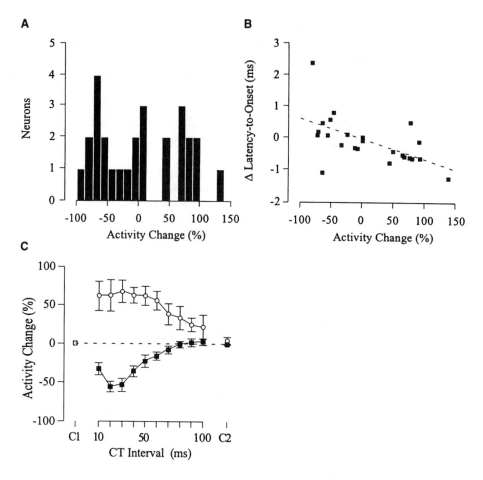

FIGURE 31.1
Afferent control of trigeminal sensory neurons in the awake cat as indicated by effects homosynaptic conditioning (HC) stimuli applied to one branch on excitability of another branch of the trigeminal nerve. Test stimuli (threshold intensity) were applied to the maxillary (upper) tooth pulp in conjunction with HC stimuli applied to the mandibular (lower) tooth pulp, or visa versa, while postsynaptic responses were recorded from individual cells within the rostral trigeminal sensory nuclear complex (TSNC). The histogram in (**A**) illustrates the tooth pulp-evoked neuronal responses of 27 neurons after HC at a conditioning test (CT) interval of 30 ms relative to baseline responses before conditioning. Overall, the application of HC stimuli resulted in ≈15% suppression in tooth pulp-evoked activity. The graph in (**B**) illustrates the relationship between HC-induced changes in tooth pulp-evoked neuronal activity and latency-to-onset of the first action potential. Overall, HC-induced modulation of neuronal activity was reciprocally correlated ($r = -0.71$) with the HC-induced change in the latency-to-onset. The graph in (**C**) depicts the relationship between CT intervals and magnitude of tooth pulp-evoked responses of TSNC neurons. The mean relative tooth pulp-evoked activity was determined during and after (C2) the application of HC stimuli at the indicated CT intervals and plotted as a proportion of baseline responses to control stimuli applied before HC onset (C1). Note that suppression (n = 6; filled squares) of neuronal activity was evident at CT intervals between 10 and 60 ms. In contrast, enhancement (n = 6; open circles) of TSNC neuronal activity had not returned to baseline after a CT interval of 100 ms. (Error bars: ±S.E.). For further details regarding the experimental paradigm, see References 13 to 15.

pulp-evoked field potentials recorded within the rostral TSNC during quiet sleep, a state characterized by synchronized electroencephalographic activity, minimal eye movements, and moderate muscle tone, do not differ from those obtained during preceding or subsequent episodes of wakefulness.[14] On the other hand, field potentials evoked by stimulation of facial receptive fields are enhanced during quiet sleep relative to the awake state.[51] However, when cats enter active sleep, characterized by a desynchronized electroencephalographic pattern, robust eye movements, and muscle atonia, field potentials orthodromically driven from either source are tonically and markedly

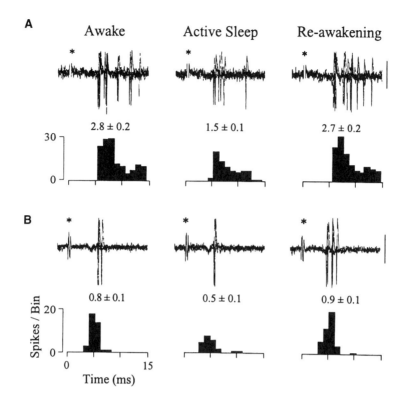

FIGURE 31.2
Active sleep-related suppression of a stimulus intensity-dependent neuron (**A**) and stimulus intensity-independent neuron (**B**) recorded in the rostral trigeminal sensory nuclear complex of an unanesthetized cat. Spike activity for each type of neuron, as illustrated in the sample oscilloscope traces, was evoked by threshold stimulation of the inferior canine tooth pulp (A: 16 µA, 0.2 ms, 1.0 Hz; B: 6.3 µA, 0.2 ms, 1.0 Hz). Each post-stimulus histogram was constructed from 50 consecutive trials during wakefulness, active sleep, and subsequent re-awakening (binwidth: 1 ms). Values above histograms denote average discharge during the 15-ms period after stimulus onset (spikes per trial; ±S.E.). Note that tooth pulp-evoked activity for the stimulus intensity-dependent (A) and -independent (B) neurons was suppressed by 45 and 41%, respectively, during active sleep relative to wakefulness (see also Reference 14). (Vertical calibration for oscilloscope traces, A: 150 µV; B: 250 µV.)

suppressed when compared to quiet sleep.[51] Moreover, the field potential is often further suppressed or abolished during phasic episodes of active sleep that are characterized by saccadic eye movements and somatic muscle twitches.[14]

In addition to overall field potential responses, the tooth pulp-evoked discharge rates of a majority of individually recorded, stimulus intensity-dependent and stimulus intensity-independent neurons located in the rostral TSNC are also suppressed during active sleep when compared to preceding or subsequent episodes of wakefulness (Figure 31.2; see also Reference 14). In this study, the population mean of tooth pulp-evoked response for rostral TSNC neurons decreased 19% during active sleep from preceding episodes of wakefulness. In addition, during phasic, rapid-eye-movement events of active sleep, the tooth pulp-evoked field potentials and unitary responses further decreased or completely disappeared.[14] These data indicate that during active sleep certain TSNC neurons are subjected to dynamic forms of suppression.

The majority of TGT neurons located in the main sensory nucleus and nucleus oralis pars$_\gamma$ are spontaneously active during wakefulness and quiet sleep.[16] Although the group mean firing rate for the recorded population as a whole does not differ between wakefulness and active sleep, individual TGT neurons undergo considerable changes in firing patterns during active sleep. Furthermore, during phasic, rapid-eye-movement events, most TGT neurons display pauses in their ongoing spike activity followed by rebound bursts, whereas other cells exhibit enhancement of ongoing firing rates.[73]

In addition to ongoing activity, peripheral nerve-evoked responses of individual TGT neurons have been monitored across the sleep/wakefulness cycle.[16] Responses evoked by stimulation of the inferior alveolar nerve (which contains large- and small-diameter afferents) increase or decrease during active sleep when compared to wakefulness, while the population mean appears unaffected. The dualistic nature of TGT cell responses to stimulation of the inferior alveolar nerve may, in part, be attributable to differences in the properties of their afferent input with respect to sensory modality and/or fiber size. When the smaller diameter fibers of the tooth pulp are used to activate TGT neurons, the tooth pulp-evoked responses recorded from each cell are markedly suppressed.[16] In contrast, when natural forms of stimuli (solenoid-controlled air puffs) are used to activate large diameter afferent fibers, i.e., those innervating perioral or facial hair mechanoreceptors, the response magnitude of TGT neurons is markedly enhanced during active sleep.[16] These data suggest the presence of a gating mechanism which is engaged during active sleep and may function in accordance with the gate-control theory proposed by Melzack and Wall[88] in that it operates in a fiber diameter-specific manner.

In summary, studies of TSNC field potential and single-unit responses to peripheral nerve or receptive field stimulation along with recordings of ongoing activity of identified cells demonstrate that trigeminal sensory neurons are subjected to profound state-dependent regulation. Moreover, the nature of state-specific modulation differs between TSNC cells and may depend on fiber size and/or modality of afferent input as well as the identity of individual neurons, including the targets of their axonal projections. The underlying physiological mechanisms are just beginning to be elucidated, as described in the next part.

31.2.6 Mechanisms Modulating Trigeminal Sensory Nuclear Complex Neuron Function During Active Sleep

31.2.6.1 Presynaptic Inhibition of Primary Afferent Terminals

Sensory transmission through the TSNC may be controlled by presynaptic inhibitory processes at the primary afferent level.[4,37] Experiments performed in acute, anesthetized cats have provided a valuable foundation for examining whether presynaptic control mechanisms operate on primary afferent terminals during active sleep. One methodological approach employs stimuli applied to rostral or caudal portions of the TSNC to elicit compound or single antidromic unit potentials in the ipsilateral mandibular or maxillary tooth pulps.[36,37,77,79,125] Conditioning stimuli applied to branches of the trigeminal nerve or to brainstem nuclei such as the nucleus raphe magnus result in depolarization of tooth pulp afferent terminals which suggests that the terminal arbors of tooth pulp afferents become less responsive to an incoming volley conducted along the conditioned axon. Such primary afferent depolarization (PAD) in the TSNC produced by other afferents of peripheral and central origins is believed to operate in a manner similar to that described for primary afferents in the spinal cord.[98] The exact biophysical mechanism linking PAD with a reduction in transmitter release is still unclear; however, substantial evidence supports the idea that the inhibitory neurotransmitter, GABA, is responsible for PAD in spinal cord Ia afferents.[32,76,98]

During active sleep, the majority of individual tooth pulp afferent terminals depolarize as evidenced by leftward shifts in their antidromic excitability curves (see also Section 31.2.6.3).[13] Hyperpolarization occurs in a minority of tooth pulp afferent terminals. Moreover, suppression of tooth pulp-evoked TSNC neuron activity during active sleep is inversely correlated with active sleep-related depolarization of tooth pulp afferent terminals.[13] This finding suggests that disfacilitatory presynaptic mechanisms are engaged on individual tooth pulp afferent terminals during active sleep and may contribute to suppression of trigeminal sensory transmission during this state.

The pharmacological basis for active sleep-related depolarization and hyperpolarization of tooth pulp afferents is not known. At the level of the trigeminal nucleus, GABA might be the principal neurotransmitter mediating tooth pulp afferent PAD.[81] However, it should be pointed out that the effects of locally applied GABA on the terminal excitability of individual tooth pulp afferents have never been reported. A valuable step towards elucidating this issue will be to determine whether each tooth pulp terminal that undergoes active sleep-related PAD also is subjected to PAD mediated by GABA.

31.2.6.2 Postsynaptic Inhibition

Trigemino-thalamic tract cell sensory transmission may be impeded by direct, postsynaptic inhibition, in addition to the effects of disfacilitation of primary afferent terminals and second-order terminals in the thalamus (see Section 31.2.6.3). The most direct approach to testing whether individual TSNC neurons undergo a process of postsynaptic inhibition during active sleep would be to perform intracellular recording experiments to compare synaptic activity and electrical properties of these cells as a function of sleep and wakefulness. Detection of an active sleep-related increase in inhibitory synaptic activity in conjunction with changes in membrane properties consistent with an increase in conductance would constitute strong evidence for the presence of postsynaptic inhibition.[21,109]

Labeling studies have indicated that TGT neurons are rather small,[16] making it exceedingly difficult to obtain intracellular records in the awake animal. Extracellular recording methods can, however, provide indirect evaluation of the existence of postsynaptic inhibition, and one approach that can be used is to measure the security of antidromic invasion of TGT neurons.[105] The security of antidromic invasion can be assessed by determining the antidromic firing index, i.e., the probability of evoking an antidromic spike over a defined number of trials, across a range of stimulus intensities. Noncompetitive decreases in the firing index are thought to reflect postsynaptic hyperpolarization.[102,105] This methodological approach has been applied to a few tooth pulp-evoked TGT neurons, of which one revealed an active sleep-specific decrease in the firing index that was not dependent on the applied stimulus intensity; another TGT cell exhibited no change in the firing index from waking levels.[16] However, such methodologies can only be applied to antidromically activated neurons.

An alternative investigative approach is to assess whether a trigeminal sensory neuronís responsiveness to excitatory neurotransmitter agonists applied juxtacellularly (e.g., via microiontophoresis) is altered during active sleep. Studies in our laboratory indicate that spike activity of trigeminal sensory neurons driven by continuous application of glutamate or N-methyl-D-aspartatate is suppressed during active sleep relative to quiet sleep or wakefulness.[12] These data, together with the antidromic firing index study results, suggest that the reduced excitability of some TSNC and TGT neurons during active sleep may be due, in part, to a process of postsynaptic inhibition.

31.2.6.3 Presynaptic Control of Trigemino-Thalamic Tract Afferent Terminals

Previous studies performed by Sessle and Dubner[101,102] in acutely prepared, anesthetized cats have demonstrated that the terminals of TGT afferents can be depolarized by conditioning stimuli applied to oro-facial receptive fields. Stimuli that result in depolarization of thalamic terminals also suppress thalamic neuronal activity.[101] These findings suggest that presynaptic inhibitory mechanisms involving a process of TGT afferent terminal depolarization may play a role in the modulation of ascending oro-facial sensory information. An important issue is whether such mechanisms are engaged during active sleep.

We recently examined this possibility in identified TGT neurons in the chronically instrumented cat preparation.[17] Changes in the excitability of TGT neuron terminals were inferred from differences between the currents required to produce an antidromic firing probability of 50% (EC_{50}) during quiet wakefulness and active sleep. Depolarization or hyperpolarization of TGT terminals was defined as an EC_{50} decrease or increase, respectively. Overall, the EC_{50} of eight TGT terminals was reduced by a mean 8.8 ± 3.6 µA during active sleep relative to quiet wakefulness.[17] This result suggests that depolarization of TGT terminals, which may suppress transfer of sensory information from the trigeminal nucleus to the thalamus, occurs during active sleep.

The neural pathways and pharmacological mechanisms mediating excitability changes of TGT afferents during active sleep are not known and probably do not involve classical PAD-like processes due to the paucity of axo-axonic synapses in the VPM nucleus.[96] These caveats notwithstanding, the axon terminals of individual TGT neurons located in the contralateral VPM apparently are subjected to depolarizing influences during active sleep when compared to wakefulness or quiet sleep. Disfacilitation of this type may provide a functional perspective for the studies of Sessle and Dubner described above[101,102] and indicates that processing of ascending sensory transmission occurs at multiple sites along the TGT during active sleep.

31.3 Studies of Spinal Cord Sensory Neurons

31.3.1 Characteristics of Dorsal Spinocerebellar Tract Neurons

Detailed surveys of the anatomical organization, electrophysiology, and pharmacology of lumbar sensory neurons comprising the spinocervical tract, spinothalamic tract (STT), and spinoreticular tract (SRT) are presented in Chapter 28. Another important sensory pathway for conveying information to higher brain centers is the dorsal spinocerebellar tract (DSCT), on which this part will focus.

The DSCT has traditionally been thought to convey proprioceptive forms of sensory input to the cerebellum and higher centers. Clarke[22] originally described two columns of cells in the spinal gray matter which have subsequently been confirmed to extend throughout the thoracic and upper lumbar spinal cord segments. Clarke's columns lie in the spinal gray matter and are located dorsal and lateral to the central canal. Neuropil in these columns consists of large, oval pyriform or stellate cells in close apposition to primary afferent terminals.[22,120] Based on soma size, Clarke's column cells can be divided into at least three types — relatively small cells (≈ 15 µm diameter), medium cells (≈ 25 µm wide $\times \approx 50$ µm long), and large cells (≈ 55 µm wide $\times <140$ µm long).[80] Intracellular horseradish-peroxidase labeling studies of individual DSCT neurons have substantiated these findings and demonstrated that DSCT neurons have intense dendritic arborizations that extend up to 3000 µm along the rostro-caudal axis of the spinal cord.[97,120]

The axons forming the DSCT ascend in the dorsolateral funiculus of the spinal cord past the lateral cervical nucleus, then course through the dorsolateral medullary brainstem just ventral to the descending root of cranial nerve V.[47,85] The tract finally terminates as mossy fibers in the anterior and posterior cerebellar lobules. In addition, collaterals of DSCT axons project to several prethalamic medullary relay sites.[67,68]

Afferents to Clarke's column are derived principally from collateral projections of primary afferent fibers that ascend in the dorsal columns.[83,118] The majority of DSCT neurons in the cat receive monosynaptic input from a variety of peripheral receptor types, including primary (group Ia) and secondary (group II) muscle spindles, Golgi tendon organs (Group Ib), as well as joint and cutaneous receptors.[44,75,118] At least three types of DSCT neurons respond to muscle stretch: (1) cells that respond at the onset of sudden passive extension of muscle, (2) cells that respond with dynamic and static components to changes in muscle length, and (3) cells that respond selectively to tendon organ activation.[65,66,91,92] Other functional classes of DSCT neurons include cells that respond to pressure applied to the footpad, cells responding to inputs from hair mechanoreceptors, and cells responding to high threshold muscle afferents or flexor reflex afferents.[45,75,93] DSCT neurons of each functional subdivision display spontaneous spike activity in the barbiturate anesthetized cat,[72,83] although their firing rates are considerably less than those of DSCT neurons recorded in the chronically instrumented cat during quiet wakefulness.[107,108]

Dorsal spinocerebellar tract cells are subjected to potent descending influences that appear to arise from higher cortical centers.[59] Both presynaptic and postsynaptic inhibitory mechanisms have been implicated in these modulatory effects.[59] Other evidence suggests the existence of a reticulospinal pathway centered in the ventro-medial aspect of the medullary reticular formation, which suppresses the responses of DSCT neurons to high- as well as low-intensity natural cutaneous stimuli.[48] This system may be tonically active.[58] The circumstances or behavioral state(s) during which these descending systems operate are not known.

A study by Walmsley and Nicol,[119] who employed elegant techniques for drug application via perfusion of the central canal combined with intracellular recording procedures, indicates that monosynaptic transmission from Ia afferents to DSCT neurons is blocked by kynurenic acid, a broad-spectrum excitatory amino acid antagonist. Earlier studies found that DSCT cells are potently suppressed by juxtacellular microiontophoretic application of inhibitory amino acids such as GABA.[31] More recently, Jankowska's group performed microiontophoretic experiments on DSCT neurons, which were situated in the more lateral and dorsal portions of the spinal gray matter and received prominent group II input. They showed that group II monosynaptic and cutaneous inputs were differentially affected by 5-HT and the 5-HT_{1A} agonist (8-OH-DPAT). Responses to group II afferent stimulation were suppressed, while responses to cutaneous stimulation of DSCT cells were facilitated, indicating that descending control of these sensory systems may be highly specific.[63,64]

Given that Clarke's column DSCT neurons also receive input from exteroceptors including high-threshold flexor reflex afferents,[83] it is possible that neurons comprising the DSCT and the more classical STT and SRT neurons located in the lumbar ventral horn share somewhat similar responses to excitatory and inhibitory neurotransmitters. However, detailed comparative electropharmacological studies of these sensory tracts have not been performed.

31.3.2 Chronically Instrumented Animal Preparations: Methodological Approaches to Spinal Cord Unit Recording

Several techniques for recording activity of lumbar sensory neurons have been described in the literature since the late 1960s. Wall[117] recorded the activity of lumbar neurons in rats that had their lumbosacral vertebral column immobilized. Bromberg and Fetz[9] adopted a similar approach of immobilizing the cervical vertebral column of monkeys that, following recovery from surgical procedures, were restrained in a primate chair during unit recording sessions.

In the mid-1980s, several groups of investigators independently obtained lumbar spinal neuron recordings in chronic, unanesthetized, adult cat preparations. In the studies of Marshall et al.,[84] Collins,[27] and Sorkin et al.,[113] cylindrical or rectangular implant chambers were positioned over the dorsal lamina of a specific lumbar vertebra. Acrylic resin was used to secure these recording chambers to vertebrae anterior and posterior to the site of interest, and wound margins were secured in close apposition to the recording chamber. Hence, several vertebrae became immobilized during the implant procedure. The recording chamber allows attachment to a stereotaxic frame that provides adequate stabilization for single-unit recording. In our laboratory, we have also implanted restraining devices and fashioned a recording chamber over specific vertebrae for subsequent unit recording.[108] Thus, lumbar restraint has been employed and considered essential for minimizing spinal cord movement and for allowing the investigator to record from lumbar sensory neurons.

Some investigators have asserted that changes in response characteristics of lumbar sensory neurons occur in chronically instrumented animals that are subjected to chronic lumbar restraint involving immobilization and fusion of several vertebrae,[52,54] although such assertions remain unsubstantiated. To circumvent this issue, Herrero et al. implanted a recording chamber on the dorsal surface of a single lumbar vertebra in the sheep, thus avoiding lumbar fusion.[52] The development of these and other novel chronic recording strategies may shed some light on this controversy in the future.

Another relevant methodological issue is the variable postoperative recovery times allowed by each laboratory group. Postoperative recovery periods have included <2 hr,[117] 2 to 3 days,[84] 2 weeks,[27] and 1 month.[108] The shorter postoperative time periods used by some investigators may affect the properties of lumbar sensory neurons as a result of the abnormal spinal cord physiology caused by "recent" surgical trauma (see Chapter 30).[23] However, there are few data in the literature to elucidate this important issue.

31.3.3 Characteristics of Lumbar Sensory Neurons in Awake Intact Preparations

A number of studies have examined the electrophysiological properties of spinal sensory tract cells in chronically instrumented animals. One approach introduced by Wall[116,117] and adopted subsequently by other investigators[9,27,52,53,113] is to characterize the responses of the recorded neuron to natural forms of stimuli (e.g., brush stimuli, moderate pinching, and/or radiant heat) applied to peripheral receptive fields (see Chapter 28). Bromberg and Fetz[9] also included electrical stimulation of peripheral nerve trunks to activate spinal cord neurons in the cervical enlargement.

A series of studies by Collins and his colleagues[24,26,29] indicate that sensory neurons in the dorsal and ventral horns of the cat can be loosely divided into two categories — namely, low-threshold neurons and wide-dynamic-range neurons. Herrero and Headley[53,55] included a third type — high-threshold mechanoreceptive nocispecific cells — in their categorization of lumbar sensory neurons in chronically instrumented sheep. Low-threshold neurons were the predominant cell type encountered by Collins in the cat. Such cells display low levels of spontaneous spike activity (<1 Hz).[24-26] This finding was substantiated by recent studies performed in the sheep.[53] Receptive fields of lumbar neurons resemble those mapped for comparable low-threshold and wide-dynamic-range types of cells in acutely prepared anesthetized animal preparations.[24,26,113,117]

Sorkin et al.[113] included electrical stimulation techniques in the battery of test stimuli they employed to distinguish between primary afferents and postsynaptic target neurons. Approximately one third of their units were postsynaptic target neurons, one third afferent fibers, and the remainder unidentified. Spontaneous activity was absent in neurons driven by cutaneous stimuli,[113] whereas cells with spike activity ranging from 5 to 40 Hz were responsive to joint movement.[113]

Unfortunately, in none of the aforementioned chronic studies were the axonal projections of the cells determined, which renders establishing the identity of the recorded cells difficult, thus making direct comparison of cell characteristics and integration of the data troublesome. Based on the published figures of recording sites, particularly within the dorsal horn, the unidentified neurons recorded by these groups may have belonged to a number of different ascending sensory pathways including the spinocervical tract and postsynaptic dorsal column pathway. Within the deeper dorsal and ventral horns, the sensory pathways may also have included the SRT, STT, and DSCT. In addition, some recorded units may have been located in spinal cord areas containing Ia and Ib interneurons, Renshaw cells,[62] and perhaps motoneurons. Nevertheless, the pioneering studies described above have provided an essential foundation for subsequent investigations of sensory processing in awake animals.

Another approach, used by our laboratory to characterize units recorded extracellularly in chronically instrumented preparations, has been to identify the cells electrophysiologically as belonging to specific sensory tracts by using classical antidromic testing procedures.[108] Clarke's column DSCT neurons located in the L_3 spinal cord segment are ideal for this purpose, as they form a compact column of neurons that can readily be located and identified. Low-intensity search stimuli (0.2 ms, 10 to 300 µA, 1.0 Hz) delivered to the anterior lobe of the cerebellum are effective in evoking a low-amplitude field potential at spinal cord depths corresponding to Clarke's column.[107] DSCT neurons can be recorded at these depths and further identified by their large-amplitude spike waveforms (300 to 1500 µV, peak-to-peak), constant antidromic latency, ability to follow high-frequency stimulus trains (0.5 to 1.0 kHz), and collision of antidromically propagated spikes with spontaneously occurring or synaptically evoked action potentials. Somatic as opposed to axonal recording of "backfired" units can be confirmed by observing facilitation and suppression of spike discharge by juxtacellular microiontophoretic ejections of glutamate and GABA (or glycine), respectively.

A characteristic feature of DSCT neurons during quiet wakefulness is their spontaneous spike activity (17.6 ± 1.3 spikes per s).[107] Most of these neurons respond to low-intensity stimuli (0.04 ms, 1.0 Hz, 50 to 500 µA) applied to cuff electrodes surrounding the sciatic nerve with short-latency (<5 ms) spikes, while a few cells respond with a longer latency discharge.[107] It is not yet clear whether the long-latency discharge is a post-inhibitory rebound and/or is due to activation of slower conducting fibers or polysynaptic pathways.

Spinoreticular tract and STT neurons can also be identified in the lower thoracic and upper lumbar segments in anesthetized cats.[1,2] These cells are located dorsolateral and ventral to Clarke's column neurons and can be identified using antidromic identification techniques in awake, chronically instrumented cats.[86] SRT cells recorded in the L_3 spinal segment of awake cats display ongoing spontaneous spike activity (16.9 ± 4.0 spikes per s) and respond to sciatic nerve stimulation similarly to Clarke's column DSCT neurons. They can also be driven by mechanical or low-intensity electrical stimuli delivered to cutaneous flank and proximal tail receptive fields (unpublished observation). However, responses of individual DSCT and SRT neurons to controlled forms of natural stimuli applied to peripheral receptive fields have not yet been reported for the chronic, awake animal preparation.

31.3.4 Modulation of Lumbar Sensory Neurons in Chronically Instrumented Animal Preparations

In contrast to the multitude of studies performed in acute, anesthetized preparations (see Chapter 28), very few investigations of afferent and descending regulation of spinal sensory neuron function have been conducted in chronically instrumented animals. In one report, Bromberg and Fetz[9] demonstrated potent suppression of sensory neurons of the cervical spinal cord in the monkey following peripheral conditioning stimuli.

We recently reported the effects of stimulation of the nucleus reticularis gigantocellularis on spontaneous spike activity of Clarke's column cells; peristimulus histograms principally revealed reductions in spontaneous firing of DSCT neurons.[110] The effects are graded as a function of stimulus intensity (10 to 80 µA) and often include abolition of cell firing lasting 10 to 20 ms after an initial burst of action potentials at higher stimulus intensities. Hence, complex excitatory and inhibitory effects on DSCT neurons can be exerted by reticular structures in the unanesthetized animal.

The low level of discharge of wide-dynamic-range neurons in chronic animal studies[24,25,29,53,113] has been attributed to the presence of tonic supraspinal inhibition akin to that described for anesthetized animals following transection or reversible cold block of the spinal cord (see, for example, References 43, 112, 116). However, direct evidence for the existence of tonic supraspinal inhibition of individual lumbar sensory neurons in awake animal preparations is not yet available.

Clearly, more experimental data are required from other lumbar sensory tract neurons such as those comprising the STT and postsynaptic dorsal column pathway before any detailed understanding of afferent and descending controls of lumbar sensory neurons in unanesthetized animals can be realized. Such studies would provide a firm foundation for determining under which behavioral circumstances descending influences may be engaged (see Chapter 29).

31.3.5 Activity of Lumbar Sensory Neurons Across the Sleep/Wakefulness Cycle

Pompeiano and his colleagues[20,94] first reported on sleep-related changes in the activity of spinal cord sensory neurons by observing responses to evoked potentials. They found that hindlimb nerve-evoked responses recorded from ipsilateral T_{12} ventrolateral quadrants were suppressed during phasic, rapid-eye-movement episodes of active sleep relative to wakefulness, quiet sleep, and tonic periods of active sleep. Similar results have been reported for field potentials recorded in the cerebellum.[18] These early studies suggested that a number of ascending sensory pathways including the SRT, STT, as well as dorsal and ventral spinocerebellar tracts might be modulated to varying degrees during active sleep.

Evoked potential recording techniques have recently been employed to investigate whether transmission through classical SRT and STT pathways is affected by sleep states.[111] The population response of SRT/STT fibers tract located in the medullary reticular formation were recorded. Sciatic nerve-evoked field potentials were suppressed during tonic periods and even abolished during phasic, rapid-eye-movement episodes of active sleep when compared with wakefulness or quiet sleep.[111] These findings suggest that regulatory influences are engaged during active sleep to modulate lumbar sensory tract neurons such as those of the SRT and STT.

Identification of sensory tracts which are subjected to active sleep-related modulation and characterization of the nature of this state-dependent sensory processing have only recently been reported. The spontaneous activity of most identified DSCT neurons[107,108] and SRT neurons[86] is markedly suppressed during active sleep relative to quiet sleep and quiet wakefulness (Figure 31.3). Sciatic nerve and cutaneous responses evoked in certain DSCT and SRT neurons, respectively, are also suppressed during active sleep (unpublished observation). The results suggest that synaptic transmission through these two ascending pathways is modulated across the sleep/wakefulness cycle and provide single-unit evidence supporting conclusions reached from previous field potential studies.

In contrast to the previous results, neuronal responses evoked in unidentified lumbar sensory neurons by manual brushing of their peripheral receptive fields are markedly facilitated during active sleep.[71] The possibility exists that a portion of the cells recorded in the latter study by Kishikawa et al.[71] may have belonged to inhibitory pathways to motoneurons[21] or ascending sensory pathways other than the DSCT or SRT (for example, the spinocervical tract or postsynaptic dorsal columns). However, active sleep-related facilitation of low-threshold, mechanoreceptor-evoked activity has also been reported for TGT neurons in the trigeminal sensory nucleus.[16] The influence of behavioral state changes on the activity of STT neurons remains to be investigated.

31.3.6 Mechanisms Modulating Lumbar Sensory Neuron Function During Active Sleep

The suppression of DSCT and SRT neuronal spike discharge observed during active sleep may arise as a consequence of pre- and/or postsynaptic forms of inhibition.[19] Substantial evidence exists that, during active sleep, lumbar motoneurons are subjected to glycine-mediated postsynaptic

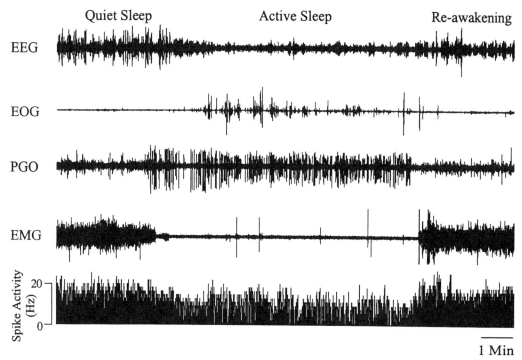

FIGURE 31.3

Continuous record of spontaneous spike activity of an antidromically identified dorsal spinocerebellar tract neuron during the transition from quiet sleep into active sleep and subsequent re-awakening. The animal's behavioral state is indicated by the electroencephalographic, electrooculographic, pedunculo-geniculo-occipital, and electromyographic activities (EEG, EOG, PGO, and EMG, respectively) in the top four traces. The bottom trace represents this cell's spike discharge plotted as a continuous rate-meter histogram (binwidth: 1 s). The average firing rate of this cell was determined for 1-min epochs in each state. During active sleep, the spike rate decreased to 8.5 Hz from 16.6 Hz during quiet sleep, and returned to the quiet sleep level upon re-awakening (17.5 Hz; see also Reference 107). These data illustrate that activity of dorsal spinocerebellar tract neurons is specifically suppressed during active sleep.

inhibition.[21,109] Lumbar sensory neurons, including those of the DSCT and SRT, may likewise be controlled by postsynaptic inhibition mediated by glycine or other neurotransmitter(s).

Recent intracellular studies performed by Chase and colleagues[122,123] suggest that classical, somatically directed postsynaptic inhibition of Clarke's column DSCT neurons is unlikely to occur during active sleep. Xi et al.[122,123] reported that in these studies of DSCT neurons recorded intracellularly in chloralose-anesthetized, artificially ventilated cats the electrical properties (e.g., membrane potential, input resistance, time constant, and rheobase) did not change during motor inhibition induced by intra-pontine microinjection of carbachol. In contrast, lumbar motoneurons during carbachol-induced motor inhibition appear to be influenced by postsynaptic inhibitory mechanisms resembling those operating during naturally occurring active sleep.[122] However, it is not immediately clear that neurons comprising ascending sensory pathways are comparably modulated by supraspinal pathways during active sleep and carbachol-induced motor inhibition. Moreover, as sensory neurons of the spinal cord dorsal horn are suppressed during chloralose anesthesia,[28,95] it is essential that the findings by Chase and colleagues are verified in the undrugged animal, as is indeed emphasized in the paper by Xi et al.[123] We have observed that glutamate-driven responses of certain SRT and DSCT neurons are reduced during active sleep which suggests that some ascending sensory tract sensory cells may be hyperpolarized during this state (Figure 31.4).

In addition to postsynaptic inhibition, other mechanisms may impede ascending sensory transmission during active sleep, including, disfacilitation arising as a result of descending influences

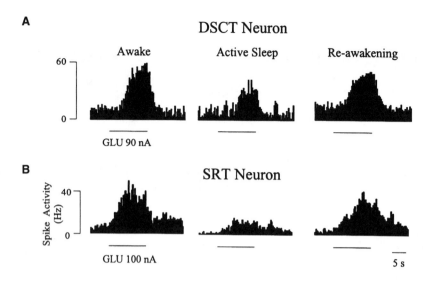

FIGURE 31.4
Effect of microiontophoretic ejections of the excitatory amino acid glutamate on an antidromically identified dorsal spinocerebellar tract (DSCT) neuron (**A**) and a spinoreticular tract (SRT) neuron (**B**) in the unanesthetized cat. The peristimulus histograms each represent cell discharge averages for six consecutive glutamate pulses (GLU; 0.1 M, pH = 8.0; indicated by the horizontal bars) delivered during wakefulness, active sleep, and re-awakening (binwidth = 0.5 s). During active sleep, average glutamate-evoked firing was suppressed by ≈52% in the DSCT cell and ≈80% in the SRT cell (determined from 5-s epochs around response maxima) relative to preceding and subsequent wakefulness. Note that the ongoing discharge rates (prior to glutamate application) also decrease during active sleep in these cells. These data indicate that during active sleep certain DSCT and SRT neurons may be hyperpolarized via a process of postsynaptic inhibition.

and/or a reduction of peripheral input (potentially via a process of primary afferent depolarization[89]). The sites of action of state-specific modulatory influences impinging on ascending sensory tracts originating in the spinal cord are not as well identified as for trigeminal sensory tracts (see Section 31.2). However, the similarity with respect to state dependency and regulatory influences invite further investigation of the extent to which presynaptic as well as postsynaptic mechanisms modulate DSCT and SRT across sleep/wakefulness states.

31.4 Conclusions

The preceding parts of this chapter summarize what is currently understood regarding trigeminal and lumbar sensory processing as a function of wakefulness and sleep. Earlier studies have indicated that peripherally evoked potentials are predominantly inhibited during active sleep. More recently, single unit recordings in the rostral TSNC and in various spinal cord segments have confirmed and expanded these observations. As a result, a number of common properties of cells in these two sensory relay areas and general principles for state-specific regulation of transmission through second-order sensory neurons are emerging.

Sensory neurons (e.g., those of the rostral TSNC and spinal cord dorsal horn) that cannot be identified as belonging to a particular ascending tract display little spontaneous activity during wakefulness. On the other hand, most neurons comprising specific ascending sensory pathways originating in the trigeminal sensory nucleus (e.g., the TGT) as well as deeper spinal gray matter (e.g., the DSCT and SRT) display regular spontaneous spike activity with discharge rates in the range of 10 to 20 spikes per second during the awake, undrugged state. The questions arise whether it is indeed a specific trait of ascending sensory tract cells that they are spontaneously active during wakefulness and what the identity of silent sensory neurons of the brainstem and spinal cord is with respect to axonal projections.

The most critical finding emerging from single-unit recording studies is that spontaneous and evoked responses of these sensory tract cells are specifically modulated during active sleep relative to other behavioral states. The modulation takes the form of suppression of ongoing spike activity when the population mean for all sampled neurons is considered.[14,16,107] Even though evoked field potential responses to certain afferents are altered during quiet sleep,[51] no overall differences between the activity of populations of individually recorded prethalamic sensory neurons during quiet sleep and wakefulness have been observed.[14] This patterning of activity across sleep/wakefulness states resembles that observed for somatic motoneurons.[21] The recent studies substantiate, at the single-cell level, the findings of evoked-potential studies conducted by Pompeiano and colleagues over 30 years ago[18–20,94] and provide experimental techniques and paradigms with which to pursue further analysis of mechanisms underlying the state-dependency of ascending sensory transmission.

Mechanisms involved in state-dependent modulation of ascending sensory tract cells appear to include both pre- and postsynaptic inhibitory processes.[13,16,17] State-specific modulation may occur at multiple presynaptic sites, at least for certain sensory systems, including at axon terminals of second-order afferents, as is the case for the TGT.[17] PAD-like processes appear to operate on primary afferent terminals in midbrain as well as spinal regions and contribute to suppression of the neural excitability, at least for the TGT, during active sleep. Additionally, ascending projection cells of the TGT (and the SRT) are apparently subjected to direct postsynaptic inhibition, whereas there is conflicting evidence regarding whether this regulatory mechanism is invoked on DSCT neurons during active sleep. Taken together, there are indications for modulation of activity and excitability at several levels of ascending sensory tracts, which may provide a required redundancy and/or avenues for precise regulation of sensory input to diencephalic regions during sleep and wakefulness.

The modulation of sensory transmission across behavioral state is, however, not uniform for all types of stimuli conveyed by a single sensory tract. Active sleep-specific gating of sensory transmission in the TGT may be dependent on sensory modality and afferent fiber diameter.[16] However, it is not certain whether such controls are common to all sensory channels or are sensory channel specific. Additional types of natural stimuli should be tested to ascertain whether principles governing the state-dependency of sensory modulation relate to the gate-control theory proposed by Melzack and Wall.[88] Nevertheless, the differential modulation of ascending sensory transmission for distinct types of stimuli may provide a diagnostic tool for studies of the mechanisms underlying state-specific sensory transmission.

The functional implications of active sleep-related modulation of sensory tract neurons remain undetermined. One possibility is that (partial) functional sensory deafferentiation contributes to the emergence and/or maintenance of active sleep. Dampening of ascending sensory input and blockade of motor output may operate in tandem to this end. Moreover, the circuitry that produces suppression of transmission through primary afferents and second-order sensory cells may be part of the brainstem systems that promote muscle atonia during active sleep. Further work is required to elucidate the pharmacological basis for these forms of modulation and to identify their neural origins, including which and to what extent descending pathways mediate state-dependent modulation of sensory transmission. Information garnered from such studies might form the scientific basis for development of analgesic agents designed to engage mechanisms that underlie state-dependent suppresser actions on sensory neurons.

In summary, it is now clear that state-dependent modulation of ascending sensory transmission occurs prior to any diencephalic processing. Sensory transmission through the brainstem and spinal cord is impeded by mechanisms impinging on multiple pre-and postsynaptic sites within ascending sensory tracts. Moreover, the activity of cells in these regions is not uniform across state and may depend on the nature of afferent input and efferent projections. These emerging properties of first-

and second-order sensory neurons provide impetus for conducting further mechanistic studies on the state specificity of sensory transmission through identified ascending tract systems in chronic, unanesthetized animal preparations.

Acknowledgments

The cited research conducted in the first author's laboratory has been supported by the U.S. National Institutes of Health (NS32306, NS34716), the Medical Research Council of Canada, and British Columbia Health Research Foundation.

References

1. Ammons, W.S., Characteristics of spinoreticular and spinothalamic neurons with renal input, *J. Neurophysiol.*, 58, 480, 1987.
2. Ammons, W.S., Renal and somatic input to spinal neurons antidromically activated from the ventrolateral medulla, *J. Neurophysiol.*, 60, 1967, 1987.
3. Azerad, J., Woda, A., and Albe-Fessard, D., Physiological properties of neurons in different parts of the cat trigeminal sensory complex, *Brain Res.*, 246, 7, 1982.
4. Baldissera, F., Broggi, G., and Mancia, M., Depolarization of trigeminal afferents induced by stimulation of brain-stem and peripheral nerves, *Exp. Brain Res.*, 4, 1, 1967.
5. Banks, D., Kuriakose, M., and Matthews, B., A technique for recording the activity of brain-stem neurones in awake, unrestrained cats using microwires and an implantable micromanipulator, *J. Neurosci. Meth.*, 46, 83, 1993.
6. Boissonade, F.M., Banks, D., and Matthews, B., Methods for recording the jaw-opening reflex to tooth-pulp stimulation in awake cats, *J. Neurosci. Meth.*, 38, 35, 1991.
7. Boissonade, F.M., Banks, D., and Matthews, B., A technique for recording from brain-stem neurones in awake, unrestrained cats, *J. Neurosci. Meth.*, 38, 41, 1991.
8. Boissonade, F.M. and Matthews, B., Responses of trigeminal brainstem neurons and the digastric muscle to tooth-pulp stimulation in awake cats, *J. Neurophysiol.*, 69, 174, 1993.
9. Bromberg, M.B. and Fetz, E.E., Responses of single units in cervical spinal cord of alert monkeys, *Exp. Neurol.*, 55, 469, 1977.
10. Burton, H. and Craig, A.D.J., Distribution of trigeminothalamic projection cells in cat and monkey, *Brain Res.*, 161, 515, 1979.
11. Bushnell, M.C., Duncan, G.H., Dubner, R., and He, L.F., Activity of trigeminothalamic neurons in medullary dorsal horn of awake monkeys trained in a thermal discrimination task, *J. Neurophysiol.*, 52, 170, 1984.
12. Cairns, B.E., Synaptic Transmission Through the Trigeminal Sensory Nuclear Complex During Wakefulness and Sleep, Ph.D. thesis, Faculty of Pharmaceutical Sciences, The University of British Columbia, Vancouver, 1996.
13. Cairns, B.E., Fragoso, M.C., and Soja, P.J., Active-sleep-related suppression of feline trigeminal sensory neurons: evidence implicating presynaptic inhibition via a process of primary afferent depolarization, *J. Neurophysiol.*, 75, 1152, 1996.
14. Cairns, B.E., Fragoso, M.C., and Soja, P.J., Activity of rostral trigeminal sensory neurons in the cat during wakefulness and sleep, *J. Neurophysiol.*, 73, 2486, 1995.
15. Cairns, B.E., Fragoso, M.C., and Soja, P.J., Homosynaptic depression of rostral trigeminal sensory neurons in the chronic cat: a role for presynaptic inhibition?, *Soc. Neuroscience Abstr.*, 21, 1199, 1995.

16. Cairns, B.E., McErlane, S.A., Fragoso, M.C., Jia, W.-G., and Soja, P.J., Spontaneous discharge and peripherally evoked orofacial responses of trigemino-thalamic tract neurons during wakefulness and sleep, *J. Neurosci.*, 16, 8149, 1996.
17. Cairns, B.E. and Soja, P.J., Active sleep-related depolarization of feline trigemino-thalamic afferent terminals, *NeuroReport*, 9, 565, 1998.
18. Carli, G., Diete-Spiff, K., and Pompeiano, O., Cerebellar responses evoked by somatic afferent volleys during sleep and waking, *Arch. Ital. Biol.*, 105, 499, 1967.
19. Carli, G., Diete-Spiff, K., and Pompeiano, O., Presynaptic and postsynaptic inhibition on transmission of cutaneous afferent volleys through the cuneate nucleus during sleep, *Experientia*, 22, 239, 1966.
20. Carli, G., Kawamura, H., and Pompeiano, O., Transmission of somatic sensory volleys through ascending spinal hindlimb pathways during sleep and wakefulness, *Pflügers Arch. für die Gesamte Physiologie des Menschen und der Tiere*, 298, 163, 1967.
21. Chase, M.H. and Morales, F.R., The atonia and myoclonia of active (REM) sleep, *Ann. Rev. Psychol.*, 41, 557, 1990.
22. Clarke, J.L., Researches into the structure of the spinal cord, *Phil. Trans. Roy. Soc.*, 607, 1851.
23. Clarke, R.W. and Matthews, B., The thresholds of the jaw-opening reflex and trigeminal brainstem neurons to tooth-pulp stimulation in acutely and chronically prepared cats, *Neuroscience*, 36, 105, 1990.
24. Collins, J.G., A descriptive study of spinal dorsal horn neurons in the physiologically intact, awake, drug-free cat, *Brain Res.*, 416, 34, 1987.
25. Collins, J.G., Inhibition of spontaneous activity of spinal dorsal horn neurons in the intact cat is naloxone-insensitive, *Brain Res.*, 401, 95, 1987.
26. Collins, J.G., Neuronal activity recorded from the spinal dorsal horn of physiologically intact, awake, drug-free, restrained cats: a preliminary report, *Brain Res.*, 322, 301, 1984.
27. Collins, J.G., A technique for chronic extracellular recording of neuronal activity in the dorsal horn of the lumbar spinal cord in drug-free, physiologically intact, cats, *J. Neurosci. Meth.*, 12, 277, 1985.
28. Collins, J.G., Kawahara, M., Homma, E., and Kitahata, L.M., Alpha-chloralose suppression of neuronal activity, *Life Sci.*, 32, 2995, 1983.
29. Collins, J.G., Ren, K., and Tang, J., Lack of spontaneous activity of cutaneous spinal dorsal horn neurons in awake, drug-free, spinally transected cats, *Exp. Neurol.*, 96, 299, 1987.
30. Contreras, D. and Steriade, M., State-dependent fluctuations of low-frequency rhythms in corticothalamic networks, *Neuroscience*, 76, 25, 1997.
31. Curtis, D.R., Gynther, B.D., and Malik, R., A pharmacological study of group I muscle afferent terminals and synaptic excitation in the intermediate nucleus and Clarke's column of the cat spinal cord, *Exp. Brain Res.*, 64, 105, 1986.
32. Curtis, D.R. and Lodge, D., The depolarization of feline ventral horn group Ia spinal afferent terminations by GABA, *Exp. Brain Res.*, 46, 215, 1982.
33. Darian-Smith, I., Neurone activity in the cat's trigeminal main sensory nucleus elicited by graded afferent stimulation, *J. Physiol.*, 153, 52, 1965.
34. Darian-Smith, I., Neurone activity in the cat's trigeminal main sensory nucleus elicited by graded afferent stimulation, *J. Physiol.*, 153, 52, 1960.
35. Darian-Smith, I., Presynaptic component in the afferent inhibition observed within the brain-stem nuclei of the cat, *J. Neurophysiol.*, 16, 634, 1965.
36. Davies, W.I., Scott, D., Vesterstrom, K., and Vyklicky, L., Depolarization of the tooth pulp afferent terminals in the brain stem of the cat, *J. Physiol.*, 218, 515, 1971.
37. Dostrovsky, J.O., Sessle, B.J., and Hu, J.W., Presynaptic excitability changes produced in brain stem endings of tooth pulp afferents by raphe and other central and peripheral influences, *Brain Res.*, 218, 141, 1981.

38. Dubner, R., Interaction of peripheral and central input in the main sensory trigeminal nucleus of the cat, *Exp. Neurol.,* 17, 186, 1967.
39. Dubner, R., Hoffman, D.S., and Hayes, R.L., Neuronal activity in medullary dorsal horn of awake monkeys trained in a thermal discrimination task. III. Task-related responses and their functional role, *J. Neurophysiol.,* 46, 444, 1981.
40. Dubner, R., Kenshalo, D.R., Maixner, W., Bushnell, M.C., and Oliveras, J.L., The correlation of monkey medullary dorsal horn neuronal activity and the perceived intensity of noxious heat stimuli, *J. Neurophysiol.,* 62, 450, 1989.
41. Dubner, R. and Sessle, B.J., Presynaptic excitability changes of primary afferent and corticofugal fibers projecting to trigeminal brain stem nuclei, *Exp. Neurol.,* 30, 223, 1971.
42. Dubner, R., Sessle, B.J. and Storey, A.T., *The Neural Basis of Oral and Facial Function,* Plenum Press, New York, 1978.
43. Duggan, A.W. and Morton, C.R., Tonic descending inhibition and spinal nociceptive transmission, *Prog. Brain Res.,* 77, 193, 1988.
44. Eccles, J.C., Oscarsson, O., and Willis, W.D., Synaptic action of group I and II afferent fibres of muscle on the cells of the dorsal spinocerebellar tract, *J. Physiol.,* 158, 517, 1961.
45. Edgley, S.A. and Jankowska, E., Information processed by dorsal horn spinocerebellar tract neurones in the cat, *J. Physiol.,* 397, 81, 1988.
46. Eisenman, J., Landgren, S., and Novin, S., Functional organization in the main sensory trigeminal nucleus and in the rostral subdivision of the nucleus of the spinal trigeminal tract in the cat, *Acta Physiol. Scand.,* 59, 3, 1963.
47. Grant, G. and Xu, Q., Routes of entry into the cerebellum of spinocerebellar axons from the lower part of the spinal cord, *Exp. Brain Res.,* 72, 543, 1988.
48. Hames, E.G. and Bloedel, J.R., Effects of descending systems on exteroceptive responses of spinocerebellar neurons to mechanical stimuli, *Physiologist,* 21, 49, 1978.
49. Hayashi, H., Sumino, R. and Sessle, B.J., Functional organization of trigeminal subnucleus interpolaris: nociceptive and innocuous afferent inputs, projections to thalamus, cerebellum, and spinal cord, and descending modulation from periaqueductal gray, *J. Neurophysiol.,* 51, 890, 1984.
50. Hayes, R.L., Dubner, R., and Hoffman, D.S., Neuronal activity in medullary dorsal horn of awake monkeys trained in a thermal discrimination task. II. Behavioral modulation of responses to thermal and mechanical stimuli, *J. Neurophysiol.,* 46, 428, 1981.
51. Hernandez-Peon, R., O'Flaherty, J.J., and Mazzuchelli-O'Flaherty, A.L., Modifications of tactile evoked potentials at the spinal trigeminal sensory nucleus during wakefulness and sleep, *Exp. Neurol.,* 13, 40, 1965.
52. Herrero, J.F., Coates, T.W., Higgins, M., Livingston, A., Waterman, A.E., and Headley, P.M., A technique for recording from spinal neurones in awake sheep, *J. Neurosci. Meth.,* 46, 225, 1993.
53. Herrero, J.F. and Headley, P.M., Cutaneous responsiveness of lumbar spinal neurons in awake and halothane-anesthetized sheep, *J. Neurophysiol.,* 74, 1549, 1995.
54. Herrero, J.F. and Headley, P.M., The dominant class of somatosensory neurone recorded in the spinal dorsal horn of awake sheep has wide dynamic range properties, *Pain,* 61, 133, 1995.
55. Herrero, J.F. and Headley, P.M., Sensitization of spinal neurons by non-noxious stimuli in the awake but not anesthetized state, *Anesthesiology,* 82, 267, 1995.
56. Hobson, J.A. and Steriade, M., Neuronal basis of behavioral state control, in *Handbook of Physiology,* Vol. 4, Mountcastle, V.B. and Bloom, F.E., Eds., Bethesda, MD, 1986, p. 701.
57. Hoffman, D.S., Dubner, R., Hayes, R.L., and Medlin, T.P., Neuronal activity in medullary dorsal horn of awake monkeys trained in a thermal discrimination task. I. Responses to innocuous and noxious thermal stimuli, *J. Neurophysiol.,* 46, 409, 1981.

58. Holmqvist, B., Lundberg, A., and Oscarsson, O., Supraspinal inhibitory control of transmission to three ascending spinal pathways influenced by the flexion reflex afferents, *Arch. Ital. Biol.*, 98, 60, 1960.
59. Hongo, T. and Okada, Y., Cortically evoked pre- and postsynaptic inhibition of impulse transmission to the dorsal spinocerebellar tract, *Exp. Brain Res.*, 3, 163, 1967.
60. Hu, J.W. and Sessle, B.J., Comparison of responses of cutaneous nociceptive and nonnociceptive brain stem neurons in trigeminal subnucleus caudalis (medullary dorsal horn) and subnucleus oralis to natural and electrical stimulation of tooth pulp, *J. Neurophysiol.*, 52, 39, 1984.
61. Jacobs, B.L., Fornal, C.A., and Wilkinson, L.O., Neurophysiological and neurochemical studies of brain serotonergic neurons in behaving animals, *Ann. N.Y. Acad. Sci.*, 600, 260, 1990.
62. Jankowska, E., Interneuronal relay in spinal pathways from proprioceptors, *Prog. Neurobiol.*, 38, 335, 1992.
63. Jankowska, E., Hammar, I., Djouhri, L., Heden, C., Szabo Lackberg, Z., and Yin, X.K., Modulation of responses of four types of feline ascending tract neurons by serotonin and noradrenaline, *Eur. J. Neurosci.*, 9, 1375, 1997.
64. Jankowska, E., Krutki, P., Lackberg, Z.S., and Hammar, I., Effects of serotonin on dorsal horn dorsal spinocerebellar tract neurons, *Neuroscience*, 67, 489, 1995.
65. Jansen, J.K., Nicolaysen, K., and Rudjord, T., Discharge pattern of neurons of the dorsal spinocerebellar tract activated by static extension of primary endings of muscle spindles, *J. Neurophysiol.*, 29, 1061, 1966.
66. Jansen, J.K., Nicolaysen, K., and Walloe, L., On the inhibition of transmission to the dorsal spinocerebellar tract by stretch of various ankle muscles of the cat, *Acta Physiol. Scand.*, 70, 362, 1967.
67. Johansson, H. and Silfvenius, H., Axon-collateral activation by dorsal spinocerebellar tract fibres of group I relay cells of nucleus Z in the cat medulla oblongata, *J. Physiol.*, 265, 341, 1977.
68. Johansson, H. and Silfvenius, H., Input from ipsilateral proprio- and exteroceptive hind limb afferents to nucleus Z of the cat medulla oblongata, *J. Physiol.*, 265, 371, 1977.
69. Khayyat, G.F., Yu, U.J., and King, R.B., Response patterns to noxious and non-noxious stimuli in rostral trigeminal relay nuclei, *Brain Res.*, 97, 47, 1975.
70. Kirkpatrick, D.B. and Kruger, L., Physiological properties of neurons in the principal sensory trigeminal nucleus of the cat, *Exp. Neurol.*, 48, 664, 1975.
71. Kishikawa, K., Uchida, H., Yamamori, Y., and Collins, J.G., Low-threshold neuronal activity of spinal dorsal horn neurons increases during REM sleep in cats: comparison with effects of anesthesia, *J. Neurophysiol.*, 74, 763, 1995.
72. Knox, C.K., Kubota, S., and Poppele, R.E., A determination of excitability changes in dorsal spinocerebellar tract neurons from spike-train analysis, *J. Neurophysiol.*, 40, 626, 1977.
73. Kristensen, M.P., Cairns, B.E., and Soja, P.J., Activity of trigemino-thalamic tract neurons during rapid-eye-movement episodes of active sleep, *Soc. Neurosci. Abstr.*, 23, 2340, 1997.
74. Kruger, L. and Michel, F., A morphological and somatotopic analysis of single unit activity in the trigeminal sensory complex of the cat, *Exp. Neurol.*, 5, 139, 1962.
75. Kuno, M., Munoz-Martinez, E.J., and Randic, M., Sensory inputs to neurones in Clarke's column from muscle, cutaneous and joint receptors, *J. Physiol.*, 228, 327, 1973.
76. Levy, R.A., The role of GABA in primary afferent depolarization, *Prog. Neurobiol.*, 9, 211, 1977.
77. Lisney, S.J., Evidence for primary afferent depolarization of single tooth-pulp afferents in the cat, *J. Physiol.*, 288, 437, 1979.
78. Lisney, S.J., Evidence for primary afferent depolarization of tooth pulp afferents following stimulation of other trigeminal afferents in the cat, *J. Physiol.*, 282, 0022, 1978.

79. Lisney, S.J. and Matthews, B., Branched afferent nerves supplying tooth-pulp in the cat, *J. Physiol.*, 279, 509, 1978.
80. Loewy, A.D., A study of neuronal types in Clarke's column in the adult cat, *J. Comp. Neurol.*, 139, 53, 1970.
81. Lovick, T.A., Primary afferent depolarization of tooth pulp afferents by stimulation in nucleus raphe magnus and the adjacent reticular formation in the cat: effects of bicuculline, *Neurosci. Lett.*, 25, 173, 1981.
82. Maixner, W., Dubner, R., Kenshalo, D.R., Bushnell, M.C., and Oliveras, J.L., Responses of monkey medullary dorsal horn neurons during the detection of noxious heat stimuli, *J. Neurophysiol.*, 62, 437, 1989.
83. Mann, M.D., Clarke's column and the dorsal spinocerebellar tract: a review, *Brain Behav. Evol.*, 7, 34, 1973.
84. Marshall, K.W., Tatton, W.G., and Bruce, I.C., A technique for recording from single neurons in the spinal cord of the awake cat, *J. Neurosci. Meth.*, 10, 249, 1984.
85. Matsushita, M. and Hosoya, Y., Spinocerebellar projections to lobules III to V of the anterior lobe in the cat, as studied by retrograde transport of horseradish peroxidase, *J. Comp. Neurol.*, 208, 127, 1982.
86. McErlane, S.A., Kristensen, M.P., and Soja, P.J., Spike activity of spinoreticular tract neurons during wakefulness and sleep, *Soc. Neuroscience Abstr.*, 23, 2341, 1997.
87. McGinty, D.J. and Harper, R.M., Dorsal raphe neurons: depression of firing during sleep in cats, *Brain Res.*, 101, 569, 1976.
88. Melzack, R. and Wall, P.D., Pain mechanisms: a new theory, *Science*, 150, 971, 1965.
89. Morrison, A.R. and Pompeiano, O., Depolarization of central terminals of group Ia muscle afferent fibres during desynchronized sleep, *Nature*, 210, 201, 1966.
90. Nord, S.G., Bilateral projection of the canine tooth pulp to bulbar trigeminal neurons, *Brain Res.*, 113, 517, 1976.
91. Osborn, C.E. and Poppele, R.E., Components of responses of a population of DSCT neurons to muscle stretch and contraction, *J. Neurophysiol.*, 61, 456, 1989.
92. Osborn, C.E. and Poppele, R.E., Cross-correlation analysis of the response of units in the dorsal spinocerebellar tract (DSCT) to muscle stretch and contraction, *Brain Res.*, 280, 339, 1983.
93. Osborn, C.E. and Poppele, R.E., Sensory integration by the dorsal spinocerebellar tract circuitry, *Neuroscience*, 54, 945, 1993.
94. Pompeiano, O., Carli, G., and Kawamura, H., Transmission of sensory information through ascending spinal hindlimb pathways during sleep and wakefulness, *Arch. Ital. Biol.*, 105, 529, 1967.
95. Radhakrishnan, V. and Henry, J.L., Excitatory amino acid receptor mediation of sensory inputs to functionally identified dorsal horn neurons in cat spinal cord, *Neuroscience*, 55, 531, 1993.
96. Ralston, H.J. and Herman, M.M., The fine structure of neurons and synapses in ventrobasal thalamus of the cat, *Brain Res.*, 14, 77, 1969.
97. Randic, M., Miletic, V., and Loewy, A.D., A morphological study of cat dorsal spinocerebellar tract neurons after intracellular injection of horseradish peroxidase, *J. Comp. Neurol.*, 198, 453, 1981.
98. Rudomin, P., Jimenez, I., and Quevedo, J., Selectivity of the presynaptic control of synaptic effectiveness of Group I afferents in the mammalian spinal cord, in *Presynaptic Inhibition and Neural Control*, Rudomin, P., Romo, R., and Mendell, L.M., Eds., Oxford University Press, New York, 1998, p. 282.
99. Satoh, T., Yamada, S., Yokota, T., Ohshima, T., and Kitayama, S., Modulation during sleep of the cat trigeminal neurons responding to tooth pulp stimulation, *Physiol. Behav.*, 39, 395, 1987.

100. Sessle, B.J., The neurobiology of facial and dental pain: present knowledge, future directions, *J. Dent. Res.,* 66, 962, 1987.

101. Sessle, B.J. and Dubner, R., Presynaptic excitability changes of trigeminothalamic and corticothalamic afferents, *Exp. Neurol.,* 30, 239, 1971.

102. Sessle, B.J. and Dubner, R., Presynaptic hyperpolarization of fibers projecting to trigeminal brain stem and thalamic nuclei, *Brain Res.,* 22, 121, 1970.

103. Sessle, B.J., Dubner, R., Hu, J.E., and Lucier, G.E., Modulation of trigeminothalamic relay and nonrelay neurones by noxious, tactile and periaqueductal gray stimuli: implications in perceptual and reflex aspects of nociception, in *Pain in the Trigeminal Region,* Anderson, D.J. and Matthews, B., Eds., Elsevier, Amsterdam, 1977, p. 285.

104. Sessle, B.J. and Greenwood, L.F., Inputs to trigeminal brain stem neurones from facial, oral, tooth pulp and pharyngolaryngeal tissues. I. Responses to innocuous and noxious stimuli, *Brain Res.,* 117, 211, 1976.

105. Sessle, B.J. and Hu, J.W., Raphe-induced suppression of the jaw-opening reflex and single neurons in trigeminal subnucleus oralis, and influence of naloxone and subnucleus caudalis, *Pain,* 10, 19, 1981.

106. Sessle, B.J., Hu, J.W., Dubner, R., and Lucier, G.E., Functional properties of neurons in cat trigeminal subnucleus caudalis (medullary dorsal horn). II. Modulation of responses to noxious and nonnoxious stimuli by periaqueductal gray, nucleus raphe magnus, cerebral cortex, and afferent influences, and effect of naloxone, *J. Neurophysiol.,* 45, 193, 1981.

107. Soja, P.J., Fragoso, M.C., Cairns, B.E., and Jia, W.G., Dorsal spinocerebellar tract neurons in the chronic intact cat during wakefulness and sleep: analysis of spontaneous spike activity, *J. Neurosci.,* 16, 1260, 1996.

108. Soja, P.J., Fragoso, M.C., Cairns, B.E., and Oka, J.-I., Dorsal spinocerebellar tract neuronal activity in the intact chronic cat, *J. Neurosci. Meth.,* 60, 227, 1995.

109. Soja, P.J., Lopez-Rodriguez, F., Morales, F.R., and Chase, M.H., The postsynaptic inhibitory control of lumbar motoneurons during the atonia of active sleep: effect of strychnine on motoneuron properties, *J. Neurosci.,* 11, 2804, 1991.

110. Soja, P.J., Nixon, G., and Fragoso, M.C., Medullary reticular control of dorsal spinocerebellar tract neurons in the cat, *Soc. Neurosci. Abstr.,* 21, 1199, 1995.

111. Soja, P.J., Oka, J.-I., and Fragoso, M., Synaptic transmission through cat lumbar ascending sensory pathways is suppressed during active sleep, *J. Neurophysiol.,* 70, 1708, 1993.

112. Soja, P.J. and Sinclair, J.G., Tonic descending influences on cat spinal cord dorsal horn neurons, *Somatosensory Res.,* 1, 83, 1983.

113. Sorkin, L.S., Morrow, T.J., and Casey, K.L., Physiological identification of afferent fibers and postsynaptic sensory neurons in the spinal cord of the intact, awake cat, *Exp. Neurol.,* 99, 412, 1988.

114. Steriade, M. and McCarley, R.W., *Brainstem Control of Sleep and Wakefulness,* Plenum Press, New York, 1990.

115. Sunada, T., Kurasawa, I., Hirose, Y., and Nakamura, Y., Intracellular response properties of neurons in the spinal trigeminal nucleus to peripheral and cortical stimulation in the cat, *Brain Res.,* 514, 189, 1990.

116. Wall, P.D., The laminar organization of dorsal horn and effects of descending impulses, *J. Physiol.,* 188, 403, 1967.

117. Wall, P.D., Freeman, J., and Major, D., Dorsal horn cells in spinal and in freely moving rats, *Exp. Neurol.,* 19, 519, 1967.

118. Walmsley, B., Central synaptic transmission: studies at the connection between primary afferent fibres and dorsal spinocerebellar tract (DSCT) neurones in Clarke's column of the spinal cord, *Prog. Neurobiol.,* 36, 391, 1991.

119. Walmsley, B. and Nicol, M.J., The effects of Ca^{2+}, Mg^{2+} and kynurenate on primary afferent synaptic potentials evoked in cat spinal cord neurones *in vivo*, *J. Physiol.*, 433, 409, 1991.

120. Walmsley, B. and Nicol, M.J., Location and morphology of dorsal spinocerebellar tract neurons that receive monosynaptic afferent input from ankle extensor muscles in cat hindlimb, *J. Neurophysiol.*, 63, 286, 1990.

121. Woda, A., Azerad, J., and Albe-Fessard, D., The properties of cells in the cat trigeminal main sensory and spinal subnuclei activated by mechanical stimulation of the periodontium, *Arch. Oral Biol.*, 28, 419, 1983.

122. Xi, M.-C., Liu, R.-H., Yamuy, J., Morales, F.R., and Chase, M.H., Electrophysiological properties of lumbar motoneurons in the α-chloralose-anesthetized cat during carbachol-induced motor inhibition, *J. Neurophysiol.*, 78, 129, 1997.

123. Xi, M.-C., Yamuy, J., Liu, R.-H., Morales, F.R., and Chase, M.H., Dorsal spinocerebellar tract neurons are not subjected to postsynaptic inhibition during carbachol-induced motor inhibition, *J. Neurophysiol.*, 78, 137, 1997.

124. Young, R.F., Effect of trigeminal tractotomy on dental sensation in humans, *J. Neurosurg.*, 56, 812, 1982.

125. Young, R.F. and King, R.B., Excitability changes in trigeminal primary afferent fibres elicited by dental pulp stimulation in the cat, *Arch. Oral Biol.*, 17, 1649, 1972.

126. Young, R.F., Oleson, T.D., and Perryman, K.M., Effect of trigeminal tractotomy on behavioral response to dental pulp stimulation in the monkey, *J. Neurosurg.*, 55, 420, 1981.

Section VII

Pain and Anesthesia

Jean-Marie Besson, Section Editor

Research in the field of pain has expanded considerably over the past 25 years, and an enormous amount of information has been published. These studies examine the modalities of activation of peripheral nociceptors, the peripheral or central release of neurotransmitters, the gene expression of various peptides, the plasticity of the nervous system, and the pharmacology of peripherally and centrally acting analgesics or putative analgesics.

As reviewed in Chapter 32, there are many strategic approaches to the development of new analgesic compounds. Several tens of thousands of molecules have been synthesized as possible analgesics. Unfortunately, we cannot be proud of our recent activity in the discovery of original analgesic substances. Indeed, the bark of the willow tree (which contains salicin) and the poppy were used before Christ, and paracetamol (acetaminophen) was discovered more than 100 years ago. To give another example, some anticonvulsants and antidepressants are now used to try to suppress particular types of pain, notably those arising from lesions of the peripheral and central nervous systems. Here again, the use of these substances does not result from rational research but simply from accidental observations by some particular clinicians.

Today, systematic and rational approaches have replaced empirical studies due to the development of more and more refined techniques. It must be recognized, however, that the rift between research hopes and clinical application has been disappointing. Clinicians make use of a limited number of analgesics which have been available for many years. The problems encountered in research are numerous; notably, these include the multiplicity of receptors and the co-localization of more than one neurotransmitter in a single neuron. By using the techniques of molecular biology, a large number of receptors have been cloned. This approach, together with the biological prediction of the structure of macromolecules, will allow the three-dimensional structure of receptors to be elucidated, which in turn will lead to the rational development of agonists and antagonists with great specificity and low side-effect liability.

The relevance (or even the lack of relevance) of the major pharmacological screening tests for clinical pain states has been discussed repeatedly. Tests used to evaluate antinociceptive activity include noxious heat, pressure to the tail or paw, colorectal distension, intraperitoneal chemical irritants, and subcutaneous administration of formalin. In most cases, these stimuli are applied to normal animals in the absence of hyperalgesia, allodynia, and hyperesthesia. Some of these tests depend on spinal mechanisms, while others also involve supraspinal structures. Some of the tests are extraordinarily sensitive to a particular class of analgesics; others are too optimistic and lead to false positive interpretations. In addition, in numerous behavioral experiments only one nociceptive test is used, and the exact method can vary from author to author. These few remarks indicate why, with a few exceptions, there are often controversies in the literature relevant to the pharmacology of pain.

More importantly, the models mentioned above are limited in that they do not mimic chronic pain states. Chronic pain states differ markedly from acute pain by the nature of their persistence and by the adaptive changes well illustrated by the phenomenon of plasticity, which has been described at various levels of the nervous system. Such limitations have led to the use and development of more appropriate chronic pain models over the past ten years. These newer models include inflammatory pain and neuropathic pain. The development of these experimental models is essential not only for the detection of new analgesics, but also for a better understanding of pain syndromes that are poorly managed clinically. Another difficulty encountered in the design of safe analgesics is the fact that some chemicals and receptors putatively involved in the transmission or modulation of pain are widely distributed throughout the nervous system. This is especially the case for peptides and excitatory or inhibitory amino acids. Most of these neuroactive substances are involved in multiple physiological functions, and developing analysis to target these systems could lead to undesirable side effects.

Major difficulties regarding the development of new analgesics are related to the multitude of neuroactive chemicals at both peripheral and spinal levels, as well as involvement of numerous supraspinal structures in the modulation and perception of pain. At the periphery, various transmitters, mediators, and other factors are involved in the generation of nociception at the peripheral endings of nociceptive C-fibers. They include immune cell factors, vascular agents, tissue injury, nerve growth factor, neurogenic factors, and sympathetic influences. Thus, a myriad of chemical sensitizers and activators have been found at the periphery, making it unlikely that a single analgesic substance will be effective against all types of pain. In addition, in animals in which most of the pharmacological investigations have been performed, recording from nociceptors requires surgery. Thus, even with the best "microdissector," one is never totally sure about conditions in the microenvironment of the nociceptors (i.e., CO_2 and O_2 partial pressures, pH, microcirculation, sympathetic tone, ionic composition of the extracellular fluid). In order to circumvent such difficulties, *in vitro* preparations have been designed. These models provide better control of the concentration of tested agents, but they are far removed from clinical situations.

The jungle of neurochemical neuroanatomy also is well illustrated by the spinal cord. Indeed, most of the classical neurotransmitters and a considerable number of neuropeptides are present in the dorsal horn. Theoretically, these substances could be released at the level of the spinal cord and intervene in the transmission of nociceptive processes. Additional complexity is provided by the co-existence of several peptides and classical neurotransmitters in the same neuron. For instance, substance P (SP) and glutamate co-exist in some primary afferent fibers. Because several compounds are co-released from the same central terminals, the problem is extraordinarily complex. If two compounds such as SP and glutamate are capable of producing excitation alone, then the action of both may have to be blocked to produce transmission failure. Thus, in this situation it is difficult to predict the effects of antagonists. This possibility may explain why the effects of systemic administration of NMDA or SP antagonists are weaker than expected.

A further complexity of the network is that some peptides — for instance SP, cholecystokinin (CCK), and somatostatin (SOM) — are localized not only in primary afferent neurons but also in intrinsic neurons and descending fibers. Interestingly, others such as calcitonin gene-related peptide (CGRP) seem to be located exclusively in primary efferent fibers. Here, again, we must be extremely careful in considering the data and not yield to the temptation of becoming infatuated with the molecule of the moment. For instance, 20 years ago SP was considered to be *the* pain neurotransmitter; now glutamate has many supporters, and in 1993 nitric oxide was the molecule of the year.

Finally, from a general point of view, the various neurotransmitters involved in pain mechanisms bind to different receptor classes so that some neurotransmitters interact synergistically and others interact antagonistically. Thus, we are faced with multifactorial interactions, and the use of cocktails to address cooperative interactions between drugs or between several modalities of action of the same drug could be highly promising. Another reason for hope is that there are multiple types of pain due to various pathophysiological mechanisms, and one could speculate that specific new chemicals would be selectively effective against certain types of pain.

In Chapter 33, Villanueva and Bernard discuss the multiplicity of ascending pain pathways. The complexity of the pain pathways, and therefore of the brain structures involved in pain processing, has been clearly emphasized by the use of powerful, modern anatomical tracing techniques. These techniques have allowed not only better visualization of the "classical" ascending pathways (origins and terminals), but also have revealed numerous "novel" ascending tracts which participate in pain integration. Presently, it is clear from the current literature that nociception (and pain) are not exclusively related to a unique system of pathways, relay nuclei, or "centers".

Based on human clinical observation, the contralateral ascending nociceptive systems located mainly in the anterolateral quadrant of the spinal cord have classically been considered as part of the spinothalamic (STT) tract. The electrophysiological characteristics of neurons at the origin of the STT in the spinal cord, as in the lateral thalamus and somatosensory cortical areas (well-defined receptive fields and encoding properties of the parameters of peripheral stimulation), show that this spinothalamic cortical tract is mainly involved in the sensory discriminative aspects of pain and nociception.

The functional roles of other ascending pathways, such as the spino-cervico-thalamic tract, the postsynaptic dorsal column fibers, and the various components of the spino-reticular and spino-reticulo-thalamic tracts, are poorly understood. Recent studies provide evidence of "new" pain pathways, such as the spino-hypothalamic tract, the spino-parabrachial-amygdaloid tract, the spino-parabrachial-hypothalamic tract, and other pathways which relay in the caudal medulla. Villanueva and Bernard's chapter focuses primarily on these newly described pathways.

In Chapter 34, Gybels addresses the question: Is there still a place in the neurosurgical treatment of pain for making lesions in the nociceptive pathways? For many years, neurosurgeons were virtually the only clinicians to take charge of patients for whom all other treatments against pain had failed and who died in distress, suffering, and misery. Fortunately, the resources we now have at our disposal are better than those available at the beginning of this century, but patients still exist for whom neurosurgery is unavoidable. In 1912, Spiller and Martin proposed the anterolateral cordotomy to relieve intractable pain. Numerous other surgical interventions have since been proposed, consisting mainly of interrupting pain pathways from the periphery to the cerebral cortex. Today, a number of these surgical approaches have been abandoned, leaving a place for more functional neurosurgery. Most of the neurosurgeons involved in the fight against pain are acquainted with advanced basic research. Moreover, against intractable pain refractory to any other therapeutic treatment, surgeons now propose less invasive techniques, with a weak risk of morbidity and presenting the best comfort for the patients. Age, general medical state, and life expectancy of these patients are now considered.

In 1989, Gybels and Sweet published a 400-page book devoted to the neurosurgical treatment of persistent pain. They described and analyzed in detail multiple interventions, and there is no doubt that some neurosurgical techniques have reached a high degree of sophistication and specificity, mainly with the help of the operating microscope and of intraoperative electrophysiological recordings. However, the major problem remains in evaluating the effects of these various surgical approaches. Thanks to his vast knowledge and experience in the field, Gybels overviews the main neurosurgical interventions which are undertaken in an attempt to relieve intractable pain.

Chapter 32

Future Pharmacological Treatments of Pain

Anthony Dickenson, Victoria Chapman, and Alison Reeve

Contents

32.1 Introduction 549
32.2 Opioids 550
    32.2.1 Novel Opioid Peptides and their Receptors 551
    32.2.2 Delta Opioids 552
    32.2.3 Endogenous Opioids 553
32.3 Neuropathic Pain — Potential Targets for Treatment 553
    32.3.1 Models of Nerve Injury 553
    32.3.2 Neurophysiology of Nerve Injury 554
32.4 Anti-Opioid Peptides 555
32.5 Central Hypersensitivity 556
    32.5.1 Excitatory Aminoacids and Central Hypersensitivity 556
    32.5.2 Controlling Excitatory Amino-Acid Events 558
    32.5.3 Inhibitory Systems 559
    32.5.4 Adenosine 559
    32.5.5 Nitric Oxide 561
32.6 Conclusions 561
References 561

32.1 Introduction

The last decade has seen an ever-increasing proliferation of data on the receptors and channels involved in the transmission and control of noxious messages, and from this a number of potential new targets for analgesic therapy have arisen.[1] The increase in the number of available experimental drugs has faciliated the study of the roles of transmitter and receptors in physiological events, and, alongside this, the development of a number of animal models for clinical pain states such as inflammation and neuropathies has clearly shown that several transmitter systems which have minor actions in acute pains have important roles in more persistent pains. Thus, we now believe that where there is pathology, a plasticity in physiological and pharmacological functions can occur. This plasticity, the capacity of the pain signaling and modulating systems to alter in different circumstances, has changed our ways of thinking about pain control. Signaling events are not fixed and are not the same in all situations but are subject to alteration. The net result is that there is now much potential for replacement of morphine by new analgesic agents, some with opioid actions, others not — a quest that has lasted for decades. The need for an improvement in the treatment of pain is well supported by a simple consideration of the number of individuals who suffer pain. Several million surgical interventions per year in the U.K. occur with, on average, 3 days of postoperative pain. Long-term chronic pain, ranging from backache to arthritis, afflicts about 5 million people in Britain, and about 10% of this pain does not respond to the currently used therapy.

The long development time for novel analgesics from within the pharmaceutical industry poses severe problems for the many patients with intractable pain who need relief now. Of people with cancer, 20% have neuropathic pain which is difficult to treat, and one third of visits to pain clinics are from patients with chronic pain. It can be estimated that about 10% of the population suffer from chronic pain, and if this is unrelieved in 20%, a conservative figure of 365 million pain days per annum is produced. This degree of disability has a huge economic toll in terms of loss of employment and disability payments, but quality of life is equally compromised.

There is also a huge number of patients for whom pain control, as a result of hospital admission for medical or surgical reasons, leaves much to be desired. Of patients who suffer pain, 33% had pain all or most of the time, and 87% of those with pain rated it as moderate to severe. If postoperative pain relief improved either as a result of pre-emptive approaches or for other reasons, the cost of 9 million pain days (3 million operations with general anesthetics per annum in the U.K. with an average 3-day stay in hospital) would be reduced.

In addition to the social and health benefits of better control of pain, the advent of new technologies within the pharmaceutical industry should aid the steps that lead to the clinical use of a new drug. For example, high-throughput screening can be used to speed up the process of identification of useful agents, which combined with improved combinatorial chemistry can lead to fast and efficient production of novel agents with good affinity for particular targets. Finally, genomic means can be used to identify targets related to specific disease and, in the field of pain and analgesia, to identify targets associated with particular pathological processes within this area.

32.2 Opioids

Opioid drugs, in particular, morphine, are the gold standards of analgesic therapy for moderate to severe pains. Thus, new analgesics must have comparable efficacy and/or reduced side-effect liability. Ever since 1807, when morphine was first identified from the juice of the opium poppy, there has been a continual hunt for opioids which are good analgesics but lack the typical morphine-like side effects. The emphasis in this chapter will be on opioids, as trying to cover all potential targets for new drugs is impossible and the opioids provide an excellent series of examples of how modern techniques can be used to seek new drugs. However, would new opioid analgesics be more useful than the existing agents? Although opioids have good effectiveness in many situations of acute and persistent pain, when the pain arises from neuropathic processes that involve damage or dysfunction to a peripheral nerve or as a result of central nervous system damage, there can be reduced opioid sensitivity. Thus, another issue that will be covered is whether non-opioids can be used in nerve injury pains and how the effectiveness of opioids can be improved.

The ideal of an opioid drug with low side-effects can only be achieved if there are opioid receptors on which morphine, fentanyl, and other clinically used opioids had low or no affinity. This is important, as the large majority of opioid drugs produced so far share the same general mu-receptor-mediated effects. The identification of endogenous opioids nearly 30 years ago prompted the discovery of multiple opioid receptors. The latter finding lent impetus to this search, because, as is the case with any transmitter, if there are many receptors the potential for separation of the desirable effects from the unwanted effects can be achieved, at least in theory, by receptor-selective drugs. The more recent isolation of the opioid receptors will further facilitate this task.[2] The production of anti-sera to parts of the receptor sequence provides tools for the unequivocal proof of location of the particular receptor. The ability to express the receptors is an approach that can give insights into the mechanisms of opioid actions in particular cell lines, but caution is needed when extrapolating from, say, an oocyte or cell-line to a neuron within the CNS. In broad terms, the sequences of opioid receptors in animals are remarkably similar in both structure and pharmacology to the human receptors. Thus, there is an obvious relevance of animal studies on opioids to clinical practice.

Until recently, three opioid receptors — the mu receptor (the receptor for morphine), the delta receptor, and the kappa receptor — were known to exist from molecular studies. A further division of the receptors into sub-types had been proposed by pharmacological studies based on the finding of differential effects and potencies of agonists, and different abilities of antagonists to reverse effects. The mu receptor was suggested to be comprised of a mu 1 and a mu 2 subtype;[3] the delta receptor was also subdivided, and the kappa receptor has been divided into three sub-types.[4,5] It still is unclear whether these subtypes have functional consequences that will translate into the clinic. The delta subtypes do appear to generate differential effects in functional studies related to pain and analgesia,[4] although at the present time there is no evidence from the cloning studies for receptor subtypes: the receptors isolated (whether mu, delta, or kappa) have been single identical species. It may be that there can be different processing in local neuronal tissue environments, allowing the subtypes to be expressed, or that the tissues used so far in these studies only contain certain sub-types so that at present we may underestimate the extent of the opioid receptor family.[2]

32.2.1 Novel Opioid Peptides and Their Receptors

A problem that did not assist investigation of the mu receptor was the lack of a ligand for the receptor in many areas of the brain and, in particular, within the spinal cord, a key site in the relay of noxious messages form the periphery to the brain. The recently discovered peptides (endomorphin-1 and endomorphin-2), isolated from bovine brain, have high affinity and selectivity for mu-opioid receptors and may be the natural endogenous ligands for mu-opioid receptors.[6] *In vitro* studies have showed endomorphin-1 to be a more potent mu-opioid agonist than the prototypical ligands previously used, and *in vivo* studies report that endomorphin-1 is equipotent with morphine in producing analgesia following intracerebroventricular (i.c.v.) administration.[6] Recently the presence of endogenous endomorphin-1 or endomorphin-2 at the level of the spinal cord has been shown, and it appears to be present in zones expected by the location of the mu receptor.[7] This bears out the idea that peptides are the sought after endogenous ligands for the abundant spinal mu-opioid receptors. In support of this concept is the behavioral analgesia seen after spinal application of these peptides and the electrophysiological evidence that spinal endomorphin-1 and endomorphin-2 reduce afferent transmission of noxious information through the spinal cord.[8] Analgesic effects were reversed by mu-opioid receptor antagonism in both studies,[6] and, overall, the findings indicate that novel drugs based on these new peptides are unlikely to produce analgesics any different from existing mu opioids such as morphine.

However, the cloning of the mu, delta, and kappa opioid receptors has led to the discovery of a new opioid receptor, the orphan or opioid receptor-like-1 (ORL1) receptor.[9,10] The heptadecapeptide, orphanin FQ/ nociceptin, is an endogenous ligand for the ORL1 receptor.[9,10] What is very interesting is that behavioral studies report that intracerebroventricular nociceptin, unlike other opioids, is pro-nociceptive.[9,10] More recent studies indicate that this supraspinal effect of the peptide may not be a direct hyperalgesia as such, but rather an anti-opioid action, in that there is a physiological antagonism of endogenous (and also exogenous) opioid actions within the brain.[11] Thus, the elevations in response to a given stimulus produced by the peptide may result from a lifting of endogenous opioid controls.

The transcript encoding the ORL1 receptor is present in the spinal cord,[12] a major site of action of other opioid receptor agonists.[1] Thus, the effect of spinal administration of nociceptin on sensory processing in the dorsal horn can be compared to that of other opioids. In complete contrast to the supraspinal hyperalgesia, spinal application of nociceptin selectively modulates spinal nociceptive events by preferentially reducing C-fiber-evoked wind-up and post-discharge of the neurons and is antinociceptive in behavioral studies by the same route.[13] These results are in keeping with *in vitro* studies demonstrating inhibitory effects of nociceptin on dorsal raphe nucleus neurons,[14] where the receptor appears to share a similar ionic mechanism of action to morphine. There is therefore little evidence that the actions of nociceptin are anything other than predominantly inhibitory at the spinal level. Although the potency of nociceptin is similar to other opioid peptides, naloxone sensitivity is much lower or entirely lacking, in keeping with the low potency of naloxone at the ORL1 receptor.[15]

Thus, this receptor could conceivably be a target for novel analgesics. The consequences of a systemic application are hard to predict due to the opposing actions of the receptor at supraspinal and spinal sites. In addition, a non-peptide antagonist would not only aid characterization of the receptor but also the functional roles of nociceptin and finally be an important drug in clinical evaluations of drugs based on this novel peptide.

32.2.2 Delta Opioids

Other novel opioids with actions differing from those of the morphine-like drugs could also be agonists at the kappa or delta receptors. Many behavioral and *in vivo* electrophysiological studies with selective agonists and antagonists for these opioid receptors have demonstrated their independence of the mu-opioid receptor in the production of antinociception.[16,17] In the case of the kappa receptor, the degree of analgesia that can be produced is still unclear, as it has often been necessary to reduce the intensity of the test stimulus in a particular model in order to demonstrate kappa receptor agonist effects, whereas full analgesia with morphine and delta opioids can be seen in many different models. Additionally, whereas mu and delta opioids are reinforcing in behavioral tests, kappa opioids appear to be aversive, which would appear to present a major problem for the development of a clinically useful agent.[18] This latter finding could be a basis to suggest that kappa opioids will have low abuse potential, but it is equally clear that dependence is not a critical issue in the clinical use of opioids. As a result of the weaker analgesia produced by kappa opioids, combined with the potential for psychotomimetic effects, this target may have less potential that first envisaged.

By contrast, the delta receptor could be an important target for novel analgesics acting via opioid mechanisms. A large number of studies have shown that selective delta receptor opioids can produce analgesia in a number of nociceptive tests and that these actions are equivalent to the analgesic effects of morphine. In addition, nociceptive transmission can be controlled by these drugs with actions at both spinal and supraspinal sites.[17,19] The endogenous opioids are peptides and so many of the early ligands used were peptides and so lacked systemic effectiveness. There are now potent and selective non-peptide delta opioids which have been thoroughly tested in a number of animal models. Thus, a highly selective delta opioid, SNC80, has been shown to be effective via central and systemic (including oral administration) routes. Furthermore, it is clear that the resultant analgesic effects are reversed by delta but not mu-opioid receptor antagonists.[20] The next stages of assessment of delta opioids will be comparisons with morphine, both in terms of analgesia in a number of models and in terms of side-effect profile.

32.2.3 Endogenous Opioids

The potential for any drug based on the endogenous opioids, whether the enkephalins or the endomorphins, is highly limited by the peptide nature of these transmitters, so that poor penetration into the CNS and rapid breakdown compromise the utility of these as drugs. Based on the therapeutic effectiveness of monoamine oxidase inhibitors and anti-cholinesterases, the tactic of prevention of transmitter degradation can be a highly effective strategy. This approach has been used as a basis for the synthesis of mixed peptidase inhibitors such as kelatorphan, which binds to the active site and blocks the actions of at least two of the important breakdown enzymes for the enkephalins.[21] By merit of this action, the drug provides almost complete protection to the enkephalins, and spinal application of the inhibitor produces a reduction of nociceptive responses of cells which is reversed by a selective delta antagonist.[22] A number of other approaches have clearly shown that the analgesic effects of the drug are accompanied by few side-effects. The more recent description of a systemically active mixed peptidase inhibitor, RB101, allows the analgesic and side-effect profiles of these agents to be compared to other direct opioid receptor ligands after systemic administration. Thus far, the side-effect profile of RB101 appears to be unlike that of morphine, as there is no evidence for physical and psychological dependence with the peptidase inhibitor.[23] If the therapeutic profile is satisfactory, this will be a major step toward the clinical application of this novel approach to pain relief, whereby delta opioid receptor activation is produced by the natural transmitter. The efficacy of these agents in animal models of more severe pain states remains to be determined.

32.3 Neuropathic Pain — Potential Targets for Treatment

Pain caused by inflammation and tissue damage and pain from nerve injury (neuropathic pain) are common clinical problems. The mechanisms of inflammatory and neuropathic pain are different from acute pain, and there is considerable plasticity in both the transmission and modulating systems in these prolonged pain states.[24] This is of great importance in consideration of a rational basis for the treatment of neuropathic pain where the pathology leads to alterations in both peripheral and central pain systems. In particular, nerve injury may well be accompanied by a loss of spinal inhibitions, including opioid controls, which leads to difficulties in the management of neuropathic pain.[16,17]

Neuropathic pain has a reduced sensitivity to opioids, indicating a loss of inhibitory function. There are a number of pathological and physiological events that can lead to this.[16,17] The level of morphine analgesia is partly controlled by cholecystokinin (CCK), which interferes with opioid inhibitory mechanisms in the normal spinal cord. After nerve damage, CCK levels rise and are at least partly responsible for reduced opioid effects.[26] In addition, allodynia is relayed via Aβ-fiber pathways which lack opioid receptors.[17] Furthermore, given the potential loss of opioid receptors after nerve damage, other measures such as hyperalgesia may also respond poorly to opioids.[1] Thus, at this stage, I will consider the neurobiological basis for neuropathic pain, an area which illustrates a pressing clinical need for novel analgesics.

32.3.1 Models of Nerve Injury

Until recently, study of the mechanisms of clinical pain syndromes has relied on data from animal studies based on the application of acute stimuli. The symptoms of pain arising from nerve injury, such as allodynia (touch-evoked pain), spontaneous pain, hyperalgesia (enhanced responses to a given noxious stimulus), sensory deficits, and in some cases, a sympathetic component are just not observed in animals when a simple acute stimulus is applied. One of the major steps towards new drug development has been the production of animal models which replicate some of the symptoms of clinical pain states. This has been achieved for both inflammation and neuropathy.

The models can be used to ascertain mechanisms and effects of drugs on the various components of this syndrome by the use of behavioral and electrophysiological approaches. There is a clear need for this multidisciplinary attack on the problem of neuropathic pain, as there is a danger of nonspecific drug effects in animal behavioral studies. More importantly, sites and modes of action of drugs are difficult to ascertain with behavioral studies. Electrophysiological studies have the potential to provide answers to these questions and also allow comparison of the effects of a drug on measures of allodynia, hyperalgesia, spontaneous pain, and nociception under controlled conditions. However, anesthesia can be a confounding factor. Due to the variable causes and constellation of symptoms presented in the clinic it will be almost impossible to gather this information from human studies.[27]

Early animal studies related to neuropathic pain used complete nerve section. More recent animal models have used a restricted partial denervation of the hindlimb following sciatic nerve injury.[28] Two of the models involve constriction of the sciatic nerve distal to the spinal cord, either a tight ligation of a portion of the sciatic nerve (partial ligation[29]) or the loose ligation of the entire nerve (chronic constriction injury, CCI[30]). The most recent model uses the tight ligation of two of the three spinal nerves which form the sciatic nerve, the spinal nerve ligation (SNL) model.[31] The behavioral consequences of these models mimic some aspects of the human symptoms, although the extent and location of the injury differ.

32.3.2 Neurophysiology of Nerve Injury

There is evidence that the aberrations in somatosensory processing which follow partial nerve injury are the culmination of a number of changes in the peripheral nervous system. Studies after nerve section suggest that the generation of ectopic discharges within the neuroma and the dorsal root ganglia (DRG) contributes to these changes.[32,33] After partial denervation (CCI model), high-frequency spontaneous activity originating in the dorsal root ganglion[34] targets spinal neurons via injured A-fibers.[35] Both fibers and DRGs contribute to the behavioral responses, which include allodynia in the SNL model,[36] where a novel mechanoreceptor now innervates the partially deafferented foot.[37] Furthermore, reorganization of Aβ-fiber termination in the spinal cord has been reported after nerve injury.[38] As a result of these changes in the peripheral nerve, there is a rational basis for the treatment of neuropathic pain with anti-convulsant drugs. The use of anti-convulsants and other agents in neuropathic pain arises primarily from the poor opioid sensitivity of this condition. There is a rational basis for the use of systemic local anesthetics, and indeed, anti-convulsants in neuropathic pain, as damaged nerves may be highly sensitive to sodium channel blockers.[39] This is probably part of the basis for the effects of those anti-convulsants with sodium channel-blocking properties (carbamezapine, phenytoin, etc.), but central effects of these compounds are also likely to be important. Calcium channels (L-, N-, and P-type) also should be studied, as the use of blockers reveals selective roles in different pain models[40] and an N-type blocker is claimed to be an effective treatment of neuropathic pain.[41]

In the future, as agents become available, these models can then be used as a basis for the investigation of the roles of a number of receptors or channels. For example, if the effect of an agonist or antagonist at a particularly receptor is greatly increased after nerve injury, as compared to normal conditions, we could suspect an up-regulation of the receptor which could be revealed with anatomical studies. Anatomical approaches with antibodies to receptors and channels can be used to compare channels and receptors in pathological states with the situation in normal animals. For example, the description of a novel TTX-resistant sodium channel in C-fibers may lead to useful research tools to probe the potential changes in this channel in the various models.[42] This may be of great importance, as this channel could be a basis for the synthesis of a drug with selective actions on C-fibers.

These peripheral changes after nerve injury have consequences for the central nervous system. Although it might be preferable to have a drug with peripheral actions, as it will lack centrally mediated side-effects, if the peripheral changes are multi-factorial a single agent with central actions may be able to block a final common pathway. In the CCI model, a high percentage of spinal neurons had abnormal levels of spontaneous activity,[43,44] despite absent somatic receptive fields. Neurons also have increased afterdischarges[43] and sensitivity to tapping of the nerve injury site;[44] however, both the number of neurons sensitive to low-intensity mechanical stimuli and the magnitude of the evoked responses were reduced. Only subtle and late changes in spinal neuronal responses have been reported after partial ligation, which were unrelated to the behavioral allodynia.[43-45]

Thus, there is little consensus on the extent of the peripheral and central changes that contribute to the behavioral changes after nerve injury and to what extent these may replicate the clinical aspects. However, the use of models of different pain states has led to the hypothesis that changes in certain peptide systems can occur after inflammation and neuropathy with important consequences for the degree of opioid analgesia in these states.

32.4 Anti-Opioid Peptides

Many factors can influence morphine analgesia, including the pathological loss of opioid receptors that can occur after nerve injury but also activity in other transmitter systems.[17] The levels in the spinal cord of the non-opioid peptide, cholecystokinin (CCK), has consequences for the level of opioid analgesic mechanisms. Exogenous application of CCK reduces the analgesic effects of morphine not only in the spinal cord but also at supraspinal sites, and the levels of the peptide have also been implicated in the development of opioid tolerance.[46] The effects are restricted to the analgesia produced by mu and kappa opioids, as spinal and supraspinal delta opioid-mediated analgesias are not altered by CCK.[17,47] Although yet untested, in physiological situations where CCK reduces mu opioid actions, delta opioid receptor agonists could be effective. CCK can therefore cause a physiological modulation of morphine analgesia in the absence of any change in opioid receptor number, etc. The receptors for CCK are found both pre- and postsynaptic to the primary afferent fibers, a distribution not dissimilar to distribution of the mu opiate receptor in the rat spinal cord.[48,49] The postsynaptic CCK receptors are mainly of the CCK_B type in the rat but of the A-type in the primate. However, the presynaptic receptors are of the CCK_B type in all species,[49] and it is likely that these receptors are those that interfere with opioid actions. CCK mobilizes calcium from intracellular stores, and if this happens on the spinal terminals of C-fibers, it will counter the opioid inhibition of transmitter release.[50] Thus, CCK_B receptor antagonists will be critical in testing whether CCK influences morphine analgesia. As would be predicted by this idea, the ability of morphine to inhibit spinal nociceptive processing is enhanced in the presence of selective CCK_B antagonists in normal animals. This, therefore, demonstrates a tonic physiological antagonism of morphine antinociception by endogenous CCK under conditions of acute nociception. More importantly, the nature of this interaction in animal models of inflammatory and neuropathic pain is altered.[25]

In neuropathies, morphine tends to have a reduced effectiveness, whereas after inflammation, morphine has enhanced actions. In fact, a few hours after carrageenan inflammation there are increases in the effects of morphine so that spinal morphine is almost 20-fold more potent than in normal rats.[51] The most likely basis for these results is a decreased availability of CCK within the spinal cord following carrageenan inflammation, either due to a decreased release of CCK or reduced content within the dorsal horn. This reduced functional activity of CCK in inflammation is a major factor in the enhanced potency of spinal morphine seen in these animals.[51]

Neuropathic models reveal that CCK produces effects entirely opposite to those seen in inflammation. There is little or no genuine CCK found in nociceptive C-fibers in normal animals;

consequently, endogenous CCK in the dorsal horn under non-pathological conditions is found in intrinsic neurons in superficial laminae of the dorsal horn and in descending fibers. This location of CCK in the spinal cord, a critical site for opioid analgesia, changes after nerve section. Intriguingly, induction of the peptide in afferents occurs after pathological nerve damage such as could occur in neuropathic pain.[26] This increase in spinal CCK leads to a reduction in the potency of spinal morphine in a rat model of neuropathic pain following peripheral nerve injury. Opioid responsiveness of this model was restored by CCK_B antagonism. Different pain states may then lead to changes in the levels and synthesis of CCK that can shift opioid sensitivity in either direction,[25] and in this regard it has been shown that CCK mediates the impact of the environment around the animal on morphine analgesia.[52]

The release of CCK is not, therefore, the same under all circumstances but varies from its normal state depending on different pain states. CCK antagonists being developed by the pharmaceutical industry may well enhance morphine analgesia in non-pathological pain states and restore morphine analgesia in humans with neuropathic pains. The possiblility of anxiolytic effects of these antagonists still remains to be determined[53] but would be a useful action when used as analgesic adjuncts especially in states of chronic pain where anxiety commonly accompanies pain. Thus, there is potential for novel opioids and opioid analgesia that can be manipulated by preventing the effects of CCK. However, in the absence of any clinically useful drugs acting on CCK at the present time, other approaches have explored the nature of the central mechanisms contributing to neuropathic pains should not be underestimated. The N-methyl-D-aspartate (NMDA) receptor for glutamate is of importance in the induction and maintenance of spinal nociceptive events following tissue damage and nerve dysfunction. Consideration of this target reveals some of the benefits of a rational approach to drug discovery but also some of the problems that can arise from drugs based on targets in the CNS.[54]

32.5 Central Hypersensitivity

Nearly 30 years ago it was reported that a burst of high-frequency C-fiber stimulation results in an marked and prolonged increase in responses of spinal neurons in the cat spinal cord. This repetition of a constant intensity, C-fiber stimulus induces the phenomenon of wind-up, whereby the responses of certain dorsal horn nociceptive neurons suddenly increase markedly (both in terms of magnitude and duration) despite the constant input into the spinal cord.[55] Wind-up would therefore seem to be a pivotal process in states where low levels of afferent input impinge upon central excitatory systems which then generate the exaggerated responses. Then, a decade ago, the same form of stimulation was shown to markedly enhance the flexion withdrawal reflex in rats recorded from motoneurons in spinal animals.[56] This demonstrated that brief noxious stimuli can enhance spinal excitatory events. Volatile general anesthesia such as with halothane fails to prevent this, thus having implications for post-operative pain states. Wind-up can be demonstrated in human subjects by the use of repeated electrical or heat stimulation.[57,58] As the pharmacology of these altered states of pain is unraveled, new analgesic targets are being revealed.

Investigation of the transmitter and receptor events behind these changes has focused on the excitatory amino-acids, because as these physiological events were being reported, tools to probe the function of the amino-acid receptors were being produced.

32.5.1 Excitatory Amino Acids and Central Hypersensitivity

Both small and large sensory fibers contain the excitatory amino-acids, glutamate and aspartate.[59] In C-fibers, the co-existence of glutamate with peptides[1,24] means that a noxious stimulus releases both peptides and excitatory amino acids into the spinal cord and so will lead to the activation of receptors for the excitatory amino acids on nociceptive neurons. The development of selective agents, especially antagonists,[60] for NMDA, metabotropic, and alpha-amino-3-hydroxy-5-methylisoxazole-4-propionic acid (AMPA) receptors has enabled their roles in the spinal processing of pain to be studied.[1,61-65]

The metabotropic receptor has a poorly defined role in pain states, but it may well contribute to nociception by acting to enhance NMDA and AMPA receptor function via intracellular actions such as making the channel opening by these receptors more effective. Furthermore, the availability of agents that influence the receptor is still somewhat limited, but it is clear that sub-types of receptor exist and that both excitatory and inhibitory actions can result from activation of the receptors. The roles of the AMPA receptor is more clear, as selective antagonists are available. This type of study has shown that not only acute noxious but also innocuous stimuli seem to be transmitted via AMPA receptor activation.[66] This low selectivity for noxious transmission may well preclude any use of AMPA antagonists in acute pain states, but this dual action of antagonists could be helpful in conditions where allodynia and pain co-exist. However, a major proviso is that AMPA receptors mediate fast synaptic transmission in many CNS functions, so blocking the receptor may well cause marked side-effects. The NMDA receptor does appear to be an important target site as evidence accumulates for a role of the receptor in more prolonged pain states involving hypersensitivity where functional alterations in central transmission processes may occur,[1,61-65] paralleling roles in other long-term events in the brain.[60,67,68]

In the hippocampus, long-term potentiation has long been viewed as a substrate for the first stages of memory. This event is also an NMDA-induced persistent enhancement in synaptic efficacy, but it endures for hours to days even in the absence of afferent input.[67] Furthermore, long-term depression can oppose these enhancements of activity.[68] Not only are spinal events considerably shorter than those in the hippocampus, but there is little evidence that central hypersensitivity states can remain active in the absence of peripheral inputs. In fact, animal and clinical studies have shown that the states of hypersensitivity are entirely dependent on peripheral inputs for maintenance.[69] In the context of neuropathic pain, it can be noted that systemic local anesthetics used in this condition can have selective effects on ectopic foci in a damaged peripheral nerve at doses that do not alter conduction in the nerve.[70] Additional sites of action of local anesthetics may well include a spinal reduction in excitability and so less wind-up as a consequence of sodium channel blockade.[71]

The complex that makes up the NMDA receptor-channel contains a number of sites that are available for modulation: the receptor for glutamate, the channel, and a co-agonist site for glycine. In addition, the interplay between chemicals acting at these sites establishes the particular conditions that have to be met in order for neuronal depolarization to occur. These include the release and binding of the co-agonists for the receptor, glycine and glutamate, and a non-NMDA-induced depolarization to remove the resting magnesium block of the channel.[1] Neurokinin receptor antagonists can reduce NMDA-mediated responses, and so it may be that C-fiber-induced release of excitatory peptides may provide the required depolarization to remove the block.[61] A summation of activity is necessary for the block to be lifted, so short-lasting, acute stimuli are unable to activate the receptor complex. Thus, the NMDA receptor-channel complex is not involved in low levels of synaptic transmission; when an input is sufficiently prolonged and/or intense, the complex will suddenly become active and add a powerful depolarizing excitatory drive to transmission of pain in the spinal cord. This is likely to be a key event in the central hypersensitivity as seen with longer term painful messages.[1,57,62-65] The use of formalin as a model of inflammation illustrates some of these principles. Studies, both electrophysiological and behavioral, indicate that only the delayed and prolonged response to formalin is reduced by NMDA antagonism. At this stage of the response to formalin, the C-fiber inputs to the cord are relatively low, but the neuronal and behavioral reactions are vigorous. Thus, the spinal NMDA receptors are amplifying and prolonging a low level of afferent input.[72] The acute activity generated is AMPA-receptor mediated. Other approaches have revealed roles of the NMDA receptor in spinal pain processes, including a number of other models of inflammatory states, and with regard to visceral pain states bladder inflammation elicits spinal hypersensitivity which is mediated by NMDA receptors. Thus, there is also evidence for an involvement of the NMDA receptor in ischaemic pain.[24,65]

NMDA-receptor-mediated events are relevant to pain evoked by nerve injury. Initial discharges from the original nerve injury involve spinal NMDA receptor activation which contributes to the subsequent pain states.[73] Here, blocking the receptor during the act of damaging the nerve has beneficial effects on the subsequent pain states. Behavioral studies in the CCI model have shown that NMDA receptor activation is required for both the induction and the maintenance of the pain-related behaviors.[74–76] There is therefore a strong association between the NMDA receptor and neuropathic hyperalgesia/allodynia, as they can both be reduced or abolished by NMDA antagonists. In addition to the potential of the use of these agents alone, the combination of an NMDA antagonist with an opioid may restore the effects of the opioid.[17,77,78] There have been a number of preliminary clinical studies suggesting that licensed drugs with NMDA antagonist properties can be effective in neuropathic pains.[1] There appears to be little difference between the effects of receptor blockers and channel blockers such as ketamine, MK-801, and memantine. However, MK-801 has been shown to be less effective when given two weeks after axotomy where the lesion is more extreme, than compared to earlie stages of the syndrome.[79] It has been shown that NMDA receptor antagonism relieves behavioral symptoms following spinal nerve ligation at a number of post-lesion time points, indicating that, in general, effciacy is maintained.[54]

32.5.2 Controlling Excitatory Amino Acid Events

The research that has implicated the NMDA receptor in these various models of different pain states has been facilitated by the fact that there are many experimental drugs which effectively block the receptor, channel, or associated sites. Furthermore, translation of the basic research to the clinic has been rapid, as the analgesic/anesthetic ketamine blocks the channel associated with the NMDA receptor. Thus, the role of the receptor can be gauged in patients. Clinical studies in patients with a number of different pain states have been reported, but it is noteworthy that a controlled study showed that ketamine can reduce allodynias and hyperalgesias and can cause pain relief in circumstances where opioids have poor or restricted efficacy.[80] The cough suppressants dextrophan and dextromethorphan are also antagonists at the NMDA channel but are relatively weak. However, they have been shown to reduce wind-up itself,[81] to reduce hyperalgesia in the CCI model of neuropathic pain after spinal application,[58] and to reduce wind-up in human psychophysical studies.[58] Another addition to the list is the anti-Parkinson drug, memantine, which has been shown to be an effective NMDA antagonist, although the extent of its efficacy has been less studied. Thus, there is great potential for NMDA antagonists as novel analgesics, although it has to be recognized that they would only be effective in reducing manifestations of central hypersensitivity such as allodynia and hyperalgesia but would not be expected to abolish background pain.[62]

The studies above lend impressive weight to support the idea that this NMDA receptor complex could be a very useful target. However, the side-effects of drugs acting on the complex are likely to be a problem. The psychotomimetic effects of ketamine are well recognized, and the predicted side-effects of sedation and potential amnesia illustrate the somewhat gloomy prognosis for these agents. However, the psychotomimetic effects gauged by pre-clinical studies do not generalize to all NMDA antagonists because of the different sites on the receptor complex.[82,83] Glycine site antagonists are highly effective at reducing mediated NMDA events, and it has been shown that motor impairment with these agents is less than with antagonists at other sites on the NMDA complex. Glycine site antagonists, which are effective in the various animal models,[84] therefore, have considerable potential. Likewise, the profile of dextrophan and memantine from clinical studies is not the same as ketamine, so there are further directions for the production of clinically useful molecules.

Reorganization of afferent termination sites of Aβ-fibers in the dorsal horn after nerve injury could result in low-threshold inputs activating projection neurons, thus targeting inappropriate higher brain centers.[38,85] Clinically, allodynia is mediated by low-threshold myelinated fibers, such that abnormal central processing of these inputs occurs.[39,86] Allodynia is a good example of inhibitions setting the level of transmission, as blocking the $GABA_A$ receptor in normal animals produces NMDA-receptor-mediated allodynia. NMDA receptor antagonists are effective against tactile-evoked nociception, whereas morphine is not.[86] Here the pain is likely to be resulting from pathways where opioid receptors do not control activity, namely low-threshold pathways. Further evidence for how levels can alter the roles of a transmitter is the finding that spinal GABA levels are enhanced after inflammation yet reduced in neuropathic states.[87] Thus, in neuropathy, the increased levels of CCK contribute to the reduced opioid sensitivity, and the decreased levels of GABA may facilitate NMDA-mediated excitations.

32.5.3 Inhibitory Systems

Thus, other than agents that reduce excitations, an alternative approach is to increase inhibitions. The extent of the reduced ability of morphine to control the various components of the neuropathic pain syndrome and the relative effectiveness of the opioid by spinal and systemic routes may still be unclear, but it is quite clear that a number of patients need other agents. A number of anti-convulsants influence GABAergic inhibitions, so the use of agents such as felbamate may shed light on the role of this transmitter after nerve injury. Agents that more directly increase GABA function, such as benzodiazepines are not going to be novel. The anti-convulsant, gabapentin, which despite its name appears not to act directly on GABA systems, has been shown to be effective in models of inflammation[88] and has been used in patients with neuropathic pain. Gabapentin produces a total lack of inhibitory effects over the same dose range in normal animals yet is effective both after the induction of carrageenan inflammation and for neuropathic pain. Clearly, then, the activity or pathology caused by neuropathy and inflammation has induced or revealed a novel action of the drug. Thus, gabapentin has clear antinociceptive effects but by an as yet unknown mechanism which is rapidly induced by inflammation and which also is revealed or induced by nerve damage. The site of action is unlikely to be a peripheral mechanism, and possibilities include channels (calcium, etc.), intracellular mediators (protein kinases, etc.), or other sites.[88] Identification of the mode of action of this drug would produce a potentially useful target.

32.5.4 Adenosine

In addition to the as yet unknown mechanisms behind the effects of gabapentin, there are more rational targets for inhibitory agents. There are several reasons why the purinergic system may be one such suitable target. The main receptor for adenosine found in the dorsal horn, the inhibitory A_1-receptor, has a postsynaptic location.[89] This would mean that unlike opioids, A_1-receptor function will not be compromised by damage to and loss of primary afferent terminals. Also, behavioral studies have indicated that spinal administration can result in antinociception.[90] From acute electrophysiological studies it is clear that the spinal activation of the A_1-receptor controls noxious inputs into the dorsal horn.[91] Electrophysiological studies have shown that the NMDA-receptor-mediated responses of neurons such as wind-up are reduced to the greatest extent by A_1-receptor agonists.[91] Unlike NMDA-receptor antagonists and mu-receptor agonists, A_1-agonists have a dual effect on wind-up, reducing both the baseline responses and increased excitability. Neuropathic pain in patients can be controlled by adenosine, and humans report a slow build-up of noxious sensations over a period of many hours after adenosine infusion has ceased,[92] suggesting that activation of the NMDA receptor must first overcome a purinergic inhibitory threshold, as has been suggested in other areas of the brain.[93]

The inhibition of these experimentally produced, enhanced responses, a reduction in C-fiber evoked responses, along with a lack of effect on the Aβ-fiber-evoked responses, provides a sound scientific basis for the use of adenosine or analogs acting at the A_1-receptor in the treatment of more persistent, and altered pain states. As with any drug, there will always be associated side-effects; adenosine also has cardiovascular effects and is clinically the drug of choice in the treatment of supra-ventricular tachycardia (SVT). However, it would appear that the clinical doses of adenosine required to inhibit noxious inputs are much lower than those required to treat SVT, and cardiovascular side-effects can be contained.[92,94,95]

The role of endogenous adenosine in inflammation is very complex, as in the periphery at least it can have a variety of effects. ATP has been shown to be released from sensory neurons.[96] The adenosine converted from ATP can add to adenosine that may be released from surrounding tissues during damage and ischemia. The consequences of release, regardless of source, will depend on the concentration of adenosine at any one time, as low concentrations will activate the A_1-receptor, and higher concentrations activate the excitatory A_2-receptor.[97] Adenosine can also add to the inflammation by causing mast cells to degranulate via activation of the A_3-receptor.[98] The effects of adenosine at the peripheral site of damage can therefore be antinociceptive or pronociceptive, depending on which receptors are activated.

It has been found that spinal A_1-receptor activation is effective at controlling both the first and second phase of the formalin evoked response.[91,99,100] These results would imply that A_1-receptor activation can control phasic as well as tonic activity at central levels during inflammation.

It has also been found that A_1-receptor activation is effective at reducing and controlling allodynia in animals and in clinical studies with adenosine and analogs in patients with allodynia resulting from nerve damage[92,95] and from allodynia induced by mustard oil.[101]

The other approach to this mediator has been to use adenosine kinase inhibitors to protect endogenously released adenosine, prolonging its half-life. When given systemically (s.c.) or intrathecally, adenosine kinase inhibitors produce antinociception, and these enzyme inhibitors are also effective after carrageenan and formalin are induced. In the formalin response, both the first and the second phase were inhibited, with a greater inhibitory effect on the second phase. This would suggest that there is a greater release of adenosine during the tonic phase of this response. The predominant inhibitory effects on wind-up and post-discharge with protected adenosine, along with a greater inhibition of the second phase of the formalin-evoked response, would suggest that adenosine (or ATP) is released in response to the phenomenon, known to be driven by activation of the NMDA-receptor (see Section 32.5.1). It is likely that adenosine is released and converted from released ATP in a calcium-dependent manner in response to NMDA-receptor activation and also AMPA-receptor activation, as seen in other areas of the CNS.[93]

It has been well documented that endogenous adenosine is released in response to ischemia and under conditions of increased excitotoxicity, such as in cases of stroke.[102] Adenosine then acts as a negative feedback mechanism to control further NMDA-receptor activity. For activation, the NMDA receptor requires membrane depolarization to remove the Mg^{++} block from the ion channel, and the A_1-receptor hyperpolarizes membranes via increasing K^+ channel activity. Thus, the NMDA receptor may be physiologically antagonized by the hyperpolarization caused by A_1-receptor activity. If this mechanism also applies to the spinal cord it could account for the enhanced duration of action of adenosine seen in the clinic,[92] as adenosine may provide an inhibitory threshold which has to be overcome before transmission can maximally proceed again.

The use of adenosine or analogs at the spinal level may therefore be of clinical use both in acute and more persistent states. If given under acute but potentially traumatic circumstances, adenosine may prevent the pain state from developing into a more damaging state. Thus, it can be seen that a number of drugs that have differing mechanisms and sites of action may have similar effects on spinal excitability after systemic administration, and a number of these can directly or indirectly control states of central hypersensitivity.

32.5.5 Nitric Oxide

There is another "neuromodulator" that can increase the responses to glutamate: the gas nitric oxide (NO).[103] The enzyme responsible for the synthesis of NO, NO synthase (NOS), has been located in the dorsal horn of the spinal cord as well as in DRG cells.[104] It has been demonstrated that NO is synthesized on demand in response to increased activity in intrinsic cells. An increase in NMDA-receptor activation can result in the increase in calcium influx.[103] This rise in intracellular calcium can result in many processes being switched on, such as the regulation of genes (namely, c-*fos* and c-*jun*) and the activation of calcium-dependent processes. One such calcium-dependent process is the production of NO. The enzyme NOS is activated by a calmodulin-sensitive site on the enzyme, which has previously been activated by the increase in intracellular calcium. NOS catalyzes the conversion of L-arginine and molecular oxygen to NO and L-citrulline. Once produced, NO can diffuse into neighboring cells or can have effects in the cell from which it is produced. One target is the primary afferent terminals, where NO activates soluble guanylate cyclase to increase cGMP which in turn can cause the further release of glutamate. For this reason NO is a possible candidate for retrograde transmission in the dorsal horn.[105] Thus, when released, a positive feedback may be established by NO. Inhibitors of nitric oxide synthase have been shown to decrease the enhanced responses of spinal neurons in both inflammation and neuropathy,[105] and the ability of certain drugs to selectively block the neuronal isoform of the enzyme lends hope to the idea that blocking the actions of a gas could provide novel analgesia without cardiovascular changes.[106]

32.6 Conclusions

Much of the target identification has come from basic research identifying receptors and enzymes as targets. A rational approach to the production of novel analgesics has therefore been produced by the identification of receptor systems in a number of different animal models. However, we can still be surprised. An example is the very recent report of potent analgesia produced by a nicotinic receptor agonist in models of thermal acute and chemical and neuropathic persistent pain states.[107] Epibatidine is a frog skin alkaloid with analgesic effects but a poor side-effect profile. This is due to a lack of any selectivity for the central vs. peripheral nicotinic receptors. The novel compound was designed to be highly selective for the neuronal receptor and consequently has improved analgesia and less adverse effects than the amphibian lead agent.[107] Until this agent was tested in a number of models, it would not have been predicted by our knowledge of the pharmacology of pain that this cholinergic system would have been a useful target.

When we add on the monoamine systems, the roles of peptides in afferent transmission, and all the possible peripheral targets, especially those revealed in inflammation, there is an impressive list of targets. Space does not permit coverage of these latter systems, but for further information, these have been recently reviewed in detail.[1]

The future pharmacological treatment of pain in the next decade or so may be very different from the past 20 years; however, for this to happen, the pharmaceutical industry has to take the next step to develop drugs with actions on the targets described in this account.

References

1. Dickenson, A.H. and Besson, J.-M., *The Pharmacology of Pain,* Springer-Verlag, Berlin, 1997.
2. Uhl, G.R., Childers, S., and Pasternak, G., An opiate receptor gene family reunion, *Trends Neurosci.,* 17, 89, 1994.

3. Pasternak G.W. and Wood, P.J., Multiple opioid receptors, *Life Sci.,* 28, 1889, 1986.
4. Jiang, Q.A.E., Takemori, A.E., and Sultana, M., Differential antagonism of opioid delta antinociception by [D-Ala2, Leu5, Cys6] enkephalin and naltrindole 5′-isothiocynanate: evidence for delta receptor subtypes, *J. Pharmacol. Exp. Ther.,* 257, 1069, 1991.
5. Traynor, J., Subtypes of the κ opioid receptor: fact or fiction?, *Trends Pharmacol. Sci.,* 10, 52–53, 1991.
6. Zadina, J.E., Hackler, L., Ge, L.J., and Kastin, A.J., A potent and selective endogenous agonist for the μ-opiate receptor, *Nature,* 386, 499, 1997.
7. Martin-Schild, S., Zadina, J.E., Gerall, A.A., Vigh, S., and Kastin, A.J., Localization of endomorphin-2-like immunoreactivity in the rat medulla and spinal cord, *Peptides,* 18, 1641, 1997.
8. Chapman, V., Diaz, A., and Dickenson, A.H., Distinct inhibitory effects of spinal endomorphin-1 and endomorphin-1 on evoked dorsal horn neuronal responses in the rat, *Br. J. Pharmacol.,* 122, 1537, 1997.
9. Meunier, J.-C., Mollereau, C., Toll, L., Suaudeau, C., Moisand, C., Alvinerie, P., Butour, J.-L., Guillemot, J.-C., Ferrara, P., Monsarrat, B., Mazarguil, H., Vassart, G., Parmentier, M., and Costentin, J., Isolation and structure of the endogenous agonist of opioid receptor-like ORL1 receptor, *Nature,* 377, 532, 1995.
10. Reinscheid, R.K., Nothacker, H-P., Bourson, A., Ardati, A., Henningsen, R.A., Bunzow, J.R., Grandy, D.K., Langen, H., Monsma, F.J., and Civelli, O., Orphanin FQ: a neuropeptide that activates an opioid like G protein-coupled receptor, *Science,* 270, 792, 1995.
11. Grisel, J.E., Mogil, J.S., Belnap, J.K., and Grandy, D.K., Orphanin FQ acts as a supraspinal, but not a spinal, anti-opioid peptide, *NeuroReport,* 7, 2125, 1996.
12. Wick, M.J., Minnerath, S.R., Lin, X., Elde, R., Law, P-Y., and Loh, H.H., Isolation of a novel cDNA encoding a putative membrane receptor with high homology to the cloned μ, δ and κ opioid receptors, *Mol. Brain Res.,* 27, 37, 1994.
13. Vaughan, C.W. and Christie, M.J., Increase by the ORL1 receptor (opioid receptor-like1) ligand, nociceptin, of inwardly rectifying K conductance in dorsal raphe nucleus neurons, *Br. J. Pharmacol.,* 117, 1609, 1996.
14. Stanfa, L.C., Chapman, V., Kerr, N., and Dickenson, A.H., Inhibitory action of nociceptin on spinal dorsal horn neurons of the rat *in vivo*, *Br. J. Pharmacol.,* 118, 1875, 1996.
15. Zhang, S. and Yu, L., Identification of dynorphins as endogenous ligands for an opioid receptor-like orphan receptor, *J. Biol. Chem.,* 270, 22772, 1995.
16. Dickenson, A.H., Mechanisms of the analgesic actions of opiates and opioids, *Br. Med. Bull.,* 47, 690, 1991.
17. Dickenson, A.H., Where and how opioids act, in *Proc. 7th World Congress on Pain, Progress in Pain Research and Management,* Vol. 2, Gebhart, G.F., Hammond, D.L., and Jensen, T., Eds., IASP Press, Seattle, 1994, p. 525.
18. Millan, M.J., κ opioid receptors and analgesia, *Trends Pharm. Sci.,* 11, 70, 1990.
19. Sullivan A.F., Dickenson, A.H., and Roques, B.P., δ-opioid mediated inhibitions of acute and prolonged noxious-evoked responses in rat dorsal horn neurons, *Br. J. Pharmacol.,* 98, 1039, 1989.
20. Calderon, S.N., Rothman, R.B., and Porreca, F., Probes for narcotic receptor mediated phenonema. 19 1 synthesis of (+)-4-[(aR)-a-(2S,5R)-4-allyl-2-5-dimethyl-1-piperazinyl0-3-methoxy]-N,N-dibethylbenz amide (SNC80): a highly selective nonpeptide δ opioid receptor agonist, *J. Med. Chem.,* 37, 2125, 1994.
21. Roques, B.P., Noble, F., Dauge, V., Fournie-Zaluski, M.C., and Beaumont, A., Neutral endopeptidase 24.11: structure, inhibition and experimental and clinical pharmacology, *Pharmacol. Rev.,* 45, 88, 1993.

22. Dickenson, A..H., Sullivan, A.F., Fournie-Zaluski, M.-C., and Roques, B., Prevention of degradation of endogenous enkephalins produces inhibition of nociceptive neurons in rat spinal cord, *Brain Res.*, 408, 185, 1986.

23. Ruiz, F., Fournie-Zaluski, M.C., Roques, B.P., and Maldonado, R., Similar decrease in spontaneous morphine abstinence by methadone and RB101, an inhibitor of enkephalin catabolism, *Br. J. Pharmacol.*, 119, 174, 1996.

24. Dray, A., Urban, L., and Dickenson, A.H., Pharmacology of chronic pain, *Trends Pharmacol. Sci.*, 15, 190, 1994.

25. Stanfa, L.C, Dickenson, A.H, Xu, X.-J., and Wiesenfeld-Hallin, Z., Cholecystokinin and morphine analgesia: variations on a theme, *Trends Pharmacol. Sci.*, 15, 65, 1995.

26. Xu, X.J, Puke, M.J.C, Verge, V.M.K, Wiesenfeld-Hallin, Z., Hughes, J., and Hokfelt, T., Up-regulation of cholecystokinin in primary sensory neurons is associated with morphine insensitivity in experimental neuropathic pain in the rat, *Neurosci. Lett.*, 152, 129, 1993.

27. McQuay, H., Carroll, D., Jadad, A.R., Wiffen, P., and Moore, A., Anticonvulsant drugs for the management of pain: a systematic review, *Br. Med. J.*, 311, 1047, 1995.

28. Bennett, G.J., Animal models of neuropathic pain, in *Proc. 7th World Congress on Pain, Progress in Pain Research and Management,* Vol. 2, Gebhart, G.F., Hammond, D.L., and Jensen, T., Eds., IASP Press, Seattle, 1994.

29. Seltzer, Z., Dubner, R., and Shir, Y., A novel behavioral model of neuropathic pain disorders produced in rats by partial sciatic nerve injury, *Pain*, 43, 205, 1990.

30. Bennett, G.J. and Xie, Y-K., A peripheral mononeuropathy in rat produces disorders of pain sensation like those seen in man, *Pain,* 33, 87, 1988.

31. Kim, S.H. and Chung, J.M., An experimental model for peripheral neuropathy produced by segmental spinal nerve ligation in the rat, *Pain*, 50, 355, 1992.

32. Wall, P.D. and Devor, M., Sensory afferent impulses originate from dorsal root ganglia as well as from the periphery in normal and nerve injured rats, *Pain,* 17, 321, 1983.

33. Tal, M. and Eliav, E., Abnormal discharge originates at the site of nerve injury in experimental constriction neuropathy (CCI) in the rat, *Pain,* 64, 511, 1996

34. Kajander, K.C., Wakisaka, S., and Bennett, G.J., Spontaneous discharge originates in the dorsal root ganglion at the onset of a painful peripheral neuropathy in the rat, *Neurosci. Lett.*, 138, 225, 1992.

35. Kajander, K.C. and Bennett, G.J., Onset of a painful peripheral neuropathy in rat: a partial and differential deafferentation and spontaneous discharge in Aβ and Aδ primary afferent neurons, *J. Neurophysiol.*, 68, 734, 1992.

36. Yoon,Y.W., Heung, S.N., and Chung, J.M., Contributions of injured and intact afferents to neuropathic pain in an experimental rat model, *Pain,* 64, 27, 1996.

37. Na, H.S., Leem, J.W., and Chung, J.M., Abnormalities of mechanoreceptors in a rat model of neuropathic pain: possible involvement in mediating mechanical allodynia, *J. Neurophysiol.*, 70, 522, 1993.

38. Lekan, H.A., Carlton, S.M., and Coggeshall, R.E., Sprouting of Aβ fibers into lamina II of the rat dorsal horn in peripheral neuropathy, *Neurosci. Lett.*, 208, 147, 1996.

39. Devor, M., Pain mechanisms and pain syndromes, in *Pain 1996: An Updated Review,* Campbell, J., Ed., IASP Press, Seattle, 1996, p. 103.

40. Diaz, A. and Dickenson, A.H., Blockade of spinal N- and P-type, but not L-type, calcium channels inhibits the excitability of rat dorsal horn neurons produced by subcutaneous formalin inflammation, *Pain,* 69, 93, 1997.

41. Miljanich, G.P. and Ramachandran, J., Antagonists of neuronal calcium channels: structure, function and therapeutic implications, *Ann. Rev. Pharmacol. Toxicol.,* 35, 707, 1995.

42. Akopian, A.N., Sivilotti, L., and Wood, J.N., A tetrodotoxin-resistant voltage-gated sodium channel expressed by sodium channels, *Nature,* 379, 257, 1996.
43. Palecek, J., Paleckova, V., Dougherty, P.M., Carlton, S.M., and Willis, W.D., Responses of spinothalamic tract cells to mechanical and thermal stimulation of skin in rats with experimental peripheral neuropathy, *J. Neurophysiol.,* 67, 1562, 1992.
44. Laird, J.M.A. and Bennett, G.J., An electrophysiological study of dorsal horn neurons in the spinal cord of rats with an experimental peripheral neuropathy, *J. Neurophysiol.,* 69, 1, 1993.
45. Takaishi, K., Eisele, J.H., and Carstens, E., Behavioral and electrophysiological assessment of hyperalgesia and changes in dorsal horn responses following partial sciatic nerve ligation in rats, *Pain,* 66, 297, 1996.
46. Baber, N.S., Dourish, C.T., and Hill, D.R., The role of CCK, caerulein, and CCK antagonists in nociception, *Pain,* 39, 307, 1989.
47. Zhou, Y., Sun, Y.-H., Zhang, Z.-W., Han, J.-S., Increased release of immunoreactive cholecystokinin octapeptide by morphine and potentiation of μ-opioid analgesia by CCK_B receptor antagonist L-365,260 in rat spinal cord, *Eur. J. Pharmacol.*, 234, 147, 1993.
48. Besse, D., Lombard, M.C., Zakac, J.M., Roques, B.P., and Besson, J.M., Pre- and postsynaptic distribution of mu, delta and kappa opioid receptors in the superficial layers of the cervical dorsal horn of the rat spinal cord, *Brain Res.,* 521,15, 1990.
49. Ghilardi, J.R., Allen, C.J., Vigna, S.R., McVey, D.C., and Mantyh, P.W., Trigeminal and dorsal root ganglion neurons express CCK receptor bindingsites in the rat, rabbit, and monkey: possible site of opiate-CCK analgesic interactions, *J. Neurosci.,* 12, 4854, 1992.
50. Wang, J., Ren, M., and Han, J., Mobilization of calcium from intracellular stores as one of the mechanisms underlying the antiopioid effects of cholecystokinin octapeptide, *Peptides,* 13, 947, 1992.
51. Stanfa, L.C. and Dickenson, A.H., Cholecystokinin as a factor in the enhanced potency of spinal morphine following carrageenan inflammation, *Br. J. Pharmacol.,* 108, 967, 1993.
52. Wiertelak, E.P., Maier, S.F., and Watkins, L.R., Cholecystokinin antianalgesia: safety cues abolish morphine analgesia, *Science,* 256, 830, 1992.
53. Rodgers, R.J and Johnson, N.J., Cholecystokinin and anxiety: promises and pitfalls, *Crit. Rev. Neurobiol.,* 9, 345, 1995.
54. Dickenson, A.H., NMDA receptor antagonists as analgesics, in *Progress in Pain Research Management,* Fields, H.L. and Liebeskind, J.C., Eds., IASP Press, Seattle, 1994, 173.
55. Dickenson, A.H., Spinal cord pharmacology of pain, *Br. J. Anaesth.,* 75, 132, 1995.
56. Woolf, C.J., Evidence for a central component of post-injury pain hypersensitivity, *Nature,* 221, 313, 1983.
57. Price, D.D., Mao, J., and Mayer, D.J., Central neural mechanisms of normal and abnormal pain states, in *Progress in Pain Research Management,* Fields, H.L. and Liebeskind, J.C., Eds., IASP Press, Seattle, 1994, p. 61.
58. Price, D.D., Mao, J., Frenk, H., and Mayer, D.J., The N-methyl-D-aspartate antagonist dextromethorphan selectively reduces temporal summation of second pain, *Pain,* 59, 165, 1994.
59. Battaglia, G. and Rustioni, A., Coexistence of glutamate and substance P in dorsal root ganglion cells of the rat and monkey, *J. Comp. Neurol.,* 277, 302, 1988.
60. Collingridge, G. and Singer, W. Excitatory amino acid receptors and synaptic plasticity, *Trends Pharmacol. Sci.,* 11, 290, 1990.
61. Urban, L., Thompson, S.W.N., and Dray, A., Modulation of spinal excitability: co-operation between neurokinin and excitatory amino acid neurotransmitters, *Trends Neurosci.,* 17, 432, 1994.
62. Dickenson, A.H., A cure for wind-up: NMDA receptor antagonists as potential analgesics, *Trends Pharmacol. Sci.,* 11, 307, 1990.

63. Daw, N.W, Stein, P.S.G., and Fox, K., The role of NMDA receptors in information processing, *Ann. Rev. Neurosci.,* 16, 207, 1993.

64. Dubner, R. and Ruda, M.A., Activity-dependent neuronal plasticity following tissue injury and inflammation, *Trends Neurosci.,* 15, 96, 1992.

65. McMahon, S.B., Lewin, G.R., and Wall, P.D., Central excitability triggered by noxious inputs, *Curr. Opin. Neurobiol.,* 3, 602, 1993.

66. Schouenborg, J. and Sjolund, B.H., First-order nociceptive synapses in rat dorsal horn are blocked by an amino acid antagonist, *Brain Res.,* 379, 394, 1986.

67. Bliss, T.V.P. and Collingridge, G.L., A synaptic model of memory: long-term potentiation in the hippocampus, *Nature,* 361, 31, 1993.

68. Dudek, S.M. and Bear, M.F., Homosynaptic long-term depression in area CA1 of the hippocampus and effects of NMDA receptor blockade, *Proc. Nat. Acad. Sci. USA,* 89, 4363, 1992.

69. Gracely, R., Lynch, S.A., and Bennett, G.J., Painful neuropathy: altered central processing maintained dynamically by peripheral input, *Pain,* 52, 251, 1993.

70. Devor, M., Wall, P.D., and Catalan, N., Systemic lidocaine silences ectopic neuroma and DRG discharge without blocking nerve conduction, *Pain,* 48, 26, 1992.

71. Fraser, H., Chapman, V., and Dickenson, A.H., Spinal local anaesthetic actions on afferent evoked responses and wind-up of nociceptive neurons in the rat spinal cord: combination with morphine produces marked potentiation of antinociception, *Pain,* 49, 33, 1992.

72. Haley, J.E, Sullivan, A.F., and Dickenson, A.H., Evidence for spinal N-methyl-D-aspartate receptor involvement in prolonged chemical nociception in the rat, *Brain Res.,* 518, 218, 1990.

73. Seltzer, Z., Cohn, S., Ginzburg, R., and Beilin, B.Z., Modulation of neuropathic pain in rats by spinal disinhibition and NMDA receptor blockade of injury discharge, *Pain,* 45, 69, 1991.

74. Qian, J., Brown, S.D., and Carlton, S.M., Systemic ketamine attenuates nociceptive behaviors in a rat model of peripheral neuropathy. *Brain Res.,* 715, 51, 1996.

75. Chaplan, S.R., Malmberg, A.B., and Yaksh, T.L., Efficacy of spinal NMDA receptor antagonism in formalin hyperalgesia and nerve injury evoked allodynia in the rat, *J. Pharmacol. Exp. Ther.,* 280, 829, 1977.

76. Wegert, S., Ossipov, M.H., Nichols, M.L., Bian, D., Vanderah, T.W., Malan, T.P., and Porreca, F., Differential activities of intrathecal MK801 or morphine to alter responses to thermal and mechanical stimuli in normal or nerve injured rats, *Pain,* 7, 57, 1977.

77. Chapman, V. and Dickenson, A.H., The combination of NMDA antagonism and morphine produces profound antinociception in the rat dorsal horn, *Brain Res.,* 573, 321, 1992.

78. Yamamoto, T. and Yaksh, T.L., Studies on the spinal interaction of morphine and the NMDA antagonist MK-801 on the hyperesthesia observed in a rat model of sciatic mononeuropathy, *Neurosci. Lett.,* 135, 67, 1992.

79. Xu, X.-J., Zhang, X., Hokfelt, T., and Wiesenfeld-Hallin, Z., Plasticity in spinal nociception after peripheral nerve section: reduced effectiveness of the NMDA receptor antagonist MK-801 in blocking wind-up and central sensitization of the flexor reflex, *Brain Res.,* 670, 342, 1995.

80. Eide, P.K., Stubhaug, A., Oye, I., and Breivik, H., Continuous subcutaneous administration of the N-methyl-D-aspartic acid (NMDA) receptor antagonist ketamine in the treatment of post-herpetic neuralgia, *Pain,* 61, 221, 1995.

81. Tal, M. and Bennett, G.J., Dextrophan relieves neuropathic heat-evoked hyperalgesia in the rat, *Neurosci. Lett.,* 151, 107, 1993.

82. Rogawski, M.A., Therapeutic potential of excitatory amino-acid antagonists; channel blockers and 2,3-benzodiazepines, *Trends Pharm. Sci.,* 14, 325, 1993.

83. Willetts, J., Balster, R.L., and Leander, D., The behavioral pharmacology of NMDA receptor antagonists, *Trends Pharm. Sci.,* 11, 423, 1993.
84. Dickenson, A.H. and Aydar, E., Antagonism at the glycine site on the NMDA receptor reduces spinal nociception in the rat, *Neurosci. Lett.,* 121, 263, 1990.
85. Woolf, C.J. and Doubell, T., The pathophysiology of chronic pain — increased sensitivity to low-threshold Aβ-fiber inputs, *Curr. Opin. Neurobiol.,* 4, 525, 1994.
86. Yaksh, T.L., Behavioral and autonomic correlates of the tactile evoked allodynia produced by spinal glycine inhibition: effects of modulatory receptor systems and excitatory amino acid antagonists, *Pain,* 37, 111, 1989.
87. Castro-Lopes, J.M., Tavares, I., and Coimbra, A., GABA decreases in the spinal cord dorsal horn after peripheral neurectomy, *Brain Res.,* 620, 287, 1993.
88. Stanfa, L.C., Singh, L., Williams, R.G., and Dickenson, A.H., Gabapentin, ineffective in normal rats, markedly reduces C-fiber evoked responses after inflammation, *NeuroReport,* 8, 587, 1977.
89. Choca, J.I., Green, R.D., and Proudfit, H.K., Adenosine A_1 and A_2 receptors of the substantia gelatinosa are located predominantly on intrinsic neurons: an autoradiography study, *J. Pharmacol. Exp. Ther.,* 247, 757, 1988.
90. Sosnowski, M., Stevens, C.W., and Yaksh, T.L., Assessment of the role of A_1/A_2 adenosine receptors mediating the purine antinociception, motor and autonomic function in the rat spinal cord, *J. Pharmacol. Exp. Ther.,* 250, 915, 1989.
91. Reeve, A.J. and Dickenson, A.H., The roles of adenosine in the control of acute and more persistent nociceptive responses of dorsal horn neurons in the anaesthetized rat, *Br. J. Pharmacol.,* 116, 2221, 1995.
92. Sollevi, A., Belfrage, M., Lundeberg, T., Segerdahl, M., and Hansson, P., Systemic adenosine infusion: a new treatment modality to alleviate neuropathic pain, *Pain,* 61, 155, 1995.
93. Craig, C.G. and White, T.D. Low-level *N*-methyl-D-aspartate receptor activation provides a purinergic inhibitory threshold against further *N*-methyl-D-aspartate-mediated neurotransmission in the cortex, *J. Pharmacol. Exp. Ther.,* 260, 1278, 1992.
94. Belfrage, M., Sollevi, A., Segerdahl, M., Sjolund, K.-F., and Hansson, P., Systemic adenosine infusion alleviates spontaneous and stimulus evoked pain in patients with peripheral neuropathic pain, *Anesth. Analg.,* 81, 713, 1995.
95. Karlsten, R. and Gordh, Jr., T., An A_1-selective adenosine agonist abolishes allodynia elicited by vibration and touch after intrathecal injection, *Anesth. Analg.,* 80, 844, 1995.
96. Holton, P., The liberation of adenosine triphosphate on antidromic stimulation of sensory nerves, *J. Physiol.,* 145, 494, 1959.
97. Fredholm, B.B., Abbracchio, M.P., Burnstock, G., Daly, J.W., Harden, T.K., Jacobson, K.A., Leff, P., and Williams, M., Nomenclature and classification of purinoreceptors, *Pharmacol. Rev.,* 46, 143, 1994.
98. Linden, J., Cloned adenosine A_3 receptors: pharmacological properties, species differences and receptor functions, *Trends Pharmacol. Sci.,* 15, 298, 1994.
99. Malmberg, A.B. and Yaksh, T.L., Pharmacology of the spinal action of ketorolac, morphine, ST-91, U50488H, and L-PIA on the formalin test and isobolographic analysis of the NSAID interaction, *Anesthesiology,* 79, 270, 1993.
100. Poon, A. and Sawynok, J., Antinociception by adenosine analogs and an adenosine kinase inhibitor: dependence on formalin concentration, *Eur. J. Pharmacol.,* 286, 177, 1995.
101. Segerdahl, M., Ekblom, A., Sjolund, K.-F., Belfrage, M., Forsberg, C., and Sollevi, A., Systemic adenosine attenuates touch evoked allodynia induced by mustard oil in humans, *NeuroReport,* 6, 753, 1995.

102. Meghji, P. and Newby, A.C., Sites of adenosine formation, action and inactivation in the brain, *Neurochem. Int.,* 16, 227, 1990.
103. Garthwaite, J., Garthwaite, G., Palmer, R.M.J., and Moncada, S., NMDA receptor activation induces nitric oxide synthesis from arginine in rat brain slices, *Eur. J. Pharmacol.,* 172, 413, 1989.
104. Morris, R., Southam, E., Braid, D.J., and Garthwaite, J., Nitric oxide may act as a messenger between dorsal root ganglion neurons and their satellite cells, *Neurosci. Lett.,* 137, 29, 1992.
105. Meller, S.T. and Gebhart, G.F., Nitric oxide (NO) and nociceptive processing in the spinal cord, *Pain*, 52, 127, 1993.
106. Moore, P.K., Wallace, P., Gaffen, Z.A., Hart, S.L., and Babbedge, R., Characterisation of the novel nitric oxide synthase inhibitor nitro indazole and related indazoles: anti-nociceptive and cardiovascular effects, *Br. J. Pharmacol.*, 110, 219, 1993.
107. Bannon, A.W., Decker, M.W., Holladay, M.W., Curzon, P., Donnelly-Roberts, D., Puttfarcken, P.S., Bitner, R.S., Diaz, A., Dickenson, A.H., Porsolt, R.D., Williams, M., and Arneric, S.P., Broad-spectrum, non-opioid analgesic activity by selective modulation of neuronal nicotinic acetylcholine receptors, *Science,* 279, 77, 1998.

Chapter 33

The Multiplicity of Ascending Pain Pathways

Luis Villanueva and Jean-François Bernard

Contents

33.1	Introduction	569
33.2	Some Anatomical Features of Primary Afferents Carrying Nociceptive Information to the Dorsal Horn	571
33.3	Some Anatomical and Functional Aspects of the Superficial Dorsal Horn	571
33.4	The Deep Dorsal Horn	574
33.5	Other Putative Pain Pathways in the Cord White Matter	577
	33.5.1 The Dorsolateral Funiculus	577
	33.5.2 The Dorsal Columns	578
33.6	Concluding Remarks	579
Acknowledgments		580
Abbreviations		580
References		581

33.1 Introduction

Although anatomical, electrophysiological, and, more recently, imaging techniques have helped to elucidate the roles of several systems that mediate sensory, autonomic, motor, and cognitive functions, the brain structures which mediate pain are multiple and still a matter of controversy. Indeed, recent data concerning ascending spinal pathways which project to different areas of the diencephalon,[1,2] the pons, and the mesencephalon,[3] as well as to the caudal medulla,[4,5] have emphasized the participation of these pathways in the transmission of nociceptive information. Although such data reveal a multiplicity of putative ascending "pain" pathways, the precise contribution of each of these pathways to pain processing remains obscure.

The purpose of this chapter is to analyze, with an emphasis on recent findings, some aspects of the anatomical and functional organization of spinal pathways that convey nociceptive information.

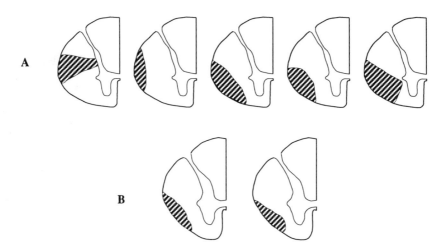

FIGURE 33.1
The importance of the ventrolateral quadrant (VLQ) in pain processing. Examples of lesions of the (**A**) thoracic or (**B**) cervical cord that produced long-lasting contralateral analgesia in humans. (Adapted from Nathan, P.W. and Smith, M.C., in *Advances in Pain Research and Therapy,* Bonica, J.J., Liebeskind, J.C., and Albe-Fessard, D.G., Eds., Raven Press, New York, 1979, p. 921.)

However, it is probably helpful to start with some comments about one of the old controversies in this field, namely the role of the ventrolateral tract (VLQ; also called the anterolateral tract) in pain transmission in humans.

It is well known that lesions produced in the VLQ by trauma as in the Brown-Séquard syndrome or by surgical transection produce an inability to feel pain contralaterally.[6] However, some authors claim that "widely scattered neurological lesions have uniformly failed to abolish pain"[7] and argue against attributing a key role in pain conduction to the VLQ, notably by citing the facts that pain can return some weeks or months following anterolateral cordotomies and that some forms of pain can still be evoked from body areas contralateral to such lesions. In fact, early and recent data have shown that the VLQ plays a key role in pain conduction. Indeed, the few clinicians who have evaluated systematically, in a large number of patients, histological sections of the spinal cord to determine the extent and location of the lesion and then correlated this information with the degree of pain alleviation, have shown clearly that when the area of the VLQ is lesioned, all forms of pathological pain on the opposite side of the body are no longer felt, and the loss of pain and thermal sensibility is lasting.[8,9] As illustrated in Figure 33.1, these lesions excluded the dorsal-most and ventromedial aspect of the white matter of the cord. A detailed analysis of the effects of VLQ lesions performed by several groups is presented in Chapter 34.

The existence of a restricted area within the VLQ which contains the main pathways responsible for the transmission of nociceptive information responsible for pain has also been demonstrated in both electrophysiological and behavioral studies in animals.[10] Moreover, anatomical studies employing *Phaseolus vulgaris* leuccoaglutinin (PHA-L) as an anterograde tracer have helped to elucidate the organization of ascending projections from different spinal areas involved in pain processing. The advantage of the PHA-L approach in comparison with previous retrograde/anterograde tracing studies lies in the fact that it provides a global view of the distribution of brain projections from spinal laminae containing nociceptive neurons. This is due mainly to the fact that this technique allows the discrimination of areas with terminal projections from those containing fibers of passage. As shown below, although these studies are still incomplete, they have provided detailed information about the organization of brain projections from spinal laminae containing nociceptive neurons.

33.2 Some Anatomical Features of Primary Afferents Carrying Nociceptive Information to the Dorsal Horn

The great majority of primary afferent fibers enter the CNS via the dorsal roots. In the radiculo-medullary junction, large fibers become separated from thin fibers. At this level, Aδ- and C-fibers occupy the ventrolateral part of the dorsal roots; this position has allowed some surgeons to destroy this area selectively to alleviate some forms of chronic pain (see Chapter 34).[11] Thin fibers then give off both ascending and descending collaterals within the dorsolateral white matter (Lissauer's tract) and terminate mainly in the superficial dorsal horn (laminae I and IIo).[12,13] Furthermore, Aδ-fibers terminate in lamina V, and C-fibers of visceral origin terminate also in laminae V–VII and X, sometimes bilaterally. Thus, it is possible to conclude that there exists an important anatomical convergence of nociceptive afferents, mainly in laminae I and V of the dorsal horn; both these areas contain neurons activated by noxious stimuli of different origins.

Indeed, several electrophysiological studies have shown that these regions contain a great number of neurons activated either specifically by noxious inputs (noxious specific cells, or NS) or by noxious and innocuous inputs (noxious non-specific cells, or NNS).[14–16] Although there is no strict segregation of these two populations, it has been shown in several species that NS cells are found mainly in the superficial dorsal horn (laminae I–II), whereas NNS cells are located mainly in deep laminae (V–VI). Moreover, there is also another population of neurons in the ventral horn (laminae VII–VIII) that respond to noxious inputs from widespread areas of the body. We will briefly analyze some features of the pathways that convey nociceptive signals from these areas.

33.3 Some Anatomical and Functional Aspects of the Superficial Dorsal Horn

Noxious specific cell neurons are located mainly in lamina I at both the spinal and the trigeminal levels;[17] however, lamina I also contains NNS cells and neurons that respond specifically to cold. NS cells have narrow receptive fields and are activated by Aδ- and C-fibers from different origins, which is consistent with the fact that many of these neurons show viscero-somatic convergence. The restricted receptive fields and the somatotopic organization indicate that NS neurons are suitable for signaling spatial and temporal features of nociceptive information. This was also found in awake, unrestrained monkeys. These neurons are able to encode the intensity of both thermal and mechanical stimuli, but within a narrower range of responses than is found for NNS neurons of the deep dorsal horn.[18,19]

It is not known what proportions of lamina I neurons are local interneurons as opposed to being cells which participate in reflex arcs or propriospinal circuits or project to higher centers. However it seems possible that lamina I cells contribute to relay systems for pain and thermal sensations, as these seem clearly associated at this level.[20] Interestingly, lesions of the VLQ that abolish pathological pain are associated with loss of thermal sensations at approximately the same levels at which analgesia occurs.

In this regard, recent studies which labeled ascending axons following injections in LI in different species showed that they tend to concentrate contralaterally in the middle aspect of the lateral funiculus.[21] It has been claimed that some of the axons of lamina I neurons ascend in the dorsal-most part of the lateral funiculus, an area not affected by the VLQ lesioning which produces relief of pain.[22] In fact there is some variability between species and spinal cord levels, with respect to the location of ascending axons from lamina I. As stated above, although these axons are concentrated in the middle part of the lateral funiculus there is some dorsoventral dispersion and some are located quite dorsally.[21] It would be interesting to compare the location of NS axons with those of thermoreceptive cells, as both early and recent studies have suggested the possibility that thermoreceptive VLQ axons are located more dorsally than nociceptive axons originating from various dermatomes. Thus, it seems that thermal and pain sensations are close from both anatomical and functional viewpoints which one might relate to the fact that both sensations are useful for homeostasis.

As illustrated in Figure 33.2, lamina I neurons terminate in several areas of the CNS which are important for processing signals relevant for homeostasis. For example, they establish propriospinal connections with the sympathetic thoraco-lumbar system, which provides the basis for somato-sympathetic reflexes. At the medullary level, lamina I neurons establish connections with neurons in the ventrolateral medulla and in the caudal portion of the nucleus of the solitary tract, two regions which are noted for their roles in cardio-respiratory regulation.

The most dense projections from lamina I neurons are at the ponto-mesencephalic level, mainly in the lateral parabrachial area (PB) and, to a lesser extent, in the ventrolateral periaqueductal grey matter (PAG).[3,20] Electrophysiological studies have shown that a high proportion of both lamina I spino-PB and PB neurons are driven by Aδ- and C-fibers, respond specifically to noxious stimuli, and encode thermal as well as mechanical stimuli within noxious ranges.[3] A smaller proportion of these neurons is also responsive to cooling. The receptive fields of spino-PB neurons are generally small (one or two toes), whereas those for PB neurons are larger (a limb to the whole body), probably indicating the heterotopic convergence of lamina I inputs onto this region. Although less systematically studied, nociceptive neurons have also been described in the PAG.

The nociceptive (lateral) parabrachial area projects densely to (1) the central nucleus of the amygdala and the bed nucleus of the stria terminalis, which are probably involved in anxiety and reactions to fear; and (2) the hypothalamic ventromedial nucleus, which participates in defensive/aggressive behavior and the regulation of energy metabolism.[3,23] In addition, the parabrachial nociceptive area receives a major visceral/autonomic input from the nucleus of the solitary tract,[24] indicating that PB is not likely to be involved in the topographical discrimination of pain but would participate in autonomic and emotional aspects of pain.

Several authors have shown that the PAG cells can be subdivided into different columns on the basis of behavioral and autonomic reactions to specific stimulation of their cell bodies. In this respect, the lateral and ventrolateral columns of the PAG contain different groups of neurons which, when activated, produce well-defined cardiovascular and defensive reactions such as decreases in blood pressure, hyporeactive immobility, avoidance behavior, and vocalization, as well as a more general emotional state of fear and anxiety.[25,26] In addition, electrical stimulation of the ventral and lateral PAG in the rat can induce powerful antinociceptive effects which seem to be mediated through its projections to the rostral ventral medulla.[27] Thus, the spino-PAG pathway could participate in feedback mechanisms involved in autonomic, aversive, and antinociceptive responses to strong nociceptive stimulation.

In rats, lamina I projections to the diencephalon terminate mainly in the ventroposterolateral, ventroposteromedial, and posterior thalamic complex. A spino-hypothalamic pathway originating from both the superficial and the deep dorsal horn was recently described.[2] However, the importance of this pathway is still unclear, as the existence of direct spinal projections to the hypothalamus is supported mainly by retrograde tracing and antidromic stimulation studies of dorsal horn neurons. However, anterograde tracing from the spinal cord and electrophysiological characterization of their hypothalamic targets are basic requirements before the existence of a nociceptive spino-hypothalamic system can be confirmed.

In monkeys, lamina I afferents terminate in several ventral posterior thalamic areas: lateral (VPL), medial (VPM), and inferior (VPI).[1,16,28,29] Dense lamina I labeling was recently described both in the ventral caudal part of the medial dorsal nucleus (MDvc) and in an area within the suprageniculate/posterior complex (SG/PO), named the posterior part of the ventromedial thalamic nucleus (VMpo).[1,20] Although lateral, medial, and posterior thalamic areas precisely encode different intensities of noxious stimuli, recordings in anesthetized and awake monkeys have revealed

The Multiplicity of Ascending Pain Pathways

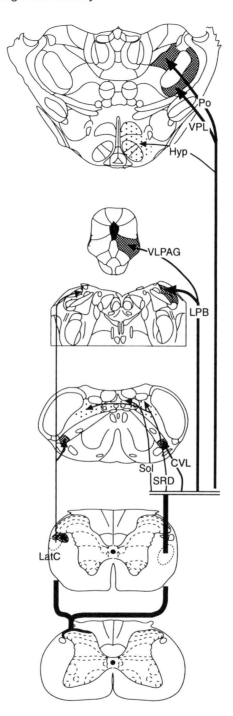

FIGURE 33.2
Main supraspinal projections from lamina I neurons. Lamina I neurons send signals from noxious and thermal inputs to spinal, bulbar, and telencephalic regions implicated in autonomic, emotional, and somatosensory processing necessary for the integrity of the body. It has been proposed that, rather than only subserving pain processing, these circuits could contribute to the creation of a human "body picture" which sustains basic emotional and motivational states. Arrow sizes and hatched areas indicate the relative density of lamina I projections. (Coronal sections diagrams adapted from Paxinos, G. and Watson, C., *The Rat Brain in Stereotaxic Coordinates*, compact 3rd. ed., Academic Press, New York, 1996.)

important differences between these areas. Interestingly, a great number of neurons in MDvc and VMpo are modality specific, showing either nociceptive or thermal responses.[1] The receptive fields of parafascicular (Pf)/MDvc cells are often very large, and both their borders and the magnitudes of their evoked responses change with the monkey's behavioral state.[30] It has been suggested that these features may reflect these cells being better suited for a behavioral reaction, thus strongly implicating these regions in the affective-emotional aspects of pain. This assertion is supported by their cortical connectivity and by positron emission tomography (PET) studies. VMpo cells project to the mid/anterior insular cortex, an area activated by both innocuous and noxious thermal stimuli in humans[31-34] and implicated in the affective components of pain on the basis of its projections to various limbic structures such as the amygdala and perirhinal cortex. Pf/MDvc cells project to area 24 of the cingulate cortex, the activity of which appears to be more selectively modulated by noxious stimuli.[34] This has led some authors to suggest that it may participate in the emotional/motivational component of pain. Interestingly, this is a functionally heterogeneous area constituted by adjacent zones which have been implicated in attentional, motor, and autonomic reactions which might allow it to elicit various behavioral reactions.[35]

By contrast, in ventroposterior thalamic areas (VPL/VPM/VPI), the majority of neurons are wide dynamic range (WDR), have receptive fields which are not modified by the behavioral state, and are smaller than those of spinal or medullary dorsal horn projecting neurons.[30] This suggests that ventral posterior areas may subserve spatial discrimination. This is also supported by behavioral evidence in monkeys showing that bilateral ablation of the main cortical target of VPL/VPM, the primary somatosensory cortex (SI), disrupts their ability to discriminate intensities of noxious heat.[30] Also, PET studies have shown that noxious and innocuous stimuli similarly activate the contralateral SI, indicating a co-existence of pain and tactile representation in this area.[32,33]

Furthermore, single-unit recordings from a caudal thalamic area in humans showed neurons which could be activated by noxious stimuli. Furthermore, stimulation of this region induced thermal and/or painful sensations.[36] Comparison of these data with electrophysiological findings in monkeys led to the conclusion that the human's thalamic region could correspond to the monkey's VPI area which contains neurons responding to both cutaneous and visceral noxious stimuli.[37]

However, taken together, these data show that the fine encoding properties of noxious stimuli applied to the skin by thalamic neurons alone cannot account for the sensory-discriminative aspects of pain, as such encoding is shared by all (thalamic and also lower CNS) regions implicated in pain processing. Moreover, the ventral posterior thalamic areas which are the best candidates for discrimination (see above), such as the VPM, can discharge with higher instantaneous frequency to innocuous than to noxious stimulation in awake monkeys.[30] Thus, it is possible that modulatory mechanisms and/or the concomitant activity of multiple neuronal populations could determine the final pain perception.

33.4 The Deep Dorsal Horn

The deep dorsal horn contains a great number of NNS. These are concentrated mainly in lamina V, although NNS cells have also been found in lamina IV and VI. Similarly, in trigeminal nucleus caudalis, these cells have been frequently encountered in the deep layers of the magnocellular region. Studies in both anesthetized and awake animals have illustrated that NNS neurons have a greater ability to encode noxious stimuli with a wide range of response levels than have NS neurons.[14-16,18,19,38] NNS neurons are characterized by their capacity to respond to a large range of mechanical stimuli, from innocuous up to strong nociceptive stimuli. They also respond to a variety of other stimuli (innocuous thermal and/or noxious and chemical stimuli), and show viscero-somatic convergence. NNS neurons usually have excitatory peripheral fields larger than NS cells, although they are still compatible with a reasonable degree of stimulus location. These cells receive Aα-, Aδ-, and C-fiber inputs and form a high proportion of neurons involved in several ascending nociceptive systems. Neurons presenting larger receptive fields and activated by noxious inputs have been found also in the ventral horn, notably in laminae VII and X at thoracic, lumbar, and sacral levels. Many of these cells receive viscero-somatic convergence.[14,16,39]

The precise sites of termination of laminae V–VII nociceptive neurons are still largely unknown, as most of the available data come from retrograde tracing. However, it is known that these areas contribute to spino-reticular, spino-mesencephalic, and spino-thalamic systems, and below the upper cervical segments deep dorsal horn cells constitute the great majority of spinal afferents to the reticular formation.[16]

As illustrated in Figure 33.3, laminae V–VII neurons establish connections with several brainstem reticular areas: the lateral reticular nucleus, subnucleus reticularis dorsalis (SRD), gigantocellular reticular nucleus (Gi), lateral paragigantocellular nucleus, and pontine and deep mesencephalic reticular nuclei. In the rostral brainstem, laminae V–VII neurons also establish connections with the internal lateral parabrachial (PBil) and the Kölliker-Fuse areas, the ventrolateral periaqueductal grey matter, the intercollicular nucleus, the peripeduncular tegmental area, and the anterior pretectal nucleus. Except for some brainstem reticular areas (see below), the contribution of most of these areas to nociception is still largely unknown. However, some of the reticular areas send dense projections to midline, intralaminar, and paralaminar thalamic nuclei.[40] The thalamic nuclei that receive these reticular afferents send dense projections notably to the striatum and the rostral premotor and associative cortical areas,[41] although these have not been determined precisely. Furthermore, we have found recently that both paracentral and parafascicular thalamic nuclei receive strong projections from the PBil.

The role of the caudal brainstem as a relay for nociceptive signals has been suggested for a long time, as the majority of VLQ ascending axons, in both animals and man, terminate within the medullary reticular formation.[42] Several groups have shown that brainstem reticular areas, most notably the Gi and surrounding regions, contain neurons responsive to noxious stimuli and that focal stimulation of some of these areas can elicit escape behavior.[43–45] However, as most nociceptive reticular units recorded in these older studies showed irregular responses and changes in excitability and presented some degree of heterosensory convergence, it was concluded that the reticular formation did not play a specific role in the processing of pain. This proposal has been challenged by data obtained in the rat showing that neurons within the medullary SRD respond exclusively to the activation of peripheral Aδ-fibers or Aδ- and C-fibers from the whole body surface, and either exclusively or preferentially to noxious stimuli.[4] By comparison with other brainstem reticular neurons, this population does not respond to visual, auditory, or proprioceptive stimuli, but does encode the intensity of thermal, mechanical, and visceral noxious stimuli and is activated via spinal pathways ascending in the ventro-lateral funiculi. Neurons with similar properties have also been recorded in the monkey's SRD.

The largest spinal afferent input to the SRD originates from the ipsilateral cervical cord, while the weakest arises bilaterally from caudal spinal regions. This is in contrast to the whole body receptive fields with a contralateral dominance of SRD neurons which suggests that some of the spinal inputs do not reach the nucleus directly. It is possible that the ascending information may relay at upper cervical levels, as these areas contain both the majority of spino-reticular afferents and neurons with heterosegmental, widespread receptive fields.[46,47] Interestingly, the largest numbers of retrogradely labeled cells in the spino-thalamic, spino-reticular, and spino-mesencephalic tracts were also found to be in the upper cervical cords of different species,[16] thus suggesting a common functional organization of several ascending somatosensory pathways. Within the framework of this hypothesis, one could envisage that at least some inputs have relays in the upper cervical cord. Such an organization could explain the widespread relief of pain, including pain from caudal segments of the body, following commissural myelotomies of the upper cervical spinal cord in human patients.[48]

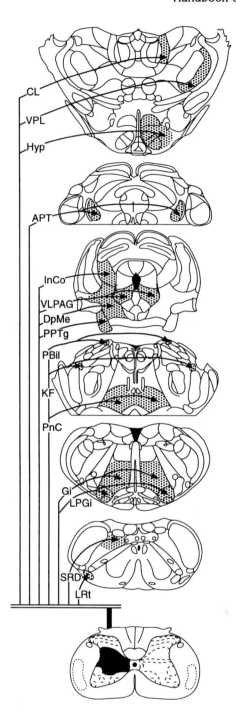

FIGURE 33.3

Main supraspinal projections from deep dorsal horn neurons. Deep dorsal horn neurons are able to capture a variety of signals originating either from the external environment through the skin or from the internal organs. Thus, they can generate a "basic somaesthetic activity". They send inputs to several regions implicated in somatosensory, motor, arousal, attentional, and perhaps motivational (aversive) processing of nociceptive inputs. Like lamina I cells, deep dorsal horn neurons appear to be implicated not only in pain processing but also in creating a "body picture" which is necessary for homeostatic regulation. (Coronal sections diagrams adapted from Paxinos, G. and Watson, C., *The Rat Brain in Stereotaxic Coordinates*, compact 3rd. ed., Academic Press, New York, 1996.)

One of the main thalamic targets of the SRD is the lateral VM area, with less dense terminals diffusing more laterally as a band immediately ventral to the VPL.[49] Some electrophysiological studies in rats have recorded units responsive to noxious stimulation in both the VM and the area immediately ventral to the VPL. Units located ventral to the VPL showed a greater amount of cutaneous and visceral nociceptive convergence than did those recorded within the VPL.[50]

In the rat, VM projections, among their various cortical targets, terminate in area 24 of the anterior cingulate cortex, a region which contains neurons activated by noxious stimulation of widespread areas of skin in rabbits.[35] Lateral VM areas which receive SRD afferents in the rat could be involved in motor reactions following noxious stimulation, as combined electrophysiological and anatomical experiments have shown that these VM areas project to cingular cortical regions involved in motor control of the forelimbs and head. As stated above, several PET studies in humans have shown an activation in the midcingulate cortex following painful cutaneous stimulation.[31-34,51] The structures activated notably include area 24.[51]

Another thalamic region that receives both SRD and Gi afferents is the lateral Pf area. Although a systematic electrophysiological exploration of the whole Pf area has not yet been performed, a recent study showed the existence of Pf units responding to both cutaneous and visceral noxious stimuli in the rat.[52] By comparison with VPL units, these neurons were driven from larger cutaneous receptive fields and responded to more intense, frankly noxious cutaneous and visceral stimuli. The lateral Pf projects to the rostral-most premotor cortex, the dorsolateral striatum, and the lateral subthalamic nucleus.[41,53] Thus, as these projections are related to forebrain structures involved in motor processing, SRD-Pf connections could mediate some emotional features of motor reactions following noxious stimulation.

In summary, brainstem nociceptive reticular areas project to several "motor", premotor, and association areas of the forebrain which strongly implicates them as links in a system involved in motor, arousal, attentional, and perhaps motivational (aversive) processing of noxious inputs from widespread origins. Furthermore, as almost all of the reticular regions considered above send projections to the medullary and spinal dorsal horns, these areas probably have an important role in the descending modulation of pain processing. These data shed new light on old hypotheses suggesting that, in addition to spino-thalamic, spino-reticulo-thalamic pathways may also play an important role in the distribution of pain signals to the forebrain.

Although not clearly delimited, direct diencephalic projections from laminae V–VII terminate in several thalamic areas, including the VPL, where a somatotopic organization has been found in different species with the exception of cats, where they end in a shell region surrounding the VPL. Other terminations include the posterior complex, the intralaminar complex (notably the central lateral area), the nucleus submedius, and other medial thalamic areas.[16]

33.5 Other Putative Pain Pathways in the White Matter of the Cord

33.5.1 The Dorsolateral Funiculus

Some authors have reported that, in various species, a great number of ascending axons from lamina I neurons are located in the dorsolateral white matter. In fact, it has been claimed that some of the axons from lamina I neurons ascend in the dorsal-most part of the lateral funiculus,[22] an area not concerned in the relief of pain.[54] On the contrary, though, electrophysiological and behavioral data show that dorsolateral lesions produce an increase in nociceptive reactions, probably by disrupting tonically active descending antinociceptive pathways.[55] The question of whether a "dorsolateral" ascending pain pathway exists may be merely a semantic problem, as both anterograde and

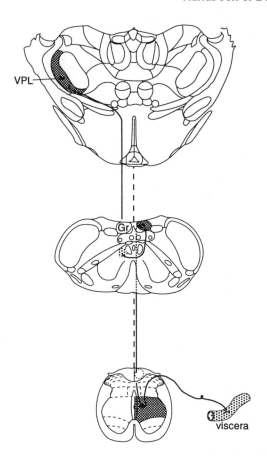

FIGURE 33.4
The putative visceral pain pathway in the dorsal columns. It has been proposed that noxious inputs from viscera are relayed in laminae VII and X neurons, whose ascending axons are confined near the midline of the dorsal columns.[5] These axons terminate in the medullary nucleus gracilis, which conveys these inputs to the thalamic VPL nucleus. (Coronal sections diagrams adapted from Paxinos, G. and Watson, C., *The Rat Brain in Stereotaxic Coordinates*, compact 3rd. ed., Academic Press, New York, 1996.)

retrograde tracing studies in monkeys have shown that lamina I spinothalamic axons are located more laterally than those of deep dorsal horn neurons, but are still within the VLQ.[16,21,29,56] Moreover, an anterograde tracing study in monkeys showed a different distribution of the thalamic projections of dorsal as opposed to ventral spinothalamic axons within the VLQ. Dorsal spinothalamic axons lie at the level of the denticulate ligament and terminate preferentially in the caudal-ventral thalamic areas known as the suprageniculate/posterior complex (SG/PO). On the other hand, spinothalamic axons ventral to the denticulate ligament terminate predominantly in the main portion of the ventroposterolateral thalamus (VPLc).[29] These data are in keeping with recent studies in different species, which have shown that anterogradely labeled lamina I axons travel in the middle of the contralateral white matter which, in monkeys, corresponds to an area ventral to the region occupied by the descending corticospinal tract.[16,21]

33.5.2 The Dorsal Columns

Two recent studies in rats have shown a visceral nociceptive pathway that activates gracile neurons.[5,58] As illustrated in Figure 33.4, the ascending axons that activate these neurons following either stimulation of reproductive pelvic viscera or colorectal distention are confined to the fasciculus gracilis. Moreover, on the basis of clinical findings showing a relief of pelvic cancer following restricted lesions, it has been proposed that an important visceral nociceptive pathway in humans is confined to the medial aspect of the dorsal columns.[5] It was also suggested that this finding could explain the widespread pain relief obtained following midline lesions and commissural myelotomies. In addition, the suppression of VPL responses to colorectal distention following dorsal column lesions in rats led to the statement that this pathway is more important than the VLQ for transmitting visceral nociceptive signals to VPL, whereas the VLQ may be more important for transmitting cutaneous nociceptive information.[58] It was concluded that the visceral inputs traveling in the dorsal columns could either cooperate with direct spinal pathways to produce the perceptions of touch and pain or serve as an alternative nociceptive pathway to the spinothalamic tract.

These data undoubtedly challenge traditional concepts that have regarded the dorsal columns as a pathway carrying only tactile inputs and kinaesthesia. However, its role in the processing of visceral pain is difficult to interpret for both clinical and experimental reasons. For instance, in man, lesions confined to the VLQ are not only able to remove cutaneous pain and thermal sensations, but also pain from various viscera. This is supported by experimental evidence showing that other brainstem structures (e.g., PB,[3] SRD[4]) that distribute noxious visceral inputs to diencephalic areas other than the VPL are activated by axons traveling in the VLQ. However, some reports of VLQ lesions relieving visceral pain in man claim that cordotomies should extend medially to the gray matter and should be bilateral, probably because many visceral structures have a bilateral innervation.[10]

It is possible that some of the difficulties in attributing an accurate role to the dorsal columns or VLQ fibers in the transmission of visceral pain lie in the fact that visceral inputs converge on a variety of spinal neurons. This provides an anatomical basis not only for referred visceral pain, but also for the activation of multiple pathways. It is supported by recent data in monkeys showing that both dorsal and ventral lesions are able to reduce VPL responses to noxious visceral stimulation.[59]

33.6 Concluding Remarks

There remains the question of whether one can label a particular area of spinal white matter as an "ascending pain pathway". For several reasons, we believe that this description is not adequate. Perhaps the simplest reason lies in the fact that a common feature of the different ascending spinal tracts is that they convey inputs from a variety of origins, including those activated by noxious and non-noxious events. Thus, they are not as neatly "individualized" as would be convenient for scientists and neurosurgeons alike. In addition to this convergence phenomenon, the description of "ascending pain pathways" based on the modification of pain following lesioning is probably complicated by the possibility that lesioning could produce a simultaneous modification of both pain signaling and pain modulatory mechanisms.

Indeed it is difficult to separate pain transmission from pain modulatory systems, as some of the latter are activated when a noxious stimulus occurs and are sustained by pathways classically involved in pain transmission. An illustrative example of this lies in diffuse noxious inhibitory controls (DNIC), the neural substrates of which have been determined in both animals and man. These controls are specifically elicited by noxious stimulation, induce strong inhibitory influences in spinal nociceptive processing, and are sustained by ascending and descending spinal pathways confined to the VLQ and dorsolateral funiculus, respectively.[60] DNIC are probably involved in analgesia elicited by some practices in popular medicine founded in old observations that "one pain can mask another". These interactions between different pain sources could explain some of the difficulties encountered in pain semiology. In some pathologies, the potentially painful sources can be varied and might interact with each other such that the resulting pain perception and localization are not strictly related to the location of tissue damage. The alteration of sensory modulatory mechanisms could contribute to the dysesthesias and pain that sometimes appear following cordotomies and more often following mesencephalic tractotomies.[61]

It is tempting to speculate that lesions confined to the dorsal columns not only disrupt an ascending pathway but might also interfere with modulatory mechanisms. This could explain how the relief of pain following midline lesions may not occur immediately but take some days to come on. In fact, the dorsal columns not only convey ascending inputs to the brain but also contain direct descending spinal projections originating from the dorsal column nuclei (DCN) and terminating in dorsal horn areas containing nociceptive neurons (laminae I and V).[62] In addition, the DCN region, where the majority of spinally projecting cells are located, is an important target for corticobulbar projections. Thus, the spinal output could be modulated not only directly by the DCN spinally projecting neurons but indirectly through cortico-DCN influences.[63] The possibility that the information traveling in the DCN could be under the influence of inputs from several levels including the cerebral cortex suggests that DCN neurons may integrate information from several ascending and descending systems. Perhaps an improved knowledge of these influences might help us understand the complex modifications of sensory perception that have been reported to follow lesioning of the dorsal columns in humans, including changes in tactile sensations, tactile and postural hallucinations, and increases in sensations of pain, tickle, warmth, and cold.[64]

From a more general point of view, the data summarized briefly here support the idea that pain is not a unique consequence of impulses in specific, unidirectional hard-wired lines which originate in the periphery and terminate in the CNS. This conclusion is in accord with some pathological situations showing that pain can be elicited by inputs conveyed by large, low-threshold fibers and in normal subjects by the simultaneous application of innocuous hot and cold stimuli.[34] It is also supported by experimental data showing a lack of areas in the CNS which are restricted to a particular pain modality and the absence of a point-to-point "pain" representation. In this respect it is tempting to speculate that the lemniscal system might also participate in accurate pain localization, as it is worth noticing that large-diameter fibers are absent in areas where a painful focus lacks precise topographical discrimination, such as the visceral territories and the dental pulp.

It seems that the sensory pathways that convey pain signals are simultaneously collecting information from many sources and not only from nociceptors. This basic somaesthetic activity could have a role in the elaboration of a "body representation" in terms of a continual transmission of information relevant to the integrity of the body. This available information is being constantly selected, filtered, and modulated in the context of an appropriate response. We believe that one of the outstanding challenges today remains the study of both the anatomical and functional organization of the different systems that convey nociceptive information and how these systems come to be disorganized during chronic pain. This will certainly contribute to a better understanding of the mechanisms underlying the generation of painful sensations and their associated reactions.

Acknowledgments

This study was supported by l'Institut National de la Santé et de la Recherche Médicale (INSERM), and l'Institut UPSA de la douleur. The authors are very grateful to Drs. J.M. Besson, S.W. Cadden, and P.W. Nathan for advice in the preparation of the manuscript.

Abbreviations

APT   anterior pretectal nucleus
CL   centrolateral thalamic nucleus

CNS	central nervous system
CVL	caudoventrolateral reticular nucleus
DCN	dorsal column nuclei
DNIC	diffuse noxious inhibitory controls
DpMe	deep mesencephalic nucleus
Gi	gigantocellular reticular nucleus
Gr	gracile nucleus
Hyp	hypothalamus
InCo	intercollicular nucleus
KF	Kölliker-Fuse nucleus
LatC	lateral cervical nucleus
LPB	lateral parabrachial nucleus
LPGi	lateral paragigantocellular nucleus
LRt	lateral reticular nucleus
MDvc	medial dorsal thalamic nucleus
NNS	noxious non-specific cells
NS	noxious specific cells
PAG	periaqueductal grey
PB	parabrachial nucleus
PBil	internal lateral parabrachial nucleus
PET	positron emission tomography
Pf	parafascicular thalamic nucleus
PHA-L	phaseolus vulgaris leuccoaglutinin
PnC	pontine reticular nucleus, caudal part
Po	posterior thalamic nuclear group
PPTg	pedunculopontine tegmental nucleus
SG/PO	suprageniculate/posterior thalamic complex
SI	primary somatosensory cortex
Sol	nucleus of the solitary tract
SRD	subnucleus reticularis dorsalis
VLPAG	ventrolateral periaqueductal gray
VLQ	ventrolateral quadrant
VM	ventromedial thalamic nucleus
VMpo	posterior part of the ventromedial thalamic nucleus
VPI	ventral posterior inferior thalamic nucleus
VPL	ventral posterolateral thalamic nucleus
VPM	ventral posteromedial thalamic nucleus
WDR	wide dynamic range neuron

References

1. Craig, A.D., Supraspinal projections of lamina I neurons, in *Forebrain Areas Involved in Pain Processing*, Besson, J.M., Guilbaud G., and Ollat, H., Eds., John Libbey Eurotext, Paris, 1995, p. 13.
2. Giesler, Jr., G.J., Katter J.T., and Dado, R.J., Direct spinal pathways to the limbic system for nociceptive information, *Trends Neurosci.*, 17, 244, 1994.
3. Bernard, J.F., Bester, H., and Besson, J.M., Involvement of the spino-parabrachio-amygdaloid and -hypothalamic pathways in the autonomic and affective emotional aspects of pain, in *The Emotional Motor System,* Holstege, G., Bandler, R., and Saper, C.B., Eds., Elsevier, Amsterdam, 1997, p. 243.

4. Villanueva, L., Bouhassira, D., and Le Bars, D., The medullary subnucleus reticularis dorsalis (SRD) as a key link in both the transmission and modulation of pain signals, *Pain*, 67, 231, 1996.

5. Hirshberg, R.M., Al-Chaer, E.D., Lawand, N.B., Westlund, K.N., and Willis W.D., Is there a pathway in the posterior funiculus that signals visceral pain?, *Pain*, 67, 291, 1996.

6. Gybels, J.M. and Sweet, W.H., Neurosurgical treatment of persistent pain, in *Pain and Headache*, Gildenberg, Ph.L., Ed., Karger, Basel, 1989, p. 293.

7. Wall, P.D., Independent mechanisms converge on pain, *Nature Med.*, 1, 740, 1995.

8. Nathan, P.W. and Smith, M.C., Clinico-anatomical correlation in anterolateral cordotomy, in *Advances in Pain Research and Therapy*, Bonica, J.J., Liebeskind, J.C., and Albe-Fessard, D.G., Eds., Raven Press, New York, 1979, p. 921.

9. Lahuerta, J., Bowsher, D., Lipton, S., and Buxton, P.H., Percutaneous cervical cordotomy: a review of 181 operations on 146 patients with a study on the location of "pain fibers" in the C-2 spinal cord segment of 29 cases, *J. Neurosurg.*, 80, 975, 1994.

10. Vierck, C.J., Greenspan, J.D., Ritz, L.A., and Yeomans, D.C., The spinal pathways contributing to the ascending conduction and the descending modulation of pain sensations and reactions, in *Spinal Afferent Processing*, Yaksh, T.L., Ed., Plenum Press, New York, 1986, p. 75.

11. Sindou, M., Quoex, C., and Baleydier, C., Fiber organization at the posterior spinal cord-rootlet junction in man, *J. Comp. Neurol.*, 153, 15, 1974.

12. Maxwell, D.J. and Rhethely, M., Ultrastructure and synaptic connections of cutaneous afferent fibers in the spinal cord, *Trends Neurosci.*, 10, 117, 1987.

13. Sugiura, Y. and Tonosaki, Y., Spinal organization of unmyelinated visceral afferent fibers in comparison with somatic afferent fibers, in *Progress in Pain Research and Management.* Vol 5. *Visceral Pain*, Gebhart, G.F., Ed., IASP Press, Seattle, 1995, p. 41.

14. Besson, J. M. and Chaouch, A., Peripheral and spinal mechanisms of nociception, *Physiol. Rev.*, 67, 67, 1987.

15. Wall, P.D., The dorsal horn, in *Textbook of Pain*, Wall, P.D. and Melzack, R., Eds., Churchill Livingstone, Edinburgh, 1989, p. 102.

16. Willis, W.D., and Coggeshall, R.E., *Sensory Mechanisms of the Spinal Cord*, 2nd ed., Plenum Press, New York, 1991.

17. Christensen, B.N. and Perl, E.R., Spinal neurons specifically excited by noxious or thermal stimuli: marginal zone of the dorsal horn, *J. Neurophysiol.*, 33, 293, 1970.

18. Hoffman, D.S., Dubner, R., Hayes, R.L., and Medlin, T.P., Neuronal activity in medullary dorsal horn of awake monkeys trained in a thermal discrimination task. I. Response to innocuous and noxious thermal stimuli, *J. Neurophysiol.*, 46, 409, 1981.

19. Maixner, W., Dubner, R., Kenshalo, D.R., Bushnell, M.C., and Oliveras, J.L., Responses of monkey medullary dorsal horn neurons during the detection of noxious heat stimuli, *J. Neurophysiol.*, 62, 437, 1989.

20. Craig, A.D., Pain, temperature, and the sense of the body, in *Somesthesis and the Neurobiology of the Somatosensory Cortex*, Franzen, O., Johansson, O., and Terenius, L., Eds., Birkhäuser, Basel, 1996, p. 27.

21. Craig, A.D., Spinal distribution of ascending lamina I axons anterogradely labeled with *Phaseolus vulgaris* leucoagglutinin (PHA-L) in the cat, *J. Comp. Neurol.*, 313, 377, 1991.

22. McMahon, S.B. and Wall, P.D., The significance of plastic changes in lamina I systems, in *Processing of Sensory Information in the Superficial Dorsal Horn of the Spinal Cord*, Cervero, F., Bennett, G.J., and Headley, P.M., Eds., NATO ASI Series A: Life Sciences, Plenum Press, New York, 1988, p. 249.

23. Bester, H., Besson, J.M., and Bernard, J.F., Organization of efferent projections from the parabrachial area to the hypothalamus: a *Phaseolus vulgaris* leucoagglutinin study in the rat, *J. Comp. Neurol.*, 383, 245, 1997.

24. Herbert, H., Moga, M., and Saper, C.B., Connections of the parabrachial nucleus with the nucleus of the solitary tract and medullary reticular formation in the rat, *J. Comp. Neurol.*, 293, 540, 1990.
25. Bandler, R. and Depaulis, A., Midbrain periaqueductal gray control of defensive behavior in the cat and the rat, in *The Midbrain Periaqueductal Gray Matter. Functional, Anatomical, and Neurochemical Organization*, Depaulis, A. and Bandler, R., Eds., NATO ASI Series A: Life Sciences, Plenum Press, New York, 1991, p. 175.
26. Lovick, T.A., Central nervous system integration of pain control and autonomic function, *News Physiol. Sci.*, 6, 82, 1991.
27. Besson, J.M., Fardin, V., and Olivéras, J.L., Analgesia produced by stimulation of the periaqueductal gray matter: True antinociceptive effects versus stress effects, in *The Midbrain Periaqueductal Gray Matter. Functional, Anatomical, and Neurochemical Organization*, Depaulis, A. and Bandler, R., Eds., NATO ASI Series A: Life Sciences, Plenum Press, New York, 1991, p. 121.
28. Apkarian, A.V. and Hodge, C.J., Primate spinothalamic pathways. III. Thalamic terminations of the dorsolateral and ventral spinothalamic pathways, *J. Comp. Neurol.*, 288, 493, 1989.
29. Ralston, H.J. and Ralston, D.D., The primate dorsal spinothalamic tract: evidence for a specific termination in the posterior nuclei (Po/SG) of the thalamus, *Pain*, 48, 107, 1992.
30. Bushnell, M.C., Thalamic processing of sensory-discriminative and affective-motivational dimensions of pain, in *Forebrain Areas Involved in Pain Processing*, Besson, J.M., Guilbaud G., and Ollat, H., Eds., John Libbey Eurotext, Paris, 1995, p. 63.
31. Jones, A.K.P., Brown, W.D., Friston, K.J., Qi, L.Y., and Frackowiak, R.S.J., Cortical and subcortical localization of responses to pain in man using positron emission tomography, *Proc. Roy. Soc. Lond.*, 244, 39, 1991.
32. Coghill, R.C., Talbot, J.D., Evans, A.C., Meyer, E., Gjedde, A., Bushnell, M.C., and Duncan, G.H., Distributed processing of pain and vibration by the human brain, *J. Neurosci.*, 14, 4095, 1994.
33. Casey, K.L., Minoshima, S., Berger, K.L., Koeppe, R.A., Morrow T.J., and Frey, K.A., Positron emission tomographic analysis of cerebral structures activated specifically by repetitive noxious heat stimuli, *J. Neurophysiol.*, 71, 802, 1994.
34. Craig, A.D., Reiman, E.M., Evans, A., and Bushnell, M.C., Functional imaging of an illusion of pain, *Nature*, 384, 258, 1996.
35. Devinsky, O., Morrell, M.J., and Vogt, B.A., Contributions of anterior cingulate cortex to behavior, *Brain*, 118, 279, 1995.
36. Lenz, F.A., Gracely, R.H., Zirh, A.T., Romanoski, A.J., and Dougherty, P.M., The sensory-limbic model of pain memory. Connections from the thalamus to the limbic system mediate the learned component of the affective dimension of pain, *Pain Forum*, 6, 22, 1997.
37. Apkarian, A.V. and Shi, T., Squirrel monkey lateral thalamus. I. Somatic nociresponsive neurons and their relation to spinothalamic terminals, *J. Neurosci.*, 14, 6779, 1994.
38. Le Bars, D., Dickenson, A.H., Besson, J.M., and Villanueva, L., Aspects of sensory processing through convergent neurons, in *Spinal Afferent Processing*, Yaksh, T.L., Ed., Plenum Press, New York, 1986, p. 467.
39. Cervero, F., Visceral pain, in *Pain Research and Clinical Management: Proc. Fifth World Congress on Pain,* Bond, M., Woolf, C., and Charlton, J.E., Eds., Elsevier, Amsterdam, 1991, p. 216.
40. Jones, B.E., Reticular formation: cytoarchitecture, transmitters, and projections, in *The Rat Nervous System,* 2nd ed., Paxinos, G., Ed., Academic Press, San Diego, 1995, p. 155.
41. Price, J.L., Thalamus, in *The Rat Nervous System,* 2nd ed., Paxinos, G., Ed., Academic Press, San Diego, 1995, p. 629.
42. Mehler, W.R., Feferman, M.E., and Nauta, W.J.H., Ascending axon degeneration following anterolateral corodotomy, an experimental study in the monkey, *Brain*, 83, 718, 1960.

43. Casey, K.L., Somatosensory responses of bulboreticular units in the awake cat: relation to escape producing stimuli, *Science*, 173, 77, 1969.
44. Bowsher, D., Role of the reticular formation in responses to noxious stimulation, *Pain*, 2, 361, 1976.
45. Gebhart, G.F., Opiate and opioid peptide effects on brain stem neurons: relevance to nociception and antinociceptive mechanisms, *Pain*, 12, 93, 1982.
46. Smith, M.V., Apkarian, A.V., and Hodge, C.J., Somatosensory response properties of contralaterally projecting spinothalamic and non-spinothalamic neurons in the second cervical segment of the cat, *J. Neurophysiol.*, 66, 83, 1991.
47. Yeziersky, R.P. and Broton, J.G., Functional properties of spino-mesencephalic tract (SMT) cells in the upper cervical spinal cord of the cat, *Pain*, 45, 187, 1991.
48. Cook, A.W., Nathan, P.W., and Smith, M.C., Sensory consequences of commissural myelotomy. A challenge to traditional anatomical concepts, *Brain*, 107, 547, 1984.
49. Villanueva, L., Desbois, C., Le Bars, D., and Bernard, J.F., Organization of diencephalic projections from the medullary subnucleus reticularis dorsalis and the adjacent cuneate nucleus: a retrograde and anterograde tracer study in the rat, *J. Comp. Neurol.*, in press.
50. Berkley, K.J., Guilbaud, G., Benoist, J.M., and Gautron, M., Responses of neurons in and near the thalamic ventrobasal complex of the rat to stimulation of uterus, cervix, vagina, colon, and skin, *J. Neurophysiol*, 69, 557, 1993.
51. Vogt, B.A., Derbyshire, S., and Jones, A.K.P., Pain processing in four regions of human cingulate cortex localised with co-registered PET and MR imaging, *Eur. J. Neurosci.* 8, 1461, 1996.
52. Berkley, K., Benoist, J.M., Gautron, M., and Guilbaud, G., Responses of neurons in the caudal intralaminar thalamic complex of the rat to stimulation of the uterus, vagina, cervix, colon and skin, *Brain Res.*, 695, 92, 1995.
53. Groenewegen, H.J. and Berendse, H.W., The specifity of the "non-specific" midline and intralaminar thalamic nuclei, *Trends Neurosci.*, 17, 52, 1994.
54. Nathan, P.W., Comments on "a dorsolateral spinothalamic tract in macaque monkey" by Apkarian and Hodge, *Pain*, 40, 239, 1990.
55. Villanueva, L., Chitour, D., and Le Bars, D., Involvement of the dorsolateral funiculus in the descending spinal projections responsible for diffuse noxious inhibitory controls in the rat, *J. Neurophysiol.*, 56, 1185, 1986.
56. Apkarian, A.V. and Hodge, C.J., A dorsolateral spinothalamic tract in macaque monkey, *Pain*, 37, 323, 1989.
57. Berkley, K. and Hubscher, C.H., Are there separate central nervous system pathways for touch and pain?, *Nature Med.*, 1, 766, 1995.
58. Al-Chaer, E., Lawand, N., Westlund, K., and Willis, W.D., Visceral nociceptive input into the ventral posterolateral nucleus of the thalamus: a new function for the dorsal column pathway, *J. Neurophysiol.*, 76, 2661, 1996.
59. Al-Chaer, E., Feng, Y., Westlund, K., and Willis, W.D., The dorsal column: a role in nociceptive viscerosensory processing in the primate, *Soc. Neurosci. Abstr.*, 915(6), 2350, 1997.
60. Le Bars, D. and Villanueva, L., Electrophysiological evidence for the activation of descending inhibitory controls by nociceptive afferent pathways, in *Pain Modulation, Progress in Brain Research*, Fields, H.L. and Besson, J.M., Eds., Elsevier, Amsterdam, 1988, p. 275.
61. Nathan, P.W., and Smith, M.C., Dysesthésie après cordotomie, *Med. Hygiène*, 42, 1788, 1984.
62. Villanueva, L., Bernard, J.F., and Le Bars, D., Distribution of spinal cord projections from the medullary subnucleus reticularis dorsalis and the adjacent cuneate nucleus: a *Phaseolus vulgaris* leucoagglutinin (PHA-L) study in the rat, *J. Comp. Neurol.*, 352, 11, 1995.

63. Towe, A., Somatosensory cortex: descending influences on ascending systems, in *Somatosensory System, Handbook of Sensory Physiology, II*, Iggo, A., Ed., Springer, Berlin, 1973, p. 701.
64. Nathan, P.W., Smith, M.C., and Cook, A.W., Sensory effects in man of lesions of the posterior columns and of some other afferent pathways, *Brain*, 109, 1003, 1986.
65. Paxinos, G. and Watson, C., *The Rat Brain in Stereotaxic Coordinates*, compact 3rd. ed., Academic Press, New York, 1996.

Chapter 34

Is There Still Room in the Neurosurgical Treatment of Pain for Making Lesions in Nociceptive Pathways?

Jan Gybels

Contents

34.1	Introduction	588
34.2	Operations on the Cranial Nerves	589
	34.2.1 The Trigeminal Nerve	589
	34.2.2 Cranial Nerves VII, IX, X, XI	591
34.3	Neurotomies and Spinal Rhizotomies	591
	34.3.1 Neurotomy and Other Direct Operations on Peripheral Nerves	591
	34.3.2 Spinal Rhizotomies	591
34.4	Lesions in the Spinal Cord	593
	34.4.1 Lesions in the Dorsal Root Entry Zone	593
	34.4.2 Anterolateral Cordotomy	593
	34.4.2.1 Historical Note	593
	34.4.2.2 General Anatomical and Physiological Considerations	593
	34.4.2.3 Results	595
	34.4.2.4 Complications	595
	34.4.2.5 Conclusion	596
	34.4.3 A Note on Commissural Myelotomy, Stereotactic C1 Central Myelotomy, and Limited (Punctate) Midline Myelotomy	596
34.5	Lesions at Supra-Spinal Levels	597
	34.5.1 The Different Supra-Spinal Procedures	597
	34.5.2 Clinical Data Until 1989	597
	34.5.3 Data from a 1994 Poll	600
	34.5.4 A Case History and Two Different Views	601
	34.5.5 A New Look at an Old Destructive Supra-Spinal Procedure for the Relief of Persistent Pain	602
	35.5.6 Conclusions	602

34.6	Lesions in the Autonomic Nervous System	603
34.7	A Cautious Answer to a Timely Question	603
References		604

34.1 Introduction

Faced with the failure of medical treatment of intractable, intolerable, and persistent pain, the physician may consider resorting to a neurosurgical intervention with analgesic consequences. For many, surgery means the surgeon's knife. But in fact, the needle, the catheter, and the electrode often replace the knife in the operating theater.

The concepts of pain surgery have evolved considerably over the last few years. Previous open operations have been supplemented by the percutaneous introduction of electrodes under very brief general or local anesthesia with the opportunity of verifying by physiological means the nervous structures one aims to destroy. In addition, the operating microscope has disclosed subtle compressive lesions on cranial nerves which may give rise to paroxysmal neuralgias and possibly can be cured by removing the compression.

But, perhaps more importantly, as described in the previous chapters, new knowledge of pain mechanisms includes the existence of both pain suppressor and pain inducer pathways and of an astonishing variety of physiological chemical mediators at the synapses in these pathways, as well as the pharmacology of the less readily destroyed analogs of the pain suppressors and antagonists of the pain inducers. This knowledge, although expanding rapidly, still remains in a state of flux.

Since the time of Foerster, neurosurgeons have learned that pain often seems to "run in front of the knife", i.e., the nervous system possesses a remarkable capacity, following destructive procedures on it, to return toward the *status quo ante* and to develop little-used or new mechanisms either to cause recurrence of the original pain or to replace it by other types of pain. Although often appearing at a slower pace than the analogous development of tolerance to chemical analgesics, it is clear that both physician and surgeon are confronted by this major problem in pain control.

This limited efficacy and significant risk of complications associated with neuro-ablative procedures have led neurosurgeons to take considerable interest in the evidence that a complex system of pain inhibitory mechanisms exists and to investigate whether pain control could be achieved by activation of these mechanisms. This activation can be realized by both electrical stimulation at specific sites of the nervous system and introduction into the cerebrospinal fluid of chemicals to bind to either opiate or many nonopiate types of pain suppressor receptors. Clinical utilization of these new concepts dates from the 1960s for electrical stimulation and the 1970s for intrathecal opoid administration and is still incomplete. However, these "neuroaugmentative" procedures, as they are called, have already been shown to have in certain persistent pain syndromes a greater efficacy than the classical destructive operations and this with extreme low morbidity and with the great advantage of being reversible. For this reason, we will examine in this chapter the question of whether there is still a place for the surgical interruption of nociceptive pathways in the neurosurgical treatment of pain. It would be tedious and impossible in the limits of the pages allowed by the editors to try to evaluate all the neurosurgical procedures that have been devised as a treatment for pain. We rather will give a general evaluation emphasizing those aspects that are of particular interest in a handbook of behavioral state control.

To be able to answer the question of whether there is still room in the neurosurgical treatment of pain for making lesions in nociceptive pathways, one must know the results obtained by this surgery, and here a major difficulty arises. We have indeed been struck by the difficulty of evaluating results reported in the world literature.[1] Reports of results range from anecdotal accounts, through studies reporting the percentages of patient responses in several response categories, to evaluations taking into account on a semi-quantitative basis the consumption of analgesics and

TABLE 34.1
Landmarks in the Surgical Treatment of *Tic douloureux*

1	First complete publication of a case	Wepfer, 1727
2	Unsuccessful peripheral section of a branch of the trigeminal nerve	Schlichting (according to Rose, 1892)
3	Description of the disorder, coining of term *Tic douloureux*, chemical destruction of peripheral branch	André, 1756
4	Gasserian ganglion resection	Krause, 1896
5	Retrogasserian rhizotomy	Horsley et al., 1891
		Spiller and Frazier, 1901
6	Alcohol injection in a branch of the trigeminal nerve	Schloesser, 1907
7	Alcohol injection in Gasserian ganglion	Harris, 1912
8	Report of 298 consecutive Gasserian ganglionectomies without mortality	Cushing, 1920
9	Partial section of V root at the pons	Dandy, 1929
10	Electrocoagulation of Gasserian ganglion	Kirschner, 1931
11	Decompression of the trigeminal root	Taarnhoj, 1952
12	Compression of the trigeminal root	Shelden et al., 1955
13	Vascular decompression	Gardner and Miklos, 1959
14	Temperature-controlled coagulation of Gasserian ganglion and rootlets	Sweet, 1968
15	Microvascular decompression	Jannetta, 1976
16	Retrogasserian glycerol injection	Hàkanson, 1981
17	Percutaneous compression of Gasserian ganglion	Mullan and Lichtor, 1983

Source: From Gybels, J.M. and Sweet, W.H., *Neurosurgical Treatment of Persistent Pain*, Karger, Basel, 1989. With permission.

activity levels and also making a distinction between short-term and long-term results. Almost all these reports can be described as presenting retrospective, uncontrolled, and incomplete data from a group of selected patients. It is only in recent years, because of the difficulty of conducting these studies in surgery, that prospective randomized cross-over studies in pain surgery have been initiated[2] and evaluation of results of pain surgery have become a topic in surgical textbooks.[3] In the absence of prospective rigorously conducted outcome studies, our subjective judgment as to how to report the results in the literature has been guided by the degree of critical assessment of the original authors, the number of their cases, the availability of thoughtful reviews, the reputation of the authors, and our own experience.

34.2   Operations on the Cranial Nerves

34.2.1   The Trigeminal Nerve

Operations in the trigeminal nerve are frequently performed for trigeminal neuralgia, also called *Tic douloureux*, a frequent disorder which is possibly one of the most painful of all human afflictions. In Table 34.1 are reproduced landmarks in the surgical treatment of *Tic douloureux*. Surgical intervention started with total section of branches of the trigeminal nerve, which was then followed by partial lesions of the nerve, the ganglion of Gasser, and the root; decompression of the root; microvascular decompression; and percutaneous compression of the Gasserian ganglion. For many years, temperature-controlled coagulation of the Gasserian ganglion and rootlets has been the standard procedure (Figure 34.1.)

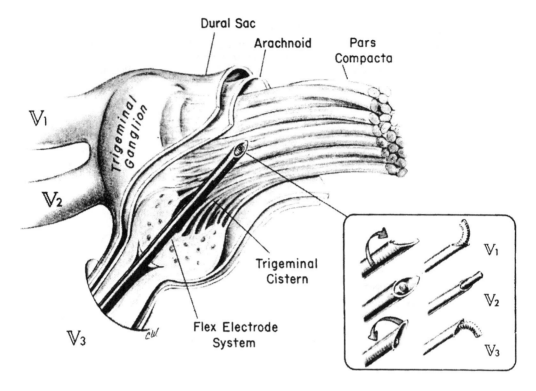

FIGURE 34.1
The trigeminal rootlets and the curved electrode with thermocouple capable of producing lesions in any of the three divisions from a single position. (From Tew, J.M. and van Loveren, H.R., in *Operative Neurosurgical Techniques*, Schmidek, H.H. and Sweet, W.H., Eds., W.B. Saunders, Philadelphia, 1988. With permission.)

Microvascular decompression is a procedure designed not to lesion nervous tissue but to eliminate the structural cause of the pain — a vascular compression by an artery or a vein at the junctional area of the trigeminal root in the cerebellopontine angle adjacent to the brainstem — rather than to replace the paroxystic pain attacks by numbness resulting from a lesion of the nerve or the root. Microvascular decompression is a major intervention, and the most puzzling set of facts continues to be the similarity of results between minor maneuvers confined to the middle fossa, such as the percutaneous compression of the Gasserian ganglion, and those confined to the posterior fossa. From a neurosurgical viewpoint, the mechanism responsible for essential trigeminal neuralgia is often mysteriously labile.

Many hypotheses have been formulated concerning the pathophysiology of *Tic douloureux*, but generally they can be divided into the group of those favoring a peripheral mechanism (compression of the nerve or root by different causes leading to demyelinization, ectopic electrogenesis, etc.) and a central mechanism. For the central mechanism, much of the evidence is speculative.[5]

In the decision process regarding which surgical strategy to follow, many factors have to be considered, and there is a continuing debate between proponents of a more invasive procedure, such as microvascular decompression in the cerebello-pontine angle, and proponents of a more conservative operation, such as a percutaneous approach. It is important to realize that in cases of trigeminal neuralgia not responding to appropriate medication, neurosurgery is very effective; indeed, the patient must to the fullest be informed, as in other therapeutic modalities, of the cost-benefit of the different procedures.

34.2.2 Cranial Nerves VII, IX, X, XI

Pain related to tumor in the rhinopharynx and essential vagoglossopharyngeal neuralgia, a disorder which displays a much richer range of symptoms than does trigeminal neuralgia, can probably be treated by percutaneous radiofrequency (RF) rhizotomy at the jugular foramen with avoidance of dysphagia and hoarseness by the use of small increments of heating and appropriate testing of the patient during and between lesions. In patients in whom this fails, open posterior fossa operations should lead to IX and partial X rhizotomies regardless of neurovascular relations because of the absence of dysesthesias after rhizotomy of these roots and the greater likelihood of recurrence of pain after microvascular decompression.

34.3 Neurotomies and Spinal Rhizotomies

34.3.1 Neurotomy and Other Direct Operations on Peripheral Nerves

Neurotomy has been used in a number of conditions such as painful neuroma, thoracotomy pains, etc. Owing to poor results, however, it has been almost entirely abandoned, except in a few conditions such as occipital neuralgia and particularly pseudoradicular pain, originating from the facets joints.

Nerve repair, though, may be one of the most important means of preventing neuroma formation following physical trauma. When nerve repair is not an option, as for instance after an amputation, controlling neuroma formation has been the goal. Many ingenious methods have been tried, such as the use of mechanical barriers, chemical destruction of regenerating axons, rerouting the cut nerve into the muscle or into bone marrow, etc. In a recent publication concerning neurosurgical procedures of the peripheral nerves, Burchiel[6] stated that a neuroma-excision, neurectomy, and nerve release for injury-related pain of peripheral nerve origin yielded substantial subjective improvement in a minority of patients; surgically proving the existence of a neuroma, with confirmed excision, was preferable to simple proximal neurectomy. The presence of a discrete nerve syndrome and mechanical hyperalgesia were positive prognostic indicators, ongoing litigation was the strongest predictor of failure, and change in work status was not a likely outcome.

34.3.2 Spinal Rhizotomies

Several variations of spinal root surgery have been developed over the years, but on the whole, after a period of rapid expansion, the method has gone through a long decline, initiated by spinothalamic cordotomy, which was introduced in 1912. Selective posterior rhizotomy merits a special mention. It was introduced by Sindou in 1972,[7] and the aim of this intervention was to interrupt selectively small-diameter nociceptive fibers at their radicular entrance into the spinal cord. Sindou reasoned that such an intervention would suppress nociceptive pathways and at the same time favor the pain inhibitory mechanism as proposed in the gate control theory of Melzack and Wall.[8] From animal work, it is known that the large and small axons in a dorsal root are randomly dispersed but that in some species such as the monkey[9] there is a segregation of fine afferents in the lateral parts of dorsal rootlets.

In man, at the spinal cord-rootlet junction, large and small fibers are anatomically dissociated. On entering the spinal cord, most of the small fibers regroup at the ventrolateral part of the junction, before penetrating into the tract of Lissauer. The large fibers run dorsomedially between the dorsal colums and the posterior horn (Figure 34.2).

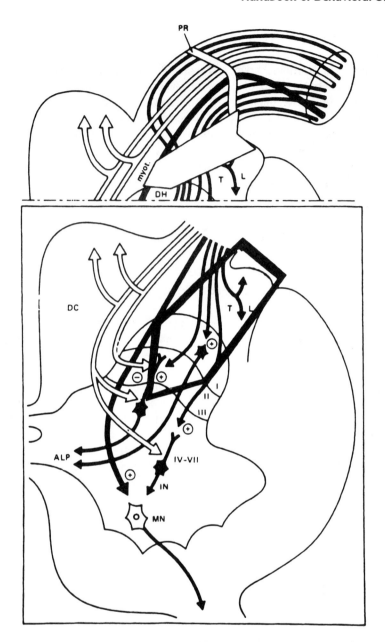

FIGURE 34.2

Schematic representation of the DREZ area and target of microsurgical DREZotomy (MDT). **(Upper)** Each rootlet can be divided, owing to the transition of its glial support, into a peripheral and a central segment. The transition between the two segments is at the pial ring (PR), which is located approximately 1 mm outside the penetration of the rootlet into the dorsolateral sulcus. Peripherally, the fibers are mixed together. As they approach PR, the fine fibers, considered nociceptive, move toward the rootlet surfaces. In the central segment, they group in the ventrolateral portion of the DREZ to enter the dorsal horn (DH) through the tract of Lissauer (TL). The large myotatic fibers (myot) are situated in the middle of the DREZ, whereas the large lemniscal fibers are located dorsomedially. **(Lower)** Schematic data on DH circuitry. Note the monosynaptic excitatory arc reflex, the lemniscal influence on a DH cell and an interneuron (IN), the fine fiber excitatory input onto DH cells and the IN, the origins in layer I and layers IV to VII of the anterolateral pathways (ALP), and the projection of the IN onto the motoneuron (MN). DC = dorsal column. Rexed laminae are marked from I to VI. MDT (arrowhead) cuts most of the fine and myotatic fibers and enters the medial (excitatory) portion of LT and the apex of the dorsal horn. It should preserve most lemniscal presynaptic fibers, the lateral (inhibitory) portion of TL, and most of the DH. (From Sindou, M., Etude de la jonction radiculo-médullaire postérieure, *La radicellotomie postérieure sélective dans la chirurgie de la douleur*, Thèse, Lyon, 1972; Sindou, M. et al., *J. Comp. Neurol.*, 153, 15, 1974. With permission.)

The major advantage of selective posterior rhizotomy is that by preservation of the lemniscal fibers, more extensive rhizotomies can be performed with significantly less impairment of the functional capacity of the limbs than is the case with total posterior rhizotomies. Sparing the lemniscal fibers may also avoid the secondary appearance of pain at the cutaneous boundary zone of sensory loss and leaves intact the substrate for subsequent electrostimulation procedures.[11] Sindou, in the *Handbook of Operative Neurosurgical Technique*,[12] now calls the procedure microsurgical DREZotomy (dorsal root entry zone), because it is situated at an anatomic entity that includes the central portion of the dorsal root, the tract of Lissauer, and the dorsal-most layers of the dorsal horn.

34.4 Lesions in the Spinal Cord

34.4.1 Lesions in the Dorsal Root Entry Zone

Nashold and Ostdahl, in 1979,[13] published the method and the results of an operative procedure in which RF heat lesions were placed in the DREZ. Among a total experience of 387 personal cases operated since 1972, Sindou[12] cites as the best indication for the procedure topographically limited cancer pain such as in Pancoast syndrome, neurogenic pain resulting from brachial plexus injuries, pain corresponding to segmental spinal cord lesions, peripheral nerve injuries, amputation, herpes zoster, and hyperspasticity with pain. Nashold,[13] with a personal experience of over 600 patients, cites the same indications for neuropathic pain but states that patients with peripheral pain do less well than patients with central pain. The success of the DREZ operation for pain related to brachial plexus avulsion is well documented, ranging in different series from 60 to 80% patients experiencing significant pain relief and has to be emphasized. In this condition, both steady and paroxystic pain are equally well relieved. Provided that selection of patients and techniques are rigorous, the intervention can be carried out with a low rate of corticospinal and dorsal columns deficit.

34.4.2 Anterolateral Cordotomy

34.4.2.1 Historical Note

In 1911, Martin, at the suggestion of Spiller, performed the operation of anterolateral tractotomy for pain, after Spiller correlated clinical with pathological evidence that lesions of the anterior quadrant of the spinal cord were associated with contralateral loss of pain and temperature sensation.[14] While the operation has remained in essence the same as that introduced by Spiller and Martin almost 90 years ago, much effort has been devoted to producing lesions in the anterolateral quadrant of the cord with minimal morbidity and maximal pain relief. A key role in this effort has been the correlation of anatomically verified lesions at autopsy with the results of detailed clinical observation.

Open anterolateral cordotomy has been performed until the present time, but gradually it has been replaced by a percutaneous method, a technique introduced by Mullan et al. in 1963.[16] The most frequently used technique is one in which an electrode is introduced laterally at the C1-C2 interspace; a radiofrequency electrode allows the lesion to be made promptly in a fractional manner with the patient awake so that the development of analgesia can be monitored and the size and the localization of the lesion adjusted.

34.4.2.2 General Anatomical and Physiological Considerations

Neuroanatomical and physiological aspects of nociceptive pathways have been extensively studied in animals and are reviewed in detail in Chapter 33. In view of the multiplicity of the ascending

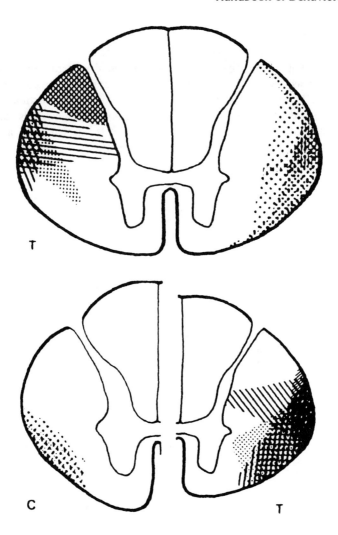

FIGURE 34.3
(**Upper**) A diagram of cordotomy lesions that were insufficient to cause adequate or dense sensory loss contralaterally below the lesion. (**Lower**) A diagram of cordotomy incisions that cause dense analgesia in a limited region of the body. Left, cervical cord; right, thoracic cord. Every case of cordotomy in both illustrations has the lesion illustrated by a different form of cross-hatching. (From Nathan, P.W. and Smith, M.C., *Adv. Pain Res. Ther.*, 3, 921, 1979. With permission.)

pathways carrying nociceptive information, it is indeed astonishing that an appropriately placed lesion (see Figure 34.3) can bring about total long-term pain relief, together with pinprick and warm and cold sensation loss. Noordenbos and Wall[17] have described in detail a patient who suffered the opposite of an anterolateral tractotomy, as her spinal cord had been completely cut across at the Th3 segment except for part of one anterolateral quadrant extending from slightly dorsal to the dentate ligament to the midline. In accordance with classical teaching, temperature and pinprick sensations on the side opposite the intact quadrant were preserved. Against classical expectations were the following findings: preservation of touch and pressure localization and pain appreciation bilaterally, ability to detect passive movement on the ipsilateral side, ability to identify von Frey hair stimuli bilaterally (sometimes at normal thresholds, usually at increased thresholds), and after repeated subthreshold stimulation, a lowering of the threshold for detection of von Frey hairs.

In human studies, the term "spinothalamic tract" has been often used very loosely and refers to any ascending fibers in the anterolateral part of the spinal cord which have to do with the ability to

TABLE 34.2
Reported Percent Pain Relief After Unilateral Percutaneous Cordotomy

	Complete (%)	Significant (%)
Tasker	72	84.2
Meglio and Cioni	79	—
Lahuerta et al.	64	87
Kühner	—	59
Siegfried et al.	—	75
Lipton	75	—
Ischia et al.	—	81.1
Lorenz	75	96
Amano et al.	64	82
Ischia et al.	—	71
Farcot et al.	—	89

Source: From Tasker, R.R. and North, R., in *Neurosurgical Management of Pain*, North, R.B. and Levy, R.M., Eds., Springer, New York, 1997, chap. 13. With permission.

feel pain and temperature. There is indeed a vast but often controversial clinical literature on the functional anatomy of the anterior quadrant of the spinal cord, but from the point of view of applied anatomy (an important aspect for the operating surgeon), the features represented in Figure 34.4 are particularly relevant.

34.4.2.3 Results

For decades, unilateral or bilateral cordotomy has been one of the most successful operations for pain control in the neurosurgical armamentarium. In many of the published series, it is difficult to evaluate fully the precise results of these cordotomies, but a few representative examples of big series allow one to trace general trends. Table 34.2 is such a representative example of results of pain relief after unilateral percutaneous cordotomy. The interested reader can find many detailed tables of open cordotomy results and complications in Sweet et al.[19] Kanpolat et al.[20] recently reported complete pain control in 97% of the cases in a series of 67 c.t.-guided percutaneous cordotomies for localized cancer pain.

34.4.2.4 Complications

Obviously, making a cut in the spinal cord can produce many complications, most of which can be deduced from the schema of applied anatomy as depicted in Figure 34.4. These complications include death; sleep-induced apnea (known as Ondine's curse); motor impairment; bladder, bowel, and sexual dysfunction; hypotension; dysesthesias; postoperative girdle pain at the level of incision; Horner's syndrome; and referred pain, known as allachesthesia. The dysesthesias are often of major clinical importance because of their severity. They arise in the areas rendered analgesic to pinprick. These complaints usually appear only when analgesia has given way to hypoalgesia and therefore are more likely to affect patients who live longer. They are considered to be a form of central pain. With present techniques, significant risks of unilateral cordotomy are postcordotomy dysesthesia and impairment of bladder function, estimated, respectively, to be 5 and 3%. Bilateral cordotomy carries an increased mortality because of respiratory risks and impaired bladder function, ranging, according to different authors, from 12 to 58%.[18]

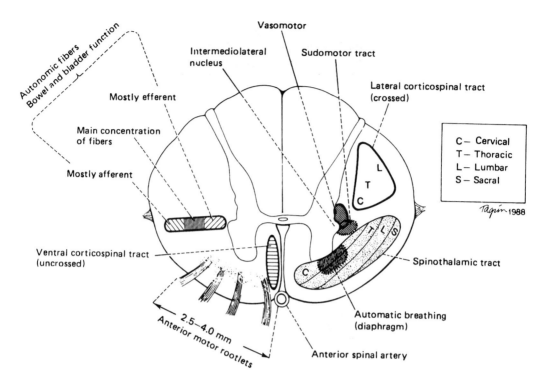

FIGURE 34.4

Diagram of spinal cord at approximately T2-T3 level to show structures affected by cordotomy of anterior quadrant. Note that, although dentate ligament is shown arising from equator of cord, it may arise farther posteriorly, leading to inadvertent incision into lateral corticospinal tract if used as an infallible landmark. The ventral rootlets do not emerge from a single sulcus as do the posterior rootlets but are scattered over an extensive mediolateral area as indicated in the diagram. At any one cross-sectional level, however, there is only one emerging rootlet — not a whole row, as presented in diagram. (From Gybels, J.M. and Sweet, W.H., *Neurosurgical Treatment of Persistent Pain*, Karger, Basel, 1989. With permission.)

34.4.2.5 Conclusion

Although cordotomy is now carried out far less often than it used to be, mainly due to better knowledge of opiate therapy and current enthusiasm for infusion techniques such as intraspinal morphine, its capability of producing a completely pain-free state in the appropriately selected and fortunate patient should not be forgotten. The extent to which anterolateral cordotomy should be utilized at present is unclear, but what is clear is that there are many pathways in the spinal cord carrying nociceptive information to suprasegmental levels (see References 21 and 22 for reviews) and that in humans pain often seems to "run in front of the knife".

Although no hard data are available, neurosurgeons report in private discussions an increase in recent years of patients referred to them by algologists, particularly anesthesiologists, as candidates for anterolateral cordotomy.

34.4.3 A Note on Commissural Myelotomy, Stereotactic C_1 Central Myelotomy, and Limited (Punctate) Midline Myelotomy

These interventions are very rarely performed at present and are mentioned here for the sake of completeness and also because we have learned from the critical observations of precise surgical therapy in man that there is a central pathway for pain in the human spinal cord, the destruction of which diminishes or eliminates the conscious awareness of clinical persistent pain with modest, if any, loss of pain from discrete noxious stimuli.

34.5 Lesions at Supra-Spinal Levels

34.5.1 The Different Supra-Spinal Procedures

In Table 4.3 some important landmarks in the surgical treatment of pain by a lesion at supra-spinal levels are listed. To answer the question of whether there is still a place for lesions at supra-spinal levels, we will first examine the data available up to 1989, at which time an exhaustive study of the problem was made.[1] We will then look at the answers to a questionnaire which was sent to all members of the European Society for Stereotactic and Functional Neurosurgery in 1994. The questionnaire was designed to find out what the actual neurosurgical practice was in the treatment of pain in malignancy and in the treatment of neuropathic pain. More particularly, the last item of this questionnaire was formulated as follows: "Do you think there are still indications for destructive neurosurgery at supra-spinal levels for the relief of painful syndromes? If yes, for which painful syndromes?" In a third section, we will comment on a case history which was published in 1996 in *Controversies in Neurosurgery* and underscores that an answer to the above-formulated question has to be approached with subtlety. Finally, we will have a new look at some old destructive interventions in the thalamus for neuropathic pain.

34.5.2 Clinical Data Until 1989

As already noted in the introduction, reporting results of surgical interventions represents a major difficulty, because by their very nature, ablative neurosurgical procedures are not amenable to double-blind, placebo-controlled prospective studies which are the standard for pharmacological studies. We reported[1] the results of the world literature guided by the degree of critical assessment of the original authors, the number of their cases, the availability of thoughtful reviews, the reputation of the authors, and our own experience. Because of space limitation here, we have to summarize these results, and we have done so by providing in Table 34.4 the conclusion we reached for each intervention. This, of course, has a subjective character but is the best we can do. For rationale and results of the different supra-spinal lesion procedures, we refer to Reference 1.

A commentary on frontal lobe lesions and particularly cingulotomy is in order. These interventions, known as "psychosurgery" were initially developed during the 1940s as a last-resort treatment in some psychiatric disorders, but it was noted that some patients, who in addition to their psychiatric disorder before the operation also complained of severe pain, no longer did so after the lobotomy. In the Western world, during the 1970s, the debate around psychosurgery became inflamed, prompting a prominent neurosurgeon to ask in a respected medical journal, "Treatment of medically intractable mental disease by limited leucotomy — justifiable?"[23] But, already in 1955, in order to reduce undesirable effects and operative risks, Foltz and White[24] began to make electrical lesions stereotactically in the white matter of the anterior cingulum in patients with a primary complaint of severe pain but with superimposed major anxiety, depression, or emotional lability. In the 23 years ending in September 1987, Ballantine et al.[25] carried out bilateral cingulotomy in 139 patients for the treatment of chronic pain. Of the 95 patients with non-malignant pain followed for 1 to 21 years (an average of 7 years) after operation, the pain in 65 of them was related to "failed low-back surgery syndrome". Complete or marked pain relief was sustained in 26%, and moderate relief continued in another 36%. There was an unequivocal absence of anything approaching significant deficit of basic function. Ballantine et al.[26] performed a very detailed prospective study of therapeutic

TABLE 34.3
Landmarks in the Surgical Treatment of Pain by a Lesion at Supraspinal Levels

Lesions in the brainstem

1	Introduction of bulbar trigeminal tractotomy	Sjöqvist (1938)
2	Introduction of spinal V nucleolysis	Hitchcock and Schwarz (1972)
3	Multiple lesions of descending cephalic pain tract and n. caudalis-DREZ lesions	Bernard, Nashold et al. (1987)

Bulbar and pontine spinothalamic tractotomy

4	First section of the spinothalamic tract in the medulla oblongata	Schwartz and O'Leary (1941)
5	Stereotactic pontine spinothalamic tractotomy	Hitchcock (1973)

Stereotactic mesencephalotomy

6	Stereotactic interruption of both spinothalamic and quintothalamic tracts, and spinoreticular tract leading to the destruction of nociceptive pathways from one half of the body	Wycis and Spiegel (1962)
7	Extensive description of results of stimulation in the midbrain of man	Nashold and Wilson (1966)

Lesions in the diencephalon

8	In 1947, a stereotactic lesion in the n. dorsomedialis thalami is added to a mesencephalotomy in order to influence the psychic components of a patient's pain	Wycis and Spiegel (1962)
9	First thalamic lesions aimed at:	
	Ventrocaudal nuclei (VPL-VPM)	Hécaen, Talairach et al. (1949)
	VPM-VPL + centrum medianum	Hécaen, Talairach et al. (1949)
	Intralaminar nuclei	Mark and Hackett (1959)
	Pulvinar	Kudo, Yoshii et al. (1968)
	Anterior nuclei	Andy (1973)
10	Publication of results of stereotactic hypothalamotomy in 54 cancer patients	Fairman (1973)

Pituitary destruction

11	First hypophysectomy in patients with advanced carcinoma of the breast and the prostate	Luft and Olivocrona (1953)
12	First description of neuroadenolysis by injection of alcohol in the pituitary gland	Greco, Sbaragli et al. (1957)

Lesions in the telencephalon:

Frontal lobe lesions

13	Freeman and Watts, in the second edition of their monograph on prefrontal lobotomies in *Psychosurgery*, added the further title "in the treatment of mental disorders and intractable pain"	Freeman and Watts (1950)
14	Results of a series of 139 chronc pain patients, treated by bilateral cingolotomy and extensively studied by independent observers are published	Ballantine, Bouckoms et al. (1987)
15	Results of a series of 35 cancer patients, treated by subcaudate tractotomy	Sweet et al. (1982)

Pre- and postcentral gyrectomy

16	First description of removal of post-central gyrus in a painful phantom limb patient	de Gutiérrez-Mahoney (1944)

Source: From Gybels, J.M. and Sweet, W.H., *Neurosurgical Treatment of Persistent Pain*, Karger, Basel, 1989. With permission.

TABLE 34.4
Clinical Data Until 1989

Lesions in the brainstem	
Bulbar trigeminal tractotomy and nucleotomy	"Trigeminal nucleotomy either at one or many levels needs further critical evaluation before its place can be decided. Intractable facial post-herpetic neuralgia is one of the most promising disorders to explore."
Bulbar and pontine spinothalamic tractotomy	"Published experience with this procedure is limited, but it may offer deep sustained high-level analgesia at a smaller risk than many other procedures."
Stereotactic mesencephalotomy	"Two recent long-term follow-ups from Tokyo and Duke University describe satisfactory relief of central neurogenic pain by mesencephalotomies in about two thirds of 55 patients. The most widespread use of the procedure is pain of malignancies involving the head and neck. It competes with brain stimulation and intraventricular morphine."
Lesions in the diencephalon	
Thalamotomy	"Tasker's conclusion, reasonable in our view, was that the risks of failure and complications are significantly greater for these procedures than for more peripheral operations so that these should usually be reserved for patients with cancer in the head and neck and certain other selected cases."
Hypothalamotomy	"Such important functions are concentrated in the hypothalamus that this has little appeal to most neurosurgeons as a stereotactic target, despite the low complication rate described today."
Lesions in the telencephalon	
Frontal lobe lesions	"Earlier conclusions were that the patients leukotomized for pain seem to suffer much more deficit than those leukotomized for psychosis. This turns out to be wrong for appropriately placed and circumscribed lesions, which may be gratifyingly beneficial."
Pre- and post-central gyrectomy	"Three cases of pre- and postcentral gyrectomy for pain relief are encouraging but too few on which to base a recommendation, especially when the successful surgeons have themselves described no more."

Source: Adapted from Gybels, J.M. and Sweet, W.H., *Neurosurgical Treatment of Persistent Pain*, Karger, Basel, 1989. With permission.

outcome, neurologic status, and behavioral test performance in 15 patients with persistent pain, 10 with back and leg pain, 3 with abdominal pain, 1 with thalamic syndrome, and 1 with amputation stump pain. The therapeutic outcome was evaluated two times, a year apart. The incidence of improvement was high at the first assessment and remained so at the second evaluation. There was no evidence of lasting behavioral deficit after cingulotomy. Based on the experience from frontal lobe lesions and, more specifically, cingulotomies for the treatment of psychiatric disorders, as far as we are able to assess, cingulotomies for persistent pain have always been bilateral. This is not without interest when looking at recent data generated by positron emission tomography (PET) studies. In one such study, Hsieh et al.[27] showed that the posterior section of the right anterior cingulate cortex (ACC), corresponding to Brodmann area 24, was activated in ongoing neuropathic pain, regardless of the side of the painful mononeuropathy (Figure 34.5); the experiments confirmed that the ACC participates in the sensorial/affectional aspect of the pain experience, but there was also a strong suggestion — and this is the new, exciting finding — that there is a possible right hemispheric lateralization of the ACC for affective processing in chronic ongoing neuropathic pain.

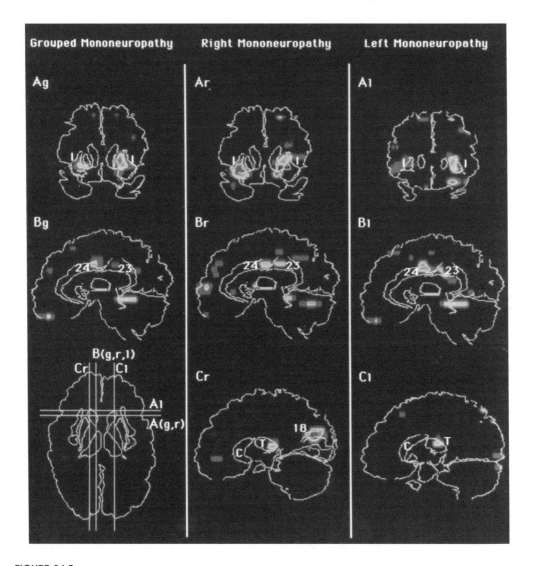

FIGURE 34.5
Omnibus significance maps of rCBF changes during chronic ongoing neuropathic pain. The left panel (Ag, Bg) depicts the general representation of the grouped PMN cases; the middle (Ar, Br, Cr) and the right (Al, Bl, Cl) panels depict the results of the laterality analysis in the right and left PMN patients, respectively. The right hemisphere in the figure is on the reader's left. The coronal views of A (g, r, l) show the activation of the anterior insula. The sagittal views of B (g, r, l) show the preferential activation of the right posterior sector of the ACC (BA 24), regardless of the laterality of the painful mononeuropathy. The sagittal views of C (r, l) show the reduced rCBF in the contralateral post thalamus. Numbers refer to the Brodmann areas. C = caudate; T = thalamus; I = insula. (From Hsieh, J.C. et al., *Pain*, 63, 225, 1995. With permission.)

34.5.3 Data from a 1994 Poll

Out of 215 questionnaires sent to the members of the European Society of Stereotactic and Functional Neurosurgery in 1994, the number of useful replies was 51. In the questionnaire it was asked whether a certain operation for pain in a given neurosurgical department was occasionally or often performed and what the actual numbers were in 1993. Table 34.5 gives the data for persistent cancer pain and neuropathic pain. This poll further revealed that for the 51 responders to the questionnaire the number of anterolateral cordotomies for cancer pain in 1993 was 86; for

TABLE 34.5
Results of a 1994 Poll on the Frequency of Destructive Neurosurgery at Supra-Spinal Levels for Relief of Cancer Pain and Neuropathic Pain

Cancer Pain / Neuropathic Pain	Occasionally	Often	Number in 1993
Telencephalic operations			
Cingulotomy	5 / 3	—	0 / 5
Postcentral gyrectomy	1 / 0	—	—
Hypothalamotomy	1 / 1	—	5 / 4
Thalamotomy			
Ventrocaudal thalamotomy (VPL-VPM)	3 / 6	1 / 1	15 / 11
Medial thalamotomy	3 / 3	2 / 2	10 / 11
Pulvinarotomy	3 / 3	1 / 0	8 / 5
Dorsomedian nucleus	0 / 2	1 / 0	8 / 3
Anterior thalamic nuclei	0 / 2	1 / 0	4 / 3
Operations in the brainstem			
Stereotactic bulbar trigeminal tractotomy and nucleotomy	2 / 3	1 / 0	3 / 0
Stereotactic bulbar and pontine spinothalamic tractotomy	4 / 2		2 / 0
Stereotactic mesencephalic tractotomy	8 / 4	0 / 1	7 / 6
Others			
Bulbar trigeminal tractotomy	1 / 0	—	—
Open trigeminal tractotomy	0 / 1	—	0 / 3
Not specified	1 / 0	—	1 / 0

Note: In the second and third column are indicated the total number of neurosurgical departments which performed a certain operation occasionally or often. In the last column is given the total number of interventions performed in the 51 departments in 1993.

Source: From Gybels, J.M. and Nuttin, B., in *Forebrain Areas Involved in Pain Processing*, Besson, J.M., Guilbaud, G., and Ollat, H., Eds., John Libbey Eurotext, Paris, 1995. With permission.

neuropathic pain, 6; for spinal cord stimulation for cancer pain, 44; for neuropathic pain, 529; for intrathecal morphine administration in cancer pain, 102; for neuropathic pain, 69. It is probably fair to state that supra-spinal destructive procedures for relief of painful syndromes are rarely performed.

In response to the question "Do you think that there are still indications for destructive neurosurgery at supra-spinal levels for the relief of painful syndromes?," 26 saw no indications for destructive neurosurgery at supra-spinal levels, while 25 did. Of these 25, four indicated cranial nerve surgery in the posterior fossa which is not really the question we sought to answer. Therefore, the negative answers were the majority (26 no, 21 yes). The majority of the indications were pain in the face in cancer; rare indications were phantom pain, brachial plexus avulsion, and other "deafferentation" pains.

34.5.4 A Case History and Two Different Views

The case:[29] a 50-year-old man underwent resection of a petroclival meningeoma. Three months after surgery, he began to have facial pain, which was persistent and increasing, associated with dysesthesia and intractable to all forms of medical treatment. This pain deeply affected his life and performance.

Surgeon A (Schvarcz):[30] What should not be done in this case? No further peripheral denervation. My first choice of neurosurgical treatment would be radiofrequency lesions stereotactically placed in the nucleus caudalis of the trigeminal system. I have performed over 200 nucleotomies. In a consecutive series of 141 patients with deafferentation pain, with follow-up from 1 to 16 years, abolition of the allodynia and a significant reduction or, less frequently, a complete abolition of the deep background pain was obtained in 72.6% of the cases without lasting side effects.

Surgeon B (Burchiel):[31] It is very unlikely that this pain will respond to further peripheral denervation. Overall, the chance of a tractotomy or caudalis nucleotomy producing satisfactory relief is about 75%. Almost 100% of the patients will experience an undesired deficit of some sort, with debilitating limb ataxia in 10% and disabling contralateral limb sensory loss in 15% of the patients. One of the guiding principles for the management of difficult pain problems has to do with the Hippocratic oath, *"primum non nocere"*. For all these reasons, this man should have a trial of deep brain stimulation, prior to consideration of an ablative operation. The probability that deep brain stimulation will help his pain is 50%, and the risk to him is about 15%.

34.5.5 A New Look at an Old Destructive Supra-Spinal Procedure for the Relief of Persistent Pain

As early as 1949, Hécaen et al.[32] suggested for the treatment of pain a lesion which interrupts, in addition to the ventrobasal complex, part of the diffuse thalamic projection system. Medial thalamotomy (essentially a lesion in the intralaminar nuclei) can bring pain relief without provoking a sensory deficit. The absence of severe complications made laminotomy at one time a popular intervention, but due to all too frequent recurrences of the original pain and the appearance of other treatment modalities such as neurostimulation and the intrathecal administration of drugs the procedure was abandoned. The procedure has been taken up again by Jeanmonod and his colleagues.[33] These authors placed a lesion in the n. centralis lateralis (CL) in 45 patients with neurogenic pain. They report that, after a medial thalamotomy, 67% of the patients reached a 50 to 100% relief of pain, without somatosensory deficits. Follow-up was short, ranging from 2 weeks to 38 months, with a mean of 14 months. Pain was assessed pre- and postoperatively using three to seven visual analog determinations of pain intensity, the patient's estimation in percent of the global postoperative improvement, activities of daily living, and pain drug intake.

However, the data given in the publications of these authors do not allow the reader to assess for himself the amount of pain relief. Interestingly, microelectrode recordings of a total of 318 units in CL showed that half of the units exhibited low-threshold calcium spike bursts. From their results, the authors propose a theory which states that neurogenic pain is due to an imbalance between central and ventroposterior nuclei, resulting in an over-inhibition of both these nuclei by the thalamic reticular nucleus.

35.5.6 Conclusions

With the data at our disposal we are of the opinion that for the time being there are only very few indications for destructive neurosurgery at supra-spinal levels for the relief of pain. However, in neurological disease, nature demonstrates that long-standing "thalamic pain" caused by a stroke can subsequently and permanently be abolished by a new stroke.[34] We agree therefore, for instance, with a respondent to the above-mentioned questionnaire when he replied to the question whether there were still indications for destructive supra-spinal procedures: "Yes, median thalamotomy, if the work by Jeanmonod and colleagues in Zurich is confirmed."

TABLE 34.6
Diagnostic Criteria of Complex Regional Pain Syndromes (CRPS)

CRPS, Type I (Reflex Sympathetic Dystrophy)[a]	CRPS, Type II (Causalgia)[b]
1. The presence of an initiating noxious event, or a cause of immobilization	1. The presence of continuing pain, allodynia, or hyperalgesia after a nerve injury, not necessarily limited to the distribution of the injured nerve
2. Continuing pain, allodynia, or hyperalgesia with which the pain is disproportionate to any inciting event	2. Evidence at some time of edema, changes in skin blood flow, or abnormal sudomotor activity in the region of the pain
3. Evidence at some time of edema, changes in skin blood flow, or abnormal sudomotor activity in the region of the pain	3. This diagnosis is excluded by the existence of conditions that would otherwise account for the degree of pain and dysfunction
4. This diagnosis is excluded by the existence of conditions that would otherwise account for the degree of pain and dysfunction	

[a] Criteria 2-4 must be satisfied.
[b] All three criteria must be satisfied.

Source: From Merskey, H. and Bogduk, N., *Classification of Chronic Pain: Descriptions of Chronic Pain Syndromes and Definitions of Pain Terms,* prepared by the International Association for the Study of Pain, Task Force on Taxonomy, IASP Press, Seattle, 1994, p. 222. With permission.

34.6 Lesions in the Autonomic Nervous System

A number of conditions, such as causalgia, Sudeck's atrophy, algodystrophy, reflex-sympathetic dystrophy, complex regional pain syndromes, etc., have in common the fact that the pain can be relieved, at least temporarily, by a sympathetic block, although this statement is not accepted by everybody. Table 34.6 summarizes the clinical characteristics of these different syndromes as they have recently been regrouped and redefined by the International Association for the Study of Pain.[35] Sympathetically maintained pain is a pain that is maintained by sympathetic efferent activity or neurochemical or circulating catecholamine action, as determined by pharmacological or sympathetic nerve blockade.

A success rate of sympathectomies carried out for sympathetically mediated disorders, characterized principally by pain, with or without dystrophic features, has by most authors been reported as being on the order of 60 to 70%. The indications for open sympathectomy have decreased steadily, consequent upon the availability of increasingly valuable sympatholytic drugs.

The ease of execution of endoscopic approaches has led to a somewhat cavalier assessment of the completeness of the denervation they achieve, and the success rate of these new methods still requires critical evaluation.

34.7 A Cautious Answer to a Timely Question

In the treatment of persistent pain, neurosurgery is moving rapidly towards reversible methods with reasonably predictable and testable outcomes, and away from interrupting nociceptive pathways. However, as discussed in the previous pages, with the data at our disposal (and needless to say these data are far from allowing us to make a scientifically founded statement) there is still room — and we think, indeed, a need in selected cases — to place a lesion in a nociceptive pathway. A typical example might be *Tic douloureux* not responding satisfactorily to medication in a patient who is not a candidate for microvascular decompression, a major intervention in the fossa posterior, or a patient with neoplasic invasion of the brachial or lumbosacral plexus not responding sufficiently to well-managed opioid therapy. It is good clinical practice to offer such a patient anterolateral cordotomy, which may provide instantaneous and complete pain relief provided the pain is strictly unilateral.

Besides establishing rigourous outcome criteria of pain surgery there is a need to develop algorithms or guidelines of treatment of specific pain syndromes. The neurosurgical community is adressing these issues by means of consensus conferences, and examples for the role of placing lesions in nociceptive pathways can be found for facial pain, cancer pain, neuropathic pain syndromes, and chronic low-back and failed back surgery syndrome in North and Levy.[36]

References

1. Gybels, J.M. and Sweet, W.H., *Neurosurgical Treatment of Persistent Pain*, Karger, Basel, 1989.
2. North, R.B., Kidd, D.H., Lee, M.S., and Piantodasi, S., A prospective, randomized study of spinal cord stimulation versus reoperation for failed back surgery syndrome: initial results, *Stereotact. Funct. Neurosurg.*, 62, 267, 1994.
3. Kupers, R.C. and Gybels, J.M., Evaluation of results in pain surgery, in *Textbook of Stereotactic and Functional Neurosurgery*, Gildenberg, P.L. and Tasker, R.R., Eds., McGraw-Hill, New York, 1997, chap. 135.
4. Tew, J.M. and van Loveren, H.R., Percutaneous rhizotomy in the treatment of intractable facial pain (trigeminal, glossopharyngeal, and vagal nerves), in *Operative Neurosurgical Techniques*, Schmidek, H.H. and Sweet, W.H., Eds., W.B. Saunders, Philadelphia, 1988, chap. 97.
5. Rappaport, Z.H. and Devor, M., Trigeminal neuralgia: the role of self-sustaining discharge in the trigeminal ganglion, *Pain*, 56, 127, 1994.
6. Burchiel, K.J., Neurosurgical procedures of the peripheral nerves, *Neurosurgical Management of Pain*, North, R.B. and Levy, R.M., Eds., Springer, New York, 1997, chap. 2.
7. Sindou, M., Etude de la jonction radiculo-médullaire postérieure, *La radicellotomie postérieure sélective dans la chirurgie de la douleur,* Thèse, Lyon, 1972.
8. Melzack, R. and Wall, P.D., Pain mechanisms: a new theory, *Science*, 150, 971, 1965.
9. Kerr, F.W.L., Neuroanatomical substrates of nociception in the spinal cord, *Pain*, 1, 325, 1975.
10. Sindou, M., Quoex, C., and Baleydier, C., Fiber organization at the posterior spinal cord-rootlet junction in man, *J. Comp. Neurol.*, 153, 15, 1974.
11. Sindou, M. and Keravel, Y., Analgésie par la méthode d'électrostimulation transcutanée. Résultats dans les douleurs d'origine neurologique. A propos de 180 cas, *Neurochirurgie*, 26, 153, 1980.
12. Sindou, M.P., Microsurgical DREZotomy, in *Operative Neurosurgical Techniques*, Schmidek, H.H. and Sweet, W.H., Eds., W.B. Saunders, Philadelphia, 1995, chap. 130.
13. Nashold, J.R. and Nashold, B.S., Microsurgical DREZotomy in treatment of deafferentation pain, *Operative Neurosurgical Techniques*, Schmidek, H.H., and Sweet, W.H., Eds., W.B. Saunders, Philadelphia, 1995, chap. 131.
14. Spiller, W.G., Martin, E., The treatment of persistent pain of organic origin in the lower part of the body by division of the anterolateral column of the spinal cord, *JAMA,* 58, 1489, 1912.
15. Nathan, P.W. and Smith, M.C., Clinico-anatomical correlation in anterolateral cordotomy, *Adv. Pain Res. Ther.*, 3, 921, 1979.
16. Mullan, S., Harper, P.V., Hekmatpanah, J., Torres, H., and Dobben, G., Percutaneous interruption of spinal-pain tracts by means of a strontium needle, *J. Neurosurg.*, 20, 931, 1963.

17. Noordenbos, W. and Wall, P.D., Diverse sensory functions with an almost totally divided spinal cord. A case of spinal cord transection with preservation of one anterolateral quadrant, *Pain*, 2, 185, 1976.
18. Tasker, R.R. and North, R., Cordotomy and myelotomy, in *Neurosurgical Management of Pain*, North, R.B. and Levy, R.M., Eds., Springer, New York, 1997, chap. 13.
19. Sweet, W.H., Poletti, C.E., and Gybels, J.M., Operations in the brainstem and spinal canal, with an appendix on the relationship of open percutaneous cordotomy, in *Textbook of Pain*, Wall, P.D. and Melzack, R., Eds., Churchill Livingstone, Edinburgh, 1994, chap. 59.
20. Kanpolat, Y., Cagler, S., Akyar, S., and Temiz, C., CT-guided pain procedures for intractable pain in malignancy, *Acta Neurochir.*, 64, 88, 1995.
21. Besson, J.M., Guilbaud, G., and Ollat, H., Eds., *Forebrain Areas Involved in Pain Processing*, John Libbey Eurotext, Paris, 1995, p. 276.
22. Gebhart, G.F., *Visceral Pain*, IASP Press, Seattle, 1995, p. 516.
23. Sweet, W.H., Treatment of medically intractable mental disease by limited frontal leucotomy — justifiable?, *New Engl. J. Med.*, 289, 1117, 1973.
24. Foltz, E.L. and White, L.E., Pain "relief" by frontal cingulotomy, *J. Neurosurg.*, 19, 89, 1962.
25. Ballantine, H.T., Bouckoms, A.J., Thomas, E.K., and Giriunas, I.E., Treatment of psychiatric illness by stereotactic cingulotomy, *Biol. Psychiatry*, 22, 807, 1987.
26. Corkin, S., Twitchell, T.E., and Sullivan, E.V., Safety and efficacy of cingulotomy for pain and psychiatric disorder, in *Modern Concepts in Psychiatric Surgery*, Hitchcock, E.R., Ballantine, H.T., and Meyerson, B.A., Eds., Elsevier/North Holland, Amsterdam, 1979, p. 253.
27. Hsieh, J.C., Belfrage, M., Stone-Elander, S., Hansson, P., and Ingvar, M., Central representation of chronic ongoing neuropathic pain studied by positron emission tomography, *Pain*, 63, 225, 1995.
28. Gybels, J.M. and Nuttin, B., Are there still indications for destructive neurosurgery at supra-spinal levels for the relief of painful syndromes?, in *Forebrain Areas Involved in Pain Processing*, Besson, J.M., Guilbaud, G., and Ollat, H., Eds., John Libbey Eurotext, Paris, 1995, chap. 18.
29. Gybels, J.M., Treatment of facial pain: descending trigeminal nucleotomy vs. deep brain stimulation, in *Controversies in Neurosurgery*, Al-Mefty, O., Origitano, T.C., and Harkey, H., Eds., Thieme, New York, 1996, p. 355.
30. Schvarcz, J.R., Descending trigeminal nucleotomy for dysesthetic facial pain, in *Controversies in Neurosurgery*, Al-Mefty, O., Origitano, T.C., and Harkey, H., Eds., Thieme, New York, 1996, chap. 36.
31. Burchiel, K.J., Deep brain stimulation for facial pain, in *Controversies in Neurosurgery*, Al-Mefty, O., Origitano, T.C., and Harkey, H., Eds., Thieme, New York, 1996.
32. Hécaen, H., Talairach, T., David, M., and Dell, M.B., Mémoires originaux. Coagulations limitées du thalamus dans les algies du syndrome thalamique, *Rev. Neurol.*, 81, 917, 1949.
33. Jeanmonod, D., Magnin, M., and Morel, A., Thalamus and neurogenic pain: physiological, anatomical and clinical data, *NeuroReport*, 4, 475, 1993.
34. Soria, E. and Fine, E., Disappearance of a thalamic pain after parietal subcortical stroke, *Pain*, 44, 258, 1991.
35. Merskey, H. and Bogduk, N., *Classification of Chronic Pain: Descriptions of Chronic Pain Syndromes and Definitions of Pain Terms,* prepared by the International Association for the Study of Pain, Task Force on Taxonomy, IASP Press, Seattle, 1994, p. 222.
36. North, R.B. and Levy, R.M., *Neurosurgical Management of Pain*, Springer, New York, 1997.

Section VIII

Immunological Alterations in Arousal States

James M. Krueger, Section Editor

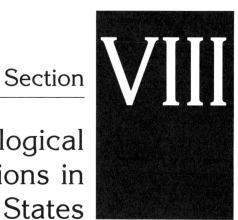

Chapter 35

Cytokines and Sleep Regulation

James M. Krueger and Jidong Fang

Contents

35.1	Tumor Necrosis Factor Is a Sleep Regulatory Substance	611
35.2	IL-1β Is a Sleep Regulatory Substance	614
35.3	Other Cytokines May Also Be Involved in Sleep Regulation	615
35.4	The Sleep Humoral Regulatory Network	615
35.5	Perspectives	616
Acknowledgment		617
References		617

Sleep is often modeled as being regulated by two processes. One of these, the homeostatic process, is posited to be a reflection of the common experience of sleepiness and sleep rebound occurring after prolonged wakefulness. The homeostatic process is thus wake dependent. The other process encompasses the common observation that it is easier to fall asleep at some times of the day than at other times; this circadian rhythm-dependent process is thought to set a threshold for the onset of sleep (see Reference 4 for a review). Both these processes result from the interactions of humoral and neuronal mechanisms.

Historically, there was considerable debate as to the importance of the humoral vs. neuronal relative contributions to physiological regulation. These soup vs. spark debates waxed and waned with the ever-improving technical abilities of the biochemists vs. electrophysiologists. It is now generally recognized that these mechanisms are inseparable from each other. Indeed, the concept of humoral modulation of neuronal circuits for the generation of rhythmic outputs has been developed for several brain functions.[26,44-47] Humoral inputs affect many aspects of neuronal circuits including their synaptic connectivity, responsiveness to stimuli, and neuronal composition, as well as the intrinsic properties of neurons within the affected circuit. Conversely, the electrical activity of neurons affects their production of a variety of substances. CNS activity during wakefulness is thought to result in the production of sleep regulatory substances (SRS). Indeed, over the past 80 years, many laboratories have demonstrated the accumulation of sleep-promoting substances in cerebrospinal fluid during prolonged wakefulness; for example, the transfer of cerebrospinal fluid

TABLE 35.1
Criteria a Putative Sleep Regulatory Substance (SRS) Should Fulfill

1. The SRS should promote physiological sleep (see Table 35.3).
2. The SRS receptors should be in areas of brain involved in sleep regulation.
3. Levels of the SRS or its receptor should change with the sleep-wake cycle.
4. Induction of increased production of the SRS should enhance sleep.
5. Inhibition of the SRS or its receptors should reduce spontaneous sleep.
6. Removal of the SRS or its receptor (e.g., in knockout animals) should reduce sleep.
7. The SRS should be part of a larger biochemical cascade involved in sleep regulation.

from sleep-deprived animals to normal animals induces excess sleep in the recipients.[4,44] It is likely that SRS-induced circuit dynamics are a fundamental mechanism of sleep.

For many years, there has been an unwritten postulate within the sleep research community that the "true" SRS would have biological actions specific to sleep and act on the CNS executive sleep regulatory network, which was also assumed to be primarily concerned with sleep regulation. However, all the SRSs identified to date have multiple biological actions (see Reference 44 for a review). Furthermore, a single neural circuit, center, or network necessary for sleep has not been demonstrated. A related issue is whether changes in molecular or electrical cellular activity are, in fact, directly tied to sleep. This is very difficult to address because any manipulation of sleep (e.g., sleep deprivation) is associated with other changes in physiology (e.g., brain temperature, caloric intake). As a consequence of such considerations, lists of criteria have been developed that an SRS (Table 35.1) or neural circuits (Table 35.2) should fulfill before they are viable candidates for being involved in sleep regulation. This review focuses on SRSs and addresses the issue of whether cytokines, as SRSs, fulfill the criteria in Table 35.1. The issue of whether specific neural circuits are involved in sleep regulation is dealt with elsewhere in this volume.

The list of endogenous substances reported to promote sleep is now lengthy, comprising some 40 to 60 substances;[4,44] however, there is convincing evidence for only a few of these for their involvement in sleep regulation. For non-rapid-eye-movement sleep (NREMS) the list is limited to tumor necrosis factor (TNF), interleukin-1 (IL-1), growth hormone releasing hormone (GHRH), prostaglandin D2, and adenosine. For REMS, they are vasoactive intestinal peptide and prolactin. To those outside the field of SRSs, even these few substances, with their well-known biological actions independent of sleep, may seem daunting. We hope to clarify this by first discussing the

TABLE 35.2
Criteria a Putative Sleep Regulatory Circuit (SRC) Should Fulfill

1. Stimulation of the SRC should promote physiological sleep.
2. The SRC should contain receptors for SRSs.
3. Firing patterns or intrinsic properties of neurons within the SRC should vary with the sleep/wake cycle.
4. Stimulation of excitatory inputs to the SRC should enhance sleep.
5. Stimulation of inhibitory inputs to the SRC should inhibit sleep.
6. Removal of the SRC (e.g., lesion) should reduce sleep.
7. The SRC should be well-connected to other SRCs.

Cytokines and Sleep Regulation

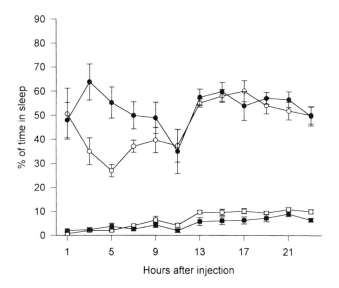

FIGURE 35.1
Murine TNFα induces increases in non-rapid-eye-movement sleep in mice. TNFα, 3 μg, was injected intraperitoneally at time 0. Mice were kept on a 12-hr light/dark cycle with lights out at 0 hr. Circles are NREMS values + S.E. Squares are REMS values + S.E. Closed symbols are experimental values; open symbols are control values.

evidence for the involvement of one of these substances, TNF, within the context of Table 35.1. Second, we will emphasize that these SRSs are linked to each other within biochemical cascades in sleep regulation. Finally, we briefly discuss how such knowledge has led to new views of the functional organization of brain as it applies to sleep.

35.1 Tumor Necrosis Factor Is a Sleep Regulatory Substance

Administration of TNFα into rabbits,[34,77] mice (Figure 35.1),[19] or rats[62] induces increased duration of NREMS (Criterion 1, Table 35.1). For example, mice given 3 μg of TNFα, intraperitoneally, have about 80 minutes of extra NREMS during the first 9 hr postinjection.[19] The excess NREMS induced by TNFα appears physiological in the sense that the criteria for SRS-induced sleep in Table 35.3 are met. TNFα induces an increase in the number and duration of individual NREMS episodes. After low somnogenic doses, sleep architecture remains normal in the sense that animals continue to cycle through stages of sleep and wakefulness, and REM-to-REM intervals remain normal (although within each cycle there may be more NREMS). After TNFα treatment, sleep remains readily reversible, e.g., the entry of the experimenter into the room quickly awakens TNFα-treated rabbits. After such a disturbance, the return to sleep is generally more rapid after TNFα treatment than in control animals. After somnogenic doses of TNFα, there are no gross behavior abnormalities; sleep postures remain normal, and no motor dysfunction is evident. Changes in sleep-coupled autonomic function also remain intact. Changes in brain temperature associated with sleep states persist in TNFα-treated rabbits; for example, during entry into NREMS there is a regulated decrease in brain temperature, whereas after entry into REM brain temperature rapidly rises if rabbits are at room temperature. Both of these brain temperature changes persist in TNF-treated rabbits, whether animals are febrile (after high doses) or not (after low somnogenic doses). After central administration of TNFα to rabbits or rats, there are increases in the amplitudes of EEG slow-waves during

TABLE 35.3
Criteria for SRSs or SRCs Regarding Physiological Sleep

1. SRS treatment or SRC stimulation should induce increased incidence or duration of sleep and/or enhance sleep intensity.
2. After SRS treatment or SRC stimulation, normal sleep architecture should be maintained, although altered duration/incidence of states may occur.
3. SRS- or SRC-induced sleep should be readily reversible by appropriate stimuli.
4. Low somnogenic doses of the SRS or mild stimulation of the SRC should not induce abnormal behavior.
5. Autonomic changes (e.g., change in brain temperature) that are coupled to sleep should persist after SRS treatment of SRC stimulation.
6. A dose-response relationship for the SRS or stimuli strength-response relationship for the SRC should persist with certain ranges.

NREMS episodes. These "supranormal" EEG slow waves also occur during the deep sleep after sleep deprivation[70] and are thus thought to be indicative of the intensity of NREMS.[4] However, there are experimental circumstances in which the duration of NREMS is independent of the amplitude of EEG slow waves, e.g., after lesions of basal forebrain cholinergic neurons.[35] Indeed, the enhanced NREMS associated with increased food intake is associated with a decrease in EEG slow wave amplitudes during NREMS.[25] Interestingly, intraperitoneal injection of TNFα also induces increases in duration of NREMS and decreases in EEG slow-wave amplitude during NREMS;[19] both these effects are dose dependent. In summary, the somnogenic actions of TNFα fulfill all the criteria in Table 35.3 for SRS-induced physiological sleep.

TNF protein and TNFα mRNA are found in many areas of normal brain, including the anterior hypothalamus and hippocampus (Criterion 2 in Table 35.1).[1,5,7,21,22,28,44] TNF receptors are also found in many areas of brain. TNF seems to be in neurons, glia, and brain endothelium. Brain production of TNF increases after treatment of animals with endotoxin.[53] Levels of TNF in brain vary with the sleep/wake cycle (Criterion 3 in Table 35.1).[7,22] Although the regulation of TNF production in brain is not very well understood, there are diurnal variations of both TNFα mRNA and TNF protein levels (Figure 35.2). In rats, highest levels occur during periods of greatest NREMS (just after lights are turned on).[7,22] Further, there is a circadian rhythm in plasma levels of TNFα, and plasma levels of TNFα correlate with EEG slow-wave activity.[11] Plasma rhythms of TNF are disrupted in sleep apnea patients[17,93] and HIV seropositive patients;[11] both sets of patients have disrupted sleep patterns. Further, TNF plasma levels increase after sleep deprivation,[27] and the ability of circulating white blood cells to produce TNFα increases after sleep deprivation.[92,97] Despite these sleep-linked changes in levels of TNF, the relative importance of circulating vs. central TNF in sleep regulation remains unknown. Further, no information is available concerning change in TNF receptors during the sleep/wake cycle.

Substances that induce TNF production, such as bacterial cell products (e.g., muramyl peptides) or viral double-stranded RNA, also induce increases in duration of NREMS (Criterion 4 in Table 35.1; see Reference 44 for a review). In contrast, administration of anti-TNFα antibodies[82,87,88] or the soluble TNF receptor[85] inhibit spontaneous NREMS (Criterion 5 in Table 35.1). In addition, the expected sleep rebound, which normally follows sleep deprivation, is attenuated if animals are pretreated with TNF inhibitors.[86] Further, inhibition of TNF also attenuates the increase in NREMS associated with acute mild increases in ambient temperatures[87] and NREMS responses induced by muramyl dipeptide.[85]

TNFα belongs to a family of molecules which includes TNFβ and two TNF receptors, the 55-kD and 75-kD receptors. TNFβ is also somnogenic,[36] but very little information is available concerning its distribution in brain or its sleep-related properties. There is also a TNF-soluble

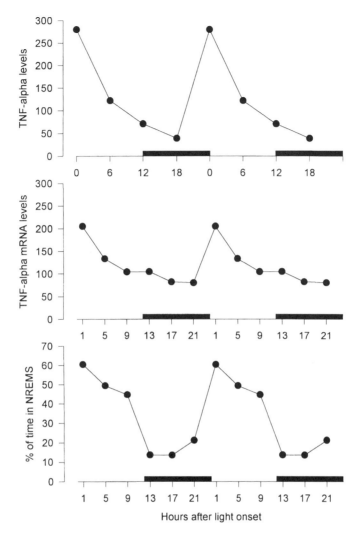

FIGURE 35.2
Diurnal changes in TNF bioactivity (top), TNFα mRNA (middle), and rat NREMS (bottom). Values were taken from References 7, 22, and 66, respectively. All three variables peak at the beginning of daylight hours.

receptor which represents the extracellular domain of the TNF receptor; it is a constituent of normal cerebrospinal fluid.[74] The TNF 55-kD receptor seems to be involved in sleep regulation. Knockout mice lacking this receptor do not have sleep responses if given TNF, although they do respond to other somnogens.[19] Further, the TNF 55-kD receptor knockout mice sleep less than other strains of mice (Criterion 6 in Table 35.1); for example, the TNF 55-kD receptor knockout mice sleep about 90 minutes less during daylight hours than strain controls.[19]

TNFα is a component of a cascade of biochemical events leading to NREMS (Criterion 7 in Table 35.1; Figure 35.3). For example, TNF induces production of interleukin-1 (IL-1);[2,12] IL-1, in turn, induces TNF.[71] IL-1 also is somnogenic (see below). If rabbits are given TNFα and pretreated with a soluble IL-1 receptor, the expected TNFα-induced NREMS responses are attenuated.[83] Conversely, if rabbits are pretreated with a TNF blocker then given IL-1β, IL-1β-induced NREMS responses are inhibited.[83] Such results suggest the somnogenic actions of IL-1 and TNF are closely linked to each other (Figure 35.3). However, their somnogenic actions can be separated under some

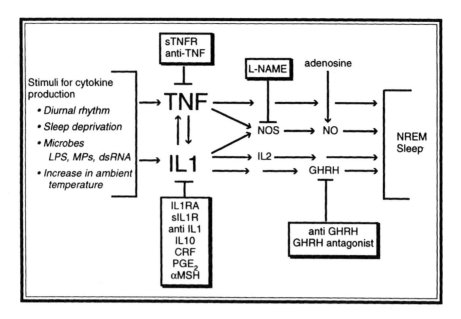

FIGURE 35.3
TNFα and IL-1β interact with each other via parallel interacting pathways to enhance NREMS. Substances in boxes inhibit NREMS and inhibit the actions or production of IL-1, TNF, NOS, or GHRH. Inhibition of any one step does not result in complete sleep loss. It is posited that animals compensate for the loss of any one step by relying on the parallel pathways shown. Such redundant pathways provide stability to the sleep regulatory system as well as a mechanism whereby the variety of known sleep-promoting stimuli may affect sleep. A major challenge to sleep research is to define how and where these molecular steps interact with known SRC mechanisms.

conditions. Thus, TNF 55-kD receptor knockout mice lacking the ability to respond to TNF exhibit robust NREMS responses if given IL-1.[19] In conclusion, TNFα meets all the criteria for a SRS outlined in Tables 35.1 and 35.3.

35.2 IL-1β Is a Sleep Regulatory Substance

Evidence, similar to that described above for TNF, also implicates IL-1 in sleep regulation (see Reference 44 for a review). Exogenous IL-1β administered intracerebroventricularly, intravenously, or intraperitoneally enhances NREMS. IL-1 was somnogenic in all species tested to date: cats, rats, mice, rabbits, and monkeys.[19,23,44,48,52,65,77,79,91] The sleep induced by somnogenic doses of IL-1 is physiological in that it meets the criteria in Table 35.3. At low somnogenic doses in rats and mice IL-1β induces excess NREMS; as doses are increased, there is a greater increase in NREMS, although after the higher doses REMS is inhibited.[65] Very high doses disrupt sleep.

IL-1-containing neurons are found in the hypothalamus.[6] IL-1β mRNA can be detected in many areas of brain (see Reference 80 for a review), and there is a diurnal variation of IL-1β mRNA in these areas.[80] Highest levels of IL-1β mRNA are found at the onset of daylight hours; this is the peak NREMS period in rats. Further, if rats are deprived of sleep levels of IL-1β mRNA increase.[55] In cats, IL-1 cerebrospinal fluid levels vary with the sleep/wake cycle,[54] and in humans plasma levels of IL-1 peak at the onset of sleep.[59]

Substances that induce IL-1 production, e.g., muramyl peptides, induce excess NREMS (see Reference 44 for a review). In contrast, inhibition of IL-1 using anti-IL-1β antibodies, the IL-1 receptor antagonist, or soluble IL-1 receptors inhibits spontaneous NREMS. These inhibitors of IL-1 also attenuate NREMS rebound after sleep deprivation[66,67,81] and the NREMS that occurs in response to acute mild increases in ambient temperature and to muramyl dipeptide.[30,84]

IL-1β is one member of a larger family of molecules which includes IL-1α and the IL-1 receptor antagonist. All these ligands bind to two IL-1 receptors; the type I receptor is the signal transduction receptor which forms a complex with an IL-1 receptor accessory protein.[24] The accessory protein also seems to be necessary for signal transduction.[95] However, the accessory protein mRNA does not have a diurnal rhythm in brain nor are its levels affected by sleep deprivation (Taishi et al., unpublished data). In contrast, the type II receptor is a decoy receptor having a truncated intracellular domain.[9,10] There is also an IL-1-converting enzyme that cleaves pre-IL-1β (but not IL-1α), thereby forming the mature 17-kD IL-1β. Finally, there is an IL-1 receptor-associated kinase. All members of this IL-1 family are found in brain.[3,10,18,96] Mice lacking the IL-1 type I receptor do not exhibit NREMS responses if given IL-1β, although they are responsive to TNFα.[20] These IL-1 type I receptor knockout mice also seem to sleep less than strain controls, although this effect is confined to the dark hours and is lesser in magnitude than that observed in TNF 55-kD receptor knockout mice.

As already mentioned, IL-1β also seems to be an import component of the cascade of biochemical events leading to NREMS (Figure 35.3). Of special note is that one step in IL-1-induced NREMS seems to involve GHRH. If rats are pretreated with anti-GHRH antibodies, then given IL-1β, the expected sleep responses fail to materialize.[63] Although not the subject of this review, there is also a wealth of data similar to those described above for IL-1 and TNF implicating GHRH in NREMS regulation.[8,16,46,64,98] In conclusion, IL-1β meets all the criteria for a SRS outlined in Table 35.1.

35.3 Other Cytokines May Also Be Involved in Sleep Regulation

Several other cytokines have also been assayed for sleep-promoting activity. The anti-inflammatory cytokines IL4 and IL-10 inhibit production of IL-1 and TNF and both inhibit sleep.[51,68] Several other cytokines have been shown to be somnogenic; the list includes interferon-α,[41] acidic fibroblast growth factor,[42] interleukin-2,[62] nerve growth factor,[89] and epidermal growth factor. However, the sleep-inhibitory or sleep-promoting properties of these cytokines have not been studied beyond these initial observations. It is thus not possible to know if they are involved in physiological sleep regulation. Two other cytokines, IL-6[69] and basic fibroblast growth factor,[42] failed to affect sleep. These latter results suggest that the cascade of cytokines involved in sleep is distinct from those cascades involved in fever or mitogenic responses, although some cytokines, e.g., IL-1 and TNF, are shared by several of these pathways.

35.4 The Sleep Humoral Regulatory Network

Several additional substances are likely involved in the cascade of events initiated by IL-1 or TNF that lead to sleep. As already mentioned, IL-1 induces GHRH, which in turn is somnogenic. Further, IL-1 induces IL-2, which is also reported to be somnogenic.[62] Both IL-1 and TNF induce nitric oxide synthase (NOS) expression and increases in NO production. NO may also be involved in the somnogenic actions of GHRH, which induces GH release via a NO mechanism.[90] Other somnogens also induce NO production, including adenosine.[29,32] Adenosine also augments IL-1-induced NOS expression.[75] If inhibitors of NOS (e.g., arginine analogs) are given to rats[37] or rabbits,[38] sleep is inhibited. In contrast, NO-donor substances greatly enhance sleep.[39] Inhibition of NOS also blocks IL-1-induced sleep responses.[38]

Several neurotransmitter systems are also likely involved in cytokine-altered sleep (see Chapter 38). For example, GABAergic mechanisms are thought to be important in thalamo-cortical EEG synchronization events (see Reference 78 for a review). The $GABA_A$ receptor is also posited to be involved in diazepam-mediated sleep responses. The IL-1β-IL-1 receptor complex affects the $GABA_A$ receptor; this interaction leads to an enhanced Cl permeability and membrane hyperpolarization. Further, the IL-1 receptor antagonist-IL-1 receptor complex inhibits $GABA_A$ receptor-induced Cl permeability.[57,58] If these interactions take place within thalamo-cortical circuits, perhaps they are responsible, in part, for IL-1-sleep relationships. A theoretical model has been developed which shows how these IL-1-IL-1 receptor-$GABA_A$ interactions would act within a population of GABAergic circuits to induce state shifts in neuronal groups.[45,49]

Histaminergic neurons may also be involved in cytokine sleep mechanisms. Neurotoxic lesions of histaminergic neurons decrease TNFα in the posterior hypothalamus while enhancing TNF production in the hippocampus.[1] Similarly, administration of a histamine blocker, alpha-fluoremethylhistidine, increased hippocampal TNF and decreased hypothalamic TNF.[21] These data suggest that neuronal histamine is involved, in part, in the regulation of the brain TNF system. Data from other tissues suggest that histamine inhibits TNFα gene expression and TNF synthesis. Some of the actions of IL-1 also seem to be mediated via histamine. The inhibitory action of IL-1β on gastric acid secretion is likely mediated via IL-1β inhibition of histamine secretion.[43] IL-1β also affects neural histamine.[61] In contrast to low doses of IL-1, high doses of IL-1 (e.g., 10 μg given centrally to rats) promote wakefulness.[65] This dose of IL-1 in rats also induces histamine release in the hypothalamus,[61] and hypothalamic histamine has been implicated in other responses induced by IL-1β.[33] Although antihistamines are popular over-the-counter somnogenic agents, the exact role of histamine in cytokine-mediated sleep remains to be determined.

TNFα and IL-1β affect many other neurotransmitter systems. For example, IL-1 activates dopaminergic, adrenergic, cholinergic, and serotinergic systems.[13,14,31] TNF also has a variety of direct effects on hypothalmic neural activity.[76] However, whether any of these actions are involved in cytokine-mediated sleep remains unknown.

35.5 Perspectives

The data summarized in this essay strongly implicate IL-1β and TNFα in NREMS regulatory mechanisms. Further, their somnogenic actions are linked to each other by affecting each other's production and linked to substances that are also thought to be part of a biochemical network regulating sleep. It seems likely that no single somnogenic substance only has sleep-related biological activity or is even necessary for sleep and yet, in concert with other biochemical steps, results in appropriate adaptive sleep responses. Such issues of specificity are germane to all areas of biological regulation; the issue of how multiple redundant pleiotropic pathways act in concert to elicit appropriate context-specific responses is a central problem in biology. These considerations in conjunction with the neuronal group theory of Edelman[15] have led to a new model of brain organization in regard to sleep.[45,47] In this model, it is proposed that sleep is a fundamental property of neuronal groups and that sleep begins at the neuronal group level being induced by substances the levels of which increase as a result of activity within the neuronal group. If sufficient numbers of neuronal groups are in a "sleep mode", then sleep as defined by EEG and behavioral criteria occurs. The coordination of sleep between neuronal groups is brought about by the neural networks previously implicated in sleep regulation. Sleep can thus be envisioned to be regulated by "bottom-up" mechanisms rather than by "top down" state imposition by sleep centers or circuits.

If, indeed, sleep is initiated within local circuits, then one would anticipate that the biochemical events responsible would be relatively widespread in brain and wakefulness dependent. In this regard, the changes in TNF and IL-1 mRNAs associated with the sleep/wake cycle and found in each area of brain examined support this view of sleep brain organization. Other data also support this notion. Unilateral stimulation of the somatosensory cortex during waking induces greater amplitudes of EEG slow waves during subsequent sleep on the stimulated side compare to the contralateral side.[40] The occurrence of NREMS in only half of the brain at any given time in dolphins[60] and fronto-occipital EEG power gradients during sleep in man[94] also support the notion that sleep is a localized process. This concept is also supported by Pigarev's findings that individual association area neurons shift from a firing patterns characteristic of wakefulness to one characteristic of sleep asychrononuslly as the animal falls asleep.[72] There are also many clinical observations consistent with the idea that it is possible for part of the brain to be asleep while part is awake (see Reference 56 for a review). These theoretical considerations, now supported by data, are important, for they lead to new suggestions for the function of sleep, e.g., that sleep serves a synaptic growth and maintenance function.[46] It seems likely that satisfactory explanations of important brain phenomena (e.g., memory, thought, emotion) will not be possible until sleep function is defined at cellular and molecular levels. We will first need to understand the biochemical mechanisms responsible for sleep before any such explanation is possible. We are beginning to do so as briefly reviewed herein.

Acknowledgment

This work was supported by grants from NIH-NINDS (NS25378, NS27250, NS31453).

References

1. Alvarez, X.A., Franco, A., Fernandez-Novoa, L., and Cacabelos, R., Effects of neurotoxic lesions in histaminergic neurons on brain tumor necrosis factor levels, *Agents Actions,* 41, C70, 1994.
2. Bachwich, P.R., Chensue, S.W., Larric, J.W., and Kunkel, S.L., Tumor necrosis factor stimulates interleukin-1 and prostaglandin E2 production in resting macrophages, *Biochem. Biophys. Res.,* 136, 94, 1986.
3. Ban, E.M., Interleukin-1 receptors in the brain: characterization by quantitative in situ autoradiography, *Immunomethods,* 5, 31, 1994.
4. Borbély, A.A. and Tobler, I., Endogenous sleep-promoting substances and sleep regulation, *Physiological Rev.,* 69, 605, 1989.
5. Breder, C.D., Tsujimoto, M., Terano, Y., Scott, D.W., and Saper, C.B., Distribution and characterization of tumor necrosis factor-α-like immuno-reactivity in the murine central nervous system, *J. Comp. Neurol.,* 337, 543, 1993.
6. Breder, C.C., Dinarello, C.A., and Saper, C.B., Interleukin-1 immuno reactive innervation of the human hypothalamus, *Science,* 240, 321, 1988.
7. Bredow, S., Taishi, P., Guha-Thakurta, N., Obál, Jr., F., and Krueger, J.M., Diurnal variations of tumor necrosis factor-α mRNA and α-tubulin mRNA in rat brain, *J. Neuroimmunomodulation,* 4, 84, 1997.
8. Bredow, S., Taishi, P., Obál, Jr., F., Guha-Thakurta, N., and Krueger, J.M., Hypothalamic growth hormone-releasing hormone mRNA varies across the day in rat, *NeuroReport,* 7, 2501, 1996.
9. Colotta, F., Dower, S.K., Sims, J.E., and Mantovani, A., The type II "decoy" receptor. A novel regulatory pathway for interleukin-1, *Immunol. Today,* 15, 652, 1994.
10. Cunningham, Jr., E.T. and DeSouza, E.B., Interleukin-1 receptors in the brain and endocrine tissues, *Immunol. Today,* 14, 171, 1993.

11. Darko, D.F., Miller, J.C., Gallen, C., White, W., Koziol, J., Brown, S.J., Hayduk, R., Atkinson, J.H., Assmus, J., Munnell, D.T., Naitoh, P., McCutchen, J.A., and Mitler, M.M., Sleep electroencophalogram delta frequency amplitude, night plasma levels of tumor necrosis factor-α and human immunodeficiency virus infection, *Proc. Natl. Acad. Sci. USA,* 92, 12080, 1995.
12. Dinarello, C.A., Cannon, J G., Wolff, S.M., Bernheim, H.A., Beutler, B., Cerami, A., Figari, I.S., Palladino, Jr., M.A., and O'Connor, J.V., Tumor necrosis factor (cachectin) is an endogenous pyrogen and induces production of interleukin 1, *J. Exp. Med.,* 163, 1433, 1986.
13. Dunn, A.J., Endotoxin-induced activation of cerebral catecholamine and serotonin metabolism: comparison with interleukin-1, *J. Pharmacol. Exp. Ther.,* 261, 964, 1992.
14. Dunn, A.J., Systemic interleukin-1 administration stimulates hypothalamic norepinephrine metabolism paralleling the increased plasma corticosterone, *Life Sci.,* 43, 429, 1988.
15. Edelman, G.M., *Neural Darwinism,* Basic Books, New York, 1987.
16. Ehlers, C.L., Reed, T.K., and Henriksen, S.J., Effects of corticotropin-releasing factor and growth hormone-releasing factor on sleep and activity in rats, *Neuroendocrinology,* 11, 467, 1986.
17. Entzian, P., Linnemann, K., Schlaak, M., and Zabel, P., Obstructive sleep apnea syndrome and circadian rhythms of hormones and cytokines, *Am. J. Respir. Crit. Care Med.,* 153, 1080, 1996.
18. Ericsson, A., Liu, C., Hart, R.P., and Sawchenko, P.E., Type I interleukin-1 receptor in the rat brain: distribution, regulation, and relationships to sites of IL-1-induced cellular activation, *J. Comp. Neurol.,* 361, 681, 1995.
19. Fang, J., Wang, Y., and Krueger, J.M., Mice lacking the TNF 55-kD receptor fail to sleep more after TNFα treatment, *J. Neurosci.,* 17, 5949, 1997.
20. Fang, J., Wang, Y., and Krueger, J.M., The affects of interleukin-1β on sleep are mediated by the type I receptor, *Am. J. Phyiol.,* 274, R655, 1998.
21. Fernandez-Novoa, L., Franco-Maside, A., Alvarez, X.A., and Cacabelos, R., Effects of histamine and alpha-fluoromethyl-histidine on brain tumor necrosis factor levels in rats, *Inflamm. Res.,* 44, 55, 1995.
22. Floyd, R.A. and Krueger, J.M., Diurnal variations of TNFα in the rat brain, *NeuroReport,* 8, 915, 1997.
23. Friedman, E.M., Boinski, S., and Coe, C.L., Interleukin-1 induces sleep-like behavior and alters call structure in juvenile rhesus macaques, *Am. J. Primatol.,* 35, 145, 1995.
24. Greenfeder, S.A., Nunes, P., Kwee, K., Labow, M., Chizzonite, R.A., and Ju, G., Molecular cloning and characterization of a second subunit of the interleukin-1 receptor complex, *J. Biol. Chem.,* 270, 13757, 1995.
25. Hansen, M., Kapas, L., Fang, J., and Krueger, J.M., Cafeteria diet-induced sleep is blocked by subdiaphragmatic vagotomy in rats, *Am. J. Physiol.,* 274, R168, 1998.
26. Harris-Warrick, R.M., Chemical modulation of central pattern generators, in *Neural Control of Rythmic Movements in Vertebrates,* Cohen, A.H., Rossignol, S., and Griller, S., Eds., John Wiley & Sons, New York, 1988, p.285.
27. Hohagen, F., Timmer, J., Weyerbrock, A., Fritsch-Montero, R., Ganter, U., Krieger, S., Berger, M., and Bauer, J., Cytokine production during sleep and wakefulness and its relationship to cortisol in healthy humans, *Neuropsychobiology,* 28, 9, 1993.
28. Hunt, J.S., Chen, H.L., Hu, X.L., Chen, T.Y., and Morrison, D.C., Tumor necrosis factor-alpha gene expression in the tissue of normal mice, *Cytokine,* 4, 340, 1992.
29. Ikeda, U., Kurosaki, K., Shimpo, M., Okada, K., Saito, T., and Shimada, K., Adenosine stimulates nitric oxide synthesis in rat cardiac myocytes, *Am. J. Physiol.,* 273, H59, 1997.
30. Imeri, L., Opp, M.R., and Krueger, J.M., An IL-1 receptor and an IL-1 receptor antagonist attenuate muramyl dipeptide- and IL-1-induced sleep and fever, *Am. J. Physiol.,* 265, R907, 1993.

31. Imeri, L., Bianchi, S., and Mancia, M., Muramyl dipeptide and IL-1 effects on sleep and brain temperature following inhibition of serotonin synthesis, *Am. J. Physiol.,*273, R1663, 1997.
32. Janigro, D., Wender, R., Ransom, G., Tinklepaugh, D.L., and Winn, H.R., Andenosine-induced release of nitric oxide from cortical astrocytes, *NeuroReport,* 7, 1640, 1996.
33. Kang, M., Yoshimatsu, H., Chiba, S., Kurokawa, M., Ogawa, R., Tamari, Y., Tatsukawa, M., and Sakata, T., Hypothalamic neuronal histamine modulates physiological responses induced by interleukin-1β, *Am. J. Physiol.,* 269, R1308, 1995.
34. Kapás, L., Hong, L., Cady, A.B., Opp, M.R., Postlethwaite, A.E., Seyer, J.M., and Krueger, J.M., Somnogenic, pyrogenic, and anorectic activities of tumor necrosis factor-α and TNF-α fragments, *Am. J. Physiol.,* 263, R708, 1992.
35. Kapás, L., Obál, Jr., F., Book, A.A., Schweitzer, J.B., Wiley, R.G., and Krueger, J.M., The effects of immunolesions of nerve growth factor-receptive neurons by 192-IgG saporin on sleep, *Brain Res.,* 712, 53, 1996.
36. Kapás, L. and Krueger, J.M., Tumor necrosis factor-α induces sleep, fever and anorexia, *Am. J. Physiol.,* 263, R703, 1992.
37. Kapás, L., Fang, J., and Krueger, J.M., Inhibition of nitric oxide synthesis inhibits rat sleep, *Brain Res.,* 664, 189, 1994.
38. Kapás, L., Shibata, M., Kimura, M., and Krueger, J.M., Inhibition of nitric oxide synthesis suppresses sleep in rabbits, *Am. J. Physiol.,* 266, R151, 1994.
39. Kapás, L. and Krueger, J.M., Nitric oxide donors SIN-1 and SNAP promote non-rapid eye movement sleep in rats, *Brain Res. Bull.,* 41, 293, 1996.
40. Kattler, H., Dijk, D.J., and Borbély, A.A., Effect of unilateral somatosensory stimulation prior to sleep on the sleep EEG in humans, *J. Sleep Res.,* 3, 1599, 1994.
41. Kimura, M., Majde, J.A., Toth, L.A., Opp, M.R., and Krueger, J.M., Somnogenic effects of rabbit and human recombinant interferons in rabbits, *Am. J. Physiol.,* 267, R53, 1994.
42. Knefati, M., Somogyi, C., Kapás, L., Bourcier, T., and Krueger, J.M., Acidic fibroblast growth factor (FGF) but not basic FGF induces sleep and fever in rabbits, *Am. J. Physiol.,* 269, R87, 1995.
43. Kondo, S., Shinomura, Y., Kanayama, S., Kawabata, S., Miyazaki, Y., Imamura, I., Fukui, H., and Matsuzawa, Y., Interleukin-1 beta inhibits gastric histamine secretion and synthesis in the rat, *Am. J. Physiol.,* 267, G966, 1994.
44. Krueger, J.M. and Majde, J.A., Microbial products and cytokines in sleep and fever regulation, *Crit. Rev. Immunol.,* 14, 355, 1994.
45. Krueger, J.M. and Obál, Jr., F., A neuronal group theory of sleep function, *J. Sleep Res.,* 2, 63, 1993.
46. Krueger, J.M. and Obál, Jr., F., Growth hormone releasing hormone and interleukin-1 in sleep regulation, *FASEB J.,* 7, 645, 1993.
47. Krueger, J.M., Obál, Jr., F., Kapás, L., and Fang, J., Brain organization and sleep function, *Behav. Brain Res.,* 69, 177, 1995.
48. Krueger, J.M., Walter, J., Dinarello, C.A., Wolff, S.M., and Chedid, L., Sleep-promoting effects of endogenous pyrogen (interleukin-1), *Am. J. Physiol.,* 246, R994, 1984.
49. Krueger, J.M., Obál, Jr., F., Opp, M.R., Toth, L., Johannsen, L., and Cady, A.B., Somnogenic cytokines and models concerning their effects on sleep, *Yale J. Biol. Med.,* 63, 157, 1990.
50. Krueger, J.M. and Fang, J., Cytokines in sleep regulation, in *Sleep and Sleep Disorders: From Molecule to Behavior,* Hayaishi, O. and Inoue, S., Eds., Academic Press, New York, 1997, pp. 261–277.
51. Kushikata, T., Fang, J., Wang, Y., and Krueger, J.M., Interleukin-4 inhibits spontaneous sleep in rabbits, *Am. J. Physiol.,* in press.

52. Lancel, M., Mathias, S., Faulhaber, J., and Schiffelholz, T., Effect of interleukin-1 beta on EEG power density during sleep depends on circadian phase, *Am. J. Physiol.,* 270, R830, 1996.
53. Lieberman, A.P., Pitha, P.M., Shin, H.S., and Shin, M.L., Production of tumor necrosis factor and other cytokines by astrocytes stimulated with lipopolysaccharide or a neutrotopic virus, *Proc. Natl. Acad. Sci. USA,* 86, 6348, 1989.
54. Lue, F.A., Bail, M., Jephthah-Ocholo, J., Carayanniotis, K., Gorczynski, R., and Moldofsky, H., Sleep and cerebrospinal fluid interleukin-1-like activity in the cat, *Int. J. Neurosci.,* 42, 179, 1988.
55. Mackiewicz, M., Sollars, P.J., Ogilvie, M.D., and Pack, A.I., Modulation of IL-1β gene expression in the rat CNS during sleep deprivation, *NeuroReport,* 7, 529, 1996.
56. Mahowald, M.W. and Schenck, C.H., What is the minimal component of the brain that is capable of sleep?, *World Fed. Sleep Res. Newslett.,* 5, 12, 1997.
57. Miller, L.G. and Fahey, J.M., Interleukin-1 modulates GABAergic and glutamergic function in brain, *Ann. N.Y. Acad. Sci.,* 31, 292, 1994.
58. Miller, L.G., Galpern, W.R., Dunlap, K., Dinarello, C.A., and Turner, T.J., Interleukin-1 augments gamma-aminobutyric acid A receptor function in brain, *Mol. Pharmacol.,* 39, 105, 1991.
59. Moldofsky, H., Lue, F.A., Eisen, J., Keystone, E., and Gorczynski, R.M., The relationship of interleukin-1 and immune functions to sleep in humans, *Psychosom. Med.,* 48, 309, 1986.
60. Mukhametov, L.M., Sleep in marine mammals, *Exp. Brain Res.,* 8, 227, 1984.
61. Niimi, M., Mochizuki, T., Yamamoto, Y., and Yamatodani, A., Interleukin-1 beta induces histamine release in the rat hypothalamus in vivo, *Neurosci. Lett.,* 181, 87, 1994.
62. Nistico, G., DeSarro, G., and Rotiroti, D., Behavioral and electrocortical spectrum power changes of interleukins and tumor necrosis factor after microinjection into different areas of the brain, in *Sleep, Hormones, and Immunological System,* Smirne, S., Francesch, M., Ferini-Strambi, L., and Zuclowi, M., Eds., Smirne, Masson, Milan, 1992, p.11.
63. Obál, Jr., F., Fang, J., Payne, L., and Krueger, J.M., Growth hormone-releasing hormone (GHRH) mediates the sleep-promoting activity of interleukin-1 (IL-1) in rats, *Neuroendocrinology,* 61, 559, 1995.
64. Obál, Jr., F., Payne, L., Kapás, L., Opp, M.R., and Krueger, J.M., Inhibition of growth hormone-releasing factor suppresses both sleep and growth hormone secretion in the rat, *Brain Res.,* 557, 149, 1991.
65. Opp, M R., Obál, Jr., F., and Krueger, J.M., Interleukin-1 alters rat sleep: temporal and dose-related effects, *Am. J. Physiol.,* 260, R52, 1991.
66. Opp, M.R. and Krueger, J.M., Anti-interleukin-1β reduces sleep and sleep rebound after sleep deprivation in rats, *Am. J. Physiol.,* 266, R688, 1994.
67. Opp, M.R. and Krueger, J.M., Interleukin-1 is involved in responses to sleep deprivation in the rabbit, *Brain Res.,* 639, 57, 1994.
68. Opp, M.R., Smith, E.M., and Hughes, T.K., Interleukin-10 acts in the central nervous system of rats to reduce sleep, *J. Neuroimmunol.,* 60, 165, 1995.
69. Opp, M.R., Obál, Jr., F., Cady, A.B., Johannsen, L., and Krueger, J.M., Interleukin-6 is pyrogenic but not somnogenic, *Physiol. Behav.,* 45, 1069, 1989.
70. Pappenheimer, J.R., Koski, G., Fencl, V., Karnovsky, M.L., and Krueger, J.M., Extractions of sleep-promoting factors from cerebrospinal fluid and from brains of sleep-deprived animals, *J. Neurophysiol.,* 38, 1299, 1991.
71. Philip, R. and Epstein, L.B., Tumor necrosis factor as immunomodulator and mediator of monocyte cytoxicity induced by itself, gamma-interferon and interleukin-1, *Nature,* 323, 86, 1986.
72. Pigarev, I., Partial sleep in cortical areas, *World Fed. Sleep Res. Soc. Newslett.,* 5, 7, 1997.
73. Plata-Salaman, C.R., Immunoregulators in the nervous system, *Neurosci. Biobehav. Rev.,* 15, 185, 1991.

74. Puccioni-Sohler, M., Rieckmann, P., Kitze, B., Lange, P., Albrecht, M., and Flegenhauer, K., A soluble form of tumor necrosis factor receptor in cerebrospinal fluid and serum of HTLV-1-associated myelopathy and other neurological diseases, *Neurology,* 242, 239, 1995.
75. Seo, H.G., Fujii, J., Asahi, M., Okado, A., Fujiwara, N., and Taniguchi, N., Roles of purine nucleotides and adenosine in enhancing NOS II gene expression in interleukin-1 beta-stimulated rat vascular smooth muscle cells, *Free Radical Res.,* 26, 409, 1997.
76. Shibata, M. and Blatteis, C.M., Differential effects of cytokines or thermosensitive neurons in guinea pig preoptic area slices, *Am. J. Physiol.,* 261, R1096, 1991.
77. Shoham, S., Davenne, D., Cady, A.B., Dinarello, C.A., and Krueger, J.M., Recombinant tumor necrosis factor and interelukin-1 enhances slow-wave sleep, *Am. J. Physiol.,* 253, R142, 1987.
78. Steriade, M., Curro Dossi, R., and Nunez, A., Network modulation of a slow intrinsic oscillation of cat thalamocortical neurons implicated in sleep delta waves: cortically induced synchronization and brainstem cholinergic suppression, *J. Neurosci.,* 11, 3200, 1991.
79. Susic, V. and Totic, S., "Recovery" function of sleep: effects of purified human interleukin-1 on the sleep and febrile response of cats, *Met. Brain Dis.,* 4, 73, 1989.
80. Taishi, P., Bredow, S., Guha-Thakurta, N., Obál, Jr., F., and Krueger, J.M., Diurnal variations of interleukin-1β mRNA and β-actin mRNA in rat brain, *J. Neuroimmunol.,* 75, 79, 1997.
81. Takahashi, S., Fang, J., Kapás, L., Wang, Y., and Krueger, J.M., Inhibition of brain interleukin-1 attenuates sleep rebound after sleep deprivation in rabbits, *Am. J. Physiol.,* 42, R677, 1997.
82. Takahashi, S., Kapás, L., Fang, J., and Krueger, J.M., An anti-tumor necrosis factor antibody suppresses sleep in rats and rabbits, *Brain Res.,* 690, 241, 1995.
83. Takahashi, S., Kapás, L., Fang, J., Seyer, J.M., and Krueger, J.M., Somnogenic relationships between interleukin-1 and tumor necrosis factor, *Sleep Res.,* 25, 31, 1996.
84. Takahashi, S., Kapás, L., Fang, J., Wang, Y., Seyer, J.M., and Krueger, J.M., An interleukin-1 receptor fragment inhibits spontaneous sleep and muramyl dipeptide-induced sleep in rabbits, *Am. J. Physiol.,* 271, R101, 1996.
85. Takahashi, S., Kapás, L., and Krueger, J.M., A tumor necrosis factor (TNF) receptor fragment attenuates TNFα and muramyl dipeptide-induced sleep and fever in rabbits, *J. Sleep Res.,* 5, 106, 1996.
86. Takahashi, S., Kapás, L., Seyer, J.M., Wang, Y., and Krueger, J.M., Inhibition of tumor necrosis factor attenuates physiological sleep in rabbits, *NeuroReport,* 7, 642, 1996.
87. Takahashi, S. and Krueger, J.M., Inhibition of tumor necrosis factor prevents warming-induced sleep responses in rabbits, *Am. J. Physiol.,* 41, R1325, 1997.
88. Takahashi, S., Tooley, D.D., Kapás, L., Fang, J., Seyer, J.M., and Krueger, J.M., Inhibition of tumor necrosis factor in the brain suppresses rabbit sleep, *Pflügers Arch.,* 431, 155, 1995.
89. Takahashi, S., Gala, S., Kapás, L., and Krueger, J.M., Nerve growth factor enhances sleep in rabbits, *Soc. Neurosci. Abstr.,* 22, 147, 1996.
90. Tene-Sempere, M., Pinilla, C., Gonzalez, D., and Aguilar, E., Involvement of endogenous nitric oxide in the control of pituitary responsiveness to different elicitors of growth hormone release in prepubertal rats, *Neuroendocrinology,* 64, 146, 1996.
91. Tobler, I., Borbély, A.A., Schwyzer, M., and Fontana, A., Interleukin-1 derived from astrocytes enhances slow-wave activity in sleep EEG of the rat, *Eur. J. Pharmacol.,* 104, 191, 1984.
92. Uthgenannt, D., Schoolmann, D., Pietrowsky, R., Fehm, H.L., and Born, J., Effects of sleep on production of cytokines in humans, *Psychosom. Med.,* 57, 97, 1994.
93. Vgontzas, A.N., Papanicolaou, D.A., Bixler, E.O., Kales, A., Tyson, K., and Chrousos, G.P., Elevation of plasma cytokines in disorders of excessive daytime sleepiness: role of sleep disturbance and obesity, *J. Clin. Endocrinol. Metab.,* 82, 1313, 1997.

94. Werth, E., Achermann, P., and Borbély, A.A., Front-occipital EEG power gradients in human sleep, *J. Sleep Res.,* 6, 102, 1997.
95. Wesche, H., Neumann, D., Resch, R., and Martin, M.U., Co-expression of mRNA for type I and type II interleukin-1 receptors and the IL-1 receptor accessory protein correlates to IL-1 responsiveness, *FEBS Lett.,* 391, 104, 1996.
96. Yabuuchi, K., Minami, M., Katsumata, S., and Satoh, M., Localization of type I interleukin-1 receptor mRNA in the rat brain, *Mol. Brain Res.,* 27, 27, 1994.
97. Yamasu, K., Shimada, Y., Sakaizumi, M., Soma, G., and Mizumo, D., Activation of the systemic production of tumor necrosis factor after exposure to acute stress, *Eur. Cytokine Netw.,* 3, 391, 1992.
98. Zhang, J., Obál, Jr., F., Fang, J., Collins, B.J., and Krueger, J.M., Sleep is suppressed in transgenic mice with a deficiency in the somatotropic system, *Neurosci. Lett.,* 220, 97, 1996.

Chapter 36

Fever, Body Temperature, and Levels of Arousal

Mark R. Opp

Contents

36.1	Introduction		623
36.2	State-Dependent Alterations in Thermoregulation		624
	36.2.1	There Are Two Components of Brain Temperature Rhythms	625
	36.2.2	Thermoregulatory Effector Mechanisms Are Altered During Sleep	626
36.3	Fever		626
	36.3.1	Fever Is Beneficial to the Host	626
	36.3.2	There Are Several Possible Mechanisms by which Fever Increases Survival	627
	36.3.3	Generating a Fever Is an Active, Energetically Expensive Process	628
	36.3.4	Fevers Are (Must Be) Regulated	629
	36.3.5	Fevers Are Modulated by Circadian Factors	629
36.4	Fever and Sleep		631
	36.4.1	Acute Infections Induce Fever and Biphasic Alterations in Sleep	631
	36.4.2	Somnogenic Cytokines Alter Sleep Architecture During Fever	632
	36.4.3	IL-1-Induced Fever and Alterations in Sleep Are Antagonized by α-MSH and CRH	633
36.5	Conclusions		634
Acknowledgments			635
References			635

36.1 Introduction

Fever and alterations in sleep are hallmarks of infectious disease. During acute infection, changes in sleep characteristically include an initial increase in non-rapid-eye-movement (NREM) sleep that generally coincides with fever, followed by a period in which sleep is suppressed. The temporal association between fever and the biphasic somnogenic responses to acute infection is the topic of this chapter.

This chapter is written from the viewpoint that both fever and sleep serve an adaptive function. Although phylogenetic data, and perhaps conventional wisdom, suggest this to be the case (see below), direct evidence supporting this assertion has only recently begun to accumulate. For both fever and sleep, the fundamental question is one of function. Although not universal in acceptance, the pervading sentiment supported by numerous studies is that fever serves to alter the host environment such that conditions are less optimal for pathogen growth (reviewed in References 37, 52, 53, 68). Perhaps one reason this idea has fostered such a high degree of acceptance is that true fever normally occurs only when the host is invaded by a pathogen (but see Reference 52). Therefore, the necessity to produce fever is a relatively infrequent event for the host, and fever in response to a pathogen is clearly a pathologic condition. Sleep, on the other hand, is neither a pathologic condition nor is it an infrequent behavior. Sleep is a fundamental central nervous system process for which, in contrast to fever, there is no direct measure. Sleep is inferred from multiple behavioral, physiological, and electrophysiological parameters. These parameters vary with vigilance state, within and between phyla. Thus, the way in which sleep is inferred across phyla ranges from the observation of behavior, posture, and/or eye state without additional measures to an almost total reliance on the electroencephalogram. In addition, although the characteristics of sleep are similar within phyla, the initiation, maintenance, patterning, and duration of sleep vary dramatically across phyla. Finally, sleep exhibits distinct seasonal, circadian, and ultradian patterns. This variability in normal, physiological sleep and the way in which sleep is inferred across phyla make it difficult, if not impossible, to ascribe a unifying function to sleep for all organisms under all conditions in which sleep occurs.

For these, and many other reasons, the function of sleep is the subject of many reviews, symposia, and conferences (e.g., see References 5, 6, 56, 67, 108). Currently in the field of sleep research the only functional concept that approaches anything close to consensus is the idea that sleep is somehow essential. The studies of Rechtschaffen[89] and Everson[26] indicate that in rats, chronic deprivation of sleep is fatal, although the exact reasons for such fatality are still not clear. It is not the purpose of this chapter to posit a functional explanation for normal physiological sleep or to suggest that sleep, as a component of the acute phase response during infectious disease, serves as an adaptive immune response; this hypothesis has been reiterated many times.[55,57,59] During the course of an acute infectious challenge, sleep is altered in a manner that supports the generation of fever. Because survival of an acute infectious challenge increases when the organism is allowed to fever, the adaptive value of the alterations in sleep that occur during such a challenge may be due, in part, to this supportive role. In this chapter, we briefly review several aspects of the normal associations between sleep and thermoregulation, the costs and benefits of developing a fever, and the alterations in sleep that occur during fevers. We conclude by suggesting that the alterations in sleep that occur during acute infection are exquisitely tailored to provide not only a means by which energy reserves may be directed to the generation of fever, but also to serve as an effector mechanism for increasing body temperature.

36.2 State-Dependent Alterations in Thermoregulation*

The close association between sleep and thermoregulation has long been known and has been the subject of much research (reviewed in References 30, 39, 73, 83, 86). Several aspects of the normal association between sleep and thermoregulation are relevant to the alterations in sleep that occur throughout the course of an immune challenge.

---

* This section focuses primarily on sleep/wake state-dependent changes in brain temperature; however, state-dependent changes are also evident in core body temperature.

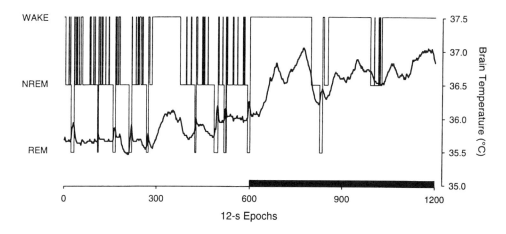

FIGURE 36.1
Representative hypnogram from an individual male Sprague-Dawley rat. Sleep-wake behavior (thin line) and brain temperature (thick line) were determined during discrete 12-s epochs for the 2 hr before and the 2 hr after the transition from the light period to the dark period of the light/dark cycle. Sleep/wake state-dependent and circadian changes in brain temperature are apparent. Brain temperature increases during rapid-eye-movement (REM) sleep and decreases prior to and during entry into non-rapid-eye-movement (NREM) sleep. Brain temperature is lower during the light period (the rest period for the rat) and higher during the dark period (the active period for the rat). The dark bar on the abscissa indicates the dark period of the light/dark cycle.

36.2.1 There Are Two Components of Brain Temperature Rhythms

There are two components of brain temperature (Tbr) changes that are tightly coupled to sleep/wake behavior: a circadian component and an ultradian component. The circadian component is composed of relatively large oscillations in Tbr that accompany rest/activity cycles with a period of approximately 24 hr. Ultradian changes in Tbr are those changes superimposed upon the circadian rhythm that are sleep/wake state dependent. An example of the association between these two components is presented in Figure 36.1. This figure depicts a 4-hr period encompassing the transition from the light period to the dark period of the light/dark cycle and the associated time course of changes for Tbr and sleep/wake behavior from an individual male Sprague-Dawley rat. The circadian rhythm of Tbr is clearly apparent; the average Tbr of this rat for the 2-hr period prior to dark onset was 35.8°C, whereas Tbr increased to an average of 36.6°C during the 2 hr following dark onset. This 0.8°C change is typical of the normal amplitude of circadian rhythm for Tbr of the rat, which may range from about 0.7 to 1.5°C (see, for example, References 74 and 76). Superimposed upon the circadian variation in Tbr are state-dependent changes. These state-dependent changes in Tbr consist of a regulated decline of Tbr upon entry into NREM sleep and the relatively rapid increase in Tbr at the onset of REM sleep. The decrease in Tbr upon entry into NREM sleep begins prior to the transition from waking to sleep (Figure 36.1).[74] Although not fully explained, this "anticipatory" reduction in Tbr prior to the initiation of NREM sleep episodes occurs when the animals are still active, suggesting that reduced locomotor behavior may not be the sole explanation. It is thus possible that the heat dissipation that occurs during NREM sleep actually begins prior to the initiation of a NREM sleep bout.

The increase in Tbr at REM sleep onset is apparent during even brief REM sleep periods and may range in magnitude from about 0.1 to 0.5°C (Figure 36.1).[1,74] The increase in Tbr at REMS onset was initially thought to result from increased heat conservation due to peripheral vasoconstriction.[2] Subsequent studies, however, indicate that vasomotion and, in fact, several other thermoregulatory effector mechanisms are impaired during REM sleep (see below).[2,28,86] Therefore, it is likely that the increase in Tbr at the transition from NREM to REM sleep is the result of processes localized to brain, e.g., a selective increase in brain metabolism, rather than peripheral vasoconstriction.[1] The magnitude of the increase in Tbr at the transition from NREM to REM sleep is a function of ambient temperature, with larger increases at cold (10°C) temperatures than at hot (29°C) temperatures,[1] suggesting that these state-dependent changes in Tbr are not regulated in the same manner as those that occur upon entry into NREM sleep.

The changes in Tbr at transitions from one state to another occur during both light and dark periods of the light/dark cycle. These changes are evident for both the regulated decrease in Tbr upon entry into NREM sleep and the rapid increase in Tbr at the transition from NREM to REM sleep (Figure 36.1).[74] There are two features of these changes that are relevant to the alterations in sleep that occur during the course of an immune challenge. First, generally speaking, the longer and more consolidated the NREM episode, the greater the decrease in Tbr until it reaches a regulated asymptote (Figure 36.1).[74] The normal occurrence of heat dissipation during NREM sleep is impeded if the NREM episodes are very brief or fragmented. Second, Tbr increases at the NREM to REM transition during the dark period, even though the overall Tbr is elevated, i.e., these changes are observable at elevated brain temperatures.

36.2.2 Thermoregulatory Effector Mechanisms Are Altered in a Sleep-State Dependent Manner

Thermoregulatory effector mechanisms are altered in a sleep/wake state-dependent fashion (see References 30, 83, 86 for reviews). There is generally a cessation of active thermoregulation (shivering, panting, etc.) during REM sleep, such that this stage of sleep renders the animal poikilothermic.[86] For example, at high ambient temperatures, panting increases during NREM sleep,[40,86] but ceases completely during REM sleep. Similarly, at low ambient temperatures, shivering occurs during NREM sleep, but ceases during REM sleep.[72,84-86,94,106] Direct stimulation of the preoptic area of the hypothalamus by means of water-perfused thermodes indicates that metabolic responses to thermal manipulation of this region are absent during REM sleep and provides an explanation for the inhibition of thermoregulatory effector mechanisms during REM sleep.[29] The REM sleep-dependent inhibition of thermoregulatory effector mechanisms has functional consequences for animals living in harsh environments. For example, species of ground-nesting gulls must remain on the nest during the breeding season to protect their eggs from predation and from excessive thermal loading that would addle them. When subjected to increasing thermal loads under these conditions, adult gulls alter their sleep/wake patterns such that active sleep (the avian equivalent to mammalian REM sleep) decreases.[77] Panting occurs during waking and quiet sleep of gulls (equivalent to mammalian NREM sleep), but not during active sleep.

36.3 Fever

Fever has long been a hallmark of disease. The role of fever in disease is complex and far from defined, although a beneficial role for fever has been postulated for centuries. Only recently have investigations of the evolution and adaptive value of fever been conducted (see References 52 and 68 for reviews).

36.3.1 Fever Is Beneficial to the Host

Both endothermic and ectothermic vertebrates, as well as invertebrates, develop fevers in response to administration of pyrogenic substances (e.g., endotoxin; reviewed in References 52 and 68).

These observations indicate an ancient phylogenetic history for fever. Observations of the phylogeny of fever are often used to support the suggestion that fever has an adaptive function, as it is frequently stated that such a costly physiologic process would not have been retained throughout the evolution of vertebrates and invertebrates if it was maladaptive (but see Reference 8). The stability of a trait throughout evolution per se need not necessarily indicate an adaptive function; however, evidence that fever serves an adaptive function may be derived from sources other than phylogenetic observations.

Several studies indicate increased survival when the host develops a moderate fever during bacterial or viral infections. Many of these studies have used ectotherms, as the fever may be manipulated by altering the environmental temperature. For example, Kluger et al.[54] found that the desert iguana *Dipsosaurus dorsalis* selects a temperature of 40°C, a temperature 2°C above the nonfebrile preferred temperature, after injection with *Aeromonas hydrophila,* a bacterium pathogenic for reptiles. In the *A. hydrophila*-injected animals, there was a 50% mortality rate 24 hr after injection if the animals were kept at 40 and 42°C, respectively. Similar experiments using goldfish,[15] grasshoppers,[9] and crickets[63] resulted in significant correlations between fever and survival. An interesting observation in these studies of ectotherms is that the absolute temperature of the febrile response is relatively unimportant. What is critical is the elevation of body temperature above the normothermic temperature. Some studies of endotherms have also demonstrated an association between fever and survival. For example, suppression of fever in rabbits subjected to bacterial challenge and ferrets infected with influenza virus reduces survival.[44,107] Furthermore, rabbits infected with rinderpest virus not only survive in greater numbers if they are not treated with antipyretics, but the rate of recovery is retarded in surviving animals that are treated with antipyretics. In humans, several retrospective clinical studies report positive correlations between fever and survival; patients with a moderate fever (100 to 101°F) survive bacterial peritonitis and bacteremia in greater numbers than those that fail to develop fever.[10,42,65,113]

36.3.2 There Are Several Possible Mechanisms by which Fever Increases Survival

If fever is adaptive and imparts survival value, and phylogenetic and survival data suggest this may indeed be the case, by what mechanisms does fever benefit the organism? There are two major avenues of defense by which the host organism may combat infection: (1) host defense (immunological) responses may be potentiated, and (2) the host environment may be directly altered such that conditions for the replication of the pathogen become less optimal. Both approaches produce an environment less conducive to pathogen growth, and fever appears, in many instances, to mediate both of these strategies. Numerous studies indicate that moderate elevations in body temperature enhance immune responses (see References 52, 68, 90, 91 for reviews). Fever may enhance nonspecific aspects of immune responses by increasing bacterial killing by neutrophils.[97] Fever may also potentiate specific immune responses by accelerating and enhancing lymphocyte proliferative responses[24,100] and enhancing antibody synthesis.[3,46] Interleukin (IL)-1, for example, stimulates T lymphocyte activation. The resulting T-cell proliferation is facilitated by fever, with T-cell proliferative responses reported to be tenfold greater at 39°C than at 37°C . Interferons (IFNs) possess potent antiviral, antibacterial, and antitumor properties. The ability of IFNs to augment killer cell function as well as the antiviral properties of IFN are potentiated at 40°C.[41]

The second approach by which the host organism may combat the pathogen is by directly altering the host environment. The optimal temperature for growth of some pathogens, for example, is at or below normal body temperature. Small et al.[99] report that allowing a febrile response to develop in rabbits during pneumococcal meningitis increases the doubling time of the pathogen in the cerebrospinal fluid to twice that of animals in which the fever was blocked. Their *in vitro* studies indicate that the pneumococci grew well at 37°C, whereas growth was nonexistent at a temperature of 41°C, a typical febrile temperature. Fever may also alter the sensitivity of pathogens to the cytolytic actions of the complement system. Gram-negative bacteria possess an outer envelope layer composed of lipopolysaccharide (LPS). LPS is the primary bacterial defense against perforation and killing by the host complement system. *In vitro* studies indicate that elevated temperatures inhibit the synthesis of LPS in many enterobacterial strains, including *Klebsiella, Enterobacter, Serratia, Salmonella, Proteus,* and multiple *Escherischia coli* serotypes.[34] Another aspect of alteration of the host environment by fever is the availability of iron in serum; plasma iron concentrations are reduced during most infections accompanied by fever. Since virtually all bacterial pathogens have a nutritional requirement for iron, the hypoferremia that occurs during infection results in reduced growth of the pathogen and has long been considered a host defense.[96,111] Although the reductions in iron are independent of fever[35,102] and are a direct result of increased IL-1 concentrations,[17,18,112] there is a synergy between fever and hypoferremia in reducing bacterial growth rates (see References 37, 52, and 68 for reviews).

36.3.3 Generating a Fever Is an Active, Energetically Expensive Process

The mechanisms by which signals from the peripheral immune system are conveyed to the central nervous system and transduced to fever have been extensively studied (see References 14, 52, 68, and 95 for reviews). Once the thermoregulatory set point is elevated, raising body temperature to this new set point involves an orchestrated reduction in heat loss and an increase in heat production. The precise mechanisms by which heat loss is reduced and heat production increased vary across species, but the major behavioral and physiological components are the same. For example, during the chill phase of fever when body temperature is below the new thermoregulatory set point, heat loss is reduced behaviorally by adopting a more compact (curled) body posture, thus reducing the surface area exposed to the environment, and by moving to a warmer place. Insulation of the body is increased by erecting hair, fur, or feathers, or by additional clothing or blankets. During this phase of fever, sweating in humans and other mammals that sweat and panting in birds and mammals that utilize this effector mechanism ceases. This reduction in heat loss coincides with increased heat production, generally accomplished by shivering and non-shivering thermogenesis.

The earliest attempts to estimate the amount of heat production necessary to generate and maintain a fever were made from human subjects suffering from malaria or injected with typhoid vaccine.[4] In these studies,[19,20] calorimetric calculations were made after the chill phase of fever had passed (i.e., shivering had ceased and the elevated temperature was stable). Under various conditions and with different etiologic agents used to induce fever, metabolism was increased by 30 to 50%, with the overall mean increase in metabolism estimated to be 13% per 1°C fever.[51] These increased energetic costs result from both the increase in metabolism necessary to raise body temperature (e.g., shivering) and from the accelerated metabolism associated with increased temperature, the Q_{10} effect. Although the 13% increase in metabolism per 1°C fever is a generally accepted figure,[51] Hart[37] points out that estimating the increase in metabolism to generate a 1°C fever across species is difficult, as multiple factors are involved. For example, body size differences result in different surface-to-body mass ratios, and hair coat density and/or hair or skin coloration influence heat loss and heat gain. In addition, differences in habitat (e.g., arctic vs. tropical) will dramatically influence the amount of energy required to generate fever, as will the efficiency of heat conservation mechanisms; less energy is required to increase body temperature if heat conservation mechanisms are efficient. In spite of the differences across species in the factors influencing the amount of energy required to generate fever, it is clear that this process is energetically expensive.

36.3.4 Fevers Are (Must Be) Regulated

Although fever is adaptive and appears to have survival value, there are consequences to fever, particularly if the fevers are of large magnitude (e.g., >3.0 to 4.0°C) or of inordinately long duration. For example, brain damage in the form of polyribosome disaggregation begins at temperatures of about 40 to 41°C, and above 41°C irreversible destruction of mitochondria occurs.[12,69] As such, there are elaborate regulatory mechanisms that normally ensure that the magnitude and duration of the fever are moderated before they become detrimental to the host.

The elucidation of regulatory mechanisms for fever began with the work of Kasting and colleagues, when they reported that fever responses to endotoxin were reduced in pregnant sheep during the 5 days preceding birth, and that no fever developed the day of birth.[50] After parturition, the fever response to endotoxin gradually returned and was indistinguishable from the fevers developed by nonpregnant ewes about 3 days after birth. Kasting also observed that newborn lambs were fever resistant and that the time to which a full fever developed in the lamb was approximately the same as that required for the ewe to regain her capability of fevering. These observations suggested that a circulating antipyretic, common to both ewe and lamb, was elevated near term. Subsequent work by several groups resulted in the identification of arginine vasopressin (AVP) as an endogenous antipyretic.[52,68] Other substances that fulfill at least some of the criteria proposed to define antipyretics[52] include α-melanocyte-stimulating hormone (α-MSH), corticotropin-releasing hormone (CRH), the IL-1 receptor antagonist (IL-1ra), some urinary factor(s) such as uromodulin,[52,68] and perhaps tumor necrosis factor.[52]

The peptides α-MSH and CRH are of particular relevance to this discussion because both have been the focus of studies on IL-1-induced alterations in sleep (see below). In the arcuate nucleus and the nucleus of the solitary tract, and in the intermediate lobe of the pituitary, α-MSH is cleaved from adrenocorticotrophic hormone (ACTH), which in turn is derived from the precursor molecule proopiomelanocortin (POMC). α-MSH inhibits fever when administered intracerebroventricularly (ICV),[32,79] when injected directly into the septal region[31] or the preoptic area of the anterior hypothalamus,[27] or when injected systemically, either intravenously[32,33] or intragastrically.[70] α-MSH administered ICV is reported to be 25,000 times more potent as an antipyretic than acetaminophen.[71] The effectiveness of α-MSH as an antipyretic has been demonstrated in antagonizing fevers induced by several pyrogens, including IL-1, IL-6, TNF, and LPS, in species as varied as the rabbit, mouse, guinea pig, squirrel monkey,[62] and rat.[11,43,109] The role of CRH as an endogenous antipyretic is more complicated than that of α-MSH due to the fact that CRH is thermogenic and many of the thermogenic responses to cytokines are mediated by CRH.[92,93] However, CRH injected centrally reduces "endogenous pyrogen"- and IL-1-induced fever[7,80] in the rabbit in a dose-related manner. Although CRH may not play a major role in modulating fever by direct antipyretic actions, it is clearly pivotal in orchestrating the hypothalamic-pituitary-adrenal (HPA) axis responses to cytokines.[25,64,82] IL-1, for example, stimulates CRH release from the hypothalamus, which subsequently induces ACTH secretion from pituitary corticotrophs. ACTH then stimulates secretion of endogenous steroids from the adrenal cortex. These endogenous steroids, released in response to IL-1, act within the CNS[13] to limit further production of IL-1.[101] With regard to fever for example, administration of corticosterone into the rat prior to LPS[88] or IL-1β[48] injection inhibits febrile responses, whereas blocking glucocorticoid receptors prior to injection of LPS prolongs the resulting fever.[66] This negative feedback mechanism normally ensures that the responses to cytokines and/or immune challenge are maintained at levels that are not detrimental to the host.

36.3.5 Fevers Are Modulated by Circadian Factors

We, and others, have previously reported that somnogenic[61,78] and pyrogenic[78] responses to IL-1, muramyl dipeptide, or LPS (Opp, M.R. and Toth, L.A., unpublished observations) are modulated

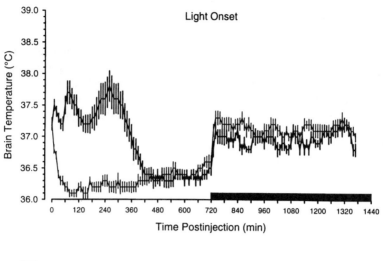
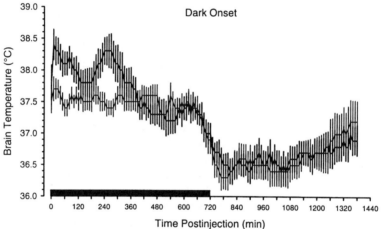

FIGURE 36.2
Circadian modulation of interleukin (IL)-1β-induced fever. Male Sprague-Dawley rats were injected intraperitoneally with either pyrogen-free saline (vehicle; thin line) or with 2.0 μg/kg IL-1β (thick line). Injections were given at either light onset (n = 12) or dark onset (n = 11), and brain temperature was determined every 10 min for the next 23 hr. The magnitude and the duration of the febrile response was greater when the IL-1β was administered at light onset. The dark bars on the abscissa indicate the dark period of the light/dark cycle.

by circadian influences. However, the influence of circadian phase on responses to immune challenge is one facet of host defense that has generally received little investigation. As indicated in Figure 36.2, the magnitude of the febrile response may be modulated by circadian factors. This figure depicts the febrile response of male Sprague-Dawley rats to intraperitoneal injection of either pyrogen-free saline (vehicle) or 2.0 μg/kg recombinant human IL-1β. These rats were maintained on a 12:12-hr, light/dark cycle at an ambient temperature of 23 ± 1°C. Brain temperatures from these animals were recorded using calibrated thermistors implanted between the surface of the dura mater and the skull above the parietal cortex. All injections were given either at the beginning of the light (n = 12) or dark (n = 11) period. When administered at light onset, this dose of IL-1β induced a fever with a peak magnitude of about 1.6°C. The same dose administered at dark onset, however, induced a maximal febrile response of about 0.8°C.

There are at least two potential explanations for these circadian differences in febrile responsiveness of rats to IL-1β. First, under these conditions an absolute Tbr of about 38.0 to 38.5°C may represent a physiological maximum, i.e., it may be impossible for rats to reduce heat loss and increase heat production to the extent required to increase Tbr beyond these values. In this case, the absolute fever would essentially be the same, regardless of the time of IL-1 administration, and the difference in magnitude would be accounted for solely on the basis of normal circadian fluctuation in brain temperature; i.e., the absolute elevation of Tbr is merely superimposed upon a higher basal brain temperature. However, rats are capable of increasing brain temperature beyond these absolute values of 38.0 to 38.5°C. By way of example, we have found that rats with chronic inflammation around an intracerebral ventricular guide cannula produce fevers in excess of 40°C when these sites are disturbed by ICV injection procedures (Opp, M.R., unpublished observations). Therefore, fevers that peak at about 38.0 to 38.5°C represent increases in brain temperature that are regulated/maintained below the physiological maximum brain temperatures that the rat is capable of producing.

Given that the rat is capable of increasing Tbr above these absolute values, a second potential explanation for the circadian difference in the magnitude of fevers concerns the HPA axis and negative feedback mechanisms for IL-1β actions (see above). The circadian pattern of HPA axis activity in the rat, as evidenced by circulating concentrations of corticosterone, is particularly striking; maximal values are observed shortly after the beginning of the dark (active) period, whereas minimum HPA axis activity occurs early in the light (rest) period.[16,36] Because the HPA axis, via actions of glucocorticoids within the CNS, is a major regulator of IL-1-induced fevers, it follows that fevers induced at a time when HPA axis activity is already maximal (e.g., dark period) would be of less magnitude than those induced when HPA axis activity is at its lowest (e.g., the light period). Definitive experiments have not been conducted to determine which of these, or other, explanations for the circadian modulation of fever are correct.

36.4 Fever and Sleep

It is a common experience that sleep is altered during the course of mild infectious diseases. For example, bouts with the "flu" include moderate fevers, in addition to feelings of sleepiness, lethargy, and malaise.

36.4.1 Acute Infections Induce Fever and Biphasic Alterations in Sleep

To date, alterations in the sleep of rabbits, rats, mice, cats, and humans have been described through the course of an infectious challenge in response to bacterial, viral, fungal, or parasitic pathogens (see Chapter 37).[103] Although the precise temporal responses to infection vary with pathogen and route of infection, acute infections are generally characterized by an initial increase in NREM sleep, followed subsequently by reductions in NREM sleep below levels observed prior to infection. These biphasic alterations in NREM sleep may be temporally dissociated from febrile responses to infection. Rabbits inoculated with *Staphylococcus aureus* during the light period exhibit enhanced NREM sleep from hr 6 to 18 after inoculation and NREM sleep suppression that lasts from about postinjection hr 26 to 32.[103,105] The febrile response of these animals persists for the entire 48-hr postinjection period. REM sleep is suppressed by this challenge from postinjection hr 6 to 42. Therefore, fever and reduction in REM sleep exhibit the same time course and persist well beyond the period of NREM sleep enhancement.

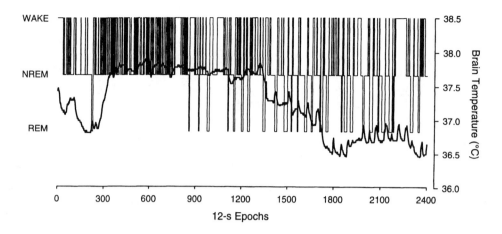

FIGURE 36.3
Representative hypnogram from an individual male Sprague-Dawley rat following an intracerebroventricular injection of 5.0 ng IL-1β. IL-1β was injected at the beginning of the dark period, and sleep/wake behavior (thin line) and brain temperature (thick line) were determined during discrete 12-s epochs for the next 8 hr. During the chill phase and first half of the plateau phase of the ensuing fever, rapid-eye-movement (REM) sleep was completely abolished, and non-rapid-eye-movement (NREM) sleep was fragmented. REM sleep began to reappear during the second half of the plateau phase and during defervescence, in conjunction with an increase in NREM episode duration.

36.4.2 Somnogenic Cytokines Alter Sleep Architecture During Fever

True infections involve replication of the pathogen in the host and induce a myriad of behavioral, physiological, and immune responses.[37] Many of the multiple responses to infection, including fever and alterations in sleep, may be mimicked by either central or systemic administration of cytokines, particularly IL-1 (see Chapter 35).[61,78–80] For example, Figure 36.3 depicts the pyrogenic and somnogenic responses of an individual male Sprague-Dawley rat to the ICV administration of 5.0 ng IL-1β at the beginning of the dark period. Initially, Tbr is increased due to handling but gradually subsides. About 1 hr after injection of IL-1, Tbr of this rat begins to increase and rapidly reaches the peak febrile temperature of about 37.8°C. The fever is then maintained at this level for about 4 hr, after which defervescence begins.

There are several key features of IL-1-induced alterations in sleep and temperature. First, REM sleep is completely abolished during the rising (chill) phase of the fever and, in this animal, for about the first 2 hr of the plateau phase. REM sleep gradually begins to reappear during the second half of the plateau phase and occurs frequently during the defervescence phase of the fever. Second, during the chill and plateau phases of the fever, NREM sleep is dramatically fragmented. During defervescence, the number transitions between waking and NREM sleep begin to diminish and NREM sleep episodes become longer. Finally, when REM sleep or NREM sleep occur during IL-1-induced fever, particularly during the plateau phase and during defervescence, the state-coupled changes in brain temperature are apparent, an observation previously reported in rabbits.[110]

Given the association of IL-1-induced fevers and alterations in sleep previously described (and see Figure 36.3), it is frequently suggested that IL-1-induced alterations in sleep are merely byproducts of fever. However, this is not the case, as this association between fever and altered sleep may be separated. For example, fevers in rabbits inoculated with *S. aureus* persist after the enhancement of NREM sleep has subsided (see above); protein synthesis inhibitors block IL-1-induced fevers, but not alterations in sleep;[60] and inhibitors of nitric oxide synthase inhibit IL-1-

induced alterations in sleep, but not fevers.[47] Furthermore, doses of CRH that do not antagonize IL-1-induced fevers completely block IL-1-induced increases in NREM sleep (see below).[80] Additional evidence that fever and sleep are separable has been recently reviewed by Krueger.[58]

36.4.3 IL-1-Induced Fever and Alterations in Sleep Are Antagonized by α-MSH and CRH

The negative feedback mechanisms that regulate/modulate fever may also play a role in the biphasic sleep responses to acute infection. Although to our knowledge there have been no sleep studies in which antipyretics have been administered in conjunction with a true infection, it is likely that the biphasic sleep response to acute infection is mediated, in part, by endogenous antipyretics. Evidence to support this assertion is derived from studies in which the somnogenic and pyrogenic effects of IL-1 have been antagonized by administration of known antipyretics, specifically α-MSH and CRH. As previously stated, rabbits injected ICV with IL-1 develop fevers in conjunction with enhanced NREM sleep and suppressed REM sleep. When either α-MSH[79] or CRH[80] is administered ICV after central administration of IL-1, the somnogenic responses are either blocked or attenuated in a dose-related manner. The time courses of the interactions of these two substances with IL-1 differ however, providing evidence that these two endogenous antipyretics antagonize IL-1 actions by different mechanisms.

α-MSH administered ICV into otherwise normal rabbits induces dose-related hypothermic responses.[32,62,79] At effective doses, the onset of α-MSH-induced hypothermia is rapid, being apparent within 10-min post-injection.[79] The antipyretic actions of α-MSH also exhibit a rapid onset. When α-MSH is administered to rabbits 30-min after they have been treated with IL-1, the reduction in IL-1-induced fever is apparent within 10 min. Depending on the dose of α-MSH, IL-1-induced fever is then attenuated or blocked for several hours, after which time Tbr gradually increases to the level achieved when the animals are injected with IL-1 alone.[79] In addition to attenuating IL-1-induced fever, this study also demonstrated that α-MSH inhibits IL-1-induced increases in NREM sleep with a similar time course, i.e., within 1 hr post-injection.

The ability of CRH to antagonize IL-1 actions diverges somewhat with respect to sleep and fever. When administered ICV into otherwise normal rabbits, CRH induces hyperthermia in a dose-related fashion.[80,93] CRH-induced increases in temperature should not be considered true fevers, as these actions are due to an increase in thermogenesis without a resetting of the thermoregulatory set point.[92] When low doses (0.02, 0.1 nmol) of CRH that do not greatly affect Tbr themselves are administered centrally 10 min after ICV IL-1, the IL-1-induced fever is attenuated. However, this effect is delayed until after the chill phase of the fever is over, some 2 hr after the IL-1 + CRH injection.[80] Higher CRH doses (0.25 nmol) that induce hyperthermia in otherwise normal rabbits are not effective in antagonizing IL-1-induced fevers; however, each of these CRH doses completely blocks IL-1-induced increases in NREM sleep, regardless of actions on Tbr.[80]

The differences in the time course between α-MSH and CRH in antagonizing the pyrogenic and somnogenic responses to IL-1 support the hypothesis that these peptides use different mechanisms to antagonize IL-1 actions. The attenuation of IL-1-induced fever by α-MSH is rapid and of relatively short duration, being evident for the initial 1 to 2 hr postinjection time period. Similarly, IL-1-induced enhancement of NREM sleep is suppressed during the first post-injection hour following α-MSH. These observations are entirely consistent with the suggestion that α-MSH has direct antipyretic actions within the CNS.[31,32] Low doses of CRH, on the other hand, antagonize IL-1-induced fever only after the chill phase of the fever is over, consistent with the idea that CRH modulates fever indirectly via actions on the HPA axis. CRH attenuates or completely abolishes IL-1-induced increases in NREM sleep within the first postinjection hour, regardless of its actions on IL-1-induced fever. These findings indicate that this peptide also acts centrally as a modulator/regulator of at least some IL-1 actions. Indeed, in addition to it hypophysiotrophic actions, CRH is a neurotransmitter within the CNS (see Reference 87 for a review) and mediates many of the autonomic and behavioral responses to a variety of stressors. It is also hypothesized that CRH is involved in the regulation of physiological sleep and waking.[75,81] With regard to the biphasic alterations in sleep throughout acute infection, however, it is likely that the major involvement of CRH is through the HPA axis. Glucocorticoids act as antipyretics within the CNS[13] and are also capable of modulating bacterially induced alterations in sleep.[104] For example, rabbits inoculated with *Escherichia coli* or *S. aureus* exhibit biphasic sleep responses (see above). Treatment with cortisone prior to bacterial inoculation attenuates the sleep suppression phase of this biphasic response.[104] Although experiments to block CRH actions prior to acute infection have not been conducted, these data support the hypothesis that CRH, via the HPA axis, is involved in sleep alterations during acute infection.

36.5 Conclusions

Why should an organism sleep more during the early stages of acute infection? At the outset of this chapter we suggested that the alterations in sleep that occur during the course of an acute infection serve to support the generation of fever. There is empirical evidence indicating that fever increases survival of the host. Therefore, anything the animal can do to support the generation of fever increases the survivability of the host. Fever is an energetically costly undertaking. Sleep, by its very nature, conserves energy, as the overall level of activity is reduced and sleep is inherently associated with reduced metabolic costs. By sleeping instead of being active, the animal makes available additional energy reserves for fever production. In addition to "just sleeping", the architecture of sleep is altered during fever. The suppression of REM sleep is critical during the chill phase of fever because shivering is abolished during this sleep stage. Shivering is the primary, but not only means, by which heat production is increased during the generation of fever. By eliminating REM sleep, the period when shivering can occur is effectively increased. In addition to suppression of REM sleep, during the chill phase of fever NREM sleep becomes fragmented. During NREM sleep, heat dissipation occurs exponentially until reaching a regulated asymptote. Fragmenting NREM sleep reduces heat dissipation because the NREM episodes are of such short duration. Therefore, the combined effect of REM sleep suppression and NREM sleep fragmentation during the chill phase of fever is to increase heat production and reduce heat loss, i.e., facilitate the production of fever.

Given the potentially detrimental effects of large or very prolonged fevers, regulation of this response to infection is vital, and multiple feedback mechanisms for moderating fever have been described. Sleep may also function in this capacity. Once the plateau stage of the fever is reached, gradually REM sleep begins to reappear and NREM sleep episodes become longer. Now shivering will be periodically abolished (due to the occurrence of REM sleep) and heat dissipation will be increased (due to progressively longer NREM sleep episodes). The combined effects of these alterations in sleep architecture are to reduce heat production and increase heat loss, i.e., facilitate defervescence and the return to normal nonfebrile temperatures.

Is there any significance to, or functional consequence of, a biphasic sleep response to acute infection? During the production of fever, the metabolic rate of the organism has increased by approximately 30 to 50%. The energy reserves that have been depleted need to be replenished. During the second phase of the biphasic sleep response, sleep is suppressed. This reduction in sleep provides additional time for the animal to forage/feed in an effort to restore energy reserves. Under some conditions, for example during fevers of longer duration, it is necessary that the animal eat even when the fever is still manifest. Such may be the case of rabbits subjected to *S. aureus* challenge described above; the febrile response (and REM sleep suppression) lasts about 46 to 48 hr, while the period of enhanced NREM sleep ends after about 20 hr. With increased waking, the animal has more time available to eat, yet when rabbits do sleep during this stage of infection, REM sleep remains suppressed, thus allowing shivering to occur more frequently to maintain body temperature at the febrile level.

The responses of the host to an infectious challenge are complex and include multiple behavioral, physiological, and immunological mechanisms. Fever appears to have evolved to increase host survival in response to infection. Although it is not likely that sleep evolved solely to support the generation of fever during an infectious challenge, the alterations in sleep throughout the course of infection are exquisitely designed to fulfill this role.

Acknowledgment

The author was supported by MH-52275 and MH-54976 while writing this chapter.

References

1. Alföldi, P., Rubicsek, G., Cserni, G., and Obál, Jr., F., Brain and core temperatures and peripheral vasomotion during sleep and wakefulness at various ambient temperatures in the rat, *Pflügers Arch. Eur. J. Physiol.,* 417, 336, 1990.
2. Baker, M.A. and Hayward, J.N., Autonomic basis for the rise in brain temperature during paradoxical sleep, *Science,* 157, 1586, 1967.
3. Banet, M., Fisher, D., Hartmann, K.U., Hentel, H., and Hilling, U., The effect of whole body heat exposure and of cooling the hypothalamus on antibody titre in the rat, *Pflügers Arch.,* 391, 25, 1981.
4. Barr, D.P., Russell, C.L., and Dubois, E.F., Temperature regulation after the intravenous injection of protease and typhoid vaccine, *Arch. Intern. Med.,* 29, 608, 1922.
5. Benington, J.H. and Heller, H.C., Restoration of brain energy metabolism as the function of sleep, *Prog. Neurobiol.,* 45, 347, 1995.
6. Berger, R.J. and Phillips, N.H., Sleep and energy conservation, *News Physiol. Sci.,* 8, 276, 1993.
7. Bernardini, G.L., Richards, D.B., and Lipton, J.M., Antipyretic effect of centrally administered CRF, *Peptides,* 5, 57, 1984.
8. Blatteis, C.M., Fever: Is it beneficial?, *Yale J. Bio. Med.,* 59, 107, 1986.
9. Boorstein, S.M. and Ewald, P.W., Costs and benefits of behavioral fever in *Melanoplus sanguinipes* infected by *Nosema acridophagus, Physiol. Zool.,* 60, 586, 1987.
10. Bryant, R.E., Hood, A.F., Hood, C.E., and Loenig, M.G., Factors affecting mortality of gram-negative rod bacteremia, *Arch. Intern. Med.,* 127, 120, 1971.
11. Bull, D.F., King, M.G., Pfister, H.P., and Singer, G., α-melanocyte-stimulating hormone conditioned suppression of a lipopolysaccharide-induced fever, *Peptides,* 11, 1027, 1990.
12. Caputa, M., Selective brain cooling: an important component of thermal physiology, in *Satellite of 28th Int.Congress of Physiological Sciences,* 1980.
13. Coelho, M.M., Luheshi, G., Hopkins, S.J., Pelá, I.R., and Rothwell, N.J., Multiple mechanisms mediate antipyretic action of glucocorticoids, *Am. J. Physiol.,* 269, R527, 1995.
14. Cooper, K.E., *Fever and Antipyresis: The Role of the Nervous System,* Cambridge University Press, New York, 1995.
15. Covert, J.B. and Reynolds, W.W., Survival value of fever in fish, *Nature,* 267, 43, 1977.
16. Dhabhar, F.S., McEwen, B.S., and Spencer, R.L., Stress response, adrenal steroid receptor levels and corticosteroid-binding globulin levels — a comparison between Sprague-Dawley, Fischer 344 and Lewis rats, *Brain Res.,* 616, 89, 1993.

17. Dinarello, C.A., An update on human interleukin-1: from molecular biology to clinical relevance, *J. Clin. Immunol.*, 5, 287, 1985.
18. Dinarello, C.A., Cannon, J.G., and Wolff, S.M., New concepts in pathogenesis of fever, *Rev. Infect. Dis.*, 10, 168, 1988.
19. Dubois, E.F., *Basal Metabolism in Health and Disease*, Lea & Febiger, Philadelphia, 1936.
20. Dubois, E.F., *Fever and Regulation of Body Temperature*, Charles C Thomas, Springfield, IL, 1948.
21. Duff, G.W., Is fever beneficial to the host: a clinical perspective, *Yale J. Biol. Med.*, 59, 125, 1986.
22. Duff, G.W. and Durum, S.K., Fever and immunoregulation: hyperthermia, interleukins 1 and 2 and T-cell proliferation, *Yale J. Biol. Med.*, 55, 437, 1982.
23. Duff, G.W. and Durum, S.K., T cell proliferation induced by interleukin 1 is greatly increased by hyperthermia, *Clin. Res.*, 30, 694A, 1982.
24. Duff, G.W. and Durum, S.K., The pyrogenic and mitogenic actions of interleukin-1 are related, *Nature*, 304, 449, 1983.
25. Dunn, A.J. and Berridge, C.W., Physiological and behavioral responses to corticotropin-releasing factor administration: is CRF a mediator of anxiety or stress responses?, *Brain Res. Rev.*, 15, 71, 1990.
26. Everson, C.A., Sustained sleep deprivation impairs host defense, *Am. J. Physiol.*, 265, R1148, 1993.
27. Feng, J.D., Dao, T., and Lipton, J.M., Effects of peroptic microinjections of a-MSH on fever and normal temperature control in rabbits, *Brain Res. Bull.*, 18, 473, 1987.
28. Franzini, C., Cianci, T., Lenzi, P., and Guidalotti, P.L., Neural control of vasomotion in rabbit ear is impaired during desynchronized sleep, *Am. J. Physiol.*, 243, R142, 1982.
29. Glotzbach, S.F. and Heller, H.C., Central nervous regulation of body temperature during sleep, *Science*, 194, 537, 1976.
30. Glotzbach, S.F. and Heller, H.C., Thermoregulation, in *Principles and Practice of Sleep Medicine*, Kryger, M.H., Roth, T., and Dement, W.C., Eds., W.B. Saunders Company, Philadelphia, 1989.
31. Glyn-Ballinger, J.R., Bernardini, G.L., and Lipton, J.M., α-MSH injected into the septal region reduces fever in rabbits, *Peptides*, 4, 199, 1983.
32. Glyn, J.R. and Lipton, J.M., Hypothermic and antipyretic effects of centrally administered ACTH (1-24) and α-melanotropin, *Peptides*, 2, 177, 1981.
33. Goelst, K., Mitchell, D., and Laburn, H., Effects of α-melanocyte stimulating hormone on fever caused by endotoxin in rabbits, *J. Physiol. (Lond.)*, 441, 469, 1991.
34. Green, M.H. and Vermeulen, C.W., Fever and the control of gram-negative bacteria, *Res. Microbiol.*, 145, 269, 1994.
35. Grieger, T.S. and Kluger, M.J., Fever and survival: the role serum iron, *J. Physiol. (Lond.)*, 279, 187, 1978.
36. Griffin, A.C. and Whitacre, C.C., Sex and strain differences in the circadian rhythm fluctuation of endocrine and immune function in the rat: implications for rodent models of autoimmune disease, *J. Neuroimmunol.*, 35, 53, 1991.
37. Hart, B.L., Biological basis of the behavior of sick animals, *Neurosci. Biobehav. Rev.*, 12, 123, 1988.
38. Heller, H.C., Thermoregulation during sleep and hibernation, *Int. Rev. Physiol.*, 15, 147, 1977.
39. Heller, H.C., Glotzbach, S., Grahn, D., and Radeke, C., Sleep-dependent changes in the thermoregulatory system, in *Clinical Physiology of Sleep*, Lydic, R. and Biebuyck, J.F., Eds., American Physiological Society, Bethesda, MD, 1988.
40. Heller, H.C., Graf, R., and Rautenberg, W., Circadian and arousal state influences on thermal regulation in the pigeon, *Am. J. Physiol.*, 245, R321, 1983.
41. Hirai, N., Hill, N.O., and Osther, K., Temperature influences on different human alpha interferon activities, *J. Interferon Res.*, 4, 507, 1984.

42. Hoefs, J., Sapico, F.L., Canawati, H.N., and Montgomerie, J.Z., The relationship of white blood cell (WBC) and pyrogenic response to survival in spontaneous bacterial peritonitis, *Gastroenterology*, 78, 1308, 1980.
43. Huang, Q.-H., Entwistle, M.L., Alvaro, J.D., Duman, R.S., Hruby, V.J., and Tatro, J.B., Antipyretic role of endogenous melanocortins mediated by central melanocortin receptors during endotoxin-induced fever, *J. Neurosci.*, 17, 3343, 1997.
44. Husseini, R.H., Sweet, C., Collie, M.H., and Smith, H., Elevation of nasal virus levels by suppression of fever in ferrets infected with influenza viruses of differeing virulence, *J. Infect. Dis.*, 145, 520, 1982.
45. Imeri, L., Bianchi, M., and Mancia, M., Muramyl dipeptide and interleukin-1 effects on sleep and brain temperature following inhibition of serotonin synthesis, *Am. J. Physiol.*, 273, R1663, 1997.
46. Janpel, H.D., Duff, G.W., Gershon, R.K., Atkins, E., and Durum, S.K., Hyperthermia augments the primary *in vitro* humoral immune response, *J. Exp. Med.*, 157, 1229, 1983.
47. Kapás, L., Shibata, M., Kimura, M., and Krueger, J.M., Inhibition of nitric oxide synthesis suppresses sleep in rabbits, *Am. J. Physiol.*, 266, R151, 1994.
48. Kapcala, L.P., Chautard, T., and Eskay, R.L., The protective role of the hypothalamic-pituitary-adrenal axis against lethality produced by immune, infectious, and inflammatory stress, *Ann. N.Y. Acad. Sci.*, 771, 419, 1995.
49. Kasting, N.W., Veale, W.L., and Cooper, K.E., Evidence for a centrally active endogenous antipyretic near parturition, in *Current Studies of Hypothalamic Function*, Lederis, K. and Veale, W.L., Eds., Karger, Basel, 1978.
50. Kasting, N.W., Veale, W.L., and Cooper, K.E., Suppression of fever at term of pregnancy, *Nature*, 271, 245, 1978.
51. Kluger, M.J., *Fever: Its Biology, Evolution and Function*, Princeton University Press, Princeton, 1979.
52. Kluger, M.J., Fever: role of pyrogens and cryogens, *Physiol. Rev.*, 71, 93, 1991.
53. Kluger, M.J., Kozak, W., Conn, C.A., Leon, L.R., and Soszynski, D., The adaptive value of fever, *Infect. Dis. Clin. N. Am.*, 10, 1, 1996.
54. Kluger, M.J., Ringler, D.H., and Anver, M.R., Fever and survival, *Science*, 188, 166, 1975.
55. Krueger, J.M. and Majde, J.A., Sleep as a host defense: its regulation by microbial products and cytokines, *Clin. Immunol. Immunopathol.*, 57, 188, 1990.
56. Krueger, J.M. and Obál, Jr., F., A neuronal group theory of sleep function, *J. Sleep Res.*, 2, 63, 1993.
57. Krueger, J.M., Opp, M.R., Toth, L., and Kapás, L., Immune regulation, hormones and sleep, in *Sleep '90*, Horne, J., Ed., Pontenagel Press, Bochum, 1990.
58. Krueger, J.M. and Takahashi, S., Thermoregulation and sleep: closely linked but separable, in *Annals of the New York Academy of Sciences. Vol. 813. Thermoregulation: Proc. 10th International Symposium on the Pharmacology of Thermoregulation*, Blatteis, C.M., Ed., New York Academy of Sciences, New York, 1997.
59. Krueger, J.M., Toth, L.A., Obál, Jr., F., Opp, M.R., Kimura-Takeuchi, M., and Kapás, L., Infections, cytokines and sleep, in *Sleep, Hormones and Immunological System*, Smirne, S., Franceschi, M., Ferini-Strambi, L., and Zucconi, M., Eds., Masson, Milano, 1991.
60. Krueger, J.M., Walter, J., Dinarello, C.A., Wolff, S.M., and Chedid, L., Sleep-promoting effects of endogenous pyrogen (interleukin-1), *Am. J. Physiol.*, 246, R994, 1984.
61. Lancel, M., Mathias, S., Faulhaber, J., and Schiffelholz, T., Effect of interleukin-1β on EEG power density during sleep depends on circadian phase, *Am. J. Physiol.*, 270, R830, 1996.
62. Lipton, J.M., Modulation of host defense by the neuropeptide α-MSH, *Yale J. Biol. Med.*, 63, 173, 1990.

63. Louis, C., Jourdan, M., and Cabanac, M., Behavioral fever and therapy in the orthoptera *Gryllus bimaculatus* during infection by an intracellular pathogenic procaryotic *Rickettsiella grylli*, *Am. J. Physiol.*, 250, R991, 1986.
64. Lumpkin, M.D., The regulation of ACTH secretion by IL-1, *Science*, 238, 452, 1987.
65. Mackowiak, P.A., Browne, R.G., Southern, P.M., and Smith, J.W., Polymicrobial sepsis: an analysis of 184 cases using log linear models, *Am. J. Med. Sci.*, 280, 73, 1980.
66. McClellan, J.L., Klir, J.J., Morrow, L.E., and Kluger, M.J., Central effects of glucocorticoid receptor antagonist RU-38486 on lipopolysaccharide and stress-induced fever, *Am. J. Physiol.*, 267, R705, 1994.
67. McGinty, D. and Szymusiak, R., Keeping cool: a hypothesis about the mechanisms and functions of slow-wave sleep, *Trends Neurol. Sci.*, 13, 480, 1990.
68. Moltz, H., Fever: causes and consequences, *Neurosci. Biobehav. Rev.*, 17, 237, 1993.
69. Murdock, L.L., Berlow, S., Colwell, R.E., and Siegel, F.L., The effects of hyperthermia on polyribosomes and amino acid levels in infant rat brain, *Neuroscience*, 3, 349, 1978.
70. Murphy, M.T. and Lipton, J.M., Peripheral administration of α-MSH reduces fever in older and younger rabbits, *Peptides*, 3, 775, 1980.
71. Murphy, M.T., Richards, D.B., and Lipton, J.M., Antipyretic potency of centrally administered α-melanocyte stimulating hormone, *Science*, 221, 192, 1983.
72. Nicol, S.C. and Maskrey, M., Thermoregulation, respiration, and sleep in the Tasmanian Devil, *Sarcophilus harrisii* (Marsupialia: Dasyuridae), *J. Comp. Physiol.*, 140, 241, 1980.
73. Obál, Jr., .F., Thermoregulation and sleep, *Exp. Brain Res.*, Suppl. 8, 157, 1984.
74. Obál, Jr., F., Rubicsek, G., Alföldi, P., Sáry, G., and Obál, F., Changes in the brain and core body temperatures in relation to the various arousal states in rats in the light and dark periods of the day, *Pflügers Arch.*, 404, 73, 1985.
75. Opp, M.R., Corticotropin-releasing hormone involvement in stressor-induced alterations in sleep and in the regulation of waking, *Adv. Neuroimmunol.*, 5, 127, 1995.
76. Opp, M.R., Rat strain differences suggest a role for corticotropin-releasing hormone in modulating sleep, *Physiol. Beh.*, 63, 67, 1997.
77. Opp, M.R., Ball, N.J., Miller, D.E., and Amlaner, Jr., C.J., Thermoregulation and sleep: effects of thermal stress on sleep patterns of glaucous-winged gulls (*Larus glaucescens*), *J. Thermal Biol.*, 12, 199, 1987.
78. Opp, M.R., Obál, Jr., F., and Krueger, J.M., Interleukin-1 alters rat sleep: temporal and dose-related effects, *Am. J. Physiol.*, 260, R52, 1991.
79. Opp, M.R., Obál, Jr., F., and Krueger, J.M., Effects of α-MSH on sleep, behavior, and brain temperature: interactions with IL-1, *Am. J. Physiol.*, 255, R914, 1988.
80. Opp, M.R., Obál, Jr., F., and Krueger, J.M., Corticotropin-releasing factor attenuates interleukin-1 induced sleep and fever in rabbits, *Am. J. Physiol.*, 257, R528, 1989.
81. Opp, M.R., Toth, L.A., and Tolley, E.A., EEG delta power and auditory arousal in rested and sleep-deprived rabbits, *Am. J. Physiol.*, 41, R648, 1997.
82. Owens, M.J. and Nemeroff, C.B., Physiology and pharmacology of corticotropin-releasing factor, *Pharmacol. Rev.*, 43, 425, 1991.
83. Parmeggiani, P.L., Thermoregulation during sleep from the viewpoint of homeostasis, in *Clinical Physiology of Sleep*, Lydic, R. and Biebuyck, J.F., Eds., American Physiological Society, Bethesda, MD, 1988.
84. Parmeggiani, P.L. and Rabini, C., Shivering and panting during sleep, *Brain Res.*, 6, 789, 1967.
85. Parmeggiani, P.L. and Rabini, C., Sleep and environmental temperature, *Arch. Ital. Biol.*, 108, 369, 1970.

86. Parmeggiani, P.L., Zamboni, G., Cianci, T., and Calasso, M., Absence of thermoregulatory vasomotor responses during fast wave sleep in cats, *Electroencephalogr. Clin. Neurophysiol.*, 42, 372, 1977.
87. Petrusz, P. and Merchenthaler, I., The corticotropin-releasing factor system, in *Neuroendocrinology*, Nemeroff, C.B., Ed., CRC Press, Boca Raton, FL, 1992.
88. Pezeshki, G., Pohl, T., and Schöbitz, B., Corticosterone controls interleukin-1β expression and sickness behavior in the rat, *J. Neuroendocrinol.*, 8, 129, 1996.
89. Rechtschaffen, A. and Bergmann, B.M., Sleep deprivation in the rat by the disk-over-water method, *Behav. Brain Res.*, 69, 55, 1995.
90. Roberts, Jr., N.J., Temperature and host defense, *Microbiol. Rev.*, 43, 241, 1979.
91. Roberts, Jr., N.J., Impact of temperature elevation on immunologic defenses, *Rev. Infect. Dis.*, 13, 462, 1991.
92. Rothwell, N.J., CRF is involved in the pyrogenic and thermogenic effects of interleukin-1β in the rat, *Am. J. Physiol.*, 256, E111, 1989.
93. Rothwell, N.J., CNS regulation of thermogenesis, *Crit. Rev. Neurobiol.*, 8, 1, 1994.
94. Roussel, B. and Bittel, J., Thermogenesis and thermolysis during sleeping and waking in the rat, *Pflügers Arch.*, 382, 225, 1979.
95. Saper, C.B. and Breder, C.D., Endogenous pyrogens in the CNS: role of the febrile response, in *Progress in Brain Research*, Swaab, D.F., Hofman, M.A., Mirmiran, M., Ravid, R., and van Leeuwen, F.W., Eds., Elsevier Science, Amsterdam, 1992.
96. Schade, A.L. and Caroline, L., An iron-binding component in human blood plasma, *Science*, 104, 340, 1946.
97. Sebag, J., Reed, W.P., and Williams, R.C., Effect of temperature on bacterial killing by serum and by polymorphonuclear leukocytes, *Infect. Immun.*, 10, 947, 1977.
98. Shoham, S. and Krueger, J.M., Muramyl dipeptide-induced sleep and fever: effects of ambient temperature and time of injection, *Am. J. Physiol.*, 255, R157, 1988.
99. Small, P.M., Tauber, M.G., Hackbarth, C.J., and Sandie, M.A., Influence of body temperature on bacterial growth rates in experimental pneumococcal meningitis in rabbits, *Infect. Immun.*, 52, 484, 1986.
100. Smith, J.B., Knowlton, R.P., and Agarwal, S.S., Human lymphocyte responses are enhanced by culture at 40°C, *J. Immunol.*, 121, 691, 1978.
101. Snyder, D.S. and Unanue, E.R., Corticosteroids inhibit murine macrophage Iα expresion and interleukin 1 production, *J. Immunol.*, 129, 1803, 1982.
102. Tocco, R.J., Kahn, L.L., Kluger, M.J., and Vander, A.J., Relationship of trace metals to fever during infection: are prostaglandins involved?, *Am. J. Physiol.*, 244, R368, 1983.
103. Toth, L.A., Sleep, sleep deprivation and infectious disease: studies in animals, *Adv. Neuroimmunol.*, 5, 79, 1995.
104. Toth, L.A., Gardiner, T.W., and Krueger, J.M., Modulation of sleep by cortisone in normal and bacterially infected rabbits, *Am. J. Physiol.*, 263, R1339, 1992.
105. Toth, L.A. and Krueger, J.M., Alterations of sleep in rabbits by *Staphylococus aureus* infection, *Infect. Immun.*, 56, 1785, 1988.
106. van Twyver, H. and Allison, T., Sleep in the armadillo *Dasypus novemcincus* at moderate and low ambient temperature, *Brain Behav. Evol.*, 9, 107, 1974.
107. Vaughn, L.K., Veale, W.L., and Cooper, K.E., Antipyresis: its effect on mortality rate of bacterially infected rabbits, *Brain Res. Bull.*, 5, 69, 1980.
108. Vertes, R.P., A life-sustaining function of REM sleep: a theory, *Neurosci. Biobehav. Rev.*, 10, 371, 1986.

109. Villar, M., Perassi, N., and Celis, M.E., Central and peripheral actions of α-MSH in the thermoregulation of rats, *Peptides,* 12, 1441, 1991.
110. Walter, J., Davenne, D., Shoham, S., Dinarello, C.A., and Krueger, J.M., Brain temperature changes coupled to sleep states persist during interleukin 1-enhanced sleep, *Am. J. Physiol.,* 250, R96, 1986.
111. Weinberg, E.D., Roles of iron in host-parasite interactions, *J. Infect. Dis.,* 124, 401, 1971.
112. Weinberg, E.D., Iron withholding: a defense against infection and neoplasia, *Physiol. Rev.,* 64, 65, 1984.
113. Weinstein, M.D., Jannini, P.B., Stratton, C.W., and Eickhoff, T., Spontaneous bacterial peritonitis: a review of 28 cases with emphasis on improved survival and factors influencing prognosis, *Am. J. Med.,* 64, 592, 1978.

Chapter 37

Microbial Modulation of Arousal

Linda A. Toth

Contents

37.1	Bacterial Infections and Sleep	643
37.2	Viral Infections and Sleep	644
37.3	Sleep Alterations Induced by Other Organisms	646
37.4	Mechanisms of Infection-Induced Alterations in Sleep	647
37.5	Clinical Relevance of Microbial Modulation of Arousal	649
37.6	Summary and Conclusions	650
Acknowledgments		651
Abbreviations		651
References		651

Common perceptions that the desire for sleep is increased during mild infectious diseases such as colds and the "flu" have fostered beliefs that sleep promotes recovery from infectious disease and that lack of sleep increases susceptibility to infections. Although the relationship between infectious disease and vigilance has historically received little scientific attention, several model systems now provide evidence that infectious disease is accompanied by alterations in sleep. Indeed, increased sleepiness, like fever and anorexia, may be viewed as a facet of the acute phase response to infectious challenge, and components of the immune response appear to provide the physiologic link between sleep and infection.

An interaction between infectious disease and sleep was initially observed by von Economo, whose study of the central nervous system lesions produced during viral infections led him to propose that sleep was an active process mediated by specific brain regions.[116] At present, the influence of microbial infections on sleep has been studied in a variety of models. These model systems include (see Table 37.1):

1. Rabbits inoculated with the Gram-positive bacteria *Staphylococcus aureus* and *Streptococcus pyogenes*; the Gram-negative bacteria *Escherichia coli* and *Pasteurella multocida;* the fungal organism *Candida albicans*; the protozoan *Trypanosoma brucei brucei*; and influenza virus[56,101,108–110]

TABLE 37.1
Infection-Induced Alterations in Sleep

Infectious Organism	Host Species	Route	Selected References
Bacteria			
Staphylococcus aureus	Rabbit	IV	108
Streptococcus pyogenes	Rabbit	IV	109
Escherichia coli	Rabbit	IV	104,109
Pasteurella multocida	Rabbit	IV, SC, IM, IN	110
Viruses			
Human immunodeficiency virus	Human	Natural infection	17, 66, 75
Feline immunodeficiency virus	Cat	IV	85, 87
Influenza virus	Rabbit	IV	56
	Mouse	IN, IP	26, 105
	Human	IN	94
Rhinovirus	Human	IN	94
Newcastle disease virus	Mouse	IP	106
Feline herpesvirus	Cat	IN	107
Rabies virus	Mouse	IC, IM	37, 39
Fungi			
Candida albicans	Rabbit	IV	109, 111
Brewer's yeast	Rat	SC	53
Protozoa			
Trypanosoma brucei	Rabbit	SC	101
	Rat	IP	40, 74
	Human	Natural infection	10
Prions			
Scrapie	Rat	IC, intra-sciatic	2, 3
Creutzfeld-Jakob agent	Cat	IC	38
Fatal familial insomnia agent	Human	Natural infection	73

Note: Routes of inoculation: IC = intracerebral; IM = intramuscular; IN = intranasal; IP = intraperitoneal; IV = intravenous; SC = subcutaneous.

2. Rats inoculated with the scrapie agent,[2,3] live brewer's yeast,[53] or trypanosomes[40,74]
3. Mice inoculated with influenza virus,[26,105] rabies virus,[37,39] or Newcastle disease virus (NDV)[106]
4. Cats infected with feline immunodeficiency virus (FIV),[85,87] feline herpesvirus,[107] or the Creutzfeldt-Jakob agent[38]
5. Humans with naturally acquired trypanosome[10] or human immunodeficiency virus (HIV)[17,66,75] infections, or with experimental inoculations of influenza virus or rhinovirus[94]

Prion-mediated processes are also linked to alterations in sleep in mice[97] and humans,[73] and viral infections have been linked to disorders in sleep and arousability in conditions such as chronic fatigue syndrome, mononucleosis, and sudden infant death syndrome.[20,22,41,46,47,59] A number of microbial components, such as the bacterial cell wall component lipopolysaccharide (LPS) and viral double-stranded RNA (dsRNA), also elicit alterations in sleep (Table 37.2).[65]

TABLE 36.2
Microbial Products that Alter Sleep

Substance	Source	Selected References
Lipopolysaccharide or lipid A	Gram-negative bacterial cell walls	62
Muramyl peptides	Bacterial cell walls	63
Double-stranded RNA	RNA viruses	71
GP120	Human immunodeficiency virus	82
SU-Env	Feline immunodeficiency virus	88

37.1 Bacterial Infections and Sleep

At present, sleep alterations during bacterial infections have been most extensively characterized in rabbits. Rabbits that are inoculated intravenously with viable bacteria typically demonstrate an initial increase and a subsequent decrease in the amount of time spent in slow-wave sleep (SWS) (Figure 37.1); rapid-eye-movement sleep (REMS) is generally reduced or absent.[108,109] The changes in the amount of SWS are typically paralleled by increases and decreases in the EEG delta-wave amplitudes (DWA) during SWS (Figure 37.1), suggesting that the depth or intensity of SWS also changes in a biphasic manner during infection. Bacterially infected rabbits typically develop sleep enhancement concurrent with fever and other pathophysiologic signs of infectious disease, but fever generally persists beyond the period of enhanced sleep,[108,109] indicating that these parameters can be dissociated. In addition, killed bacteria and isolated bacterial components are capable of eliciting enhanced sleep, although increased doses may be required.[49,61,62,72,108,109] Treatment of infected animals with bacteriocidal antibiotics attenuates the development of sleep alterations.[108] Thus, bacterial replication may amplify sleep enhancement, but it is not essential for somnogenesis.

The precise temporal pattern of bacterially induced changes in sleep varies substantially depending on the infecting microorganism and the route of administration. For example, Gram-negative bacteria induce enhanced sleep more quickly and for a much shorter duration than do Gram-positive bacteria.[109] The variations in sleep alteration that accompany infectious disease caused by various microorganisms may be related to structural or biologic differences in the organisms. For example, lipid A, which is a component of endotoxin from Gram-negative bacteria, increases sleep in rabbits within an hour after intravenous injection, and the increase persists for 3 to 4 hr.[62] In contrast, muramyl dipeptide, a synthetic analog of the monomeric muramyl peptide component of bacterial cell wall peptidoglycan, increases sleep after a longer latency but for a longer duration.[91] Variations in sleep patterns may also reflect differences in the disease process induced by or the physiologic response to the challenge organism. For example, inoculation of rabbits with *Pasteurella multocida*, a natural pathogen in that species, causes altered sleep as well as clinical illness.[110] However, intravenous administration, which results in septicemia, induces a different pattern of changes in sleep than does intranasal administration, which causes pneumonia.[110] Temporal and directional variations in sleep propensity after microbial challenge may explain the apparently inconsistent changes in sleep reported in some studies of humans with spontaneous infections.[36,69]

Bacterial components may influence sleep even in healthy animals. Muramyl peptides (MPs), which are components of bacterial cell wall peptidoglycans, promote sleep after their administration to experimental animals.[65] MPs have been detected in the brain and cerebrospinal fluid of some mammals,[61,83,90] which do not synthesize MPs *de novo*[52] but may instead obtain them from exogenous sources, as they do with vitamins.[1] MPs can pass from the intestinal lumen into blood[84] and from blood into brain, liver, and heart.[64] Furthermore, serum from humans, rats, and rabbits contains enzymes capable of degrading bacterial peptidoglycan into its MP constituents.[33,44,113,114]

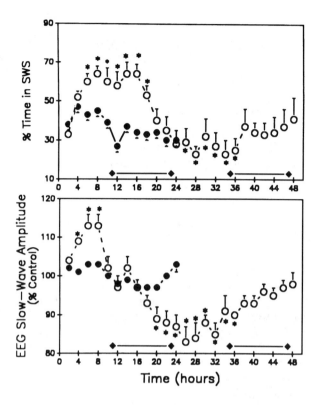

FIGURE 37.1
Effects of inoculation with *Staphylococcus aureus* on SWS in rabbits. Panels indicate the percentage of time spent in SWS (top) and EEG slow-wave (delta-wave) amplitudes during SWS in rabbits (n = 16) for 24 hr prior to (filled circles) and 48 hr after (open circles) the intravenous administration of 10^7 to 10^8 colony-forming units of viable *Staphylococcus aureus*. Individual data points represent the mean ± S.E.M. of values obtained from each rabbit during the preceding 2-hr period. Lines on the abscissa indicate the lights-off period. *$p < 0.03$ relative to the corresponding baseline values. (Adapted from Toth, L.A. and Krueger, J.M., *Infect. Immun.*, 56, 1785, 1988. With permission.)

and mammalian macrophages and glia can degrade bacteria, bacterial cell walls, and bacterial peptidoglycan to produce somnogenically active substances.[30,50,96,115,119] Because the body is continually exposed to bacteria at mucosal surfaces, especially the gastrointestinal tract, and because normal intestinal microorganisms can cross the gut epithelial barrier,[4,120] mammalian processing of bacterial cell walls is likely to be a normal daily occurrence, as well as an early event in the initiation and amplification of the immune response. Indeed, macrophages have high- and low-affinity MP binding sites that have K_a values within the physiologic range, and macrophage activation can alter the binding characteristics of the high-affinity site.[92,93] In addition, rats that are maintained on enteric antibiotic regimens to reduce intestinal bacteria demonstrate reduced sleep[9] and decreased body temperature.[58] Bacterial products may therefore participate in the regulation of normal sleep. In addition, these substances would likely exert a greater effect during bacterial infections, when large numbers of invading organisms would provide a ready source of MPs.

37.2 Viral Infections and Sleep

The somnogenic effects of several viral infections have also been characterized. The two pathogens that have been studied most extensively are influenza virus and immunodeficiency viruses (HIV in humans and FIV in cats). Polysomnographic studies of asymptomatic HIV-infected men reveal several alterations in normal sleep patterns. Most prominent is a significant increase in the percentage of SWS during the second half of the night; frequent nighttime awakenings and abnormal REMS architecture are also common.[18,75,76,123] These altered sleep patterns are not related to psychological, social, or medical etiologies, and they precede the onset of secondary infections or overt neurologic involvement.[18,75,123] As the HIV infection progresses to AIDS, patients tend to develop severe reductions in SWS, marked fragmentation of sleep, and an extreme disruption of normal sleep architecture.[66,124] These severe changes may be related either to HIV-induced encephalitis, or to the development of opportunistic infections or aberrant immune responses. Like HIV-infected humans, cats infected with FIV (a retrovirus closely related to HIV that is a natural pathogen of cats) develop an unusual temporal pattern of SWS, an increased frequency of arousal from sleep, and a decrease in the amount of REMS.[85] Intracerebroventricular administration of GP120 (the coat glycoprotein of HIV) or SU-Env (the envelope glycoprotein of FIV) also modulates sleep patterns in rats.[82,88]

Like infection with immunodeficiency virus, infection with influenza virus also alters normal sleep patterns in humans and in animals. Human volunteers experimentally infected with rhinovirus, influenza virus, or both sleep less during the incubation period but sleep longer during the symptomatic period; sleep quality and the number of awakenings are not affected.[94] In animal models, mice inoculated intranasally with influenza virus demonstrate marked sleep enhancement that is most apparent during the dark portion of the circadian cycle, when mice are normally most active (Figure 37.2); however, sleep enhancement does not develop after intraperitoneal challenge or after intranasal challenge that induces only upper respiratory tract infection.[26,27,105] In rabbits, intravenous administration of influenza virus induces a brief but consistent period of sleep enhancement.[56] Mice and rabbits differ in that influenza virus can replicate completely in mice, but the virus undergoes only partial replication in rabbits. Mice inoculated intraperitoneally with the avian paramyxovirus NDV develop transient sleep enhancement similar to that of influenza-inoculated rabbits.[106] Like influenza virus in rabbits, NDV cannot replicate in mice. This inability to induce active infection may contribute to the relative brevity of the sleep and temperature effects induced by such challenges. Moreover, killed influenza virus does not promote sleep in rabbits or mice.[56,105] These observations suggest that viral replication may be essential to virus-induced potentiation of sleep, thus contrasting with bacterial infection models. Double-stranded RNA (dsRNA), which is produced in the infected host cells as part of the viral replication process, may mediate at least some of the somnogenic properties of viral infections. Intracerebroventricular administration of dsRNA extracted from the lungs of influenza-infected mice induces sleep in rabbits, and synthetic dsRNA (polyinosinic:polycytidilic acid, or poly I:C) induces increased sleep in both rabbits and mice.[55,71,106]

Several other viruses have been studied in terms of their ability to modify sleep. Intranasal infection of kittens with feline herpesvirus, which induces upper respiratory tract infection and sometimes pneumonia, causes an age-dependent increase in auditory arousability from sleep and a reduction in DWA.[107] Mice inoculated with rabies virus, which targets neuronal tissue and causes fatal encephalitis, develop sleep abnormalities that are characterized by generalized EEG slowing, flattening of cortical activity, and replacement of normal sleep/waking stages with abnormal sleep.[37,39] These alterations are somewhat similar to those observed in animals that become moribund subsequent to microbial infection or chronic sleep deprivation.[89,100] However, an important consideration in evaluating sleep changes that develop after challenge with neurotropic agents is the recognition that encephalitis is associated with marked alterations in normal EEG patterns.[122] These encephalitis-induced changes can complicate the determination of sleep stages based on standard EEG criteria.[37] Similar caveats may apply to evaluation of sleep patterns in moribund animals.

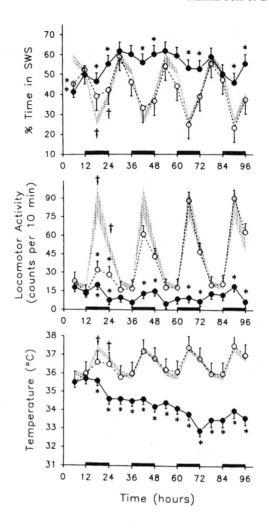

FIGURE 37.2
Sleep, locomotor activity, and body temperature in influenza- and vehicle-inoculated C57BL/6 mice. The percentage of time spent in SWS, locomotor activity, and core temperature was monitored before (shaded area) and after intranasal inoculation of C57BL/6 mice with influenza virus (strain H3N2-x31; 4000 hemagglutinating units in 0.1 ml of allantoic fluid) (n = 15) (filled circles); or with an equivalent volume of uninfected allantoic fluid (n = 5) (open circles). The shaded area represents the mean ± S.E.M. of values collected during 2 days of baseline recording; these data are replicated over the 4-day postinoculation period to facilitate visual comparisons. For baseline measurements, † denotes $p < 0.03$ relative to the baseline values at the 6-hr time point. Postinoculation data points represent the mean ± S.E.M. of values collected during the preceding 6-hr period. *$p < 0.03$ relative to baseline values at the comparable circadian time point. Bars on the abscissa indicate lights-off periods. (Adapted from Toth, L.A. et al., *J. Neuroimmunol.*, 58, 89, 1995. With permission.)

37.3 Sleep Alterations Induced by Other Organisms

Some subspecies of the protozoan *Trypanosoma brucei* cause the chronic clinical condition known as "sleeping sickness" in humans. Sleep patterns have been evaluated in humans undergoing natural trypanosomal infections and in rats and rabbits with experimentally induced infections. In rabbits, subcutaneous inoculation with *Trypanosoma brucei brucei*, which somewhat mimics the intradermal inoculation that normally occurs via the bite of an infected tsetse fly, increases sleep after a latency of several days, coincident with the onset of fever and other signs of clinical illness.[101] This initial increase gradually abates, although periods of increased somnolence continue to occur in association with the episodic recrudescence of parasitemia.[101] In addition to such episodes of hypersomnolence, loss of the normal circadian organization of sleep also gradually develops during chronic trypanosomiasis in rabbits, rats, and humans.[10,40,74,101]

Fungal challenge also alters sleep patterns in animals. Subcutaneous administration of live brewer's yeast to rats significantly increases the time spent in SWS during both light and dark phases of the circadian cycle.[53] Intravenous inoculation of rabbits with *Candida albicans* induces sleep alterations similar to those elicited by Gram-positive bacteria.[109] *Candida*-infected rabbits demonstrate an initial increase and a subsequent decrease in the amount of time spent in SWS, with parallel changes in the amplitude of EEG delta waves during sleep.[109]

Prion-related conditions are also associated with alterations in sleep. Rats inoculated intracerebrally or in the sciatic nerve with brain homogenate from scrapie-infected animals demonstrate unusual spiking patterns in the EEG during periods of quiet wakefulness.[2,3] These changes begin about 4 months after inoculation; later, SWS and active wakefulness are reduced, and drowsiness is increased.[2,3] Cats inoculated intracerebrally with brain homogenate from a human with Creutzfeldt-Jakob disease demonstrate increased SWS time, reduced wakefulness, and abnormal EEG and physiological correlates of REMS approximately 20 months after inoculation.[38] In humans, the condition known as fatal familial insomnia is associated with prion-related thalamic neurodegeneration.[31,73] Mutations in the prion protein, a glycoprotein on neuronal membranes and in astrocytes, may underlie the pathological changes that accompany this condition.[13] Mice that genetically lack the prion protein gene demonstrate alterations in both sleep and circadian rhythms.[97,98]

37.4 Mechanisms of Infection-Induced Alterations in Sleep

The somnogenic effects of viable microorganisms, killed or non-replicating microbes, and isolated microbial components appear to be related to their ability to initiate an immune response and to trigger the production and release of endogenous immune modulators that promote sleep. Data collected in infectious disease models suggest that immune processes contribute to microbially induced sleep enhancement and that alteration of the immune response capabilities of the host modifies the magnitude and temporal pattern of the sleep responses elicited by microbial challenge.[99,103,105] For example, treatment of rabbits with immunosuppressive doses of cortisone or cyclosporine modulates the sleep alterations associated with bacterial or fungal inoculations, although the immune-stimulatory treatments of immunization or adjuvant administration do not.[99,102] However, mice immunized against influenza virus demonstrate an abbreviated period of sleep enhancement after viral challenge as compared to virus-naive mice.[105] The rapid recovery of normal sleep patterns in immunized mice may reflect augmented viral clearance due to the anamnestic immune response. In trypanosome-infected rabbits, parasitemic recrudescences, which reflect proliferation of new antigenic variants and subsequent stimulation of a host immune response, occur concurrent with periods of enhanced sleep.[101]

Several endogenous cytokines, including interleukin-1 (IL-1), tumor necrosis factor, and interferon-α (IFNα), induce sleep when administered exogenously to animals and humans,[29,80] suggesting that these modulators may contribute to sleep enhancement during microbial infections. These cytokines are produced *in vivo* after infection with viable microbes or injection of microbial components such as MPs or LPS.[6,11,12,15,117] In rabbits, the somnogenic effects of MPs are attenuated by the IL-1 receptor antagonist,[48] suggesting a role for IL-1 in mediating the somnogenic effects of this treatment. Similarly, plasma IFN concentrations increase in rabbits that demonstrate sleep enhancement in response to IFN-inducing substances, but the increase is attenuated in tolerant rabbits that do not show increased sleep.[55,56] Among different strains of mice, sleep enhancement in response to poly I:C and some viral challenges is correlated with alleles for high or low IFN production.[106] The types and amounts of cytokines produced after infectious challenge vary with time postinoculation,[6,11,15] suggesting that variations in the cytokine milieu could account for the biphasic sleep changes that occur after microbial infection in rabbits.

The mechanisms underlying microbially induced sleep suppression are more conjectural. Rabbits prevented from sleeping during the postinoculation interval associated with *E. coli*-induced sleep enhancement show a subsequent sleep rebound similar to that of uninfected sleep-deprived rabbits; however, after a latency comparable to that of infected rabbits permitted spontaneous sleep, a period of decreased sleep occurs.[104] These data support two conclusions. First, stimuli that promote sleep enhancement appear transiently after microbial challenge. Microbially induced somnogens may be rapidly cleared from their effector sites, or the milieu of somnogenic substances may change as the infectious disease state progresses, resulting in dissipation of the stimuli that elicit sleep. Second, infection-induced sleep suppression appears to be actively induced, rather than a simple "rebound" effect due to previous hypersomnolence. Thus, microbial inoculation may induce a temporal progression from the production of sleep-promoting to sleep-reducing substances, with the latter eventually dominating the modulation of sleep. Endogenous substances that reduce sleep, including adrenocorticotrophic hormone, glucocorticoids, corticotropin-releasing hormone, α-melanocyte stimulating hormone, the IL-1 receptor antagonist, and interleukin-10,[14,35,48,77,78,81,99] are produced during infectious or inflammatory conditions[32,70,95,109] and may contribute to microbially induced sleep suppression.

In addition to immune and hormonal influences, other factors can also modify infection-induced alterations in sleep. The somnogenic and pyrogenic effects of IL-1, poly I:C, and LPS are influenced by the circadian phase in which they are administered (Opp and Toth, unpublished observations).[67,79,106] In rats, IL-1-induced sleep alterations include increases in the amount of sleep, which predominate during the behaviorally active (dark) portion of the circadian cycle, and increases in DWA during sleep, which predominate during the somnolent (light) phase.[67,79] Circadian variation in the sensitivity and susceptibility of rodents to challenge with various microbes and endotoxin has also been reported.[28,42,125] Patterns of light exposure influence infection-induced sleep responses. In rabbits, exposure to constant light potentiates microbially induced sleep enhancement, whereas the sleep response is reduced during exposure to constant darkness.[111] Environmental temperature is also modulatory.[53,91] Indeed, in rhesus monkeys, the development of IL-1-induced somnolence appears to be context dependent and requires a permissive environment; thus, IL-1-treated monkeys demonstrate sleep-like inactivity if they are maintained in a quiet environment, but not if they are placed in situations requiring interactions with humans or other primates or the performance of a learned task.[34] Finally, observations that subdiaphragmatic vagotomy prevents or attenuates fever and other behavioral and physiologic responses to systemic administration of IL-1 or LPS[16,118] have led to suggestions that the vagus nerve serves as an important conduit for communication between the peripheral immune system and the CNS, thereby potentially contributing to the generation of clinical signs associated with clinical disease. Recent findings indicate that vagotomy also attenuates some of the sleep alterations induced by intraperitoneal administration of LPS and IL-1 in rats.[43,51,112]

The distinctive immunologic sequelae of microbial infections induced by different pathogens support the probability that specific classes of microbial organisms may induce distinctive alterations in sleep. For example, the acute phase responses and the sleep alterations elicited by the viral dsRNA analog poly I:C differ from those induced by bacterially derived LPS.[54] However, generalized conclusions regarding qualitative differences in the sleep responses elicited by different classes of microorganims are not currently possible because appropriate studies have not been reported to date. For example, rabbits infected intravenously with bacterial or fungal organisms show sequential increases and decreases in DWA during SWS over the course of the infection;[109] in contrast, influenza-infected mice demonstrate only decreases in this parameter.[105] These observations might suggest a qualitative difference in the somnogenic properties of bacterial and viral infections.

However, intranasal administration of the bacteria *Pasteurella multocida* to rabbits or of the feline herpesvirus to cats fails to cause marked or prolonged increases in DWA during sleep.[107,109] These observations may indicate that increased DWA during sleep reflects septicemic but not pneumonic conditions, as opposed to characterizing viral or bacterial etiologies. In addition, species-specific variations in host responses are an important consideration. For example, mice respond to influenza infection by developing hypothermia,[57,105] whereas fever is more common in other species. Valid comparisons of the somnogenic properties of various classes of microorganisms would minimally require evaluation of two different classes of pathogen (e.g., a virus and a bacterium) administered to the same host species by the same route.

37.5 Clinical Relevance of Microbial Modulation of Arousal

Little information is available regarding the clinical relevance of altered arousal during microbial infections. However, the occurrence of prolonged periods of increased sleep during various infectious states in animals and humans implies an adaptive function for infection-induced sleep. Such advantages could be related to the general physiologic and behavioral alterations that accompany sleep.[45] For example, because sleep is associated with a decreased metabolic rate and muscular inactivity, increased sleep may permit the animal to conserve metabolic energy. Such conservation could be particularly important during febrile conditions when overall energy requirements are high and food intake may be low. Similarly, under natural conditions, animals generally sleep in relatively protected environments. Infection-induced increases in sleep propensity may therefore promote the animal's remaining in comparatively secure surroundings while physically debilitated, thereby potentially reducing the possibility of predation or injury during illness.

In addition to such general advantages, sleep may also directly promote recovery from infectious disease. Animals that respond to microbial challenge with a robust enhancement of SWS and of DWA during SWS have a greater probability of surviving than do animals that show little or no increase in SWS (Figure 37.3); analogously, animals that die after microbial challenge demonstrate less SWS than do animals that survive.[100,108] Infection-induced sleep alterations correlate not only with mortality but also with the severity of various clinicopathologic indices.[100] A similar situation may occur in HIV-infected humans. Individuals who are seropositive for HIV but otherwise healthy have excessive SWS,[18,75,76,123] but sleep deteriorates and becomes disrupted as the disease progresses.[66,124] Other studies in humans report correlations between insomnia or unusually short nighttime sleep durations and decreased life expectancy.[60,86] Associations between absent or diminished sleep, reduced EEG amplitude, and imminent death also occur in aged mice prior to spontaneous death,[121] in mice with fatal experimental rabies infections,[37,39] and in rats that die subsequent to chronic sleep deprivation and septicemia.[23,24,89] SWS and DWA during SWS also gradually decline in rabbits with trypanosome infections,[101] which are eventually fatal.

These observations may indicate that impaired sleep reflects a deteriorating clinical condition during illness. Thus, sleep may simply provide a general index of well-being during microbial infections, such that less severely ill animals are likely to sleep better. However, an alternative hypothesis is that dynamic changes in sleep actively contribute to the recovery process. This possibility is indirectly supported by reports that immune competence is adversely affected by sleep deprivation. For example, sleep-deprived mice immunized against influenza virus fail to clear the virus from the lungs after challenge, although immunized animals that are not deprived clear the infection.[7] A related study demonstrated that sleep-deprived rats have a suppressed secondary antibody response to sheep red blood cells;[8] however, these intriguing observations have not been replicated. Rats that undergo chronic sleep deprivation eventually die,[23,89] often in association with septicemia[24] that may develop secondary to increased microbial translocation from the gastrointestinal

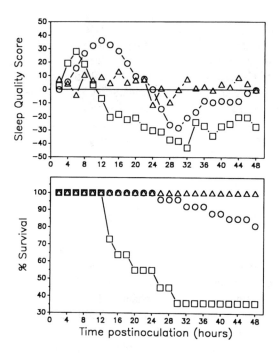

FIGURE 37.3
Sleep and survival in rabbits inoculated with *Staphylococcus aureus*. **(Top panel)** Sleep quality score during the 48-hr postinoculation interval, expressed as a difference from baseline values. **(Lower panel)** Percentage of animals that survived at given postinoculation intervals. Sleep quality score is calculated as (% time in SWS) × (% baseline DWA during SWS)/100, and thus reflects both the intensity and amount of SWS. Rabbits were grouped based on the duration of the sleep quality score increase, as follows: triangles, minimal sleep change, n = 8; circles, ≥16 hr of increased sleep, n = 26; squares, ≤14 hr of increased sleep, n = 11. Rabbits that became moribund were euthanized and considered to be non-survivors. (Adapted from Toth, L.A. et al., *Proc. Soc. Exp. Biol. Med.*, 203, 179, 1993. With permission.)

tract.[25] In addition to possible immunosuppressive effects, however, sleep deprivation augments some immune indices in humans[21] and retards tumor growth in rats.[5]

37.6 Summary and Conclusions

Intriguing data support significant interactions among sleep, immune function, and infectious disease. These findings also indicate important directions for future work, such as identifying the mechanisms by which infectious disease alters sleep, determining whether sleep directly promotes recovery from infectious illness, and considering whether infectious disease should represent a contraindication for performing activities that require a high degree of vigilance. Resolution of these issues is critical not only for advancing our basic understanding of sleep but also for addressing significant problems of human welfare. Sleep and sleepiness are major economic and public health considerations. Circadian periods characterized by sleepiness are associated with greater numbers of accidents, from individual mishaps to catastrophic tragedies.[68] In addition, poor sleep or excessive sleepiness causes significant economic loss through reduced productivity and compromises the quality of life of many people who suffer from sleep disorders.[19] An improved understanding of the factors that mediate sleep and sleepiness may contribute to the alleviation of these important public health concerns.

Acknowledgments

This work was supported in part by NIH grants NS-26429 and CA-21765, and by the American Lebanese Syrian Associated Charities (ALSAC). The author thanks Dr. Amy L.B. Frazier for editorial review of this manuscript.

Abbreviations

CNS	central nervous systemds
RNA	double-stranded RNA
DWA	delta-wave amplitude(s)
EEG	electroencephalogram
FIV	feline immunodeficiency virus
HIV	human immunodeficiency virus
IFN	interferon
IL-1	interleukin-1
LPS	lipopolysaccharide
MP	muramyl peptide
NDV	Newcastle disease virus
poly I:C	polyinosinic:polycytidilic acid
REMS	rapid-eye-movement sleep
SWS	slow-wave sleep

References

1. Adam, A. and Lederer, E., Muramyl peptides: immunomodulators, sleep factors, and vitamins, *Med. Res. Rev.*, 4, 111, 1984.
2. Bassant, M., Cathala, F., Court, L., Gourmelon, P., and Hauw, J.J., Experimental scrapie in rats: first electrophysiological observations, *Electroencephalogr. Clin. Neurophysiol.*, 57, 541, 1984.
3. Bassant, M., Baron, H., Gumpel, M., Cathala, F., and Court, L., Spread of scrapie to the central nervous system: study of a rat model, *Brain Res.*, 383, 397, 1986.
4. Berg, R.D. and Garlington, A.W., Translocation of certain indigenous bacteria from the gastrointestinal tract to the mesenteric lymph nodes and other organs in a gnotobiotic mouse model, *Infect. Immun.*, 23, 403, 1979.
5. Bergmann, B.M., Rechtschaffen, A., Gilliland, M.A., and Quintans, J., Effect of extended sleep deprivation on tumor growth in rats, *Am. J. Physiol.*, 40, R1460, 1996.
6. Bjork, L., Andersson, J., Cesaka, M., and Andersson, U., Endotoxin and *Staphylococcus aureus* enterotoxin A induce different patterns of cytokines, *Cytokine*, 4, 513, 1992.
7. Brown, R., Pang, G., Husband, A.J., and King, M.G., Suppression of immunity to influenza virus infection in the respiratory tract following sleep disturbance, *Reg. Immunol.*, 2, 321, 1989a.
8. Brown, R., Price, R.J., King, M.G., and Husband, A.J., Interleukin-1-α and muramyl dipeptide can prevent decreased antibody response associated with sleep deprivation, *Brain Behav. Immun.*, 3, 320, 1989b.
9. Brown, R., Price, R.J., King, M.G., and Husband, A.J., Are antibiotic effects on sleep behavior in the rat due to modulation of gut bacteria?, *Physiol. Behav.*, 48, 561, 1990.

10. Buguet, A., Bert, J., Tapie, P., Tabaraud, F., Doua, F., Lonsdorfer, J., Bogui, P., and Dumas, M., Sleep-wake cycle in human African trypanosomiasis, *J. Clin. Neurophysiol.*, 10, 190, 1993.

11. Chensue, S.W., Terebuh, P.D., Remick, D.G., Scales, W.E., and Kunkle, S.L., *In vivo* biologic and immunohistochemical analysis of interleukin-1 alpha, beta and tumor necrosis factor during experimental endotoxemia, *Am. J. Pathol.*, 138, 395, 1991.

12. Clark, B.D., Bedrosian, I., Schindler, R., Cominelli, F., Cannon, J.G., Shaw, A.R., and Dinarello, C.A., Detection of interleukin 1α and 1β in rabbit tissues during endotoxemia using sensitive radioimmunoassay, *J. Appl. Physiol.*, 71, 2412, 1991.

13. Collinge, J., Palmer, M.S., Sidle, K.C., Gowland, I., Medori, R., Ironside, J., and Lantos, P., Transmission of fatal familial insomnia to laboratory animals, *Lancet*, 346, 569, 1995.

14. Concu, A., Ferrari, W., Gessa, G.L., Mereu, G.P., and Tagliamonte, A., EEG changes induced by the intraventricular injection of ACTH in cats, in *Sleep 1974*, Levin, P. and Koella W.P., Eds., Karger, Basel, 1975, p. 321.

15. Cross, A.S., Opal, S.M., Sadoff, J.C., and Gemski, P., Choice of bacteria in animal models of sepsis, *Infect. Immun.*, 61, 2741, 1993.

16. Dantzer, R. Cytokines and behavior, *Physiologist*, 37, A4, 1994.

17. Darko, D.F., McCutchan, J.A., Kripke, D.F., Gillin, J.C., and Golshan, S., Fatigue, sleep disturbance, disability, and indices of progression of HIV infection, *Am. J. Psychiat.*, 149, 514, 1992.

18. Darko, D.F., Mitler, M.M., and Henriksen, S.J., Lentiviral infection, immune response peptides and sleep, *Adv. Neuroimmunol.* 5, 57, 1995.

19. Dement, W.C. and Gelb, M., Somnolence: its importance in society, *Neurophysiol. Clin.* 23, 5, 1993.

20. Dickinson, C.J., Chronic fatigue syndrome — aetiological aspects, *Eur. J. Clin. Invest.*, 27, 257, 1997.

21. Dinges, D.F., Douglas, S.D., Zaugg, L., Campbell, D.E., McMann, J.M., Whitehouse, W.G., Orne, E.C., Kapoor, S.C., Icaza, E., and Orne, M.T., Leukocytosis and natural killer cell function parallel neurobehavioral fatigue induced by 64 hours of sleep deprivation, *J. Clin. Invest.*, 93, 1930, 1994.

22. Drucker, D.B. and Sayers, N.M., Sudden infant death syndrome: a microbial aetiology, *Biomed. Lett.* 48, 255, 1993.

23. Everson, C.A., Bergmann, B.M., and Rechtschaffen, A., Sleep deprivation in the rat. III. Total sleep deprivation, *Sleep*, 12, 13, 1989.

24. Everson, C.A., Sustained sleep deprivation impairs host defense, *Am. J. Physiol.*, 265, R1148, 1993.

25. Everson, C.A. and Toth, L.A. Abnormal control of viable bacteria in body tissues during sleep deprivation in rats, *Sleep Res.*, 26, 613, 1997.

26. Fang, J., Sanborn, C.K., Renegar, K.B., Majde, J.A., and Krueger, J.M., Influenza viral infections enhance sleep in mice, *Proc. Soc. Exp. Biol. Med.*, 210, 242, 1995.

27. Fang, J., Tooley, D., Gatewood, C., Renegar, K.B., Majde, J.A., and Krueger, J.M., Differential effects of total and upper airway influenza viral infection on sleep in mice, *Sleep*, 19, 337, 1996.

28. Feigin, R.D., Middelkamp, J.N., and Reed, C., Circadian rhythmicity in susceptibility of mice to sublethal Coxsackie B3 infection, *Nature (New Biol.)*, 240, 57, 1972.

29. Fent, K. and Zbinden, G., Toxicity of interferon and interleukin, *Trends Pharmacol. Sci.*, 8, 100, 1987.

30. Fincher, E.F., Johannsen, L., Kápas, L., Takahashi, S., and Krueger, J.M. Microglia digest *Staphylococcus aureus* into low molecular weight biologically active compounds, *Am. J. Physiol.*, 271, R149, 1996.

31. Fiorino, A.S., Sleep, genes and death: fatal familial insomnia, *Brain Res. Rev.*, 22, 258, 1996.

32. Fischer, E., van Zee, K.J., Marano, M.A., Rock, C.S., Kenney, J.S., Poutsiaka, D.D., Dinarello, C.A., Lowry, S.F., and Moldawer, L.L., Interleukin-1 receptor antagonist circulates in experimental inflammation and in human disease, *Blood*, 79, 2196, 1994.

33. Fox, A. and Fox, K., Rapid elimination of a synthetic adjuvant peptide from the circulation after systemic administration and absence of detectable natural muramyl peptides in normal serum at current analytical levels, *Infect. Immun.*, 59, 1202, 1991.
34. Friedman, E.M., Reyes, T.M., and Coe, C.L., Context-dependent behavioral effects of interleukin-1 in the Rhesus monkey (*Mucacca mulatta*), *Psychoneuroendocrinology*, 21, 455, 1996.
35. Gillin, J.C., Jacobs, L.S., Fram, D.H., and Snyder, F., Acute effect of a glucocorticoid on normal human sleep, *Nature*, 237, 398, 1972.
36. Gould, J.B., Lee, A.F.S., Cook, P., and Morelock, S., Apnea and sleep state in infants with nasopharyngitis, *Pediatrics*, 65, 713, 1980.
37. Gourmelon, P., Briet, D., Court, L., and Tsiang, H., Electrophysiological and sleep alterations in experimental mouse rabies, *Brain Res.*, 398, 128, 1986.
38. Gourmelon, P., Amyx, H.L., Baron, H., Lemercier, G., Court, L., and Gibbs, C.J., Sleep abnormalities with REM disorder in experimental Creutzfeldt-Jakob disease in cats: a new pathological feature, *Brain Res.*, 411, 391, 1987.
39. Gourmelon, P., Briet, D., Clarencon, D., Court, L., and Tsiang, H., Sleep alterations in experimental street rabies virus infection occur in the absence of major EEG abnormalities, *Brain Res.*, 554, 159, 1991.
40. Grassi-Zucconi, G., Harris, J.A., Mohammed, A.H., Ambrosini, M.V., Kristensson, K., and Bentivoglio, M., Sleep fragmentation, and changes in locomotor activity and body temperature in trypanosome-infected rats, *Brain Res. Bull.*, 37, 123, 1995.
41. Guilleminault, C. and Mondini, S., Mononucleosis and chronic daytime sleepiness: a long-term follow-up study, *Arch. Intern. Med.*, 146, 1333, 1986.
42. Halberg, F., Johnson, E.A., Brown, B.W., and Bittner, J.J., Susceptibility rhythm to *E. coli* endotoxin and bioassay, *Proc. Soc. Exp. Biol. Med.*, 103, 142, 1960.
43. Hansen, M.K. and Krueger, J.M., Subdiaphragmatic vagotomy attenuates interleukin-1-induced sleep, *Sleep Res.*, 26, 75, 1997.
44. Harrison, J. and Fox, A., Degradation of muramyl dipeptide by mammalian serum, *Infect. Immun.*, 50, 320, 1985.
45. Hart, B.L., Biological basis of the behavior of sick animals, *Neurosci. Biobehav. Rev.*, 12, 123, 1988.
46. Hoffman, H.J., Damus, K., Hillman, L., and Krongrad, E., Risk factors for SIDS: results of the National Institute of Child Health and Human Development SIDS cooperative epidemiological study, *Ann. N.Y. Acad. Sci.*, 533, 13, 1988.
47. Holmes, G.P., Kaplan, J.E., Gantz, N.M., Komaroff, A.L., Schonberger, L.B., Straus, S.E., Jones, J.F., Dubois, R.E., Cunningham-Rundles, C., Pahwa, S., Tosato, G., Zegans, L.S., Purtilo, D.T., Brown, N., Schooley, R.T., and Brus, I., Chronic fatigue syndrome: a working case definition, *Ann. Intern. Med.*, 108, 387, 1988.
48. Imeri, L., Opp, M.R., and Krueger, J.M., An IL-1 receptor and an IL-1 receptor antagonist attenuate muramyl dipeptide- and IL-1-induced sleep and fever, *Am. J. Physiol.*, 265, R907, 1993.
49. Johannsen, L., Toth, L.A., Rosenthal, R.S., Opp, M.R., Obál, F., Cady, A.B., and Krueger, J.M., Somnogenic, pyrogenic, and hematologic effects of bacterial peptidoglycan, *Am. J. Physiol.*, 259, R182, 1990.
50. Johannsen, L., Wecke, J., Obál, F., and Krueger, J.M., Macrophages produce somnogenic and pyrogenic muramyl peptides during digestion of staphylococci, *Am. J. Physiol.*, 260, R126, 1991.
51. Kapás, L., Hansen, M.K., Chang, H., and Krueger, J.M., The somnogenic effects of lipopolysaccharide are attenuated in vagotomized rats, *Sleep Res.*, 26, 77, 1997.
52. Karnovsky, M.L., Muramyl peptides in mammalian tissues and their effects at the cellular level, *Fed. Proc.*, 42, 2556, 1986.

53. Kent, S., Price, M., and Satinoff, E., Fever alters characteristics of sleep in rats, *Physiol. Behav.*, 44, 709, 1988.
54. Kimura, M., Toth, L.A., Agostini, H., Cady, A.B., Majde, J.A., and Krueger, J.M., Comparison of acute phase responses induced in rabbits by lipopolysaccharide and double-stranded RNA, *Am. J. Physiol.*, 267, R1596, 1994.
55. Kimura-Takeuchi, M., Majde, J.A., Toth, L.A., and Krueger, J.M., The role of double-stranded RNA in induction of the acute-phase response in an abortive influenza virus infection model, *J. Infect. Dis.*, 166, 1266, 1992a.
56. Kimura-Takeuchi, M., Majde, J.A., Toth, L.A., and Krueger, J.M., Influenza virus-induced changes in rabbit sleep and acute phase responses, *Am. J. Physiol.*, 263, R1115, 1992b.
57. Klein, M.S., Conn, C.A., and Kluger, M.J., Behavioral thermoregulation in mice inoculated with influenza virus, *Physiol. Behav.*, 52, 1133, 1992.
58. Kluger, M.J., Conn, C.A., Franklin, B., Freter, R., and Abrams, G.D., Effect of gastrointestinal flora on body temperature of rats and mice, *Am. J. Physiol.*, 258, R552, 1990.
59. Komaroff, A.L., Chronic fatigue syndromes: relationship to chronic viral infections, *J. Virol. Meth.*, 21, 3, 1988.
60. Kripke, D.F., Simons, R.N., Garfinkel, L., and Hammond, C., Short and long sleep and sleeping pills: is increased mortality associated? *Arch. Gen. Psychiatry*, 36, 103, 1991.
61. Krueger, J.M., Pappenheimer, J.R., and Karnovsky, M.L., Sleep-promoting effects of muramyl peptides, *Proc. Nat. Acad. Sci. USA*, 79, 6102, 1982.
62. Krueger, J.M., Kubillus, S., Shoham, S., and Davenne, D., Enhancement of slow-wave sleep by endotoxin and lipid A, *Am. J. Physiol.*, 251, R591, 1986.
63. Krueger, J.M., Davenne, D., Walter, J., Shoham, S., Kubillus, S.L., Rosenthal, R.S., Martin, S.A., and Biemann, K., Bacterial peptidoglycans as modulators of sleep. II. Effects of muramyl peptides on the structure of rabbit sleep, *Brain Res.*, 403, 258, 1987.
64. Krueger, J.M., Toth, L.A., Cady, A.B., Johannsen, L., and Obál, F., Immunomodulation and sleep, in *Sleep Peptides: Basic and Clinical Approaches*, Inoue S. and Schneider-Helmert D., Eds., Springer-Verlag, Berlin, 1988, p. 95.
65. Krueger, J.M. and Majde, J.A., Microbial products and cytokines in sleep and fever regulation, *Crit. Rev. Immunol.*, 14, 355, 1994.
66. Kubicki, S., Henkes, H., Terstegge, K., and Ruf, B., AIDS related sleep disturbances — a preliminary report, in *HIV and the Nervous System*, Kubicki, S., Henkes, H., Bienzle, U., and Pohle, H.D., Gustav Fischer, New York, 1988, p. 97.
67. Lancel, M., Mathias, S., Faulhaber, J., and Schiffelholz, T., Effect of interleukin-1β on EEG power density during sleep depends on circadian phase, *Am. J. Physiol.*, 270, R830, 1996.
68. Leger, D., The cost of sleep-related accidents: a report for the National Commission on Sleep Disorders Research, *Sleep*, 17, 84, 1994.
69. Lipshitz, A., Lopez, M., Fiorello, S., Medina, E., Osuna, G., and Halabe, J., The electroencephalogram in adult patients with fever, *Clin. Electroencephalogr.*, 18, 85, 1987.
70. Lipton, J.M., Modulation of host defense by the neuropeptide α-MSH, *Yale J. Biol. Med.*, 63, 173, 1990.
71. Majde, J.A., Brown, R.K., Jones, M.W., Dieffenbach, C.W., Maitra, N., Krueger, J.M., Cady, A.B., Smitka, C.W., and Maassab, H.F., Detection of toxic viral-associated double-stranded RNA (dsRNA) in influenza-infected lung, *Microb. Pathogen.*, 10, 105, 1991.
72. Masek, K., Kadlecová, O., and Petrovicky, P., The effect of some bacterial products on temperature and sleep in rat, *Z. Immun.-Forsch. Bd.*, 149, 273, 1975.

73. Monari, L., Chen, S.G., Brown, P., Parchi, P., Petersen, R.B., Mikol, J., Gray, F., Cortelli, P., Montagna, P., Ghetti, B., Goldfarb, L.G., Gajdusek, D.C., Lugaresi, E., Gambetti, P., and Autilio-Gambetti, L., Fatal familial insomnia and familial Creutzfeld-Jakob disease: different prion proteins determined by DNA polymorphism, *Proc. Nat. Acad. Sci. USA*, 91, 2839, 1994.

74. Montmayeur, A. and Buguet, A., Time-related changes in the sleep-wake cycle of rats infected with *Trypanosoma brucei brucei*, *Neurosci. Lett.*, 168, 172, 1994.

75. Norman, S.E., Chediak, H.D., Kiel, M., and Cohn, M.A., Sleep disturbances in HIV-infected homosexual men, *AIDS*, 4, 775, 1990.

76. Norman, S.E., Chediak, A.D., Freeman, C., Liel, M., Mendez, A., Duncan, R., Simoneau, J., and Nolan, B., Sleep disturbances in men with asymptomatic human immunodeficiency (HIV) infection, *Sleep*, 15, 150, 1992.

77. Opp, M., Obál, F., and Krueger, J.M., CRF attenuates interleukin-1 induced sleep and fever in rabbits, *Am. J. Physiol.*, 257, R528, 1989.

78. Opp, M.R., Obál, F., and Krueger, J.M., Effects of α-MSH on sleep, behavior, and brain temperature: interactions with IL-1, *Am. J. Physiol.*, 255, R914, 1988.

79. Opp, M.R., Obál, F., and Krueger, J.M., Interleukin 1 alters rat sleep: temporal and dose-related effects, *Am. J. Physiol.*, 260, R52, 1991.

80. Opp, M.R., Kapás, L., and Toth, L.A., Cytokine involvement in the regulation of sleep, *Proc. Soc. Exp. Biol. Med.*, 201, 16, 1992.

81. Opp, M.R., Smith, E.M., and Hughes, T.K., Interleukin-10 (cytokine synthesis inhibitory factor) acts in the central nervous system of rats to reduce sleep, *J. Neuroimmunol.*, 60, 165, 1995.

82. Opp, M.R., Rady, P.L., Hughes, T.K., Cadet, P., Tyring, S.K., and Smith, E.M., Human immunodeficiency virus envelope glycoprotein 120 alters sleep and induces cytokine mRNA expression in rats, *Am. J. Physiol.*, 270, R963; 1996.

83. Pappenheimer, J.R., Koski, G., Fencl, V., Karnovsky, M.L., and Krueger, J., Extraction of sleep-promoting factor S from cerebrospinal fluid and from brains of sleep-deprived animals, *J. Neurophysiol.*, 38, 1299, 1975.

84. Pappenheimer, J.R., and Zich, K.E., Absorption of hydrophilic solutes from the small intestine, *J. Physiol.*, 371, 138P, 1986.

85. Phillips, T.R., Prospero-Garcia, O., Puaoi, D.L., Lerner, D.L., Fox, H.S., Olmsted, R.A., Bloom, F.E., Henriksen, S.J., and Elder, J.H., Neurological abnormalities associated with feline immunodeficiency virus infection, *J. Gen. Virol.*, 75, 979, 1994.

86. Pollack, C.P., Perlick, D., Linsner, J.P., Wenston, J., and Hsieh, F., Sleep problems in the community elderly as predictors of death and nursing home placement, *J. Commun. Health*, 15, 123, 1990.

87. Prospero-Garcia, O., Herold, N., Phillips, T.R., Elder, J.H., Bloom, F.E., and Henriksen, S.J., Sleep patterns are disturbed in cats infected with feline immunodeficiency virus, *Proc. Nat. Acad. Sci. USA*, 91, 12947, 1994.

88. Prospero-Garcia, O., Herold, N., Waters, A.K., Phillips, T.R., Elder, J.H., and Henriksen, S.J., Intraventricular administration of a FIV-envelope protein induces sleep architecture changes in rats, *Brain Res.*, 659, 254, 1994.

89. Rechtschaffen, A., Gilliland, M.A., Bergmann, B.M., and Winter, J.B., Physiological correlates of prolonged sleep deprivation in rats, *Science*, 221, 182, 1983.

90. Sen, Z. and Karnovsky, M.L., Qualitative detection of muramic acid in normal mammalian tissues, *Infect. Immun.*, 43, 937, 1984.

91. Shoham, S. and Krueger, J.M., Muramyl dipeptide-induced sleep and fever: effects of ambient temperature and time of injections, *Am. J. Physiol.*, 255, R157, 1988.

92. Silverman, D.H.S., Wu, H., and Karnovsky, M.L., Muramyl peptides and serotonin interact at specific binding sites on macrophages and enhance superoxide release, *Biochem. Biophys. Res. Commun.*, 131, 1160, 1985.
93. Silverman, D.H.S., Krueger, J.M., and Karnovsky, M.L., Specific binding site for muramyl peptides on murine macrophages, *J. Immunol.*, 136, 2195, 1986.
94. Smith, A., Sleep, colds, and performance, *Sleep, Arousal and Performance*, Broughton, R.J. and Ogilvie, R.D., Eds., Birkhauser, Boston, 1992, p. 233.
95. Smith, E.M., Meyer, W.J., and Blalock, J.E., Virus-induced corticosterone in hypophysectomized mice: a possible lymphoid adrenal axis, *Science*, 218, 1311, 1982.
96. Smith, H., Microbial surfaces in relation to pathogenicity, *Bact. Rev.*, 41, 475, 1977.
97. Tobler, I., Gaus, S.E., Deboer, T., Achermann, P., Fischer, M., Rülicke, T., Moser, M., Oesch, B., McBride, P.A., and Manson, J.C., Altered circadian rhythms and sleep in mice devoid of prion protein, *Nature*, 380, 639, 1996.
98. Tobler, I., Deboer, T., and Fischer, M., Sleep and sleep regulation in normal and prion protein-deficient mice, *J. Neurosci.*, 17, 1869, 1997.
99. Toth, L.A., Gardiner, T.W., and Krueger, J.M., Modulation of sleep by cortisone in normal and bacterially infected rabbits, *Am. J. Physiol.*, 263, R1339, 1992.
100. Toth, L.A., Tolley, E.A., and Krueger, J.M., Sleep as a prognostic indicator during infectious disease in rabbits, *Proc. Soc. Exp. Biol. Med.*, 203, 179, 1993.
101. Toth, L.A., Tolley, E.A., Broady, R., Blakely, B., and Krueger, J.M., Sleep during experimental trypanosomiasis in rabbits, *Proc. Soc. Exp. Biol. Med.*, 205, 174, 1994.
102. Toth, L.A., Immune-modulatory drugs alter *Candida albicans*-induced sleep patterns in rabbits, *Pharmacol. Biochem. Behav.*, 5, 877, 1995.
103. Toth, L.A., Immune-modulatory drugs alter *Candida albicans*-induced sleep patterns in rabbits, *Pharmacol. Physiol. Behav.*, 51, 877, 1995.
104. Toth, L.A., Opp, M.R., and Mao, L., Somnogenic effects of sleep deprivation and *Escherichia coli* inoculation in rabbits, *J. Sleep Res.*, 4, 30, 1995.
105. Toth, L.A., Rehg, J.E., and Webster, R.G., Strain differences in sleep and other pathophysiological sequelae of influenza virus infection in naive and immunized mice, *J. Neuroimmunol.*, 58, 89, 1995.
106. Toth, L.A., Strain differences in the somnogenic effects of interferon inducers in mice, *J. Interferon Cytokine Res.*, 16, 1065, 1996.
107. Toth, L.A. and Chaudhary, M.A., Developmental alterations in auditory arousal from sleep in healthy and virus-infected neonatal cats, *Sleep*, 21, 143–152, 1998.
108. Toth, L.A. and Krueger, J.M., Alteration of sleep in rabbits by *Staphylococcus aureus* infection, *Infect. Immun.*, 56, 1785, 1988.
109. Toth, L.A. and Krueger, J.M., Effects of microbial challenge on sleep in rabbits, *FASEB J.*, 3, 2062, 1989.
110. Toth, L.A. and Krueger, J.M., Somnogenic, pyrogenic and hematologic effects of experimental pasteurellosis in rabbits, *Am. J. Physiol.*, 258, R536, 1990.
111. Toth, L.A. and Krueger, J.M., Lighting conditions alter *Candida albicans*-induced sleep responses in rabbits, *Am. J. Physiol.*, 269, R1441, 1995.
112. Toth, L.A. and Opp, M.R., Pyrogenic and somnogenic effects of intraperitoneal interleukin-1 in intact and vagotomized rats, *Sleep Res.*, 26, 217, 1997.
113. Valinger, Z., Ladesic, B., and Tomasic, J., Partial purification and characterization of *N*-acetylmuramyl-L-alanine amidase from human and mouse serum, *Biochim. Biophys. Acta*, 701, 63, 1982.

114. Vanderwinkel, E., De Vlieghere, M., De Pauw, P., Cattalini, N., Ledoux, V., Gigot, D., and Have, J.-P.T., Purification and characterization of *N*-acetylmuramyl-L-alanine amidase from human serum, *Biochim. Biophys. Acta*, 1039, 331, 1990.
115. Vermeulon, M.W., and Grey, G.R., Processing of *Bacillus subtilis* peptidoglycan by a mouse macrophage cell line, *Infect. Immun.*, 46, 476, 1984.
116. von Economo, C., Sleep as a problem of localization, *J. Nerv. Ment. Dis.*, 71, 249, 1930.
117. Wakabayashi, G., Gelfand, J.A., Burke, J.F., Thompson, R.C., and Dinarello, C.A., A specific receptor antagonist for interleukin 1 prevents *Escherichia coli*-induced shock in rabbits, *FASEB J.*, 5, 338, 1991.
118. Watkins, L.R., Maier, S.F., and Goehler, L.E., Cytokine-to-brain communication: a review and analysis of alternative mechanisms, *Life Sci.*, 57, 1011, 1995.
119. Wecke, J., Johannsen, L., and Giesbrecht, P., Reduction of wall degradability of clindamycin-treated staphylococci within macrophages, *Infect. Immun.*, 58, 197, 1990.
120. Wells, C.L., Maddaus, M.A., and Simmons, R.L., Proposed mechanisms for the translocation of intestinal bacteria, *Rev. Infect. Dis.*, 10, 958, 1988.
121. Welsh, D.K., Richardson, G.S., and Dement, W.C., Effect of age on the circadian pattern of sleep and wakefulness in the mouse, *J. Gerontol.*, 41, 579, 1986.
122. Westmorland, B.F., The EEG in cerebral inflammatory processes, in *Electroencephalography: Basic Principles, Clinical Applications and Related Fields*, 2nd ed., Niedermeyer, E. and Lopes da Silva, F., Eds., Urban & Schwarzenberg, Baltimore, MD, 1991, p. 259.
123. White, J.L., Darko, D.F., Brown, S.J., Miller, J.C., Hayduk, R., Kelly, T., and Mitler, M.M., Early central nervous system response to HIV infection: sleep distortion and cognitive-motor decrements, *AIDS*, 9, 1043, 1995.
124. Wiegand, M., Müller, A.A., Schreiber, W., Krieg, J.C., Fuchs, D., Wachter, H., and Holsboer, F., Nocturnal sleep EEG in patients with HIV infection, *Eur. Arch. Psychiatr. Clin. Neurosci.*, 240, 153, 1991.
125. Wongwiwat, M., Sukapanit, S., Triyanond, C., and Sawyer, W.D., Circadian rhythm of the resistance of mice to acute pneumococcal infection, *Infect. Immun.* 5, 442, 1972.

Chapter 38

Immune Alterations in Neurotransmission

Luca Imeri and Maria Grazia de Simoni

Contents

- 38.1 Introduction 659
- 38.2 Interactions Between Immune Molecules and Neurotransmitters 660
    - 38.2.1 Noradrenaline 660
    - 38.2.2 Dopamine 661
    - 38.2.3 Serotonin 661
    - 38.2.4 Acetylcholine 664
    - 38.2.5 GABA 664
    - 38.2.6 Glutamic Acid 665
    - 38.2.7 Histamine 665
    - 38.2.8 Opioids 665
    - 38.2.9 Nitric Oxide 665
    - 38.2.10 Adenosine 666
- 38.3 Discussion 666
- References 668

38.1 Introduction

The reciprocal interaction between neurotransmitters and immune-active molecules (such as infectious agents and their components and cytokines) represents one possible mechanism mediating the bi-directional communication between the central nervous system (CNS) and the immune system. Data reviewed in this chapter show that the activation of the immune system induces profound alterations in the CNS and in several neurotransmitters at this level, as well as modifications in food and water intake, locomotion, sexual behavior, social interactions, and sleep.[36,53,56] These behavioral effects may be achieved through, and accounted for by, the changes induced in central neurotransmitters by immune challenges. The present review will summarize the available evidence of the reciprocal interactions between immune-active molecules and neurotransmitters in the CNS and will discuss the possible relevance of these interactions in the regulation of the sleep/wake cycle.

38.2 Interactions Between Immune Molecules and Neurotransmitters

38.2.1 Noradrenaline

An early report by Besedovsky[9] showed that injection of antigens such as sheep red blood cells (SRBC) or of supernatant from activated immunological cells affects noradrenaline (NA) content in the hypothalamus of rats. Since then, the observations that the febrile response induced by polyI:polyC (a synthetic double-stranded RNA, an interferon-inducer mimicking viral symptoms)[70] and adrenocortical activation induced by herpes virus[4] are prevented by NA depletion indicate that the noradrenergic system is involved in mediating the central effect of different immune challenges. A variety of immune challenges have been shown to activate the brain NA system. Besides SRBC,[9,110,127] these challenges include viruses such as Newcastle disease virus,[22] bacteria and their constituents such as lipopolysaccharide (LPS, the biologically active component of Gram-negative bacterial cell wall endotoxin), and cytokines. LPS is a potent stimulator of the immune response, eliciting many of the symptoms (including cytokine synthesis and sleep responses) associated with Gram-negative bacterial infections.[23]

Among immune challenges that affect noradrenergic activity, LPS and interleukin-1 (IL-1) have been the most thoroughly investigated. Dunn[21] showed that LPS activates the noradrenergic system in different brain areas, although the response is largest in the hypothalamus. Studies by Linthorst et al.[67,68] indicate that NA release in the preoptic area and hippocampus increases when LPS (100 µg) is given intraperitoneally (IP). Peripheral administration of IL-1α or β also activates the noradrenergic system in different brain areas such as the hypothalamus and hippocampus,[19,46,77,80,113,126] 10 ng being the minimal effective dose.[21]

As for central administrations, 10 ng of IL-1 injected intracerebroventricularly (ICV) into freely moving rats increases NA release in the hypothalamic paraventricular nucleus (PVN).[118] Local injections in the anterior hypothalamus (1 ng per rat) are also effective, suggesting an action on noradrenergic terminals in this area.[111]

Lipopolysaccharide action on the noradrenergic system is partially reduced by indomethacin, an inhibitor of the cyclo-oxygenase pathway and of prostaglandin (PG) synthesis, suggesting that PGs are also involved in this effect. On the other hand, as both peripheral and central IL-1 actions on the noradrenergic system are completely prevented by indomethacin,[116,118] it is likely that IL-1 action on this neurotransmitter system is mediated by PG, specifically PGD_2.[116]

Interleukin-1-enhanced secretion of NA is thought to mediate cytokine-induced activation of the hypothalamus-pituitary-adrenal (HPA) axis.[19,76,118] In fact, the hypothalamic NA system plays a major role in the control of the HPA axis.[93,119] In particular, corticotropin-releasing hormone (CRH) neurons located in the hypothalamic paraventricular nucleus (PVN) receive a dense innervation originating from the noradrenergic groups in the brainstem. NA release at this level triggers HPA axis activation, ultimately leading to adrenal glucocorticoid release. Accordingly, neurotoxic lesions of the ventral noradrenergic ascending bundle prevent the activation of the HPA axis induced by IP IL-1β,[92] and the depletion of cerebral NA substantially reduces the IL-1-induced Fos increase in the PVN.[115] Taken together, these data suggest that catecholaminergic innervation of the PVN plays a major role in the HPA axis activation induced by peripheral immune stimuli. Antisera against PGE_2, PGE_1, and $PGF_2\alpha$ prevent IL-1-induced ACTH activation; however, whether these PGs activate the NA system is not yet clear.[116,122]

Few data are available regarding action of other cytokines on noradrenergic system. Tumor necrosis factor (TNF) exerts a presynaptic inhibitory control on NA release in rat hippocampal brain slices[40] and in the rat isolated median eminence, an action thought to be involved in the regulation of CRH release.[24] However, no change in NA turnover in the rat hypothalamus is detected after either IP or ICV TNFα[117] or in the mouse brain after IV TNFα or β.[20] Interleukin-2 (IL-2) inhibits

K+-evoked NA release in hypothalamic slices,[61] whereas an increase in NA metabolite MHPG (3-methoxy-4-hydroxyphenylglycol) in response to IL-2 has been observed *in vivo*.[126]

The interactions between cytokines and NA have a second facet, as not only can cytokines act on NA, but the converse is also true. Tringali et al.[121] have shown that IL-1β can be released by NA from neurons in hypothalamic explants, an effect antagonized by the β-blocker propranolol. NA also induces both interleukin-6 (IL-6) mRNA and protein in primary neonatal rat astrocytes and synergizes with IL-1β and TNFα in inducing this effect.[73,88] Stimulation of NAα$_2$ presynaptic receptors decreases constitutive TNFα in different brain areas.[40]

38.2.2 Dopamine

Dopamine (DA) release in medial hypothalamus and prefrontal cortex of freely moving rats increases in response to IP LPS, with a peak response at about 2 hr.[62] In contrast to LPS, inconsistent or no effects of systemic IL-1 are reported on DA release or the levels of the DA metabolite DOPAC (dihydroxyphenilacetic acid) in different brain areas,[19,23,124] although DA increases in the prefrontal cortex of mice following IP IL-1.[126] On the other hand, central IL-1 is effective in releasing DA. In fact, IL-1 enhances DA release when infused through a push-pull cannula into the medial basal hypothalamus[81] or injected in small amounts (1 ng per rat) into the anterior hypothalamus.[111] Thus, it is possible that the effects of IP LPS are mediated by LPS-induced IL-1 in the brain; careful time-course studies are needed to verify this possibility.

Several reports have evaluated the effect of IL-2 on DA release *in vitro*, but data available are not consistent (see Reference 33 for a review). *In vivo*, IL-2 decreases DA efflux in the nucleus accumbens and reduces responding for rewarding lateral hypothalamic stimulation.[2] IL-6[126] and TNF[18] increase DOPAC concentration in prefrontal cortex and tuberculus olfactorium, indicating an activation of the dopaminergic system in these areas.

38.2.3 Serotonin

Muramyl peptides (MPs) are the monomeric building blocks of bacterial cell wall peptidoglycans and are released by mammalian macrophages during the digestion of bacterial cell walls.[43] MPs are well-known somnogenic substances (see Reference 56 for a review); the administration of MPs tailored from bacterial cell walls induces slow-wave sleep (SWS) and fever.[43,58] The same effects are induced by the synthetic muramyl dipeptide (MDP; *N*-acetylmuramyl-L-alanyl-D-isoglutamine), which was originally characterized as the minimal component capable of replacing the mycobacteria of Freund's complete adjuvant.[25] Sleep factors isolated from brain tissue of sleep-deprived animals and urine have been identified as MPs.[59] There is evidence suggesting that MPs/MDP interact with the serotonergic system at multiple levels: (1) serotonin (5-HT) turnover is increased in response to MPs,[75] (2) MPs have specific binding sites on macrophages and glial cells and this binding can be competed for by 5-HT, and (3) MDP inhibits 5-HT uptake by human platelets, whose serotonergic system is commonly considered to resemble that of the CNS.[94] Different MPs/MDP effects may be mediated through interactions with the serotonergic system. The finding that MDP does not alter sleep/wake activity and brain cortical temperature when given to rats pretreated with the 5-HT-depleting drug *para*-chlorophenylalanine (PCPA) suggests that 5-HT is essential for MDP to exert these effects (Figure 38.1).[41] In addition, fever induced by bacterial pyrogens is attenuated if 5-HT synthesis is prevented, if the raphe nuclei are lesioned, if animals are pretreated with the 5-HT receptor antagonist cyproheptadine, or if whole brain 5-HT is depleted (see Reference 123 for a review).

Substantial increases in the 5-hydroxyindolacetic acid (5-HIAA, the 5-HT metabolite)/5-HT ratio occur in different brain areas in response to IP LPS.[21] Endotoxin increases 5-HT release in the

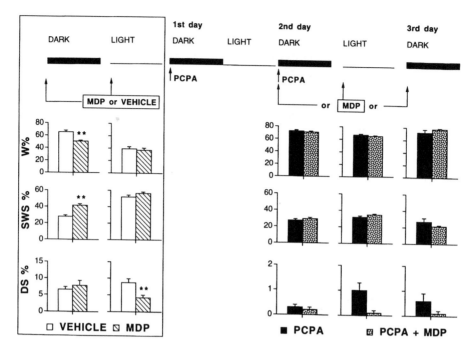

FIGURE 38.1
Histograms in the left box show the effects of the ICV administration of vehicle (pyrogen-free saline) and of muramyl dipeptide (MDP) at either dark (n = 6) or light (n = 5) onset on wakefulness (W), slow-wave sleep (SWS), and desynchronized sleep (DS). Data are expressed as percent of total recording time during 12-hr recordings. Control and test substances administered at time 0. Histograms in the same column refer to data from the same 12-hr time block of the dark/light cycle. Histograms on the right side of the figure show the effects of the administration of *para*-chlorophenylalanine (PCPA) and of MDP in PCPA-pretreated animals (PCPA + MDP). Substances were given according to the treatment protocol shown in the upper part of the figure. In particular, animals were given MDP 5 min after the second PCPA administration (2nd day, dark onset, n = 5), 12 hr after the second PCPA administration (2nd day, light onset, n = 6), or 24 hr after the second PCPA administration (3rd day, dark onset, n = 6). Histograms in the same column refer to data from the same 12-hr time block of the dark/light cycle. **$p < 0.01$ vs. vehicle.

hippocampus, an effect significantly attenuated by the IL-1 receptor antagonist and mimicked by ICV IL-1, indicating that this LPS effect is mediated, in part, by central IL-1 receptors.[69]

IL-1 activates the serotonergic system (Figure 38.2) by enhancing 5-HT release, metabolism,[28,29,66,111] and turnover,[21,77,126] as well as by raising brain tryptophan levels,[21] suggesting multiple levels of interaction. A wide range of IL-1β doses has been injected IP or ICV; the dose-response curve is U-shaped: low doses (5 to 50 ng/rat ICV) increase 5-HIAA,[28] whereas mid-range doses (200 to 400 ng) are not effective. However, higher doses (1 µg/rat ICV) do elicit this response. IL-1-induced 5-HIAA increase is completely abolished by the IL-1 receptor antagonist (40 µg/rat ICV) showing that even with this high dose the action is specific and is mediated by the IL-1 receptor.[29] Local infusions of IL-1β by microdialysis probe in the rat anterior hypothalamus[111] or hippocampus[66] increase 5-HT release, indicating that these areas may be important for IL-1-mediated effects on the serotonergic system.

Central IL-1 alters sleep and induces fever, anorexia, and activation of the HPA axis; 5-HT plays a role in each of these processes. These observations suggest that this cytokine may exert some of its effects by interacting with the serotonergic system.[18] Consistent with this hypothesis, the first phase (but not the second one) of the biphasic SWS increase induced in rats by ICV IL-1 (2.5 ng) is associated with an early increase in serotonergic activity in the medial preoptic area (MPA),[30] and

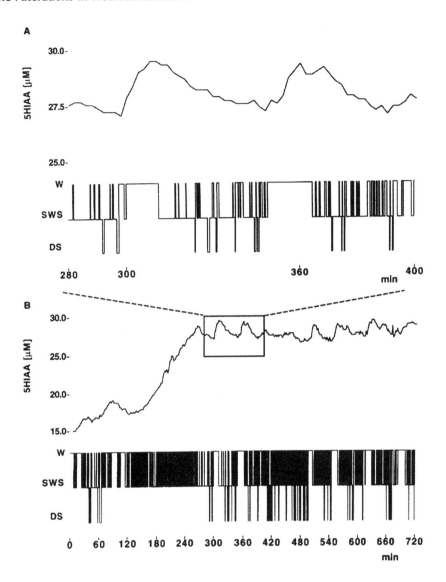

FIGURE 38.2
(A), (B) Typical changes in 5-HIAA concentrations (upper trace) from a single representative freely behaving rat induced by ICV injection of IL-1 (2.5 ng) and the simultaneous determination of the sleep/wake cycle (lower trace). Injection performed at time 0. (A) shows in detail a 2-hr period from the entire recording. Please note the phasic, state-specific changes in 5-HIAA levels, which are superimposed on the overall, tonic increase in 5-HIAA activity. W = wakefulness, SWS = slow wave sleep, DS = desynchronized sleep, 5HIAA: 5-hydroxyindolacetic acid. (From Gemma C. et al., *Am. J. Physiol.*, 272, R601–R606, 1997. With permission.)

it is abolished in PCPA-pretreated animals.[41] These findings also suggest that the mechanisms mediating the first and second phases of IL-1-induced SWS excess are different: 5-HT could be involved in the first phase, but not in the second one. On the other hand, PCPA pretreatment does not modify the febrile response to IL-1, suggesting that 5-HT is not involved in IL-1-induced fever.[41]

In freely behaving animals, serotonergic activity exhibits state-specific changes; it increases during wakefulness and decreases during sleep.[15,42,96] The close correlation between 5-HT levels and behavioral activity appears to be disrupted in LPS- and IL-1-treated animals, which show a decrease in behavioral activity associated with increased hippocampal 5-HT release.[66,69] However, when a detailed time-course of 5-HT activity is obtained in the MPA every 2 min and the animal's activity is polygraphically defined, 5-HT phasic state-specific changes are superimposed on the tonic, overall increase in 5-HT activity induced by IL-1 (Figure 38.2). The persistence of these phasic fluctuations suggests that IL-1, at least in the MPA, does not alter the normal physiology of the serotonergic system, whose tonic activity, in this condition, is just set at a higher level.[30]

The interactions between IL-1 and 5-HT have also been explored from other points of view. At variance with all the data showing that IL-1 increases 5-HT release and activity, an upregulation of the 5-HT transporter gene expression by IL-1 has also been observed.[103] This would result in an enhanced clearance of 5-HT at the synaptic cleft. Kugaia has shown that IL-1 (as well as LPS) inhibits intracellular Ca^{++} mobilization induced by $5-HT_2$ receptor stimulation, showing the existence of an inhibitory interaction between IL-1 and the $5-HT_2$ pathway.[60] Taken together, studies of IL-1 effects on the serotonergic system indicate that, although the cytokine enhances 5-HT release, interactions between IL-1 and 5-HT also occur at other levels, resulting in a different and more complicated net effect.

IL-6 has been reported to affect the serotonergic system and in particular to increase brain tryptophan in different brain areas.[23,126] Very few studies have evaluated TNFα or interferon-α (IFNα) action on the serotonergic system,[20,77] and no significant effects have been reported.

Little consideration has been given to the influence of the serotonergic system on brain cytokines. 5-HT stimulates IL-1 release in fresh rat brain preparations.[112] 5-HT does not affect IL-1α mRNA expression in hippocampal astrocytes but is effective in stimulating IL-6 and TNFα mRNA expression.[97]

38.2.4 Acetylcholine

As far as sleep is concerned, ACh has been shown to play a role in cortical desynchronization and in the orchestration of REM sleep.[44,120] Systemically injected IL-1β reduces ACh release in the rat hippocampus in a dose-related manner.[102] IL-2 also appears to affect this transmitter; K^+-stimulated, although not basal, release of ACh decreases in hippocampal and cortical, but not striatal, slices incubated with nanomolar concentrations of IL-2, showing a selective action depending on the area.[3,34] However, at very low, picomolar, concentrations, IL-2 acts as a stimulant of hippocampal ACh release.[34]

The central cholinergic system inhibits the peripheral immune response. Inhibition of brain ACh biosynthesis enhances, whereas inhibition of acetylcholinesterase (AChE) suppresses, the antibody response of rats challenged with SRBC. During the period when antibody responses to SRBC are diminished (i.e., 3 to 6 days after the challenge), AChE activity is reduced in the hypothalamus and hippocampus.[99]

38.2.5 GABA

Interleukin-1, but not TNF, increases $GABA_A$ receptor function in synaptosomes derived from mouse cortex by augmenting Cl^- uptake.[79] It has been suggested that the inhibitory effects of IL-1 on supraoptic neurons are mediated by activation of local GABAergic interneurons.[64] IL-1 also enhances GABAergic inhibitory postsynaptic potentials in hippocampal neurons.[128] Altogether, these data indicate that IL-1 could enhance GABAergic-induced hyperpolarization and inhibition. IL-1 reduces the $GABA_A$ effect on Purkinje cells in rat cerebellar slices.[98] Although also in this case the net result is in an increase in the inhibitory output from the cerebellar cortex, it shows that the action of IL-1 on $GABA_A$ may be different in different brain areas.

38.2.6 Glutamic Acid

The brain glutamatergic system is affected by peripheral immune stimuli. IP LPS and IL-1β enhance glutamic acid (GLU) release in the nucleus tractus solitarius, which receives vagal afferents, supporting the hypothesis that vagal sensitive afferent fibers may convey immune stimuli from the periphery to the CNS.[74] GLU, glutamine, and GABA levels decrease in hippocampal tissue following IP IL-1β but not IL-2.[10] This action may be due to an increased amino-acid release and consequent metabolization.

Many studies suggest an interaction of cytokines with the different glutamate receptors.[17,32,100,114] This may be important in brain injury and neurodegeneration. Depending on the experimental model used, either protective or toxic actions of cytokines are reported. Some studies indicate that inflammatory cytokines, including IL-1β, TNFα, and IFNγ, attenuate astrocytic glutamate uptake, resulting in an accumulation of extracellular GLU at excitotoxic concentrations. This effect involves interactions with NMDA (N-methyl-D-aspartic acid) receptor and the activation of a pathway that induces nitric oxide (NO).[16,125] As for a neuroprotective role, GLU dehydrogenase mRNA expression is increased by TNFα and IL-1α, indicating that these cytokines may contribute to a reduction in elevated extracellular GLU levels, thus preventing excitatory amino acid neurotoxicity.[35] IL-1 protective action can be also mediated by nerve growth factor.[114]

38.2.7 Histamine

Histaminergic neurons, located in the tuberomamillary nucleus and adjacent areas of the posterior hypothalamus and innervating several brain regions, are a major determinant of the waking state and modulate REM sleep.[65,83] Intravenous administration of LPS produces a rapid increase in the amount of histamine in the median eminence. This action seems to be involved in the early phase (<60 min) of LPS-induced activation of the HPA axis.[31] Central administration of IL-1β increases hypothalamic histamine.[48] Moreover, IL-1β-induced alterations in food and water intake are significantly modulated by depletion of neuronal histamine.[47] Changes in the histaminergic system may, in turn, influence TNFα production in the brain, in an area-specific fashion.[26]

38.2.8 Opiods

A few reports indicate that LPS and cytokines may increase the production of endogenous opioids in the neuroendocrine system. LPS, for instance, increases β-endorphin concentration in cerebrospinal fluid.[13] IL-1 enhances proopiomelanocortin gene expression in pituitary cells[12] and μ-opioid receptor mRNA expression in astrocytes.[105] TNFα increases proenkephalin expression in the same cells.[71] IFNα binds to central μ-opioid receptors, an action that may mediate central IFN effects such as inhibition of adrenocortical secretion[11,78,107] or induction of the early phase of fever.[86]

Brain opioids, in turn, modulate the immune response by regulating a variety of immune parameters, including cytokines. In particular, the different opioid receptor subtypes in the brain control and finely tune specific aspects of the immune response.[6,7,14]

38.2.9 Nitric Oxide

There is increasing evidence that nitric oxide (NO), a free radical that can act both as a signaling molecule and a neurotoxin, is involved in multiple aspects of pathophysiology in the brain.[37,39] Because LPS, IL-1, TNFα, and IFNγ induce NO synthase (NOS),[54,55,63,101,104,106] the rate-limiting enzyme in NO production, it has been suggested that enhanced NO activity may be one of the mechanisms mediating either physiological or pathological cytokine actions in the brain. We shall only mention here that most studies focus on NO as a possible mediator of neurotoxic effects induced by enhanced cytokine production in neuropathological conditions. However, cytokine-NO interactions may play a role in regulating normal brain functions, such as sleep.[49,51,52] This issue will be addressed in the discussion. Different lines of evidence suggest that NO plays an important function in mediating the HPA axis response to an immune challenge: (1) NOS has been described in neurons that regulate ACTH secretion,[63] (2) systemic LPS increases mRNA levels for this enzyme in the PVN,[63] and (3) CRH release induced in an *in vitro* hypothalamic preparation can be antagonized by NO inhibitors.[106] IL-2-induced CRH release from the amygdala is also mediated by NO.[101]

38.2.10 Adenosine

Adenosine has recently been proposed as a physiological sleep factor that accumulates in brain during wakefulness (due to increased metabolic activity) and promotes subsequent sleep.[5,95] Although interactions between adenosine and cytokines have been described outside the CNS, data available on these interactions in the CNS are very scanty. Adenosine has been reported to increase IL-6 in astroglioma cells.[27]

38.3 Discussion

The data reviewed in this chapter indicate that important and reciprocal interactions occur within the CNS between neurotransmitters and immune-active molecules. The activity of several neurochemical systems is altered in response to immune challenge, and these alterations may contribute to the behavioral and sleep changes in response to immune activation. This discussion will focus on sleep as a behavioral endpoint and a model for the study of CNS-immune interactions.

The pattern of neurochemical changes induced in the CNS by an immune challenge is complex. The behavioral consequences of these neurochemical changes result from the integration of changes in multiple systems. These neurochemical changes may be subdivided into two different types: those that, in fact, directly drive the observed alterations in behavior and those that counteract the same effects and may mediate a return to homeostasis. The interactions between 5-HT and IL-1 may be useful to illustrate this hypothesis. There are also other reasons to start the discussion of this chapter from 5-HT/IL-1 interactions. 5-HT is one of the most extensively studied neurotransmitters with regard to sleep regulation.[1,44] Similarly, IL-1 has received the most attention of the cytokines in terms of its involvement in sleep regulation and in the alterations in sleep that occur during an immune challenge.[56]

Interleukin-1 induces SWS excess in several animal species,[56] and, as reviewed in this chapter, increases serotonergic activity. It has been proposed that 5-HT, which is released during wakefulness, is necessary for sleep, as it might induce the synthesis and/or release of hypnogenic factor(s) that would be secondarily responsible for sleep induction.[45] Consistent with this hypothesis[30] is the finding that in rats the first phase of the biphasic SWS excess induced by IL-1 is associated with an early increase in serotonergic activity in the MPA and this first phase is abolished following brain serotonin depletion.[41] On the other hand, neurophysiological, neurochemical, and pharmacological data suggest that 5-HT promotes wakefulness per se and may be considered a sleep-inhibiting agent; for instance, the spontaneous activity of the 5-HT system increases with wakefulness and decreases with sleep, the electrical stimulation of serotonergic dorsal raphe cells enhances wakefulness, and blockade of 5-HT$_2$ receptors increases SWS in both rats and humans (see Reference 18 for a review). If this is indeed the case, 5-HT could play a more complex role in mediating IL-1 effects on sleep/

wake activity, and it could also be considered part of those negative feed-back/feed-forward mechanisms which are activated by IL-1 to counteract its own effects and maintain homeostasis. For instance, IL-1 activates the HPA axis, and different molecules of this system inhibit IL-1 production (feed-back mechanism) and/or actions (feed-forward mechanism).[8,57,89] The IL-1-induced activation of the serotonergic system, which is long lasting and continues through the second phase of IL-1-induced SWS enhancement,[30] could be regarded as part of these homeostatic mechanisms. According to this hypothesis, part of the IL-1-induced SWS excess (i.e., the second phase) would occur not through the activation of the serotonergic system, but in spite of it, and SWS enhancement would be the result of the balance between different (in some cases, opposing) stimuli. Different mechanisms could promote the second phase of IL-1-induced SWS excess.[30,41,56,89] As discussed below, these mechanisms may include some of the other complex changes induced by IL-1 in different brain neurochemical systems.

The observations that IL-1, while activating the serotonergic system, also enhances 5-HT clearance from the synaptic cleft[103] and antagonizes some effects induced by the stimulation of the 5-HT$_2$ receptors[60] indicate that IL-1 affects the serotonergic system at multiple levels. IL-1, through these mechanisms (which could prevail at different time points or in different brain structures) could regulate (in time or extent) its own-induced activation of the serotonergic system.

Interactions between IL-1 and 5-HT also have a second facet, as not only does IL-1 affect 5-HT, but 5-HT also stimulates IL-1 release *in vitro*.[112] IL-1 could therefore be considered one of the sleep factors through which it has been proposed that 5-HT exerts its hypnogenic effects.[45] It is worth noting that IL-1 and 5-HT concentrations in the CNS appear to alternate during the different phases of the sleep/wake cycle, IL-1 levels being high during sleep and low during W, and 5-HT activity following the opposite pattern.[72,82,92]

The observation that some cytokines, such as IL-6, do not affect the sleep/wake cycle[91] but activate the serotonergic system, whereas others, such as TNF, promote SWS[50] but do not activate the serotonergic system, indicates that behavioral and neurochemical modifications induced by cytokines may be dissociated. This dissociation underlines the fact that the net behavioral effects of an immune challenge are not dependent on the activation of a single neurochemical system.

Because catecholamines (DA, NA), as well as histamine, mainly promote wakefulness[18,44,83] but are activated by many immune challenges which facilitate SWS, they could be considered part of those mechanisms aimed to maintain homeostasis. Other neurochemical changes induced by an immune challenge can directly account for the sleep modifications observed in this situation. For instance, an attractive suggestion about the possible mechanisms mediating the hypnogenic effects of IL-1 concerns the involvement of GABA. As the activation of central GABAergic inhibitory neurotransmission is thought to represent the cellular mechanisms of action of hypnotics of the benzodiazepine class,[85] IL-1-induced activation of the GABAergic system and potentiation of GABA effects could explain IL-1 effects on sleep, although *in vivo* data are needed to establish the importance of this action. Also, in this case, the observation that TNF does not affect GABA transmission indicates that hypnogenic cytokines act through different mechanisms. The inhibitory effects on ACh release by IL-1 (which have thus far been studied only in the hippocampus) could be relevant in sleep regulation. IL-1 inhibits desynchronized sleep. Because desynchronized sleep is facilitated by the cholinergic system,[18,44] this action of IL-1 may reflect a decrease in ACh transmission. The observation that IL-2 (whose somnogenic properties are controversial,[87,89,90]) may increase or decrease ACh, depending on the concentration used, suggests that different doses of each cytokine should be tested before any conclusion are drawn. There is evidence that NO (which is activated by different cytokines) could be involved in normal sleep regulation and in mediating sleep changes induced by IL-1. In fact, inhibition of NO production inhibits normal sleep and IL-1-induced sleep enhancement.[49,51,52]

Most studies of the interactions between neurotransmitters and cytokines (and other immune-active molecules) have used acute exogenous administration of cytokines in amounts and through routes of administration that could be far from normal and may not reflect a physiological condition. These manipulations may modify the normal interplay among the molecules involved.

The central interactions between immune-active molecules and neurotransmitters can be relevant to many functions regulated by the CNS and under many pathological conditions. These interactions should be thoroughly investigated. Moreover, because at least some cytokines are constitutively expressed in normal brain[38,56,108] and muramyl peptides are isolated from normal brain tissue of sleep-deprived animals,[59] the possible role and relevance of interactions between these immune-active compounds and central neurotransmitters in the physiology of CNS functioning and behavioral state control deserve further investigation. There are many questions to be addressed by future studies. For example, although IL-1 mRNA, IL-1 receptors, and neurons immunoreactive for IL-1 have been described in different brain areas and in different animal species,[57,89,109] to our knowledge there is no information on possible co-localization with classical neurotransmitters and/or their receptors. While the study of the effects of cytokines and other immune-active molecules on brain neurotransmitters has begun, very little is known about the reciprocal interaction of neurotransmitters on cytokines (especially in the brain) and the immune response. The brain levels of many neurotransmitters show state-specific changes (i.e., they change in relation to the different phases of the sleep/wake cycle). Although it has been shown that the pharmacological manipulation of some central neurochemical systems may affect the immune response, the possible relevance of physiological state-dependent neurochemical changes in modulating the immune system is unknown. Different and specific effects can be achieved through the interaction of the same neurotransmitter with multiple receptor subtypes specifically distributed in different brain areas, leading to a fine tuning of physiological functions. Whether and how receptor subtypes mediate the interactions of neurotransmitters with cytokines has not been investigated.

References

1. Adrien, J.,The serotoninergic system and sleep/wakefulness regulation, in *The Pharmacology of Sleep*, Kales, A., Ed., Springer-Verlag, Berlin, 1995, p. 91.
2. Anisman, H., Kokkinidis, L., and Merali, Z., Interleukin-2 decreases accumbal dopamine efflux and responding for rewarding lateral hypothalamic stimulation, *Brain Res.*, 731, 1, 1996.
3. Araujo, D.M., Lapchak, P.A., Collier, B., and Quirion, R., Localization of interleukin-2 immunoreactivity and interleukin-2 receptors in the rat brain: interaction with the cholinergic system, *Brain Res.*, 498, 257, 1989.
4. Ben Hur, T., Rosenthal, J., Itzik, A., and Weidenfeld, J., Adrenocortical activation by herpes virus: involvement of IL-1β and central noradrenergic system, *NeuroReport*, 7, 927, 1996.
5. Benington, J.H. and Heller, H.C., Restoration of brain energy metabolism as the function of sleep, *Prog. Neurobiol.*, 45, 347, 1995.
6. Bertolucci, M., Perego, C., and de Simoni, M.G., Central opiate modulation of peripheral IL-6 in rats, *NeuroReport*, 7, 1181, 1996.
7. Bertolucci, M., Perego, C., and de Simoni, M.G., Interleukin-6 is differently modulated by central opioid receptor subtypes, *Am. J. Physiol.*, 273, R95, 1998.
8. Besedovsky, H., Del Rey, A., Sorkin, E., and Dinarello, C.A., Immunoregulatory feedback between interleukin-1 and glucocorticoid hormones, *Science*, 233, 652, 1986.
9. Besedovsky, H.O., Del Rey, A., Sorkin, E., Da Prada, M., Burri, R., and Honegger, C., The immune response evokes changes in brain noradrenergic neurons, *Science*, 221, 564, 1983.
10. Bianchi, M., Ferrario, P., Zonta, N., and Panerai, A.E., Effects of interleukin-1β and interleukin-2 on amino acids levels in mouse cortex and hippocampus, *NeuroReport*, 6, 1689, 1995.

11. Blalock, J.E. and Smith, E.M., Human leukocyte interferon (HuIFN-α):potent endorphin-like opioid activity, *Biochem. Biophys. Res. Comm.*, 101, 472, 1981.
12. Brown, S.L., Smith, L.R., and Blalock, J.E., Interleukin-1 and interleukin-2 enhance proopiomelanocortin gene expression in pituitary cells, *J. Immunol.*, 139, 3181, 1987.
13. Carr, D.B., Bergland, R., Hamilton, A., Blume, H., Kasting, N., Arnold, M., Martin, J.B., and Rosenblatt, M., Endotoxin-stimulated opioid peptide secretion: two secretory pulse and feedback control *in vivo, Science*, 217, 845, 1982.
14. Carr, D.J.J., The role of endogenous opioids and their receptors in the immune system, *Proc. Soc. Exp. Biol. Med.*, 198, 710, 1991.
15. Cespuglio, R., Sarda, N., Gharib, A., Chastrette, N., Houdouin, F., Rampin, C., and Jouvet, M., Voltammetric detection of the release of 5-hydroxyindole compounds throughout the sleep/waking cycle of the rat, *Exp. Brain Res.*, 80, 121, 1990.
16. Chao, C.C., Hu, S., Ehrlich, L., and Peterson, P.K., Interleukin-1 and tumor necrosi factor-alpha synergistically mediate neurotoxicity: involvement of nitric oxide and *N*-methyl-D-aspartate receptors, *Brain Behav. Immun.*, 9, 355, 1995.
17. Cheng, B., Christakos, S., and Mattson, M.P., Tumor necrosis factors protect neurons against metabolic-excitotoxic insults and promote maintenance of calcium homeostasis, *Neuron*, 12, 139, 1994.
18. de Simoni, M.G., Imeri, L., De Matteo, W., Perego, C., Simard, S., and Terrazzino, S., Sleep regulation: interactions among cytokines and classical neurotransmitters, *Adv. Neuroimmunol.*, 5, 189, 1995.
19. Dunn, A.J., Systemic interleukin-1 administration stimulates hypothalamic norepinephrine metabolism paralleling the increased plasma corticosterone, *Life Sci.*, 43, 429, 1988.
20. Dunn, A.J., The role of interleukin-1 and tumor necrosis factor α in the neurochemical and neuroendocrine responses to endotoxin, *Brain Res. Bull.*, 29, 807, 1992.
21. Dunn, A.J., Endotoxin-induced activation of cerebral catecholamine and serotonin metabolism: comparison with interleukin-1, *J. Pharmacol. Exp. Ther.*, 261, 964, 1992.
22. Dunn, A.J. and Vickers, S.L., Neurochemical and neuroendocrine responses to Newcastle dsease virus administration in mice, *Brain Res.*, 645, 103, 1994.
23. Dunn, A.J. and Wang, J., Cytokine effects on CNS biogenic amines, *Neuroimmunomodulation*, 2, 319, 1996.
24. Elenkov, I.J., Kovaks, K., Duda, E., Stark, E., and Vizi, E.S., Presynaptic inhibitory effect of TNF-α on the release of noradrenaline in isolated median eminence, *J. Neuroimmunol.*, 41, 117, 1992.
25. Ellouz, F., Adam, A., Ciorbaru, R., and Lederer, E., Minimal structural requirements for adjuvant activity of bacterial peptidoglycan derivatives, *Biochem. Biophys. Res. Commun.*, 59, 1217, 1987.
26. Fernandez-Novoa, L., Franco-Maside, A., Alvarez, X., and Cacabelos, R., Effects of histamine and alpha-fluoromethylhistidine on brain tumor necrosis factor levels in rats, *Inflamm. Res.*, 44, 55, 1995.
27. Fiebich, B.L., Biber, K., Gyufko, K., Berger, M., Bauer, J., and van Calker, D., Adenosine A_{2B} receptors mediate an increase in interleukin (IL)-6 mRNA and IL-6 protein synthesis in human astroglioma cells, *J. Neurochem.*, 66, 1426, 1996.
28. Gemma, C., De Luigi, A., and De Simoni, M.G., Permissive role of glucocorticoids on interleukin-1 activation of the hypothalamic serotonergic system, *Brain Res.*, 651, 169, 1994.
29. Gemma, C., Ghezzi, P., and De Simoni, M.G., Activation of the hypothalamic serotoninergic system by central interleukin-1, *Eur. J. Pharmacol.*, 209, 139, 1991.
30. Gemma, C., Imeri, L., de Simoni, M.G., and Mancia, M., Interleukin-1 induces changes in sleep, brain temperature and serotonergic metabolism, *Am. J. Physiol.*, 272, R601, 1997.

31. Givalois, L., Siaud, P., Mekaouche, M., Ixart, G., Malaval, F., Assenmacher, I., and Barbanel, G., Involvement of central histamine in the early phase of ACTH and corticosterone responses to endotoxin in rats, *Neuroendocrinology*, 63, 219, 1996.
32. Hagan, P., Poole, S., Bristow, A.F., Tilders, F., and Silverstein, F.S., Intracerebral NMDA injection stimulates production of interleukin-1 beta in perinatal rat brain, *J. Neurochem.*, 67, 2215, 1996.
33. Hanisch, R. and Quirion, R., Interleukin-2 as a neuroregulatory cytokine, *Brain Res. Rev.*, 21, 246, 1996.
34. Hanisch, U.K., Seto, D., and Quirion, R., Modulation of hippocampal acetylcholine release: a potent central action of interleukin-2, *J. Neurosci.*, 13(8), 3368, 1993.
35. Hardin-Pouzet, H., Giraudon, P., Bernard, A., Derrington, E., Belin, M.F., and Didier-Bazes, M., Cytokines are increased in the rat hippocampus after serotonergic neuron degeneration and upregulate the expression of GDH, an enzyme involved in glutamate detoxification, *J. Neuroimmunol.*, 69, 117, 1996.
36. Hart, B.L., Biological basis of the behavior of sick animals, *Neurosci. Biobehav. Rev.*, 12, 123, 1988.
37. Holscher, C., Nitric oxide, the enigmatic neuronal messanger: its role in synaptic plasticity, *Trends Neurosci.*, 20, 298, 1997.
38. Hopkins, S.J. and Rothwell, N.J., Cytokines and the nervous system I: expression and recognition, *Trends Neurol. Sci.*, 18, 83, 1995.
39. Iadecola, C., Bright and dark sides of nitric oxide in ischemic brain injury, *Trends Neurosci.*, 20, 132, 1997.
40. Ignatowski, T.A., Chou, R.C., and Spengler, R.N., Changes in noradrenergic sensitivity to tumor necrosis factor-α in brains of rats administered clonidine, *J. Neuroimmunol.*, 70, 55, 1996.
41. Imeri, L., Bianchi, S., and Mancia, M., Muramyl dipeptide and IL-1 effects on sleep and brain temperature following inhibition of serotonin synthesis, *Am. J. Physiol.*, 273, R166, 1998.
42. Imeri, L., De Simoni, M.G., Giglio, R., Clavenna, A., and Mancia, M., Changes in the serotonergic system during the sleep/wake cycle: simultaneous polygraphic and voltammetric recordings in hypothalamus using a telemetry system, *Neuroscience*, 58, 353, 1994.
43. Johannsen, L., Wecke, J., Obál, Jr., F., and Krueger, J.M., Macrophages produce somnogenic and pyrogenic muramyl peptides during digestion of staphylococci, *Am. J. Physiol.*, 260, R126, 1991.
44. Jones, B.E., Basic mechanisms of sleep/wake states, in *Principles and Practice of Sleep Medicine*, Kryger, M.H., Roth, T., and Dement, W.C., Eds., W.B. Saunders, Philadelphia, 1989, p. 121.
45. Jouvet, M., Sallanon, M., Petitjean, F., and Bobillier, P., Serotonergic and non-serotonergic mechanisms in sleep, in *Sleep Disorders: Basic and Clinical Research*, Gibson, C.J. and Chase, M.H., Eds., Spectrum, New York, 1983, p. 557.
46. Kabiersch, A., Del Rey, A., Honegger, C.G., and Besedovsky, H., Interleukin-1 induces changes in norepinephrine metabolism in the rat brain, *Brain Behav. Imm.*, 2, 267, 1988.
47. Kang, M., Yoshimatsu, H., Chiba, S., Kurokawa, M., Ogawa, R., Tamari, Y., Tatsukawa, M., and Sakata, T., Hypothalamic neuronal histamine modulates physiological responses induced by interleukin-1β, *Am. J. Physiol.*, 269, R1308, 1995.
48. Kang, M., Yoshimatsu, H., Ogawa, R., Kurokawa, M., Oohara, Y., Tamari, Y., and Sakata, T., Thermoregulation and hypothalamic histamine turnover modulated by interleukin-1β in rats, *Brain Res. Bull.*, 35, 299, 1994.
49. Kapás, L., Fang, J., and Krueger, J.M., Inhibition of nitric oxide synthesis inhibits rat sleep, *Brain Res.*, 664, 189, 1994.
50. Kapás, L., Hong, L., Cady, A.B., Opp, M.R., Postlethwaite, A.E., Seyer, J.M., and Krueger, J.M., Somnogenic, pyrogenic, and anorectic activities of tumor necrosis factor-α and TNF-α fragments, *Am. J. Physiol.*, 263, R708, 1992.

51. Kapás, L. and Krueger, J.M., Nitric oxide donors SIN-1 and SNAP promote nonrapid-eye-movement sleep in rats, *Brain Res. Bull.*, 41, 293, 1996.
52. Kapás, L., Shibata, M., Kimura, M., and Krueger, J.M., Inhibition of nitric oxide synthesis suppresses sleep in rabbits, *Am. J. Physiol.*, 266, R, 1994.
53. Kent, S., Bluthé, R., Kelley, K.W., and Dantzer, R., Sickness behavior as a new target for drug development, *Trends Neurosci.*, 13, 24, 1992.
54. Kifle, Y., Monnier, J., Chesrown, S.E., Raizada, M.K., and Nick, H.S., Regulation of the manganese superoxide dismutase an inducible nitric oxide synthase gene in rat neuronal and glial cells, *J. Neurochem.*, 66, 2128, 1996.
55. Korytko, P.J. and Boje, K.M., Pharmacological characterization of nitric oxide production in a rat model of meningitis, *Neuropharmacology*, 35, 231, 1996.
56. Krueger, J.M. and Majde, J.A., Microbial products and cytokines in sleep and fever regulation, *Crit. Rev. Immunol.*, 14, 355, 1994.
57. Krueger, J.M. and Obál, Jr., F., Sleep factors, in *Sleep and Breathing*, Saunders, N.A. and Sullivan, C.E., Eds., Marcel Dekker, New York, 1994, p. 79.
58. Krueger, J.M., Pappenheimer, J.R., and Karnovsky, M.L., Sleep-promoting effects of muramyl peptides, *Proc. Natl. Acad. Sci. USA*, 79, 6102, 1982.
59. Krueger, J.M., Pappenheimer, J.R., and Karnovsky, M.L., The composition of sleep-promoting factor isolated from human urine, *J. Biol. Chem.*, 257, 1664, 1982.
60. Kugaya, A., Kagaya, A., Uchitomi, Y., Motohashi, N., and Yamawaki, S., Inhibition of serotonin-induced Ca^{++} mobilization by interleukin-1β in rat C6BU-1 glioma cells, *Brain Res.*, 682, 151, 1995.
61. Lapchak, P.A. and Araujo, D.M., Hippocampal interleukin-2 regulates monoamine and opioid peptide release from the hippocampus, *NeuroReport*, 4, 303, 1993.
62. Lavicky, J. and Dunn, A.J., Endotoxin administration stimulates cerebral catecholamine release in freely moving rats as assessed by microdialysis, *J. Neurosci. Res.*, 40, 407, 1995.
63. Lee, S., Barbanel, G., and Rivier, C., Systemic endotoxin increases steady-state gene expression of hypothalamic nitric oxide synthase: comparison with corticotropin-releasing factor and vasopressin gene transcripts, *Brain Res.*, 705, 136, 1995.
64. Li, Z., Inenaga, K., and Yamashita, H., GABA-ergic inputs modulate effects of interleukin-1β on supraoptic neurones *in vitro*, *NeuroReport*, 5, 181, 1993.
65. Lin, J., Hou, W., Sakai, K., and Jouvet, M., Histaminergic descending inputs to the mesopontine tegmentum and their role in the control of cortical activation and wakefulness in the cat, *J. Neurosci.*, 16, 1523, 1996.
66. Linthorst, A.C.E., Flachskamm, C., Holsboer, F., and Reul, J.M.H.M., Local administration of recombinant human interleukin-1β in the rat hippocampus increases serotonergic neurotransmission, hypothalamic-pituitary-adrenocortical axis activity, and body temperature, *Endocrinology*, 135(2), 520, 1994.
67. Linthorst, A.C.E., Flachskamm, C., Holsboer, F., and Reul, J.M.H.M., Activation of serotonergic and noradrenergic neurotransmission in the rat hippocampus after peripheral administration of bacterial endotoxin: involvement of the cyclo-oxygenase pathway, *Neuroscience*, 72, 989, 1996.
68. Linthorst, A.C.E., Flachskamm, C., Holsboer, F., and Reul, M.H.M., Intraperitoneal administration of bacterial endotoxin enhances noradrenergic neurotransmission in the rat preoptic area: relationship with body temperature and hypothalamic-pituitary-adrenocortical axis activity, *Eur. J. Neurosci.*, 7, 2418, 1995.
69. Linthorst, A.C.E., Flachskamm, C., Müller-Preuss, P., Holsboer, F., and Reul, J.M.H., Effect of bacterial endotoxin and interleukin-1β on hippocampal serotonergic neurotransmission, behavioral activity, and free corticosterone levels: an *in vivo* microdialysis study, *J. Neurosci.*, 15, 2920, 1995.

70. Liu, H.J., Young, C.M., and Lin, M.T., Depletion of hypothalamic norepinephrine reduces the fever induced by polyriboinosinic acid:polyribocytidylic acid (Poly I:Poly C) in rats, *Experientia*, 45, 720, 1989.
71. Low, K.J., Allen, R.G., and Melner, R.H., Differential regulation of proenkephalin expression in astocytes by cytokines, *Endocrinology*, 131, 1908, 1992.
72. Lue, F.A., Bail, M., Jephenta-Ocholo, J., Carayanniotis, K., Gorczynski, R.M., and Moldofsky, H., Sleep and cerebrospinal fluid interleukin 1-like activity in the cat, *Int. J. Neurosci.*, 42, 179, 1988.
73. Maimone, D., Cioni, C., Rosa, S., Macchia, G., Aloisi, F., and Annunziata, P., Norepinephrine and vasoactive intestinal peptide induce IL-6 secretion by astrocytes: synergism with IL-1β and TNFα, *J. Neuroimmunol.*, 47, 73, 1993.
74. Mascarucci, P., Perego, C., Terrazzino, S., and de Simoni, M.G., Glutamic acid release in the nucleus tractus solitarius induced by peripheral lipopolysac and interleukin-1β, *Neuroscience*, in press.
75. Masek, K. and Kadlec, O., Sleep factor, muramyl peptides, and the serotonergic system, *Lancet*, i, 1277, 1983.
76. Matta, S.G., Singh, J., Newton, R., and Sharp, B., The adrenocorticotropin response to interleukin-1β instilled into the rat median eminence depends on the local release of catecholamines, *Endocrinology*, 127(5), 2175, 1990.
77. Mefford, I.N. and Heyes, M.P., Increased biogenic amine release in mouse hypothalamus following immunological challenge: antagonism by indomethacin, *J. Neuroimmunol.*, 27, 55, 1990.
78. Menziers, R.A., Patel, A., Hall, N.R.S., O'Grady, M.P., and Rier, S.E., Human recombinant interferon alpha inhibits naloxone binding to rat brain membranes, *Life Sci.*, 50, 227, 1992.
79. Miller, L.G., Galpern, W.G., Lumpkin, M., Chesley, S.F., and Dinarello, C.A., Interleukin-1 (IL-1) augments gamma-aminobutyric acid receptor function in the brain, *Mol. Pharmacol.*, 39, 105, 1991.
80. Mohankumar, P.S. and Quadri, S.K., Systemic administration of interleukin-1 stimulates norepinephrine release in the paraventricular nucleus, *Life Sci.*, 52, 1961, 1993.
81. Mohankumar, P.S., Thyagarajan, S., and Quadri, S.K., Interleukin-1 stimulates the release of dopamine and dihydroxyphenylacetic acid from the hypothalamus *in vivo*, *Life Sci.*, 48, 925, 1991.
82. Moldofsky, H., Central nervous system and peripheral immune functions and the sleep/wake system, *J. Psychiatr. Neurosci.*, 19, 368, 1994.
83. Monti, J.M., Pharmacology of the histaminergic system, in *Pharmacology of Sleep*, Kales, A., Ed., Springer, Berlin, 1995, p. 117.
84. Monti, J.M. and Jantos, H., Dose-dependent effects of the 5-HT$_{1A}$ receptor agonist 8-OH-DPAT on sleep and wakefulness in the rat, *J. Sleep Res.*, 1, 169, 1992.
85. Muller, W.E., Pharmacology of the GABAergic/benzodiazepine system, in *Pharmacology of Sleep*, Kales, A., Ed., Springer, Berlin, 1995, p. 211.
86. Nakasghima, T., Murakami, T., Murai, Y., Hori, T., Miyata, S., and Kiyohara, T., Naloxone suppresses the rising phase of fever induced by interferon-alpha, *Brain Res. Bull.*, 37, 61, 1995.
87. Nisticò, G., and De Sarro, G.B., Is interleukin-2 a neuromodulator in the brain?, *Trends Neurosci.*, 14, 146, 1991.
88. Norris, J.G. and Benveniste, E.N., Interleukin-6 production by astrocytes: induction by the neurotransmitter norepinephrine, *J. Neuroimmunol.*, 45, 137, 1993.
89. Opp, M.R., Kapás, L., and Toth, L.A., Cytokine involvement in the regulation of sleep, *Proc. Soc. Exp. Biol. Med.*, 201, 16, 1992.
90. Opp, M.R. and Krueger, J.M., Somnogenic actions of interleukin-2: real or artifact?, *Abstract*, 1995.
91. Opp, M.R., Obál, Jr., F., Cady, A.B., Johannsen, L., and Krueger, J.M., Interleukin-6 is pyrogenic but not somnogenic, *Physiol. Behav.*, 45, 1069, 1989.

92. Parsadaniantz, S.M., Gaillet, S., Malaval, F., Lenoir, V., Batsché, E., Barbanel, G., Gardier, A., Terlain, B., Jacquot, C., Szafarczyk, A., Assenmacher, I., and Kerdelhué, B., Lesions of the afferent catecholaminergic pathways inhibit the temporal activation of the CRH and POMC gene expression and ACTH release induced by human interleukin-1β in the male rat, *Neuroendocrinology*, 62, 586, 1995.

93. Plotsky, P.M., Cunningham, Jr., E.T., and Widmaier, E.P., Catecholaminergic modulation of corticotropin-releasing factor and adrenocorticotropin secretion, *Endocr. Rev.*, 10, 437, 1989.

94. Polanski, M. and Karnovsky, M.L., Serotonergic aspects of the response of human platelets to immune-adjuvant muramyl dipeptide, *J. Neuroimmunol.*, 37, 149, 1992.

95. Porkka-Heiskanen, T., Strecker, R.E., Thakkar, M., Biørkum, A.A., Greene, R.W., and McCarley, R.B., Adenosine: a mediator of the sleep-inducing effects of prolonges wakefulness, *Science*, 276, 1265, 1997.

96. Portas, C.M. and McCarley, R.W., Behavioral state-related changes of extracellular serotonin concentration in the dorsal raphe nucleus: a microdialysis study in the freely moving cat, *Brain Res.*, 648, 306, 1994.

97. Pousset, F., Fournier, J., Legoux, P., Keane, P., Shire, D., and Soubrie, P., Effect of serotonin on cytokine mRNA expression in rat hippocampal astrocytes, *Brain Res. Mol. Brain Res.*, 38, 54, 1996.

98. Pringle, A.K., Gardner, C.R., and Walker, R.J., Reduction of cerebellar $GABA_A$ responses by interleukin-1 (IL-1) through an indomethacin insensitive mechanism, *Neuropharmacology*, 35, 147, 1996.

99. Qiu, Y., Peng, Y., and Wang, J., Immunoregulatory role of neurotransmitters, *Adv. Neuroimmunol.*, 6, 223, 1996.

100. Qiu, Z., Parsons, K.L., and Groul, D.L., Interleukin-6 selectively enhances the intracellular calcium responses to NMDA in developing CNS neurons, *J. Neurosci.*, 15, 6688, 1995.

101. Raber, J., Koob, G.F., and Bloom, F.E., Interleukin-2 (IL-2) induces corticotropin-releasing factor (CRF) release from the amygdala and involves a nitric oxide-mediated signaling; comparison with the hypothalamic response, *J. Pharmacol. Exp. Ther.*, 272, 815, 1995.

102. Rada, P., Mark, G.P., Vitek, M.P., Mangano, R.M., Blume, A.J., Beer, B., and Hoebel, B.G., Interleukin-1β decreases acetylcholine measured by microdialysis in the hippocampus of freely moving rats, *Brain Res.*, 550, 287, 1991.

103. Ramamoorthy, S., Ramamoorthy, J.D., Prasad, P.D., Bhat, G.K., Mahesh, V.B., Leibach, F.H., and Ganapathy, V., Regulation of the human serotonin transporter by interleukin-1β, *Biochem. Biophys. Res. Commun.*, 216, 560, 1995.

104. Romero, L.I., Tatro, J.B., Field, J.A., and Reichlin, S., Roles of IL-1 and TNF alpha in endotoxin-induced activation of nitric oxide synthase in cultured rat brain cells, *Am. J. Physiol.*, 270, R326, 1996.

105. Ruzicka, B.B., Thomson, R.C., Watson, S.J., and Akil, H., Interleukin-1β mediated regulation of m-opioid receptors mRNA in primary astrocytes-enriched cultures, *J. Neurochem.*, 66, 425, 1996.

106. Sandi, C. and Guaza, C., Evidence for a role of nitric oxide in the corticotropin-releasing factor release induced by interleukin-1β, *Eur. J. Pharmacol.*, 274, 17, 1995.

107. Saphier, D., Welch, J.E., and Chuluyan, E., α-Interferon inhibits adrenocortical secretion via $μ_1$-opioid receptors in the rat, *Eur. J. Pharmacol.*, 236, 183, 1993.

108. Schöbitz, B., Holsboer, F., and De Kloet, E.R., Cytokines in the healty and diseased brain, *News Physiol. Sci.*, 9, 138, 1994.

109. Schöbitz, B., Ron De Kloet, E., and Holsboer, F., Gene expression and function of interleukin-1, interleukin-6 and tumor necrosis factor in the brain, *Prog. Neurobiol.*, 44, 397, 1994.

110. Shanks, N., Francis, D., Zalcman, S., Meaney, M.J., and Anisman, H., Alterations in central catecholamines associated with immune responding in adult and aged mice, *Brain Res.*, 666, 77, 1994.

111. Shintani, F., Kanba, S., Nakaki, T., Nibuya, M., Kinoshita, M., Suzuki, E., Yagi, G., Kato, R., and Asai, M., Interleukin-1β augments release of norepinephrine, dopamine, and serotonin in the rat anterior hypothalamus, *J. Neurosci.*, 13(8), 3574, 1993.

112. Silverman, D.H., Imam, K., and Karnovsky, M.L., Muramyl peptide/serotonin receptors in brain-derived preparations, *Pept. Res.*, 2(5), 338, 1989.

113. Smagin, G.N., Swiergiel, A.H., and Dunn, A.J., Peripheral administration of interleukin-1 increases extracellular concentrations of norepinephrine in rat hypothalamus: comparison with plasma corticosterone, *Psychoneuroendocrinol.*, 21, 83, 1996.

114. Strijbos, P.J.L.M. and Rothwell, N.J., Interleukin-1β attenuates excitatory amino acid-induced neurodegenration *in vitro:* involvement of nerve growth factor, *J. Neurosci.*, 15, 3468, 1995.

115. Swiergiel, A.H., Dunn, A.J., and Stone, E.A., The role of cerebral noradrenergic system in the Fos response to interleukin-1, *Brain Res. Bull.*, 41, 61, 1996.

116. Terao, A., Kitamura, H., Asano, A., Kobayashi, M., and Saito, M., Roles of prostaglandins D_2 and E_2 in interleukin-1-induced activation of norepinephrine turnover in the brain and peripheral organs of rats, *J. Neurochem.*, 65, 2742, 1995.

117. Terao, A., Oikawa, M., and Saito, M., Cytokine-induced change in hypothalamic norepinephrine turnover: involvement of corticotropin-releasing hormone and prostaglandins, *Brain Res.*, 662, 257, 1993.

118. Terrazzino, S., Perego, C., and De Simoni, M.G., Noradrenaline release in hypothalamus and ACTH secretion induced by central interleukin-1β, *NeuroReport*, 6, 2465, 1995.

119. Terrazzino, S., Perego, C., and De Simoni, M.G., Effect of development of habituation to restraint stress on hypothalamic noradrenaline release and adrenocorticortopic secretion, *J. Neurochem.*, 65, 263, 1995.

120. Tononi, G. and Pompeiano, O., Pharmacology of the cholinergic system, in *Pharmacology of Sleep*, Kales, A., Ed., Springer, Berlin, 1995, p. 143.

121. Tringali, G., Mancuso, C., Mirtella, A., Pozzoli, G., Parente, L., Preziosi, P., and Navarra, P., Evidence for the neuronal origin of immunoreactive interleukin-1β released by rat hypothalamic explants, *Neurosci. Lett.*, 219, 143, 1996.

122. Watanobe, H., Sasaki, S., and Takebe, K., Role of prostaglandins E_1, E_2 and $F_2\alpha$ in the brain interleukin-1β-induced adrenocorticotropin secretion in the rat, *Cytokine*, 7, 710, 1995.

123. Wilkinson, L.O., Auerbach, S.B., and Jacobs, B.L., Extracellular serotonin levels change with behavioral state but not with pyrogen-induced hyperthermia, *J. Neurosci.*, 11, 2732, 1991.

124. Yang, Z.J. and Meguid, M.M., Continuous systemic interleukin-1 alpha infusion suppresses food intake without increasing lateral hypothalamic dopamine activity, *Brain Res. Bull.*, 36, 417, 1996.

125. Ye, Z.C. and Sontheimer, H., Cytokine modulation of glial glutamate uptake: a possible involvement of nitric oxide, *NeuroReport*, 7, 2181, 1996.

126. Zalcman, S., Green-Johnson, J.M., Murray, L., Nance, D.M., Dyck, D., Anisman, H., and Greenberg, A.H., Cytocine-specific central monoamine alterations induced by interleukin-1, -2, and -6, *Brain Res.*, 643, 40, 1994.

127. Zalcman, S., Shanks, N., and Anisman, H., Time-dependent variations of central norepinephrine and dopamine following antigen administration, *Brain Res.*, 557, 69, 1991.

128. Zeise, M.L., Madamba, S., and Siggins, G.R., Interleukin-1β increases synaptic inhibition in rat hippocampal pyramidal neurons *in vitro*, *Regul. Peptides*, 39, 1, 1992.

Index

Index

A

AANAT, 46, 49–52
Aβ-fibers, 553, 554
Abstinence syndromes, 378, 379, 429, 491
N-Acetyl-aspartyl-glutamate (NAAG), 32, 46, 49–52
Acetylcholine, 312. *See also* Cholinergic neurons; Mesopontine cholinergic system
   basalo-cortical system cholinergic interactions, 222
   cortical activation function, 214
   cortical secretion in response to basal forebrain stimulation, 224
   cytokine and immune-active substance interactions, 664, 667
   input to REM sleep induction zone of pontine reticular formation, 162
   locus coeruleus response, 259
   mesopontine (LDT/PPT) neurons and, 169. *See also* Mesopontine cholinergic system
   monoaminergic neuron response, 196
   non-cholinergic mesopontine neurons and, 171
   pontine cholinergic system electrophysiological interactions, 284–285
   REM sleep and, 392. *See also* REM sleep; specific cholinergic systems
   state-related thalamic and reticular formation activity, 172–173
   suprachiasmatic nucleus afferents, 36
Acetylcholinesterase (AChE), 214, 392, 664
N-Acetyltransferase, 8
   pineal melatonin rhythm-generating enzyme (AANAT), 46, 49–52
Acidic fibroblast growth factor, 615
Activation-synthesis (A-S) hypothesis, 102, 109, 114
Activator protein-1 (AP-1), 236
Active sleep, 88, 89
   lumbar sensory neuron function, 534–537
   trigeminal sensory nuclear complex function, 526–530, 537
Acupuncture, 455, 456
A current, 167, 170
Addiction, 365–379. *See also* Drug addiction
Aδ-fibers, 476, 571, 572, 575
Adenosine, 610
   central hypersensitivity and, 559–561
   cytokine-induced expression, 615
   cytokine and immune-active substance interactions, 666
   endogenous somnogen, 311, 316–318
   hypothalamic model of state control, 318
   inflammation and, 560
   locus coeruleus response, 259
   medial preoptic area effects, 416
   mesopontine cholinergic system and, 170, 285, 290–291
   monoaminergic neuron inhibition, 206
   receptor pharmacology and distribution, 316–317
   spinal nociceptive function, 559–560
Adenosine kinase inhibitors, 560
Adenylate cyclase, 398
α_1-Adrenoceptors, raphe electrophysiology, 270
Afterhyperpolarization
   basal forebrain, 299, 302
   dorsal raphe nucleus, 265–266, 270
   locus coeruleus noradrenergic neurons, 259–260, 262
   pontine cholinergic system, 279–280
AIDS, 645
AIM model of sleep mentation, 109
Alcoholism. *See* Drug addiction; Ethanol
Alertness, drug effects
   caffeine, 424–425
   ethanol, 423, 428
   nicotine, 426–427, 428
Algodystrophy, 603
Allachesthesia, 595
Allodynia, 472, 506. *See also* Nociceptive pathways; Pain control
   adenosine receptor and, 560
   NMDA antagonists and, 558–559
   sensitization, 506, 510–512, 416
Alpha EEG oscillations
   dream recall and, 113
   marijuana effects, 435
   nicotine effects, 426
Alpha-fluoromethylhistidine, 616
Alpha herpesviruses, 18
Alphadolone, 509, 511
Alphaxalone, 509, 511

Alzheimer's disease, 239
Amino-3-hydroxy-5-methylisoxazole-4-propionic acid. *See* AMPA
4-Aminopyridine, 259
AMPA, 164, 220, 224, 262, 282, 557
Amphetamine, 369–370, 377
Amplified restriction fragment length polymorphism (AFLP), 79
Amygdala
   benzodiazepine effects, 411
   entorhinal sharp waves and, 353–354
   fear-associated antinociceptive response, 496
   lateral pontine projections, 393, 394–395
   REM sleep modulation, 396–398, 400
      pontine-amygdala interactions, 393–395, 400
   state-specific IEG expression, 247
Analgesia. *See* Pain control; specific analgesics
Analgesia center, 489. *See* Rostral ventromedial medulla
Anandamide, 434, 436–437
Anesthesia. *See also* Pain control
   GABA interactions, 512
   preparative effects on spinal sensory processing, 505–517. *See also* Spinal somatosensory systems: surgical/anesthetic electrophysiological effects
   rostral ventromedial medulla nociception-modulating cell response, 491, 495
   rostral ventromedial medulla bulbospinal cell response, 491
Angiotensin, 259
Angiotensin II, 16
Anterior cingulate cortex (ACC), 599, 600
Anterolateral cordotomy, 593–596, 600–601, 604
Antheraea pernyi, 65
Anti-convulsants, 554
Anti-opioid peptides, 552, 555–556
Antidepressants
   penile tumescence, 149
   REM sleep abnormality induction, 148
   sleep disorder treatment, 132
Antigravity muscle tone, 186
Antipyretics, 629, 633–634
   serotonin, 661
Anxiolytic effects of drugs, 372
Apamin, 260, 266, 270
Apex dreaming, 105, 108
Aplysia, 64
Arachidonylethanolamide, 436–437
Archetypal dreaming, 105
Arcuate nucleus, 629
Arginine vasopressin, 7, 36–37
   endogenous antipyretic function, 629
Arousal. *See* Waking state
Arousal disorders, state dissociation and NREM sleep variations, 148
Arousal threshold
   defining sleep states, 89
   monotremes, 94
   reptile sleep, 95–96
Arylalkylamine *N*-acetyltransferase (AANAT), 46, 49–52

Ascending reticular activating system. *See* specific neuroanatomical areas
Astrocytes
   mesopontine cholinergic neurons, 166
   neurotransmitter-cytokine interactions, 665
Atonia, 89
   cataplexy episodes in narcolepsy, 130
   non-atonic REM sleep, 394
   persistence of REM state (sleep paralysis), 130–131, 147
   REM sleep behavior disorder, 147–148
   serotonergic neurons and, 186
ATP, locus coeruleus response, 259
Atriopeptin, 165
Atropine, 164
Attention
   dorsal raphe serotonergic response, 187
   fast oscillations, 340
   marijuana effects, 435
   pain sensation and, 487
Autoimmunity, narcolepsy and, 135, 137
Automatic behaviors, 144, 400
Autonomic functions, 400
Autonomic nervous system, neurosurgical pain intervention, 603
Avian models
   circadian systems, 6
   melatonin rhythm-generating system, 54–55
   sleep, 87, 95

B

Bacterial infection, 627, 631, 660. *See also* Infection
   model systems, 641–642
Barbiturates, 372, 408, 410, 489, 491
   mechanisms of action, 409
   preoptic area lesioning study, 413–414
Basal forebrain, 218, 297–298
   adenosine and, 170, 317
   afterhyperpolarizations, 299, 302
   benzodiazepine action (medial preoptic area), 409, 410, 412–416
   c-Fos expression, 313
   cholinergic neurons, 1, 214–230, 297–298, 312
      adenosinergic interactions, 291
      brainstem afferents, 217–218
      cortical activation role, 214, 223–227
      discharge properties, 223
      distribution, 215
      electrophysiology, 297–305
      in vitro electrophysiology, 218–220
      monoaminergic interactions, 221–222
      pharmacological modulation, 220–222
      projections, 215–217
      stimulation and REM sleep abnormalities, 132
   drug reinforcement effects, 378–379
   efferent projections, 298
   monoaminergic afferents, 218
   non-cholinergic (possible GABAergic) neurons, 214–230, 297, 298

Index

electrophysiology, 302–305
reward system, 368
SCN afferents, 36
sleep-active and wake-active cells, 223, 312–313
state-specific IEG expression, 24
Basal forebrain-preoptic area. *See also* Preoptic area
   adenosine-induced responses, 317
   benzodiazepine action, 409, 410, 412–416
   state regulation model, 312–318
Basal gene expression, 237–239
Basalo-cortical system, 214–230. *See also* Basal forebrain
Basic fibroblast growth factor, 615
Basic helix-loop-helix (bHLH) transcription factors, 65, 77
Basket cells, 170
B-CCT, 416
BDNF, 36
Bed nucleus of stria terminalis, 203
Beetle model, 68
Behavioral state. *See* Active sleep; Non-REM sleep; REM sleep; Slow-wave sleep; Waking state
   conceptual development, 145–146
   dissociation. *See* Dissociated states
   multiple component perspective, 151
   normal state determination, 145
   ontogeny of, 150–151
Benzodiazepines, 34–35, 196, 372, 407–416
   functional neuroanatomy of state-altering effects, 409–412
      amygdala basomedial nucleus, 411
      diagonal band of broca, 412
      dorsal raphe nuclei, 411
      drug design implications, 416
      gigantocellular tegmental field, 411
      lateral preoptic area, 412
      locus coeruleus, 409, 411
      medial preoptic area, 409, 410, 412–416
      recovery sleep following preoptic area lesioning, 413–414
   mechanism of action, 408–409
   receptors, 408–409, 412, 414, 416
   wakefulness-enhancing enantiomer, 416
Beta EEG oscillations, nicotine effects, 426
Bicuculline, 34, 183, 196–197, 199–200, 271
Birds. *See* Avian models
Bizarreness ratings, 112, 115–116
Bladder distention, 514
Bladder inflammation, 558
Blind animal models, 8
Blood flow, cerebral, correlated with REM and dream recall, 400
Blood pressure, descending pain-modulating system and, 496–497
Body temperature. *See also* Brain temperature
   fever effects, 623–635. *See also* Fever
   preoptic area response and state alterations, 313
   regulation. *See* Thermoregulation
Bombesin, 166
Brachial plexus avulsion, 601
Brachium conjuntivum (BC), 393

Brain temperature. *See also* Fever; Thermoregulation
   rhythms, 625–626
   state-dependent alterations in thermoregulation, 624–626
   TNFα and, 611
Brainstem. *See also* specific areas
   basal forebrain afferents from, 217–218
   basal IEG expression, 238, 239
   cholinergic stimulation and REM sleep abnormalities, 132
   circadian IEG fluctuations, 240
   descending nociceptive modulatory systems, 487–497. *See also* Rostral ventromedial medulla
   glycinergic/GABAergic projections to monoaminergic neurons, 201–207
   nociceptive projections (to reticular areas), 575
   pontine cholinergic system (LDT/PPT). *See* Mesopontine cholinergic system
   SCN afferents, 36
   state dissociation induction, 146
   trigeminal sensory system. *See* Trigeminal sensory nuclear complex
Brewer's yeast, 642
Brown-Séquard syndrome, 570
Buprenorphine, 508

C

CA1, 349–353, 355. *See also* Hippocampal formation; Hippocampus
Caffeine, 270, 290–291, 316, 317, 421, 424–425, 429
Caimans, 96
Calbindin, 17
Calcitonin gene-related peptide (CGRP), 451, 470
Calcium-binding proteins, SCN neurons, 17
Calcium-dependent action potentials, locus coeruleus neurons, 262
Calcium channel-blocking analgesia, 554
Calcium channels, basal IEG expression and, 239
Calcium low-voltage current, basal forebrain activity, 299
Calretinin, 17
canarc-1, 136–137
Cancer pain treatment, 579, 601
Candida albicans, 641–642, 647
Canine model of narcolepsy. *See* Narcolepsy
Cannabinoids, 373–374, 435–436
   receptors, 434, 436–437
Capsaicin, lumbosacral neuron responses, 472, 473
Carbachol (and associated REM-like sleep states), 162, 165, 172, 392
   amygdala-cholinergic pons effects, 395–398
   cortical response to basal forebrain stimulation, 225
   mesopontine (LDT/PPT) electrophysiological response, 284–285
   muscarinic receptors and, 398
   PGO waves and, 392–393, 397–398
   postsynaptic inhibition of lumbar motoneurons, 535

rostral ventromedial medulla c-*fos* expression, 495
state-related motor effects, 187
state-specific cholinergic neuron response, 167
state-specific IEG expression patterns, 248
suprachiasmatic nucleus electrophysiology and, 36
Carbenoxolone, 264
5-Carboxamidotryptamine (5-CT), 283
Cardiovascular function, descending pain-modulating system and, 496–497
Cat models
chronic implant approach to brainstem unit recording, 523–524
REM sleep, 88
viruses, 642, 645
Cataplexy, 130, 147. *See also* Narcolepsy
treatment, 132
Causalgia, 603
Central hypersensitivity, targets for treatment of, 556–561
adenosine, 559–561
excitatory amino acids, 556–559
inhibitory systems, 559
nitric oxide, 561
Centralis superior, 182, 183
Cerebral cortex
activation. *See* Cortical activation
benzodiazepine receptors, 408
circadian IEG fluctuations, 240
descending modulation of spinothalamic tract, 476
descending somatosensory projections to trigeminal brainstem complex, 454
state-specific IEG expression, 241, 243–245, 247–248
Cerebrospinal fluid, accumulation of sleep-promoting substance, 609–610
C-fibers, 450, 476, 554, 571
c-*fos* (and c-Fos expression), 236–249, 311. *See also* Immediate early genes
anesthetic effects, 507
light-induced expression, 8, 32
sleep deprivation and, 313–314
state-specific expression and neuroanatomy, 237–249, 312–314
basal expression, 237–239
circadian fluctuations, 239–241
expression during sleep, 247–249
expression during spontaneous wakefulness, 241–247
neuronal activation marker, 236
null mouse model, REM sleep, 318
open questions, 249
rostral ventromedial medulla, 495
sleep deprivation and, 247
species specificity, 239
Chameleon, 95
Chelonians, 96
Chicken, 54–55, 64
Children's dreaming, 107, 109
α-Chloralose, 509, 511–512
Chloramphenicol, 147–148
Cholecystokinin (CCK), 393, 553, 555–556

Cholera toxin (CT), 18, 215–217, 219–220
Cholera-toxin B (CTb), 201–204
Cholinergic neurons. *See also* Acetylcholine; Basal forebrain; Mesopontine cholinergic system
abnormal REM sleep (narcolepsy) and, 132
basal forebrain and basalo-cortical system, 214–230. *See also* under Basal forebrain
behavioral state and, 278–279
intrinsic electrophysiology, 218–220
pharmacological modulation, 220–222
pontine (LDT/PPT) system. *See* Mesopontine cholinergic system
suprachiasmatic nucleus and, 36
Choral hydrate, 200
Chronic constriction injury, 554
Chronic FRAs, 378
Chronic implant systems, 506, 507–509, 523–525, 531–534
Chronic pain. *See* Neuropathic pain; specific types of pain
Chronic pain treatment, 550. *See* Neurosurgical pain treatment; Pain control
Cingulotomy, 597, 599, 600
Circadian rhythms, 609
brain temperature rhythms, 625–626
Drosophila model, 9, 61–71
feedback regulation, 62–65, 76–77
electrophysiological processes, 37–38
endogenous neuropeptides, 7
feline model of serotonergic neuron activity, 82
fever modulation and immune response, 629–631
forebrain extracellular serotonin and, 190
functional neuroanatomy. *See* Suprachiasmatic nucleus
functions, 3–4
future prospects, 39
gene analysis approaches
Drosophila model, 62–71
functional analysis (reverse genetics), 80–81
molecular pathway analysis, 79–80
gene identification approaches
expression, 79
genetic, 61–62, 77–78
molecular, 78–79
gene expression, 79, 239–241. *See also* Immediate early genes
Drosophila model, 9, 61–71, 76–77
feedback regulation, 62–65, 76–77
light-induced, 8, 32, 33, 37
mouse model, 75–81
hypothalamic model of state control, 315
IL-1 levels and, 614
infection-induced sleep alterations and, 648
input/output entrainment pathways, 4, 7–8
light-induced gene expression, 8, 37, 32, 33
non-photic, 5, 8, 33, 34–35, 36
phase-response curve, 4, 8, 33
pineal melatonin system. *See* Melatonin rhythm-generating systems
restoration of age-related deterioration, 38–39
serotonergic effects, 82, 191

sleep mentation (dreaming), 114
suprachiasmatic nucleus function, 5–7. *See also*
Suprachiasmatic nucleus
tissue-autonomous oscillators, 64
TNF levels, 612
transcriptional feedback regulation, 62–67
visual system, 8. *See also* Retinohypothalamic
tract
c-Jun, 236
Clarke's column, 533
Claustrum, 247
Clock gene, 69, 70, 75, 77, 78, 80
CLOCK protein binding partner, 80
Clonazepam, 149
CNQX, 465
Cocaine, 268, 269, 369–371, 376, 378, 379
Cochlear nuclei, GABA antagonist effects, 200
Cognitive activation, cortical activation and, 113–114
Cognitive activity during sleep. *See* Dreaming; REM/
NREM sleep mentation differences
Cold stress response, 185–186
Collateralization, mesopontine (cholinergic) efferent
projections, 164–165
Commissural connections, suprachiasmatic nucleus,
14, 24
Commissural myelotomy, 575, 596
Complex regional pain syndrome, 603
Confusional arousal, 148
Cordotomies, 593–596
Cortical activation. *See also* Corticothalamic system;
specific neuroanatomical systems
basal forebrain and. *See also* Basal forebrain
discharge properties, 223
IEG expression and, 242
modulation by stimulation/inactivation,
223–227
monoaminergic interactions, 221–222
cholinergic interactions, 163–164, 169, 214, 278.
See also Acetylcholine; Basal forebrain;
Mesopontine cholinergic system
EEG synchronization/desynchronization patterns,
145
locus coeruleus and, 264. *See* Locus coeruleus
sleep mentation cognitive activation and, 113–114
glutamate effects, 169
nicotine effects, 426
Corticosteroid rhythms, 24
Corticothalamic system, 327–344. *See also* Thalamus
functional outcome, 342, 344
intrinsic properties of thalamic reticular neurons,
332–333
long-lasting hyperpolarization, 332
paroxysmal oscillations (seizure patterns), 337–339
pontine cholinergic interactions, 163, 278. *See
also* Mesopontine cholinergic system
slow-oscillation and grouping of sleep rhythms
into complex wave sequences, 329–339
state-associated fast oscillations, 340–342
Corticotropin-releasing factor (CRF, also
corticotropin-releasing hormone, CRH), 17,
166, 259, 376, 629, 633–634

Cortisone, 647
Cough suppressants, 558
Counter-irritation, 456
CPA, 285, 291
CPP, 511
Cranial nerve operations, 589–591
Craniofacial innervation, 446. *See also* Trigeminal
sensory nuclear complex
Creutzfeld-Jakob disease, 149, 642, 647
Cyclic alternating pattern (CAP), 148–149
Cyclic AMP (cAMP)
cholinergic REM sleep induction interaction, 398
drug dependence and, 376
LC electrophysiology, 261–262
melatonin rhythm-generating system enzyme
regulation, 52, 53, 54, 55
Cyclic AMP response element (CRE), 53
Cyclic AMP response element-binding protein
(CREB), 238, 378
Cyclic GMP (cGMP), 33, 561
Cyclopentyltheophylline, 285
Cyclosporine, 647
Cyproheptadine, 661–662
Cytokines, 609–617. *See also* Interleukin-1; Tumor
necrosis factor; other specific substances
fever regulation, 629
induced alterations in sleep and temperature,
632–634
mechanisms of infection-induced sleep alterations,
647
neurotransmitter interactions, 616, 659–667
sleep humoral regulatory network, 615–616

D

Dantrolene, 260
Deafferentation pains, 601–602
Deep brain stimulation, 455
Delta EEG oscillations
bacterial infection and, 643
basal forebrain activity, 298
cortical response to stimulation, 225
extracellular adenosine and, 170
dream recall and, 113
growth hormone and, 314
infection-induced sleep alterations and, 648–649
mesopontine cholinergic system and, 163
thalamocortical system, 329–331, 333–334
δ^9-Tetrahydrocannabinol (THC), 435–437
Delta-opioid receptor, 551, 555
Delta sleep-inducing peptide, 316
Dentate spikes, 355
Depression, sleep deprivation effects, 183
Developmental stage
IEG expression and, 237
sleep-related electrophysiological differences, 88
Dextromethorphan, 558
Dextrophan, 558
Diagonal band of Broca, 297–298, 312, 317, 318. *See
also* Basal forebrain

benzodiazepine effects, 412
cholinergic/GABAergic basal forebrain
  projections, 215
state-specific IEG expression, 241
Dialysis, *in vivo* microtechnique, 132, 172, 189
Diazepam, 408, 616
Diencephalon
  circadian IEG fluctuations, 240
  nociceptive afferents, 572
  state-specific IEG expression, 241, 245–246, 248
Diffuse noxious inhibitory controls (DNIC), 456, 579–580
Dissociated states, 143–151
  clinical neurologic model, 147
    nocturnal penile tumescence, 149
    NREM sleep variations, 148–149
    REM sleep variations, 147–148
    status dissociatus, 144, 149
    wakefulness variations, 147
  CNS plasticity, 151
  conceptual development, 145–146
  experimental animal models, 146
  locomotor center activation/inhibition, 149
  narcolepsy, 147. See Narcolepsy
  naturally occurring in animals, 146
  normal state determination, 145
  psychogenic model, 149–150
DMNAT, 51
Dolphins, 89, 95, 146, 617
Dopamine
  cytokine and immune-active substance
    interactions, 661
  drug reinforcement actions, 370–374
  narcolepsy and, 134
Dorsal column pathways, 464–468, 578–580
Dorsal funiculus, 464, 488
Dorsal horn. *See also* Spinal somatosensory
    systems
  nitric oxide synthase, 561
  nociceptive pathways, 488, 571–577. See Rostral
    ventromedial medulla
    deep dorsal horn, 574–577
    primary afferents, 571
    superficial dorsal horn, 571–574
  noxious non-specific neurons, 571, 574
  surgical/anesthetic-preparation effects, 507, 508, 510
  trigeminal brainstem complex similarities, 451–452
  trigeminal projections, 446
Dorsal raphe nucleus (DRN, especially serotonergic
    neurons), 182–189, 196–197, 200
  benzodiazepine effects, 409, 411
    medial preoptic area projections, 410
  challenges/stressors, 185–186
  cholinergic input/interactions, 196, 218, 221–222, 229, 278
  circadian cycles, 182
  cytokine-stimulated immune response and sleep
    alterations, 666–667
  feline CNS function, 182–189

  glycinergic/GABAergic-induced inhibition during
    sleep, 196–197, 200, 204–207
    afferent projections, 203–204, 207
  hippocampal theta, 189
  intrinsic membrane properties, 265, 271
    action potentials/afterhyperpolarizations, 265–266, 270
    voltage clamp experiments, 266–267
  medullary neuron response comparison, 187
  motor activity, 186–187
  noradrenergic interactions, 265, 269–270
  PGO waves, 188
  preoptic area state-regulatory network, 318
  SCN afferents, 20
  state-specific activity, 182–185, 258, 270–271, 312
  state-specific IEG expression, 246–248
  stress response, 190
  structure, 265
  synaptic potentials, 267
    5-HT, 265, 268–269
    GABA, 265, 268
    glutamate, 265, 267
    norepinephrine, 265, 269–270
Dorsal root entry zone lesioning, 593
Dorsal root ganglia (DRG), 554
Dorsal spinocerebellar tract, 530–531, 533–536
Dorsal striatum, basal IEG expression, 238
Dorsolateral funiculus, 530, 577–578
Double-stranded RNA (dRNA), 642, 645, 648
Double immunostaining, 201–204, 215
Dreaming, 101–119
  bizarreness ratings, 112, 115–116
  children, 107, 109
  cognitive activation, 108, 113–115
  cognitive activity and, 105
  dissociated states and motor behavior, 149
  H-reflex, 148
  limbic interactions, 400
  lucid, 105, 148
  NREM/REM distinction, 101–119. See REM/
    NREM sleep mentation differences
  one-generator model, 102, 103, 108–109
  phantom REM sleep model, 102, 112, 117
  recall, 106–107, 113, 114, 118
  REM and quality of, 102
  REM sleep/dreaming perspective, 102
  REM sleep behavior disorder, 147–148
  REM sleep insufficient for, 109–110, 116
  two-generator model, 102, 109–110, 116
  types of, 105
  vivid form (apex), 108
  wakeful hallucinations, 107, 108, 147
DREZotomy, 593
Drosophila model, 9, 61–71, 75–80
  arylalkylamine *N*-acetyltransferase, 51
  circadian feedback loop, 62–65
  circadian mRNA cycling, 65–66
  extraocular photoreceptors, 70
  gene expression regulation (*per* and *tim*), 62–71, 76–77

Index

light entrainment, 69–70
regulation of protein cycling, 66–68
subcellular localization of gene products, 68
Drug addiction, 365–379
  animal models, 365–366
  extended amygdala and, 368, 378–379
  motivational (reinforcement) model, 366–367, 427–428
  negative reinforcement (neuronal substrates), 374
    cAMP upregulation, 376–377
    neurochemical substrates, 374–376
    reward, 374
    sensitization, 377–378
    tolerance, 367, 377, 379, 425
    withdrawal, 374, 379, 425
  neuroadaptation and sensitization, 367, 374–376
  neurochemical substrates, 374–376
  opponent process theory, 374, 428–429
  positive reinforcement (neuronal mechanisms), 367–368, 428
    nicotine, 373
    opiates, 371–372
    opponent process theory, 374, 428–429
    sedative-hypnotics, 372–373
    stimulants, 369–371
    THC, 373–374
  protracted abstinence, 378, 379
  relapse, 378, 422
  transcription factors, 378
Drug reinforcement. *See* Drug addiction
Drug sensitization, 367, 374, 377–378
Drug tolerance, 367, 377, 555
  caffeine, 424, 425
Drug withdrawal, 374, 379, 425
Dysesthesia, 595

E

Ecdysone-inducible system, 81
Echidna, 90–95
EEG. *See* specific oscillation patterns and electrophysiological phenomena
  caffeine effects, 424
  cholinergic activation of thalamus and basal forebrain structures, 163–164
  cortical response to basal forebrain stimulation/inactivation, 224–229
  corticothalamic fast oscillations, 340–342
  corticothalamic slow-oscillation and grouping of sleep rhythms into complex wave sequences, 329–339
  desynchronization, 145
  determining sleep function, 328
  fast oscillations. *See* Fast rhythms
  feline sleep-related electrophysiology, 88
  functional significance, 342, 344, 355–357
  local phase shifts, 357
  nicotine effects, 426
  REM sleep-associated low voltage patterns, 89
    evolutionary development, 97
  monotremes, 91–93
  reptilian sleep, 95–97
  sleep mentation cortical activation, 113
  slow waves. *See* Delta EEG oscillations; Slow-waves
  state-dependent network dynamics of hippocampal formation, 349–358. *See also* Hippocampal formation
  state-specific corticothalamic activity, 327–344. *See also* Corticothalamic system
  viral infection and, 645
egr-1, 236
Electrotonic coupling locus coeruleus noradrenergic neurons, 263–264
Encephalitis-induced sleep alterations, 645
Encephalitis lethargica, 410
Endogenous opioids. *See also* Mu-opioid receptor; Opioids; specific substances
  cytokine/immune-active substance interactions, 665
  endogenous mu-opioid agonists, 551
  geniculohypothalamic tract, 20
  locus coeruleus response, 259
  mixed peptidase inhibitors, 553
  monoaminergic neuron inhibition, 206
  trigeminal brainstem neurons, 448
Endomorphins, 551, 553
β-Endorphin, 665
Endotoxin, 629, 661
Enkephalins
  geniculohypothalamic tract, 20
  locus coeruleus response, 259
  mixed peptidase inhibitors, 553
  monoaminergic neuron inhibition, 206
Entorhinal-hippocampal system, 349–358. *See* Hippocampal formation
Epidermal growth factor (EGF), 615
Epilepsy, 239, 355, 400
Escherichia coli, 641–642
Essential vagoglossopharyngeal neuralgia, 591
Ethanol, 372–373, 376, 421, 422–424, 429
  alcoholism relapse, 378, 422
  $GABA_A$-benzodiazepine receptor complex and, 414
  reinforcement mechanisms. *See* Drug addiction
  state-altering effects, 422–423, 428
    medial preoptic area, 416
Etomidate, 414
Event-related potentials, 111
Evolution
  circadian rhythms, 9, 70
  REM sleep, 91, 94–95, 97
Excessive daytime somnolence (EDS), 129, 130. *See also* Narcolepsy
  genetic predisposition, 134
  treatment, 131–132
Excitatory amino acids. *See also* Glutamate
  central hypersensitivity, 556–559
  cholinergic interactions, 220
  cortical response to basal forebrain stimulation, 224
  dorsal spinocerebellar tract afferent input and, 531

locus coeruleus synaptic potentials, 262
serotonergic neuron response to phasic sensory input and, 183
spinothalamic tract, 474
trigeminal brainstem complex somatosensory function, 451
Excitatory post-synaptic potential (EPSP)
hippocampal-entorhinal system functional significance, 355–356
hippocampal sharp waves and, 353
pontine cholinergic system, 282
thalamocortical system, 329, 332
Existential dreaming, 105
Extended amygdala, drug reinforcement and, 368, 378–379

F

Facial pain, 601
Failed low-back surgery syndrome, 597
Fasciculus gracilis, 579
Fast rhythms
corticothalamic systems, 340–342
entorhinal-hippocampal system dynamics, 350
functional significance, 350, 355–357
functional considerations, 344, 355–357
Fast prepotentials (FPPs), 332
Fatal familial insomnia, 149, 642, 647
Fatty acid amide hydrolase (FAAH), 437
Fear-related responses, 495–496
Felbamate, 559
Feline herpesvirus, 642, 645
Feline immunodeficiency virus, 642, 645
Feline models. *See* Cat models
Fentanyl, 509, 550
Fetal tissue grafts, suprachiasmatic nucleus, 7, 38–39
Fever, 185, 623–635. *See also* Brain temperature
antipyretic-induced antagonism of IL-1 effects, 633–634
biphasic sleep alterations, 631–635
circadian factors, 629–631
function, 624, 627, 634
heat loss/heat production mechanisms, 628
host benefits, 626–628
host detrimental effects, 629
regulation, 629, 634
serotonergic interactions, 661
Fine touch, 449
Fish, melatonin rhythm-generating system, 55
5-HT, 265, 268–269. *See also* Serotonin
Flunixin, 508
Fluoxetine, 268, 269, 373
Flurazepam, 409
Forebrain, basal. *See* Basal forebrain
Forskolin, 261–262
Forward genetics, 77–78
Fos B, 236, 239
Fos expression, 236–249. *See also* c-*fos*; Immediate early genes
basal expression, 238

light-induced expression, 20
state-associated neuroanatomy, 312–314
Fos-lacZ, 239
Fos-related antigen-2 (Fra-2), 53
Fos-related antigens (Fras), 236
basal expression, 238
chronic drug action and, 378
Free-running, 4, 64
Frontal lobe lesioning, 597, 599, 601
frq, 62, 69, 70, 76
Fruit fly. *See Drosophila* model
Fungal infections, 641–642
Fungal model, 9
Funiculus, 464, 488, 530

G

GABA (and GABAergic neurons). *See also* Glutamic acid decarboxylase (GAD)
anesthetic interactions, 512
anxiolytic drug effects and, 372
basalis non-cholinergic neurons, 302
basalo-cortical system, 214–230. *See* Basal forebrain
cytokine and immune-active substance interactions, 664, 667
dorsal raphe nucleus synaptic potentials, 265, 268
double immunostaining, 201–205
ethanol effects and, 422
input to serotonergic neurons, 183
hippocampal gamma/theta oscillations and, 352
hypothalamic model of state regulation, 314–315
locus coeruleus response, 259
synaptic potentials, 262
state-associated inhibition of monoaminergic neurons, 264–265
NMDA-mediated pain mechanisms, 559
pontine (LDT/PPT) cholinergic system input, 169
postsynaptic potentials, 282
reticular thalamocortical electrophysiology and, 338
state-associated sensory afferent depolarization, 528–529
state-specific monoaminergic inhibition, 195–207, 264–265
antagonist iontophoresis, 196, 197, 199–200
suprachiasmatic nucleus circadian function and, 9, 17, 20, 25, 34–36
trigeminal brainstem neurons, 448
tuberomammillary histaminergic interactions, 318
GABA-benzodiazepine receptor, 408, 414, 416
Gabapentin, 559
$GABA_A$ receptor, 169. *See also* GABA
benzodiazepine receptor complex, 408, 414, 416
cytokine-associated state altering mechanism and, 616
$GABA_B$ receptor, 35, 169
GAD. *See* Glutamic acid decarboxylase
Galanin, 170, 259, 314, 318
Gamma EEG oscillations, 145

cortical response to basal forebrain stimulation, 225, 229
entorhinal-hippocampal system dynamics, 350–353, 358
Gasserian ganglion, 589
Gastric acid secretion, 616
Gastrin-releasing peptide (GRP), 36–37
Gender differences, IEG expression, 237
Gene expression, 235–249. *See also* c-*fos*; Immediate early genes; specific genes, products
 basal conditions, 237–239
 circadian rhythms. *See under* Circadian rhythms
  Drosophila model (*tim* and *per*), 9, 61–71, 76–77
  mouse model, 75–81
 light-induced, 8, 32, 33, 37
 photoreceptor conserved elements (PCEs), 49–50
 pineal/retinal enzymes, 49
 transcriptional feedback regulation, 62–67, 76–77
Gene identification
 expression approach, 79
 genetic approach, 61–62, 77–78
 molecular approach, 78–79
Gene knockouts, 80–81
Geniculohypothalamic afferents to SCN, 20, 33–35
Gigantocellular tegmental field, benzodiazepine effects, 411
Glia, SCN cellular communication and, 39
Globus pallidus, 215
Glucocorticoids, drug dependence neuroadaptive model, 376
Glutamate. *See also* Excitatory amino acids; NMDA
 basalo-cortical projections, 217
 cholinergic interactions, 220, 228
 central hypersensitivity and, 557
 dorsal raphe nucleus synaptic potentials, 265, 267
 drug sensitization effects, 378
 immune function interactions, 665
 locus coeruleus response, 259, 262
 postsynaptic dorsal column, 465
 pontine cholinergic interactions, 165, 169, 170–171, 290
 retinohypothalamic tract, 18, 32
 spinocervical tract, 469
 suprachiasmatic nucleus and, 32–33, 35
  retinohypothalamic afferents, 18, 32
 trigeminal brainstem complex somatosensory function, 451
Glutamate receptors. *See also* NMDA receptors
 cholinergic activation of basal forebrain, 164
 hyperalgesia and, 506
 spinothalamic tract, 471
 trigeminal brainstem neurons, 448
Glutamatergic neurons. *See* Glutamate; specific neuroanatomical systems
Glutamic acid, cytokine and immune-active substance interactions, 665
Glutamic acid decarboxylase (GAD), 318. *See also* GABA
 double immunostaining
  basalo-cortical projections, 215, 218
  locus coeruleus, 203

dorsal raphe nucleus, 204
SCN localization, 17
Glycine
 locus coeruleus response, 259, 262
 state-specific monoaminergic inhibition, 195–207
Gonadotrophin-releasing hormone, 22
G protein
 cholinergic REM sleep induction interaction, 398
 opioid receptor uncoupling and drug tolerance effects, 377
Growth hormone, 314
Growth hormone releasing hormone (GHRH), 314, 610, 615
Gypsy moth, 64

H

Hallucinations, 394
 hypnagogic, narcolepsy and, 131, 147
 sleep-like mentation while awake, 107, 108
Hallucinogenic drugs, 433–438. *See* specific drugs
Halothane, 200, 489, 491, 507, 512, 516
Heat stress. *See* Fever; Thermal stress response
Heroin, 371
Herpesviruses, 18
 feline herpesvirus, 642, 645
5-HIAA, 661–662
Hibernation
 hypothermia, 9
 state dissociation, 149
HIOMT, 47, 49
Hippocampal formation (hippocampus-subiculum-entorhinal cortex), 349–358
 dendritic field stratification, 354–355
 dentate spikes, 355
 memory function, 349–350, 358
 monoaminergic input, 352
 punctate sharp waves, 350, 353–355
 serotonergic modulation of theta activity, 189
 state-dependent network dynamics, 350, 352–353
  functional significance of fast-frequency dynamics, 350, 355–357
  local phase shifts, 357
  theta/gamma ensemble, 352–353, 358
  two-state model of memory formation, 358
Hippocampus
 benzodiazepine receptors, 408
 cholinergic/GABAergic basal forebrain projections, 215
 circadian IEG fluctuations, 239–240
 long-term potentiation and memory, 557
 memory function, 349–350, 358, 557
 state-specific IEG expression, 241–242, 247
 TNF, 612, 616
Histamine
 basal forebrain state activation, 312
 cytokine and immune-active substance interactions, 616, 665
 mesopontine cholinergic neurons and, 169, 285
 tuberomammillary nucleus neurons, 312, 314, 318

HIV, 612, 642, 644–645, 649
HLA DQA1*0102, 135, 136
HLA DQB1*0602, 135–136
HLA DR2, 135
3-HMC, 409
Horner's syndrome, 595
Horseradish peroxidase, 18
hper, 79
H-reflex, 148
5-HT receptors (5-HT$_{1A}$), 168, 183, 188, 271. *See also* Serotonin
   cannabinoid interactions, 436
   cholinergic interactions, 168
   hippocampal theta and, 188
   sleep deprivation and, 183
   suprachiasmatic nucleus activity, 35
5-HT-1B/D receptor, 269
Human immunodeficiency virus (HIV), 612, 642, 644–645, 649
Human *per* gene, 79
8-Hydroxy-dipropylaminotetralin. *See* 8-OH-DPAT
6-Hydroxydopamine (6-OHDA), 146, 370
5-Hydroxyindoleacetic acid (5-HIAA), 661–662
3-Hydroxymethyl-β-carboline (3-HMC), 409
5-Hydroxytryptophan, 188. *See also* Serotonin
Hyperalgesia, 491. *See also* Central hypersensitivity; Neuropathic pain; Nociceptive pathways; Pain control
   NMDA antagonists and, 558
   NMDA receptors and chronic nociceptive input, 506, 510–512, 516
Hypersensitivity, central. *See* Central hypersensitivity
Hypersomnias, 131
Hypnagogic hallucinations, 131
Hypnotic drugs. *See* Antidepressants; Barbiturates; Benzodiazepines; Ethanol
Hypoferremia, 628
Hypothalamic-pituitary-adrenal axis
   fever regulation function, 629, 631
   noradrenergic immune response and, 660
Hypothalamotomy, 599, 601
Hypothalamus. *See also* specific subanatomy and systems
   basal IEG expression, 238
   GABA antagonist effects, 200
   GABAergic/glycinergic projections to monoaminergic neurons, 201–205
   IL-1, 614
   mesopontine cholinergic projections, 164
   noradrenergic immune response and, 660
   retinohypothalamic tract, 8, 18–19, 31–33, 35
   spinal afferents, 572
   state dissociation induction, 146
   state-specific IEG expression, 245–248
   TNF, 612, 616
Hypothermia of hibernation, 9

I

Ibotenic acid, 413–414
Idazoxan, 284
Idiopathic hypersomnia, 118
IL-1. *See* Interleukin-1
Immediate early genes (IEGs), 235–239, 311. *See also* c-*fos*; Gene expression; specific genes and gene products
   basal expression, 237–239
      biological clock, 239–241
      species specificity, 239
   circadian fluctuations, 239–241. *See under* Circadian rhythms
   expression during sleep, 247–249
   expression during spontaneous wakefulness, 241–247
      brainstem, 246
      cortex, 243–245
      diencephalon, 245–246
      noradrenergic interactions, 246
      spontaneous wakefulness, 246–247
   light-induced expression in SCN, 8, 32, 33, 37, 241, 247
   melatonin rhythm-generating system enzyme regulation, 53
   neural lesions-induced state dissociation and, 146
   neuroanatomical substrates, 237–249
   neuronal activation markers, 236
   noradrenergic interactions, 246
   open questions, 249
   sleep deprivation and, 247, 313–314
   suprachiasmatic nucleus expression, 8, 32, 33, 37, 241, 247
Immune response. *See also* Fever; Infection; Inflammation
   body temperature (fever) and, 627–628. *See also* Fever
   circadian factors, 630
   mechanisms of infection-induced sleep alterations, 647
   narcolepsy and, 135, 137
   neurotransmitter interactions with cytokines and immune-active substances, 659, 666–667
      acetylcholine, 664
      adenosine, 666
      dopamine, 661
      GABA, 664
      glutamic acid, 665
      histamine, 665
      nitric oxide, 665–666
      norepinephrine, 660–661
      opioids, 665
      serotonin, 661–664
   sleep deprivation and, 649–650
   sleep function, 624, 634
Implant systems, 506, 507–509, 523–525, 531–534
In vivo microdialysis, 132, 172, 189
Indomethacin, 660
Inducible transgenics, 81
Infection. *See also* Immune response
   fever and immune response, 627–628, 631–635. *See also* Fever
   noradrenergic immune response and, 660
   sleep alterations, 623, 641–651

bacterial infections, 643–644
clinical relevance, 649–650
fever and, 631–635
fungal infections, 647
mechanisms, 647–649
model systems, 641–642
prion-related conditions, 642, 647
trypanosomiasis ("sleeping sickness"), 241, 646–647
viral infections, 644–645, 649
Inflammation, 557–558
adenosine and, 560
anti-convulsant (gabapentin) and, 559
morphine and, 555
nitric oxide and, 561
Inflammatory pain, 553, 554
Influenza virus, 642, 644–645
Inhibitory postsynaptic potential (IPSP), 200
hippocampal gamma oscillations, 351
pontine cholinergic system, 282
stimulus-induced locus coeruleus response, 263
thalamocortical system, 329, 332–333
In situ, in vitro mouse model of spinal function, 507, 509, 514–516
Insomnia. *See also* Sleep disorders
ethanol self-administration and, 428
fatal familial insomnia, 149, 642, 647
Insulin sensitivity, serotonergic raphe neurons, 186
Intercalatus, state-specific IEG expression, 248
Interferon-α (IFN-α), 615, 647, 664, 665
Intergeniculate leaflet (IGL)
basal IEG expression, 238
state-specific IEG expression, 248
suprachiasmatic nucleus connections, 20, 24
Interleukin-1 (IL-1 or IL-1β), 316, 610, 614–615
diurnal variation, 614
fever regulation, 629
sleep-altering effects, 631, 632–634
T-cell proliferation, 627
infection-induced sleep alterations and, 647–648
neurotransmitter interactions, 616, 660–668
acetylcholine, 664
dopamine, 661
GABA, 664, 667
glutamate, 665
histamine, 665
noradrenaline, 660
opioid, 665
serotonin, 662–664, 666–667
receptor antagonist (IL-1ra), 629
sleep alterations, 631–634, 647–648, 666–667
TNF interactions, 613
Interleukin-2 (IL-2), 615, 629, 661, 664, 665, 666, 667
Interleukin-4 (IL-4), 615
Interleukin-10 (IL-10), 615, 648
Intramodule scanning, 78
Ionotropic effectors, pontine cholinergic system, 282
Ipsapirone, 188
Iron deficiency, 628
Ischaemic pain, 558

J

Jaw, 450, 451, 506
jun gene family
basal expression, 238, 239
circadian oscillations, 240
Jun B, 236, 239
Jun D, 236

K

Kainate, 220
Kappa-opioid receptor, 551, 552
Kelatorphan, 553
Ketamine, 509, 511, 558
Knockout genes, 80–81
Kölliker-Fuse area, 201–202, 575
Krox-24, 236, 234, 239
Kyneurenate, 271
Kynurenic acid, 164, 183, 531

L

Laminectomy, 508, 509
Lateral funiculus, 571
Lateral geniculate nucleus, 163
basal IEG expression, 238
GABA antagonist effects, 200
PGO activity, 188, 286, 392–393
Lateral habenula, glycinergic/GABAergic projections to dorsal raphe nucleus, 203
Lateral preoptic area. *See* Preoptic area
benzodiazepine effects, 412
mesopontine projections, 164
Lateral septum, 22
Laterodorsal tegmental nucleus (LDT), 161–173, 278, 312, 392. *See also* Mesopontine cholinergic system
amygdala connections, 393–395
basal forebrain connections, 218
cellular factors affecting excitability, 278–285
intrinsic and extrinsic properties and impact on behavioral state, 286–291
preoptic area state-regulatory network, 318
REM sleep modulation, 392–393
amygdala interactions, 393–395
dual role in REM sleep and wakefulness, 161–173. *See also* Mesopontine cholinergic system
proposed mechanism, 398–400
Lennox-Gastaut syndrome, 338
Lesion-based pain control. *See* Neurosurgical pain treatment
Ligand-gated currents, in pontine cholinergic neurons, 282–286, 289
Light/dark cycle
brain temperature transitions, 626
IEG expression patterns, 240
infection-induced sleep alterations and, 648

photic entrainment physiology and circadian rhythmicity, 4
  gene expression, 8, 32, 33, 37, 69–70, 76, 241
  phase-response curve (PRC), 4, 5, 8, 33
  pineal melatonin rhythm-generating system, 48–49. *See also* Melatonin rhythm-generating systems
  suprachiasmatic nucleus and associated structures, 7–8, 16, 18–20, 25, 32–35
  serotonergic interactions, 182, 190
Limited (punctate) midline myelotomy, 596
Lipopolysaccharide (LPS), 628, 629, 642, 648, 660, 661, 663, 665
Lissauer's tract, 571, 591
Lizards. *See* Reptiles
L-NAME, 33
Local phase shifts, 357
Locus coeruleus (and associated noradrenergic neurons), 196, 258–265
  acetylcholine input, 196
  afferents, 258
  afterhyperpolarization, 262
  basal IEG expression, 239
  basalo-cortical cholinergic system interactions, 218, 221–222, 229
  benzodiazepine effects, 411
  cAMP-dependent electrophysiology, 261–262
  cortical activation, 264
  electrotonic coupling, 263–264
  endogenous neurochemicals, 258
  glycinergic/GABAergic-induced inhibition during sleep, 195–207, 264–265
    afferent projections, 201–203, 207
    anesthesia effects, 200
    glycine/GABA antagonists and, 197–200
  intrinsic membrane properties, 259, 271
    action potentials, 259
    afterhyperpolarization, 259–260
    voltage clamp experiments, 260–262
  mesopontine cholinergic system connections, 168
  neurochemical effects (table), 259
  preoptic area state-regulatory network, 318
  reticular formation connections, 162
  sleep/wake cycle, 264–265, 271, 312
  spontaneous activity during sleep/wake cycle, 197
  state dissociation and, 146
  state-specific IEG expression, 246–248
  structure, 258
  synaptic potentials, 262–263
Low-threshold calcium current. *See also* PGO waves
  basalis cholinergic neurons, 299
  pontine cholinergic neuron activity, 167, 169, 287–289
Low-threshold mechanoreceptive (LTM) neurons, 449–450, 455
Low-threshold neurons, state-associated lumbar sensory system, 532
LSD, 434, 438
Lucid dreaming, 105, 148
Lumbosacral sensory system, 463–477, 530. *See also* Spinal somatosensory systems
  postsynaptic dorsal column pathway organization and control, 464–468
  sensory neuron function during active sleep, 534–536
  spinocerebellar tract, 530–531
  spinocervical tract, 468–470, 530
  spinohypothalamic pathways, 572
  spinomesencephalic tracts, 476–477, 489
  spinoreticular tract, 476–477, 489, 530, 533–536
  spinothalamic tract, 470–474, 530, 533, 534, 572–574, 578–579, 594–595
    cordotomy, 591
    tractotomy, 599, 600
Lysergic acid diethylamide (LSD), 434, 438

M

Magnocellular basal nucleus, 214. *See* Basal forebrain
Magnocellular preoptic area, 215, 220, 312–313. *See also* Preoptic area
Magnocellular reticular nucleus, 205
Main sensory nucleus, 522, 527. *See* Trigeminal sensory nuclear complex
Manoalide, 270
Marijuana, 373–374, 434, 435–437
  receptors, 434, 436–437
Marine mammals, 89, 95, 146, 617
Marsupials, 90
Mecamylamine, 36
Mechanoreceptive somatosensory system, 449–450, 465
  dorsal spinocerebellar tract, 530
Medial forebrain bundle, reward/reinforcement system, 367–368
Medial prefrontal cortex, 169
Medial preoptic. *See also* Preoptic area
  benzodiazepine action, 409, 410, 412–416
  state-specific IEG expression, 246
Medial septal nucleus, 297–299, 302. *See also* Basal forebrain
Medial thalamotomy, 599, 601, 602
Median raphe nucleus, 20, 35
Medulla, descending pathways modulating nociception, 488. *See also* Rostral ventromedial medulla
Medullary dorsal horn, 452
Medullary reticular formation. *See* Reticular formation
Medullary serotonergic neurons, 187
α-Melanocyte-stimulating hormone (α-MSH), 629, 633
Melatonin rhythm-generating systems, 8, 45–46
  AANAT (rhythm-generating enzyme), 46, 49–52
  biosynthetic site, 46–47
  cAMP and, 52
  chicken, 54–55, 64
  endogenous clock (suprachiasmatic nucleus), 46–47, 52–53
  fish, 55
  mammals, 52–54

paraventricular thalamic SCN circadian function, 21, 23
photodetection site, 48
retinal synthesis, 47
species-specific functionality, 46, 52–56
Memantine, 558
Membrane polarization. *See* Afterhyperpolarization; EEG; specific electrophysiological phenomena and neuroanatomical substrates
Memory
    activation model for sleep mentation differences, 108, 113–116
    hippocampal network activity, 349–350, 358, 557
    marijuana effects, 435
    sources for sleep mentation, 110–111
    thalamocortical function, 344
Mescaline, 434, 438
Mesencephalotomy, 599, 601
Mesocorticolimbic dopaminergic system, drug dependency and, 370–374, 378, 428
Mesopontine cholinergic system, (pedunculopontine tegmental (PPT) and laterodorsal tegmental (LDT) nuclei), 161–173, 272–292, 312
    acetylcholine inputs/effects, 162, 169, 284–285
    adenosine inputs/interactions, 170, 285, 290–291
    afterhyperpolarization, 279–280
    amygdala connections, 393, 394–395
    basal forebrain connections, 218
    cellular factors affecting neuronal excitability
        ionotropic effectors, 282
        ligand-gated currents, 282–286, 289
        metabotropic effectors, 282–285
        voltage-gated currents, 279–282
    cholinergic effects on behavioral state, 278–279
    glutamate and GABA inputs, 169–171
    histamine-stimulated response, 285
    monoaminergic input/interactions, 162, 168, 278, 284
        serotonin-activated metabotropic system, 282–284
        state-specific effects, 289–291
    neuron subtypes, 167–168, 279
    NOS inputs, 166, 170
    PGO wave generation, 392–393
    preoptic area state-regulatory network, 318
    REM sleep modulation, 162–163, 392–393
        amygdala interactions, 393–395
        proposed mechanism, 398–400
    reticular formation afferents, 162
    role in REM sleep and wakefulness, 161–173
        basal forebrain activation, 163–164
        collateralization, 165
        efferents, 164–165
        electrophysiology, 167–168
        neurochemical inputs, 162, 168–171
        neurotransmitter colocalization, 165–166
        proposed mechanisms, 172–173
        thalamic connections, 163, 164–165
        ultrastructure, 166–167
        wakefulness, 163–164
    state-specific IEG expression, 246, 247, 248, 313
    thalamic activation, 172–173

Metabotropic effects, 282–285
Metabotropic receptor, 557
Methacholine, 285
Methohexitone, 509, 511
3-Methoxy-4-hydroxyphenylglycol (MHPG), 661
N-Methyl-D-aspartate. *See* NMDA
Methylphenidate, 424
MHPG, 661
Microbial modulation of arousal, 641–651
Microdialysis, 132, 172, 189
Microsleep episodes, 130
Microsurgical DREZotomy, 593
Microvascular decompression, 589–590
Midcingulate cortex, 577
Mifepristone-inducible system, 81
MK-801, 36, 558
Monkey, chronic implant approach to brainstem unit recording, 523–524
Monoamine oxidase inhibitors, 149
Monoaminergic neurons, state-associated synaptic and intrinsic properties, 257–271. *See also* Norepinephrine; Serotonergic neurons; specific monoamines, neuroanatomical systems
Monoamines. *See* Norepinephrine; Serotonin; specific receptors
Monotremes, 90–95
Morphine, 185, 371, 489, 491–492, 550, 555. *See also* Opioids
    studies of descending pain-modulating pathways, 489, 491–492, 496
Motor activity. *See also* Atonia
    feline model of serotonergic neuron activity, 186–187
    monotremes, 94
    nociceptive stimulus response and sympathetic output, 514–516
    REM sleep-associated atonia, 88, 89. *See also* Atonia
Mouse
    circadian genetics model (*Clock* and *mper* genes), 77–78
    in situ, in vitro model of spinal function, 507, 509, 514–516
*mper*1, 75, 77, 78–79, 80
*mper*2, 75, 77, 79, 80
*mper*3, 79
mRNA, circadian cycling, 65–66
Multiple Sleep Latency Test (MSLT), 118–119, 131, 423
Mundane dreaming, 105
Mu-opioid receptor, 371, 550–551, 665
    endogenous ligands, 551
Muramyl dipeptide, 612, 629, 643
Muramyl peptides, 316, 647, 661, 668
Murine models. *See* Mouse
Muscarinic receptors
    acetylcholine-stimulated pontine cholinergic response, 285
    basalo-cortical system cholinergic interactions, 222
    carbachol-induced interactions, 162, 398

mesopontine cholinergic neurons expression, 169
suprachiasmatic nucleus, 36
Muscimol, 205, 313
Muscle tone loss. *See* Atonia

N

NAAG, 32
N-Acetyl-aspartyl-glutamate (NAAG), 32
N-Acetyltransferase, 8
  pineal melatonin rhythm-generating enzyme (AANAT), 46, 49–52
NADPH-diaphorase, 18
Naloxone, 373, 491, 552
Naltrexone, 373, 378
NAN-190, 268
Narcolepsy, 129–137, 147
  associated disorders, 131
  autoimmunity, 135, 137
  cataplexy, 130, 147
  clinical aspects, 129–131
  diagnostic procedures, 131
  excessive daytime somnolence (EDS), 129, 130
  genetic aspects, 134–137
  hypnagogic hallucinations, 131, 147
  neurochemistry, 132–134
  sleep paralysis, 130–131, 147
  spectrum, 131
  state dissociation, 147
  treatment, 131–132
NBTI, 291
N^6-cyclohexyladenosine (CPA), 285, 291
NDV, 645
NECA, 318
Neocortex
  basal IEG expression, 238
  cholinergic/GABAergic basal forebrain projections, 215
  spike-wave seizure generation, 338–339
Neonatal sleep, 88, 89. *See also* Active sleep
Neostigmine, 392
Nerve growth factor (NGF), 36, 615
Nerve growth factor-induced A. *See* NGFI-A
Nerve injury, neuropathic pain model, 553–555
Network dynamics, state-dependent hippocampal formation activity, 349–358. *See* Hippocampal formation
  functional significance of fast-frequency dynamics, 355–357
  local phase shifts, 357
  theta/gamma ensemble, 352–353, 358
  two-state model of memory formation, 358
Neuroma, 591
Neuronal group theory, 616
Neuropathic pain, 474, 550
  adenosine and, 559
  anti-convulsant (gabapentin) and, 559
  hemispheric lateralization of the anterior cingulate cortex, 599
  microsurgical DREZotomy, 593

nitric oxide and, 561
NMDA receptor interactions, 557–558
spinothalamic tract stimulus responses, 474
treatment. *See* Neurosurgical pain treatment; Pain control
Neuropeptide Y (NPY), 17, 20, 34, 259
Neurophysin, 14
Neurospora model, 9, 62, 69, 70
Neurosurgical pain treatment, 587–604
  autonomic nervous system, 603
  cranial nerves, 589–591
  medial thalamotomy, 602
  neurosurgeon's perspectives on indications and use, 600–603
  neurotomy/peripheral nerve procedures, 591
  outcome evaluation issues, 588–589
  psychosurgical interventions (lobotomies), 597
  spinal cord, 593–596
    dorsal root entry zone, 593
    potential complications, 595
    rhizotomies, 591–593
  supraspinal lesions, 597–602
  sympathectomy, 603
Neurotensin, 17, 299, 302
Neurotomy, 591
Neurotransmitter colocalization, mesopontine (cholinergic) efferent projections, 165–166
Neurotropic virus, 18, 23
Newcastle disease virus, 642, 660
NGFI-A, 236, 239, 247
Nicotine, 373, 376, 421, 425–427, 429
Nicotinic receptors, 373
  basalo-cortical system cholinergic interactions, 222
  nicotine pharmacodynamics, 426
  suprachiasmatic nucleus, 36
Nitric oxide (NO)
  central hypersensitivity and, 561
  cytokine and immune-active substance interactions, 665–666
  suprachiasmatic nucleus function, 18, 33
Nitric oxide synthase (NOS), 561
  cytokine interactions, 615, 665–666
  mesopontine cholinergic system and, 166, 170
  suprachiasmatic nucleus, 33
Nitrobenzylthioionosine (NBTI), 291
NMDA, 557–558. *See also* Glutamate; NMDA receptors
  analgesic activity of antagonist, 557–559
  cholinergic interactions, 220, 222
  cortical response to basal forebrain stimulation, 224, 227
  depolarization of non-cholinergic mesopontine neurons, 171
  ethanol effects and, 372–373, 422
  locus coeruleus synaptic potentials, 262
  prolonged pain states and, 557
  suprachiasmatic nucleus photic response and, 32
NMDA receptors. *See also* Glutamate receptors
  chronic nociceptive input and hyperalgesic response, 506, 510–512, 516
  ethanol effects and, 422

inflammation/hypersensitivity model, 557–559
pontine cholinergic system activation, 169, 170
   basal forebrain stimulation, 164
   postsynaptic potentials, 282
spinal nociceptive mechanisms, 556–559
trigeminal brainstem complex somatosensory function, 451
Nociceptin, 551–552
Nociceptive pathways, 446, 487–497, 571–577. *See also* Somatosensory systems; Spinal somatosensory systems
   analgesia center model, 489
   animal preparation effects (surgery/anesthesia), 506–517. *See* Surgical and anesthetic preparation effects
   anterolateral cordotomy effects, 570
   ascending pathways, 571–577. *See also* Spinal somatosensory systems; specific systems
      dorsal columns, 578–580
      dorsal horn, 571–577
      dorsolateral funiculus, 577–578
      functional localization problems, 579–580
      lumbosacral systems, 465, 466, 471–475
      noxious non-specific cells, 571, 574
      noxious specific cells, 571
      ventrolateral tract, 570
   descending modulatory systems
      brainstem reticular areas, 577
      integration of somatomotor/autonomic control with nociceptive modulation, 496–497
      on- and off-cell functional relationships, 489–492, 496
      rostral ventromedial medulla, 487–497
      serotonergic interactions, 493–495
      state-specific neuronal recruitment, 495–496
   excitatory amino acids and central hypersensitivity, 556–559
   lumbosacral systems, 465, 466, 471–475
   neurosurgical treatment of pain, 587–604. *See also* Neurosurgical pain treatment
   neurotransmitter mechanisms of hyperalgesic/allodynic sensitization, 506, 510–512, 516
   problematic interpretation of lesioning data, 579–580
   spino-reticulo-thalamic pathways, 575, 577
   spinothalamic tract, 471, 472, 474–476
      descending control, 475–476
      non-localizable pain response, 474
      wide-dynamic range (WDR) neurons, 452, 471, 512
   trigeminal brainstem, 451–454
      neuroplastic changes, 453–454
   visceral sensation, 465, 466, 471, 514, 558, 578–579
   wide-dynamic range neurons, 452, 471
Nocturnal penile tumescence, 117, 149
Non-REM sleep (also called resting sleep, quiet sleep). *See also* Slow-wave sleep
   adenosine effects, 317, 318
   arousal disorders and dissociated states, 148
   benzodiazepine effects, functional neuroanatomy, 411–413
brain temperature rhythms, 625–626
cognitive activity, 102–108. *See also* REM/NREM sleep mentation differences
cyclic alternating pattern (CAP), 148–149
defining REM sleep, 89
fever and, 631–634
functional neuroanatomy, 313–314
hypothalamic model of state regulation, 313–316
infection and, 623
IL-1 and, 614–615
   induced sleep and temperature alterations, 632–633
PGO waves and REM sleep deprivation, 146
sleep regulatory substances, 610
stages 3 and 4. *See* Slow-wave sleep
state dissociation, 148–149
stimulus tagging, 105–106
TNFα and, 611–614
unihemispheric (dolphin), 89, 146, 617
Noradrenergic neurons. *See* Locus coeruleus; Norepinephrine
Norepinephrine (and noradrenergic neurons), 312
   basalo-cortical cholinergic system interactions, 221–222, 229
   basalo-cortical system noncholinergic (potentially GABAergic) interactions, 222
   cortical response to basal forebrain stimulation, 224–225
   cytokine and immune-active substance interactions, 660–661
   dorsal raphe nucleus synaptic potentials, 265, 269–270
   drug tolerance effects, 377
   locus coeruleus response, 259, 262, 271. *See also* Locus coeruleus
   non-cholinergic mesopontine neurons and, 171
   pontine cholinergic interactions, 168, 284
   raphe electrophysiology, 269–270
   SCN-pineal rhythmic function and, 53
   state-associated immediate early gene expression and, 246
Noxious non-specific (NNS) cells, 571, 574
Noxious specific (NS) cells, 571
Nucleus abducens, state-specific IEG expression, 248
Nucleus accumbens, drug reinforcement and, 370–372, 376–377, 379
Nucleus basalis, 214–215, 299–302. *See also* Basal forebrain
   state-associated c-Fos expression, 313
Nucleus basalis magnocellularis, 297–298. *See also* Basal forebrain; Nucleus basalis
Nucleus gigantocellular alpha, 205
Nucleus of the solitary tract, 572, 629, 665
   GABA antagonist effects, 200
Nucleus oralis, 522, 527. *See also* Trigeminal sensory nuclear complex
Nucleus raphe magnus, 183, 185
   analgesic function and spinothalamic tract inhibition, 475–476
   descending nociception-modulating pathways, 488, 493. *See* Rostral ventromedial medulla

descending somatosensory projections to
   trigeminal brainstem complex, 454–455
GABAergic/glycinergic projections to
   monoaminergic neurons, 203, 205
stimulus-induced tooth pulp afferent
   depolarization, 528
Nucleus raphe obscurus, 182, 183, 185, 186
Nucleus raphe pallidus, 182, 183, 185, 186

O

Obstructive sleep apnea syndrome (OSAS), 118, 129, 131
Occipital neuralgia, 591
OFQ, 259
8-OH-DPAT, 35, 183, 185, 188, 531
6-OHDA, 146
Oleamide, 437
Olfactory bulb
   basal IEG expression, 239
   cholinergic/GABAergic basal forebrain
      projections, 215
   state-specific IEG expression, 247
Omega 1, 408
One-generator (1–gen) model, 102, 103, 108–109, 113, 116, 117
Ontogeny of state development, 150–151
Opioid antagonists, 373, 378, 491, 552
Opioid receptor-like-1 (ORL1) receptor, 551–552
Opioid receptors, 550–551, 665. *See also* specific
      receptors
   drug addiction relationships, 371, 373
   endogenous ligands, 551
   G protein uncoupling and drug tolerance effects, 377
   ORL1, 551–552
Opioids, 371–372
   cytokine and immune-active substance
      interactions, 665
   endogenous, 551, 553. *See also* specific substance
   mesopontine cholinergic neurons and, 170
   new targets for analgesic therapy, 550–553
      anti-opioid peptides, 552, 553–555
      delta opioids, 552
      endogenous opioids, 553
      mu receptor ligand, 551
      orphanin FQ/nociceptin ligand for ORL1
         receptor, 551–552
   NMDA antagonists and, 558
   nonpeptide (SCN80), 552
   receptors. *See* Opioid receptors; specific receptors
   rostral ventromedial medulla function, 488–489
   serotonergic response, 185
   spinal response to preparative surgery/anesthesia, 509
   tolerance, 555
   trigeminal brainstem nociceptive system, 455
Opponent process theory, 374, 428–429
Oral-buccal responses, 187
ORL1 receptor, 551–552

Orofacial somatosensory transmission. *See* Trigeminal
   sensory nuclear complex
Orphanin FQ, 551–552
Out-of-body experiences, 148

P

PACAP, 33
Pacemakers. *See* Circadian rhythms
Pachmorphas exguttata, 68
Pain. *See* Neuropathic pain; Nociceptive pathways;
   Pain control
Pain control. *See also* Anesthesia; Nociceptive
      pathways
   analgesia center perspective, 489
   descending controls of spinothalamic tract, 475–476
   descending influences on trigeminal brainstem, 455
   diffuse noxious inhibitory controls (DNIC), 456, 579–580
   dorsal column lesions, 579
   pharmacotherapeutic targets, 549–561
   neurosurgical treatment. *See* Neurosurgical pain
      treatment
   new treatment targets, 553–555
      adenosine, 559–561
      anti-opioid peptides, 552, 555–556
      central hypersensitivity, 552, 555–556
      excitatory amino acids, 556–559
      inhibitory systems, 559
      neuropathic pain, 553–555
      nitric oxide, 561
      NMDA antagonists, 557–559
      opioids, 550–553
      plasticity concept, 550
   rostral ventromedial medulla function, 488–492, 496
   serotonin and, 185, 494
Pain surgery. *See* Neurosurgical pain treatment
Pancoast syndrome, 593
Panting, 626
Parabrachial areas
   autonomic function, 400
   descending somatosensory projections to
      trigeminal brainstem complex, 454–455
   GABAergic/glycinergic projections to dorsal raphe
      nucleus, 203
   glycinergic/GABAergic projections to locus
      coeruleus, 203
   nociceptive spinal pathways, 572, 575
   state-specific IEG expression, 247
   trigeminal projections, 446
Parachlorophenylalanine (PCPA), 312, 434–435, 661, 663
Paradoxical sleep, 88. *See* REM sleep
Paragigantocellular nucleus, 201–204, 206, 207, 262, 575
Parasomnias, 89, 106
Paraventricular nucleus, 21–23

Index

Paroxysmal oscillations, 337–339
PAS domain, 77
Pasteurella multocida, 641–642, 643, 649
Pavor nocturnas, 148
PCPA, 312, 434–435, 661, 663
Peduncular hallucinosis, 147
Pedunculopontine tegmental (PPT) nucleus, 161–173, 272–292, 312, 393. *See also* Mesopontine cholinergic system
    amygdala connections, 393
    cellular factors affecting excitability, 278–285
    dual role in REM sleep and wakefulness, 161–173
    intrinsic and extrinsic properties and impact on behavioral state, 286–291
    low-threshold calcium spike, 218
    preoptic area state-regulatory network, 318
Pemoline, 424
Penile tumescence, 117, 149
Pentobarbital, 409, 411, 413
Peptidase inhibitors, 553
Peptide-histidine isoleucine (PHI), 17, 37
per, 62–66, 75–77
    mammalian homologue, 70, 75, 77–79
    subcellular localization, 68
Percutaneous radiofrequency rhizotomy, 591
Periaqueductal gray (PAG)
    analgesic function and descending control of spinothalamic tract, 475
    basal IEG expression, 239
    descending somatosensory projections to trigeminal brainstem complex, 454–455
    GABA antagonist effects, 200
    GABAergic/glycinergic projections to monoaminergic neurons, 202–204, 206
    nociceptive spinal afferents, 572
    rostral ventromedial medulla and analgesic function, 488
    state-specific IEG expression, 241
    trigeminal projections, 446
Pericellular baskets, 170
Period gene. *See per*
Periodic leg movements, 131
Peripheral nerve injury, and neuropathic pain treatment, 553–555
Peripheral nerve surgery, 591
Pertussis toxin, 263
Peyote, 437
PGO waves, 391
    autonomic/somatomotor function, 400
    carbachol effects, 397–398
    feline model of serotonergic neuron activity, 188
    IGE expression patterns, 248–249
    lateral pontine origin, 392–393
    limbic activity, 394
    pontine cholinergic neuron activity, 286–289
    REM sleep deprivation and, 146
    serotonergic interactions, 434–435
    slow-wave sleep (SPHOL) episodes, 393. *See also* SPHOL
    state-specific cholinergic neuron response, 167
PHAL, 201–202, 570

Phantom pain, 601
Phantom REM sleep, 112, 117–119, 151
Phase-response curve (PRC), for circadian system
    inputs, 4, 5, 8, 33
    non-photic, 5, 8, 33, 34–35, 36
Phasic motor activation (twitching), 88, 89
Phenylephrine, 183, 269, 270, 271
Phospholipase C (PLC), 270, 398
Photoreceptor conserved elements (PCEs), 49
Photoreceptors, 48
    Drosophila model, 70
    pineal, 54
Pigment-dispersing hormone (PDH), 71
Pike, 55
Pindobind, 268
Pineal gland
    chicken, 54–55, 64
    circadian system and, 8
    fish, 55
    melatonin activity, 8, 45–56. *See* Melatonin rhythm-generating systems
    photoreceptors, 54
    SCN network, 23, 52–53
Pirenzepine, 285
Pituitary-adrenal stress axis, drug dependency and, 376, 378
Pituitary adenylate cyclase activating peptide (PACAP), 33
Platypus, 91–95
Poly I:C, 645, 648, 660
Pontine cholinergic neurons. *See* Mesopontine cholinergic system
Pontine reticular formation. *See* Reticular formation
Ponto-geniculo-occipital spikes. *See* PGO waves
Pontomesencephalic tegmental nuclei. *See* Mesopontine cholinergic system
Positron emission tomography, 400, 574, 599, 600
Postsynaptic dorsal column sensory system, 464–468
Prazosin, 269
Pre- and post-central gyrectomy, 599, 600
Preoptic area
    benzodiazepine effects, 409, 410, 412–416
    cholinergic/GABAergic basal forebrain projections, 215, 220
    GABAergic/glycinergic projections to monoaminergic neurons, 201–203, 206, 207
    mesopontine projections, 164
    recovery sleep following ibotenic acid lesions, 413–414
    SCN efferents, 22
    state regulation model, 312–318
    state-specific IEG expression, 246, 247, 248, 313
Prepositus hypoglossi, 248, 264
Pretectal nucleus, 164, 476
Prion-related sleep alterations, 642, 647
Prolactin, 610
Proopiomelanocortin, 665
Propofol, 414
Prostaglandins
    cannabinoid receptor ligands, 436

noradrenergic immune response and, 660
somnogenic effects, 316, 610
Protein cycling regulation, *Drosophila* model, 66–68
Protein kinase A (PKA), 51, 398
Protein kinase C (PKC), 53, 270
Protein kinase G, 33
Proto-oncogenes. *See c-fos*; Immediate early genes
Protracted abstinence, 378, 379, 429
Psilocin, 434, 438
Psilocybin, 434, 438
Psychedelic mushrooms, 434, 437
Psychobiology, 143
Psychogenic model of state dissociation, 149–150
Psychomimetic drugs, 433–438. *See also* specific drugs
Psychomotor stimulants, 369–371. *See also* Cocaine
Psychosurgical interventions, 597
Pulvinarotomy, 601
Pyrogens, 185, 626. *See also* Fever; specific pyrogens

R

Rabies virus, 642
Radiculo-medullary junction, 571
Raphe. *See also* Dorsal raphe nucleus (DRN); Nucleus raphe magnus
basal IEG expression, 239
mesopontine cholinergic system connections, 168
reticular formation connections, 162
serotonergic neurons. *See* Dorsal raphe nucleus
suprachiasmatic nucleus afferents, 20, 35
Rapid eye movement sleep. *See* REM sleep
PGO waves and, 392
vivid dreaming and, 102
RB101, 553
Recall, of sleep mentation content, 106–107, 113, 114
Recessive screens, 77
Reciprocal interaction model, 315
Reflex-sympathetic dystrophy, 603
Reinforcement models of drug dependence, 366–367, 427–429. *See under* Drug addiction
Relapse, 378, 422
REM-active basal forebrain cells, 312
REM/NREM mentation differences, 101–119
circadian factors, 114
event-related potentials, 111
evidence for cognitive activity in NREM sleep, 105–107
insufficiency of REM sleep for dreaming, 109–110, 116
locomotor center activation/inhibition, 149
memory activation, 113–116
memory consolidation, 111
memory sources, 110–111
"off" neurons, 88, 248
one-generator model, 103, 108–109
"on" neurons, 88, 167, 248
phantom REM sleep, 112, 117, 151
qualitative differences, 103, 108–109, 112, 115–116

recall, 106–107, 114, 118
report content interrelationships, 111, 112
report length, 103, 110, 111, 112, 113, 115–116
residual stage differences, 112
sleep onset, 106, 111, 118–119
stages 3 and 4 sleep, 107
stimulation effects, 105–106, 112
subject differences, 112–113
two-generator model, 102, 109–110, 116
wakefulness, 107, 108, 147
mesopontine cholinergic system, 172–173
REM sleep (also called paradoxical sleep), 87–97, 145, 162, 391. *See also* Active sleep
abnormalities
atonia persistence (sleep paralysis), 130–131, 147
dissociated states and, 147–148
genetics, 134–137
neurochemistry, 132–134
state dissociation, 147–148
arousal threshold. *See* Arousal threshold
avian, 87, 95
bacterial infection and, 643
benzodiazepine effects, functional neuroanatomy, 411–413
brain temperature rhythms, 625
carbachol-induced state. *See* Carbachol
cat model, 88–89
c-fos-null mice, 318
cholinergic modulation, 162–173, 391–400. *See* Mesopontine cholinergic system
amygdala and, 396–398, 400
lateral pons and, 392–393. *See also* Mesopontine cholinergic system
pons-amygdala interactions, 393–395, 400
proposed mechanisms, 398–400
thalamocortical activation, 278
cognitive activity (dreaming), 101–119. *See also* Dreaming; REM/NREM sleep mentation differences
conceptual evolution of state, 145
cortical activation, 214. *See* Cortical activation
definition of, 87–90
developmental differences in electrophysiology, 88
dissociated states and motor behavior, 149
dreaming. *See* REM/NREM sleep mentation differences
echidna, 90–95
EEG desynchronization, 145
efficiency concept, 119
ethanol effects, 422
evolutionary speculations, 91, 94–95, 97
fever and, 634
function, 328. *See also* Sleep function
H-reflex, 148
hypothalamic model of sleep control, 315
IEG expression patterns, 248–249
locomotor center activation/inhibition, 149
locus coeruleus electrophysiology, 264
mammals, 90
monoamine effects, 168

effects on mesopontine cholinergic function, 289–290
glycinergic/GABAergic inhibition of monoaminergic neurons, 196–200, 205–207
motoneuron inhibition, 186. *See also* Atonia
narcolepsy-associated abnormalities, 130–132. *See also* Narcolepsy
non-atonic, 186, 394
non-cholinergic mesopontine neurons and, 171
penile tumescence, 117
PGO waves and, 188, 286, 391. *See also* PGO waves
associated autonomic/somatomotor responses, 400
phantom REM sleep, 112, 117–119, 151
phasic motor activation (twitching), 88, 89
monotremes, 94, 97
platypus, 91–95
prion-related alterations, 647
reciprocal interactions model, 196
regional cerebral blood flow and, 400
REM sleep/dreaming perspective, 102
reptilian sleep and, 95–97
serotonergic interactions, 183
hippocampal theta, 188
sleep deprivation effects, 146. *See also* Sleep deprivation
sleep onset, 118–119, 131
sleep regulatory substances, 610
species differences in sleep times, 90
species differences of neuroanatomical substrates, 88
temporal lobe epileptic subjects, 400
viral infection and, 645
REM sleep atonia. *See* Atonia
REM sleep behavior disorder (RBD), 117, 131, 147, 149
Reptiles
circadian systems, 6
sleep states, 95–97
temperature effects on immune function, 627
Respiration
benzodiazepine properties, 416
REM sleep and PGO wave association, 400
marine mammal sleep, 146
Reticular formation
carbachol-induced REM sleep-like state, 162, 165
cholinergic-monoaminergic-glutamatergic input and state-specific activity, 162–163
descending pathways modulating nociception, 488, 496. *See also* Rostral ventromedial medulla
EEG synchronization/desynchronization patterns, 145
glutamatergic projections, 217
glycinergic/GABAergic projections to locus coeruleus, 201–203
projections to basal forebrain, 217–218
spino-reticulo-thalamic nociceptive afferents, 575, 577

state-specific activation by mesopontine cholinergic neurons, 172–173
state-specific IEG expression, 248
trigeminal projections, 446
Reticular thalamic neurons. *See* Corticothalamic system; Thalamic reticular neurons
Retinal melatonin, 47, 49
Retinohypothalamic tract, 8, 18–19, 31–33, 35
Reverse genetics, 80–81
Reverse tetracycline-controlled transactivator (rtTA) system, 81
Reward, neuronal elements of, 374
Rhinovirus, 642, 645
Ripple oscillations, 350
Rostral ventromedial medulla
anesthetic effects, 491
cardiovascular regulation and, 496–497
descending nociceptive modulatory system, 487–497
integration of somatomotor/autonomic control with nociceptive modulation, 496–497
on- and off-cell functional relationships, 489–492
serotonergic interactions, 493–495
state-specific neuronal recruitment, 495–496
Ryanodine, 260, 270

S

SCN. *See* Suprachiasmatic nucleus
SCN80, 552
Scrapie, 642
Sedative-hypnotics, 372–373. *See also* Antidepressants; Barbiturates; Benzodiazepines; Ethanol
Seizures, 149, 239
naturally synchronized sleep oscillations, 337–338
Selective posterior rhizotomy, 591–593
Sensitization, drug, 367, 374, 377–378
Sensitization, hyperalgesic/allodynic, 506, 510–512, 516
Sensory pathways. *See* Pain; Somatosensory systems; specific neuroanatomy
Septum
basal IEG expression, 238
cholinergic/GABAergic basal forebrain projections, 215
state-specific IEG expression, 241, 247
Serial analysis of gene expression (SAGE), 79
Serotonergic neurons, 181–191. *See also* Dorsal raphe nucleus; 5-HT receptors; Serotonin
analgesia and, 185
circadian effects, 191
feline CNS function, 182–189
challenges/stressors, 185–186
circadian cycles, 182
hippocampal theta, 189
motor activity, 186–187
PGO waves, 188
sleep/wake cycles, 182–185
GABAergic input, 183

medullary activity, compared to pontine, 187
sleep deprivation effects, 183, 185, 191
state-related function, 182–185
suprachiasmatic nucleus afferents, 35
sympathetic activation, 186
Serotonin, 312, 434–435. *See also* Dorsal raphe nucleus
   antinociceptive function, 494
   antipyretic function, 661
   basalo-cortical cholinergic system interactions, 221–222, 224–225, 229
   basalo-cortical noncholinergic (potentially GABAergic) interactions, 222
   cytokine and immune-active substance interactions, 661–664, 666–667
   descending nociceptive modulation, 493–495
   dorsal raphe nucleus. *See* Dorsal raphe nucleus
   dorsal spinocerebellar tract afferent input and, 531
   drug tolerance effects, 377
   locus coeruleus response, 259
   PGO waves and, 188, 434–435
   pontine cholinergic electrophysiological response, 282–284
   psychomimetic drug action and, 434–435, 438
   SCN afferents, 20, 25
   sensory consciousness effects, 434
   state-specific activation/inhibition of mesopontine cholinergic neurons, 168
   synaptic potentials, 265, 268–269
Serotonin *N*-acetyltransferase, 46, 49–52. *See also* AANAT
Serotonin receptors. *See* 5-HT receptors
Serotonin-specific reuptake inhibitors (SSRIs), 148
Sharp waves, 350, 353–355
Sheep model of spinal electrophysiology, 508, 511, 512–514
Shivering, 626
Silk moth, 65, 68
single-minded gene, 77
Single-unit recording
   chronically-implanted animal models of sensory transmission, 523–524, 531–534
      brainstem, 523–525
      spinal cord, 531–534
   iontophoretic application of GABA/glycine antagonists and state-specific effects on monoaminergic neurons, 197–200
   mesopontine activity during REM sleep, 162
   serotonergic effects on PGO wave generation, 188
   state-specific mesopontine cholinergic system function, 163, 167, 172
   surgical/anesthetic effects on spinal sensory processing, 507–514
   SCN electrophysiology, 9
   validity for assessing serotonergic function, 190
Sleep
   active. *See* Active sleep
   non-REM. *See* Non-REM sleep
   paradoxical. *See* REM sleep
   REM. *See* REM sleep
   slow-wave. *See* Slow-wave sleep

state dissociation (in animals), 146. *See also* Dissociated states
   unihemispheric in marine mammals, 89, 146, 617
Sleep-active neurons, basal forebrain preoptic area, 313–314
Sleep apnea syndrome, 118, 129, 131, 612
Sleep attacks, 130
Sleep deprivation
   antidepressant effects, 183
   fatal in rats, 624
   IL-1β and, 615
   immune response, 649–650
   NREM functional neuroanatomy, 313–314
   plasma TNF levels and, 612
   serotonergic response, 183, 185, 191
Sleep disorders. *See also* Dissociated states
   arousal disorders NREM sleep variations, 148–149
   fatal familial insomnia, 149, 642, 647
   narcolepsy, 129–137. *See* Narcolepsy
   REM sleep behavior disorder (RBD), 117, 131, 147, 149
   sleeping sickness (trypanosomiasis), 241, 646–647
   sleep onset REM and, 118
   treatment, 131–132
Sleep drunkenness, 148
Sleep function, 328, 624
   infection-induced alterations and, 649
   immune response, 634
   oscillation patterns, 342, 344
   synaptic growth and maintenance, 617
Sleep-inducing drugs, 407–416. *See* Barbiturates; Benzodiazepines; Ethanol
Sleeping sickness (trypanosomiasis), 241, 646–647
Sleep lipid, 316
Sleep mentation. *See* Dreaming; REM/NREM sleep mentation differences
Sleep-onset mentation, 106, 111
Sleep onset REM (SOREM), 118–119, 131
Sleep paralysis, 130–131, 147
Sleep-regulatory substances, 316, 609–617. *See also* Adenosine; Cytokines; Muramyl peptides; Vasoactive intestinal peptide; specific endogenous somnogens
   criteria for circuits and substances, 610
   infection-induced sleep alterations and, 647–648
   sleep humoral regulatory network, 615–616
   suprachiasmatic nucleus peptides, 36–37
Sleep talking, 105
Sleep terrors, 148
Sleep/wake cycle. *See* specific states, modulatory factors and systems
Sleep walking, 89, 148
Slow-waves
   caffeine effects, 424
   corticothalamic substrates, 328–339
      grouping of sleep rhythms into complex wave sequences, 329–339
   functional considerations, 342
   human sleep, 334–335
   localized stimulus response, 617

Index

long-lasting hyperpolarization, 332
neuronal synchronization, 332
paroxysmal oscillations (seizure patterns), 337–339
TNFα and, 611–612
Slow-wave sleep (SWS), 391
  bacterial infection and, 643
  basalo-cortical system, 223
  cellular substrates. *See* Corticothalamic system;
    Slow-waves
  cerebral function, 328. *See* Sleep function
  cholinergic influence, 163
  glycinergic/GABAergic inhibition of
    monoaminergic neurons, 196–200,
    205–207
  IL-1–induced alterations, 666–667
  marijuana effects, 435
  monoaminergic effects on mesopontine cholinergic
    function, 289–290
  ponto-geniculo-occipital waves. *See* PGO waves
  prion-related alterations, 647
  serotonin effects, 434
  sharp waves, 350
  viral infection and, 645, 649
Sodium channel-blocking anti-convulsants/analgesics,
  554–555
Somatosensory cortex, nociceptive spinal afferents,
  574
Somatosensory systems. *See also* Spinal
  somatosensory systems
  low-threshold mechanoreceptive (LTM) neurons,
    449–450, 455, 456
  lumbosacral system, 463–477. *See also*
    Lumbosacral sensory system; Spinal
    somatosensory systems
  pain transmission. *See* Nociceptive pathways
  preparative surgical/anesthetic effects, 505–517,
    523. *See also* Surgical and anesthetic
    preparation effects
  trigeminal complex, 446–457, 522–529. *See also*
    Trigeminal sensory nuclear complex
  state-specific variation, 495
  wide-dynamic range (WDR) neurons, 452, 471,
    512, 532, 533, 574
Somatostatin, 16, 36–37, 393, 400, 259, 400
Somnambulism, 89, 148
SPHOL, 393, 396–398
Spike-wave (SW) seizures, 337–339
Spinal cord
  benzodiazepine receptors, 408
  dorsal horn. *See* Dorsal horn
  GABA antagonist effects, 200
  glycinergic afferents, 207
Spinal cord surgical interventions, 593–596
  anterolateral cordotomy, 593–596, 600–601, 604
  dorsal root entry zone (DREZ), 593
  rarely used procedures, 596
Spinal nerve ligation, 554
Spinal rhizotomies, 591–593
Spinal shock, 514, 517
Spinal somatosensory systems, 464, 530–538. *See
  also* specific tracts

adenosine receptor function, 559–560
anti-opioid peptide (CCK), 555–556, 559
ascending sensory and nociceptive pathways,
    464–477, 530–536, 569–577. *See also*
    Nociceptive pathways; specific tracts
  dorsal columns, 464, 578–579, 580
  dorsal horn, 571–577
  dorsolateral funiculus, 577–578
  functional localization problems, 579–580
  noxious non-specific cells, 571, 574
  noxious specific cells, 571
  post-synaptic dorsal column, 464–477. *See also*
    specific spinal tracts
  problematic interpretation of lesioning data,
    579–580
chronically instrumented animal approach to unit
  recording, 531–534
neurosurgical intervention for pain, 593–596
  dorsal root entry zone, 593
  potential complications, 595
  rhizotomies, 591–593
nitric oxide and, 561
opioid receptors and ligands, 551, 552
preparative surgical/anesthetic electrophysiological
  effects in animals, 505–517
  anesthetic effects, 508, 511–514, 516
  conscious sheep model single neuron
    recording, 508, 511, 512–514
  coupled sympathetic/somatic motor
    responsiveness, 514–516
  dorsal horn activity, 508, 510–511
  in situ, in vitro mouse model, 507, 509,
    514–516
  NMDA receptors and hyperalgesic response to
    chronic nociceptive input, 506,
    510–512, 516
  rat model single unit recording, 507, 509–512
  receptive field size, 512–514, 516
  surgery effects, 507, 509–522, 526
postsynaptic inhibition, 534–536
trigeminal projections, 446
rostral ventromedial medulla nociception-
  modulating projections, 489, 492, 496
state-associated electrophysiology, 534–537
  active sleep, 534–537
Spindle oscillations, 298, 329–331, 333
Spinocervical tract, 468–470, 530
Spinohypothalamic pathways, 572
Spinomesencephalic tract, 476–477, 489
Spinoreticular tract, 476–477, 489, 530, 533–536
Spinothalamic cordotomy, 591
Spinothalamic tract, 470–474, 530, 533, 534,
    572–574, 578–579, 594–595
Spinothalmic tractotomy, 599, 600
Spiperone, 268
Stages 3 and 4 sleep. *See* Slow-wave sleep
Staphylococcus aureus, 641–642
State dissociation. *See* Dissociated states
Status dissociatus, 144, 149
Stereotactic C1 central myelotomy, 596
Stereotactic mesencephalotomy, 599, 600

Stimulants, 369–371. *See* Caffeine; Cocaine; Nicotine
  benzodiazepine enantiomer, 416
Streptococcus pyogenes, 641–642
Stress response. *See also* Fever
  drug sensitization effects, 376, 378
  feline model of serotonergic neuron activity,
    185–186
  forebrain serotonin and, 190
  rostral ventromedial medulla antinociceptive
    function and, 487, 495–496
  serotonergic neurons (raphe), 185–186
  surgical/anesthetic preparation effects on spinal
    somatosensory processing, 505–517
Striatum, IEG expression, 239, 240, 247
Strychnine, 34, 196–197, 199–200
Subiculum, 349. *See also* Hippocampal formation
Subincertal nucleus, 203
Subnucleus reticularis dorsalis (SRD), 575, 577
Substance dependence. *See* Drug addiction
Substance P, 33, 166, 170, 259, 451, 470, 471, 474
Substantia gelatinosa, 448
Substantia innominata, 215, 220, 312, 313, 379
Substantia nigra, mesopontine projections, 164
Sudeck's atrophy, 603
Superior colliculus
  mesopontine projections, 164
  IEG expression, 241, 246, 247
Suprachiasmatic nucleus (SCN)
  afferents, 8, 21–22, 33
    cholinergic, 36
    geniculohypothalamic, 20, 33–35
    nerve growth factor, 36
    raphe, 20
    retinohypothalamic tract, 8, 18–19, 31–33, 35
    serotonergic, 20, 25
  cellular ionic mechanisms, 37–38
  circadian pacemaker function, 5–7. *See also*
    Circadian rhythms
  cytoarchitecture, 14
  efferents, 8, 24, 25, 38–39
    commissural connections, 14, 24
    dorsomedial nucleus, 24
    preoptic area, 22
    raphe, 35
    supraventricular zone and paraventricular
      nucleus, 22–23
  endogenous neuropeptides, 7, 36–37
  fetal transplants, 38–39
  functional neuroanatomy, 13–25
  future studies, 39
  hypothalamic model of state control, 315
  in vitro electrophysiology, 7, 9
  light-induced electrophysiology, 32–35
  light-induced immediate early gene expression, 8,
    32, 33, 37, 241, 247
  neural grafts of fetal tissue, 7
  neurochemical organization, 14–18
  photoreceptors, 8, 48
  pineal melatonin production rhythms and, 47, 52–53
  retinohypothalamic innervation, 8, 18–19, 31–33,
    35

serotonergic inputs, 182, 191
species-specific phenotypic differences, 17
Suprageniculate/posterior complex, 578
Supramammillary nucleus, 246
Supraventricular zone, 22–23
Surgical and anesthetic preparation effects, 505–517
  anesthetic effects, 508, 511–514, 516
  conscious sheep model single neuron recording,
    508, 511, 512–514
  coupled sympathetic/somatic motor
    responsiveness, 514–516
  dorsal horn activity, 508, 510–511
  in situ, in vitro mouse model, 507, 509, 514–516
  rat model single unit recording, 507, 509–512
  receptive field size, 512–514, 516
  surgery effects, 507, 509–522, 526
  NMDA receptors and hyperalgesic response to
    chronic nociceptive input, 506, 510–512,
    516
Surgical treatment of pain. *See* Neurosurgical pain
  treatment
Sympathectomy, 603
Sympathetic activation, stress-induced serotonergic
  response, 186
Sympathetic output, nociceptive stimulus response,
  514–516
Synaptic potentials
  dorsal raphe nucleus serotonergic neurons, 267
  GABA, 265, 268
  glutamate, 265, 267
  5-HT, 265, 268–269
  norepinephrine, 269–270
  locus coeruleus noradrenergic neurons, 262–263
  pontine cholinergic system, 282

T

t-ACPD, 220
Tactile sensation
  postsynaptic dorsal column system, 465
  spinocervical tract function, 469
  spinothalamic tract function, 474–475
  low-threshold mechanoreceptive neurons, 449–450
Tail flick reflex, 489
T-cells, 627
Tectal state-specific IEG expression, 247
Tegmental nucleus of Castaldi, 204
Temperature compensation, 61
Temperature regulation. *See* Thermoregulation
Temporal lobe epilepsy, 355, 400
Temporomandibular joint (TMJ), 450, 451
TENS, 456
Teonanacatl, 437
Tetrahydrocannabinol (THC), 373–374, 435–437
Tetrodotoxin (TTX), 34, 36, 38, 261, 262
Thalamic pain, 602
Thalamic reticular neurons, 338. *See also*
  Corticothalamic system
  intrinsic electrophysiology, 329, 332–333
Thalamotomy, 599, 601, 602

Thalamus, 278. *See also* Corticothalamic system; Spinothalamic tract
  descending modulation of spinothalamic tract, 476. *See also* Spinothalamic tract
  mesopontine cholinergic connections, 163, 164–165, 172–173, 278
  state dissociation induction, 146
  state-specific IEG expression, 245–248
  trigeminal projections, 446
Thalidomide, 137
Thapsigargin, 270
THC, 373–374, 435–437
Theophylline, 425
Thermal stress response, 185–186, 465, 472. *See also* Brain temperature; Fever; Thermoregulation
Thermoregulation
  ascending pathways, 571–572
  fever heat loss/heat production mechanisms, 628
  lumbosacral system neurons, 465, 472, 475
  state-dependent alterations, 624–626
Thermosensitive neurons, 450–451
Theta EEG oscillations, 350–351
  basal forebrain activity, 298
  entorhinal-hippocampal system dynamics, 350–353
  state-dependent modulation, 352–353, 358
  feline model of serotonergic neuron activity, 189
Tic douloureux, 589–590, 604
tim, 62–71. *See also Drosophila* model
  Drosophila, 75–77
  mammalian, 80
  subcellular localization, 68
Timeless. *See tim*
tis-8, 236
Titanic dreaming, 105
Tobacco, 425. *See* Nicotine
Tolerance, 367, 377
  caffeine, 424, 425
  opioid, 555
Tooth pulp-evoked responses, 523, 524–528
Transcendental dreaming, 105
Transcriptional feedback regulation, 62–67, 76–77
Transcription factors, 236. *See also* Gene expression; Immediate early genes; specific genes and gene products
  chronic drug action and, 378
  light induced-expression (suprachiasmatic nucleus), 8
Transgenic approach to circadian gene rhythm, 80–81
Triazolam, 34–35, 409–411
Tricyclic antidepressants, 148, 149. *See also* Antidepressants
Trigeminal nerve surgery, 589
Trigeminal neuralgia, 589–590, 604
Trigeminal sensory nuclear complex, 446–457, 522–529
  analgesic mechanisms and, 455
  ascending pathways, 446
  chronically implanted animal approach to brainstem unit recording, 523–525
  descending and afferent modulation, 454–456, 525
  modulation of somatosensory transmission, 454
  neuronal types, 448
  nociceptive inputs, 451–453
    neuroplastic changes, 453–454
    spinal nociceptive afferents, 574
  non-nociceptive inputs, 449–450
  presynaptic control of afferent terminals, 528–530
  primary afferents and organization, 446–448
  similarity to spinal dorsal horn, 451–452
  state-associated modulation, 525–528, 536–537
    presynaptic/postsynaptic inhibition during active sleep, 528–530
  structure-function relationships, 448
Trigeminal tractotomy/nucleotomy, 599, 600
Trigemino-thalamic tract, 524, 529–530
Trout, 55
Trypanosoma brucei brucei, 241, 641–642, 646–647
Trypanosomiasis, 241, 646–647
Tryptophan hydroxylase, 46, 49
Tuber cinereum, 203
Tuberomammillary nucleus (TMN), 169, 203, 312, 318, 665
Tumor necrosis factor (TNF, TNFα), 316, 610–614, 667
  fever regulation, 629
  IL-1 interactions, 613
  infection-induced sleep alterations and, 647
  diurnal variations, 612
  neurotransmitter interactions, 616
    dopamine, 661
    norepinephrine, 660
    serotonin, 664
Twitching, 88, 89, 94, 97
Two-generator (2–gen)model, 102, 109–110, 116, 117

U

Unihemispheric sleep, 89, 146, 617
Unit recording. *See* Single-unit recording
Uromodulin, 629

V

Vagotomy, 648
Vasoactive intestinal peptide (VIP), 393, 400, 610
  locus coeruleus response, 259
  suprachiasmatic nucleus, 16–17, 22, 23, 36–37
Vasopressin
  locus coeruleus response, 259
  suprachiasmatic nucleus, 14, 16–17, 23
V brainstem complex, 446. *See also* Trigeminal sensory nuclear complex
Venlafaxine, 148
Ventral lateral preoptic area (VLPO), 313–318
Ventral posterior thalamic areas, spinal afferents, 572–574. *See also* Spinothalamic tract
Ventral tegmental area (VTA)
  abnormal REM sleep and, 134
  drug dependency/reinforcement and, 368, 371, 377, 428

Ventrolateral tract, 570
VGF, 26
Vibrissae, 447
Viral double-stranded RNA (dsRNA), 645
Viral infection, 627, 660. *See also* Infection
    model systems, 641–642
Visceral sensation, 465, 466, 471, 514, 558, 578–579
Visual cortex, IEG expression, 237
Voltage-gated channels
    ethanol effects, 372–373
    pontine cholinergic neurons, 279–282
Voltage clamp experiments
    dorsal raphe nucleus, 266–267
    locus coeruleus, 260–262

W

Wake-active basal forebrain cells, 312
Waking state, 145
    benzodiazepine enantiomer enhancement of, 416
    cholinergic neurons and, 161–173. *See*
    Mesopontine cholinergic system
        influence on thalamocortical activation, 278
    cortical activation, 214, 278. *See* Cortical
        activation
    dreaming/hallucination, 107, 108, 147
    drug effects on alertness, 427–428
        caffeine, 424–425
        ethanol, 423, 428
        nicotine, 426–427, 428
    EEG desynchronization, 145
    functional neuroanatomy, 312–313
    IEG expression, 241–247
        brainstem, 246
        cortex, 243–245
        diencephalon, 245–246
        noradrenergic interactions, 246
        spontaneous wakefulness, 246–247
    monoaminergic effects on mesopontine cholinergic
        function, 289–290
    wide dynamic range neurons, 512
Warming-induced state alterations, 313
W cells, 18
Wheel-running rhythms, 35
Wide dynamic range (WDR) neurons, 452, 471, 512, 532, 533, 574
Wind-up, 506, 552, 556, 559, 560
Withdrawal effects, 374
    caffeine, 425
W/REM-on neurons, 167, 170, 172–173

X

X area, 393
Xenopus, 64

Y

Yeast infections, 641–642, 647

Z

Zebrafish, 55
zif/268, 236, 239
Zolpidem, 408, 414
Zona incerta, 203